INDEX to RECORD and TAPE REVIEWS

A Classical Music Buying Guide

1976

Antoinette O. Maleady

Chulainn Press

San Anselmo, California • 1977

ML156
.9
.M28

ISBN 0-917600-02-9
Library of Congress Catalog Card No. 72-3355
Manufactured in the United States of America
Copyright © 1977 by Antoinette O. Maleady

CHULAINN PRESS
Post Office Box 770
San Anselmo, California 94960

S.S.O.

To Tom and Sarah

CONTENTS

v

This Index brings together in one volume a listing of all classical music recordings reviewed in 1976 in the major reviewing media of the United States, England and Canada.

The Index has four sections. Section I is a straight listing by composer. Section II, "Music in Collections", lists records or tapes with several composers on one disc or tape. Section II main entries are arranged alphabetically by name of the manufacturer, and then serially by manufacturer's number within each main entry. The work of any composer in each of these collections appears in Section I under the name of that composer, with a reference to the Section II entry. Section III lists, alphabetically by title, anonymous works, with a reference to their location in Section I or Section II. Section IV is a Performer Index to recordings listed in Sections I and II, with reference by citation number to its location within the section. Each citation gives, if available, the disc label and number; variant labels and numbers; tape cassette, cartridge or reel numbers; quadraphonic disc or tape numbers; reissues; location of the reviews and the reviewers evaluation of the recording. The main entry for each recording is in upper case letters. Tapes or discs reviewed in 1976 that were also reviewed in 1975, 1974, 1973, 1972 and/or 1971 have all the reviews brought together in the 1976 Index.

A key is provided for understanding the entries of the four sections. The entries are fictitious entries to show the various possibilities of form.

Section I

MACONCHY, Elizabeth
1219 Ariadne, soprano and orchestra. WALTON: Songs: Daphne; Old Sir
[entry   Faulk; A song for the Lord Mayor's table; Through gilded
   no.]   trellises. Heather Harper, s [Heather Harper, soprano]; Paul
          Hamburger, pno [Paul Hamburger, piano]; ECO [English Chamber
          Orchestra]; Raymond Leppard [conductor]. Columbia M 30443
          [disc number] (2) [number of discs in set]. Tape (c) MT 30443
          [cassette number] (ct) MA 30443 [cartridge number] (r) L 30443
          [reel number] (Q) MQ 30443 [quadraphonic disc number] Tape (ct)
          MAQ 30443 [quadraphonic tape cartridge number]. (also CBS
          72941) [recording also available on CBS label]
              ++Gr 9-74 p1025 [evaluation excellent; review in Gramo-
                          phone, September 9, 1974, page 1025]
               +Gr 11-76 p667 tape [evaluation good, review in Gramo-
                          phone, November 1976, page 667 for tape]
               /MJ 5-73 p73 [evaluation fair, review in Music Journal
                          May, 1973, page 73]
               -NR 4-72 p4 [evaluation poor, review in New Records,
                          April, 1972, page 4]

Section II

LONDON

OS 36578 (also Decca SXL 3315) [label number. Also available on Decca
     label] (Reissue from OS 3716) [reissue from London OS 3716]
2395 Baroque flute sonatas. BLAVET: Sonata, flute, no. 2, F major.
[entry   GUALTIER: Suite, G minor. HANDEL: Sonata, flute, op. 1, no.
  no.]   5, G major. LOEILLET: Sonata, flute, F major. VINCI: Sonata,
     flute, D major. Jean-Pierre Rampal, flt [Jean-Pierre Rampal,
     flute]; Raymond Leppard, hpd [Raymond Leppard, harpsichord];
     Claude Viala, vlc [Claude Viala, violoncello]; AMF [Academy
     of St. Martin-in-the-Fields]; Raymond Leppard [conductor]
     +-NYT 4-30-75 pD32 [evaluation mixed, review in New
           York Times, April 30, 1975, page D32]
     +STL 1-17-76 p30 [evaluation good, review in Sunday Times,
           London, January 17, 1976, page 30]

Section III

Kyrie trope: Orbis factor. cf TALLIS: Missa salve intemerata. [Anony-
     mous work, appears in Section I under entry for Tallis]
Deo gracias Anglia, Agincourt carol. cf BASF 25 22286-1. [Anonymous
     work, appears in Section II under entry BASF 25 22286-1]

Section IV

Baker, Janet, mezzo-soprano (contralto) 231, 494, 939, 1376, 1776,
     2513 [performer, listed in reviews as mezzo-soprano or
     contralto. Citation numbers in Section I and II, where artist
     has performed]

ABBREVIATIONS

Periodicals Indexed

AR	American Recorder
ARG	American Record Guide
ARSC	Association for Recorded Sound Collections Journal
Audio	Audio
CJ	Choral Journal
CL	Clavier
Gr	Gramophone
GTR	Guitar Review
Ha	Harpers
HF	High Fidelity
HFN	Hi-Fi News & Record Review
HPD	Harpsichord
IN	The Instrumentalist
LJ	Library Journal/School Library Journal Previews
MJ	Music Journal
MM	Music & Musicians
MQ	Musical Quarterly
MR	Music Review
MT	Musical Times
MU	Music/American Guild of Organists
NR	New Records
NYT	New York Times
OC	Opera Canada
ON	Opera News
Op	Opera, London
PNM	Perspectives of New Music
PRO	Pro Musica
RR	Records and Recording
SFC	San Francisco Examiner and Chronicle (This World)
SR	Saturday Review/World
St	Stereo Review
ST	Strad
STL	Sunday Times, London
Te	Tempo

Performers

Orchestral

AMF	Academy of St. Martin-in-the Fields
BBO	Berlin Bach Orchestra
BeSO	Berlin Symphony Orchestra
BPhO	Berlin Philharmonic Orchestra

BPO	Boston "Pops" Orchestra
Brno PO	Brno State Philharmonic Orchestra
BRSO	Berlin Radio Symphony Orchestra
BSO	Boston Symphony Orchestra
CnSO	Cincinnati Symphony Orchestra
CO	Cleveland Orchestra
COA	Concertgebouw Orchestra, Amsterdam
CPhO	Czech Philharmonic Orchestra
CSO	Chicago Symphony Orchestra
DBS	Deutsche Bach Solisten
ECO	English Chamber Orchestra
FK	Frankfurt Kantorei
HCO	Hungarian Chamber Orchestra
HRT	Hungarian Radio and Television
HSO	Hungarian State Symphony Orchestra
LAPO	Los Angeles Philharmonic Orchestra
LOL	Little Orchestra, London
LPO	London Philharmonic Orchestra
LSO	London Symphony Orchestra
MB	Munich Bach Orchestra
MPAC	Munich Pro Arte Chamber Orchestra
MPO	Moscow Philharmonic Orchestra
MRSO	Moscow Radio Symphony Orchestra
NPhO	New Philharmonia Orchestra
NSL	New Symphony, London
NWE	Netherlands Wind Ensemble
NYP	New York Philharmonic Orchestra
ORTF	O.R.T.F. Philharmonic Orchestra
OSCCP	Orchestre de la Société des Concerts du Conservatoire de Paris
OSR	l'Orchestre de la Suisse Romande
PCO	Prague Chamber Orchestra
PH	Philharmonia Hungarica
PhO	Philharmonia Orchestra, London
PO	Philadelphia Orchestra
PSO	Prague Symphony Orchestra
ROHO	Royal Opera House Orchestra, Covent Garden
RPO	Royal Philharmonic Orchestra
SBC	Stuttgart Bach Collegium
SCO	Stuttgart Chamber Orchestra
SDR	Stuttgart S.D.R. Symphony Orchestra
SPO	Stuttgart Philharmonic Orchestra
SSO	Sydney Symphony Orchestra
VCM	Vienna Concentus Musicus
VPM	Vienna Pro Musica
VPO	Vienna Philharmonic Orchestra
VSO	Vienna Symphony Orchestra
VSOO	Vienna State Opera Orchestra

Instrumental

acc	accordion	cld	clavichord
bal	balalaika	clt	clarinet
bs	bass	cor	cornet
bsn	bassoon	dr	drums
c	celesta	Eh	English horn
cimb	cimbalon	flt	flute

Fr hn	French horn		pic	piccolo
gtr	guitar		pno	piano
harm	harmonica		rec	recorder
hn	horn		sax	saxophone
hp	harp		sit	sitar
hpd	harpsichord		tpt	trumpet
lt	lute		trom	trombone
mand	mandolin		vib	vibraphone
mar	marimba		v	viol
ob	oboe		vla	viola
ond	ondes martenot		vlc	violoncello
org	organ		vln	violin
perc	percussion		z	zither

Vocal

bar	baritone		ms	mezzo-soprano
bs	bass		s	soprano
con	contralto		t	tenor
c-t	countertenor			

Qualitative Evaluation of Recordings

++	excellent or very good		/	fair
+	good		–	poor
+–	mixed		*	no evaluation

ABEL
      Sonata, G major. cf BIS LP 22.
ABSIL, Jean
      Sonata, op. 37. cf Orion ORS 76235.
      Suite on a Rumanian folk tune, op. 90. cf Transatlantic TRA 308.
      Suite on popular Rumanian themes. cf Golden Crest CRS 4143.
ACHRON, Joseph
      Hebrew dance. cf RCS ARM 4-0942/7.
      Hebrew lullaby, op. 33. cf Da Camera Magna SM 93399.
      Hebrew lullaby, op. 33. cf RCA ARM 4-0942/7.
      Hebrew melody, op. 33 (2). cf RCA ARM 4-0942/7.
      Scher. cf Da Camera Magna SM 93399.
      Stimmung. cf RCA ARM 4-0942/7.
ADAM
      O holy night. cf CBS 61771.
      Songs: Comrades in arms. cf Columbia SCX 6585.
ADAM, Adolphe-Charles
      Cantique Noël. cf Decca SXL 6781
1   Giselle (arr. Busser). Si j'étais roi: Overture. OSCCP; Albert
      Wolff, Jean Martinon. Decca SPA 384. Tape (c) KCSP 15384.
      (Reissues from RCA SB 2018, SXL 2008).
            +Gr 4-75 p1871   +RR 5-75 p26
            -+HFN 7-75 p90 tape   +RR 8-76 p83 tape
2   Giselle. DELIBES: Coppélia. Sylvia. TCHAIKOVSKY: The nut-
      cracker, excerpts. Sleeping beauty, op. 66, excerpts. Swan
      Lake, op. 20, excerpts. London Concert Orchestra; Marcus Dods.
      Polydor 2383 356. Tape (c) 3170 256.
            +Gr 1-76 p1244
3   Giselle: Pas de deux, Act 1; Grand pas de deux and finale (arr.
      Busser). MINKUS: Don Quixote: Pas de deux. TCHAIKOVSKY: The
      nutcracker, op. 71: Pas de deux, Act 2. Sleeping beauty,
      op. 66: Blue bird pas de deux. Swan Lake, op. 20: Pas de deux,
      Act 3. LSO, OSR, OSCCP; Richard Bonynge, Ernest Ansermet,
      Jean Martinon. Decca SPA 487.
            ++RR 12-76 p62
      Si j'étais roi: Overture. cf Giselle.
      Si j'étais roi: Overture. cf Decca GOSD 674/6.
      Si j'étais roi: Un regard de ses yeux. cf Club 99-98.
      Variations on a theme of Mozart's "Ah, vous dirai-je Maman".
        cf BIS LP 45.
      Variations on a theme of Mozart's "Ah, vous dirai-je Maman".
        cf Richmond R 23197.
      Variations on a theme of Mozart's "Ah, vous dirai-je Maman".
      cf Westminister WGS 8268.

ADAM DE LA HALLE
        Fines amouretes ai. cf Nonesuch H 71326.
        Li mans d'amer. cf Vanguard VCS 10049.
        Songs: De cueur pensieu; En mai quant rosier. cf Candide
            CE 31095.
        Tant con je vivrai. cf Nonesuch H 71326.
ADAMS
        Songs: The holy city. cf Argo ZFB 95/6.
ADAMS, John
4   American standard. BRYARS: 1, 2, 1-2-3-4. HOBBS: Aran.
            McCrimmon will never return. Christopher Hobbs, John White,
            Gavin Bryars; San Francisco Conservatory of Music New Music
            Ensemble. Obscure no. 2.
                +RR 3-76 p18
ADDINSELL, Richard
        Warsaw concerto. cf Rediffusion 15-46.
ADDISON, John
5   Concerto, trumpet, strings and percussion. ARNOLD: Concerto,
            2 violins and strings, op. 77. CROSSE: Some marches on a
            ground, op. 28. SEIBER: Concertino, clarinet and strings.
            Leon Rapier, tpt; Paul Kling, Peter McHugh, vln; James
            Livingston, clt; Louisville Orchestra; Jorge Mester.
            RCA GL 25018.
                +Gr 10-76 p577   +RR 11-76 p49
                +-HFN 12-76 p136
6   Divertimento, op. 9. DODGSON: Sonata, brass. BENNETT: Com-
            media IV. GARDNER: Theme and variations, op. 7. Philip Jones
            Brass Ensemble. Argo ZRG 813.
                ++Gr 12-75 p1066   ++NR 8-76 p6
                ++HFN 12-75 p156   ++RR 12-75 p35
        Divertimento, op. 9: Valse. cf Argo SPA 464.
ADLER
        Songs: A kiss; Strings in the earth. cf Orion ORS 75025.
ADRIAENSSEN, Emanuel
        Branle Englese. cf DG Archive 2533 302.
        Branle simple de Poictou. cf DG Archive 2533 302.
        Courante. cf DG Archive 2533 302.
        Fantasia. cf DG Archive 2533 302.
        Io vo gridando. cf DG Archive 2533 323.
        Madonna mia pietá. cf DG Archive 2533 323.
AGRICOLA, Alexander
        Comme femme. cf HMV SLS 5049.
        Oublier veul. cf Argo ZRG 823.
AIN, Noa
        Used to call me sadness. cf Folkways FTS 33901/4.
AITKEN, Hugh
        Montages, solo bassoon. cf Crystal S 351.
ALABIEFF (Alabiev), Alexander
        Die Nachtigall. cf Rococo 5379.
        The nightingale. cf Richmond R 23197.
        The Russian nightingale. cf BIS LP 45.
ALAIN, Jehan
        Litanies, op. 79. cf Audio 2.
        Litanies, op. 79. cf Decca SDD 499.
ALBENIZ, Isaac
        Capricho catalan. cf RCA ARL 1-1323.

Cantos de Espãna, op. 232: Cordoba. cf Angel S 36094.

7   Espãna, op. 165, no. 2: Tango (arr. Godowsky). CHOPIN: Ballade,
no. 3, op. 47, A flat major. Etudes, op. 10, no. 4, C sharp
minor; op. 25, no. 7, C sharp minor; op. 25, no. 10, B minor.
SCHUBERT: Moment musical, op. 94, no. 3, D 780, F minor (arr.
Godowsky). SCHUMANN: Symphonic etudes, op. 13. Shura
Cherkassky, pno. L'Oiseau-Lyre DSLO 15.
     +-Gr 11-76 p843   +-RR 12-76 p86

España, op. 165, no. 2: Tango. cf Angel S 36094.

España, op. 165, no. 2: Tango. cf CBS 30063.

Espãņa, op. 165, no. 2: Tango. cf London STS 15239.

España, op. 165, no. 2: Tango. cf L'Oiseau-Lyre DSLO 7.

España, op. 165, no. 2: Tango; no. 3: Malagueña. cf London
CS 6953.

España, op. 165, no. 3: Malagueña. cf DG 3335 182.

8   Iberia. TURINA: Danzas fantásticas. Blanca Uribe, pno.
Orion ORS 75202/3 (2).
     +NR 11-76 p14

Liebesleid. cf Tango.

Pavana-capricho, op. 12. cf London CS 6953.

Piezas características, no. 12: Torre bermeja. cf DG 3335 182.

Recuerdos de viaje, op. 71, no. 5: Puerta de tierra; no. 6:
Rumores de la caleta. cf London CS 6953.

Recuerdas de viaje, op. 71, no. 6: Rumores de la caleta. cf
DG 3335 182.

Recuerdos de viaje, op. 71: Rumores de la caleta. cf London
CS 7015.

Suite española, excerpts. cf Opus unnumbereed.

9   Suite española, no. 3: Sevillanas; no. 4: Cadiz. GRANADOS:
Spanish dances, op. 37: Villanesco, Andaluza, Danza triste.
RODRIGO: Sonata giacosa. TARREGA: Danza mora. Preludio. Sueño.
Vincenzo Macaluso, gtr. Klavier KS 552.
     +NR 7-76 p13

Suite española, no. 3: Sevillanas. cf London CS 6953.

Suite española, no. 3: Sevillanas. cf RCA ARM 4-0942/7.

Suite española, no. 5: Asturias. cf DG 3335 182.

Tango. cf Desmar DSM 1005.

10   Tango (arr. Kreisler). DVORAK: Slavonic dance, no. 3, G major
(arr. Kreisler). FALLA: La vida breve: Danse espagnole (arr.
Kreisler). KREISLER: Allegretto in the style of Boccherini.
Andantino in the style of Martini. Caprice viennois, op. 2.
Chanson Louis XIII and Pavane in the style of Couperin. La
gitana. Liebesfreud. Liebesleid. Recitativo and scherzo
capriccio, op. 6. Rondino on a theme by Beethoven. Schön
Rosmarin. Tambourin chinois, op. 3. Itzhak Perlman, vln;
Samuel Sanders pno. HMV ASD 3258 (also Angel S 37171).
     +Gr 9-76 p478   +RR 9-76 p82
     ++HFN 9-76 p123   ++SFC 7-25-76 P29
     +NR 8-76 p13

Zambra granadina. cf London CS 7015.

ALBENIZ, Mateo

Sonata, guitar. cf RCA SB 6891.

Sonata, piano, D major. cf London CG 6953.

d'ALBERT

Zur Drossel sprach der Fink. cf Rococo 5370.

ALBINONI, Tommaso
    Adagio, G minor. cf Works, selections (Erato STU 70231).
    Adagio, G minor. cf DG 2548 219.
    Adagio, G minor. cf Angel S 37157.
    Adagio, G minor. cf HMV ESD 7011.
    Adagio, G minor. cf Vanguard SRV 344.
    Adagio, organ and strings. cf Abbey LPB 754.
    Adagio, organ and strings, G minor. cf HMV (Q) ASD 3131.
11   Concerto, oboe, op. 7, no. 3, B flat major. Concerto, oboe,
       op. 7, no. 6, D major. BELLINI: Concerto, oboe, E flat major.
       VIVALDI: Concerto, oboe, op. 8, no. 9, D minor. Concerto,
       oboe, op. 9, no. 2, A minor. Eugene Nepalov, ob; Moscow
       Chamber Orchestra; Rudolf Barshai. Westminster WGS 8323
       (Reissues from Melodiya).
           +NR 11-76 p5
    Concerto, oboe, op. 7, no. 6, D major. cf Concerto, oboe, op.7,
       no. 3, B flat major.
    Concerto, oboe, op. 9, no. 2, D minor. cf Works, selections
       (Erato STU 70231).
12   Concerto, oboe and strings, op. 9, no. 8, G minor. TELEMANN:
       Pimpinone. TESSARINI: Concerto, violin, no. 8, op. 1, no. 7,
       B major. VIVALDI: Concerto, op. 7, no. 2. Uta Spreckelsen,
       s; Siegmund Nimsgern, bs; Florilegium Musicum Ensemble;
       Hans Ludwig Hirsch. Telefunken ER 6-35285 (2).
           +RR 4-76 p30
    Concerto, trumpet, B flat major. cf Erato STU 70871.
    Concerto, trumpet, B flat major. cf RCA CRL 3-1430.
13   Concerto, trumpet, C major. HERTEL: Concerto, trumpet, 2 oboes
       and 2 bassoons. TELEMANN: Concerto, trumpet, D major. Gerald
       Schwarz, tpt; Ronald Roseman, Susan Weiner, Virginia Brewer,
       ob; Donald MacCourt, William Scribner, bsn; Edward Brewer, hpd.
       Desto DC 6438 (also Peerless PRCM 209).
           +Gr 1-76 p1187   +HFN 12-75 p148
           +HF 1-76 p97   +St 11-75 p140
    Concerto, trumpet (oboe), op. 7, no. 6, D major. cf Works,
       selections (Erato STU 70231).
    Sonata, C minor. cf Works, selections (Erato STU 70231).
14   Sonata à 5, op. 2, no. 6, G minor. LOCATELLI: Concerto grosso,
       op. 7, no. 12, F major. VIVALDI: Concerto, violin, E flat
       major. Collegium Aureum. BASF HB 29364 (also BASF BAC 3090
       Tape (c) KBACC 3090).
           +Gr 5-76 p1771   ++NR 9-73 p7
           +HFN 7-76 p105 tape   ++RR 5-76 p59
           ++HFN 8-76 p87   ++SFC 11-11-73 p32
           +MQ 4-74 p312
    Sonata à 5, op. 5, no. 9, E minor. cf DG 2548 219.
    Symphony, B flat major. cf Works, selections (Erato STU 70231).
15   Works selections: Adagio, G Minor (arr. Giazotto). Concerto,
       oboe, op. 9, no. 2, D minor. Concerto, trumpet, op. 7, no. 6,
       D major. Symphony, B flat major. Sonata, C minor. Maurice
       André, tpt; Jacques Chambon, ob; Georg-Friedrich Hendel, vln;
       Saar Radio Chamber Orchestra; Karl Ristenpart. Erato STU
       70231.
           +Gr 4-76 p1587   +NR 3-74 p14
           +-HFN 5-76 p91   +RR 4-76 p35

ALBRECHTSBERGER, Johann
    Fugue, op. 16, no. 5, G major. cf Pelca PRS 40577.
    Fugue on B-A-C-H, G minor. cf Pelca PRSRK 41017/20.
16   Partita, harp, F major. HOFFMEISTER: Concerto, flute, D major.
       Janos Szebenyi, flt; Anna Lelkes, hp; Györ Philharmonic Orch-
       estra; János Sándor. Hungaroton LPX 11454.
       +SFC 7-11-76 p13
    Prelude and fugue, B flat major. cf Pelca PRS 40581.
ALCOCK, John
    Voluntary, D major. cf Cambridge CRS 2540.
ALEXANDROV, Boris
    Volga bargehaulers song. cf HMV ASD 3200.
ALFONSO X, El Sabio
    Rosa das rosas. cf DG 2530 504.
    Santa Maria. cf DG 2530 504.
ALFORD, Harry
    The mad major. cf Mercury SRI 75055.
    Purple carnival. cf Columbia 33513.
17   Works, selections: By land and sea; Colonel Bogey; Colonel Bogey
       on parade; Dunedin; H. M. Jollies; The great little army;
       The middy; On the quarter deck; The standard of St. George;
       The thin red line; The voice of the guns. HM Royal Marines
       Band, Plymouth Division; Major F. J. Ricketts. EMI One Up
       OUM 2104.
       +Gr 4-76 p1665
18   Works, selections: By land and sea: Slow march. Eagle squadron:
       March. The lightning switch: Fantasia. Mac and Mac, 2 xylo-
       phones. The mad major: March. A musical switch: Fantasia.
       The smithy: Pastoral fantasia. Thoughts: Waltz. The two imps,
       2 xylophones. The vedette: March. I. Wilder, A. Flook, xylo-
       phone; HM Royal Marines Naval Home Command Band; J. R. Mason.
       Polydor 2383 362. Tape (c) 3170 262.
       +Gr 3-76 p1513
ALFVEN, Hugo
    Aftonen. cf Swedish Society SLT 33188.
19   Festspel, op. 25. Swedish rhapsody, no. 1, op. 19. Swedish
       rhapsody, no. 3, op. 48. Stockholm Philharmonic Orchestra,
       Kungl Hovkapellet; Stig Westerberg, Hugo Alfvén. Swedish
       Society SLT 33145.
       +RR 7-76 p42
20   King Gustav Adolf II, op. 49. Swedish Radio Symphony Orchestra;
       Stig Westerberg. Swedish Society Discofil SLT 33173.
       +Gr 4-76 p1587   +HFN 1-76 p101
21   The mountain king. The prodigal son. Royal Swedish Orchestra;
       Hugo Alfvén. Swedish Discofil SLT 33182.
       +Gr 4-76 p1587   +St 6-75 P48
       +/RR 2-76 p24
    The prodigal son. cf The mountain king.
22   En Skärgårdssägen, op. 20. Symphony, no. 5, op. 55, A minor:
       1st movement. Swedish Radio Symphony Orchestra; Stig Wester-
       berg. Swedish Society SLT 33174.
       +-RR 5-76 p33
    Swedish rhapsody, no. 1, op. 19. cf Festspel, op. 25.
    Swedish rhapsody, no. 3, op. 48. cf Festspel, op. 25.
23   Symphony, no. 1, op. 7, F minor. Swedish Radio Symphony Orch-
       estra; Stig Westerberg. Swedish Society Discofil SLT 33213.
       ++Gr 3-76 p1457   ++HFN 1-76 p101

24   Symphony, no. 2, op. 11, D major. Stockholm Philharmonic Orches-
      tra; Leif Segerstam. Swedish Society SLT 33211.
         +HFN 10-76 p163   +RR 10-76 p38
25   Symphony, no. 3, op. 23, E major. Stockholm Philharmonic Orches-
      tra; Nils Grevillius. Swedish Society SLT 33161.
         +-RR 5-76 p33
26   Symphony, no. 4, op. 39, C minor. Sven Erik Vikström, t; Gunilla
      af Malmborg, s; Stockholm Philharmonic Orchestra; Nils
      Grevillius. Swedish Society Discofil SLT 33186.
         +-RR 2-75 p29   +-RR 5-76 p33
      Symphony, no. 5, op. 55, A minor: 1st movement. cf En Skärgårds-
      sägen, op. 20.
ALISON, Richard
      Dolorosa pavan. cf L'Oiseau-Lyre 12BB 203/6.
ALKAN, Charles-Henri
27   Trio, no. 1, op. 30, G minor. HENSELT: Trio, op. 24, A minor.
      LITOLFF: Trio, no. 1, op. 47, D minor. THALBERG: Trio, op. 69,
      A major. Mirecourt Trio. Genesis GS 1058/9 (2).
         +NR 4-76 p7
ALLISON
      Mancini magic. cf Pye TB 3006.
      Silver threads among the gold. cf Pye TB 3006.
ALLITSEN
      The lute player. cf HMV EMD 5528.
ALNAES, Eyvind
      Songs (4). cf Caprice CAP 1107.
ALONSO (16th century Spain)
      La tricotea Samartin. cf Nonesuch H 71326.
ALTENBURG, Johann
      Concerto, 7 trumpets and timpani: Allegro. cf Argo SPA 464.
AMALIA, Princess of Prussia
      Regimental marches (3). cf Gemini Hall RAP 1010.
      AMBROSIAN CHANT. cf Anonymous Works.
d'AMBROSIO, Alfredo
      Serenade. cf RCA ARM 4-0942/7.
      AMERICA SINGS: THE FOUNDING YEARS, 1620-1800. cf Vox SVBX 5350.
      AMERICA SINGS, THE GREAT SENTIMENTAL AGE--STEPHEN FOSTER TO
      CHARLES IVES. cf Vox SVBX 5304.
      AMERICAN SAMPLER. cf Personal Touch 88.
      AMERICAN SAMPLER. cf Washington University Press OLY 104.
      AMERICAN SONGS FOR A CAPELLA CHOIR. cf Orion ORS 75205.
      THE AMERICAN STRING QUARTET IN U.S.A., 1900-1950. cf Vox
      SVBX 5305.
ANCHIETTA, Juan de
      Con amores, la mi madre. cf DG 2530 598.
ANDERSON
      Clarinet candy. cf Pye GH 603.
      Summer skies. cf Stentorian SP 1735.
ANDERSON, Leroy
      Belle of the ball. cf Pye TB 3004.
ANDERSON, Robert
      Canticle of praise: Te deum. cf Delos FY 025.
ANDERSON, Thomas Jefferson, Jr.
28   Squares. HAKIM: Visions of Ishwara. WILSON: Akwan. Richard
      Bunger, pno; Baltimore Symphony Orchestra; Paul Freeman.
      Columbia M 33434.

```
        +Audio 12-75 p106        ++NR 9-75 p5
        +HF 10-75 p65            +St 5-76 p127
```
Variations on a theme by M. B. Tolson. cf Nonesuch H 71302/2.
ANDREE, Elfride
        Quintet, E major: Allegro molto vivace. cf Gemini Hall RAP 1010.
ANDRIESSEN, Hendrik
        Theme and variations. cf Vista VPS 1035.
ANDROZZO
        If I can help somebody. cf Canon VAR 5968.
d'ANGLEBERT, Jean-Henri
        Chaconne, D major. cf Harmonia Mundi HMU 334.
29   Pièces de clavecin. Kenneth Gilbert, hpd. Harmonia Mundi
        HMU 941.
                +Gr 1-76 p1221   +St 6-76 p100
                ++HFN 1-76 p105
        Le tombeau de M. de Chambonnieres. cf Harmonia Mundi HMU 334.
d'ANGLETERRA, Gallot
30   Carillon, G major. CAMPIAN: Pieces, D major. VISEE: Suite,
        D minor. WEISS: Chaconne, E flat major. Suite, C minor.
        David Rhodes, gtr, lt. Titanic TI 5.
                +NR 7-76 p14
ANNA AMALIA, Duchess of Saxe-Weimar
        Erwin and Elmore, excerpt. cf Gemini Hall RAP 1010.
ANTES, John
        How beautiful upon the mountains. cf Vox SVBX 5350.
        ANTHOLOGY OF ENGLISH SONG. cf Saga 5213.
        THE ANTIPHONAL ORGANS OF THE CATHEDRAL OF FREIBURG. cf Columbia
        M 33514.
APPLEBAUM
        All's well that ends well: The Stratford fanfares; Suite of
        dances. cf Citadel CT 6007.
APPLEBY, Thomas
31   Magnificat. DAVY: Ah mine heart. MASON: Quales sumus; Vae
        nobis miserere. PRESTON: Beatus Laurentius. Magdalen College
        Choir; Ian Crabbe, org; Bernard Rose. Argo ZRG 846.
                +Gr 6-76 p77   +RR 4-76 p67
                +HFN 5-76 p103
ARBEAU, Thoinot
        Basse danse "Jouyssance vous donneray". cf Angel SFO 36895.
        Belle qui tient ma vie. cf HMV CSD 3740.
ARCADELT, Jacques
        Margo labourez les vignes. cf Argo ZRG 667.
ARDITI, Luigi
        Il bacio   cf Rubini GV 68.
        Parla. cf Court Opera Classics CO 342.
        Se saran rose. cf HMV RLS 719.
AREL, Bulent
        Mimiana II: Frieze. cf CRI SD 300.
        Stereo electronic music, no. 2. cf Finnadar QD 9010.
ARENSKY, Anton
32   Concerto, piano, no. 2. PADEREWSKI: Fantaisie polonaise, op. 19.
        Felicja Blumenthal, pno; Innsbruck Symphony Orchestra; Robert
        Wagner. Everest SDBR 3376.
                +-Gr 1-76 p1187   +NR 5-75 p7
        Concerto, violin, A minor: Tempo di valse. cf RCA ARM 4-0942/7.
33   Concerto, violin, op. 54, A minor. RIMSKY-KORSAKOV: Fantasy on
```
```

Russian themes, op. 33. WIENIAWSKI: Polonaise, violin. Aaron
Rosand, vln; Luxembourg Radio Orchestra; Louis de Froment.
Turnabout QTV 34629.
          +NR 12-76 p7
Concerto waltz. cf Odyssey Y 33825.
Suite, 2 pianos, op. 15: Waltz. cf HMV SXLP 30181.

ARGENTO, Dominick
34   To be sung upon the water. ROREM: King Midas. Sandra Walker,
      ms; John Stewart, t; Charles Russo, clt; Donald Hassard, Ann
      Schein, pno. Desto DC 6443.
          +St 7-76 p106

ARKANGELSKII, Alexandr
      Nunc Dimittis; Polveleos. cf Ikon IKO 3
      Outoli bolzni. cf Harmonia Mundi HMU 105.
      Songs: Brajen Mouj; Pomichliaiou dien strachniy. cf Harmonia
      Mundi HMU 137.

ARMSTRONG, Thomas
      Christ whose glory fills the skies. cf Vista VPS 1023.

ARNE, Thomas
      Artaxerxe: The soldier tir'd of war's alarms. cf Westminister
      WGS 8268.
      Artaxerxes: Water parted from the sea. cf Smithsonian Collection
      N 002.
      As you like it: Under the greenwood tree. cf HMV SLS 5022 (2).
      Concerto, organ: Con spirito. cf Wealden WS 142.
35   Concerti, organ, nos. 4-6. Jean Guillou, org; Berlin Brandenberg
      Orchestra; René Klopfenstein. Philips 6581 016. (Reissue
      from Philips 4FO 7009).
          +-Gr 4-76 p1587   +-RR 9-76 p42
          ++HFN 3-76 p109
      Flute tune. cf Stentorian SC 1724.
36   Overtures, nos. 1-8. Academy of Ancient Music; Christopher
      Hogwood, cond and hpd. L'Oiseau-Lyre Florilegium DSLO 503.
          +-Audio 9-76 p85   +NR 6-75 p6
          +Gr 11-74 p883   +RR 11-74 p17
          ++HF 8-75 p79   +SFC 11-30-75 p34
      Songs: Blow, blow thou winter wind; Come away death. cf National
      Trust NT 002.
      Songs: Where the bee sucks. cf HMV ESD 7002.
37   Symphony, no. 1, C major. Symphony, no. 2, F major. Symphony,
      no. 3, E flat major. Symphony, no. 4, C minor. WESLEY, S.:
      Symphony, D major. Bournemouth Sinfonietta Orchestra;
      Kenneth Montgomery. HMV SQ CSD 3767.
          ++Gr 3-76 p1457   +MM 9-76 p36
          +HFN 6-76 p79   +RR 3-76 p30
      Symphony, no. 2, F major. cf Symphony, no. 1, C major
      Symphony, no. 3, E flat major. cf Symphony, no. 1, C major.
      Symphony, no. 4, C minor. cf Symphony, no. 1, C major.

ARNOLD
      Songs: Sara. cf Columbia SCX 6585.

ARNOLD, Malcolm
      Concerto, guitar, op. 67. cf RCA ARL 3-0997.
      Concerto, 2 violins and strings, op. 77. cf ADDISON: Concerto,
      trumpet, strings and percussion.
      Quintet, brass: Con brio. cf Argo SPA 464.
      Scottish dances, op. 59. cf RCA GL 25006.

ARRIAGA Y BALZOLA, Juan de
38   Quartets, strings, nos. 1-3. Chilingirian Quartet. CRD CRD
         1012/3 (2).
                  +Gr 6-75 p56   +-St 3-76 p108
                  +HFN 9-75 p91E   ++STL 1-4-76 P36
                  +RR 6-75 p57
39   Symphony, D major. SCHMIDT: Variations on a Hussar's song.
         NPhO; Hans Bauer. HMV CSD 3769.
                  +-Gr 4-76 p1588   +-MT 10-76 p831
                  ++HFN 4-76 p99   +RR 4-76 p35
ARS ANTIQUA
         Motets from the Bamberg and Montpellier Codices. Hoqueti: Instru-
            mental pieces. cf LEONIN: Organa.
ARS NOVA
         Motets from the Roman de Fauvel and the Ivrea and Chantilly
            Codices. cf LEONIN: Organa.
         THE ART OF COURTLY LOVE. cf HMV SLS 863.
         THE ART OF SHERRILL MILNES, OPERATIC ARIAS, DUETS AND SONGS.
            cf RCA LRL 2-7531
         THE ART OF THE NETHERLANDS. cf HMV SLS 5049.
         THE ART OF THE RECORDER. cf HMV SLS 5022 (2).
         THE ART SONG IN AMERICA. cf Duke University Press DWR 7306.
ARUTYUNIAN, Aleksander
40   Concerto, trumpet and orchestra, A flat major. KRIUKOV: Concerto-
            poem, trumpet and orchestra, op. 59. VAINBERG: Concerto,
            trumpet and orchestra, op. 94, B flat major. Timofey Dok-
            schitser, tpt; Bolshoi Theatre Orchestra; Gennady Rozhdestvensky,
            Algis Zuraitis. Melodiya/Angel SR 40149. (also HMV Melodiya
            ASD 3236).
                  +ARG 7-72 p552   +HFN 8-76 p75
                  +Gr 10-76 p615   ++NR 1-72 p5
                  ++HF 2-72 p110   +RR 8-76 p26
ASENCIO
         Dipso. cf RCA ARL 1-0864.
ASMUSSEN
         So sorry. cf Swedish Society SLT 33197.
ASTAFEV
         Paschal Troparion (Christ is risen). cf Ikon IKO 2.
ASTON, Hugh
         Hornpype. cf Argo SPA 464.
ATKINS, Ivor
         Psalm, no. 107, O give thanks unto the Lord. cf HMV SQ CSD 3768.
ATTAINGNANT, Pierre
         Au joly bois. cf Hungaroton SLPX 11669/70.
41   Basse dance. Tant que vivray. BALLARD: Branles de village.
            Courante. Entrée de Luth, nos. 1-3. BESARD: Air de cour.
            Allemande. Ballet. Branle. Branle gay (2). Chorea rustica.
            Gagliarda. Gagliarda vulgo dolorata. Gesamtzeit mit Pausen.
            Guillemette. Pass'e mezo. Volte. Le ROY: Basse dance.
            Branle gay. Destre amoureux. Haulberroys. Passemeze. Konrad
            Ragossnig, renaissance lute. DG Archive 2533 304.
                  +Gr 6-76 p70   +NR 7-76 p14
                  ++HFN 5-76 p103   +RR 7-76 p69
         Basse danse La brosse and recoupe. cf Turnabout TV 34137.
         Bransle. cf CBS 76183.
         C'est grand plaisir, phantasia. cf Hungaroton SLPX 11669/70.

Chanson. cf Turnabout TV 34137.
Content desire. cf Hungaroton SLPX 11669/70.
Content desir basse danse. cf L'Oiseau-Lyre 12BB 203/6.
Gaillarde. cf Turnabout TV 34137.
Pavane. cf Turnabout TV 34137.
Pavane et galliarde. cf Harmonia Mundi HMU 610.
Tant que vivray. cf Basse dance.
Tant que vivray. cf Turnabout TV 34137.
Tordion. cf Harmonia Mundi HMU 610.
Tordion. cf Turnabout TV 34137.

ATTERBERG, Kurt
42   De favitska jungfrurna. LINDBERG: Leksandssvit. Stockholm
          Philharmonic Orchestra; Nils Grevillius. Swedish Society
          SLT 33192.
              +-RR 2-76 p24
43   Suite, violin, viola and orchestra, op. 19, no. 1. FRUMERIE:
          Pastoral suite, flute, harp and strings. WIREN: Sinfonietta,
          op. 7a, C major. Mircea Saulesco, vln; Gideon Roehr, vla;
          Börge Mårelius, flt; Swedish Radio Symphony Orchestra; Stig
          Westerberg. Swedish Society SLT 33167.
              +RR 5-76 p33
44   Suite Barocco, op. 23. Suite pastorale, op. 34. Swedish Radio
          Symphony Orchestra; Kurt Atterberg. Swedish Society SLT 33175.
              -RR 12-76 p50
      Suite pastorale, op. 34. cf Suite Barocco, op. 23.
45   Symphony, no. 2, op. 6, F major. Swedish Radio Symphony Orchestra;
          Stig Westerberg. Swedish Society SLT 33179.
              +Gr 1-76 p1187   +-RR 2-75 p29
              +HFN 11-75 p148

AUBER, Daniel
      Fra Diavolo: Romanze. cf Rococo 5379.
      Manon Lescaut: C'est l'histoire amoureuse. cf Decca GOSD 674/6.
46   Marco spada. LSO; Richard Bonynge. Decca SXL 6707 (also London
          6923).
              +Gr 7-75 p173   +RR 7-75 p21
              +HF 1-76 p81   ++SFC 4-4-76 p34
              +HFN 7-75 p71   +SR 1-24-76 p53
              +NR 2-76 p3

AUGMENTED HISTORY OF THE VIOLIN ON RECORDS, 1920-1950. cf
      Thomas L. Clear TC 2580.

AURIC, Georges
      Adieu, New York. cf Supraphon 111 1721/2.

AZEVADO
      Delicada. cf Transatlantic XTRA 1160.

AZZAIOLO, Filippo
      Al di dolce. cf Hungaroton SLPX 11669/70.
      Chi passa. cf Hungaroton SLPX 11669/70.
      Quando le sera; Sentomi la formicula. cf L'Oiseau-Lyre 12BB 203/6.

BABBITT, Milton
      All set. cf Nonesuch H 71302/3.
      Ensembles for synthesizer. cf Finnadar QD 9010.
47   Quartet, strings, no. 3. WUORINEN: Quartet, strings. Fine Arts
          Quartet. Turnabout TV 34515.
              +Gr 9-74 p529   +RR 7-74 p48
              ++HF 5-73 p73   ++SFC 7-1-73 p31
              ++NR 3-73 p7   ++St 5-73 p107
              +NYT 4-5-73 pC54   ++St 7-76 p75
              +NYT 6-17-73 pD30

Sextets. cf Desto DC 6435/7.
BACARISSE, Salvador
  Concertino, guitar and orchestra, op. 72, A minor. cf HALFFTER:
    Concerto, guitar.
BACEWICZ, Grazyna
  Sonata, piano, no. 2. cf Avant AV 1012.
BACH, Carl Philipp Emanuel
  Adagio affettuoso e sostenuto, A flat major. cf Works, selections
    (Orion ORS 76223).
  Adagio assai, B flat minor. cf Works, selections (Orion ORS 76223).
  Allegro di molto, F minor. cf Works, selections (Orion ORS 76223).
48 Concerto 4 harpsichords, F major. BACH, J. S.: Concerto, 3 harpsi-
    chords, S 1063, D minor. Concerto, 4 harpsichords, S 1065,
    A minor. MALCOLM: Variations on a theme by Mozart for 4 harpsi-
    chords. George Malcolm, Valda Aveling, Geoffrey Parsons, hpd;
    Simon Preston, hpd continuo; ECO; Raymond Leppard. Decca
    SDD 451. (Reissue from SXL 6318).
          +Gr 4-76 p1588   +RR 3-76 p33
          +HFN 3-76 p109
49 Concerto, oboe, W 164, B flat major. Concerto, oboe, W 165, E flat
    major. BACH, J. S.: Cantata, no. 12, Weinen, Klagen, Sorgen,
    Zagen: Sinfonia. Cantata, no. 21, Ich hatte viel Bekümmernis:
    Sinfonia. Heinz Hollinger, ob; ECO; Raymond Leppard. Philips
    6500 830.
          +Gr 2-76 p1331   ++NR 12-75 p7
          ++HF 2-76 p87   +RR 2-76 p24
          ++HFN 3-76 p87   ++St 12-75 p116
  Concerto, oboe, E flat major. cf Concerto, oboe, B flat major.
50 Concerto, piano, A major. SOLER: Fandango, D minor. Sonata,
    harpsichord, D major. Doyle White, pno; NPhO; Paul Freemand.
    Orion ORS 76225.
          -NR 11-76 p5
  Fantasias, F major, C major, C minor. cf Works, selections
    (Orion ORS 76223).
51 Fantasias, W 58/6, E flat; W 59/5, F major. Rondos, W 56/1, C major;
    W 57/1, E major; W 61/4, D minor. Sonatas, fortepiano, W 55/1,
    C major; W 57/4, D minor. Huguette Dreyfus, fortepiano.
    DG Archive 2533 327.
          +RR 12-76 p76
52 Minuets, Wq 189 (2). Polonaises, Wq 190 (5). RAMEAU: Zoroastre:
    Dances (7). STARZER: Diane et Endimione: Dances. Roger et
    Bradamante: Dances. Gli Orazi e Gli Curiazi: Dances. Eduard
    Melkus Ensemble; Eduard Melkus. DG Archive 2533 303.
          +Audio 4-76 p88   +NR 2-76 p7
          +Gr 8-76 p306   +RR 8-76 p27
          +HFN 7-76 p94   +St 8-76 p106
          +MJ 1-76 p39
  Polonaises, WQ 190. cf Minuets, Wq 189.
  Rondos, W 56/1, C major; W 57/1, E major; W 61/4, D minor. cf
    Fantasias, W 58/6, E flat; W 59/5, F major.
  Sonatas, fortepiano, W 55/1, C major; W 57/4, D minor. cf
    Fantasias, W 58/6, E flat; W 59/5, F major.
  Sonata, harpsichord, no. 1. cf Works, selections (Orion ORS 76223).
  Sonata, organ, F major. cf Pelca PRSRK 41017/20.
53 Sonatas, organ, nos. 1-6. Xavier Darasse, org. Arion ARN 236
    006 (2).
          +HFN 1-76 p101

54   Trio sonatas (4). Ars Rediviva Ensemble; Milan Munclinger.
        Supraphon 111 0640.
            ++SFC 8-1-76 p27
55   Works, selections: Adagio affettuoso e sostenuto, A flat major.
        Adagio assai, B flat minor. Allegro di molto, F minor. Fan-
        tasias, F major, C major, C minor. Sonata, harpsichord, no. 1.
        Joan Benson, pno and cld. Orion ORS 76223.
            +-NR 11-76 p12
BACH, Johann Bernhard
56   Du Friedensfürst, Herr Jesu Christ, partita.** Passacaglia
        (Chaconne), B major.** BACH, Johann Christoph: Aus meines
        Herzens Grunde. Prelude and fugue, E flat major.** Wach auf,
        mein Herz und singe.** Warum betrübst du dich, mein Herz.**
        BACH, Johann Ernst: Fantasia and fugue, F major.** BACH, Johann
        Lorenz: Prelude and fugue, D major. BACH, Johann Michael:
        Chorale preludes: Allein Gott in der Höh sei Ehr;** Wenn
        wir in höchsten Nöten sein. BACH, Johann Sebastian: Capriccio,
        S 993, E major.** Fantasia and fugue, S 904, A minor.* Chorale
        prelude, Wenn wir in höchsten Nöten sein, S Anh 78.** Prelude
        and fugue, S 542, G minor.* Prelude and fugue, S 547, C major.*
        Toccata and fugue, S 565, D minor.* Wilhelm Krumbach, org.
        Telefunken 6-35273 (2). (*Reissue from SAWT 9503; **Reissue
        from SAWT 9551).
            +-Gr 2-76 p1359
        Passacaglia (Chaconne), B major. cf Du Fiedensfürst, Herr Jesu
        Christ, partita.
BACH, Johann Christian
        Aria with 15 variations, A minor. cf Pelca PRSRK 41017/20.
57   Concerto, harpsichord, A major. HAYDN: Concerto, harpsichord,
        op. 21, D major. George Malcolm, hpd; AMF; Neville Marriner.
        London STS 15172.
            ++NR 6-75 p6   ++St 6-76 p103
            ++SFC 2-1-76 p29
58   Concerti, keyboard, op. 7, nos. 1-3; op. 13, no. 4. Ingrid Haebler,
        fortepiano; Vienna Capella Academica; Eduard Melkus. Philips
        6500 846.
            ++Gr 7-75 p173   ++MT 2-76 p139
            ++HFN 10-75 p135   ++RR 6-75 p29
        Concerto, keyboard, op. 13, no. 4. cf Concerti, keyboard, op. 7,
        nos. 1-3.
        Duets, A major, F major. cf Sonatas, flute and piano, op. 18,
        nos. 1-4.
        Quartet, 2 flutes, viola and violoncello, D major. cf Quartet,
        2 flutes, violin and violoncello, C major.
59   Quartet, 2 flutes, violin and violoncello, C major. Quartet,
        2 flutes, viola and violoncello, D major. BACH, W. F.: Duet,
        flute, F major. BACH, W. F. E.: Trio, 2 flutes and viola,
        G major. Jean-Pierre Rampal, Eugenia Zukerman, flt; Pinchas
        Zukerman, vln, vla; Charles Tunnell, vlc. Columbia M 33310
        (also CBS 73497).
            +Gr 8-76 p306   ++NR 8-75 p8
            ++HF 11-75 p96   +RR 5-76 p60
            +HFN 6-76 p79   +St 2-76 p98
60   Sinfonia, op. 6, no. 6, C minor. HAYDN, M.: Symphony, G major
        (with introduction by Mozart, K 444). MOZART: Cassation
        (serenade), no. 1, K 62a, D major (with March K 62, D major).

St. Paul Chamber Orchestra; Dennis Russell Davies. Nonesuch
    H 71323.
        +-Gr 11-76 p775   +-NR 8-76 p3
        +HF 8-76 p91   +RR 11-76 p69
        +HFN 11-76 p155   +SFC 7-18-76 p31
        +IN 12-76 p20
61   Sinfonias, op. 18, nos. 2, 4, 6. TELEMANN: Don Quichotte.
        Stuttgart Chamber Orchestra; Karl Münchinger. Decca SXL 6755.
        (also London 6988).
        +-Gr 10-76 p577   +-RR 9-76 p43
        +-HFN 10-76 p163
62   Sonatas, flute and piano, op. 18, nos. 1-4. Duets, A major, F
        major. Nicholas McGegan, flt; Christopher Hogwood, Colin
        Tilney, fortepiano. L'Oiseau-Lyre DSLO 516.
        +Gr 10-76 p616   ++RR 10-76 p85
        +HFN 11-76 p155
63   Sonatas, flute and fortepiano, op. 19, nos. 1-6. Ingrid Haebler,
        fortepiano; Kurt Redel, flt. Philips 6500 849.
        +-Gr 5-76 p1772   +-RR 5-76 p63
        +HFN 5-76 p92   +STL 7-4-76 p36
     Sonata, harpsichord, op. 18, no. 2, A major. cf Smithsonian
        Collection N 002.
64   Symphonies, winds, nos. 1-6. Camden Wind Ensemble. Pye GSGC 15029.
        +-Gr 8-76 p278   +-RR 8-76 p27
     Symphony, op. 3, no. 1, D major. cf Philips 6580 114.
BACH, Johann Christoph
     Aus meines Herzens Grunde. cf BACH, J. B.: Du Friedensfürst,
        Herr Jesu Christ, partita.
     Prelude and fugue, E flat major. cf BACH, J. B.: Du Friedens-
        fürst, Herr Jesu Christ, partita.
     Songs: Ach, dass ich Wassers genug hätte. cf Telefunken 6-41929.
     Wach auf, mein Herz und singe. cf BACH, J. B.: Du Friedensfürst,
        Herr Jesu Christ, partita.
     Warum betrübst du dich, mein Herz. cf BACH, J. B.: Du Friedens-
        fürst, Herr Jesu Christ, partita.
BACH, Johann Ernst
     Fantasia and fugue, F major. cf BACH, J. B.: Du Friedensfürst,
        Herr Jesu Christ, partita.
BACH, Johann Lorenz
     Prelude and fugue, D major. cf BACH, J. B.: Du Friedensfürst,
        Herr Jesu Christ, partita.
BACH, Johann Michael
     Chorale preludes: Allein Gott in der Höh sei Ehr; Wenn wir in
        höchsten Noten sein. cf BACH, J. B.: Du Friedensfürst, Herr
        Jesu Christ, partita.
BACH, Johann Sebastian
65   Allabreve, S 589, D major. Canzona, S 588, D minor. Fugues,
        S 575, 577/9. Pastorale, S 590, F major. Lionel Rogg, org.
        Oryx 1002.
        +-St 12-76 p72
     Allabreve, S 589, D major. cf Organ works (DG 2533 140).
     Allabreve, S 589, D major. cf Organ works (DG 2722 014).
     Anna Magdalena notebook, S 508, excerpts. cf RCA ARL 1-1323.
     Anna Magdalena notebook, S 508: 3 pieces. cf Philips 6581 017.
     Anna Magdalena notebook, S 508: Bist du bei mir. cf Works,
        selections (HMV SQ ASD 3265).

Anna Magdalena notebook, S 508: Bist du bei mir. cf Stentorian
SC 1724.

Anna Magdalena notebook, S 508: Gib dich zufrieden und sei still:
Wie wohl ist mir, O Freund der Seelen. cf Songs (DG Archive
2533 299).

Anna Magdalena notebook, S 508: Musette, March. cf Saga 5426.

Aria, flute and strings. cf Brandenburg concerto, no. 4, S 1049,
G major.

Aria variata alla maniera Italiana, S 989, A minor. cf Harpsichord
works (DG 2722 020).

Arioso. cf London STS 15239.

Ave Maria. cf Decca SXL 6781.

Ave Maria. cf HMV RLS 719.

66   Brandenburg concerti, nos. 1-6, S 1046-51. ECO; Benjamin Britten.
Decca SXL 6774/5 (2). Tape (c) KSXC 6774/5. (Reissues from
SET 410/1).
      +Gr 8-76 p277   +RR 6-76 p32
      +HFN 7-76 p103   ++RR 8-76 p83 tape
      +HFN 8-76 p94

67   Brandenburg concerti, nos. 1-6, S 1046-51. Jean-François Paillard
Chamber Orchestra; Jean-François Paillard. Erato STU 70801/2 (2).
      ++Gr 7-74 P190   +RR 12-73 p53
      +-Gr 7-76 p167   +-RR 6-76 p32
      +-HFN 8-76 p93

68   Brandenburg concerti, nos. 1-6, S 1046-1051. Boyd Neel Chamber
Orchestra; Boyd Neel. Olympic 8131/2 (2) (Reissue).
      +Audio 6-76 p97

69   Brandenburg concerti, nos. 1-6. S 1046-1051. ECO; Raymond
Leppard. Philips 6747 166 (2). Tape (c) 7699 006.
      ++Gr 4-76 p1587   ++RR 3-76 p30
      +-Gr 12-76 p1066 tape   +-RR 11-76 p108 tape
      ++HF 6-76 p75   +SFC 8-1-76 p27
      +-HFN 4-76 p99   ++St 7-76 p106
      +MJ 10-76 p24   ++STL 6-6-76 p37
      +NR 6-76 p5

70   Brandenburg concerti, nos. 1-6, S 1046-51. Virtuosi of England;
Arthur Davison. Vanguard SRV 313/4 (2). (Also Classics for
Pleasure CFP 40010/1. Tape (c) TC CFP 40010/1).
      -Gr 10-72 p679   +NR 7-75 p3
      +HF 9-75 p81   +RR 10-72 p53
      +-HFN 9-72 p1659   ++St 7-76 p106
      +HFN 12-76 p153 tape

71   Brandenburg concerti, nos. 1-6, S 1046-51. ECO; Johannes Somary.
Vanguard VSD 71208/9 (2). (Q) VSQ 30049/50 (2).
      ++HF 3-76 p79   ++SFC 11-30-75 p34
      +NR 2-76 p3   ++St 4-76 p110

72   Brandenburg concerti, nos. 1-3, S 1046-48. BSO; Charles Munch.
Camden CCV 5007. (Reissue).
      -ST 2-76 p741

Brandenburg concerto, no. 2, S 1047, F major. cf Works, selections
(Decca DPA 535/6).

Brandenburg concerto, no. 3, S 1048, G major. cf Philips 6580 114.

73   Brandenburg concerto, no. 4, S 1049, G major. Aria, flute and
strings. Suite, flute, no. 2, S 1067, B minor. Jean-Pierre
Rampal, flt; Baroque Chamber Ensemble. Everest SDBR 3383.
      +NR 4-76 p6

74   Brandenburg concerto, no. 5, S 1050, D major. Chorale preludes:
        Ich ruf zu dir, Herr Jesu Christ, S 639; Nun komm der Heiden
        Heiland, S 699; Wir glauben all an einen Gott, S 680. PO;
        Leopold Stokowski. CBS 30061.
                +-Gr 1-76 p1187   +-RR 12-75 p35
                +HFN 1-76 p123
75   Brandenburg concerto, no. 5, S 1050, D major. HAYDN: Concerto,
        harpsichord, D major.* Igor Kipnis, Colin Tilney, hpd; Hans-
        Martin Linde, flt; Neville Marriner, vln; London Strings;
        Neville Marriner. CBS 61694. (*Reissue from 61067).
                +-Gr 8-76 p277   +RR 7-76 p49
                +-HFN 7-76 p83
76   Canonic variations, S 769. Fantasias, S 562, S 572, C minor and
        G major. Toccata and fugue, S 566, E major. Trio, S 583,
        D minor. Lionel Rogg, org. Oryx 1003.
                +-St 12-76 p72
77   Cantatas, S 1, 4, 6, 11 (Ascension oratorio), 12, 44, 61, 64-65,
        67, 104, 111, 121, 124, 158, 171, 182. Christmas oratorio,
        S 248. Magnificat, S 243, D major. Edith Mathis, s; Anna
        Reynolds, ms; Peter Schreier, t; Dietrich Fischer-Dieskau, bar;
        Munich Bach Orchestras and Choirs; Karl Richter. DG Archive
        2722 018 (11). (Reissues from SAPM 198 353/5, 198197, 2722005,
        198465).
                +Gr 1-76 p1222   +-MT 1-76 p39
                +-HFN 12-75 p147   +-RR 1-76 p53
                +-MM 7-76 p31
        Cantata, no. 6, Bleib bei uns: Hochgelobter Gottessohn. cf Works,
        selections (HMV SQ ASD 3265).
78   Cantata, no. 10, Meine Seele erhebt den Herren. Cantata, no. 24,
        Ein ungefärbt Gemüte. Cantata, no 135, Ach Herr, mich armen
        Sünder. Edith Mathis, s; Anna Reynolds, con; Peter Schreier, t;
        Kurt Moll, Dietrich Fischer-Dieskau, bs; Munich Bach Orchestra
        and Choir; Karl Richter. DG Archive 2533 329.
                +Gr 12-76 p1027   +-RR 12-76 p88
        Cantata, no. 11, Lobet Gott: Ach, bleibe doch. cf Works, selec-
        tions (HMV SQ ASD 3265).
        Cantata, no. 12, Weinen, Klagen, Sorgen, Zagen: Sinfonia. cf
        BACH C. P. E: Concerto, oboe, B flat major.
79   Cantatas (Advent and Christmas), nos. 13, 28, 58, 61, 63-65, 81-82,
        111, 121, 124, 132, 171. Lotte Schädle, Edith Mathis, Sheila
        Armstrong, s; Hertha Töpper, Anna Reynolds, alto; Peter
        Schreier, Ernst Häfliger, t; Dietrich Fischer-Dieskau, Theo
        Adam, bs; MB and Choir; Karl Richter. DG Archive 2722 005 (6).
        (Reissues).
                +Gr 11-72 p935   +RR 11-72 p93
                +-HF 2-73 p74   +RR 10-76 p92
                +HFN 12-72 p2436   +St 2-73 p111
                ++NR 2-73 p8
80   Cantata, no. 18: Sinfonia. Cantata, no. 21: Sinfonia. Cantata,
        no. 29: Sinfonia. Cantata, no. 31: Sonata. Cantata, no. 35:
        Sinfonia I and II. Cantata, no. 42: Sinfonia. Cantata, no. 49:
        Sinfonia. Cantata, no. 248: Sinfonia. VCM; Nikolaus Harnoncourt.
        Telefunken AF 6-41970.
                ++RR 7-76 p42
81   Cantatas, nos. 21, 34, 39, 51, 55-56, 60, 68, 76, 93, 106, 129,
        175, 189, 201, 211-212. Edith Mathis, Adele Stolte, Elizabeth

Speiser, s; Anna Reynolds, Hertha Töpper, Eva Fleischer,
Ingeborg Springer, ms; Peter Schreier, Ernst Häfliger, Hans-
Joachim Rotzsch, t; Dietrich Fischer-Dieskau, Kurt Moll, Theo
Adam, Gunter Leib, Siegfried Lorenz, bs; Munich Bach Orchestra
and Choir; Leipzig Gewandhaus Orchestra; Berlin Solisten-
vereinigung; Berlin Chamber Orchestra; Kurt Thomas, Helmut Koch.
DG Archive 2722 019 (11).

    +Gr 12-75 p1086   +MM 7-76 p31
    +-HFN 1-76 p102   +-RR 12-75 p87

Cantata, no. 21, Ich hatte viel Bekümmernis: Sinfonia. cf BACH,
C. P. E.: Concerto, oboe, B flat major.

Cantata, no. 21: Sinfonia. cf Cantata, no. 18: Sinfonia.

82   Cantata, no. 23, Du wahrer Gott and Davids Sohn. Cantata, no. 87,
Bisher habt ihr nichts gebeten in meinem Namen. Edith Mathis,
s; Anna Reynolds, con; Peter Schreier, t; Dietrich Fischer-
Dieskau, bs; Munich Bach Orchestra and Choir; Karl Richter.
DG Archive 2533 313.

    -Gr 12-76 p1027   -RR 12-76 p88

Cantata, no. 24, Ein ungefärbt Gemüte. cf Cantata, no. 10, Meine
Seele erhebt den Herren.

Cantata, no. 29, Wir danken dir, Gott: Sinfonia. cf Works,
selections (DG 2530 647).

Cantata, no. 29: Sinfonia. cf Cantata, no. 18: Sinfonia.

Cantata, no. 29, Wir danken dir, Gott: Sinfonia. cf Polydor
2460 262.

Cantata, no. 31: Sonata. cf Cantata, no. 18: Sinfonia.

83   Cantata, no. 34, O ewiges Feuer, O Ursprung der Liebe. Cantata,
no. 68, Also hat Gott die Welt geliebt. Cantata, no. 175, Er
rufet seinen Schafen mit Namen. Edith Mathis, s; Anna Reynolds,
con; Peter Schreier, t; Dietrich Fischer-Dieskau, bs; Munich
Bach Orchestra and Choir; Karl Richter. DG Archive 2533 306
(Reissue from 1711 019).

    +Gr 12-76 p1028   -RR 12-76 p88

Cantata, no. 34, O ewiger Feuer: Wohl euch, ir auserwählten Seelen.
cf Works, selections (HMV SQ ASD 3265).

Cantata, no. 35: Sinfonia I and II. cf Cantata, no. 18: Sinfonia.

84   Cantatas, nos. 39-42. René Jacobs, Paul Esswood, c-t; Marius von
Altena, Kurt Equiluz, t; Max von Egmont, Ruud van der Meer, bs;
Hanover Boys' Choir, Leonhardt Consort, Vienna Boys' Choir;
Chorus Viennensis; VCM: Gustav Leonhardt, Hans Gillesberger,
Nikolaus Harnoncourt. Telefunken EX 6-35269 (2). (also SKW 11).

    +AR 5-76 p27   ++NR 10-75 p6
    ++Gr 8-75 p351   +RR 9-75 p60
    ++HF 2-76 p88   ++STL 3-7-76 p37
    ++HFN 7-75 p72

Cantata, no. 42: Sinfonia. cf Cantata, no. 18: Sinfonia.

85   Cantatas, nos. 43-46. Peter Jelosits, treble; Paul Esswood, René
Jacobs, c-t; Kurt Equiluz, t; Ruud van der Meer, Hanns-Friedrich
Kunz, bs; Vienna Boys' Choir, Chorus Viennensis, Hanover Boys'
Choir; VCM, Leonhardt Consort; Nikolaus Harnoncourt, Gustav
Leonhardt. Telefunken 6-35283 (2). (also SWK 12).

    +Gr 4-76 p1645   ++NR 12-75 p10
    ++HFN 3-76 p88   +RR 3-76 p64
    ++MJ 10-76 p24   ++STL 3-7-76 p37

86   Cantatas, nos. 47-50. Soloists; VCM; Chorus Viennensis; Vienna
Boys' Choir; Nikolaus Harnoncourt. Telefunken 6-35284.

    +Gr 7-76 p199   ++NR 2-76
    +HFN 6-76 p79   +RR 7-76 p73
    ++MJ 10-76 p24
Cantata, no. 49: Sinfonia. cf Cantata, no. 18: Sinfonia.

87   Cantatas, nos. 51-52, 54-56. Marianne Kweksilber, s; Seppi
    Kronwitter, boy soprano; Paul Esswood, c-t; Kurt Equiluz, t;
    Michael Schopper, bs; Hanover Boys' Choir; Leonhardt Consort;
    Gustav Leonhardt. Telefunken 6-35304. (also SKW 14) (2).
        +Gr 12-76 p1028   ++MJ 10-76 p24
        +-HF 8-76 p79   ++NR 9-76 p8
        +HFN 12-76 p136   +-RR 11-76 p92

88   Cantata, no. 51, Jauchzet Gott in allen Landon. SCARLATTI, A.:
    Su le sponde del Tebro. Carole Bogard, s; Armando Ghitalla,
    tpt; Copenhagen Chamber Orchestra; John Moriarty. Cambridge
    CRS 2710.
        ++AR 5-76 p27

89   Cantata, no. 67, Halt, halt im Gedächtnis Jesum Christ. Cantata,
    no. 101, Nimm von uns, Herr, du treuer Gott. Cantata, no. 130,
    Herr Gott, dich loben alle wir. Elly Ameling, s; Helen Watts,
    con; Werner Krenn, t; Tom Krause, bar; Lausanne Pro Arte Chorus;
    OSR; Ernest Ansermet. Decca ECS 790. (Reissue from SXL 6392).
        +-Gr 11-76 p844   +-RR 10-76 p92
        +-HFN 11-76 p173
  Cantata, no. 67, Halt im Gedächtnis Jesum Christ. cf Vista VPS
    1023.
  Cantata, no. 68, Also hat Gott die Welt geliebt. cf Cantata, no.
    34, O ewiges Feuer, O Ursprung der Liebe.

90   Cantata, no. 69, Lobe den Herren, meine Seele. Cantata, no. 120,
    Gott, man lobet dich in der Stille. Louisville Bach Society;
    Melvin Dickinson. Rivergate LP 1002.
        ++MU 10-76 p14

91   Cantata, no. 79, Gott, der Herr, ist Sonn und Schild. Cantata,
    no. 80, Ein feste Burg ist unser Gott. Elly Ameling, s;
    Janet Baker, alto; Theo Altmeyer, t; Hans Sotin, bs; South
    German Madrigal Choir; Consortium Musicum; Wolfgang Gönnenwein.
    Seraphim S 60248.
        +NR 3-76 p6   +St 1-76 p102
  Cantata, no. 80, Ein feste Burg ist unser Gott. cf Cantata, no.
    79, Gott, der Herr, ist Sonn und Schild.
  Cantata, no. 87, Bisher habt ihr nichts gebeten in meinem Namen.
    cf Cantata, no. 23, Du wahrer Gott und Davids Sohn.

92   Cantata, no. 92, Ich hab in Gottes Herz und Sinn. Cantata, no.
    126, Erhalt uns, Herr, bei deinem Wort. Edith Mathis, s;
    Anna Reynolds, ms; Peter Schreier, t; Dietrich Fischer-Dieskau,
    Theo Adam, bs; Munich Bach Choir; Munich Bach Orchestra; Karl
    Richter. DG Archive 2533 312.
        ++DG 7-76 p199   +RR 8-76 p72
        +HFN 8-76 p75
  Cantata, no. 99: Opening chorus. cf Coronet LPS 3031.
  Cantata, no. 101, Nimm von uns, Herr, du treuer Gott. cf Cantata,
    no. 67, Halt, halt im Gedächtnis Jesum Christ.
  Cantata, no. 106, Sonatina, Magnificat, S 243, D major: Esurientes.
    cf HMV SLS 5022.
  Cantata, no. 120, Gott man lobet dich in der Stille. cf Cantata,
    no. 69, Lobe den Herren, meine Seele.
  Cantata, no. 126, Erhalt uns, Herr, bei deinem Wort. cf Cantata,
    no. 92, Ich hab in Gottes Herz und Sinn.

Cantata, no. 129, Gelobet sei der Herr. cf Works, selections
(HMV SQ ASD 3265)

Cantata, no. 130, Herr Gott, dich loben alle wir. cf Cantata,
no. 67, Halt, halt im Gedächtnis Jesum Christ.

Cantata, no. 135, Ach Herr, mich armen Sünder. cf Cantata, no.
10, Meine Seele erhebt den Herren.

Cantata, no. 140, Wachet auf, ruft uns die Stimme. cf Works,
selections (DG 2530 647)

Cantata, no. 140, Wachet auf, ruft uns die Stimme. cf RCA CRL
3-1430.

93   Cantata, no. 142, Uns ist ein kind geboren. Magnificat, S 243,
D major. Sena Jurinac, s; Eugenia Zareska, con; Theo Altmeyer,
t; Heinz Rehfuss, bs; Munich Pro Arte Orchestra and Choir;
Kurt Redel. Philips 6581 014. (Reissue)
     +-Gr 3-76 p1488   +-RR 2-76 p54
     +-HFN 4-76 p100

94   Cantata, no. 147, Herz und Mund und Tat und Leben. Motets: Der
Geist hilft unsrer Schwachheit auf, S 226; Fürchte dich nicht,
S 228; Lobet den Herrn, alle Heiden, S 230. Elly Ameling, s;
Janet Baker, ms; Ian Partridge, t; John Shirley-Quirk, bs;
Cambridge, King's College Choir; AMF; Neville Marriner, David
Willcocks. Angel S 36804. (also HMV HQS 1254 Tape (c) TCEXE
189)
     ++AR 8-73 100   +HFN 5-76 p117 tape
     +ARG 8-72 p598   ++NR 7-72 p9
     +Gr 5-72 p1915   +-RR 6-76 p86 tape
     +HF 10-72 p84   +SFC 5-21-72 p46
     +-HFN 5-72 p913   ++St 11-72 p103

Cantata, no. 147, Jesu, joy of man's desiring. cf Organ works
(Philips 6500 925)

Cantata, no. 147, Jesu, joy of man's desiring. cf Works,
selections (Decca DPA 535/6).

Cantata, no. 147, Jesu bleibt meine Freude. cf Works, selections
(DG 2530 647).

Cantata, no. 147, Jesu, joy of man's desiring. cf Works,
selections (DG 2584 001).

Cantata, no. 147, Jesu, joy of man's desiring. cf Abbey LPB 754.

Cantata, no. 147, Jesu, joy of man's desiring. cf CBS 73487.

Cantata, no. 147, Jesu, joy of man's desiring. cf HMV HLM 7093.

Cantata, no. 147, Jesu, joy of man's desiring. cf Vanguard VSD
707/8.

Cantata, no. 156: Arioso. cf Works, selections (RCA ARL 1-0880).

Cantata, no. 156: Sinfonia. cf Klavier KS 551.

Cantata, no. 161, Komm, du süsse Todesstunde. cf Works, selections
(HMV SQ ASD 3265).

Cantata, no. 175, Er rufet seinen Schafen mit Namen. cf Cantata,
no. 34, O ewiges Feuer, O Ursprung der Liebe.

Cantata, no. 190, Lobe, Zion, deinen Gott. cf Works, selections
(HMV SQ ASD 3265).

Cantata, no. 208, Schafe können sicher weiden. cf HMV SLS 5022.

Cantata, no. 208, Sheep may safely graze (arr. Walton). cf Works,
selections (DG 2584 001).

Cantata, no. 208, Sheep may safely graze. cf HMV SXLP 30181.

Cantata, no. 208, Sheep may safely graze. cf Vanguard VSD 707/8.

95   Cantata, no. 209: Sinfonia. Concerto, E minor (reconstr. Radeke,
based on S 1059 & S 35). Suite, orchestra, S 1067, B minor.

Jean-Pierre Rampal, flt; Jean-François Paillard Chamber Orchestra; Jean-François Paillard.   RCA FRL 5693.
         ++NR 11-76 p5
96   Cantata, no. 211, Schweigt stille, plaudert nicht. Cantata, no.
      no. 212, Mer hahn en neue Oberkeet. Edith Mathis, s; Peter
      Schreier, t; Theo Adam, bs; Berlin Chamber Orchestra; Peter
      Schreier. DG Archive 2533 269.
         +Gr 12-76 p1028   +-RR 11-76 p94
         +HFN 11-76 p155
      Cantata, no. 212, Mer hahn en neue Oberkeet. cf Cantata, no. 211,
      Schweigt stille, plaudert nicht.
      Cantata, no. 248: Sinfonia. cf Cantata, no. 18: Sinfonia.
      Canzona, S 588, D minor. cf Allabreve, S 589, D major.
      Canzona, S 588, D minor. cf Organ works (DG 2533 140)
      Canzona, S 588, D minor. cf Organ works (DG 2722 014)
      Capriccio, S 993, E major. cf BACH, J. B.: Du Friedensfürst,
      Herr Jesu Christ, partita.
      Capriccio sopra la lontananze del suo fratello dilettissimo,
      S 992, B flat major. cf Harpsichord works (DG 2722 020).
      Chaconne. cf Columbia/Melodiya M 33593.
97   Chorales, S 645/50, 690/1, 696/9, 701, 703/4, 706, 709, 711, 714,
      731, 738. Lionel Rogg, org. Oryx 1006.
         +St 12-76 p72
98   Chorales, S 651/68. Lionel Rogg, org. Oryx 1017/8 (2).
         +St 12-76 p72
99   Chorales, S 653b, 720, 727, 733/4, 736/7. Trio sonatas, nos.
      1-6, S 525/30. Lionel Rogg, org. Oryx 1009/10 (2).
         +St 12-76 p72
100   Chorales, S 695, 710, 712/3, 717/8, 740. Preludes and fugues,
      S 534, S 539, F minor and D minor. Lionel Rogg, org. Oryx
      1004.
         +-St 12-76 p72
101   Chorale partitas, S 766/8. Lionel Rogg, org. Oryx 1016.
         +St 12-76 p72
      Chorale preludes of diverse kinds (18). cf Organ works (DG
      Archive 2722 016).
      Chorale prelude: Herr Gott nun Schleuss den Himmel auf. cf
      Westminster WGS 8116.
      Chorale prelude: In dulci jubilo. cf Vista VPS 1022.
      Chorale prelude: Jesu Christus, unser Heiland. cf Pelca PRSRK
      41017/20.
      Chorale prelude: Liebster, Jesu, wir sind hier. cf Audio 16.
      Chorale prelude: Wenn wir in hochsten Noten sein, S Anh 78. cf
      BACH, J. B.: Du Friedensfürst, Herr Jesu Christ, partita.
      Chorale prelude: Singet dem Herrn, S 225. cf HMV SLS 5047.
      Chorale preludes: Bist du bei mir, S 508; Herzlich tut mich
      verlangen, S 161, S 248, S 727; Nun komm der Heiden Heiland,
      S 659; O Gott, du frommer Gott, S 64, S 129; Vom Himmel hoch,
      da komm ich her, S 248, S 606, S 728; Wir glauben all' an
      einem Gott, S 740. cf Organ works (Philips 6500 925)
102   Chorale preludes (Das Orgelbüchlein). Robert Noehren, org.
      Orion ORS 75200/1.
         +-HF 4-76 p97   ++NR 1-76 p14
         ++MU 10-76 p12
103   Chorale preludes (Orgelbüchlein), S 599-644. Lionel Rogg, org.
      Oryx 1013/4 (2).
         +St 12-76 p72

104 Chorale preludes (Orgelbüchlein), S 599-644. Michel Chapuis, org.
     Telefunken EK 6-35085 (2).
         +-Gr 11-76 p837   -RR 10-76 p82
         +-HFN 10-76 p163   ++SFC 10-10-76 p32
105 Chorale preludes (Orgelbüchlein), S 599-644. Robert Noehren,
     org. Orion ORS 75200 (2).
         +-HF 4-76 p97   ++NR 1-76 p14
         ++MU 10-76 p12
     Chorale preludes: In dulce jubilo, S 608; Liebster Jesu, wir sind
     hier, S 731. cf Grosvenor GRS 1041.
106 Chorale preludes: Das alte Jahr vergangen ist, S 614; In dir ist
     Freude, S 615; In dulci jubilo (2 settings); Nun komm, der
     Heiden Heiland (3 settings); Wachet auf, ruft uns die Stimme,
     S 645. KAUFFMANN: O Jesulein süss. Vom Himmel hoch. OLEY:
     Jesus meine Zuversicht. Wir Christenleut (2 versions).
     TELEMANN: Wie schön leuchtet der Morgenstern (2 settings).
     James Dalton, org. FAS 16.
         +HFN 5-76 p67   +RR 5-76 p67
     Chorale prelude: Das alte Jahr vergangen ist, S 614. cf Wealden
     WS 137.
107 Chorale preludes: In dir ist Freude, S 615; Nun komm der Heiden
     Heiland, S 599. Prelude and fugue, S 532, D major. Trio
     sonata, no. 1, S 525, E flat major. BOHM: Jesu, du bist allzu
     schöne, partita. SCHEIDEMANN: Toccata Mixolydisch, G major.
     Herfried Mencke, org. Pelca PSR 40597.
         +NR 7-76 p10
     Chorale preludes (Orgelbüchlein): Erstandin ist der Heilige
     Christ, S 628. cf Visto VPS 1023.
     Chorale preludes: Ich ruf zu dir, Herr Jesu Christ, S 639; Nun
     komm der Heiden Heiland, S 699; Wir glauben all an einen Gott,
     S 680. cf Brandenburg concerto, no. 5, S 1050, D major.
     Chorale preludes (Schübler), S 645-50. cf Organ works (DG Archive
     2722 016).
     Chorale preludes (Schübler), S 645-50. cf Argo 5BBA 1013/5.
     Chorale preludes, S 645-650, 684, 688. cf Organ works (Saga
     5428/9).
     Chorale prelude: Wachet auf, S 645. cf Works, selections (Decca
     DPA 535/6).
     Chorale prelude: Wachet auf, S 645. cf Decca DPA 523/4.
     Chorale prelude: Wachet auf, ruft uns die Stimme, S 645. cf
     London STS 15222.
     Chorale prelude: Wachet auf, ruft uns die Stimme, S 645. cf
     Telefunken AF 6-41872.
     Chorale prelude: Ach bleib bei uns, S 649. cf London STS 15222.
     Chorale prelude: Kommst du Nun, Jesu, S 650. cf Wealden WSQ 134.
108 Chorale preludes, S 651-668. Peter Hurford, org; Alban Singers.
     Argo ZRG 843/4 (2).
         +ARG 12-76 p10
109 Chorale preludes, S 651-S 688a. Michael Chapuis, org. Telefunken
     EK 6-35083.
         ++Gr 10-75 p652   -MU 10-76 p18
         +-HFN 10-75 p136   +RR 10-75 p69
     Chorale prelude: Nun komm, der Heiden Heiland, S 659. cf RCA VH
     020.
110 Chorale preludes: Allein Gott, S 662-664; An Wasserflussen Baby-
     lon, S 653; Herr Jesu Christ, S 655; Jesus Christus, S 665,

S 666; Komm Gott, Schöpfer, S 667; Komm heilige Geist, S 651,
S 652; O Lamm Gottes, unschuldig, S 656; Nun danket alle Gott,
S 657; Nun komm er Heiden Heiland, S 659-661; Schmucke dich,
O liebe, S 654; Von Gott will ich nicht lassen, S 658; Wenn
wir in höchsten, S 668a. Alban Singers; Peter Hurford, org.
Argo ZRG 843/4 (2).
       ++Gr 6-76 p71   +RR 9-76 p78
       +HFN 8-76 p76

Chorale preludes: Allein Gott in der Hoh sei Ehr, S 675; Aus
tiefer Not, S 686; Christ unser Herr, S 685; Dies sind die
heil'gen zehn Gebot, S 678; Jesus Christus, unser Heiland,
S 688; Kyrie, Gott Vater in Ewigkeit, S 669; Vater unser im
Himmelreich, S 683; Wir glauben all einen Gott, S 680. cf
Organ works (Telefunken 6-41977).

Chorale preludes: Sei gegrüsset Jesus gütig, S 678; Vom Himmel
hoch variations, S 679. cf Organ works (DG Archive 2722 016).

111   Chorale prelude: Aus tiefer Not schrei ich zu dir, S 687. Prelude and fugue, S 545, C major. Toccata and fugue, S 538, D
minor. HAYDN: Pieces, mechanical clock (4). MENDELSSOHN:
Sonata, organ, op. 65, no. 2, C minor. TORRES: Saeta, no. 4.
Robert Munns, org. Wealden WS 120.
       +RR 8-76 p66

Chorale prelude: Christ lag in Todesbanden, S 695. cf Vista
VPS 1023.

112   Chorale preludes: Jesu meine Freude, S 713: Fantasia; Vom Himmel
hoch, S 701. Magnificat, S 243: Vom Himmel hoch; Freut euch
und jubiliert. Motets: Fürchte dich nicht, S 228; Der Geist
hilft unsrer Schwachheit auf, S 226; Jesu, meine Freude,
S 227; Komm, Jesu, komm, S 231; Lobet den Herrn, alle Heiden,
S 230; Singet dem Herrn, S 225. David Lumsden, org; Louis
Halsey Singers; Louis Halsey. L'Oiseau-Lyre SOL 340/1 (2).
       +HF 7-76 p76   +NR 5-76 p8
       /MJ 10-76 p24   ++St 9-76 p116

Chorale preludes: Ein feste Burg, S 720; Komm süsser Tod; Wachet
auf, S 645. cf Works, selections (RCA ARL 1-0880).

Chorale preludes: Herzlich tut mich verlangen (Befiehl du deine
Wege), S 727; Ich ruf zu dir, Herr Jesu Christ, S 639; In
dulci jubilo; Nun komm der Heiden Heiland, S 659. cf Works,
selections (DG 2530 647).

Chorale prelude: Liebster Jesu, wir sind hier, S 731. cf Wealden
WS 121.

Chorale preludes: Liebster Jesu, wir sind hier, S 731; Valet will
ich dir geben, S 736. cf Wealden WS 142.

Chorale prelude: Rejoice, beloved Christians, S 734 (arr. Busoni).
cf International Piano Library IPL 104.

113   Chorale preludes: O Gott, du frommer Gott, S 767. BULL: Revenant.
BUXTEHUDE: Chorale prelude: Wie schön leuchtet der Morgenstern.
Trauermusic. GRIGNY: Veni creator. Marion Rowlatt, s; Peter
Hurford, org. Argo ZRG 835.
       +Gr 10-76 p633   +-RR 10-76 p86
       +HFN 10-76 p167

114   Chorale preludes: Sei gegrüsset Jesus gütig, S 768; Von Gott will
ich nicht lassen, S 658. Fantasia, S 562, C minor. Toccata
and fugue, S 540, F major. Alice Jucker-Baumann, org. Pelca
PSR 40588.
       -NR 5-76 p13

115   Chorale prelude: Sei gegrüsset, Jesus gütig, S 768. SCHEIDT:
      Warum betrübst du dich. WALTHER: Jesu, meine Freude. Dieter
      Wellmann, org. Pelca PSR 40596.
          +NR 5-76 p15
116   Christmas oratorio, S 248. Hans Buchhierl, treble; Andreas
      Stein, alto; Theo Altmeyer, t; Barry McDaniel, bar; Tölzer
      Boys' Choir; Collegium Aureum; Gerhard Schmidt-Gaden. BASF
      59 21749-3 (3).
          +Gr 3-76 p418   +RR 1-76 p54
          +-HFN 1-76 p102
      Christmas oratorio, S 248. cf Cantatas (DG Archive 2722 018).
      Christmas oratorio, S 248, excerpts. cf Works, selections
      (Decca DPA 535/6).
      Christmas oratorio, S 248: Bereite dich, Zion. cf Works, selec-
      tions (HMV SQ ASD 3265).
      Christmas oratorio, S 248: Sinfonia. cf Decca SDD 411.
117   Chromatic fantasia and fugue, S 903, D minor. Concerto, harpsi-
      chord, S 971, F major. French suite, no. 5, S 816, G major.
      Toccata, S 912, D major. George Malcolm, hpd. Decca ECS 788.
      Tape (c) KECC 788. (Reissue from SXL 2259)
          +Gr 10-76 p622   +HFN 12-76 p155 tape
          +HFN 11-76 p171   +-RR 10-76 p82
118   Chromatic fantasia and fugue, S 903, D minor. Fantasia, S 922,
      A minor. Toccatas, S 910-916. Blandine Verlet, hpd.
      Philips 6780 500 (2).
          +Gr 4-76 p1628   +RR 4-76 p62
          ++HFN 4-76 p100
119   Chromatic fantasia and fugue, S 903, D minor. Fantasia, S 922,
      A minor. Toccatas, S 912, D major; S 914, E minor; S 916,
      G major. Blandine Verlet, hpd. Philips 6833 184.
          ++ARG 11-76 p13   +NR 12-76 p15
          ++MJ 11-76 p60
      Chromatic fantasia and fugue, S 903, D minor. cf Concerto,
      harpsichord, S 971, F major.
      Chromatic fantasia and fugue, S 903, D minor. cf Harpsichord
      works (DG 2722 020).
120   Clavierübung, Pt 3. Lionel Rogg, org. Oryx 1007/8 (2).
          +St 12-76 p72
121   Clavierübung, Pt 3. Michel Chapuis, org. Telefunken 6-35084 (2).
          +Gr 4-76 p1628   +RR 5-76 p63
          +HFN 2-76 p91   +-St 12-76 p72
      Clavierübung, Pt 3. cf Organ works (DG Archive 2722 016).
      Concerto, E minor (reconstr. Radeke, based on S 1059 and S 35).
      cf Cantata, no. 209: Sinfonia.
122   Concerto, flute and strings, G minor (arr. from S 1056). Concerto,
      oboe, violin and strings, D minor (arr. from S 1060). Con-
      certo, 3 violins and strings, D major (arr. from S 1064).
      Carmel Kaine, Ronald Thomas, Richard Studt, vln; Neil Black,
      ob; William Bennett, flt; Christopher Hogwood, Nicholas Krämer,
      hpd; AMF; Neville Marriner. Argo ZRG 820. Tape (c) KZRC 820.
          ++Gr 12-75 p1022   ++MT 5-76 p408
          +HF 12-76 p97   +NR 5-76 p5
          ++HFN 12-75 p147   +RR 12-75 p36
          ++HFN 2-76 p116 tape
123   Concerto, flute, violin and harpsichord, S 1044, A minor. Suite,
      orchestra (flute and strings), S 1067, B minor. Peter Lukas
      Graf, flt; José Luis Garcia, vln; Jörg Ewald Dahler, hpd; ECO;

Peter Lukas Graf. CRD Claves 30-324.
>   +Gr 10-76 p577   +-RR 6-76 p34
124 Concerto, flute, violin and harpsichord, S 1044, A minor. Concerto, 2 harpsichords, S 1061, C major. Frans Brüggen, baroque flt; Marie Leonhardt, baroque vln; Gustav Leonhardt, Anneke
Uittenbosch, hpd; Leonhardt Consort. Telefunken SAWT 9552.
>   +AR 8-76 p77
125 Concerto, harpsichord, S 971, F major (Italian). Partita, harpsichord, S 831, B minor. Igor Kipnis, hpd. Angel S 36096.
>   +ARG 12-76 p12   +NR 12-76 p15
>   +MJ 4-76 p31   +-SFC 10-31-76 p35
126 Concerto, harpsichord, S 971, F major. Fantasia, S 906, C minor.
Inventions, 2 part, S 772-786. Suite, lute (harpsichord),
S 997, C minor. Lionel Party, hpd. Desmar DSM 1008.
>   +MJ 3-76 p25   +RR 4-76 p62
127 Concerto, harpsichord, S 971, F major. Chromatic fantasia and
fugue, S 903, D minor. Partita, S 831, B minor. Elizabeth
de la Porte, hpd. Saga 5424.
>   +-Gr 3-76 p1482   +RR 3-76 p57
>   +-MT 7-76 p575
Concerto, harpsichord, S 971, F major. cf Chromatic fantasia and
fugue, S 903, D major.
Concerto, harpsichord, S 971, F major. cf Harpsichord works
(DG 2722 020).
Concerto, harpsichord, S 971, F major. cf Works, selections
(Decca DPA 535/6).
128 Concerti, harpsichord, nos. 1-7, S 1052-1058. Concerto, 2 harp-
sichords, S 1060, C minor. Concerto, 2 harpsichords, S 1061,
C major. George Malcolm, Simon Preston, hpd; Menuhin Festival
Orchestra; Yehudi Menuhin. HMV SLS 5039 (3). (Reissues from
ASD 3007, 2713, 2647)
>   ++Gr 5-76 p1743   ++RR 4-76 p36
>   +-MT 5-76 p408
129 Concerti, harpsichord, nos. 1-2, 4-5, S 1052-3, 1055-6. Concerti,
2 harpsichords, S 1060-61. Concerti, 3 harpsichords, S 1063-64.
Concerto, 4 harpsichords, S 1065, A minor. Raymond Leppard,
Andrew Davis, Philip Ledger, Blandine Verlet, hpd; ECO;
Raymond Leppard. Philips 6747 194 (3).
>   +HF 12-76 p97   +St 12-76 p134
>   ++NR 8-76 p5
130 Concerti, harpsichord, nos. 1, 14-5, S 1052, 1055, 1056. Concerti,
3 harpsichords, S 1063-64. Concerto, 4 harpsichords, S 1065,
A minor. Concerti, violin and strings, S 1041-42. Concerto,
2 violins and strings, S 1043, D minor. Karl Richter, hpd;
Eduard Melkus, vln; Munich Bach Orchestra; Eduard Melkus. DG
Tape (c) 3376 006.
>   +HF 9-76 p84 tape
131 Concerto, harpsichord, no. 1, S 1052, D minor. Concerto, harpsichord, no. 5, S 1056, F minor. Concerto, violin and strings,
S 1042, E major. William Armon, vln; Harold Lester, hpd; LOL;
Leslie Jones. Oryx ORPS 7.
>   +-Gr 8-76 p277
132 Concerto, harpsichord, no. 1, S 1052, D minor. Concerto, harpsi-
chord, no. 4, S 1055, A major. Concerto, harpsichord, no. 5,
S 1056, F minor. Anton Heiller, hpd; VSOO; Miltiades Caridis.
Vanguard Bach Guild HM 44.
>   +-Gr 8-76 p277   +-RR 6-76 p34
>   +HFN 7-76 p103

BACH, J. S. (cont.)   24

Concerto, harpsichord, no. 2, S 1053, F major (arr.). cf Con-
    certo, 3 harpsichords, S 1063, D minor (arr).
Concerto, harpsichord, no. 4, S 1055, A major. cf Concerto,
    harpsichord, no. 1, S 1052, D minor.
Concerto, harpsichord, no. 4, S 1055, A major (arr.). cf Concerto,
    3 harpsichords, S 1063, D minor (arr.).
Concerto, harpsichord, no. 5, S 1056, F minor. cf Concerto,
    harpsichord, no. 1, S 1052, D minor (Vanguard Bach Guild 44).
Concerto, harpsichord, no. 5, S 1056, F minor. cf Concerto,
    harpsichord, no. 1, S 1052, D minor (Oryx 7).
Concerto, harpsichord, no. 5, S 1056, F minor: Largo. cf Works
    selections (DG 2530 647).
Concerto, 2 harpsichords and strings, C minor: Adagio. cf CBS
    61684.

133 Concerti, 2 harpsichords, S 1060-1. Concerti, 2 harpsichords,
     S 1063-4. Concerto, 4 harpsichords, S 1065, A minor. Anton
     Heiller, Kurt Rapf, Christa Landon, hpd; I Solisti di Zagreb;
     Antonio Janigro. Bach Guild HM 52/53. (also Bach 70659/60)
          ++ARG 11-76 p25
Concerti, 2 harpsichords, S 1060-61. cf Concerti, harpsichord,
    nos. 1-2, 4-5, S 1052-3, 1055-6.
Concerto, 2 harpsichords, S 1060, C minor. cf Concerti, harpsi-
    chord, nos. 1-7, S 1052-1058.
Concerto, 2 harpsichords, S 1061, C major. cf Concerto, flute,
    violin and harpsichord, S 1044, A minor.
Concerto, 2 harpsichords, S 1061, C major. cf Concerti, harpsi-
    chord, nos. 1-7, S 1052-1058.
Concerti, 3 harpsichords, S 1063-64. cf Concerti, harpsichord,
    nos. 1, 4-5, S 1052, 1055, 1056.
Concerti, 3 harpsichords, S 1063-64. cf Concerti, harpsichord,
    nos. 1-2, 4-5, S 1052-3, 1055-6.
Concerti, 3 harpsichords, S 1063-4. cf Concerti, 2 harpsichords,
    S 1060-1.

134 Concerto, 3 harpsichords, S 1063, D minor (arr.). Concerto,
     harpsichord, no. 4, S 1055, A major (arr.). Concerto, harpsi-
     chord, no. 2, S 1053, F major (arr.). Carmel Kaine, vln;
     Neil Black, ob, oboe d'amore; William Bennett, flt; AMF;
     Neville Marriner. Argo ZRG 821.
          +-Gr 5-76 p1743   ++RR 4-76 p36
          +-HFN 4-76 p100   +STL 4-11-76 p36
          ++NR 6-76 p6
Concerto, 3 harpsichords, S 1063, D minor. cf BACH, C. P. E.:
    Concerto, 4 harpsichords, F major.
Concerto, 4 harpsichords, S 1065, A minor. cf Concerti, harpsi-
    chord, nos. 1, 4-5, S 1052, 1055, 1056.
Concerto, 4 harpsichords, S 1065, A minor. cf Concerti, harpsi-
    chord, nos. 1-2, 4-5, S 1052-3, 1055-6.
Concerto, 4 harpsichords, S 1065, A minor. cf Concerti, 2 harpsi-
    chords, S 1060-1.
Concerto, 4 harpsichords, S 1065, A minor. cf BACH, C. P. E.:
    Concerto, 4 harpsichords, F major.

135 Concerti, keyboard, S 972-987. Sophie Svirsky, pno. Monitor MCS
     2149/51 (3).
          +NR 12-76 p14
Concerto, oboe, violin and strings, D minor (arr. from S 1060).
    cf Concerto, flute and strings, G minor (arr. from S 1056).

Concerto, organ, A minor. cf Klavier KS 551.

136   Concerti, organ, S 592-7 (after Vivaldi, Johann Ernst of Sachsen-
       Weimar and others). Karl Richter, org. DG Archive 2533 170.
       Tape (c) 3310 128.
          ++St 12-76 p71

Concerto, organ, no. 1, S 592, G major. cf Audio 2.

Concerto, organ, no. 2, S 593, A minor. cf Organ works (Classics
for Pleasure CFP 40241).

Concerto, organ, no. 2, S 593, A minor. cf Toccata and fugue,
S 565, D minor.

Concerto, organ, no. 2, S 593, A minor. cf Creative Record
Service R 9115.

Concerto, organ, no. 3, S 594, C major: Recitative. cf CBS
Classics 61579.

137   Concerto, organ, no. 5, S 596, D minor (after Vivaldi). Preludes
       and fugues, S 553-60 (Little). E. Power Biggs, org. Columbia
       M 33975.
          ++NR 5-76 p13   +St 12-76 p72

Concerto, trumpet, D minor. cf RCA CRL 3-1430.

Comcerto, violin and strings, C minor. cf CBS 61684.

Concerti, violin and strings, S 1041-42. cf Concerti, harpsichord, S 1052, 1055, 1056.

138   Concerto, violin and strings, S 1041, A minor. Concerto, violin
       and strings, S 1052, D minor (after Concerto, harpsichord).
       Concerto, violin, oboe and strings, S 1060, C minor. Itzhak
       Perlman, vln; Neil Black, ob; ECO; Daniel Barenboim. Angel
       S 37076. (also HMV ASD 3076)
          +-Gr 6-75 p31   +NR 5-75 p7
          +HF 6-75 p83   +RR 6-75 p26
          +HFN 6-75 p83   ++St 3-76 p108
          +-MT 8-75 p713

139   Concerto, violin and strings, S 1041, A minor. MOZART: Concerto,
       violin, no. 3, K 216, G major. Jaime Laredo, vln; Washington
       National Symphony Orchestra; Howard Mitchell. Camden CCV 5041.
          +HFN 4-76 p121   +RR 5-76 p34

140   Concerto, violin and strings, S 1041, A minor. Concerto, violin
       and strings, S 1042, E major. Concerto, 2 violins and strings,
       S 1043, D minor. Kenneth Sillito, Hugh Bean, vln; Virtuosi
       of England; Arthur Davison. Classics for Pleasure CFP 40244.
          +Gr 8-76 p277   +RR 6-76 p34
          +-HFN 6-76 p80

141   Concerto, violin and strings, S 1041, A minor. Concerto, violin
       and strings, S 1042, E major. Concerto, 2 violins and strings,
       S 1043, D minor. Josef Suk, Ladislav Jašek, vln; Prague Sym-
       phony Orchestra; Václav Smetáček. Supraphon 50672. Tape (c)
       4-50672.
          +-HFN 2-76 p116 tape   +RR 1-76 p65 tape

142   Concerto, violin and strings, S 1041, A minor. Concerto, violin
       and strings, S 1042, E major. Concerto, violin and strings,
       S 1052, D minor. Ruggiero Ricci, vln; London City Ensemble;
       Ruggiero Ricci. Unicorn UNS 202.
          ++St 3-76 p108

Concerto, violin and strings, S 1041, A minor. cf HMV RLS 718.

Concerto, violin and strings, S 1042, E major. cf Concerto,
harpsichord, no. 1, S 1052, D minor.

Concerto, violin and strings, S 1042, E major. cf Concerto, violin
and strings, S 1041, A minor (Classics for Pleasure CFP 40244).

Concerto, violin and strings, S 1042, E major. cf Concerto,
  violin and strings, S 1041, A minor (Supraphon 50672).
Concerto, violin and strings, S 1042, E major. cf Concerto,
  violin and strings, S 1041, A minor (Unicorn UNS 202).
Concerto, violin and strings, S 1042, E major. cf HMV SEOM 22.
Concerto, violin and strings, S 1052, D minor (after Concerto,
  harpsichord). cf Concerto, violin, no. 1, S 1041, A minor
  (Angel 37076).
Concerto, violin and strings, S 1052, D minor. cf Concerto,
  violin and strings, S 1041, A minor (Unicorn UNS 202).
143 Concerto, 2 violins and strings, S 1043, D minor. Sonata, solo
  flute, S 1013, A minor. Suite, orchestra, S 1067, B minor.
  Okko Kamu, Leif Segerstam, vln; Gunilla von Gahr, flt;
  Nationalmusei Kammarorkester; Claude Genetay. BIS LP 21.
     +-HFN 9-76 p117   +-RR 9-76 p43
144 Concerto, 2 violins and strings, S 1043, D minor. Concerto,
  violin, oboe and strings, S 1060, C minor. Isaac Stern,
  Alexander Schneider, vln; Marcel Tabuteau, ob; Prades Festival
  Orchestra; Pablo Casals. CBS 61679.
     -Gr 4-76 p1588   -RR 3-76 p33
     -HFN 4-76 p99
Concerto, 2 violins and strings, S 1043, D minor. cf Concerto,
  violin and strings, S 1041, A minor (Classics for Pleasure
  CFP 40244).
Concerto, 2 violins and strings, S 1043, D minor. cf Concerto,
  violin and strings, S 1041, A minor (Supraphon 50672).
Concerto, 2 violins and strings, S 1043, D minor. cf Concerti,
  harpsichord, nos. 1, 4-5, S 1052, 1055, 1056.
Concerto, 2 violins and strings, S 1043, D minor. cf CBS SQ 79200.
Concerto, 2 violins and strings, S 1043, D minor. cf HMV RLS 718.
Concerto, 2 violins and strings, S 1043, D minor. cf RCA ARM
  4-0942/7.
Concerto, 3 violins and strings, D major (arr. from S 1064). cf
  Concerto, flute and strings, G minor (arr. from S 1056).
145 Concerto, violin, oboe and strings, S 1060, D minor. HANDEL:
  Concerto, oboe, G minor. MARCELLO: Concerto, oboe, D minor.
  VIVALDI: Concerto, oboe, A minor. Emilia Csanky, ob; Janos
  Rolla, vln; Liszt Ferenc Academy Chamber Orchestra; Frigyes
  Sándor. Hungaroton SLPX 11740.
     +NR 5-76 p5   +RR 6-76 p34
Concerto, violin, oboe and strings, S 1060, C minor. cf Concerto,
  2 violins and strings, S 1043, D minor.
Concerto, violin, oboe and strings, S 1060, C minor. cf Concerto,
  violin and strings, no. 1, S 1041, A minor.
Contrapunctus VII. cf Crystal S 206.
Duets, S 802-805. cf Organ works (DG 2533 140).
Duet, S 803, F major. cf Organ works (Telefunken 6-41977).
Easter oratorio, S 249: Saget, saget mir geschwinde. cf Works,
  selections (HMV SQ ASD 3265).
Easter oratorio, S 249: Sinfonia. cf Works, selections (DG 2584
  001).
146 English suites, nos. 1-6, S 806-11. Huguette Dreyfus, hpd. DG
  Archive 2533 164/166 (3). (Reissue from 2722 020)
     +Gr 9-76 p443   ++RR 9-76 p79
     +-HF 11-74 p99   +St 12-74 p129
     +HFN 9-76 p117

English suites, nos. 1-6, S 806-811. cf Harpsichord works
    (DG 2722 020).
English suite, no. 3, S 808, G minor: Sarabande, Gavotte, Musette.
    cf RCA ARM 4-0942/7.
English suite, no. 6, S 811: Gavottes, nos. 1 and 2. cf RCA ARM
    4-0942/7.
Fantasia, S 562, C minor. cf Canonic variations, S 769.
Fantasia, S 562, C minor. cf Chorale preludes: Sei gegrüsset
    Jesus gütig, S 768; Von Gott will ich nicht lassen, S 658.
Fantasia, S 562, C minor. cf Organ works (DG 2722 014).
Fantasia, S 562, C minor. cf Organ works (Saga 5428/9).
Fantasias, S 563, B minor; S 570, C major. cf Organ works
    (Telefunken BC 25098-T/1-2).
Fantasia, S 572, G major. cf Canonic variations, S 769.
Fantasia, S 572, G major. cf Organ works (DG 2722 014).
Fantasia, S 572, G major. cf Grosvenor GRS 1041.
147   Fantasia, S 906, C minor. Toccata and fugue, S 910, F sharp minor.
    MOZART: Fantasia, K 396, C minor. Sonata, piano, no. 17,
    K 576, D major. Variations on a minuet by Duport, K 573.
    Grant Johannesen, pno. Golden Crest CRSQ 4142.
        +HF 9-76 p105
Fantasia, S 906, C minor. cf Concerto, harpsichord, S 971, F
    major.
Fantasia, S 906, C minor. cf Harpsichord works (DG 2722 020).
Fantasia, S 922, A minor. cf Chromatic fantasia and fugue,
    S 903, D minor (Philips 6780 500).
Fantasia, S 922, A minor. cf Chromatic fantasia and fugue, S 903,
    D minor (Philips 6833 184).
148   Fantasia and fugue, S 537, C minor. Fantasia and fugue, S 542,
    G minor. Passacaglia and fugue, S 582, C minor. Toccata and
    fugue, S 565, D minor. Lionel Rogg, org. Oryx EXP 21.
        +-St 12-76 p72
Fantasia and fugue, S 537, C minor. cf Organ works (DG 2722 014).
149   Fantasia and fugue, S 542, G minor. Passacaglia and fugue, S 582,
    C minor. Prelude and fugue, S 543, A minor. Toccata and fugue,
    S 565, D minor. Karl Richter, org. Decca Tape (c) KSDC 258.
        +-RR 2-76 p70 tape
150   Fantasia and fugue, S 542, G minor. Passacaglia and fugue, S 582,
    C minor. Prelude and fugue, S 548, E minor. Toccata and fugue,
    S 565, D minor. Karl Richter, org. Decca SPA 459. (Reissue
    from SXL 2219).
        +-Gr 12-76 p1015   +-RR 12-76 p75
Fantasia and fugue, S 542, G minor. cf Fantasia and fugue, S 537,
    C minor
Fantasia and fugue, S 542, G minor. cf Organ works (Classics for
    Pleasure CFP 40241).
Fantasia and fugue, S 542, G minor. cf Organ works (DG 2722 014).
Fantasia and fugue, S 542, G minor. cf Decca DPA 523/4.
Fantasia and fugue, S 542, G minor. cf London STS 15222.
Fantasia and fugue, S 904, A minor. cf BACH, J. B.: Du Friedens-
    fürst, Herr Jesu Christ, partita.
Fantasia and fugue, S 944, A minor. cf Harpsichord works (DG
    2722 020).
151   French suites, nos. 1-6, S 812-7. Kenneth Gilbert, hpd. Harmonia
    Mundi HMU 438 (2).
        +-Gr 8-76 p318   +STL 8-8-76 p29

French suites, nos. 1-6, S 812-817. cf Harpsichord works
(DG 2722 020).

French suite, no. 5, S 816, G major. cf Chromatic fantasia and
fugue, S 903, D minor.

Fugue, S 574, C minor. cf Organ works (DG 2722 014).

Fugues, S 575, 577-9. cf Allabreve, S 589, D major.

Fugues, S 575, C minor; S 578, G minor. cf Organ works (Telefunken BC 25098-T/1-2).

Fugue, S 577, G major. cf Organ works (Classics for Pleasure CFP
40241).

Fugue, S 577, G major. cf Organ works (Saga 5428/9).

Fugue, S 577, G major (arr. Holst). cf Works, selections (DG
2584 001).

Fugue, S 578, G minor. cf Organ works (DG 2722 014).

Fugue, S 578, G minor. cf Works, selections (RCA ARL 1-0880).

Fugue, S 578, G minor. cf Wealden WSQ 134.

Fugue, S 579, B minor. cf Organ works (DG 2722 014).

Fugue, S 1080, D minor. cf Organ works (DG 2722 014).

152   Fugue, lute, S 1000, G minor. Suite, lute, S 995, G minor. Suite,
lute, S 1006a, E major. Narciso Yepes, gtr. DG 2530 461.
    +Gr 4-76 p1628   +NR 3-75 p14
    +-HFN 4-76 p100   +-RR 3-76 p57

153   Fugue, lute, S 1000, G minor. Partita, violin, no. 1, S 1002,
B minor: Sarabande. Prelude, fugue and allegro, S 998, E flat
major. Suite, lute, S 996, E minor. Prelude, S 999, C minor.
Vladimir Mikulka, gtr. Supraphon 111 1585.
    +Gr 10-76 p622   +-RR 8-76 p67
    -NR 12-76 p14

Fugue, lute, S 1000, G minor. cf Lute works (Columbia M2 33510).

Fugue, lute, S 1000, G minor. cf Works, selections (Erato STU
70885).

Gavotte, nos. 1 and 2. cf London CS 7015.

God liveth still. cf Vista VPS 1023.

154   Harpsichord works: Aria variata alla maniera Italiana, S 989, A
minor. Capriccio sopra la lontananze del suo fratello dilettis-
simo, S 992, B flat major. Chromatic fantasia and fugue,
S 903, D minor. English suites, nos. 1-6, S 806-811. Fantasia,
S 906, C minor. French suites, nos. 1-6, S 812-817. Fantasia
and fugue, S 944, A minor. Concerto, S 971, F major. Partitas,
harpsichord, nos. 1-6, S 825-830. Partita, harpsichord, S 831,
B minor. Huguette Dreyfus, Ralph Kirkpatrick, hpd. DG 2722 020
(10). (Reissues from 2533 138/9, SAPM 198003/5, 198655, 198183,
198185).
    ++Gr 1-76 p1216   +RR 11-75 p67

In dulci jubilo in canone doppio al ottava. cf Coronet LPS 3032.

In thee is joy. cf Coronet LPS 1722.

Inventions, 2 part, S 772-786. cf Concerto, harpsichord, S 971,
F major.

Inventions, 2 parts, no. 1, S 772, C major. cf Angel S 36095.

Komm süsser Tod, S 42. cf Club 99-99.

155   Die Kunst der Fuge (The art of the fugue), S 1080. AMF: Neville
Marriner. Philips 6747 172 (2). Tape (c) 7699 007.
    +Gr 10-75 p601   +-MT 1-76 p39
    +-Gr 12-76 p1066 tape   ++NR 12-75 p4
    +-HF 1-76 p76   ++RR 9-75 p26
    ++HFN 9-75 p92   +RR 11-76 p108 tape

Die Kunst der Fuge, S 1080. cf Organ works (DG Archive 2722 016).
156   Lute works: Fugue, S 1000, G minor. Prelude, S 999, C minor.
Prelude, fugue and allegro, S 998, E major. Suite, S 996,
E minor. Suite, S 997, G minor. Suite, S 995, G minor.
Suite, S 1006a, E major. John Williams, gtr. Columbia M2
33510 (2). (also CBS 79203. Tape (c) 40-79203)
    ++Gr 1-75 p652   +RR 11-75 p68
    ++HFN 11-75 p148   +RR 1-76 p65 tape
    ++HFN 12-75 p173 tape ++St 12-75 p116
    +NR 10-75 p13   ++STL 10-5-75 p36
    +-SFC 10-27-75 p24
157   Magnificat, S 243, D major (with Christmas interpolations).
Helen Donath, Gundula Bernat-Klein, s; Birgit Finnilä, alto;
Peter Schreier, t; Barry McDaniel, bs; South German Madrigal
Choir; German Bach Soloists; Wolfgang Gönnenwein. Sine Qua
Non SQN 7739.
    /St 1-76 p102
Magnificat, S 243, D major. cf Cantatas (DG Archive 2722 018).
Magnificat, S 243, D major. cf Cantata, no. 142, Uns ist ein
kind geboren.
Magnificat, S 243, D major: Esurientes. cf HMV SLS 5022.
Magnificat, S 243, D major: Et exultavit. cf Works, selections
(HMV SQ ASD 3265).
Magnificat, S 243: Freut euch und jubiliert; Vom Himmel hoch. cf
Chorale preludes (L'Oiseau-Lyre SOL 340/1).
Mass, S 232, B minor: Sanctus and Osanna. cf Works, selections
(Decca SDD 535/6).
158   Mass, S 235, G minor. Mass, S 236, G major. Wendy Eathorne, s;
Paul Esswood, c-t; Philip Langridge, t; Stephen Roberts, bs;
Richard Hickox Singers and Orchestra; Richard Hickox. Argo
ZRG 829.
    +Gr 2-76 p1364   +-MT 10-76 p831
    +HFN 5-76 p92   +RR 2-76 p54
Mass, S 236, G major. cf Mass, S 235, G minor.
Minuet, G major. cf Connoisseur Society (Q) CSQ 2066.
Motets: Der Geist hilft unsrer Schwachheit auf, S 226; Fürchte
dich nicht, S 228; Lobet den Herrn, alle Heiden, S 230. cf
Cantata, no. 147, Herz und Mund und Tat und Leben.
Motets: Fürchte dich nicht, S 228; Der Geist hilft unsrer Schwach-
heit auf, S 226; Jesu, meine Freude, S 227; Komm, Jesu, komm,
S 231; Lobet den Herrn, alle Heiden, S 230; Singet dem Herrn,
S 225. cf Chorale preludes (L'Oiseau-Lyre SOL 340/1).
Musette, D major. cf Connoisseur Society (Q) CSQ 2066.
159   A musical offering, S 1079. Leonhard Consort. ABC L 67007.
    +SFC 12-12-76 p55
160   Organ works: Concerto, organ, no. 2, S 593, A minor (after Vivaldi).
Fantasia and fugue, S 542, G minor. Fugue, S 577, G major.
Prelude and fugue, S 544, B minor. Toccata and fugue, S 540,
F major. Nicolas Kynaston, org. Classics for Pleasure CFP
40241.
    +Gr 11-76 p837   +-HFN 8-76 p75
161   Organ works: Allabreve, S 589, D major. Canzona, S 588, D minor.
Duets, S 802-805. Trio sonata, no. 1, S 525, E flat major.
Trio sonata, no. 6, S 530, G major. Helmut Walcha, org. DG
2533 140.
    +NR 8-75 p13   *St 12-76 p71

162   Organ works: Allabreve, S 589, D major. Canzona, S 588, D minor.
       Fantasias, S 562, C minor; S 572, G major. Fantasias and
       fugues, S 537, C minor; S 542, G minor. Fugues, S 574, C minor;
       S 578, G minor; S 579, B minor; S 1080, D minor. Passacaglia
       and fugue, S 582, C minor. Pastroale, S 590, F major. Preludes
       and fugues, S 531-536, S 539, S 541, S 543-552. Trio sonatas,
       nos. 1-6, S 525-530. Toccata, adagio and fugue, S 564, C major.
       Toccatas and fugues, S 538, D minor; S 540, F major; S 565,
       D minor. Helmut Walcha, org. DG 2722 014 (8). Tape (c) 3376
       004. (Reissue from 2722 002/1)
             *Gr 8-75 p347   +-HFN 7-75 p72
             +HF 6-76 p97 tape
163   Organ works: Chorale preludes (Schübler), S 645-50. Chorale
       preludes of diverse kinds (18). Chorale preludes: Sei gegrüs-
       set Jesus gütig, D 678; Vom Himmel hoch variations, S 679.
       Clavierübung, Pt 3. Die Kunst der Fuge, S 1080. Helmut Walcha,
       org. DG Archive 2722 016 (8). (Reissue from 2722 003)
             ++Gr 4-76 p1631   ++RR 12-75 p77
164   Organ works (arranged for trumpets, organ and ensemble): Cantata,
       no. 147: Jesu, joy of man's desiring. Chorale preludes: Bist
       du bei mir, S 508; Herzlich tut mich verlangen, S 161, S 248,
       S 727; Nun komm der Heiden Heiland, S 659; O Gott, du frommer
       Gott, S 64, S 129; Vom Himmel hoch, da komm ich her, S 248,
       S 606, S 738; Wir glauben all' an einem Gott, S 740. Prelude
       and fugue, S 532, D major. REICHE: Abblasen. Don Smithers,
       clarino trumpet, piccolo trumpet, cornetto, trumpet; William
       Neil, org; Clarion Consort. Philips 6500 925.
             +Audio 11-76 p108   ++HFN 7-76 p104 tape
             ++Audio 12-76 p92   +MJ 11-76 p60
             ++Gr 4-76 p1612   +NR 9-76 p15
             +HF 11-76 p140   ++RR 3-76 p55
             +-HFN 3-76 p88   ++St 10-76 p120
165   Organ works: Chorale preludes, S 645-650, 684, 688. Fugue, S 577,
       G major. Fantasia, S 562, C minor. Prelude and fugue, S 541,
       G major. Prelude and fugue, S 536, A major. Passacaglia and
       fugue, S 582, C minor. Toccata and fugue, S 565, D minor.
       Toccata and fugue, S 566, E major. David Sanger, org. Saga
       5428/9 (2).
             +Gr 7-76 p188   +-RR 10-76 p84
166   Organ works: Fantasias, S 563, B minor; S 570, C major. Fugues,
       S 578, G minor; S 575, C minor. Preludes, S 569, A minor;
       S 568, G major. Trio, S 584, G minor. Trio sonatas, nos.
       1-6, S 525-530. Michel Chapuis, org. Telefunken BC 25098-T/1-2
       (2).
             +AR 5-75 p51   +-NR 3-74 p14
             +Gr 4-74 p1876   ++PRO 1/2-76 p20
             +HF 9-74 p83   +RR 3-74 p51
             ++HFN 3-74 p22   ++SFC 1-6-74 p32
             ++MJ 5-74 p52   +-St 6-74 p115
             +MQ 10-74 p685
167   Organ works: Chorale preludes: Allein Gott in der Hoh sei Ehr,
       S 675; Aus tiefer Not, S 686; Christ unser Herr, S 685; Dies
       sind die heil'gen zehn Gebot, S 678; Jesus Christus, unser
       Heiland, S 688; Kyrie, Gott Vater in Ewigkeit, S 669; Vater
       unser im Himmelreich, S 683; Wir glauben all einen Gott, S
       680. Duet, S 803, F major. Prelude and fugue, S 552, E flat

major. Fritz Heitmann, org. Telefunken 6-41977. (Reissues)
      *HFN 10-76 p183
168   Partita, flute, S 1013, A minor. Sonatas, flute and harpsichord,
      S 1030, S 1034, S 1035, B minor, E minor, E major. Alexander
      Murray, baroque flute; Martha Goldstein, hpd. Pandora PC 176.
      -St 4-76 p110
      Partita, flute, S 1013, A minor. cf Sonatas, flute, S 1030-5.
      Partitas, harpsichord, nos. 1-6, S 825-830. cf Harpsichord
      works (DG 2722 020).
169   Partita, harpsichord, no. 1, S 825, B flat major. BEETHOVEN:
      Concerto, piano, no. 4, op. 58, G major. CHOPIN: Etude, op.
      25, no. 1, A flat major. Etude, op. 25, no. 2, F minor.
      Nocturne, op. 9, no. 3, B major. DEBUSSY: Images: Reflets dans
      l'eau. RAVEL: Jeux d'eau. Walter Gieseking, pno; Saxon State
      Orchestra; Karl Böhm. Bruno Walter Society RR 415.
         +NR 10-76 p4
      Partita, harpsichord, S 831, B minor. cf Concerto, harpsichord,
      S 971, F major (Angel 36096)
      Partita, harpsichord, S 831, B minor. cf Concerto, harpsichord,
      S 971, F major (Saga 5424)
      Partita, harpsichord, S 831, B minor. cf Harpsichord works
      (DG 2722 020).
      Partita, violin, no. 1, S 1002, B minor: Sarabande. cf Fugue,
      lute, S 1000, G minor.
170   Partita, violin, no. 2, S 1004, D minor. Sonata, violin, no. 3,
      S 1005, C major. Kyung-Wha Chung, vln. Decca SXL 6721. Tape
      (c) KSXC 6721. (also London CS 6940)
         +-Gr 11-75 p859   +-RR 11-75 p67
         +HF 10-76 p102   +St 11-76 p136
         ++HFN 11-75 p149   +STL 11-2-75 p38
         ++NR 8-76 p13
171   Partita, violin, no. 2, S 1004, D minor: Chaconne. MARTINU:
      Sonata, violin and piano. PAGANINI: Le Streghe, variations on
      a theme by Süssmayr, op. 8. SLUKA: Sonata, violin and piano.
      Petr Messiereur, vln; Jarmila Kozderkova, pno; Milan Zelenka,
      gtr. Panton 110 373.
         +-RR 7-76 p69
172   Partita, violin, no. 2, S 1004, D minor. Partita, violin, no. 3,
      S 1006, E major. Nathan Milstein, vln. DG 2530 730. (Reissue
      from 2709 047)
         +Gr 11-76 p837   ++RR 11-76 p88
         +HFN 12-76 p151
      Partita, violin, no. 2, S 1004, D minor: Chaconne. cf Works,
      selections (RCA ARL 1-0880).
      Partita, violin, no. 2, S 1004, D minor: Chaconne. cf Columbia/
      Melodiya M 33593.
      Partita, violin, no. 2, S 1004, D minor: Chaconne (arr. Busoni).
      cf International Piano Library IPL 104.
      Partita, violin, no. 3, S 1006, E major. cf Partita, violin, no.
      2, S 1004, D minor.
      Partita, violin, no. 3, S 1006, E major: Gavotte and rondo. cf
      HMV HLM 7093.
      Partita, violin, no. 3, S 1006, E minor: Minuets, nos. 1 and 2.
      cf RCA ARM 4-0942/7.
      Partita, violin, no. 3, S 1006, E major: Preludio. cf Works,
      selections (RCA ARL 1-0880).

Partita, violin, no. 3, S 1006, E major: Praeludium, Gavotte. cf
    Works, selections (DG 2584 001).
Partita, violin, no. 3, S 1006, E major: Prelude, Loure, Gigue.
    cf CBS 76420.
Partita, violin, no. 3, S 1006, E major: Prelude, Loure, Gigue.
    cf Columbia M2 33444.
Passacaglia and fugue, S 582, C minor. cf Fantasia and fugue,
    S 537, C minor.
Passacaglia and fugue, S 582, C minor. cf Fantasia and fugue,
    S 542, G minor (Decca KSDC 258).
Passacaglia and fugue, S 582, C minor. cf Fantasia and fugue,
    S 542, G minor (Decca SPA 459).
Passacaglia and fugue, S 582, C minor. cf Organ works (DG 2722
    014).
Passacaglia and fugue, S 582, C minor. cf Organ works (Saga
    5428/9).
Passacaglia and fugue, S 582, C minor. cf Argo ZRG 806.
Pastorale, S 590, F major. cf Allabreve, S 589, D major.
Pastorale, S 590, F major. cf Organ works (DG 2722 014).
Pastorale, S 590, F major: Aria. cf CBS 61579.
Prelude, A minor. cf London CS 7015.
Prelude, C minor. cf Connoisseur Society (Q) CSQ 2065.
Preludes, S 568, G major; S 569, A minor. cf Organ works (Tele-
    funken BC 25098-T/1-2).
173   Prelude, S 999, C minor. Prelude, fugue and allegro, S 998, E
        flat major. Suite, lute, S 996, E minor. Suite, lute, S 997,
        C minor. Narciso Yepes, gtr. DG 2530 462.
            +Gr 4-76 p1628   +NR 3-75 p14
            +-HFN 4-76 p100   +-RR 3-76 p57
    Prelude, S 999, C minor. cf Fugue, lute, S 1000, G minor.
    Prelude, S 999, C minor. cf Lute works (Columbia M2 33510).
    Prelude, S 999, C minor. cf Works, selections (Erato STU 70885).
    Prelude, S 999, C minor. cf Saga 5426.
    Prelude and fugue. cf Pye TB 3006.
174   Preludes and fugues, S 531-3, 535, 549-50. Lionel Rogg, org.
        Oryx 1015.
            ++St 12-76 p72
    Preludes and fugues, S 531-536, S 539, S 541, S 543-552. cf
        Organ works (DG 2722 014).
175   Prelude and fugue, S 532, D major. Prelude and fugue, S 544, B
        minor. Toccata, adagio and fugue, S 564, C major. Trio sonata,
        no. 5, S 529, C major. Hans Heintze, org. Nonesuch H 71321.
            +-NR 7-76 p10   +-St 12-76 p72
    Prelude and fugue, S 532, D major. cf Chorale preludes (Pelca
        PRS 40597).
    Prelude and fugue, S 532, D major. cf Organ works (Philips 6500
        925).
    Prelude and fugue, S 532, D major. cf Vista VPS 1030.
    Prelude and fugue, S 533, E minor. cf Wealden WS 121.
    Prelude and fugue, S 534, F minor. cf Chorales, S 695, 710,
        712-3, 717-8, 740.
    Prelude and fugue, S 535, G minor. cf Works, selections (DG
        2584 001).
176   Preludes and fugues, S 536, 541, 543-548. Lionel Rogg, org.
        Oryx 1011/2 (2).
            +-St 12-76 p72

Prelude and fugue, S 536, A major.   cf Organ works (Saga 5428/9).
Prelude and fugue, S 539, D minor.   cf Chorales, S 695, 710,
712-3, 717-8, 740.
Prelude and fugue, S 539, D minor.   cf Works, selections (Tele-
funken AS 6-41111).
Prelude and fugue, S 541, G major.   cf Organ works (Saga 5428/9).
Prelude and fugue, S 541, G major.   cf Klavier KS 551.
Prelude and fugue, S 542, G minor.   cf BACH, J. B.: Du Friedens-
fürst, Herr Jesu Christ, partita.
Prelude and fugue, S 543, A minor.   cf Fantasia and fugue, S 542,
G minor.
Prelude and fugue, S 543, A minor.   cf Telefunken AF 6-41872.
Prelude and fugue, S 544, B minor.   cf Organ works (Classics for
Pleasure CFP 40241).
Prelude and fugue, S 544, B minor.   cf Prelude and fugue, S 532,
D major.
Prelude and fugue, S 545, C major.   cf Chorale prelude: Aus tiefer
Not schrei ich zu dir, S 687.
Prelude and fugue, S 545, C major.   cf Audio EAS 16.
Prelude and fugue, S 545, C major.   cf International Piano Library
IPL 104.
177   Prelude and fugue, S 546, C minor.   FRANCK: Prière, op. 20.
SALLINEN: Chaconne.   STENIUS: Partita on a Finnish folk melody.
Folke Forsmann, org.   BIS LP 12.
        +RR 1-76 p42
Prelude and fugue, S 547, C major.   cf BACH, J. B.: Du Friedens-
fürst, Herr Jesu Christ, partita.
Prelude and fugue, S 548, E minor.   cf Fantasia and fugue, S 542,
G minor.
Prelude and fugue, S 552, E flat major.   cf Organ works (Tele-
funken 6-41977).
Preludes and fugues, S 553-60 (Little).   cf Concerto, organ, no.
5, S 596, D minor.
Prelude, fugue and allegro, S 998, E flat major.   cf Fugue, lute,
S 1000, G minor.
Prelude, fugue and allegro, S 998, E flat major.   cf Lute works
(Columbia M2 33510).
Prelude, fugue and allegro, S 998, E flat major.   cf Prelude,
S 999, C minor.
Prelude, fugue and allegro, S 998, E flat major.   cf Works,
selections (Erato STU 70885).
Presto, A minor.   cf London CS 7015.
178   St. John Passion S 245.   Marianne Koehnlein-Goebel, Elly Ameling,
s; Julia Hamari, con; Dieter Ellenbeck, Wolfgang Isenhardt,
Werner Hollweg, t; Walter Berry, Allan Ahrans, Manfred Acker-
mann, Hermann Prey, bs; Stuttgart Hymnus Boys' Choir; Stuttgart
Chamber Orchestra; Karl Munchinger.   Decca SET 590/2 (3).
        +-Gr 4-75 p1844   +RR 4-75 p58
        +-MT 2-76 p139
129   St. John Passion, 245, excerpts.   Irgard Seefreid, Antonia Fah-
berg, Evelyn Lear, s; Hertha Töpper, con; Ernst Häfliger, t;
Dietrich Fischer-Dieskau, bar; Keith Engen, Max Proebstl,
Hermann Prey, bs; Munich Bach Orchestra and Choir; Munich Boys'
Choir; Karl Richter.   DG 2535 152.   (Reissue from DG Archive
2722 010)
        +HFN 3-76 p111   +RR 3-76 p65

St. John Passion, S 245: Es ist vollbracht. cf Works, selections
(HMV SQ ASD 3265).

180   St. Matthew Passion, S 244. Jo Vincent, s; Ilona Durigo, ms;
      Karl Erb, Lous van Tulder, t; Willem Ravelli, Hermann Schey,
      bs; Piet van Egmond, org; Johannes den Hertog, hpd; Amsterdam
      Toonkunst Choir; Zanglust Boys' Choir; COA; Willem Mengelberg.
      Philips 6747 168 (3). (Reissue from Columbia SL 179).
          +Audio 6-76 p97   ++NYT 8-10-75 pD14
          +HF 8-75 p76   +St 8-75 p94
          ++NR 7-75 p9

181   St. Matthew Passion, S 244. Adele Stolte, Erika Wustmann, Eva
      Hassbecker, s; Annelies Burmeister, Gerda Schriever, con;
      Peter Scheier, Hans-Joachim Rotzsch, t; Theo Adam, bs-bar;
      Johannes Künzel, Siegfried Vogel, Hans Martin Nau, Hermann
      Christian Polster, Günther Leib, bs; Dresden Kreuzchor, St.
      Thomas's Choir; Leipzig Gewandhaus Orchestra; Rudolf Mauers-
      berger. RCA LRL 4-5098 (4).
          +-Gr 2-76 p1363   +-RR 2-76 p54
          +-HFN 3-76 p87

182   St. Matthew Passion, S 244. Elisabeth Feuge, s; Margarethe Klose,
      con; Koloman von Pataky, t; Paul Schöffler, bar; Kurt Böhme,
      bs; Leipzig University Choir, St. Peter's Boys' Choir;
      Leipzig Gewandhaus Orchestra; Hans Weisbach; Karl Meyer, org.
      Friedbert Sammler, hpd. Rococo 1010 (3).
          +-NR 10-76 p9

183   St. Matthew Passion, S 244: Kommt, ihr Töchter, helft mir klagen;
      Du lieber Heiland du; Buss und Reu; Blute nur, du liebes Herz;
      Ich will bei meinem Jesu wachen; Was mein Gott will, das
      g'scheh allzeit; Erbarme dich, mein Gott; Wenn ich einmal soll
      scheiden; Mache dich, mein Herze, rein, Nun ist der Herr zur
      Ruh gebracht; Wir setzen uns mit Tränen nieder. Gundula Jano-
      witz, s; Christa Ludwig, con; Horst Laubenthal, t; Walter
      Berry, bs; Vienna Singverein; BPhO; Herbert von Karajan. DG
      2530 631. (Reissue from 2720 070).
          +-Gr 6-76 p70   ++RR 5-76 p67
          +-HFN 6-76 p80

      St. Matthew Passion, S 244: Peter's denial. cf Works, selections
      (Decca DPA 535/6).

184   Sonata, no. 3. HOY: The DeMazio quintet. Lament. Carol Stein,
      vln; Bonnee Hoy, pno; The Amado Quartet. Encore EN 3003.
          +NR 2-76 p6
      Sonata, solo flute, S 1013, A minor. cf Concerto, 2 violins and
      strings, S 1043, D minor.

185   Sonatas, flute and harpsichord, S 1030-5. Partita, flute, S 1013,
      A minor. Stephen Preston, flt; Trevor Pinnock, hpd; Jordi
      Saval, vla da gamba. CRD CRD 1014/5 (2).
          +Gr 8-75 p336   +St 4-76 p110
          +RR 8-75 p54   +STL 11-2-75 p38

186   Sonatas, flute and harpsichord, S 1030, S 1032, S 1034-5. Son-
      atas, 2 flutes and harpsichord, S 1038-9. Leopold Stastny,
      Frans Brüggen, flt; Alice Harnoncourt, vln; Nikolaus Harnon-
      court, vlc; Herbert Tachezi, hpd. Telefunken EX 6-35339 (2).
          ++RR 12-76 p75
      Sonata, flute and harpsichord, S 1030, B minor. cf Partita, solo
      flute, S 1013, A minor.
      Sonata, flute and harpsichord, S 1031, E flat major: Siciliano.

cf Works, selections (DG 2530 647).
Sonata, flute and harpsichord, S 1031, E flat major: Siciliano.
   cf HMV HQS 1360.
Sonata, flute and harpsichord, S 1033, C minor: Allegro. cf
   RCA LRL 1-5131.
Sonatas, flute and harpsichord, S 1034, S 1035, E minor, E
   major. cf Partita, solo flute, S 1013, A minor.
Sonatas, 2 flutes and harpsichord, S 1038-9. cf Sonatas, flute
   and harpsichord, S 1030, S 1032, S 1034-5.
Sonata, harpsichord, D minor (transcription of Sonata, solo violin,
   S 1003). cf Works, selections (Telefunken AS 6-41111).
Sonata, harpsichord, G major (Adagio, S 968; 2nd, 3rd, 4th move-
   ments reconstructed by Leonhardt after Sonata, solo violin,
   S 1005, C major). cf Works, selections (Telefunken AS 6-41111).
187 Sonatas, viola da gamba and harpsichord, nos. 1-3, S 1027-9.
        Leonard Rose, vlc; Glenn Gould, pno. Columbia M 32934. (also
        CBS 76373)
            -Gr 5-75 p1980   +RR 5-75 p51
            +-HF 12-74 p97   +St 3-75 p96
            +-HFN 5-75 p124   +St 2-76 p741
            +-NR 11-74 p7
188 Sonata, violin, no. 1, S 1001, G minor. LISZT: Etudes d'execution
        transcendente, no. 11, G 139: Harmonies du soir. RACHMANINOFF:
        Etude tableaux, op. 33, no. 3, E flat major. TCHAIKOVSKY: The
        seasons, op. 37, no. 5: May. Theme and variations, op. 19,
        no. 6, F major. Valse scherzo, violin and piano, op. 34.
        Cenek Pavlik, vln; Alfréd Holecek, Ivan Klánský, pno. Panton
        110 515.
            +-RR 10-76 p85
Sonata, violin, no. 1, S 1001, G minor. cf RCA ARM 4-0942/7.
Sonata, violin, no. 2, S 1003, A minor: Andante. cf HMV RLS 723.
Sonata, violin, no. 3, S 1005, C major. cf Partita, violin,
   no. 2, S 1004, D minor.
Sonata, violin, no. 3, S 1005, C major. cf RCA ARM 4-0942/7.
Sonata, violin, no. 3, S 1005, C major: Adagio. cf Works,
   selections (DG 2584 001).
189 Sonata, violin and harpsichord, A minor. DEBUSSY: Sonata, violin
        and piano. MOZART: Divertimento, no. 15, K 287, B flat major.
        Joseph Szigeti, vln; Andor Földes, pno; Goberman Chamber Orch-
        estra; Max Goberman. Bruno Walter Society WSA 704. (Reissue
        from Columbia X-2, CX 242, CM 322).
            +ARSC Vol VIII, no. 2-3, p92
            ++NR 11-76 p7
190 Sonatas, violin and harpsichord, nos. 1-6, S 1014-1019. Eduard
        Melkus, vln; Huguette Dreyfus, hpd. DG Archive 2708 032 (2).
        Tape (c) 3375 002.
            +Gr 8-76 p311   +NR 8-76 p5
            ++Gr 11-76 p887 tape   ++RR 9-76 p79
            +-HF 9-76 p88   ++St 9-76 p116
            +HFN 9-76 p112
191 Sonatas, violin and harpsichord, nos. 1-6, S 1014-1019. Alice
        Harnoncourt, vln; Nikolaus Harnoncourt, vla da gamba; Herbert
        Tachezi, hpd. Telefunken EX 6-35310.
            +-Gr 9-76 p436   ++RR 10-76 p84
            +HFN 9-76 p112
192 Sonata, violin and harpsichord, no. 6, S 1019, G major. BUSONI:

Sonata, violin and piano, no. 2, op. 36a, E minor. Sergiu
Luca, vln; David Golub, pno; Albert Fuller, hpd. Delos DEL
25404.
  +HF 6-76 p77   +St 3-76 p109
  ++NR 8-76 p5
Sonata, 2 violins and harpsichord, S 1037, C major: Gigue. cf
CBS 73487.

193 Sonatas and partitas, solo violin, S 1001-6. Yehudi Menuhin, vln.
Angel SC 3817 (3). (also HMV SLS 5045)
  +-Gr 4-76 p1617   ++NR 7-76 p12
  +-HF 10-76 p102   +-RR 7-76 p69
  +-HFN 6-76 p80   +-St 11-76 p136
  +-MT 9-76 p744

194 Sonatas and partitas, solo violin, S 1001-6. Nathan Milstein,
vln. DG 2709 047 (3).
  ++Audio 6-76 p100   ++RR 4-75 p47
  ++Gr 4-75 p1830   ++SFC 10-26-75 p24
  ++HF 1-76 p81   ++St 2-76 p104
  ++NR 12-75 p15   ++STL 5-4-75 p37

195 Sonatas and partitas, solo violin, S 1001-6. Paul Zukofsky, vln.
Vanguard VSD 71194/6 (3).
  +-HF 3-75 p76   -SFC 1-19-75 p28
  ++MJ 2-75 p40   +SR 1-11-75 p51
  +-NR 2-75 p14   +-St 2-76 p104
Sonatas and partitas, violin, S 1002: Sarabande; S 1003: Andante.
cf Saga 5426.

196 Songs (from Schemellis Gesangbuch): Ach, dass nicht die letzte
Stunde; Die bittre Leidenszeit beginnet abermal, S 450;
Brich entzwei, mein armes Herze; Dir, Dir, Jehova, will ich
singen, S 299; Eins ist not, Ach Herr, dies eine; Es kostet
viel, ein Christ zu sein, S 459; Gott lebet noch; Gott, wie
gross ist deine Güte; Die güldne Sonne, voll Freud und Wonne;
Ich lass dich nicht, du musst mein Jesus bleiben; Ich steh an
deiner Krippen hier; Ihr Gestirn, ihr hohen Lüfte; Komm,
süsser Tod, S 478;; Kommt, Seelen, dieser Tag muss heilig sein
besungen; Kommt wieder aus der finstern Gruft; Der lieben Sonne
Licht und Pracht; Liebster Herr Jesu, wo bleibst du so lange;
Mein Jesu, was für Seelenweh; O Jesulein süss, O Jesulein mild,
S 493; So gehst du nun, mein Jesu, hin; So gibst du nun, mein
Jesu, gute Nacht; Steh ich bei meinem Gott; Vergiss mein nicht,
mein allerliebster Gott, S 505; Wo ist mein Schäflein, das ich
liebe; Anna Magdalena notebook, S 508: Gib dich zufrieden und
sei still; Wie wohl ist mir, O Freund der Seelen. Elisabeth
Speiser, s; Peter Schreier, t; Hedwig Bilgram, org. DG Archive
2533 299.
  +NR 4-76 p10   +St 9-76 p116
  ++SFC 8-1-76 p27
Suite, no. 3: Aria. cf HMV RLS 723.
Suite, flute, no. 2, S 1067, B minor. cf Brandenburg concerto,
no. 4, S 1049, G major.
Suite, flute, no. 2, S 1067, B minor. cf RCA CRL 3-1430.
Suite, lute, S 995, G minor. cf Fugue, lute, S 1000, G minor.
Suite, lute, S 995, G minor. cf Lute works (Columbia M2 33510).
Suite, lute, S 995, G minor: Prelude; Presto. cf Works, selec-
tions (Erato STU 70885).
Suite, lute, S 996, E minor. cf Fugue, lute, S 1000, G minor.

Suite, lute, S 996, E minor. cf Lute works (Columbia M2 33510).
Suite, lute, S 996, E minor. cf Prelude, S 999, C minor.
Suite, lute, S 996, E minor. cf Works, selections (Erato STU
     70885).
Suite, lute, S 996, E minor. cf Swedish Society SLT 33189.
Suite, lute (harpsichord), S 997, C minor. cf Concerto,
     harpsichord, S 971, F major.
Suite, lute, S 997, C minor. cf Lute works (Columbia M2 33510).
Suite, lute, S 997, C minor. cf Prelude, S 999, C minor.
Suite, lute, S 997, C minor: Prelude and fugue. cf Philips
     6581 017.
Suite, lute, S 1006a, E major. cf Fugue, lute, S 1000, G minor.
Suite, lute, S 1006a, E major. cf Lute works (Columbia M2 33510).
Suite, lute, S 1006a, E major. cf CBS 30066.
197 Suites, orchestra, S 1066-9. Collegium Aureum. BASF BHM 23
     20332 (2).
               +AR 2-76 p133   +NR 1-74 p5
               /Gr 10-72 p679   -RR 10-72 p58
               +HFN 5-73 p978   ++SFC 3-10-74 p26
198 Suites, orchestra, S 1066-9. Ars Rediviva Orchestra; Milan
     Munclinger. Supraphon 110 1361/2 (2).
               +-HFN 8-75 p71   +RR 7-75 p22
               +NR 4-75 p5   +St 2-76 p98
199 Suite, orchestra, S 1067, B minor. Suite, orchestra, S 1068, D
     major. BPhO; Herbert von Karajan. DG 2535 138. (Reissue
     from SLPM 138 978)
               +-Gr 4-76 p1588   +-RR 3-76 p33
               +HFN 3-76 p109
Suite, orchestra, S 1067, B minor. cf Cantata, no. 209: Sinfonia.
Suite, orchestra (flute and strings), S 1067, B minor. cf Con-
     certo, flute, violin and harpsichord, S 1044, A minor.
Suite, orchestra, S 1067, B minor. cf Concerto, 2 violins and
     strings, S 1043, D minor.
Suite, orchestra, S 1067, B minor: Minuet; Badinerie. cf RCA
     LRL 1-5094.
Suite, orchestra, S 1067, B minor: Minuet; Badinerie. cf RCA
     LRL 1-5127.
Suite, orchestra, S 1067, B minor: Rondeau; Minuet; Badinerie.
     cf Decca SPA 394.
Suite, orchestra, S 1068, D major. cf Suite, orchestra, S 1067,
     B minor.
Suite, orchestra, S 1068, D major. cf Works, selections (Decca
     DPA 535/6).
Suite, orchestra, S 1068, D major. cf Bruno Walter Society RR 443.
Suite, orchestra, S 1068, D major: Air. cf Works, selections
     (RCA ARL 1-0880).
Suite, orchestra, S 1068, D major: Air on the G string. cf
     Brassworks Unlimited BULP 2.
Suite, orchestra, S 1068, D major: Air on the G string. cf Golden
     Crest CRS 4143.
Suite, solo violoncello, no. 1, S 1007, G major: 3 movements. cf
     RCA ARL 1-0864.
200 Suite, solo violoncello, no. 3, S 1009, C major. KODALY: Sonata,
     solo violoncello, op. 8. Frans Helmerson, vlc. BIS LP 25.
               ++Gr 6-76 p62   ++St 8-76 p92
               +RR 8-76 p67

Suite, solo violoncello, no. 3, S 1009, C major: Bourrée. cf
CBS 73487.

201   Suite, solo violoncello, no. 5, S 1011, C minor. BRITTEN: Suite,
violoncello, no. 1, op. 72. Frans Helmerson, vlc. BIS LP 5.
  +Gr 10-76 p622   ++RR 11-75 p68
  ++HFN 11-75 p148

Toccata, C major: Adagio. cf HMV SQ ASD 3283.

Toccatas, S 910-916. cf Chromatic fantasia and fugue, S 903, D
minor.

202   Toccatas, S 911-915. Zola Shaulis, pno. DG 2530 697.
  ++Gr 10-76 p621   -RR 9-76 p79
  +HFN 9-76 p117

Toccata, S 912, D major. cf Chromatic fantasia and fugue, S 903,
D minor (Decca ECS 788).

Toccata, S 912, D major. cf Chromatic fantasia and fugue, S 903,
D minor (Philips 6833 184).

Toccata, S 912, D major. cf Saga 5402.

Toccata, S 914, E minor. cf Chromatic fantasia and fugue, S 903,
D minor.

Toccata, S 916, C major. cf Chromatic fantasia and fugue, S 903,
D minor.

Toccata, adagio and fugue, S 564, C major. cf Organ works (DG
2722 014).

Toccata, adagio and fugue, S 564, C major. cf Prelude and fugue,
S 532, D major.

Toccata, adagio and fugue, S 564, C major. cf Toccata and fugue,
S 538, C minor.

203   Toccata and fugue, S 538, C minor. Toccata and fugue, S 540, F
major. Toccata, adagio and fugue, S 564, C major. Lionel
Rogg, org. Oryx 1005.
  +-St 12-76 p72

Toccata and fugue, S 538, D minor. cf Chorale prelude: Aus tiefer
Not schrei ich zu dir, S 687.

Toccata and fugue, S 538, D minor. cf Organ works (DG 2722 014).

Toccata and fugue, S 540, F major. cf Chorale preludes: Sei
gegrüsset Jesus gütig, S 768; Von Gott will ich nicht lassen,
S 658.

Toccata and fugue, S 540, F major. cf Organ works (Classics for
Pleasure CFP 40241).

Toccata and fugue, S 540, F major. cf Organ works (DG 2722 014).

Toccata and fugue, S 540, F major. cf Toccata and fugue, S 538,
C minor.

204   Toccata and fugue, S 565, D minor (orch. Ormandy). BIZET:
L'Arlésienne: Suites nos. 1 and 2: Farandole. CHABRIER: España.
TCHAIKOVSKY: Capriccio italien, op. 45. PO; Eugene Ormandy.
CBS 30071. Tape (c) 40-30071. (Reissues from SPR 43, 72527)
  +-Gr 11-76 p888   +-RR 8-76 p27
  +-HFN 10-76 p183   +-RR 12-76 p105 tape
  +-HFN 10-76 p185 tape

205   Toccata and fugue, S 565, D minor (orch. Brun). LISZT: Hungarian
rhapsody, no. 2, G 244, G sharp minor (orch. Brun). ROSSINI:
La gazza ladra: Overture (trans. Richard). SUPPE: Poet and
peasant: Overture (trans. Richard). French Garde Republicaine
Band; François Julien Brun. Connoisseur CS 2112.
  ++NR 12-76 p15

206   Toccata and fugue, S 565, D minor. Concerto, organ, no. 2, S 593,

A minor. BOELLMANN: Suite gothique, op. 25. MULET: Byzantian
sketches: Wiehnachtslied. Du bist der Fels. Kirchenfenster.
Klaus Germann, org. Pelca PRS 40576.
    -NR 7-76 p10
207   Toccata and fugue, S 565, D minor. Trio sonata, no. 6, G 530,
    G major. MOZART: Adagio and allegro, K 594, F minor. SWEELINCK:
    Mein Junges Leben hat ein End, variations. Gustav Leonhardt,
    Erich Piasetzki, Daniel Chorzempa, Gabor Lehotka, org. Philips
    6833 141.
    +Gr 4-76 p1644   +RR 8-76 p67
Toccata and fugue, S 565, D minor. cf Fantasia and fugue, S 537,
   C minor.
Toccata and fugue, S 565, D minor. cf Fantasia and fugue, S 542,
   D minor (Decca KSDC 258).
Toccata and fugue, S 565, D minor. cf Fantasia and fugue, S 542,
   G minor (Decca SPA 459).
Toccata and fugue, S 565, D minor. cf Organ works (DG 2722 014).
Toccata and fugue, S 565, D minor. cf Organ works (Saga 5428/9).
Toccata and fugue, S 565, D minor. cf Works, selections (Decca
   DPA 535/6).
Toccata and fugue, S 565, D minor (arr. Stokowski). cf Works,
   selections (DG 2584 001).
Toccata and fugue, S 565, D minor. cf BACH, J. B.: Du Friedens-
   fürst, Herr Jesu Christ, partita.
Toccata and fugue, S 565, D minor. cf Decca SDD 499.
Toccata and fugue, S 565, D minor. cf Decca DPA 523/4.
Toccata and fugue, S 565, D minor. cf HMV SQ HQS 1356.
Toccata and fugue, S 565, D minor. cf Telefunken AF 6-41872.
Toccata and fugue, S 566, E major. cf Canonic variations, S 769.
Toccata and fugue, S 566, E major. cf Organ works (Saga 5428/9).
Toccata and fugue, S 910, F sharp minor. cf Fantasia, S 906, C
   minor.
Trio, S 583, D minor. cf Canonic variations, S 769.
Trio, S 583, D minor. cf Telefunken AF 6-41872.
Trio, S 584, G minor. cf Organ works (Telefunken BC 25098-T/1-2).
208   Trio sonatas, nos. 1-6, S 525-530. Gunther Pohl, flt; Waldemar
    Doling, hpd. Barenreiter-Musicaphon BM 30 SL 1913/4 (2).
    ++AR 5-76 p27
Trio sonatas, nos. 1-6, S 525-530. cf Chorales, S 653b, 720, 727,
   733-4, 736-7.
Trio sonatas, nos. 1-6, S 525-530. cf Organ works (DG 2722 014).
Trio sonatas, nos. 1-6, S 525-530. cf Organ works (Telefunken
   BC 25098-T/1-2).
Trio sonata, no. 1, S 525, E flat major. cf Chorale preludes
   (Pelca PSR 40597).
Trio sonata, no. 1, S 525, E flat major. cf Organ works (DG
   2533 140).
209   Trio sonatas, nos. 2, 4-6, S 526, S 528-30. Paul Hersh, vla;
    Laurette Goldberg, hpd. 1750 Arch S 1756.
    +NR 12-75 p8   +PRO 1/2-76 p20
Trio sonata, no. 5, S 529, C major. cf Prelude and fugue, S 532,
   D major.
Trio sonata, no. 5, S 529, C major. cf Abbey LPB 752.
Trio sonata, no. 6, S 530, G major. cf Organ works (DG 2533 140).
Trio sonata, no. 6, G 530, G major. cf Toccata and fugue, S 565,
   D minor.

210   Variations, harpsichord, S 988 (Goldberg). David Sanger, hpd.
       Saga 5395.
            +-Gr 7-75 p213   /MT 12-76 p1003
            +HFN 7-75 p72   +RR 6-75 p58
       Das wohltemperierte Klavier: Fugues, nos. 2, 7, 17. cf Coronet
       LPS 3032.
       Das wohltemperierte Klavier: Prelude, no. 1, S 846, C major; Pre-
       lude and fugue, no. 2, S 847, C minor.
211   The well-tempered clavier, Bk II, S 870-893: Prelude, no. 22.
       BRAHMS: Variations and fugue on a theme by Handel, op. 24.
       MOZART: Minuet, K 355, D major. Variations on "Ah, vous
       dirai-je, Maman, K 265", C major. Clarion Wind Quintet.
       Golden Crest CRS 4146.
            +IN 8-76 p22
212   Works, selections: Brandenburg concerto, no. 2, S 1047, F major.
       Cantata, no. 147, Jesu, joy of man's desiring. Chorale pre-
       lude: Wachet auf, S 645. Concerto, harpsichord, S 971, F
       major. Christmas oratorio, S 248, excerpts. Mass, S 232, B
       minor: Sanctus and Osanna. St. Matthew Passion, S 244: Peter's
       denial. Suite, orchestra, S 1068, D major. Toccata and fugue,
       S 565, D minor. George Malcolm, hpd; Karl Richter, Simon
       Preston, org; St. John's College Chapel Choir; Stuttgart Orch-
       estra; George Guest, Karl Münchinger. Decca DPA 535/6. Tape
       (c) KDPC 535/6.
            +-Gr 6-76 p92   +HFN 6-76 p103
            +Gr 9-76 p497 tape   +-RR 6-76 p32
213   Works, selections: Cantata, no. 29, Wir danken dir Gott: Sinfonia.
       Cantata, no. 140, Wachet auf, ruft uns die Stimme. Cantata,
       no. 147, Jesu bleibt meine Freude. Chorale preludes: Herz-
       lich tut mich verlangen (Befiehl du deine Wege), S 727; Ich
       ruf zu dir, Herr Jesu Christ, S 639; In dulci jubilo; Nun
       komm der Heiden Heiland, S 659. Concerto, harpsichord, no. 5,
       S 1056, F minor: Largo. Sonata, flute and harpsichord, S 1031,
       E flat major: Siciliano. GLUCK: Orfeo ed Euridice: Ballet
       music. HANDEL: Minuet, G minor. (Piano transcriptions by
       Kempff). Wilhelm Kempff, pno. DG 2530 647.
            -Gr 5-76 p1788   +RR 5-76 p63
            +-HFN 5-76 p92
214   Works, selections: Cantata, no. 147, Jesu, joy of man's desiring.
       Cantata, no. 208, Sheep may safely graze (arr. Walton). Pre-
       lude and fugue, S 535, G minor (arr. Caillet). Fugue, S 577,
       G major (arr. Holst). Easter oratorio, S 249: Sinfonia (arr.
       Whittaker). Partita, violin, no. 3, S 1006, E major: Praeludium,
       Gavotte (arr. Bachrich). Sonata, violin, no. 3, S 1005, C
       major: Adagio (arr. Bachrich). Toccata and fugue, S 565, D
       minor (arr. Stokowski). BPO; Arthur Fiedler. DG 2584 001.
            +-Gr 5-76 p1807   +-RR 6-76 p33
            +-HFN 6-76 p80
215   Works, selections: Fugue, lute, S 1000, G minor. Prelude, S 999,
       C minor. Prelude, fugue and allegro, S 998, E flat major.
       Suite, lute, S 996, E minor. Suite, lute, S 995, G minor:
       Prelude, Presto. Turibio Santos, gtr. Erato STU 70885.
            +-Gr 12-75 p1075   +RR 12-75 p77
            ++HFN 1-76 p102
216   Works, selections: Anna Magdalena's notebook, S 508: Bist du bei
       mir (arr. Connah). Cantata, no. 6, Bleib bei uns: Hochgelobter
       Gottessohn. Cantata, no. 11, Lobet Gott: Ach, bleibe doch.

Cantata, no. 34, O ewiger Feuer: Wohl euch, ihr auserwählten
Seelen. Cantata, no. 129, Gelobet sei der Herr. Cantata, no.
161, Komm, du süsse Todesstunde. Cantata, no. 190, Lobe,
Zion, deinen Gott. Christmas oratorio, S 248: Bereite dich,
Zion. Easter oratorio, S 249: Saget, saget mir geschwinde.
Magnificat, S 243, D major: Et exultavit. St. John Passion,
S 245: Es ist vollbracht. Janet Baker, s; AMF; Neville Marri-
ner. HMV SQ ASD 3265.
    +Gr 10-76 p633   ++RR 10-76 p92

217 Works, selections: Cantata, no. 156: Arioso. Chorale preludes:
Ein feste Burg, S 720; Komm süsser Tod; Wachet auf, S 645.
Fugue, S 578, G minor. Partita, violin, no. 2, S 1004, D
minor: Chaconne. Partita, violin, no. 3, S 1006, E major:
Preludio. Suite, orchestra, S 1068, D major: Air. LSO;
Leopold Stokowski. RCA ARL 1-0880. Tape (c) ARK 1-0880 (ct)
ARS 1-0880 (Q) ARD 1-0880.
    +Audio 1-76 p69   +NR 7-75 p3
    +HF 2-76 p130 tape   +RR 9-75 p26
    +HFN 5-76 p92   +SFC 6-15-75 p24

218 Works, selections (Works for solo violin in versions for keyboard
instruments): Prelude and fugue, S 539, D minor. Sonata,
harpsichord (transcription of Sonata, solo violin, S 1003,
attr. to Bach). Sonata, harpsichord, G major (Adagio, S 968;
2nd, 3rd, 4th movements reconstructed by Leonhardt after
Sonata, solo violin, S 1005, C major). Gustav Leonhardt,
hpd, org. Telefunken AS 6-41111.
    +-Gr 4-76 p1628   +-RR 2-76 p47
    +-HFN 2-76 p91

BACH, Wilhelm Friedemann
    Duet, flute, F major. cf BACH, J. C.: Quartet, 2 flutes, violin
      and violoncello, C major.

BACH, Wilhelm Friedrich Ernst
    Das Dreyblatt. cf Desmar DSM 1005.
    Trio, 2 flutes and viola, G major. cf BACH, J. C.: Quartet,
      2 flutes, violin and violoncello, C major.

BACK, Sven-Erik
    Neither nor. cf Caprice RIKS 59.

BADARZEWSKA-BARANOWSKA, Tekla
    The maiden's prayer. cf Decca SPA 473.

BADINGS, Henk
219 Intermezzo. HOLLER: Fantasie, op. 49. SCHROEDER: Praeludium,
Kanzone and Rondo. WALTHER: Sonata, G major. Richard Prosser,
vln; John Burke, org. Available from John Burke, First Con-
gregational Church, Berkeley, Calif.
    +MU 10-76 p14

BAGLEY, E.
    National emblem. cf Department of Defense Bicentennial Edition
      50-1776.
    National emblem. cf Michigan University SM 0002.
    National emblem. cf RCA AGL 1-1334.

BAINTON, Edgar
    And I saw a new heaven. cf Argo ZRG 789.

BAIRSTOW, Edward
    Blessed city, heavenly Salem. cf Abbey LPB 757.
    The lamentation. cf Abbey LPB 739.
    Let all mortal flesh keep silence. cf HMV HQS 1350.

Songs: Jesu, the very thought of Thee; Lord, I call upon thee.
cf Audio 1.

BAKER, David
220   Le chat qui pêche. GOULD: Symphonette, no. 2. Linda Andersons,
s; Jamey Aebersold, alto and tenor sax; Dan Haerle, pno;
John Clayton, bs; Charlie Craig, drums; Louisville Orchestra;
Jorge Mester. Louisville LS 751.
+MJ 7-76 p57   +-St 7-76 p106
221   Sonata, violoncello and piano. WHITE: Concerto, violin, F sharp
minor (ed. Glass and Moore). Aaron Rosand, vln; Janos Starker,
vlc; Alain Plaines, pno; LSO; Paul Freeman. Columbia M 33432.
+HF 10-75 p65   +NYT 4-11-76 pD23
++NR 9-75 p9

BAKER, George
Far-West toccata. cf Delos FY 025.

BAKFARK, Balint
Fantasia (after "D'amour me plains" by Roger). cf Hungaroton
SLPX 11721.

BALADA, Leonardo
Cuatris, 4 instruments. cf Serenus SRS 12064.
Música en cuatro tiempos, piano. cf Serenus SRS 12064.

BALAI, Leonid
222   Divertimento, wind quintet and harp, op. 7. HINDEMITH: Sonata,
4 horns. POOT: Concertino, wind quartet. VILLA-LOBOS: Chôros,
no. 4. Leningrad State Philharmonic Wind Quintet. ABC Westminster WGS 8322.
+NR 10-76 p7   ++SFC 6-6-76 p33

BALAKIREV, Mily
223   Russia. BORODIN: Symphony, no. 2, B minor. RIMSKY-KORSAKOV:
Skazka, op. 29. Bournemouth Symphony Orchestra; Anshel Brusi-
low. HMV SQ ASD 3193. Tape (c) TC ASD 3193.
+Gr 7-76 p167   +HFN 11-76 p175 tape
+-Gr 11-76 p887 tape   +-RR 7-76 p48
+HFN 7-76 p83   +STL 7-4-76 p36
224   Symphony, no. 1, C major.* BORODIN (orch. Rimsky-Korsakov):
Prince Igor: Polovtsian dances. · RPO; Beecham Choral Society;
Thomas Beecham. HMV SXLP 30171. Tape (c) TC EXE 193. (*Re-
issue from SXLP 30002)
++Gr 10-74 p676   ++RR 10-74 p33
+HFN 6-76 p105 tape

BALASSA, Sándor
225   Cantata V, op. 21. Legend, op. 12. Requiem for Lajos Kossák,
op. 15. Erika Sziklay, s; Sándor Palcso, t; Endre Utö, bar;
HRT Orchestra and Chorus; György Lehel, Ferenc Sapszon.
Hungaroton SLPX 11681.
++HFN 11-75 p149   ++RR 9-75 p67
+NR 7-75 p8   +Te 3-76 p26
226   Iris, op. 22. Lupercalia, op. 24. Tabulae, op. 25. Xenia, op.
20. Budapest Chamber Ensemble, Budapest Symphony Orchestra;
György Lehel, András Mihály. Hungaroton SLPX 11732.
+NR 9-76 p5
Legend, op. 12. cf Cantata V, op. 21.
Lupercalia, op. 24. cf Iris, op. 22.
Requiem for Lajos Kossák, op. 15. cf Cantata V, op. 21.
Tabulae, op. 25. cf Iris, op. 22.
Xenia, op. 20. cf Iris, op. 22.

BALAZS
    Songs: Borsod song; The song about melody. cf Hungaroton SLPX
        11696.
    Songs: Deceptive enticing; Song about the song. cf Hungaroton
        SLPX 11762.
BALBASTRE, Claude
    La bellaud. cf Works, selections (Philips 6581 013).
    La castelmore. cf Works, selections (Philips 6581 013).
    La d'héricourt. cf Works, selections (Philips 6581 013).
    Fugue et duo. cf Calliope CAL 1917.
    La genty. cf Works, selections (Philips 6581 013).
    La laporte. cf Works, selections (Philips 6581 013).
    La lugeac. cf Works, selections (Philips 6581 013).
    La malesherbe. cf Works, selections (Philips 6581 013).
    La segur. cf Works, selections (Philips 6581 013).
227 Works, selections: La bellaud. La castelmore. La d'héricourt.
        La genty. La laporte. La malesherbe. La lugeac. La segur.
        DUPHLY: La félix. La de Drummond. La de vatre. La forqueray.
        La millettina. Médée. Rondeau. La victoire à Madame Vic-
        toire de France. Blandine Verlet, hpd. Philips 6581 013.
            ++Gr 4-76 p1643   ++MT 8-76 p666
            +-HFN 3-76 p101   +RR 3-76 p59
BALFE, Michael
    Killarney. cf Rococo 5397.
    Songs: The bohemian girl; The heart bowed down. cf Library of
        Congress OMP 101/2.
BALL, Eric
    Cornish festival overture. cf Pye TB 3007.
    Holiday overture. cf Pye TB 3007.
    Journey into freedom. cf HMV OU 2105.
BALLARD, Robert
    Allemande. cf BIS LP 22.
    Branles de village. cf ATTAINGNANT: Basse dance.
    Courante. cf ATTAINGNANT: Basse dance.
    Courante. cf BIS LP 22.
    Entrée de Luth, nos. 1-3. cf ATTAINGNANT: Basse dance.
    Prelude. cf BIS LP 22.
    Recontins. cf BIS LP 22.
    Tourdion. cf CBS 76183.
BALLET, William
    Sweet was the song the virgin sang. cf Abbey LPB 761.
BANCHIERI, Adriano
    Contrapunto bestiale. cf HMV CSD 3756.
    Dialogo per organo. cf Columbia M 33514.
    Rostiva i corni. cf Hungaroton SLPX 11669/70.
BANKS, Don
228 Concerto, horn. MUSGRAVE: Concerto, clarinet. SEARLE: Aubade,
        op. 28. Barry Tuckwell, hn; Gervase de Peyer, clt; NPhO, LSO;
        Norman Del Mar. Argo ZRG 726.
            ++Gr 8-75 p315   ++NR 2-76 p4
            +HF 6-76 p90   ++RR 9-75 p40
            +-HFN 11-75 p149   +SR 1-24-76 p53
            ++MT 3-76 p235   +-Te 3-76 p31
BANTOCK, Granville
    The frogs of Aristophanes overture. cf Decca SB 322.

BARBER, Samuel
229   Adagio, strings. Medea, op. 23: Suite. Overture to School for
        scandal, op. 5. Symphony, no. 1, op. 9. Eastman Rochester
        Symphony Orchestra; Howard Hanson. Mercury SRI 75012. (Reissues)
              +NYT 7-4-76 pD1   +SFC 10-6-74 p26
230   Antony and Cleopatra, excerpts. Knoxville, summer of 1915.
        Leontyne Price, s; NPhO; Thomas Schippers. RCA LSC 3062.
              +St 7-76 p74
231   Concerto, piano, op. 38. Concerto, violin, op. 14. John Browning,
        pno; Isaac Stern, vln; CO, NYP; Georg Szell, Leonard Bernstein.
        CBS 61621. (Reissues from CBS SBRG 72345, Columbia SAX 2575)
              +Gr 3-75 p1639   +-RR 4-75 p21
              +-MT 11-76 p912   +-St 2-76 p739
232   Concerto, violin, op. 14. Symphony, no. 1, op. 9. Wolfgang
        Stavonhagen, vln; Imperial Philharmonic Symphony, Japan
        Philharmonic Symphony Orchestra; William Strickland. CRI
        SD 137.
              +-Gr 2-76 p1331   /RR 1-76 p29
      Concerto, violin, op. 14. cf Concerto, piano, op. 38.
233   Essays, orchestra, no. 1, op. 12. Essays, orchestra, no. 2, op.
        17. Night flight, op. 19a. Symphony, no. 1, op. 9. LSO;
        David Measham. Unicorn RHS 342.
              ++Gr 9-76 p409   +RR 8-76 p27
              +HFN 9-76 p117
      Essays, orchestra, no. 2, op. 17. cf Essays, orchestra, no. 1,
        op. 12.
      Knoxville, summer of 1915. cf Antony and Cleopatra, excerpts.
      Medea, op. 23: Suite. cf Adagio, strings.
      Mutations from Bach. cf Unicorn RHS 339.
      Night flight, op. 19a. cf Essays, orchestra, no. 1, op. 12.
      Overture to School for scandal, op. 5. cf Adagio, strings.
234   Quartet, strings, op. 11, B minor. IVES: Quartet, strings, no. 2.
        Scherzo, string quartet. Cleveland Quartet. RCA ARL 1-1599 (Q)
        ARD 1-1599.
              ++HF 10-76 p110   +NR 9-76 p6
              -MJ 11-76 p44   ++St 11-76 p136
      Quartet, strings, op. 11, B minor: Adagio. cf Argo ZRG 845.
      Reincarnation, op. 16. cf Orion ORS 75205.
235   Sonata, piano, op. 26. PROKOFIEV: Sonata, piano, no. 7, op. 83,
        B flat major. SCRIABIN: Etude, op. 8, no. 7, B flat minor.
        Etude, op. 42, no. 5, C sharp minor. Prelude, op. 11, no. 5,
        D major. Prelude, op. 22, no. 1, G sharp minor. Sonata, piano,
        no. 9, op. 68, F major. Vladimir Horowitz, pno. RCA VH 014
        (Reissues from RB 6555, RB 16019, HMV ALP 1431).
              +-Gr 1-76 p1221
      Songs: The daisies; Monks and raisins; Nocturne; Songs to poems
        from chamber music (by James Joyce); Sure on this shining night;
        With rue my heart is laden. cf Cambridge CRS 2715.
      Souvenirs: Pas de deux. cf Personal Touch 88.
      Symphony, no. 1, op. 9. cf Adagio, strings.
      Symphony, no. 1, op. 9. cf Concerto, violin, op. 14.
      Symphony, no. 1, op. 9. cf Essays, orchestra, no. 1, op. 12.
      Vanessa: Intermezzo. cf Columbia MG 33728.
BARBERIIS (16th century)
      Madonna, qual certezza. cf L'Oiseau-Lyre 12BB 203/6.
BARBIERI, Francisco
      Songs: El barberillo de lavapiés; Canción de paloma; Jugar con

         fuego; Romanza de la Duquesa. cf London OS 26435.
BARBIREAU, Jacques.
      Songs: Een vrolic Wesen. cf HMV SLS 5022 (2)
      Songs: Ein fröhlich wesen. cf HMV SLS 5049.
BARDOS
      Songs: Gay melodies; Uszküdárá; The youth of March. cf Hungaroton
         SLPX 11696.
      Songs: Jocose marrying off; Hey put it right. cf Hungaroton SLPX
         11762.
BARGONI
      Autumn concerto. cf Rediffusion 15-46.
BARKIN, Elaine
236   Quartet, strings. BOYKAN: Quartet, strings, no. 1. Contemporary
         Quartet, American Quartet. CRI SD 338.
            +MQ 10-76 p616   +NR 9-75 p8
BARNARD, Charlotte
      Come back to Erin. cf HMV RLS 719.
BARNBY, Joseph
      Sweet and low. cf Prelude PRS 2501.
      THE BAROQUE SOUND OF THE TRUMPET. cf Argo ZDA 203.
BARRIOS, Agustin
      Danza paraguaya. cf Saga 5412.
BARSOTTI
      Sundown. cf Canon VAR 5968.
BARTA, Lubor
237   Quintet, winds, no. 2. FELD: Quintet, winds, no. 2. FLOSMAN:
         Sonata, wind quintet and piano. KALABIS: Little chamber music,
         wind quintet, op. 27. Prague Wind Quintet. Supraphon 11 1426.
            +NR 7-76 p5   +RR 8-76 p62
238   Sonata, guitar. BURGHAUSER: Czech dances (3). Sonata, guitar,
         E minor. OBROVSKA: Preludes (6). TRUHLAR: Impromptu. Milan
         Zelenka, gtr. Supraphon 111 0969.
            +SFC 4-25-76 p30
239   Symphony, no. 3. VALEK: Symphony, no. 7. Janáček Philharmonic
         Orchestra; Otakar Trhlik. Panton 110 393.
            ++RR 3-76 p34
BARTLETT
      Praise the Lord with joyful cry. cf Grosvenor GRS 1039.
BARTOK, Bela
      Bagatelles (3). cf Swedish Society SLT 33200.
240   Bluebeard's castle, op. 11. Tatiana Troyanos, ms; Siegmund
         Nimsgern, bs; BBC Symphony Orchestra; Pierre Boulez. CBS 76518.
         Tape (c) 40-76518 (also Columbia M 34217).
            ++Gr 9-76 p466   ++HFN 11-76 p175 tape
            ++HFN 9-76 p114   ++RR 9-76 p26
      Concerto, piano, no. 3, E major. cf Concerto, piano, no. 1,
         A major.
241   Concerto, orchestra. BPhO; Herbert von Karajan. DG 2535 202.
         (Reissue from SLPM 139 003).
            +-Gr 12-76 p989   +RR 12-76 p51
            ++HFN 12-76 p151
242   Concerto, orchestra. The miraculous Mandarin, op. 19: Suite.
         Strasbourg Philharmonic Orchestra; Alain Lombard. Erato STU
         70835.
            -Gr 12-75 p1022   +-RR 1-76 p30
            +-MM 4-76 p29

243   Concerto, orchestra. Hungarian sketches. Israel Philharmonic
       Orchestra; Zubin Mehta. London CS 6949. Tape (c) 5-6949 (also
       Decca SXL 6730 Tape (c) KSXC 6730).
            ++ARG 12-76 p13   +HFN 12-76 p136
             +Gr 12-76 p989   /NR 12-76 p3
             /HF 12-76 p98   +RR 12-76 p51
244   Concerto, piano, no. 1, A major. Concerto, piano, no. 3, E major.
       Stephen Bishop-Kovacevich, pno; LSO; Colin Davis. Philips
       9500 043.
             +Gr 7-76 p167   +NR 10-76 p6
             +-HFN 7-76 p84   ++RR 7-76 p47
245   Concerto, piano, no. 2, G major. Sonata, 2 pianos and percussion.
       György Sándor, Rolf Reinhardt, pno; VSO; Michael Gielen. Turn-
       about TV 34036. Tape (c) KTVC 34036.
            +HFN 5-76 p117 tape   +-RR 5-76 p77 tape
246   Concerto, piano, no. 3, E major. Concerto, viola. Géza Németh,
       vla; Dezsö Ránki, pno; Budapest Philharmonic Orchestra,
       Hungarian State Symphony Orchestra; András Kóródy, János
       Feréncsik. Hungaroton SLPX 11421.
             +Gr 3-76 p1458   +RR 2-76 p24
             +-HF 8-76 p79   ++SFC 2-8-76 p26
             +NR 3-76 p3
       Concerto, viola. cf Concerto, piano, no. 3, E major.
       Concerto, violin, no. 1. cf HMV SLS 5058.
       Contrasts. cf Hungarian folksongs, violin and piano.
247   Dance suite. KODALY: Variations on a Hungarian folk song
       (Peacock). Hungarian Radio and Television Orchestra; György
       Lehel. Hungaroton SHLX 90043.
             -Gr 1-76 p1188   +-NR 11-75 p3
             +-HFN 1-76 p102   -RR 12-75 p36
248   Dance suite. Music, strings, percussion and celesta. PH; Antal
       Dorati. Philips 6500 931.
             +Audio 10-76 p151   +MJ 5-76 p29
             +Gr 2-76 p1457   +NR 5-76 p4
             +HF 10-76 p102   +RR 3-76 p34
            ++HFN 3-76 p88   +-SFC 4-11-76 p38
249   Divertimento, string orchestra. RAVEL: Ma Mère l'oye. Pavane
       pour une infante défunte. CPhO; Jiří Bělohlávek. Panton
       110 375.
             +-RR 8-76 p49
250   Hungarian folksongs, violin and piano (trans. Országh and Bartok).
       Rhapsody, violoncello and piano, no. 1. Rhapsodies, violin
       and piano (2). Contrasts. Zoltán Székely, Miklós Szenthelyi,
       Mihály Szücs, vln; Lászlo Mezõ, vlc; Kálmán Berkes, clt;
       Isobel Moore, Zoltán Kocsis. Erzsébet Tusa, pno. Hungaroton
       SLPX 11357.
             +-Gr 5-76 p1772   +RR 6-76 p61
             +NR 6-76 p10
251   Hungarian folksongs (1906, 1907-17). Songs, op. 15-16. Eszter
       Kovács, s; Adám Fellegi, pno. Hungaroton SLPX 11603.
             +-Gr 8-76 p324   +NR 6-76 p14
             +-HF 8-76 p79   +-RR 6-76 p77
252   Hungarian peasant suite. DEBUSSY: Syrinx, flute. POULENC: Sonata,
       flute and piano. PROKOFIEV: Sonata, flute and piano, op. 94,
       D major. Jean-Pierre Rampal, flt; Robert Veyron-Lacroix, pno.
       Musical Heritage MHS 906 (also Odyssey Y 33905).
             +AR 2-73 p25   +NR 4-76 p6

Hungarian sketches. cf Concerto, orchestra.
Mikrokosmos: 4 songs. cf Songs (Hungaroton SLPX 11610).
Mikrokosmos, nos. 1, 2, 5-6: Bulgarian dances. cf Desmar DSM 1005.
253 Mikrokosmos, Bks 1-6: Nos. 13, 25, 35, 39-40, 45, 47, 49, 52, 54,
61-63, 69-70, 72, 82-85, 87-88, 91-92, 97, 100-102, 107-109,
124, 126, 128, 133, 135, 140, 143-144, 146-147, 149, 153.
Dagmar Baloghová, pno. Supraphon 111 1615.
/Gr 7-76 p188   +RR 6-76 p67
+-NR 7-76 p11
254 The miraculous Mandarin, op. 19: Suite. PROKOFIEV: The love for
three oranges, op. 33: Suite. Scythian suite, op. 20. LSO,
BBC Symphony Orchestra; Antal Dorati. Philips 6582 011. Tape
(c) 7321 021. (Reissues from SAL 3569, Mercury AMS 16009).
+Gr 2-76 p1331   +-HFN 1-76 p123
+Gr 9-76 p494   +-RR 1-76 p37
The miraculous Mandarin, op. 19: Suite. cf Concerto, orchestra.
Music, strings, percussion and celesta. cf Dance suite.
255 Out of doors. BERG: Sonata, piano, op. 1. SCHONBERG: Klavierstück,
op. 33a and b. STRAVINSKY: Petrouchka. Adám Fellegi, pno.
Hungaroton SLPX 11529.
+-HF 5-76 p101   +-NR 11-75 p12
+HFN 2-76 p92   +-3-76 p58
256 Quartet, strings, no. 4, C major. ROSENBERG: Quartet, strings,
no. 4. Fresk Quartet. Caprice RIKS LP 51. (also CAP 1051).
+-RR 7-74 p49   ++St 5-76 p112
Rhapsodies, violin and piano (2). cf Hungarian folksongs, violin
and piano.
257 Rhapsody, violin and piano, no. 1. Rumanian folk dances (2).
Sonata, violin and piano, no. 1. Denés Zsigmondy, vln;
Anneliese Nissen, pno. Klavier KS 535.
+Audio 9-75 p70   ++NR 5-75 p9
+-HF 2-76 p87   +St 10-75 p104
Rhapsody, violoncello and piano, no. 1. cf Hungarian folksongs,
violin and piano.
258 Rumanian dances, op. 8a (2). RAVEL: Ma mère l'oye. STRAVINSKY:
Pulcinella: Suite. St. John's Symphony Orchestra; John
Lubbock. Pye Nixa QS PCNHX 3. Tape (c) ZCPCNHX 3.
+-Gr 7-76 p167   +-RR 7-76 p63
+HFN 7-76 p84
Rumanian folk dances. cf Connoisseur Society CS 2070.
Rumanian folk dances (2). cf Rhapsody, violin and piano, no. 1
Rumanian folk dances, nos. 1-6. cf HMV HLM 7077.
259 Sonata, 2 pianos and percussion. Sonata, solo violin. Dezsö Ránki,
Zoltán Kocsis, pno; Ferenc Petz, József Marton, perc; Dénes
Kovács, vln. Hungaroton SLPX 11479.
++Gr 3-75 p1668   ++SFC 2-8-76 p26
++RR 4-75 p48
Sonata, 2 pianos and percussion. cf Concerto, piano, no. 2,
G major.
260 Sonata, solo violin. Sonata, violin and piano, no. 2. Dénes
Zsigmondy, vln; Anneliese Nissen, pno. Klavier KS 542.
+-HF 2-76 p87   +St 10-75 p104
+NR 8-75 p7
Sonata, solo violin. cf Sonata, 2 pianos and percussion.
261 Sonata, violin and piano, no. 1. BRAHMS: Sonata, violin and piano,
no. 2, op. 100, A major. David Oistrakh, vln; Sviatoslav Richter,

pno. Rococo 2111.
          ++NR 9-76 p6
Sonata, violin and piano, no. 1. cf Rhapsody, violin and piano,
     no. 1.
Sonata, violin and piano, no. 2. cf Sonata, solo violin.
262   Songs, op. 16 (5). Songs, op. 15 (5). Eszter Kovacs, s; Adám
          Fellegi, pno. Hungaroton SLPX 11603.
              +-Gr 8-76 p324   +NR 6-76 p14
              +-HF 8-76 p79   +-RR 6-76 p77
263   Songs: Hungarian folk songs. Mikrokosmos: 4 songs. Village
          scenes. Erika Sziklay, s; István Lantos, pno. Hungaroton
          SLPX 11610.
              +NR 9-76 p10
Songs: Girls mocking song; Lads mocking song; Letter to the folks
     at home. cf Hungaroton SLPX 11696.
Songs: Bread breaking; Pillow dance. cf Hungaroton SLPX 11762.
Village scenes. cf Songs (Hungaroton SLPX 11610).
BASSANO, Jerome
264   Hodie Christus natus est. GABRIELI, G.: Motets: Audite principes;
          Angelus ad pastores; O magnum mysterium; Salvator noster; Sol
          sol la sol; Quem vidistis pastores. Sonata, pian'e forte.
          MONTEVERDI: Exultent caeli. Monteverdi Choir; Philip Jones
          Brass Ensemble; John Eliot Gardiner. London STS 15256.
              ++NR 2-76 p9
265   Ricercare. NOORT: Les petit branles. TELEMANN: Fantasias. VOIS:
          Pavane de Spanje. Scott-Martin Kosofsky, rec. Titanic TI 7.
              +NR 2-76 p16
BASSETT, Leslie
266   Sextet. EDWARDS: Kreuz and Quer. ERB: Pieces for brass quintet
          and piano (3). MacDOUGALL: Anacoluthon: A confluence. John
          Graham, vla; Gilbert Kalish, James Smolko, pno; Concord String
          Quartet, New York Brass Quintet, Boston Musica Viva, Contemporary
          Music Ensemble; Matthias Bamert, Richard Pittman, Arthur Weisberg.
          CRI SD 323.
              +NR 1-76 p7   +-St 8-76 p100
BASTON, John
     Concerto, D major. cf HMV SLS 5022 (2).
BATTEN, Adrian
     O praise the Lord. cf Argo ZRG 789.
BAUER, Robert
     Sokasodik. cf Golden Crest CRS 4143.
     Willy rag. cf Golden Crest CRS 4143.
BAX, Arnold
     A Christmas carol song. cf Grosvenor GRS 1034.
     Fanfare for the wedding of Princess Elizabeth, 1948. cf Decca 419.
267   The garden of Fand. Mediterranean. Northern ballad, no. 1.
          Tintagel. LPO; Adrian Boult. Musical Heritage Society MHS 1769.
          Tape (c) MHC 21204.
              +HF 11-74 p99   +HF 1-76 p111 tape
     Mediterranean. cf The garden of Fand.
     Mediterranean. cf RCA ARM 4-0942/7.
     Northern ballad, no. 1. cf The garden of Fand.
     Tintagel. cf The garden of Fand.
BAYCO
     Royal Windsor. cf Pye GH 603.
BAYLY, Thomas
     Long long ago. cf Club 99-99.

BAZELON, Irwin
268   Chamber concerto, no. 2. Symphony, no. 5. Indianapolis Symphony
       Orchestra Chamber Ensemble; Solomon Izler, Irwin Bazelon. CRI
       SD 287.
            +-ARG 7-72 p554   ++SFC 4-23-72 p47
            -NR 7-72 p3   ++St 10-72 p120
            *RR 8-76 p28
       Duo, viola and piano. cf CRI SD 342.
       Symphony, no. 5. cf Chamber concerto, no. 2.
BAZZINI, Antonio
       La ronde des lutins, op. 25. cf London STS 15239.
       La ronde des lutins, op. 25. cf RCA ARL 1-1172.
       La ronde des lutins, op. 25 (2). cf RCA ARM 2-0942/7.
BEAT, Duncan
       Royal mile. cf Philips 6308 246.
BEAUVARLET-CHARPENTIER (18th century France)
       Fugue, G minor. cf L'Oiseau-Lyre SOL 343.
BECKER, John J.
269   Concerto arabesque, piano. IVES: Holidays: Thanksgiving. RIEGGER:
       Canon and fugue, op. 33, D minor. Jan Henrik Kayser, pno; Oslo
       Philharmonic Orchestra, Members, Iceland State Radio Orchestra
       and Chorus; William Strickland. CRI SD 117.
            +-RR 10-76 p65
BEDFORD, David
270   The rime of the ancient mariner. David Bedford, keyboards,
       recorders, percussion; Mike Oldfield, gtr; Robert Powell,
       narrator; Queen's College Girls Choir, London. Virgin V 2038.
       Tape (c) TCV 2038.
            -Gr 2-76 p1331   ++RR 12-75 p71
            +-HFN 11-75 p150
BEASON, Jack
271   Captain Jinks of the horse marines. Carol Wilcox, s; Carolyne
       James, ms; Robert Owen Jones, t; Ronald Highley, Walter Hook,
       bar; Eugene Green, Brian Steele, bs; Kansas City Philharmonic
       Orchestra; Russell Patterson. RCS ARL 2-1727 (2).
            +MJ 12-76 p28   +ON 11-76 p98
            +NR 12-76 p9   +St 10-76 p130
BEETHOVEN, Ludwig van
272   Adagio, E flat major. Sonatinas, mandolin, C major, C minor.
       Variations, mandolin, D major. SCHLICK: Divertimento, 2 mando-
       lins and continuo, D major. Maria Hinterleitner, hpd; Elfriede
       Hunschak, Vinzenz Hladky, mand. Turnabout TV 34110. Tape (c)
       KTVC 34110.
            -Gr 8-72 p351   +HFN 7-76 p104 tape
            +-HFN 8-72 p1456   /RR 9-76 p93 tape
273   Andante favori, F major. Bagatelle, op. 126, no. 4, B minor.
       Polonaise, op. 89, C major. SCHUBERT: Sonata, piano, no. 16,
       op. 42, D 845, A minor: Finale. SCHUMANN: Concerto, piano,
       op. 54, A minor. Artur Schnabel, pno; Orchestra; Alfred
       Wallenstein. Bruno Walter Society BWS 724.
            +-ARSC Vol VIII, no. 2-3 +-NR 10-76 p4
              p89
274   Andante favori, F major. Sonata, piano, no. 21, op. 53, C major.
       Sonata, piano, no. 31, op. 110, A flat major. Alfred Brendel,
       pno. Philips 6500 762. Tape (c) 7300 351.
            +Gr 11-75 p859   ++MJ 10-76 p52
            +HF 8-76 p80   +NR 6-76 p11

```
            ++HFN 7-75 p90 tape        +-RR 11-75 p74
            +-HFN 10-75 p137           ++St 12-76 p134
```

Bagatelle, no. 25, A minor. cf Works, selections (Philips 6833 179).
Bagatelle, no. 25, A minor. cf CBS 30064
Bagatelle, no. 25, A minor. cf Philips 6747 204.

275 Bagatelles, opp. 33, 119, 126. Stephen Bishop-Kovacevich, pno.
     Philips 6500 930.

```
            +-Augio 9-76 p85           +MM 6-76 p42
            +-Gr 1-76 p1216            ++NR 3-76 p12
            +-HF 5-76 p79              +-RR 11-75 p73
            ++HFN 11-75 p150           +-SFC 2-15-76 p38
            ++MJ 4-76 p30              ++St 5-76 p112
```

276 Bagatelles, op. 33 (6). Bagatelles, op. 126 (6). Glenn Gould,
     pno. Columbia M 33265. (also CBS 76424).

```
            +-Gr 1-76 p1216            ++NR 8-75 p12
            +-HF 8-75 p80              +-RR 11-75 p73
            ++HFN 11-75 p150           ++St 8-75 p94
            -MM 6-76 p42
```

Bagatelles, op. 126 (6). cf Bagatelles, op. 33 (6).
Bagatelle, op. 126, no. 4, B minor. cf Andante favori, F major.
Bundeslied, op. 122. cf Works, selections (CBS 76404).

277 Concerti, piano, nos. 1-5. Wilhelm Kempff, pno; BPhO; Ferdinand
     Leitner. DG 2740 131 (4). Tape (c) 3371 010 (3). (Reissues
     from SLPM 138774/77).

```
            +Gr 11-75 p909             +-RR 11-75 p36
            +-HF 1-76 p111 tape
```

278 Concerti, piano, nos. 1-5. Vladimir Ashkenazy, pno; CSO; Georg
     Solti. London CSA 2404 (4). Tape (c) D 10270 (r) W 480270.
     (also Decca SXLG 6594/7 Tape (c) KSXCG 7019/21).

```
            ++Gr 9-73 p477             +RR 9-73 p57
            +HF 10-73 p98              +-RR 2-74 p71 tape
            ++HF 5-74 p122 tape        ++SFC 8-12-73 p32
            ++HF 1-76 p111 tape        ++St 11-73 p74
            +HFN 12-73 p2621 tape
```

279 Concerti, piano, nos. 1-5. Stephen Bishop, pno; LSO; BBC Symphony
     Orchestra; Colin Davis. Philips 6747 104 (4). (Nos. 1, 3, 5
     reissues from 6500 179, 6500 315, SAL 3787).

```
            +Gr 10-75 p607             +RR 11-75 p36
            +-HFN 9-75 p96             ++SFC 10-19-75 p33
            +NR 12-75 p6               +-St 4-76 p111
            ++NYT 1-18-76 pD1
```

280 Concerti, piano, nos. 1-5. Arthur Rubinstein, pno; LPO; Daniel
     Barenboim. RCS CRL 5-1415 (5). Tape (c) ARK 1-1416/20
     (ct) ARS 1-1416/20 (Q) ARD 5-1415 (5).

```
            +Gr 5-76 p1743             +MJ 10-76 p25
            ++HF 5-76 p79              -NR 5-76 p4
            +HF 9-76 p84 tape          +-RR 5-76 p34
            +-HFN 7-76 p84             +-SFC 8-22-76 p38
```

281 Concerto, piano, no. 1, op. 15, C major. MOZART: Concerto, piano,
     no. 9, K 271, E flat major. Walter Gieseking, pno; Berlin
     State Opera Orchestra; Hans Rosbaud. Bruno Walter Society RR 411.

```
            +-ARSC Vol VIII, no. 2-3  ++NR 9-75 p7
            p91
```

282 Concerto, piano, no. 1, op. 15, C major. Fantasia, op. 80, C minor.
     Julius Katchen, pno; LSO and Chorus; Pierino Gamba. Decca
     SDD 227. Tape (c) KSDC 227. (also London CS 6451).

```
            +MJ 10-76 p25              +-RR 2-75 p74 tape
```

283   Concerto, piano, no. 1, op. 15, C major. Concerto, piano, no. 2,
        op. 19, B flat major. Wilhelm Backhaus, pno; VPO; Hans Schmidt-
        Isserstedt. Decca SPA 401. Tape (c) KCSP 401. (Reissue from
        SXL 2178).
                    -Gr 9-76 p409   +-HFN 10-76 p181
                    +Gr 11-76 p887 tape   +-RR 11-76 p49
284   Concerto, piano, no. 1, op. 15, C major. Sonata, piano, no. 8,
        op. 13, C minor. Vladimir Ashkenazy, pno; CSO; Georg Solti.
        Decca SXL 6651. Tape (c) KSXC 6651. (Reissue from SXLF 6594/7,
        SXL 6706).
                    +Gr 2-76 p1332   +RR 2-76 p25
                    +HFN 2-76 p115   +-RR 5-76 p77 tape
                    ++HFN 4-76 p125 tape
285   Concerto, piano, no. 1, op. 15, C major. Sonata, piano, no. 6,
        op. 10, no. 2, F major. Claudio Arrau, pno; COA; Bernard
        Haitink. Philips 6580 122. (Reissue from SABL 20, SAL 3550)
        (also Philips 839 749. Tape (c) 18242 CAA).
                    ++Gr 4-76 p1588   +RR 3-76 p35
                    +-HFN 3-76 p109
286   Concerto, piano, no. 1, op. 15, C major. HUMMEL: Sonata, piano,
        op. 13, E flat major. Dino Ciani, pno; Orchestra; Bruno
        Bartoletti. Rococo 2134.
                    +NR 8-76 p5
287   Concerto, piano, no. 2, op. 19, B flat major. Sonata, piano, no.
        21, op. 53, C major. Vladimir Ashkenazy, pno; CSO; Georg Solti.
        Decca SXL 6652. Tape (c) KSXC 6652. (Reissues from SXL 6594-7,
        SXL 6706).
                    +Gr 1-76 p1188   +RR 2-76 p24
                    +HFN 12-75 p171
288   Concerto, piano, no. 2, op. 19, B flat major. Concerto, piano,
        no. 4, op. 58, G major. Wilhelm Kempff, pno; BPhO; Ferdinand
        Leitner. DG 138775. Tape (c) 3300 485.
                    +HFN 12-75 p173 tape   +RR 2-76 p70 tape
289   Concerto, piano, no. 2, op. 19, B flat major. Concerto, piano,
        no. 4, op. 58, G major. Stephan Bishop-Kovacevich, pno; BBC
        Symphony Orchestra; Colin Davis. Philips 6500 975. Tape (c)
        7300 454. (Reissue from 6747 104).
                    +-Audio 11-76 p111   -MJ 4-76 p30
                    +-Gr 1-76 p1337   +NR 3-76 p5
                    +-Gr 10-76 p658 tape   +RR 2-76 p25
                    +-HF 2-76 p87   +SFC 2-15-76 p38
                    +HFN 2-76 p115
290   Concerto, piano, no. 2, op. 19, B flat major. Sonata, piano, no. 1,
        op. 2, no. 1, F minor. Claudio Arrau, pno; COA; Bernard
        Haitink. Philips 6580 123. (Reissues from SABL 20, SAL 3568).
                    +Gr 6-76 p38   +-RR 8-76 p28
                    +-HFN 6-76 p102
      Concerto, piano, no. 2, op. 19, B flat major. cf Concerto,
        piano, no. 1, op. 15, C major.
291   Concerto, piano, no. 3, op. 37, C minor. Sonata, piano, no. 14,
        op. 27, no. 2, C sharp minor. Wilhelm Backhaus, pno; VPO;
        Hans Schmidt-Isserstedt. Decca SPA 402. Tape (c) KCSP 402.
        (Reissue from 2190).
                    +-Gr 6-76 p38   +-HFN 4-76 p121
                    +Gr 11-76 p887 tape   +RR 8-76 p28
292   Concerto, piano, no. 3, op. 37, C minor. Sviatoslav Richter, pno;
        VSO; Kurt Sanderling. DG 2535 107. Tape (c) 3335 107.

(Reissue from 138848).
          -Gr 1-76 p1188   +-RR 9-75 p77 tape
     +-HFN 11-75 p173   +RR 1-76 p30
293   Concerto, piano, no. 3, op. 37, C minor. MOZART: Rondos, piano,
     K 382, D major; K 386, A major. Annie Fischer, pno; Bavarian
     State Orchestra; Ferenc Fricsay. DG 2548 238. (Reissue from
     SLPM 138 087).
          ++Gr 8-76 p278   +HFN 8-76 p93
294   Concerto, piano, no. 3, op. 37, C minor. Sonata, piano, no. 26,
     op. 81a, E flat major. Vladimir Ashkenazy, pno; CSO; Georg
     Solti. London CS 6855. (also Decca SXL 6653).
          +-NR 12-76 p6   +RR 1-76 p30
295   Concerto, piano, no. 4, op. 58, G major. HAYDN: Symphony, no. 104,
     D major. Pietro Scarpini, pno; RAI Rome Symphony Orchestra,
     BPhO; Wilhelm Furtwängler. Bruno Walter Society RR 441.
          +-NR 11-76 p4
296   Concerto, piano, no. 4, op. 58, G major. Sonata, piano, no. 8,
     op. 13, C minor. Wilhelm Backhaus, pno; VPO; Hans Schmidt-
     Isserstedt. Decca SPA 403. Tape (c) KCSP 403.
          -Gr 5-76 p1744   +-HFN 4-76 p121
          +Gr 11-76 p887 tape   +-RR 8-76 p28
297   Concerto, piano, no. 4, op. 58, G major. SCHUBERT: Impromptu,
     op. 142, no. 3, D 935, B flat major. Clifford Curzon, pno;
     VPO; Hans Knappertsbusch. Decca ECS 752. (Reissues from LXT
     2948, 2781).
          +-Gr 8-76 p278   +-RR 8-76 p28
          -HFN 5-76 p115
298   Concerto, piano, no. 4, op. 58, G major. Leonore overture, no. 3,
     op. 72. Vladimir Ashkenazy, pno; CSO; Georg Solti. London
     CS 6856. (also Decca SXL 6654 Tape (c) KSXC 6654. Reissues
     from SXLG 6594/7, SXLP 6684).
          +Gr 5-75 p1962   +NR 5-76 p5
          ++HFN 6-75 p85   ++RR 6-75 p30
          +HFN 11-76 p175
299   Concerto, piano, no. 4, op. 58, G major. Sonata, piano, no. 15,
     op. 28, D major. Sonata, piano, no. 26, op. 81a, E flat major.
     Wilhelm Backhaus, pno. Rococo 2085.
          +-NR 9-76 p5
     Concerto, piano, no. 4, op. 58, G major. cf Concerto, piano,
     no. 2, op. 19, B flat major (DG 138775).
     Concerto, piano, no. 4, op. 58, G major. cf Concerto, piano,
     no. 2, op. 19, B flat major (Philips 6500 975).
     Concerto, piano, no. 4, op. 58, G major. cf BACH: Partita,
     harpsichord, no. 1, S 825, B flat major.
300   Concerto, piano, no. 5, op. 73, E flat major. DELIUS: Concerto,
     piano, C minor. Benno Moiseiwitsch, pno; Orchestra; Malcolm
     Sargent. Bruno Walter Society SWS 725.
          +-NR 10-76 p4
301   Concerto, piano, no. 5, op. 73, E flat major. Wilhelm Backhaus,
     pno; VPO; Hans Schmidt-Isserstedt. Decca SPA 452. Tape (c)
     KCSP 452. (Reissue from SXL 2179).
          +Gr 10-76 p578   +-HFN 10-76 p181
          +-Gr 11-76 p887 tape   +-RR 11-76 p49
302   Concerto, piano, no. 5, op. 73, E flat major. Egmont, op. 84:
     Overture. Vladimir Ashkenazy, pno; CSO; George Solti. Decca
     SLX 6655. Tape (c) KSXC 16655. (Reissue from SXLG 6594/7,
     SXLP 6684). (also London 6857).

```
        +Audio 11-76 p111           +HFN 7-75 p90 tape
         +Gr 3-75 p1640             -NR 6-75 p7
        ++Gr 6-75 p106 tape         +RR 4-75 p21
        ++HFN 5-75 p124             -RR 7-75 p69 tape
```
303   Concerto, piano, no. 5, op. 73, E flat major. Sonata, piano,
      no. 25, op. 79, G major. Andor Foldes, pno; BPhO; Ferdinand
      Leitner. DG 2548 206. Tape (c) 3348 206 (Reissue from
      138019).
```
        +-Gr 11-75 p840             +-RR 12-75 p37
        +-HFN 10-75 p152            +RR 9-76 p92 tape
```
304   Concerto, piano, no. 5, op. 73, E flat major. Emil Gilels, pno;
      CO; Georg Szell. HMV SXLP 30223. Tape (c) TC SXLP 30223.
      (Reissue from World Records SM 156/60).
```
        +Gr 11-76 p775             +-HFN 11-76 p171
        +Gr 11-76 p887 tape        +HFN 12-76 p155
```
305   Concerto, piano, no. 5, op. 73, E flat major. Rudolf Firkusny,
      pno; RPO; Rudolf Kempe. RCA GL 25014. Tape (c) GK 25014.
      (Previously issued by Reader's Digest).
```
        +-Gr 11-76 p775            +-RR 11-76 p50
        +-HFN 12-76 p137           +-RR 12-76 p104 tape
        +-HFN 12-76 p153 tape
```
      Concerto, piano, no. 5, op. 73, E flat major. cf Works, selec-
      tions (Decca DPA 529/30).
      Concerto, piano, no. 5, op. 73, E flat major: 1st movement. cf
      HMV SEOM 24.
306   Concerto, violin, op. 61, D major. Isaac Stern, vln; NYP;
      Leonard Bernstein. CBS 61598. Tape (c) 40-61598. (Reissue
      from Fontana SCFL 120).
```
        ++Gr 2-75 p1473            +RR 1-75 p26
         -Gr 12-75 p1121 tape      +-RR 2-76 p70 tape
        +-HFN 12-75 p173 tape
```
307   Concerto, violin, op. 61, D major. Isaac Stern, vln; NYP; Daniel
      Barenboim. Columbia M 33587. Tape (c) MT 33587 (Q) MQ 33587.
      (also CBS 76477).
```
        +-Gr 3-76 p1458            +NR 5-76 p5
        +HF 6-76 p75               +-RR 2-76 p25
        +-HFN 3-76 p89             +-St 7-76 p107
```
308   Concerto, violin, op. 61, D major. BRAHMS: Concerto, violin, op.
      77, D major. SIBELIUS: Concerto, violin, op. 47, D minor.
      TCHAIKOVSKY: Concerto, violin, op. 35, D major. Christian
      Ferras, vln; BPhO; Herbert von Karajan. DG 2740 137. Tape
      (c) 3371 021. (Reissues from 139 021, 138 930, 138 961,
      SKL 922/8).
```
        +-Gr 6-76 p61             +HFN 6-76 p102
        +HF 12-76 p147 tape       +-RR 5-76 p38
```
309   Concerto, violin, op. 61, D major. Coriolan overture, op. 62.
      Josef Suk, vln; NPhO; Adrian Boult. HMV ASD 2667. Tape (c)
      TC ASD 2667.
```
        ++Gr 3-71 p1459           -RR 5-76 p77 tape
        +-HFN 4-76 p125 tape
```
310   Concerto, violin, op. 61, D major. David Oistrakh, vln; French
      National Radio Orchestra; André Cluytens. HMV SXLP 30168.
      Tape (c) TCEXE 197. (also Angel S 35783). (Reissue from
      Columbia SAX 2315).
```
        +Gr 11-74 p889            +RR 10-74 p36
        +HFN 8-76 p94 tape
```
311   Concerto, violin, op. 61, D major. Hermann Krebbers, vln; COA;

Bernard Haitink. Philips 6580 115.
   +Gr 2-76 p1337   +-RR 2-76 p25
   +-HFN 2-76 p92

312   Concerto, violin, op. 61, D major. Igor Oistrakh, vln; VSO;
     David Oistrakh. RCA GL 25005. Tape (c) GK 25005.
       ++Gr 10-76 p578   +-RR 10-76 p40
       +-HFN 12-76 p137   +-RR 12-76 p104 tape
       +-HFN 12-76 p153 tape

313   Concerto, violin, op. 61, D major. Josef Suk, vln; NPhO; Adrian
     Boult. Vanguard SRV 353 SD.
       +ARG 11-76 p13   ++NR 11-76 p4
       +HF 10-76 p94   +-SFC 9-12-76 p31
   Concerto, violin, op. 61, D major. cf RCA ARM 4-0942/7.

314   Concerto, volin, violoncello and piano, op. 56, C major. Rudolf
     Serkin, pno; Jaime Laredo, vln; Leslie Parnas, vlc; Marlboro
     Festival Orchestra; Alexander Schneider. CBS 61663. (Reissue
     from 72202).
       +-Gr 2-76 p1332   +RR 1-76 p31
       +-HFN 2-76 p115

315   Concerto, violin, violoncello and piano, op. 56, C major. Géza
     Anda, pno; Wolfgang Schneiderhan, vln; Pierre Fournier, vlc;
     Berlin Radio Symphony Orchestra; Ferenc Fricsay. DG 2535 153.
     (Reissue from 136 236).
       +-Gr 2-76 p1332   -RR 1-76 p31

316   Concerto, violin, violoncello and piano, op. 56, C major. Josef
     Suk, vln; Josef Chuchro, vlc; Jan Panenka, pno; CPhO: Kurt
     Masur. Supraphon 110 1558.
       -Gr 11-75 p802   +-NR 5-76 p5
       +HF 10-76 p94   +-RR 10-75 p37
       +-HFN 10-75 p136
   Consecration of the house, op. 124: Overture. cf Overtures
     (DG 2548 138).
   Consecration of the house, op. 124: Overture. cf Symphonies,
     nos. 1-9 (HMV SLS 788).
   Contradances (Country dances), G 141. cf DG Archive 2533 182.

317   Country dances, G 141. German dances G 140 (12). March, C major.
     Minuet, E flat major. PH; Hans Ludwig Hirsch. Telefunken
     AW 6-41996.
       +-Gr 8-76 p278   +RR 8-76 p29
       +-HFN 8-76 p76
   Country dances, op. 141 (12). cf Nonesuch H 71141.

318   Coriolan overture, op. 62. Fidelio, op. 72: Overture. Egmont,
     op. 84: Overture. Leonore overture, no. 3, op. 72. LPO;
     Andrew Davis. Classics for Pleasure CFP 40227.
       +HFN 3-76 p89   /RR 2-76 p25
   Coriolan overture, op. 62. cf Concerto, violin, op. 61, D major.
   Coriolan overture, op. 62. cf Overtures (DG 2535 135).
   Coriolan overture, op. 62. cf Symphonies, nos. 1-9 (HMV SLS 788).
   Coriolan overture, op. 62. cf Symphonies, nos. 1-9 (HMV SLS 5053).
   Coriolan overture, op. 62. cf Symphonies, nos. 1-9 (London CPS9).
   Coriolan overture, op. 62. cf Symphony, no. 3, op. 55, E flat major.
   Coriolan overture, op. 62. cf Symphony, no. 7, op. 92, A major
     (DG 2535 147).
   Coriolan overture, op. 62. cf Symphony, no. 7, op. 92, A major
     (Decca SXL 6764).
   Coriolan overture, op. 62. cf Bruno Walter Society RR 443.

Duo, clarinet and bassoon, no. 1, C major. cf HMV SLS 5046.

Ecossaises (arr. Busoni). cf International Piano Library IPL 104.

319 Egmont, op. 84: Incidental music. Gundula Janowitz, s; BPhO;
     Herbert von Karajan. DG 2530 301. (Reissue from 643628/30,
     2720 011).
         +Gr 5-73 p2098   +-ON 2-7-76 p34
         +HF 9-73 p87   +RR 5-73 p45
         +HFN 6-73 p1171   +SFC 7-15-73 p32
         +NR 5-73 p11   -St 8-73 p98
     Egmont, op. 84: Incidental music. cf Symphonies, nos. 1-9
     (HMV SLS 788).

320 Egmont, op. 84: Overture. Fidelio, op. 72: Overture. König
     Stefan, op. 117: Overture. Leonore overtures, nos. 1-3, opp.
     72, 138. CO; Georg Szell. CBS 61580. Tape (c) 40-61580.
         +RR 2-76 p70 tape

321 Egmont, op. 84: Overture. LISZT: Les préludes, G 97. MOZART:
     Serenade, no. 13, K 525, G major. SMETANA: Ma Vlast: Vltava.
     Berlin Radio Symphony Orchestra; Ferenc Fricsay. DG Tape (c)
     3318 061.
         +RR 3-76 p76 tape
     Egmont, op. 84: Overture. cf Concerto, piano, no. 5, op. 73,
     E flat major.
     Egmont, op. 84: Overture. cf Coriolan overture, op. 62.
     Egmont, op. 84: Overture. cf Overtures (DG 2535 135).
     Egmont, op. 84: Overture. cf Overtures (DG 2548 138).
     Egmont, op. 84: Overture. cf Symphonies, nos. 1-9 (HMV SLS 5053).
     Egmont, op. 84: Overture. cf Symphonies, nos. 1-9 (London CPS 9).
     Egmont, op. 84: Overture. cf Symphonies, nos. 1-9 (Seraphim SIH
     6093).
     Egmont, op. 84: Overture. cf Symphony, no. 2, op. 36, D major.
     Egmont, op. 84: Overture. cf Symphony, no. 4, op. 60, B flat
     major.
     Egmont, op, 84: Overture. cf Symphony, no. 5, op. 67, C minor.
     Egmont, op. 84: Overture. cf Symphony, no. 7, op. 92, A major
     (Decca 4342).
     Egmont, op. 84: Overture. cf Symphony, no. 7, op. 92, A major
     (Decca SXLN 6673).
     Egmont, op. 84: Overture. cf Symphony, no. 7, op. 92, A major
     (Decca KSXC 6780).
     Egmont, op. 84: Overture. cf Works, selections (Decca DPA 529/30).
     Egmont, op, 84: Overture. cf Works, selections (Philips 6833 179).
     Egmont, op. 84: Overture. cf Decca SPA 409.
     Egmont, op. 84: Overture. cf Philips 6747 204.
     Egmont, op. 84: Overture; Die Trommel gerühret; Freudvoll und
     Leidvoll; Klärchens Tod; Sieges Symphonies. cf Symphony, no.
     9, op. 125, D minor.

Equali, 4 trombones, no. 1: Andante. cf Argo SPA 464.

Fantasia, op. 77, G minor. cf Sterling 1001/2.

322 Fantasia, op. 80, C minor. The ruins of Athens, op. 113: Inci-
     dental music. Jörg Demus, pno; Arleen Auger, s; Klaus Hirte,
     bar; Franz Crass, bs; Vienna Singverein, RIAS Chamber Choir;
     VSO, BPhO; Ferdinand Leitner, Bernhard Klee. DG 2535 151.
     (Reissue from 2720 017, 643628/30).
         +Gr 2-76 p1332   +RR 1-76 p54
         +HFN 2-76 p115
     Fantasia, op. 80, C minor. cf Concerto, piano, no. 1, op. 15, C
     major.

323   Fidelio, op. 72 (arr. Sedlak and ed. Hellyer). MOZART: Die
       Entführung aus dem Serail, K 384 (arr. Wendt and ed. Brown).
       London Wind Soloists; Jack Brymer. Decca SDD 485.
           +-Gr 10-76 p616   +RR 10-76 p82
           +HFN 10-76 p165
       Fidelio, op. 72: Ha, Welch ein Augenblick. cf Decca SPA 449.
       Fidelio, op. 72: Ha, Welch ein Augenblick; Abscheulicher. cf
       Decca GOSE 677/79.
       Fidelio, op. 72: Hat man nicht auch Gold beineban; Ha, Welch
       ein Augenblick. cf Preiser LV 192.
324   Fidelio, op. 72: Overtures (4). PhO; Otto Klemperer. Seraphim
       S 60261. (Reissue)
           +NR 1-76 p3   ++SFC 12-28-75 p30
       Fidelio, op. 72: Overture. cf Coriolan overture, op. 62.
       Fidelio, op. 72: Overture. cf Egmont, op. 84: Overture.
       Fidelio, op. 72: Overture. cf Overtures (DG 2535 135).
       Fidelio, op. 72: Overture. cf Overtures (DG 2548 138).
       Fidelio, op. 72: Overture. cf Symphonies, nos. 1-9.
       Fidelio, op. 72: Prisoners chorus. cf DG 2548 212.
       Fugue, C major. cf Organ works (Musical Heritage Society MHS
       1517).
       Fugue cycle on themes by J. S. Bach, D minor. cf Organ works
       (Musical Heritage Society MHS 1517).
       German dances, G 140 (12). cf Country dances, G 141.
       German dance, no. 6. cf RCA ARM 4-0942/7.
       Die Geschöpfe des Prometheus (The creatures of Prometheus), op.
       43: Overture. cf Overtures (DG 2535 135).
       The creatures of Prometheus, op. 43: Overture. cf Symphony, no.
       2, op. 36, D major.
       The creatures of Prometheus, op. 43: Overture. cf Symphonies,
       nos. 1-9 (HMV SLS 788).
       The creatures of Prometheus, op. 43: Overture. cf Symphonies,
       nos. 1-9 (Seraphim SIH 6093).
       The creatures of Prometheus, op. 43: Overture. cf Symphony,
       no. 2, op. 36, D major.
       The creatures of Prometheus, op. 43: Overture. cf Symphony,
       no. 6, op. 68, F major.
       The creatures of Prometheus, op. 43: Overture. cf Decca SXL 6782.
       The creatures of Prometheus, op. 43: Overture. cf HMV ESD 7011.
       Grosse Fuge, op. 133, B flat major. cf Quartets, strings, nos.
       1-6, op. 18, nos. 1-6.
       Grosse Fuge, op. 133, B flat major. cf Quartets, strings, nos.
       1-16.
       Grosse Fuge, op. 133, B flat major. cf Quartets, strings,
       nos. 12-16 (DG 2734 006).
       Grosse Fuge, op. 133, B flat major. cf Quartets, strings, nos.
       12-16 (Supraphon 111 1151/4).
       Grosse Fuge, op. 133, B flat major. cf Symphony, no. 4, op. 60,
       B flat major.
       The heavens resound. cf Coronet LPS 3031.
325   König Stefan (King Stephen), op. 117: Incidental music. Songs:
       Elegiac song, op. 118; Opferlied, op. 121b; Bundeslied, op.
       122; Calm sea and prosperous voyage, op. 112. Lorna Haywood,
       s; Ambrosian Opera Chorus; LSO; Michael Tilson Thomas. Colum-
       bia M 33509. (Q) MQ 33509. (also CBS 76404 Tape (c) 40-76404).
           +-Audio 11-76 p111   +NR 6-76 p12

              +Gr 11-75 p871   +NYT 9-21-75 pD18
              +HF 1-76 p82   +ON 2-7-76 p28
              +HFN 11-75 p150   +-RR 12-75 p87
              +-HFN 2-76 p116 tape   +-SFC 10-19-75 p33
              +MT 7-76 p575   ++St 2-76 p98
King Stephen, op. 117: Incidental music. cf Works, selections
    (CBS 76404).
König Stefan, op. 117: Overture. cf Egmont, op. 84: Overture.
King Stephen, op. 117: Overture. cf Symphonies, nos. 1-9.
Leonore overtures, nos. 1-3, opp. 72, 138. cf Egmont, op. 84:
    Overture.
Leonore overtures, nos. 1-3, opp. 72, 138. cf Symphonies, nos.
    1-9.
Leonore overture, no. 2, op. 72. cf HMV RLS 717.
Leonore overtures, nos. 2 and 3, op. 72. cf Overtures (DG 2548
    138).
Leonore overture, no. 3, op. 72. cf Concerto, piano, no. 4,
    op. 58, G major.
Leonore overture, no. 3, op. 72. cf Coriolan overture, op. 62.
Leonore overture, no. 3, op. 72. cf Overtures (DG 2535 135).
Leonore overture, no. 3, op. 72. cf Symphonies, nos. 1-9 (HMV
    SLS 5053).
Leonore overture, no. 3, op. 72. cf Symphonies, nos. 1-9 (London
    CPS 9).
Leonore overture, no. 3, op. 72. cf Symphonies, nos. 1-9 (Sera-
    phim SIH 6093).
Leonore overture, no. 3, op. 72. cf Symphony, no. 5, op. 67, C
    minor.
Leonore overture, no. 3, op. 72. cf Works, selections (Decca
    DPA 529/30).
Leonore overture, no. 3, op. 72. cf CBS SQ 79200.
Leonore overture, no. 3, op. 72. cf Pye PCNHX 6.
March, C major. cf Country dances, G 141.
326   Mass, op. 123, D major. Gundula Janowitz, s; Agnes Baltsa, con;
          Peter Schreier, t; José van Dam, bs; Vienna Singverein; BPhO;
          Herbert von Karajan. Angel SB 3821 (2). (also HMV SLS 979)
              +-Gr 7-75 p224   -NYT 9-21-75 pD18
              +HF 8-75 p80   +ON 2-7-76 p34
              +HFN 7-75 p75   +-RR 7-75 p55
              +MT 1-76 p39   ++SFC 7-6-75 p16
              +NR 7-75 p9   +-St 12-75 p116
327   Mass, op. 123, D major. Heather Harper, s; Janet Baker, ms;
          Robert Tear, t; Hans Sotin, bs; New Philharmonia Chorus; LPO;
          Carlo Maria Giulini. Angel SB 3836 (2). (also HMV SQ SLS 989
          Tape (c) TC SLS 989).
              +-ARG 12-76 p13   +HFN 10-76 p185 tape
              +Gr 7-76 p200   +RR 8-76 p72
              -Gr 10-76 p658 tape   +RR 10-76 p104 tape
              +-HF 12-76 p98   +STL 7-4-76 p36
              +-HFN 9-76 p118
328   Mass, op. 123, D major. Margaret Price, s; Christa Ludwig, ms;
          Wieslaw Ochman, t; Martti Talvela, bs; Vienna State Opera Chorus;
          VPO; Karl Böhm. DG 2707 080 (2).
              +Audio 3-76 p68   +-NYT 9-21-75 pD18
              ++Gr 6-75 p75   +ON 2-7-76 p34
              +-HF 11-75 p96   +RR 7-75 p55

```
              +HFN 7-75 p75              +SFC 10-19-75 p33
              +MT 1-76 p39              +St 12-75 p116
              +NR 1-76 p9
```
329   Mass, op. 123, D major. Gundula Janowitz, s; Christa Ludwig,
        ms; Fritz Wunderlich, t; Walter Berry, bs; Wiener Singverein;
        VPO; Herbert von Karajan. DG 2726 048 (2). (Reissue from
        SKL 95/6).
```
              +Gr 11-76 p844           +-RR 12-76 p89
              +-HFN 11-76 p173
```
      Mass, op. 123, D major: Credo. cf HMV SEOM 25.
      Meerestille und Gluckliche Fahrt (Calm sea and prosperous
        journey), op. 112. cf Works, selections (CBS 76404).
      Minuet, E flat major. cf Country dances, G 141.
330   Minuets, WoO 7 (12). Minuets, WoO 10 (6). Minuets, 2 violins
        and double bass, WoO 9 (6). PH; Hans Ludwig Hirsch. Tele-
        funken AW 6-41935.
```
              +-Gr 3-76 p1458
```
      Minuets, WoO 10 (6). cf Minuets, WoO 7 (12).
      Minuets, 2 violins and double bass, WoO 9 (6). cf Minuets,
        WoO 7 (12).
      Mödlinger dances, nos. 1-4, 6, 8. cf Saga 5411.
      Mödlinger dances: Waltzes, nos. 3, 10-11. cf Saga 5421.
331   Namensfeier, op. 115: Overture. BRAHMS: Symphony, no. 4, op. 98,
        E minor. Lamoureux Orchestra; Igor Markevitch. DG 2548 217.
        (Reissue from SLPM 138 032).
```
               -Gr 4-76 p1617            -RR 4-76 p38            -
               -HFN 4-76 p121
```
      Namensfeier, op. 115: Overture. cf Overtures (DG 2548 138).
332   Octet, op. 103, E flat major. Quintet, piano, op. 16, E flat
        major. Rudolf Serkin, pno; Marlboro Festival Winds. Columbia
        M 33527. (Reissue from MS 6116).
```
              +-HF 2-76 p89             ++SFC 11-23-75 p26
              ++NR 8-75 p7
```
      Opferlied, op. 121b. cf Works, selections (CBS 76404).
333   Organ works: Fugue, C major. Fugue cycle on themes by J. S.
        Bach, D minor. Prelude, F minor. Prelude in every key, op.
        39, no. 1. Suites, mechanical organ, nos. 1-3. Trio, E
        minor. Wilhelm Krumbach, org. Musical Heritage Society MHS
        1517. (also Schwann AMS 2592).
```
              +MU 10-76 p14             +St 10-73 p133
```
334   Overtures: Coriolan, op. 62. The creatures of Prometheus, op. 43.
        Egmont, op. 84. Fidelio, op. 72. Leonore, no. 3, op. 72.
        VPO, Dresden Staatskapelle; Karl Böhm. DG 2535 135. (Reissues
        from 2720 045, 643614/6).
```
              +-Gr 3-76 p1458           +RR 3-76 p34
              +HFN 3-76 p109
```
335   Overtures: Consecration of the house, op. 124. Egmont, op. 84.
        Fidelio, op. 72. Leonore, nos. 2 and 3, op. 72. Namensfeier,
        op. 115. The ruins of Athens, op. 113. BPhO, Lamoureux
        Orchestra, Bavarian Radio Symphony Orchestra; Eugen Jochum,
        Igor Markevitch. DG 2548 138. Tape (c) 3348 138. (Reissues
        from SLPM 138039, 138038, 138694, 138032, 136019).
```
              +-Gr 10-75 p601           /RR 9-75 p28
              +HFN 9-75 p108           +-RR 9-76 p92 tape
```
      Pariser Einzugsmarsche (attrib.). cf DG 2721 077.
      Polonaise, op. 89, C major. cf Andante favori, F major.
      Prelude, F minor. cf Organ works (Musical Heritage Society MHS
        1517).

Prelude in every key, op. 39, no. 1. cf Organ works (Musical
    Heritage Society MHS 1517).

336   Quartets, strings, nos. 1-16. Grosse Fuge, op. 133, B flat
    major. Quartetto Italiano. Philips 6747 272 (10).
        ++Gr 9-76 p477   +RR 9-76 p75
        ++HFN 10-76 p183

337   Quartets, strings, nos. 1-6, op. 18. Juilliard Quartet. CBS
    77362 (3). (also Columbia M3 30084).
        +Gr 12-76 p1010   +-RR 12-76 p68

338   Quartets, strings, nos. 1-6, op. 18. Quartets, strings, nos.
    7-9, op. 59. Quartet, strings, no. 10, op. 74, E flat major.
    Quartet, strings, no. 11, op. 95, F minor. Quartet, strings,
    no. 12, op. 127, E flat major. Quartet, strings, no. 13, op.
    130, B flat major. Quartet, strings, no. 14, op. 131, C sharp
    minor. Quartet, strings, no. 15, op. 132, A minor. Quartet,
    strings, no. 16, op. 135, F major. Grosse Fuge, op. 133, B
    flat major. Amadeus Quartet. DG 2721 071 (10). (Reissues
    from SLPM 138531/3, 138534/6, 138537/40).
        +-Gr 6-74 p66   +-RR 6-74 p57
        +-NR 9-76 p7

339   Quartets, strings, nos. 1-6, op. 18. Végh Quartet. Telefunken
    6-35042 (3).
        ++HF 3-76 p79

340   Quartet, strings, no. 1, op. 18, no. 1, F major. Quartet, strings,
    no. 2, op. 18, no. 2, G major. Gabrieli Quartet. Decca SDD
    478. Tape (c) KSDC 478.
        +Gr 4-76 p1623   +-RR 3-76 p52
        +Gr 6-76 p102 tape   ++STL 5-9-76 p38
        +HFN 3-76 p89

341   Quartet, strings, no. 2, op. 18, no. 2, G major. Quartet, strings,
    no. 4, op. 18, no. 4, C minor. Quartetto Italiano. Philips
    6500 646.
        ++Audio 12-76 p91   /MJ 10-76 p25
        +Gr 6-76 p61   +NR 8-76 p6
        /HF 9-76 p89   ++RR 4-76 p60
        ++HFN 5-76 p95
    Quartet, strings, no. 2, op. 18, no. 2, G major. cf Quartet,
    strings, no. 1, op. 18, no. 1, F major.
    Quartet, strings, no. 4, op. 18, no. 4, C minor. cf Quartet,
    strings, no. 2, op. 18, no. 2, G major.

342   Quartets, strings, nos. 7-9, op. 59. Quartetto Italiano. Philips
    6747 139 (2).
        +Gr 12-75 p1062   ++NR 2-76 p6
        +-HF 3-76 p79   +RR 11-75 p58
        ++HFN 12-75 p148   +St 11-75 p64
        +MJ 1-76 p24
    Quartets, strings, nos. 7-9, op. 59. cf Quartets, strings, nos.
    1-6, op. 18.

343   Quartets, strings, nos. 7-11, opp. 59, 74, 95. Juilliard Quartet.
    Columbia D3M 34094 (3). (Reissues).
        ++MJ 10-76 p25   ++SFC 7-18-76 p31
        ++NR 7-76 p5

344   Quartets, strings, nos. 7-11, opp. 59, 74, 95. Végh Quartet.
    Telefunken EX 6-35041 (3).
        +Gr 8-76 p311
    Quartet, strings, no. 10, op. 74, E flat major. cf Quartets,
    strings, nos. 1-6, op. 18.

Quartet, strings, no. 11, op. 95, F minor. cf Quartets, strings,
nos. 1-6, op. 18.

345 Quartets, strings, nos. 12-16. Grosse Fuge, op. 133, B flat major.
Amadeus Quartet. DG 2734 006 (4).
+HFN 10-76 p183   +-RR 9-76 p75

346 Quartets, strings, nos. 12-16. Grosse Fuge, op. 133, B flat major.
Smetana Quartet. Supraphon 111 1151/4 (4). (no. 13, reissue
from SUAST 50604, no. 15, reissue from SUAST 50885).
+-Gr 5-76 p1775   +-RR 6-76 p62
+-NR 5-76 p7

347 Quartet, strings, no. 12, op. 127, E flat major. HAYDN: Quartet,
strings, op. 103, B flat major. Weller Quartet. Decca SDD
442. (Reissue from SXL 6423).
++Gr 6-76 p61   +-RR 7-76 p66
++HFN 6-76 p102

348 Quartet, strings, no. 12, op. 127, E flat major. Quartet, strings,
no. 14, op. 131, C minor. Quartet, strings, no. 16, op. 135,
F major. Busch Quartet. World Records SHB 38 (2). (Reissue
from 4410/2, DB 3044/8, 2810/4, 2113/6).
++Gr 11-76 p824   ++RR 9-76 p75
+HFN 10-76 p164   ++STL 10-10-76 p37

Quartet, strings, no. 12, op. 127, E flat major. cf Quartets,
strings, nos. 1-6, op. 18.

Quartet, strings, no. 13, op. 130, B flat major. cf Quartets,
strings, nos. 1-6, op. 18.

Quartet, strings, no. 13, op. 130, B flat major: Lento assai
(orchestral version). cf Symphony, no. 7, op. 92, A major.

Quartet, strings, no. 14, op. 131, C sharp minor. cf Quartets,
strings, nos. 1-6, op. 18.

Quartet, strings, no. 14, op. 131, C minor. cf Quartet, strings,
no. 12, op. 127, E flat major.

Quartet, strings, no. 15, op. 132, A minor. cf Quartets, strings,
nos. 1-6, op. 18.

Quartet, strings, no. 16, op. 135, F major. cf Quartets, strings,
nos. 1-6, op. 18.

Quartet, strings, no. 16, op. 135, F major. cf Quartet, strings,
no. 12, op. 127, E flat major.

Quartet, strings, no. 16, op. 135, F major: Lento (orchestral
version). cf Symphony, no. 5, op. 67, C minor.

Quintet, piano, op. 16, E flat major. cf Octet, op. 103, E flat
major.

349 Romances, nos. 1 and 2, opp. 40, 50. SIBELIUS: Concerto, violin,
op. 47, D minor. Pinchas Zukerman, vln; LPO; Daniel Barenboim.
DG 2530 552. Tape (c) 3300 496.
+-Gr 10-75 p633   +-NYT 1-18-76 pD1
-HF 7-76 p92   +RR 10-75 p55
+HFN 11-75 p151   ++RR 10-76 p106 tape
+-MJ 5-76 p29   +-St 6-76 p108
+NR 2-76 p4

Romance, no. 1, op. 40, G major. cf RCA ARM 4-0942/7.
Romance, no. 2, op. 50, F major. cf Works, selections (Philips
6833 179).
Romance, no. 2, op. 50, F major. cf RCA ARM 4-0942/7.
Rondo, G major. cf HMV HLM 7077.

350 Rondo a capriccio, op. 129, G major. The ruins of Athens, op. 113:
Turkish march. Sonata, piano, no. 3, op. 2, no. 3, C major.
Sonata, piano, no. 14, op. 27, no. 2, C sharp minor. Josef

Hofmann, pno. Saga 5392. (From Duo-Art piano rolls).
          +-Gr 7-75 p214   +-RR 7-75 p46
          +-HFN 8-76 p76
The ruins of Athens, op. 113: Chorus of dervishes Turkish march.
    cf RCA ARM 4-0942/7.
The ruins of Athens, op. 113: Incidental music. cf Fantasia,
    op. 80, C minor.
The ruins of Athens, op. 113: Overture. cf Overtures (DG 2548 138).
The ruins of Athens, op. 113: Overture. cf Symphony, no. 2,
    op. 36, D major.
Die Ruinen von Athens, op. 113: Overture. cf HMV RLS 717.
The ruins of Athens, op. 113: Turkish march. cf Rondo a capriccio,
    op. 129, G major.
The ruins of Athens, op. 113: Turkish march. cf Classics for
    Pleasure CFP 40254.
The ruins of Athens, op. 113: Turkish march. cf Desmar DSM 1005.
The ruins of Athens, op. 113: Turkish march. cf International
    Piano Library IPL 5001/2.
Septet, op. 20, E flat major. cf HMV SLS 5046.
Serenade, flute, violin and viola, op. 25, D major: Adagio; Allegro.
    cf Decca SPA 394.
Serenade, flute, violin and viola, op. 25, D major: Entrata. cf
    Works, selections (Philips 6833 179).
351 Serenade, string trio, op. 8. Trios, strings, op. 3, no. 1;
    op. 9, nos. 1-3. Jascha Heifetz, vln; William Primrose, vla;
    Gregor Piatigorsky, vlc. RCA SER 5729/31 (3).
          -Gr 10-76 p616   +RR 11-76 p86
352 Sketchbooks: Symphonies, nos. 1-2, 5-7. Denis Matthews. Dis-
    courses All About Music ABM 4-5.
          +Gr 1-76 p1240   +RR 12-75 p37
          +HFN 12-75 p149
353 Sketchbooks: Symphony, no. 8, op. 93, F major. Symphony, no. 9,
    op. 125, D minor. Presented and illustrated by Denis Matthews.
    Joan Sutherland, s; Marilyn Horne, con; James King, t; Martti
    Talvela, bs; VPO; Hans Schmidt-Isserstedt. Discourses All
    About Music ABM 6.
          +Gr 9-76 p474   +RR 5-76 p38
          +-HFN 5-76 p95
354 Sketchbooks: The piano sonatas. Presented and illustrated by
    Denis Matthews. Discourses All About Music ABM 7.
          +Gr 9-76 p474   +RR 6-76 p67
          +HFN 6-76 p81
355 Sketchbooks: The string quartets. Presented by Denis Matthews.
    Rocca String Quartet. Discourses All About Music ABM 8.
          +Gr 8-76 p474   +RR 9-76 p76
          +-HFN 10-76 p164
356 Sonatas, piano, nos. 1-2, 5-6, 11, 13, 15, 17, 19-21, 24-25, 28,
    30-32. Alfred Brendel, pno. Turnabout TV 34111, 34113, 34115,
    34117, 34119, 34120. Tape (c) KTVC 34111, 34113, 34115, 34117,
    34119, 34120.
          +Gr 10-76 p658 tape
357 Sonatas, piano, nos. 1, 3, 7-8, 13-14, 17-18, 21-23, 26-32. Solo-
    mon, pno. HMV RLS 722 (7). (Reissues from ALP 1573, C 3847/9,
    ALP 1062, 1900, BPL 1051, ALP 1303, 1160, 1546, 1272, 294,
    1141).
          +-Gr 11-76 p838   +-HFN 11-76 p155

Sonata, piano, no. 1, op. 2, no. 1, F minor. cf Concerto, piano, no. 2, op. 19, B flat major.

358   Sonata, piano, no. 3, op. 2, no. 3, C major. Sonata, piano, no. 14, op. 27, no. 2, C sharp minor. Sonata, piano, no. 20, op. 49, no. 2, G major. Roger Woodward, pno. RCA LRL 1-5097.
        -Gr 7-76 p193   +-HFN 7-76 p84
Sonata, piano, no. 3, op. 2, no. 3, C major. cf Rondo a capriccio, op. 129, G major.

359   Sonata, piano, no. 6, op. 10, no. 2, F major. Sonata, piano, no. 27, op. 90, E minor. Sonata, piano, no. 30, op. 109, E major. Alfred Brendel, pno. Philips 9500 076.
        +Gr 11-76 p837   +-RR 11-76 p88
        +HFN 11-76 p156
Sonata, piano, no. 6, op. 10, no. 2, F major. cf Concerto, piano, no. 1, op. 15, C major.

360   Sonatas, piano, nos. 8, 13, 23. Bruno-Leonardo Gelber, pno. Connoisseur Society CSQ 2113.
        +MJ 12-76 p28

361   Sonatas, piano, nos. 8, 21, 26. Vladimir Ashkenazy, pno. Decca SXL 6706. (also London CS 6921).
        +Gr 5-75 p1992   +HFN 7-75 p90 tape
        +-HF 10-76 p103   ++NR 11-76 p12
        ++HFN 6-75 p87   +-RR 5-75 p52

362   Sonata, piano, no. 8, op. 13, C minor. Sonata, piano, no. 17, op. 31, no. 2, D minor. Sonata, piano, no. 26, op. 81a, E flat major. Andor Földes, pno. DG 2548 210. (Reissues from SLPM 138 671, SLPM 138 784).
        +-Gr 4-76 p1617   /RR 8-76 p67
        +HFN 4-76 p121

363   Sonata, piano, no. 8, op. 13, C minor. Sonata, piano, no. 14, op. 27, no. 2, C sharp minor. Sonata, piano, no. 23, op. 57, F minor. John Lill, pno. Enigma VAR 1001.
        +Gr 12-76 p1015   +RR 12-76 p76

364   Sonata, piano, no. 8, op. 13, C minor. Sonata, piano, no. 18, op. 31, no. 3, E flat major. Sonata, piano, no. 19, op. 49, no. 1, G minor. Alfred Brendel, pno. Philips 9500 077.
        ++Gr 12-76 p1016   +-RR 12-76 p76
        +-HFN 12-76 p137
Sonata, piano, no. 8, op. 13, C minor. cf Concerto, piano, no. 1, op. 15, C major.
Sonata, piano, no. 8, op. 13, C minor. cf Concerto, piano, no. 4, op. 58, G major.
Sonata, piano, no. 8, op. 13, C minor: Adagio cantabile. cf Decca SPA 473.

365   Sonatas, piano, nos. 9, 10, 28. Alfred Brendel, pno. Philips 9500 041. Tape (c) 7300 475.
        +-ARG 12-76 p18   +-HFN 10-76 p165
        +Gr 10-76 p622   +-RR 10-76 p86
        +-Gr 10-76 p658 tape

366   Sonata, piano, no. 12, op. 26, A flat major. Sonata, piano, no. 16, op. 31, no. 1, G major. Emil Gilels, pno. DG 2530 654. Tape (c) 3300 654.
        +-Gr 6-76 p62   +HFN 6-76 p81
        +-Gr 10-76 p658 tape   +RR 6-76 p67

367   Sonata, piano, no. 14, op. 27, no. 2, C sharp minor. SCHUBERT: Impromptus, op. 90, nos. 2, 4, D 899. Impromptus, op. 142, nos. 1, 2, D 935. Vladimir Horowitz, pno. Columbia M 32342.

Tape (c) MT 32342 (ct) MA 32342. (also CBS 73173).
    -Gr 1-76 p1217   +-NR 11-73 p12
   +-HF 12-73 p99   +NYT 9-2-73 pD18
   +HF 5-74 p122 tape   -RR 1-76 p44
   -HFN 1-76 p103   /-St 1-74 p105

368   Sonata, piano, no. 14, op. 27, no. 2, C sharp minor. CHOPIN:
Fantaisie-Impromptu, op. posth., C sharp minor. SCARLATTI:
Pastorale and capriccio (arr. Tausig). TCHAIKOVSKY: The
sleeping beauty, op. 66: Coda prologue. Charles Lynch, pno.
Cork Dance Company SN 55.
   +Gr 4-76 p1644

Sonata, piano, no. 14, op. 27, no. 2, C sharp minor. cf Concerto,
piano, no. 3, op. 37, C minor.
Sonata, piano, no. 14, op. 27, no. 2, C sharp minor. cf Rondo a
capriccio, op. 129, G major.
Sonata, piano, no. 14, op. 27, no. 2, C sharp minor. cf Sonata,
piano, no. 3, op. 2, no. 3, C major.
Sonata, piano, no. 14, op. 27, no. 2, C sharp minor. cf Sonata,
piano, no. 8, op. 13, C minor.
Sonata, piano, no. 14, op. 27, no. 2, C sharp minor. cf Works,
selections (Decca DPA 529/30).
Sonata, piano, no. 14, op. 27, no. 2, C sharp minor: 1st movement.
cf Works, selections (Philips 6833 179).
Sonata, piano, no. 14, op. 27, no. 2, C sharp minor: 1st movement.
cf CBS 30064.

369   Sonata, piano, no. 15, op. 28, D major. Sonata, piano, no. 21,
op. 53, C major. Edwin Fischer, pno. Bruno Walter Society
RR 435.
     +-NR 10-76 p14

Sonata, piano, no. 15, op. 28, D major. cf Concerto, piano, no. 4,
op. 58, G major.
Sonata, piano, no. 16, op. 31, no. 1, G major. cf Sonata, piano,
no. 12, op. 26, A flat major.
Sonata, piano, no. 17, op. 31, no. 2, D minor. cf Sonata, piano,
no. 8, op. 13, C minor.

370   Sonata, piano, no. 18, op. 31, no. 3, E flat major. Sonata, piano,
no. 23, op. 57, F minor. Lazar Berman, pno. CBS 76533. (also
Columbia M 34218. Tape (c) MT 34218 (ct) MA 34218.
   /Gr 12-76 p1016   +-RR 12-76 p76

Sonata, piano, no. 18, op. 31, no. 3, E flat major. cf Sonata,
piano, no. 8, op. 13, C minor.
Sonata, piano, no. 19, op. 49, no. 1, G minor. cf Sonata, piano,
no. 8, op. 13, C minor.
Sonata, piano, no. 20, op. 49, no. 2, G major. cf Sonata, piano,
no. 3, op. 2, no. 3, C major.
Sonata, piano, no. 20, op. 49, no. 2, G major. cf Connoisseur
Society (Q) CSQ 2065.
Sonata, piano, no. 21, op. 53, C major. cf Andante favori, F
major.
Sonata, piano, no. 21, op. 53, C major. cf Concerto, piano, no. 2,
op. 19, B flat major.
Sonata, piano, no. 21, op. 53, C major. cf Sonata, piano, no. 15,
op. 28, D major.
Sonata, piano, no. 21, op. 53, C major. cf International Piano
Archives IPA 5007/8 (2).

371   Sonata, piano, no. 23, op. 57, F minor. Sonata, piano, no. 32,

op. 111, C minor. Roger Woodward, pno. RCA Victor LRL 1-5016.
            +-Gr 10-74 p724   +-RR 10-74 p75
            +-MT 1-76 p39
372   Sonata, piano, no. 23, op. 57, F minor. LISZT: Sonata, piano,
        G 178, B minor. Lazar Berman, pno. Saga 5430.
            +-ARG 11-76 p14   -RR 5-76 p66
            +-Gr 6-76 p69   +-St 9-76 p117
      Sonata, piano, no. 23, op. 57, F minor. cf Sonata, piano, no. 8
        op. 13, C minor.
      Sonata, piano, no. 23, op. 57, F minor. cf Sonata, piano, no. 18,
        op. 31, no. 3, E flat major.
373   Sonata, piano, no. 24, op. 78, F sharp major. Sonata, piano,
        no. 29, op. 106, B flat major. Claudio Arrau, pno. Philips
        6833 145.
            +NR 6-76 p11
374   Sonatas, piano, nos. 25-27. Emil Gilels, pno. DG 2530 589.
            +Gr 12-75 p1075   +-NR 3-76 p12
            +HF 6-76 p75   ++RR 12-75 p78
            +-HFN 12-75 p149   +SFC 2-15-76 p38
      Sonata, piano, no. 25, op. 79, G major. cf Concerto, piano,
        no. 5, op. 73, E flat major.
      Sonata, piano, no. 26, op. 81a, E flat major. cf Concerto, piano,
        no. 3, op. 37, C minor.
      Sonata, piano, no. 26, op. 81a, E flat major. cf Concerto, piano,
        no. 4, op. 58, G major.
      Sonata, piano, no. 26, op. 81a, E flat major. cf Sonata, piano,
        no. 8, op. 13, C minor.
375   Sonata, piano, no. 27, op. 90, E minor. BUSONI: All Italia.
        Indian diary. Perpetuum mobile. LISZT: Concerto, piano, no.
        2, G 125, A major. Années de pelerinage, 2nd year, G 161.
        Egon Petri, pno; Orchestra; Alfred Wallenstein. Bruno Walter
        Society RR 430.
            +-NR 10-76 p14
      Sonata, piano, no. 27, op. 90, E minor. cf Sonata, piano, no. 6,
        op. 10, no. 2, F major.
376   Sonata, piano, no. 28, op. 101, A major. Sonata, piano, no. 30,
        op. 109, E major. Sviatoslav Richter, pno. Rococo 2115.
            +-NR 10-76 p15
377   Sonata, piano, no. 29, op. 106, B flat major. Istvan Antal, pno.
        Hungarian SHLX 90033.
            +-RR 3-76 p58
      Sonata, piano, no. 29, op. 106, B flat major. cf Sonata, piano,
        no. 24, op. 78, F sharp major.
378   Sonata, piano, no. 30, op. 109, E major. Sonata, piano, no. 31,
        op. 110, A flat major. Maurizio Pollini, pno. DG 2530 645.
        Tape (c) 3300 645.
            ++Gr 5-76 p1782   +-RR 5-76 p64
            ++HFN 5-76 p95   +-RR 9-76 p93 tape
            +-MJ 12-76 p28   ++SFC 9-12-76 p31
            +NR 10-76 p14   ++St 12-76 p134
379   Sonata, piano, no. 30, op. 109, E major. Sonata, piano, no. 32,
        op. 111, C minor. Artur Schnabel, pno. RCA AVM 1-1410. (also
        RCA SMA 7013).
            +Gr 7-76 p188   +-NR 7-76 p11
            -MJ 10-76 p52   +RR 6-76 p68
      Sonata, piano, no. 30, op. 109, E major. cf Sonata, piano, no. 6,
        op. 10, no. 2, F major.

Sonata, piano, no. 30, op. 109, E major. cf Sonata, piano,
   no. 28, op. 101, A major.
380   Sonata, piano, no. 31, op. 110, A flat major. Sonata, piano, no.
      32, op. 111, C minor. Gary Graffman, pno. Columbia M 33890.
         ++HF 9-76 p89   +-SFC 9-12-76 p31
         +-MJ 10-76 p52   +St 12-76 p134
         +-NR 7-76 p11
381   Sonata, piano, no. 31, op. 110, A flat major. Sonata, piano, no.
      32, op. 111, C minor. Stephen Bishop, pno. Philips 6500 764.
         +Gr 6-75 p67   ++NR 4-75 p12
         +HFN 8-75 p72   +RR 6-75 p59
         ++MT 9-76 p744   ++SFC 3-2-75 p24
Sonata, piano, no. 31, op. 110, A flat major. cf Andante favori,
   F major.
Sonata, piano, no. 31, op. 110, A flat major. cf Sonata, piano,
   no. 30, op. 109, E major.
Sonata, piano, no. 32, op. 111, C minor. cf Sonata, piano, no.
   23, op. 57, F minor.
Sonata, piano, no. 32, op. 111, C minor. cf Sonata, piano, no.
   30, op. 109, E major.
Sonata, piano, no. 32, op. 111, C minor. cf Sonata, piano, no.
   31, op. 110, A flat major (Columbia M 33890).
Sonata, piano, no. 32, op. 111, C minor. cf Sonata, piano, no.
   31, op. 110, A flat major (Philips 6500 764).
382   Sonatas, violin and piano, nos. 1-10. Josef Suk, vln; Jan Panenka,
      pno. Supraphon 111 0561/5 (5). (Some reissues from SUAST
      50905/7).
         +-Gr 11-75 p847   ++NR 6-76 p8
         +HF 10-76 p94   +RR 10-75 p69
         +-HFN 10-75 p136
383   Sonata, violin and piano, no. 1, op. 12, no. 1, D major. Sonata,
      violin and piano, no. 5, op. 24, F major. Arthur Grumiaux, vln;
      Claudio Arrau, pno. Philips 9500 055. Tape 7300 473.
         +-Gr 5-76 p1776   ++HFN 5-76 p95
         +Gr 10-76 p658 tape   +-RR 5-76 p64
384   Sonata, violin and piano, no. 3, op. 12, E flat major. Sonata,
      violin and piano, no. 8, op. 30, no. 3, G major. Itzhak Perl-
      man, vln; Vladimir Ashkenazy, pno. Decca SXL 6789.
         +Gr 12-76 p1010   +RR 12-76 p77
385   Sonata, violin and piano, no. 4, op. 23, A minor. Sonata, violin
      and piano, no. 5, op. 24, F major. Itzhak Perlman, vln; Vladi-
      mir Ashkenazy, pno. Decca SXL 6736. Tape (c) KSXC 6736.
         +Gr 7-76 p187   +HFN 7-76 p104 tape
         +Gr 7-76 p230 tape   +-RR 5-76 p64
         +HFN 5-76 p95
386   Sonata, violin and piano, no. 4, op. 23, A minor. Sonata, violin
      and piano, no. 5, op. 24, F major. Oleg Kagaan, vln; Sviatos-
      lav Richter, pno. HMV ASD 3295.
         ++HFN 12-76 p137   +RR 12-76 p77
387   Sonata, violin and piano, no. 5, op. 24, F major. BRAHMS: Sonatas,
      violin and piano, nos. 1-3, opp. 78, 100, 108. Wanda Wilko-
      mirska, vln; Antonio Barbosa, pno. Connoisseur Society CS
      2079, 2080 (2).
         +-HF 12-75 p88   ++St 3-76 p109
         ++NR 3-76 p5
388   Sonata, violin and piano, no. 5, op. 24, F major. Sonata, violin
      and piano, no. 9, op. 47, A major. Adolf Busch, vln; Rudolf

BEETHOVEN (cont.)   66

Serkin, pno. Rococo 2079.
        +-NR 8-76 p6
Sonata, violin and piano, no. 5, op. 24, F major. cf Sonata,
    violin and piano, no. 1, op. 12, no. 1, D major.
Sonata, violin and piano, no. 5, op. 24, F major. cf Sonata,
    violin and piano, no. 4, op. 23, A minor (Decca SXL 6736).
Sonata, violin and piano, no. 5, op. 24, F major. cf Sonata,
    violin and piano, no. 4, op. 23, A minor (HMV 3295).
Sonata, violin and piano, no. 8, op. 30, no. 3, G major. cf
    Sonata, violin and piano, no. 3, op. 12, E flat major.
Sonata, violin and piano, no. 9, op. 47, A major. cf Sonata,
    violin and piano, no. 5, op. 24, F major.
Sonata, violin and piano, no. 9, op. 47, A major. cf RCA ARM
    4-0942/7.
389 Sonatas, violoncello and piano, no. 1-5. Variations on Mozart's
    "Bei Männern" (7). Variations on Mozart's "Ein Mädchen oder
    Weibchen" (12). Variations on a theme from Handel's "Judas
    Maccabeus" (12). Jacqueline du Pre, vlc; Daniel Barenboim,
    pno. Angel SCB 3823 (3). (also HMV SLS 5042).
        +-Gr 5-76 p1775   +NR 6-76 p8
        -HF 10-76 p95   +RR 3-76 p20
        +HFN 5-76 p92   ++St 9-76 p117
        +HFN 10-76 p165   +STL 6-6-76 p37
        +MT 6-76 p495
390 Sonatas, violoncello and piano, nos. 1-5. Pablo Casals, vlc;
    Rudolf Serkin, pno. CBS 78291 (2). (Reissues from Philips
    ABL 3131, 3223; Columbia 33CX 1093).
        +-Gr 5-76 p1775   +RR 3-72 p20
        +HFN 4-76 p121
Sonatas, violoncello, nos. 1-5. cf Works, selections (DG 2733 009).
Sonata, violoncello, no. 3, op. 69, A major. cf Philips 6780 751.
Sonatina, G major. cf Connoisseur Society (Q) CSQ 2066.
Sonatinas, mandolin, C major, C minor. cf Adagio, E flat major.
391 Songs (arr.): Bonny laddie, Highland laddie; Cease your funning;
    Faithful Johnie; Polly Stewart; The sweetest lad was Jamie.
    HAYDN: Songs (arr.): The Birks of Abergeldie; The brisk young
    lad; Cumbernauld house; Duncan Gray; Green grow the rashes; I
    have lost my love; I'm o'er young to marry yet; John Anderson;
    Love will find out the way; My ain kind dearie; My boy Tammy;
    O bonny lass; O can ye sew cushions; The ploughman; Shepherds,
    I have lost my love; Sleepy Bodie; Up in the morning early; The
    white cockade. Janet Baker, ms; Yehudi Menuhin, vln; George
    Malcolm, hpd, pno; Ross Pople, vlc. Angel (Q) S 37172. (also
    HMV (Q) SQ ASD 3167).
        +Gr 3-76 p1500   +ON 5-76 p48
        +HF 7-76 p81   +-RR 3-76 p71
        +HFN 3-76 p103   ++St 8-76 p102
        +NR 6-76 p14
Songs: Elegischer Gesang, op. 118. cf Works, selections (CBS
    76404).
Songs: Elegiac song, op. 118; Opferlied, op. 121b,; Bundeslied,
    op. 122; Calm sea and prosperous voyage, op. 112. cf King
    Stephen, op. 117: Incidental music.
Songs: An die Hoffnung, op. 94; Ich liebe dich, G 235. cf Desto
    DC 7118/9.
Songs from sonatas and symphonies arranged as songs by Friedrich
    Silcher. cf Symphony, no. 8, op. 93, F major.

Suites, mechanical organ, nos. 1-3. cf Organ works (Musical
   Heritage Society MHS 1517).
392   Symphonies, nos. 1-9. Helen Donath, s; Teresa Berganza, ms;
   Wieslaw Ochman, Thomas Stewart, bar; LSO, COA, BPhO, Israel
   Philharmonic Orchestra, BSO, Orchestre de Paris, VPO, CO,
   Bavarian Radio Symphony Orchestra and Chorus; Rafael Kubelik.
   DG 2740 155 (8).
         +-ARG 12-76 p16   +-NR 10-76 p2
         +Gr 10-76 p583   +RR 9-76 p44
         +-HF 11-76 p104   +SFC 9-19-76 p33
         +HFN 10-76 p163
393   Symphonies, nos. 1-9. Overtures: Consecration of the house,
   op. 124. Coriolan, op. 62. Fidelio, op. 72. King Stephen, op.
   117. Leonore, nos. 1-3, opp. 72, 138. The creatures of Pro-
   metheus, op. 43. Egmont, op. 84: Incidental music. Birgit
   Nilsson, s; Christa Ludwig, con; Hans Hotter, bar; Waldemar
   Kmentt, t; PhO and Chorus; Otto Klemperer. HMV SLS 788 (5).
   Tape (c) TC SLS 788 (5).
         +-Gr 9-76 p497 tape   +-RR 11-76 p109 tape
         +-HFN 8-76 p94 tape
394   Symphonies, nos. 1-9. Overtures: Coriolan, op. 64. Egmont, op.
   84. Leonore, no. 3, op. 72. Elisabeth Schwarzkopf, s; Marga
   Höffgen, alto; Ernest Häfliger, t; Otto Edelmann, bs-bar;
   Vienna Singverein; PhO; Herbert von Karajan. HMV SLS 5053 (7).
   (Reissues from Angel originals)
         +Gr 5-76 p1744   +-HFN 6-76 p81
         +-HF 11-76 p104   +RR 6-76 p35
395   Symphonies, nos. 1-9. Coriolan overture, op. 62. Egmont, op. 84:
   Overture. Leonore overture, no. 3, op. 72. CSO; Georg Solti.
   London CPS 9 (5). Tape (c) CSPS 9 (6). (also Decca 11BB 188/9
   (9). Reissues from SXL 6655, 6684, 6BB 121/2).
         ++Gr 9-75 p443   +MJ 10-76 p25
         +-HF 2-76 p90   ++NR 5-76 p1
         +HF 9-76 p84   +-RR 9-75 p28
         +HFN 9-75 p95   ++SFC 11-23-75 p27
396   Symphonies, nos. 1-9. LSO; Josef Krips. Olympic Tape (c) 3065/5
   (5).
         +-HF 11-76 p153 tape
397   Symphonies, nos. 1-9. Overtures: Egmont, op. 84. Leonore, no. 3,
   op. 72. The creatures of Prometheus, op. 43. Ursula Koszut,
   s; Brigitte Fassbaender, ms; Nicolai Gedda, t; Donald McIntyre,
   bar; Munich Motet Choir; Munich Philharmonic Orchestra and
   Chorus; Rudolf Kempe. Seraphim SIH 6093 (8).
         +-HF 8-76 p73   +NR 6-76 p1
398   Symphony, no. 1, op. 21, C major. Symphony, no. 2, op. 36, D
   major. NYP; Leonard Bernstein. CBS 72694.
         +-Gr 1-76 p1188   +-RR 1-76 p30
         +-HFN 1-76 p103
399   Symphony, no. 1, op. 21, C major. Symphony, no. 8, op. 93, F
   major. CSO; Georg Solti. Decca SXL 6760. Tape (c) KSXC 6760.
   (Reissue from 11BB 188/96).
         -Gr 9-76 p409   +-RR 10-76 p39
         ++HFN 10-76 p181
400   Symphony, no. 1, op. 21, C major. Symphony, no. 8, op. 93, F
   major. Bavarian Radio Symphony Orchestra, BPhO; Eugen Jochum.
   DG 2548 224. Tape (c) 3348 224. (Reissue from SLPM 138 037).

```
                 +Gr 4-76 p1617              +RR 4-76 p36
                ++HFN 4-76 p121             +RR 9-76 p92 tape
```
401   Symphony, no. 1, op. 21, C major. Symphony, no. 2, op. 36, D
      major. AMF; Neville Marriner. Philips 6500 113. Tape (c)
      7300 087. (Reissue from 6707 013).
```
                 +Gr 12-72 p1214 tape       -NR 3-75 p2
                 -HF 5-75 p72               +RR 1-73 p88
                 +HF 2-76 p130 tape        ++SFC 4-20-75 p23
                 +HFN 5-72 p913            ++St 5-75 p94
```
402   Symphony, no. 2, op. 36, D major. Egmont, op. 84: Overture.
      CSO; Georg Solti. Decca SXL 6761. (Reissue from 11BB 188/196).
```
                 +Gr 11-76 p989
```
403   Symphony, no. 2, op. 36, D major. The creatures of Prometheus,
      op. 43: Overture. VPO; Karl Böhm. DG 2530 448. (Reissue
      from 2720 045).
```
                 +Gr 10-76 p602             +NR 1-76 p3
                 +HFN 8-75 p87              +RR 8-75 p28
```
404   Symphony, no. 2, op. 36, D major. The creatures of Prometheus,
      op. 43: Overture. The ruins of Athens, op. 113: Overture.
      BPhO; Herbert von Karajan. DG Tape (c) 3300 456.
```
                 +-HFN 2-76 p117 tape
```
405   Symphony, no. 2, op. 36, D major. CPhO; Paul Kletzki. Supraphon
      50792.
```
                 +Gr 10-76 p578
```
      Symphony, no. 2, op. 36, D major. cf Symphony, no. 1, op. 21,
      C major (CBS 72694).
      Symphony, no. 2, op. 36, D major. cf Symphony, no. 1, op. 21,
      C major (Philips 6500 113).
406   Symphony, no. 3, op. 55, E flat major. BERLIOZ: Roman carnival,
      op. 9. SIBELIUS: Kuolema, op. 44: Valse triste. En Saga, op.
      9. WAGNER: Die Walküre: Ride of the Valkyries. LPO; Victor
      de Sabata. Decca 6BB 236/7 (2). (Reissues from Decca K 1507/13,
      K 1552, K 1504/6, K 1562).
```
                 +Gr 10-76 p612             +RR 11-76 p50
                 +HFN 11-76 p155
```
407   Symphony, no. 3, op. 55, E flat major. Scottish National Orches-
      tra; Carlos Paita. Decca SXL 4367. (also London SPC 21152
      Tape (c) 5-21152).
```
                 +-ARG 12-76 p13            +-RR 8-76 p28
                 +-HFN 8-76 p76             +STL 8-8-76 p29
```
408   Symphony, no. 3, op. 55, E flat major. BPhO; Ferenc Fricsay. DG
      2548 088. Tape (c) 3348 088. (Reissue from DG 138 038).
```
                 +Gr 4-75 p1867             +-RR 9-76 p92 tape
                 +-RR 4-75 p21
```
409   Symphony, no. 3, op. 55, E flat major. BBC Symphony Orchestra;
      John Barbirolli. HMV SXLP 30209. (Reissue from ASD 2348).
      (also S 36461 Tape (c) 4XS 36461 (ct) 8XS 36461).
```
                 +Gr 7-76 p168              +RR 5-76 p34
                 +-HFN 6-76 p102
```
410   Symphony, no. 3, op. 55, E flat major. Columbia Symphony Orches-
      tra; Bruno Walter. Odyssey Y 33925.
```
                 +MJ 10-76 p25
```
411   Symphony, no. 3, op. 55, E flat major. COA; Eugene Jochum. Phil-
      ips 6580 137. (Reissue from AXS 9000/1-9).
```
                 +Gr 11-76 p775      /      +-RR 11-76 p50
                 -HFN 11-76 p170
```

412   Symphony, no. 3, op. 55, E flat major. San Francisco Symphony
       Orchestra; Seiji Ozawa. Philips 9500 002. Tape (c) 7300 420.
              -NR 3-76 p2   -SFC 12-28-75 p30
              +-NYT 1-18-76 pD1
413   Symphony, no. 3, op. 55, E flat major. Coriolan overture, op. 62.
       LSO; Leopold Stokowski. RCA ARL 1-0600. Tape (c) ARK 1-0600
       (ct) ARS 1-0600. (also RCA Tape (c) RK 11710).
              +-Gr 10-75 p602   +NR 7-75 p4
              -Gr 7-76 p230 tape   +-RR 9-75 p30
              +-HFN 11-75 p150   +-SFC 7-13-75 p21
              +-HFN 7-76 p104 tape   +St 9-75 p106
414   Symphony, no. 3, op. 55, E flat major. VPO; Erich Kleiber.
       Richmond R 23202.
              +-ARG 12-76 p13
       Symphony, no. 3, op. 55, E flat major. cf HMV RLS 717.
415   Symphony, no. 4, op. 60, B flat major. Egmont, op. 84: Overture.
       BBC Symphony Orchestra; Colin Davis. Philips 9500 032. Tape
       (c) 7300 455.
              +-Gr 4-76 p1591   ++NR 5-76 p2
              +HF 6-76 p76   +RR 4-76 p37
              +-HFN 4-76 p101   -SFC 7-11-76 p13
              +MJ 10-76 p25
416   Symphony, no. 4, op. 60, B flat major. Grosse Fuge, op. 133,
       B flat major. AMF; Neville Marriner. Philips 9500 033. Tape
       (c) 7300 456.
              +-Gr 4-76 p1591   +NR 12-76 p3
              +HF 12-76 p147 tape   ++RR 3-76 p35
              +HFN 3-76 p88   +SFC 9-19-76 p33
              +HFN 7-76 p104 tape   ++St 12-76 p134
417   Symphony, no. 5, op. 67, C minor. HANDEL: Berenice: Overture.
       Berlin State Opera Orchestra; Erich Kleiber. Bruno Walter
       Society IGI 330.
              +-ARSC Vol VIII, no. 2-3 +-NR 10-76 p4
                 p93
418   Symphony, no. 5, op. 67, C minor. Egmont, op. 84: Overture.
       Symphony, no. 9, op. 125, D minor: Finale. Philharmonic
       Orchestra, LAPO, Danish State Radio Symphony Orchestra and
       Chorus; Fritz Busch. Bruno Walter Society RR 396.
              +-ARSC Vol VIII, no. 2-3 p92
419   Symphony, no. 5, op. 67, C minor. Symphony, no. 8, op. 93, F
       major. BPhO; André Cluytens. Classics for Pleasure CFP 40007.
       Tape (c) TC CFP 40007. (Reissue)
              +-HFN 12-76 p153 tape
420   Symphony, no. 5, op. 67, C minor. MENDELSSOHN: Symphony, no. 4,
       op. 90, A major. MOZART: Symphony, no. 40, K 550, G minor.
       SCHUMANN: Symphony, no. 1, op. 38, B flat major. OSR, NPhO,
       LSO; Ernest Ansermet, Carlo Maria Giulini, Josef Krips. Decca
       DPA 527/8 (2). (Reissues from SXL 2003, 6166, 6225, 2223).
              +-Gr 3-76 p1513   +-RR 3-76 p47
              +-HFN 3-76 p109
421   Symphony, no. 5, op. 67, C minor. VPO; Carlos Kleiber. DG 2530
       516. Tape (c) 3300 472.
              ++Audio 3-76 p65   ++NYT 8-10-75 pD14
              ++Gr 6-75 p32   ++RR 6-75 p32
              ++HF 11-75 p91   +RR 7-75 p68 tape
              +HF 9-76 p84 tape   ++RR 10-75 p96 tape

```
              ++HFN 6-75 p85              +SR 11-1-75 p45
              ++MT 10-75 p885            ++St 12-75 p81
              +NR 12-75 p3
```

422   Symphony, no. 5, op. 67, C minor. RPO; Antal Dorati. DG 2535
      216. (Reissue from Contour 2870 482)
              +Gr 10-76 p578   /RR 11-76 p51
              +-HFN 11-76 p170

423   Symphony, no. 5, op. 67, C minor. BPhO; Ferenc Fricsay. DG 2548
      028. Tape (c) 3348 028. (Reissue from 138813).
              +-Gr 11-75 p840   +-RR 10-75 p37
              -HFN 10-75 p152   +-RR 9-76 p92 tape

424   Symphony, no. 5, op. 67, C minor. BRAHMS: Tragic overture, op.
      81. SCHUBERT: Symphony, no. 8, D 759, B minor. BPhO; Lorin
      Maazel. DG Tape (c) 3335 103.
              +-RR 2-76 p71 tape

425   Symphony, no. 5, op. 67, C minor. Leonore overture, no. 3, op.
      72. CSO; Georg Solti. London CS 6930. Tape (c) 5-6930.
              +-ARG 12-76 p18

426   Symphony, no. 5, op. 67, C minor. MOZART: Symphony, no. 40,
      K 550, G minor. Così fan tutte, K 588: Overture. COA, LSO;
      Georg Szell, Colin Davis. Philips Tape (c) 7317 133.
              +Gr 9-76 p494 tape

427   Symphony, no. 5, op. 67, C minor. Quartet, strings, no. 16, op.
      135, F major: Lento (orchestral version). GLUCK: Alceste:
      Overture. Salzburg Mozarteum Orchestra; Leopold Hager. Pye
      TPLS 13068. Tape (c) ZCTPL 13068.
              -Gr 5-76 p1749

428   Symphony, no. 5, op. 67, C minor. SCHUBERT: Symphony, no. 8,
      D 759, B minor. BSO; Seiji Ozawa. RCA GL 25002.
              +-Gr 10-76 p578   +-RR 11-76 p51
              +-HFN 12-76 p137
      Symphony, no. 5, op. 67, C minor: 1st movement. cf Works,
      selections (Philips 6833 179).

429   Symphony, no. 6, op. 68, F major. The creatures of Prometheus,
      op. 43: Overture. OSR; Ernest Ansermet. Decca ECS 781.
      (Reissue from SXL 2193).
              -Gr 9-76 p410   +RR 6-76 p36
              +-HFN 6-76 p102

430   Symphony, no. 6, op. 68, F major. CSO; Georg Solti. Decca SXL
      6763. Tape (c) KSXC 16763. (Reissue from 11BB 188/96).
              ++Gr 4-76 p1591   +-RR 3-76 p35
              +HFN 3-76 p109   +-RR 8-76 p83 tape
              +HFN 7-76 p104 tape

431   Symphony, no. 6, op. 68, F major. BPhO; Lorin Maazel. DG 2548
      205. Tape (c) 3348 205. (Reissue from 138642).
              +Gr 11-75 p840   +RR 10-75 p37
              +HFN 10-75 p152   -RR 9-76 p92 tape

432   Symphony, no. 6, op. 68, F major. Munich Philharmonic Orchestra;
      Rudolf Kempe. HMV ESD 7004. Tape (c) TC ESD 7004.
              +-Gr 10-76 p658 tape   +RR 10-76 p39
              +-HFN 11-76 p175 tape

433   Symphony, no. 6, op. 68, F major. Hungarian State Symphony Orch-
      estra; János Feréncsik. Hungaroton SLPX 11790.
              +ARG 12-76 p15

434   Symphony, no. 6, op. 68, F major. Columbia Symphony Orchestra;
      Bruno Walter. Odyssey Y 33924.
              +MJ 10-76 p25

435   Symphony, no. 6, op. 68, F major. BBC Symphony Orchestra; Colin
      Davis. Philips 6500 463. Tape (c) 7300 361.
            ++ARG 11-76 p14   -NR 12-76 p3
             +Gr 5-76 p1749   -RR 5-76 p35
             +-HFN 5-76 p92   +SFC 9-19-76 p33
             +-HFN 2-76 p116 tape
436   Symphony, no. 6, op. 68, F major. Orchestra dell'Augusteo;
      Victor de Sabata. World Records SH 235. (Reissue from
      HMV DB 6473/7).
             +Gr 12-76 p989   +RR 11-76 p50
             +HFN 10-76 p165
      Symphony, no. 6, op. 68, F major. cf Works, selections
      (Decca DPA 529/30)
437   Symphony, no. 7, op. 92, A major. Marlboro Festival Orchestra;
      Pablo Casals. Columbia M 33788. (also CBS 61671).
             +-Gr 2-76 p1331   +RR 1-76 p30
             +-HFN 1-76 p103   +St 3-76 p109
             +NR 12-75 p3
438   Symphony, no. 7, op. 92, A major. Egmont, op. 84: Overture.
      NPhO; Leopold Stokowski, Decca PFS 4342. (also London 21139
      Tape (c) 5-1139).
             +Gr 2-76 p1331   +-MM 6-76 p43
             +-HFN 3-76 p88
439   Symphony, no. 7, op. 92, A major. Egmont, op. 84: Overture.
      LAPO; Zubin Mehta. Decca SXLN 6673. Tape (c) KSXLN 6673.
      (also London CS 6870 Tape (c) 056870 (ct) 086870).
             +-Gr 10-74 p681   +-NYT 1-18-76 pD1
              +Gr 1-75 p1403 tape   +RR 10-74 p34
              -HF 8-75 p82   +SFC 8-10-75 p26
              -NR 9-75 p5   +-St 3-76 p109
440   Symphony, no. 7, op. 92, A major. Coriolan overture, op. 62.
      CSO; Georg Solti. Decca SXL 6764. (Reissue from 11BB 188/96).
             ++Gr 6-76 p38   +-RR 5-76 p35
             ++HFN 6-76 p102
441   Symphony, no. 7, op. 92, A major. Egmont, op. 84: Overture.
      CSO; Georg Solti. Decca Tape KSXC 6780.
             +HFN 8-76 p94 tape   +-RR 10-76 p105 tape
442   Symphony, no. 7, op. 92, A major. VPO; Carlos Kleiber.
      DG 2530 706. Tape (c) 3300 706.
             +-Gr 9-76 p409   ++RR 9-76 p45
             ++HFN 10-76 p163   +RR 10-76 p105 tape
443   Symphony, no. 7, op. 92, A major. Coriolan overture, op. 62.
      BPhO; Karl Böhm. DG 2535 147. (Reissue from SLPM 138 018).
             -Gr 2-76 p1331   +-RR 3-76 p36
             +HFN 5-76 p115
444   Symphony, no. 7, op. 92, A major. LSO; André Previn. HMV ASD
      3119. Tape (c) TC ASD 3119. (also Angel S 37116).
             +-Gr 11-75 p802   -NR 1-76 p3
             ++HFN 11-75 p150   +-NYT 1-18-76 pD1
             ++HFN 3-76 p113 tape   -RR 11-75 p41
             +-MM 6-76 p43   ++SFC 12-28-75 p30
445   Symphony, no. 7, op. 92, A major. Hungarian State Symphony
      Orchestra; János Feréncsik. Hungaroton SLPX 11791.
             /ARG 12-76 p15   -NR 12-76 p3
             +HF 12-76 p99   -RR 11-76 p51
             +-HFN 10-76 p164   ++SFC 9-19-76 p33

446   Symphony, no. 7, op. 92, A major.   Quartet, strings, no. 13, op.
      130, B flat major: Lento assai (orchestral version).   GLUCK:
      Iphigénie en Aulide: Overture.   Salzburg Mozarteum Orchestra;
      Leopold Hager.   Pye TPLS 13067.   Tape (c) ZCTPL 13067.
            -Gr 5-76 p1749   -RR 3-76 p36
            -HFN 3-76 p89
447   Symphony, no. 7, op. 92, A major.   CPhO; Paul Kletzki.   Supraphon
      50797.
            +Gr 10-76 p578
448   Symphony, no. 8, op. 93, F major.   WAGNER: Tannhäuser: Overture.
      WEBER: Euryanthe: Overture.   BPhO; Wilhelm Furtwängler.
      Bruno Walter Society RR 413.
            +-NR 10-76 p2
449   Symphony, no. 8, op. 93, F major (arr. Liszt).   Songs from
      sonatas and symphonies arranged as songs by Friedrich Silcher.
      Hermann Prey, bar; Leonard Hokanson, pno.   DG Archive 2533 121.
            +Audio 8-75 p82   +NYT 8-10-75 pD14
            +HF 5-75 p73   -ON 2-7-76 p34
            +NR 5-75 p12   ++SFC 2-16-75 p24
450   Symphony, no. 8, op. 93, F major.   Symphony, no. 9, op. 125,
      D minor.   Carole Farley, s; Alfreda Hodgson, con; Stuart Burrows,
      t; Norman Bailey, bs; Brighton Festival Chorus; RPO; Antal
      Dorati.   DG 2726 073 (2).
            +-Gr 12-76 p989   +-HFN 12-76 p137
      Symphony, no. 8, op. 93, F major.   cf Symphony, no. 1, op. 21,
      C major (Decca SXL 6760).
      Symphony, no. 8, op. 93, F major.   cf Symphony, no. 1, op. 21,
      C major (DG 2548 224).
      Symphony, no. 8, op. 93, F major.   cf Symphony, no. 5, op. 67,
      C minor.
451   Symphony, no. 9, op. 125, D minor.   BSO; Charles Munch.   Camden
      CCV 5021.   (Reissue).
            +ST 2-76 p741
452   Symphony, no. 9, op. 125, D minor.   Lucine Amara, s; Lili Chookasian,
      con; John Alexander, t; John Macurdy, bs; Mormon Tabernacle
      Choir; PO; Eugene Ormandy.   CBS 61747.   (also Columbia M 31818
      Tape (c) MT 31818 (ct) MA 31818).
            ++Gr 11-76 p776   -RR 11-76 p51
            +HFN 11-76 p170
453   Symphony, no. 9, op. 125, D minor.   CSO and Chorus; Georg Solti.
      Decca 6BB 121/2.   Tape KBB 27041.   (also London CSP 8 Tape
      (c) 5-8).
            +HFN 7-76 p104 tape   +-RR 6-76 p86 tape
454   Symphony, no. 9, op. 125, D minor.   Egmont, op. 84: Overture;
      Die Trommel gerühret; Freudvoll and Leidvoll; Klärchens Tod;
      Sieges symphonie.   Eva Andor, s; Márta Szirmay, ms; György Kor-
      ondi, t; Sándor Nagy, bar; Budapest Chorus; HSO; János Feréncsik.
      Hungaroton SPLX 11736/7 (2).
            +-HF 10-75 p70   +-RR 3-76 p36
            -NR 9-75 p3
455   Symphony, no. 9, op. 125, D minor.   VPO; Erich Kleiber.   London
      R 23201.   (Reissue).
            +SFC 9-19-76 p33
456   Symphony, no. 9, op. 125, D minor.   cf Symphony, no. 8, op. 93,
      F major.
      Symphony, no. 9, op. 125, D minor: 3rd movement.   cf Philips
      6747 204.

Symphony, no. 9, op. 125, D minor: 4th movement excerpt. cf
    HMV SEOM 24.
Symphony, no. 9, op. 125, D minor: Finale. cf Symphony, no. 5,
    op. 67, C minor.
Symphony, no. 9, op. 125, D minor: Presto; Allegro assai. cf
    Works, selections (Philips 6833 179).
Trio, E minor. cf Organ works (Musical Heritage Society MHS 1517).
456 Trio, clarinet, op. 11, B flat major. BRAHMS: Trio, clarinet,
    violoncello and piano, op. 114, A minor. David Glazer, clt;
    Frank Glazer, pno; David Soyer, vlc. Turnabout TV 34108.
    Tape (c) KTVC 34108.
        +HFN 7-76 p104 tape
457 Trios, piano, nos. 1-6. Trio in one movement, B flat major.
    Wilhelm Kempff, pno; Henryk Szeryng, vln; Pierre Fournier, vlc.
    DG 2734 003 (4). (Reissue from 2720 016).
        +-Gr 4-76 p1623   ++RR 4-76 p60
458 Trio, piano, no. 3, op. 1, no. 3, C minor. MARTINU: Trio, piano,
    no. 3, C major. New Prague Trio. Panton 110 378.
        +Gr 9-76 p435   +RR 10-76 p75
459 Trio, piano, no. 6, op. 97, B flat major. Variations on "Ich bin
    der Schneider Kakadu", op. 121a, G major.* Jacques Thibaud, vln;
    Pablo Casals, vlc; Alfred Cortot, pno. World Records SH 230.
    (*Reissue from HMV DB 1223/7).
        +-Gr 1-76 p1210   +RR 12-75 p72
        +HFN 12-75 p149
460 Trio, piano, clarinet and violoncello, op. 38, E flat major.
    GOUNOD: Petite symphonie. HAYDN: Trio, piano, flute and
    violoncello. MOZART: Serenade, no. 11, K 375, E flat major.
    Rudolf Serkin, pno; Michel Dubost, flt; Richard Stoltzman,
    clt; Peter Wiley, Alain Meunier, vlc. Marlboro Recording
    Society MRS 7/8 (2).
        +HF 10-75 p80   +NYT 5-25-75 pD14
        +HF 2-76 p89
Trios, strings, op. 3, no. 1; op. 9, nos. 1-3. cf Serenade,
    string trio, op. 8.
461 Trio, strings, op. 9, no. 1, G major. Trio, strings, op. 9,
    no. 3, C minor. Leonid Kogan, vln; Rudolf Barshai, vla;
    Mstislav Rostropovitch, vlc. Saga 5396. (Reissue from Artia
    ALP 164).
        -Gr 1-76 p1210   +RR 11-75 p59
Trio, strings, op. 9, no. 3, C minor. cf Trio, strings, op. 9,
    no. 1, G major.
Trio in one movement, B flat major. cf Trios, piano, nos. 1-6.
462 Variations, G 191 (32). Variations, op. 34, F major (6). Varia-
    tions and fugue, op. 35, E flat major. Glenn Gould, pno.
    CBS 72882. (also Columbia M 30080).
        +-Gr 3-76 p1482   +-RR 1-76 p44
        +-HFN 1-76 p103
Variations, op. 34, F major (6). cf Variations, G 191 (32).
Variations, mandolin, D major. cf Adagio, E flat major.
Variations and fugue, op. 35, E flat major. cf Variations, G 191
    (32).
463 Variations on a theme by Diabelli, op. 120. Daniel Barenboim, pno.
    ABC Westminster WGA 8272. (Reissue from HMV SXLP 20083).
        +-Gr 12-76 p1016
Variations on a theme from Handel's "Judas Maccabeus" (12). cf
    Sonatas, violoncello and piano, nos. 1-5.

Variations on Handel's "See the conquering hero comes" (12).
cf Works, selections (DG 2733 009).

Variations on "Ich bin der Schneider Kakadu", op. 121a, G major.
cf Trio, piano, no. 6, op. 97, B flat major.

Variations on "Kind, willst du ruhig schlafen". cf International
Piano Library LPL 5003/4.

Variations on Mozart's "Bei Männern" (7). cf Sonatas, violoncello
and piano, nos. 1-5.

Variations on Mozart's "Bei Männern, welche Liebe fühlen" (7).
cf Works, selections (DG 2733 009).

Variations on Mozart's "Bei Männern" (7). cf HMV RLS 723.

Variations on Mozart's "Ein Mädchen oder Weibchen" (12). cf Works,
selections (DG 2733 009).

Variations on Mozart's "Ein Mädchen oder Weibchen" (12). cf
Sonatas, violoncello and piano, nos. 1-5.

York'scher Marsch. cf DG 2721 077 (2).

464   Works, selections: Bundeslied, op. 122. Elegischer Gesang, op.
       118. King Stephen, op. 117. Meerestille und Gluckliche Fahrt,
       op. 112. Opferlied, op. 121b. Lorna Haywood, s; Ambrosian
       Singers; LSO; Michael Tilson Thomas. CBS 76404. (also
       Columbia M 33509).

              +-Audio 11-76 p111   +NR 6-76 p12
              +-Gr 11-75 p33   +NYT 9-21-75 pD18
              +HF 1-76 p82   +ON 2-7-76 p28
              +HFN 11-75 p150   +-RR 12-75 p87
              +-HFN 2-76 p116 tape   +-SFC 10-19-75 p33
              +MT 7-76 p575   ++St 2-76 p98

465   Works, selections: Concerto, piano, no. 5, op. 73, E flat major.
       Egmont, op. 84: Overture. Leonora overture, no. 3, op. 72.
       Sonata, piano, no. 14, op. 27, no. 2, C sharp minor. Symphony,
       no. 6, op. 68, F major. Julius Katchen, Friedrich Gulda, pno;
       LSO, OSR; Pierino Gulda, Lorin Maazel. Decca DPA 529/30.
       Tape (c) KDPC 529/30.

              +-Gr 6-76 p92   +-HFN 6-76 p103
              +Gr 9-76 p497 tape   +-RR 6-76 p35

466   Works, selections: Sonatas, violoncello, nos. 1-5. Variations on
       Mozart's "Ein Mädchen oder Weibchen" (12). Variations on
       Mozart's "Bei Männern, welche Liebe fühlen" (7). Variations
       on Handel's "See the conquering hero comes" (12). Pierre
       Fournier, vlc; Wilhelm Kempff, pno. DG 2733 009 (3). (Reissue
       from 138 993/5).

              +-Gr 5-76 p1775   +RR 3-76 p20
              +HFN 3-76 p121

467   Works, selections: Bagatelle, no. 25, A minor (Für Elise). Egmont,
       op. 84: Overture. Romance, no. 2, op. 50, F major. Serenade,
       flute, violin and viola, op. 25, D major: Entrata. Sonata,
       piano, no. 14, op. 27, no. 2, C sharp minor: 1st movement.
       Symphony no. 5, op. 67, C minor: 1st movement. Symphony, no. 9,
       op. 125, D minor: Presto; Allegro assai. Ingeborg Wenglor, s;
       Ursula Zollenkopf, con; Hans-Joachim Rotzsch, t; Theo Adam, bs;
       Claudio Arrau, Hans Richter-Haaser, pno; Arthur Grumiaux, vln;
       COA, NPhO, VSO, Leipzig Gewandhaus Orchestra; Georg Szell,
       Edo de Waart, Christoph von Dohnányi, Franz Konwitschny.
       Philips 6833 179.

              +-Gr 5-76 p1808   +-RR 4-76 p37
              +HFN 5-76 p115

BEHREND
   Auntie. cf Transatlantic XTRA 1159.
   Eichstatter Hofmühltanz. cf DG Archive 2533 184.
   Reidenburger Tanz. cf DG Archive 2533 184.
   Tanz im Aicholdinger Schloss. cf DG Archive 2533 184.
BELKNAP
   The seasons. cf Vox SVBX 5350.
BELL, Daniel
   Grass. cf Caprice RIKS 59.
BELLINI, Vicenzo
   Adelson e Salvini: Io provo un palpito per quel dimora. cf
      English National Opera ENO 1001.
   Almen se non poss'io. cf L'Oiseau-Lyre SOL 345.
468   I Capuleti ed i Montecchi. Beverly Sills, s; Janet Baker, ms;
         Nicolai Gedda, t; Robert Lloyd, bs; Raimund Herincx, bar; John
         Alldis Choir; NPhO; Giuseppe Patanè. Angel SCLX 3824 (3).
         Tape (c) 4X3S 3824. (also HMV SQ SLS 986).
               -Gr 4-76 p1654   +OC 5-76 p49
            +-HF 4-76 p89   +ON 2-28-76 p53
            +HF 10-76 p147 tape   +-RR 4-76 p28
            +-HFN 4-76 p101   ++SFC 4-4-76 p34
             +MM 9-76 p35   +-SR 5-29-76 p52
            +-MT 12-76 p1003   ++St 4-76 p78
            +-NR 5-76 p10   +-STL 5-9-76 p38
            +-NYT 6-20-76 pD27
469   I Capuletti ed i Montecchi: O, quante volte, O, quante. La
         sonnambula: Ah, non credea mirarti. DONIZETTI: Lucia di
         Lammermoor: Ardon gli incensi. ROSSINI: Il barbiere di Siviglia:
         Una voce poco fa. La gazza ladra: Di piacer mi balza il cor.
         VERDI: Rigoletto: Caro nome. Mady Mesplé, s; Paris Opera
         Orchestra; Gianfranco Masini. Angel S 37095.
               -HF 3-76 p102   +OC 2-76 p4
            +-NR 12-75 p11   +St 3-76 p124
   Concerto, oboe, E flat major. cf ALBINONI: Concerto, oboe, op. 7,
      no. 3, B flat major.
   Norma: Casta diva. cf English National Opera ENO 1001.
   Norma: Casta diva...Ah, bello, a me ritorno. cf Rubini GV 63.
   Norma: Casta diva. cf Rubini GV 68.
   Norma: Dormono entrambi...Teneri, teneri figlia, In mia man alfin
      tu sei. cf Court Opera Classics CO 347.
   Norma: Sediziose voci...Casta diva. cf HMV SLS 5057.
   Il pirata: Oh, s'io potessi...Col sorriso d'innocenza. cf
      HMV SLS 5057.
470   I puritani. Joan Sutherland, s; Anita Caminada, ms; Luciano
         Pavarotti, Renato Cazzaniga, t; Piero Cappuccilli, bar; Nicolai
         Ghiaurov, Giancarlo Luccardi, bs; ROHO Chorus; LSO; Richard
         Bonynge. London OSA 13111 (3). (also Decca SET 587/9.
               +Gr 7-75 p237   ++ON 6-75 p10
            ++HF 6-75 p84   +RR 7-75 p16
             +HFN 7-75 p76   ++SFC 2-23-75 p22
             +MT 2-76 p139   ++St 5-75 p73
             +NYT 4-6-75 pD18   +STL 10-5-75 p36
             +ON 4-19-75 p54
471   I puritani: Fini...me lassa. DONIZETTI: L'Elisir d'amore: Chiedi
         all'aura lusinghiera. La fille du régiment: Depuis l'instant.
         Lucia di Lammermoor: Sulla tomba...Verranno a te. VERDI:

Rigoletto: E il sol dell'anima. Joan Sutherland, s; Luciano
Pavarotti, t; Various orchestras; Richard Bonynge. London
OS 26437.
　　+NR 1-76 p10　　　　　　　　++St 9-76 p130
I puritani: O amato zio...Sai com arde.　cf Rococo 5386.

472　I puritani: O rendetemi la speme.　DONIZETTI: Lucia di Lammermoor:
Mad scene.　MEYERBEER: Dinorah: Shadow song.　THOMAS: Hamlet:
Mad scene.　Joan Sutherland, s; Orchestral accompaniments.
London OS 26436.
　　+NR 1-76 p10
I puritani: O rendetemi la speme...Qui la voce.　cf HMV SLS 5057.
I puritani: Qui la voce...Vien, diletto.　cf Richmond R 23197.
I puritani: Son vergin vezzosa.　cf Rubini GV 68.

473　Songs: Dolente immagine; Il fervido desidereo; Vaga luna, che
inargenti.　CHOPIN: Songs: Dumka; The maiden's wish, op. 74,
no. 1; Spring, op. 74, no. 13; The trooper, op. 74, no. 10.
DONIZETTI: Songs: La corrispondenza amorosa; A messanotte;
La mère et l'enfant.　ROSSINI: Songs: La danza; L'orgia;
La promessa.　Leyla Gencer, s; Marcello Guerrini, pno.
Cetra LPO 2003.
　　-HF 9-76 p105

474　Songs: Dolente immagine di fille mia; Malinconia, ninfa gentile;
Per pietà, bell'idol mio; Vaga luna, che inargenti.　DONIZETTI:
La corrispondenza amorosa; Una lagrima; La mère et l'enfant;
Ne ornerà la bruna chioma.　ROSSINI: Giovanna d'Arco (Cantata):
La danza.　VERDI: Brindisi; Lo spazzacamino; Stornello.　Renata
Scotto, s; Walter Baracchi, pno.　RCA AGL 1-1341.
　　+-HF 7-76 p102　　　　　　　+NR 5-76 p11
　　+ON 5-76 p48　　　　　　　　+-St 8-76 p108

475　La sonnambula: Ah, se un volta sola...Ah, non credea mirarti...
Ah, non giunge.　CILEA: Adriana Lecouvreur: Io sono l'umile
ancella.　PUCCINI: Suor Angelica: Senza mamma.　VERDI: Un ballo
in maschera: Morrò, ma prima in grazia.　Rigoletto: Caro nome.
Il trovatore: D'amor sull'ali rosee.　I vespri siciliani:
Arrigo, Ah parli a un core; Mercè, dilette amiche.　Montserrat
Caballé, s; Barcelona Symphony Orchestra; Gianfranco Masini.
London OS 26424.
　　++HF 1-76 p98　　　　　　　　++NR 1-76 p10
　　++OC 5-76 p49　　　　　　　　+-SFC 11-16-75 p32
　　+ON 1-17-76 p32
La sonnambula: Ah, non credea mirarti.　cf Capuletti ed i
Montecchi: O quante volte, O, quante.
La sonnambula: Care compagne...Come per me sereno.　cf HMV SLS 5057.
La sonnambula: Come per me sereno...Sovra il sen.　cf Richmond
R 23197.
La sonnambula: Lisa, mendace anch essa.　cf English National
Opera ENO 1001.

BELSAYAGA, Cristóbal de
Magnificat a 8.　cf Eldorado S-1.

BELSTERLING
March of the steel men.　cf Michigan University SM 0002.

BEMBERG, Herman
Songs: Les anges pleurent; Un ange est venu; Chant hindou; Chant
venétien; Elaine: L'amour est pur; Nymphs et Sylvains; Sur le
lac.　cf HMV RLS 719.

BENDA, Jiří (Georg)
Sonatina, D major.　cf RCA ARL 1-0864.

Sonatina, D minor. cf RCA ARL 1-0864.
BENEDICT, Julius
    La capinera. cf BIS LP 45.
    The gypsy and the bird. cf Decca KCSP 367.
    The gypsy and the moth. cf Decca SPA 394.
    Songs: The rose of Erin; Hungers' chorus. cf Library of Congress
       OMP 101/2.
BEN HAIM, Paul
    Sephardic lullaby. cf Da Camera Magna SM 93399.
BENJAMIN, Arthur
    Jamaican rumba. cf CBS 30073.
    Jamaican rumba. cf HMV SXLP 30181.
    Mattie rag. cf HMV SXLP 30181.
BENN
    Ricercar a 3. cf Pelca PRSRK 41017/20.
BENNET, John
    Weep, o mine eyes. cf HMV CSD 3756.
BENNETT, Richard Rodney
    Commedia IV. cf ADDISON: Divertimento, op. 9.
476 Concerto, guitar and chamber orchestra. BERKELEY: Theme and
    variations, guitar. RAWSTHORNE: Elegy, guitar. WALTON: Baga-
    telles, guitar (5). Julian Bream, gtr; Melos Ensmeble; David
    Atherton. RCA ARL 1-0049. Tape (c) ARK 1-0049 (ct) ARS 1-0049.
    (also RCA SB 6876 Tape (c) RK 11709).
        +Gr 10-73 p690   ++RR 10-73 p59
        -Gr 7-76 p230 tape   +-RR 9-76 p93 tape
        +HFN 7-76 p105 tape   +SR 5-18-74 p6 tape
        ++MJ 5-74 p52   ++St 7-74 p114
        +NR 2-74 p14
    Concerto, guitar and chamber ensemble. cf RCA ARL 3-0997.
BENNETT, Robert Russell
477 Concerto, violin, A major. Hexapoda. A song sonata. Louis
    Kaufman, vln; Annette Kaufman, pno; LSO; Bernard Herrmann.
    Citadel CT 6005.
        +NR 11-76 p5   +-St 12-76 p134
    Hexapoda. cf Concerto, violin, A major.
    A song sonata. cf Concerto, violin, A major.
BENNETT, William Sterndale
478 January, op. 36, no. 1. BURTON: Sonata, piano, no. 1, D major.
    FIELD: Rondo, E flat major. Sonata, piano, no. 1, E flat
    major: Rondo. PINTO: Sonata, piano, op. 3, no. 1, E flat minor.
    Alan Cuckston, pno. RCA 1-5101.
        +Gr 3-76 p1487   +MT 9-76 p744
        +HFN 5-76 p109   ++RR 3-76 p16
        +MM 9-76 p36
479 Sonata, piano, no. 1, op. 13, F minor. SCHUMANN, C.: Mazurka,
    op. 6, G major. Romances, op. 21, nos. 2-3. Variations on
    a theme by Robert Schumann, op. 20. James Sykes, pno. Orion
    ORS 75182.
        +-HF 7-76 p91   +-NR 4-76 p12
480 Songs: Song set, op. 23: Forget-me-not; Gentle zephyr; Maienthau;
    Musing on the roaring ocean; The past; To Chloe in sickness.
    Song set, op. 35: Castle Gordon; Dawn gentle flower; Indian
    love; Sing maiden sing; Waldeinsamkeit; Winter's gone. Sacred
    duets: And who is he that will harm you; Remember now thy
    creator. Meryl Drower, s; Antony Brahms, bar; Gary Peacock,

        pno. Rare Recorded Editions SRRE 165.
            +Gr 8-76 p324   +HFN 7-76 p97
        Songs: The carol singers. cf Argo ZFB 95/6.
BENSHOOF, Kenneth
        Songs: The cow; Dinky; The fox; John Brown's body; The waking.
        cf Washington University Press OLY 104.
BEREZOVSKY, Maximus
        Do not reject me in my old age. cf HMV Melodiya ASD 3102.
BERG, Alban
481 Concerto, violin. MOZART: Concerto, violin, no. 5, K 219, A major.
        Joseph Szigeti, vln; Orchestras; Eugene Ormandy, Dmitri
        Mitropoulos. Bruno Walter Society WSA 701.
            +-NR 11-76 p6
482 Lulu: Suite. STRAUSS, R.: Salome, op. 54: Final scene. Anja
        Silja, s; VPO; Christoph von Dohnanyi. Decca SXL 6657. (also
        London OS 26397).
            +-Audio 10-75 p118   -ON 1-10-76 p33
            +-Gr 10-74 p742   +-Op 11-74 p990
            +-HF 6-75 p87   -RR 9-74 p38
            +NR 6-75 p6   +St 9-75 p106
483 Lulu: Suite. Wozzeck: Suite. Helga Pilarczyk, s; LSO; Antal
        Dorati. Mercury SRI 75065. (Reissue from SR 90278).
            +MJ 5-76 p29
484 Lyric suite: Pieces (3). Orchestral pieces, op. 6 (3). SCHOEN-
        BERG: Pelleas und Melisande, op. 5. Verklärte Nacht, op. 4.
        Variations, orchestra, op. 31. WEBERN: Movements, string
        quartet, op. 5 (15). Pieces, op. 6 (6). Passacaglia, op. 1.
        Symphony, op. 21. BPhO; Herbert von Karajan. DG 2711 014 (4).
        (also DG 2530 485/8).
            +Gr 3-75 p1640   +RR 3-75 p34
            +-HF 7-75 p61   ++SFC 5-25-75 p17
            +-MJ 2-76 p32   +-St 9-75 p120
            ++NR 6-75 p3
        Orchestral pieces, op. 6. cf Lyric suite: Pieces.
485 Quartet, op. 3. SHOSTAKOVICH: Quartet, no. 10, op. 118. Weller
        Quartet. London STS 15287.
            +NR 6-76 p10
        Sonata, piano, op. 1. cf BARTOK: Out of doors.
        Songs: Frühe Lieder (7); Lieder, op. 2 (4). cf Hungaroton SLPX
            11713.
        Wozzeck: Suite. cf Lulu: Suite.
BERGER, Arthur
        Duo, no. 2. cf Desto DC 6435/7.
BERGER, Jean
        I lift up my eyes. cf Columbia M 34134.
        Songs: Snake baked a hoe-cake; The Frisco whale. cf Orion ORS
            75205.
BERGHMANS
        La femme à barbe. cf Boston Brass BB 1001.
BERGSMA, William
486 The fortunate island. LUENING: A poem in cycles and bells. King
        Lear: Suite. USSACHEVSKY: Piece, tape recorder. Royal Danish
        Radio Orchestra, Rome Santa Cecilia Orchestra; Alfredo Antonini,
        Otto Luening. CRI SD 112. (Reissue).
            +-NR 8-76 p14

BERIO, Luciano
487   Allelujah II. Concerto, 2 pianos. Nones. Bruno Canino, Antonio
       Ballista, pno; LSO, BBC Symphony Orchestra; Pierre Boulez. RCA
       ARL 1-1674.
          ++SFC 12-19-76 p50
488   A-Ronne. Cries of London. Swingle II; Luciano Berio. Decca
       HEAD 15.
          ++HFN 12-76 p137   ++RR 12-76 p89
       Concerto, 2 pianos. cf Allelujah II.
       Cries of London. cf A-Ronne.
       Nones. cf Allelujah II.
       O King. cf Delos DEL 25406.
489   Opera: Agnus; Air; E vo; El mar la mar; Melodrama; O King. Elise
       Ross, Mary Thomas, s; Gerald English, t; Alide Maria Salvetta;
       London Sinfonietta; Luciano Berio. RCA ARL 1-0037.
          +HF 11-75 p98   +SR 9-6-75 p40
          +NR 9-75 p11   +-St 6-76 p100
          +-NYT 12-21-75 pD18
       Sequenza. cf Nonesuch HB 73028.
BERKELEY, Lennox
       Aria. cf Wealden WS 149.
       Aubade. cf Wealden WS 149.
490   Concerto, guitar. RODRIGO: Concierto de Aranjuez, guitar and
       orchestra. Julian Bream, gtr; Monteverdi Orchestra; John
       Eliot Gardiner. RCA ARL 1-1181. Tape (c) ARK 1-1181 (ct)
       ARS 1-1181.
          +Gr 11-75 p807   +NR 12-75 p14
          +HF 2-76 p130 tape   ++RR 12-75 p60
          ++HFN 12-75 p149   +SR 11-29-75 p50
          +MM 8-76 p35
491   Concerto, guitar. RODRIGO: Concierto de Aranjuez, guitar and
       orchestra. Julian Bream, gtr. RCA Tape (c) RK 11734.
          +Gr 11-76 p887 tape   ++HFN 11-76 p175 tape
       Sonatina, op. 13. cf RCA LRL 1-5127.
       Sonatina, op. 51. cf RCA SB 6891
492   Songs: Automne; Chinese songs; D'un vanneur de blé; Tant que mex
       yeux. CROSSE: The new world, op. 25. DICKINSON: Extravangan-
       zas. Meriel Dickinson, ms; Peter Dickinson, pno. Argo ZRG 788.
          +MT 8-76 p659   +STL 11-2-75 p38
          ++RR 9-75 p68
       Theme and variations, guitar. cf BENNETT: Concerto, guitar and
       chamber orchestra.
       Toccata. cf Wealden WS 149.
BERLIN, Irving
       God bless America. cf Department of Defense Bicentennial Edition
       50-1776.
BERLINSKI, Herman
       The burning bush. cf Delos FY 025.
       Kol nidre. cf Serenus SRS 12039.
BERLIOZ, Hector
       Béatrice et Bénédict: Dieu, Que viens-je d'entre...Il m'en souvient.
       cf CBS 76522.
       Béatrice et Bénédict. cf Overtures (Angel 37170).
       Béatrice and Bénédict. cf Overtures (Classics for Pleasure CFP
       40097).
       Béatrice et Bénédict: Vous soupirez, madame. cf Decca DPA 517/8.

Béatrice et Bénédict: Vous soupirez, madame. cf Decca GOSD 674/6.
Benvenuto Cellini, op. 23. cf Overtures (Angel S 37170).
Benvenuto Cellini, op. 23. cf Overtures (Classics for Pleasure
    CFP 40097).
Benvenuto Cellini, op. 23. cf Works, selections (Klavier KS 553).
493   Le carnaval romain (The Roman carnival), op. 9. Le Corsaire, op.
        2. RESPIGHI: The pines of Rome. THOMAS: Mignon: Overture.
        World Youth Symphony Orchestra, Interlochen; George Wilson.
        Golden Crest GCIN 401.
            ++IN 1-76 p14
494   Le carnaval romain, op. 9. La damnation de Faust, op. 24: Menuet
        des follets; Danse des sylphes; Marche hongrois. Romeo et
        Juliette, op. 17: Queen Mab scherzo; Love scene. Les Troyens:
        Royal hunt and storm. John Alldis Choir; LSO and Chorus, ROHO;
        Colin Davis. Philips 6580 116.
            +NR 7-76 p3   +St 10-76 p120
Le carnaval romain, op. 9. cf Overtures (Angel S 37170).
Roman carnival, op. 9. cf Overtures (Classics for Pleasure CFP
    40097).
Le carnaval romain, op. 9. cf Overtures (Philips 7300 080).
Roman carnival, op. 9. cf Works, selections (Klavier KS 553).
Roman carnival, op. 9. cf BEETHOVEN: Symphony, no. 3, op. 55,
    E flat major.
Roman carnival, op. 9. cf Decca SXL 6782.
Le carnaval romain, op. 9. cf HMV SLS 5019.
Roman carnival, op. 9. cf Pye PCNHX 6.
Le Corsaire, op. 21. cf Le carnaval romain, op. 9.
Le Corsaire, op. 21. cf Overtures (Angel S 37170).
The Corsair, op. 21. cf Overtures (Classics for Pleasure CFP
    40097).
Le Corsaire, op. 21. cf Overtures (Philips 7300 080).
The damnation of Faust, op. 24, excerpts. cf GOUNOD: Faust.
La damnation de Faust, op. 24: D'amour l'ardente flamme. cf CBS
    76522.
The damnation of Faust, op. 24: Dance of the sylphs. cf Decca
    DPA 519/20.
La damnation de Faust, op. 24: Hungarian march. cf Angel S 37231.
La damnation de Faust, op. 24: Hungarian march. cf Classics for
    Pleasure CFP 40254.
La damnation de Faust, op. 24: Hungarian march. cf HMV SLS 5019.
La damnation de Faust, op. 24: Hungarian march. cf Pye QS PCNH 4.
La damnation de Faust, op. 24: Menuet des follets; Danse des sylphs;
    Marche hongrois. cf Le carnaval romain, op. 9.
La damnation de Faust, op. 24: Minuet of the will-o-the-wisps;
    Dance of the sylphs; Hungarian march. cf Works, selections
    (Klavier KS 553).
La damnation de Faust, op. 24: Serenade de Mephistopheles. cf
    Club 99-101.
La damnation de Faust, op. 24: Soldiers chorus. cf DG 2548 212.
Les Francs Juges, op. 3. cf Overtures (Angel S 37170).
Les Francs Juges, op. 3. cf Overtures (Philips 7300 080).
Les Francs Juges, op. 3. cf Philips 6780 030.
Hamlet, op. 18: Funeral march. cf Works, selections (Klavier KS
    553).
Hamlet, op. 18: Funeral march. cf Works, selections (Philips
    6747 271).

495 Harold in Italy, op. 16. Daniel Benyamini, vla; Israel Philharmonic
     Orchestra; Zubin Mehta. Decca SXL 6732. (also London CS
     6951).
     +-Gr 8-75 p316   +NR 8-76 p2
     +-HF 4-76 p97   +-RR 8-75 p29
     +HFN 8-75 p73   +-St 6-76 p100
496 Harold in Italy, op. 16. Nobuko Imai, vla; LSO; Colin Davis
     Philips 9500 026. Tape (c) 7300 441.
     +Gr 3-76 p1463   +RR 2-76 p26
     +-HF 10-76 p103   ++SFC 5-23-76 p36
     +HFN 2-76 p92   ++St 9-76 p118
     ++NR 6-76 p4
     Harold in Italy, op. 16. cf Works, selections (Philips 6747 271).
     King Lear (Le Roi Lear), op. 4. cf Overtures (Classics for
     Pleasure CFP 40097).
     Le Roi Lear, op. 4. cf Overtures (Philips 7300 080).
497 La mort d'Ophelie, op. 18, no. 2. CHERUBINI: Ave Maria. HANDEL:
     Giulio Cesare: Piangero la sorte mia. Rodelinda: Mio caro bene.
     MOZART: Exultate jubilate, K 165. Or che il cielo, K 374.
     Maria Kurenko, s; Orchestral accompaniments. Collectors Guild
     668.
     +NR 3-76 p9
498 Les nuits d'été, op. 7. DEBUSSY: La damoiselle élue. Victoria de
     los Angeles, s; BSO; Charles Munch. RCA AVM 1-1412.
     +ARSC Vol VIII, no. 2-3, p98
     Les nuits d'été, op. 7: Le spectre de la rose; Absence. cf
     HMV RLS 716.
     Les nuits d'été, op. 7: Villanelle. cf HMV SEOM 24.
499 Overtures: Béatrice et Bénédict. Benvenuto Cellini, op. 23. Le
     carnival romain, op. 9. Le Corsaire, op. 21. Les Francs
     Juges, op. 3. LSO; André Previn. Angel (Q) S 37170. (also
     HMV SQ ASD 3212 Tape (c) TC ASD 3212).
     +Gr 6-76 p38   +NR 6-76 p7
     +-Gr 10-76 p658 tape   +RR 6-76 p36
     +-HF 9-76 p89   ++RR 7-76 p47
     ++HFN 6-76 p81   -SFC 5-23-76 p36
     ++HFN 11-76 p175   ++St 10-76 p120
500 Overtures: Béatrice and Bénédict. Benvenuto Cellini, op. 23.
     The Corsair, op. 21. King Lear, op. 4. Roman carnival, op. 9.
     Hallé Orchestra; James Loughran. Classics for Pleasure
     CFP 40097.
     +-Gr 12-76 p990   +RR 12-76 p52
     +-HFN 12-76 p139
501 Overtures: Le carnaval romain, op. 9. Le Corsaire, op. 21. Les
     Francs Juges, op. 3. Le Roi Lear, op. 4. Waverly, op. 2b.
     LSO; Colin Davis. Philips Tape (c) 7300 080.
     +Gr 11-76 p887 tape
     Rakoczy march. cf Stentorina SP 1735.
502 Requiem (Grande messe des morts), op. 5. Stuart Burrows, t;
     French National Radio Orchestra and Chorus; Leonard Bernstein.
     CBS 79205 (2). (also Columbia M2 34202).
     +Gr 11-76 p844   ++RR 11-76 p94
     ++HFN 11-76 p153   ++SFC 10-24-76 p35
     +NR 12-76 p9
503 Requiem, op. 5. Robert Tear, t; Birmingham City Symphony Chorus
     and Orchestra; Louis Frémaux. HMV SLS 982 (2). Tape (c) TC SLS
     982. (also Angel SB 3814).

```
              +Gr 9-75 p499              /NR 11-76 p10
              +Gr 12-75 p1121 tape      ++RR 10-75 p81
              +HFN 10-75 p138            +RR 3-76 p76 tape
              +-HFN 12-75 p173 tape     ++STL 9-7-75 p37
              +-MT 2-76 p139
```

504   Requiem, op. 5. LSO and Chorus; Colin Davis. Philips Tape (c)
      7699 008.
```
              /Gr 12-76 p1066 tape       +RR 11-76 p108
```
      Requiem, op. 5: Sanctus. cf Decca SXL 6781.
      Rêverie et caprice, op. 8. cf Philips 6780 030.
505   Roméo et Juliette, op. 17. Julia Hamari, ms; Jean Dupouy, t;
      José van Dam, bs-bar; New England Conservatory Chorus; BSO;
      Seiji Ozawa. DG 2707 089 (2).
```
              +-Gr 11-76 p844           +-RR 11-76 p94
              +-HFN 11-76 p156
```
      Roméo et Juliette, op. 17. cf Works, selections (Philips 6747 271).
      Roméo and Juliet, op. 17: Queen Mab scherzo. cf RCA CRM 5-1900.
      Roméo et Juliette, op. 17: Queen Mab scherzo; Love scene. cf Le
      carnaval romain, op. 9.
506   Roméo and Juliet, op. 17: Romeo alone; Festival of the Capulets;
      Love scene; Queen Mab: Scherzo. TCHAIKOVSKY: Romeo and Juliet
      overture.* NYP; Leonard Bernstein. CBS 61682. (*Reissue
      from 77391).
```
              +-Gr 6-76 p43             -RR 5-76 p38
              -HFN 4-76 p101
```
507   Roméo et Juliette, op. 17: Suite. Les Troyens: Royal hunt and
      storm. Orchestre de Paris; Daniel Barenboim. CBS 76524.
```
              -Gr 10-76 p584            +-RR 11-76 p51
              +-HFN 11-76 p156
```
      Songs: Absence. cf Desto DC 7118/9.
508   Symphonie fantastique, op. 14. BSO; Georges Prêtre. Camden
      CCV 5048. (Reissue from RCA 1646).
```
              +-HFN 4-76 p121           -RR 4-76 p37
```
509   Symphonie fantastique, op. 14. NPhO; Leopold Stokowski. Decca
      SDD 495. (Reissue from PFS 4160).
```
              +Gr 8-76 p278             +RR 8-76 p29
              +HFN 8-76 p93
```
510   Symphonie fantastique, op. 14. BPhO; Herbert von Karajan.
      DG 2530 597. Tape (c) 3300 498.
```
              ++Gr 3-76 p1463           +-MJ 1-76 p39
              ++Gr 10-76 p658 tape      +-NR 1-76 p5
              +HF 1-76 p883             +-RR 3-76 p36
              +HF 9-76 p84 tape         +-SFC 1-4-76 p27
              +-HFN 3-76 p89            ++STL 2-8-76 p36
              +HFN 6-76 p105 tape
```
511   Symphonie fantastique, op. 14. Lamoureux Orchestra; Igor
      Markevitch. DG 2548 172. Tape (c) 3318 034. (Reissue from
      138712).
```
              /Gr 11-75 p840            +-RR 11-75 p41
              ++HFN 10-75 p152          +RR 9-76 p92 tape
              +HFN 2-76 p117
```
512   Symphonie fantastique, op. 14. ORTF; Jean Martinon. HMV (Q)
      Q4ASD 2945. (also Angel (Q) S31738).
```
              ++Gr 6-74 p40             +-RR 2-74 p24
              +MJ 11-76 p45             ++St 12-76 p135
              ++NR 10-76 p4
```

513 Symphonie fantastique, op. 14. French National Radio Orchestra;
     Jean Martinon. HMV SQ ASD 3263. (Reissue from Q4 ASD 2945).
          ++Gr 10-76 p584   +-RR 10-76 p40
          +-HFN 11-76 p170
514 Symphonie fantastique, op. 14. Budapest Symphony Orchestra;
     Charles Munch. Hungaroton SLPX 11842.
          +HFN 12-76 p139   -RR 12-76 p52
515 Symphonie fantastique, op. 14. LSO; Colin Davis. Philips 6580
     127. (Reissue from SAL 3441).
          +Gr 10-76 p584   +RR 10-76 p40
          +HFN 11-76 p170
516 Symphonie fantastique, op. 14. Sydney Symphony Orchestra; Willem
     van Otterloo. RCA GL 25012.
          +Gr 10-76 p584   -RR 10-76 p40
          +-HFN 12-76 p139
     Symphonie fantastique, op. 14. cf Works, selections (Philips
     6747 271).
     Symphonie funèbre et triomphale, op. 15. cf Works, selections
     (Philips 6747 271).
517 Les Troyens, excerpts. Régine Crespin, Jane Berbié, s; Guy
     Chauvet, Gerard Dunan, t; Paris Opera Orchestra and Chorus;
     Georges Prêtre. Seraphim S 60263.
          ++St 7-76 p108
     Les Troyens: Je vais mourir. cf Decca GOSD 674/6.
518 Les Troyens: Overture, March Troyenne. MASSENET: La Vièrge: Le
     dernier sommeil de la vierge (The last sleep of the virgin).
     RIMSKY-KORSAKOV: Le coq d'or: Wedding march. SIBELIUS: Karelia
     suite, op. 11, no. 3: March. The tempest, op. 109: Miranda,
     The Naiads, The storm. RPO; Thomas Beecham. Odyssey Y 33288.
     (also CBS 61655. Reissue from Philips SBR 6215, Columbia
     33CX 1087).
          +Gr 7-75 p199   +-RR 7-75 p27
          +HF 7-75 p89   +St 7-75 p110
          +-HFN 9-75 p108   +ST 2-76 p737
     Les Troyens: Prelude. cf Works, selections (Philips 6747 271).
     Les Troyens: Royal hunt and storm. cf Le carnaval romain, op. 9.
     Les Troyens: Royal hunt and storm. cf Roméo et Juliette, op. 17:
     Suite.
     Les Troyens: Royal hunt and storm. cf HMV SLS 5019.
     Les Troyens: Royal hunt and storm, Trojan march. cf Works,
     selections (Klavier KS 553).
     Les Troyens: Trojan march. cf Classics for Pleasure CFP 40254.
     Les Troyens: Trojan march. cf HMV RLS 717.
     The Trojans: Trojan march. cf Pye TB 3004.
     Waverly, op. 2b. cf Overtures (Philips 7300 080).
519 Works, selections: Benvenuto Cellini, op. 23. La damnation de
     Faust, op. 24: Minuet of the will-o-the-wisps; Dance of the
     sylphs; Hungarian march. Hamlet, op. 18: Funeral march. Roman
     carnival, op. 9. Les Troyens: Royal hunt and storm, Trojan
     march. Birmingham City Symphony Orchestra; Louis Frémaux.
     Klavier KS 553.
          +NR 7-76 p3
520 Works, selections: Hamlet, op. 18: Funeral march. Harold in
     Italy, op. 16. Roméo et Juliette, op. 17. Symphonie fantas-
     tique, op. 14. Symphonie funèbre et triomphale, op. 15.
     Les Troyens: Prelude. Patricia Kern, con; Robert Tear, t;

John Shirley-Quirk, bar; John Alldis Choir; LSO, COA; Colin
Davis. Philips 6747 271 (5)
    +-Gr 9-76 p474   +RR 9-76 p45
    ++HFN 10-76 p181

BERNEVILLE, Gilebert de
De moi doleros vos chant. cf Telefunken 6-41275.

BERNHARD, Christoph
Songs: Was betrübst du dich, meine Seele. cf Telefunken 6-41929.

BERNSTEIN
Fanfare for Bima. cf Crystal S 204.

BERNSTEIN, Leonard
521   Candide: Overture. COPLAND: Appalachian spring. GERSHWIN: An
    American in Paris. LAPO; Zubin Mehta. Decca SXL 6811. Tape
    (c) KSXC 6811.
        +Gr 9-76 p410   ++HFN 12-76 p155
        +HFN 9-76 p119   ++RR 10-76 p40
Candide: Overture. cf Works, selections (Columbia MG 32174).
Candide: Overture. cf London CSA 2246.
522   Dybbuk. David Johnson, bar; John Ostendorf, bs; New York City
    Ballet Orchestra; Leonard Bernstein. Columbia M 33082. Tape
    (c) MT 33082 (ct) MA 33082 (Q) MQ 33082 Tape (ct) MAQ 33082.
    (also CBS 76486).
        +Gr 8-76 p278   ++RR 7-76 p47
        +HF 11-74 p100   +SFC 10-6-74 p26
        +HF 3-75 p108 Quad, tape +SR 11-30-74 p40
        +HFN 6-76 p81   ++St 11-74 p122
        +NYT 8-4-74 pD20   +Te 12-76 p34
Facsimile. cf Works, selections (Columbia MG 32174).
Fancy free. cf Works, selections (Columbia MG 32174).
523   Jeremiah symphony. HARRIS: Symphony, no. 3. Jennie Tourel, ms;
    NYP; Leonard Bernstein. Columbia MS 6303.
        +St 7-76 p74
524   Mass, excerpts. Alan Titus, celebrant; Norman Scribner Choir,
    Berkshire Boys' Choir; Rock band, blues band and orchestra;
    Leonard Bernstein. CBS 73541. (Reissue from 77256).
        +Gr 8-76 p324   +RR 6-76 p77
        +HFN 6-76 p102
Mass: A simple song. cf RCA LRL 2-7531.
Mass: Two meditations. cf Works, selections (Columbia MG 32174).
On the town: Ballet music. cf Works, selections (Columbia MG
32174).
525   Seven anniversaries. COPLAND: Sonata, piano. Leonard Bernstein,
    pno. RCA SMA 7015.
        +-Gr 6-76 p69   +RR 6-76 p69
West side story: Maria. cf RCA LRL 2-7531.
West side story: Symphonic dances. cf Works, selections (Columbia
MG 32174).
526   Works, selections: Candide: Overture. Facsimile. Fancy free.
    Mass: Two meditations. On the town: Ballet music. West side
    story: Symphonic dances. NYP; Kennedy Center Theater Orchestra;
    Leonard Bernstein. Columbia MG 32174 (2). Tape (c) MGT 32174.
        ++NR 10-73 p2   *St 7-76 p73
        ++St 10-73 p134

BERTOLI
Sonata prima. cf Cambridge CRS 2826.

BERTRAND, Antoine de
Je suis un demi-dieu. cf Argo ZRG 667.

527   Songs (Les amours de Ronsard): Beaulté qui, sans pareille; Ce ris
        plus doulx que l'oeuvre d'une abeille; Certes mon oeil fut trop
        avantureux; Ces deux yeulx bruns; Le coeur loyal qui n'a
        l'ocasion; Je ne suis seulement amoureux de Marie; Je sui
        tellement amoureus; Je vy ma nymphe entre cent damoyselles;
        Las, pour vous trop aymer je ne puis vous aymer; Mon dieu,
        que ma am maistresse est belle; Nature ornant la dame qui
        devoyt; O doux plaisir, o mon plaisant dommage; Oeil, qui
        mes pleurs de tes rayons essuye; Prenés mon coeur, dame, prenés
        mon coeur. Jocelyn Chamonin, s; Joseph Sage, c-t; André Meurant,
        t; Georges Abdoun, bs; Paris Polyphonic Ensemble; Charles
        Ravier. Telefunken AW 6-41916. (Reissue from Valois MB 973).
              +-Gr 3-76 p1488   +RR 4-76 p67
              +-HFN 1-76 p103
BERWALD, Franz
528   Symphony, no. 1, G minor. Symphony, no. 3, C major. Stockholm
        Philharmonic Orchestra; Hans Schmidt-Isserstedt. Swedish
        Society TRS 11037. (Reissue from Nonesuch H 71087)
              +-Gr 4-76 p1591   +RR 2-76 p26
529   Symphony, no. 2, D major. BLOMDAHL: Sisyphos. ROSENBERG: Voyage
        to America: Intermezzo; The railway fugue. Stockholm Phil-
        harmonic Orchestra; Antal Dorati. RCA VICS 1319.
              +Gr 1-76 p1188
      Symphony, no. 3, C major. cf Symphony, no. 1, G minor.
530   Trio, piano, no. 1, E major. Trio, piano, no. 3, D minor.
        Berwald Trio. Swedish Society TRS 11038.
              +-Gr 3-76 p1478   +RR 2-76 p44
              +-HFN 2-76 p92
      Trio, piano, no. 3, D minor. cf Trio, piano, no. 1, E major.
BESARD, Jean
      Air de cour. cf ATTAINGNANT: Basse dance.
      Allemande. cf ATTAINGNANT: Basse dance.
      Ballet. cf ATTAINGNANT: Basse dance.
      Branle. cf ATTAINGNANT: Basse dance.
      Branle gay (2). cf ATTAINGNANT: Basse dance.
      Chorea rustica. cf ATTAINGNANT: Basse dance.
      Gagliarda. cf ATTAINGNANT: Basse dance.
      Gagliarda vulgo dolorata. cf ATTAINGNANT: Basse dance.
      Gesamtzeit mit Pausen. cf ATTAINGNANT: Basse dance.
      Guillemette. cf ATTAINGNANT: Basse dance.
      Pass'e mezo. cf ATTAINGNANT: Basse dance.
      Volte. cf ATTAINGNANT: Basse dance.
      THE BEST OF LEOPOLD STOKOWSKI. cf Vanguard VSD 707/8.
BETHUNE
      The battle of Manassas. cf Columbia M 34129.
BIBER, Carl Heinrich
      Sonata, 2 choirs. cf Nonesuch H 71301.
      Sonata, trumpet, C major. cf Nonesuch H 71301.
      Sonata a 7, C major. cf Nonesuch H 71301.
      Suite, 2 clarino trumpets. cf Philips 6500 926.
BICCI-BILLI
      E canta il grillo. cf Rubini RS 301.
      A BICENTENNIAL CELEBRATION, 200 YEARS OF AMERICAN MUSIC. cf
        Columbia M 33838.
BIGAGLIA, Drogenio
      Sonata, recorder, A minor. cf Telefunken DT 6-48075.

Sonata, recorder, B flat major.   cf Telefunken DT 6-48075.
BILLINGS, William
    America.   cf Columbia M 33838.
    Chester.   cf Columbia MS 6161.
    Chester.   cf Columbia M 34129.
    Chester.   cf Department of Defense Bicentennial Edition 50-1776.
    Chester.   cf Michigan University SM 0002.
531   Songs: Be glad than America; The bird; Boston; Chester; Cobham;
       Connection; Consonance; Creation; David's lamentaion; Hopkinton;
       I am the rose of Sharon; Jargon; Kittery; The Lord is risen;
       Modern music; Morpheus; The shepherd's carol; Swift as an
       Indian arrow flies; A virgin unspotted; When Jesus wept.   Gregg
       Smith Singers; Gregg Smith.   Columbia MS 7277.
         +-St 7-76 p72
    Songs: I am come into my garden; I am the rose of Sharon; I charge
       you; An anthem for Thanksgiving: O praise the Lord.   cf
       Nonesuch H 71276.
    Songs: The hart panteth; Consonance; Thus saith the high, the loft
       one.   cf Vox SVBX 5350.
BIMBONI
    Sospiri miei andant ove vi mando.   cf Pye Ember GVC 51.
BINCHOIS, Gilles de
    Amoreux suy.   cf HMV SLS 863.
    Bien puist.   cf HMV SLS 863.
    Dueil angoisseus.   cf Telefunken ER 6-35257 (2).
    Files a marier.   cf HMV SLS 863.
    Gloria, laud et honor.   cf Telefunken ER 6-35257 (2).
    Je ne fai toujours.   cf HMB SLS 863.
    Jeloymors.   cf HMV SLS 863.
    Votre trés doulz regart.   cf HMV SLS 863.
BINGE, Ronald
532   Concerto, saxophone, E flat major.   Saturday symphony.   Aage Voss,
       sax; South German Radio Orchestra; Ronald Binge.   Rediffusion
       ZS 75.
         +Gr 1-76 p1188
    Elizabethan serenade.   cf Music for Pleasure SPR 90086.
    Elizabethan serenade.   cf RCS GL 25006.
    Saturday symphony.   cf Concerto, saxophone, E flat major.
    The watermill.   cf Virtuosi VR 7506.
BINKERD, Gordon
533   Sonata, violin and piano.   COPLAND: Sonata, violin and piano.
       IVES: Sonata, violin and piano, no. 4.   Jaime Laredo, vln;
       Ann Schein, pno.   Desto DC 6439.   (also Peerless PRCM 210).
         +Gr 1-76 p1210   +-HFN 2-76 p92
         ++HF 11-75 p129   +SFC 5-2-76 p38
BIRD, Arthur
    Carnival scene.   cf Louisville LS 753/4.
    BIRDS IN MUSIC.   cf Decca KCSP 367.
BIRTWISTLE, Harrison
    Chronometer.   cf The triumph of time.
534   Tragoedia.   CROSSE: Concerto da camera.   WOOD: Pieces, piano,
       op. 6 (3).   Manoug Parikian, vln; Susan McGaw, pno; Melos
       Ensemble; Edward Downes, Lawrence Foster.   Argo ZRG 759.
       (Reissues from HMV ASD 2333).
         +Gr 5-76 p1776   +NR 6-76 p8
         +HFN 4-76 p123   +-RR 12-75 p72
535   The triumph of time.   Chronometer.   BBC Symphony Orchestra;
       Pierre Boulez.   Argo ZRG 790.

```
  +Gr 7-75 p174          ++NR 1-76 p4
 ++HF 3-76 p80           +RR 6-75 p32
 +-HFN 10-75 p137        +ST 2-76 p741
 +MT 7-76 p580           +Te 12-75 p43
```

BISHOP, Henry
  Forrester sound the cheerful horn. cf Prelude PRS 2501.
  Grand march, E major. cf Saga 5417.
  Songs: Bid me discourse; Home, sweet home; Lo, hear the gentle
    lark. cf HMV RLS 719.
  Songs: Coming through the rye; Should he upbraid. cf Rococo 5397.
  Songs: Home, sweet home. cf Collectors Guild 611.
  Songs: Home, sweet home. cf HMV EMD 5528.
  Songs: Home, sweet home. cf Prelude PRS 2505.
  Songs: Lo, here the gentle lark. cf BIS LP 45.
  Songs: Lo, here the gentle lark. cf Westminster WGS 8268.

BITTNER, Jacques
  Allemande. cf Turnabout TV 34137.
  Courante. cf Turnabout TV 34137.
  Passacaglia. cf Turnabout TV 34137.
  Praeludium. cf Turnabout TV 34137.
  Sarabande. cf Turnabout TV 34137.

BIZET, Georges
  Agnus Dei. cf Decca SXL 6781.
  L'Arlésienne: Suites, nos. 1 and 2. cf Works, selections (Decca
    DPA 559/60).
536 L'Arlésienne: Suite, no. 1; Suite, no. 2: Farandole. GRIEG: Peer
       Gynt, op. 46: Suite no. 1. Peer Gynt, op. 55: Suite no. 2:
       Solveig's song. CO; Georg Szell. CBS 61303.
         +Gr 4-76 p1665   +-RR 4-76 p41
     L'Arlésienne: Suite, no. 1. Suite, no. 2: Farandole. cf BACH:
       Toccata and fugue, S 565, D minor.
     L'Arlésienne: Suite, no. 1. Suite, no. 2: Intermezzo; Farandole.
       cf HMV SLS 5019.
537 L'Arlésienne: Suite, no. 1. Carmen: Suites, nos. 1 and 2. NPhO;
       Charles Munch. Decca SDD 492. (Reissue from PFS 4127).
         +-Gr 9-76 p410   +RR 8-76 p29
         +-HFN 8-76 p93
     L'Arlésienne: Suite, no. 1: Minuet. cf HMV SXLP 30181.
538 L'Arlésienne: Suite, no. 2. GRIEG: Peer Gynt, op. 55: Suite no. 2.
       SMETANA: Ma Vlast: The Moldau. Budapest Philharmonic Orchestra;
       Adam Medveczky. Hungaroton SLPX 11813.
         +NR 12-76 p4
     L'Arlésienne: Suite, no. 2, excerpts. cf Philips 6780 030.
539 Carmen. Risë Stevens, s; Metropolitan Opera Orchestra. CBS 30069.
         +-HFN 4-76 p123
540 Carmen. Régine Crespin, Jeannette Pilou, Maria Rosa Carminati,
       Nadine Denize, s; Gilbert Py, Rémy Corazza, t; José van Dam,
       Jacques Trigeau, bar; Pierre Thau, Paul Guige, bs; St. Maurice
       Children's Chorus, Rhine Opera Chorus; Strasbourg Philharmonic
       Orchestra; Alain Lombard. Erato STU 70900/2 (3).
         +-Gr 8-75 p359   +-NR 2-76 p10
         +-HF 12-75 p85   +-ON 12-13-75 p48
         +MJ 4-76 p31   +-St 2-76 p99
541 Carmen. Victoria de los Angeles, Janine Micheau, Denise Monteil,
       Marcelle Croisier, Monique Linval, s; Michel Hamel, Nicolai
       Gedda, t; Ernest Blanc, Jean-Christophe Benoit, Bernard Plantey,
       bar; Xavier Depraz, bs; Les Petits Chanteurs de Versailles;
```
```

French National Radio Orchestra and Chorus; Thomas Beecham.
HMV SLS 5021 (3). Tape (c) TC SLS 5021. (Reissue from ASD
331/3). (also Angel S 3613).
  ++Gr 2-76 p1374   +RR 1-76 p24
  +-HFN 2-76 p92   +RR 4-76 p80 tape

542   Carmen. Kiri Te Kanawa, Norma Burrowes, s; Tatiana Troyanos,
Jane Berbié, ms; Placido Domingo, Michel Sénéchal, t; José
van Dam, Michel Roux, Thomas Allen, bar; Pierre Thau, bs; John
Alldis Choir; LPO; Georg Solti. London OSA 13115 (3).
Tape (c) 5-13115. (also Decca D 11D 3 Tape (c) K11K 33).
  ++ARG 12-76 p19   +HFN 10-76 p161
  +Gr 10-76 p640   +RR 11-76 p34
  /HF 12-76 p89   +St 12-76 p97

Carmen: L'amour est un oiseau rebelle (Habañera). cf Decca
GOSD 674/6.
Carmen: C'est toi, c'est toi. cf Decca DPA 517/8.
Carmen: Con vol ber; Si tu m'aimes. cf Discophilia DIS KGA 2.
Carmen: Euren Toast kann ich wohl erwidern. cf Preiser LV 192.
Carmen: Finale. cf Club 99-98.
Carmen: La fleur que vous. cf Cantilena 6238.
Carmen: La fleur que tu m'avais jetée. cf RCA CRM 1-1749.
Carmen: Flower Song. cf Decca SXL 6649.
Carmen: Habañera; Chanson bohème; Séguidille. cf Rubini RS 301.
Carmen: Habañera, Kartenarie. cf Discophilia KIS KGG 3.
Carmen: Habañera; Seguidilla; Toreador song; Flower song; Card
trio. cf Works, selections (Decca DPA 559/60).
Carmen: Je dis que rien ne m'épouvante. cf Club 99-99.

543   Carmen: March and prelude; Chorus of street boys; Entry of Carmen
and habañera; Séguidilla; Entr'acte (acts 2, 3, 4); Flower
song; Micaëla's aria (arr. Howarth); Toreador song. (arr.
Civil). Kingsway Symphony Orchestra; Alan Civil. Decca PFS
4348.
  +-Gr 11-76 p893   -RR 11-76 p52
  +HFN 11-76 p156

Carmen: Micaëla's aria. cf Bongiovanni GB 1.
Carmen: Micaëla's aria. cf CBS 30072.
Carmen: Micaëla's air. cf Discophilia DIS 13 KHH 1.
Carmen: Prelude, Act 1. cf HMV SEOM 25.

544   Carmen: Prelude, habañery; Séguidilla and duet; Gypsy song; Toreador
songs; Flower song; Card song; Micaëla's aria; Final duet.
Risë Stevens, ms; Raoul Jobin, t; Robert Weede, bar; Metropolitan Opera Chorus and Orchestra; George Sebastian. CBS
30069. (Reissue from Fontana CFE 15002/3; Philips GBL 5641).
  +-Gr 4-76 p1654   +-RR 5-76 p26

Carmen: Séguidilla. cf Rococo 5377.
Carmen: Suite, no. 1. cf Philips 6747 204.
Carmen: Suites, nos. 1 and 2. cf L'Arlésienne: Suite, no. 1.
Carmen: Toreador chorus. cf Decca DPA 525/6.
Carmen fantasy, op. 25. cf HMV HQS 1360.
Carmen fantasy, op. 25 (arr. Busoni). cf International Piano
Library IPL 104.
Ivan le Terrible: Ouvre ton coeur. cf Cantilena 6239.

545   Jeux d'enfants, op. 22. IBERT: Divertissement. SAINT-SAENS:
Danse macabre, op. 40. Le rouet d'omphale, op. 31. OSCCP;
Jean Martinon. Decca ECS 782. (Reissue from SXL 2252).
  +-Gr 8-76 p279   +RR 6-76 p36
  +HFN 6-76 p102

Jeux d'enfants, op. 22. cf Works, selections (Decca DPA 559/60).
La jolie fille de Perth: Suite. cf Symphony, C major.
La jolie fille de Perth: Suite. cf Works, selections (Decca DPA
  559/60).
Pastorale. cf HMV RLS 719.
Patrie overture. cf Symphony, C major.
Les pêcheurs de perles (The pearl fishers): Au fond du temple.
  cf Cantilena 6238.
Les pêcheurs de perles: Au fond du temple saint. cf RCA LRL 2-7531.
Les pêcheurs de perles: C'est toi...Au fond du temple saint.
  cf Decca DPA 517/8.
The pearl fishers: C'est toi...Au fond du temple saint. cf
  Decca GOSD 674/6.
Les pêcheurs de perles: Leila, Leila, Dieu puissant le voilà. cf
  Angel S 37143.
I pescatori di perle: Mi par d'udir ancora. cf Everest SDBR 3382.
546 Roma suite, C major. Symphony, C major. Birmingham Symphony
     Orchestra; Louis Frémaux. HMV ASD 3039. Tape (c) TC ASD 3039.
     (also Klavier KS 546).
          *Audio 9-76 p85   ++NR 1-76 p4
          ++Gr 1-75 p1341   +RR 12-74 p25
          +Gr 9-76 p497 tape   +RR 12-76 p105 tape
          +HFN 8-76 p94 tape
     Songs: Ouvre ton coeur. cf Columbia M 33933.
547 Symphony, C major. La jolie fille de Perth: Suite. Patrie over-
     ture. Orchestre de Paris; Daniel Barenboim. HMV SQ ASD 3277.
          +Gr 10-76 p584   +-RR 11-76 p52
          +-HFN 10-76 p165
548 Symphony, C major. PROKOFIEV: Symphony no. 1, op. 25, D major.
     AMF; Neville Marriner. Argo ZRG 719. Tape (c) KZRC 719.
          +Gr 12-73 p1192   ++RR 8-76 p84 tape
          +-HF 7-74 p88   +-SFC 9-1-74 p20
          +HFN 7-76 p104 tape   +St 8-74 p107
          ++NR 6-74 p3
     Symphony, C major. cf Roma suite, C major.
     Symphony, C major. cf Works, selections (Decca DPA 559/60).
549 Works, selections: L'Arlésienne: Suites, nos. 1 and 2. Carmen:
     Habañera; Séguidilla; Toreador song; Flower song; Card trio.
     Jeux d'enfants, op. 22. La jolie fille de Perth: Suite.
     Symphony, C major. Georgette Spanellys, s; Regina Resnik,
     Yvonne Minton, con; Mario del Monaco, t; Tom Krause, Claude
     Cales, bar; Robert Geay, bs; OSR; Ernest Ansermet. Decca
     DPA 559/60. Tape (c) KDPC 559/60.
          +-HFN 12-76 p152   -RR 12-76 p52
          +-HFN 12-76 p155 tape
BJORKANDER, Nils
     Sketches from the Archipelago: Idyll. cf Swedish Society SLT
     33229.
BLACHER, Boris
     Ornaments. cf Sonata, solo violin, op. 40.
550 Sonata, solo violin, op. 40. Ornaments (4). LINKE: Violencia.
     MADERNA: Dedication. Pièce pour Ivry. Christiane Edinger,
     vln. Orion ORS 75171.
          ++HF 9-75 p82   ++NR 10-75 p11
          +IN 11-76 p18

BLAKE, Howard
551   Songs: Dressing up; Home; I don't want to be a number; Cultivating
       the land; My guru's gone away; People are saying; One of the
       things; Start a new society; We were watching the birds. The
       Scholars. Firecrest FEU 1002.
              +-RR 1-76 p55
BLANK, Allan
552   Rotation. JENNI: Musique printanière. KARLINS: Variations on
       'Obiter dictum". STOCK: Quintet, clarinet and strings. Allen
       Blustine, clt; Joel Krosnick, vlc; Elizabeth Buccheri, Gilbert
       Kalish, pno; Thomas Siwe, perc; Betty Bang Mather, flt; Con-
       temporary String Quartet; John Simms. CRI SD 329.
              +-HFN 1-76 p104   +-RR 7-75 p44
              ++NR 1-75 p7
BLASCO, Manuel
       Versos con duo para chirimias. cf Eldorado S-1.
BLAVET, Michel
553   Sonatas, flute and harpsichord, op. 2 (6). Ransom Wilson, flt;
       Lionel Party, hpd. Musical Heritage Society MHS 1861.
              +AR 2-76 p133
554   Sonatas, flute and harpsichord, op. 2, nos. 1-2, 4-5. Gabriel
       Fumet, flt; Jean-Louis Petit, hpd. Society Francaise du
       Son SXL 20 140.
              +AR 2-76 p133
BLEWITT
       They don't propose. cf Transatlantic XTRA 1159.
BLEY, Carla
555   3/4, piano and orchestra. MANTLER: 13, piano and 2 orchestras.
       Carla Bley, pno; Orchestra; Carla Bley, Michael Mantler. Watt 3.
              ++St 4-76 p111
BLISS, Arthur
       An age of kings. cf Pye GH 603.
       Antiphonal fanfare, 3 brass choirs. cf Argo SPA 464.
       Antiphonal fanfare, 3 brass choirs. cf Decca 419.
556   Belmont variations (arr. Wright). Kenilworth. IRELAND: A comedy
       overture. Downland suite. Gus (Kettering) Band; Geoffrey
       Brand. Studio Two TWOX 1053.
              +Gr 9-76 p410
       Bliss. cf Polydor 2383 391.
       Checkmate: The ceremony of the red bishops; Finale checkmate.
       cf Transatlantic XTRA 1160.
557   Concerto, violin. Introduction and allegro. Theme and cadenza,
       violin and orchestra. Alfredo Campoli, vln; LSO, LPO; Arthur
       Bliss. Decca ECS 783. (Reissues from LXT 5166 5170).
              +-Gr 4-76 p1591   +-RR 3-76 p37
              +-HFN 3-76 p109
       Introduction and allegro. cf Concerto, violin.
       Kenilworth. cf Belmont variations.
       Kenilworth. cf Virtuosi VR 7506.
       The rout trot. cf Polydor 2383 391.
       Theme and cadenza, violin and orchestra. cf Concerto, violin.
       Welcome the Queen. cf HMV HLM 7093.
BLITZSTEIN, Marc
558   The airborne symphony. Orson Welles, narrator; Andrea Velis, t;
       David Watson, bar; Choral Art Society; NYP; Leonard Bernstein.
       Columbia M 34136.

         +NR 9-76 p3   +-St 11-76 p136
         +ON 11-76 p98
559 The cradle will rock. Theatre Four cast; Howard Da Silva,
         Gershon Kingsley. CRI SD 266 (2).
         +-HF 3-71 p100   -SFC 7-11-71 p36
         +MJ 3-71 p79   ++St 5-71 p79
         +NR 3-71 p6   +St 7-76 p73
         +ON 2-20-71 p34

BLOCH, Augustyn
560 Dialogues, violin and orchestra. Gilgamesh. Wanda Wilkomirska,
         vln; Warsaw National Philharmonic Choir; Warsaw National
         Philharmonic Orchestra; Andrzej Markowski. Muza SX 1208.
         +NR 9-76 p3
     Gilgamesh. cf Dialogues, violin and orchestra.

BLOCH, Ernest
561 America, an epic rhapsody. American Concert Choir; Symphony of
         the Air; Leopold Stokowski. Introductory comments by Ernest
         Bloch. Vanguard SRV 346 SD. (Reissue from VSD 2065).
         +-HF 6-76 p70   -SFC 2-15-76 p38
         *MJ 3-76 p24   +St 6-76 p100
     Baal Shem. cf Da Camera Magna SM 93399.
     Baal Shem: Ningun. cf CBS 76420.
     Baal Shem: Ningun. cf Columbia M2 33444.
     Baal Shem: Ningun. cf RCA ARL 1-1172.
562 Nocturnes (3). HOPKINS: Diferencias sobre una tema original.
         PISTON: Trio, piano. SCHWANTNER: Autumn cantiles. Western
         Arts Trio. Laurel LR 104.
         ++NR 10-76 p6   +SFC 12-19-76 p50
563 Schelomo. ELGAR: Concerto, violoncello, op. 85, E minor. Pierre
         Fournier, vlc; BPhO; Alfred Wallenstein. DG 2535 201. (Reissue
         from SLPM 139 128).
         +-Gr 12-76 p990   +RR 12-76 p54
         +-HFN 12-76 p151
564 Schelomo. Suite, viola and orchestra. Laszlo Varga, vlc;
         Milton Katims, vla; Westphalian Symphony Orchestra, Seattle
         Symphony Orchestra; Siegfried Landau, Henry Siegl. Turnabout
         TVS 34622.
         +NR 12-76 p5   +SFC 10-76 p32
     Sonata, violin, nos. 1 and 2. cf RCA ARM 4-0942/7.
     Suite, viola and orchestra. cf Schelomo.
565 Suite symphonique. Symphony, trombone and orchestra. Howard
         Prince, trom; Portland Junior Symphony Orchestra; Jacob
         Avshalomov. CRI SD 351.
         +ARG 12-76 p22   +MJ 11-76 p45
         +HF 12-76 p99   +NR 11-76 p3
     Symphony, trombone and orchestra. cf Suite symphonique.

BLOCKLEY
     List to the convent bells. HMV EMD 5528.

BLOMDAHL, Karl-Birger
566 In the hall of mirrors (I speglarnas sal). Margareta Hallin, s;
         Barbro Ericson, con; Sven-Erik Vikström, t; Anders Näslund, bar
         and speaker; Bengt Rundgren, bs; Eric Ericson Choir; Stockholm
         Philharmonic Orchestra; Sixten Ehrling. Caprice LP 6. (also
         CAP 1006).
         +Gr 7-74 p249   ++St 9-76 p118
         ++RR 4-74 p73
     Sisyphos. cf BERWALD: Symphony, no. 2, D major.

BLOW, John
567   Coronation and symphony anthems: Blessed is the man. Cry aloud,
        and spare not. God spake sometime in visions. I was glad. O
        sing unto the Lord. Charles Brett, c-t; Philip Langridge, t;
        James Lancelot, org; Kenneth Heath, vlc; Cambridge, King's
        College Chapel Choir; AMF; David Willcocks. Argo ZRG 767.
                +Gr 4-75 p1844   +NR 8-75 p8
                +-HF 1-76 p85   ++RR 4-75 p58
                +MT 2-76 p140   +St 11-75 p126
        Fugue, F major: Vers. cf Philips 6500 926.
        Verse for cornet and single organ. cf Wealden WS 121.
        Voluntary for a single organ. cf Wealden WS 121.
BOCCHERINI, Luigi
        Concerto, violoncello, B flat major. cf HMV RLS 723.
568   Quintet, strings, C major. Sextet, strings, op. 24, no. 3,
        D major. Sestetto Chigiano. Musical Heritage Society MHS
        3122.
                +St 6-76 p101
        Sextet, strings, op. 24, no. 3, D major. cf Quintet, strings,
        C major.
569   Sonatas, flute and harpsichord, op. 5 (6). Sheridan Stokes, flt;
        Bess Karp, hpd. Orion ORS 75173.
                /IN 3-76 p12   +NR 5-75 p8
        Sonata, violoncello, no. 6, A major: Adagio and allegro. cf
        HMV RLS 723.
BODIN, Lars-Gunnar
        Dedicated to you II. cf Caprice RIKS LP 35.
        Place of plays. cf Caprice RIKS LP 35.
BOELLMAN, Leon
        Ronde français. cf Stentorian SP 1735.
        Suite gothique, op. 25. cf BACH: Toccata and fugue, S 565,
        D minor.
        Suite gothique, op. 25: Toccata. cf Decca DPA 523/4.
BOHAC, Josef
570   Elegie in memory of Ludvíka Poděste. JIRKO: Elegie on the death
        of a friend. PODEST: Partita, strings, guitar and percussion.
        Zdeněk Pitr, gtr; Jaroslav Chovanec, vlc; CPhO; Jiří Bělohlávek,
        Vladimir Válek. Panton 110 357.
                +RR 4-76 p51
571   Fragment. BORKOVEC: Il symphonietta. DVORACEK: Ex post. EBEN:
        Vox clamantis. CPhO; Václav Neumann, Zdenek Kosler. Panton
        110 300.
                +HFN 2-76 p92   ++RR 1-76 p32
BOHM, Georg
572   Ach wie nichtig, ach wie flüchtig, chorale partita. Prelude and
        fugue, D minor. BUXTEHUDE: Chaconne, E minor. Prelude, fugue
        and chaconne, C major. WALTHER: Concerto del Signore Torelli,
        appropriato all'organo. Jesu, meine Freude, partita. Kenneth
        Gilbert, org. Saga 5403.
                +GR 1-76 p1218   +RR 5-76 p65
                +MT 6-76 p497
        Jesu, du bist allzu schöne, partita. cf BACH: Chorale preludes
        (Pelca PSR 40597).
        Prelude and fugue, D minor. cf Ach wie nichtig, ach wie flüchtig,
        chorale partita.
        Präludium, Fuge und Postludium, G minor. cf Saga 5402.
        Still as the night. cf Prelude PRS 2505.

BOIELDIEU, François
573   Concerto, harp, C major. Concerto, piano, F major. Marie-Claire
       Jamet, hp; Martin Galling, pno; Paris Chamber Orchestra,
       Innsbruck Symphony Orchestra. Turnabout TV 34148. Tape (c)
       KTVC 34148.
             ++HFN 5-76 p117 tape   +-RR 6-76 p86 tape
       Concerto, piano, F major. cf Concerto, harp, C major.
       La dame blanche: Maintenant, observons..Viens gentille dame. cf
       Decca GOSD 674/6.
       Weisse Dame (La dame blanche): Duet. cf Rococo 5377.
574   Jean de Paris: Overture. GODARD: Fragments poetiques, op. 13.
       Impressions de campagne au printemps, op. 123. Suite, flute
       and orchestra, op. 116. Ian Reynolds, flt; Orchestra; Graham
       Nash. Rare Recorded Editions SRRE 162.
             +-HFN 7-76 p87
BOISMORTIER, Joseph Bodin de
575   Concerto, 5 recorders without bass, D minor. HEINICHEN: Concerto,
       4 recorders and strings. SCARLATTI, A.: Concerto, recorder
       and strings, A minor. TELEMANN: Concerto, recorder and strings,
       F major. Concerto, 2 recorders and strings, B minor. Clas
       Pehrsson, rec; Musica Dolce; Drottingholm Baroque Ensemble.
       BIS LP 8.
             ++St 11-76 p166
       Rondeau, A minor. cf Cambridge CRS 2826.
BOITO, Arrigo
       Luna fedel. cf Everest SDBR 3382.
576   Mefistofele: L'altra notte in fondo al mare. CHARPENTIER: Louise:
       Da quel di. CILEA: Adriana Lecouvreur: Poveri fiori; Io son
       l'umile ancella. PUCCINI: La boheme: Si mi chiamano Mimi.
       Tosca: Vissi d'arte. Turandot: Tu che di gel sei cinta;
       Signore ascolta. Magda Olivero, s; Italian Radiotelevision
       Symphony Orchestra; Franco Ghione, Ugo Tansini, Armando La
       Rosa Parodi, Alfredo Simonetto. Pye Ember GVC 53.
             +Gr 3-76 p1508   ++RR 3-76 p29
             ++HFN 5-76 p105
       Mefistofele: L'altra notte. cf VERDI: Il trovatore.
       Mefistofele: L'altra notte. cf Club 99-100.
       Mefistofele: L'altra notte. cf Rococo 5386.
       Mefistofele: L'altra notte in fondo al mare. cf Telefunken
       AG 6-41947.
577   Mefistofele: Dio de pièta; Dai campi; L'altra notte in fondo al mare.*
       PUCCINI: Manon Lescaut: Tu, tu, amore, tu; Donna non vidi mai;
       In quelle trine morbide.* VERDI: Aida: Pur ti riveggo, mia
       dolce Aida; La fatal pietra; Se guerrier...Celeste Aida; Ritorna
       vincitor; O patria mia. Don Carlo: Io vengo a domandar, E
       dessa; Fontainebleau; Non pianger, mia, compagna; Tu che le
       vanità.* Montserrat Caballé, s; Fiorenza Cossotto, ms; Placido
       Domingo, t; Ruggero Raimondi, Giovanni Foiani, Simon Estes, bs;
       NPhO, LSO, ROHO and Chorus; Bruno Bartoletti, Riccardo Muti,
       Carlo Maria Giulini, Julius Rudel. HMV SLS 5051 (2). (*Reissues
       from SLS 962, 956, 973).
             +Gr 7-76 p212   +RR 8-76 p23
             +-HFN 7-76 p83
       Mefistofele: Giunto sul passo estremo, Dai campi dai prati. cf
       Everest SDBR 3382.
       Mefitofele: Lontano, lontano. cf Rubini GV 63.

Mefistofele: Lontano, lontano. cf Rubini RS 301.
578   Mefistofele: Prologue. VERDI: Pezzi sacri, no. 4: Te Deum.
       Nicola Mascona, bs; Columbus Boys' Choir, Robert Shaw Chorale;
       NBC Symphony Orchestra; Arturo Toscanini. RCA AT 131.
       (Reissues HMV ALP 1363).
              +Gr 8-75 p336   +ST 2-76 p737
              +RR 6-75 p22
       Mefistofele: Spunta l'aurora palida. cf Rubini GV 58.
BOLCOM, William
579   Commedia. Open house. Paul Sperry, t; Saint Paul Chamber
       Orchestra; Dennis Russell Davies. Nonesuch H 71324.
              +ON 6-76 p52   +NYT 4-11-76 pD23
              +NR 7-76 p9   ++St 11-76 p136
       Open house. cf Commedia.
       Whisper moon. cf Folkways FTS 33901/4.
BOLLE, James
580   Oleum canis (Oil of dog). Neva Pilgrim, s; Jan Curtis, ms;
       Jerrold Siena, Richard Fithian, t; Donald Miller, bs; New
       Hampshire Sinfonietta; The Committee Male Quartet; James
       Bolle. Serenus SRS 12060.
              +ARG 11-76 p15   -St 7-76 p108
              +NR 5-76 p7
BOLLING, Claude
581   Concerto, guitar and piano. Alexandre Lagoya, gtr; Claude
       Bolling, pno; Michel Gaudry, bs; Marcel Sabiani, drum.
       RCA FRL 1-0149.
              +-ARG 12-76 p23   +-NR 12-76 p8
582   Suite, flute and jazz piano. Jean-Pierre Rampal, flt; Claude
       Bolling, pno; Max Hédiguer, double bs; Marcel Sabiani, drums.
       Columbia M 33233.
              +NR 10-75 p4   ++St 2-76 p99
BOND
       Sonata, violoncello. cf Laurel-Protone LP 13.
BOND, Carrie Jacobs
       Songs: A perfect day. cf Argo ZFB 95/6.
       Songs: I love you truly; A perfect day. cf Washington University
       Press OLY 104.
BONDON, Jacques
583   Quartet, strings, no. 1. MILHAUD: Quartet, strings, no. 12.
       TAILLEFERRE: Quartet, strings. Provence Quartet. Calliope
       CAL 1803.
              +Gr 9-76 p436
BONNET, Pierre
       Francion vint l'autre jour. cf Argo ZRG 667.
BONONCINI, Giovanni
584   Divertimenti da camera, A minor, G minor, E minor, G major,
       F major, D minor, C minor, B flat major. Hans-Martin Linde,
       rec; Eduard Müller, hpd; Konrad Ragossnig, lt; Josef Ulsamer,
       vla da gamba. DG Archive 2533 167.
              ++Audio 3-76 p63   ++NR 5-75 p8
              +Gr 12-74 p1161   ++RR 12-74 p45
              +HF 7-75 p72   +SFC 7-27-75 p22
       Songs: Deh più a me non vascondete. cf Decca SXL 6629.
BOONE, J. W.
       A grand fantasy, pianoforte. cf Transatlantic XTRA 1159.
BOREL
       Fugato, F major. cf Coronet LPS 3030.

BORETZ, Benjamin
    Group variations. cf CRI SD 300.
BORKOVEC, Pavel
    Songs on poems by V. Nezval, op. 15 (7). cf Panton 110 385.
    Symphonic allegro. cf Symphony, no. 3.
    Il symphonietta. cf BOHAC: Fragment.
585   Symphony, no. 3. Symphonic allegro. CPhO; Zdenek Kosler.
      Panton 110 367.
        +-RR 7-76 p48
BORLET (14th century)
    Hé trés doulz roussignol. cf HMV SLS 863.
    Ma tredol rosignol. cf HMV SLS 863.
BORODIN, Alexander
586   In the steppes of Central Asia. Symphony, no. 2, B minor.
      TCHAIKOVSKY: Romeo and Juliet: Overture. Dresden Staatskapelle;
      Kurt Sanderling. DG 2548 226. (Reissue from SLPM 138 686).
        +Gr 4-76 p1617   -RR 4-76 p38
        ++HFN 4-76 p121
    In the Steppes of Central Asia. cf Columbia M 34127.
    In the Steppes of Central Asia. cf Philips 6780 755.
    Prince Igor: Ah, I bitterly weep. cf Rubini GV 63.
    Prince Igor: Dance of the Polovtsian maidens; Polovtsian dances.
      cf Angel S 37232.
    Prince Igor: How goes it, Prince. cf Hungaroton SLPX 11444.
587   Prince Igor: Overture. GLINKA: Russlan and Ludmila: Overture.
      MUSSORGSKY (orch. Rimsky-Korsakov): A night on the bare
      mountain. Khovanschina: Prelude and Persian dance. BPhO;
      Georg Solti. Decca SPA 257. (also London CS 6944).
        +Gr 7-73 p193   +RR 7-73 p55
        +-HF 12-75 p111   +-SFC 8-24-75 p28
        +NR 10-75 p2   +St 3-76 p118
588   Prince Igor: Overture; Polovtsian dances. RIMSKY-KORSAKOV: Le
      coq d'or: Suite. LSO; Antal Dorati. Philips 6582 012. Tape (c)
      7321 020. (Reissues from Pye MRL 2537; EMI Mercury AMS 16102).
        -Gr 4-76 p1591   +-RR 12-76 p36
        +-HFN 2-76 p115
589   Prince Igor: Polovtsian dances (orch. Rimsky-Korsakov/Glazunov).
      MUSSORGSKY: A night on the bare mountain (orch. Rimsky-Korsakov).
      RIMSKY-KORSAKOV: Capriccio espagnol, op. 34. Russian Easter
      festival overture, op. 36. Orchestre de Paris; Gennady
      Rozhdestvensky. HMV ESD 7006. Tape TC ESD 7006. (Reissue
      from Columbia TWO 395).
        ++Gr 9-76 p435   +RR 9-76 p46
590   Prince Igor: Polovtsian dances. ROZYCKI: Orchestra prelude,
      op. 31. SCHUMANN: Manfred overture, op. 115. VERDI: I vespri
      siciliani: Overture. Silesian Philharmonic Orchestra; Karol
      Stryja. Musa SXL 1103.
        +NR 9-76 p4
    Prince Igor: Polovtsian dances. cf BALAKIREV: Symphony, no. 1,
      C major.
    Prince Igor: Polovtsian dances. cf Columbia M 34127.
    Prince Igor: Polovtsian dances. cf Decca DPA 525/6.
    Prince Igor: Polovtsian dances, Act 2. cf HMV SLS 5019.
    Prince Igor: Polovtsian dance, no. 2. cf CBS 30072.
591   Quartet, strings, no. 1, A major. Borodin Quartet. Odyssey/
      Melodiya Y 33827.

```
                ++HF 3-76 p81              ++SR 3-6-76 p41
                +NR 4-76 p8               ++St 7-76 p108
```
      Quartet, strings, no. 2, D major: Nocturne. cf CBS 30072.
      Quartet, strings, no. 2, D major: Nocturne. cf HMV HQS 1360.
      Quartet, strings, no. 2, D major: Nocturne. cf RCA CRL 3-2026.
      Quartet, strings, no. 2, D major: Nocturne. cf Vanguard SRV 344.
      Songs: Listen, maidens, to my song. cf Musical Heritage Society
            MHS 3276.
592   Symphony, no. 2, B minor. RIMSKY-KORSAKOV: The tale of Tsar
            Sultan: Suite; Flight of the bumblebee. Monte Carlo Opera
            Orchestra; Roberto Benzi. Philips 6580 130.
                +-Gr 8-76 p279   +RR 6-76 p40
                +HFN 7-76 p103
      Symphony, no. 2, B minor. cf In the steppes of Central Asia.
      Symphony, no. 2, B minor. cf BALAKIREV: Russia.
BORTNIANSKY (BORTNYANSKY), Dmitri
      Cherubim hymn, no. 7. cf HMV Melodiya ASD 3102.
      Dostoino est; Slava vo vichnih Bogou. cf Harmonia Mundi HMU 105.
      I will lift up my eyes to the hills. cf HMV Melodiya ASD 3102.
      Nativity Kontakion. cf Ikon IKO 3.
      Songs: I shall praise the name of God; O come, let us bless the
            ever-memorable Joseph. cf Ikon IKO 2.
BOSKERCK
      Semper paratus. cf Department of Defense Bicentennial Edition
            50-1776.
BOSSI, Marco
      Divertimento in forma de giga. cf Vista VPS 1035.
      Etude symphonique. cf Argo ZRG 807.
BOTTEGARI
      Mi stare pone Totesche. cf L'Oiseau-Lyre 12BB 203/6.
BOUIN (17th century, France)
      La montauban. cf DG Archive 2533 172.
BOULANGER, Lili
      Clairières dans le ciel, excerpts. cf Gemini Hall RAP 1010.
      Nocturne. cf Gemini Hall RAP 1010.
BOULANGER, Nadia
      Cortege. cf Avant AV 1012.
      Cortege. cf RCA ARM 4-0942/7.
      D'un vieux jardin. cf Avant AV 1012.
      Nocturne, F major. cf RCA ARM 4-0942/7.
BOULEZ, Pierre
593   Le soleil des eaux. KOECHLIN: Les Bandar-log, op. 176. MESSIAEN:
            Chronochromie. Josephine Nendick, s; Barry McDaniel, t; Louis
            Devos, bs; BBC Symphony Orchestra and Chorus; Pierre Boulez,
            Antal Dorati. Argo ZRG 756. (Reissue from HMV ASD 639).
                +Gr 4-76 p1592   +NR 5-76 p8
                +HFN 4-76 p123   +RR 1-76 p34
                +MT 7-76 p580
BOUTRY
      Capriccio. cf Boston Brass BB 1001.
BOWERS
      Because I love you. cf Rococo 5397.
BOYCE, William
      Heart of oak. cf HMV ESD 7002.
      Trio sonata, D major. cf National Trust NT 002.
      Voluntaries, D major, G minor. cf Cambridge CRS 2540.

Voluntary, D major. cf Abbey LPB 739.
Voluntary, A minor. cf Wealden WS 121.
BOYDELL, Brian
594   Symphonic inscapes. VICTORY: Jonathan Swift. Radio Telefis
      Eireann Symphony Orchestra; Albert Rosen. CRD New Irish
      Recording NIR 011.
         +Gr 5-76 p1749
BOYKAN, Martin
      Quartet, strings, no. 1. cf BARKIN: Quartet, strings.
BOZIC, Darijan
595   Audiogemi I-IV. DETONI: Dokument 75. DEVCIC: Entre nous.
      HORVAT: Träumerei. RUZDJAK: Piste. Zagreb String Quartet,
      Acezantez Ensemble; RTZ Symphony Orchestra; Dubravko Detoni,
      Kuljeric, Lajovic, conductors. Jugoton LSY 61198.
         +Te 9-76 p33
BOZZA, Eugene
      Sonatina, brass quintet. cf Crystal S 204.
      Sonatina, flute and bassoon. cf Crystal S 351.
BRAGA, Gaetano
      La serenata. cf Rococo 5397.
BRAHE
      Songs: Bless this house. cf Argo ZFB 95/6.
BRAHMS, Johannes
596   Academic festival overture, op. 80. Serenade, no. 2, op. 16, A
      major. Variations on a theme by Haydn, op. 56a. BPhO; Dresden
      Staatskapelle; Claudio Abbado. DG Tape (c) 3300 508.
         +Gr 7-76 p302 tape
597   Academic festival overture, op. 80. Alto rhapsody,op. 53. Tragic
      overture, op. 81. Variations on a theme by Haydn, op. 56a.
      Monica Sinclair, con; Croydon Philharmonic Choir; LPO; Adrian
      Boult. Pye GSGC 15021. (Reissues from Nixa NCL 16001,
      NCL 16003, NCL 16000, NCL 16002).
         +-Gr 1-76 p1191
      Academic festival overture, op. 80. cf Symphonies, nos. 1-4.
      Akademische Festouverture, op. 80. cf Symphony, no. 3, op. 90,
      F major.
      Academic festival overture, op. 80. cf Symphony no. 4, op. 98,
      E minor.
      Academic festival overture, op. 80. cf Works, selections (Decca
      DPA 553/4).
      Academic festival overture, op. 80. cf Works, selections
      (Philips 6747 270).
      Academic festival overture, op. 80. cf Decca SXL 6782.
      Academic festival overture, op. 80. cf International Piano
      Library IPL 5001/2.
598   Alto rhapsody, op. 53.* STRAUSS, R.: Liebeshymnus, op. 32, no. 3;
      Muttertändelei, op. 43, no. 2; Das Rosenband, op. 36, no. 1;
      Ruhe, meine Seele, op. 27, no. 1. WAGNER: Wesendonk Lieder.
      Janet Baker, ms; LPO; John Alldis Choir; Adrian Boult. HMV
      SQ ASD 3260. Tape (c) TC ASD 3260. (*Reissue from ASD 2749).
         +Gr 9-76 p454   +MT 12-76 p1007
         ++HFN 10-76 p166   +RR 9-76 p87
         ++HFN 12-76 p155
      Alto rhapsody, op. 53. cf Academic festival overture, op. 80.
      Alto rhapsody, op. 53. cf Symphonies, nos. 1-4.
      Alto rhapsody, op. 53. cf Works, selections (Decca DPA 553/4).

599   Ballades, op. 10. Fantasias, op. 116. Emil Gilels, pno.
         DG 2530 655. Tape (c) 3300 655.
                  +Gr 7-76 p193   +-NR 12-76 p14
                  +-Gr 10-76 p658   ++RR 6-76 p68
                  +HFN 7-76 p85
       Ballade, no. 1, op. 10, D minor. cf Piano works (Vanguard
         VSD 71213).
       Chorale preludes, op. 122, nos. 1, 4, 8, 10. cf Argo 5BB
         1013-5.
       Chorale prelude, op. 122, no. 5: Schmücke dich. cf Wealden
         WSQ 134.
       Chorale prelude, op. 122, no. 8: Es ist ein Ros entsprungen.
         cf Audio EAS 16.
       Chorale prelude, op. 122, no. 8: Es ist ein Ros entsprungen.
         cf HMV SQ HQS 1356.
       Chorale prelude, op. 122, no. 8: Es ist ein Ros entsprungen.
         cf Wealden WSQ 134.
       Chorale preludes, op. 122, no. 8: Es ist ein Ros entsprungen;
         no. 9, Herzlich tut mich verlangen; no. 11, O Welt, ich muss
         dich lassen. cf Wealden WS 121.
600   Concerto, piano, no. 1, op. 15, D minor. Bruno-Leonardo Gelber,
         pno; Munich Philharmonic Orchestra; Franz Paul Decker.
         Connoisseur Society CS 2102.
                  +-ARG 11-76 p17   ++SFC 10-10-76 p32
                  +NR 9-76 p6   +St 11-76 p138
601   Concerto, piano, no. 1, op. 15, D minor. Radu Lupu, pno; LPO;
         Edo de Waart. Decca SXL 6728. Tape (c) KSXC 6728. (also
         London CS 6947 Tape (c) 5-6947).
                  -Gr 11-75 p808   +NR 9-76 p6
                  +-HF 6-76 p76   +-RR 10-75 p38
                  +-HFN 11-75 p152   +SFC 5-30-76 p24
                  +HFN 12-75 p173 tape   +-St 11-76 p138
602   Concerto, piano, no. 1, op. 15, D minor. Artur Rubinstein, pno;
         Israel Philharmonic Orchestra; Zubin Mehta. Decca SXL 6797.
         Tape (c) KSXC 6797. (also London 7018 Tape (c) 5-7018).
                  -ARG 11-76 p17   +HFN 11-76 p175 tape
                  +-Gr 10-76 p589   +-NR 11-76 p4
                  +HFN 10-76 p166   +RR 9-76 p46
603   Concerto, piano, no. 1, op. 15, D minor. Julius Katchen, pno;
         LSO; Pierre Monteux. London STS 15209. (Reissue).
                  ++SFC 4-11-76 p38
       Concerto, piano, no. 1, op. 15, D minor. cf Works, selections
         (Philips 6747 270).
604   Concerto, piano, no. 2, op. 83, B flat major. Emil Gilels, pno;
         CSO; Fritz Reiner. Camden CCV 5042.
                  +HFN 4-76 p123   +RR 4-76 p38
605   Concerto, piano, no. 2, op. 83, B flat major. Bruno-Leonardo
         Gelber, pno; RPO; Rudolf Kempe. Connoisseur Society 2088 (Q)
         CSQ 2088.
                  +HF 6-76 p77   ++SFC 11-16-75 p32
                  +-MJ 4-76 p30   +St 7-76 p108 Quad
                  +-NR 3-76 p4
606   Concerto, piano, no. 2, op. 83, B flat major. Vladimir Horowitz,
         pno; NBC Symphony Orchestra; Arturo Toscanini. RCA VH 0019.
         (Reissue from DB 5861/6).
                  +Gr 6-76 p43   +-HFN 4-76 p123

607   Concerto, piano, no. 2, op. 83, B flat major. Sviatoslav Richter,
      pno; CSO; Erich Leinsdorf. RCA AGL 1-1267. (Reissue from
      LSC 2466).
             +MJ 2-76 p47   -SFC 4-11-76 p38
      Concerto, piano, no. 2, op. 83, B flat major. cf Works,
      selections (Philips 6747 270).
608   Concerto, violin, op. 77, D major. Isaac Stern, vln; PO; Eugene
      Ormandy. CBS 61325. (Reissue from Fontana SCFL 129).
             +Gr 8-76 p279   +-RR 6-76 p40
             +-HFN 7-76 p103
609   Concerto, violin, op. 77, D major. Nathan Milstein, vln; VPO;
      Eugen Jochum. DG 2530 592. Tape (c) 3300 592.
             +Audio 11-76 p111   +RR 12-75 p22
             +-Gr 12-75 p1031   +RR 4-76 p81 tape
             +-HF 5-76 p80   ++SFC 2-29-76 p25
             +HFN 3-76 p113 tape   +St 6-76 p101
             ++MJ 5-76 p29   +STL 1-11-76 p36
             +-NR 4-76 p3
610   Concerto, violin, op. 77, D major. Gidon Kremer, vln; BPhO;
      Herbert von Karajan. HMV SQ ASD 3261. (also Angel 37226).
             -Gr 10-76 p590   +-RR 9-76 p46
             -HFN 10-76 p166   -SFC 12-26-76 p34
611   Concerto, violin, op. 77, D major. Efrem Zimbalist, vln; BSO;
      Serge Koussevitzky. Rococo 2100.
             +-NR 4-76 p6
612   Concerto, violin, op. 77, D major. Nathan Milstein, vln; PhO;
      Anatole Fistoulari. Seraphim S 60265.
             +-NR 6-76 p6
      Concerto, violin, op. 77, D major. cf Works, selections
      (Philips 6747 270).
      Concerto, violin, op. 77, D major. cf BEETHOVEN: Concerto, violin,
      op. 61, D major.
      Concerto, violin, op. 77, D major. cf RCA ARM 4-0942/7.
613   Concerto, violin and violoncello, op. 102, A minor. Tragic
      overture, op. 81. Wolfgang Schneiderhan, vln; Janos Starker,
      vlc; Berlin Radio Symphony Orchestra; BPhO; Ferenc Fricsay,
      Lorin Maazel. DG 2535 140. (Reissues from 133 237, 138 022).
             +-Gr 3-76 p1463   -RR 1-76 p31
             +-HFN 2-76 p115
614   Concerto, violin and violoncello, op. 102, A minor. SCHUMANN:
      Fantasy, violin and orchestra, op. 131, C major. Ruggiero
      Ricci, vln; Giorgio Ricci, vlc; NPhO; Leipzig Gewandhaus
      Orchestra; Kurt Masur. Turnabout TVS 34593.
             +Gr 9-76 p415   +-NR 10-75 p3
             +-HF 1-76 p85   +-RR 8-76 p29
             -HFN 8-76 p76   ++SFC 8-24-75 p28
      Concerto, violin and violoncello, op. 102, A minor. cf Works,
      selections (Philips 6747 270).
      Concerto, violin and violoncello, op. 102, A minor. cf HMV
      RLS 723.
      Concerto, violin and violoncello, op. 102, A minor. cf RCA ARM
      4-0942/7.
      Ein Deutsches Requiem (German requiem), op. 45: How lovely are
      thy dwellings fair. cf Abbey LPB 757.
      German requiem, op. 45: How lovely are thy dwellings fair. cf
      HMV SXLP 50017.

F-A-E sonata: Scherzo. cf Works, selections (DG 2709 058).
Fantasia, op. 116. cf Ballades, op. 10.
615 Hungarian dances (21). Michel Beroff, Jean-Philippe Collard, pno.
     Connoisseur Society CS 2083 (Q) CSQ 2083.
          +HF 6-76 p77   ++SFC 10-12-75 p22
          +NR 3-76 p12   ++St 5-76 p78
     Hungarian dances (7). cf Works, selections (Decca DPA 553/4).
616 Hungarian dances, nos. 1-21. Alfons and Aloys Kontarsky, pno.
     DG 2530 710. Tape (c) 3300 710.
          ++Gr 10-76 p622   ++RR 9-76 p80
          +HFN 10-76 p166
     Hungarian dances, nos. 1, 3, 19. cf Symphony, nos. 3, op. 90,
     F major.
617 Hungarian dances, nos. 1, 5-7, 12-13, 19, 21. DVORAK: Slavonic
     dances, op. 46, nos. 1, 3, 8; op. 72, nos. 1 and 2. VPO;
     Fritz Reiner. Decca SPA 377. Tape (c) KCSP 377. (Reissue
     from SXL 2249).
          ++Gr 5-76 p1807   ++RR 4-76 p40
          +-HFN 4-76 p121   +RR 10-76 p105 tape
     Hungarian dance, no. 1, G minor.. cf HMV SXLP 30181.
     Hungarian dance, no. 1, G minor. cf Pye QS PCNH 4.
     Hungarian dance, no. 1, G minor. cf RCA ARM 4-0942/7.
     Hungarian dance, no. 5, G minor. cf HMV HQS 1360.
618 Piano works: Ballade, no. 1, op. 10, D minor. Pieces, piano, op.
     76, nos. 2 and 6. Pieces, piano, op. 117, no. 2, B flat major.
     Pieces, piano, op. 118, nos. 1-6. Rhapsodies, op. 79 (2).
     Bruce Hungerford, pno. Vanguard VSD 71213.
          +ARG 11-76 p15   +NR 11-76 p13
          +-MJ 11-76 p45   ++SFC 10-31-76 p35
     Pieces, piano, op. 76, nos. 2 and 6. cf Piano works (Vanguard
     VSD 71213).
619 Pieces, piano, op. 117 (3). Pieces, piano, op. 119 (4). Varia-
     tions and fugue on a theme by Handel, op. 24. John McCabe,
     pno. Oryx ORPS 78.
          -Gr 9-76 p443
620 Pieces, piano, op. 117 (3). Trio, piano, no. 1, op. 8, B major.
     Gyula Kiss, pno; Eszler Perényi, vln; Miklos Perényi, vlc.
     Hungaroton SLPX 11796.
          +NR 11-76 p6   ++SFC 10-10-76 p32
     Pieces, piano, op. 117, no. 1, E flat major. cf Saga 5427.
     Pieces, piano, op. 117, no. 1, E flat major. cf Sonata, piano,
     no. 3, op. 5, F minor.
621 Pieces, piano, op. 117, no. 2, B flat minor. Sonata, violin and
     piano, no. 3, op. 108, D minor. Waltz, op. 39, no. 15, A flat
     major. MUSSORGSKY: Pictures at an exhibition (edit. Horowitz).
     Vladimir Horowitz, pno; Nathan Milstein, vln. RCA VH 017.
     (Reissue from RB 16019).
          +Gr 4-76 p1631   +-RR 3-76 p59
          +-HFN 4-76 p123
     Pieces, piano, op. 117, no. 2, B flat minor. cf Piano works
     (Vanguard VSD 71213).
     Pieces, piano, op. 117, no. 2, B flat minor. cf HMV HQS 1354.
     Pieces, piano, op. 118, nos. 1-6. cf Piano works (Vanguard
     VSD 71213).
     Pieces, piano, op. 119 (4). cf Pieces, piano, op. 117.
     Quartets, piano and strings, nos. 1-3, opp. 25, 26, 60. cf Works,
     selections (DG 2740 117).

Quartets, strings, nos. 1-3, op. 51, nos. 1 and 2, op. 67. cf
   Works, selections (DG 2740 117).
622   Quartet, strings, no. 1, op. 51, no. 1, C minor. Quartet, strings,
      no. 2, op. 51, no. 2, A minor. Weller String Quartet. London
      STS 15245. (also Decca SDD 322. Reissue from SXL 6151).
               /HFN 4-72 p705   +MJ 12-76 p44
   Quartet, strings, no. 2, op. 51, no. 2, A minor. cf Quartet,
      strings, no. 1, op. 51, no. 1, C minor.
   Quintet, clarinet, op. 115, B minor. cf Works, selections
      (DG 2740 117).
   Quintet, clarinet, op. 115, B minor. cf HMV SLS 5046.
   Quintet, clarinet, op. 115, B minor: 3rd movement. cf Decca
      SPA 395.
623   Quintet, piano, op. 34, F minor. Werner Haas, pno; Berlin
      Philharmonic Octet Members. Philips 6500 705.
               +Gr 1-76 p1210   +RR 2-76 p44
               +-HFN 1-76 p104
   Quintet, piano, op. 34, F minor. cf Works, selections (DG 2740
      117).
   Quintets, strings, nos. 1 and 2, opp 88, 111. cf Works, selections
      (DG 2740 117).
   Rhapsodies, op. 79 (2). cf Piano works (Vanguard VSD 71213).
   Rhapsody, op. 79, no. 2, G minor. cf Sonata, piano, no. 3, op. 5,
      F minor.
   Rhapsody, op. 79, no. 2, G minor. cf Works, selections (Decca
      DPA 553/4).
   Serenade, no. 2, op. 16, A major. cf Academic festival overture,
      op. 80.
   Sextets, strings, nos. 1 and 2, opp. 18 and 36. cf Works, selec-
      tions (DG 2740 117).
   Sonatas, clarinet, nos. 1 and 2, op. 120, nos. 1 and 2. cf Works,
      selections (DG 2740 117).
624   Sonata, clarinet, no. 2, op. 120, E flat major (trans. Brahms for
      violin). Sonata, violin and piano, no. 1, op. 78, G major.*
      Sonata, violin and piano, no. 2, op. 100, A major. Sonata,
      violin and piano, no. 3, op. 108, D minor.* Isaac Stern, vln;
      Alexander Zakin, pno. CBS 78232 (2). (*Reissues from SBRG
      72157).
               +Gr 9-76 p436   +-RR 12-76 p77
               +HFN 9-76 p119
625   Sonata, piano, no. 2, op. 2, F sharp minor. Variations on a theme
      by Paganini, op. 35. Claudio Arrau, pno. Philips 9500 066.
               +-ARG 11-76 p18   +MJ 12-76 p28
               ++Gr 6-76 p69   +NR 12-76 p14
               +HF 12-76 p99   +RR 5-76 p64
               ++HFN 8-76 p97
626   Sonata, piano, no. 3, op. 5, F minor. Rhapsody, op. 79, no. 2,
      G minor. Bruno-Leonardo Gelber, pno. Connoisseur Society
      2084 (Q) CSQ 2084.
               ++Audio 5-76 p77   ++NR 11-75 p14 Quad
               +-Audio 8-76 p75   ++SFC 11-16-75 p32 Quad
               -HF 9-76 p85   ++ St 2-76 p99 Quad
               +MJ 4-76 p30
627   Sonata, piano, no. 3, op. 5, F minor. Pieces, piano, op. 117,
      no. 1, E flat major; op. 119, no. 3, C major. Clifford
      Curzon, pno. London STS 15272. (Reissue from CS 6341).

BRAHMS (cont.)   102

        (also Decca SDD 498 Reissue from SXL 6041).
           +Gr 12-76 p1016   ++St 7-75 p95
           ++HF 6-75 p87
628   Sonata, viola, no. 1, op. 120, F minor. Sonata, viola, no. 2,
        op. 120, E flat major. Pinchas Zukerman, vla; Daniel Barenboim,
        pno. DG 2530 722. (Reissue from 2709 058).
           +Gr 11-76 p829   +-RR 12-76 p77
           ++HFN 11-76 p171
      Sonatas, viola, nos. 1-2, op. 120. cf Sonatas, violin and
        piano, nos. 1-3.
      Sonata, viola, no. 1, op. 120, F minor. cf Works, selections
        (DG 2709 058).
      Sonata, viola, no. 2, op. 120, E flat major. cf Sonata, viola,
        no. 1, op. 120, F minor.
      Sonata, viola, no. 2, op. 120, E flat major. cf Works, selections
        (DG 2709 058).
629   Sonatas, violin and piano, nos. 1-3. Sonatas, viola (clarinet),
        nos. 1-2, op. 120. Pinchas Zukerman, vln and vla; Daniel
        Barenboim, pno. DG 2709 058 (3).
           +-Gr 11-75 p847   +-MJ 1-76 p24
           +-HF 3-76 p82   +NR 1-76 p9
           ++HFN 11-75 p152   +-RR 11-75 p74
      Sonatas, violin and piano, nos. 1-3, opp. 78, 100, 108. cf
        Works, selections (DG 2740 117).
      Sonatas, violin and piano, nos. 1-3, opp. 78, 100, 108. cf
        BEETHOVEN: Sonata, violin and piano, no. 5, op. 24, F major.
630   Sonata, violin and piano, no. 1, op. 78, G major. Sonata, violin
        and piano, no. 3, op. 108, D minor. Miklós Szenthelyi, vln;
        András Schiff, pno. Hungaroton SLPX 11731.
           +-NR 3-76 p5   -RR 1-76 p47
      Sonata, violin and piano, no. 1, op. 78, G major. cf Sonata,
        clarinet, no. 2, op. 120, E flat major.
      Sonata, violin and piano, no. 1, op. 78, G major. cf Works,
        selections (DG 2709 058).
631   Sonata, violin and piano, no. 2, op. 100, A major. PROKOFIEV:
        Sonata, violin and piano, no. 1, op. 80, F minor. David
        Oistrakh, vln; Sviatoslav Richter, pno. Melodiya/Angel
        SR 40268.
           +HF 8-76 p92   ++NR 7-76 p7
632   Sonata, violin and piano, no. 2, op. 100, A major. Sonata,
        violin and piano, no. 3, op. 108, D minor. Arthur Grumiaux,
        vln; György Sebok, pno. Philips 9500 108.
           +Gr 11-76 p829   ++RR 11-76 p88
           ++HFN 11-76 p156
      Sonata, violin and piano, no. 2, op. 100, A major. cf Sonata,
        clarinet, no. 2, op. 120, E flat major.
      Sonata, violin and piano, no. 2, op. 100, A major. cf Works,
        selections (DG 2709 058).
      Sonata, violin and piano, no. 2, op. 100, A major. cf BARTOK:
        Sonata, violin and piano, no. 1.
      Sonata, violin and piano, no. 2, op. 100, A major. cf RCA ARM
        4-0942/7.
      Sonata, violin and piano, no. 3, op. 108, D minor. cf Pieces,
        piano, op. 117, no. 2, B flat minor.
      Sonata, violin and piano, no. 3, op. 108, D minor. cf Sonata,
        clarinet, no. 2, op. 120, E flat major.

Sonata, violin and piano, no. 3, op. 108, D minor. cf Sonata,
    violin and piano, no. 1, op. 78, G major.
Sonata, violin and piano, no. 3, op. 108, D minor. cf Sonata,
    violin and piano, no. 2, op. 100, A major.
Sonata, violin and piano, no. 3, op. 108, D minor. cf Works,
    selections (DG 2709 058).
Sonata, violin and piano, no. 3, op. 108, D minor. cf RCA
    ARM 4-0942/7.
633 Sonata, violoncello and piano, no. 1, op. 38, E minor. Sonata,
    violoncello and piano, no. 2, op. 99, F major. Pierre Fournier,
    vlc; Wilhelm Backhaus, pno. Decca ECS 785. (Reissue from
    LXT 5077).
        +-Gr 11-76 p824   +-RR 12-76 p77
        +-HFN 11-76 p171
Sonatas, violoncello and piano, nos. 1 and 2, opp. 38, 99. cf
    Works, selections (DG 2740 117).
634 Sonata, violoncello and piano, no. 1, op. 38, E minor. GRIEG:
    Sonata, violoncello and piano, op. 36, A minor. Mstislav
    Rostropovich. vlc; Sviatoslav Richter, pno. Olympic 8140.
        +-NR 7-76 p7
Sonata, violoncello and piano, no. 2, op. 99, F major. cf Sonata,
    violoncello and piano, no. 1, op. 38, E minor.
635 Songs: Auf dem Kirchhofe, op. 105, no. 4; Mit vierzig Jahren ist
    der Berg erstiegen, op. 94, no. 1; Ruhe, Süssliebchen, op. 33,
    no. 9. LOEWE: Songs: Hinkenden Jamben, op. 62, no. 5; Hoch-
    zeitlied, op. 20, no. 1; Odins Meeres-Ritt, op. 118; Die
    Wandelnde Glocken, op. 20, no. 3. STRAUSS, R.: Songs: Ach
    weh mir unglückhaftem Mann, op. 21, no. 4; All mein Gedanken,
    mein Herz und mein Sinn, op. 21, no. 1; Du meines Herzens
    Krönelein, op. 21, no. 2; Gefunden, op. 56, no. 1; Himmels-
    boten, op. 32, no. 5; Nachtgang, op. 29, no. 3. WOLF: Songs:
    Anakreons Grab; Der Musikant; Der Soldat I; Der verzweifelte
    Liebhafer; Wenn du zu Blumen gehst; Wer sein holdes Lieb
    verloren. Hans Hotter, bar; Geoffrey Parsons, pno. Decca
    SXL 6738.
        +-Gr 3-76 p1503   +-RR 3-76 p68
        +HFN 3-76 p95
636 Songs: Children's folk songs: Beim Ritt auf dem Knie; Dornröschen;
    Heidenröslein; Die Henne; Der Jäger im Walde; Das Mädchen und
    der Hasel; Der Mann; Marienwürmchen; Die Nachtigall; Sandmänn-
    chen; Das Schlaraffenland; Dem Schutzengel; Wiehnachten;
    Wiegenlied. German folk songs: Des Abends kann ich; Ach
    englische Schäferin; Ach Gott, wie weh tut Scheiden; Ach könnt
    ich diesen Abend; All mein Gedanken; Da unten im Tale; Dort
    in den Weiden; Du mein en einzig Licht; Erlaube mir, feins
    Mädchen; Es ging ein Maidlein zarte; Es reit' ein Herr und
    auch sein Knecht; Es ritt ein Ritter; Es steht ein Lind; Es
    war ein Markgraf über Rhein; Es war eine schöne Jüdin; Es
    wohnet ein Fiedler; Feinsliebchen; Gar lieblich hat sich gesellet;
    Gunhilde; Guten Abend; Ich stand auf hohem Berge; Ich wiess
    mir'n Maidlein; In stiller Nacht; Jungfräulein, soll ich mit
    euch gehn; Maria ging aus wandern; Mein Mädel hat einem Rosen-
    mund; Mir ist ein schöns brauns Madelein; Nur ein Gesicht auf
    Erden lebt; Och Moder, ich well; Der Reiter spreitet seinen;
    Sagt mir, o schönste Schäf'rin mein; Schöner Augen; Schönster
    Schatz, mein Engel; Schwesterlein; So will ich frisch; So

wünsch ich ihr ein' gute Nacht; Soll ich der Mond; Die Sonne
scheint nicht mehr; Wach' auf mein Herzenschöne; Wach' auf,
mein Hort; Wie komm' ich; Wo gehst du hin, du Stolze. German
folk songs for chorus: Abschiedslied; Bei nächtlicher Weil; Der
englische Jäger; In stiller Nacht; Mit Lust tät ich ausreiten;
Morgengesang; Schnitter Tod; Von edler Art; Die Wollust in den
Maien. Edith Mathis, s; Peter Schreier, t; Karl Engel, pno;
Leipzig Radio Chorus; Horst Neumann. DG 2709 057 (3).
> +-Audio 6-76 p98   ++NR 1-76 p12
> +Gr 10-75 p662   +ON 12-20-75 p38
> +HF 2-76 p93   +RR 10-75 p81
> +HFN 10-75 p139   +SR 1-24-76 p53
> ++MJ 1-76 p24   +St 6-76 p101

637   Songs: Es ist das Heil uns kommen her, op. 29, no. 1; Schaffe in
      mir, Gott, op. 29, no. 2; Warum ist das Licht gegeben dem
      Mühseligen, op. 74, no. 1. SCHUBERT: Songs: Christ ist erstan-
      den, D 440; Gebet, D 815; Gott im Ungewitter, D 985; Psalm 23,
      Gott ist mein Hirt, D 706. Cambridge, King's College Chapel
      Choir; Philip Ledger, pno; Philip Ledger. HMV ASD 3091.
> -MT 2-76 p143   +-RR 9-75 p71

638   Songs: Ave Maria, op. 12; Psalm 13, op. 27; Sacred choruses, op.
      37 (3); Songs, op. 17 (4); Songs and romances, op. 44 (12).
      Zoltán Kodály Women's Choir; Ilona Andor. Hungaroton SLPX
      11691.
> ++HF 8-76 p83   +NR 6-76 p12

639   Songs: Das Mädchen spricht, op. 107, no. 3; Die Mainacht, op. 43,
      no. 2; Nachtigall, op. 97, no. 1; Von ewiger Liebe, op. 43,
      no. 1. SCHUBERT: Songs: Die abgeblühte Linde, D 514; Heim-
      liches Lieben, D 922; Minnelied, D 429; Der Musensohn, D 764.
      SCHUMANN: Frauenliebe und Leben, op. 42. Janet Baker, ms;
      Martin Isepp, pno. Saga 5277. Tape (c) CA 5277.
> +HF 11-76 p124   +HFN 10-76 p185 tape

      Songs: Klänge II, op. 66, no. 2; Klosterfräulein, op. 61, no. 2;
      Phänomen, op. 61, no. 3; Walpurgisnacht, op. 75, no. 4; Weg der
      Liebe I and II, op. 20, nos. 1 and 2. cf CBS 76476.
      Songs: Auf dem Kirchhofe, op. 105, no. 4; Mainacht, op. 43, no. 2;
      Sappische Ode, op. 94, no. 4. cf Club 99-99.
      Songs: Sändmannchen. cf HMV SEOM 25.
      Songs: An eine Aeolscharfe, op. 19, no. 5; Meine Liebe ist grün,
      op. 63, no. 5; O wüsst ich doch den Weg zurück, op. 63; Der
      Tod, das ist die kühle Nacht, op. 96, no. 1; Verzagen, op. 72,
      no. 4. cf Prelude PRS 2505.
      Songs: Am Wildbach die Weiden; Die Berge sind spitz; Nun stehn
      die Rosen in Blüte; Und gehst du über den Kirchhof. cf RCA
      PRL 1-9034.
      Songs: Liebestreu, op. 3, no. 1; Nicht mehr zu dir zu gehen. cf
      Rococo 5370.

640   Symphonies, nos. 1-4. Academic festival overture, op. 80. Tragic
      overture, op. 81. Variations on a theme by Haydn, op. 56a.
      OSR; Ernest Ansermet. Decca SPA 378/81 (4). (Reissues from
      SXL 6059, 6060, 6061, 6062).
> +-Gr 2-76 p1337

641   Symphonies, nos. 1-4. BPhO; Herbert von Karajan. DG 2721 075
      (4). Tape (c) 3371 015. (Reissue from SKL 133/9).
> +Gr 7-74 p199   +RR 7-74 p35
> ++Gr 12-76 p1066 tape   ++RR 2-75 p75 tape

642   Symphonies, nos. 1-4. VPO; Karl Böhm. DG 2740 154 (4). (also
      DG 2711 017 Tape (c) 3371 023).
              +-Gr 10-76 p589   +RR 10-76 p45
              -Gr 12-76 p1066 tape   ++RR 10-76 p105 tape
              +HFN 10-76 p165
643   Symphonies, nos. 1-4. Alto rhapsody, op. 53. Janet Baker, s;
      John Alldis Choir; LPO, LSO; Adrian Boult. HMV Tape (c)
      TC SLS 5009 (2).
              +-Gr 9-76 p497 tape   +RR 11-76 p109 tape
      Symphonies, nos. 1-4. cf Works, selections (Philips 6747 270).
644   Symphony, no. 1, op. 68, C minor. Munich Philharmonic Orchestra;
      Rudolf Kempe. BASF BAC 3083. Tape (c) KBACC 3083.
              +-Gr 5-76 p1749   ++RR 3-76 p37
              +-HFN 3-76 p91   ++RR 5-76 p77 tape
              +-HFN 4-76 p125 tape
645   Symphony, no. 1, op. 68, C minor. BSO; Charles Munch. Camden
      CCV 5018. (Reissue)
              +St 2-76 p741
646   Symphony, no. 1, op. 68, C minor. Hallé Orchestra; James Loughran.
      Classics for Pleasure CFP 40096. Tape (c) TC CFP 40096.
              +Gr 3-75 p1648   +RR 3-75 p24
              ++HFN 12-76 p153 tape
647   Symphony, no. 1, op. 68, C minor. CO; Lorin Maazel. Decca SXL
      6783. Tape (c) KSXC 6783. (also London 7007 Tape (c) 5-7007).
              +-Gr 11-76 p776   +-RR 10-76 p45
              +-HFN 10-76 p165   ++SFC 12-5-76 p58
648   Symphony, no. 1, op. 68, C minor. BPhO; Wilhelm Furtwängler.
      DG 2530 744. (also DG 2535 162).
              +ARG 11-76 p16   +-NR 10-76 p2
              +Gr 5-76 p1771   +RR 5-76 p23
              +-HF 11-76 p105
649   Symphony, no. 1, op. 68, C minor. Symphony, no. 3, op. 90, F
      major. Variations on a theme by Haydn, op. 56a. BRUCKNER:
      Symphony, no. 7, E major. HINDEMITH: Symphonic metamorphoses
      on a theme by Carl Maria von Weber. TCHAIKOVSKY: Symphony, no.
      6, op. 74, B minor. BPhO; Wilhelm Furtwängler. DG 2535 161
      (5).
              +HFN 5-76 p100
650   Symphony, no. 1, op. 68, C minor. PhO; Otto Klemperer. HMV SXLP
      30217. (Reissue from Columbia SAX 2262).
              ++Gr 12-76 p990   +RR 12-76 p53
651   Symphony, no. 1, op. 68, C minor. LSO; Jascha Horenstein. RCA
      GL 25001. Tape (c) GK 25001. (Previously issued by Reader's
      Digest).
              ++Gr 10-76 p589   +RR 11-76 p52
              +HFN 12-76 p139   ++RR 12-76 p104 tape
              +HFN 12-76 p153 tape
652   Symphony, no. 1, op. 68, C minor. CSO; James Levine. RCA ARL
      1-1326. Tape (c) ARK 1-1326 (ct) ARS 1-1326 (Q) ARD 1-1326.
              ++Gr 6-76 p43   +-NR 4-76 p3
              ++HF 5-76 p82   +-RR 5-76 p38
              ++HF 9-76 p84 tape   +SFC 3-7-76 p27
              +HFN 5-76 p95   +SR 4-17-76 p51
              ++MJ 4-76 p31   +St 7-76 p109
      Symphony, no. 1, op. 68, C minor. cf Works, selections (Decca
      DPA 553/4).

653   Symphony, no. 2, op. 73, D major. Hallé Orchestra; James Loughran.
      Classics for Pleasure CFP 40219. Tape (c) TC CFP 40219.
                +Gr 11-75 p808   -HFN 12-76 p153 tape
                +HFN 11-75 p152   +RR 12-75 p45
654   Symphony, no. 2, op. 73, D major. PhO; Otto Klemperer. HMV Tape
      (c) TC EXE 196.
                +HFN 9-76 p133 tape
655   Symphony, no. 2, op. 73, D major. VPO; Pierre Monteux. London
      STS 15192.
                +-St 10-76 p120
656   Symphony, no. 2, op. 73, D major. VPO; Wilhelm Furtwängler.
      Olympic 8141.
                -NR 8-76 p4
657   Symphony, no. 2, op. 73, D major. Variations on a theme by
      Haydn, op. 56a. COA; Bernard Haitink. Philips 6500 375. Tape
      (c) 7300 375.
                +Gr 7-75 p175   ++NR 11-75 p2
                +-HF 12-75 p88   +RR 7-75 p23
                +-HFN 7-75 p77   ++St 10-76 p120
658   Symphony, no. 2, op. 73, D major. Danish State Radio Symphony
      Orchestra; Jascha Horenstein. Unicorn UNS 236.
                +Gr 11-76 p776
      Symphony, no. 2, op. 73, D major. cf HMV RLS 717.
659   Symphony, no. 3, op. 90, F major. Hungarian dances, nos. 1, 3,
      19. Hallé Orchestra; James Loughran. Classics for Pleasure
      CFP 40237. Tape (c) TC CFP 40237.
                +Gr 7-76 p168   +-HFN 12-76 p153 tape
                +-HFN 3-76 p91   +RR 2-76 p26
660   Symphony, no. 3, op. 90, F major. BPhO; Wilhelm Furtwangler.
      DG 2535 163.
                +-Gr 5-76 p1772   +RR 5-76 p23
661   Symphony, no. 3, op. 90, F major. Symphony, no. 4, op. 98, E
      minor. Tragic overture, op. 81. BPhO, Lamoureux Orchestra;
      Lorin Maazel, Igor Markevitch. DG Tape (c) 3348 217.
                +-RR 9-76 p92 tape
662   Symphony, no. 3, op. 90, F major. Akademische Festouverture,
      op. 80. HRT Orchestra; Tamás Pál. Hungaroton SLPX 11734.
                -NR 11-75 p2   -RR 3-76 p37
      Symphony, no. 3, op. 90, F major. cf Symphony, no. 1, op. 68, C
      minor.
663   Symphony, no. 4, op. 98, E minor. PhO; Otto Klemperer. HMV SXLP
      30214. (Reissue from Columbia SAX 2350).
                +Gr 8-76 p279   ++RR 7-76 p48
                +HFN 8-76 p93
664   Symphony, no. 4, op. 98, E minor. VPO; István Kertész. London CS
      6838. (also Decca SXL 6678. Reissue from SXLH 6610/3).
                +Gr 1-76 p1191   +NR 6-75 p4
                +HF 6-75 p88   +-RR 8-75 p31
                +HFN 8-75 p87   +St 6-75 p95
665   Symphony, no. 4, op. 98, E minor. Academic festival oveture, op.
      80. NPhO; Leopold Stokowski. RCA ARL 1-0719. Tape (c) 1-0719
      (ct) ARS 1-0719 (Q) ARD 1-0719 Tape (ct) ART 1-0719.
                ++Audio 3-76 p67   +NR 11-75 p2
                +Gr 12-75 p1031   +-RR 11-75 p42
                +HF 12-73 p88   +-St 3-76 p109
                +-HFN 11-75 p152

666   Symphony, no. 4, op. 98, E minor.   RPO; Fritz Reiner.   RCA AGL
       1-1961
              +SFC 11-21-76 p35
       Symphony, no. 4, op. 98, E minor.   cf Symphony, no. 3, op. 90,
       F major.
       Symphony, no. 4, op. 98, E minor.   cf BEETHOVEN: Namensfeier, op.
       115: Overture.
       Tragic overture, op. 81.   cf Concerto, violin and violoncello,
       op. 102, A minor.
       Tragic overture, op. 81.   cf Symphonies, nos. 1-4.
       Tragic overture, op. 81.   cf Symphony, no. 3, op. 90, F major.
       Tragic overture, op. 81.   cf Works, selections (Philips 6747 270).
       Tragic overture, op. 81.   cf BEETHOVEN: Symphony, no. 5, op. 67,
       C minor.
       Tragic overture, op. 81.   cf Bruno Walter Society RR 443.
       Trio, clarinet, violoncello and piano, op. 114, A minor.   cf
       Works, selections (DG 2740 117).
       Trio, clarinet, violoncello and piano, op. 114, A minor.   cf
       BEETHOVEN: Trio, clarinet, op. 11, B flat major.
       Trio, horn, violin and piano, op. 40, E flat major.   cf Works,
       selections (DG 2740 117).
       Trio, piano, no. 1, op. 8, B major.   cf Pieces, piano, op. 117.
       Trios, piano, nos. 1-3, opp. 8, 87, 101.   cf Works, selections
       (DG 2740 117).
       Variations and fugue on a theme by Handel, op. 24.   cf Pieces,
       piano, op. 117.
       Variations and fugue on a theme by Handel, op. 24.   cf BACH: The
       well-tempered clavier, Bk II, S 870-893: Prelude, no. 22.
667   Variations on a theme by Haydn, op. 56a.   ELGAR: Enigma variations,
       op. 36.   LSO; Eugen Jochum.   DG 2530 586.   Tape (c) 3300 586.
              +Gr 12-75 p1031   -NR 6-76 p3
              +HF 8-76 p83   +RR 12-75 p48
              +-HFN 12-75 p151   ++St 8-76 p93
              +HFN 2-76 p116 tape
668   Variations on a theme by Haydn, op. 56a.   HINDEMITH: Symphonic
       metamorphoses on themes by Carl Maria von Weber.   BPhO; Wilhelm
       Furtwängler.   DG 2535 164.
              +-Gr 5-76 p1772   +RR 5-76 p22
669   Variations on a theme by Haydn, op. 56a.   MOZART: Sonata, 2 pianos,
       K 448 (375a), D major.   RAVEL: Ma mère l'oye.   Dezsö Ránki,
       Zoltán Kocsis, pno.   Hungaroton SLPX 11646.
              +-Gr 2-75 p1515   +-RR 1-75 p46
              +-HFN 1-76 p104
       Variations on a theme by Haydn, op. 56a.   cf Academic festival
       overture, op. 80 (DG 3300 508).
       Variations on a theme by Haydn, op. 56a.   cf Academic festival
       overture, op. 80 (Pye 15021).
       Variations on a theme by Haydn, op. 56a.   cf Symphonies, nos. 1-4.
       Variations on a theme by Haydn, op. 56a.   cf Symphony, no. 1, op.
       68, C minor.
       Variations on a theme by Haydn, op. 56a.   cf Symphony, no. 2,
       op. 73, D major.
       Variations on a theme by Haydn, op. 56a.   cf Works, selections
       (Decca DPA 553/4).
       Variations on a theme by Haydn, op. 56a.   cf Works, selections
       (Philips 6747 270).

Variations on a theme by Paganini, op. 35. cf Sonata, piano,
  no. 2, op. 2, F sharp minor.
Vier ernste Gesänge, op. 121. cf RCA LRL 2-7531.
Waltz, op. 39, no. 15, A flat major. cf Pieces, piano, op. 117,
  no. 2, B flat minor.
Waltz, op. 39, no. 15, A flat major. cf Works, selections (Decca
  DPA 553/4).
Waltz, op. 39, no. 15, A flat major. cf London STS 15239.

670   Works, selections: Academic festival overture, op. 80. Alto
      rhapsody, op. 53. Hungarian dances (7). Rhapsody, op. 79, no.
      2, G minor. Symphony, no. 1, op. 68, C minor. Variations on
      a theme by Haydn, op. 56a. Waltz, op. 39, no. 15, A flat major.
      Helen Watts, con; Julius Katchen, pno; OSR, LSO, VPO; Ernest
      Ansermet, Pierre Monteux, Fritz Reiner, Josef Krips. Decca
      DPA 553/4 (2). Tape (c) KDPG 553/4. (Reissues)
            +-HFN 10-76 p183   +HFN 11-76 p175 tape

671   Works, selections: Sonata, viola, no. 1, op. 120, F minor. Sonata,
      viola, no. 2, op. 120, E flat major. Sonata, violin and piano,
      no. 1, op. 78, G major. Sonata, violin and piano, no. 2, op.
      100, A major. Sonata, violin and piano, no. 3, op. 108, D minor.
      F-A-E sonata: Scherzo. Pinchas Zukerman, vla, vln; Daniel
      Barenboim, pno. DG 2709 058 (3).
            ++St 4-76 p112

672   Works, selections: Quartets, piano, nos. 1-3, opp. 25, 26, 60.
      Quartets, strings, nos. 1-3, op. 51, nos. 1-2; op. 67. Quintet,
      clarinet, op. 115, B minor. Quintet, piano, op. 34, F minor.
      Quintets, strings, nos. 1 and 2, opp. 88, 111. Sextets,
      strings, nos. 1 and 2, opp. 18 and 36. Sonatas, clarinet. nos.
      1 and 2, opp. 120, nos. 1 and 2. Sonatas, violin, nos. 1-3,
      opp. 78, 100, 108. Sonatas, violoncello and piano, nos. 1 and 2,
      opp. 38, 99. Trio, clarinet, violoncello and piano, op. 114,
      A minor. Trio, horn, violin and piano, op. 40, E flat major.
      Trios, piano, nos. 1-3, opp. 8, 87, 101. Cecil Aronowitz,
      Stefano Passagggio, vla; William Pleeth, Georg Donderer, Pierre
      Fournier, vlc; Christoph Eschenbach, Jörg Demus, Pierre Barbi-
      zet, Rudolf Firkusny, pno; Karl Leister, clt; Eduard Drolc
      Christian Ferras, vln; Amadeus Quartet, Trieste Trio. DG 2740
      117 (15). (Reissue from 104973/87).
            +Gr 1-76 p1211   +-RR 8-75 p46
            +-HFN 9-75 p109

673   Works, selections: Academic festival overture, op. 80. Concerto,
      piano, no. 1, op. 15, D minor. Concerto, piano, no. 2, op. 83,
      B flat major. Concerto, violin, op. 77, D major. Concerto,
      violin and violoncello, op. 102, A minor. Symphonies, nos. 1-4.
      Tragic overture, op. 81. Variations on a theme by Haydn, op.
      56a. Claudio Arrau, pno; Henryk Szeryng, vln; Janos Starker,
      vlc; COA; Bernard Haitink. Philips 6747 270 (8).
            +Gr 9-76 p474   +-RR 10-76 p46
            +HFN 10-76 p183

BRANT, Henry
      Quombex, viola d'amore, music box and organ. cf Desto DC 6435/7.
BRASSART, Jean
      O flos fragrans. cf Telefunken ER 6-35257.
BRAUN, Yehezekiel
      Psalm 98. cf Serenus SRS 12039.
BRIAN, Havergal
674   Symphony, no. 10, C minor. Symphony, no. 21, E flat major.

Leicestershire Schools Symphony Orchestra; James Loughran,
Eric Pinkett. Unicorn RHS 313.
     ++Gr 5-73 p2042   ++RR 5-73 p47
     +HFN 5-73 p980   +-St 2-76 p99

BRIDGE, Frank
    Adagio, E major. cf Wealden WS 149.
    Allegro con spirito, B flat major. cf Wealden WS 149.
    Cherry ripe. cf Works, selections (HMV SQ ASD 3190).
    Enter spring. cf Works, selections (HMV SQ ASD 3190).
    Lament. cf Works, selections (HMV SQ ASD 3190).
    The sea. cf Works, selections (HMV SQ ASD 3190).
    Summer. cf Works, selections (HMV SQ ASD 3190).
675   Works, selections: Cherry ripe. Enter spring. Lament. The sea.
      Summer. Royal Liverpool Philharmonic Orchestra; Charles Groves.
      HMV SQ ASD 3190.
         +Gr 5-76 p1749   ++RR 5-76 p39
         +HFN 5-76 p97   ++Te 9-76 p31
         ++MT 9-76 p744

BRIDGES
    All my hope on God is founded. cf Polydor 2460 250.
BRIGHAM, Erl
    City of Madrid. cf Philips 6308 246.
BRIGHT, Houston
    Rainsong. cf Columbia M 34134.
BRINDLE, Reginald Smith
676   Orion 42. JONES: Sonata, 3 timpani. SIMPSON: Expose. STOCK-
      HAUSEN: Zyklus. Tristan Fry Percussion. Classics for Pleasure
      CFP 40307.
         +-HFN 8-76 p87   /*RR 2-76 p52
BRISCIALDI
    Carnival of Venice. cf RCA LRL 1-5131.
BRITTEN, Benjamin
677   A birthday Hansel, op. 92. Canticle: The death of Saint Narcissus,
      op. 89. Gloriana, op. 53: 2nd lute song. Folk songs: O can
      ye sew cushions; Ca' the yowes. Suite, harp, op. 83. Peter
      Pears, t; Osian Ellis, hp. Decca SXL 6788.
         +Gr 7-76 p200   +RR 7-76 p73
         ++HFN 7-76 p85   +STL 7-4-76 p36
         +MT 10-76 p831
    Canticle: The death of Saint Narcissus, op. 89. cf A birthday
      Hansel, op. 92.
678   Canticles: My beloved is mine, op. 40; Abraham and Isaac, op. 51;
      Still falls the rain, op. 55. Robert Tear, t; James Bowman, c-t;
      Alan Civil, hn; Philip Ledger, pno. HMV CSD 3773.
         ++Gr 10-76 p633   +RR 10-76 p95
         ++HFN 10-76 p166
679   A ceremony of carols op. 28. The golden vanity, op. 78. Missa
      brevis, op. 63, D major. Marisa Robles, hp; Robert Bottone,
      pno; Clement McWilliam, org; Winchester Cathedral Choir; Martin
      Neary. Pye TPLS 13065.
         +Gr 1-76 p1225   +RR 1-76 p55
         +HFN 1-76 p104
680   A charm of lullabies, op. 41. Folksong arrangements: The Ash
      grove; The bonny Earl o' Moray; Come you not from Newcastle;
      O can ye sew cushions; O waly, waly; Oliver Cromwell; The
      Salley gardens; There's none to soothe; Sweet Polly Oliver; The

trees they grow so high. Bernadette Greevy, con; Paul Hamburger, pno. London STS 15166.
    +NR 6-76 p14   +St 9-76 p120

681   Concerto, violin, op. 15, D minor. Serenade, op. 31. Ian Partridge, t; Rodney Friend, vln; Nicholas Busch, hn; LPO; John
Pritchard. Classics for Pleasure CFP 40250.
    +-HFN 12-76 p139

682   Death in Venice. Penelope Mackay, Iris Saunders, s; Peter Pears,
Kenneth Bowen, t; James Bowman, c-t; John Shirley-Quirk, bar;
Peter Leeming, bs; Neville Williams, bs-bar; English Opera
Group Chorus; ECO; Steuart Bedford. Decca SET 581/3 (3).
(also London 13109 Tape (c) Q 513109 (r) R 413109).
    +Gr 11-74 p955   +RR 11-74 p20
    ++HF 2-75 p79   +SR 11-30-74 p40
    +MJ 2-75 p30   +-St 1-75 p104
    +NR 2-75 p8   +Te 6-76 p45
    +ON 12-14-74 p46

Fanfare for St. Edmundsbury. cf Argo SPA 464.
Fanfare for St. Edmondsbury. cf Decca 419.
Festival Te Deum. cf Argo ZRG 789.
Folk songs: O can ye sew cushions; Ca' the yowes. cf A birthday
Hansel, op. 92.
Folksong arrangements: The Ash grove; The bonny Earl o' Moray;
Come you not from Newcastle; O can ye sew cushions; O waly,
waly; Oliver Cromwell; The Salley gardens; There's none to
soothe; Sweet Polly Oliver; The trees they grow so high. cf A
charm of lullabies, op. 41.

683   Friday afternoons, op. 7. Gemini variations, op. 73. Psalm 150,
voices and orchestra. Zoltán Jeney, flt, pno; Gabriel Jeney,
vln, pno; Parley Downside School Choir; Viola Tunnard, pno;
Benjamin Britten. London STS 15173.
    +Audio 11-76 p110   ++St 9-76 p120
    +NR 10-76 p8

Gemini variations, op. 73. cf Friday afternoons, op. 7.
Gloriana, op. 53: The courtly dances. cf RCA ARL 3-0997.
Gloriana, op. 53: 2nd lute song. cf A birthday Hansel, op. 92.
The golden vanity, op. 78. cf A ceremony of carols, op. 28.
Hymn to St. Cecilia, op. 27. cf HMV SLS 5047.
A hymn to the virgin. cf HMV CSD 3774.

684   Les illuminations, op. 18. Serenade, op. 31. Heather Harper, s;
Robert Tear, t; Alan Civil, hn; Northern Sinfonia Orchestra;
Neville Marriner. HMV SXLP 30194. (Reissue from CSD 3684).
    ++Gr 1-76 p1222   +RR 1-76 p55

Metamorphoses after Ovid, oboe, op. 49 (6). cf Phantasy, oboe
and strings, op. 2.
Missa brevis, op. 63, D major. cf A ceremony of carols.
Missa brevis, op. 63, D major. cf HMV SLS 5047.

685   Noye's fludde. Trevor Anthony, speaker; Caroline Clark, Marie-
Thérèse Pinto, Eileen O'Donovan, Patricia Garrod, Margaret
Hawes, Kathleen Petch, Gillian Saunders, girl sopranos; David
Pinto, Darien Angadi, Stephen Alexander, boy trebles; Sheila
Rex, con; Owen Brannigan, bs-bar; Children's Chorus; English
Opera Group Orchestra, Suffolk Children's Orchestra; Norman
Del Mar. Argo ZK 1. (Reissue from ZNF 1).
    +Gr 12-76 p1045

686   Peter Grimes: Four sea interludes and passacaglia, op. 33 a and b.

Sinfonia da requiem, op. 20. LSO; André Previn. Angel S
37142. (also HMV SQ ASD 3154)
+Gr 3-76 p1463   +-RR 3-76 p37
+HFN 3-76 p91   ++SFC 2-22-76 p29
+MJ 4-76 p31   +St 7-76 p109
++NR 3-76 p2

687   Phantasy, oboe and strings, op. 2. Metamorphoses after Ovid, oboe,
op. 49 (6). CROSSE: Ariadne, op. 31. Sarah Francis, ob;
Emanuel Hurwitz, vln; Margaret Major, vla; Derek Simpson, vlc;
LSO; Michael Lankester. Argo ZRG 842.
+Gr 10-76 p590   +RR 8-76 p30
++HFN 8-76 p77

688   Prelude and fugue, 18 solo strings, op. 29. LUYTENS: O saisons,
O chateaux. SCHOENBERG: Suite, string orchestra. RPO; Norman
Del Mar. Argo ZRG 754. (Reissue from HMV ASD 612).
/Gr 10-74 p682   +-RR 10-74 p56
++MQ 1-76 p139

Prelude and fugue on a theme by Vittoria. cf Argo 5BBA 1013/5.
Psalm 150, voices and orchestra. cf Friday afternoons, op. 7.

689   Quartet, strings, no. 1, op. 25, D major. Quartet, strings, no.
2, op. 36, C major. Allegri Quartet. Decca SXL 6564. (also
London STS 15303).
+Gr 6-73 p67   +NR 6-76 p10
+HFN 6-73 p1173   +RR 6-73 p65
++MJ 12-76 p44   ++St 9-76 p120

Quartet, strings, no. 2, op. 36, C major. cf Quartet, strings,
no. 1, op. 25, D major.
Scherzo. cf HMV SLS 5022.
Serenade, op. 31. cf Concerto, violin, op. 15, D minor.
Serenade, op. 31. cf Les illuminations, op. 18.
A shepherd's carol. cf Grosvenor GRS 1034.

690   Simple symphony, strings, op. 4. Young person's guide to the
orchestra, op. 34. ELGAR: Cockaigne overture, op. 40. Enigma
variations, op. 36. Pomp and circumstances march, op. 39, no.
1, D major. WALTON: Crown imperial. LSO, COA, I Musici,
Eastman Wind Ensemble; Frederick Fennell, Colin Davis, Antal
Dorati, Bernard Haitink. Philips 6780 753 (2).
+-Gr 1-76 p1244   +RR 1-76 p32
+-HFN 12-75 p171

Simple symphony, strings, op. 4. cf Vanguard SRV 344.
Sinfonia da requiem, op. 20. cf Peter Grimes: Four sea interludes
and passacaglia, op. 33 a and b.

691   Sinfonietta, op. 1. HINDEMITH: Octet. Vienna Octet, Members.
London STS 15288.
+NR 8-76 p7

692   Songs: Folk songs(4); Holy sonnets of John Donne, op. 35; Winter
words, op. 53. Raymond Gilvan, t; Frédéric Capon, pno. Oryx
1925.
+-ARG 11-76 p18

Suite, harp, op. 83. cf A birthday Hansel, op. 92.
Suite, violoncello, no. 1, op. 72. cf BACH: Suite, solo violoncello, no. 5, S 1011, C minor.
Variations and fugue on a theme by Purcell, op. 34. cf RCA GL
25006.

693   Variations on a theme by Frank Bridge, op. 10. BUTTERWORTH: The
banks of green willow. English idylls (2). A Shropshire lad.

AMF; Neville Marriner. Argo ZRG 860.
+-Gr 11-76 p776   +-RR 11-76 p52
++HFN 11-76 p157

694 War requiem, op. 66. Jeanine Altmeyer, s; Douglas Lawrence, bs-
     bar; Michael Sells, t; William Hall Chorale, Columbus Boys'
     Choir; Vienna Festival Orchestra; William Hall. Klavier KS 544
     (2).
          ++Audio 11-75 p96   +NR 8-75 p9
          -CL 4-76 p30   +St 1-76 p102
          +-Gr 4-76 p1645

695 War requiem, op. 66. Galina Vishnevskaya, s; Peter Pears, t;
     Dietrich Fischer-Dieskau, bar; Simon Preston, org; Bach
     Choir, Highgate School Choir; LSO and Chorus, Melos Ensemble;
     Benjamin Britten. London A 4255.
          ++CJ 4-76 p30

696 Winter words, op. 53. MARX: Songs: Hat dich die Liebe berüht;
     Japanisches Regenlied; Nocturne; Selige Nacht; Waldseligkeit;
     Windräder. NYSTROEM: Sjal och landskap. Dorothy Irving, s;
     Erik Werba, pno. Caprice RIKS LP 61.
          +-RR 1-76 p59

697 Young person's guide to the orchestra (Variations and fugue on a
     theme by Purcell), op. 34. PROKOFIEV: Cinderella, op. 87:
     Introduction; Quarrel; The dancing lesson; Spring fairy; Summer
     fairy; Grasshoppers dance; Winter fairy; The interrupted de-
     parture; Clock scene; Cinderella's arrival at the ball; Grande
     valse; Cinderella's waltz; Midnight; Apotheosis. LSO; Andrew
     Davis. CBS 76453. Tape (c) 40-76453. (also Columbia M 33891).
          +Gr 12-75 p1031   ++NR 4-76 p2
          +-HF 3-76 p92   /-RR 11-75 p42
          +HFN 12-75 p151   ++St 5-76 p120
          +HFN 2-76 p116 tape

698 Young person's guide to the orchestra, op. 34. PROKOFIEV: Peter
     and the wolf, op. 67. Richard Baker, narrator; NPhO; Raymond
     Leppard. Classics for Pleasure CFP 185. Tape (c) TC CFP 185.
          +Gr 1-72 p1208   +HFN 12-76 p153 tape
          ++HFN 2-72 p305

699 Young person's guide to the orchestra, op. 34. PROKOFIEV: Peter
     and the wolf, op. 67. Will Geer, narrator; ECO; Johannes Som-
     ary. Vanguard VSD 71189. (Q) VSQ 30033.
          +Audio 3-76 p64   +NR 4-75 p5
          +HF 11-75 p108   +St 8-75 p104

     Young person's guide to the orchestra, op. 34. cf Simple symphony,
     strings, op. 4.

     BROAD STRIPES/BRIGHT STARS. cf Department of Defense Bicentennial
     Edition 50-1776.

BROADBENT, Derek
     Centaur. cf Pye TB 3009.

BROCKLESS, Brian
     Songs: Christ is now rysen agayne. cf Abbey LPB 750.

BRODSZKY
     Be my love. cf DG 2530 700.

BRONSART, Ingeborg von
     Jery und Bätely: Lied and duet. cf Gemini Hall RAP 1010.

BROUGHTON
     The immortal hour: Faery song. cf Decca SKL 5208.

BROUGHTON, Bruce
     My country 'tis of thee. cf Brassworks Unlimited BULP 2.

BROUWER, Leo
     Micro-piezas (4).   cf Delos DEL FY 008.
BROWN
     Rondo, G major.   cf Columbia MS 6161.
BROWN, Earle
     Music, violoncello and piano.   cf DG 2530 562.
BROWNE, John
     Woefully array'd.   cf BASF 25 22286-1.
BROWNSON
     Salisbury.   cf Vox SVBX 5350.
BRUCH, Max
700   Concerto, violin, no. 1, op. 26, G minor.   Concerto, violin, no.
          2, op. 44, D minor.   Yehudi Menuhin, vln; LSO; Adrian Boult.
          Angel S 36920.   (also HMV ASD 2852 Tape (c) TC ASD 2852).
               ++Gr 4-73 p1871   +-NR 4-73 p6
               +Gr 7-76 p230 tape   +RR 3-73 p51
               +-HF 5-73 p76   +St 8-73 p98
               +-HFN 6-76 p105 tape
701   Concerto, violin, no. 1, op. 26, G minor.   Scottish fantasia, op.
          46.   Maurice Hasson, vln; Scottish National Orchestra; Alex-
          ander Gibson.   Classics for Pleasure CFP 40248.
               -Gr 11-76 p781   +-RR 7-76 p49
               +-HFN 8-76 p77
702   Concerto, violin, no. 1, op. 26, G minor.   Scottish fantasia, op.
          46.   Arthur Grumiaux, vln; NPhO; Heinz Wallberg.   Philips 6500
          780.   Tape (c) 7300 291
               +-Gr 5-76 p1750   ++NR 11-75 p5
               ++HF 12-75 p90   +RR 5-76 p39
               +-HFN 8-76 p97   ++St 5-76 p112
       Concerto, violin, no. 1, op. 26, G minor.   cf MENDELSSOHN: Concerto,
          violin, op. 64, E minor.
       Concerto, violin, no. 1, op. 26, G minor.   cf HMV RLS 718.
       Concerto, violin, no. 1, op. 26, G minor.   cf HMV SLS 5068.
       Concerto, violin, no. 1, op. 26, G minor.   cf RCA ARM 4-0942/7.
       Concerto, violin, no. 2, op. 44, D minor.   cf Concerto, violin,
          no. 1, op. 26, G minor.
703   Kol Nidrei, op. 47.   LALO: Concerto, violoncello, D minor.   SAINT-
          SAENS: Concerto, violoncello, no. 1, op. 33, A minor.   Pierre
          Fournier, vlc; Lamoureux Orchestra; Jean Martinon.   DG 2535 157.
          (Reissue from SLPM 138669).
               +-Gr 4-76 p1592   +-RR 4-76 p48
               +HFN 4-76 p123
704   Kol Nidrei, op. 47.   MEDINS: Suite concertante.   STRAUSS, R.:
          Sonata, violoncello and piano, op. 6.   Ingus Naruns, vlc; Ana-
          tol Berzkalns, pno.   Kaibala 60F 01.
               ++Audio 8-76 p75
705   Kol Nidrei, op. 47.   DVORAK: Concerto, violoncello, op. 104, B
          minor.   Pablo Casals, vlc; CPhO, LSO; Georg Szell, Landon Ron-
          ald.   Seraphim 60240.
               +St 9-76 p119
       Kol Nidrei, op. 47.   cf HMV HLM 7093.
706   Scottish fantasia, op. 46.   HINDEMITH: Concerto, violin, D flat
          major.   David Oistrakh, vln; LSO; Jascha Horenstein, Paul Hinde-
          mith.   Decca SDD 465.   (Reissue from SXL 6035).
               ++Gr 6-76 p43   ++RR 5-76 p39
               ++HFN 5-76 p115

BRUCH (cont.)   114

707   Scottish fantasia, op. 46. MENDELSSOHN: Concerto, violin, op.
       64, E minor. Alfredo Campoli, vln; LPO; Adrian Boult. Decca
       ECS 775. (Reissue from SXL 2026).
            +Gr 8-76 p280   +RR 6-76 p44
            +-HFN 6-76 p102
       Scottish fantasia, op. 46. cf Concerto, violin, no. 1, op. 26,
       G minor (Classics for Pleasure CFP 40248).
       Socttish fantasia, op. 46. cf Concerto, violin, no. 1, op. 26,
       G minor (Philips 6500 780).
       Scottish fantasia, op. 46. cf RCA ARM 4-0942/7.
708   Symphony, no. 2, op. 36, F minor. RIETZ: Concert overture, op. 7.
       Louisville Orchestra; Jorge Mester. RCA GL 25017.
            +Gr 10-76 p590   /RR 12-76 p53
            +-HFN 12-76 p139
709   Trios, op. 83, nos. 2, 6, 7. MOZART: Trio, clarinet, K 498, E
       flat major. SCHUMANN: Märchenerzählungen, op. 132. Boris
       Kroyt, vla; Harold Wright, clt; Murray Perahia, pno. Turnabout
       TVS 34615.
            +-HF 11-76 p139
BRUCKNER, Anton
710   Intermezzo, string quintet, D minor. Quintet, strings, F major.
       Vienna Philharmonia Quintet. Decca SDD 490.
            +Gr 9-76 p436   +RR 8-76 p61
            +HFN 8-76 p77
711   Mass, no. 1, D minor. Edith Mathis, s; Marga Schiml, con; Martin
       Ochman, t; Karl Ridderbusch, bs; Bavarian Radio Symphony Orch-
       estra and Chorus; Eugen Jochum. DG 2530 314. (Reissue from
       2720 054).
            +Gr 7-73 p241   ++RR 7-73 p74
            ++NR 6-75 p10   ++SFC 9-14-75 p28
            +NYT 9-21-75 pD18   ++St 4-76 p112
712   Mass, no. 1, D minor. Pieces, orchestra (4). Barbara Yates, s;
       Sylvia Swan, con; John Steel, t; Colin Wheatley, bs; Alexandra
       Choir; LPO; Hans-Hubert Schönzeler. Unicorn UNS 210.
            +NR 8-75 p8   +-St 4-76 p112
713   Mass, no. 2, E minor. John Alldis Choir; ECO; Daniel Barenboim.
       HMV ASD 3079. (also Angel S 37112).
            +-Gr 10-75 p669   +NR 8-76 p10
            +-HF 8-76 p83   +RR 9-75 p67
            +-HFN 10-75 p140
714   Mass, no. 2, E minor. Gächinger Kantorei, Spandauer Kantorei;
       Bach Collegium Wind Ensemble; Helmuth Rilling. Musical Heritage
       Society MHS 1801. (also Oryx 3C 320).
            /HF 12-74 p95   +St 4-76 p102
            +St 10-74 p125
       Pieces, orchestra (4). cf Mass, no. 1, D minor.
       Quadrilles. cf Sterling 1001/2.
       Quintet, strings, F major. cf Intermezzo, string quintet, D minor.
715   Requiem, D minor. Herrad Wehrung, s; Hildegard Laurich, ms;
       Friedreich Melzer, t; Günter Reich, bs; Laubacher Kantorei;
       Werner Keltsch Instrumental Ensemble; Hans Michael Beuerle.
       Nonesuch H 71327.
            -ARG 11-76 p19   +NR 9-76 p8
            +HF 11-76 p106   +SFC 12-5-76 p58
716   Symphonies, nos. 1-9. Te Deum. BPhO; Bavarian Radio Symphony
       Orchestra; Eugen Jochum. DG 2740 136 (11).
            +HFN 4-76 p121   +-RR 4-76 p38

717   Symphony, no. 2, C minor. VSO; Carlo Maria Giulini. HMV (Q) ASD
       3146.
              ++Gr 12-75 p1032   -RR 12-75 p45
              +-HFN 12-75 p151   +STL 3-7-76 p37
              +-MT 9-76 p744
718   Symphony, no. 4, E flat major. LSO; István Kertész. Decca SDD
       464. (Reissue from SXL 6227). (also London STS 15289)
              +Gr 4-76 p1592   +-RR 3-76 p38
              +HFN 3-76 p109
719   Symphony, no. 4, E flat major. VPO; Karl Böhm. DG Tape (c) KBB2
       7039 (2).
              +HFN 5-76 p117 tape   +RR 5-76 p77
720   Symphony, no. 4, E flat major. BPhO; Herbert von Karajan. DG
       2530 674.
              +Gr 10-76 p590   ++RR 10-76 p46
              +-HFN 10-76 p166   ++STL 10-10-76 p37
721   Symphony, no. 4, E flat major. VPO; Karl Böhm. London CSA 2240 (2).
              ++HF 6-76 p77   ++SFC 2-29-76 p25
              +-NR 5-76 p3
722   Symphony, no. 6, A major. VPO; Horst Stein. Decca SXL 6682.
       (also London 6880).
              +-Gr 4-75 p1802   ++NR 2-76 p2
              +HF 2-76 p94   +-RR 4-75 p26
              +MJ 2-76 p47
723   Symphony, no. 6, A major. BSO; William Steinberg. RCA GL 25009.
              -Gr 11-76 p781   +-RR 10-76 p48
              +-HFN 12-76 p140
724   Symphony, no. 7, E major (orig. version). BPhO; Wilhelm Furt-
       wängler. Bruno Walter Society RR 416.
              +NR 10-76 p2
725   Symphony, no. 7, E major (orig. version). BPhO; Wilhelm Furt-
       wängler. DG 2535 161.
              +-Gr 5-76 p1772   +RR 5-76 p22
726   Symphony, no. 7, E major. BPhO; Jascha Horenstein. Unicorn UN
       111. (Reissue from Polydor 66802/8).
              +-Gr 12-76 p990
       Symphony, no. 7, E major. cf BRAHMS: Symphony, no. 1, op. 68,
       C minor.
727   Symphony, no. 8, C minor. BPhO; Herbert von Karajan. DG 2707 085
       (2).
              +Gr 5-76 p1750   ++SFC 12-26-76 p34
              +HFN 8-76 p77   +STL 7-4-76 p36
              ++RR 8-76 p29
728   Symphony, no. 8, C minor. BPhO; Herbert von Karajan. HMV SXDW
       3024 (2). (Reissue from Columbia 33 CX 1586/7).
              +Gr 5-76 p1750   +RR 8-76 p30
729   Symphony, no. 9, D minor. CSO; Daniel Barenboim. DG 2530 639.
              +Gr 4-76 p1592   +-NR 8-76 p2
              +-HF 8-76 p83   +RR 4-76 p39
              +HFN 4-76 p101   +St 12-76 p135
       Te Deum. cf Symphonies, nos. 1-9.
BRULE, Gace
       Biaus m'est estez. cf Telefunken 6-41275.
BRUMEL, Antoine
       Missa et ecce terrae motus: Gloria. HMV SLS 5049.
       Songs: Du tout plongiet; Fors seulement, l'attente. cf HMV SLS 5049.

BRUMEL (cont.)   116

        Vray dieu d'amours. cf HMV SLS 5049.
BRUNEAU, Alfred
        L'attaque du Moulin: Adieu forêt profonde. cf Club 99-98.
BRYARS, Gavin
730   Jesus' blood never failed me yet. The sinking of the Titanic.
        Cockpit Ensemble and other musicians; San Francisco Conserva-
        tory of Music New Music Ensemble; Gavin Gryars, John Adams.
        Obscure no. 1.
            +RR 3-76 p18
        1, 2, 1-2-3-4. cf ADAMS: American standard.
        The sinking of the Titanic. cf Jesus' blood never failed me yet.
BUCHT, Gunnar
731   Dramma per musica. Hund skenar glad. Symphony, no. 7. Dorothy
        Dorow, s; Swedish Radio Women's Chorus; Swedish Radio Symphony
        Orchestra, Nörrkoping Symphony Orchestra; Stig Westerberg, Okko
        Kamu. Caprice CAP 1076. (also RIKS 75).
            ++HF 8-76 p86   +RR 10-76 p48
            +MT 10-76 p831   *Te 6-76 p50
        Hund skenar glad. cf Dramma per musica.
        Symphony, no. 7. cf Dramma per musica.
BUCK, Dudley
        Concert variations on "The star-spangled banner". cf Columbia
        M 34129.
BULL, John
        English toy. cf BASF BAC 3075.
        Fantasia, D minor. cf BASF BAC 3075.
        Fantasia, D minor. cf Hungaroton SLPX 11741.
        The herd girl's Sunday. cf Swedish Society SLT 33229.
        The King's hunt. cf BASF BAC 3075.
        The King's hunt. cf Saga 5402.
        Prelude and In nomine. cf Turnabout TV 34017.
        Revenant. cf BACH: Chorale prelude: O Gott, du frommer Gott,
        S 767.
        Variations on the Dutch chorale "Laet ons met herten reijne".
        cf Philips 6500 926.
BURGE, David
        Sources IV. cf CRI SD 345.
BURGHAUSER
        Czech dances (3). cf BARTA: Sonata, guitar.
        Sonata, guitar, E minor. cf BARTA: Sonata, guitar.
BURIAN, Karel
        American suite. cf Supraphon 111 1721/2.
BURKHARD
        Serenade, op. 71, no. 3. cf Claves LP 30-406.
BURKHART
        Tre advetntssanger. cf BIS LP 2.
BURLEIGH, Cecil
        Little mother of mine. cf Pye Ember GVC 51.
BURTON, Eldin
732   Sonatina, flute and piano. COPLAND: Duo, flute and piano.
        PISTON: Sonata, flute and piano. VAN VACTOR: Sonatina, flute
        and piano. Keith Bryan, flt; Karen Keys, pno. Orion ORS
        76242.
            ++NR 10-76 p6
BURTON, John
        Sonata, piano, no. 1, D major. cf BENNETT, W.: January, op. 36,
        no. 1.

BUSNOIS, Antoine
    Songs: Fortuna desperata. cf HMV SLS 5049.
BUSONI, Ferruccio
    All Italia. cf BEETHOVEN: Sonata, piano, no. 27, op. 90, E minor.
733   Berceuse elégiaque, op. 42. DALLAPICCOLA: Piccola musica nottur-
        na. Preghiere. Sex carmina alcaei. WOLPE: Piece in two parts,
        6 players. Henry Datyner, vln; Colin Bradbury, clt; David
        Mason, tpt; Charles Tunnell, vlc; Michael Jeffries, hp; Kather-
        ine Wolpe, pno; Heather Harper, s; Barry McDaniel, bar; NPhO,
        ECO; Frederick Prausnitz. Argo ZRG 757. (Reissue from HMV
        ASD 2388).
                +Gr 5-76 p1771   +NR 6-76 p14
                +HFN 4-76 p123   +RR 12-75 p45
                +MT 7-76 p580
734   Concerto, violin, op. 35a, D major. Sonata, violin and piano,
        no. 2, op. 36a, E minor. Joseph Szigeti, vln; Clara Haskil,
        pno; RAI Orchestra; Fernando Previtali. Bruno Walter Society
        WSA 700.
                ++NR 11-76 p6
735   Duettino concertante. Fantasia contrappuntistica. Fantasia per
        un Orgelwalze. Improvisation on Bach's chorale "Wie wohl ist
        mir". Gino Giorini, Sergio Lorenzi, pno. Harmonia Mundi HM
        314.
                +-HFN 5-76 p97
    Indian diary. cf BEETHOVEN: Sonata, piano, no. 27, op. 90, E
        minor.
    Indian diary, Bk 1. cf International Piano Library IPL 104.
    Fantasia contrappuntistica. cf Duettino concertante.
    Fantasia per un Orgelwalze. cf Duettino concertante.
    Improvisation on Bach's chorale "Wie wohl ist mir". cf Duettino
        concertante.
    Perpetuum mobile. cf BEETHOVEN: Sonata, piano, no. 27, op. 90,
        E minor.
736   Sonata, violin and piano, no. 2, op. 36a, E minor. PADEREWSKI:
        Sonata, violin and piano, op. 13, A minor. Endre Granat, vln;
        Harold Gray, pno. Desmar DSM 1004.
                +-Gr 2-76 p1355   ++NR 12-75 p9
                +HF 6-76 p78   +RR 1-76 p51
                +-HFN 2-76 p93   +St 1-76 p106
                ++MJ 12-75 p38
    Sonata, violin and piano, no. 2, op. 36a, E minor. cf Concerto,
        violin, op. 35a, D major.
    Sonata, violin and piano, no. 2, op. 36a, E minor. cf BACH: Son-
        ata, violin and harpsichord, no. 6, S 1019, G major.
    Sonatina, no. 2. cf International Piano Library IPL 102.
BUSSOTTI, Sylvano
    Ultima rara. cf DG 2530 561.
BUTTERFIELD
    When you and I were young, Maggie. cf Pye Ember GVC 51.
BUTTERLY, Nigel
    The white throated warbler. cf HMV SLS 5022.
BUTTERWORTH, George
737   The banks of green willow. English idylls (2). A Shropshire
        lad. HOWELLS: Elegy, op. 15. Merry eye, op. 20b. Music for
        a prince. Herbert Downes, vla; Desmond Bradley, Gillian East-
        wood, vln; Albert Cayzer, vla; Norman Jones, vlc; NPhO, LPO;

Adrian Boult. Lyrita SRCS 69.
    +Gr 1-76 p1211   +RR 12-75 p50
    ++HFN 11-75 p153
The banks of green willow. cf BRITTEN: Variations on a theme
    by Frank Bridge, op. 10
English idylls (2). cf BRITTEN: Variations on a theme by Frank
    Birdge, op. 10.
English idylls. cf The banks of green willow.
738 A Shropshire lad. SOMERVELL: Maud. John Carol Case, bar; Daphne
     Ibbott, pno. Pearl SHE 527.
         ++Gr 5-76 p1788   +STL 4-11-76 p36
         +RR 5-76 p68
739 A Shropshire lad. GURNEY: Songs: An epitaph; Black Stichel; De-
     sire in spring; Down by the salley gardens. IRELAND: Songs:
     The land of lost content. WARLOCK: Songs: Pretty ring-time;
     My own country; Passing by; A prayer to St. Anthony of Padua;
     The sick heart. Anthony Rolfe Johnson, t; David Willison, pno.
     Polydor 2460 258.
         +-Gr 1-76 p1229   +-RR 5-76 p67
         +HFN 4-76 p101
A Shropshire lad. cf The banks of green willow.
A Shropshire lad. cf BRITTEN: Variations on a theme by Frank
    Bridge, op. 10.
740 Songs: Bredon Hill: O fair enough are sky and plain, On the idle
     hill of summer, When the lad for longing sighs, With rue my
     heart is laden; A Shropshire lad. FINZI: Earth and air and rain,
     op. 15. Benjamin Luxon, bar; David Willison, pno. Argo ZRG
     838.
         ++Gr 4-76 p1650   +MT 8-76 p658
         +HFN 4-76 p103   ++RR 4-76 p68
BUXTEHUDE, Dietrich
     Chaconne, E major. cf Organ works (Titanic TI 11).
     Chaconne, E minor. cf BOHM: Ach wie nichtig, ach wie flüchtig,
        chorale partita.
741 Chorale preludes (6). Preludes and fugues, G major, G minor, E
     minor, C major. Toccatas, D minor, G major. Michel Chapuis,
     org. Telefunken AF 6-42001.
         +HFN 10-76 p167   +-RR 10-76 p86
     Chorale preludes: Nun bitten wir den heiligen Gott; Jesus Christus,
        unser Heiland. cf Telefunken AF 6-41872.
     Chorale prelude: Nun komm, der Heiden Heiland. cf Audio EAS 16.
     Chorale prelude: Wie schön leuchtet der Morgenstern. cf BACH:
        Choral prelude: O Gott, du frommer Gott, S 767.
     Est ist das Heil uns zu kommen. cf Organ works (Titanic TI 11).
     Fanfare and chorus. cf Unicorn RHS 339.
     Fugue, C major. cf Telefunken AF 6-41872.
     Fugue à la gigue, C major. cf Westminster WGS 8116.
     Ich ruf zu dir. cf Organ works (Titanic TI 11).
     Jubilate Domino. cf Telefunken 6-41929.
742 Das Jüngste Gericht (The last judgment). Annemarie Grünewald,
     Margarethe Lerche, Ingrid Rattunde-Würtz, s; Sabine Kirchner,
     alto; Raimund Gilvan, t; Traugott Schmohl, bs; Mannheim Bach
     Choir; Heidelberg Chamber Orchestra; Heinz Markus Göttsche.
     Musical Heritage Society MHS 1579/90 (2). (also Oryx 1702/3)
         +St 1-74 p107   +St 4-76 p112
     Lobt Gott, ihr Christen. cf Organ works (Titanic TI 11).

743 Organ works: Chaconne, E major. Est ist das Heil uns zu kommen.
        Lobt Gott, ihr Christen. Ich ruf zu dir. Preludes and fugues,
        F major, E major, G minor. Toccata, F major. Mireille Lagacé,
        org. Titanic TI 11.
            ++NR 11-76 p11
     Passacaglia, D minor. cf Argo ZRG 806.
     Prelude on "Ein feste Burg". cf Grosvenor GRS 1039.
     Prelude on the Te Deum. cf Grosvenor GRS 1039.
     Preludes and fugues, G major, G minor, E minor, C major. cf
        Chorale preludes (6).
     Preludes and fugues, F major, E major, G minor. cf Organ works
        (Titanic TI 11).
     Prelude and fugue, D minor. cf London STS 15222.
     Prelude, fugue and chaconne, C major. cf BOHM: Ach wie nichtig,
        ach wie flüchtig, chorale partita.
     Prelude, fugue and chaconne, C major. cf Argo ZRG 806.
     Toccatas, D minor, G major. cf Chorale preludes (6).
     Toccata, F major. cf Organ works (Titanic TI 11).
     Toccata, F major. cf Telefunken AF 6-41872.
     Toccata and fugue, F major. cf Columbia M 33514.
     Trauermusic. cf BACH: Chorale prelude: O Gott, du frommer Gott,
        S 767.
BYRD, William
     Earle of Oxford's march. cf Argo SPA 464.
     Earle of Oxford's march. cf Argo ZRG 823.
     The Earl of Salisbury pavan. cf Decca PFS 4351.
     Elegy on the death of Thomas Tallis. cf Turnabout TV 34017.
     Exsurge Domine. cf Argo ZRG 789.
744 Fantasias, nos. 2 and 3. Miserere. Pavan and galliard. LAWES:
        Sonata, no. 7, D minor. Suite, no. 2, F major. Suite, no. 3,
        B flat major, excerpts. PURCELL: Fantasia. Ground, D minor.
        A new ground, E minor. Overtures, D minor, G minor. Overture
        and suite, G major. Parts on a ground, D major (3). Pavans,
        B flat major, A minor. Pavan of four parts, G minor. Sefauchi's
        farewell, D minor. Sonata, A minor. Suite, D major. TOMKINS:
        A sad pavan for these distracted times. Leonhardt Consort;
        Gustav Leonhardt. Telefunken 6-35286 (2).
            ++HFN 5-76 p99   +RR 5-76 p61
            ++NR 8-76 p7
     Galliard (after Francis Tregian). cf Decca PFS 4351.
     Haec dies. cf HMV SLS 5047.
     Haec dies a 5. cf Harmonia Mundi HMU 473.
     Haec dies a 6. cf Vista VPS 1023.
     In nomine a 5. cf Turnabout TV 34017.
     The leaves be green. cf HMV SLS 5022.
     Lord Salisbury's pavan. cf Turnabout TV 34017.
     Miserere. cf Fantasias, nos. 2 and 3.
     O Lord, how vain. cf Turnabout TV 34017.
     Pavan and galliard. cf Fantasias, nos. 2 and 3.
     Pavan and galliard of Mr. Peter. cf BASF BAC 3075.
745 Songs (Psalms and anthems): Ave verum corpus; Magnificat and nunc
        dimittis (The great service); Prevent us O Lord; Praise our
        Lord, all ye Gentiles; Psalm, no. 47, O clap your hands; Psalm,
        no. 54, Save me, O God; Psalm, no. 55, Hear my prayer; Psalm,
        no. 114, When Israel came out of Egypt; Psalm, no. 119, Teach
        me, O Lord; Bow thine ear; Sing joyfully. New College Choir,

Oxford; Patrick Russill, Paul Trepte, org; David Lumsden.
Abbey LPB 751.
    +-Gr 11-76 p851   ++RR 10-76 p96
    +HFN 11-76 p157
Songs: Make ye joy to God; Justorum animae. cf Abbey LPB 750.
Songs: Constant Penelope; What pleasures have great princes. cf
    Prelude PRS 2501.
Ut, Re. cf Argo ZRG 806.
Ut, Re mee fa sol la. cf Abbey LPB 752.
Walsingham variations. cf BASF BAC 3075.
Wolsey's wilde. cf Saga 5402.
The woods so wild. cf RCA RK 11708.
CABALLERO
    Chateau Margaux: Romanza de Angelita. cf London OS 26435.
    Gigantes y cabezudos: Romanza de Pilar. cf Decca SXLR 6792.
    El señor Joaquin: Balada y alborada. cf London OS 26435.
CABANILLES, Juan
    Passacalles du 1er mode. cf Argo ZRG 806.
CABEZON, Antonio de
    Diferencias sobre "La dama le demanda". cf Nonesuch H 71326.
    La gamba, pavan. cf Hungaroton SLPX 11669/70.
CABLE, Howard
    Newfoundland rhapsody. cf Citadel CT 6007.
CACAVAS
    Burnished brass. cf Canon VAR 5968.
CACCINI, Francesca
    La liberazione di Ruggiero dall'Isola d'Alcina, excerpts. cf
    Gemini Hall RAP 1010.
CACCINI, Giulio
    Songs: Amarilli mia bella; Belle rose porporine; Perfidissimo
    volto; Udite amante. cf DG Archive 2533 305.
CADMAN, Charles
    At dawning. cf Pye Ember GVC 51.
CAGE, John
746 Concerto, prepared piano and chamber orchestra. FOSS: Baroque
    variations. Yuji Takahashi, prepared pno; Buffalo Philharmonic
    Orchestra; Lukas Foss. Nonesuch H 71202. Tape (c) N5 71202.
        +St 7-76 p74
747 Experiences, nos. 1 and 2. Forever and sunsmell. In a landscape.
    The wonderful widow of 18 springs. STEELE: All day. Distant
    saxophones. Rhapsody spaniel. Jan Steele, Janet Sherbourne,
    Richard Bernas, pno; Robert Wyatt, Carla Bley, voice; Richard
    Bernas; Various musicians. Obscure no. 5.
        +-RR 12-76 p95
    Forever and sunsmell. cf Experiences, nos. 1 and 2.
    In a landscape. cf Experiences, nos. 1 and 2.
    Nocturne. cf Desto DC 6435/7.
748 Sonatas and interludes, prepared piano. John Tilbury, prepared
    piano. Decca HEAD 9.
        +Gr 11-76 p838   +RR 7-76 p70
        +HFN 7-76 p85
    The wonderful widow of 18 springs. cf Experiences, nos. 1 and 2.
CAHUZAC, Louis
    Cantilene. cf Grenadilla GS 1006.
CAIMO, Giuseppe
    Mentre il cuculo. cf HMV CSD 3756.

CAIX d'HERVELOIS, Louis de
    Suite, A major. cf BIS LP 22.
CALDARA, Antonio
    La rosa. cf Westminster WGS 8268.
CALESTANI, Vincenzo
    Songs: Damigella tutta bella. cf DG Archive 2533 305.
CALLEJA
    Canción triste. cf London CS 7015.
CALVI, Carlo
    The Medici court. cf Saga 5420.
CALVIERE
    Pièce. cf Calliope CAL 1917.
CAMBRAI (13th century)
    Retrowange novelle. cf Telefunken 6-41275.
CAMIDGE, Matthew
    Concerto, organ, op. 13, no. 3, A minor: Gavotte. cf Wealden
        WS 121.
CAMPBELL
    Capital City suite: River by night and confusion square. cf
        Citadel CT 6007.
CAMPBELL, Sidney
    Jubilate Deo. cf Argo ZRG 789.
CAMPIAN, Thomas
    Never weather-beaten sail. cf Philips 6500 926.
    Pieces, D major. cf d'ANGLETERRA: Carillon, G major.
    Songs: It fell on a summer's day; The cypress curtain of the
        night; Shall I come, sweet love, to thee. cf Philips 6500 282.
CAMPRA, André
    Rigaudon, A major. cf Columbia M 33514.
CAMSEY
    Songs: Benediction; Sing hallelujah, shout hallelujah. cf HMV
        SXLP 50017.
CANBY, Edward
    The interminable farewell. cf CRI SD 102.
CANN, Richard
    Bonnylee. cf Odyssey Y 34139.
CANTELOUBE, Joseph
749   Chants d'Auvergne: La pastoura als camps; L'antouèno; La past-
        rouletta è lou chibalié; Lo calhé; Lou boussu; Malurous qu'o
        uno fenno; Oï ayaï; Pour l'enfant; Pastorale; Lou coucut; Obal,
        din lo coumbelo; La haut, sur le rocher; Hé, beyla-z-y dáu
        fè; Tè, l'co, tè; Uno jionto postouro. Victoria de los Angeles,
        s; Lamoureux Orchestra; Jean-Pierre Jacquillat. HMV (Q) ASD
        3134. Tape (c) TC ASD 3134. (also Angel S 36898).
                -Gr 11-75 p875   +-ON 5-76 p48
                -HF 9-76 p89   +-RR 12-75 p93
                +HFN 12-75 p151   +St 3-76 p109
                +NR 12-75 p12
CAPEL
    Love, could I only tell thee. cf Pearl SHE 528.
CARA, Marchetto
    Non e tempo. cf Nonesuch H 71326.
CARBONETTI
    Tu non mi vuoi piu ben. cf Everest SDBR 3382.
CARDILLO
    Core 'ngrato. cf DG 2530 700.

CARISSIMI, Giacomo
750   Choral works: Mass for 8 voices. The story of Abraham and Isaac.
       The story of Hezekiah. Tolle sponsa. Karine Rosat, Jennifer
       Smith, s; Hanna Schaer, con; John Elwes, Fernando Serafim, t;
       Philippe Huttenlocher, bar; Michel Brodard, bs; Gulbenkian
       Foundation Symphonic Choir, Chamber Choir and Instrumental
       Ensemble; Michel Corbőz. Erato STU 70762. (also Musical
       Heritage Society MHS 3091).
               +-Gr 7-73 p220   +St 1-76 p103
               +RR 6-73 p86
751   Historia di Jonas. CAVALLI: Missa pro defunctis. Louis Halsey
       Singers; Louis Halsey. L'Oiseau-Lyre SOL 347.
               ++HFN 12-76 p140   +-RR 12-76 p90
       Mass, 8 voices. cf Choral works (Erato STU 70762).
       The story of Abraham and Isaac. cf Choral works (Erato STU 70762).
       The story of Hezekiah. cf Choral works (Erato STU 70762).
       Tolle sponsa. cf Choral works (Erato STU 70762).
CARLSTEDT, Jan
752   Sinfonietta, 5 wind instruments. HOLMBOE: Notturno, op. 19.
       MORTENSEN: Quintet, winds, op. 4. SALMENHAARA: Quintet, winds.
       Göteborg Wind Quintet. BIS LP 24.
               ++RR 12-76 p72
       CARMINA BURANA. cf Harmonia Mundi HMU 337.
       CARMINA BURANA: Songs of drinking and eating; Songs of unhappy
       love. cf Harmonia Mundi HMU 335.
CAROUBEL (16th century Spain)
       Courante (2). cf DG Archive 2533 184.
       Volte (2). cf DG Archive 2533 184.
CARPENTER
       Jazz boys; Crying' blues. cf Club 99-101.
       The player queen. cf Washington University Press OLY 104.
CARPENTER, John Alden
753   Adventures in a perambulator. PHILLIPS: Selections from McGuffy's
       readers. Eastman-Rochester Symphony Orchestra; Howard Hanson.
       ERA 1009.
               *NYT 7-4-76 pD1
CARR, Gordon
       Prism for brass. cf Unicorn RHS 339.
CARTER, Elliot
754   Concerto, harpsichord, piano and 2 chamber orchestras. Duo,
       violin and piano. Paul Jacobs, hpd; Gilbert Kalish, pno;
       Contemporary Chamber Ensemble; Arthur Weisberg. Nonesuch H
       71314.
               ++Gr 4-76 p1597   +NR 12-75 p7
               +HF 3-76 p82   ++St 8-76 p92
               +MM 7-76 p32   +Te 3-76 p29
               ++MT 9-76 p745
       Duo, violin and piano. cf Concerto, harpsichord, piano and 2
       chamber orchestra.
755   Eight pieces, 4 timpani. Fantasy about Purcell's Fantasia upon
       one note. Quintet, brass. American Brass Quintet; Morris Lang,
       timpani. Odyssey Y 34137.
               +SR 11-13-76 p52
756   Etudes and a fantasy, woodwind quartet (8). PORTER: Quartet,
       strings, no. 8. Stanley Quartet, University of Michigan;
       New York Woodwind Quintet, Members. CRI SD 118.
               +-NR 7-76 p6   *RR 2-76 p46

Fantasy about Purcell's Fantasia upon one note. cf Eight pieces,
    4 timpani.
757 Quartets, strings, nos. 1 and 2. Composers Quartet. Nonesuch H
    71249.
            +ARG 7-71 p756   +NR 1-71 p6
            +Gr 3-72 p1545   +NYT 2-7-71 pD27
            ++HF 2-71 p76   ++SFC 11-29-70 p34
            ++HF 4-71 p64   +St 2-71 p89
            ++HFN 3-72 p501   -St 7-76 p74
            ++MJ 3-71 p77
    Quintet, brass. cf Eight pieces, 4 timpani.
CARULLI, Fernando
    Serenade, C major. cf Claves LP 30-406.
CASALS, Pablo
    Nigra sum, sed formosa. cf Abbey LPB 754.
CASELLA, Alfredo
    Contrasts, op. 31 (2). cf International Piano Library IPL 102.
CASERTA, Antonellus de
    Amour m'a le cuer mis. cf HMV SLS 863.
CASSADO, Gaspar
    Requiebros. cf Laurel-Protone LP 13.
CASTELNUOVO-TEDESCO, Mario
758 Concertino, harp, op. 93. RODRIGO: Fantasia para un gentilhombre.
    VILLA-LOBOS: Concerto, harp. Catherine Michel, hp; Monte Carlo
    Opera Orchestra; Antonio de Almeida. Philips 6500 812.
            +Gr 7-75 p175   +NR 3-76 p3
            +HFN 7-75 p79   +RR 6-75 p55
            +MJ 7-76 p57   ++St 6-76 p114
759 Concertino, harp, op. 93. KOUTZEN: Sonata, solo violin. NELHYBEL:
    Arco and pizzicato. Miniatures, strings (3). WEIGL: Songs,
    contralto and string quartet (3). Pearl Chertok, hp; Maureen
    Forrester, con; Aaron Rosand, vln; Orchestra da Camera Romana,
    Phoenix String Trio; Nicolas Flagello. Serenus SRS 12062.
            ++NR 4-76 p15
    Coplas. cf GOLD: Songs of love and parting.
    Etudes d'ondes: Sea murmurs. cf CBS 76420.
    Etudes d'ondes: Sea murmurs. cf Columbia M2 33444.
    Etudes d'ondes: Sea murmurs (2). cf RCA ARM 4-0942/7.
    La guarda cuydadosa. cf DG 2530 561.
760 Sonata, violoncello and harp. MONDELLO: Poem, flute and harp.
    MOURANT: Elegy, flute and harp. RIBARI: Pezzi, violoncello
    and harp (4). Pearl Chertok, hp; Nathan Stutch, vlc; Harold
    Jones, flt. Orion ORS 76227.
            +NR 7-76 p5
    Sonatina, op. 205. cf BIS LP 30.
    Tango. cf RCA ARM 4-0942/7.
    Tarantella. cf DG 2530 561.
CASTRO, Jean de
    Enfans a laborder. cf Hungaroton SLPX 11669/70.
CATALANI, Alfredo
    Loreley: Non fui da um padre mai bendetta, Dove son, d'onde vengo
    ...O forze recondite. cf Court Opera Classics CO 347.
761 La Wally: Ebben, Ne andrò lontana. DONIZETTI: Catarina Cornaro:
    Torna all'ospitetto...Vieni o tu, che ognor io chiamo. Luc-
    rezia Borgia: Com'e bello. Maria Stuarda: O nube, che lieve.
    Roberto Devereux: E Sara in questi orribili momenti...Vivi,

ingrato, a lei d'accanto...Quel sangue versato. VERDI: La
forza del destino: Pace, pace, mio Dio. Aida: O cieli azzuri.
La traviata: Addio del passato. Il trovatore: Timor di me...
D'amor sull'ali rosee. Leyla Gencer, s; RAI Torino Symphony
Orchestra, Turin Symphony Orchestra; Arturo Basile, Gianandrea
Gavazzeni. Cetra LPL 69001.
    +-HF 9-76 p105
762 La Wally: Ebben, Ne andrò lontana. CILEA: Adriana Lecouvreur: Io
     son l'amile ancella; Poveri fiori. MASCAGNI: Lodoletta: Flammen,
     perdonami. Iris: Un dì (ero piccina) al tempio. PUCCINI: La
     bohème: Quando m'en vo' soletta. Gianni Schicchi: O mio babbino
     caro. Manon Lescaut: In quelle trine morbide; Sola, perduta,
     abbandonata. La rondine: Ch' il bel sogno di Doretta. Suor
     Angelica: Senza mamma, o bimbo. Le Villi: Non ti scordar di
     me. Renata Scotto, s; LSO; Gianandrea Gavazzeni. Columbia M
     33435. (also CBS 76407 Tape (c) 40-76407).
          +Gr 11-75 p904   +NYT 7-6-75 pD11
          +HF 9-75 p101   +-ON 8-75 p23
          +HFN 12-75 p169   +RR 11-75 p34
          +HFN 3-76 p113 tape   ++SFC 8-3-75 p30
          +-NR 8-75 p9   +St 9-75 p116
     La Wally: Ebben, Ne andrò lontana. cf VERDI: Il trovatore.
     La Wally: Ebben, Ne andrò lontana. cf Court Opera Classics CO 347.
CATELINET
     Jolson memories. cf Pye TB 3006.
     The star. cf Pye TB 3006.
CATO, Diomedes
     Fantasia. cf Hungaroton SLPX 11721.
     Favorito. cf Hungaroton SLPX 11721.
     Villanella. cf Hungaroton SLPX 11721.
CAURROY, Eustache de
     Fantasia; Prince la France te veut. cf L'Oiseau-Lyre 12BB 203/6.
CAVALIERI, Emilio de
     O che nuovo. cf Hungaroton SLPX 11669/70.
CAVALLI, Pietro
763 Messa concertata. Munich Vocal Soloists; Karl Heinz Klein, hpd;
     Franz Lehrndorfer, org; Bavarian State Orchestra Chamber En-
     semble; Hans Ludwig Hirsch. Telefunken AW 6-41931.
          /Gr 8-76 p324   +-RR 7-76 p74
          +-HFN 8-76 p79
     Missa pro defunctis. cf CARISSIMI: Historia di Jonas.
CAVENDISH, Michael
     Sly thief, if so you will believe. cf Harmonia Mundi 593.
     Wand'ring in this place. cf L'Oiseau-Lyre 12BB 203/6.
CEELY, Robert
764. Elegia. Mitsyn music. DEL MONACO: Electronic study, no. 2.
     Metagrama. RANDALL: Music for the film 'Eakins'. Princeton
     University Computer Center; Milan, Studio di Fonologia; Boston
     Experimental Electronic-music Projects; Columbia-Princeton
     Music Center. CRI SD 328.
          +-HFN 10-75 p140   +NR 2-75 p15
          +MQ 1-76 p152   +-RR 7-75 p60
     Mitsyn music. cf Elegia.
CENNICK ·
     Happy in the Lord. cf Columbia M 33838.
     CENTRAL EUROPEAN LUTE MUSIC, 16th AND 17th CENTURIES. cf Hungaro-
     ton SLPX 11721.

CEREMUGA, Joseph
    Lasské pastorale. cf Panton 110 440.
CERTON, Pierre
    Lá, lá, la je ne l'oise dire. cf HMV CSD 3740.
    Psalms and nunc Dimittis (3). cf BIS LP 2.
    Que n'est elle aupres de moi. cf Argo ZRG 667.
CESARIS, Johannes
    Bonté biauté. cf Telefunken ER 6-35257.
CHABRIER, Emanuel
    Bourrée fantasque. cf Works, selections (Mercury SRI 75078).
    Bourrée fantasque. cf International Piano Library IPL 102.
    Danse slave. cf Works, selections (Mercury SRI 75078).
765 España. DEBUSSY: Prelude à l'aprés-midi d'un faune. IBERT:
    Escales. RAVEL: Daphnis et Chloé: Suite, no. 2. Orchestra
    de Paris; Daniel Barenboim. CBS 76523. (also Columbia M
    34500 Tape (c) MT 34500).
        +Gr 10-76 p615   +-RR 11-76 p53
        +HFN 10-76 p167
766 España. PROKOFIEV: Symphony, no. 1, op. 25, D major.* RAVEL:
    Boléro. STRAVINSKY: Circus polka. Fireworks, op. 4.* NPhO;
    Rafael Frühbeck de Burgos. HMV ESD 7019. Tape (c) TC ESD
    7019. (*Reissues from ASD 2315, Columbia TWO 239).
        -Gr 12-76 p1071   +-RR 12-76 p53
        +HFN 12-76 p152
767 España. DUKAS: L'Apprenti sorcier. FALLA: El amor brujo: Dances.
    RAVEL: Boléro. Hermann Scherchen, pno; Vienna State Opera
    Orchestra. Westminster Tape (c) GRT 8131.
        +HF 7-76 p114 tape
    España. cf Works, selections (Mercury SRI 75078).
    España. cf BACH: Toccata and fugue, S 565, D minor.
    España. cf Decca DPA 519/20.
    España: Rhapsody. cf Pye QS PCNH 4.
    Gwendoline overture. cf Works, selections (Mercury SRI 75078).
    Marche joyeuse. cf Works, selections (Mercury SRI 75078).
    Marche joyeuse. cf Classics for Pleasure CFP 40254.
    Marche joyeuse. cf HMV SLS 5019.
768 Le Roi malgrè lui: Fête polonaise. LALO: Rapsodie norvégienne,
    op. 21. MASSENET: Scènes hongroises. Luxembourg Radio Orch-
    estra; Pierre Cao. Turnabout (Q) QTV 34570.
        +Audio 2-76 p95   +-NR 3-75 p3
    Le Roi malgrè lui: Fête polonaise. cf Works, selections (Mercury
    SRI 75078).
769 Trois valse romantiques. SAINT-SAENS: The carnival of the animals.
    SEVERAC: Le soldat de Plomb. Marylène Dosse, Annie Petit, pno;
    Württemberg Chamber Orchestra; Jörg Faerber. Turnabout TVS
    34586.
        +NR 2-76 p14   -St 6-76 p108
    Villanelle des petits canards. cf HMV ASD 2929.
770 Works, selections: Bourrée fantasque. España. Danse slave.
    Gwendoline overture. Marche joyeuse. Le Roi malgrè lui: Fête
    polonaise. MASSENET: Phedre: Overture. Detroit Symphony Orch-
    estra; Paul Paray. Mercury SRI 75078. (Reissue).
        +MJ 12-76 p28   ++SFC 11-28-76 p45
CHADABE, Joel
    Echoes. cf Folkways FTS 33901/4.

CHADWICK, George Whitefield
    Euterpe. cf Louisville LS 753/4.
    Quartet, strings, no. 4, E minor. cf Vox SVBX 5301.
CHAITKIN, David
    Etudes, piano (3). cf CRI SD 345.
CHAJES, Julius
    Hechassid, op. 24, no. 1. cf Da Camera Magna SM 93399.
CHAMBERS
    The boys of the old brigade. cf Michigan University SM 0002.
CHAMBONNIERES, Jacques
    Chaconne, F major. cf Harmonia Mundi HMU 334.
    Chaconne, G major. cf Argo ZRG 806.
    Rondeau. cf Harmonia Mundi HMU 334.
CHAMINADE, Cecile
    Autrefois, op. 87, A minor. cf L'Oiseau-Lyre DSLO 7.
    Caprice espagnole. cf Gemini Hall RAP 1010.
    Chanson slave. cf Discophilia KIS KGG 3.
    Scarf dance. cf Connoisseur Society (Q) CSQ 2065.
    CHANSON DER TROUVERES, 13th CENTURY NORTHERN FRANCE. cf Telefunken
        6-41275.
CHAPI Y LORENTE, Ruperto
    Las hijas del Zebedeo: Carceleras. cf London OS 26435.
    La patria chica: Canción de pastora. cf London OS 26435.
CHAPIN, Lucius
    Rockbridge. cf Nonesuch H 71276.
CHARLES, Ernest
    My lady walks in loveliness. cf Cambridge CRS 2715.
CHARPENTIER, Gustave
771   Louise. Ileana Cotrubas, Jane Berbié, s; Lyliane Guitton, ms;
        Placido Domingo, Michel Sénéchal, t; Gabriel Bacquier, bs-
        bar; Ambrosian Opera Chorus; NPhO; Georges Prêtre. CBS 79302
        (3). (also Columbia M3 34207).
            +-Gr 10-76 p643   ++SFC 11-14-76 p30
            +HFN 12-76 p140   ++STL 10-10-76 p37
            ++RR 11-76 p34
    Louise: Berceuse. cf Club 99-101.
    Louise: Da quel di. cf BOITO: Mefistofele: L'altra notte in fondo
        al mare.
    Louise: Depuis le jour. cf Club 99-99.
    Louise: Depuis le jour. cf Decca GOSD 674/6.
CHARPENTIER, Marc-Antoine
    Messe de minuet. cf HMV SLS 5047.
    Te deum: Prelude. cf Polydor 2460 262.
CHAUSSON, Ernest
    Chanson perpetuelle, op. 37. cf CBS 76476.
    Concerto, violin, piano and string quartet, op. 21, D major. cf
        RCA ARM 4-0942/7.
772   Poème, op. 25. RAVEL: Tzigane.* SAINT-SAENS: Introduction and
        rondo capriccioso, op. 28. Havanaise, op. 83. Itzhak Perlman,
        vln; Orchestre de Paris; Jean Martinon. Angel (Q) S 37118.
        Tape (c) 4XS 37118 (ct) 8XS 37118. (also HMV ASD 3125). (*Re-
        issue from SLS 5016).
            ++Gr 1-76 p1191   ++NR 3-76 p13
            +HF 12-75 p110   +RR 12-75 p61
            ++HF 2-76 p130 tape   +St 1-76 p110
            +HFN 12-75 p152

Poème, op. 25. cf CBS 30063.
Poème, op. 25. cf HMV RLS 718.
Songs: Le colibri, op. 2. cf Club 99-97.
Songs: Le colibri, op. 2, no. 7; Les papillons, op. 2, no. 3;
    Poème de l'amour et la mer, op. 19; Les temps des lilas. cf
    HMV RLS 716.
Songs: Le temps des lilas. cf HMV RLS 719.
773 Symphony, op. 20, B flat major. FRANCK: Les Eolides. OSR;
    Ernest Ansermet. London STS 15294. (Reissue).
      +Audio 11-76 p111   +SFC 5-23-76 p36

CHAVEZ, Carlos
774 Concerto, piano. Eugene List, pno; VSOO; Carlos Chávez. West-
    minister WGS 8324. (Reissue).
      +NR 10-76 p6
775 Symphonies, nos. 1, 2, 4. New York Stadium Symphony Orchestra;
    Carlos Chávez. Peerless Everest 3029.
      +Gr 10-76 p590

CHAYNES, Charles
M'zab. cf Tarquinia.
776 Tarquinia. M'zab. Trio Deslogères; Odette Chaynes-Decaux, pno.
    Calliope CAL 1847.
      ++RR 10-76 p76

CHEDVILLE, Nicolas
Musette. cf DG Archive 2533 172.

CHERRY
Will-o-the-wisp. cf HMV EMD 5528.

CHERTOK
Around the clock. cf Orion ORS 76231.

CHERUBINI, Luigi
Ave Maria. cf BERLIOZ: La mort d'Ophelie, op. 18, no. 2.
Medea: Dei tuoi figli. cf HMV SLS 5057.
777 Quartets, strings, nos. 1-6. Melos Quartet. DG Archive 2723 044
    (3).
      ++Gr 8-76 p311   +MT 9-76 p745
      +HFN 8-76 p83   +RR 7-76 p67
      ++MJ 12-76 p44   +SR 11-13-76 p52
778 Requiem, male voices and orchestra, D minor. Ambrosian Singers;
    NPhO; Riccardo Muti. HMV ASD 3073 (Q) SQ 4 ASD 3073. (also
    Angel S 37096).
      +Gr 5-75 p2000 Quad   +NR 8-75 p8
      ++HF 12-75 p90   +ON 1-17-76 p32
      +HFN 9-75 p97   +RR 4-75 p58
      +MT 11-75 p977   ++SFC 10-19-75 p33
      ++NYT 9-21-75 pD18
Sonata, horn, no. 1, F major. cf BASF BAC 3085.
Sonata, horn, no. 2, F major. cf BASF BAC 3085.

CHESNOKOV, Pavel
Songs: Paschal canon (Hymns to the mother of God); Troparia;
    Great compline (God is with us). cf Ikon IKO 2.
CHICHESTER CATHEDRAL, 900 YEARS. cf HMV HQS 1350.

CHILDS, Barney
779 Trio, clarinet, violoncello and piano. LENTZ: Songs of the sirens.
    NØRGARD: Spell. Montagnana Trio. ABC/Command COMS 9005.
      ++NR 1-76 p8   +St 6-76 p116

CHOPIN, Frederic
780 Andante spianato and grande polonaise, op. 22, E flat major.

Scherzo, no. 2, op. 31, B flat minor. Sonata, piano, no. 2,
op. 35, B flat minor. Martha Argerich, pno. DG 2530 530.
Tape (c) 3300 474.

+Gr 6-75 p68   +-MM 8-76 p33
+HF 7-75 p74   ++NR 7-75 p15
+HF 7-76 p114 tape   +-RR 6-75 p65
++HFN 6-75 p87

781   Andante spianato and grande polonaise, op. 22, E flat major.
      SCHUMANN: Concerto, violoncello, op. 129, A minor. TCHAIKOVSKY:
      Variations on a rococo theme, op. 33. Mstislav Rostropovich,
      vlc; MRSO; Kiril Kondrashin. Everest SDBR 3391.
      -NR 5-76 p7

782   Andante spianato and grande polonaise, op. 22, E flat major. Noc-
      turne, op. 62, no. 1, B major. Polonaise-Fantaisie, op. 61,
      A flat major. Scherzo, no. 4, op. 54, E major. Emanuel Ax,
      pno. RCA ARL 1-1569.

++Gr 9-76 p443   +RR 9-76 p80
++HF 9-76 p90   ++St 10-76 p121
+MJ 11-76 p45   +STL 10-10-76 p37
+NR 9-76 p12

783   Andante spianato and grande polonaise, op. 22, E flat major.
      LISZT: Hungarian fantasia, G 123. Totentanz, G 126. Sviatoslav
      Richter, Shura Cherkassky, pno; LSO, ORTF; Kiril Kondrashin,
      Milan Horvat. Rococo 2092.
      +-NR 9-76 p11
      Andante spianato and grande polonaise, op. 22, E flat major. cf
      Piano works (Decca ECS 768/70).
      Andante spianato and grande polonaise, op. 22, E flat major. cf
      International Piano Library IPL 5001/2.

784   Ballades, nos. 1-3, opp. 23, 38, 47, 52. Impromptus, nos. 1-3,
      opp. 29, 36, 51. Fantasie-Impromptu, op. 66, C sharp minor.
      Nikita Magaloff, pno. Philips 6580 117.
      -Gr 8-76 p318   ++RR 7-76 p70
      +-HFN 7-76 p85

785   Ballades, no. 1, op. 23, G minor; no. 4, op. 52, F minor. Etude,
      op. 10, no. 12, C minor. Nocturne, op. 9, no. 2, E flat major.
      Waltzes, op. 18, E flat major; op. 64, no. 2, C sharp minor;
      op. 70, no. 1, G flat major; op. 70, no. 2, F minor; op. posth.,
      E minor. Philippe Entremont, pno. CBS 30087.
      +-HFN 11-76 p171   +-RR 11-76 p88
      Ballade, no. 1, op. 23, G minor. cf Piano works (Oryx ORPS 55).
      Ballade, no. 1, op. 23, G minor. cf International Piano Library
      IPL 5001/2.

786   Ballade, no. 2, op. 38, F major. Etude, op. 10, no. 3, E major.
      Etude, op. 25, no. 5, E minor. Polonaise, op. 26, no. 1, C
      sharp minor. SCHUBERT: Impromptu, op. 142, no. 2, D 935, A
      flat major. Moments musicaux, op. 94, nos. 1, 3, 6, D 780.
      Sviatoslav Richter, pno. Columbia/Melodiya M 33826.
      ++HF 4-76 p98   +-NR 2-76 p13
      Ballade, no. 3, op. 47, A flat major. cf Piano works (Decca ECS
      768/70).
      Ballade, no. 3, op. 47, A flat major. cf Piano works (DG 2548 215).
      Ballade, no. 3, op. 47, A flat major. cf Piano works (RCA VH 018).
      Ballade, no. 3, op. 47, A flat major. cf ALBENIZ: España, op. 165,
      no. 2: Tango.
      Ballade, no. 4, op. 52, F minor. cf Ballade, no. 1, op. 23, G
      minor.

Ballade, no. 4, op. 52, F minor. cf Piano works (Decca PFS 4313).
Ballade, no. 4, op. 52, F minor. cf Piano works (Saga 5394).
Ballade, no. 4, op. 52, F minor. cf International Piano Archives
    IPA 5007/8.
Barcarolle, op. 60, F sharp major. cf Piano works (Decca ECS
    768/70).
Barcarolle, op. 60, F sharp major. cf RCA ARL 1-1176.
787 Berceuse, op. 57, D flat major. Ecossaises, op. 72. Etudes, opp.
    10 and posth. (15). Ruth Sleczynska, pno. Musical Heritage
    Society MHS 3216.
        +HF 5-76 p80   +MJ 4-76 p30
Berceuse, op. 57, D flat major. cf Piano works (Decca ECS 768/70).
Berceuse, op. 57, D flat major. cf Piano works (Oryx ORPS 55).
Berceuse, op. 57, D flat major. cf Preludes, op. 28, nos. 1-24.
Berceuse, op. 57, D flat major. cf International Piano Library
    IPL 5001/2.
788 Concerto, piano, no. 1, op. 11, E minor. Concerto, piano, no. 2,
    op. 21, F minor. Fantasy on Polish airs, op. 13. Krakowiak,
    op. 14, F major. Garrick Ohlsson, pno; Polish Radio Symphony
    Orchestra; Jerzy Maksymiuk. Angel S 37179/80. (also HMV SQ
    SLS 5043).
        +Gr 4-76 p1597   ++NR 8-76 p4
        +HF 10-76 p104   +-RR 3-76 p38
        +HFN 4-76 p103   ++St 11-76 p138
        +-MJ 11-76 p45
789 Concerto, piano, no. 1, op. 11, E minor. Mazurkas, op. 7, no. 1,
    B flat major; op. 67, no. 3, C major; op. 68, no. 2, A minor;
    op. posth., D major. Tamás Vásáry, pno; BPhO; Jerzy Semkov.
    DG 2535 206. (Reissue).
        ++RR 12-76 p53
790 Concerto, piano, no. 1, op. 11, E minor. Krakowiak, op. 14, F
    major. Stefan Askenase, pno; Hague Philharmonic Orchestra;
    Willem van Otterloo. DG 2548 066. Tape (c) 3348 066. (Re-
    issue from SLPM 138085).
        ++Gr 10-75 p608   +RR 9-75 p34
        +HFN 9-75 p108   +RR 9-76 p92 tape
791 Concerto, piano, no. 1, op. 11, E minor. Abbey Simon, pno; Ham-
    burg Symphony Orchestra; Heribert Beissel. Turnabout TV 37083.
        -Gr 5-76 p1757
Concerto, piano, no. 1, op. 11, E minor: Romanza. cf Desmar DSM
    1005.
Concerto, piano, no. 2, op. 21, F minor. cf Concerto, piano, no.
    1, op. 11, E minor (Angel S 37179/80).
Concerto, piano, no. 2, op. 21, F minor: 2nd movement. cf HMV
    SEOM 24.
Ecossaises, op. 72. cf Berceuse, op. 57, D flat major.
Etude, C minor. cf CBS 30064.
792 Etudes, opp. 10 and 25. Sylvia Kersenbaum, pno. Classics for
    Pleasure CFP 40239.
        -Gr 8-76 p318   +RR 6-76 p69
        +HFN 8-76 p83
793 Etudes, opp. 10 and 25. Ilana Vered, pno. Connoisseur Society
    CS 2045. Tape (c) Advent E 1018 (Q) CSQ 2045.
        +-HF 10-73 p101   +LJ 2-74 p50
        +HF 7-75 p102 tape   +-MJ 9-73 p44
        +HF 7-76 p114 tape   ++NR 7-73 p11

794   Etudes, opp. 10 and 25. Maurizio Pollini, pno. DG 2530 291.
       Tape (c) 3300 287.
              +Gr 11-72 p928   ++RR 11-72 p80
              +HF 3-73 p82   ++RR 1-74 p77 tape
              +HF 7-76 p114 tape   ++SFC 3-4-73 p33
              ++HFN 4-73 p792 tape   ++SFC 7-15-73 p32
              ++NR 6-73 p12   ++St 8-73 p99
              +NYT 4-1-73 pD28
795   Etudes, opp. 10 and 25. Vladimir Ashkenazy, pno. Everest SDBR
       3387. (Reissue).
              -NR 5-76 p12
796   Etudes, opp. 10 and 25. Vladimir Ashkenazy, pno. London CS 6844.
       Tape (c) CS5 6844. (also Decca SXL 6710 Tape (c) KSXC 6710).
              +Gr 12-75 p1076   +NR 11-75 p12
              +-HF 11-75 p102   +-RR 12-75 p79
              ++HF 7-76 p114   +-RR 9-76 p94 tape
              +HFN 2-76 p93   ++SFC 8-24-75 p28
              ++HFN 4-76 p125 tape   ++St 12-75 p117
       Etudes, opp. 10 and posth. (15). cf Berceuse, op. 57, D flat
       major.
       Etudes, op. 10, nos. 3, 4, 12. cf Piano works (Columbia M 32932).
797   Etude, op. 10, no. 3, E major. LISZT: Etudes d'exécution trans-
       cendente, no. 5, G 139: Feux follets; no. 11, G 139: Harmonies
       du soir. Valses oubliées, no. 1, G 215, F sharp major; no. 2,
       G 215, A flat major. MUSSORGSKY: Pictures at an exhibition.
       SCHUBERT: Impromptu, op. 90. no. 2, D 899, E flat major. Im-
       promptu, op. 90, no. 4, D 899, A flat major. Moments musicaux,
       op. 94, no. 1, D 780, C major. Sviatoslav Richter, pno.
       Philips 6780 502 (2). (Reissues from ABL 3314, ABL 3301).
              +Gr 3-76 p1487   +-RR 3-76 p61
              +-HFN 3-76 p111
       Etude, op. 10, no. 3, E major. cf Ballade, no. 2, op. 38, F major.
       Etude, op. 10, no. 3, E major. cf Saga 5427.
       Etude, op. 10, no. 4, C sharp minor. cf ALBENIZ: España, op. 165,
       no. 2: Tango.
       Etude, op. 10, no. 5, G flat major (2 versions). cf International
       Piano Library IPL 104.
       Etude, op. 10, no. 12, C minor. cf Ballade, no. 1, op. 23, G
       minor.
       Etude, op. 25, no. 1, A flat major. cf BACH: Partita, harpsichord,
       no. 1, S 825, B flat major.
       Etude, op. 25, no. 2, F minor. cf Piano works (Vanguard VSD
       71214).
       Etude, op. 25, no. 2, F minor. cf BACH: Partita, harpsichord,
       no. 1, S 825, B flat major.
       Etude, op. 25, no. 5, E minor. cf Ballade, no. 2, op. 38, F major.
       Etude, op. 25, no. 5, E minor. cf International Piano Library
       IPL 104.
       Etude, op. 25, no. 6, G flat major. cf International Piano Library
       IPL 5001/2.
       Etude, op. 25, no. 7, C sharp minor. cf ALBENIZ: España, op. 165,
       no. 2: Tango.
       Etude, op. 25, no. 10, B minor. cf ALBENIZ: España, op. 165,
       no. 2: Tango.
798   Fantasia, op. 49, F minor. TCHAIKOVSKY: Concerto, piano, no. 1,
       op. 23, B flat minor. Solomon, pno; Hallé Orchestra; Hamilton

Harty. Bruno Walter Society IGI 333. (Reissue from Columbia
originals).
+HF 11-76 p130
799   Fantasia, op. 49, F minor. Nocturnes, op. 9, nos. 1 and 2.
Polonaise, op. 26, nos. 1 and 2; op. 71, nos. 1 and 3. Bruno
Rigutto, pno. French Decca 7293.
+RR 3-76 p59
Fantasia, op. 49, F minor. cf Piano works (Decca ECS 768/70).
Fantasia, op. 49, F minor. cf Scherzi, nos. 1-4.
Fantasie-Impromptu. cf CBS 30064.
Fantasie-Impromptu, op. 66, C sharp minor. cf Ballades, nos.
1-4, opp. 23, 38, 47, 52.
Fantasie-Impromptu, op. 66, C sharp minor. cf Piano works
(Decca ECS 768/70).
Fantasie-Impromptu, op. 66, C sharp minor. cf Piano works
(DG 2548 215).
Fantasie-Impromptu, op. 66, C sharp minor. cf Piano works
(Oryx ORPS 55).
Fantasie-Impromptu, op. posth., C sharp minor. cf BEETHOVEN:
Sonata, piano, no. 14, op. 27, no. 2, C sharp minor.
Fantasy on Polish airs, op. 13. cf Concerto, piano, no. 1, op.
11, E minor.
Impromptus, nos. 1-3, opp. 29, 36, 51. cf Ballades, nos. 1-4,
opp. 23, 38, 47, 52.
Impromptus, nos. 1-3, opp. 29, 36, 51. cf Piano works (Decca ECS
768/70).
Impromptus, nos. 1-3, opp. 29, 36, 51. cf Piano works (DG 2548
215).
Impromptu, no. 1, op. 29, A flat major. cf Piano works (Saga 5394).
Krakowiak, op. 14, F major. cf Concerto, piano, no. 1, op. 11,
E minor (Angel 37179/80).
Krakowiak, op. 14, F major. cf Concerto, piano, no. 1, op. 11,
E minor (DG 2548 066).
800   Mazurkas, complete. Ronald Smith, pno. HMV SLS 5014 (3).
+-Gr 5-75 p1992   +-MM 8-76 p33
++HFN 5-75 p125   +-RR 6-75 p60
Mazurka, op. 7, no. 1, B flat major. cf Concerto, piano, no. 1,
op. 11, E minor.
Mazurka, op. 7, no. 1, B flat major. cf Piano works (Decca PFS
4313).
Mazurka, op. 7, no. 1, B flat major. cf Piano works (Oryx ORPS 55).
Mazurka, op. 7, no. 3, F minor. cf Piano works (Columbia M 32932).
Mazurka, op. 17, no. 4, A minor. cf Piano works (Decca PFS 4313).
Mazurka, op. 17, no. 4, A minor. cf Decca PFS 4351.
Mazurka, op. 24, no. 4, B flat minor. cf Piano works (RCA VH 018).
Mazurka, op. 30, no. 3, D flat major. cf Piano works (Columbia M
32932).
Mazurka, op. 33, no. 2, D major. cf Piano works (Columbia M 32932).
Mazurkas, op. 41 (4). cf Piano works (DG 2548 215).
Mazurka, op. 41, no. 1, F minor. cf Piano works (Columbia M 32932).
Mazurka, op. 41, no. 2, B major. cf Piano works (Oryx ORPS 55).
Mazurka, op. 50, no. 3, C sharp minor. cf Piano works (Columbia
M 32932).
Mazurka, op. 50, no. 3, C sharp minor. cf Piano works (Vanguard
VSD 71214).
Mazurka, op. 59, no. 3, F sharp minor. cf Piano works (Columbia
M 32932).

Mazurka, op. 59, no. 3, F sharp minor. cf Piano works (Oryx
ORPS 55).
Mazurka, op. 59, no. 3, F sharp minor. cf International Piano
Library IPL 5003/4.
Mazurka, op. 67, no. 3, C major. cf Concerto, piano, no. 1, op.
11, E minor.
Mazurka, op. 68, no. 1, C major. cf Piano works (Oryx ORPS 55).
Mazurka, op. 68, no. 2, A minor. cf Concerto, piano, no. 1, op.
11, E minor.
Mazurka, op. posth., D major. cf Concerto, piano, no. 1, op. 11,
E minor.
Minuet. cf Piano works (Everest SDBR 3377).
Nocturne, D flat major. cf RCA ARM 4-0942/7.
Nocturne, E minor. cf RCA ARM 4-0942/7.
Nocturne, E flat major. cf RCA ARM 4-0942/7.
Nocturnes, op. 9, nos. 1 and 2. cf Fantasia, op. 49, F minor.
Nocturne, op. 9, no. 2, E flat major. cf Ballade, no. 1, op. 23,
G minor.
Nocturne, op. 9, no. 2, E flat major. cf Piano works (Oryx ORPS
55).
Nocturne, op. 9, no. 2, E flat major. cf International Piano
Library IPL 5001/2.
Nocturne, op. 9, no. 2, E flat major. cf Saga 5427.
Nocturne, op. 9, no. 3, B major. cf Piano works (Decca ECS 768/70).
Nocturne, op. 9, no. 3, B major. cf Piano works (RCA VH 018).
Nocturne, op. 9, no. 3, B major. cf BACH: Partita, harpsichord,
no. 1, S 825, B flat major.
Nocturne, op. 9, no. 3, B major. cf International Piano Library
IPL 5003/4.
Nocturne, op. 9, no. 3, B major. cf International Piano Archives
IPA 5007/8.
Nocturne, op. 15, no. 1, F major. cf Piano works (Everest SDBR
3377).
Nocturne, op. 15, no. 1, F major. cf Piano works (Oryx ORPS 55).
Nocturne, op. 15, no. 1, F major. cf Piano works (RCA VH 018).
Nocturne, op. 15, no. 2, F sharp major. cf HMV HLM 7093.
Nocturne, op. 15, no. 2, F sharp major. cf International Piano
Library IPL 104.
Nocturne, op. 15, no. 2, F sharp major. cf International Piano
Library IPL 5001/2.
Nocturne, op. 27, no. 2, D flat major. cf Piano works (Vanguard
VSD 71214).
Nocturne, op. 37, no. 1, G minor. cf Piano works (Saga 5394).
Nocturne, op. 48, no. 1, C minor. cf Piano works (Oryx ORPS 55).
Nocturne, op. 55, no. 1, F minor. cf Piano works (Decca PFS 4313).
Nocturne, op. 55, no. 2, E flat major. cf Piano works (Vanguard
VSD 71214).
Nocturne, op. 62, no. 1, B major. cf Andante spianato and grande
polonaise, op. 22, E flat major.
Nocturne, op. 72, E minor. cf Piano works (RCA VH 018).
Nocturne, op. posth., C sharp minor. cf Piano works (Oryx ORPS
55).
Nouvelles etudes (3). cf Preludes, op. 28, nos. 1-24.
Nouvelles etudes, no. 1, F minor. cf Piano works (Vanguard VSD
71214).
801   Piano works: Etudes, op. 10, nos. 3, 4, 12. Mazurkas, op. 7, no.

3; op. 30, no. 3; op. 33, no. 2; op. 41, no. 2; op. 50, no. 3;
op. 59, no. 3. Polonaise, op. 40, no. 1, A major. Prelude,
op. 38, no. 6, B minor. Waltz, op. 64, no. 2, C sharp minor.
Vladimir Horowitz, pno. Columbia M 32932. Tape (c) MT 32932
(ct) MA 32932. (also CBS 76307).

+Gr 2-75 p1516   ++RR 2-75 p51
+HF 2-75 p90   +SFC 10-20-74 p26
+MM 1-76 p41   +-St 4-75 p98
+-NR 4-75 p11

802   Piano works: Andante spianato and grande polonaise, op. 22, E
      flat major. Ballade, no. 3, op. 47, A flat major. Barcarolle,
      op. 60, F sharp major. Berceuse, op. 57, D flat major.
      Fantasia, op. 49, F minor. Fantasie-Impromptu, op. 66, C sharp
      minor. Impromptus, nos. 1-3, opp. 29, 36, 51. Nocturne, op. 9,
      no. 3, B major. Polonaise-Fantaisie, op. 61, A flat major.
      Scherzo, no. 3, op. 39, C sharp minor. Sonata, piano, no. 2,
      op. 35, B flat minor. Sonata, piano, no. 3, op. 58, B minor.
      Wilhelm Kempff, pno. Decca ECS 768/70 (3). (Reissues from
      SXL 2081, 2024, 2025).

      -Gr 4-76 p1632   +-RR 2-76 p48
      +-HFN 2-76 p115

803   Piano works: Ballade, no. 4, op. 52, F minor. Mazurka, op. 17,
      no. 4, A minor. Mazurka, op. 7, no. 1, B flat major. Nocturne,
      op. 55, no. 1, F minor. Polonaise, op. 40, no. 1, A major.
      Sonata, piano, no. 3, op. 58, B minor. Waltz, op. 34, no. 2,
      A minor. Ilana Vered, pno. Decca PFS 4313. Tape (c) KPFC
      4313. (also London SPC 21119 Tape (c) 052119 (ct) 082119 (r)
      E42119).

      ++Gr 11-74 p922   +-RR 1-76 p65 tape
      +Gr 2-75 p1562 tape   ++SFC 11-17-74 p32
      +HF 7-75 p102 tape   +-SR 1-25-75 p50
      -RR 11-74 p78

804   Piano works (Ballade, no. 3, op. 47, A flat major. Fantasie-
      Impromptu, op. 66, C sharp minor. Impromptus, nos. 1-3, opp.
      29, 36, 51. Mazurkas, op. 41 (4). Scherzo, no. 3, op. 39,
      C sharp minor. Stefan Askenase, pno. DG 2548 215. (Reissue
      from 2538 078).

      -Gr 4-76 p1617   +-RR 4-76 p62
      +HFN 4-76 p123

805   Piano works: Minuet. Nocturne, op. 15, no. 1, F major. Polish
      songs, op. 74, no. 1: Maiden's wish. Scherzo, no. 2, op. 31,
      B flat minor. Waltz, op. 18, E flat major. RACHMANINOFF:
      Barcarolle, op. 10, no. 3, G minor. Elegie, op. 3, no. 1.
      Mélodie, op. 3, no. 3, E major. Polichinelle. op. 3, no. 4,
      F sharp minor. Polka de W. R. Prelude, op. 3, no. 2, C sharp
      minor. Prelude, op. 23, no. 5, G minor. Sergei Rachmaninoff,
      pno. Everest SDBR 3377.

      +-Gr 9-76 p453   +NR 11-75 p15

806   Piano works: Ballade, no. 1, op. 23, G minor. Berceuse, op. 57,
      D flat major. Fantasie-Impromptu, op. 66, C sharp minor.
      Nocturnes, op. 9, no. 2, E flat major; op. 15, no. 1, F major;
      op. 48, no. 1, C minor; op. posth., C sharp minor. Mazurkas,
      op. 7, no. 1, B flat major; op. 41, no. 2, B major; op. 68,
      no. 1, C major; op. 59, no. 3, F sharp minor. Polonaise, op.
      53, A flat major. Anthony Goldstone, pno. Oryx ORPS 55.

      +-Gr 9-76 p443

807   Piano works: Ballade, no. 3, op. 47, A flat major. Mazurka, op.
       24, no. 4, B flat minor. Nocturnes, op. 9, no. 3, B major;
       op. 72, E minor; op. 15, no. 1, F major. Scherzo, no. 2, op.
       31, B flat minor. Scherzo, no. 3, op. 39, C sharp minor.
       Vladimir Horowitz, pno. RCA VH 018.
             ++Gr 4-76 p1631   +-RR 12-75 p80
             +-HFN 2-76 p117
808   Piano works: Ballade, no. 4, op. 52, F minor. Impromptu, no. 1,
       op. 29, A flat major. Nocturne, op. 37, no. 1, G minor. Noc-
       turne, op. 62, no. 1, B major. Polonaise, op. 71, no. 2, B
       flat major. Waltz, op. 18, E flat major. Waltz, op. 64, no.
       1, D flat major. Ignaz Friedman, pno. Saga 5394. (Reissue
       from Piano Rolls).
             +HFN 8-76 p74   +RR 7-75 p47
             +MT 2-76 p140
809   Piano works: Etude, op. 25, no. 2, F minor. Mazurka, op. 50, no.
       3, C sharp minor. Nocturne, op. 27, no. 2, D flat major.
       Nocturne, op. 55, no. 2, E flat major. Nouvelles etudes,
       no. 1, F minor. Sonata, piano, no. 3, op. 58, B minor. Waltz,
       op. 34, no. 1, A flat major. Bruce Hungerford, pno. Vanguard
       VSD 71214.
             +-ARG 11-76 p20   +-MJ 11-76 p45
             +HF 12-76 p99   +-NR 9-76 p11
       Polish songs (Chants polonaise), op. 74, no. 1: The maiden's wish.
       cf Piano works (Everest SDBR 3377).
       Chants polonaise, op. 74, no. 1: The maiden's wish; no. 2, My joy.
       cf Scherzi, nos. 1-4.
       Chants polonaise, op. 74, no. 5. cf Decca SPA 473.
       Polonaise, A flat major. cf CBS 30064.
810   Polonaise, opp. 26, 40, 44, 53, 61, nos. 1-7. Maurizio Pollini,
       pno. DG 2530 659. Tape (c) 3300 659.
             ++Gr 12-76 p1021   ++RR 12-76 p78
             ++HFN 12-76 p140
       Polonaise, op. 26, nos. 1 and 2. cf Fantasia, op. 49, F minor.
       Polonaise, op. 26, no. 1, C sharp minor. cf Ballade, no. 2, op.
       38, F major.
       Polonaise, op. 26, no. 2, E flat minor. cf International Piano
       Archives IPA 5007/8.
       Polonaise, op. 40, no. 1, A major. cf Piano works (Columbia
       32932).
       Polonaise, op. 40, no. 1, A major. cf Piano works (Decca PFS 4313).
       Polonaise, op. 53, A flat major. cf Piano works (Oryx ORPS 55).
       Polonaise, op. 71, nos. 1 and 3. cf Fantasia, op. 49, F minor.
       Polonaise, op. 71, no. 2, B flat major. cf Piano works (Saga 5394).
       Polonaise-Fantaisie, op. 61, A flat major. cf Andante spianato
       and grande polonaise, op. 22, E flat major.
       Polonaise-Fantaisie, op. 61, A flat major. cf Piano works (Decca
       ECA 768/70).
811   Preludes, op. 28, nos. 1-24. Prelude, op. 45, C sharp minor. Pre-
       lude, op. posth., A flat major. Murray Perahia, pno. Columbia
       M 33507. (also CBS 76422 Tape (c) 40-76422).
             +-Gr 1-76 p1217   +MJ 10-76 p25
             ++HF 11-76 p106   +-NR 5-76 p12
             +HFN 2-76 p93   +-RR 1-76 p47
             +-HFN 2-76 p116 tape   +St 6-76 p102
812   Preludes, op. 28, nos. 1-24. Prelude, op. 45, C sharp minor.
       Prelude, op. posth., A flat major. Daniel Barenboim, pno.

HMV ASD 3254.
     -Gr 10-76 p627   -RR 10-76 p87

813   Preludes, op. 28, nos. 1-24. Berceuse, op. 57, D flat major.
      Alicia de Larrocha, pno. Decca SXL 6733. (also London CS
      6952).
          +Gr 11-75 p860   -NR 5-76 p12
          +-HF 11-76 p106   +-RR 11-75 p76
          +-HFN 11-75 p152   +-St 6-76 p102

814   Preludes, op. 28, nos. 1-24. Maurizio Pollini, pno. DG 2530 550.
      Tape (c) 3300 550.
          +Gr 12-75 p1076   +NR 5-76 p13
          ++HF 11-76 p106   ++RR 12-75 p80
          +HFN 3-76 p113 tape   ++RR 3-76 p76
          +MJ 10-76 p25   ++St 8-76 p92

815   Preludes, op. 28, nos. 1-24. Prelude, op. 45, C sharp minor.
      Prelude, op. posth., A flat major. Claudio Arrau, pno.
      Philips 6500 622. Tape (c) 7300 335.
          +Gr 6-75 p68   -MM 8-76 p33
          +-Gr 10-75 p721 tape ++NR 9-75 p12
          ++HF 10-75 p72   ++RR 6-75 p65
          ++HFN 6-75 p87   ++SFC 8-10-75 p26
          +HFN 7-75 p90   ++St 11-75 p126

816   Preludes, op. 28, nos. 1-24. Prelude, op. 45, C sharp minor.
      Prelude, op. posth., A flat major. Nouvelles etudes (3).
      Nikita Magaloff, pno. Philips 6580 118.
          +-RR 11-76 p89

Prelude, op. 28, no. 6, B minor. cf Piano works (Columbia M
    32932).
Prelude, op. 28, no. 7, A major. cf International Piano Library
    IPL 104.
Prelude, op. 45, C sharp minor. cf Preludes, op. 28, nos. 1-24
    (Philips 6580 118).
Prelude, op. 45, C sharp minor. cf Preludes, op. 28, nos. 1-24
    (Columbia M 33507).
Prelude, op. 45, C sharp minor. cf Preludes, op. 28, nos. 1-24
    (Philips 6500 622).
Prelude, op. posth., A flat major. cf Preludes, op. 28, nos. 1-24
    (Philips 6580 118).
Prelude, op. posth., A flat major. cf Preludes, op. 28, nos. 1-24
    (Columbia M 33507).
Prelude, op. posth., A flat major. cf Preludes, op. 28, nos. 1-24
    (Philips 6500 622).

817   Scherzi, nos. 1-4. Chants polonaise, op. 74, no. 1: The maiden's
      wish; no. 12, My joy. Antonio Barbosa, pno. Connoisseur Society
      CS 2071 (Q) CSQ 2071. Tape (c) Advent E 1045.
          +HF 3-75 p77   ++SFC 7-27-75 p22
          +HF 7-76 p114 tape   +St 3-75 p98 Quad
          +-NR 10-75 p10

818   Scherzi, nos. 1-4. Fantasia, op. 49, F minor. François Duchâble,
      pno. Connoisseur Society CSQ 2086.
          +Audio 12-76 p86   +-NR 3-76 p12
          -HF 11-76 p108   -SFC 11-16-75 p32
          -MJ 4-76 p30

Scherzo, no. 2, op. 31, B flat minor. cf Andante spianato and
    grande polonaise, op. 22, E flat major.
Scherzo, no. 2, op. 31, B flat minor. cf Piano works (Everest
    SDBR 3377).

Scherzo, no. 2, op. 31, B flat minor. cf Piano works (RCA VH 018).
Scherzo, no. 3, op. 39, C sharp minor. cf Piano works (Decca
    ECS 768/70).
Scherzo, no. 3, op. 39, C sharp minor. cf Piano works (DG 2548
    215).
Scherzo, no. 3, op. 39, C sharp minor. cf Piano works (RCA VH 018).
819 Scherzo, no. 4, op. 54, E major. PROKOFIEV: Cinderella, op. 87:
    Gavotte. Visions fugitives, op. 22 (5). RACHMANINOFF: Pre-
    ludes (3). RAVEL: Jeu d'eaux. Miroirs: La vallee des cloches.
    Sviatoslav Richter, pno. RCA AGL 1-1279. (Reissue).
        +MJ 1-76 p25   +SFC 11-16-75 p32
Scherzo, no. 4, op. 54, E major. cf Andante spianato and grande
    polonaise, op. 22, E flat major.
820 Sonata, piano, no. 2, op. 35, B flat minor. Sonata, piano, no. 3,
    op. 58, B minor. Witold Malcuzynski, pno. Classics for
    Pleasure CFP 40095.
        +RR 9-76 p80
821 Sonata, piano, no. 2, op. 35, B flat minor. Sonata, piano, no. 3,
    op. 58, B minor. Daniel Barenboim, pno. HMV ASD 3064.
        +Gr 8-75 p347   +-MM 8-76 p33
        ++HFN 8-75 p74   ++RR 8-75 p56
Sonata, piano, no. 2, op. 35, B flat minor. cf Andante spianato
    and grande polonaise, op. 22, E flat major.
Sonata, piano, no. 2, op. 35, B flat minor. cf Piano works (Decca
    ECS 768/70).
822 Sonata, piano, no. 3, op. 58, B minor. LISZT: Sonata, piano,
    G 178, B minor. Paul Badura-Skoda, pno. Bruno Walter Society
    AUDAX 761.
        +NR 11-76 p12
823 Sonata, piano, no. 3, op. 58, B minor. LISZT: Gnomenreigen, G 145.
    Etude d'exécution transcendente d'après Paganini, no. 6, G 140,
    A minor. Paraphrases, SCHUBERT: Horch, Horch, Die Lerch, G
    558/9; Der Muller und der Bach; Liebesbotschaft; Das Wandern.
    Emanuel Ax, pno. RCA ARL 1-1030. Tape (ct) ARS 1-1030.
        +Gr 12-75 p1085   ++RR 12-75 p79
        +HF 11-75 p128   +SR 9-6-75 p40
        +HFN 2-76 p93   +St 10-75 p72
        +NR 9-75 p11
824 Sonata, piano, no. 3, op. 58, B minor. LISZT: Sonata, piano,
    G 178, B minor. Alfred Cortot, pno. Seraphim 60241.
        +ARSC Vol III, no. 2-3, p87
Sonata, piano, no. 3, op. 58, B minor. cf Piano works (Decca ECS
    768/70).
Sonata, piano, no. 3, op. 58, B minor. cf Piano works (Decca
    PFS 4313).
Sonata, piano, no. 3, op. 58, B minor. cf Sonata, piano, no. 2,
    op. 35, B flat minor (Classics for Pleasure CFP 40095).
Sonata, piano, no. 3, op. 58, B minor. cf Sonata, piano, no. 2,
    op. 35, B flat minor (HMV 3064).
Sonata, piano, no. 3, op. 58, B minor. cf Piano works (Vanguard
    VSD 71214).
Sonata, piano, no. 3, op. 58, B minor: 2nd and 3rd movements.
    cf CBS 30066.
825 Sonata, violoncello, op. 65, G minor. SHOSTAKOVICH: Sonata,
    violoncello, op. 40, D minor. Mstislav Rostropovich, vlc;
    Aleksander Dediukhin, pno. Bruno Walter Society IGI 321.
        +NR 11-76 p7

826   Songs: Polish songs, op. 74: Melodya; Narzeczony; Spiew grobowy;
      Moja pieszczotka; Wiosna; Nie ma czego trzeba; Hulanka. LISZT:
      Songs: Angiolin dal biondo crin, G 269; Es rauschen die Winde,
      G 294; Es muss ein Wunderbares sein, G 314; In Liebeslust,
      G 318; Kling leise, mein Lied, G 301; Schwebe, schwebe,
      blaues Auge, G 305; Die Vätergruft, G 281; Wie singt die Lerche
      schön, G 312. Robert Tear, t; Philip Ledger, pno. Argo ZRG
      814.
            +-Gr 3-76 p1493   +-RR 2-76 p60
            +HFN 2-76 p93   +-STL 5-9-76 p38
      Songs: Dumka; The maiden's wish, op. 74, no. 1; Spring, op. 72,
      no. 2; The trooper, op. 74, no. 10. cf BELLINI: Songs (Cetra
      LPO 2003).
827   Les sylphides (orch. Douglas). DELIBES: Coppelia: Suite. BPhO;
      Herbert von Karajan. DG 2535 189. (also DG 136257 Tape (ct)
      86-257).
            -RR 12-76 p54
      Les sylphides: Waltz. cf CBS 30072.
      Variations brillantes on a melody from the opera Ludovic, op. 12.
      cf International Piano Library IPL 5003/4.
      Variations on a theme from Rossini's "La Cenerentola". cf RCA LRL
      1-5131.
828   Waltzes (16). Peter Katin, pno. Decca SDD 353. (also London
      STS 15305).
            ++Gr 1-73 p1344   +NR 5-76 p12
            +HFN 2-73 p336   +RR 1-73 p60
829   Waltzes (18). Aldo Ciccolini, pno. Seraphim S 60252.
            ++HF 1-76 p85   /NR 12-75 p13
830   Waltzes (19). Abbey Simon, pno. Turnabout TVS 34580.
            +-Gr 5-76 p1782   +NR 9-75 p12
            +HF 11-75 p102   -RR 5-76 p65
            +-HFN 4-76 p103
831   Waltzes, nos. 1-14. Stefan Askenase, pno. DG 2548 146. Tape
      (c) 3348 146. (Reissue from 136396).
            +Gr 4-75 p1867   +-RR 4-75 p50
            +HFN 10-75 p152   +RR 9-76 p92 tape
832   Waltzes, nos. 1-14. Arthur Rubinstein, pno. RCA Tape (c) RK 11705.
            ++HFN 7-76 p104 tape   -RR 10-76 p106 tape
      Waltz, A flat major. cf CBS 30064.
      Waltz, C sharp minor. cf CBS 30064.
      Waltz, op. 18, E flat major. cf Ballade, no. 1, op. 23, G minor.
      Waltz, op. 18, E flat major. cf Piano works (Everest SDBR 3377).
      Valse, op. 18, E flat major. cf Piano works (Saga 5394).
      Waltz, op. 18, E flat major. cf International Piano Archives
      IPA 5007/8.
      Waltz, op. 18, E flat major. cf Philips 6747 204.
      Waltz, op. 34, no. 1, A flat major. cf Piano works (Vanguard VSD
      71214).
      Valse, op. 34, no. 1, A flat major. cf Rococo 2049.
      Waltz, op. 34, no. 2, A minor. cf Piano works (Decca PFS 4313).
      Waltz, op. 34, no. 2, A minor. cf Connoisseur Society (Q) CSQ
      2066.
      Waltz, op. 42, A flat major. cf International Piano Library IPL
      5001/2.
      Waltz, op. 64, no. 1, D flat major. cf Piano works (Saga 5394).
      Waltz, op. 64, no. 1, D flat major. cf Coronet LPS 1722.

Waltz, op. 64, no. 1, D flat major. cf International Piano
Archives IPA 5007/8.
Waltz, op. 64, no. 1, D flat major (arr. and orch. Gerhardt). cf
RCA LRL 1-5094.
Waltz, op. 64, no. 2, C sharp minor. cf Ballade, no. 1, op. 23,
G minor.
Waltz, op. 64, no. 2, C sharp minor. cf Piano works (Columbia
M 32932).
Waltz, op. 64, no. 2, C sharp minor. cf International Piano
Library IPL 5001/2).
Waltz, op. 64, no. 2, C sharp minor. cf Saga 5427.
Waltz, op. 69, no. 1, A flat major. cf Connoisseur Society (Q)
CSQ 2065.
Waltz, op. 70, no. 1, G flat major. cf Ballade, no. 1, op. 23,
G minor.
Waltz, op. 70, no. 2, F minor. cf Ballade, no. 1, op. 23, G
minor.
Waltz, op. posth., E minor. cf Ballade, no. 1, op. 23, G minor.
CHORBAJIAN, John
Songs: Bitter for sweet. cf Orion ORS 75205.
CHRISTINE
Phi-Phi: Ah, Cher Monsieur, excusez-moi. cf London OS 26248.
CHRISTMAS EVE AT THE CATHEDRAL OF ST. JOHN THE DIVINE. cf
Vanguard SVD 71212.
CHRISTOV, Dobri
Hvalite imia Gospodne. cf Harmonia Mundi HMU 105.
Songs: Ije heruvimi, lako da tsaria; Tebe poem; Vo tsarstiviy
tvoiem. cf Harmonia Mundi HMU 137.
CILEA, Francesco
Adriana Lecouvreur: Ecco, respiro appena...Io son l'umile ancella.
cf Telefunken AG 6-41947.
Adriana Lecouvreur: Io son l'umile ancella. cf BELLINI: La
sonnambula: Ah, se una volta sola...Ah, non credea mirarti...
Ah, non giunge.
Adriana Lecouvreur: Io son l'umile ancella; Poveri fiori. cf
VERDI: Il trovatore.
Adriana Lecouvreur: Io son l'umile ancella; Poveri fiori. cf
CATALANI: La Wally: Ebben, Ne andrò lontana.
Adriana Lecouvreur: Non piu nobile. cf Everest SDBR 3382.
Adriana Lecouvreur: Poveri fiori. cf Rococo 5386.
Adriana Lecouvreur: Poveri fiori; Io son l'umile ancella. cf
BOITO: Mefistofele: L'altra notte in fondo al mare.
Adriana Lecouvreur: Respiro appena...Io son l'umile ancella.
cf HMV SLS 5057.
CIMAROSA, Domenico
Arias: Apri il timpano sonoro; A mme sto vico 'nfaccio. cf Li due
Baroni di Rocca Azzurra: Overture.
Concerto, 2 flutes and orchestra, G major: Allegro. cf Decca
SPA 394.
833   Li due Baroni di Rocca Azzurra: Overture. Il maestro di Capella:
Intermezzo. Arias: Apri il timpano sonoro; A mme sto vico
'nfaccio. Gastone Sarti, bar; I Solisti di Milano; Angelo
Ephrikian. Hungaroton SLPX 11585.
+-HFN 12-76 p141   +RR 11-76 p95
+NR 12-76 p11   +SFC 11-14-76 p30
Il maestro di Capella: Intermezzo. cf Li due Baroni di Rozza
Azzurra: Overture.

Sonatas, guitar, C sharp minor, A major. cf RCA SB 6891

CIVIL, Alan
834   Tarantango. HOROWITZ, J.: Music hall suite. HOWARTH: Berne
           patrol. Basel march. The cuckoo. Lucerne song. The old
           chalet. Variations on "Carnival of Venice". Zurich march.
           KOESTSIER: Petite suite. Philip Jones Brass Ensemble; Elgar
           Howarth. CRD Claves DFP 600.
               +Gr 4-76 p1664   +RR 3-76 p55

CLAFLIN, Avery
       Songs: Lament for April 15; The quangle wangle-s hat; Design for
           the atomic age. cf CRI SD 102.

CLARKE
       Songs: The blind ploughman. cf Argo ZFB 95/6.

CLARKE, Jeremiah
       Come, my way, my truth, my life. cf Grosvenor GRS 1039.
       Interlude. cf CBS 61648.
       King William's march. cf CBS 61648.
       Prince of Denmark's march. cf CBS 61648.
       Prince of Denmark's march. cf Klavier KS 551.
       Prince of Denmark's march. cf Saga 5417.
       Suite, D major. cf Argo ZDA 203.
       Trumpet voluntary. cf Decca DPA 523/4.
       Trumpet voluntary. cf Decca PFS 4351.
       Trumpet voluntary. cf Philips 6747 204.
       Trumpet voluntary. cf Polydor 2460 262.
       CLASSIC FAVORITES FOR STRINGS. cf Vanguard SRV 344.

CLAY, Frederic
       I'll sing three songs of Araby. cf HMV EMD 5528.

CLEMENS NON PAPA
       Au joly bois. cf Hungaroton SLPX 11669/70.
       La belle margarite. cf HMV CSD 3766.
       Je prens en gré, à 3. cf Hungaroton SLPX 11669/70.

CLEMENTI, Muzio
       Sonata, flute and piano, op. 2, no. 3, G major. cf Smithsonian
           Collection N 002.
       Sonata, piano, op. 47, no. 2, B flat major: Rondo. cf RCA VH 013.
835   Sonatinas, op. 36, nos. 1-6. GRIEG: Concerto, piano, op. 16, A
           minor. Marie-Aimée Varro, pno; Moravian Philharmonic Orchestra;
           Ludovic Rajter. Orion ORS 75196.
               -HF 12-75 p94   +NR 9-75 p7
               +MJ 4-76 p31

CLERAMBAULT, Louis
       Basse et dessus de trompette. cf Telefunken AF 6-41872.
       Dialogue sur les grande jeux. cf Telefunken AF 6-41872.
       Largo on the G string. cf RCA ARM 4-0942/7.
       Récits de cromorne et cornet séparé. cf Telefunken AF 6-41872.
       Suite du deuxième ton. cf Wealden WS 121.

COATES, Eric
       Calling all workers. cf The jester at the wedding: March; Valse.
       The dambusters march. cf Decca 419.
836   The jester at the wedding: March (The Princess arrives); Valse
           (Dance of the orange blossoms). Calling all workers. The
           merrymakers overture. Miniature suite: Children's dance; Inter-
           mezzo; Scene du bal. The three Elizabeths suite. Felix Kok,
           vln; Birmingham City Symphony Orchestra; Reginald Kilbey. HMV
           ESD 7005. Tape (c) TC ESD 7005. (Reissue from TWO 361).

COATES (cont.)   140

                ++Gr 9-76 p478   +HFN 11-76 p175 tape
                ++Gr 10-76 p658 tape   +RR 9-76 p46
                ++HFN 10-76 p181
        Knightsbridge. cf Philips 6308 246.
        The merrymakers overture. cf The jester at the wedding: March;
            Valse.
        Miniature suite: Children's dance; Intermezzo; Scene du bal. cf
            The jester at the wedding: March; Valse.
        The three Elizabeths suite. cf The jester at the wedding: March;
            Valse.
COCKER, Norman
        Tuba tune. cf Wealden WS 137.
COHAN, George
        Songs: Over there; You're a grand old flag; Yankee doodle boy.
            cf Department of Defense Bicentennial Edition 50-1776.
COLE
        The reason. cf HMV SXLP 50017.
COLERIDGE-TAYLOR, Samuel
        Big lady moon. cf Prelude PRS 2505.
        Hiawatha's wedding feast: Onaway, awake beloved. cf Decca SKL
            5208.
COLLIER
        Arc de Triomphe. cf Philips 6308 246.
COMPERE, Louis
        O bone Jesu, motet. cf HMV SLS 5049.
        Virgo celesti. cf L'Oiseau-Lyre 12BB 203/6.
        A COMPUTER-PERFORMED ORGAN RECITAL. cf Creative Record Service
            R 9115.
CONFREY
        Kitten on the keys. cf Stentorian SP 1735.
CONVERSE, Frederick Shepherd
        Endymion's narrative, op. 10. cf Louisville LS 753/4.
        Flivver ten million. cf Louisville LS 753/4.
COOLIDGE, Peggy Stuart
837   New England autumn. Rhapsody, harp and orchestra. Pioneer dances.
            Spirituals in sunshine and shadow. Westphalian Symphony Orch-
            estra; Siegfried Landau. Turnabout QTV 34635.
                +NR 12-76 p5
        Pioneer dances. cf New England autumn.
        Rhapsody, harp and orchestra. cf New England autumn.
        Spirituals in sunshine and shadow. cf New England autumn.
COOPER, Paul
838   Symphony, no. 4. JONES: Elegy, string orchestra. Let us now
            praise famous men. Houston Symphony Orchestra; Samuel Jones.
            CRI SD 347.
                +MJ 7-76 p57   +SFC 6-6-76 p33
                +NR 7-76 p2
COPLAND, Aaron
839   Appalachian spring. Billy the kid: Suite. Music for the theatre.
            Rodeo. El salón México. NYP; Leonard Bernstein. Columbia MG
            30071 (2). Tape (c) MGT 30071.
                ++HF 6-72 p112 tape   +St 7-76 p74
                +St 10-72 p123 tape
840   Appalachian spring. Lincoln portrait. El salón México. Melvyn
            Douglas, speaker; BSO; Serge Koussevitzky. RCA AVM 1-1739.
            (Reissues from RCA originals).

+ARSC Vol III, no. 2-3   +-HF 11-76 p110
      p85   ++SFC 8-8-76 p38
Appalachian spring. cf Works, selections (Columbia D3M 33720).
Appalachian spring. cf BERNSTEIN: Candide: Overture.
Appalachian spring. cf London CSA 2246.
841   Billy the kid. GERSHWIN: An American in Paris.* RCA Victor
      Symphony Orchestra; Leonard Bernstein. RCA SMA 7016. (*Reissue
      from HMV C 3881/2).
            +Gr 6-76 p43   +-RR 6-76 p42
Billy the kid. cf Works, selections (Columbia D3M 33720).
Billy the kid: The open prairie; Celebration dance; Billy's de-
      mise; The open prairie again. cf Personal Touch 88.
Billy the kid: Suite. cf Appalachian spring.
The city: New England countryside; Sunday traffic. cf Works,
      selections (CBS 61672).
842   Dance panels. Danzón cubano. Latin American sketches (3). El
      salón México. LSO, NPhO; Aaron Copland. Columbia M 33269.
      (also CBS 73451).
            +Audio 4-76 p88   +RR 9-75 p34
            +Gr 9-75 p452   ++SFC 6-15-75 p24
            +HF 8-75 p82   +St 7-75 p95
            +HFN 9-75 p97   +Te 12-75 p40
            +NR 6-75 p2
Dance panels. cf Works, selections (Columbia D3M 33720).
843   Dance symphony. STEVENS: Symphony, no. 1. Japan Philharmonic
      Symphony Orchestra; Akeo Watanabe. CRI SD 129.
            +-Gr 1-76 p1191   +-RR 12-75 p46
            +-HFN 1-76 p105
Danzón cubano. cf Dance panels.
844   Down a country lane. John Henry. Letter from home. Music for
      movies. The red pony. NPhO; Aaron Copland. Columbia M 33586.
            +HF 4-76 p100   ++SFC 12-21-75 p39
            +NR 1-76 p2
Down a country lane. cf Works, selections (CBS 61672).
Duo, flute and piano. cf BURTON: Sonatina, flute and piano.
Fanfare for the common man. cf Works, selections (Columbia D3M
      33720).
John Henry. cf Down a country lane.
John Henry. cf Works, selections (CBS 61672).
Latin American sketches (3). cf Dance panels.
Letter from home. cf Down a country lane.
Letter from home. cf Works, selections (CBS 61672).
845   Lincoln portrait. Our town. Outdoor overture. Quiet city.
      Charlton Heston, narrator; Utah Symphony Orchestra; Maurice
      Abravanel. Vanguard SRV 348. (Reissue from VSD 2115).
            +MJ 3-76 p24   +SFC 5-2-76 p38
Lincoln portrait. cf Appalachian spring.
Lincoln portrait. cf Works, selections (Columbia D3M 33720).
Music for movies. cf Down a country lane.
Music for the theatre. cf Appalachian spring.
Of mice and men: Barley wagons, Threshing machines. cf Works,
      selections (CBS 61672).
Our town. cf Lincoln portrait.
Our town. cf Works, selections (Columbia D3M 33720).
Our town: Grovers corner. cf Works, selections (CBS 61672).
Outdoor overture. cf Lincoln portrait.

Piano blues (4). cf Supraphon 111 1721/2.
Pieces, string orchestra (2). cf Vox SVBX 5305.
Quiet city. cf Lincoln portrait.
Quiet city. cf Argo ZRG 845.
The red pony. cf Down a country lane.
The red pony. cf Works, selections (CBS 61672).
Rodeo. cf Appalachian spring.
Rodeo. cf Works, selections (Columbia D3M 33720).

846   El salón México. GERSHWIN: An American in Paris. Porgy and
      Bess (arr. Russell Bennett). LPO; John Pritchard. Classics
      for CFP 40240. Tape (c) TC CFP 40240.
         +HFN 3-76 p91   -RR 3-76 p39
         ++HFN 12-76 p153 tape

El salón México. cf Appalachian spring (Columbia 30071).
El salón México. cf Appalachian spring (RCA 1-1739).
El salón México. cf Dance panels.
El salón México. cf Works, selections (Columbia D3M 33720).
Sonata, piano. cf BERNSTEIN: Seven anniversaries.
Sonata, violin and piano. cf BINKERD: Sonata, violin and piano.
Songs: Ching-a-ring chaw. cf Hungaroton SLPX 11696.

847   Symphony, no. 3. HARRIS: Symphony, no. 3, NYP; Leonard Bern-
      stein. CBS 61681. (Reissues from 72559, 72399).
         +Gr 6-76 p44   +-RR 6-76 p42
         +HFN 6-76 p102

848   Works, selections: The city: New England countryside; Sunday
      traffic. Down a country lane. John Henry. Letter from home.
      Of mice and men: Barley wagons, Threshing machines. Our town:
      Grovers corner. The red pony. NPhO, LSO; Aaron Copland.
      CBS 61672.
         +Gr 3-76 p1464   +HFN 3-76 p91

849   Works, selections: Appalachian spring. Billy the kid. Dance
      panels. Fanfare for the common man. Lincoln portrait. Our
      town. Rodeo. El salón México. LSO, NPhO; Henry Fonda, nar-
      rator; Aaron Copland. Columbia D3M 33720 (3).
         ++NR 1-76 p2   +SFC 12-21-75 p39

CORDELL, Frank
    Patterns. cf Translatlantic TRA 308.

CORELLI, Arcangelo
850   Concerti grossi, op. 6, nos. 2-3, 8, 12. AMF; Neville Marriner.
      Argo Tape (c) KZRC 15016.
         +HFN 1-76 p125 tape   ++RR 1-76 p65 tape

851   Concerti grossi, op. 6, nos. 6-8. AMF; Neville Marriner. Argo
      ZRG 828. (Reissue from ZRG 773/5).
         +Gr 1-76 p1191   ++RR 12-75 p46
         +HFN 12-75 p171

852   Concerto grosso, op. 6, no. 8, G minor. GLUCK: Chaconne. PACHEL-
      BEL: Kanon. RICCIOTTI: Concertino, no. 2, G major. Stuttgart
      Chamber Orchestra; Karl Münchinger. London 6206. Tape (c)
      CS5-6206.
         +HF 7-76 p114 tape

853   Concerto grosso, op. 6, no. 8, G minor. LOCATELLI: Concerto
      grosso, op. 1, no. 8, F minor. MANFREDINI: Concerto grosso,
      op. 3, no. 12, G major. TORELLI: Concerto grosso, op. 8, no.
      6, G minor. I Musici. Philips 6580 121.
         +Gr 6-76 p44   +HFN 6-76 p102
         +HFN 5-76 p115   ++RR 4-76 p39

Concerto grosso, op. 6, no. 8, G minor. cf Decca SDD 411.
Sonata, violin, no. 12, D minor. cf HMV HLM 7077.
854   Sonatas, violin and continuo, op. 5 (11). Eduard Melkus, vln;
        Huguette Dreyfus, hpd, org; Karl Scheit, lt; Garo Altmacayan,
        vlc; Vienna Capella Academica. DG 2533 132/3 (2).
          +Gr 10-75 p640   ++RR 9-75 p58
          +-HF 6-73 p74   ++St 6-73 p118
          ++HFN 9-75 p98   +STL 10-5-75 p36
          +PRO 9/10-75 p18
855   Sonata, violin and continuo, op. 5, no. 5, G minor. FORSTER:
        Concerto, waldhorn, E flat major. HANDEL: Concerto grosso,
        op. 29, F major. TELEMANN: Concerto, horn, D major. Hermann
        Baumann, hn; Herbert Tachezi, org. Telefunken 6-41932.
          *MU 10-76 p14   +NR 2-76 p15
          +-HFN 5-76 p101   -RR 4-76 p63
CORNYSSHE, William
      Ah, Robin. cf BASF 25 22286-1.
      Ah, Robin; Hoyda jolly Rutterkin. cf Harmonia Mundi 204.
      Blow thy horn, hunter. cf BASF 25 22286-1.
      Hoyda, jolly Rutterkin. cf BASF 25 22286-1.
      Part songs (2). cf Argo ZRG 833.
CAROSO, Fabrizio
      Laura soave: Gagliarda, Saltarello (Balleto). cf DG 2530 561.
CORRETTE, Michel
      Magnificat du 8ème ton. cf Calliope CAL 1917.
      Menuets, nos. 1, 2. cf DG Archive 2533 172.
CORTECCIA, Francesco
      St. John Passion. Arnold Foà, speaker; Schola Cantorum Francesco
        Cordini; Fosco Corti. DG Archive 2533 301.
          +-Gr 4-76 p1645   +RR 5-76 p69
          ++HFN 5-76 p99
COSTELEY, Guillaume
      Hèlas, hèlas, que de mal. cf L'Oiseau-Lyre 12BB 203/6.
      Mignonne, allons voir. cf Argo ZRG 667.
COUPERIN, François
      L'apothéose de Corelli. cf Works, selections (Philips 6747 174).
      L'apothéose de Lully. cf Works, selections (Philips 6747 174).
      L'art de toucher le clavecin: Preludes (8). cf Livres de clavecin
        (Telefunken EK 6-35276).
857   Concerts royaux. Nouveaux concerts. Thomas Brandis, vln; Heinz
        Holliger, ob; Aurèle Nicolet, flt; Josef Ulsamer, Laurenzius
        Strehl, viola da gamba; Manfred Sax, bsn; Christianne Jaccottet,
        hpd, and others. DG Archive 2723 046 (4). (also DG 2712 003).
          +Gr 10-76 p616   ++RR 10-76 p48
          +HFN 10-76 p167
      Concerts royaux, nos. 1-4. cf Works, selections (Philips 6747 174).
858   Livres de clavecin, Bk I: Ordres, nos, 3, 4; Bk II, Ordres, nos.
        6, 11; Bk III, Ordres nos. 13-18; Bk IV, Ordres, nos. 24, 26-27.
      L'art de toucher le clavecin: Preludes (8). Huguette Dreyfus,
        hpd. Telefunken EK 6-35276 (4).
          +Audio 9-76 p85   +NR 4-76 p13
          +-Gr 1-76 p1217   +-PRO 3/4-76 p17
      Livres de clavecin, Bk I, Ordre, no. 3: La favorite; Bk I, Ordre,
        no. 4: Le reveil matin. cf Angel S 36095.
      Livres de clavecin, Bk III, Ordre, no. 14: La linote efarouchee;
        Le rossignol vainqueur. cf Orion ORS 76216.

COUPERIN, F. (cont.)   144

Livres de clavecin, Bk IV, Ordre, no. 17: Les petits Moulins à
    vent. cf RCA ARM 4-0942/7.
Messe pour les paroisses. cf Pelca PRSRK 41017/20.
Musête de choisi. cf HMV SLS 5022.
Musête de taverni. cf HMV SLS 5022.
La nations: L'Espagnola. cf Telefunken DT 6-48075.
Nouveaux concerts. cf Concerts royaux.
Nouveaux concerts, nos. 5-7, 9-12, 14. cf Works selections
    (Philips 6747 174).
La steinquerque. cf Works, selections (Philips 6747 174).
La sultane, D minor. cf Works, selections (Philips 6747 174).
La superbe. cf Works, selections (Philips 6747 174).
859   Works, selections: L'apothéose de Corelli. L'apothéose de Lully.
      Concerts royaux, nos. 1-4. Nouveaux concerts, nos. 5-7, 9-12,
      14. La sultane, D minor. La superbe. La steinquerque.
      Sigiswald Kuijken, baroque violin, treble viol, bass viola da
      gamba; Lucy van Dael, Janine Rubinlicht, baroque violins;
      Wieland Kuijken, baroque violoncello, bass viola da gamba;
      Adelheid Glatt, bass viola da gamba; Barthold Kuijken, German
      flute, recorder; Oswald van Olmen, Frans Brüggen, German flute;
      Bruce Haynes, Jürg Schaeftlein, Paul Dombrecht, baroque oboe;
      Hans Jürg Lange, Milan Turkovíc, baroque bassoon; Robert
      Kohnen, hpd, speaker. Philips 6747 174 (6).
          +-Gr 10-75 p645   ++MT 3-76 p235
          +HFN 10-75 p141   +RR 9-75 p58
COUPERIN, Louis
Chaconne, C major. cf Vista VPS 1030.
Passacaille, C major. cf Harmonia Mundi HMU 334.
La Piémontaise, A minor. cf Harmonia Mundi HMU 334.
Suite, A minor. cf Hungaroton SLPX 11741.
860   Suites, harpsichord, G minor, D major, A minor, F major. Alan
      Curtis, hpd. DG Archive 2533 325.
          +Gr 12-76 p1021   ++RR 12-76 p78
Le tombeau de M. Blancrocher. cf Saga 5402.
COURTS AND CHAPELS OF RENAISSANCE FRANCE. cf Titanic TI 4.
COWELL, Henry
Advertisment. cf Works, selections (CRI SD 109).
Aeolian harp. cf Works, selections (CRI SD 109).
The banshee. cf Works, selections (CRI SD 109).
Hymn and fuguing tune, no. 10. cf Argo ZRG 845.
Lilt of the reel. cf Works, selections (CRI SD 109).
Prelude, violin and harpsichord. cf Works, selections (CRI SD 109).
Sinister resonance. cf Works, selections (CRI SD 109).
The tides of Manaunaun. cf Works, selections (CRI SD 109).
Twilight: Texas. cf Columbia MG 33728.
861   Works, selections: Advertisement. The banshee. Aeolian harp.
      Lilt of the reel. Prelude, violin and harpsichord. Sinister
      resonance. The tides of Manaunaun. PINKHAM: Cantilena.
      Capriccio. Concerto, celeste and harpsichord. HOVHANESS: Duet
      violin and harpsichord. Henry Cowell, pno; Robert Brink, vln;
      Edward Low, celeste; Daniel Pinkham, hpd. CRI SD 109.
          +NR 8-76 p14
CRAUS, Stephan
Chorea, Auff und nider. cf Hungaroton SLPX 11721.
Tantz, Hupff auff. cf Hungaroton SLPX 11721.
Die trunke pinter. cf Hungaroton SLPX 11721.

CRAWFORD
      The Air Force song.   cf Department of Defense Bicentennial Edition
            50-1776.
CRECQUILLON, Thomas
      Content desire.   cf Hungaroton SLPX 11669/70.
      Un gay bergier.   cf Hungaroton SLPX 11669/70.
CRESPO
      Norteña, Homenaje a Julian Aguirre.   cf Saga 5412.
CRESTON, Paul
      Midnight: Mexico.   cf Columbia MG 33728.
      A rumor.   cf Argo ZRG 845.
CROFT, William
862   By purling streams.   A hymn on divine musick.   Sonatas, violin,
            G minor, B minor.   Suites, harpsichord, C minor, E minor, E
            flat major, C minor.   Honor Sheppard, s; Marjorie Lavers, vln;
            Michael Dobson, ob; Jane Ryan, vla da gamba; Robert Elliott,
            hpd.   Oryx 1730.
                  +St 8-76 p92
      A hymn on divine musick.   cf By purling streams.
      O worship the King.   cf Saga 5225.
      Sonata, recorder, G major.   cf Orion ORS 76216.
      Sonatas, violin, G minor, B minor.   cf By purling streams.
      Suites, harpsichord, C minor, E minor, E flat major, C minor.   cf
            By purling streams.
      Trumpet tune, D major.   cf CBS 61648.
      Voluntary, organ and trumpets.   cf CBS 61648.
CROIX, De la
      S'amours eust point de poer.   cf Candide CE 31095.
CROSS, Lowell
      Etudes, magnetic tape.   cf CRI SD 342.
CROSSE
      Unter den Linden.   cf Philips 6308 246.
CROSSE, Gordon
      Ariadne, op. 31.   cf BRITTEN: Phantasy, oboe and strings, op. 2.
      Concerto da camera.   cf BIRTWISTLE: Tragoedia.
      The new world, op. 25.   cf BERKELEY: Songs (Argo ZRG 788).
863   Purgatory.   Glenville Hargreaves, bar; Peter Bodenham, t; Royal
            Northern School of Music Orchestra and Chorus; Michael Lan-
            kester.   Argo ZRG 810.
                  +Gr 12-75 p1093   +NR 2-76 p12
                  +HFN 12-75 p152   ++RR 11-75 p32
                  +MJ 7-76 p56   +SR 4-17-76 p51
                  +MM 7-76 p37
      Some marches on a ground, op. 28.   cf ADDISON: Concerto, trumpet,
            strings and percussion.
CROUCH, Frederick
      Kathleen Mavourneen.   cf CBS 61746.
CRUMB, George
864   Ancient voices of children.   Jan DeGaetani, ms; Michael Dash,
            boy soprano; Contemporary Chamber Ensemble; Arthur Weisberg.
            Nonesuch H 71255.   Tape Advent (c) F 1035.
                  +ARG 8-71 p807   +NYT 7-4-71 pD17
                  +HF 8-71 p78   *SR 7-31-71 p44
                  ++HF 4-76 p148 tape   ++St 9-71 p80
                  +NR 8-71 p10   +St 7-76 p75
865   Black angels, electric string quartet.   JONES: Quartet, strings,

          no. 6. Sonatina. Gilbert Kalish, pno; New York String Quartet.
          CRI SD 283.
                    +-Gr 4-76 p1624   +NR 6-72 p8
                    ++HF 10-72 p80   +-RR 1-76 p40
                    +LJ 10-72 p44   ++St 11-72 p104
866   Black angels, electric string quartet. RAXACH: Quartet, strings,
          no. 2. LEEUW: Quartet, strings, no. 2. Gaudeamus Quartet.
          Philips 6500 881.
                    +-HF 6-75 p106   ++St 11-75 p127
                    ++NR 6-75 p8   +Te 3-76 p29
                    +-SR 6-14-75 p46
867   Music for a summer evening (Makrokosmos III). Gibert Kalish,
          James Freeman, pno; Raymond DesRoches, Richard Fitz, perc.
          Nonesuch H 71311.
                    +-Gr 2-76 p1355   +-RR 10-75 p75
                    ++HF 10-75 p73   ++SFC 2-15-76 p38
                    ++HFN 12-75 p152   ++St 12-75 p117
                    ++MQ 4-76 p293   +Te 3-76 p29
                    +NR 10-75 p14
       Night music, II. cf Desto DC 6435/7.
868   Sonata, solo violoncello. HINDEMITH: Sonata, solo violoncello,
          op. 25, no. 3. WELLESZ: Sonata, solo violoncello, op. 30.
          YSAYE: Sonata, solo violoncello, op. 28. Robert Sylvester, vlc.
          Desto DC 7169.
                    ++HF 6-75 p108   ++St 10-75 p120
                    +SFC 6-6-76 p33
869   Sonata, solo violoncello. DALLAPICCOLA: Ciaconna, intermezzo e
          adagio. HINDEMITH: Sonata, solo violoncello, op. 25, no. 3.
          SCHULLER: Fantasy, op. 19. Roy Christensen, vlc. Gasparo GS
          101.
                    ++IN 5-76 p12
870   Twelve fantasy pieces after the Zodiac (Makrokosmos II). Robert
          Miller, pno. Odyssey Y 34135.
                    ++ARG 12-76 p25   ++NR 12-76 p13
CRUSELL, Bernard
871   Quartet, clarinet, no. 2, op. 4, C minor. HUMMEL: Quartet, clari-
          net, E flat major. The Music Party. L'Oisea-Lyre Florilegium
          DSLO 501.
                    ++Audio 9-76 p85   ++NR 5-75 p8
                     +Gr 11-74 p915   +RR 11-74 p17
                    +-HF 11-75 p104   +SFC 12-28-75 p30
CSERMAK, Antal
872   Hungarian dances (6). The threatening dance or the love of the
          fatherland. ROZSAVOLGYI: Czardas. First Hungarian round dance.
          Hungarian Chamber Orchestra; Vilmos Tátrai. Hungaroton SLPX
          11698.
                    +NR 9-76 p5
       The threatening dance or the love of the fatherland. cf Hungarian
          dances.
CUMMING, Richard
       Songs: Go lovely rose; The little black boy; Memory, hither come.
          cf Duke University Press DWR 7306.
CUNDICK
       The west wind. cf Columbia M 34134.
CURTIS, E. de
       Non ti scordar di me. cf DG 2530 700.

CURTIS-SMITH, Curtis
    Rhapsodies, piano. cf CRI SD 345.
873   Sonorous inventions (5). MORRIS: Phases. Gerald Fischbach, vln;
        Robert Morris, William Albright, Curtis Curtis-Smith, pno.
        CRI SD 346.
            +CL 10-76 p7   +NR 6-76 p7
            ++MJ 7-76 p57   *St 8-76 p100
CUSTER, Arthur
874   Found objects. Number 3, contrabass and tape. Number 5, five
        instruments and tape. Number 6, flute and tape. Number 8,
        violin and tape. Barbara Poularikas, Daniel Kobialka, vln;
        Philips McClintock, clt; John Pellegrino, tpt; Louis Pezzullo,
        trom; Victor Preston, Bertram Turetzky, contrabass; Marjorie
        Shansky, flt; Arthur Custer, pno and conductor. Serenus SRS
        12045.
            +ARG 12-76 p27   +NR 4-76 p15
    Number 3, contrabass and tape. cf Found objects.
    Number 5, five instruments and tape. cf Found objects.
    Number 6, flute and tape. cf Found objects.
    Number 8, violin and tape. cf Found objects.
CUTTING, Francis
    Galliard, G minor. cf Philips 6500 282.
    Greensleeves. cf Saga 5426.
    Packington's pound. cf RCA RK 11708.
CUTTS
    As the bridegroom to his chosen. cf Grosvenor GRS 1039.
CZERNY, Karl
    Variations on La ricordanza, op. 33. cf RCA VH 013.
DAGERE
    Downberry down. cf HMV CSD 3766
DAHL, Ingolf
875   Hymn. Sonata seria, piano. Sonata pastorale. Charles Fierro,
        pno. Orion ORS 76209.
            ++MJ 7-76 p57   +NR 4-76 p12
    Sonata pastorale. cf Hymn.
    Sonata seria, piano. cf Hymn.
DALLAPICCOLA, Luigi
    Ciaconna, intermezzo e adagio. cf CRUMB: Sonata, solo violoncello.
    Piccola musica notturna. cf BUSONI: Berceuse elégiaque, op. 42.
    Preghiere. cf BUSONI: Berceuse elégiaque, op. 42.
876   Il Prigionero. Giulia Barrera, con; Maurizio Mazzieri, bar;
        Romano Emile, Gabor Carelli, Ray Harrell, t; University of
        Maryland Chorus; Washington National Symphony Orchestra; Antal
        Dorati. Decca HEAD 10. (also London OSA 1166).
            +Gr 5-75 p2010   ++NYT 12-21-75 pD18
            +HF 1-76 p78   +ON 1-3-76 p33
            +HFN 7-75 p79   +RR 5-75 p23
            +-MJ 5-76 p29   ++SFC 11-9-75 p22
            +-MM 10-76 p40   +-St 5-76 p113
            ++MT 10-75 p885   +Te 6-75 p48
            +NR 2-76 p10
    Sex carmina alcaei. cf BUSONI: Berceuse elégiaque, op. 42.
    Sicut umbra. cf Tempus destruendi, Tempus aedificandi.
877   Tempus destruendi, Tempus aedificandi. Sicut umbra. SHAW: A
        lesson from Ecclesiastes. Music when soft voices die. Peter
        and the lame man. To the Bandusian spring. Sybil Michelow,

ms; BBC Singers; London Sinfonietta Orchestra; Gary Bertini.
Argo ZRG 791.

+Gr 12-75 p1086	+NR 2-76 p8
+-HF 7-76 p92	+NYT 12-21-75 pD18
+HFN 4-76 p103	+-RR 11-75 p84
+MM 10-76 p40	+Te 3-76 p33
+MT 7-76 p580	

DALZA, Joan Ambrosio
Calata ala Spagnola. cf DG Archive 2533 184.
Recercar; Suite ferrarese; Tastar de corde. cf L'Oiseau-Lyre
12BB 203/6.
DAME NELLI MELBA, THE LONDON RECORDINGS 1904-1926. cf HMV RLS 719.
DANCE MUSIC OF THE HIGH BAROQUE. cf DG Archive 2533 172.
DANCE MUSIC OF THE RENAISSANCE. cf CBS 76183.

DANDRIEU, Jean
Allons voir de divin Gage. cf Telefunken AF 6-41872.
Armes, amours. cf Telefunken ER 6-35257.
Armes, amours: O flour des flours. cf HMV SLS 863.

DANKS
Silver threads among the gold. cf Pye Ember GVC 51.

DANZI, Franz
878   Quintet, op. 68, no. 2, F major. Quintet, op. 68, no. 3, D minor.
Soni Ventorum Wind Quintet. Crystal S 251.
+NR 6-76 p9
Quintet, op. 68, no. 3, D minor. cf Quintet, op. 68, no. 2,
F major.

DAQUIN, Louis
Le coucou. cf CBS 73487.
The cuckoo. cf Decca DCSP 367.
879   Noëls (12). MARCHAND: Basse de cromorne. Duo. Dialogue (2).
Grand jeu. Fugue. Plein jeu. Quatuor. Tierce en taille.
Arthur Wills, org. Saga 5433/4 (2).
+Gr 12-76 p1021   +-RR 12-76 p83
Noël étranger. cf Telefunken AF 6-41872
Noël no. VII en trio et en dialogue. cf Abbey LPB 752.

DARE, Elkanah Kelsey
Babylonian captivity. cf Nonesuch H 71276.

DARGOMIZHSKY, Alexander
Russalka: Natasha's aria, Act 4. cf Rubini GV 63.
Songs: Romance. cf Desto DC 7118/9.

DARWALL
Ye holy angels bright. cf Saga 5225.

DAVID, Johann
880   Toccata and fugue, F minor. DEMESSIEUX: Prelude and fugue, C
major. KARL-ELERT: Music for organ. McCABE: Le poisson
magique (Meditation after Paul Klee). Graham Barber, org.
Vista VPS 1025.
+Gr 1-76 p1221   +RR 3-76 p60
+HFN 2-76 p101   +STL 1-11-76 p36

DAVIDOV
Night, love and moon. cf Rubini RS 301.

DAVIDOVSKY, Mario
Electronic study, no. 3. cf Finnadar QD 9010.
Junctures. cf Nonesuch HB 73028.
Synchronisms, no. 3. cf Delos DEL 25406.

DAVIES
Creep-mouse. cf HMV EMD 5528.

DAVIES, Peter Maxwell
881   Antechrist. From stone to thorn. L'homme armé. Hymnos. Mary
         Thomas, s; Vanessa Redgrave, speaker; Alan Hacker, clt;
         Stephen Pruslin, pno; The Fires of London; Peter Maxwell
         Davies. L'Oiseau-Lyre DSLO 2.
                +-Gr 12-72 p1161   *NYT 3-4-73 pD31
                ++HF 8-73 p79   ++RR 12-72 p30
                +HFN 12-72 p2441   +SR 4-73 p72
                ++MQ 4-76 p302   ++St 8-73 p101
                +NR 7-73 p13
      Ave Maria. cf HMV HQS 1350.
      Fantasia on John Taverner's 'In Nomine', no. 2. cf Taverner:
         Points and dances.
      From stone to thorn. cf Antechrist.
      L'homme armé. cf Antechrist.
882   Hymn to St. Magnus. Psalm, no. 124. Renaissance Scottish dances.
         Mary Thomas, s; The Fires of London; Peter Maxwell Davies.
         L'Oiseau-Lyre DSLO 12.
                ++HFN 12-76 p141   +RR 12-76 p90
      Hymnos. cf Antechrist.
      Organ fantasia on "O magnum mysterium". cf Argo 5BB 1013/5.
      Psalm, no. 124. cf Hymn to St. Magnus.
      Renaissance Scottish dances. cf Hymn to St. Magnus.
883   Taverner: Points and dances. Fantasia on John Taverner's 'In
         Nomine', no. 2. The Fires of London; Maxwell Davies, NPhO;
         Charles Groves. Argo ZRG 712.
                +-Gr 5-73 p2048   ++MQ 1-76 p139
                +HF 6-75 p89   +RR 5-73 p25
                +HFN 5-73 p983
DAVIES, Walford
      The Lord is my shepherd, Psalm, no. 23. cf Abbey LPB 761.
      Out of the deep, Psalm, no. 130. cf HMV SQ CSD 3768.
      Solemn melody. cf Decca DPA 523/4.
      Songs: God be in my head. cf Abbey LPB 739.
      Songs: God be in my head; Psalm, no. 121. cf Audio 3.
DAVIS
      Little drummer boy. cf RCA PRL 1-8020.
      Songs: God will watch over you. cf Argo ZFB 95/6.
DAVY, Richard
      Ah mine heart. cf APPLEBY: Magnificat.
      The Bay of Biscay. cf HMV ESD 7002.
DAWSON
      Ain'a that good news. cf Grosvenor GRS 1034
DAY, Edgar
      Psalm 84: O how amiable. cf Polydor 2460 250.
      When I survey the wondrous cross. cf Abbey LPB 757.
DEBIASY
      Dance. cf Saga 5421.
DE BO
      Sonatina, D major. cf Orion ORS 76235
DE BOIS (Du Boys), F.
      Je suis deshéritée. cf Argo ZRG 667.
DEBUSSY, Claude
      Arabesques (2). cf Piano works (DG 2535 158).
      Arabesques (2). cf Piano works (Muza SX 1049).
      Arabesques (2). cf Piano works (Seraphim S 60253).

Arabesques, nos. 1 and 2. cf Stentorian SC 1724.
Ballade. cf Piano works (Seraphim S 60253).
884 La boîte à joujoux. Printemps. French National Radio Orchestra;
       Jean Martinon. Angel S 37124.
           +-HF 1-76 p86   ++SFC 9-21-75 p34
           ++NR 12-75 p4
Chanson de Bilitis (3). cf Desto DC 7118/9.
Chanson de Bilitis: La chevelure. cf RCA ARM 4-0942/7.
Children's corner suite. cf HMV HQS 1364.
Children's corner suite: Golliwog's cakewalk. cf Connoisseur
       Society (Q) CSQ 2065.
Children's corner suite: Golliwog's cakewalk. cf Saga 5427.
Children's corner suite: Golliwog's cakewalk. cf Supraphon
       111 1721/2.
Children's corner suite: Serenade for the doll. cf RCA VH 020.
Cortège et air de danse. cf Stentorian SC 1724.
La damoiselle élui. cf BERLIOZ: Les nuits d'été, op. 7.
Danse. cf Piano works (DG 2535 158).
Danse. cf Piano works (Seraphim S 60253).
En blanc et noir. cf Piano works (Telefunken DX 6-35272).
L'Enfant prodigue: Air de Lia. cf Discophilia KIS KGG 3.
L'Enfant prodigue: Prelude. cf RCA ARM 4-0942/7.
Epigraphes antiques (6). cf Piano works (Telefunken DX 6-35272).
885 Estampes: Jardins sous la pluie. Images: Reflets dans l'eau;
       Poissons d'or. L'isle joyeuse. LISZT: Années de pelerinage,
       1st year, G 160: Au lac de Wallenstadt. Années de pelerinage,
       3rd year, G 161: Les jeux d'eau a la Villa d'Este. RAVEL:
       Gaspard de la nuit: Ondine. Miroirs: Une barque sur l'océan.
       Jeux d'eau. SMETANA: On the seashore, op. 17. Albert Ferber,
       pno. Saga 5422.
           -HFN 9-76 p127   +-RR 4-76 p63
886 Etudes. Anthony di Bonaventura, pno. Connoisseur Society (Q)
       CSQ 2074.
           +Gr 8-76 p319   +NR 11-75 p13
           ++HF 11-75 p102   ++SFC 4-20-75 p23
           +-MJ 4-76 p30   +St 9-75 p107
887 Etudes. Paul Jacobs, pno. Nonesuch H 71322.
           ++Gr 9-76 p443   +-RR 9-76 p80
           ++HF 10-76 p104   +-SFC 10-31-76 p35
           +HFN 9-76 p119   +St 10-76 p121
           ++NR 8-76 p12
Fêtes galantes, Set I: Clair de lune. cf RCA LRL 1-5127.
Images. cf Works, selections (RCA FVL 3-7276).
888 Images, Bk I. RAVEL: Gaspard de la nuit. REZAC: Sisyfova nedele
       (The Sunday of Sisyfos). Boris Krajný, pno. Supraphon 110 227.
           +-HFN 2-76 p95   +-RR 6-75 p72
Images: Cloches à travers les feuilles; Et la lune descend sur le
       temple qui fut; Hommage à Rameau; Mouvement; Reflets dans l'eau;
       Poissons d'or. cf Piano works (Muza SX 1049).
Images: Reflets dans l'eau. cf BACH: Partita, harpsichord, no. 1,
       S 825, B flat major.
Images: Reflets dans l'eau; Poissons d'or. cf Estampes: Jardins
       sous la pluie.
889 Images pour orchestra: Ibéria. Nocturnes, nos. 1-3. Washington
       Oratorio Society; Washington National Symphony Orchestra; Antal
       Dorati. Decca SXL 6742. Tape (c) KSXC 6742. (also London CS
       6968 Tape (c) 5-6968).

```
        +-Gr 11-76 p782            +-RR 11-76 p53
        ++HFN 10-76 p167           ++SFC 12-26-76 p34
         +HFN 12-76 p155 tape
```
Images pour orchestra: Ibéria. cf RCA CRM 5-1900.
Intermezzo. cf Sonata, violoncello and piano, no. 1, D minor.
L'isle joyeuse. cf Piano works (DG 2535 158).
L'isle joyeuse. cf Piano works (Muza SX 1049).
L'isle joyeuse. cf RCA ARL 1-1176.
Lindaraja. cf Piano works (Telefunken DX 6-35272).
The little shepherd. cf RCA LRL 1-5127.
Marche écossaise. cf Piano works (Telefunken DX 6-35272).
Le martyre de Saint-Sebastian. cf Works, selections (RCA FVL
    3-7276).
Masques. cf Piano works (DG 2535 158).
890   La mer. RAVEL: Rapsodie espagnole. BSO; Charles Munch. Camden
      CCV 5039.
          +HFN 4-76 p121   +-RR 4-76 p40
891   La mer. Prélude à l'après-midi d'un faune. FAURE: Masques et
      bergamasques, op. 112. LPO; Vernon Handley. Classics for
      Pleasure CFP 40231.
          +HFN 4-76 p103   +RR 2-76 p29
892   La mer. HOLST: The planets, op. 32. Interlochen Arts Academy
      Orchestra, World Youth Symphony Orchestra, Interlochen; Thor
      Johnson. Golden Crest GCIN 404 (2).
          ++IN 2-76 p8
893   La mer. Nocturnes. PhO; Carlo Maria Giulini. HMV Tape (c) TC
      EXE 185.
          +HFN 4-76 p125 tape
894   La mer. RAVEL: Daphnis et Chloé: Suite, no. 2. La valse.
      Hungarian Radio and Television Orchestra; György Lehel. Hun-
      garoton SLPX 11761.
          +NR 6-76 p2   ++SFC 5-23-76 p36
La mer. cf Piano works (Telefunken DX 35272).
La mer. cf Works, selections (RCA FVL 3-7276).
La mer. cf Philips 6780 030.
La mer. cf RCA CRM 5-1900.
Nocturnes. cf La mer.
Nocturnes, nos. 1-3. cf Images pour orchestra: Ibéria.
Nocturnes: Fêtes (3). cf Desmar DSM 1005.
Nocturnes: Nuages; Fêtes; Printemps. cf Works, selections (RCA
    FVL 3-7276).
Pelléas et Mélisande: C'est le dernier soir. cf Decca GOSD 674/6.
Pelleas et Mélisande: Voici ce qui'il écrit; Tu ne sais pas
    pourquoi. cf HMV RLS 716.
Le petite nègre. cf Golden Crest CRS 4143.
Petite pièce, B flat major. cf Decca SPA 395.
Petite pièce, B flat major. cf Hungaroton SLPX 11748.
Petite suite. cf Piano works (Telefunken DX 6-35272).
Petite suite: En bateau. cf Connoisseur Society CS 2070.
895   Piano works: Arabesques (2). L'isle joyeuse. Danse. Masques.
      La plus que lente. Pour le piano. Suite bergamasque. Tamás
      Vásáry, pno. DG 2535 158. (Reissue from 139458).
          +-Gr 1-76 p1217   +-RR 1-76 p48
          +HFN 2-76 p117
896   Piano works: Arabesques (2). Images: Cloches à travers les feuilles;
      Et la lune descend sur le temple qui fut; Hommage à Rameau;

Mouvement; Reflets dans l'eau; Poissons d'or. L'isle joyeuse.
Bernard Ringeissen, pno. Muza SX 1049.
       +-NR 10-76 p13
897 Piano works: Arabesques (2). Ballade. Danse. Pour le piano.
     Rêverie. Suite bergamasque. Aldo Diccolini, pno. Seraphim
     S 60253.
       ++HF 3-76 p84   ++NR 1-76 p13
898 Piano works: Epigraphes antiques (6). En blanc et noir. Linda-
     raja. Marche écossaise. La mer. Petite suite. Prélude à
     l'après-midi d'un faune. Noël Lee, Bernard Ringeissen, pno.
     Telefunken DX 6-35272 (2).
          -Gr 8-76 p312   +RR 7-76 p71
       +-HFN 8-76 p79
     La plus que lente. cf Piano works (DG 2535 158).
     La plus que lente. cf CBS 76420.
     La plus que lente. cf Columbia M2 33444.
     La plus que lente. cf Connoisseur Society CS 2070.
     La plus que lente. cf Odyssey Y 33825.
     La plus que lente. cf RCA ARM 4-0942/7.
     Pour le piano. cf Piano works (DG 2535 158).
     Pour le piano. cf Piano works (Seraphim S 60253).
899 Preludes, Bk I. Livia Rev, pno. Saga 5391. Tape (c) CA 5391
       +Gr 3-75 p1678   +-RR 4-75 p50
       +HFN 12-76 p155 tape
     Prelude, Bk I, no. 8: La fille aux cheveux de lin. cf Decca
     SPA 473.
     Prelude, Bk I, no. 8: La fille aux cheveux de lin. cf HMV HLM
     7077.
     Prelude, Bk I, no. 8: La fille aux cheveux de lin. cf London
     CS 7015.
     Prelude, Bk I, no. 8: The girl with the flaxen hair. cf Pye TB
     3004.
     Prelude, Bk I, no. 8: La fille aux cheveux de lin (2). cf RCA
     ARM 4-0942/7.
     Prelude, Bk I, no. 8: La fille aux cheveux de lin. cf Transat-
     lantic XTRA 1160.
     Prelude, Bk I, no. 12: Minstrels. cf HMV SQ ASD 3283.
900 Prélude à l'après-midi d'un faune (Prelude to the afternoon of a
     faun). FAURE: Pélleas et Mélisande, op. 80: Incidental music.
     RAVEL: Pavane pour une infante défunte. ROUSSEL: Bacchus et
     Ariane, op. 43: Suite, no. 2. Strasbourg Philharmonic Orches-
     tra; Alain Lombard. Erato STU 70889.
       +Gr 1-76 p1209   +MM 4-76 p30
       +HFN 1-76 p105   -RR 1-76 p31
901 Prelude to the afternoon of a faun. MUSSORGSKY: A night on the
     bare mountain. STRAVINSKY: The firebird. Isao Tomita, syn-
     thesizers. RCA ARL 1-1312.
       -NR 3-76 p1   +-St 8-76 p99
     Prélude à l'après-midi d'un faune. cf Piano works (Telefunken DX
     6-35272).
     Prélude à l'après midi d'un faune. cf Works, selections (RCA FVL
     3-7276).
     Prélude à l'après-midi d'un faune. cf CHABRIER: España.
     Prélude à l'après-midi d'un faune. cf La mer.
     Prélude à l'après-midi d'un faune. cf CBS 30066.
     Prélude à l'après-midi d'un faune. cf Decca DPA 519/20.
     Printemps. cf La boîte à joujoux.

902   Quartet, strings, op. 10, G minor. RAVEL: Quartet, strings, F
        major. Parrenin Quartet. Connoisseur Society CS 2103.
            +-HF 11-76 p110
      Rêverie. cf Piano works (Seraphim S 60253).
      Rêverie. cf CBS 30072.
      Rhapsody, no. 1. cf Grenadilla GS 1006.
      Rhapsody, clarinet and orchestra. cf Hungaroton SLPX 11728.
      Sonata, flute, viola and harp. cf Decca SPA 394.
903   Sonata, violin and piano. FAURE: Sonata, violin and piano, no. 1,
        op. 13, A major. Maurice Hasson, vln; Michael Isador, pno.
        Classics for Pleasure CFP 40210.
            ++Gr 7-75 p200   +RR 5-75 p52
            +HFN 9-75 p98   +-St 2-76 p739
904   Sonata, violin and piano. FAURE: Sonata, violin and piano, no. 1,
        op. 13, A major. RAVEL: Sonata, violin and piano. Jean-Pierre
        Wallez, vln; Bruno Rigutto, pno. French Decca QS 7174.
            +Gr 8-76 p318   +-RR 6-76 p71
            +HFN 8-76 p79
      Sonata, violin and piano. cf BACH: Sonata, violin and harpsi-
        chord, A minor.
905   Sonata, violoncello and piano, no. 1, D minor. Intermezzo.
        HONEGGER: Sonatina. SAINT-SAENS: Sonata, violoncello and
        piano, no. 1, op. 32, A minor. Jeffrey Solow, vlc; Irma
        Vallecillo, pno. Desmar DSM 1006.
            +HF 4-76 p127   ++NR 2-76 p6
            +MJ 3-76 p25   +-RR 4-76 p65
906   Sonata, violoncello and piano, no. 1, D minor. PROKOFIEV: Sonata,
        violoncello and piano, op. 119. WEBERN: Kleine Stücke, violon-
        cello and piano, op. 11 (3). Lynn Harrell, vlc; James Levine,
        pno. RCA ARL 1-1262.
            ++HF 4-76 p127   ++SFC 3-28-76 p28
            ++MJ 4-76 p31   ++SR 11-1-75 p46
            ++NR 2-76 p5
907   Songs: Le promenoir des deux amants. RESPIGHI: Il tramonto.
        SANTOLIQUIDO: Poesi persiane (3). WOLF-FERRARI: Quattro rispetti,
        op. 11, no. 4. Howard Thain, t; Epsilon Quartet; Hubert Dawkes,
        Magdaleine Panzera-Baillot, pno; Yannis Daris. Pearl SHE 521.
            /Gr 1-76 p1229   -RR 12-75 p91
      Songs: Ariettes oubliées, no. 2: Il pleure dans mon coeur. cf
        Command COMS 9006.
      Songs: Ariettes oubliées, no. 2: Il pleure dans mon coeur. cf
        RCA ARM 4-0942/7.
      Songs: Ariettes oubliées: Beau soir. cf Seraphim S 60251.
      Songs: Ballade des femmes de Paris; Chanson de Bilitis (3); Fêtes
        galantes, I and II; Green. cf HMV RLS 716.
      Songs: Noël des enfants qui n'ont plus de maisons. cf Hungaroton
        SLPX 11762.
      Suite bergamasque. cf Piano works (DG 2535 158).
      Suite bergamasque. cf Piano works (Seraphim S 60253).
      Suite bergamasque: Clair de lune. cf CBS 30064.
      Suite bergamasque: Clair de lune. cf CBS 30072.
      Suite bergamasque: Clair de lune. cf CBS 30073.
      Suite bergamasque: Clair de lune. cf Connoiseur Society (Q) CSQ 2066.
      Suite bergamasque: Clair de lune. cf Music for Pleasure SPR 90086.
      Suite bergamasque: Clair de lune. cf Seraphim S 60259.
      Syrinx, flute. cf BARTOK: Hungarian peasant suite.
      Syrinx, flute. cf RCA LRL 1-5127.
908   Works, selections: Images. Le martyre de Saint Sebastian. La mer.
        Nocturnes: Nuages; Fêtes; Printemps. Prélude à l'après-midi d'un

faune. BSO; Charles Munch. RCA FVL 3-7276 (3). (Reissues
from VICS 1041, 1162, 1227, SB 6540).
+-Gr 11-76 p782   +-RR 11-76 p53
+HFN 12-76 p151

DEFAYE
Danses (2). cf Boston Brass BB 1001.
DEGTIAREV
Preslavnaia dnies. cf Harmonia Mundi HMU 137.
DELIBES, Leo
Arioso. cf Club 99-97.
909 Coppélia. Minneapolis Symphony Orchestra; Antal Dorati. Philips
6780 253 (2). (Reissue from Mercury AMS 16018/9).
+-Gr 11-75 p817   +RR 12-75 p46
+-HFN 1-76 p123
Coppélia. cf ADAM: Giselle.
910 Coppélia: Suite. Sylvia: Suite. LPO; Stanley Black. Decca PFS
4358. Tape (c) KPFC 4358. (also London 21147 Tape (c) 5-21147).
+Gr 8-76 p342   +RR 8-76 p35
+HFN 8-76 p79
Coppélia: Suite. cf CHOPIN: Les sylphides.
Les filles de Cadiz. cf Columbia M 33933.
Lakmé: Dovè l'Indiana bruna. cf VERDI: Rigoletto.
Lakmé: Dovè l'Indiana bruna. cf HMV SLS 5057.
Lakmé: Fantaisie aux divins mensonges. cf Club 99-98.
911 Lakmé: Fantasien nebelhafte Träume. GOUNOD: Faust: Was is der Gott
...O gib junges Blut; Es ist schön spat...Verweile doch ein
Augenblick...Er liebt mich. VERDI: La forza del destino; Unsonst
Alvarez. WOLF: Songs: Heimweh; Verschwiegene Liebe; Waldwander-
ung. Helge Roswänge, t; Instrumental accompaniment. Rococo 5375.
+NR 3-76 p7
Lakmé: Où va la jeune Indoue. cf Rubini GV 70.
Lakmé: Prendre le dessin d'un bijou...Fantasie aux divins men-
songes; Où va la jeune Indoue. cf Decca GOSD 674/6.
Sylvia: Pizzicato polka. cf CBS 30073.
Sylvia: Suite. cf Coppélia: Suite.
DELISLE
La Marseillaise. cf Everest 3360.
DELIUS, Frederick
912 Aquarelles (2) (orch Fenby). Fennimore and Gerda: Intermezzo.
On hearing the first cuckoo in spring. Summer night on the
river. VAUGHAN WILLIAMS: Fantasia on "Greensleeves". The
lark ascending. WALTON: Henry V: Death of Falstaff, Passa-
caglia, Touch her soft lips and part. Pinchas Zukerman, vln;
ECO; Daniel Barenboim. DG 2530 505. Tape (c) 3300 500.
+Audio 11-75 p95   +-NYT 1-18-76 pD1
+Gr 4-75 p1807   ++RR 4-75 p37
+HF 8-75 p102   ++RR 5-76 p78 tape
+HF 10-76 p147 tape   ++SFC 10-5-75 p38
++NR 8-75 p2   +STL 6-8-76 p36
913 Brigg Fair. Dance rhapsody, no. 2. In a summer garden. On
hearing the first cuckoo in spring. PO; Eugene Ormandy. CBS
61426. Tape (c) 40-61426. (Reissue from SBRG 72086).
+Gr 6-74 p46   -RR 5-74 p32
+-HFN 12-75 p173 tape   -RR 6-76 p86 tape
914 Brigg Fair. On hearing the first cuckoo in spring. A song of
summer. A village Romeo and Juliet: The walk to the paradise
garden. LSO; Anthony Collins. Decca ECS 633. Tape (c) KECC 633.
/HFN 4-76 p125 tape   -RR 4-76 p80 tape
Brigg Fair. cf Works, selections (World Records SHB 32).
915 Caprice and elegy. Concerto, piano. Concerto, violin. Albert
Sammons, vln; Benno Moiseiwitsch, pno; Beatrice Harrison, vlc;

155   DELIUS (cont.)

Liverpool Philharmonic Orchestra, PhO, CO; Malcolm Sargent,
Constant Lambert. World Records SH 224. (Reissues from
Columbia DX 1160/2, ITMV C 3533/5, HMV B 3721).
   +Gr 8-75 p321   +RR 7-75 p24
   +HFN 7-75 p79   +ST 2-76 p737
Concerto, piano. cf Caprice and elegy.
Concerto, piano, C minor. cf BEETHOVEN: Concerto, piano, no. 5,
op. 73, E flat major.
Concerto, violin. cf Caprice and elegy.
Dance rhapsody, no. 2. cf Brigg Fair.
Dance rhapsody, harpsichord. cf Angel S 36095.
Eventyr. cf Works, selections (World Records SHB 32).
916   Fennimore and Gerda. Elisabeth Söderström, Kirsted Buhl-Møller,
Bodil Kongsted, Ingeborg Junghans, s; Hedvig Rummel, ms;
Robert Tear, Anthony Rolfe Johnson, Michael Hansen, t; Brian
Rayner Cook, Peter Fog, Hans Christian Hansen, bar; Birger
Brandt, Mogens Berg, bs; Danish State Radio Symphony Chorus;
Danish Radio Symphony Orchestra; Meredith Davies. HMV SQ
SLS 991 (2). (also Angel SX 3835).
   +-Gr 12-76 p1045   +-RR 12-76 p42
Fennimore and Gerda: Intermezzo. cf Aquarelles.
Fennimore and Gerda: Intermezzo. cf Works, selections (World Records).
Florida suite: La Calinda. cf Works, selections (World Records
SHB 32).
Hassan: Intermezzo and serenade. cf Works, selections (HMV TC
ASD 2477).
Hassan: Intermezzo, serenade, closing scene. cf Works, selec-
tions (World Records SHB 32).
In a summer garden. cf Brigg Fair.
In a summer garden. cf Works, selections (HMV TC ASD 2477).
In a summer garden. cf Works, selections (World Records SHB 32).
Irmelin: Prelude. cf Works, selections (World Records SHB 32).
Irmelin: Prelude. cf HMV HLM 7093.
Koanga: La Calinda. cf Works, selections (HMV TC ASD 2477).
Koanga: La Calinda; Final scene. cf Works, selections (World
Records SHB 32).
917   A late lark. HERMANN: The fantasticks. For the fallen.
WARLOCK: Motets. Gillian Humphreys, s; Meriel Dickinson, con;
John Amis, t; Michael Rippon, bs; Stephen Hicks, org; Thames
Chamber Choir; National Philharmonic Orchestra; Bernard
Herrmann, Louis Halsey. Unicorn RHS 340.
   +Gr 9-76 p454   +-RR 8-76 p75
   +-HF 12-76 p120   ++St 11-76 p166
   +-HFN 8-76 p79
918   Life's dance. North country sketches. A song of summer. RPO;
Charles Groves. HMV (Q) ASD 3139. (also Angel S 37140 (Q)
SQ 37140).
   ++Gr 1-76 p1192   +MT 7-76 p575
   +HF 1-76 p86   ++NR 12-75 p3
   +HF 1-76 p86 Quad   ++RR 12-75 p47
   +HFN 11-75 p154   ++St 1-76 p103 Quad
North country sketches. cf Life's dance.
North country sketches, no. 2. cf HMV SEOM 24.
On hearing the first cuckoo in spring. cf Aquarelles.
On hearing the first cuckoo in spring. cf Brigg Fair (CBS 61426).
On hearing the first cuckoo in spring. cf Brigg Fair (Decca ECS
633).

On hearing the first cuckoo in spring. cf Works, selections
(HMV TC ASD 2477).

On hearing the first cuckoo in spring. cf Works, selections
(World Records SHB 32).

Over the hills and far away. cf Works, selections (World Records
SHB 32).

Quartet, strings: 3rd movement (Late swallows). cf Works,
selections (HMV TC ASD 2477).

919   Sonata, violin and piano, no. 1. Sonata, violin and piano, B
major. David Stone, vln; Robert Threlfall, pno. Pearl SHE 522.
+Gr 10-75 p645   +-RR 12-75 p81
+-HFN 12-75 p153   +Te 6-76 p48
+MT 7-76 p575

Sonata, violin and piano, B major. cf Sonata, violin and piano,
no. 1.

A song before sunrise. cf Works, selections (HMV TC ASD 2477).

A song of summer. cf Brigg Fair.

A song of summer. cf Life's dance.

Songs: Appalachia; Le ciel est pardessus le toit (2); Cradle
song; Evening voices; I Brasil; Iremlin Rose (2); Klein
Venevil; Love's philosophy; La luna blanche; Mass of life:
Prelude, part 2; The nightingale; Sea drift; So white, so
soft; To the queen of the heart; The violet (2); Whither.
cf Works, selections (World Records SHB 32).

Songs: To be sung on a summer night on the water. cf Argo ZRG 833.

Summer night on the river. cf Aquarelles.

Summer night on the river. cf Works, selections (HMV TC ASD 2477).

Summer night on the river. cf Works, selections (World Records
SHB 32).

920   To be sung on a summer night on the water. HADLEY: The hills.
Felicity Palmer, s; Robert Tear, t; Robert Lloyd, bs; Cambridge,
King's College Choir, Cambridge University Musical Society
Chorus; LPO; Philip Ledger. HMV SQ SAN 393.
+Gr 6-76 p71   +-MT 9-76 p746
+-HFN 7-76 p85   +RR 7-76 p75

A village Romeo and Juliet: The walk to the paradise garden. cf
Brigg Fair.

A village Romeo and Juliet: The walk to the paradise garden. cf
Works, selections (World Records SHB 32).

921   Works, selections: In a summer garden. Quartet, strings: 3rd
movement (Late swallows). On hearing the first cuckoo in
spring. A song before sunrise. Summer night on the river.
Hassan: Intermezzo and serenade. Koanga: La Calinda. Hallé
Orchestra; John Barbirolli. HMV Tape (c) TC ASD 2477.
+-HFN 4-76 p125 tape

922   Works, selections: Brigg Fair. Eventyr. Florida suite: La
Calinda. Hassan: Intermezzo, serenade, closing scene. In
a summer garden. On hearing the first cuckoo in spring. Over
the hills and far away. Summer night on the river. Fennimore
and Gerda: Intermezzo. Irmelin: Prelude. Koanga: La Calinda;
Final scene. A village Romeo and Juliet: The walk to the
paradise garden. Songs: Appalachia; Le ciel est pardessus le
toit (2); Cradle song; Evening voices; I Brasil; Irmelin Rose
(2); Klein Venevil; Love's philosophy; La luna blanche; Mass
of life: Prelude, part 2; The nightingale; Sea drift; So white,
so soft; To the queen of the heart; The violet (2); Whither.
Dora Labbette, s; Heddle Nash, t; Gerald Moore, pno; LPO, RPO,

Symphony Orchestra; Thomas Beecham. World Records SHB 32 (5).
(Reissues).
       ++ Gr 11-76 p781

DELL'ACQUA
    Villanelle. cf BIS LP 45.
    Villanelle. cf Columbia M 33933.
    Villanelle. cf Court Opera Classics CO 342.
    Villanelle. cf Rococo 5397.
DELLO JOIO, Norman
    A jubilant song. cf Columbia M 34134.
    Notes from Tom Paine. cf Columbia M 33838.
923 Satiric dances. MUDGETT (arr.): Colonial songs. Musical history
        of the national anthem. GRUNDEMAN: The spirit of '76 (arr.).
        SCHUMAN: Chester overture. Concord Band; William Toland.
        Concord Band unnumbered.
           *IN 6-76 p22
DEL MONACO, Alfredo
    Electronic study, no. 2. cf CEELY: Elegia.
    Metagrama. cf CEELY: Elegia.
DEL RIEGO
    Thank God for a garden. cf Pye Ember GVC 51.
DEMANTIUS, Johann
    Polnischer Tanz und Galliarda. cf Harmonia Mundi HMU 610.
DEMESSIEUX, Jeanne
    Prelude and fugue, C major. cf DAVID: Toccata and fugue, F minor.
DENCKE
    O, be glad, ye daughters of His people. cf Vox SVBX 5350.
DENZA, Luigi
    Non t'amo piu. cf Everest SDBR 3382.
    Songs: Si vous l'aviez compris. cf Musical Heritage Society
        MHS 3276.
DERING, Richard
    Jesu, dulcis memoria. cf Harmonia Mundi HMU 473.
DESMARETS, Henri
    Menuet. cf DG Archive 2533 172.
    Passepied. cf DG Archive 2533 172.
DETONI, Dubravko
    Dokument 75. cf BOZIC: Audiogemi I-IV.
DETT, Robert Nathaniel
924 In the bottoms. GRIFFES: Sonata, piano, F major. IVES: Three-
        page sonata. Clive Lythgoe, pno. Philips 9500 096.
           +-ARG 11-76 p22   +-MJ 12-76 p28
           +Gr 10-76 p627   +NR 12-76 p13
           +-HF 12-76 p122   +RR 10-76 p87
           +HFN 10-76 p167   +St 12-76 p144
    Juba dance. cf Washington University Press OLY 104.
DEVCIC, Natko
    Entre nous. cf BOZIC: Audiogemi I-IV.
925 Igra rječi II. Non nova. Prolog. Panta rei. Sonata. V. Kovačić,
        reciter; RTZ Chorus; Acezantez Ensemble; V. Krpan, pno; Zagreb
        Philharmonic Orchestra; RTZ Symphony Orchestra; Milan Horvat,
        Kuljeric. Jugoton LSY 61202.
           +Te 9-76 p33
    Non nova. cf Igra rječi II.
    Prolog. cf Igra rječi II.
    Panta rei. cf Igra rječi II.

        Sonata, piano. cf Igra rječi II.
DEVIENNE, François
926   Concerto, flute, no. 2, D major. IBERT: Concerto, flute. Peter-
       Lukas Graf, flt; ECO; Raymond Leppard. Claves P 501.
            +Gr 7-76 p168   +RR 6-76 p40
            +HFN 2-76 p95
927   Concerto, flute, no. 4, G major. ROSETTI: Concerto, flute, G
       major. János Szebenyi, flt; Györ Philharmonic Orchestra;
       János Sándor. Hungaroton SLPX 11694.
            +NR 5-76 p5   +RR 6-76 p40
DIAMOND, David
928   Romeo and Juliet. THORNE: Burlesque overture. Rhapsodic
       variations. Francis Thorne, pno; Polish Radio and Television
       Orchestra; Jan Krenz, William Strickland. CRI SD 216.
            +-RR 10-76 p49
DICKINSON, Peter
       Extravaganzas. cf BERKELEY: Songs (Argo ZRG 788).
       Recorder music. cf HMV SLS 5022 (2).
DIEMER, Louis
       Valse de concerto. cf Rococo 2049.
DIJON, Guiot de
       Chanterai por mon coraige. cf Telefunken 6-41275.
DILETZKY, Nikolai
       Glorify the name of the Lord. cf HMV Melodiya ASD 3102.
DINICU, Dmitri
       Hora staccato. cf Odyssey Y 33825.
       Hora staccato (arr. and orch. Gerhardt). cf RCA LRL 1-5094.
       Hora staccato. cf RCA ARM 4-0942/7.
DITTERSDORF, Karl
929   Sinfonia concertante, viola, double bass and orchestra, D major.
       HAYDN, M.: Divertimento, viola, violoncello and double bass.
       KEYPER: Romance and rondo, double bass and orchestra. ROSSINI:
       Duet, violoncello and double bass. Rodney Slatford, double
       bass; Stephen Shingles, vla; Kenneth Heath, vlc; AMF; Neville
       Marriner. HMV SQ ASD 3264.
            +Gr 11-76 p824   +-RR 11-76 p54
            +HFN 11-76 p164
       LA DIVINA: THE ART OF MARIA CALLAS. cf HMV SLS 5057.
       DIVISIONS ON A GROUND: AN INTRODUCTION TO THE RECORDER AND ITS
       MUSIC. cf Transatlantic TRA 292.
DLUGORAI, Adalbert
       Chorea polonica. cf Hungaroton SLPX 11721.
       Fantasia. cf Hungaroton SLPX 11721.
       Finale. cf Hungaroton SLPX 11721.
       Villanella polonica. cf Hungaroton SLPX 11721.
DLUGOSZEWSKI, Lucia
       Angeles of the utmost heavens. cf Folkways FTS 33901/4.
DOBIAS, Václav
930   Prague, my only one. Quintet, winds and timpani. Sonata, piano
       and string orchestra. Jiří Pokorný, pno; Jindřich Jindrák, bar;
       CPhO, Czech Radio Orchestra, Czech Wind Quintet; Alois Klíma,
       Václav Neumann. Panton 110 458.
            +RR 9-76 p47
       Quintet, winds and timpani. cf Prague, my only one.
       Sonata, piano and string orchestra. cf Prague, my only one.
DODGE, Charles
       Extensions, trumpet and tape. cf CRI SD 300.

Folia, chamber orchestra. cf CRI SD 300.
931   In celebration. Speech songs. The story of our lives.
      Synthesized voice music realized at Columbia University Center
      for Computing Activities, Nevis Laboratories, Bell Telephone
      Laboratories. CRI SD 348.
       *MJ 7-76 p57   +NYT 4-11-76 pD23
       +NR 9-76 p14   -St 8-76 p100
      Speech songs. cf In celebration.
      The story of our lives. cf In celebration.
DODGSON, Stephen
      Sonata, brass. cf ADDISON: Divertimento, op. 9.
      Sonata, brass: Poco adagio. cf Argo SPA 464.
DOHNANYI, Ernst von
      Rhapsody, op. 11, no. 3, C major. cf Decca SPA 473.
      Ruralia Hungarica: Gypsy andante. cf RCA ARM 4-0942/7.
      Suite, op. 19, F sharp minor. cf Variations on a nursery song,
      op. 25.
932   Variations on a nursery song, op. 25. RACHMANINOFF: Rhapsody on
      a theme by Paganini, op. 43. Christina Ortiz, pno; NPhO;
      Kazuhiro Koizumi. Angel (Q) S 37178. (also HMV SQ ASD 3197
      Tape (c) TC ASD 3197).
       +Gr 6-76 p44   ++NR 7-76 p11
       +-HF 9-76 p90   +-RR 7-76 p62
       +HFN 7-76 p87   +St 9-76 p119
       ++HFN 11-76 p175
933   Variations on a nursery song, op. 25. Suite, op. 19, F sharp
      minor. Béla Siki, pno; Seattle Symphony Orchestra; Milton
      Katims. Turnabout TVS 34623.
       +HF 9-76 p90   +St 9-76 p119
       +NR 8-76 p12
DONIZETTI, Gaetano
      Anna Bolena: Cielo, a miei lunghi spasimi. cf English National
      Opera ENO 1001.
      Anna Bolena: Piangete voi...Al dolce guidami. cf Rococo 5386.
934   Arias: Torquato Tasso: Fatal Goffredo....Trono e corona. Gemma
      di Vergy: Lascia, Guido, ch'io possa vendicare...Una voce al
      cor...Egli riede. Belisario: Plauso, Voce di gioia...Sin la
      tomba e a me negata. Parisina: No piu salir non ponno...
      Ciel, sei tu che in tal momento...Ugo e spento. Montserrat
      Caballé, Ambrosian Opera Chorus; LSO; Carlo Cillario. RCA
      LSC 3164. (also RCA SER 5591).
       +ARG 1-71 p279   +NYT 5-16-71 pD24
       +OC 5-76 p49   +St 1-71 p87
       +-ON 2-20-71 p34
      Belisario: Plauso, Voce di gioia...Sin la tomba e a me negata.
      cf Arias (RCA LSC 3164).
      Catarina Cornaro: Torna all'ospitetto...Vieni o tu, che ognor io
      chiamo. cf CATALANI: La Wally: Ebben, ne andrò lontana.
      La conocchia. cf L'Oiseau-Lyre SOL 345.
      Don Pasquale: Com'e gentil. cf Decca SKL 5208.
      Don Pasquale: O summer night. cf Library of Congress OMP 101/2.
      Don Pasquale: So anch'io la virtu; Pronto io son. cf Rubini GV 68.
      L'Elisir d'amore: Chiedi all'aura lusinghiera. cf BELLINI: I
      puritani: Fini...me lassa.
      L'Elisir d'amore: Con se va contento...Quanto amore. cf Rococo
      5386.

L'Elisir d'amore: Una furtiva lagrima. cf Decca SPA 449.
L'Elisir d'amore: Una furtiva lagrima. cf Decca SKL 5208.
L'Elisir d'amore: Una furtiva lagrima. cf English National Opera
ENO 1001.
L'Elisir d'amore: Una furtiva lagrima (2 versions). cf Everest
SDBR 3382.
L'Elisir d'amore: Una furtiva lagrima. cf RCA CRM 1-1749.

935   La favorita: Spirto gentil; Una vergine, un angel di Dio. Poliuto:
Lasciando la terra. MEYERBEER: Gli Ugonotti: Bianca al par di
neve alpina. ROSSINI: Guglielmo Tell: O muto asil del pianto.
VERDI: I Lombardi: La mia Letizia infondere. Il trovatore:
Di quella pira; Ah, si ben mio coll'essere; Miserere...Ah, Che
la morte. Caterina Mancini, Margherita Benetti, s; Giacomo
Lauri-Volpi, t; Rome Symphony Orchestra, Turin Radio Orchestra
and Chorus, Rome Radio Symphony Orchestra, Milan Symphony
Orchestra; Gennaro d'Angelo, Tullio Serafin. Ember GVC 47.
(Reissues from Cetra LPV 45017, OLPC 50153, EPO 0344).
   +-Gr 8-76 p337

La favorita: Ah l'alto ardor. cf Discophilia DIS KGA 2.
La favorita: O mio Fernando. cf Columbia/Melodiya M 33931.
La favorita: Pour tant d'amour. cf Rubini GV 65.
La figlia del reggimento: Convien partir. cf Bongiovanni GB 1.
La fille du régiment: Depuis l'instant. cf BELLINI: I puritani:
Fini...me lassa.   .
La figlia del reggimento: Heil dir mein Vaterland. cf Discophilia
DIS 13 KHH 1.
La figlia del reggimento: Weiss nicht die Welt. cf Rococo 5379.
Gemma di Vergy: Lascia, Guido, ch'io possa vendicare...Una voce
al cor...Egli riede. cf Arias (RCA LSC 3164).

936   Linda di Chamounix: Ah tardai troppo...O luce di quest'anima.
Lucia di Lammermoor: Il doce suono mi corpi sua voce...Ardon
gl'incensi; Ancor non giunse...Regnava nel silenzio. VERDI:
Ernani: Surta è la notte...Ernani, Ernani, involami. I vespri
Siciliani: Mercé, diletti amiche. Joan Sutherland, Nadine
Sautereau, s; Paris Opera Chorus: OSCCP; Nello Santi. Decca
SDD 146. Tape (c) KSDC 146. (Reissues).
   +-RR 3-75 p74 tape   +-RR 5-76 p77 tape
Linda di Chamounix: O luce di quest anima; Rondo. cf Rubini
GV 68.

937   Lucia di Lammermoor. Joan Sutherland, Ana-Raquel Sartre, s;
Renato Cioni, Kenneth MacDonald, Rinaldo Pelizzoni, t; Robert
Merrill, bar; Cesare Siepi, bs; Rome, Santa Cecilia Orchestra
and Chorus; John Pritchard. Decca GOS 663/5 (3). (Reissue
from SET 212/4). (also London 1327).
   +-Gr 1-76 p1230   +RR 1-76 p24
   +HFN 1-76 p105

938   Lucia di Lammermoor. Joan Sutherland, s; Huguette Tourangeau, ms;
Luciano Pavarotti, Ryland Davies, t; Sherrill Milnes, bar;
Nicolai Ghiaurov, bs; ROHO and Chorus; Richard Bonynge. London
OSA 130103 (3). Tape (c) D 31210 (r) 90210. (also Decca
SET 258/30 Tape (c) K2L 22).
    +ARG 10-72 p680   ++ON 11-72 p43
   ++Gr 5-72 p1928   +-Op 7-72 p636
    +Gr 8-76 p341 tape   +-RR 8-76 p82 tape
   +-HF 9-72 p78   ++SFC 8-6-72 p31
   ++HFN 5-72 p920   +SR 8-12-72 p38
    +HFN 8-76 p94 tape   ++St 10-72 p81
   ++NR 8-72 p12   +STL 6-11-72 p38

939   Lucia di Lammermoor, excerpts. Korola Agay, s; Olga Szönyi, ms;
      József Simándy, József Réti, t; László Jambor, bar; Gyula
      Veress, bs; Hungarian State Opera House Choir; Budapest
      Philharmonic Orchestra; András Kórody. Hungaroton SLPX 11648.
          +NR 5-76 p9   +-RR 3-76 p28
      Lucia di Lammermor: Ardon gli incensi. cf BELLINI: Capuletti ed
      i Montecchi: O, quante volte, O, quante.
      Lucia di Lammermoor: Il dolce suono mi corpi sua voce...Ardon
      gl'incensi; Ancor non giunse...Regnava nel silenzio. cf
      Linda di Chamounix: Ah, tardai troppo...O luce di quest'
      anima.
      Lucia di Lammermoor: Fra poco a me, Tu che a Dio. cf Cantilena 6239.
      Lucia di Lammermoor: Mad scene. cf BELLINI: I puritani: O ren-
      detemi la speme.
      Lucia di Lammermoor: Mad scene. cf HMV RLS 719.
      Lucia di Lammermoor: Mad scene. cf Richmond R 23197.
      Lucia di Lammermoor: Oh giusto cielo; Ardon gl'incensi. cf
      HMV SLS 5057.
      Lucia di Lammermoor: Regnava nel silenzio; Sulla tomba; Verrano a
      te. cf Rubini GV 68.
      Lucia di Lammermoor: Tombe degliavi miei. cf Muza SXL 1170.
      Lucia di Lammermoor: Tombe degli avi miei...Fra poco a me. cf
      Rococo 5383.
      Lucrezia Borgia: Brindisi. cf Collectors Guild 611.
      Lucrezia Borgia: Com'e bello. cf CATALANI: La Wally: Ebben, ne
      andrò lontana.
      Lucrezia Borgia: T'amo qual s'ama un angelo. cf English National
      Opera ENO 1001.
      Maria di Rohan: Rival, se tu spaersi...Cupa fatal mestizia. cf
      Rococo 5386.
940   Maria Stuarda. Joan Sutherland, s; Huguette Tourangeau, Margreta
      Elkins, ms; Luciano Pavarotti, t; James Morris, Roger Soyer,
      bs; Bologna Teatro Comunale Orchestra and Chorus; Richard
      Bonynge. London OSA 13117 (3). (also Decca D2D 3 Tape (c)
      K2A 33).
          +-Gr 6-76 p78   +NYT 6-20-76 pD27
          +-HF 8-76 p86   +-ON 9-76 p70
           +HFN 6-76 p84   +RR 6-76 p20
          +-MT 11-76 p913   +SFC 5-16-76 p28
          +-NR 12-76 p10   +St 8-76 p67
      Maria Stuarda: O nube, che lieve. cf CATALANI: La Wally: Ebben,
      ne andrò lontana.
      Parisina: No, piu salir non ponno...Ciel, sei tu che in tal momento
      ...Ugo e spento. cf Arias (RCA LSC 3164).
      Poliuto: Lasciando la terra. cf La favorita: Spirto gentil; Una
      vergine, un angel di Dio.
941   Quartet, strings, D major. ROSSINI: Sonatas, strings, nos. 2 and
      4. AMF; Neville Marriner. Argo ZRG 603. Tape (c) KRC 603.
          +HFN 7-76 p104 tape
      Roberto Devereux: E Sara in questi orribili momenti...Vivi, ingrato,
      a lei d'accanto...Quel sangue versato. cf CATALANI: La Wally:
      Ebben, ne andrò lontana.
      Songs: La corrispondenza amorosa; A mezzanotte; La mère et l'enfant.
      cf BELLINI: Songs (Cetra LPO 2003).
      Songs: La corrispondenza amorosa; Una lagrima; La mère et l'enfant;
      Ne ornerà la bruna chioma. cf BELLINI: Songs (RCA AGL 1-1341).

DONIZETTI (cont.)   162

Torquato Tasso: Fatal Goffredo...Trono e corona. cf Arias
    (RCA LSC 3164).
DOPPLER, Franz
Fantaisie pastorale hongroise, op. 26. cf RCA LRL 1-5094.
DORNEL, Louis-Antoine
942 Suite, no. 1, C minor. GAUTIER: Suite, E minor. REBEL: Sonata,
    recorder, no. 3, D major. Sonata, recorder, no. 6, B minor.
    Quadro Hotteterre. Telefunken AW 6-41927.
            +Gr 7-76 p188   +PRO 9/10-76 p19
            +-HFN 8-76 p85   +RR 6-76 p62
            +NR 2-76 p7
DOROW, Dorothy
Songs: Dream; Pastourelles, pastoureux. cf BIS LP 45.
DOUBRAVA, Jaroslav
943 Pastorale. FISER: Double pro orchestr. JANECEK: Legenda o Praze.
    KREJCI: Malý balet. CPhO, Studio Orchestra; Václav Smetáček,
    Jerí Kout. Panton 110 363.
            +RR 11-76 p67
DOUGHTY
Grandfather's clock. cf Decca SB 322.
DOULCE MEMOIRE. cf Argo ZRG 667.
DOWLAND, John
Awake, sweet love. cf Vista VPS 1022.
Bonnie sweet robin. cf RCA RK 11708.
Captain Digorie Piper's galliard. cf Works, selections (RCA
    ARL 1-1491).
A fancy (2). cf Works, selections (RCA ARL 1-1491).
Farewell. cf Works, selections (RCA ARL 1-1491).
Fine knacks for ladies. cf Enigma VAR 1017.
944 First booke of songes, 1597. Consort of Musicke; Anthony Rooley.
    L'Oiseau-Lyre DSLO 508/9 (2).
            +Gr 11-76 p851   +RR 10-76 p20
            +HFN 11-76 p157
Flow my teares. cf Hungaroton LSPX 11669/70.
Forlorn hope fancy. cf Works, selections (RCA ARL 1-1491).
Galliard to lachrimae. cf Works, selections (RCA ARL 1-1491).
Go from my window. cf RCA RK 11708.
The King of Denmark's galliard. cf Hungaroton SLPX 11669/70.
945 Lachrimae 1604. Consort of Musicke; Anthony Rooley. L'Oiseau-
    Lyre DSLO 517.
            ++Gr 11-76 p851   +RR 10-76 p20
            +HFN 10-76 p169
Lacrimae: Antiquae pavan. cf Philips 6500 926.
Loth to depart. cf RCA RK 11708.
Melancholy galliard. cf Saga 5426.
Mr. George Whitehead his almand. cf DG Archive 2533 323.
Mr. Langton's galliard. cf Works, selections (RCA ARL 1-1491).
Mistress Winter's jump. cf Saga 5426.
My Lady Hunsdon's puffe. cf Saga 5426.
My Lord Chamberlain, his galliard. cf Works, selections (RCA
    ARL 1-1491).
My Lord Chamberlain's galliard. cf Angel S 36851.
My Lord Willoughby's welcome home. cf Works, selections (RCA
    ARL 1-1491).
My Lord Willoughby's welcome home. cf DG Archive 2533 323.
Piper's pavan. cf Works, selections (RCA ARL 1-1491).
Precious moment. cf Hungaroton SLPX 11762.

Resolution. cf Works, selections (RCA ARL 1-1491).
Resolution. cf BIS LP 22.
Sir John Souche's galliard. cf Works, selections (RCA ARL 1-1491).
Songs: Can she excuse my wrongs; Come again sweet love. cf Argo
    ZRG 833.
Songs: If that a sinner's sighs; In this trembling shadow cast;
    Now, o now I needs must part; My Lord Willoughby's welcome
    home. cf Harmonia Mundi HMD 223.
Songs: Awake, sweet love thou art return'd; Away with these self-
    loving lads; Come again, sweet love doth now invite; Galliard,
    D major; Fine knacks for ladies; In darkness let me dwell; I
    saw my lady weep; Shall I sue; Tarletones risurrectione; What
    if I never speed. cf Philips 6500 282.
Songs: Mrs. White's nothinge; My Lady Hunsdon's puffe. cf Pearl
    SHE 525.
Walsingham. cf RCA RK 11708.
946   Works, selections: Captain Digorie Piper's galliard. A fancy (2).
    Forlorn hope fancy. Farewell. Galliard to lachrimae. Mr.
    Langton's galliard. My Lord Chamberlain, his galliard. My
    Lord Willoughby's welcome home. Piper's pavan. Resolution.
    Sir John Souche's galliard. Julian Bream, lt. RCA ARL 1-1491.
        +NR 12-76 p15
DOWNING
    Free and easy. cf Library of Congress OMP 101/2.
DRECHSLER
    Der Bauer als Millionär: Brüderlein fein. cf RCA PRL 1-9034.
DREJSL
947   Concerto, piano. PAUER: Concerto, trumpet. Mirka Pokorna, pno;
    Stanislav Sejpal, tpt; Czech Radio Symphony Orchestra, CPhO;
    Eduard Fischer, Václav Neumann. Panton 110 487.
        +RR 12-76 p54
DRIGO, Richard
    Les millions d'Arléquin: Serenade (arr. and orch. Gamley). cf
    RCA LRL 1-5094.
    Valse bluette (2). cf RCA ARM 4-0942/7.
DUARTE, John
    Friendships (6). cf Delos DEL FY 008.
DUBENSKI
    Otchne nach. cf Harmonia Mundi HMU 105.
DUBOIS, Pierre Max
    Quatuor pour saxophones. cf Transatlantic TRA 308.
DUBROVAY
948   Quartet, strings. KAROLYI: Quartet, strings, no. 2. PAPP: Quintet,
    cimbalom. SZOKOLAY: Quartet, strings, no. 1. Kodály Quartet.
    Hungaroton SLPX 11754.
        ++NR 8-76 p7
DUFAY, Guillaume
    L'alta bellezza. cf Telefunken ER 6-35257 (2).
    Ave virgo. cf Telefunken ER 6-35257 (2).
    La belle se siet. cf HMV SLS 863.
    Bien veignes vous. cf Telefunken ER 6-35257 (2).
    Bon jour, bon mois. cf Telefunken ER 6-35257 (2).
    Ce moys de may. cf HMV SLS 863.
    C'est bien raison. cf Telefunken ER 6-35257 (2).
    Credo. cf Telefunken ER 6-35257 (2).
    La dolce vista. cf Telefunken ER 6-35257 (2).

DUFAY (cont.)   164

Dona i ardente ray. cf Telefunken ER 6-35257.
Donnés l'assault. cf HMV SLS 863.
949   Ecclesiae militantis. Missa Sancti Jacobi. Rite Majorem.
      Capella Cordina; Alejandro Planchart. Lyrichord LLST 7275.
           +ARG 12-76 p27   +NR 7-75 p8
      Ecclesiae militantis. cf Telefunken ER 6-35257.
950   Franc cuer gentil. Missa sine nomine. MACHAUT: Ma chière dame.
      ANON. (13th century): Donc le rieu de la fontaine. (14th
      century): Ben ch'io; Saltarello; Trotto. (15th century):
      Bagpipes solo; La danse de clèves; L'Espérance de Bourbon;
      Filles à marier; Sans faire. SCHOOL OF NOTRE DAME: Benedi-
      camus domino. Clemencíc Consort; René Clemencíc. Harmonia
      Mundi HMU 939.
           +-RR 4-76 p69
      Gloria. cf Telefunken ER 6-35257.
951   Gloria ad modum tubae. Se la face ay pale, chanson. Se la
      face ay pale, keyboard (2 versions). Se la face ay pale,
      four-part instrumental version. Se la face ay pale, mass.
      Early Music Consort; David Munrow. HMV CSD 3751. (also
      Seraphim S 60267).
           ++Gr 5-74 p2049   +RR 7-74 p74
           +NR 11-76 p10   ++St 11-76 p138
           +NYT 8-15-76 pD15
      Helas mon dueil. cf HMV SLS 863.
      Helas mon dueil. cf Telefunken ER 6-35257.
      J'attendray tant qu'il vous playra; J'ay grant doleur; Gloria
      ad modum tubai. cf Titanic TI 4.
      J'ay mis mon cuer. cf Telefunken ER 6-35257.
      Je vous pri. cf Telefunken ER 6-35257.
      Kyrie. cf Telefunken ER 6-35257.
      Lamentatio Sanctae matris ecclesiae. cf HMV SLS 863.
      Lamentatio Sanctae matris ecclesiae Constantinopolitanae. cf
      Telefunken ER 6-35257.
952   Las, que feray. Ma bella dame souveraine. Missa se la face ay
      pale. Vergine bella. Capella Cordina; Alejandro Planchart.
      Lyrichord LLST 7274.
           +-ARG 12-76 p27   +NR 7-75 p8
      Ma bella dame souveraine. cf Las, que feray.
953   Missa Ave regina coelorum. Clemencíc Consort; René Clemencíc.
      Harmonia Mundi HMU 985.
           ++St 6-76 p102
      Missa Sancti Jacobi. cf Ecclesiae militantis.
      Missa se la face ay pale. cf Las, que feray.
      Missa sine nomine. cf Franc cuer gentil.
      Mon chier amy. cf Telefunken ER 6-35257.
      Moribus et genere Christo. cf Telefunken ER 6-35257.
954   Motets: Supremum est mortalibus. Flos forum. Ave virgo quae
      de caelis. Vasilissa, ergo gaude. Alma redemptoris mater
      (II). DUNSTABLE: Motets: Beata mater. Preco proheminencie.
      Salve regina misericordie. Veni sancti spiritus. Pro Can-
      tione Antiqua, Hamburg Wind Ensemble, Members; Bruno Turner.
      DG Archive 2533 291.
           ++Gr 8-75 p352   +NR 3-76 p6
           ++HF 3-76 p81   ++St 1-76 p103
           ++HFN 8-75 p75

Navré je suis. cf HMV SLS 863.
Par droit je puis. cf HMV SLS 863.
Qui latuit. cf Telefunken ER 6-35257 (2).
Rite Majorem. cf Ecclesiae militantis.
Sanctus. cf Telefunken ER 6-35257 (2).
Se la face ay pale, chanson. cf Gloria ad modum tubae.
Se la face ay pale, four-part instrumental version. cf Gloria
    ad modum tubae.
Se la face ay pale, keyboard (2 versions). cf Gloria ad modum
    tubae.
Se la face ay pale, mass. cf Gloria ad modum tubae.

955   Songs: Alons ent, resvelons vous; Ce jour de l'an; Les douleurs;
        En triumphant; Entre vous gentils amoureux; Dona i ardenti
        ray; Helas mon dueil; J'ay mis mon cuer; Je ne suy plus; Je
        sui povere de leesse; Malheureulx cueur; Ma tres douce,
        Tant que mon argent, Je vous pri; Puisque vous estez campieur;
        Mon bien m'amour; Par le regart; Resveilliés vous. Musica
        Mundana; David Fallows. 1750 Arch Records 1751.
            +St 1-76 p103
      Songs: C'est bien raison de devoir essaucier; Je me complains
        piteusement; Invidia nimica; Malheureux cuer que veux to faire;
        Par droit je puis bien complaindre. cf 1750 Arch S 1753.
      Veni creator spiritus. cf Telefunken ER 5-35257 (2).
      Vergine bella. cf Las, que feray.
      Vergine bella. cf BIS LP 2.
      Vergine bella. cf HMV SLS 863.
      Vergine bella. cf Nonesuch H 71326.

DUKAS, Paul
956   La péri. ROUSSEL: Symphony, no. 3, op. 42, G minor. NYP;
        Pierre Boulez. Columbia M 34201. Tape (c) MT 34201.
            ++SFC 12-19-76 p50
957   La péri: Fanfare. Polyeucte. The sorcerer's apprentice. CPhO;
        Antonio de Almeida. Supraphon 110 1560.
            +Gr 7-76 p167   +-RR 4-76 p40
            +NR 4-76 p3
      La péri: Fanfare. cf Argo SPA 464.
958   La plainte au loin du faune. Prélude elegiaque sur le nom de
        Haydn. ROUSSEL: L'Accueil des muses. Prélude et fugue sur
        le nom de Bach, op. 46. SCHMITT: Chaine brisée, op. 87: La
        tragique chevauchée. Mirages, op. 70. Annie d'Arco, pno.
        Calliope CAL 1813.
            +RR 11-76 p92
      Polyeucte. cf La péri: Fanfare.
      Prélude elegiaque sur le nom de Haydn. cf La plainte au loin
        du faune.
959   The sorcerer's apprentice (L'Apprenti sorcier). Symphony, C major.
        LPO; Walter Weller. London CS 6995. (also Decca SXL 6770).
            +-Audio 10-76 p152   +MT 11-76 p913
            +Audio 12-76 p92   +NR 9-76 p4
            +Gr 6-76 p44   +-RR 3-76 p38
            ++HF 9-76 p90   +SFC 5-23-76 p36
            +HFN 3-76 p92
960   The sorcerer's apprentice. RESPIGHI: La boutique fantasque.
        Israel Philharmonic Orchestra; Georg Solti. London STS 15005.
        Tape (c) A 30605. (also Decca SPA 376 Tape (c) KCSP 376).
            +Gr 6-75 p90   ++RR 6-75 p46
            ++HFN 7-75 p91 tape   +RR 10-76 p106 tape
            +HFN 8-75 p89

DUKAS (cont.)   166

L'Apprenti sorcier. cf CHABRIER: España.
The sorcerer's apprentice. cf La péri: Fanfare.
The sorcerer's apprentice. cf Decca DPA 519/20.
The sorcerer's apprentice. cf DG 2584 004.
L'Apprenti sorcier. cf HMV (Q) ASD 3131.
Symphony, C major. cf The sorcerer's apprentice.
DUKE, John
Luke Havergal. cf Cambridge CRS 2715.
Songs: I carry your heart; In just spring; The mountains are
    dancing. cf Duke University Press DWR 7306.
DUMONT, Henry
Pavane, D minor. cf Harmonia Mundi HMU 334.
DUNCALF
My God, my King, thy various praise. cf Grosvenor GRS 1039.
DUNHILL, Thomas
Songs: The cloths of heaven. cf HMV HLM 7093.
Songs: The cloths of heaven: To the Queen of heaven. cf Saga
    5213.
DUNSTABLE, John
Motets: Beata mater. cf Telefunken ER 6-35257 (2).
Motets: Beata mater. Preco proheminencie. Salve regina
    misericordie. Veni sancti spiritus. cf DUFAY: Motets
    (DG 2533 291).
Songs: O rosa bella; Hastu mir. cf L'Oiseau-Lyre 12BB 203/6.
DUPARC, Henri
961   Songs: Chanson triste; Extase; Phidylé; Testament. FAURE: Songs:
    Adieu, op. 21, no. 3; Après un rêve, op. 7, no. 1; Barcarolle,
    op. 7, no. 3; Clair de lune, op. 46, no. 2; Chanson d'amour,
    op. 27, no. 1; En prière; Nell, op. 18, no. 1; Poème d'un
    jour, op. 21; Le secret, op. 23, no. 3. Ian Partridge, t;
    Jennifer Partridge, pno. Pearl SHE 514.
        ++Gr 2-76 p1364   +STL 1-11-76 p36
        +-RR 3-76 p67
Songs: Chanson triste. cf HMV RLS 719.
Songs: Chanson triste; Extase; L'invitation au voyage; Phidylé.
    cf HMV RLS 716.
Songs: Extase. cf Decca PFS 4351.
Songs: L'Invitation au voyage; Phidylé. cf Seraphim S 60251.
DUPHLY, Jacques
Chaconne. cf La félix.
962   La félix. La forqueray. Chaconne. FORQUERAY: La Laborde. La
    Bellemont. La Couperin. MARCHAND: Suite. Kenneth Gilbert,
    hpd. Harmonia Mundi HMU 940.
        +Gr 1-76 p1221   ++RR 11-75 p76
La félix. cf BALBASTRE: Works, selections (Philips 6581 013).
La forqueray. cf La félix.
La forqueray. cf BALBASTRE: Works, selections (Philips 6581 013).
La de Drummond. cf BALBASTRE: Works, selections (Philips 6581
    013).
La de vatre. cf BALBASTRE: Works, selections (Philips 6581 013).
Médée. cf BALBASTRE: Works, selections (Philips 6581 013).
La millettina. cf BALBASTRE: Works, selections (Philips 6581 013).
Rondeau. cf BALBASTRE: Works, selections (Philips 6581 013).
La victoire à Madame Victoire de France. cf BALBASTRE: Works,
    selections (Philips 6581 013).

DUPRE, Marcel
963   Choral and fugue, op. 57. Cortège et litanie, op. 19. Elevation,
       op. 32, no. 1. Final, op. 27, no. 7. I am black but comely,
       op. 18. Musette, op. 51. Prelude and fugue, op. 7, no. 3,
       G minor. Michael Murray, org. Advent S 5014.
           +NR 11-76 p12
964   Cortège et litanie, op. 19. Symphonie-Passion, op. 23. Preludes
       and fugues, op. 7. Evocation, op. 37. Variations on an old
       noël, op. 20. Pierre Cochereau, org. Delos DEL FY 020-021 (2).
           +-MU 10-76 p17   ++NR 9-76 p12
       Cortège et litanie, op. 19. cf Choral and fugue, op. 57.
       Elévation, op. 32, no. 1. cf Choral and fugue, op. 57.
       Evocation, op. 37. cf Cortège et litanie, op. 19.
       Fileuse. cf Vista VPS 1029.
       Final, op. 27, no. 7. cf Choral and fugue, op. 57.
       I am black but comely, op. 18. cf Choral and fugue, op. 57.
       Musette, op. 51. cf Choral and fugue, op. 57.
       Preludes and fugues, op. 7. cf Cortège et litanie, op. 19.
       Prelude and fugue, op. 7, no. 3, G minor. cf Choral and fugue,
       op. 57.
       Symphonie-Passion, op. 23. cf Cortège et litanie, op. 19.
       Triptyque, op. 51. cf Gaudeamus XSH 101.
       Variations on an old noël, op. 20. cf Cortège et litanie, op. 19.
DURAN DE LA MOTA, Antonio
       Laudate pueri. cf Eldorado S-1.
DURKO, Zsolt
965   Burial prayer. Chamber music. Erzsébet Tusa, István Lantos, pno;
       Atilla Fülöp, t; Endre Utö, bs; Hungarian Radio and Television
       Chorus; Budapest Symphony Orchestra; György Lehel. Hungaroton
       SLPX 11803.
           +NR 10-76 p7   +RR 12-76 p91
       Chamber music. cf Burial prayer.
DURUFLE, Maurice
966   Prelude et fugue sur le nom d'Alain, op. 7. Requiem, op. 9.
       Robert King, treble; Christopher Keyte, bar; Stephen Cleobury,
       org; St. John's College Chapel Choir, Cambridge; George Guest.
       Argo ZRG 787.
           +-Gr 5-75 p2000   +NR 10-75 p6
           +HF 4-76 p102   +RR 5-75 p58
           ++HFN 6-75 p88   -St 2-76 p100
           +MT 12-75 p1070
       Prelude and fugue on the name Alain, op. 7. cf Westminster
       WGS 8116.
       Requiem, op. 9. cf Prelude et fugue sur le nom d'Alain, op. 7.
       Suite, op. 5: Toccata, B minor. cf Vista VPS 1029.
DUSSEK, Johann Ladislaus (also Dusik or Dessek)
       La chasse. cf International Piano Library IPL 102.
967   Concerto, 2 pianos, no. 10, op. 63, B flat major. SCHUMANN:
       Andante and variations, op. 46, B flat major. Toni and Rosi
       Grünschlag, pno; Richard Harand, Günther Weiss, vlc; Walter
       Tombock, hn; Vienna Volksoper Orchestra; Paul Angerer.
       Turnabout 34204. Tape (c) KTVC 34204.
           +-HFN 7-76 p104 tape   +RR 9-76 p94 tape
968   La consolation, op. 62. HUMMEL: Fantasie and rondo, op. 19,
       E major. Sonata, piano, op. 20, F minor. VORISEK: Sonata,
       op. 20, B flat minor. Vladimir Pleshakov, pno. Orion ORS
       75178.

DUSSEK (cont.)   168

        +-HF 10-76 p134   ++NR 4-76 p11
969   Sonatas, harp, op. 2, nos. 1-3. Sonatas, harp, op. 34, nos. 1,
      2. Susann McDonald, hp. Saga 5413.
          +Gr 3-76 p1482   +RR 2-76 p48
          +HFN 3-76 p109
      Sonatas, harp, op. 34, nos. 1, 2. cf Sonatas, harp, op. 2, nos.
      1-3.
DUTERTRE
      Pavane and galliarde. cf CBS 76183.
DUTILLEUX, Henri
970   Concerto, violoncello. LUTOSLAWSKI: Concerto, violoncello.
      Mstislav Rostropovitch, vlc; Orchestre de Paris; Serge Baudo,
      Witold Lutoslawski. HMV (Q) ASD 3145. (also Angel S 37146).
          ++Gr 2-76 p1336   ++NR 5-76 p4
          ++HF 6-76 p78   ++RR 12-75 p47
          ++HFN 12-75 p153   ++STL 3-7-76 p37
          ++MJ 7-76 p56
971   Sonatine, flute and piano. MULLER-ZURICH: Capriccio. QUANTZ:
      Sonata, flute, op. 1, no. 1, A minor. Sonata, flute, op. 1,
      no. 2, B flat major. Anne Diener Giles, flt; Allen Giles,
      hpd and pno. Crystal S 312.
          +ARG 11-76 p35   ++NR 11-76 p7
972   Symphony, no. 1. MARTINON: Symphony, no. 2. ORTF; Jean
      Martinon. Barclay Inédits 995 028.
          +Gr 9-73 p478   ++St 12-76 p135
DVORACEK, Jiri
      Dialogue. cf Panton 110 440.
      Ex post. cf BOHAC: Fragment.
      Music, harp. cf Panton 110 380.
DVORAK, Antonin
973   Carnival overture, op. 92. Slavonic dances, opp. 46 and 72.
      SMETANA: Má Vlast: Vltava. CO; Georg Szell. CBS 61709/10 (2).
      (Reissues from Columbia SCX 6053, 6054, SAX 2539).
          +Gr 8-76 p280
974   Carnival overture, op. 92. Slavonic dances, opp. 46, 72. CO;
      Georg Szell. Odyssey Y2 33524 (2). (Reissue from Columbia
      M2S 726).
          +HF 3-76 p77   +SFC 8-31-75 p20
          +-NR 10-75 p10
      Carnival overture, op. 92. cf Slavonic dances, opp. 46 and 72.
      Carnival overture, op. 92. cf Symphony, no. 7, op. 70, D minor.
      Carnival overture, op. 92. cf Symphony, no. 9, op. 95, E minor
      (Classics for Pleasure CFP 104).
      Carnival overture, op. 92. cf Symphony, no. 9, op. 95, E minor
      (Decca SXL 6751).
      Carnival overture, op. 92. cf Symphony, no. 9, op. 95, E minor
      (Philips 9500 001).
      Carnival overture, op. 92. cf Works, selections (Decca DPA
      539/40).
975   Concerto, piano, op. 33, G minor. Justus Frantz, pno; NYP;
      Leonard Bernstein. Columbia M 33889. (also CBS 76480 Tape (c)
      40-76480).
          +Gr 7-76 p168   +NR 4-76 p6
          -Gr 9-76 p497 tape   ++RR 6-76 p41
          +-HF 5-76 p81   -SFC 3-7-76 p27
          +-HFN 8-76 p84   /SR 5-29-76 p52
          +-HFN 10-76 p185 tape   ++St 8-76 p93

976   Concerto, piano, op. 33, G minor. Bruno Rigutto, pno; French
        Radio Orchestra; Zdeněk Mácal. French Decca 7352.
            +RR 10-76 p49
977   Concerto, piano, op. 33, G minor. Sviatoslav Richter, pno; LSO;
        Kyril Kondrashin. Rococo 2118.
            ++NR 8-76 p5
      Concerto, piano, op. 33, G minor. cf Works, selections (Vox
        QSVBX 5135).
978   Concerto, violin, op. 53, A minor. SIBELIUS: Concerto, violin,
        op. 47, D minor. Ruggiero Ricci, vln; LSO; Malcolm Sargent,
        Øivin Fjeldstad. Decca SPA 398. Tape (c) KCSP 398. (Reissues
        from SXL 2279, SXL 2077).
            +-Gr 7-75 p176   +RR 7-75 p35
            +-HFN 9-75 p108   +RR 3-76 p76 tape
979   Concerto, violin, op. 53, A minor. Romance, op. 11, F minor.
        Itzhak Perlman, vln; LPO; Daniel Barenboim. HMV ASD 3120.
        (also Angel S 37069).
            +Gr 10-75 p611   +-NYT 1-18-76 pD1
            -HF 5-76 p81   +RR 10-75 p38
            +HFN 10-75 p141   ++SFC 11-16-75 p32
            ++NR 12-75 p7   ++St 9-76 p119
      Concerto, violin, op. 53, A minor. cf Works, selections (Vox
        QSVBX 5135).
      Concerto, violin, op. 53, A minor: 3rd movement. cf Works,
        selections (HMV tape (c) TC EXES 5023).
980   Concerto, violoncello, op. 104, B minor. Paul Tortelier, vlc;
        PhO; Malcolm Sargent. HMV SXLP 30018. Tape (c) TC EXE 158.
            +-HFN 12-75 p173 tape   +-RR 1-76 p65 tape
981   Concerto, violoncello, op. 104, B minor. Lynn Harrell, vlc; LSO;
        James Levine. RCA ARL 1-1155. Tape (c) ARK 1-1155 (ct) ARS
        1-1155 (Q) ARD 1-1155 (ct) ART 1-1155. (also RK 11713).
            +-Audio 4-76 p89   -HFN 7-76 p104 tape
            ++Gr 2-76 p1337   ++MJ 4-76 p31
            -Gr 7-76 p230 tape   ++NR 11-75 p7
            +-HF 1-76 p111 tape   +-RR 2-76 p29
            ++HF 2-76 p94   -RR 6-76 p86 tape
            +-HF 2-76 p94 Quad   +SR 11-1-75 p46
            +-HFN 2-76 p97   ++St 1-76 p103
982   Concerto, violoncello, op. 104, B minor. Zara Nelsova, vlc;
        St. Louis Symphony Orchestra; Walter Susskind. SMG Vox Tape (c)
        8T 152 (ct) CT 152.
            +HF 1-76 p111 tape
      Concerto, violoncello, op. 104, B minor. cf Works, selections
        (HMV Tape (c) TC EXES 5023).
      Concerto, violoncello, op. 104, B minor. cf Works, selections
        (Vox QSVBX 5135).
      Concerto, violoncello, op. 104, B minor. cf BRUCH: Kol Nidrei,
        op. 47.
983   The golden spinning wheel, op. 109. The wood dove, op. 110.
        Bavarian Radio Symphony Orchestra; Rafael Kubelik. DG 2530 713.
            +Gr 11-76 p782   +RR 11-76 p63
            +HFN 11-76 p157
      Gypsy songs, op. 55, no. 4: Songs my mother taught me. cf HMV
        HLM 7077.
      Humoresque, op. 101. cf Connoisseur Society (Q) CSQ 2066.
      Humoresque, op. 101, no. 7, G flat major. cf HMV SXLP 30188.

Humoresque, op. 101, no. 7, G flat major. cf RCA LRL 1-5131.
Husitska overture, op. 67. cf Symphony, no. 8, op. 88, G major.
984   Legends, op. 59. Brno State Philharmonic Orchestra; Jiří Pinkas.
       Supraphon 110 1395.
           ++Gr 4-75 p1807   +RR 3-75 p25
           +HF 3-76 p77   +SFC 2-23-75 p23
           +NR 8-75 p2
985   Legends, op. 59, nos. 1-4. Slavonic dances, op. 46, nos. 1-8;
       op. 72, nos. 9-16. Vlastimil Lejsek, Věra Lejsková, pno.
       Supraphon 111 1301/2 (2).
           -Gr 7-75 p205   +NR 10-75 p10
           +HF 3-76 p77   +RR 7-75 p24
           -HFN 7-75 p80
986   Mass, op. 86, D major. Christ Church Cathedral Choir; Simon
       Preston. Argo ZRG 781.
           +Gr 12-74 p1200   +-RR 12-74 p71
           +HF 7-75 p74   +-SFC 7-6-75 p16
           ++MQ 1-76 p149   ++St 7-75 p96
           +-NR 6-75 p10
       Mazurek, violin, op. 49, E minor. cf Works, selections (Vox
       QSVBX 5135).
987   My home overture, op. 62. Slavonic dances, op. 72. Bavarian
       Radio Symphony Orchestra; Rafael Kubelik. DG 2530 593.
       Tape (c) 3300 593.
           ++Gr 12-75 p1032   ++NR 6-76 p4
           +HF 3-76 p77   ++RR 12-75 p47
           ++HFN 1-76 p107   +RR 6-76 p87 tape
       Nocturne, op. 40, B major. cf HMV SQ ESD 7001.
988   The noonday witch, op. 108. Symphonic variations, op. 78. The
       water goblin, op. 107. Bavarian Radio Symphony Orchestra;
       Rafael Kubelik. DG 2530 712.
           +Gr 11-76 p782   +RR 11-76 p63
           +HFN 11-76 p157
       Quartet, strings, op. 80 (27), E major. cf Quartet, strings,
       no. 3, op. 51, E flat major.
989   Quartet, strings, no. 2, op. 34, D minor. Quartet, strings,
       no. 6, op. 96, F major. Janáček Quartet. London STS 15207.
           ++NR 2-76 p2   ++SFC 2-22-76 p29
990   Quartet, strings, no. 3, op. 51, E flat major. Quartet, strings,
       no. 7, op. 105, A flat major. Gabrieli Quartet. Decca SDD
       479.
           +Gr 1-76 p1211   +MM 5-76 p33
           +HFN 2-76 p97
991   Quartet, strings, no. 3, op. 51, E flat major. Quartet, strings,
       op 80 (27), E major. Prague String Quartet. DG 2530 719.
           +Gr 11-76 p829   +RR 10-76 p76
           +HFN 10-76 p169
992   Quartet, strings, no. 6, op. 96, F major. Quartet, strings, no. 7,
       op. 105, A flat major. Prague Quartet. DG 2530 632.
           +Gr 4-76 p1623   ++HFN 4-76 p103
           +-HF 12-76 p100   +RR 4-76 p61
993   Quartet, strings, no. 6, op. 96, F major. Quintet, strings, no. 3,
       op. 97, E flat major. Guarneri Quartet; Walter Trampler, vla.
       RCA ARL 1-1791. Tape (c) ARK 1-1791 (ct) ARS 1-1791.
           +ARG 11-76 p20   +MJ 12-76 p44
           +-HF 12-76 p100   ++NR 11-76 p6

Quartet, strings, no. 6, op. 96, F major. cf Quartet, strings,
    no. 2, op. 34, D minor.
Quartet, strings, no. 7, op. 105, A flat major. cf Quartet,
    strings, no. 3, op. 51, E flat major.
Quartet, strings, no. 7, op. 105, A flat major. cf Quartet,
    strings, no. 6, op. 96, F major.
994   Quartet, strings, no. 13, op. 106, G major. Alban Berg Quartet.
    Telefunken 6-41933. Tape (c) 4 41933.
        ++Gr 5-76 p1776   +NR 6-76 p10
        +HF 8-76 p88   ++RR 5-76 p60
        +HFN 5-76 p99   ++St 11-76 p140
Quintet, strings, no. 3, op. 97, E flat major. cf Quartet,
    strings, no. 6, op. 96, F major.
Romance, op. 11, F minor. cf Concerto, violin, op. 53, A minor.
Romance, op. 11, F minor. cf Works, selections (Vox QSVBX 5135).
Rondo, op. 94, G minor. cf Works, selections (HMV Tape (c)
    TC EXES 5023).
Rondo, op. 94, G minor. cf Works, selections (Vox QSVBX 5135).
995   Rusalka, op. 114. Alena Míková, Milada Subrtová, s; Maria
    Ovcacikova, con; Ivo Zidek, t; Jiři Joran, bar; Eduard Haken,
    bs; Prague National Theatre Orchestra; Zdenek Chalabala.
    Supraphon ST 50440/3 (4).
        +NR 11-76 p9
Rusalka, op. 114: Gods of the lake. cf HMV SXLP 30205.
Rusalka, op. 114: O lovely moon. cf Muza SX 1144.
Rusalka, op. 114: O silver moon. cf Works, selections (Decca
    DPA 539/40).
Rusalka, op. 114: O silver moon. cf Decca SPA 449.
996   Scherzo capriccioso, op. 66. Slavonic dances, op. 46. Bavarian
    Radio Symphony Orchestra; Rafael Kubelik. DG 2530 466. Tape
    (c) 3300 422.
        +HF 3-76 p77   +-NR 2-76 p2
        +HF 8-76 p70 tape   +RR 6-76 p87 tape
        +MJ 1-76 p39   +SR 4-17-76 p51
Scherzo capriccioso, op. 66, D flat major. cf Works, selections
    (Decca DPA 539/40).
997   Serenade, op. 44, D minor. Serenade, strings, op. 22, E major.
    Czech Chamber Orchestra; Martin Turnovský, Josef Vlach.
    Supraphon 50760. Tape (c) 4-50760.
        +HFN 2-76 p116 tape   +-RR 1-76 p65 tape
998   Serenade, strings, op. 22, E major. GRIEG: Holberg suite, op. 40.
    AMF; Neville Marriner. Argo ZRG 670. Tape (c) KZRC 670.
        *Gr 11-70 p787   +-RR 7-72 p92 tape
        ++HF 11-76 p153 tape   ++SFC 10-31-71 p34
        ++HFN 5-73 p997 tape   ++St 1-72 p92 tape
        +NR 10-71 p4
999   Serenade, strings, op. 22, E major. TCHAIKOVSKY: Serenade,
    strings, op. 48, C major. AMF; Neville Marriner. Argo ZRG
    848. Tape (c) KZRC 848. (Reissues from ZRG 670, 584).
        ++Gr 7-76 p175   +-RR 5-76 p58
        +HFN 6-76 p102   +RR 11-76 p110 tape
        +HFN 7-76 p104 tape   ++SFC 10-10-76 p32
1000 Serenade, strings, op. 22, E major. SUK: Serenade, strings, op.
    6, E flat major. CSO; Karl Münchinger. Decca SXL 6758. (Reissue
    from SXL 6533).
        +-Gr 11-76 p782   +-RR 11-76 p63
        +HFN 11-76 p159

1001 Serenade, strings, op. 22, E major. TCHAIKOVSKY: Serenade,
        strings, op. 48, C major. Hamburg Radio Orchestra, Dresden
        Staatskapelle Orchestra; Hans Schmidt-Isserstedt, Otmar
        Suitner. DG Heliodor 2548 121. Tape (c) 3348 121. (Reissue
        from DG 136 481, 135 109).
              +Gr 4-75 p1867   +-RR 9-76 p92 tape
              +RR 4-75 p36
1002 Serenade, strings, op. 22, E major. TCHAIKOVSKY: Serenade,
        strings, op. 48, C major. ECO; Daniel Barenboim. HMV ASD
        3036. (also Angel S 37045).
              +Gr 11-74 p890   +-NYT 1-18-76 pD1
              +HF 11-75 p102   +-RR 11-74 p36
              -NR 8-75 p2
1003 Serenade, strings, op. 22, E major. JANACEK: Idyll, string
        orchestra. South West German Chamber Orchestra; Paul Angerer.
        Turnabout TV 34532.
              +-Gr 7-75 p176   +MT 2-76 p141
              +HF 10-74 p95   +NR 9-74 p4
              +HFN 5-75 p126   +RR 5-75 p33
      Serenade, strings, op. 22, E major. cf Serenade, op. 44, D
        minor.
      Serenade, strings, op. 22, E major. cf Works, selections (Decca
        DPA 539/40).
      Serenade, strings, op. 22, E major. cf Works, selections (HMV
        Tape (c) TC EXES 5023).
      Silent woods, op. 68. cf Works, selections (Vox QSVBX 5135).
      Slavonic dance, no. 2. cf RCA ARM 4-0942/7.
      Slavonic dance, no. 3, G major. cf ALBENIZ: Tango.
1004 Slavonic dances, opp. 46 and 72. CO; Georg Szell. CBS 61709/10.
        Reissues. (also Columbia MS 7208).
              +-HFN 8-76 p93
1005 Slavonic dances, opp. 46 and 72. Carnival overture, op. 92.
        SMETANA: Má Vlast: Vltava. CO; Georg Szell. CBS 78299 (2).
              +-RR 8-76 p35
1006 Slavonic dances, opp. 46 and 72. SMETANA: The bartered bride:
        Overture, Polka, Furiant, Dance of the comedians. Minneapolis
        Symphony Orchestra; Antal Dorati. Mercury SRI 2-77001 (2).
        (Reissue from SR 2-9007).
              +HF 3-76 p77
1007 Slavonic dances, opp. 46 and 72. Alfred Brendel, Walter Klien,
        pno. Turnabout TV 34060. Tape (c) KTVC 34060.
              +HFN 5-76 p117 tape   +-RR 6-76 p87 tape
      Slavonic dances, opp. 46 and 72. cf Carnival overture, op. 92.
        (CBS 61709/10).
      Slavonic dances, opp. 46 and 72. cf Carnival overture, op. 92.
        (Odyssey 33524).
1008 Slavonic dances, op. 46 (8). Bamberg Symphony Orchestra; Antal
        Dorati. Turnabout (Q) QTV 34582.
              -Gr 8-76 p280   +NR 11-75 p1
              +HF 3-76 p77   +RR 8-76 p35
              +-HFN 8-76 p79
      Slavonic dances, op. 46. cf Scherzo capriccioso, op. 66.
1009 Slavonic dances, op. 46, nos. 1-6. GLINKA: Russlan and Ludmila:
        Overture. SMETANA: Má Vlast: Vltava. COA; Bernard Haitink.
        Philips Tape (c) 7317 096.
              +-RR 3-76 p78 tape
      Slavonic dance, op. 46, no. 1, C major. cf Decca DPA 519/20.

Slavonic dances, op. 46, nos. 1, 8. cf Works, selections (Decca
    DPA 539/40).
Slavonic dances, op. 46, nos. 1, 3, 8. cf BRAHMS: Hungarian
    dances, nos. 1, 5-7, 12-13, 19, 21.
Slavonic dances, op. 46, nos. 2, 4, 6. cf Symphony, no. 8, op.
    88, G major.
1010 Slavonic dances, op. 72. Bamberg Symphony Orchestra; Antal
    Dorati. Turnabout (Q) QTV S 34583.
        +HF 3-76 p77
Slavonic dances, op. 72. cf My home overture, op. 62.
Slavonic dances, op. 72, nos. 1 and 2. cf BRAHMS: Hungarian
    dances, nos. 1, 5-7, 12-13, 19, 21.
Slavonic dance, op. 72, no. 1, B major. cf HMV (Q) ASD 3131.
1011 Slavonic dance, op. 72, no. 2, E minor. RIMSKY-KORSAKOV:
    Capriccio espagnol, op. 34. SCRIABIN: Poème de l'extase,
    op. 54. CPhO, NPhO; Leopold Stokowski. Decca PFS 4333.
    (also London 40-21117).
        +Gr 7-75 p176   ++RR 6-75 p48
        +HF 11-75 p124   +SFC 9-21-75 p34
        ++HFN 6-75 p102   ++St 2-76 p109
        +NR 11-75 p2
Slavonic dance, op. 72, no. 2, E minor. cf Works, selections
    (Decca DPA 539/40).
Slavonic dance, op. 72, no. 2, E minor. cf Decca PFS 4351.
Slavonic dance, op. 72, no. 2, E minor. cf RCA ARM 4-0942/7.
Slavonic dance, op. 72, no. 8. cf RCA ARM 4-0942/7.
Songs: Die Bescheidene, op. 32, no. 8; Die Gefangene, op. 32,
    no. 11; Scheiden ohne leiden, op. 32, no. 4; Die verlassene,
    op. 32, no. 6; Die Zuversicht, op. 32, no. 10. cf BIS LP 17.
Songs: op. 55: Songs my mother taught me. cf Works, selections
    (HMV Tape (c) TC EXES 5023).
Songs: op. 55: Songs my mother taught me. cf HMV RLS 723.
Symphonic variations, op. 78. cf The noonday witch, op. 108.
1012 Symphonies, nos. 1-9. LSO; István Kertész. Decca D6D7 (7).
        +Gr 9-76 p477   +-RR 9-76 p47
        ++HFN 10-76 p181
1013 Symphony, no. 6, op. 60, D major. RPO; Charles Groves. HMV SQ
    ASD 3169.
        +Gr 5-76 p1757   -RR 5-76 p40
        +-HFN 8-76 p99
1014 Symphony, no. 6, op. 60, D major. CPhO; Václav Neumann.
    Supraphon 110 1833. (Reissue from 110 1621/8).
        +-Gr 12-76 p995   +-RR 9-76 p47
        ++NR 12-76 p6
1015 Symphonies, nos. 7-9. CO; Georg Szell. CBS 78304 (3). (also
    Columbia D3S 814).
        +Gr 9-76 p477   +RR 9-76 p48
        +HFN 9-76 p119
1016 Symphony, no. 7, op. 70, D minor. LSO; Pierre Monteux. Decca
    ECS 779. Tape (c) KECC 779. (Reissue from RCA SB2155).
    (also London STS 15157).
        +Gr 7-76 p175   +HFN 8-76 p94 tape
        +Gr 10-76 p658 tape   +RR 6-76 p41
        +HFN 7-76 p103   +RR 9-76 p94 tape
1017 Symphony, no. 7, op. 70, D minor. Carnival overture, op. 92.
    LSO; Witold Rowicki. Philips Tape (c) 7300 407.
        +-RR 12-76 p106 tape

1018   Symphony, no. 7, op. 70, D minor.   CPhO; Václav Neumann.
       Vanguard SU 7.
              +HF 1-76 p86   +NR 11-75 p1
1019   Symphony, no. 8, op. 88, G major.   Symphony, no. 9, op. 95, E
       minor.   CPhO; Frantisek Stupka.   Panton 110 447/48.
              +-RR 1-76 p31
1020   Symphony, no. 8, op. 88, G major.   Slavonic dances, op. 46,
       nos. 2, 4, 6.   COA; Bernard Haitink.   Philips 6580 126.
       (Reissue from SAL 3451).
              /Gr 11-76 p785   +-RR 10-76 p50
              +HFN 11-76 p170
1021   Symphony, no. 8, op. 88, G major.   Husitska overture, op. 67.
       LSO; Witold Rowicki.   Philips Tape (c) 7300 408.
              +RR 6-76 p87
1022   Symphony, no. 9, op. 95, E minor.   Carnival overture, op. 92.
       PhO; Wolfgang Sawallisch.   Classics for Pleasure Tape (c)
       TC CFP 104.
              +-HFN 8-76 p94 tape
1023   Symphony, no. 9, op. 95, E minor.   VPO; István Kertész.   Decca
       SPA 87.   Tape (c) KCSP 87.   (also London STS 15101).
              +RR 1-76 p65 tape
1024   Symphony, no. 9, op. 95, E minor.   Carnival overture, op. 92.
       LAPO; Zubin Mehta.   Decca SXL 6751.   (also London 6980 Tape (c)
       5-6980).
              +-Gr 11-76 p785   +RR 11-76 p63
              +HFN 11-76 p157   ++SFC 10-10-76 p32
1025   Symphony, no. 9, op. 95, E minor.   SMETANA: Má Vlast: Vltava.
       BPhO; Herbert von Karajan.   HMV ASD 2863.   Tape (c) TC ASD
       2863.   (Reissue from Columbia SAX 2275).
              +Gr 8-73 p376   +RR 10-76 p106 tape
              +RR 8-73 p40
1026   Symphony, no. 9, op. 95, E minor.   Budapest Philharmonic
       Orchestra; Adám Medveczky.   Hungaroton SLPX 11785.
              +-ARG 12-76 p28   /NR 8-76 p4
1027   Symphony, no. 9, op. 95, E minor.   Carnival overture, op. 92.
       San Francisco Symphony Orchestra; Seiji Ozawa.   Philips 9500
       001.   Tape (c) 7300 419.
              -Gr 3-76 p1464   -NR 6-76 p4
              -HFN 3-76 p92   -RR 3-76 p39
              -HFN 7-76 p104 tape   /St 8-76 p93
              /MJ 5-76 p29
1028   Symphony, no. 9, op. 95, E minor.   CPhO; Václav Neumann.   Vanguard
       SU 8.
              +-HF 1-76 p86   ++NR 11-75 p1
       Symphony, no. 9, op. 95, E minor.   cf Symphony, no. 8, op. 88,
       G major.
       Symphony, no. 9, op. 95, E minor.   cf Works, selections (Decca
       DPA 539/40).
       Symphony, no. 9, op. 95, E minor.   cf Pye TB 3009.
       Symphony, no. 9, op. 95, E minor: Largo.   cf Works, selections
       (HMV Tape (c) TC EXES 5023).
1029   Trio, piano, op. 65, F minor.   GLIERE: Duo, violin and violoncello,
       op. 39.   HANDEL: Passacaglia.   STRAVINSKY: Suite italienne.
       Jascha Heifetz, vln; Gregor Piatigorsky, vlc; Leonard Pennario,
       pno.   Columbia M 33447.
              ++SFC 12-76 p34

1030 Trio, piano, op. 90, E minor. SMETANA: Trio, piano, op. 15,
      G minor. Yuval Trio. DG 2530 594.
          +Gr 3-76 p1478   +NR 6-76 p10
          +-HF 8-76 p88   +-RR 3-76 p52
          +HFN 3-76 p92   ++St 8-76 p93
      The water goblin, op. 107. cf The noonday witch, op. 108.
      The wood dove, op. 110. cf The golden spinning wheel, op. 109.
1031 Works, selections: Carnival overture, op. 92. Rusalka, op. 114:
      O silver moon. Scherzo Capriccioso, op. 66, D flat major.
      Serenade, strings, op. 22, E major. Slavonic dances, op. 46,
      nos. 1, 8; op. 72, no. 2. Symphony, no. 9, op. 95, E minor.
      Pilar Lorengar, s; LSO, VPO, Rome, Santa Cecilia Orchestra,
      Israel Philharmonic Orchestra; István Kertész, Giuseppe Patané,
      Rafael Kubelik, Fritz Reiner. Decca DPA 539/40. Tape (c)
      KDPC 539/40.
          +Gr 6-76 p92   +HFN 6-76 p103
          +Gr 9-76 p497 tape   ++RR 6-76 p41
1032 Works, selections: Concerto, violin, op. 53, A minor: 3rd move-
      ment. Concerto, violoncello, op. 104, B minor. Rondo, op. 94,
      G minor. Serenade, strings, op. 22, E major. Songs, op. 55:
      Songs my mother taught me. Symphony, no. 9, op. 95, E minor:
      Largo. Elisabeth Schwarzkopf, s; Hermann Krebbers, vln; Paul
      Tortelier, Mstislav Rostropovich, vlc; Gerald Moore, Shuku
      Iwasaki, pno; PhO, ECO, RPO, Amsterdam Philharmonic Orchestra;
      István Kertész, Adrian Boult, Carlo Maria Giulini, Daniel
      Barenboim. HMV Tape (c) TC EXES 5023 (ct) 8X EXES 5023.
          +HFN 9-76 p133 tape
1033 Works, selections: Concerto, piano, op. 33, G minor. Concerto,
      violoncello, op. 104, B minor. Concerto, violin, op. 53, A
      minor. Mazurek, violin, op. 49, E minor. Romance, op. 11, F
      minor. Rondo, op. 94, G minor. Silent woods, op. 68. Ruggiero
      Ricci, vln; Rudolf Firkusny, pno; Zara Nelsova, vlc; St. Louis
      Symphony Orchestra; Walter Susskind. Vox (Q) QSVBX 5135 (3).
          +HF 5-76 p81 Quad   ++SFC 2-22-76 p29
          ++NR 3-76 p3   ++St 9-76 p119 Quad
DVORKIN
      Maurice. cf CRI SD 102.
DYKES
      Songs: Eternal father; The Marine's hymn. cf Department of
      Defense Bicentennial Edition 50-1776.
      Songs: The King of love my shepherd is; Eternal Father, strong to
      save. cf Saga 5225.
EARLS, Paul
      Songs: Arise my love; Entreat me not to leave you. cf Duke
      University Press DWR 7306.
      EARLY AMERICAN VOCAL MUSIC: NEW ENGLAND ANTHEMS AND SOUTHERN
      FOLK HYMNS. cf Nonesuch H 71276.
      EARLY STRING QUARTET IN U. S. A. cf VOX SVBX 5301.
EAST, Michael
      Peccavi. cf Crystal S 206.
EASTLEY, Max
1034 The centriphone. Elastic aerophone/centriphone. Hydrophone.
      Metallophone. TOOP: The chairs story. The divination of the
      bowhead whale. Do the bathosphere. David Toop; The Cetaceans;
      David Toop, Frank Perry, Paul Burwell, Brian Eno, Hugh Davies.
      Obscure no. 4.
          +RR 3-76 p18

Elastic aerophone/centriphone. cf The centriphone.
Hydrophone. cf The centriphone.
Metallophone. cf The centriphone.

EBEN, Petr
Vox clamantis. cf BOHAC: Fragment.
EDINBURG FESTIVAL SAMPLER. cf HMV SEOM 25.

EDWARDES
My colleague Caruso. cf Rococo 5379.
AN EDWARDIAN MUSICAL EVENING. cf Pearl SHE 528.

EDWARDS, George
Kreuz und Quer. cf BASSETT: Sextet.
1035 Quartet, strings. GRIFFITH: Quartet, one string. POLLOCK:
Movement and variations. THIMMIG: Seven profiles. Composers
String Quartet. CRI SD 265.
    +Gr 9-76 p437   +-RR 8-76 p63

EISMA, Will
1036 Le gibet. Little lane. RAXACH: Imaginary landscape. Paraphrase.
Ileana Melita, con; Rien de Reede, flt; Willy Goudswaard, perc;
Cor Coppens, ob; Radio Chamber Orchestra; Pro-Hontra Ensemble;
Paul Hupperts. Donemus Audio-Visual DAVS 7475/2.
    +RR 10-75 p67   *Te 9-76 p28
Little lane. cf Le gibet.

ELGAR, Edward
Adieu. cf Piano works (Prelude PRS 2503).
Adieu. cf Works, selections (Polydor 2383 359).
1037 The apostles, op. 49. Sheila Armstrong, s; Helen Watts, con;
Robert Tear, t; Benjamin Luxon, bar; Clifford Grant, John
Carol Case, bs; Downe House School Choir; LPO and Choir; Adrian
Boult. HMV SLS 976 (3). Tape (c) TC SLS 976. (also
Connoisseur Society CS 3-2094 (3)).
    +Gr 11-74 p936   +RR 11-74 p18
    +Gr 3-75 p1712 tape   +RR 6-75 p91 tape
    +HF 7-76 p76   ++St 12-76 p136
    +MJ 11-76 p45
Bavarian dances, op. 27, nos. 1-3. cf Works, selections (RCA
LRL 1-5133).
Beau Brummel: Minuet. cf Works, selections (Polydor 2383 359).
Cantique, op. 3, no. 1. cf Organ works (RCA LRL 2-5120).
La capricieuse, op. 17 (2). cf RCA ARM 4-0942/7.
Caractacus, op. 35: Woodland interlude. cf Works, selections
(RCA LRL 1-5133).
Carillon, op. 75. cf Organ works (RCA LRL 2-5120).
Carissima. cf Works, selections (CBS 76423).
Carissima. cf Works, selections (HMV ESD 7009).
Chanson de matin, op. 15, no. 2. cf Works, selections (CBS 76423).
Chanson de matin, op. 15, no. 2. cf Works, selections (Classics
for Pleasure CFP 40235).
Chanson de matin, op. 15, no. 2. cf Works, selections (Decca
DPA 537/8).
Chanson de matin, op. 15, no. 2. cf Works, selections (RCA LRL
1-5133).
Chanson de matin, op. 15, no. 2. cf RCA GL 25006.
Chanson de nuit, op. 15, no. 1. cf Works, selections (CBS 76423).
Chanson de nuit, op. 15, no. 1. cf Works, selections (Classics
for Pleasure CFP 40235).
Chanson de nuit, op. 15, no. 1. cf Works, selections (Decca
DPA 537/8).

Chanson de nuit, op. 15, no. 1. cf Works, selections (RCA LRL
     1-5133).
Chantant. cf Piano works (Prelude PRS 2503).
1038 Cockaigne overture, op. 40. Enigma variations, op. 36. Serenade,
     strings, op. 20, E minor. RPO; Thomas Beecham. CBS 61660.
     (Reissue from Philips ABL 3053).
          +Gr 9-75 p457   +RR 10-75 p43
          -HFN 11-75 p153   +ST 2-76 p737
1039 Cockaigne overture, op. 40. Enigma variations, op. 36. CSO,
     LPO; Georg Solti. Decca SXL 6795.
          +Gr 11-76 p786   +RR 11-76 p65
          +-HFN 12-76 p141
1040 Cockaigne overture, op. 40. Froissart overture, op. 19. In the
     south overture, op. 50. HANDEL (arr. Elgar): Overture, D
     minor. LPO; Adrian Boult. HMV ASD 2822. Tape (c) TC ASD 2822.
          +Gr 9-72 p485   +RR 9-72 p54
          +Gr 7-76 p230 tape   ++RR 8-76 p83 tape
          +-HFN 10-72 p1906   +STL 10-1-72 p38;
          +-HFN 6-76 p105 tape   12-10-72 p35
     Cockaigne overture, op. 40. cf Works, selections (Classics for
     Pleasure CFP 40235).
     Cockaigne overture, op. 40. cf Works, selections (HMV SLS 5030).
     Cockaigne overture, op. 40. cf BRITTEN: Simple symphony, strings,
     op. 4.
     Concert allegro, op. 41. cf Piano works (Prelude PRS 2503).
1041 Concerto, violin, op. 61, B minor. Pinchas Zukerman, vln; LPO;
     Daniel Barenboim. CBS 76528.
          ++Gr 11-76 p786   ++RR 11-76 p65
          +HFN 11-76 p159
     Concerto, violin, op. 61, B minor. cf RCA ARM 4-0942/7.
1042 Concerto, violoncello, op. 85, E minor. Enigma variations, op.
     36. Jacqueline du Pré, vlc; PO, LPO; Daniel Barenboim.
     CBS 76529.
          ++Gr 11-76 p785   +RR 11-76 p64
          +-HFN 11-76 p159
1043 Concerto, violoncello, op. 85, E minor. Sea pictures. Janet
     Baker, s; Jacqueline du Pré, vlc; LSO; John Barbirolli. HMV
     Tape (c) TC ASD 655.
          +-HFN 4-76 p125 tape   +-RR 8-76 p83 tape
     Concerto, violoncello, op. 85, E minor. cf BLOCH: Schelomo.
     Concerto, violoncello, op. 85, E minor. cf HMV SLS 5068.
     Contrasts, op. 10, no. 3. cf Works, selections (HMV ESD 7009).
     Contrasts, op. 10, no. 3. cf Works, selections (RCA LRL 1-5133).
     Crown of India suite, op. 66. cf Imperial march, op. 32.
     Dream children, op. 43. cf Works, selections (RCA LRL 1-5133).
1044 The dream of Gerontius. Helen Watts, con; Nicolai Gedda, t;
     Robert Lloyd, bs; John Alldis Choir; LPO, NPhO; Adrian Boult.
     HMV 987 (2). Tape (c) TC SLS 987.
          ++Audio 12-76 p92   ++MT 8-76 p659
          +Gr 5-76 p1788   ++RR 4-76 p20
          ++Gr 8-76 p341 tape   ++RR 8-76 p83 tape
          +HFN 8-76 p84   ++STL 4-11-76 p36
          +MM 10-76 p41
1045 The dream of Gerontius, op. 38: Angel's farewell (arr. Ball).
     Enigma variations, op. 36: Nimrod (arr. Wright). Imperial
     march, op. 32 (arr. Ball). HOLST: Suite, no. 2, op. 28, F
     major. HOWELLS: Three figures. Virtuosi Brass Band; Maurice
     Handford. Virtuosi VR 7507.
          +Gr 9-76 p483

The dream of Gerontius, op. 38: Prelude; Angel's farewell. cf
    Organ works (RCA LRL 2-5120).
1046 Elegy, strings, op. 58. Introduction and allegro, op. 47. Sere-
    nade, strings, op. 20, E minor. Sospiri, op. 70. The Spanish
    lady, op. 89: Suite. AMF; Neville Marriner. Argo ZRG 573.
    Tape (c) KZRC 573.
        ++HF 10-76 p147 tape   +HFN 5-75 p142 tape
        +HFN 2-74 p347 tape   +RR 1-74 p77 tape
1047 Elegy, strings, op. 58. Froissart overture, op. 19. Pomp and
    circumstance marches, op. 39. Sospiri, op. 70. PhO, NPhO;
    John Barbirolli. HMV ASD 2292. Tape (c) TC ASD 2292.
        +-Gr 6-76 p102 tape   +-HFN 5-76 p117
    Elegy, strings, op. 58. cf Enigma variations, op. 36.
    Elegy, strings, op. 58. cf Works, selections (CBS 76423).
    Elegy, strings, op. 58. cf HMV SEOM 24.
1048 Enigma variations, op. 36. Elegy, strings, op. 58. Serenade,
    strings, op. 20, E minor. CPhO, LSO; Leopold Stokowski, Ains-
    lee Cox. Decca PFS 4338. (also London SPC 21136).
        +-Gr 8-75 p322   +NR 12-76 p3
        +HF 12-76 p100   +-RR 8-75 p32
        +-HFN 8-75 p75
1049 Enigma variations, op. 36. Pomp and circumstance marches, op. 39,
    nos. 1-5. RPO; Norman del Mar. DG 2535 217. (Reissue from
    Contour 2870 440).
        +Gr 10-76 p595   +-RR 11-76 p65
        +HFN 11-76 p170
1050 Enigma variations, op. 36. Falstaff, op. 68. NPhO; Andrew Davis.
    Lyrita SRCS 77.
        +Gr 9-75 p457   +RR 9-75 p36
        +-HFN 9-75 p98   +St 8-76 p93
    Enigma variations, op. 36. cf Cockaigne overture, op. 40 (CBS
    61660).
    Enigma variations, op. 36. cf Cockaigne overture, op. 40 (Decca
    SXL 6795).
    Enigma variations, op. 36. cf Concerto, violoncello, op. 85, E
    minor.
    Enigma variations, op. 36. cf Works, selections (Decca DPA 537/8).
    Enigma variations, op. 36. cf Works, selections (HMV SLS 5030).
    Enigma variations, op. 36. cf BRAHMS: Varitaions on a theme by
    Haydn, op. 56a.
    Enigma variations, op. 36. cf BRITTEN: Simple symphony, strings,
    op. 4.
    Enigma variations, op. 36: Nimrod. cf The dream of Gerontius,
    op. 38: Angel's farewell.
    Enigma variations, op. 36: Nimrod. cf Decca PFS 4351.
    Enigma variations, op. 36: Nimrod. cf RCA GL 25006.
1051 Falstaff, op. 68. The sanguine fan, op. 8. Fantasia and fugue,
    op. 86, C minor. (Bach trans. Elgar). LPO; Adrian Boult. HMV
    ASD 2970. Tape (c) TC ASD 2970.
        +Gr 3-74 p1694   ++RR 5-74 p33
        +Gr 8-75 p376 tape   +RR 12-76 p106 tape
        +-HFN 9-75 p110 tape
    Falstaff, op. 68. cf Enigma variations, op. 36.
    Falstaff, op. 68. cf Works, selections (HMV SLS 5030).
    Falstaff, op. 68: Interludes (2). cf Works, selections (RCA LRL
    1-5133).

Fantasia and fugue, op. 86, C minor (Bach trans. Elgar). cf
  Falstaff, op. 68.
For the fallen, op. 80, no. 3: Prelude. cf Organ works (RCA
  LRL 2-5120).
Froissart overture, op. 19. cf Cockaigne overture, op. 40.
Froissart overture, op. 19. cf Elegy, strings, op. 58.
Griffinesque. cf Piano works (Prelude PRS 2503).
1052 Imperial march, op. 32. Pomp and circumstance marches, op. 39,
  nos. 1-5. Crown of India suite, op. 66. LPO; Daniel
  Barenboim. Columbia M 32936. Tape (c) MT 32926 (ct) MA 32926
  (Q) MQ 32936 Tape (ct) MAQ 32936. (also CBS 76248 Tape (c)
  40-76248).
            -Gr 9-74 p495   +-HFN 12-75 p173 tape
         +-Gr 12-75 p1121 tape   +-RR 10-74 p45
           -HF 4-75 p112 tape   +-RR 6-76 p87 tape
         +-HF 4-75 p112 Quad   ++SFC 9-22-74 p22
Imperial march, op. 32. cf The dream of Gerontius, op. 38: Angel's
  farewell.
Imperial march, op. 32. cf Organ works (RCA LRL 2-5120).
Imperial march, op. 32. cf Decca 419.
In Smyrna. cf Piano works (Prelude PRS 2503).
1053 In the south overture, op. 50. VAUGHAN WILLIAMS: Fantasia on a
  theme by Thomas Tallis. The wasps: Overture. Bournemouth
  Symphony Orchestra; Constantin Silvestri. HMV ESD 7013.
  Tape (c) TC ESD 7013. (Reissue from ASD 2370).
            +Gr 10-76 p595   +RR 10-76 p74
            +HFN 11-76 p170
In the south overture, op. 50. cf Cockaigne overture, op. 40.
1054 Introduction and allegro, op. 47. Serenade, strings, op. 20,
  E minor. TIPPETT: Fantasia concertante on a theme by Corelli.
  Little music, strings. St. John's Symphony Orchestra; John
  Lubbock. Pye (Q) TPLS 13069.
            -Gr 12-75 p1032   +-RR 5-76 p40
Introduction and allegro, op. 47. cf Elegy, strings, op. 58.
Introduction and allegro, op. 47. cf Works, selections (Decca
  DPA 537/8).
Introduction and allegro, op. 47. cf Works, selections (HMV
  SLS 5030).
1055 The kingdom, op. 51. Margaret Price, s; Yvonne Minton, con;
  Alexander Young, t; John Shirley-Quirk, bs; LPO and Choir;
  Adrian Boult. Connoisseur Society CS 2089 (2).
            +Audio 11-76 p106   ++SFC 4-11-76 p38
            +HF 7-76 p76   ++St 4-76 p77
            ++MJ 3-76 p25
The kingdom, op. 51: Prelude. cf Organ works (RCA LRL 2-5120).
May song. cf Piano works (Prelude PRS 2503).
May song. cf Works, selections (HMV ESD 7009).
Mazurka, op. 10, no. 1. cf Works, selections (HMV ESD 7009).
Mina. cf Works, selections (HMV ESD 7009).
Minuet. cf Piano works (Prelude PRS 2503).
Minuet, op. 21. cf Works, selections (HMV ESD 7009).
1056 Organ works: Cantique, op. 3, no. 1. Carillon, op. 75 (arr.
  Blair). The dream of Gerontius, op. 38: Prelude; Angel's
  farewell (arr. Brewer). For the fallen, op. 80, no. 3:
  Prelude (arr. Grace). Imperial march, op. 32 (arr. Martin).
  The kingdom, op. 51: Prelude (arr. Brewer). Sonata, organ,

no. 1, op. 28, G major. Sonata, piano, no. 2, op. 87a (arr.
Atkins). Vespers, op. 14. Donald Hunt, Christopher Robinson,
org. RCA LRL 2-5120 (2).
+-Gr 5-76 p1782   +MT 8-76 p659
+-HFN 5-76 p99   +-RR 6-76 p70

1057 Piano works: Adieu. Chantant. Concert allegro, op. 41. Griffin-
esque. In Smyrna. May song. Minuet. Serenade. Skizze.
Sonatina. John McCabe, pno. Prelude PRS 2503.
+Gr 10-76 p627   +MT 12-76 p1003
+-HFN 10-76 p169   +RR 10-76 p87

Pomp and circumstance marches, op. 39. cf Elegy, strings, op. 58.
Pomp and circumstance marches, op. 39. cf Works, selections
(Decca DPA 537/8).
Pomp and circumstance marches, op. 39. cf Decca 419.
Pomp and circumstance marches, op. 39, nos. 1-5. cf Enigma
variations, op. 36.
Pomp and circumstance marches, op. 39, nos. 1-5. cf Imperial
march, op. 32.
Pomp and circumstance marches, op. 39, nos. 1-5. cf Works,
selections (HMV SLS 5030).
Pomp and circumstance marches, op. 39, nos. 1, 4. cf Works,
selections (Classics for Pleasure CFP 40235).
Pomp and circumstance march, op. 39, no. 1, D major. cf BRITTEN:
Simple symphony, strings, op. 4.
Pomp and circumstance march, op. 39, no. 1, D major. cf BBC
REB 228.
Pomp and circumstance march, op. 39, no. 1, D major. cf HMV
SEOM 24.
Pomp and circumstance march, op. 39, no. 4, G major. cf HMV
HLM 7093.
Romance, bassoon, op. 62. cf Works, selections (CBS 76423).
Romance, bassoon, op. 62. cf Works, selections (HMV ESD 7009).
Romance, bassoon, op. 62. cf Washington University RAVE 761.
Rosemary. cf Works, selections (CBS 76423).
Rosemary. cf Works, selections (HMV ESD 7009).
Salut d'amour, op. 12. cf Works, selections (CBS 76423).
Salut d'amour, op. 12. cf Works, selections (RCA LRL 1-5133).
Salut d'amour, op. 12. cf Pearl SHE 528.
The sanguine fan, op. 8. cf Falstaff, op. 68.
Sea pictures. cf Concerto, violoncello, op. 85, E minor.
Serenade. cf Piano works (Prelude PRS 2503).
Serenade, strings, op. 20, E minor. cf Cockaigne overture, op. 40.
Serenade, strings, op. 20, E minor. cf Elegy, strings, op. 58.
Serenade, strings, op. 20, E minor. cf Enigma variations, op. 36.
Serenade, strings, op. 20, E minor. cf Introduction and allegro,
op. 47.
Serenade, strings, op. 20, E minor. cf Works, selections (CBS
76423).
Serenade, strings, op. 20, E minor. cf Works, selections (Classics
for Pleasure CFP 40235).
Serenade, strings, op. 20, E minor. cf Works, selections (Decca
DPA 537/8).
Serenade lyrique. cf Works, selections (HMV ESD 7009).
Serenade lyrique. cf Works, selections (RCA LRL 1-5133).
Serenade mauresque, op. 10, no. 2. cf Works, selections (HMV
ESD 7009).

Sevillana, op. 7. cf Works, selections (HMV ESD 7009).

Skizze. cf Piano works (Prelude PRS 2503).

Soliloquy, oboe (orch. Jacob). cf Works, selections (RCA LRL 1-5133).

Sonata, organ, G major: Last movement (organ solo). cf Audio 3.

Sonata, organ, no. 1, op. 28, G major. cf Organ works (RCA LRL 2-5120).

Sonata, organ, no. 1, op. 28, G major. cf Argo 5BBA 1013-5.

Sonata, piano, no. 2, op. 87a. cf Organ works (RCA LRL 2-5120).

1058   Sonata, violin, op. 82, E minor. WALTON: Sonata, violin. Sidney
       Weiss, vln; Jeanne Weiss, pno. Unicorn RHS 341.
          +Gr 7-76 p187   +-HFN 7-76 p87

Sonatina. cf Piano works (Prelude PRS 2503).

1059   Songs: After, op. 31, no. 1; As I laye a-thynkynge; Come, gentle
       night; In the dawn, op. 41, no. 1; Is she not passing fair;
       The language of flowers; The pipes of Pan; Pleading, op. 48,
       no. 1; Rondel, op. 16, no. 3; The shepherds' song, op. 16,
       no. 1; A song of autumn; Speak, music, op. 41, no. 2; A song
       of flight, op. 31, no. 2; Through the long days, op. 16, no. 2;
       The wind at dawn. Brian Rayner Cook, bar; Roger Vignoles, pno.
       Pearl SHE 526.
          +-Gr 7-76 p205   +RR 7-76 p74
          +MT 12-76 p1003

1060   Songs: Caractacus: O'er-arch'd by leaves; Clapham town end;
       Grania and Diarmid: There are seven that pull the thread;
       Like to the damask rose; O salutaris hostia; Oh, soft was the
       song, op. 59, no. 3; Pleading, op 48, no. 1; Queen Mary's
       lute song; The river, op. 60, no. 2; Rondel, op. 16, no. 3;
       The shepherd's song, op. 16, no. 1; Spanish lady: Modest and
       fair; The starlight express, op. 78; The blue-eyed fairy;
       Still to be neat; The torch, op. 60, no. 1; Twilight, op. 59,
       no. 6. Mary Thomas s; John Carol Case, bar; Daphne Ibbott,
       pno. Saga 5304. (Reissue from Alpha SPHA 3017).
          -Gr 9-73 p518   +-St 12-76 p139

Sospiri, op. 70. cf Elegy, strings, op. 58 (Argo ZRG 573).

Sospiri, op. 70. cf Elegy, strings, op. 58 (HMV 2292).

Sospiri, op. 70. cf Works, selections (CBS 76423).

Sospiri, op. 70. cf Works, selections (Polydor 2383 359).

The Spanish lady, op. 89: Burlesco. cf Works, selections
   (Polydor 2383 359).

The Spanish lady, op. 89: Suite. cf Elegy, strings, op. 58.

1061   The starlight express, op. 78. Valerie Masterson, s; Derek
       Hammond-Stroud, bar; LPO; Vernon Handley. HMV SQ SLS 5036 (2).
          +Gr 5-76 p1757   +RR 4-76 p69
          ++HFN 4-76 p104   +-STL 5-9-76 p38
          +MT 7-76 p575

The starlight express, op. 78: Waltz. cf Works, selections
   (Polydor 2383 359).

Sursum corda, op. 11. cf Works, selections (Polydor 2383 359).

1062   Symphony, no. 1, op. 55, A flat major. Symphony, no. 2, op. 63,
       E flat major. LPO; Daniel Barenboim. CBS 78289 (2). (Reissues
       from 76247, 73094).
          +-Gr 1-76 p1192   -RR 12-75 p48
          +HFN 2-76 p115

1063   Symphony, no. 1, op. 55, A flat major. Hallé Orchestra; John
       Barbirolli. Pye GSGC 15022. (Reissue from CCL 30102/3).

                    +Gr 1-76 p1192   -RR 1-76 p33
                    +-HFN 1-76 p123
1064 Symphony, no. 1, op. 55, A flat major. Scottish National
        Orchestra; Alexander Gibson. RCA LRL 1-5130.
                    +Gr 9-76 p415   +-RR 9-76 p48
        Symphony, no. 1, op. 55, A flat major. cf Works, selections (HMV
          SLS 5030).
1065 Symphony, no. 2, op. 63, E flat major. LPO; Georg Solti. Decca
        SXL 6723. Tape (c) KSXC 16723. (also London CS 6941).
                    ++Audio 11-76 p111   +NR 6-76 p3
                    +Gr 6-75 p36   ++RR 6-75 p35
                    ++HF 5-76 p83   +RR 12-75 p173 tape
                    ++HFN 6-75 p88   ++SFC 4-25-76 p30
                    +-MJ 5-76 p29   ++St 6-76 p102
                    +-MT 9-75 p798   +STL 9-7-75 p37
1066 Symphony, no. 2, op. 63, E flat major. LPO; Adrian Boult.
        HMV SQ ASD 3266. (also Angel S 37218).
                    +Gr 10-76 p595   ++RR 10-76 p50
                    +-HFN 10-76 p169
        Symphony, no. 2, op. 63, E flat major. cf Symphony, no. 1,
          op. 55, A major.
        Symphony, no. 2, op. 63, E flat major. cf Works, selections
          (HMV SLS 5030).
        Vals capricho. cf Piano works (RCA TRL 1-7073).
        Vespers, op. 14. cf Organ works (RCA LRL 2-5120).
1067 ·Works, selections: Carissima. Chanson de matin, op. 15, no. 2.
        Chanson de nuit, op. 15, no. 1. Elegy, strings, op. 58.
        Romance, bassoon, op. 62. Rosemary. Salut d'amour, op. 12.
        Serenade, strings, op. 20, E minor. Sospiri, op. 70. Martin
        Gatt, bsn; ECO; Daniel Barenboim. CBS 76423. Tape (c)
        40-76423. (also Columbia M 33584).
                    +-Gr 11-75 p818   +-NR 1-76 p4
                    +-HFN 11-75 p153   +-NYT 1-18-76 pD1
                    +-HFN 2-76 p116 tape   +RR 11-75 p43
                    +MT 7-76 p575   +St 5-76 p113
1068 Works, selections: Chanson de matin, op. 15, no. 2. Chanson de
        nuit, op. 15, no. 1. Cockaigne overture, op. 40. Pomp and
        circumstance marches, op. 39, nos. 1, 4. Serenade, strings,
        op. 20, E minor. RPO, Pro Arte Orchestra; George Weldon.
        Classics for Pleasure CFP 40235. (Reissue from World Record
        ST 296).
                    ++Gr 11-76 p786   +RR 8-76 p35
                    +HFN 10-76 p183
1069 Works, selections: Chanson de matin, op. 15, no. 2. Chanson de
        nuit, op. 15, no. 1. Engima variations, op. 36. Intro-
        duction and allegro, op. 47. Pomp and circumstance marches,
        op. 39. Serenade, strings, op. 20, E minor. LSO, AMF, LPO,
        RPO; Neville Marriner, Adrian Boult, Ainslee Cox, Pierre
        Monteux, Arthur Bliss. Decca DPA 537/8. Tape (c) KDPC 537/8.
                    +Gr 6-76 p92   +-HFN 6-76 p103
                    +Gr 9-76 p408 tape   +-RR 6-76 p41
1070 Works, selections: Cockaigne overture, op. 40. Enigma varia-
        tions, op. 36. Falstaff, op. 68. Introduction and allegro,
        op. 47. Pomp and circumstance marches, op. 39, nos. 1-5.
        Symphony, no. 1, op. 55, A flat major. Symphony, no. 2,
        op. 63, E flat major. PhO, Hallé Orchestra, Sinfonia of

London, NPhO; John Barbirolli. HMV SLS 5030 (4). (Reissues
from ASD 548, ASD 610/1, ASD 521, RES 4310, ASD 2292, ASD 540).
      +Gr 1-76 p1192   +-RR 1-76 p32
      +HFN 2-76 p115

1071 Works, selections: Carissima. Contrasts, op. 10, no. 3 (The
     gavotte, AD 1700 and 1900). May song. Mazurka, op. 10,
     no. 1. Minuet, op. 21. Mina. Romance, bassoon, op. 62.
     Rosemary (That's for remembrance). Serenade lyrique. Sere-
     nade mauresque, op. 10, no. 2. Sevillana, op. 7. Michael
     Chapman, bsn; Northern Sinfonia Orchestra; Neville Marriner.
     HMV ESD 7009. Tape (c) TC ESD 7009. (Reissue from ASD 2638).
      +Gr 11-76 p786   +HFN 11-76 p175
      +Gr 11-76 p887 tape   +RR 11-76 p64

1072 Works, selections: Adieu (orch. Geehl). Beau Brummel: Minuet.
     Sospiri, op. 70. The starlight express, op. 78: Waltz. The
     Spanish lady, op. 89: Burlesco (arr. and ed. Young). Sursum
     corda, op. 11. VAUGHAN WILLIAMS: Hymn tune preludes, no. 1,
     Eventide; no. 2, Dominus regit me. The poisoned kiss:
     Overture. The running set. Sea songs: Quick march.
     Bournemouth Sinfonietta; George Hurst. Polydor 2383 359.
     Tape (c) 3170 259.
      +Gr 4-76 p1597   +-HFN 4-76 p104

1073 Works, selections: Bavarian dances, op. 27, nos. 1-3. Caractacus,
     op. 35: Woodland interlude. Chanson de matin, op. 15, no. 2.
     Chanson de nuit, op. 15, no. 1. Contrasts, op. 10. no. 3.
     Dream children, op. 43. Falstaff, op. 68: Interludes (2).
     Serenade lyrique. Salut d'amour, op. 12. Soliloquy, oboe
     (orch. Jacob). Leon Goossens, ob; Bournemouth Sinfonietta;
     Norman del Mar. RCA LRL 1-5133.
      +Gr 11-76 p786   +RR 11-76 p64

     ELIZABETHAN AND JACOBEAN MADRIGALS. cf Enigma VAR 1017.
     ELIZABETHAN LUTE SONGS AND SOLOS. cf Philips 6500 282.

ELLINGTON, Edward (Duke)
1074 Latin American suite. Duke Ellington and His Orchestra.
     Fantasy 8419.
      +St 7-76 p73

ELMS
     Wembley Way. cf Canon VAR 5968.

ELWYN-EDWARDS, Dilys
     Caneuom y tri aderun. cf Argo ZRG 769.

EMMETT, Daniel
     Dixie. cf CBS 61746.
     Dixie. cf Department of Defense Bicentennial Edition 50-1776.
     Dixie. cf RCA AGL 1-1334.

ENCINA, Juan del
     Ay triste que vengo. cf Nonesuch H 71326.
     Romerico. cf DG 2530 504.

ENESCO, Georges
     Rumanian rhapsody, op. 11, no. 1, A minor. cf HMV ESD 7011.
1075 Songs, op. 15 (7). ROUSSEL: Songs: Adieu; Odes anacréontiques,
     nos. 1, 5; Odelette; A flower given to my daughter; Jazz
     dans la nuit; Light; Mélodies, op. 20 (2); Poèmes chinois,
     op. 12 (2); Poèmes chinois, op. 35 (2). Yolanda Marcoulescou,
     s; Katja Phillabaum, pno. Orion ORS 75184. (also Saga 5416).
      ++Gr 5-76 p1788   +NYT 6-8-75 pD19
      ++HF 10-75 p78   +RR 4-76 p70

ENESCO (cont.)

    +-MT 8-76 p660   +-St 5-76 p114
    +NR 7-75 p12
ENGEL
    Chabad melody and Freilachs, op. 20, nos. 1, 2. cf Da Camera
      Magna SM 93399.
    ENGLISH FOLK SONGS. cf Anonymous works (Prelude PMS 1502).
    ENGLISH MADRIGALS AND FOLKSONGS. cf Harmonia Mundi 593.
    ENGLISH VIRGINALISTS. cf BASF BAC 3075.
1076 Discreet music. Variations on a Canon in D major by Johann
      Pachelbel. Cockpit Ensemble; Gavin Bryars. Obscure no. 3.
      +-RR 3-76 p18
ENO, Brian
    Variation on a Canon in D major by Johann Pachelbel. cf Discreet
      music.
EPHROS, Gershon
1077 The priestly benediction. PERGAMENT: Kol nidre. ROSENBLUTH:
      Y'hi ratson; R'tseh; Psalm 116; Tavo l'Fanecha; M'loch;
      V'Hagen. Leo Rosenblüth, cantor; Gunilla von Bahr, flt;
      Maria Thyrsesson, org; Andris Vitolins, org; Stockholm Royal
      Conservatory Chamber Choir; Eric Ericson. BIS LP 1.
      +St 2-76 p112
ERB, Donald
    Pieces for brass quintet and piano (3). cf BASSETT: Sextet.
ERBACH, Christian
    Canzon a 4 del quarto tono. cf Pelca PRSRK 41017/20.
ERICKSON, Robert
1078 End of the mime. FERRITTO: Oggi, op. 9. IVEY: Hera, hung from
      the sky. RANDALL: Improvisation. Bethany Beardslee, s;
      Neva Pilgrim, Elaine Bonazzi, ms; Allen Blustine, clt; Ursula
      Oppens, pno; New Music Choral Ensemble; Instrumental Ensemble;
      Kenneth Gaburo, David Gilbert, Andrew Thomas. CRI SD 325.
      +-HFN 1-76 p107   +NR 2-75 p10
      +MQ 10-75 p636   +St 8-76 p101
ERTL, D.
    Hoch-und-Deutschmeister-Marsch. cf DG 2721 077 (2).
L'ESCUREL (14th century)
    Amours, cent mille merciz. cf Candide CE 31095.
ESTEVE Y GRIMAU, Pablo
    Songs: Alam, sintamos; Ojos, llorar. cf DG 2530 598.
EVANS, Gil
    Bluefish. cf Folkways FTS 33901/4.
EVANS, Merle
    Symphonia. cf Columbia 33513.
EYBLER, Joseph
    Polonaise. cf DG Archive 2533 182.
EYCK, Jacob van
    Engels Nachtegaeltje. cf Telefunken DT 6-48075.
1079 Der Fluyten Lust-Hof, excerpts. Scott-Martin Kosofsky, rec.
      Titanic TI 1
          +NR 2-76 p7
    Variations on 'Amarilli mia bella'. cf Transatlantic TRA 292.
EYSER, Eberhard
1080 Hjärter Kung (The king of hearts). Sista Resan (The last voyage).
      Helge Lannerbäck. Annika Falk, Siw Sjöberg, Ingeborg Nordenfelt,
      Torbjörn Lilliequist, Catharina Olsson, Göran Fransson, Lars
      Erland Holmlund, Tom Sandberg, soloists; Chorus and Chamber

Ensemble; Arnold Ostman, Connie Kjäll, clt; Electronic tape
realised in EMS, Stockholm. Caprice CAP 1080.
    +RR 9-76 p26
Sista Resan (The last voyage). cf Hjärter Kung.

FALL, Leo
Der Fidele Bauer: O frag mich nicht. cf Angel S 37108.
Die Rose von Stambul: Zwei Augen; Ihr stillen, süssen Frau'n.
cf Angel S 37108.

FALLA, Manuel de
1081 El amor brujo. GRANADOS: Goyescas: Intermezzo. RAVEL: Miroirs:
Alborado del gracioso. Pavane pour une infante défunte.
Nati Mistral, ms; NPhO; Rafael Frühbeck de Burgos. London
STS 15358.
    ++St 10-76 p121
El amor brujo: El círculo magico; Canción del fuego fatuo. cf
DG 3335 182.
El amor brujo: Dances. cf CHABRIER: España.
El amor brujo: Pantomime. cf RCA ARM 4-0942/7.
El amor brujo: Ritual fire dance. cf CBS 30073.
El amor brujo: Ritual fire dance. cf Connoisseur Society (Q)
CSQ 2066.
El amor brujo: Ritual fire dance. cf Opus unnumbered.
El amor brujo: Ritual fire dance. cf Stentorian SP 1735.
1082 El amor brujo: Suite. Fantasia baetica. Piezas españolas:
Aragonese, Cubana, Montanesca, Andaluza. El sombrero de tres
picos: Three dances. Alicia de Larrocha, pno. Decca SXL
6683. (also London CS 6881).
    ++Gr 4-75 p1836   ++RR 4-75 p50
    ++HF 12-75 p92   ++SFC 6-29-75 p26
    +MM 4-76 p29   ++St 9-75 p112
1083 Concerto, harpsichord, flute, oboe, clarinet, violin and violoncello. The three-cornered hat. Jan DeGaetani, ms; Igor
Kipnis, hpd; Paige Brook, flt; Harold Gomberg, ob; Stanley
Drucker, clt; Eliot Chap, vln; Lorne Munroe, vlc; NYP; Pierre
Boulez. Columbia M 33970. Tape (c) MT 33970 (Q) MQ 33970.
(also CBS 76500 Tape (c) 40-76500).
    +-Audio 10-76 p148   ++NR 7-76 p2
    +Gr 8-76 p280   +RR 8-76 p36
    +HF 8-76 p90   ++SFC 4-25-76 p30
    +HFN 8-76 p79   ++St 8-76 p94
    +-HFN 12-76 p155 tape
Fantasia baetica. cf El amor brujo: Suite.
Fantasia baetica. cf Piano works (RCA TRL 1-7073).
Homenaje a Paul Dukas. cf Piano works (RCA TRL 1-7073).
1084 Mélodies (3). Spanish popular songs (7). GRANADOS: Tonadilla:
La maja dolorosa. Tonadillas al estilo antiguo: Amor y odio;
El maja discreto; El majo timido; El mirar de la maja; El tra
la la y el punteado. TURINA: Poema en forma de canciones,
op. 19. Jill Gomez, s; John Constable, pno. Saga 5409.
Tape (c) CA 5409.
    +Gr 2-76 p1373   ++RR 12-75 p88
    +HFN 1-76 p107   +STL 2-8-76 p36
    +HFN 10-76 p185 tape
Nocturno. cf Piano works (RCA TRL 1-7073).
1085 Piano works: Fantasia baetica. Homenaje a Paul Dukas. Nocturno.
Piezas españolas (4). Serenata andaluza. Vals capricho.

Joaquin Achucarro, pno. RCA TRL 1-7073.
        +Gr 12-76 p1021   +RR 12-76 p83
Piezas españolas (4). cf Piano works (RCA TRL 1-7073).
Piezas españolas: Aragonese, Cubana, Montanesa, Andaluza. cf
    El amor brujo: Suite.
Serenata andaluza. cf Piano works (RCA TRL 1-7073).
Spanish popular songs (Canciones populares españolas) (7). cf
    Melodies.
Spanish popular songs: Jota. cf RCA ARM 4-0942/7.
Canciones populares españolas: Jota. cf Seraphim S 60259.
Spanish popular songs: Nana. cf CBS 61579.
Spanish popular songs: Nana. cf CBS 76420.
Spanish popular songs: Nana. cf Columbia M2 33444.
1086 The three-cornered hat (El sombrero de tres picos). Barbara
        Howitt, s; LSO; Enrique Jorda. Everest 3057. Tape (c) 3057.
            +HF 11-76 p153 tape
1087 The three-cornered hat. Victoria de los Angeles, s; PhO; Rafael
        Frühbeck de Burgos. HMV SXLP 30187. Tape (c) TC EXE 188.
        (Reissue from ASD 608). (also Angel S 36235).
            +Gr 2-76 p1338   ++HFN 5-76 p117 tape
The three-cornered hat. cf Concerto, harpsichord, flute, oboe,
    clarinet, violin and violoncello.
El sombrero de tres picos: Three dances. cf El amor brujo: Suite.
El sombrero de tres picos: Danza del molinero. cf DG 3335 182.
The three-cornered hat: Jota. cf HMV SEOM 25.
La vida breve: Danse espagnole. cf ALBENIZ: Tango.
La vida breve: Danza, no. 1. cf RCA ARM 4-0942/7.
La vida breve: Vivan los que rien. cf Caprice CAP 1107.
FANTINI, Girolamo
1088 Sonatas, organ (3). FRESCOBALDI: Corrente (4). Toccata per
        l'elevatione. TELEMANN: Melante: Heroic music (trans.).
        Douglas Butler, org. Ars Forma 4001.
            ++MU 10-76 p14   ++NR 9-76 p15
Sonata, 2 trumpets, B flat major. cf Philips 6500 926.
FARBERMAN, Harold
Alea, 6 percussion. cf Serenus SRS 12064.
FARKAS, Ferenc
1089 Concertino all'antica. LUTOSLAWSKI: Concerto, violoncello.
        MARTIN: Ballade. Miklós Perényi, vlc; Budapest Symphony Orch-
        estra; György Lehel. Hungaroton SLPX 11749.
            +HFN 12-76 p141   +RR 11-76 p68
Tillio-lio. cf Hungaroton SLPX 11762.
FARMER, Henry
Moonbeam waltzes. cf Library of Congress OMP 101/2.
FARMER, John
A little pretty bonny lass. cf Harmonia Mundi 593.
A little pretty bonny lass. cf HMV CSD 3766.
Songs: Fair Phyllis I saw sitting all alone. cf Coronet LPS 3032.
Songs: Fair Phyllis I saw sitting all alone. cf HMV CSD 3756.
FARNABY, Giles
Fantasia. cf Hungaroton SLPX 11741.
Giles Farnaby's dream. cf Argo ZRG 823.
His rest. cf Argo ZRG 823.
Loath to depart. cf Saga 5402.
Maske, G minor. cf BASF BAC 3075.
The new Sa-Hoo. cf Argo ZRG 823.

The old spagnoletta. cf Argo ZRG 823.
The old spagnoletta. cf National Trust NT 002.
Tell me, Daphne. cf Argo ZRG 823.
A toye. cf Argo SPA 464.
A toye. cf Argo ZRG 823.
FARNON, Robert
    Colditz march. cf Pye TB 3004.
    Concorde march. cf Pye TB 3004.
    State occasion. cf Pye GH 603.
    Une vie de matelot. cf Transatlantic XTRA 1160.
    Westminster waltz. cf Pye TB 3004.
FARRANT, Richard
    Call to remembrance. cf Argo ZRG 789.
FARRAR
    Bombasto. cf Michigan University SM 0002.
FARRENC, Louise
    Quintet: Scherzo. cf Gemini Hall RAP 1010.
FASCH, Johann Friedrich
1090 Concerto, guitar and strings, D minor. KREBS: Concerto, guitar
        and strings, G major. VIVALDI: Concerto, guitar and strings,
        D major. Konrad Ragossnig, gtr; Southwest German Chamber
        Orchestra; Paul Angerer. Turnabout TV 34547.
            +Gr 8-76 p280   +RR 7-76 p65
            /HFN 8-76 p76
    Concerto, trumpet, 2 oboes and strings, D major. cf PACHELBEL:
        Canon, D major.
    Symphony, A major. cf PACHELBEL: Canon, D major.
    Symphony, G major. cf PACHELBEL: Canon, D major.
FAURE, Gabriel
1091 Ballade, op. 19, F sharp major. FRANCK: Symphonic variations,
        piano and orchestra. d'INDY: Symphonie sur un chant montagnard
        français, op. 25. Marie Françoise Bucquet, pno; Monte Carlo
        Opera Orchestra; Paul Capolongo. Philips 6580 140.
            -Gr 11-76 p782   +-RR 11-76 p54
            +HFN 12-76 p151
1092 Barcarolles, nos. 1-13. Jean-Philippe Collard, pno. Connoisseur
        Society CS 2078.
            ++HF 3-76 p84   *MJ 4-76 p30
1093 Cantique de Jean Racine, op. 11. Requiem, op. 48. Benjamin Luxon,
        bar; Jonathon Bond, treble; Stephen Cleobury, org; St. John's
        College Chapel Choir, Cambridge; AMF; George Guest. Argo
        ARG 841. Tape (c) 841.
            -Gr 4-76 p1646   +MT 8-76 p660
            +Gr 6-76 p102 tape   +-RR 4-76 p70
            +HFN 5-76 p100   +RR 8-76 p84 tape
            +HFN 7-76 p104 tape   /St 12-76 p139
1094 Dolly, op. 56. Masques et bergamasques, op. 112. Pelléas et
        Mélisande, op. 80. Orchestre de Paris; Serge Baudo. Sera-
        phim S 60273.
            +HF 11-76 p112   +SFC 8-15-76 p38
            ++NR 10-76 p4   +SR 9-18-76 p50
1095 Elégie, op. 24, C minor. Sonata, violoncello, no. 1, op. 109,
        D minor. Sonata, violoncello, no. 2, op. 117, G minor.
        Sicilienne, op. 78. Thomas Igloi, vlc; Clifford Benson, pno.
        CRD 1016.
            +-Gr 12-75 p1061   +St 10-76 p121
            +-HFN 9-75 p98

Elégie, op. 24, C minor. cf Works, selections (HMV SQ ASD 3153).
Elégie, op. 24, C minor. cf Command COMS 9006.
Masques et bergamasques, op. 112. cf Dolly, op. 56.
Masques et bergamasques, op. 112. cf DEBUSSY: La mer.

1096 Nell, op. 18, no. 1 (arr. Grainger). GERSHWIN: Love walked in;
      The man I love (arr. Grainger). GRAINGER: Country gardens.
      Irish tune from County Derry. Eastern intermezzo. Handel
      in the Strand. Knight and shepherd's daughter. Molly on the
      shore. To a Nordic princess. Over the hills and far away.
      Sailor's song. Shepherd's hey. Tribute to Foster: Lullaby.
      Walking tune. Daniel Adni, pno. HMV SQ HQS 1363.
           +Gr 11-76 p843   +RR 11-76 p90
           +HFN 12-76 p143

1097 Nocturnes, op. 33, nos. 1-3; op. 36; op. 37; op. 63; op. 74;
      op. 84, no. 8; op. 97; op. 99; op. 104; op. 107; op. 119.
      Theme and variations, op. 73, C minor. Jean-Philippe Collard,
      pno. Connoisseur Society/Pathe Marconi CS 2072 (2).
           +HF 6-75 p94   ++SFC 4-6-75 p22
           +MJ 4-76 p30   ++St 7-75 p69

1098 Nocturne, op. 74, C sharp minor. Nocturne, op. 99, no. 10, E
      minor. FRANCK: Prélude, aria et final. Prélude, chorale et
      fugue. Paul Crossley, pno. L'Oiseau-Lyre DSLO 8.
           +Gr 5-76 p1787   +RR 5-76 p65
           +HFN 5-76 p100

Nocturne, op. 99, no. 10, E minor. cf Nocturne, op. 74, no. 7,
C sharp minor.
Papillon, op. 77. cf Works, selections (HMV SQ ASD 3153).
Papillon, op. 77. cf Command COMS 9006.
Pavane, op. 50. cf Requiem, op. 48.
Pavane, op. 50. cf Decca DPA 519/20.
Pavane, op. 50. cf Philips 6780 030.

1099 Pelléas et Mélisande, op. 80. FRANCK: Symphony, D minor. NPhO;
      Andrew Davis. CBS 76526. (also Columbia M 34506 Tape (c) MT
      34506).
           +Gr 10-76 p596   +RR 11-76 p66
           +HFN 10-76 p169

Pelléas et Mélisande, op. 80. cf Dolly, op. 56.
Pelléas et Mélisande, op. 80: Incidental music. cf DEBUSSY: Prélude à l'après-midi d'un faune.

1100 Requiem, op. 48. Victoria de los Angeles, s; Dietrich Fischer-
      Dieskau, bar; Elisabeth Brasseur Chorale; Henriette Puig-
      Roget, org; OSCCP; André Cluytens. Classics for Pleasure CFP
      40234. (Reissue from HMV SAN 107).
           +Gr 6-76 p71   +-RR 5-76 p69
           +-HFN 3-76 p111

1101 Requiem, op. 48. Suzanne Danco, s; Gérard Souzay, bar; OSR;
      Ernest Ansermet. Decca SDD 154. Tape (c) KSDC 154.
           +-HFN 7-75 p90 tape   +-RR 8-76 p83 tape

1102 Requiem, op. 48. Pavane, op. 50. Elly Ameling, s; Bernard
      Druysen, bar; Daniel Chorzempa, org; Netherlands Radio Chorus;
      Rotterdam Philharmonic Orchestra; Jean Fournet. Philips 6500
      968. Tape (c) 7300 417.
           +-Gr 3-76 p1493   +RR 2-76 p55
           +HF 7-76 p77   ++St 8-76 p94
           +HF 11-76 p153   +STL 2-8-76 p36
           +NR 6-76 p12

Requiem, op. 48. cf Cantique de Jean Racine, op. 11.
Requiem, op. 48: Pie Jesu. cf Abbey LPB 761.
Sérénade, op. 98. cf Works, selections (HMV SQ ASD 3153).
Sicilienne, op. 78. cf Elégie, op. 24, C minor.
Sicilienne, op. 78. cf Command COMS 9006.
Sicilienne, op. 78. cf HMV SQ ASD 3283.
Sicilienne, op. 78. cf Pearl SHE 528.
Sonata, violin and piano, no. 1, op. 13, A major. cf DEBUSSY:
    Sonata, violin and piano (Classics for Pleasure CFP 40210).
Sonata, violin and piano, no. 1, op. 13, A major. cf DEBUSSY:
    Sonata, violin and piano (Connoisseur Society QS 7174).
Sonata, violin and piano, no. 1, op. 13, A major. cf RCA ARM
    4-0942/7.
Sonata, violoncello, no. 1, op. 109, D minor. cf Elégie, op.
    24, C minor.
Sonata, violoncello, no. 1, op. 109, D minor. cf Works, selec-
    tions (HMV SQ ASD 3153).
Sonata, violoncello, no. 2, op. 117, G minor. cf Elégie, op. 24,
    C minor.
Sonata, violoncello, no. 2, op. 117, G minor. cf Works, selec-
    tions (HMV SQ ASD 3153).
1103   Songs: La bonne chanson, op. 61. Mélodies de Venise, op. 58 (5).
       Songs, op. 23 (3). Songs, op. 39: Fleur jetée; Les roses
       d'Ispahan. Felicity Palmer, s; John Constable, pno. Argo
       ZRG 815.
            ++Gr 8-76 p324   +RR 8-76 p74
            ++HFN 8-76 p80
1104   Songs: L'Absent, op. 5, no. 3; Après un rêve, op. 7, no. 1; Au
       bord de l'eau, op. 8, no. 1; Aubade, op. 6, no. 1; Aurore, op.
       39, no. 4; Barcarolle, op. 7, no. 3; Chanson du pêcheur, op.
       4, no. 1; Chant d'automne, op. 5, no. 1; Dans les ruins d'un
       abbaye; Hymne, op. 7, no. 2; Ici-bas, op. 8, no. 3; Lydia, op.
       4, no. 2; Mai, op. 1, no. 2; Les matelots, op. 2, no. 2; Le
       papillon et la fleur, op. 2, no. 1; Le rançon, op. 8, no. 2;
       Rêve d'amour, op. 5, no. 2; Sylvia, op. 6, no. 3; Tristesse,
       op. 6, no. 2. Jacques Herbillon, bar; Theodore Paraskivesco,
       pno. Calliope CAL 1841.
            -Gr 5-76 p1794   +RR 10-75 p24
            +MT 8-76 p660   +RR 3-76 p67
1105   Songs: Aurore, op. 39, no. 4; Automne, op. 18, no. 3; Les berceaux,
       op. 23, no. 1; Chanson d'amour, op. 27, no. 1; La fée aux
       chansons, op. 27, no. 2; Fleur jetée, op. 39, no. 2; Nell, op.
       18, no. 1; Nocturne, op. 43, no. 2; Noël, op. 43, no. 1; Le
       pays des rêves, op. 39, no. 3; Poème d'un jour, op. 21: Ren-
       contre, Tourjours, Adieu; Les roses d'Ispahan, op. 39, no. 4;
       Secret, op. 23, no. 3; La voyageur, op. 18, no. 2. Jacques
       Herbillon, bar; Theodore Paraskivesco, pno. Calliope CAL 1842.
            /Gr 5-76 p1794   +RR 3-76 p67
            +MT 8-76 p660
1106   Songs: Au cimentière, op. 51, no. 2; Chanson (Extraite de Shylock),
       op. 57, no. 1; Claire de lune, op. 46, no. 2; En prière; Larmes,
       op. 51, no. 1; Madrigal, op. 35; Melodies de Venise, op. 58:
       Mandoline, En sourdine, Green, C'est l'extase, A clymène; Les
       presents, op. 46, no. 1; La rose, op. 51, no. 4; Spleen, op.
       51, no. 3. Jacques Herbillon, bar; Theodore Paraskivesco, pno.
       Calliope CAL 1843.
            /Gr 5-76 p1794   +RR 3-76 p67
            +MT 8-76 p660

1107 Songs: Accompagnement, op. 85, no. 3; Arpège, op. 76, no. 2; La
      bonne chanson, op. 6; Dans la forêt de septembre, op. 85, no.
      1; La fleur qui va sur l'eau, op. 85, no. 2; Le perfum impéris-
      sable, op. 76, no. 1; Le plus doux chemin, op. 87, no. 1;
      Prison, op. 83, no. 1; Le ramier, op. 87, no. 2; Sérénade du
      bourgeois gentilhombre; Soir, op. 83, no. 2. Jacques Herbillon,
      bar; Theodore Paraskivesco, pno. Calliope CAL 1844.
            -Gr 5-76 p1794   +RR 10-75 p24
            +MT 8-76 p660   +RR 3-76 p67
1108 Songs: Chanson, op. 94. Le don silencieux, op. 92. L'horizon
      chimérique, op. 118: La mer est infinie; Je me suis embarqué;
      Diane Séléne; Vaisseaux, nous vous aurons aimés. Le Jardin
      Clos, op. 106: Exaucement; Quand tu plonges tes yeux; La
      messagère; Je me poserai sur ton coeur; Dans la nymphée; Dans
      la pénombre; Il m'est cher, mon amour; Inscription sur le
      sable. Mirages, op. 113: Cynges sur l'eau; Reflets dans l'eau;
      Jardin nocturne; Danseuse. Sérénade toscane, op. 3, no. 2.
      Anne-Marie Rodde, s; Theodore Paraskivesco, pno. Calliope CAL
      1845.
            +Gr 5-76 p1794   +RR 3-76 p67
            +MT 8-76 p660
1109 Songs: La chanson d'Eve, op. 95: Paradis; Prima verba; Roses
      ardentes; Comme Dieu rayonne; L'Aube blanche; Eau vivante;
      Veilles-tu, ma senteur de soleil; Dans un parfum de roses
      blanches; Crépuscule, O mort, poussières d'etoiles. Mélisande's
      song, op. posth. Notre amour, op. 23, no. 2. Pleurs d'or,
      op. 72. Puisqu'ici-bas, op. 10, no. 1. Seule, op. 3, no. 1.
      Tarentelle, op. 10, no. 2. Vocalise without opus. Anne-Marie
      Rodde, Sonia Nigoghossian, s; Theodore Paraskivesco, pno.
      Calliope CAL 1846.
            +Gr 5-76 p1794   +RR 3-76 p67
            +MT 8-76 p660
1110 Songs: Après un rêve, op. 7, no. 1; Au bord de l'eau, op. 8, no. 1;
      Chanson d'amour, op. 27, no. 1; Clair de lune, op. 46, no. 2;
      Lydia, op. 4, no. 2; Nell, op. 18, no. 1; Sylvie, op. 6, no. 3.
      HAHN: Songs: D'une prison; L'heure exquise; Mai; Le rossignol
      des lilas; Offrande; Paysage; Si mes vers avaient des ailes.
      MASSENET: Songs Chant provençal; Elégie; Nuit d'espagne; Séré-
      nade d'automne; Stances; Un adieu; Vous aimerez demain. Martyn
      Hill, t; John Constable, pno. Saga 5419.
            +Gr 12-76 p1042   +-RR 12-76 p93
1111 Songs: L'Horizon chimérique, op. 118; Nocturne, op. 43, no. 2;
      Poeme d'un jour, op. 21. LULLY: Alceste: Air de Caron. POULENC:
      L'Anguille; La belle jeunesse; Priez pour paix; Serenade.
      TIERSOT: Chants de la vieille France (4). Martial Singher, bar;
      Alden Gilchrist, hpd or pno. 1750 Arch S 1754.
            +NR 12-75 p12   +St 5-76 p129
            +-ON 5-76 p48
      Songs: Adieu, op. 21, no. 3; Après un rêve, op. 7, no. 1; Barca-
      rolle, op. 7, no. 3; Clair de lune, op. 46, no. 2; Chanson
      d'amour, op. 27, no. 1; En prière; Nell, op. 18, no. 1; Poème
      d'un jour, op. 21; Le secret, op. 23, no. 3. cf DUPARC: Songs
      (Pearl SHE 524).
      Songs: L'Absent, op. 5, no. 3; Après un rêve, op. 7, no. 1; Clair
      de lune, op. 46, no. 2; Dans les ruines d'une abbaye, op. 2,
      no. 1; Ici-bas, op. 8, no. 3; Nell, op. 18, no. 1; Les roses
      d'Ispahan, op. 39, no. 4; Le secret op. 23, no. 3; Soir, op. 83,
      no. 2. cf HMV RLS 716.

Songs: Clair de lune, op. 46. cf Pearl SHE 528.
Songs: The crucifix. cf Rubini RS 301.
Songs: Les bercaux, op. 23, no. 1; La chanson du pêcheur, op. 4,
    no. 1; Mai, op. 1, no. 2. cf Seraphim S 60251.
Theme and variations, op. 73, C minor. cf Nocturnes, op. 33, nos.
    1-3 (Connoisseur Society CS 2072).
1112 Trio, piano, op. 120, D minor. SHOSTAKOVICH: Trio, piano, no. 2,
    op. 67, E flat major. Hans Pålson, pno; Arve Tellefsen, vln;
    Frans Helmerson, vlc. BIS LP 26.
        +HFN 10-76 p169   +-RR 9-76 p76
1113 Works, selections: Elégie, op. 24, C minor. Papillon, op. 77.
    Sérénade, op. 98. Sonata, violoncello, no. 1, op. 109, D
    minor. Sonata, violoncello, no. 2, op. 117, G minor. Paul
    Tortelier, vlc; Eric Heidseick, pno. HMV SQ ASD 3153.
        +Gr 3-76 p1478   +RR 2-76 p51
        ++HFN 3-76 p92
    FAVORITE AMERICAN CONCERT SONGS. cf Cambridge CRS 2715.
    FAVOURITE OPERETTA DUETS. cf HMV CSD 3748.
FAYRFAX, Robert
    I love, loved; Thatt was my woo. cf L'Oiseau-Lyre 12BB 203/6.
FEARIS
    Beautiful isle of somewhere. cf Pye Ember GVC 51.
FEATHERSTONE-CATELINET
    My Jesus I love Thee. cf HMV SXLP 50017.
FELCIANO, Richard
    Chöd. cf Crasis.
1114 Crasis. Chöd. Gravities. Spectra. Milton and Peggy Salkind,
    pno; Nancy Turetzky, contrabass; Philadephia Composers' Forum,
    Ensemble and tape; Joel Thome, Richard Felciano. CRI SD 349.
        *MJ 7-76 p57   +St 8-76 p100
        +NR 6-76 p7
    Gravities. cf Crasis.
    Spectra. cf Crasis.
FELD, Jindrich
    Quintet. cf Crystal S 206.
    Quintet, winds, no. 2. cf BARTA: Quintet, winds, no. 2.
FELDMAN, Morton
1115 For Frank O'Hara. Rothko chapel. Karen Phillips, vla; James
    Holland, perc; Gregg Smith Singers; Center of the Creative and
    Performing Arts, State University of New York at Buffalo,
    Members; Jan Williams, Gregg Smith. Odyssey Y 34138.
        ++ARG 12-76 p28   +NR 10-76 p7
        +HF 12-76 p126   *ON 11-76 p98
    Rothko chapel. cf For Frank O'Hara.
    Vertical thoughts. cf Desto DC 6435/7.
FELIX, Vaclav
    Living earth. cf Panton 110 440.
FENNELLY, Brian
1116 Evanescences. HIBBARD: Quartet, strings. Da Capo Chamber
    Players, Members; Stradivari Quartet. CRI SD 322.
        +HF 1-75 p83   ++NR 10-74 p9
        +-HFN 1-76 p107   +-RR 10-75 p62
FERGUSON, Barry
    Festival march. cf Audio EAS 16.
    Prelude on the hymn tune "Durness". cf Audio EAS 16.
    Toccata. cf Audio EAS 16.

FERNANDES, Gaspar
        Eso rigor e repente. cf Eldorado S-1.
FERNANDEZ, Oscar Lorenzo
        Brasileira, no. 2: Ponteio, Moda, Cataretè. cf Angel S 37110.
FERNANDEZ HIDALGO, Gutierre
        Salve regina a 5. cf Eldorado S-1.
FERRABOSCO, Alfonso II
        Pavane. cf Turnabout TV 34017.
FERRITTO, John
        Oggi, op. 9. cf ERICKSON: End of the mime.
FESTA, Constanza
        L'ultimo di mi maggio. cf HMV CSD 3756.
        FESTIVAL OF EARLY LATIN AMERICAN MUSIC. cf Eldorado S-1.
        A FESTIVAL OF FRENCH MUSIC. cf Philips 6780 030.
FEVRIER
        Monna Vanna: Ce n'est pas un vieillard. cf Club 99-101.
FIBICH, Zdenek
        Poem. cf Rediffusion 15-46.
        Poem, no. 14. cf HMV SXLP 30188.
1117   Quintet, violin, clarinet, horn, violoncello and piano, op. 42,
            D major. Trio, piano, F minor. Karel Dlouhý, clt; Zdeněk
            Tylsar, hn; Fibich Trio. Supraphon 111 1617.
                    +ARG 12-76 p29   +NR 11-76 p6
                    +Gr 9-76 p437   +RR 8-76 p62
        Trio, piano, F minor. cf Quintet, violin, clarinet, horn,
            violoncello and piano, op. 42, D major.
FIELD, John
        Nocturne, no. 9, E minor. cf International Piano Library
            IPL 102.
1118   Nocturne, no. 11, E flat major. Nocturne, no. 12, G major.
            Pastorale, E major. Sonata, no. 4, B major. MENDELSSOHN:
            Songs without words, op. 30, no. 6; op. 38, no. 6; op. 53,
            no. 4; op. 62, nos. 5, 6; op. 67, no. 5. Studies, op. 104 (3).
            Richard Burnett, pno. Prelude PRS 2504.
                    +-Gr 12-76 p1027   +-RR 12-76 p85
                    +HFN 12-76 p141
        Nocturne, no. 12, G major. cf Nocturne, no. 11, E flat major.
        Pastorale, E major. cf Nocturne, no. 11, E flat major.
        Sonata, piano, no. 1, E flat major: Rondo. cf BENNETT, W.:
            January, op. 36, no. 1.
        Sonata, piano, no. 4, B major. cf Nocturne, no. 11, E flat
            major.
FILLMORE, Henry
        Americans we. cf Department of Defense Bicentennial Edition
            50-1776.
        Americans we. cf Mercury SRI 75055.
        Americans we. cf Michigan University SM 0002.
        The footlifter. cf Columbia 33513.
        Miss trombone. cf Washington University Press OLY 104.
FINGER, Gottfried
        Divisions on a ground. cf Transatlantic TRA 292.
FINLAYSON
        Bright eyes. cf Canon VAR 5968.
FINZI, Gerald
        Earth and air and rain, op. 15. cf BUTTERWORTH: Songs (Argo
            ZRG 838).
        Let us garlands bring, op. 18: It was a lover and his lass. cf

HMV ASD 2929.
Let us garlands bring, op. 18: Come away, come away, death;
    It was a lover and his lass. cf Saga 5213.
FIOCCO, Joseph-Hector.
    Andante. cf Polydor 2460 262.
FISCHER, Ernst
    South of the Alps. cf Rediffusion 15-46.
FISCHER, Johann
    Bourée. cf DG Archive 2533 172.
    Gigue. cf DG Archive 2533 172.
FISER, Lubos
    Double pro orchestr. cf DOUBRAVA: Pastorale.
    FIVE CENTURIES AT ST. GEORGE'S. cf Argo ZRG 789.
FLOSMAN, Oldrich
    Sonata, wind quintet and piano. cf BARTA: Quintet, winds, no. 2.
FLOTOW, Friedrich
    Allesandro stradella: Overture. cf Supraphon 110 1637.
    Martha: Ach, so fromm. cf RCA CRM 1-1749.
    Martha: Last rose of summer; Twelve o'clock, twelve o'clock. cf
        English National Opera ENO 1001.
    Martha: M'appari. cf Decca SXL 6649.
FO (15th century Italy)
    Tua voisi esser sempre mai. cf Nonesuch H 71326.
FONTANA, Giovanni Battista
1119 Sonatas, trumpet and bassoon, nos. 1-6. FRESCOBALDI: Canzone
        per sonar (5). Gerard Schwarz, tpt. Desto 6481. Tape (c)
        X 46481.
                +HF 6-76 p97 tape
    Sonata, violin. cf L'Oiseau-Lyre 12BB 203/6.
FOOTE, Arthur
    Francesca da Rimini. cf Louisville LS 753/4.
    Quartet, strings, op. 70, D major. cf Vox SVBX 5301.
    FOOTLIFTER: A CENTURY OF AMERICAN MARCHES IN AUTHENTIC VERSIONS.
        cf Columbia 33513.
FORD
    Almighty God. cf Harmonia Mundi HMD 223.
    The pill to purge melancholy. cf National Trust NT 002.
FORQUERAY, Antoine
    La Bellemont. cf DUPHLY: La Félix.
    La Couperin. cf DUPHLY: La Félix.
    La Laborde. cf DUPHLY: La Félix.
FORSTER
    Concerto, waldhorn, E flat major. cf CORELLI: Sonata, violin
        and continuo, op. 5, no. 5, G minor.
FORSTER, Georg
    Vitrum nostrum Gloriosum. cf L'Oiseau-Lyre 12BB 203/6.
FOSCARINI, Giovanni
    Il furioso. cf Saga 5420.
FOSS, Lukas
    Baroque variations. cf CAGE: Concerto, prepared piano and
        chamber orchestra.
1120 Behold, I build an house. Psalms. SHIFRIN: Serenade for five
        instruments. Roger Wagner Chorale; Melvin Kaplan, ob; Charles
        Russo, clt; Robert Cecil, Fr hn; Ynes Lynch, vla; Harriet
        Wingreen, James MacInnes, Lukas Foss, pno; Rober Wagner. CRI 123.
            +Gr 4-76 p1645   *RR 1-76 p56

Psalms. cf Behold, I build an house.
FOSTER, Stephen
    Jeanie with the light brown hair. cf Pye GH 603.
    Medley. cf RCA LRL 2-7531.
    Old folks at home. cf HMV RLS 719.
1121 Songs: Ah, May the red rose live alway; Beautiful dreamer;
      Gentle Annie; If you've only got a moustache; I'm nothing
      but a plain old soldier; Jeanie with the light brown hair;
      Mr. and Mrs. Brown; Slumber my darling; Some folks; Sweetly
      she sleeps, my Alice fair; That's what's the matter; There's
      a good time coming; Was my brother in the battle; Wilt thou
      be gone, love. Jan DeGaetani, ms; Leslie Guinn, bar; Gilbert
      Kalish, pno, melodeon; Robert Sheldon, flt, keyed bugle; Sonya
      Monosoff, vln. Nonesuch H 71268.
          +St 7-76 p72
1122 Songs: Ah, May the red rose live alway; Beautiful dreamer; Come
      where my love lies dreaming; De camptown races; I dream of
      Jeanie with the light brown hair; Massa's in de cold, cold
      ground; Oh Susanna; Old black Joe; Old folks at home; My old
      Kentucky home. Richard Crooks, t; the Balladeers; Frank
      LaForge, pno. RCA AVM 1-1738.
         *ON 11-76 p98   +St 12-76 p149
    Songs: Ay, may the red rose live alway; Old memories; Why, no one
      to love. cf Library of Congress OMP 101/2.
    Songs: We are coming Father Abraham, 300,000 more; Willie has
      gone to war; Jenny June; Wilt thou be true; Katy Bell. cf
      Vox SVBX 5304.
    Songs: I cannot sing tonight; Some folks; Summer longings; Why,
      no one to love. cf Washington University Press OLY 104.
    FOURTEENTH CENTURY ITALIAN MONOPHONIC DANCES. cf Anonymous works
      (Decca DL 79418).
FRANCAIX, Jean
1123 Concerto, piano, D major. HAHN: Concerto, pinao, no. 1, E minor.
      LAMBERT: The Rio Grande. MILHAUD: Scaramouche. Jean Françaix,
      Magda Tagliafero, Hamilton Harty, Marcelle Meyer. Darius
      Milhaud, pno; Paris Philharmonic Orchestra, Hallé Orchestra,
      Orchestra; St. Michael's Singers; Nadia Boulanger, Reynaldo
      Hahn. World Records SH 227.
         +Gr 1-76 p1209   +RR 10-75 p85
         +HFN 10-75 p140
1124 Sonatine. PROKOFIEV: Melodies, op. 35 (5). RAVEL: Sonata,
      violin and piano. SZYMANOWSKI: La Fontaine d'Areethuse, op. 30,
      no. 1. Diana Steiner, vln: David Berfield, pno. Orion ORS
      75195.
         +IN 10-76 p18   ++NR 1-76 p13
FRANCHETTI, Alberto
    Germania: No, non chiuder gli occhi vaghi (2 versions); Studenti,
      Udite. cf Everest SDBR 3382.
FRANCHOS
    Trumpet intrada. cf Argo ZRG 823.
FRANCISCUS (Francisque), Antoine
    Phiton, Phiton. cf HMV SLS 863.
FRANCK, César
    Cantabile. cf Organ works (CRD CAL 1919/21).
    Cantabile, B major. cf Organ works (Saga 5390).

1125 Le chasseur maudit. Nocturne. Psyché. Christa Ludwig, ms;
       Orchestre de Paris; Daniel Barenboim. DG 2530 771.
       +Gr 12-76 p995   +-RR 12-76 p55
       +HFN 12-76 p143
1126 Chorales, nos. 1-3. Thomas Murray, org. Nonesuch H 71310.
       ++NR 10-75 p11   ++St 1-76 p104
     Chorales, nos. 1-3. cf Organ works (CRD CAL 1919/21).
     Chorale, no. 1. cf Argo ZRG 807.
1127 Chorale, no. 2, B minor. Pièce héroïque. LISZT: Variations on
       Bach's "Weinen, Klagen, Sorgen, Zagen", G 673. REGER: Pieces,
       op. 145, no. 2: Dankpsalm. SAINT-SAENS: Prelude and fugue,
       op. 99, no. 3, E flat major. Nicholas Danby, org. CBS 76514.
       ++Gr 9-76 p454   +HFN 9-76 p127
     Chorale, no. 2, B minor. cf Gaudeamus XSH 101.
     Chorale, no. 2, B minor. cf L'Oiseau-Lyre SOL 343.
     Chorale, no. 2, B minor. cf Vista VPS 1030.
     Chorale, no. 3, A minor. cf HMV SQ HQS 1356.
     Les Eolides. cf Symphony, D minor.
     Les Eolides. cf CHAUSSON: Symphony, op. 20, B flat major.
     Fantasia, A major. cf Organ works (CRD CAL 1919/21).
     Fantasia, A major. cf Organ works (Saga 5390).
     Fantasia, op. 16, C major. cf Organ works (CRD CAL 1919/21).
     Final, op. 21, B flat major. cf Organ works (CRD CAL 1919/21).
     Grande pièce symphonique, op. 17. cf Organ works (CRD CAL 1919/21).
     Nocturne. cf Le chasseur maudit.
1128 Organ works: Cantabile. Chorales, nos. 1-3. Fantasia, A major.
       Fantasia, op. 16, C major. Final, op. 21, B flat major. Grande
       pièce symphonique, op. 17. Pastorale, op. 19. Pièce héroïque.
       Prière, op. 20. Prelude, fugue and variations, op. 18. André
       Isoir, org. CRD Calliope CAL 1919/21 (3).
       +Gr 4-76 p1632   ++RR 8-76 p68
1129 Organ works: Cantabile, B major. Fantasia, A major. L'Organiste:
       Poco allegretto; Tres lent; Prière (quasi lento); Andantino;
       Poco allegretto; Poco lento; Quasi allegro; Non troppo lento.
       Pièce heroique. Pierre Cochereau, org. Saga 5390.
       -Gr 5-75 p1995   +-MT 1-76 p40
       /HFN 5-75 p127   +-RR 6-75 p66
     L'Organiste: Poco allegretto; Tres lent; Prière (quasi lento);
       Andantino; Poco allegretto; Poco lento; Quasi allegro; Non
       troppo lento. cf Organ works (Saga 5390).
     Panis Angelicus. cf Decca SXL 6781.
     Pastorale, op. 19. cf Organ works (CRD CAL 1919/21).
     Pièce héroïque. cf Chorale, no. 2, B minor.
     Pièce héroïque. cf Organ works (CRD CAL 1919/21).
     Pièce héroïque. cf Organ works (Saga 5390).
     Pièce héroïque. cf Argo 5BBA 1013-5.
     Pièce héroïque. cf Decca DPA 523/4.
     Prélude, aria et final. cf FAURE: Nocturne, op. 74, no. 7, C
       sharp minor.
     Prélude, chorale et fugue. cf FAURE: Nocturne, op. 74, no. 7,
       C sharp minor.
     Prelude, fugue and variations. cf Grosvenor GRS 1041.
     Prelude, fugue and variations, op. 18. cf Organ works (CRD CAL
       1919/21).
     Prelude, fugue and variations, op. 18. cf Argo 5BBA 1013-5.
     Prière, op. 20. cf Organ works (CRD CAL 1919/21).

FRANCK (cont.)   196

        Prière, op. 20. cf BACH: Prelude and fugue, S 546, C minor.
1130 Psyché. Belgian Radio Chorus; Orchestre de Liège; Paul Strauss.
        Connoisseur Society (Q) CSQ 2096. (also HMV SQ ASD 3164).
              +-ARG 11-76 p21   +NR 9-76 p4
              +Gr 3-76 p1464   +RR 2-76 p29
              +-HFN 3-76 p92   ++St 12-76 p140
        Psyché. cf Le chasseur maudit.
        Rédemption. cf Symphony, D minor.
1131 Sonata, violin and piano, A major. STRAUSS, R.: Sonata, violin
        and piano, op. 18, E flat major. Jascha Heifetz, vln; Brooks
        Smith, pno. CBS 76419. Tape (c) 40-76419.
              +Gr 3-76 p1481   +-HFN 7-76 p105 tape
              +-Gr 6-76 p102 tape   +RR 2-76 p51
              +HFN 2-76 p109
1132 Sonata, violin and piano, A major. MILHAUD: Sonatas, viola and
        piano, nos. 1 and 2. Bernard Zaslav, vla; Naomi Zaslav, pno.
        Orion ORS 75186.
              ++NR 7-76 p5
1133 Sonata, violin and piano, A major. PROKOFIEV: Sonata, flute
        and piano, op. 94, D major. James Galway, flt; Martha Argerich,
        pno. RCA LRL 1-5095. Tape (c) LRK 1-5095 (ct) LRS 1-5095.
              ++Gr 11-75 p848   ++MJ 1-76 p25
              ++HF 5-76 p89   ++NR 2-76 p5
              ++HF 6-76 p97 tape   ++RR 10-75 p78
              ++HFN 12-75 p153   ++SFC 5-23-76 p36
        Sonata, violin and piano, A major. cf Columbia M2 33444.
1134 Songs: O salutaris hostia; Psalm, no. 150. POULENC: Mass, G
        major. VILLETTE: Hymne à la vierge. WIDOR: Mass, 2 choirs
        and 2 organs. Worcester Cathedral Choir; Harry Bramma, org;
        Donald Hunt. Abbey LPB 758.
              ++Gr 11-76 p867   +-RR 11-76 p95
              +-HFN 11-76 p159
1135 Symphonic variations, violin and piano. GRIEG: Concerto, piano,
        op. 16, A minor. György Cziffra, pno; Budapest Symphony
        Orchestra; György Cziffra, Jr. Connoisseur Society CS 2090.
              +-SFC 4-25-76 p30   +-St 12-76 p141
        Symphonic variations, piano and orchestra. cf FAURE: Ballade,
        op. 19, F sharp major.
        Symphonic variations, piano and orchestra. cf HMV SLS 5033.
        Symphonic variations, piano and orchestra. cf Philips 6780 030.
1136 Symphony, D minor. Rédemption. Orchestre de Paris; Daniel
        Barenboim. DG 2530 707. Tape (c) 3300 707.
              +Gr 10-76 p596   -RR 9-76 p48
              +HFN 10-76 p169   -SFC 11-28-76 p45
1137 Symphony, D minor. BRSO; Lorin Maazel. DG 2535 156. (Reissue
        from SLPM 138 693).
              ++Gr 4-76 p1598   +-RR 3-76 p39
              +HFN 3-76 p109
1138 Symphony, D minor. Les Eolides. COA; Willem van Otterloo.
        Philips 6580 109.
              /Gr 1-76 p1192   +-RR 1-76 p33
              +-HFN 1-76 p107
1139 Symphony, D minor. RCA Victor Symphony Orchestra; Adrian Boult.
        RCA GL 25004. Tape (c) GK 25004 (Previously issued by
        Reader's Digest).
              +Gr 10-76 p596   +RR 11-76 p66

```
        +HFN 12-76 p143          +RR 12-76 p104 tape
        +HFN 12-76 p153 tape
    Symphony, D minor.  cf FAURE: Pelléas et Mélisande. op. 80
FRANCK, Melchior
    Pavanne et galliarde.  cf Harmonia Mundi HMU 610.
FRANK, Andrew
    Orpheum (Night music I).  cf CRI SD 345.
FRANKLIN, Benjamin
    Quartet, strings.  cf Vox SVBX 5301.
FRANZ, Robert
    O thank me not, op. 14, no. 1.  cf HMV RLS 716.
    Widmung, op. 14, no. 1.  cf Club 99-99.
FREDERICK II, King of Prussia
    Torgauer Marsch.  cf Polydor 2489 523.
FREDERIKSEN
    Copenhagen.  cf Philips 6308 246.
FREIRE
    Ay-ay-ay.  cf DG 2630 700.
FRENCH-COLES
    Why hang your harp on the willow.  cf HMV SXLP 50017.
    FRENCH SONGS.  cf HMV CSD 3740.
FRESCOBALDI, Girolamo
    Aria detta "La Frescobaldi".  cf RCA SB 6891.
    Canzona.  cf Coronet LPS 3032.
    Canzone per sonar (5).  cf FONTANA: Sonatas, trumpet and bassoon,
        nos. 1-6.
    Capriccio sopra un soggetto.  cf Philips 6500 926.
    Cento.  cf Argo ZRG 806.
    Corrente (4).  cf FANTINI, Sonatas, organ.
    Galliards (5).  cf Hungaroton SLPX 11741.
1140 Harpsichord works: Balleto terzo-Corrente-Passacagli.  Capriccio
        Fra Jacopino, sopra l'aria di Ruggiero, Toccata ottava.  Cento
        partite sopra Passacagli.  Capriccio sopra la Battaglia.
        Partite sopra l'aria di Monicha.  Partite 14 sopra l'aria
        della Romanesca.  Toccata duodecima and quinta.  Edward Brewer,
        hpd.  Musical Heritage Society MHS 3245.
            ++MJ 11-76 p45
    Partite sopra passacagli.  cf Argo ZRG 806.
    Toccata.  cf L'Oiseau-Lyre 12BB 203/6.
    Toccata IX.  cf Hungaroton SLPX 11741.
    Toccata per l'elevatione.  cf FANTINI: Sonatas, organ.
FRIEDERICH, G. W. E.
1141 The American brass band journal: Works composed or arranged by
        G. W. E. Friederich.  Hail Columbia; Massa's in the cold
        ground; My old Kentucky home; Old dog tray; Lilly bell quick-
        step; Pelham schottische; Signal march; The star spangled
        banner; Yankee doodle; and twelve others.  Empire Brass
        Quintet and Friends.  Columbia M 34192.
            +ARG 12-76 p30            +St 11-76 p137
            +IN 12-76 p18
    Lilly Belle quickstep.  cf Library of Congress OMP 101/2.
FRIEDMAN
    Viennese dance, no. 1.  cf Desmar DSM 1005.
FROBERGER, Johann
    Capriccio, C major.  cf Pelca PRSRK 41017/20.
    Suite, no. 19, C minor.  cf Hungaroton SLPX 11741.
    Tombeau de M. Blancrocher.  cf Angel S 36095.
```

FRUMERIE, Gunnar de
    Pastoral suite, flute, harp and strings. cf ATTERBERG: Suite,
      violin, viola and orchestra, op. 19, no. 1.
1142 Songs: Det blir vackert där du går; Det kom ett brev; Kärleckens
      visa; Låt mig gå vilse i ditt ijus; Saliga väntan. NYSTROEM:
      Bara hos dem; Vitt land; Onskan. RANGSTROM: Flickan under
      nymånen; Pan; Villemo Villemo; Notturno. STENHAMMAR: Det far
      att skepp; Flickan knyter i Johannenatten; Flickan Kom ifrån
      sin älsklings möte; I skogen. Kerstin Meyer, con; Elisabeth
      Söderström, s; Jan Eyron, pno. Swedish Society SLT 33171.
        +RR 11-76 p102
FUCIK, Julius
    Florentiner Marsch. cf DG 2721 077 (2).
    Regimentskinder. cf DG 2721 077 (2).
FUENLLANA, Miguel de
    Perdida de Antequera. cf DG 2530 504.
FUKUSHIMA, Kazuo
    Pieces from Chu-u. cf Nonesuch HB 73028 (2).
FULKERSON, James
    Patterns, no. 7. cf Folkways FTS 33901/4.
FURTWANGLER, Wilhelm
1143 Symphony, no. 2. BPhO; Wilhelm Furtwängler. DG 2707 086 (2).
        +NR 10-76 p2   +SFC 10-3-76 p33
FUX, Johann Josef
1144 Air and 30 variations. Capriccio, G minor. Ciaconna, D major.
      Harpeggio e fuga, G major. Parthie suite, G minor. Sonata
      septima. Michael Thomas, cld. Oryx 1716.
        +-Gr 4-73 p1894   +St 8-76 p94
        +RR 8-73 p64
    Capriccio, G minor. cf Air and 30 variations.
    Ciaconna, D major. cf Air and 30 variations.
    Harpeggio e fuga, G major. cf Air and 30 variations.
    Parthie suite, G minor. cf Air and 30 variations.
    Sonata, organ, no. 7. cf Pelca PRS 40577.
    Sonata septima. cf Air and 30 variations.
GABAYE, Pierre
    Sonatine, flute and bassoon. cf Crystal S 351.
GABRIELI, Andrea
    Canzona francese. cf L'Oiseau-Lyre 12BB 203/6.
    Pour ung plaisir, phantasia. cf Hungaroton SLPX 11669/70.
GABRIELI, Domenico
    Sonata, trumpet. cf Nonesuch H 71301.
GABRIELI, Giovanni
    Motets: Audite principes; Angelus ad pastores; O magnum mysterium;
      Salvator noster; Sol sol la sol; Quem vidistis pastores. cf
      BASSANO: Hodie Christus natus est.
    Sanctus Dominus Deus. cf L'Oiseau-Lyre 12BB 203/6.
    Sonata, pian'e forte. cf BASSANO: Hodie Christus natus est.
    Sonata, pian'e forte. cf Argo SPA 464.
1145 Symphoniae sacre (12). Sofia Soloists Chamber Ensemble; Vassil
      Kasandjiev. Monitor MCS 2144.
        +SFC 1-25-76 p30
GAGLIANO, Marco da
1146 La Dafne. Mary Rawcliffe, Maurita Thornburgh, s; Robert White,
      t; Dale Terbeek, ct; Chamber Chorus and Instrumental Ensemble;
      Paul Vorwerk. ABC Command COMS 9004 (2).

           +HF 2-76 p85   ++ON 12-13-75 p48
           +NR 2-76 p11   ++St 6-76 p116
1147 La Dafne. Elizabeth Humes, Christine Whittlesey, s; Daniel
         Collins, c-t; Ray DeVoll, t; New York Pro Musica Antiqua;
         George Houle. Musical Heritage Society MHS 1953/4 (2).
             +-HF 2-76 p85   +-St 7-75 p96
             +-ON 4-17-76 p42
      Songs: Valli profonde. cf DG Archive 2533 305.
GALILEI, Michelangelo
      For the Duke of Bavaria. cf Saga 5420.
GALILEI, Vicenzo
      Fuga a l'unisono. cf DG Archive 2533 323.
GARDINER, Henry Balfour
      Evening hymn. cf Polydor 2460 250.
      Evening hymn. cf Wealden WS 137.
GARDNER, John
      Songs: The old man and young wife; Sandgate girl's lament. cf
         Argo ZRG 833.
      Theme and variations, op. 7. cf ADDISON: Divertimento, op. 9.
GARRETT
      Psalm, no. 93, The Lord is King. cf HMV SQ CSD 3768.
GASTOLDI, Giovanni
1148 Balletti per cantare, sonare e ballare. Lyon Vocal Ensemble;
         Lyon Early Music Ensemble; François Castet. Musical Heritage
         Society MHS 3310.
             +St 11-76 p140
      Mascherata di Cacciatori. cf HMV CSD 3756.
GATES
      Oh, my luve's like a red, red rose. cf Columbia M 34134.
GAUTIER, Pierre
      Suite, E minor. cf DORNEL: Suite, no. 1, C minor.
GAYFER
      Royal visit. cf Citadel CT 6007.
GEEHL, Henry
      Romanza. cf HMV OU 2105.
GEIJER
      Dansen. cf BIS LP 17.
GEMINIANI, Francesco
      Concerto grosso, D minor (from Corelli's Sonata, violin, op. 5,
         no. 12). cf DG 2548 219.
1149 Sonatas, violoncello, op. 5, nos. 1-6. Anthony Pleeth, Richard
         Webb, vlc; Christopher Hogwood, hpd. L'Oiseau-Lyre DSLO 513.
             +-Gr 6-76 p61   ++RR 6-76 p71
             +HFN 6-76 p85
GENIN
      The carnival of Venice variations. cf Pearl SHE 528.
GENTIAN (15th century France)
      Je suis Robert. cf Harmonia Mundi 204.
GERHARD, Roberto
      Don Quixote: Dances. cf Symphony, no. 1.
1150 Symphony, no. 1 Don Quixote: Dances. BBC Symphony Orchestra;
         Antal Dorati. Argo ZRG 752. (Reissue from HMV ASD 613).
             ++Gr 10-74 p687   ++RR 11-74 p37
             ++MQ 1-76 p139   ++SFC 8-17-75 p22
             +NR 1-75 p4

GERMAN, Edward
    Men of Harlech. cf HMV OU 2105.
GERNSHEIM, Freidrich
1151 Sonata, violoncello and piano, no. 1, op. 12, D minor. RUBINSTEIN:
     Sonata, violoncello and piano, no. 2, op. 39, G major. Gayle
     Smith, vlc; John Jensen, pno. Genesis GS 1060.
         +NR 11-76 p14   +SFC 7-11-76 p13
GERSHWIN, George
1152 An American in Paris. Rhapsody in blue. George Gershwin, pno;
     Columbia Jazz Band, NYP; Michael Tilson Thomas. Columbia
     M 34205. Tape (c) MT 34205 (ct) MA 34205.
         +-HF 12-76 p100   +SR 11-13-76 p52
         ++NR 11-76 p13
1153 An American in Paris. Cuban overture. Rhapsody in blue. Ivan
     Davis, pno; Daniel Majeski, vln; CO; Lorin Maazel. Decca SXL
     6727. (also London CS 6946 Tape (c) 56946 (ct) 86946
     (r) 46946).
         ++Gr 9-75 p458   +NYT 1-18-76 pD1
         -HF 4-76 p102   ++RR 8-75 p32
         +HFN 8-75 p75   ++SFC 9-21-75 p34
         ++NR 12-75 p13   ++St 12-75 p125
1154 An American in Paris. GROFE: Grand Canyon suite. NBC Symphony
     Orchestra; Arturo Toscanini. RCA AT 129. (Reissues from HMV
     ALP 1107, HMV ALP 1232). (also AVM 1-1737 Reissue from
     LM 9020, LM 1004).
         +-ARSC Vol VIII, no. 2-3 +-HF 12-76 p100
              p85   +-RR 6-75 p41
         +Gr 8-75 p336
1155 An American in Paris. Preludes (3). Oh, Kay, excerpts.
     Rhapsody in blue. Tip-toes, excerpts. George Gershwin, pno;
     Paul Whiteman and His Concert Orchestra; RCA Victor Symphony
     Orchestra; Nathaniel Shilkret. RCA AVM 1-1740.
         +ARSC Vol VIII, no. 2-3   +NR 10-76 p14
              p86   +SFC 8-8-76 p38
         +HF 12-76 p100   +St 11-76 p140
         +MJ 11-76 p44
     An American in Paris. cf BERNSTEIN: Candide: Overture.
     An American in Paris. cf COPLAND: Billy the kid.
     An American in Paris. cf COPLAND: El salón Mexico.
     An American in Paris. cf London CSA 2246.
     Concerto, piano, F major. cf Columbia MG 33728.
     Cuban overture. cf An American in Paris.
1156 Gershwin's song book: Swanee, Somebody loves me; Who cares;
     I'll build a stairway to paradise; The man I love; Nobody but
     you; Do it again; 'S wonderful; O lady be good; Sweet and
     low-down; That certain feeling; Liza; I got rhythm. Preludes
     (3). Rhapsody in blue. André Watts, pno. Columbia M 34221.
     Tape (c) MT 34221 (ct) MA 34221. (also CBS 76508 Tape 40-
     76508.
         -ARG 11-76 p22   +NR 10-76 p14
         ++Gr 9-76 p444   +RR 9-76 p81
         +-HF 12-76 p100   -SFC 8-8-76 p38
         +HFN 9-76 p121   +St 11-76 p140
         +HFN 11-76 p175 tape
     Lullaby, string quartet. cf Vox SVBX 5305.
     Oh, Kay, excerpts. cf An American in Paris.

1157 Porgy and Bess. Frances Faye, Betty Roche, Sallie Blair, Mel
      Torme, George Kirby, soloists; Bethlehem Orchestra; Duke
      Ellington and His Orchestra; Australian Jazz Quintet, Pat
      Moran Quartet, Stan Levey Group; Russ Garcia. Bethlehem 3BP 1.
           +NR 1-76 p12   +-NYT 4-25-76 pD16
1158 Porgy and Bess. Willard White, Leona Mitchell, McHenry Boatwright,
      Florence Quivar, Barbara Hendricks, Barbara Conrad, François
      Clemmons, Other; CO and Chorus and Children's Chorus; Loren
      Maazel. London OSA 13116 (3). Tape (c) 5-13116 (ct) 8-13116.
      (also Decca SET 609/11 Tape (c) K3Q 28).
           ++Gr 4-76 p1654   +-NYT 4-25-76 pD16
            +Gr 7-76 p230 tape   +ON 4-10-76 p32
            +-HF 5-76 p77   +RR 4-76 p28
            +HF 8-76 p70 tape   +-RR 8-76 p82 tape
           ++HFN 4-76 p105   ++SFC 3-14-76 p27
           ++HFN 8-76 p94 tape   +-SR 6-12-76 p48
           ++MJ 7-76 p56   ++St 7-76 p73
           ++NR 6-76 p11   +Te 9-76 p30
1159 Porgy and Bess. Orchestra, soloists and chorus; Lehman Engel.
      Odyssey 32 360016 (3).
           +-NYT 4-25-76 pD16
1160 Porgy and Bess (Symphonic picture) (arr. Bennett). KERN: Showboat:
      Scenario, orchestra (arr. & orch. Miller). Utah Symphony
      Orchestra; Mauric Abravanel. Vanguard SRV 345SD.
           +NR 4-76 p3
1161 Porgy and Bess. Cleo Laine, Ray Charles. RCA CPL 2-1831 (2).
           +ON 11-76 p98
      Porgy and Bess (arr. Russell Bennett). cf COPLAND: El salón
      Mexico.
1162 Porgy and Bess, excerpts. Eleanor Steber, s; Robert Merrill,
      Lawrence Tibbett, bar; Cab Calloway, pno; RCA Orchestra;
      Alexander Smallens. RCA AVM 1-1742.
           +ARSC Vol VIII, no. 2-3   +ON 11-76 p98
             p97
1163 Porgy and Bess, excerpts. Anne Brown, s; Todd Duncan, bar;
      Orchestra and chorus; Alexander Smallens. Decca 79024. Decca
      Tape (c) 739024 (r) 6-9024.
           +NYT 4-25-76 pD16   *St 6-71 p111
1164 Porgy and Bess, excerpts. Soloists, chorus and orchestra; George
      Gershwin. Mark 56 667.
           +NYT 4-25-76 pD16
1165 Porgy and Bess, excerpts. Leontyne Price, s; William Warfield,
      bar; Orchestra and chorus; Skitch Henderson. RCA LSC 2679.
           +NYT 4-25-76 pD16
      Preludes. cf Supraphon 111 1721/2.
      Preludes (3). cf An American in Paris.
      Preludes (3). cf Gershwin's song book: Thirteen songs.
      Rhapsody in blue. cf An American in Paris (Columbia M 34205).
      Rhapsody in blue. cf An American in Paris (Decca 6727).
      Rhapsody in blue. cf An American in Paris (RCA AVM 1-1740).
      Rhapsody in blue. cf Gershwin's song book: Thirteen songs.
      Songs: Love walked in; The man I love. cf FAURE: Nell, op. 18,
      no. 1.
      Strike up the band. cf Department of Defense Bicentennial
      Edition 50-1776.
      Strike up the band. cf RCA AGL 1-1334.

Strike up the band, medley. cf Pye GH 603.
Summertime. cf Transatlantic XTRA 1160.
Tip-toes, excerpts. cf An American in Paris.
GERSTER, Robert
Bird in the spirit. cf Crystal S 351.
GERVAISE, Claude
Branle. cf Harmonia Mundi HMU 610.
Branle de Bourgogne. cf DG Archive 2533 184.
Branle de Champaigne. cf DG Archive 2533 184.
Danceries a quatre. cf CBS 76183.
Dances from the French Renaissance (4). cf Klavier KS 551.
Galliards (2). cf Angel SFO 36895.
M'amye est tant honneste. cf Argo ZRG 667.
GESUALDO, Carlo
Canzona francese; Mille volte il dir moro. cf L'Oiseau-Lyre
     12BB 203/6.
Sacrae cantiones a 5: Ave, dulcissima Maria; Ave, Regina coelorum;
     Hei mihi, Domine; O crux benedicta; O vos omnes. cf Harmonia
     Mundi HMU 473.
GHEZZO, Dinu
1166 Kanones. Music, flutes and tape. Ritualen. Thalia. Gretel
          Shanley Andrus, flt; Selene Hurford, vlc; Susanne Shapiro,
          hpd; Sever Tipei, pno; University of Michigan Contemporary
          Direction Ensemble; Uri Mayer. Orion ORS 75172.
               +-HR 9-76 p91   ++NR 1-76 p8
      Music, flutes and tape. cf Kanones.
      Ritualen. cf Kanones.
      Thalia. cf Kanones.
GHISELIN, Johannes
Songs: Ghy syt die werste boven al (Verbonnet). cf HMV SLS 5049.
GIACOBBI, Girolamo
Exultate Deo. cf HMV CSD 3766
GIANELLA, Luigi
Concerto, flute, no. 1, D minor. cf RCA CRL 3-1429.
Concerto, flute, no. 3, C major. cf RCA CRL 3-1429.
Concerto lugubre, C minor. cf RCA CRL 3-1429.
GIBBONS, Orlando
Alman (2). cf Works, selections (L'Oiseau-Lyre DSLO 515).
Alman, "The King's jewel". cf Works, selections (L'Oiseau-Lyre
     DSLO 515).
Coranto. cf Works, selections (L'Oiseau-Lyre DSLO 515).
The cries of London. cf Harmonia Mundi 204.
A fancy. cf Works, selections (L'Oiseau-Lyre DSLO 515).
Fancy, D minor. cf BASF BAC 3075.
Fantasias (2). cf Works, selections (L'Oiseau-Lyre DSLO 515).
Fantasia, D minor. cf BASF BAC 3075.
Fantasia of 4 parts. cf Works, selections (L'Oiseau-Lyre DSLO 515).
Galliard (3). cf Works, selections (L'Oiseau-Lyre DSLO 515).
Great Lord of Lords. cf Turnabout TV 34017.
Ground, A major. cf Works, selections (L'Oiseau-Lyre DSLO 515).
In nomine. cf Argo ZRG 823.
Italian ground. cf Works, selections (L'Oiseau-Lyre DSLO 515).
Jesu, grant me this, I pray. cf Saga 5225.
Lincoln's Inn masque. cf Works, selections (L'Oiseau-Lyre DSLO
     515).
Lord Salisbury's pavan. cf Turnabout TV 34017.

Now each flowery bank. cf L'Oiseau-Lyre 12BB 203/6.
O clap your hands. cf Argo ZRG 789.
O God, the King of glory. cf Turnabout TV 34017.
Pavan (2). cf Works, selections (L'Oiseau-Lyre DSLO 515).
Pavan, G minor. cf BASF BAC 3075.
Pavan "The Earl of Salisbury". cf Works selections (L'Oiseau-
    Lyre DSLO 515).
Prelude. cf Works, selections (L'Oiseau-Lyre DSLO 515).
Royal pavane. cf Argo ZRG 823.
The Queens command (2). cf Works selections (L'Oiseau-Lyre DSLO
    515).
The silver swan. cf Enigma VAR 1017.
The silver swan. cf Prelude PRS 2501.
The silver swan. cf Vista VPS 1022.

1167 Songs (Church music): Hosanna to the son of David; Hymnes and songs
    of the church, nos. 1, 3-5, 9, 13, 18, 20, 22, 24, 31, 47, 67;
    I am the resurrection; Lord we beseech Thee; O clap your hands;
    O Lord in Thy wrath; Praise the Lord, O my soul; See, see, the
    word is incarnate. The Clerkes, Oxenford; David Wulstan.
    Calliope CAL 1611.
        ++Gr 11-76 p851

1168 Songs (Madrigals and motets, Set I, 5 parts): Ah, dear heart;
    Dainty fine bird; Fair is the rose; Fair ladies that to love;
    Farewell all joys; How art thou thralled; I feign not friend-
    ship; I see ambition never pleased; I tremble not at noise of
    war; I weigh not fortune's frown; Lais now old; Mongst thousand
    good; Nay let me weep; Ne'er let the sun; Now each flowerly
    bank of May; O that the learned poets; The silver swan; Trust
    not too much fair youth; What is our life; Yet if that age.
    Consort of Musicke; Anthony Rooley. L'Oiseau-Lyre DSLO 512.
        +-Gr 1-76 p1222   +MT 4-76 p321
        ++HFN 2-76 p97   ++NR 6-76 p13
        +MJ 11-76 p45   +-RR 1-76 p56
        +-MM 5-76 p34   ++St 11-76 p142
    Verse. cf Works, selections (L'Oiseau-Lyre DSLO 515).
    Verse for ye single organ. cf Audio EAS 16.

1169 Works, selections: Alman (2). Alman, "The King's jewel". Coranto.
    A fancy. Fantasias (2). Fantasia of 4 parts. Galliard (3).
    Ground, A major. Italian ground. Lincoln's Inn masque. Pavan
    "The Earl of Salisbury". Pavan (2). Prelude. The Queen's
    command (2). Verse. Christopher Hogwood, hpd, org, spinet.
    L'Oiseau-Lyre DSLO 515.
        ++Gr 2-76 p1359   +MT 4-76 p321
        +HFN 2-76 p97   ++NR 7-76 p13
        +MJ 11-76 p45   +RR 1-76 p48
        ++MM 5-76 p34   +St 11-76 p142

GIBBS, Armstrong
    Songs: Love is a sickness; This is a sacred city. cf Saga 5213.

GIDEON, Miriam
1170 The condemned playground. Questions on nature. WEISGALL: End of
    summer. Jan DeGaetani, ms; Phyllis Bryn-Julson, s; Charles
    Bressler, Constantine Cassolas, t; New York Chamber Soloists;
    Instrumentalists. CRI SD 343.
        +NR 2-76 p12   ++St 8-76 p100
        ++NYT 12-21-75 pD18
    Questions on nature. cf The condemned playground.

GIGOUT, Eugene
    Scherzo. cf Argo ZRG 807.
GILMORE, Patrick
    When Johnny comes marching home. cf CBS 61746.
GIMENEZ
    El barbero de Sevilla: Me llaman la primorosa. cf London OS 26435.
GINASTERA, Alberto
    Canción al arbol. cf Polydor 2383 389.
1171 Concerto, piano, no. 2, op. 37. Quintet, piano and strings. A.
    Black, A. Edelberg, vln; Jacob Glick, vla; Seymour Barab, vlc;
    Hilde Somer, pno. Orion ORS 76241.
       ++NR 11-76 p5
    Estancia: Danza final. cf DG 2584 004.
    Quintet, piano and strings. cf Concerto, piano, no. 2, op. 37.
    Toccata, villancico y fuga. cf Wealden WS 149.
GIORDANO, Umberto
    Adriana Lecouvreur: Io sono l'umile ancella; Poveri fiori. cf
    Club 99-100.
    Andrea Chénier: La mamma morta. cf VERDI: Il trovatore.
    Andrea Chénier: La mamma morta. cf Telefunken AG 6-41947.
    Andrea Chénier: La mamma reggia. cf HMV SXLP 30205.
    Andrea Chénier: Un di all'azzuro spazio. cf Rococo 5383.
    Andrea Chénier: Vicino a te...La morte nostra. cf PUCCINI:
    Madama Butterfly.
    Fedora: Amor ti vieta. cf Everest SDBR 3382.
    Fedora: Amor ti vieta. cf Muza SXL 1170.
    Fedora: Morte di Fedora. cf Club 99-100.
    Siberia: Nel suo amor, Non odi la il martis. cf Club 99-100.
    Siberia: Nel suo amore rianimata la coscienza. cf Court Opera
    Classics CO 347.
    Siberia: Nel suo amore. cf Rubini GV 58.
GIULIANI, Mauro
1172 Concerto, guitar, op. 30, A major. RODRIGO: Concierto madrigal,
    2 guitars and orchestra. Pepe and Angel Romero, gtr; AMF;
    Neville Marriner. Philips 6500 918. Tape (c) 7300 369.
       +Gr 1-76 p1197   ++NR 2-76 p4
       +-HFN 1-76 p107   +-RR 12-75 p48
       +MM 8-76 p35   ++SFC 2-8-76 p26
    Concerto, guitar, op. 30, A major. cf RCA ARL 3-0997.
    Grand overture, op. 61. cf Angel S 36093.
    Grande ouverture, op. 61. cf DG 2530 571.
1173 Introduction, theme with variations and polonaise, op. 65.
    RODRIGO: Fantasia para un gentilhombre. Pepe Romero, gtr; AMF;
    Neville Marriner. Philips 9500 042. Tape (c) 7300 442.
       +-Gr 11-76 p791   +RR 11-76 p77
       +-HFN 11-76 p159
1174 Rondo, no. 1, C major. Rondo, no. 2, F major. Sonata, 2 guitars,
    op. posth. SOR: Divertimento, op. 55, G major. L'Encourage-
    ment, op. 34, E major. Hugh and Thomas Geoghegan, gtr. Orion
    ORS 76229.
       +NR 12-76 p14
    Rondo, no. 2, F major. cf Rondo, no. 1, C major.
    Sonata, flute and guitar, op. 85, A major. cf BIS LP 30.
    Sonata, 2 guitars, op. posth. cf Rondo, no. 1, C major.
1175 Sonata, violin and guitar, op. 25. PAGANINI: Cantabile, op. 17.
    Centone di sonata, op. 64, no. 1. Sonata, violin and guitar,

op. 3, no. 6, E minor. Sonata, violin and guitar, A major.
Itzhak Perlman, vln; John Williams, gtr. CBS 76525.
+-Gr 11-76 p829   ++RR 11-76 p91
+-HFN 11-76 p159
Sonata, violin and guitar, op. 25, E minor. cf Claves LP 30-406.
GLASER
    Kleine Stücke, 4 saxophones, op. 8a. cf Coronet LPS 3030.
GLASS, Philip
    Two pages. cf Folkways FTS 33901/4.
GLAZUNOV, Alexander
1176 Concerto, piano, no. 2, op. 100, B major. Concerto, violin, op.
    82, A minor. Meditation, violin. Ruggiero Ricci, vln; Michael
    Ponti, pno; PH, Westpahalian Symphony Orchestra; Reinhard
    Peters, Siegfried Landau. Turnabout QTS S 34621.
        +-NR 5-76 p6   ++SFC 7-25-76 p29
    Concerto, violin, op. 82, A minor. cf Concerto, piano, no. 2,
    op. 100, B major.
    Concerto, violin, op. 82, A minor. cf RCA ARM 4-0942/7.
    Meditation, op. 32 (@). cf RCA ARM 4-0942/7.
    Meditation, violin. cf Concerto, piano, no. 2, op. 100, B major.
1177 Quartet, strings, no. 3, op. 26b, G major. Quartet, strings, no.
    5, op. 70, D major. Dartington Quartet. Pearl SHE 536.
        +RR 12-76 p72
    Quartet, strings, no. 5, op. 70, D major. cf Quartet, strings,
    no. 3, op. 26b, G major.
    Raymonda, op. 57: Overture. cf CBS 61748.
    Raymonda, op. 57: Valse grande adagio. cf RCA ARM 4-0942/7.
1178 Song of the troubadour, op. 71. SHOSTAKOVICH: Concerto, violon-
    cello, no. 2, op. 126, G major. Mstislav Rostropovich, vlc;
    BSO; Seiji Ozawa. DG 2530 653. Tape (c) 3300 653.
        +Gr 11-76 p791   ++SFC 12-5-76 p58
        +-HFN 10-76 p171   +STL 9-19-76 p36
        +RR 9-76 p66
    Waltz, op. 42, no. 3, D major. cf L'Oiseau-Lyre DSLO 7.
GLIERE, Reinhold
    The bronze horseman, op. 89: Suite, no. 2. cf Symphony, no. 3,
    op. 42, B minor.
    Duo, violin and violoncello, op. 39. cf DVORAK: Trio, piano,
    op. 65, F minor.
    The red poppy, op. 70: Ballet suite. cf Symphony, no. 3, op.
    42, B minor.
    The red poppy, op. 70: Sailors' dance. cf Columbia M 34127.
1179 Symphony, no. 3, op. 42, B minor (Ilya Murometz). Moscow Radio
    Symphony Orchestra; Nathan Rakhlin. Columbia/Melodiya MG
    33832 (2).
        +-Audio 3-76 p68   +NR 2-76 p2
        +HF 3-76 p84   +St 6-76 p103
1180 Symphony, no. 3, op. 42, B minor. The bronze horseman, op. 89:
    Suite, no. 2. MRSO, Bolshoi Theatre Orchestra; Nathan
    Rakhlin, Algis Zuraitis. HMV Melodiya SLS 5062 (2).
        +Gr 8-76 p280   +RR 8-76 p36
        +-HFN 8-76 p80
1181 Symphony, no. 3, op. 42, B minor. The red poppy, op. 70: Ballet
    suite. Vienna State Opera Orchestra; Hermann Scherchen. West-
    minster WGD 2001 (2).
        +RR 12-76 p55
    Tarantella, op. 9, no. 2. cf Columbia/Melodiya M 33593.

GLINKA, Mikhail
 Barcarolle, G major. cf Piano works (Musical Heritage Society
  MHS 1973).
1182 Jota aragonesa. RIMSKY-KORSAKOV: Capriccio espagnol, op. 34.
  TCHAIKOVSKY: Capriccio italien, op. 45. Eugene Onegin, op. 24,
  excerpts. RPO; Pierino Gamba, Paul Kletzki, Walter Weller.
  Classics for Pleasure CFP 40242.
   /HFN 3-76 p109   +-RR 3-76 p49
 The lark. cf HMV HQS 1354.
 Mazurkas, C minor, A minor. cf Piano works (Musical Heritage
  Society MHS 1973).
 Nocturne, F minor. cf Piano works (Musical Heritage Society MHS
  1973).
1183 Piano works: Barcarolle, G major. Mazurkas, C minor, A minor.
  Variations on Alabiev's song "The nightingale". Nocturne,
  F minor. Waltz, G major. Trio pathetique, D minor. Thomas
  Hrynkiv, pno; Esther Lamneck, clt; Michael McCraw, bsn; New
  American Trio. Musical Heritage Society MHS 1973.
   ++HF 9-75 p85   ++St 7-76 p109
 Russlan and Ludmila: Overture. cf BORODIN: Prince Igor: Overture.
 Russlan and Ludmila: Overture. cf DVORAK: Slavonic dances, op. 46.
 Russlan and Ludmila: Overture. cf Decca SPA 409.
 Russlan and Ludmilla: Overture. cf Decca SXL 6782.
 Russlan and Ludmilla: Overture. cf Philips 6780 755.
 Songs: Do not tempt me needlessly. cf Rubini GV 63.
 Songs: Do not tempt me needlessly. cf Rubini RS 301.
 Songs: Doubt. cf Musical Heritage Society MHS 3276.
 Songs: Doubt; Vain temptation. cf Desto DC 7118/9.
 Songs: Travelling song. cf HMV ASD 3200.
1184 Trio pathetique, D minor. SCHUMANN: Märchenerzählungen, op. 132.
  WEBER: Duo concertant, op. 48. The Music Party. L'Oiseau-Lyre
  DSLO 524.
   -Gr 7-76 p187   +NR 12-76 p8
   +HFN 6-76 p85   +-RR 6-76 p65
   +-MT 10-76 p832   +STL 9-19-76 p36
 Trio pathetique, D minor. cf Piano works (Musical Heritage Society
  MHS 1973).
 Variations on Alabiev's song "The nightingale." cf Piano works
  (Musical Heritage Society MHS 1973).
 Waltz, G major. cf Piano works (Musical Heritage Society MHS
  1973).
GLOVER
 The gipsy countess. cf Transatlantic XTRA 1159.
 Songs: Rose of Tralee. cf Argo ZRB 95/6.
GLUCK, Christoph
1185 Arias: Alceste: Divinités du Styx. Armide: Le perfide Renaud.
  Iphigénie en Aulide: Vous essayez en vain...Par la crainte;
  Adieu consevez votre âme. Iphigénie en Tauride: Non cet
  affreux devoir. Orfeo ed Euridice: Che puro ciel; Che farò
  senza Euridice. Paride ed Elena: Spiagge amate; Oh, del mio
  dolce ardor; Le belle immagini; Di te scordarmi. La rencontre
  imprevue: Bel inconnu; Je cherche à vous faire. Janet Baker,
  ms; ECO; Raymond Leppard. Philips 9500 023. Tape (c) 7300 440.
   +Gr 10-76 p644   +RR 12-76 p43
   +HFN 10-76 p171   ++SFC 11-14-76 p30
 Alceste: Divinités du Styx. cf Arias (Philips 9500 023).

Alceste: Divinités du Styx. cf Decca SXL 6629.
Alceste: Divinités du Styx. cf English National Opera ENO 1001.
Alceste: Overture. cf BEETHOVEN: Symphony, no. 5, op. 67, C minor.
Armide: Le perfide Renaud. cf Arias (Philips 9500 023).
Chaconne. cf CORELLI: Concerto grosso, op. 6, no. 8, G minor.
Chaconne. cf Decca SDD 411.
Don Juan: Allegretto. cf DG Archive 2533 182.
1186 Iphigénie en Aulide (ed. Wagner, sung in German). Anna Moffo,
    Arleen Auger, s; Trudeliese Schmidt, ms; Ludovic Spiess, t;
    Dietrich Fischer-Dieskau, Bernd Weikl, bar; Thomas Stewart, bs;
    Bavarian Radio Chorus; Munich Radio Orchestra; Kurt Eichhorn.
    Eurodisc 86271 XR (2). (also RCA ARL 2-1104)
        +-Gr 12-76 p1046   +-SFC 1-18-76 p38
        +HF 4-73 p74   +-SR 3-6-76 p41
        +MJ 3-76 p24   ++St 3-73 p86
        +NR 2-76 p11   ++St 5-76 p114
        +ON 2-21-76 p32
Iphigénie en Aulide: Overture. cf BEETHOVEN: Symphony, no. 7,
    op. 92, A major.
Iphigénie en Aulide: Vous essayez en vain...Par la crainte; Adieu
    consevez votre âme. cf Arias (Philips 9500 023).
Iphigénie en Tauride: Dieux qui me poursuivez; Dieux protecteurs.
    cf Caprice CAP 1062.
Iphigénie en Tauride: Non cet affreux devoir. cf Arias (Philips
    9500 023).
1187 Orfeo ed Euridice. Hilde Gueden, Magda Gabory, s; Fedora Barbieri,
    con; La Scala Opera Orchestra and Chorus; Wilhelm Furtwängler.
    Bruno Walter Society RR 419 (2).
        +NR 10-76 p10
1188 Orfeo ed Euridice, Act 2. Nan Merriman, ms; Barbara Gibson, s;
    Robert Shaw Chorale; NBC Symphony Orchestra; Arturo Toscanini.
    RCA AT 127. (Reissue from HMV ALP 13507).
        +Gr 8-75 p336   +ST 2-76 p737
        +RR 6-75 p22
Orfeo ed Euridice: Ballet. cf DG Archive 2533 182
Orfeo ed Euridice: Ballet music. cf BACH: Works, selections (DG
    2530 647).
Orfeo ed Euridice: Che puro ciel; Che farò senza Euridice. cf
    Arias (Philips 9500 023).
Orfeo ed Euridice: Dance of the blessed spirits. cf Decca SPA 394.
Orfeo ed Euridice: Dance of the blessed spirits. cf RCA LRL 1-5094.
Orpheus and Eurydice: Dance of the blessed spirits. cf RCA LRL
    1-5127.
Orfeo ed Euridice: Mélodie. cf HMV SXLP 30188.
Orphée et Eurydice: Viens, viens, Eurydice, suis-moi. cf Angel
    S 37143.
Paride ed Elena: O del mio dolce ardor. cf Decca SXL 6629.
Paride ed Elena: Spiagge amate. cf Club 99-99.
Paride ed Elena: Spiagge amate; Oh, del mio dolce ardor; Le belle
    immagini; Di te scordarmi. cf Arias (Philips 9500 023).
La rencontre imprevue: Bel inconnu; Je cherche à vous faire. cf
    Arias (Philips 9500 023).
GODARD, Benjamin
    Chanson d'Estelle. cf HMV RLS 716.
    Fragments poetiques, op. 13. cf BOIELDIEU: Jean de Paris: Over-
    ture.

Impressions de campagne au printemps, op. 123. cf BOIELDIEU:
Jean de Paris; Overture.
Jocelyn: Angels guard thee. cf Pye Ember GVC 51.
Jocelyn: Berceuse. cf Decca SB 322.
Jocelyn: Berceuse. cf Musical Heritage Society MHS 3276.
Pieces, op. 116: Waltz. cf RCA LRL 1-5094.
Suite, flute and orchestra, op. 116. cf BOIELDIEU: Jean de Paris:
Overture.
Le tasse: Les regrets. cf Club 99-97.
Vivandière: Viens avec nous. cf Discophilia KIS KGG 3.
GODOWSKY, Leopold
Alt Wien. cf L'Oiseau-Lyre DSLO 7.
Alt Wien. cf RCA ARM 4-0942/7.
The gardens of Buitenzorg. cf International Piano Library IPL
102.
Passacaglia, B minor. cf MOSZKOWSKI: Concerto, piano, op. 59, E
major.
Waltz, D major. cf RCA ARM 4-0942/7.
Waltz-poem, no. 4, for the left hand. cf L'Oiseau-Lyre DSLO 7.
GOETZ, Hermann
Leichte Stücke, violin and violoncello, op. 2. cf Quartet, op.
6, E major.
1189 Quartet, op. 6, E major. Quintet, op. 16, C minor. Trio, op. 1,
G minor. Leichte Stücke, violin and violoncello, op. 2. Ger-
ald Robbins, pno; Glenn Dicterow, vln; Terry King, vlc; Dennis
Trembly, double bs; Alan de Veritch, vla. Genesis GS 1037/38
(2).
+-Audio 7-76 p73   ++NR 4-76 p7
Quintet, op. 16, C minor. cf Quartet, op. 6, E major.
Trio, op. 1, G minor. cf Quartet, op. 6, E major.
GOETZE
Still as the night. cf HMV RLS 716.
GOLD, Ernest
1190 Songs of love and parting. CASTELNUOVO-TEDESCO: Coplas. Marni
Nixon, s; Vienna Volksoper Orchestra; Ernest Gold. Crystal
S 501.
+-Audio 2-76 p97   ++St 6-75 p98
+-NR 4-75 p9
GOLDEN DANCE HITS OF 1600. cf DG Archive 2533 184.
GOLDMAN
Boy Scouts of America. cf Mercury SRI 75055.
Bugles and drums. cf Mercury SRI 75055.
Children's march. cf Mercury SRI 75055.
Illinois march. cf Mercury SRI 75055.
The Interlochen bowl. cf Mercury SRI 75055.
Onward-upward. cf Mercury SRI 75055.
GOLDMAN, Richard
1191 Sonata, violin and piano. Sonatina, 2 clarinets. PERSCHETTI:
Parable. WYLIE: Psychogram. Theodore Cole, Thomas Falcone,
clt; Berl Senofsky, vln; Ellen Mack, pno; Arthur Weisberg, bsn.
CRI SD 353.
+-NR 7-76 p5
Sonatina, 2 clarinets. cf Sonata, violin and piano.
GOLDMARK, Carl
Concerto, violin, A minor: Andante. cf RCA ARM 4-0942/7.
Königin von Saba: Magische tone. cf Rococo 5379.

GOLLAND
    Relay. cf Virtuosi VR 7506.
GOMBERT, Nicolas
    Caeciliam cantate. cf L'Oiseau-Lyre 12BB 203/6.
    Je suis trop ionette. cf Hungaroton SLPX 11669/70.
GOODMAN, Joseph
    Jadis III. cf Crystal S 351.
GOODWIN
    The headless horseman. cf HMV OU 2105.
    London serenade. cf Music for Pleasure SPR 90086.
    Prairie serenade. cf Music for Pleasure SPR 90086.
    Puppet serenade. cf Music for Pleasure SPR 90086.
GOODWIN, George
    Door latch quickstep. cf Library of Congress OMP 101/2.
GOOSSENS, Eugene
    Folk-tune. cf Polydor 2383 391.
GOSS, John
    Praise, my soul the King of heaven. cf Saga 5225.
    Psalm, no. 37, Fret not thyself: Gloria. cf HMV SQ CSD 3768.
GOSSEC, Francois
    Tambourin. cf RCA LRL 1-5131.
GOTTSCHALK, Louis
    Bamboula, op. 2. cf Works, selections (Vanguard VSD 723/4).
    Le bananier, op. 5. cf Piano works (London CS 6943).
    Le bananier, op. 5. cf Works, selections (Vanguard VSD 723/4).
    The banjo, op. 15. cf Piano works (London CS 6943).
    Le banjo, op. 15. cf Works, selections (Vanguard VSD 723/4).
    The banjo, op. 15. cf Personal Touch 88.
    Battle cry of freedom, op. 55. cf Piano works (Angel S 36090).
    Berceuse, op. 47. cf Piano works (Angel S 36090).
    Columbia, op. 34. cf Piano works (Angel S 36090).
    Concerto paraphrase on national airs, op. 48: The union. cf
      Works, selections (Turnabout TV 37034).
    The dying poet. cf Piano works (London CS 6943).
    They dying poet. cf Works, selections (Vanguard VSD 723/4).
    L'Etincelle, op. 21. cf Piano works (Vanguard VSD 71218).
    La gallina, op. 53. cf Piano works (Angel S 36090).
    La gallina, op. 53. cf Piano works (Vanguard VSD 71218).
    La gallina, op. 53. cf Works, selections (Turnabout TV 37034).
    Grand scherzo, op. 57. cf Piano works (Angel S 36090).
    Grand scherzo, op. 57. cf Piano works (London CS 6943).
    Grande fantaisie triomphale sur l'hymne nationale Brésilien, op.
      69. cf Works, selections (Turnabout TV 37034).
    Grande tarantelle, piano and orchestra. cf Works, selections
      (Turnabout TV 37034).
    Grande tarantelle, piano and orchestra, op. 67. cf Works,
      selections (Vanguard VSD 723/4).
    La jota aragonesa, op. 14. cf Piano works (Vanguard VSD 71218).
    The last hope, op. 16. cf Works, selections (Vanguard VSD 723/4).
    Maiden's blush. cf Works, selections (Vanguard VSD 723/4).
    Le mancenillier, op. 11. cf Piano works (London CS 6943).
    Manchega, op. 38. cf Piano works (London CS 6943).
    Marche de nuit, op. 17. cf Piano works (Vanguard VSD 71218).
    Marguerite. cf Piano works (Angel S 36090).
    Mazurka, F sharp minor. cf Piano works (Angel S 36090).
    O, ma charmante, epargnez-moi, op. 44. cf Piano works (Angel S
      36090).

O, ma charmante, epargnez-moi, op. 44. cf Piano works (London
   CS 6943).
Ojos criollos, op. 37. cf Works, selections (Turnabout TV 37034).
Ojos criollos, op. 37. cf Works, selections (Vanguard VSD 723/4).
Orfa, op. 71. cf Piano works (Vanguard VSD 71218).
Pasquinade, op. 59. cf Piano works (London CS 6943).
Pasquinade, op. 59. cf Works, selections (Turnabout TV 37034).
Pasquinade, op. 59. cf Works, selections (Vanguard VSD 723/4).
1192 Piano works: Battle cry of freedom, op. 55. Berceuse, op. 47.
   Columbia, op. 34. La gallina, op. 53. Grand scherzo, op. 57.
   Marguerite. Mazurka, F sharp minor. O, ma charmante, epargnez-
   moi, op. 44. Polka, B flat major. Suis-moi, op. 45. Tournament galop. Leonard Pennario, pno. Angel S 36090. Tape (c)
   4XS 36090 (ct) 8XS 36090.
          +HF 3-76 p88   +SFC 1-18-76 p38
          +NR 5-76 p12   +St 12-75 p120
1193 Piano works: Le bananier, op. 5. The banjo, op. 15. The dying
   poet. Grand scherzo, op. 57. Le mancenillier, op. 11. Man-
   chega, op. 38. O, ma charmante, epargnez-moi, op. 44. Pasquinade, op. 59. Souvenirs d'Andalousie, op. 22. Souvenir de
   Porto Rico. Suis-moi, op. 45. Tournament galop. Ivan Davis,
   pno. London CS 6943. (also Decca SXL 6725 Tape (c) KSXC 6725).
          +Gr 7-76 p193   +NR 5-76 p12
          +-HF 5-76 p83   ++RR 7-76 p24
          +HFN 7-76 p87   ++SFC 1-18-76 p38
          ++HFN 11-76 p175 tape   +St 4-76 p119
          ++MJ 11-76 p52
Polka, B flat major. cf Piano works (Angel S 36090).
Printemps d'amour, op. 40. cf Piano works (Vanguard VSD 71218).
Radieuse, op. 72. cf Piano works (Vanguard VSD 71218).
Radieuse, op. 72. cf Works, selections (Turnabout TV 37034).
Reponds-moi, op. 50. cf Piano works (Vanguard VSD 71218).
La Savane. cf Works, selections (Vanguard VSD 723/4).
Ses yeux, op. 66. cf Piano works (Vanguard VSD 71218).
Ses yeux, op. 66. cf Works, selections (Turnabout TV 37034).
Souvenir de Porto Rico. cf Piano works (London CS 6943).
Souvenir de Porto Rico. cf Works, selections (Vanguard VSD 723/4).
Souvenirs d'Andalousie, op. 22. cf Piano works (London CS 6943).
Souvenirs d'Andalousie, op. 22. cf Piano works (Vanguard VSD
   71218).
Suis-moi, op. 45. cf Piano works (Angel S 36090).
Suis-moi, op. 45. cf Piano works (London CS 6943).
Suis-moi, op. 45. cf Works, selections (Vanguard VSD 723/4).
Symphony, no. 1 (Night in the tropics). cf Works, selections
   (Vanguard VSD 723/4).
Tournament galop. cf Piano works (Angel S 36090).
Tournament galop. cf Piano works (London CS 6943).
Tournament galop. cf Works, selections (Vanguard VSD 723/4).
Tremolo, op. 58. cf Piano works (Vanguard VSD 71218).
The Union, op. 48. cf Piano works (Vanguard VSD 71218).
Variations on the Portuguese national hymn. cf Works, selections
   (Turnabout TV 37034).
1194 Works, selections: Grande fantaisie triomphale sur l'hymne nationale
   Bresilien, op. 69. Grande tarantelle, piano and orchestra.
   Variations on the Portuguese national hymn. Concerto para-
   phrase on national airs, op. 48: The Union. Pieces, piano, 4
   hands: Radieuse, op. 72. Ses yeux, op. 66. La gallina, op. 53.

Ojos criollos, op. 37. Pasquinade, op. 59. Eugene List, Cary
Lewis, Brady Millican, pno; BeSO, VSOO; Samuel Adler, Igor
Buketoff. Turnabout TV 37034. Tape (c) KTVC 37034.
  +-Gr 8-73 p333   -RR 10-76 p106 tape
  +-RR 7-73 p48

1195 Works, selections: Bamboula, op. 2. Le bananier, op. 5. Le
  banjo, op. 15. The dying poet. Grande tarantelle, piano
  and orchestra, op. 67. The last hope, op. 16. Maiden's blush.
  Ojos criollos, op. 37. Pasquinade, op. 59. La Savane.
  Souvenir de Porto Rico. Suis-moi, op. 45. Symphony, no. 1
  (Night in the tropics). Tournament galop. Eugene List, Reid
  Nibley, pno; Orchestra, Utah Symphony Orchestra; Maurice
  Abravanel. Vanguard VSD 723/4. (Reissues).
  +LJ 3-75 p33   +SFC 6-30-74 p39
  +MJ 7-74 p30   +St 7-76 p72

1196 Works, selections: L'Etincelle, op. 21. La gallina, op. 53.
  La jota aragonesa, op. 14. Marche de nuit, op. 17. Orfa,
  op. 71. Printemps d'amour, op. 40. Radieuse, op. 72.
  Reponds-moi, op. 50. Ses yeux, op. 66. Souvenirs d'Andalousie,
  op. 22. Tremolo, op. 58. The Union, op. 48. Eugene List,
  Joseph Werner, Cary Lewis, pno. Vanguard VSD 71218.
  +St 10-76 p124

GOUDIMEL
  Psalms, 77, 86, 137; Priere avant le repas. cf Titanic TI 4.
GOULD
  Songs: The curfew. cf Argo ZFB 95/6.
GOULD, Morton
  Interplay: Blues. cf Personal Touch 88
  Latin American symphonette: Guaracha. cf Personal Touch 88.
  Party rag. cf Personal Touch 88.
  Symphonette, no. 2. cf BAKER: Le chat qui pêche.
GOUNOD, Charles
  Ave Maria (Bach). cf Decca SXL 6781.
  Cinq-mars: Nuit respendissante. cf Club 99-97.
1197 Fantasy on the Russian national hymn. MASSANET: Concerto, piano.
  SAINT-SAENS: Africa, op. 89. Marylène Dosse, pno; Westphalian
  Symphony Orchestra; Siegfried Landau. Candide (Q) QCE 31088.
  +Audio 2-76 p95   /SFC 4-4-76 p34
  +NR 3-75 p3
1198 Faust. BERLIOZ: The damnation of Faust, op. 24, excerpts.
  Mireille Berthon, s; Marthe Coiffier, ms; Jeanne Montfort,
  alto; César Vezzani, t; Louis Musy, Michel Cozette, bar;
  Marcel Journet, Louis Morturier, bs; Paris Opera Orchestra and
  Chorus; St. Gervais Chorus; Paseloup Concerts Orchestra; Henri
  Busser, Piero Coppola. Club 99 OP 1000 (3). (Reissues from
  French EMI originals).
  +HF 7-76 p78
1199 Faust (sung in German). Emmy Destin, s; Maria Gotze, ms; Ida von
  Scheele-Müller, alto; Karl Jörn, t; Desider Zador, Arthur
  Neuendahn, bar; Paul Knüpfer, bs; Orchestra and Chorus; Bruno
  Seidler-Winkler. Discophilia KS 4/6 (3). (Reissue from DG
  originals, 1908).
  +-HF 7-76 p78   +NR 2-76 p11
1200 Faust (sung in German). Margarete Teschemacher, s; Elisabeth
  Waldenau, alto; Helge Roswänge, t; Hans Hermann Nissen, Alex-
  ander Welitsch, bar; Georg Hann, bs-bar; Stuttgart Radio Orch-
  estra and Chorus; Joseph Keilberth. Preiser FST 3 (3).

(Recorded from broadcast, Dec. 5, 1937).
        -HF 7-76 p78
Faust: Avant de quitter ces lieux. cf Caprice CAP 1062.
Faust: Avant de quitter ces lieux; Que vois-je la...Ah, je ris;
        Alerte, alerte. cf Decca GOSD 674/6.
1201 Faust: Ballet music; Waltz. PONCHIELLI: La gioconda: Dance of
        the hours. TCHAIKOVSKY: Eugene Onegin, op. 24: Waltz and
        polonaise. VERDI: Aida: Ballet music. Otello: Ballet music.
        Berlin Radio Symphony Orchestra; Ferenc Fricsay. DG 2548 133.
        +RR 1-76 p33
Faust: Duet, Act 1. cf Club 99-98.
Faust: Il était un roi de Thulé...O Dieu, que de bijoux. cf
        Telefunken AG 6-41945.
Faust: Faites-lui mex aveux. cf Collectors Guild 611.
Faust: Jewel song; Final trio. cf HMV RLS 719.
Faust: O Dieu, que les bijoux. cf Decca SPA 449.
Faust: Salut, demeure. cf Decca SXL 6649.
Faust: Salut, demeure chaste et pure. cf RCA CRM 1-1749.
Faust: Seigneur, daignez permettre; Prison scene. cf Club 99-97.
Faust: Soldiers chorus. cf Decca DPA 525/6.
Faust: Soldiers chorus. cf Philips 6747 204.
Faust: Le veau d'or est toujours debout, Il était temps, Vous qui
        faites l'endormie. cf Preiser LV 192.
Faust: Was ist der Gott...O gib junges Blut; Es ist schön spat...
        Verweile doch ein Augenblick...Er liebt mich. cf DELIBES:
        Lakmé: Fantasien nebelhafte Träume.
Funeral march of the marionette. cf Philips 6780 030.
1202 Messe solennelle (St. Cecilia). Pilar Lorengar, s; Heinz Hoppe,
        t; Franz Crass, bs; Henriette Puig-Roget, org; René Duclos
        Choir; OSCCP; Jean-Claude Hartemann. HMV SXLP 30206. (Re-
        issue from ASD 589). (also Angel S 36214)
        +Gr 1-76 p1225   +-RR 1-76 p57
        +HFN 2-76 p117
Mireille: Anges du paradis. cf Club 99-98.
Mireille: O légère hirondelle; Heureux petit berger. cf Richmond
        R 23197.
Mireille: Vincenette à votre âge. cf Angel S 37143.
Mireille: Waltz. cf Columbia M 33933.
Mors et vita: Judex. cf Pye TB 3007.
Petite symphonie. cf BEETHOVEN: Trio, piano, clarinet and
        violoncello, op. 38, E flat major.
1203 Petite symphonie, 9 wind instruments. MOZART: Serenade, no. 11,
        K 375, E flat major. Paul Dunkel, flt; Rudolph Vrbsky, Allan
        Vogel, Richard Woodhams, ob; Frank Cohen, John Fullam, Richard
        Stoltzman, clt; Alexander Heller, Vincent Ellin, Eric Arbiter,
        bsn; Robert Routch, E. Scott Brubaker, John Serkin, hn. Marlboro
        Recording Society MRS 8.
        ++St 2-76 p100
La Reine de Saba (Königin von Saba): Plus grande dans son obscurité.
        cf Club 99-97.
Königin von Saba: Plus grand dans son obscurité. cf Discophilia
        KIS KGG 3.
Roméo et Juliette: Depuis hier je cherche en vain. cf CBS 76522.
Roméo et Juliette: Je veux vivre dans ce reve. cf Court Opera
        Classics CO 342.
Roméo et Juliette: Je veux vivre. cf Telefunken AG 6-41945.

Roméo et Juliette: Madrigal of Juliet and Romeo. cf Angel S 37143.
Roméo et Juliette: Waltz song. cf HMV RLS 719.
Sappho: O ma lyre immortelle. cf Dicophilia KIS KGG 3.
Sappho: Où suis-je...O ma lyre immortelle. cf Decca GOSD 674/6.
Songs: Aimons-nous; Où voulez-vous aller. cf Seraphim S 60251.
Songs: Serenade. cf HMV ASD 2929.
Songs: Serenade. cf Musical Heritage Society MHS 3276.
GRAFULLA, Claudio
    Captain Finch's quickstep; Captain Shepherd's quickstep. cf
        Library of Congress OMP 101/2.
    Washington greys. cf Department of Defense Bicentennial Edition
        50-1776.
    Washington greys. cf Michigan University SM 0002.
    Washington greys. cf Philips 6308 246.
GRAINGER, Percy
    Colonial song. cf Piano works (Everest X 913).
    Country gardens. cf Piano works (Everest X 913).
    Country gardens. cf FAURE: Nell, op. 18, no. 1.
    Eastern intermezzo. cf FAURE: Nell, op. 18, no. 1.
    Handel in the Strand. cf FAURE: Nell, op. 18, no. 1.
    Handel in the Strand. cf HMV SXLP 30181.
    Irish tune from County Derry. cf FAURE: Nell, op. 18, no. 1.
    Knight and shepherd's daughter. cf FAURE: Nell, op. 18, no. 1.
    Molly on the shore. cf Piano works (Everest X 913).
    Molly on the shore. cf FAURE: Nell, op. 18, no. 1.
    Over the hills and far away. cf FAURE: Nell, op. 18, no. 1.
1204 Paraphrase on Tchaikovsky: Nutcracker: Waltz of the flowers. LISZT:
        Parapharase on Tchaikovsky: Eugen Onegin: Polonaise, G 429.
        TCHAIKOVSKY: Concerto, piano, no. 2, op. 44, G major. Michael
        Ponti, pno. Turnabout TV 34560.
            +-Gr 4-76 p1598   -RR 3-76 p50
            +HFN 3-76 p93   -SFC 10-5-75 p38
            +NR 1-75 p5
1205 Piano works: Colonial song. Country gardens. Molly on the shore.
        Shepherd's hey. Spoon River. Sussex mummers' carol. STANFORD:
        Irish dances: Reel; Leprechaun's dance (arr. Grainger). Percy
        Grainger, pno. Everest X 913. (Reissue from Duo-Art piano
        rolls).
            +Gr 8-76 p319
    Sailor's song. cf FAURE: Nell, op. 18, no. 1.
    Shepherd's hey. cf Piano works (Everest X 913).
    Shepherd's hey. cf FAURE: Nell, op. 18, no. 1.
1206 Songs: Children's march: Over the hills and far away; Colonial song;
        Country gardens; Duke of Marlborough's fanfare; Handel in the
        Strand; Harvest hymn for string orchestra; Harvest hymn, 18
        single instruments; The lonely desert man sees the tents of the
        happy tribes; Scotch Strathspey and reel; Shallow brown; Under
        En Bro; Le vallée des cloches (arr. from Ravel). Lauris Elms,
        con; Pearl Berridge, s; David Parker, t; Ronal Jackson, Chris-
        topher Field, bar; Clarence Mellor, hn; Joyce Hutchingson,
        Colin Forbes, pno; Oriana Singers, Male Voices; Sydney Symphony
        Orchestra; John Hopkins. EMI EMD 5514.
            +Gr 3-76 p1493   ++RR 2-76 p56
            +-HFN 3-76 p93
    Spoon River. cf Piano works (Everest X 913).
    Sussex mummers' carol. cf Piano works (Everest X 913).

To a Nordic princess. cf FAURE: Nell, op. 18, no. 1.
Tribute to Foster: Lullaby. cf FAURE: Nell, op. 18, no. 1.
Walking tune. cf FAURE: Nell, op. 18, no. 1.
GRANADOS, Enrique
    Allegro de concierto, C major. cf Piano works (CRD CRD 1023).
    Canciones amatorias: Gracia mia; Llorad corazón que teneis razón;
        Mañinca era; Mira que soy niña; Serranas de Cuenca. cf Songs
        (Everest 3237).
    Capricho espagnol, op. 39. cf Piano works (CRD CRD 1023).
    Carezza vals, op. 38. cf Piano works (CRD CRD 1023).
1207 Danza lenta. Escenas romanticas: Mazurka. Piezas sobre cantos
        populares españoles (6). Thomas Rajna, pno. CRD CRD 1022.
        ++Gr 8-76 p319   ++HFN 8-76 p80
    Escenas romanticas: Mazurka. cf Danza lenta.
1208 Goyescas. Francisco Aybar, pno. Connoisseur Society CS 2091.
        +MJ 10-76 p52   ++SFC 5-30-76 p24
    Goyescas: Intermezzo. cf FALLA: El amor brujo.
    Goyescas: Intermezzo. cf HMV SLS 5019.
    Goyescas: The maiden and the nightingale. cf Decca KCSP 367.
    Goyescas: The maiden and the nightingale. cf HMV HQS 1354.
    Goyescas: The maiden and the nightingale. cf HMV HQS 1360.
    Goyescas: The maja and the nightingale. cf Opus unnumbered.
    Goyescas: Quejas o la maja y el ruiseñor. cf RCA ARL 1-1176.
    Impromptus (2). cf Piano works (CRD CRD 1023).
1209 Piano works: Allegro de concierto, C major. Capricho espagnol,
        op. 39. Carezza vals, op. 38. Impromptus (2). Spanish dances,
        op. 37: Oriental. Rapsodia aragonesa. Valses poéticas.
        Thomas Rajna, pno. CRD CRD 1023.
        +Gr 10-76 p628   ++RR 10-76 p87
        +HFN 10-76 p171
    Piezas sobre cantos populares españoles (6). cf Danza lenta.
    Rapsodia aragonesa. cf Piano works (CRD CRD 1023).
1210 Songs: Las currutacas modestas; El majo olvidado. Canciones
        amatorias: Gracia mia; Llorad corazón que teneis razón;
        Mañanica era; Mira que soy niña; Serranas de Cuenca. Tonadillas
        (La maja dolorosa): Ay majo de mi vida; De quel majo amante; Oh
        muerte cruel. Tonadillas al estilo antiguo: Amor y odio;
        Callejo; La maja de Goya; El majo discreto; El majo timido; El
        mirar de la maja; El tra-la-la y el punteado. Conchita Badia,
        s; Alicia de Larrocha, pno. Everest 3237.
        +-Gr 10-76 p634
1211 Spanish dances (Danzas españolas), op. 37 (12). Gonzalo Soriano,
        pno. Connoisseur Society CS 2105.
        ++SFC 12-26-76 p34   ++St 10-76 p122
        +SR 9-18-76 p50
1212 Spanish dances, op. 37 (12). Thomas Rajna, pno. CRD 1021.
        +Gr 7-76 p193   ++RR 7-76 p71
        +HFN 8-76 p80   ++St 10-76 p122
    Danza española, op. 37: Andaluza. cf RCA ARM 4-0942/7.
    Danzas españolas, op. 37: Andaluza, Valenciana. cf London CS 6953.
    Spanish dances, op. 37: Oriental. cf Piano works (CRD CRD 1023).
    Danza española, op. 37: Villanesco. cf DG 3335 182.
    Spanish dances, op. 37: Villanesco, Andaluza, Danza triste. cf
        ALBENIZ: Suite española, no. 3: Sevillanas; no. 4: Cadiz.
    Tonadillas: La maja dolorosa, nos. 1-3; El majo discreto; El tra-
        la-la y el punteado; El majo timido. cf DG 2530 598.

Tonadilla: La maja dolorosa. cf FALLA: Mélodies.
Tonadillas (La maja dolorosa): Ay majo de mi vida; De quel majo
    amante; O muerte cruel. cf Songs (Everest 3237).
Tonadillas al estilo antiguo: Amor y odio; Callejo; La maja de
    Goya; El majo discreto; El majo timido; El mirar de la maja;
    El tra-la-la ye el punteado. cf Songs (Everest 3237).
Tonadillas al estilo antiguo: Amor y odio; El majo discreto;
    El majo timido; El mirar de la maja; El tra-la-la y el punteado.
    cf FALLA: Mélodies.
Tonadillas al estilo antiguo: La maja de Goya. cf Angel S 36094.
Valses poéticas. cf Piano works (CRD CRD 1023).

GRANDI, Alessandro
1213 Motets: Ave Regina; Dixit Dominus; Exaudi Deus; Jesu mi dulcissime;
    O quam tu pulchra es; O vos omnes; Plorabo die ac nocte;
    Vulnerasti cor meum. Paul Esswood, c-t; Edgar Fleet, Nigel
    Rogers, t; Trinity Boys' Choir; Accademia Monteverdiana; Denis
    Stevens. Nonesuch H 71329.
        +ARG 12-76 p30   +NR 9-76 p8
        +HFN 9-76 p171   +St 12-76 p140

GRANDJANY, Marcel
    Fantasia on a theme by Haydn. cf Panton 110 380.
GRAVES, Milford
    Transmutations. cf Folkways FTS 33901/4.
GRAY, Alan
    Evening service, F minor. cf Audio 1.
    GREAT AMERICAN MARCHES. cf RCA AGL 1-1334.
    GREAT HITS YOU PLAYED WHEN YOU WERE YOUNG, volume 3. cf Connoisseur
        Society (Q) CSQ 2065.
    GREAT HITS YOU PLAYED WHEN YOU WERE YOUNG, volume 4. cf Connoisseur
        Society (Q) CSQ 2066.
GREENE, Maurice
    Lord let us know mine end. cf Argo ZRG 789.
1214 Songs (Anthems): Arise, shine, O Zion; I will sing of thy pow'r,
    O God; Like as the hart; Lord, let me know mine end; Magnificat
    and nunc dimittis, C major; My God, my God, look upon me; O
    clap your hands together all ye people. St. Alban's Abbey
    Choir; Peter Hurford. Argo ZRG 832.
        +-Gr 6-76 p71   +RR 5-76 p69
        +HFN 5-76 p100
    Voluntaries, G major, C minor. cf Cambridge CRS 2540.
    GREGORIAN CHANT. cf Anonymous works.
GREGSON, Edward
    Prelude for an occasion. cf HMV OU 2105.
GRENON, Nicholaw
    La plus jolie. cf Telefunken ER 6-35257.
GRESSEL, Joel
    Points in time. cf Odyssey Y 34139.
GRETCHANINOV, Alexander
    Otpoutchtchaiechi (9). cf Harmonia Mundi HMU 137.
GRETRY, André
1215 Céphale et Procris, suite de ballet. Danses Villageoises.
    Overtures: L'Epreuve Villageoise, Les mariages Samnites, Richard
    Coeur de Lion. Orchestre de Liege; Paul Strauss. Seraphim S
    60268.
        ++NR 6-76 p2   +St 12-76 p141
    Danses Villageoises. cf Céphale et Procris, suite de ballet.

L'Epreuve Villageoise overture. cf Céphale et Procris, suite de
     ballet.
Les mariages Samnites overture. cf Céphale et Procris, suite de
     ballet.
Richard Coeur de Lion overture. cf Céphale et Procris, suite de
     ballet.
GREVER
     Songs: Magic is the moonlight; Jurame. cf DG 2530 700.
GRIEG, Edvard
     Album leaf, op. 47, no. 2. cf Lyric pieces (DG 2530 476).
     Arietta, op. 12, no. 1. cf Lyric pieces (DG 2530 476).
     At the cradle, op. 68, no. 5. cf Lyric pieces (DG 2530 476).
     At your feet, op. 68, no. 3. cf Lyric pieces (DG 2530 476).
     Ballade, op. 65, no. 5. cf Lyric pieces (DG 2530 476).
     Berceuse, op. 38, no. 1. cf Lyric pieces (DG 2530 476).
     Brooklet, op. 62, no. 4. cf Lyric pieces (DG 2530 476).
     Butterfly, op. 43, no. 1. cf Lyric pieces (DG 2530 476).
1216 Concerto, piano, op. 16, A minor. SCHUMANN: Concerto, piano, op.
     54, A minor. Solomon, pno; PhO; Herbert Menges. Classics for
     Pleasure CFP 40255. (Reissue from HMV ASD 272).
          +Gr 12-76 p995   +-RR 10-76 p50
          +-HFN 11-76 p171
1217 Concerto, piano, op. 16, A minor. RACHMANINOFF: Concerto, piano,
     no. 2, op. 18, C minor. Clifford Curzon, pno; LPO, LSO; Anatole
     Fistoulari, Adrian Boult. Decca ECS 753. (Reissues from LXT
     5165, 5178).
          /-Gr 6-75 p39   -RR 7-76 p62
1218 Concerto, piano, op. 16, A minor. SCHUMANN: Concerto, piano, op.
     54, A minor. Radu Lupu, pno; LSO; André Previn. Decca SXL
     6624. Tape (c) KSXC 6624. (also London CS 6840)
          +-Gr 2-74 p1552   ++NR 11-74 p5
          +HF 11-74 p122   +RR 2-74 p33
          +HFN 3-74 p105   ++RR 2-76 p71 tape
          +HFN 9-75 p120   +-SFC 8-11-74 p34
          +MJ 1-75 p48
1219 Concerto, piano, op. 16, A minor. SCHUMANN: Concerto, piano, op.
     54, A minor. Sviatoslav Richter, pno; Monte Carlo National
     Orchestra; Lovro von Matacic. HMV (Q) ASD 3133. (also Angel
     (Q) S 36899 Tape (c) 4XS 36899 (ct) 8XS 36899).
          +-Gr 11-75 p818   ++MJ 1-76 p25
          +-HF 12-75 p94   ++NR 11-75 p6
          +-HF 4-76 p148 tape   -RR 12-75 p49
          +-HFN 12-75 p153   ++St 12-75 p126
1220 Concerto, piano, op. 16, A minor. RAVEL: Concerto, piano, G major.
     Jenö Jandó, pno; HRT Orchestra; Antal Jancsovics. Hungaroton
     SLPX 11710.
          ++HF 6-76 p80   +-NR 11-75 p6
          /HFN 1-76 p108   +-RR 12-75 p58
1221 Concerto, piano, op. 16, A minor. Lyric pieces, op. 65, no. 6:
     Wedding day at Troldhaugen. Norwegian dances, op. 35. Grant
     Johannesen, pno; Utah Symphony Orchestra; Maurice Abravanel.
     Turnabout (Q) QTV S 34624.
          +HF 11-76 p118   +St 12-76 p141
          +NR 8-76 p5
     Concerto, piano, op. 16, A minor. cf Works, selections (Decca DPA
     567/8).

Concerto, piano, op. 16, A minor.   cf CLEMENTI: Sonatinas, op. 36,
   nos. 1-6.
Concerto, piano, op. 16, A minor.   cf FRANCK: Symphonic variations,
   violin and piano.
Concerto, piano, op. 16, A minor.   cf HMV SLS 5033.
Concerto, piano, op. 16, A minor.   cf HMV SLS 5068.
1222 Elegiac melodies, op. 34. Holberg suite, op. 40. NIELSEN: Little
   suite, op. 1, A minor. SIBELIUS: Rakastava, op. 14. LOL;
   Leslie Jones. Unicorn ZCUN 201.
        +Gr 7-72 p247   +St 3-76 p110
   Elegiac melodies, op. 34 (2).   cf Works, selections (Vox QSVBS
   5140).
   Elegiac melodies, op. 34 (2).   cf Swedish Society SLT 33229.
   Elegiac melodies, op. 34: Heart's wounds; The last spring. cf
   Works, selections (Decca SXL 6766).
   Elegiac melodies, op. 34: Heart's wounds; The last spring. cf
   CBS 30062.
   Funeral march. cf Argo SPA 464.
   Funeral march. cf Unicorn RHS 339.
   Gone, op. 71, no. 6. cf Lyric pieces (DG 2530 476).
   Grandmother's minuet, op. 68, no. 2. cf Lyric pieces (DG 2530 476).
1223 Holberg suite, op. 40. Norwegian melodies, op. 63: Cowkeeper's
   tune and country dance. Peer Gynt, op. 46: Morning mood;
   The death of Aase; Anitra's dance; In the hall of the mountain
   King; Solveig's song. Sigurd Jorsalfar, op. 56: Incidental
   music. LSO, Stuttgart Chamber Orchestra, London Proms Symphony
   Orchestra; Øivin Fjeldstad, Karl Münchinger, Charles Mackerras.
   Decca SPA 421.
        ++RR 12-76 p55
1224 Holberg suite, op. 40. TCHAIKOVSKY: Serenade, strings, op. 48,
   C major\ Netherlands Chamber Orchestra; David Zinman. Philips
   6580 102.
        +Gr 4-75 p1807   ++RR 4-75 p36
        +NR 12-75 p2   -St 3-76 p110
   Holberg suite, op. 40. cf Elegiac melodies, op. 34.
   Holberg suite, op. 40. cf Works, selections (Decca DPA 567/8).
   Holberg suite, op. 40. cf Works, selections (Decca SXL 6766).
   Holberg suite, op. 40. cf Works, selections (Vox QSVBS 5140).
   Holberg suite, op. 40. cf DVORAK: Serenade, strings, op. 22, E
   major.
   Holberg suite, op. 40. cf Coronet LPS 3031.
   Home-sickness, op. 57, no. 6. cf Lyric pieces (DG 2530 476).
   Homeward, op. 62, no. 6. cf Lyric pieces (DG 2530 476).
1225 Humoresques, op. 6, nos. 1-4. Lyric pieces, op. 43; op. 54, nos.
   3 and 4; op. 65; op. 71, nos. 2 and 3. John McCabe, pno.
   Oryx ORPS 97.
        +-Gr 8-76 p319
   In autumn, op. 11. cf Works, selections (Vox QSVBS 5140).
1226 Landkjenning, op. 31. LARSSON: The disguised God, op. 24.
   Elisabeth Söderström, s; Erik Saedén, bar; Martin Lidstam
   Vocal Ensemble; Stockholm Radio Orchestra, Stockholms Students-
   ängarförbund, Royal Swedish Orchestra; Stig Westerberg, Einar
   Ralf. Swedish Society SLT 33146.
        +-RR 4-76 p72
   Lonely wanderer, op. 43, no. 2. cf Lyric pieces (DG 2530 476).

1227 Lyric pieces: Album leaf, op. 47, no. 2. Arietta, op. 12, no. 1.
       At the cradle, op. 68, no. 5. At your feet, op. 68, no. 3.
       Ballade, op. 65, no. 5. Berceuse, op. 38, no. 1. Brooklet,
       op. 62, no. 4. Butterfly, op. 43, no. 1. Gone, op. 71, no. 6.
       Grandmother's minuet, op. 68, no. 2. Home-sickness, op. 57,
       no. 6. Homeward, op. 62, no. 6. Lonely wanderer, op. 43,
       no. 2. Melody, op. 47, no. 3. Nocturne, op. 54, no. 4.
       Norwegian dance, op. 47, no. 4. Once upon a time, op. 71,
       no. 1. Puck, op. 71, no. 3. Remembrances, op. 71, no. 7.
       Scherzo, op. 54, no. 5. Emil Gilels, pno. DG 2530 476. Tape
       (c) 3300 499.

++Audio 3-76 p63	++NR 9-75 p12
+Gr 3-75 p1681	+RR 3-75 p47
++Gr 6-76 p102 tape	++SFC 7-27-75 p22
++HF 9-75 p85	++St 12-75 p126
+HF 8-76 p70	

Lyric pieces, op. 43. cf Humoresques, op. 6, nos. 1-4.
Lyric pieces, op. 43, no. 6: To the spring. cf Connoisseur
       Society (Q) CSQ 2065.
Lyric pieces, op. 47, no. 3: Melody. cf Lyric pieces (DG 2530 476).
Lyric pieces, op. 54, nos. 3 and 4. cf Humoresques, op. 6, nos.
       1-4.
Lyric pieces, op. 54, no. 4: Nocturne. cf Lyric pieces (DG 2530
       476).
Lyric pieces, op. 54, no. 4: Nocturne. cf Saga 5427.
Lyric pieces, op. 54, no. 6: Scherzo. cf RCA ARM 4-0942/7.
Lyric pieces, op. 54: Suite. cf Works, selections (Decca DPA
       567/8).
Lyric pieces, op. 54: Suite. cf Works, selections (Vox QSVBS
       5140).
Lyric pieces, op. 65. cf Humoresques, op. 6, nos. 1-4.
Lyric pieces, op. 65, no. 6: Wedding day at Troldhaugen. cf
       Concerto, piano, op. 16, A minor.
Lyric pieces, op. 65, no. 6: Wedding day at Troldhaugen. cf
       HMV HQS 1360.
Lyric pieces, op. 65, no. 6: Wedding day at Troldhaugen. cf
       HMV SXLP 30181.
Lyric pieces, op. 65, no. 6: Wedding day at Troldhaugen. cf
       Saga 5427.
Lyric pieces, op. 68, no. 4: Evening in the mountains; no. 5:
       At the cradle. cf Works, selections (Vox QSVBS 5140).
Lyric pieces, op. 71, no. 1: Once upon a time. cf Lyric pieces
       (DG 2530 476).
Lyric pieces, op. 71, nos. 2 and 3. cf Humoresques, op. 6, nos.
       1-4.
Lyric pieces, op. 71, no. 3: Puck. cf Lyric pieces (DG 2530 476).
Lyric pieces, op. 71, no. 3: Puck. cf RCA ARM 4-0942/7.
Lyric pieces, op. 71, no. 7: Remembrances. cf Lyric pieces
       (DG 2530 476).
1228 Norwegian dances, op. 35 (4) (orch. Sitt). Peer Gynt, op. 46:
       Suite no. 1; op. 55: Suite no. 2. ECO; Raymond Leppard.
       Philips 9500 106. Tape (c) 7300 513.

+Gr 11-76 p791	++RR 11-76 p66
++HFN 11-76 p160	++SFC 12-5-76 p58

Norwegian dances, op. 35. cf Concerto, piano, op. 16, A minor.
Norwegian dance, op. 47, no. 4. cf Lyric pieces (DG 2530 476).

Norwegian melodies, op. 63: Cowkeeper's tune and country dance.
  cf Holberg suite, op. 40.
Norwegian melodies, op. 63: In popular folk style; Cowkeeper's
  tune and country dance. cf HMV SQ ESD 7001.
Norwegian melodies, op. 63: In the style of a folksong; Cowkeeper's
  tune and country dance. cf Works, selections (Decca SXL 6766).
Og jeg vil ha'. cf Cantilena 6238.
Old Norwegian melody with variations, op. 51. cf Works, selections (Vox QSVBS 5140).
Peer Gynt, op. 46: Dawn. cf CBS 30072
Peer Gynt, op. 46: Morning mood; The death of Aase; Anitra's
  dance; In the hall of the mountain king; Solveig's song.
  cf Holberg suite, op. 40.
1229 Peer Gynt, op. 46: Suite, no. 1. RAVEL: Daphnis et Chloé: Suite
  no. 2. STRAUSS, R.: Don Juan, op. 20. COA; Willem Mengelberg.
  Rococo 2066.
       +-NR 4-76 p2
Peer Gynt, op. 46: Suite, no. 1. cf Works, selections (Vox QSVBS
  5140).
Peer Gynt, op. 46: Suite, no. 1. cf BIZET: L'Arlésienne: Suite,
  no. 1; Suite, no. 2: Farandole.
1230 Peer Gynt, op. 46: Suite, no. 1; op. 55: Suite, no. 2. Songs:
  En svane, op. 25, no. 2; Fra Monte Pincio, op. 39, no. 1; Ich
  liebe dich, op. 5, no. 3; Lauf der Welt, op. 48, no. 3;
  Prinsessen. Elisabeth Söderström, s; NPhO; Andrew Davis.
  CBS 76527.
       +RR 12-76 p16
1231 Peer Gynt, op. 46: Suite, no. 1; op. 55: Suite, no. 2. Sigurd
  Jorsalfar, op. 56. Bamberg Symphony Orchestra; Richard Kraus.
  DG Tape (c) 3318 041.
       -HFN 2-76 p117
1232 Peer Gynt, op. 46: Suite, no. 1; op. 55: Suite, no. 2. Symphonic
  dances, op. 64. PhO; Walter Susskind. HMV SXLP 30105. Tape
  (c) TC EXE 190.
       +HFN 5-76 p117 tape
Peer Gynt, op. 46: Suite, no. 1; op. 55: Suite, no. 2. cf
  Norwegian dances, op. 35.
Peer Gynt, op. 46: Suite, no. 1; op. 55: Suite, no. 2. cf Works,
  selections (Decca DPA 567/8).
Peer Gynt, op. 46: Suite, no. 1; op. 55: Suite, no. 2: The death
  of Aase. cf Works, selections (Decca SXL 6766).
1233 Peer Gynt, op. 46: Suite, no. 1; op. 55: Suite, no. 2: Solveig's
  song. TCHAIKOVSKY: The nutcracker, op. 71. BPO; Arthur
  Fiedler. Decca PFS 4352. Tape (c) KPFC 4352. (also London
  SPC 21142 Tape (c) 5-21142).
       +-Gr 5-76 p1807   ++RR 10-76 p107 tape
       +-HFN 5-76 p117   ++SFC 1-4-76 p27
       /NR 2-76 p4
Peer Gynt, op. 46: Der Wingen mag scheiden. cf Court Opera
  Classics CO 342.
Peer Gynt, op. 55: Suite, no. 2. cf Works, selections (Vox
  QSVBS 5140).
Peer Gynt, op. 55: Suite, no. 2. cf BIZET: L'Arlésienne: Suite,
  no. 2.
Peer Gynt, op. 55: Suite, no. 2: Solveig's song. cf BIZET:
  L'Arlésienne: Suite, no. 1; Suite, no. 2: Farandole.

Scenes from peasant life, op. 19, no. 2: Bridal procession
    passing by. cf Works, selections (Vox QSVBS 5140).
Scherzo, op. 54, no. 5. cf Lyric pieces (DG 2530 476).
Sigurd Jorsalfar, op. 56. cf Peer Gynt, op. 46: Suite, no. 1;
    op. 55: Suite, no. 2.
Sigurd Jorsalfar, op. 56. cf Works, selections (Decca DPA 567/8).
Sigurd Jorsalfar, op. 56. cf Works, selections (Vox QSVBS 5140).
Sigurd Jorsalfar, op. 56: Hommage march. cf Works, selections
    (Decca SXL 6766).
Sigurd Jorsalfar, op. 56: Incidental music. cf Holberg suite,
    op. 40.
1234 Sonatas, violin and piano, nos. 1-3. Henri Temianka, vln;
      James Fields, pno. Orion ORS 75193.
           +-HF 5-76 p84   ++NR 4-76 p8
           +IN 4-76 p16   +St 7-76 p109
      Sonata, violin and piano, no. 2, op. 13, G minor. cf RCA ARM
      4-0942/7.
      Sonata, violin and piano, no. 3, op. 45, C minor. cf RCA ARM
      4-0942/7.
      Sonata, violoncello and piano, op. 36, A minor. cf BRAHMS:
      Sonata, violoncello and piano, no. 1, op. 38, E minor.
      Songs: Ave maris stella. cf Swedish Society SLT 33197.
      Songs: A dream, op. 48, no. 6. cf Polydor 2383 389.
      Songs: The first meeting; The goal; A hope; I love thee. cf
      Rococo 5380.
      Songs: I laid me down to slumber, op. 30, no. 1; Kvaalin's
      Halling, op. 30, no. 4; Little Thora, op. 30, no. 3; When I
      take a stroll, op. 30, no. 6. cf HMV CSD 3766.
      Songs: En svane. cf Rubini GV 65.
      Songs: En svane, op. 25, no. 2; Fra Monte Pincio, op. 39, no. 1;
      Ich liebe dich, op. 5, no. 2; Lauf der Welt, op. 48, no. 3;
      Prinsessen. cf Peer Gynt, op. 46: Suite no. 1; op. 55: Suite,
      no. 2.
      Symphonic dances, op. 64. cf Peer Gynt, op. 46: Suite, no. 1;
      op. 55: Suite, no. 2.
      Symphonic dances, op. 64. cf Works, selections (Vox SQVBS 5140).
1235 Works, selections: Concerto, piano, op. 16, A minor. Holberg
      suite, op. 40. Lyric pieces, op. 54: Suite. Peer Gynt, op.
      46: Suite, no. 1; op. 55: Suite, no. 2. Sigurd Jorsalfar,
      op. 56. Julius Katchen, pno; LSO, Israel Philharmonic
      Orchestra, Stuttgart Chamber Orchestra; Øivin Fjeldstad,
      István Kertész, Karl Münchinger, Stanley Black. Decca DPA
      567/8. Tape (c) KDPC 567/8. (Reissue).
           +HFN 11-76 p173   +-RR 11-76 p66
           +HFN 12-76 p155 tape
1236 Works, selections: Elegiac melodies, op. 34: Heart's wounds; The
      last spring. Holberg suite, op. 40. Norwegian melodies,
      op. 63: In the style of a folksong; Cowkeeper's tune and
      country dance. Peer Gynt, op. 46: Suite, no. 1; op. 55: Suite,
      no. 2: The death of Aase. Sigurd Jorsalfar, op. 56: Homage
      march. National Philharmonic Orchestra; Willi Boskovsky.
      Decca SXL 6766. Tape (c) KSXC 6766.
           +-Gr 4-76 p1598   +-RR 3-76 p40
           +HFN 3-76 p93
1237 Works, selections: Scenes from peasant life, op. 19, no. 2:
      Bridal procession passing by. Elegiac melodies, op. 34 (2).

Holberg suite, op. 40. In autumn (concert overture), op. 11.
Lyric pieces, op. 54: Suite. Lyric pieces, op. 68, no. 4:
Evening in the mountains; no. 5: At the cradle. Old Norwegian
melody with variations, op. 51. Peer Gynt, op. 46: Suite no. 1.
Peer Gynt, op. 55: Suite, no. 2. Sigurd Jorsalfar, op. 56.
Symphonic dances, op. 64. Utah Symphony Orchestra; Maurice
Abravanel. Vox QSVBS 5140 (3).
     +HF 11-76 p118   ++NR 9-76 p3
     +MJ 11-76 p45   ++SFC 8-15-76 p38

GRIFFES, Charles
1238 Bacchanale. Clouds. The pleasure dome of Kubla Khan. The white
     peacock. LOEFFLER: Memories of my childhood. Poem, orchestra.
     Eastman-Rochester Symphony Orchestra; Howard Hanson. Mercury
     SRI 75090.
        *NYT 7-4-76 pD1
    Clouds. cf Bacchanale.
    Indian sketches (2). cf Vox SVBX 5301.
    The lament of Ian the Proud. cf Cambridge CRS 2715.
1239 The pleasure dome of Kubla Khan. The white peacock. HANSON:
     Symphony, no. 2, op. 30. National Philharmonic Orchestra;
     Charles Gerhardt. RCA GL 25021.
        ++Gr 10-76 p596   +-RR 10-76 p51
        +HFN 12-76 p143
    The pleasure dome of Kubla Khan. cf Bacchanale.
    The pleasure dome of Kubla Khan. cf Columbia MG 33728.
    Sonata, piano, F major. cf DETT: In the bottoms.
    Songs: Early morning in London; The lament of Ian the Proud.
     cf Washington University Press OLY 104.
    The white peacock. cf Bacchanale.
    The white peacock. cf The pleasure dome of Kubla Khan.
    The white peacock. cf Columbia MG 33728.
    The white peacock. cf Personal Touch 88.
GRIFFITH, Peter
    Quartet, one string. cf EDWARDS: Quartet, strings.
GRIGNY, Nicolas de
    Veni creator. cf BACH: Chorale prelude: O Gott, du frommer Gott,
     S 767.
GRIMACE, Magister
    A l'arme, a l'arme. cf HMV SLS 863.
    A l'arme, a l'arme. cf 1750 Arch S 1753.
GRODSKI
    The night hovered over the earth. cf Rubini GV 63.
    The seagull's cry. cf Rubini RS 301.
GROFE, Ferde
    Grand Canyon suite. cf GERSHWIN: An American in Paris.
    Trick or treat. cf Columbia MG 33728.
GRUBER, Franz
    Silent night. cf CBS 61771.
    Stille Nacht. cf RCA PRL 1-8020.
GRUNDEMAN, Clare
    The spirit of '76. cf DELLA JOIO: Satiric dances.
GRUNFELD
    Schonen von Fogaras: Ganselied. cf Rococo 5379.
GUAMI, Gioseffo
    La brillantina. cf L'Oiseau-Lyre 12BB 203/6.
GUARNIERI, Camargo
    Dansa brasileira. cf Angel S 37110.

GUARNIERI (cont.)   222

        Dansa negra. cf Angel S 37110.
        Ponteios, nos. 24, 30. cf Angel S 37110.
GUASTAVINO, Carlos
        La rose y el sauce. cf Polydor 2383 389.
GUILMANT, Alexandre
        Concert piece, trombone. cf HMV SXLP 50017.
        Morceau symphonique, op. 88. cf Boston Brass BB 1001.
1240   Sonata, organ, no. 1, op. 42, D minor. LANGLAIS: Pasticcio.
            PEETERS: Suite modale, op. 43. Richard Galloway, org. Vista
            VPS 1024.
                +-Gr 3-76 p1488   ++RR 3-76 p61
                +-HFN 1-76 p115
        Sonata, organ, no. 1, op. 42, D minor: Finale. cf Wealden WSQ 134.
GUIMARAES
        Sounds of bells. cf London CS 7015.
        GUITAR AND LUTE CONCERTOS. cf RCA ARL 3-0997.
GULIELMUS (15th century)
        Falla con misuras. cf Nonesuch H 71326.
GURIDI, Jesus
        Songs: Llámle con el pañuelo; No quiero tus avellanas; Cómo
            quieres que advine. cf DG 2530 598.
GURLITT
        Toy symphony, op. 169, C major. cf Angel S 36080.
GURNEY, Ivor
        Songs: An epitaph; Black Stichel; Desire in spring; Down by the
            salley gardens. cf BUTTERWORTH: A Shropshire lad.
        Songs: I will go with my father a'ploughing; Sleep. cf Saga 5213.
HADLEY, Henry
        Quintet, piano, op. 50, A minor. cf Vox SVBX 5301.
HADLEY, Patrick
        The hills. cf DELIUS: To be sung on a summer night on the water.
        I sing of a maiden. cf HMV CSD 3774.
HAGEMAN, Richard
        Do not go, my love. cf Cambridge CRS 2715.
HAHN, Reynaldo
        Ciboulette: Moi je m'appelle, Y'a des arbres...C'est sa banlieue.
            cf London OS 26248.
        Concerto, piano, no. 1, E minor. cf FRANCAIX: Concerto, piano,
            D major.
        Songs: D'une prison. cf Discophilia KIS KGG 3.
        Songs: D'une prison; L'heure exquise; Mai; Le rossignol des lilas;
            Offrande; Paysage; Si mes vers avaient des ailes. cf FAURE:
            Songs (Saga 5419).
        Songs: En sourdine; L'heure exquise; L'Offrande; Si mes vers
            avaient des ailes; Mozart: Etre adoré; L'adieu. cf HMV RLS 716.
        Songs: L'heure exquise. cf HMV ASD 2929.
        Songs: If my songs were only winged. cf Polydor 2383 389.
        Songs: Je me mets en votre mercy; Offrande. cf Club 99-101.
        Songs: Si mes vers avaient des ailes. cf Decca SKL 5208.
        Songs: Si mes vers avaient des ailes. cf Desto DC 7118/9.
        Songs: Si mes vers avaient des ailes. cf HMV RLS 719.
HAINES
        Sonata, harp. cf Orion ORS 75207.
HAKIM, Talib
        Placements. cf Folkways FTS 33901/4.
        Visions of Ishwara. cf ANDERSON: Squares.

HALEVY, Jacques
    La Juive: Rachel, quand du Seigneur. cf RCA CRM 1-1749.
    La Juive: Wenn ew'ger Hass gluhende Rache. cf Preiser LV 192.
    La Juive: Wenn ew'ger Hass. cf Rococo 5397.
HALFFTER, Ernesto
1241 Concerto, guitar. BACARISSE: Concertino, guitar and orchestra,
        op. 72, A minor. Narciso Yepes, gtr; Spanish Radio Television
        Symphony Orchestra; Odón Alonso. DG 2530 326.
            +–Audio 2-76 p95   +–RR 7-73 p42
            +–Gr 6-73 p187   +SR 6-14-75 p46
            -NR 6-75 p8
    Danza de la gitana (Escriche). cf RCA ARM 4-0942/7.
    Songs: Al que linda moca. cf Polydor 2383 389.
HALL
    Officer of the day. cf Mercury SRI 75055
HANBY
    Nelly Gray. cf Vox SVBX 5304.
HABA, Alois
    Suite, bass clarinet, op. 96. cf Supraphon 111 1390.
HANDEL, Georg Friedrich
    Acis and Galatea: I rage, I melt, I burn; O ruddier than the
        cherry. cf Works, selections (Decca SPA 448).
    Acis and Galatea: Love sounds the alarm; Where shall I seek. cf
        Arias (Westminster WGS 8169).
    Acis and Galatea: O ruddier than the cherry. cf HMV SLS 5022 (2).
    Agrippina: Overture. cf Concerto a due cori, no. 1, B flat
        major.
    Alceste: Enjoy the sweet Elysian grove. cf Arias (Westminster
        WGS 8169).
    Alcina: Dream music. cf HMV RLS 717.
    Alcina: Here are the Heavens all joyful. cf Works, selections
        (CBS 61559).
    Alcina: Overture. cf Works, selections (Philips 6833 200).
    Alcina: Verdi prati. cf Decca SXL 6629.
1242 Alexander's feast. Honor Sheppard, s; Max Worthley, t; Maurice
        Bevan, bar; Oriana Concerto Orchestra and Choir; Alfred Deller.
        Vanguard Bach Guild HM 50/1 (2). (also Vanguard S 282/3).
        (Reissue from SRV 282/3).
            +ARG 11-76 p25   +–SFC 12-12-76 p55
    Alexander's feast: War is toil and trouble. cf Arias (West-
        minster WGS 8169).
    Arianna: Overture. cf Concerto a due cori, no. 1, B flat major.
1243 Arias: Acis and Galatea: Love sound the alarm; Where shall I seek.
        Alceste: Enjoy the sweet Elysian grove. Alexander's feast:
        War is toil and trouble. Atalanta: Say to Irene. Jeptha:
        Waft her, angels. Judas Maccabaeus: Sound an alarm. Radamisto:
        Sommi Dei. Samson: Total eclipse. Semele: Where'er you walk.
        Jan Peerce, t; VSOO; Hans Schweiger. ABC Westminster WGS 8169.
            +–Gr 12-76 p1046
    Atalanta: Say to Irene. cf Arias (Westminster WGS 8169).
    Athalia: The Gods, who chosen blessings shed. cf Works, selec-
        tions (CBS 61559).
    Aylesford pieces: Fugue, G major; Saraband; Impertinence. cf
        Columbia M 33514.
    Aylesford pieces: Minuets, A minor (2). cf Saga 5426.
    Belshazzar: Martial symphony. cf Works, selections (CBS 61559).
    Belshazzar: Martial symphony. cf CBS 61684.

1244 Berenice: Overture. Concerti, oboe, nos. 1-3. Concerto, oboe,
       no. 2, F major (variant). Solomon: Arrival of the Queen of
       Sheba. Roger Lord, ob; AMF; Neville Marriner. Argo ZK 2.
       (Reissue from ZRG 5442).
            +Gr 12-76 p95
      Berenice: Overture. cf BEETHOVEN: Symphony, no. 5, op. 67, C
       minor.
      Chaconne, G major. cf Harpsichord works (Erato STU 70906).
1245 Chandos anthems: I will magnify Thee; In the Lord I put my trust.
       Caroline Friend, s; Philip Langridge, t; Cambridge, King's
       College Chapel Choir; AMF; David Willcocks. Argo ZRG 766.
            ++AR 8-76 p77   ++NR 8-75 p9
            +Gr 3-75 p1688   +RR 3-75 p61
            -HF 1-76 p87   /SFC 10-26-75 p24
            +MT 5-76 p408   ++St 1-76 p104
1246 Chandos anthems: O praise the Lord with one consent; Let God
       arise. Elizabeth Vaughan, s; Alexander Young, t; Forbes
       Robinson, bs; King's College Chapel Choir; AMF; Andrew
       Davis, hpd; John Langdon, org; David Willcocks. Argo SZRG 5490.
            +AR 8-76 p77
1247 The choice of Hercules. Heather Harper, s; Helen Watts, ms; James
       Bowman c-t; Robert Tear, t; King's College Chapel Choir; AMF;
       Philip Ledger, cond; Francis Grier, hpd; James Lancelot, org.
       HMV (Q) ASD 3148.
            +-Gr 2-76 p1367   +-MT 7-76 p576
            +-HFN 2-76 p99   +RR 2-76 p57
1248 Concerto, 2 lutes, strings and recorders, op. 4, no. 6, B flat
       major. KOHAUT: Concerto, guitar (lute), F major. VIVALDI:
       Concerto, lute, D major. Julian Bream, lt; Robert Spencer,
       chitarrone; Nicholas Krämer, hpd; Marilyn Sansom, vlc; John
       Gray, violone; Monteverdi Orchestra; John Eliot Gardiner.
       RCA ARL 1-1180. Tape (c) ARK 1-1180 (ct) ARS 1-1180.
            +Gr 11-75 p839   +RR 12-75 p70
            ++HFN 12-75 p155   ++SFC 1-25-76 p30
            +MM 4-76 p28   ++St 3-76 p122
            ++NR 12-75 p7
      Concerti, oboe, nos. 1-3. cf Berenice: Overture.
      Concerto, oboe, G minor. cf BACH: Concerto, violin, oboe and
       strings, S 1060, D minor.
      Concerto, oboe, no. 3, G minor: Allegro. cf Works, selections
       (Philips 6833 200).
1249 Concerti, organ, opp. 4 and 7. Herbert Tachezi, org; VCM; Nikolaus
       Harnoncourt. Telefunken 6-35282 (3).
            +HF 3-76 p88   +PRO 7/8-76 p18
            ++NR 3-76 p4   ++SFC 1-25-76 p30
1250 Concerti, organ, op. 4, nos. 1-6; op. 7, nos. 13-18. Karl
       Richter, org; Chamber orchestra; Karl Richter. Decca SDD
       470/2 (3). (Reissues from SXL 2115, 2187, 2201).
            +Gr 11-75 p818   +RR 5-76 p40
            +HFN 10-75 p153
1251 Concerti, organ, op. 4, nos. 1-6; op. 7, nos. 13-18; nos. 19-20.
       Sonata, organ and strings (Il trionfo del tempo e del dis-
       inganno). George Malcolm, org and hpd; AMF; Neville Marriner.
            +Gr 9-76 p415   +RR 9-76 p53
            +-HFN 9-76 p121

1252 Concerti, organ, op. 4, nos. 1-6; op. 7, nos. 13-18; nos. 19-20.
      Daniel Chorzempa, org; Concerto Amsterdam; Jaap Schröder.
      Philips 6709 009 (5).
              +Gr 9-76 p415   +RR 10-76 p50
              +HFN 9-76 p121   +SFC 12-12-76 p55
1253 Concerti, organ, op. 4, nos. 1-6; op. 7, nos. 13-16, 18; nos.
      19-20. E. Power Biggs, org; LPO; Adrian Boult. Columbia D3M
      33716 (Reissues from K 25602, 604). (also CBS 77358. Reissues
      from Philips SABL 148-8, ABL 3326/7).
              +Audio 10-76 p146   +HFN 11-73 p2315
              +Gr 3-74 p1765   ++RR 11-73 p36
              +-HF 3-76 p88   +SFC 11-30-75 p34
      Concerto, organ, op. 4, no. 2, B flat major. cf Works, selections
      (Decca DPA 551/2).
      Concerto, organ, op. 4, no. 2, B flat major. cf Decca DPA 523/4.
      Concerto, organ, op. 4, no. 6, B flat major: Allegro moderato.
      cf Works, selections (Philips 6833 200).
      Concerto, organ, op. 4, no. 8, A major. cf Westminster WGS 8116.
      Concerto, organ, op. 7, no. 16, D minor. cf Decca DPA 523/4.
      Concerto, trumpet, B flat major. cf Philips 6500 926.
      Concerto, trumpet, G minor. cf Erato STU 70871.
      Concerto, trumpet, G minor. cf RCA CRL 3-1430.
1254 Concerto, viola, B minor (Casadesus). STAMITZ: Concerto, viola,
      D major. TELEMANN: Concerto, viola, G major. WEBER: Andante
      and rondo, op. 35, C minor. Pinchas Zukerman, vla; ECO;
      Pinchas Zukerman. Columbia M 33979. (also CBS 76490 Tape (c)
      40-76490).
              +Gr 5-76 p1771   +RR 6-76 p61
              +HFN 6-76 p101   +RR 12-76 p111 tape
              +-HFN 8-76 p95   +St 11-76 p159
              ++NR 7-76 p4
1255 Concerto, 2 wind choirs and strings, no. 1, B flat major. Royal
      fireworks music. Water music. La Grande Ecurie et La Chambre
      du Roy; Jean-Claude Malgoire. Columbia MG 32813. (also CBS   -J
      73172).
              +AR 5-75 p52   +-PRO 1/2-76 p20
              +-Gr 8-74 p358   -RR 8-74 p35
              +HF 11-74 p102   +-St 10-74 p129
              ++NR 11-74 p4
1256 Concerto a due cori, no. 1, B flat major. Concerto a due cori,
      no. 3, F major. Agrippina: Overture. Arianna: Overture. AMF;
      Neville Marriner. HMV SQ ASD 3182. (also Angel S 37176).
              +-Gr 5-76 p1757   ++RR 4-76 p41
              ++HFN 4-76 p107   +SFC 12-12-76 p55
1257 Concerto, 2 wind choirs and strings, no. 2, F major. Royal
      fireworks music. LOL; Leslie Jones. Oryx BRL 16.
              +-Gr 10-76 p596   +-HFN 3-74 p105
      Concerto a due cori, no. 3, F major. cf Concerto a due cori,
      no. 1, B flat major.
      Concerto grosso, C major: 1st movement. cf Coronet LPS 3031.
1258 Concerti grossi (2). Royal fireworks music. Orchestra; Arthur
      Davison. Classics for Pleasure (c) TC CFP 105.
              +-HFN 12-76 p153 tape
1259 Concerti grossi, op. 3. La Grande Ecurie et la Chambre du Roy;
      Jean-Claude Malgoire. CBS 76474. Tape (c) 40-76474.
              +-Gr 8-76 p285   -RR 5-76 p40

```
            +-HFN 6-76 p85              -RR 8-76 p84 tape
            +-HFN 8-76 p94
```

1260 Concerti grossi, op. 6 (12). South West German Chamber
     Orchestra; Paul Angerer. Vox QSVBX 558 (3).
```
            ++HF 10-76 p109             +NR 10-75 p1
```
     Concerto grosso, op. 29, F major. cf CORELLI: Sonata, violin and
       continuo, op. 5, no. 5, G minor.
1261 Delirio amoroso. Nel dolce del'oblio (Pensieri notturni di Filli).
     Magda Kalmár, s; Ferenc Liszt Academy Chamber Orchestra;
     Frigyes Sándor. Hungaroton SLPX 11653.
```
            /HF 2-76 p97               +RR 4-75 p59
            ++HFN 7-75 p81             ++SFC 6-1-75 p21
            ++NR 6-75 p12             ++St 8-75 p96
```
     Esther: Arias. cf Serenus SRS 12039.
     Ezio: March. cf Works, selections (CBS 61559).
     The faithful shepherd: Minuet. cf Works, selections (Decca DPA
       551/2).
     Fantasia, C major. cf Harpsichord works (Erato STU 70906).
     Floridante: Marches, nos. 1 and 2. cf Works, selections (CBS
       61559).
1262 Funeral anthem for Queen Caroline (Ode on the death of Queen
     Caroline). Soloists Dresden Cathedral Orchestra and Choir;
     Kurt Bauer. Everest 3227.
```
            -Gr 8-76 p327
```
     Giulio Cesare: Piangero la sorte mia. cf BERLIOZ: La mort
       d'Ophelie, op. 18, no. 2.
1263 Harpsichord works: Chaconne, G major. Fantasia, C major. Partita,
     A major. Prelude and lesson, A minor. Suite, harpsichord,
     D minor. Luciano Sgrizzi, hpd. Erato STU 70906.
```
            +-Gr 12-75 p101             +-RR 1-76 p48
            +HFN 1-76 p108
```
     Hornpipe, D major. cf Works, selections (Philips 6833 200).
1264 Israel in Egypt. Elizabeth Gale, Lilian Watson, s; James Bowman,
     alto; Ian Partridge, t; Tom McDonnell, Alan Watt, bs; Christ
     Church Cathedral Choir; ECO; Simon Preston. Argo ZRG 817/8 (2).
```
            +-ARG 11-76 p24            +RR 4-76 p70
            +-Gr 4-76 p1646           +-St 12-76 p144
            +-HFN 4-76 p104
```
     Israel in Egypt: Arias. cf Serenus SRS 12039
     Jeptha: Waft her, angels. cf Arias (Westminster WGS 8169).
     Joshua: O had I Jubel's lyre. cf Works, selections (Philips
       6833 200).
     Joshua: See the raging flames arise. cf RCA LRL 2-7531.
     Joy to the world. cf CBS 61771.
     Joy to the world. cf RCA PRL 1-8020.
     Judas Maccabaeus: Arias. cf Serenus SRS 12039.
     Judas Maccabaeus: See the conquering hero comes. cf Works,
       selections (Decca SPA 448).
     Judas Maccabaeus: See, the conquering hero comes; March no. 2.
       cf Works, selections (CBS 61559).
     Judas Maccabaeus: Sound an alarm. cf Arias (Westminster WGS 8169).
     Largo. cf HMV SXLP 30188.
1265 Messiah. Elly Ameling, s; Anna Reynolds, con; Philip Langridge,
     t; Gwynne Howell, bs; AMF; Neville Marriner. Argo D183D 3 (3).
     Tape (c) K 18 K32.
```
            +Gr 11-76 p852             +-RR 11-76 p95
            ++HFN 11-76 p160
```

1266 Messiah. Helen Donath, s; Anna Reynolds, ms; Stuart Burrows, t;
     Donald McIntyre, bar; Hedwig Bilgram, hpd; Edgar Krapp, org;
     John Alldis Choir; LPO; Karl Richter. DG 2709 045 (3). Tape
     (c) 3371 009. (also DG 2720 069 Tape (c) 3300 365/7).
          -ARG 12-76 p6   ++NR 3-74 p7
          +Gr 11-73 p965   +RR 10-73 p112
          +Gr 9-74 p592 tape   ++RR 5-74 p85 tape
          ++HF 12-73 p96   +-SFC 12-16-73 p40
          ++MJ 3-74 p49   +St 12-73 p127
1267 Messiah (arr. Mozart). Edith Mathis, s; Birgit Finnila, con;
     Peter Schreier, t; Theo Adam, bs; Austrian Radio Symphony
     Orchestra and Chorus; Charles Mackerras. DG Archive
     2723 019 (3). (also DG Archive 2710 016).
          +-ARG 12-76 p6   +NR 1-75 p8
          ++Gr 11-74 p941   ++RR 11-74 p88
          +-HF 3-75 p78   ++St 3-75 p101
1268 Messiah. Felicity Palmer, s; Helen Watts, con; Ryland Davies, t;
     John Shirley-Quirk, bs; Philip Jones, tpt; Leslie Pearson, org;
     ECO and Chorus; Raymond Leppard. Erato STLU 70921/3 (3).
     (also RCA CRL3-1426, also Musical Heritage Society MHS 3273/5).
          +ARG 12-76 p6   ++MT 7-76 p575
          +Gr 2-76 p1364   ++MU 10-76 p16
          +HF 9-76 p91   +NR 6-76 p12
          +HFN 2-76 p99   ++RR 2-76 p57
          +MJ 10-76 p24   ++SFC 4-18-76 p23
1269 Messiah. Elizabeth Harwood, s; Janet Baker, ms; Paul Esswood,
     c-t; Robert Tear, t; Raimund Herincx, bar; Ambrosian Singers;
     ECO; Charles Mackerras. HMV SLS 774. Tape (c) TC SLS 774.
     (also Angel S 3705).
          +-ARG 12-76 p6   +HFN 1-76 p125 tape
          +Gr 12-76 p1066 tape   ++RR 1-76 p66 tape
1270 Messiah. James Bowman, c-t; Robert Tear, t; Benjamin Luxon, bar;
     Boys of King's College Choir, soprano solos; King's College
     Choir, Cambridge; AMF; David Willcocks. HMV SLS 845 (3).
          +ARG 12-76 p6   ++RR 8-73 p70
          +Gr 6-73 p82
1271 Messiah. Joan Sutherland, s; Huguette Tourangeau, ms; Tom
     Krause, bs-bar; Werner Krenn, t; Ambrosian Singers; ECO;
     Richard Bonynge. London OSA 1396 (3).
          +ARG 12-70 p205   +NR 5-71 p9
          -ARG 12-76 p6   +ON 4-17-71 p34
          -HF 3-71 p84   -St 4-71 p75
1272 Messiah. Heather Harper, s; Helen Watts, con; John Wakefield, t;
     John Shirley-Quirk, bs; LSO and Chorus; Colin Davis. Philips
     6703 001. Tape (c) 769 9009. (also SC71 AX 300).
          +ARG 12-76 p6   +RR 11-76 p108 tape
          +Gr 12-76 p1066 tape
1273 Messiah. Margaret Price, s; Yvonne Minton, con; Alexander Young,
     t; Justino Diaz, bs; Amor Artis Chorale; ECO; Johannes Somary.
     Vanguard C 10090/2. Tape (c) ZCVS 10092.
          +ARG 12-76 p6   +ON 4-17-71 p34
          +Gr 11-71 p870   +RR 3-75 p74 tape
          +-HF 3-71 p58   ++SFC 3-28-71 p20
          +NR 5-71 p9
1274 Messiah. Pierette Alarie, s; Nan Merriman, con; Leopold Simoneau,
     t; Richard Standen, bs; Vienna State Opera Orchestra and

Chorus; Hermann Scherchen. Westminster 8163.
       -ARG 12-76 p6
1275 Messiah, excerpts. Jennifer Vyvyan, s; Monica Sinclair, con;
      Jon Vickers, t; Giorgio Tozzi, bs; RPO and Chorus; Thomas
      Beecham. RCA CRL 2-0192 (2).
         +-ARG 12-76 p6
1276 Messiah: Amen; And the glory of the Lord; Behold the Lamb of
      God; Come unto Him, Comfort ye; Ev'ry valley; For unto us a
      child is born; Hallelujah; He shall feed his flock; He was
      despised; I know that my Redeemer liveth; Then shall the
      eyes; Worthy is the lamb. Elsie Morison, s; Marjorie Thomas,
      con; Richard Lewis, t; Huddersfield Choral Society; Royal
      Liverpool Philharmonic Orchestra; Malcolm Sargent. Classics
      for Pleasure CFP 40020. Tape (c) TC CFP 40020. (Reissue,
      1959 SAX 2365).
         +HFN 2-73 p347   +RR 2-73 p82
         +HFN 12-76 p153 tape
1277 Messiah: And the glory of the Lord; And He shall purify; For unto
      us a child is born; Glory to God; His yoke is easy; Behold the
      lamb of God; Surely, he hath borne our griefs; And with his
      stripes; All we like sheep; He trusted in God; Lift up your
      heads; Let all the angels of God; The Lord gave the word; Let
      us break their bonds; Hallelujah; Since my man came death;
      Worthy is the lamb; Amen. King's College Choir; AMF; David
      Willcocks. HMV CSD 3778. (Reissue from SLS 845).
         +-Gr 12-76 p1028   +RR 12-76 p92
         +-HFN 12-76 p151
1278 Messiah: O thou that tellest good tidings; For unto us a child
      is born; Pifa; There were shepherds; Glory to God; He shall
      feed His flock; The Lord gave the word; Why do the nations so
      furiously rage; Hallelujah; I know that my Redeemer liveth;
      Worthy is the lamb; Amen. Helen Donath, s; Anna Reynolds, con;
      Stuart Burrows, t; Donald McIntyre, bs; John Alldis Choir;
      LPO; Karl Richter. DG 2530 643. (Reissue from 2720 069).
         +Gr 12-76 p1033   ++RR 12-76 p92
         +HFN 12-76 p151
      Messiah: Every valley shall be exalted: O thou that tellest;
         For unto us a child is born; I know that my redeemer liveth;
         The trumpet shall sound; Hallelujah. cf Works, selections
         (Decca DPA 551/2).
      Messiah: Hallelujah chorus. cf Abbey LPB 754.
      Messiah: Hallelujah chorus. cf CBS SQ 79200.
      Messiah: Hallelujah chorus. cf Vista VPA 1023.
      Messiah: He was despised; Hallelujah. cf Works, selections
         (Decca SPA 448).
      Messiah: How beautiful are the feet; I know that my redeemer
         liveth. cf Abbey LPB 761.
      Messiah: I know that my redeemer liveth; Hallelujah. cf Works,
         selections (Philips 6833 200).
      Minuet, G minor. cf BACH: Works, selections (DG 2530 647).
      Nel dolce del'oblio (Pensieri notturno di Filli). cf Delirio
         amoroso.
      Ode for Saint Cecilia's day: March. cf Works, selections
         (CBS 61559).
      Ode for Saint Cecilia's day: A trumpet voluntary. cf CBS 61648.
      Or che il cielo, K 374. cf BERLIOZ: La mort d'Ophelie, op. 18, no.

Overture, D minor (arr. Elgar).   cf ELGAR: Cockaigne overture,
   op. 40.
1279 Overtures: Admeto; Alcina; Esther; Lotario; Orlando; Ottone;
   Partenope; Poro. Leslie Pearson, hpd; ECO; Raymond Leppard.
   Philips 6500 053. Tape (c) 7317113. (also 6599 053).
      +AR 2-74 p24   +HFN 2-76 p116
      +-Gr 11-72 p909   ++RR 11-72 p49
Il Parnasso in Festa: Largo. cf Works, selections (CBS 61559).
Partita, A major. cf Harpsichord works (Erato STU 70906).
Passacaglia. cf DVORAK: Trio, piano, op. 65, F minor.
Il pastor Fido: Allegro, March. cf Works, selections (CBS 61559).
Il pensieroso: Sweet bird. cf HMV RLS 719.
Prelude and lesson, A minor. cf Harpsichord works (Erato STU
   70906).
Radamisto: Sommi Dei. cf Arias (Westminster WGS 8169).
Rinaldo: March. cf Works, selections (CBS 61559).
Rodelinda: Dove sei. cf Works, selections (Decca SPA 448).
Rodelinda: Dove sei. cf Works, selections (Decca DPA 551/2).
Rodelinda: Mio caro bene. cf BERLIOZ: La mort d'Ophelie, op. 18,
   no. 2.
Rodrigo: Passacaglia. cf Works, selections (CBS 61559).
1280 Royal fireworks music. Water music: Suite. RCA Symphony Orchestra;
   Leopold Stokowski. Camden CCV 5002.
      +ST 2-76 p743
1281 Royal fireworks music. Water music: Suite. Schola Cantorum;
   Archive Wind Ensemble; August Wenzinger. DG 2548 169. Tape (c)
   3348 169. (Reissue from Archive 198365, 198146).
      ++Gr 11-75 p840   +-RR 10-75 p43
      +HFN 10-75 p153   +-RR 9-76 p92 tape
1282 Royal fireworks music. Water music: Suite. Minnesota Orchestra;
   Stanislaw Skrowaczewski. Turnabout QTV 34632. Vox Tape (c)
   CT 2108.
      +NR 10-76 p3
1283 Royal fireworks music. MUSSORGSKY (Ravel): Pictures at an exhibi-
   tion. NPhO; Charles Mackerras. Vanguard (Q) (r) VSS 24/5.
      +HF 3-76 p116 Quad tape
Royal fireworks music. cf Concerti grossi (2).
Royal fireworks music. cf Concerto, 2 wind choirs and strings,
   no. 1, B flat major.
Royal fireworks music. cf Concerto, 2 wind choirs and strings,
   no. 2, F major.
Royal fireworks music. cf Works, selections (Decca SPA 448).
Royal fireworks music: Bourrée. cf Saga 5225.
Royal fireworks music: La paix; La rejouissance. cf Works,
   selections (Philips 6833 200).
Royal fireworks music: Suite. cf Works, selections (Decca DPA
   551/2).
Samson: Awake the trumpet's lofty sound. cf CBS 61648.
Samson: Awake the trumpet's lofty sound. cf Columbia M 33514.
Samson: Let the bright seraphim. cf Works, selections (Decca SPA
   448).
Samson: Total eclipse. cf Arias (Westminster WGS 8169).
Sarabande, D major. cf Philips 6581 017.
Saul: Dead march. cf Works, selections (CBS 61559).
Saul: Dead march. cf CBS 61684.
Scipione: March. cf Works, selections (CBS 61559).

Semele: Where'er you walk. cf Arias (Westminster WGS 8169).
Serse (Xerxes): Largo. cf Works, selections (Decca DPA 551/2).
Serse: Ombra mai fù. cf Works, selections (Decca SPA 448).
Serse: Ombra mai fù. cf Works, selections (Philips 6833 200).
Xerxes: Ombra mai fù. cf Decca SXL 6629.
Serse: Ombra mai fù. cf RCA CRM 1-1749.
Solomon: Arrival of the Queen of Sheba. cf Berenice: Overture.
Solomon: Arrival of the Queen of Sheba. cf Works,, selections
    (Decca SPA 448).
Solomon: Arrival of the Queen of Sheba. cf Works, selections
    (Decca DPA 551/2).
Solomon: Arrival of the Queen of Sheba. cf Works, selections
    (Philips 6833 200).
Solomon: Arrival of the Queen of Sheba. cf RCA LRL 1-5131.
Solomon: Entrance of the Queen of Sheba. cf Vanguard SRV 344.
Sonata, flute, op. 1, no. 5, G major. cf Decca SPA 394.
Sonata, flute, op. 1, no. 11, F major. cf Cambridge CRS 2826.
Sonata, organ and strings (Il trionfo del tempo e del disinganno).
    cf Concerti, organ, op. 4, nos. 1-6.
Sonata, recorder, op. 1, no. 4, A minor. cf Orion ORS 76216.
Sonata, recorder, op. 1, no. 11, F major. cf Transatlantic TRA
    292.
Sonata, violin, op. 1, no. 13, D major. cf RCA ARM 4-0942/7.
Songs: Jehova, to my words give ear. cf Musical Heritage Society
    MHS 3276.
Songs: Meine Seele hor im Sehen. cf Westminster WGS 8268.
Suite, harpsichord, D minor. cf Harpsichord works (Erato STU
    70906).
1284 Suites, harpsichord, nos. 1-8. Colin Tilney, hpd. DG Archive
    2533 168/9 (2).
              ++Gr 2-75 p1516   +NR 1-75 p15
              +HF 4-75 p86   +-RR 2-75 p56
              +MT 4-76 p321
Suite, harpsichord, no. 5, E major (The harmonious blacksmith).
    cf HMV HQS 1360.
Suite, harpsichord, no. 11, D minor: Sarabande (2). cf CBS 61684.
Suite, trumpet, D major. cf Argo ZDA 203.
1285 Tamerlano. Sophia Steffan, Carole Bogard, s; Gwendolyn Killebrew,
    Johanna Simon, alto; Alexander Young, t; Marius Rintzler, bs;
    Albert Fuller, hpd; Lars Holm Johansen, vlc; Copenhagen Chamber
    Orchestra; John Moriarty. Oryx 4XLC 2 (4). (also Cambridge
    2902).
              +Gr 8-76 p311   /ON 3-27-71 p34
              +-HF 5-71 p72   -SR 3-27-71 p79
              +HFN 2-73 p337   +St 4-71 p76
              +NR 3-71 p6
Terpsichore: Sarabande; Gigue. cf Coronet LPS 3032.
Tolomeo: Overture. cf Works, selections (CBS 61559).
1286 Trio sonata, op. 2, no. 1, C minor. HAYDN: Divertimenti (12).
    McGIBBON: Sonata, no. 5, G major (edit. Elliot). PURCELL:
    Chacony, G minor. TRAD.: Airs and dances of Renaissance Scot-
    land (edit. and arr. Elliot). Scottish Baroque Ensemble.
    CRD CRD 1028.
              +Gr 8-76 p306   +RR 8-76 p61
              +-HFN 8-76 p86
1287 Water music. Virtuosi of England; Arthur Davison. Classics for

Pleasure CFP 40092. Tape (c) TC CFP 40092.
  +HFN 6-76 p105 tape
1288 Water music. Wolfgang Meyer, hpd; BPhO; Rafael Kubelik. DG 2535
      137. Tape (c) 3335 137. (Reissue from 138799).
        +Gr 3-76 p1464   ++RR 1-76 p33
        +HFN 2-76 p117   +RR 6-76 p87 tape
1289 Water music. LOL; Leslie Jones. Oryx ORPS 15.
        +Gr 10-76 p601
1290 Water music (Chrysander edition). LPO; Adrian Boult. Pye GSGC
      15026. (Reissue from NCL 16017).
        +Gr 9-76 p416   -RR 8-76 p37
        -HFN 8-76 p93
      Water music. cf Concerto, 2 wind choirs and strings, no. 1, B
      flat major.
      Water music: Air; Largo. cf Decca DPA 523/4.
      Water music: Air; Hornpipe. cf Works, selections (Decca SPA 448).
      Water music: Pomposo. cf Columbia M 33514.
      Water music: Suite. cf Royal fireworks music (DG 2548 169).
      Water music: Suite. cf Royal fireworks music (Turnabout QTV 34632).
      Water music: Suite. cf Works, selections (Decca DPA 551/2).
      Water music: Suite, D major. cf Works, selections (Philips 6833
      200).
      Water music: Suite, G major: Rigaudons and minuets. cf CBS 30066.
      Water music: Suite, no. 3. cf Philips 6747 204.
1291 Works, selections: Alcina: Here are the heavens all joyful. Athalia:
      The Gods, who chosen blessings shed. Belshazzar: Martial
      symphony. Ezio: March. Floridante: Marches, nos. 1 and 2.
      Judas Maccabaeus: See the conquering hero comes; March, no. 2.
      Ode for Saint Cecilia's day: March. Il Parnasso in Festo:
      Largo. Il pastor Fido: Allegro; March. Rinaldo: March.
      Rodrigo: Passacaglia. Saul: Dead march. Scipione: March.
      Tolomeo: Overture. Derek Wickens, ob; Neville Taweel, vln;
      E. Power Biggs, org; RPO; Charles Groves. CBS 61559. (also
      Columbia M 31206 Tape (c) MT 31206 (r) MA 31206).
        +-Gr 5-76 p1758   +NR 6-72 p13
        +HF 8-72 p61   +RR 4-76 p41
        +HFN 4-76 p107   ++SFC 5-21-72 p46
1292 Works, selections: Acis and Galatea: I rage, I melt, I burn; O
      ruddier than the cherry. Judas Maccabaeus: See the conquering
      hero comes. Messiah: He was despised; Hallelujah. Rodelinda:
      Dove sei. Royal fireworks music. Serse: Ombra mai fù. Samson:
      Let the bright seraphim. Solomon: Arrival of the Queen of Sheba.
      Water music: Air; Hornpipe. Zadok the priest. Joan Sutherland,
      s; Bernadette Greevy, Kathleen Ferrier, con; Kenneth McKellar,
      t; Forbes Robinson, bs; Colin Tilney, hpd; Harry Dilley, tpt;
      Handel Opera Society Orchestra and Chorus, LSO and Chorus, ROHO,
      LPO, AMF; Adrian Boult, Francesco-Molinari-Pradelli, Neville
      Marriner. Decca SPA 448. (Reissues).
        +HFN 8-76 p93   ++RR 8-76 p74
1293 Works, selections: Concerto, organ, op. 4, no. 2, B flat major.
      The faithful shepherd: Minuet (arr. Beecham). Messiah: Every
      valley shall be exalted; O thou that tellest; For unto us a
      child is born; I know that my redeemer liveth; The trumpet shall
      sound; Hallelujah. Rodelinda: Dove sei. Royal fireworks music:
      Suite (arr. Harty). Solomon: Arrival of the Queen of Sheba.
      Xerxes: Largo (arr. Reinhard). Water music: Suite (arr. Harty-

Szell). Zadok the priest. Joan Sutherland, s; Bernadette
Greevy, Kathleen Ferrier, con; Kenneth McKellar, t; David
Ward, bs; Alan Stringer, tpt; King's College Chapel Choir;
Karl Richter, org; ECO, AMF, Chamber Orchestra, LSO and Chorus;
David Willcocks, Neville Marriner, Raymond Leppard, Karl
Richter, Georg Szell, Adrian Boult. Decca DPA 551/2 (2).
Tape (c) KDPC 551/2. (Reissues).
+--HFN 11-76 p173   +RR 10-76 p51
+HFN 11-76 p175 tape

1294 Works, selections: Alcina: Overture. Concerto, organ, op. 4, no.
      6, B flat major: Allegro moderato. Concerto, oboe, no. 3, G
      minor: Allegro. Hornpipe, D major. Joshua: O had I Jubel's
      lyre. Messiah: I know that my redeemer liveth; Hallelujah.
      Royal fireworks music: La paix; La rejouissance. Serse: Ombra
      ma fù. Solomon: Arrival of the Queen of Sheba. Water music:
      Suite, D major. Heather Harper, s; Janet Baker, ms; Heinz
      Holliger, ob; André Pépin, Jean-Claude Hermanjat, flt; Ursula
      Holliger, hp; LSO and Chorus, AMF, ECO, I Musici; Colin Davis,
      Neville Marriner, Raymond Leppard. Philips 6833 200. (Re-
      issues).
      +--HFN 11-76 p173   +RR 10-76 p51
      Zadok the priest. cf Works, selections (Decca SPA 448).
      Zadok the priest. cf Works, selections (Decca DPA 551/2).

HANDL, Jacob
      Jesu dulcis memoria. cf HMV CSD 3766.
HANDY, William
      St. Louis blues march. cf RCA AGL 1-1334.
HANSON (20th century Sweden)
      L'Inferno de Strindberg. cf Caprice RIKS LP 35.
HANSON, Howard
1295 Concerto, piano, op. 36. Mosaics. LaMONTAINE: Birds of paradise.
      Eastman-Rochester Symphony Orchestra; Howard Hanson. Era 1006.
      *NYT 7-4-76 pD1
1296 Drum taps: Songs. THOMPSON: The testament of freedom. Eastman-
      Rochester Symphony Orchestra; Howard Hanson. Era 1007.
      *NYT 7-4-76 pD1
1297 Elegy in memory of Serge Koussevitzky. Song of democracy. LANE:
      Songs (4). Eastman-Rochester Symphony Orchestra; Howard Hanson.
      Era 1010.
      *NYT 7-4-76 pD1
      Mosaics. cf Concerto, piano, op. 36.
1298 Psalms (4). THOMSON: Feast of love. Symphony on a hymn tune.
      David Clatworthy, Gene Boucher, bar; Eastman-Rochester Orches-
      tra; Howard Hanson. Mercury SRI 75063.
      ++SFC 5-2-76 p38
      Psalm, no. 150. cf Columbia M 34134.
      Quartet, op. 23. cf Vox SVBX 5305.
      Song of democracy. cf Elegy in memory of Serge Koussevitzky.
      Symphony, no. 2, op. 30. cf GRIFFES: The pleasure dome of Kubla
      Khan.
d'HARDELOT
      Because. cf DG 2530 700.
      Because. cf HMV RLS 716.
      I know a lovely garden. cf Rococo 5397.
      Three green bonnets. cf HMV RLS 719.
HARMAN
      A hymn to the Virgin. cf CRI SD 102.

HARRIS, Donald
1299 Fantasy, violin and piano. MOSS: Elegy. Timepiece. RIEGGER:
      Quartet, strings, no. 2. Paul Zukofsky, Romuard Teco, vln;
      Gilbert Kalish, pno; Jean Dupouy, vla; Raymond DesRoches, perc;
      New Music Quartet. CRI SD 307.
           *St 8-76 p100
      Ludis II. cf Delos DEL 25406.
HARRIS, Roy
1300 Symphony, no. 3. IVES: Three places in New England. PO; Eugene
      Ormandy. RCA ARL 1-1682. Tape (c) ARK 1-1682 (ct) ARS 1-1682.
           +MJ 11-76 p44   ++NR 9-76 p2
      Symphony, no. 3. cf BERNSTEIN: Jeremiah symphony.
      Symphony, no. 3. cf COPLAND: Symphony, no. 3.
1301 Symphony, no. 4. Utah Chorale; Utah Symphony Orchestra; Maurice
      Abravanel. Angel (Q) S 36091.
           +HF 12-75 p94   +SR 9-6-75 p40
           ++NR 11-75 p3   +St 1-76 p104 Quad
           *ON 11-76 p98
1302 Symphony, no. 4. American Festival Orchestra and Chorus; Vladimir
      Golschman. Vanguard SRV 347. (Reissue from VSD 2082).
           +-HF 6-76 p70   /MJ 3-76 p24
HARRIS, William
      Behold now praise the Lord. cf Argo ZRG 789.
HARRISON
      Songs: Give me a ticket to heaven. cf Argo ZFB 95/6.
HARTLEY, Walter
      Octet, saxophones. cf Coronet LPS 3031.
      Orpheus. cf Crystal S 206.
      Sonorities II. cf Crystal S 371.
HASPROIS, Jehan Simon
      Ma douce amour. cf HMV SLS 863.
      Puisque je voy. cf Telefunken ER 6-35257.
HASSE, Johann
      Sonata, recorder, F major. cf Orion ORS 76216.
HASSLER, Hans
      Canzon. cf DG Archive 2533 323.
      Canzon, G minor. cf Pelca PRSRK 41017/20.
      Intraden (3). cf Harmonia Mundi HMU 610.
HATTON, John
      The Indian maid. cf Prelude PRS 2501.
HAUBENSTOCK-RAMATI, Roman
1303 Quartet, strings, no. 1. URBANNER: Quartet, strings, no. 3.
      WEBERN: Bagatelles, string quartet, op. 9 (6). Five movements
      for string quartet, op. 5. Quartet, strings, op. 28. Alban
      Berg Quartet. Telefunken 6-41994.
           ++Gr 9-76 p437   +RR 8-76 p65
           ++HFN 9-76 p121   ++St 12-76 p149
      Ständchen sur le nom de Heinrich Strobel. cf Caprice RIKS LP 34.
HAUBIEL, Charles
1304 Metamorphoses (Variations on a theme by Stephen Foster). LEGINSKA:
      Victorian portraits. Jeanane Dowis, pno. Orion ORS 75188.
           +NR 10-75 p10   ++St 2-76 p103
HAUSSMANN, Valentin
      Catkanei. cf DG Archive 2533 184.
      Galliard. cf DG Archive 2533 184.
      Tantz. cf DG Archive 2533 184.

HAWES
     Psalm, no. 45: My heart is inditing. cf HMV SQ CSD 3768.
HAWTHORNE
     Listen to the mocking bird. cf Vox SVBX 5304.
HAYDN, Josef
     Adagio, F major. cf Piano works (Telefunken FK 6-35281).
     Adagio, F major. cf Sonatas, piano, nos. 1, 4, 20, 30, 32, 35,
        41-42, 53, 57, 59.
     Allegro molto, D major (fragment). cf Piano works (Telefunken FK
        6-35281).
1305 Allemandes (6). Concerto, trumpet, E flat major. HAYDN, M.:
        Concerto, horn, D major. Minuets (6). Alan Stringer, tpt;
        Barry Tuckwell, hn; AMF; Neville Marriner. Argo ZRG 543.
        Tape (c) KZRC 543.
               +HFN 7-76 p104 tape   +RR 6-76 p87
1306 Andante and variations, F minor. MOZART: Fantasia, K 397, D
        minor. Sonata, piano, no. 9, K 311, D major. Sonata, piano,
        no. 10, K 330, C major. Alicia da Larrocha, pno. London CS
        7008. (also Decca SXL 6784).
               +Gr 12-76 p1021   ++NYT 8-8-76 pD13
               +HFN 12-76 p144   +-RR 12-76 p84
               ++NR 9-76 p12   ++St 8-76 p96
     Capriccio, G major. cf Piano works (Telefunken FK 6-35281).
1307 Cassation, no. 6, C major. SCHUBERT: Quartet, guitar, flute,
        viola and violoncello, D 96, G major. WEBER: Trio, piano, op.
        63, G minor. Luise Walker, gtr; Gottfried Hechtl, flt; Paul
        Roczek, vln; Jurgen Geise, vla; Wilfried Tachezi, vlc. Turna-
        bout 34171. Tape (c) KTVC 34171.
               +RR 1-76 p67 tape
1308 Concerto, 2 flutes and orchestra, F major. Concerto, violin,
        harpsichord and strings. Jeanette Dwyer, Claude Legrand, flt;
        Jacques-Francis Manzone, vln; Françoise Petit, hpd; Mozart
        Society Orchestra; Orchestra; Henri-Claude Fantapie, Guido
        Bozzi. Orion ORS 75198.
               +Audio 2-76 p95   ++NR 10-75 p4
               +HF 2-76 p97
     Concerto, harpsichord, D major. cf BACH: Brandenburg concerto,
        no. 5, S 1050, D major.
     Concerto, harpsichord, op. 21, D major. cf BACH, J. C.: Concerto,
        harpsichord, A major.
1309 Concerti, organ, nos. 1-3, C major. E. Power Biggs, org; Columbia
        Symphony Orchestra; Zoltan Rozsnyai. CBS 61675. (Reissue from
        72231).
               +Gr 2-76 p1338   +-RR 9-76 p54
               +HFN 2-76 p115
1310 Concerto, piano, D major. MOZART: Concerto, piano, no. 15, K 450,
        B flat major. Arturo Benedetti Michelangeli, pno; Zurich Cham-
        ber Orchestra; Edmond de Stoutz. Rococo 2076.
               +NR 4-76 p6
1311 Concerti, piano, nos. 2-4, 9, 11. Divertimento, 2 horns, piano
        and strings, C major. Ilse von Alpenheim, pno; Bamberg Symp-
        hony Orchestra; Antal Dorati. Vox QSVBX 5136 (3).
               +NR 3-76 p4   -St 6-76 p103
               ++SFC 3-7-76 p27
     Concerto, piano, no. 4, G major. cf Concerto, piano (clavier),
        no. 11, D major.

1312 Concerto, piano (clavier), no. 11, D major. Concerto, piano, no.
      4, G major. Arturo Benedetti Michelangeli, pno; Zurich
      Chamber Orchestra; Edmond de Stoutz. HMV ASD 3128. Tape (c)
      TC ASD 3128. (also Angel S 37136).
            -Gr 9-75 p458   +NR 11-75 p5
            -HF 2-76 p97   +RR 9-75 p38
            +-HFN 9-75 p99   +-St 6-76 p103
1313 Concerto, piano, op. 21, D major. MOZART: Concerto, piano, no.
      13, K 415, C major. Arturo Benedetti Michelangeli, pno; RAI
      Radio Orchestra. Olympic 8142.
            +-NR 5-76 p6
1314 Concerto, trumpet, E flat major. L'Incontro improvviso overture.
      Sinfonia concertante, B flat major. Maurice André, tpt; Otto
      Winter, ob; Helman Jung, bsn; Walter Forchert, vln; Hans
      Haublein, vlc; Bamberg Symphony Orchestra; Theodor Guschlbauer.
      Erato STU 70652.
            +Gr 9-76 p421   +RR 9-76 p55
1315 Concerto, trumpet, E flat major. VIVALDI: Concerto, piccolo and
      strings, P 83, A minor. WEBER: Concertino, clarinet, op. 26,
      C minor. WIENIAWSKI: Polonaise, op. 4, D major. Scherzo
      tarantelle, op. 16. Thomas Stevens, tpt; Miles Zentner, pic;
      Michele Zukovsky, clt; Glenn Dicterow, vln; LAPO; Zubin Mehta.
      London CS 6967. (also Decca SXL 6737).
            +-Gr 2-76 p1338   ++NR 5-76 p7
            /HF 11-76 p139   ++RR 2-76 p30
            +HFN 2-76 p95   ++SFC 2-15-76 p38
            +MJ 5-76 p28
1316 Concerto, trumpet, E flat major. HUMMEL: Concerto, trumpet, E
      flat major. NERUDA: Concerto, trumpet, E flat major. William
      Lang, tpt; Northern Sinfonia Orchestra; Christopher Seaman.
      Unicorn RHS 337.
            +Gr 4-76 p1598   ++RR 4-76 p41
            +HFN 4-76 p107   +St 10-76 p122
      Concerto, trumpet, E flat major. cf Allemandes (6).
      Concerto, trumpet, E flat major. cf HMV SLS 5068.
1317 Concerto, violin, no. 1, C major. Sinfonia concertante, op. 84,
      B flat major. Franzjosef Maier, vln; Collegium Aureum. BASF
      KHC 21799.
            +-Gr 3-76 p1467   +NR 2-75 p5
            ++HF 9-75 p86   +RR 2-76 p30
            +-HFN 2-76 p99   -SFC 7-27-75 p22
      Concerto, violin, harpsichord and strings. cf Concerto, 2 flutes
      and orchestra, F major.
1318 Concerto, violoncello, C major. Concerto, violoncello, op. 101,
      D major. AMF; Mstislav Rostropovich, vlc and cond. Angel S
      37193. (also HMV SQ ASD 3255).
            ++Gr 9-76 p421   +NR 11-76 p5
            +HF 12-76 p102   +RR 9-76 p54
            +HFN 9-76 p123
      Concerto, violoncello, op. 101, D major. cf Concerto, violoncello,
      C major.
1319 The creation (Die Schöpfung). Helen Donath, s; Adalbert Kraus, t;
      Kurt Widmer, bs; South German Madrigal Choir; Ludwigsburger
      Schlossfestspiele Orchestra; Wolfgang Gönnenwein. Vox (Q)
      QSVBX 5414 (3).
            +NR 2-76 p8   +-St 10-76 p122

1320 The creation: Im Anfange schuf Gott Himmel und Erde; Nun schwanden
     vor dem heiligen Strahle; Rollend in schäumenden Wellen; Nun
     beaut die Flur das frische Grun; In vollem Glanze steiget jetzt;
     Die Himmel erzählen die Ehre Gottes; Und Gott sprach; Es bringe
     das Wasser; Auf starkem Fittiche; Und Gott schuf den Menschen
     nach seinem Bilde; Mit Würd und Hoheit angetan; Vollendet ist
     das grosse Werk; Von deiner Güt or Herr und Gott; Singt dem
     Herren alle Stimmen. Gundula Janowitz, s; Christa Ludwig, con;
     Fritz Wunderlich, Werner Krenn, t; Dietrich Fischer-Dieskau,
     bar; Walter Berry, bs; Vienna Singverein; BPhO; Herbert von
     Karajan. DG 2535 146. (Reissue from 2707 044).
          +-Gr 5-76 p1793   +RR 4-76 p71
          +HFN 5-76 p115
     The creation: The heavens are telling. cf HMV SXLP 50017.
     The creation: In native worth. cf Decca SKL 5208.
     Deutsche Tänze (German dances) (12). cf Piano works (Telefunken
     FK 6-35281).
     German dances, nos. 4, 10-11. cf Saga 5411.
     Deutsche Tänze und Coda (12). cf Piano works (Telefunken FK 6-
     35281).
     Divertimenti (12). cf HANDEL: Trio sonata, op. 2, no. 1, C minor.
1321 Divertimenti, op. 31 (6). LOL; Leslie Jones. Oryx 1740/41 (2).
          /SFC 5-9-76 p38
     Divertimento, 2 horns, piano and strings, C major. cf Concerti,
     piano, nos. 2-4, 9, 11.
1322 Divertimenti, 2 oboes, 2 horns and 2 bassoons, nos. 1-5. Péter
     Pongracz, Bertalan Hock, ob; András Medveczky, Dezsö Mesterházy,
     hn; Tibor Fülemile, András Nagy, bsn. Hungaroton SLPX 11719.
          +-HFN 1-76 p108   +RR 12-75 p73
          ++NR 11-75 p8
     Fantasia, C major. cf Piano works (Telefunken FK 6-35281)
     Fantasia, C major. cf Sonatas, piano, nos. 6, 10, 18, 33, 38-39,
     47, 50, 52, 60.
1323 Fantasia, C major (Capriccio). Sonatas, piano, nos. 40-42, 47-52.
     Variations, Hob XVII:5-6. Dezso Ránki, pno. Hungaroton SLPX
     11625/7 (3).
          ++ARG 12-76 p31
1324 La fedeltà premiata. Ileana Cotrubas, Kari Lövaas, s; Frederica
     von Stade, ms; Lucia Valentini, con; Tonny Landy, Luigi Alva,
     t; Alan Titus, Maurizio Mazzieri, bar; OSR Chorus; Lausanne
     Chamber Orchestra; Antal Dorati. Philips 6707 028 (4).
          +Gr 9-76 p466   +-ON 9-76 p70
          +HF 6-76 p68   +RR 9-76 p19
          +HFN 9-76 p122   +SR 9-18-76 p50
          +MJ 10-76 p24   ++St 8-76 p94
          ++NR 5-76 p11
     Feldpartita, B flat major. cf Saga 5417.
     L'Incontro improvviso overture. cf Concerto, trumpet, E flat major.
     La infedeltà delusa: Non v'e rimedio. cf Hungaroton SLPX 11444.
     March for the Prince of Wales. cf Unicorn RHS 339.
1325 Mass, no. 5, B flat major. MOZART: Mass, no. 7, K 167, C major.
     Elly Ameling, s; Peter Planyavsky, org; Vienna State Opera
     Chorus; VPO; Karl Münchinger. Decca SXL 6747. (also London
     26443).
          +-Gr 3-76 p1494   +HFN 3-76 p95
          +-HF 11-76 p212   +-RR 3-76 p69

1326 Mass, no. 7, C major. Elsie Morison, s; Marjorie Thomas, con;
      Peter Witsch, t; Karl Christian Kohn, bs; Bedrich Janácek,
      org; Bavarian Radio Orchestra and Chorus; Rafael Kubelik. DG
      2548 229. (Reissue from SLPM 138881).
          +Gr 4-76 p1617   +RR 4-76 p71
          +-HFN 5-76 p115
1327 Mass, no. 10, B flat major. Elisabeth Speiser, s; Maureen Lehane,
      con; Theo Altmeyer, t; Wolfgang Schöne, bs; Tölzer Boys'
      Choir; Collegium Aureum. BASF 22287-3.
          /Gr 8-76 p327   +RR 6-76 p78
          -HFN 6-76 p87
1328 Mass, no. 12, B flat major. Judith Blegen, s; Frederica von
      Stade, con; Kenneth Riegel, t; Simon Estes, bs; Westminster
      Symphony Choir; NYP; Leonard Bernstein. Columbia M 33267.
      (Q) MQ 32267. (also CBS 86410).
          +-Gr 12-75 p1089   ++NYT 9-21-75 pD18
          +-HF 9-75 p85   ++RR 10-75 p84
          ++HFN 10-75 p142   ++SFC 7-6-75 p16
          +-MT 2-76 p141   ++St 9-75 p84
          ++NR 8-75 p9
      Menuets de la redoute (12). cf Piano works (Telefunken FK 6-35281).
      Minuets (2). cf DG Archive 2533 182.
      Minuets (12). cf Piano works (Telefunken FK 6-35281).
1329 Minuets (24). PH; Antal Dorati. Decca HDNW 90/1 (2).
          +Gr 10-76 p601   +RR 9-76 p54
          +HFN 9-76 p122
      Minuet and trio. cf Coronet LPS 3032.
1330 Piano works: Adagio, F major. Allegro molto, D major (fragment).
      Capriccio, G major. Deutsche Tänze (12). Deutsche Tänze und
      Coda (12). Fantasia, C major. Minuets (12). Menuets de la
      redoute (12). Sonata, piano, no. 15, C major. Sonata, piano,
      no. 16, E flat major. Sonatas, piano, nos. 54-62. Variations,
      F minor, E flat major, C major, A major (plus alternative to
      variation 10), D major. Rudolf Buchbinder, pno. Telefunken
      FK 6-35281 (6).
          +-Gr 4-76 p1632   +-RR 1-76 p48
          +-HFN 11-76 p161
      Pieces, mechanical clock (4). cf BACH: Chorale prelude: Aus
      tiefer not schrei ich zu dir, S 687.
      Pieces, mechanical clock, no. 17, F major; no. 21, C major; no.
      23, C major. cf Audio 2.
      Quartet, strings, D major: Vivace. cf RCA ARM 4-0942/7.
      Quartet, strings, op. 3, no. 5, F major: Andante cantabile. cf
      Pye QS PCNH 4.
1331 Quartets, strings, op. 20, nos. 1-6. Quartets, strings, op. 64,
      nos. 1-6. Aeolian Quartet. Argo HDNT 70/5 (6).
          +-Gr 5-76 p1776   ++RR 5-76 p60
          +HFN 5-76 p100
1332 Quartet, strings, op. 20, no. 2, C major. Quartet, strings, op.
      20, no. 4, D major. Esterhazy Quartet. ABC L 67011.
          -SFC 12-12-76 p55
      Quartet, strings, op. 20, no. 4, D major. cf Quartet, strings,
      op. 20, no. 2, C major.
1333 Quartets, strings, op. 33, nos. 1-6, op. 46, D minor; op. 50, nos.
      1-6. Aeolian Quartet. Argo HDNU 76/81 (6).
          +Gr 11-76 p829   +-RR 10-76 p76
          +HFN 11-76 p160

Quartet, strings, op. 46, D minor. cf Quartets, strings, op. 33, nos. 1-6.

1334 Quartets, strings, op. 50, nos. 1-6. Tokyo Quartet. DG 2740 135 (3). (*Nos. 44, 45 reissue from 2530 440).

+Gr 7-76 p187   +MT 12-76 p1005
+HFN 6-76 p87   +RR 6-76 p63

Quartets, strings, op. 50, nos. 1-6. cf Quartets, strings, op. 33, nos. 1-6.

Quartets, strings, op. 64, nos. 1-6. cf Quartets, strings, op. 20, nos. 1-6.

1335 Quartet, strings, op. 64, no. 5, D major. Quartet, strings, op. 76, no. 2, D minor. Cleveland Quartet. RCA ARL 1-1409.

+HF 8-76 p91   +NR 6-76 p10
+MJ 10-76 p24   ++SFC 7-18-76 p31

1336 Quartets, strings, opp. 71 and 74. Aeolian Quartet. London STS 15325/7 (3).

/HF 11-76 p112   ++SFC 9-12-76 p31

1337 Quartet, strings, op. 74, no. 3, G minor. Quartet, strings, op. 76, no. 3, C major. Alban Berg Quartet. Telefunken 6-41302. Tape (c) 4-41302.

+HFN 9-75 p110 tape   ++RR 2-76 p70 tape

Quartet, strings, op. 76, no. 2, D minor. cf Quartet, strings, op. 64, no. 5, D major.

Quartet, strings, op. 76, no. 3, C major. cf Quartet, strings, op. 74, no. 3, G minor.

Quartet, strings, op. 103, B flat major. cf BEETHOVEN: Quartet, strings, no. 12, op. 127, E flat major.

Raccolta de menuetti ballabili, nos. 1, 14. cf Nonesuch H 71141.

1338 The seasons (Die Jahreszeiten). Helen Donath, s; Adalbert Kraus, t; Kurt Widmer, bs; South German Madrigal Choir; Ludwigsburger Schlossfestspiele Orchestra; Wolfgang Gönnenwein. Vox (Q) QSVBX 5215 (3).

+NR 2-76 p8   +-St 10-76 p122

Sinfonia concertante, B flat major. cf Concerto, trumpet, E flat major.

Sinfonia concertante, op. 84, B flat major. cf Concerto, violin, no. 1, C major.

Sonata, no. 1: Tempo di menuetto. cf HMV RLS 723.

1339 Sonatas, piano, nos. 1-19. Variations, A major. Variations, D major. Zsuzsa Pertis, János Sebestyén, hpd. Hungaroton SPLX 11614/7 (4).

++NR 5-76 p14   -St 11-76 p143
+RR 6-76 p72

1340 Sonatas, piano, nos. 1, 4, 20, 30, 32, 35, 41-42, 53, 57, 59. Adagio, F major. Variations, C major. John McCabe, pno. Decca 3HDN 106/8 (3).

+Gr 9-76 p444   +RR 9-76 p81
++HFN 9-76 p122

1341 Sonatas, piano, nos. 6, 10, 18, 33, 38-39, 47, 50, 52, 60. Fantasia, C major. Variations, F minor. John McCabe, pno. Decca 1 HSN 100/2 (3). (also London STS 15343/5)

+-Gr 10-75 p657   +-RR 10-75 p76
+HFN 10-75 p142   +-St 11-76 p143
+MT 11-76 p914   +STL 6-6-76 p37
+NR 12-76 p12

1342 Sonatas, piano, nos. 9, 17, 31, 36, 43-45, 46, 48, 54-56. Variations, A major. John McCabe, pno. Decca HDN 103/5 (3).

```
        +Gr 5-76 p1782            +MT 11-76 p914
        +HFN 4-76 p107            +RR 4-76 p63
```
Sonata, piano, no. 9, D major: Adagio. cf CBS 61579.
Sonata, piano, no. 15, C major. cf Piano works (Telefunken FK
    6-35281).
Sonata, piano, no. 16, E flat major. cf Piano works (Telefunken
    FK 6-35281).
1343 Sonatas, piano, nos. 19, 37, 44. Variations, F minor. Gilbert
    Kalish, pno. Nonesuch H 71328.
```
        -ARG 11-76 p26            +-NR 9-76 p12
```
1344 Sonatas, piano, nos. 23, 35, 46, 51. Vasso Devetzi, pno. Moni-
    tor MCS 2147.
```
        +-MJ 10-76 p52            +NR 9-76 p12
```
1345 Sonatas, piano, nos. 32, 34, 46, 51. Gilbert Kalish, pno.
    Nonesuch H 71318.
```
        +-HFN 9-76 p123           +-RR 10-76 p88
        ++NR 3-76 p12             ++St 4-76 p79
```
Sonatas, nos. 40-42, 47-52. cf Fantasia, C major (Capriccio).
Sonata, piano, no. 42, D major. cf International Piano Library
    IPL 5003/4.
1346 Sonata, piano, no. 46, A flat major. Sonata, piano, no. 48, C
    major. MOZART: Adagio, K 540, B minor. Andante, K 616, F
    major. Eine kleine Gigue, K 574, G major. Minuet, K 355, D
    major. Rondo, piano, K 511, A minor. Renée Sándor, pno.
    Hungaroton LXP 11638.
```
        ++St 2-76 p112
```
1347 Sonatas, piano, nos. 48-50, 53. Christoph Eschenbach, pno. DG
    2530 736.
```
        -Gr 10-76 p628            +-RR 10-76 p88
        +HFN 10-76 p171
```
Sonata, piano, no. 48, C major. cf Sonata, piano, no. 46, A flat
    major.
Sonatas, piano, nos. 54-62. cf Piano works (Telefunken FK 6-35281).
Songs: The Birks of Abergeldie; The brisk young lad; Cumbernauld
    house; Duncan Gray; Green grow the rashes; I have lost my
    love; I'm o'er young to marry yet; John Anderson; Love will
    find the way; My ain kind dearie; My boy Tammy; O bonny lass;
    O can ye sew cushions; The ploughman; Shepherds, I have lost
    my love; Sleepy Bodie; Up in the morning early; The white cock-
    ade. cf BEETHOVEN: Songs (Angel S 37172).
Stücke für die Flötenuhr. cf Pelca PRS 40577.
1348 Symphonies, nos. 6-8. Prague Chamber Orchestra; Bernhard Klee.
    DG 2530 591. Tape (c) 3300 591
```
        +Gr 2-76 p1338            +-RR 3-76 p40
        +-Gr 10-76 p658           +RR 11-76 p109 tape
        +HF 7-76 p77              ++SFC 8-1-76 p27
        +HFN 3-76 p95             ++St 8-76 p95
        +NR 6-76 p3               +STL 3-7-76 p37
```
1349 Symphonies, nos. 20-22. PH; Antal Dorati. Decca SDD 468. (Re-
    issue from HDNB 7-12).
```
        +Gr 4-76 p1598            +-RR 3-76 p40
        ++HFN 3-76 p109
```
1350 Symphonies, nos. 25-28. PH; Antal Dorati. Decca SDD 457. Tape
    (c) KSDC 457. (Reissue from HDNB 7/12).
```
        +Gr 1-76 p1197            +-RR 11-75 p43
        +HFN 11-75 p173
```

1351 Symphony, no. 48, C major. Symphony, no. 56, C major. VSOO; Max
      Goberman. CBS 61661.
            ++ARSC Vol VIII, no. 2-3   +HFN 11-76 p170
                  p87   +RR 10-76 p51
            +-Gr 11-76 p791
1352 Symphony, no. 48, C major. Symphony, no. 92, G major. PCO; Dean
      Dixon. Supraphon 110 1202.
            +HFN 11-75 p154   +-RR 9-75 p38
            +NR 6-76 p3
      Symphony, no. 56, C major. cf Symphony, no. 48, C major.
1353 Symphonies, nos. 82-87. NYP; Leonard Bernstein. CBS 77307 (3).
      (Reissues from SBRG 72240, 72529, 72641).
            +-Gr 9-76 p416   +-RR 8-76 p37
            +-HFN 8-76 p80
1354 Symphonies, nos. 82-87. PH; Antal Dorati. Decca SDD 482/4 (3).
      Tape (c) KSDC 482/4. (Reissues from HDNH 35/40).
            +Gr 8-76 p285   ++HFN 8-76 p94
            ++Gr 10-76 p658 tape   +-RR 8-76 p37
            +HFN 7-76 p103   +RR 12-76 p106 tape
1355 Symphonies, nos. 82-87. ECO; Daniel Barenboim. HMV SLS 5065 (3).
            ++Gr 10-76 p601   +-RR 9-76 p55
            +HFN 9-76 p122
1356 Symphony, no. 83, G minor. Symphony, no. 86, D major. Menuhin
      Festival Orchestra; Yehudi Menuhin. HMV ASD 3214.
            +HFN 6-76 p87   +-RR 6-76 p42
1357 Symphony, no. 85, B flat major. Symphony, no. 87, A major. Menu-
      hin Festival Orchestra; Yehudi Menuhin. HMV ASD 3186.
            +Gr 4-76 p1603   +-RR 4-76 p42
            +HFN 4-76 p107
      Symphony, no. 86, D major. cf Symphony, no. 83, G minor.
      Symphony, no. 87, A major. cf Symphony, no. 85, B flat major.
1358 Symphony, no. 88, G major. Symphony, no. 94, G major. RAI Torino
      Orchestra, VPO; Wilhelm Furtwängler. Bruno Walter Society RR
      399.
            +-NR 10-76 p2
1359 Symphony, no. 88, G major. Symphony, no. 89, F major. VPO; Karl
      Böhm. DG 2530 343. Tape (c) 3300 325.
            +-Gr 4-74 p1857   +NR 2-74 p2
            +-HF 6-74 p84   ++RR 4-74 p41
            +-HFN 2-76 p117   +-SFC 5-5-74 p28
1360 Symphony, no. 88, G major. Symphony, no. 98, B flat major. BPhO;
      Eugen Jochum. DG 2548 241. (Reissue from 138823).
            ++Gr 10-76 p601   ++RR 10-76 p52
            ++HFN 11-76 p170
1361 Symphony, no. 88, G major. Symphony, no. 100, G major. Hungarian
      State Symphony Orchestra; Adam Fischer. Hungaroton SLPX 11786.
            +-ARG 12-76 p31   -NR 12-76 p4
      Symphony, no. 89, F major. cf Symphony, no. 88, G major.
1362 Symphony, no. 91, E flat major. Symphony, no. 92, G major. VPO;
      Karl Böhm. DG 2530 524. Tape (c) 3300 470.
            +-Gr 7-75 p181   ++RR 7-75 p26
            +HF 12-75 p97   +RR 10-75 p97 tape
            ++HFN 7-75 p81   ++SFC 7-18-76 p31
            +-NR 12-75 p5
      Symphony, no. 92, G major. cf Symphony, no. 48, C major.
      Symphony, no. 92, G major. cf Symphony, no. 91, E flat major.

1363 Symphonies, nos. 93-98. PH; Antal Dorati. Decca SDD 500/2 (3).
      (Reissue from HDNJ 41/6).
           +Gr 11-76 p792
1364 Symphony, no. 93, D major. Symphony, no. 94, G major. PH; Antal
      Dorati. London STS 15319. (Reissue from STS 15319/24).
           +NR 10-76 p3
1365 Symphony, no. 94, G major. Symphony, no. 101, D major. LPO; Eugen
      Jochum. DG 2530 628. Tape (c) 3300 628. (Reissues from DG
      2545 008, 2720 064).
           ++Gr 7-76 p175   +NR 11-76 p2
           +Gr 7-76 p230 tape   +-RR 5-76 p45
           +HFN 6-76 p102   +RR 9-76 p94 tape
1366 Symphony, no. 94, G major. Symphony, no. 96, D major.* Orchestra
      Sinfonica di Roma; Antonio de Almeida. Philips 6580 124.
      (*Reissue from Unicorn UNS 236).
           +-Gr 7-76 p175   +NR 8-76 p3
           +-HFN 5-76 p115   +-RR 5-76 p45
           +-MJ 10-76 p24
      Symphony, no. 94, G major. cf Symphony, no. 88, G major.
      Symphony, no. 94, G major. cf Symphony, no. 93, D major.
      Symphony, no. 96, D major. cf Symphony, no. 94, G major.
      Symphony, no. 98, B flat major. cf Symphony, no. 88, G major.
1367 Symphony, no. 99, E flat major. Symphony, no. 100, G major.
      NYP; Leonard Bernstein. Columbia M 34126. Tape (c) 34126
           +MJ 11-76 p44   +NR 7-76 p3
1368 Symphony, no. 100, G major. Symphony, no. 103, E flat major.
      NYP; Leonard Bernstein. CBS 76507.
           +-RR 12-76 p56
1369 Symphony, no. 100, G major. Symphony, no. 101, D major. Leipzig
      Gewandhaus Orchestra, BRSO; Otmar Suitner, Rolf Kleinert.
      DG 2548 218. (Reissue from 135037).
           /Gr 4-76 p1617   -RR 4-76 p42
           +-HFN 4-76 p121
      Symphony, no. 100, G major. cf Symphony, no. 88, G major.
      Symphony, no. 100, G major. cf Symphony, no. 99, E flat major.
1370 Symphony, no. 101, D major. Symphony, no. 103, E flat major.
      NYP; Leonard Bernstein. Columbia M 33531. (also CBS 76507).
           +Gr 12-76 p996   +-NR 12-75 p5
           +HFN 12-76 p143   +St 1-76 p105
1371 Symphony, no. 101, D major. Symphony, no. 104, D major. LOL;
      Leslie Jones. Oryx ORPS 18. Tape (c) BRL 18.
           +-Gr 9-76 p416   +-RR 2-75 p75 tape
      Symphony, no. 101, D major. cf Symphony, no. 94, G major.
      Symphony, no. 101, D major. cf Symphony, no. 100, G major.
      Symphony, no. 103, E flat major. cf Symphony, no. 100, G major.
      Symphony, no. 103, E flat major. cf Symphony, no. 101, D major.
1372 Symphony, no. 104, D major. SCHUBERT: Symphony, no. 8, D 759,
      B minor. BPhO; Herbert von Karajan. HMV SQ ASD 3203. Tape
      (c) TC ASD 3203. (also Angel S 37058).
           ++Gr 9-76 p416   +-HFN 11-76 p175 tape
           +HFN 11-76 p160   +-RR 9-76 p60
      Symphony, no. 104, D major. cf Symphony, no. 101, D major.
      Symphony, no. 104, D major. cf BEETHOVEN: Concerto, piano, no. 4,
      op. 58, G major.
1373 Trios, piano, nos. 1, 13-14, 16-17. Beaux Arts Trio. Philips Tape
      (c) 7300 492/3 (2).
           +Gr 11-76 p887 tape

1374 Trio, piano, no. 14, A flat major. Trio, piano, no. 15, G major.
        Beaux Arts Trio. Philips 9500 034.
                +Gr 11-76 p830   +-RR 11-76 p86
                +HFN 11-76 p161
      Trio, piano, no. 15, G major. cf Trio, piano, no. 14, A flat major.
      Trio, piano, flute and violoncello. cf BEETHOVEN: Trio, piano,
        clarinet, violoncello, op. 38, E flat major.
      Variations, Hob XVII:5-6. cf Fantasia, C major (Capriccio).
      Variations, A major. cf Sonatas, piano, nos. 1-19.
      Variations, A major. cf Sonatas, piano, nos. 9, 17, 31, 36, 43,
        45-46, 48, 54-56.
      Variations, C major. cf Sonatas, piano, nos. 1, 4, 20, 30, 32,
        35, 41-42, 53, 57, 59.
      Variations, D major. cf Sonatas, piano, nos. 1-19.
      Variations, F minor. cf Sonatas, piano, nos. 6, 10, 18, 33, 38-39,
        50, 52, 60.
      Variations, F minor. cf Sonatas, piano, nos. 19, 37, 44.
      Variations, F minor, E flat major, C major, A major (plus alterna-
        tive to variation 10), D major. cf Piano works (Telefunken FK
        6-35281).
HAYDN, Michael
      Coburger-Marsch. cf DG 2721 077.
      Concerto, horn, D major. cf HAYDN, J.: Allemandes (6).
      Divertimento, viola, violoncello and double bass. cf DITTERSDORF:
        Sinfonia concertante, viola, double bass and orchestra.
      Minuets (6). cf HAYDN, J.: Allemandes (6).
      Pappenheimer-Marsch. cf DG 2721 077.
      Quintet, strings, F major. cf Quintet, strings, G major.
1375 Quintet, strings, G major. Quintet, strings, F major. Vienna
        Philharmonia Quintet. Decca SDD 340. (also London STS 15309).
                +Gr 2-73 p1529   +NR 6-76 p10
                +HFN 2-73 p339   ++RR 2-73 p69
                +MJ 10-76 p24   ++SFC 5-9-76 p38
      Songs: Anima nostra; In dulci jubilo. cf RCA PRL 1-8020.
      Symphony, G major (with introduction by Mozart, K 444). cf BACH,
        J. C.: Sinfonia, op. 6, no. 6, C minor.
HAYNE VON GHIZEGHEM
      Songs: De tous biens plaine; A la audienche. cf HMV SLS 5049.
HEAD, Michael
      A piper. cf Saga 5213.
HEATH
      Angel voices. cf Pye TB 3009.
      Frolic for trombones. cf Pye TB 3009.
HECKEL, Wolf (Wolfgang)
      Mille regretz; Nach willen dein. cf L'Oiseau-Lyre 12BB 203/6.
      Ein ungarischer Tantz, Proportz auff den ungarischen Tantz. cf
        Hungaroton SLPX 11721.
HEDAR, Josef
      Musik. cf Swedish Society SLT 33188.
HEIDEN, Bernhard
      Canons, 2 horns, nos. 2 and 5. cf Crystal S 371.
HEINICHEN, Johann
      Concerto, 4 recorders and strings. cf BOISMORTIER: Concerto, 5
        recorders without bass, D minor.
HEINRICH, Anthony Philip
      Barbecue divertimento. cf The dawning of music in Kentucky.

1376 The dawning of music in Kentucky: Barbecue divertimento. Epitaph
         on Joan Butt. Gipsey dance. Irradiate cause. Hail to Ken-
         tucky. The minstrel's march (or Road to Kentucky). The musi-
         cal bachelor. The voice of faithful love. The young Columbian
         midshipman. American Music Group; Neely Bruce, pno and cond.
         Vanguard SRV 349. (Reissue from VSD 71178).
                 *HF 6-76 p70   +NR 2-76 p6
                 +MJ 3-76 p24   ++SFC 1-11-76 p38
         Epitaph on JOan Butt. cf The dawning of music in Kentucky.
         Gipsey dance. cf The dawning of music in Kentucky.
         Irradiate cause. cf The dawning of music in Kentucky.
         Hail to Kentucky. cf The dawning of music in Kentucky.
         The minstrel's march (or Road to Kentucky). cf The dawning of
            music in Kentucky.
         The musical bachelor. cf The dawning of music in Kentucky.
         The voice of faithful love. cf The dawning of music in Kentucky.
         The young Columbian midshipman. cf The dawning of music in
            Kentucky.
HEKKING, Gerard
      Villageoise. cf HMV SQ ASD 3283.
HELD
      Chromatic rag. cf Washington University Press OLY 104.
HELLER, Stephen
      L'Avalanche. cf Connoisseur Socity (Q) CSQ 2066.
HELY-HUTCHINSON, Victor
1377 Carol symphony. VAUGHAN WILLIAMS: Fantasia on Christmas carols.
         WARLOCK: Adam lay y bounden. Bethlehem down. TRAD. (arr.
         Vaughan Williams): And all in the morning. Wassail song. Pro
         Arte Orchestra; Guildford Cathedral Choir; Barry Rose. HMV
         ESD 7021.
                 +RR 12-76 p97
HENDRIE, Gerald
      As I outrode this end'res night. cf Grosvenor GRS 1034.
HENNAGIN
      Songs: Crossing the Han River; Walking on the green grass. cf
         Orion ORS 75205.
HENRION, R.
      Fehrbelliner Reitermarsch. cf DG 2721 077.
      Kreuzritter-Fanfare. cf DG 2721 077.
      Kreuzritter-Fanfare. cf Polydor 2489 523.
HENRY VIII, King
      Pastime with good company, ballad. cf Angel SFO 36895.
      Pastime with good company. cf BASF 25 22286-1.
      Pastime with good company. cf HMV CSD 3766.
      Tauner naken. cf National Trust NT 002.
1378 Vocal and instrumental works: Consorts, nos. 2-5, 8, 12-16, 22,
         23: Adew madame et ma mastres; Alac, alac, what shall I do;
         Alas what shall I do for love; En vray amoure; I love now
         rynyd; Gentil prince de renom; Helas madame; Lusty youth should
         us ensue; O my heart; Pastyme with good company; The tyme of
         youth; Taunder naken; Who so that wyll all feattes optayne;
         Who so that wyll for grace sew; Thow that men do call it dotage;
         Without dyscord. St. George's Canzona; John Sothcott. Oryx
         EXP 57.
                 -Gr 7-72 p211   /St 7-76 p110
      HENRY VIII AND HIS SIX WIVES: Music from the film soundtrack. cf
         Angel SFO 36895.

HENSCHEL, George
    Spring. cf HMV RLS 719.
HENSELT, Adolph von
1379 Etudes. Daniel Graham, pno. Musical Heritage Society MHS 1836.
        +MJ 11-76 p45
    Trio, op. 24, A minor. cf ALKAN: Trio, no. 1, op. 30, G minor.
HENZE, Hans Werner
    Apollo et hyazinthus. cf Works, selections (L'Oiseau-Lyre DSLO 4).
1380 Compases para preguntas ensemismadas, viola and 22 players. Con-
        certo, violin, tape, voices and 13 instrumentalists, no. 2.
        Hirofumi Fukai, vla; Brenton Langbein, vln; London Sinfonietta;
        Hans Werner Henze. Decca HEAD 5. (also London HEAD 5).
            +Gr 11-74 p895   +NR 5-76 p4
            +HF 4-76 p102   +RR 11-74 p43
            *MJ 7-76 p56   +SR 4-17-76 p51
            +MT 8-75 p713
    Concerto, double-bass. cf Works, selections (DG 2740 150).
    Concerto, oboe, harp and strings. cf Works, selections (DG 2740
        150).
    Concerto, piano, no. 2. cf Works, selections (DG 2740 150).
    Concerto, violin, tape, voices and 13 instrumentalists, no. 2.
        cf Compases para preguntas ensemismadas, viola and 22 players.
1381 In memoriam: Die Weisse Rose. Kammermusik. Philip Langridge, t;
        Timothy Walker, gtr; London Sinfonietta; Hans Werner Henze.
        L'Oiseau-Lyre DSLO 5.
            +Gr 11-75 p875   +NR 2-76 p3
            +HF 3-76 p89   +-NYT 12-21-75 pD18
            +HFN 12-75 p155   +RR 9-75 p69
            ++MT 9-76 p746   +STL 4-11-76 p36
    Kammermusik. cf In memoriam: Die Weisse Rose.
    Labyrinth. cf Works, selections (L'Oiseau-Lyre DSLO 4).
    Symphonies, nos. 1-5. cf Works, selections (DG 2740 150).
    Symphony, no. 6, 2 chamber orchestras. cf Works, selections
        (DG 2740 150).
    L'Usignolo dell'imperatore. cf Works, selections (L'Oiseau-Lyre
        DSLO 4).
    Wiegenlied der Mutter Gottes. cf Works, selections (L'Oiseau-
        Lyre DSLO 4).
1382 Works, selections: Concerto, double-bass. Concerto, oboe, harp
        and strings. Concerto, piano, no. 2. Symphony, no. 6, 2
        chamber orchestras. Symphonies, nos. 1-5. Christoph Eschen-
        bach, pno; Heinz Holliger, ob; Ursula Holliger, hp; Gary Karr,
        double bs; BPhO, LPO, LSO, Collegium Musicum, ECO; Hans Werner
        Henze, Paul Sacher. DG 2740 150 (5). (Reissues from SLPM
        139203/4, 2530 056, 2530 261, 139395, 139456).
            +Gr 8-76 p285   +RR 7-76 p49
            +HFN 8-76 p93
1383 Works, selections: Apollo et hyazinthus. Labyrinth. L'Usignolo
        dell'imperatore. Wiegenlied der Mutter Gottes. Anna Reynolds,
        ms; Sebastian Bell, flt; John Constable, hpd; Finchley Child-
        ren's Music Group; London Sinfonietta; Hans Werner Henze.
        L'Oiseau-Lyre DSLO 4.
            -Gr 4-76 p1646   +NR 7-76 p6
            +-HF 9-76 p92   *RR 2-76 p30
            +-HFN 5-76 p101   +STL 4-11-76 p36
            ++MT 9-76 p746   +-Te 6-76 p46

HERBECK, Johann
    Songs: Angels we have heard on high: Kommet ihr Hirten; Pueri
        concinite. cf RCA PRL 1-8020.
HERBERT
    The gold bug. cf Columbia M 33838.
    March of the toys. cf RCA AGL 1-1334.
HERBST
    God was in Jesus. cf Vox SVBX 5350.
HERIKSTAD
    Heaven came down: March. cf HMV SXLP 50017.
HERITTE-VIARDOT, Louise
    Quartet: Serenade. cf Gemini Hall RAP 1010.
HEROLD, Louis Joseph
    Zampa: Overture. cf CBS 61748.
    Zampa: Overture. cf HMV ESD 7010.
HERRMANN, Bernard
1384 Echoes, string quartet. Quintet, clarinet and strings. Robert
      Hill, clt; Ariel Quartet, Amici Quartet. Unicorn RHS 332.
        +HF 6-76 p80
    The fantasticks. cf DELIUS: A late lark.
    For the fallen. cf DELIUS: A late lark.
    Quintet, clarinet and strings. cf Echoes, string quartet.
1385 Symphony. National Philharmonic Orchestra; Bernard Herrmann.
      Unicorn RHS 331.
        +Gr 2-76 p1345   ++St 7-76 p110
        ++HF 2-76 p97
HERTEL, Johann Wilhelm
    Concerto, trumpet, 2 oboes and 2 bassoons. cf ALBINONI: Concerto,
      trumpet, C major.
HERZER
    Hoch Heidecksburg. cf Polydor 2489 523.
HERZOGENBERG, Heinrich von
    Choralvorspiele. cf Pelca PRS 40577.
HESSE
    Fantasy, op. 87, D minor. cf Pelca PRS 40581.
HEUBERGER, Richard
    The opera ball: In our secluded rendezvous. cf CBS 30070.
    Der Opernball: Overture. cf Supraphon 110 1638.
HEWITT, James
    The battle of Trenton. cf Columbia MS 6161.
    The battle of Trenton. cf Columbia M 34129.
    The battle of Trenton. cf Columbia M 33838.
    The battle of Trenton. cf Department of Defense Bicentennial
      Edition 50-1776.
HIBBARD, William
    Quartet, strings. cf FENNELLY: Evanescences.
    Trio. cf CRI SD 324.
    HIDDEN MELODIES: I've got you under my skin (in the style of Men-
      delssohn); Three blind mice (Bach); Waltzing Matilda (Scar-
      latti); I saw three ships come sailing by (Schumann); When
      Johnny comes marching home (Schubert); For he's a jolly good
      fellow (Chopin); The Lambeth walk (Rachmaninoff); The London-
      derry air (Brahms); Three blind mice (Debussy). cf Decca SPA
      473.
HILLER, Lejaren
1386 Sonatas, piano, nos. 4 and 5. Frina Boldt, Kenwyn Boldt, pno.

HILLER (cont.)   246

        Orion ORS 75176.
            +-HF 2-76 p97   ++NR 10-75 p10
HIMES, William
        America the beautiful. cf Brassworks Unlimited BULP 2.
HINDEMITH, Paul
        Concerto, violin, D flat major. cf BRUCH: Scottish fantasia, op.
            46.
        Concerto, violin, D flat major. cf HMV SLS 5058.
        Konzertstück, 2 saxophones. cf Coronet LPS 3030.
1387   Ludus tonalis. Richard Tetley-Kardos, pno. Orion ORS 75189.
            +MJ 4-76 p31   ++NR 4-76 p12
1388   Mathis der Maler symphony. STRAUSS, R.: Death and transfigura-
            tion, op. 24. LSO; Jascha Horenstein. Nonesuch H 71307. Tape
            Advent (c) D 1043.
            +HF 5-76 p114 tape   ++SFC 10-5-75 p38
            ++NR 6-75 p6   ++St 10-75 p107
        Octet. cf BRITTEN: Sinfonietta, op. 1.
1389   Requiem "For those we love". Louise Parker, con; George London,
            bs-bar; NYP; Schola Cantorum; Paul Hindemith. Odyssey Y 33821.
            (Reissue from Columbia MS 6573).
            +ASRC Vol VIII, no. 2-3 +NYT 5-16-76 pD19
                p83   +SFC 6-6-76 p33
            ++MU 10-76 p16
        Sonata, bass clarinet and piano. cf Panton 110 369.
        Sonata, horn and piano. cf Works, selections (Columbia M2 33971).
        Sonata, alto horn and piano, E flat major. cf Works, selections
            (Columbia M2 33971).
        Sonata, 4 horns. cf BALAI: Divertimento, wind quintet and harp.
1390   Sonatas, organ (3). George Baker, org. Delos FY 026.
            ++MU 10-76 p16   ++NR 10-76 p15
        Sonata, organ, no. 3. cf Argo 5BBA 1013/5.
        Sonata, trombone and piano. cf Works, selections (Columbia M2
            33971).
        Sonata, trumpet and piano, B flat major. cf Works, selections
            (Columbia M2 33971).
        Sonata, bass tuba and piano. cf Works, selections (Columbia M2
            33971).
        Sonata, solo violoncello, op. 25, no. 3. cf CRUMB: Sonata, solo
            violoncello (Desto 7169).
        Sonata, solo violoncello, op. 25, no. 3. cf CRUMB: Sonata, solo
            violoncello (Gasparo GS 101).
        Suite "1922", op. 26. cf Supraphon 111 1721/2.
        Symphonic metamorphoses on a theme by Carl Maria von Weber. cf
            BRAHMS: Symphony, no. 1, op. 68, C minor.
        Symphonic metamorphoses on a theme by Carl Maria von Weber. cf
            BRAHMS: Variations on a theme by Haydn, op. 56a.
        Trio for soprano and 2 alto recorders. cf HMV SLS 5022.
1391   Works, selections: Sonata, horn and piano. Sonata, alto horn and
            piano, E flat major. Sonata, trombone and piano. Sonata,
            trumpet and piano, B flat major. Sonata, bass tuba and piano.
            Glenn Gould, pno; Philadelphia Brass Ensemble Members. Colum-
            bia M2 33971 (2).
            +-NR 7-76 p5   ++St 10-76 p123
            +SFC 6-6-76 p33
        HISTORIC AND NEW ORGANS OF THE SWISS COUNTRYSIDE. cf Pelca PRSRK
            41017/20.

HLOBIL, Emil
1392 Concerto, double bass, op. 70. KALABIS: Sonata, violoncello.
        REZAC: Torso of a Schumann statue. Vaclav Fuka, double bs;
        Musici de Prague. Panton 110 394.
            +RR 12-76 p56
     Quartet, saxophones, op. 93. cf Coronet LPS 3030.
HOBBS, Christopher
     Aran. cf ADAMS: American standard.
     McCrimmon will never return. cf ADAMS: American standard.
HODDINOTT, Alun
1393 Concertino, viola and small orchestra, op. 14. Dives and Lazarus,
        op. 39. Night music, op. 48. Sinfonietta, no. 1, op. 56.
        Felicity Palmer, s; Thomas Allen, bar; Csabo Erdélyi, vla;
        Welsh National Opera Chorus; NPhO; David Atherton. Argo ZRG
        824.
            +Gr 7-76 p175   +RR 7-76 p75
            +-HFN 7-76 p89
1394 Concerto grosso, no. 2, op. 46. Investiture dances, op. 66.
        Welsh dances, op. 64. MATHIAS: Celtic dances, op. 60. Sin-
        fonietta, op. 34. National Youth Orchestra, Wales; Arthur
        Davison. BBC Records REC 222.
            +Gr 6-76 p44   +RR 6-76 p43
1395 Divertimento, op. 32. Septet, op. 10. Sonata, violin, no. 1,
        op. 63. Clarence Myerscough, vln; Martin Jones, pno; Nash
        Ensemble. Argo ZRG 770.
            +Gr 8-75 p341   ++MT 1-76 p40
            +HFN 8-75 p76   +RR 6-75 p70
     Dives and Lazarus, op. 39. cf Concertino, viola and small orch-
        estra, op. 14.
     Investiture dances, op. 66. cf Concerto grosso, no. 2, op. 46.
     Night music, op. 48. cf Concertino, viola and small orchestra,
        op. 14.
     Septet, op. 10. cf Divertimento, op. 32.
     Sinfonietta, no. 1, op. 56. cf Concertino, viola and small orch-
        estra, op. 14.
     Sonata, violin and piano, no. 1, op. 63. cf Divertimento, op. 32.
     Welsh dances, op. 64. cf Concerto grosso, no. 2, op. 46.
HOFFMEISTER, Franz
     Concerto, flute, D major. cf ALBRECHTSBERGER: Partita, harp, F
        major.
1396 Concerto, viola, D major. PAGANINI: Sonata, viola, op. 35, C
        minor. STAMITZ: Concerto, viola, D major. Atar Arad, vla; PH;
        Reinhard Peters. Telefunken AW 6-42007.
            ++Gr 11-76 p792
HOFHEIMER, Paul
     Nach willen dein. cf L'Oiseau-Lyre 12BB 203/6.
HOFMANN, Josef
     Chromaticon, piano and orchestra. cf International Piano Library
        IPL 5001/2.
     Kaleideskop, op. 40. cf International Piano Archives IPA 5007/8.
     Kaleideskop, op. 40. cf L'Oiseau-Lyre DSLO 7.
     Penguine. cf International Piano Archives IPA 5007/8.
HOLBORNE, William
     The choice. cf HMV SLS 5022.
     The fairy round. cf RCA RK 11708.
     Heart's ease. cf RCA RK 11708.

HOLBORNE (cont.)   248

        Heigh-ho holiday. cf Argo SPA 464.
        Heigh-ho holiday. cf RCA RK 11708.
        The image of melancholly. cf International Trust NT 002.
        Muylinda. cf HMV SLS 5022.
        Pavan and galliard. cf HMV SLS 5022.
        Sic semper soleo. cf HMV SLS 5022.
HOLDEN, Smollet
        Donshier quickstep. cf Columbia M 34129.
HOLLER, Karl
1397 Ciacona, op. 54. KARG-ELERT: Passacaglia and fugue on B-A-C-H,
           op. 150. REGER: Pieces, op. 145, no. 2: Dankpsalm. Andrew
           Armstrong, org. Vista VPS 1028.
               +-Gr 7-76 p199   +-RR 10-76 p91
        Fantasie, op. 49. cf BADINGS: Intermezzo.
HOLLINS, Alfred
        Concert overture, C minor. cf Vista VPS 1035.
HOLLOWAY
        Wood-up quick-step. cf Michigan University SM 0002.
HOLMBOE, Vagn
        Notturno, op. 19. cf CARLSTEDT: Sinfonietta, 5 wind instruments.
        Quintet, 2 trumpets, horn, trombone and tuba, op. 79. cf Swedish
           Society SLT 33200.
HOLST, Gustav
1398 Choral symphony, op. 41. Felicity Palmer, s; LPO and Chorus;
           Adrian Boult. HMV SAN 354. (also Angel S 37030).
               ++Gr 10-74 p738   ++SFC 11-24-74 p31
                +HF 3-75 p80   +SR 1-11-75 p51
                +NR 1-75 p10   ++St 3-75 p104
                +ON 1-24-76 p51   ++Te 3-75 p44
               ++RR 10-74 p46
        Egdon Heath, op. 47. cf The tale of the wandering scholar, op. 50.
        Festival chorus, op. 36, no. 2: Turn back, O man. cf HMV HQS 1350.
        The perfect fool, op. 39. cf The tale of the wandering scholar,
           op. 50.
1399 The planets, op. 32. Ambrosian Singers; LSO; André Previn.
           Angel S 36991. Tape (ct) 4XS 36991. (also HMV ASD 3002 (Q)
           Q4 ASD 3002).
                +Gr 9-74 p206   +RR 7-74 p43
                +-HF 2-75 p92   -SFC 9-29-74 p28
                +-HF 10-76 p147 tape   ++St 3-75 p104
                +NR 11-74 p2
1400 The planets, op. 32. Hallé Womens Chorus; Hallé Orchestra; James
           Loughran. Classics for Pleasure CFP 40243.
               /-HFN 12-76 p144   ++RR 11-76 p67
1401 The planets, op. 32. Bournemouth Municipal Choir; Bournemouth
           Symphony Orchestra; George Hurst. Contour CN 2020. Tape (c)
           CN 42020. (Reissue from Contour 2870 367).
                +Gr 9-76 p421
1402 The planets, op. 32 (trans. Gleeson). Eu Polyphonic Synthesizer;
           Patrick Gleeson. Mercury SRI 80000.
                +NR 11-76 p14
1403 The planets, op. 32. Mendelssohn Club Chorus, Women's Voices;
           PO; Eugene Ormandy. RCA CRL 1-1921. Tape (c) CRK 1-1921 (ct)
           CRS 1-1921.
                +-HF 12-76 p104   ++NR 11-76 p2
1404 The planets, op. 32. LSO and Chorus, BBC Symphony Orchestra and
           Chorus; Gustav Holst, Malcolm Sargent. Stanyan 2SR 9017.
                +-ARG 11-76 p27

1405 The planets, op. 32. St. Louis Symphony Orchestra; Walter Suss-
          kind. Vox Turnabout (Q) QTVS 34598. Tape (c) KTVC 34598.
            +-HF 4-76 p111   ++NR 12-75 p3
            +-HFN 12-76 p155 tape   ++SFC 11-16-75 p32
      The planets, op. 32. cf DEBUSSY: La mer.
      The planets, op. 32: Jupiter, the bringer of jollity. cf HMV 7011.
      The planets, op. 32: Mars. cf Transatlantic XTRA 1160.
      The planets, op. 32: Venus. cf Angel S 37157.
1406 Songs: English folk songs: Matthew, Mark, Luke and John; I sowed
          the seed of love; I love my love; Jig. Welsh folk songs: The
          lively pair; Lisa Lan; My sweetheart's like Venus; The mother-
          in-law; 2 folksong fragments; Oh, I hae seen the roses blaw;
          The shoemaker; Adar man ymynydd (The nightingale and the linnet);
          Lliw gwyn rhosyn yr haf (White summer rose); The first love;
          Green grass; Awake, awake; The dove; O twas on a Monday morning;
          The lover's complaint. Two-part canons: If twere the time of
          lilies; Evening on the Moselle. Nocturne. Toccata. Newburn
          lads. This have I done for my true love. Keith Swallow, pno;
          BBC Northern Singers; Stephen Wilkinson. Abbey LPB 736.
            ++Gr 5-76 p1794   +RR 4-75 p60
            +-HFN 9-75 p99
      Songs, op. 34: Lullay my liking. cf Grosvenor GRS 1034.
      Suite, no. 2, op. 28, F major. cf ELGAR: The dream of Gerontius,
          op. 38: Angel's farewell.
1407 The talke of the wandering scholar, op. 50. The perfect fool,
          op. 39. Egdon Heath, op. 47. Norma Burrowes, s; Robert Tear,
          t; Michael Rippon, Michael Langdon, bs; ECO, LSO; Steuart
          Bedford, André Previn. HMV ASD 3097. (also Angel S 37152).
            +-Gr 9-75 p506   +-NYT 1-18-76 pD1
            +-HF 5-76 p84   +ON 1-24-76 p51
            +HFN 9-75 p99   +RR 9-75 p18
            +MM 1-76 p43   +SR 1-24-76 p53
            +MT 12-75 p1070   +St 5-76 p117
            +NR 1-76 p12
HOMER, Sidney
      The sick rose. cf Cambridge CRS 2715.
HOMILIUS, Gottfried
      Wer nur den lieben Gott lasst walten. cf Pelca PRSRK 41017/20.
HONEGGER, Arthur
      Choral. cf Pelca PRSRK 41017/20.
1408 Concerto da camera, flute and cor anglais. IBERT: Concerto, flute.
          JOLIVET: Concerto, flute. English Sinfonia Orchestra; Neville
          Dilkes. HMV EMD 5526.
            +-Gr 5-76 p1758   +-RR 4-76 p47
            +MM 7-76 p30
      Sonatina. cf DEBUSSY: Sonata, violoncello and piano, no. 1, C
          minor.
      Sonatina. cf Grenadilla GS 1006.
      Sonatina, clarinet and piano, A major. cf Hungaroton SLPX 11748.
1409 Symphony, no. 1. Symphony, no. 4. CPhO; Serge Baudo. Supraphon
          110 1536.
            +HF 1-76 p87   +RR 5-75 p29
            ++NR 9-75 p4   ++St 3-76 p114
      Symphony, no. 4. cf Symphony, no. 1.
HOOK, James
      The Caledonian laddy. cf Smithsonian Collection N 002.

HOPKINS, James
 Diferencias sobre una tema original. cf BLOCH: Nocturnes.
HOPKINSON, Francis
 Songs: A toast; Beneath a weeping willow's shade; Come fair Rosina;
 Enraptur'd I gaze; My days have been so wond'rous free; My
 gen'rous heart disdains; My love is gone to sea; O'er the hills
 far away; See down Maria's blushing cheek; The traveler be-
 nighted and lost. cf Vox SVBX 5350.
HORN, Charles
 Cherry ripe. cf HMV ESD 7002.
 Cherry ripe. cf Prelude PRS 2505.
HOROBIN
 Star of Erin. cf Pye TB 3007.
HOROWITZ (Horovitz), Joseph
 Majorcan pieces (2). cf Decca SPA 395.
 Music hall suite. cf CIVIL: Tarantango.
HOROWITZ, Vladimir
 Variations on themes from Bizet's Carmen. cf RCA VH 020.
HORSLEY
 There is a green hill. cf Abbey LPB 761.
HORVAT, Stanko
 Träumerei. cf BOZIC: Audiogemi I-IV.
HOTTETERRE, Jean
 Bourrée. cf DG Archive 2533 172.
 Suite, 2 blockflöten. cf Telefunken DT 6-48075.
1410 Suites, recorder and harpsichord (5). Scott-Martin Kosofsky, rec;
 John Gibbons, hpd. Titanic TI 2/3 (2).
 +NR 2-76 p7
HOVE, Joachim van den
 Galliarde. cf DG Archive 2533 302.
HOVHANESS, Alan
1411 Armenian rhapsody, no. 1. Avak, the healer, op. 65. Prayer of
 St. Gregory. Tzalkerk. Marni Nixon, s; Thomas Stevens, flt;
 Gretel Shanley, vln; Eudice Shapiro, vln; Chamber Orchestra;
 Ernest Gold. Crystal S 800.
 ++MJ 7-76 p57   +NR 9-76 p14
 Avak, the healer, op. 65. cf Armenian rhapsody, no. 1.
 Duet, violin and harpsichord. cf COWELL: Works, selections (CRI
 SD 109).
1412 Magnificat, op. 157. Audrey Nossaman, s; Elizabeth Johnson, con;
 Thomas East, t; Richard Dales, bar; Louisville University
 Chorus; Louisville Orchestra; Robert Whitney. Poseidon 1018.
 +NR 2-76 p8
1413 Meditation on Orpheus. KELLER: Symphony, no. 3. WOOD: Poem,
 orchestra. Japan Philharmonic Symphony Orchestra, Tokyo Asahi
 Orchestra; William Strickland, Richard Korn. CRI SD 134.
 +RR 12-76 p61
 Meditation on Orpheus. cf Columbia MG 33728.
 Prayer of St. Gregory. cf Armenian rhapsody, no. 1.
 Prayer of St. Gregory. cf Avant AV 1014.
1414 Requiem and resurrection, op. 224. Symphony, no. 25, op. 275.
 Polyphonia Orchestra; West Caldwell Symphonic Band; Alan Hov-
 haness. Unicorn RHS 335.
 +Gr 7-76 p175   ++RR 8-76 p40
 +HFN 7-76 p89
 Symphony, no. 25, op. 275. cf Requiem and resurrection, op. 224.

1415   Triptych. HUSA: Mosaiques. STRAIGHT: Development. Benita
          Valente, s; Bamberg Symphony Orchestra, Members, Singers;
          Stockholm Radio Orchestra, LPO; Alfredo Antonini, Karel Husa,
          Russell Stanger. CRI SD 221.
                /-HFN 1-76 p108   ++RR 6-75 p79
          Tzalkerk. cf Armenian rhapsody, no. 1.
HOWARTH, Elgar
          Basel march. cf CIVIL: Tarantango.
          Berne patrol. cf CIVIL: Tarantango.
          The cuckoo. cf CIVIL: Tarantango.
          Lucerne song. cf CIVIL: Tarantango.
          The old chalet. cf CIVIL: Tarantango.
          Variations on "Carnival of Venice". cf CIVIL: Tarantango.
          Zurich march. cf CIVIL: Tarantango.
HOWE
          Battle hymn of the republic. cf Columbia SCX 6585.
          Battle hymn of the Republic. cf Department of Defense Bicenten-
          nial Edition 50-1776.
HOWELLS, Herbert
          Elegy, op. 15. cf BUTTERWORTH: The banks of green willow.
          Gavotte. cf HMV ASD 2929.
          Magnificat. cf HMV HQS 1350.
          Magnificat and nunc dimittis, G major. cf Polydor 2460 250.
          Merry eye, op. 20b. cf BUTTERWORTH: The banks of green willow.
          Music for a prince. cf BUTTERWORTH: The banks of green willow.
          Pieces, organ: Paean. cf Audio 2.
          Rhapsody, no. 3, C sharp minor. cf Wealden WSQ 134.
          Siciliano for a high ceremony. cf Wealden WS 137.
          Songs: Come sing and dance; King David. cf Saga 5213.
          A spotless rose. cf HMV CSD 3774.
          Three figures. cf ELGAR: The dream of Gerontius, op. 38: Angel's
          farewell.
HOWET, Gregorio
          Fantasie. cf DG Archive 2533 302.
HOY, Bonnee
          The DeMazio quintet. cf BACH: Sonata, no. 3.
          Lament. cf BACH: Sonata, no. 3.
HUBAY, Jenö
1416   Csárda scenes: Comely Katy; Memory of Szalatna; On the waves of
          the Balaton; They say they don't give me; The water of the
          Maros is flowing calmly; Yellow maybeetle. Rajkó Gypsy En-
          semble. Qualiton SLPX 10136.
                +NR 5-76 p14
HUDSON, Joe
          Reflexives, piano and tape. cf CRI SD 345.
HUE, Georges
          Soir païen. cf HMV RLS 719.
HUFFER
          Black Jack. cf Michigan Univesity SM 0002.
HUFFINE, D. W.
          Them basses. cf Columbia 33513.
HUGGLER, John
          Quintet, no. 1. cf Crystal S 204
HUGHES
          Cwm Rhondda. cf HMV OU 2105.
HUHN
          Songs: Invictus. cf Argo ZFB 95/6.

HUME, Tobias
    Musick and mirth. cf L'Oiseau-Lyre 12BB 203/6.
HUMMEL, Johann
1417 Concerto, trumpet, E flat major (ed. Oubradous). MOZART, L.:
        Concerto, trumpet, D major (ed. Seiffert). TELEMANN: Concerto,
        trumpet, D major (ed. Grebe). VIVALDI: Concerto, trumpet, A
        flat major (ed. Thilde). Maurice André, tpt; BPhO; Herbert von
        Karajan. HMV ASD 3044. Tape (c) TC ASD 3044. (also Angel
        S 37063).
            +Gr 6-75 p55   ++RR 5-75 p30
            +HFN 6-75 p107   ++RR 10-76 p106 tape
            +NR 2-75 p4   ++STL 6-8-75 p36
      Concerto, trumpet, E flat major. cf HAYDN: Concerto, trumpet,
        E flat major.
      Fantasie and rondo, op. 19, E major. cf DUSSEK: La consolation,
        op. 62.
1418 Partita, E flat major. SPOHR: Notturno, op. 34, C major. LOL;
        Leslie Jones. Oryx 1830.
            +Gr 7-76 p188
      Quartet, clarinet, E flat major. cf CRUSELL: Quartet, clarinet,
        no. 2, op. 4, C minor.
      Rondo, op. 11, E flat major. cf RCA ARM 4-0942/7.
      Sonata, piano, op. 13, E flat major. cf BEETHOVEN: Concerto,
        piano, no. 1, op. 15, C major.
      Sonata, piano, op. 20, F minor. cf DUSSEK: La consolation, op. 62.
      Sonata, piano, op. 92, A flat major. cf Sterling 1001/2.
1419 Sonata, viola, E flat major. REGER: Sonata, solo viola, op. 31,
        G minor. TAUSINGER: Duetti compatibili. Ladislav Cerný, vla;
        Jarmila Kozderková, pno; Brigita Sulcova, s. Panton 110 430.
            -RR 4-76 p64
HUMPERDINCK, Engelbert
1420 Hansel and Gretel. Patricia Kern, ms; Margaret Neville, Jennifer
        Eddy, Rita Hunter, Elisabeth Robinson, s; Ann Howard, con;
        Raimund Herincx, bar; Children's Chorus; Sadler's Wells Orches-
        tra and Chorus; Mario Bernardi. HMV SXDW 3023 (2). (Reissue
        from CSD 1576/7).
            +Gr 1-76 p1230   +RR 1-76 p24
            +HFN 1-76 p123
      Hänsel und Gretel: Ein Männlein steht im Walde. cf Rubini RS 301.
      Hänsel und Gretel: Overture. cf HMV (Q) ASD 3131.
      Königskinder: Ei, ist das schwer, ein Bettler sein. cf Court
        Opera Classics CO 354.
HURFORD, Peter
    Laudate Dominum. cf Argo ZRG 807.
HUSA, Karel
    Mosaiques. cf HOVHANESS: Triptych.
HUSTON, Scott
    Lifestyles, clarinet, violoncello and piano. cf Serenus SRS
        12064.
1421 Sounds at night. Suite for our times. KAPR: Woodcuts. KUPFERMAN:
        Madrigal, brass quartet. TILLIS: Quintet, brass. Indianapolis
        Brass Ensemble, New York Brass Quintet, Edward Tarr Brass En-
        semble. Serenus SRS 12066.
            ++NR 4-76 p7
      Suite for our times. cf Sounds at night.
      HYMNS SUNG BY SHERRILL MILNES. cf RCA ARL 1-1403.

IANNACONNE, Anthony
1422 Bicinia, flute and alto saxophone. Partita, piano. Rituals,
         violin and piano. Sonatine, trumpet and tuba. Rodney Hill,
         flt; Max Plank, alto sax; Alfio Pignotti, vln; Dady Mehta,
         Joseph Gurt, pno; Carter Eggers, tpt; John Smith, tuba. Coro-
         net LPS 3038.
               +NR 10-76 p7
         Partita, piano. cf Bicinia, flute and alto saxophone.
         Rituals, violin and piano. cf Bicinia, flute and alto saxophone.
         Sonatine, trumpet and tuba. cf Bicinia, flute and alto saxophone.
IBERT, Jacques
1423 Bacchanale. Bostoniana. Louisville concerto. Symphonie marine.
         Birmingham City Symphony Orchestra; Louis Frémaux. HMV SQ ASD
         3176.
               ++Gr 5-76 p1758   +RR 4-76 p47
         Bostoniana. cf Bacchanale.
         Concerto, flute. cf DEVIENNE: Concerto, flute, no. 2, D major.
         Concerto, flute. cf HONEGGER: Concerto da camera, flute and cor
            anglais.
1424 Concerto, violoncello and wind instruments. MARTINU: Duo, violin
         and violoncello, no. 2. Concertino, violoncello, piano and
         wind instruments. Josef Suk, vln; André Navarra, vlc; Prague
         Chamber Harmony; Martin Turnovsky. Supraphon 50877.
               ++Gr 8-76 p286   +RR 7-76 p67
         Divertissement. cf BIZET: Jeux d'enfants, op. 22.
         Entr'acte. cf BIS LP 30.
         Entr'acte. cf Claves LP 30-406.
1425 Escales (Ports of call). Ouverture de fête. Tropismes pour des
         amours imaginaires. ORTF: Jean Martinon. Angel (Q) 37194.
         (also HMV (Q) ASD 3147).
               ++Gr 1-76 p1197   +RR 1-76 p34
               +-HF 11-76 p114   +SFC 5-23-76 p36
                +HFN 2-76 p101   ++St 11-76 p144
               +-NR 7-76 p2
         Escales. cf CHABRIER: España.
         Histoires. cf HMV HQS 1364.
         Louisville concerto. cf Bacchanale.
         Ouverture de fête. cf Escales.
         Symphonie marine. cf Bacchanale.
         Tropismes pour des amours imaginaires. cf Escales.
d'INDIA, Sigismondo
         Songs: Cruda amarilli; Intenerite voi, lagrime mie. cf DG
            Archive 2533 305.
d'INDY, Vincent
         Symphonie sur un chant montagnard français, op. 25. cf FAURE:
            Ballade, op. 19, F sharp major.
INGALLS, Jeremiah
         Northfield. cf Nonesuch H 71276.
         Northfield. cf Vox SVBX 5350.
         INSTRUMENTS OF THE MIDDLE AGES AND RENAISSANCE. cf HMV SLS 988.
         INTERNATIONAL PIANO LIBRARY GALA CONCERT. cf Desmar DSM 1005.
IPPOLITOV-IVANOV, Mikhail
         Caucasian sketches, op. 10: Procession of the Sardar. cf Colum-
            bia M 34127.
         Caucasian sketches, op. 10: Procession of the Sardar. cf Pye
            QS PCNH 4.

IPPOLITOV-IVANOV (cont.)   254

Caucasian sketches, op. 10: Procession of the Sardar.   cf RCA
    CRL 3-2026.
IRELAND, John
    A comedy overture.   cf BLISS: Belmont variations.
    A Downland suite.   cf BLISS: Belmont variations.
    Greater love hath no man.   cf HMV HQS 1350.
    The Salley gardens.   cf HMV ASD 2929.
    Songs: Greater love hath no man.   cf Audio 3.
    Songs: The land of lost content.   cf BUTTERWORTH: A Shropshire
        lad.
    Songs: Her song; A thanksgiving.   cf Saga 5213.
ISAAC, A.
    Ne più bella di queste; Palle, palle; Quis dabit pacem.   cf
        L'Oiseau-Lyre 12BB 203/6.
ISAAC, Heinrich
    Innsbruck, ich muss dich lassen.   cf Nonesuch H 71326.
    La la hö hö.   cf L'Oiseau-Lyre 12BB 203/6.
    Songs: Donna di dentro di tua casa; Missa la bassadanza: Agnus
        Dei; A la battaglia.   cf HMV SLS 5049.
ISALOW
    Songs: Sanctus.   cf Columbia SCX 6585.
ISTVAN, Miloslav
1426 Isle of toys.   KUCERA: The kenetic ballet: The labyrinth; The
        spiral.   VOSTRAK: Scales of light.   Electronic tape.   Supra-
        phon 111 1423.
              /HFN 11-75 p155   +-RR 7-75 p60
              +NR 3-76 p15
1427 Ritmi ed Antiritmi.   REINER: Trio, flute, bass clarinet and per-
        cussion.   RYCHLIK: Relazioni, alto flute, English horn and
        bassoon.   TAUSINGER: Concertino, viola and chamber orchestra.
        Vlastimil Lejsek, Vera Lejsková, pno; Bohumil Krška, Vladislav
        Benes, perc; Ladislav Cerný, vla; Sonatori di Praga, Musica
        Viva Pragensis, CPhO; Frantisek Vajnar.   Supraphon 111 1184.
              +NR 1-74 p8   +RR 1-76 p49
IVES, Charles
    Allegretto (Inventions).   cf Piano works (Desto DST 6458/61).
    Bad resolutions and good.   cf Piano works (Desto DST 6458/61).
1428 Browning overture.   Three places in New England.   Holidays:
        Washington's birthday.   American Symphony Orchestra, PO, NYP;
        Leopold Stokowski, Eugene Ormandy, Leonard Bernstein.   Columbia
        MS 7015.
              *St 7-76 p74
    Celestial railroad.   cf Piano works (Desto DST 6458/61).
    Country band march.   cf March Mega Lambda Chi.
    Hallowe'en.   cf Works, selections (Vox SVBX 564).
1429 Holidays.   Temple University Concert Choir; PO; Eugene Ormandy.
        RCA ARL 1-1249.   Tape (c) ARK 1-1249 (ct) ARS 1-1249 (Q) ARD
        1-1249.
              -HF 6-76 p70   ++SFC 2-29-76 p25
              ++NR 2-76 p3   +St 6-76 p104
1430 Holidays: Decoration day.   Symphony, no. 2.   Variations on
        "America" (orch. William Schuman).   LAPO; Zubin Mehta.   Decca
        SXL 6753.
              ++Gr 7-76 p176   +RR 7-76 p52
              ++HFN 7-76 p89
    Holidays: Decoration day.   cf London CSA 2246.

Holidays: Thanksgiving. cf BECKER: Concerto arabesque, piano.
1431 Holidays: Washington's birthday; Decoration day; The fourth of
      July; Thanksgiving. Tokyo Philharmonic Orchestra, Finnish
      Radio Symphony Orchestra, Gothenburg Symphony Orchestra, Ice-
      land Symphony Orchestra; Iceland State Radio Chorus; William
      Strickland. CRI S 190.
          -Gr 7-76 p176   -RR 7-76 p49
Holidays: Washington's birthday. cf Browning overture.
In re con moto et al. cf Works, selections (Vox SVBX 564).
The innate (Adagio cantabile), piano and string quartet. cf
      Works, selections (Vox SVBX 564).
Largo, violin and piano. cf Works, selections (Vox SVBX 564).
Largo, violin, clarinet and piano. cf Sonatas, violin and piano,
      nos. 1-4.
Largo, violin, clarinet and piano. cf Works, selections (Vox
      SVBX 564).
Largo, violin, clarinet and piano. cf Delos DEL 25406.
Largo risoluto I and II. cf Works, selections (Vox SVBX 564).
March in G and D, "Here's to good old Yale". cf Piano works
      (Desto DST 6458/61).
March intercollegiate. cf Columbia 33513.
1432 March Mega Lambda Chi. Country band march. Overture and march
      "1776". They are there. LOCKWOOD: Suite, band. ROSS: Con-
      certo, tuba and orchestra. STRAUSS, R.: Fanfare für die Wien
      Philharmoniker. Harvey Phillips, tuba; Cornell University Wind
      Ensemble; Cornell University Symphony Band, Members; Marice
      Stith. Cornell CUWE 17.
          +-NR 4-76 p13
March Omega Lambda Chi. cf Columbia 33513.
Overture and march "1776". cf March Mega Lambda Chi.
1433 Piano works: Allegretto (Invention). Bad resolutions and good.
      Baseball take-off. Celestial railroad. March in G and D,
      "Here's to good old Yale". Processional, anthem. Rough and
      ready. Scene episode. Seen and unseen: Processional. Son-
      atas, piano, nos. 1, 2. Song with (good) words. Storm and
      distress. Studies, nos. 2, 5-8, 9, 15, 18, 20-22. Three-
      page sonata. Varied air and variations (six protests). Waltz-
      rondo. Alan Mandel, pno. Desto DST 6458/61 (4).
          +-Gr 4-75 p1836   +St 7-76 p74
          -HFN 5-76 p129
Processional, anthem. cf Piano works (Desto DST 6458/61).
Quarter-tone pieces, 2 pianos (3). cf Works, selections (Vox
      SVBX 564).
1434 Quartets, strings, nos. 1 and 2. Concord Quartet. Nonesuch H
      71306.
          ++Gr 8-76 p312   ++NR 6-75 p10
          ++HF 8-75 p88   ++SFC 7-20-75 p27
          ++HFN 7-75 p83   ++St 10-75 p107
          +MT 11-75 p977   +St 7-76 p74
Quartet, strings, no. 2. cf BARBER: Quartet, strings, op. 11,
      B minor.
Rough and ready. cf Piano works (Desto DST 6458/61).
Scene episode. cf Piano works (Desto DST 6458/61).
Scherzo, string quartet. cf BARBER: Quartet, strings, op. 11,
      B minor.
Scherzo, string quartet. cf Vox SVBX 5305.

Seen and unseen: Processional. cf Piano works (Desto DST 6458/61).
Sonatas, piano, nos. 1, 2. cf Piano works (Desto DST 6458/61).
1435   Sonata, piano, no. 1. Noël Lee, pno. Nonesuch H 71169.
          +MT 10-76 p831
1436   Sonata, piano, no. 2. Hadassah Sahr, pno; Carl Adams, flt.
          Critics Choice CC 1705.
              +-HF 6-76 p70
1437   Sonatas, violin and piano, nos. 1-4. Largo, violin, clarinet and
          piano. Paul Zukofsky, vln; Gilbert Kalish, pno. Nonesuch HB
          73025 (2).
              ++Gr 12-74 p1162   +RR 11-74 p81
              +HF 10-74 p79   ++SFC 8-25-74 p29
              +MT 10-76 p831   ++St 10-74 p128
              ++NR 10-74 p7
       Sonatas, violin and piano, nos. 1-4. cf Works, selections (Vox
          SVBX 564).
1438   Sonata, violin and piano, no. 2. Sonata, violin and piano, no.
          3. Marilyn Dubow, vln; Marsha Cheraskin Winokur, pno. Musical
          Heritage Society MHS 3160.
              +MJ 3-76 p24   +St 1-76 p105
       Sonata, violin and piano, no. 3. cf Sonata, violin and piano,
          no. 2.
       Sonata, violin and piano, no. 4. cf BINKERD: Sonata, violin and
          piano.
       Song without (good) words. cf Piano works (Desto DST 6458/61).
1439   Songs: Abide with me; Ann Street; At the river; Autumn; The child-
          ren's hour; A Christmas carol; Disclosure; Elegie; A farewell
          to land; Feldeinsamkeit; Ich grolle nicht; In Flanders fields;
          The swimmers; Tom sails away; Two little flowers; Weil auf mir;
          West London; Where the eagle; The white gulls. Dietrich Fischer-
          Dieskau, bar; Michael Ponti, pno. DG 2530 696.
              +Gr 7-76 p205   ++RR 7-76 p76
              ++HFN 8-76 p81   +STL 8-8-76 p29
1440   Songs: Ann Street; At the river; The cage; A Christmas carol;
          The circus band; A farewell to land; From "Paracelsus"; The
          Housatonic at Stockbridge; The Indians; The innate; Like a
          sick eagle; In the mornin'; Majority; Memories (A--Very pleas-
          ant; B--Rather sad); Serenity; The things our fathers loved;
          Thoreau. Jan DeGaetani, ms; Gilbert Kalish, pno. Nonesuch H
          71325.
              +Gr 11-76 p852   +ON 6-76 p52
              ++HF 8-76 p84   +RR 11-76 p96
              +HFN 11-76 p161   +SFC 8-8-76 p38
              +NR 7-76 p9   ++St 9-76 p86
              ++NYT 7-4-76 pD1
       Songs: An old flame; Circus band; A Civil War memory; In the
          alley; Karen; Romanza di Central Park; A son of a gambolier.
          cf Vox SVBX 5304.
       Songs: Walt Whitman; The white gulls. cf Washington University
          Press OLY 104.
       Storm and distress. cf Piano works (Desto_DST 6458/61).
       Studies, nos. 2, 5-9, 15, 18, 20-22. cf Piano works (Desto DST
          6458/61).
       Symphony, no. 2. cf Holidays: Decoration day.
       Symphony, no. 2. cf London CSA 2246.
1441   Symphony, no. 3. Three places in New England. Eastman-Rochester

         Symphony Orchestra; Howard Hanson. Mercury SRI 75035.
            *NYT 7-4-76 pD1
      Symphony, no. 3. cf Argo ZRG 845.
1442 Symphony, no. 4. LPO; José Serebrier. RCA ARL 1-0589. (Q) ARD
         1-0589 Tape (Q) 1-0589.
            +Gr 10-74 p688   +NR 11-74 p2
            +HF 10-74 p81   ++NYT 10-20-74 pD26
            +HF 2-75 p110 Quad   +-RR 10-74 p47
            +MJ 12-74 p45   ++SFC 10-6-74 p26
            +MT 10-76 p831   +St 11-74 p134
      They are there. cf March Mega Lambda Chi.
      Three-page sonata. cf Piano works (Desto DST 6458/61).
      Three-page sonata. cf DETT: In the bottoms.
      Three places in New England. cf Browning overture.
      Three places in New England. cf Symphony, no. 3.
      Three places in New England. cf HARRIS: Symphony, no. 3
1443 Trio, violin, violoncello and piano. KORNGOLD: Trio, piano, op.
         1. Pacific Art Trio. Delos DEL 25402.
            +HF 5-76 p85   +-SFC 7-20-75 p27
            +NR 9-75 p8   +St 7-76 p110
1444 Trio, violin, violoncello and piano. SHOSTAKOVICH: Trio, piano,
         no. 2, op. 67, E minor. Beaux Arts Trio. Philips 6500 860.
            +-Gr 1-76 p1212   +-RR 2-76 p44
            +HF 5-76 p85   +SFC 3-14-76 p27
            +HFN 1-76 p108   ++St 7-76 p110
           ++NR 1-76 p8
      Trio, violin, violoncello and piano. cf Works, selections (Vox
         SVBX 564).
      Variations on "America" (orch. William Schuman). cf Holidays:
         Decoration day.
      Variations on "America". cf Columbia MS 6161.
      Variations on "America". cf Columbia MG 33728.
      Variations on "America". cf Creative Record Service R 9115.
      Variations on "America". cf Delos FY 025.
      Variations on "America". cf London CSA 2246.
      Variations on "America". cf Personal Touch 88.
      Varied air and variations (six protests). cf Piano works (Desto
         DST 6458/61).
      Waltz-rondo. cf Piano works (Desto DST 6458/61).
1445 Works, selections: Hallowe'en. In re con moto et al. The innate
         (Adagio cantabile), piano and string quartet. Largo, violin,
         clarinet and piano. Largo, violin and piano. Largo risoluto
         I and II. Quarter-tone pieces, 2 pianos (3). Sonatas, violin
         and piano, nos. 1-4. Trio, violin, violoncello and piano.
         Millard Taylor, John Celentano, vln; Francis Tursi, vla; Alan
         Harris, vlc; Stanley Hasty, clt; Artur Balsam, Frank Glazer,
         pno. Vox SVBX 564 (3).
            +-HF 11-76 p114   +NR 8-76 p6
            +MJ 11-76 p44   +St 12-76 p144
IVEY, Jean Eichelberger
      Hera, hung from the sky. cf ERICKSON: End of the mime.
JACKSON, Francis
      Fanfare. cf Audio 2.
JACOB
      Partita. cf Washington University RAVE 761.

JACOB, Gordon
1446   Divertimento, harmonica and string quartet. MOODY: Quintet,
       harmonica. Tommy Reilly, harmonica; Hindar Quartet. Argo
       ZDA 206.
              +Gr 6-75 p61   +RR 8-75 p52
              +NR 2-76 p5
       Quartet, saxophone. cf Transatlantic TRA 308.
JACOBI
       Niederdeutscher Tanz. cf Coronet LPS 3032.
       Skizze. cf Coronet LPS 3032.
JACOTIN, Jacques (Jacob)
       A Paris à trois fillettes. cf HMV CSD 3740.
       Voyant souffrir. cf HMV SLS 5022.
JACQUET DE LA GUERRE, Elizabeth
       Jacob et Rachel: Air. cf Gemini Hall RAP 1010.
       Suite, D minor. cf Avant AV 1012.
       Susanne: Recitative and air. cf Gemini Hall RAP 1010.
JAEGER
       Indiana polka. cf Library of Congress OMP 101/2.
JANACEK, Leos
       By overgrown tracks, Bk I. cf Works, selections (Supraphon 111
          1481/2).
       Capriccio, piano, left hand and wind ensemble. cf Works, selec-
          tions (Supraphon 111 1481/2).
       Concertino, piano and chamber ensemble. cf Works, selections
          (Supraphon 111 1481/2).
1447   Glagolitic mass. Teresa Kubiak, s; Anne Collins, ms; Robert Tear,
       t; Wolfgang Schöne, bs; John Birch, org; Brighton Festival
       Chorus; RPO; Rudolf Kempe. Decca SXL 6600. Tape (c) KSXC
       6600. (also London OS 26338)
              +-Gr 2-74 p1589   +NR 11-74 p7
              +-Gr 9-76 p497 tape   ++NYT 8-18-74 pD20
              +-HF 11-74 p106   +ON 12-21-74 p28
              +HFN 2-74 p339   +RR 2-74 p18
              ++HFN 7-76 p104 tape   +RR 8-76 p84 tape
              ++MJ 1-75 p48   ++St 1-75 p112
1448   Glagolitic mass. Soloists; CPhO; Karel Ančerl. Supraphon 50519.
       Tape (c) 4-50519.
              +HFN 2-76 p116 tape   +RR 1-76 p66 tape
1449   Idyll, string orchestra. MARTINU: Concerto, 2 orchestras, piano
       and timpani. PCO; Hans-Hubert Schönzeler. RCA LHL 1-5086.
              +Gr 3-75 p1653   +RR 3-75 p27
              +-MT 2-76 p141
       Idyll, string orchestra. cf DVORAK: Serenade, op. 22, E major.
       In the mists. cf Works, selections (Supraphon 111 1481/2).
       Sonata, piano. cf Works, selections (Supraphon 111 1481/2).
1450   Sonata, violin and piano, op. 21. PROKOFIEV: Sonata, violin and
       piano, no. 1, op. 80, F minor. David Oistrakh, vln; Frieda
       Bauer, pno. Westminster WGM 8292.
              +HF 8-76 p92   ++St 4-75 p105
1451   Songs (choral works): Divím se milému (I wonder at my love);
       Láska opravdivá (True love); Kačena divoká (Wild duck); Na
       košatej jedli dva holubi sed'á; Na prievoze (On the ferry);
       Naše píseň; Nestálost lásky; Odpočin si (Rest in peace);
       Oráni (Ploughing) Osudu neujdeš (No escape from fate); Osa-
       melá bez techy (Forsaken); Píseň v jeseni (Autumn song);
       Vínek stonylý; Zpěvná dumá. Prague Philharmonic Choir; Josef

Veselka. Panton 110 400.
     +Gr 10-76 p634   +RR 8-76 p75
1452 Songs (choral works): Hradčany songs; Kašpar Ruckẏ; Říkadia; Wolf
      tracks. Various soloists; CPhO and Chorus; Josef Veselka.
      Supraphon 112 1486.
        +Gr 8-76 p327   +NR 10-76 p7
        +HF 12-76 p104   ++RR 8-76 p75
        +-HFN 8-76 p81
1453 Works, selections: By overgrown tracks, Bk I. Capriccio, piano,
      left hand and wind ensemble. Concertino, piano and chamber
      ensemble. In the mists. Sonata, piano (October 1, 1905).
      Josef Pálenícek, pno; Czech Philharmonic Wind Ensemble. Supra-
      phon 111 1481/2 (2).
        +-Gr 11-75 p860   +RR 9-75 p58
        +-HFN 10-75 p142   ++St 3-76 p112
        ++NR 11-75 p11
JANECEK, Karel
1454 Grand symposium, 15 soloists. JAROCH: Metamorphoses, 12 woodwinds.
      JIRKO: Sonata, 14 woodwinds and timpani. Czech Philharmonic
      Wind Ensemble; Vladimir Cernẏ. Panton 110 301.
        +RR 1-76 p40
      Legenda o Praze. cf DOUBRAVA: Pastorale.
JANNEQUIN, Clement
      Au joly jeu. cf HMV CSD 3766.
      Les cris de Paris. cf L'Oiseau-Lyre 12BB 203/6.
      De son amour. cf Hungaroton SLPX 11699/70.
      La guerre, parts 1 and 2. cf HMV CSD 3740.
      J'ai trop soudainement aymé. cf Hungaroton SLPX 11669/70.
      La plus belle de la ville. cf Argo ZRG 667.
JARNEFELT, Armas
      Berceuse. cf Swedish Society SLT 33229.
      Söndag. cf Swedish Society SLT 33197.
JAROCH, Jiri
      Metamorphoses, 12 woodwinds. cf JANECEK: Grand symposium, 15
      soloists.
      JASCHA HEIFETZ, IN CONCERT. cf Columbia M2 33444.
JEANJEAN, Paul
      Arabesques. cf Grenadilla GS 1006.
JENKINS, John
      Newark siege. cf National Trust NT 002.
JENNI, Donald
      Cucumber music. cf CRI SD 324.
      Musique printanière. cf BLANK: Rotation.
      JENNIE TOUREL AT ALICE TULLY HALL. cf Desto DC 7118/9.
JENSEN, Adolf
      Erotikon, op. 44: Eros. cf International Piano Library IPL 102.
JEUNE, Claude le
      Fière cruelle. cf L'Oiseau-Lyre 12BB 203/6.
JEZEK, Jaroslav
      Children's choruses on poems by V. Nezval. cf Panton 110 385.
      Songs (6). cf Panton 110 385.
JIRASEK, Ivo
1455 Partita. MATEJ: Concerto, trumpet, French horn and trombone.
      Miroslav Kejmar, tpt; Milos Petr, hn; Zdenek Pulec, trom;
      Musici de Prague, Czech Philharmonic Wind Ensemble; Vladimir
      Valek. Panton 110 456.
        +RR 4-76 p49

JIRKO, Ivan
      Elegie on the death of a friend. cf BOHAC: Elegie in memory of
         Ludvíka Poděště.
1456 Requiem. Brigita Sulcová, s; Blanka Vitková, ms; Oldřich Spisar,
         t; René Tuček, bar; Dalibor Jedlička, bs; Pevecky Choir; CPhO;
         Bohumir Liška.   Panton 110 428.
            +RR 11-76 p96
      Sonata, 14 woodwinds and timpani. cf JANECEK: Grand symposium,
         15 soloists.
      JOHN PHILIP SOUSA CONDUCTS BAND MUSIC OF THE WORLD. cf Everest
         3360.
JOHNSON, Bengt
      Through the mirror of thirst (second passage). cf Caprice RIKS
         LP 35.
JOHNSON, David
      Trumpet tune. cf Wealden WS 142.
JOHNSON, Edward
      Eliza is the fairest queen. cf Harmonia Mundi HMD 223.
JOHNSON, Robert
      Care-charming sleep. cf Harmonia Mundi 204.
      Treble to a ground. cf DG Archive 2533 323.
JOLIVET, André
      Concertino. cf RCA CRL 3-1430.
      Concerto, flute. cf HONEGGER: Concerto da camera, flute and cor
         anglais.
      Concerto, flute. cf RCA CRL 3-1429.
      Concerto, trumpet, no. 2. cf RCA CRL 3-1430.
      Sérénade. cf Delos DEL FY 008.
      Suite en concert, flute and percussion. cf RCA CRL 3-1429.
      THE JOLLY MINSTRELS: Minstrel tunes, songs and dances of the Middle
         Ages on authentic instruments. cf Vanguard VCS 10049.
JONES
      Songs: Morte Christe. cf Columbia SCX 6585.
JONES, Charles
      Quartet, strings, no. 6. cf CRUMB: Black angels, electric
         string quartet.
      Sonatina. cf CRUMB: Black angels, electric string quartet.
JONES, Daniel
1457 Quartet, strings, no. 9. Sonata, 3 timpani. Trio, strings.
         Gabrieli Quartet; Tristan Fry, timpani. Argo ZRG 772.
            /Gr 8-76 p317   +RR 7-76 p64
            ++HFN 6-76 p83
      Sonata, 3 timpani. cf Quartet, strings, no. 9.
      Sonata, 3 timpani. cf BRINDLE: Orion 42.
      Trio, strings. cf Quartet, strings, no. 9.
JONES, Jeff
      Ambiance. cf Nonesuch H 71302/3.
JONES, M.
      Rhondda rhapsody. cf HMV OU 2105.
JONES, Samuel
      Elegy, string orchestra. cf COOPER: Symphony, no. 4.
      Let us now praise famous men. cf COOPER: Symphony, no. 4.
JONGEN, Joseph
      Petit prélude. cf Wealden WS 142.
1458 Symphonie concertante. Toccata. Rollin Smith, org. Repertoire
         Recording Society RRS 13.
            +MU 10-76 p12

Toccata. cf Symphonies concertante.
Toccata. cf L'Oiseau-Lyre SOL 343.
JONSSON, Josef
    Du käre gud fader i himmelrik. cf Swedish Society SLT 33197.
JOPLIN, Scott
    Bethena, a concert waltz. cf Piano works (Nonesuch HB 73026).
    Combination march (arr. Schuller). cf Columbia 33513.
    The easy winners. cf Personal Touch 88.
    Elite syncopation. cf Piano works (Nonesuch HB 73026).
1459 The entertainer (The prodigal son). London Festival Ballet Orch-
        estra; Michael Bassett, pno; Grant Hossack. Columbia M 33185.
        (Q) Tape (c) MAQ 33185 (ct) 33185.
            ++HF 3-76 p116   +SFC 12-22-74 p24
    The entertainer. cf Piano works (Nonesuch HB 73026).
    The entertainer. cf HMV SEOM 22.
    Eugenia. cf Piano works (Nonesuch HB 73026).
    Euphonic sounds. cf Piano works (Nonesuch HB 73026).
    Fig leaf rag. cf Piano works (Nonesuch HB 73026).
    Fig leaf rag. cf Washington University Press OLY 104.
    Gladiolus rag. cf Piano works (Nonesuch HB 73026).
    Leola, two-step. cf Piano works (Nonesuch HB 73026).
    Magnetic rag. cf Piano works (Nonesuch HB 73026).
    Maple leaf rag. cf Piano works (Nonesuch HB 73026).
    Maple leaf rag. cf Creative Record Service R 9115.
    Marching onward. cf Columbia M 34129.
    Paragon rag. cf Piano works (Nonesuch HB 73026).
1460 Piano works: Bethana, a concert waltz. Elite syncopation. The
        entertainer. Eugenia. Euphonic sounds. Fig leaf rag. Gladi-
        olus rag. Leola, two-step. Magnetic rag. Maple leaf rag.
        Paragon rag. Pineapple rag. The ragtime dance. Rose leaf
        rag, a rag time two-step. Scott Joplin's new rag. Solace,
        a Mexican serenade. Joshua Rifkin, pno. Nonesuch HB 73026.
            +St 7-76 p72
    Pineapple rag. cf Piano works (Nonesuch HB 73026).
    The ragtime dance. cf Piano works (Nonesuch HB 73026).
    Rose leaf rag, a rag time two-step. cf Piano works (Nonesuch
        HB 73026).
    Scott Joplin's new rag. cf Piano works (Nonesuch HB 73026).
    Solace, a Mexican serenade. cf Piano works (Nonesuch HB 73026).
    Something doing rag. cf Golden Crest CRS 4143.
1461 Treemonisha. Carmen Balthrop, s; Betty Allen, ms; Curtis Rayam,
        t; Willard White, bs; Orchestra and chorus; Gunther Schuller.
        DG 2707 083 (2). Tape (c) 3370 012.
            +Gr 7-76 p206   ++NR 4-76 p9
            +-Gr 12-76 p1066 tape   +ON 6-76 p52
            +-HF 7-76 p72   /RR 7-76 p28
            +HF 11-76 p153 tape   /SFC 5-16-76 p28
            +HFN 7-76 p90   +St 5-76 p116
            +-MJ 7-76 p56   +-STL 8-8-76 p29
    Treemonisha: Finale. cf Columbia M 34129.
JOSQUIN DES PRES (also Des Pres, Depres)
    La Bernadina. cf HMV SLS 5049.
    Mille regretz. cf L'Oiseau-Lyre 12BB 203/6.
    Motets: Benedicta es caelorum Regina; De profundis; Inviolata,
        integra et casta es, Maria. cf HMV SLS 5049.
1462 Songs: Adieu mes amours; Bergerette savoyenne; Coeurs désolez par

toutes nations (3); Deploration sur la mort de Johannes
Ockeghem; Faulte d'argent; El grillo; In te Domine speravi;
Mille regretz; Plaine de dueil; Recordans de mia segnora;
Regretz sans fin; Scaramella va alla guerra; Tenez-moy en vos
bras. SUSATO: Mille regretz, pavane. ANON.: Dances "In te
Domine speravi". Musica Reservata; Andrew Parrott. Argo ZRG
793.

    +Gr 2-76 p1367   ++MT 7-76 p577
    +-HF 11-76 p116   +NR 6-76 p13
    +HFN 2-76 p101   +-RR 2-76 p55
    +-MJ 11-76 p45   ++STL 3-7-76 p37

Songs: Allegez moy, doulce plaisant brunette; Adieu mes amours;
El grillo è buon cantore; Guillaume se va; Scaramella va alla
guerra. cf HMV SLS 5049.
Songs: Baisez moi; Petite camusette. cf HMV CSD 3766.
La Spagna. cf HMV SLS 5049.
Vive le roy. cf HMV SLS 5049.

JOSTEN, Werner
1463 Concerto sacro I-II. American Symphony Orchestra; Leopold Stokowski. CRI SD 200.
    +Gr 4-76 p1603   *RR 1-76 p34

JUDENKUNIG, Hans
Ellend bringt peyn. cf DG Archive 2533 302.
Hoff dantz. cf DG Archive 2533 302.

JUON, Paul
Berceuse. cf RCA ARM 4-0942/7.

JUREK
Deutschmeister Regiments Marsch. cf Polydor 2489 523.

KABALEVSKY, Dimitri
1464 Concerto, piano, no. 3, op. 50, D major. Concerto, violin, op.
48, C major. Overture pathétique, op. 64, B minor. Spring,
op. 65. Vladimir Fetsman, pno; Victor Pikaizen, vln; MPO;
Dimitri Kavalevsky, Fuat Mansurov. HMV Melodiya ASD 3078.
    +-Gr 8-75 p322   +MM 1-76 p42
    +-HFN 8-75 p76   +RR 8-75 p34
Concerto, violin, op. 48, C major. cf Concerto, piano, no. 3,
op. 50, D major.
Overture pathétique, op. 64, B minor. cf Concerto, piano, no. 3,
op. 50, D major.
1465 Sonata, violoncello, op. 71, B flat major. RACHMANINOFF: Vocalise,
op. 34, no. 14. STRAVINSKY: Suite italienne. Raphael Wall-
fisch, vlc; Richard Markham, pno. Prelude PMS 1501
    +-Gr 10-76 p621   +-RR 10-76 p25
    +-HFN 10-76 p171
Song about Russia. cf HMV ASD 3200.
Spring, op. 65. cf Concerto, piano, no. 3, op. 50, D major.
Variations on a Ukrainian folksong. cf Swedish Society SLT 33200.

KADOSA, Pal
Verseire, op. 68. cf Hungaroton SLPX 11713.

KAGEL, Mauricio
Unguis incarnatus est. cf DG 2530 562.

KALABIS, Victor
Little chamber music, wind quintet, op. 27. cf BARTA: Quintet,
winds, no. 2.
Sonata, violoncello. cf HLOBIL: Concerto, double bass, op. 70.

KALASHNIKOV, Nikolai
Concerto, 12 voice choir: Cherubic hymn. cf HMV Melodiya ASD 3102.

KALLIWODA, Johann Wenzel (also Kalivoda, Jan Václav)
  Introduction and rondo, op. 51, F major. cf BASF BAC 3085.
KALMAN, Emmerich
  Die Czardasfürstin: Tanzen mocht ich; Tausend kleine Engel singen;
    Sich verlieben kann man öfters; Mädchen gibt es underfeine. cf
    HMV CSD 3748.
  Grafin Maritza (Countess Maritza): Mein lieber Schatz. cf HMV CSD
    3748.
  Countess Maritza: Komm Zigany; Grüss mir mein Wien. cf Angel
    S 37108.
  Grafin Maritza: Tassilo's song; Maritza and Tassilo duet; Aus-
    schnitt. cf Heut nacht hab ich getraumt von dir.
1466 Heut nacht hab ich getraumt von dir. Grafin Maritza: Tassilo's
    song, Maritza and Tassilo duet, Ausschnitt. LEHAR: Friederike:
    Poet's song. Das Land des Lächelns: Su Tshong's songs.
    STRAUSS, O.: Rund und die Liebe: Hans songs. ZELLER: Der
    Vögelhandler: Adam's song. József Simándy, t; Choral and Orch-
    estral accompaniments. Qualiton SLPX 16581.
      -NR 3-76 p10
  Die Zirkusprinzessin: Zwei Märchenaugen. cf Angel S 37108.
KAMINSKI
  Recitative and dance. cf Da Camera Magna SM 93399.
KAPR, Jan
  Dialogues, flute and harp (5). cf Serenus SRS 12064.
  Studies, soprano, flute and harp. cf Hungaroton SLPX 11713.
  Woodcuts. cf HUSTON: Sounds at night.
KARAI, Jozsef
  Summer night. cf Hungaroton SLPX 11762.
KARG-ELERT, Siegfried
1467 Chorale improvisations, op. 65: Herr Jesu Christ, dich zu uns
    wend; Herzliebster Jesu, was hast du verbrochen; Allein Gott
    in der Höh sei ihr, Ach bleib mit deine Gande; Wie schön
    leuchtet der Morgenstern; Ein feste Burg ist unser Gott.
    PEETERS: Chorale fantasie "Lasst uns erfreuen", op. 81, no. 2.
    Chorale prelude, op. 81, no. 1: Stuttgart. Suite modale, op.
    43. Toccata, fugue and hymn, op. 28. Robert Husson, org.
    Wealden WS 128.
      +Gr 4-76 p1643   +RR 3-76 p62
      +-HFN 1-76 p119
  Music for organ. cf DAVID: Toccata and fugue, F minor.
  Nun danket alle Gott. cf Decca DPA 523/4.
  Passacaglia and fugue on B-A-C-H, op. 150. cf HOLLER: Ciacona,
    op. 54.
KARJINSKY
  Esquisse. cf HMV SQ ASD 3283.
KARKOFF, Maurice
1468 Figures for wind instruments and percussion. MOZART: Divertimento,
    E flat major (arr. from Serenade, no. 10, K 361, B flat major).
    SODERLUNDH: Land o'Bella: Suite. ANON.: Marches from the music
    of the Royal Södermanland Regiment (6). Musica Camerata Region-
    alis; Claes-Merithz Pettersson. Caprice CAP 1074.
      +Gr 6-76 p58   +RR 6-76 p32
1469 Symphony, no. 4, op. 69. LARSSON: Orchestral variations, op. 51.
    Swedish Radio Symphony Orchestra; Sixten Ehrling, Stig Wester-
    berg. Swedish Discofil SLT 33164.
      +-RR 12-76 p61   +St 6-75 p51

KARLINS, M. William
    Variations on 'Obiter dictum'. cf BLANK: Rotation.
KARLOWICZ, Mieczyslaw
1470 Concerto, violin. The sorrowful tale. Wanda Wilkomirska, vln;
        Warsaw Philharmonic Orchestra; Witold Rowicki. Muza XL 0179.
            *HF 3-76 p70
1471 Rhapsody, orchestra. Warsaw Philharmonic Orchestra; Stanislaw
        Wislocki. Muza XL 0290.
            *HF 3-76 p70
      The sorrowful tale. cf Concerto, violin.
1472 Symphonic poems (2). Warsaw Philharmonic Orchestra; Stanislaw
        Wislocki. Muza XL 0269.
            *HF 3-76 p70
KAROLYI
    Quartet, strings, no. 2. cf DUBROVAY: Quartet, strings.
KASTAISKY
    Songs: From the rising of the sun unto the going down thereof;
        Vespers (O joyful light). cf Ikon IKO 2.
KASTNER
    Sextet, saxophones. cf Coronet LPS 3031.
KAUFFMANN
    O Jesulein süss. cf BACH: Chorale preludes (FAS 16).
    Vom Himmel hoch. cf BACH: Chorale preludes (FAS 16).
KAUFFMANN, Leo
    Music for brass. cf Unicorn RHS 339.
KAY
    Songs: How stands the glass around; What's in a name. cf CRI SD
        102.
KAY, Ulysses
1473 Dances, string orchestra. STILL: Darker America. From the black
        belt. Westchester Symphony Orchestra, Westphalian Symphony
        Orchestra; Siegfried Landau, Paul Freeman. Turnabout TV 34546.
            *Audio 6-76 p101   +St 10-75 p118
            +-NR 9-74 p4
KELLER, Homer
    Symphony, no. 3. cf HOVHANESS: Meditation on Orpheus.
KELLEY, Edgar Stillman
    Eldorado. cf Cambridge CRS 2715.
KELLY
    Songs: Last week I took a wife; The mischievous bee. cf Vox
        SVBX 5350.
KENNEDY
    Say "Au revoir" but not "Goodbye". cf Pye Ember GVC 51.
KENNEDY, Russell
    Land of hearts desire. cf HMV RLS 716.
KENNEY, H. A.
    Concerto, horn. cf Pye TB 3007.
KERN, Jerome
    All the things you are. cf Coronet LPS 1722.
    Showboat: Ol' man river. cf RCA LRL 2-7531.
    Showboat: Scenario, orchestra. cf GERSHWIN: Porgy and Bess.
KETTING, Otto
1474 For moon light nights. A set of pieces, wind quintet. Time
        machine. Abbie de Quant, flt; Ardito Quintet, Radio Philhar-
        monic Orchestra; Otto Keeting. Donemus CV 7601.
            ++RR 7-76 p52   +Te 9-76 p28

A set of pieces, wind quintet. cf For moon light nights.
Time machine. cf For moon light nights.
KETTING, Piet
1475 Preludes and fugues, nos. 1 and 2. Prelude and fuguetta. Pre-
        lude, interlude and postlude. Gerard Hengeveld, Henk Largen-
        daal, Jan Van Der Meer, pno; Koos Verheul, flt. Donemus DAVS
        7475/4.
              +-RR 7-76 p72   +Te 9-76 p28
      Prelude and fuguetta. cf Preludes and fugues, nos. 1 and 2.
      Prelude, interlude and postlude. cf Preludes and fugues, nos. 1
        and 2.
KEURIS, Tristan
1476 Concerto, saxophone. LOEVENDIE: Scaramuccia. PORCELIJN: 10-5-
        6-5(a). Continuations. SCHAT: Thema. Hans de Vries, ob;
        Netherlands Wind Ensemble, Radio Philharmonic Orchestra, Radio
        Chamber Orchestra; Ed Bogaard, sax; Piet Honingh, clt; Peter
        Schat, Diego Masson, Roelof Krol, David Atherton. Donemus
        7374/4.
              +RR 4-75 p30   +Te 9-76 p28
KEY, Francis Scott
      Star spangled banner. cf Coronet LPS 1722.
KEYPER, Franz
      Romance and rondo, double bass and orchestra. cf DITTERSDORF:
        Sinfonia concertante, viola, double bass and orchestra.
KHACHATURIAN, Aram
1477 Concerto, flute and orchestra. Jean-Pierre Rampal, flt; ORTF;
        Jean Martinon. Odyssey Y 33906. (Reissue from Musical Heri-
        tage Society MHS 1186).
              ++HF 10-76 p112   +-SR 4-17-76 p51
              ++NR 4-76 p6   +St 7-76 p110
      Concerto, violin, D minor. cf HMV SLS 5058.
      Gayaneh: Adagio. cf HMV HQS 1364.
      Gayaneh: Sabre dance. cf CBS 30073.
      Gayaneh: Sabre dance. cf DG 2584 004.
1478 Gayaneh suite. PROKOFIEV: Symphony, no. 1, op. 25, D major. Love
        for three oranges, op. 33: March; Scherzo. LSO, London Festi-
        val Orchestra; Stanley Black. Decca PFS 4349.
              +Gr 12-76 p1071
1479 Gayaneh suite. MASSENET: El Cid: Ballet music. Netherlands
        Radio Philharmonic Orchestra, LSO; Stanley Black. London SPC
        21133.
              +SFC 10-24-76 p35
1480 Gayaneh suite. Masquerade suite. Brno State Philharmonic Orch-
        estra; Jiří Belohlávek. Supraphon 110 1226. Tape (c) 40-
        01226.
              +-HFN 2-76 p116 tape   ++RR 1-76 p66 tape
              -NR 3-75 p4
      Masquerade suite. cf Gayaneh suite.
      Pictures of childhood (4). cf HMV HQS 1364.
1481 Rhapsody concerto, violoncello and orchestra. Symphony, no. 3,
        C major. Mstislav Rostropovich, vlc; Harry Grodberg, org;
        Bolshoi Theatre Trumpeters' Ensemble; USSR Symphony Orchestra,
        MPO; Kyril Kondrashin, Yevgeny Svetlanov. HMV Melodiya ASD
        3108.
              +-Gr 9-75 p458   +-MT 12-75 p1071
              +-HFN 8-75 p77   +RR 8-75 p34
              +MM 1-76 p42

1482 Spartacus. Bolshoi Theatre Orchestra; Algis Zhuratis. Columbia
         Melodiya D4M 33493 (4). (also HMV Melodiya SQ SLS 5061).
                   +Gr 10-76 p602   -NR 8-75 p4
                   +-HF 11-75 p105   ++RR 8-76 p40
                   +-HFN 8-76 p81   +St 9-75 p107
         Spartacus: Adagio. cf Philips 6780 755.
         Symphony, no. 3, C major. cf Rhapsody concerto, violoncello and
            orchestra.
KHRENNIKOV, Tikhon
1483 Concerto, piano, no. 1, op. 1, F major. Concerto, piano, no. 2,
         op. 21, C major. Concerto, violoncello, op. 16, C major.
         Tikhon Khrennikov, pno; Michael Khomitser, vlc; MPO, MRSO,
         USSR Symphony Orchestra; Kiril Kondrashin, Yevgeny Svetlanov,
         Gennady Rozhdestvensky. HMV Melodiya ASD 3227.
                   +Gr 11-76 p792   +RR 9-76 p55
                   +MT 12-76 p1005
         Concerto, piano, no. 2, op. 21, C major. cf Concerto, piano, no.
            1, op. 1, F major.
         Concerto, violoncello, op. 16, C major. cf Concerto, piano, no. 1,
            op. 1, F major.
KIENZL, Wilhelm
         Evangelimann: Selig song. cf Rococo 5379.
         Kahn Szene, neuer Walzer. cf Rococo 2049.
KING
         Hosts of freedom. cf Michigan University SM 0002.
KIRCHNER, Theodore
         Allegro con passione, op. 30, no. 17. cf Piano works (Genesis
            GS 1032).
         Aquarellen, op. 21, no. 3. cf Piano works (Genesis GS 1032).
         Days gone by, op. 73, no. 4. cf Piano works (Genesis GS 1032).
         Elegy, op. 73, no. 16. cf Piano works (Genesis GS 1032).
         Moderato, op. 30, no. 8. cf Piano works (Genesis GS 1032).
         Moderato, op. 71, no. 100. cf Piano works (Genesis GS 1032).
         Nocturne, op. 73, no. 12. cf Piano works (Genesis GS 1032).
1484 Piano works: Allegro con passione, op. 30, no. 17. Aquarellen,
         op. 21, no. 3. Days gone by, op. 73, no. 4. Elegy, op. 73,
         no. 16. Moderato, op. 30, no. 8. Moderato, op. 71, no. 100.
         Nocturne, op. 73, no. 12. Romanze, op. 73, no. 6. Spring
         greeting, op. 73, no. 2. VOLKMANN: Sonata, piano, no. 12,
         C minor. Fantasie, op. 25a, C major. Adrian Ruiz, pno.
         Genesis GS 1032.
                   +-Gr 9-76 p445
         Romanze, op. 73, no. 6. cf Piano works (Genesis GS 1032).
         Spring greeting, op. 73, no. 2. cf Piano works (Genesis GS 1032).
KIRKPATRICK
         Away in a manger. cf Abbey LPB 761.
         Away in a manger. cf HMV CSD 3774.
         KIRSTEN FLAGSTAD MEMORIAL ALBUM. cf Rococo 5380.
KITTREDGE, Walter
         Tenting on the old camp ground. cf CBS 61746.
         Tenting on the old camp ground. cf Vox SVBX 5304.
KLEINSINGER
         Pavane for Seskia. cf Orion ORS 76231.
KLING
         Kitchen symphony, op. 445. cf Angel S 36080.

KLOHR
        The billboard. cf Michigan University SM 0002.
KLUSAK, Jan
        Reydowak. cf Supraphon 111 1390.
KNAEBLE
        General Taylor storming Monterey. cf Library of Congress OMP
            101/2.
KNIGHT
        Songs: Rocked in the cradle of the deep. cf Argo ZFB 95/6.
KNIPPER, Lev
        Cavalry of the Steppes. cf Canon VAR 5968.
        Cavalry song. cf HMV ASD 3200.
KOCH,   von
        Canto e danza. cf BIS LP 30.
KOCH, Erland von
        Dance, no. 2. cf Swedish Society SLT 33229.
1485 Nordic rhapsody, op. 26. BYSTROEM: Sinfonia concertante, violon-
            cello and orchestra. Erling Blöndal-Bengtsson, vlc; Stockholm
            Radio Symphony Orchestra; Stig Westerberg. Swedish Society
            SLT 33136.
                    +RR 4-76 p50   +St 6-75 p51
1486 Oxberg variations. STENHAMMAR: Serenade, op. 31, F major. Stock-
            holm Philharmonic Orchestra; Rafael Kubelik, Stig Westerberg.
            Swedish Society SLT 33227.
                    +RR' 10-76 p66
KOCSAR, Miklos
        The dawn awakens. cf Hungaroton SLPX 11696.
KODALY, Zoltan
1487 Ballet music. Hungarian tunes. Summer evening. Variations
            on a Hungarian folksong (Peacock). PH; Antal Dorati. Decca
            SXL 6714. (Reissue from SXLM 6665/7).
                    +Gr 12-76 p996   +RR 12-76 p61
                    +HFN 11-76 p170
1488 Chorale preludes, violoncello and piano (Bach arr. Kodaly).
            Duo, violin and violoncello, op. 7. Epigrams, voice or instru-
            ment with piano. Peter Komlos, vln; Ede Banda, vlc; Béla
            Kovács, clt; Kornél Zempléni, Adám Fellegi, pno. Hungaroton
            SLPX 11559.
                    +NR 8-76 p7
        Duo, violin and violoncello, op. 7. cf Chorale preludes, violon-
            cello and piano.
        Die Engel und die Hirten. cf RCA PRL 1-8020.
        Epigrams, voice or instrument with piano. cf Chorale preludes,
            violoncello and piano.
1489 Háry János: Suite. PROKOFIEV: Lieutenant Kijé suite, op. 60.
            Netherlands Radio Philharmonic Orchestra; Antal Dorati. Decca
            PFS 4355. Tape (c) KPFC 4355. (also London 21146-Tape (c)
            5-21146).
                    +Gr 5-76 p1758   -MM 11-76 p43
                    +HFN 5-76 p101   +-RR 4-76 p47
                    +HFN 5-76 p117 tape   +-RR 6-76 p88 tape
1490 Háry János: Suite. Minuetto serio. Symphony, C major. PH;
            John Leach, cimbalon; Antal Dorati. Decca SXL 6713. (Reissue
            from SXLM 6665/7).
                    +Gr 8-76 p286   +RR 7-76 p53
                    +HFN 8-76 p93

1491 Háry János: Suite. PROKOFIEV: Lieutenant Kijé suite, op. 60.
       PO; Eugene Ormandy. RCA ARL 1-1325.
              +MJ 4-76 p31   ++SFC 2-8-76 p29
              ++NR 3-76 p1
     Hungarian dances (3). cf Seraphim S 60259.
     Hungarian tunes. cf Ballet music.
     Minuetto serio. cf Háry János: Suite.
1492 Missa brevis. Pange lingua. Elizabeth Gale, Sally Le Sage, Hannah
       Francis, s; Alfreda Hodgson, con; Ian Caley, t; Michael Rippon,
       bs; Brighton Festival Chorus; Christopher Bowers-Broadbent,
       org; Laszlo Heltay. Decca SXL 6803.
              +Gr 12-76 p1033   +-RR 12-76 p92
              +HFN 12-76 p145
     Pange lingua. cf Missa brevis.
     Sonata, solo violoncello, op. 8. cf BACH: Suite, solo violoncello,
       no. 3, S 1009, C major.
     Songs: Csillagoknak teremtöje; Kiolvaso. cf BIS LP 17.
     Songs: Egytem Begyeterm; Gömör District folk songs (3). cf Hung-
       aroton SLPX 11762.
     Songs: Dance song; Egytem Begyetem; Italian madrigal, no. 3;
       László Lengyel; Peal of bells; Stork song. cf Hungaroton SLPX
       11696.
     Summer evening. cf Ballet music.
     Symphony, C major. cf Háry János: Suite.
     Variations on a Hungarian folksong (Peacock). cf Ballet music.
     Variations on a Hungarian folksong (Peacock). cf BARTOK: Dance
       suite.
KOECHLIN, Charles
     Les Bandar-log, op. 176. cf BOULEZ: Le soleil des eaux.
     Songs: Si tu le veux. cf Columbia M 33933.
KOESTSIER, Jan
     Petite suite. cf CIVIL: Tarantango.
KOHAUT, Carl
     Concerto, guitar (lute), F major. cf HANDEL: Concerto, 2 lutes,
       strings and recorders, op. 4, no. 6, B flat major.
KOHOUTEK, Ctirad
1493 Teatro del mondo. LIDL: Hic homo sum. Miroslav Svejda, t; Prague
       Radio Chorus; Jiří Pokorný, pno; Jaroslav Echtner, Ivo Kies-
       lich, perc; Brno State Philharmonic Orchestra; Frantisek Jílek.
       Supraphon 112 1460.
              +NR 10-76 p8   +-RR 4-76 p72
KOMADINA, Vojin
     Mikrosonate. cf KULJERIC: Omaggio à Lukačić.
KOMZAK, Karel
     Erzherzog-Albrecht-Marsch, op. 136. cf DG 2721 077.
     Erzherzog-Albrecht-Marsch, op. 136. cf Polydor 2489 523.
     Vindobona-Marsch. cf DG 2721 077.
KONTONSKI, Wlodzimerz
     Canto für kammerorchester. cf Caprice RIKS LP 34.
KORLING
     Aftonstamning. cf Rubini GV 65.
KORNGOLD, Erich
     Die Kathrin: Letter song; Prayer. cf Songs and arias (Entr'acte
       ERS 6502).
     Love letter, op. 9. cf Songs and arias (Entr'acte ERS 6502).
     Much ado about nothing, op. 11: Holzapfel und Schlehwein, Garden
       scene. cf RCA ARM 4-0942/7.

Much ado about nothing, op. 11: The page's song. cf Songs and
    arias (Entr'acte ERS 6502).
The private lives of Elizabeth and Essex: Come live with me. cf
    Songs and arias (Entr'act ERS 6502).
Der Ring des Polykrates: Diary song. cf Songs and arias (Entr'
    acte ERS 6502).
The sea hawk: Old Spanish song. cf Songs and arias (Entr'acte
    ERS 6502).
The silent serenade: Song of bliss; Without you. cf Songs
    and arias (Entr'acte ERS 6502).
1494 Songs and arias: Die Kathrin: Letter song; Prayer. Love letter,
    op. 9. Much ado about nothing, op. 11: The page's song. The
    private lives of Elizabeth and Essex: Come live with me. Der
    Ring des Polykrates: Diary song. The sea hawk: Old Spanish
    song. The silent serenade: Song of bliss; Without you. Die
    tote Stadt, op. 12: Lute song. Polly Jo Baker, s; George
    Calusdian, pno. Rediffusion Entr'acte ERS 6502.
        +Gr 12-76 p1033   +RR 11-76 p39
        +NR 7-76 p9
1495 Suite, op. 11. WEILL: Quodlibet, op. 9. Westphalian Symphony
    Orchestra; Siegfried Landau. Candide CE 31091
        +-NR 8-76 p4
1496 Die tote Stadt, op. 12. Carol Neblett, s; Rose Wagemann, ms;
    René Kollo, t; Hermann Prey, bar; Benjamin Luxon, bs; Bavarian
    Radio Chorus; Munich Radio Orchestra; Erich Leinsdorf. RCA ARL
    3-1199 (3).
        +Gr 1-76 p1230   +-ON 1-11-76 p33
        +-HF 1-76 p88   +RR 2-76 p22
        +HFN 3-76 p97   +-SR 1-24-76 p52
        +MJ 3-76 p24   +-St 3-76 p115
        +-NR 1-76 p11   +STL 2-8-76 p36
        +-NYT 11-16-75 pD1
    Die tote Stadt, op. 12: Lute song. cf Songs and arias (Entr'acte
        ERS 6502).
    Trio, piano, op. 1. cf IVES: Trio, violin, violoncello and piano.
KOUGUELL
    Berceuse. cf Da Camera Magna SM 93399.
KOUTZEN, Boris
    Sonata, solo violin. cf CASTELNUOVO-TEDESCO: Concertino, harp,
        op. 93.
KRAMAR-KROMMER, Frantisek
    Quartet, flute and strings, op. 75, D major. cf REICHA: Quartet,
        flute and strings, op. 98, no. 1, G minor.
KREBS, Johann
    Concerto, guitar and strings, G major. cf FASCH: Concerto, guitar
        and strings, D minor.
    Fugue on B-A-C-H. cf Columbia M 33514.
    Sei Lob und Ehr dem höchsten Gut. cf Pelca PRSRK 41017/20.
    Warum betrubst du dich mein Herz. cf Pelca PRSRK 41017/20.
KREISLER, Fritz
    Allegretto in the style of Boccherini. cf ALBENIZ: Tango.
    Andantino in the style of Martini. cf ALBENIZ: Tango.
    Caprice viennois, op. 2. cf ALBENIZ: Tango.
    Chanson Louis XIII and Pavane in the style of Couperin. cf
        ALBENIZ: Tango.
    La chasse. cf Columbia M2 33444.

La chasse (in the style of Cartier). cf CBS 76420.
Cradle song. cf Pye Ember GVC 51.
La gitana. cf ALBENIZ: Tango.
Liebesfreud. cf ALBENIZ: Tango.
Liebesfreud. cf CBS 30063.
Liebesfreud. cf Odyssey Y 33825.
Liebesleid. cf CBS 30063.
Liebesleid. cf Odyssey Y 33825.
Londonderry air (arr.). cf Connoisseur Society CS 2070.
Minuet. cf RCA ARM 4-0942/7.
Recitativo and scherzo capriccio, op. 6. cf ALBENIZ: Tango.
Rondino on a theme by Beethoven. cf ALBENIZ: Tango.
Schön Rosmarin. cf ALBENIZ: Tango.
Schön Rosmarin. cf HMV HLM 7093.
Schön Rosmarin. cf Odyssey Y 33825.
Schön Rosmarin. cf RCA LRL 1-5131.
Sicilienne et rigaudon. cf RCA ARM 4-0942/7.
Tambourin chinois, op. 3. cf ALBENIZ: Tango.
Tambourin chinois, op. 3. cf HMV HLM 7077.
Variations on a theme by Corelli in the style of Tartini. cf
   HMV SEOM 22.
Variations on a theme by Corelli in the style of Tartini. cf
   HMV SXLP 30188.
KREJCI, Isa
   Maly balet. cf DOUBRAVA: Pastorale.
   Songs on texts by V. Nezval (5). cf Panton 110 385.
KRENEK, Ernst
1497 Aulokithara, op. 213a. Echoes from Austria, op. 166. Sacred
   pieces, op. 210 (3). Wechselrahmen, op. 189. James Ostryniec,
   ob; Karen Lindquist, hp; Beverly Ogdon, s; Ernst Krenek, pno;
   COD Vocal Ensemble; John Norman. Orion ORS 76246.
        +-NR 10-76 p6
   Echoes from Austria, op. 166. cf Aulokithara, op. 213a.
1498 O lacrymosa (3 songs). Santa Fe timetable, chorus. Tape and
   double, 2 pianos. Toccata, accordion. Genevieve Weide, s;
   Young McMahan, accordion; John Dare, Patricia Marcus, William
   Tracy, pno; California State University Northridge Chamber
   Singers; John Alexander. Orion ORS 75204.
        +-HF 6-76 p80   +NR 2-76 p8
        ++MJ 7-76 p57
   Sacred pieces, op. 210 (3). cf Aulokithara, op. 213a.
   Santa Fe timetable, chorus. cf O lacrymosa.
   Tape and double, 2 pianos. cf O lacrymosa.
   Toccata, accordion. cf O lacrymosa.
   Wechselrahmen, op. 189. cf Aulokithara, op. 213a.
KRESTYANIN, Feodor
   Befittingly. cf HMV Melodiya ASD 3102.
KREUTZER, Conradin
   Das Nachtlager von Granada: Overture. cf Supraphon 110 1637
KRIEGER, Arthur
   Short piece. cf Odyssey Y 34139.
KRIUKOV, Vladimir
   Concerto-poem, trumpet and orchestra, op. 59. cf ARUTYUNIAN: Con-
   certo, trumpet and orchestra, A flat major.
KUBIK, Ladislav
1499 Lament of a warrior's wife. LOUDOVA: Chorale, orchestra. Spleen:

Hommage à Charles Baudelaire. RYBAR: Sonata, 12 wind instruments. Brigita Sulcová, s; Karel Rehák, vla; Josef Horák, bs
clt; Emma Kovárnová, pno; Ivo Kieslich, Oldřich Satava, Vladimir Vlasák, Jan Klousa, perc; Lai-Thuy-Hein, reciter; Czech
Philharmonic Wind Ensemble, Prague Symphony Orchestra; Vladimir
Cerný, Ladislav Slovák. Panton 110 490.
          +RR 11-76 p97

KUCERA, Vaclav
     Invariant, bass clarinet, piano and stereo tape recorder. cf
          Supraphon 111 1390.
     The kenetic ballet: The labyrinth; The spiral. cf ISTVAN: Isle
          of toys.

KUCHAR, Jan
     Fantasia, D minor. cf Panton 110 418/9.

KUHNAU, Johann
     Sonata, organ, no. 1. cf Pelca PRSRK 41017/20.

KUKUCK
     Die Brücke. cf BIS LP 2.

KULJERIC, Igor
1500 Omaggio á Lukačić. LEBIC: Meditacije za dva. KOMADINA: Mikro-
          sonate. MAKSIMOVIC: Tri Haiku-a RADICA: Extensio II. SAKAC:
          Turm-Musik. D. Thune, vla; J. Stojanović, vlc; RT Belgrade
          Choir and Orchestra; RTZ Symphony Orchestra, MBZ Ensemble; Milan
          Horvat, K. Sipus and other conductors.
               +-Te 9-76 p33

KUNST, Joe
1501 No time. KUNST/VRIEND: Elements of logic. STRAESSER: Ramassasin.
          VRIEND: Huantan. Lia Rittier, ms; Emile Biessen, flt; Nether-
          lands Percussion Ensemble; Residentie Orchestra; Ensemble;
          Radio Wind Ensemble; Hubert Soudant, Ernest Bour, Jan Vriend,
          Hans Vonk. Donemus DAVS 7475/3.
               *Te 9-76 p28

KUNST, Joe/Jan Vriend
     Elements of logic. cf KUNST: No time.

KUPFERMAN, Meyer
1502 Fantasy concerto, violoncello, piano and tape. MENDELSSOHN:
          Sonata, violoncello and piano, no. 2, op. 58, D major. Laszlo
          Varga, vlc; Paul Hersh, pno. Serenus SRS 12059.
               +ARG 11-76 p27   +NR 4-76 p7
     Madrigal, brass quartet. cf HUSTON: Sounds at night.

LABREY
     Songs: Le chant des Charpentiers; Le chant des demoisselles de
          Magasin. cf Club 99-101.

LACALLE
     Amapola. cf DG 2530 700.

LACHNER
     Introduction and fugue, op. 62, D minor. cf Pelca PRS 40581.

LALO, Edouard
1503 Concerto, violoncello, D minor. Symphonie espagnole, op. 21.
          Paul Tortelier, vlc; Yan Pascal Tortelier, vln; Birmingham
          City Symphony Orchestra; Louis Frémaux. HMV SQ ASD 3209.
               +Gr 6-76 p49   ++RR 6-76 p43
               +HFN 6-76 p89
1504 Concerto, violoncello, D minor. SCHUMANN: Concerto, violoncello,
          op. 129, A minor. Csabo Onczay, vlc; Hungarian Radio and Tele-
          vision Orchestra; Antal Jancsovics. Hungaroton SLPX 11705.
               +-NR 3-76 p4   +-RR 3-76 p48

Concerto, violoncello, D minor. cf BRUCH: Kol Nidrei, op. 47.
1505 Rapsodie norvégienne, op. 21. Le Roi d'Ys: Overture. Symphony,
      G minor. Monte Carlo Opera Orchestra; Antonio de Almeida.
      Philips 6500 927.

*Audio 8-76 p78	+NR 6-76 p2
+Gr 4-76 p1603	+-RR 3-76 p45
+-HFN 3-76 p95	+SFC 5-23-76 p36
++MT 9-76 p747	

      Rapsodie norvégienne, op. 21. cf Symphonie espagnole, op. 21.
      Rapsodie norvégienne, op. 21. cf CHABRIER: Le Roi malgré lui:
        Fête polonaise.
      Le Roi d'Ys: Aubade. cf HMV RLS 719.
      Le Roi d'Ys: Cher mylio. cf Angel S 37143.
      Le Roi d'Ys: En silence pourquoi souffrir. cf Club 99-97.
      Le Roi d'Ys: Overture. cf Rapsodie norvégienne, op. 21.
      Le Roi d'Ys: Vainement ma bien-aimée. cf Decca SKL 5208.
1506 Symphonie espagnole, op. 21. Rapsodie norvégienne, op. 21.
      Pierre Amoyal, vln; Monte Carlo Opera Orchestra; Paul Paray.
      Erato STU 70771. (also Musical Heritage Society Tape (c)
      MHC 2101).

-Gr 6-73 p55	-HFN 6-73 p1178
+-HF 2-76 p130 tape	+RR 6-73 p53

      Symphonie espagnole, op. 21. cf Concerto, violoncello, D minor.
      Symphonie espagnole, op. 21: Andante. cf RCA ARM 4-0942/7.
      Symphony, G minor. cf Rapsodie norvégienne, op. 21.
LAMB
      American beauty rag; Ragtime nightingale. cf Washington Univer-
        sity Press OLY 104.
      Madrigal, 3 saxophones. cf Coronet LPS 3030.
      Songs: The volunteer organist. cf Argo ZFB 95/6.
LAMBERT, Constant
      Concerto, piano. cf Polydor 2383 391.
      Elegiac blues. cf Polydor 2383 391.
      Elegy. cf Polydor 2383 391.
1507 The Rio Grande. WALTON: Symphony, no. 2. Scapino overture.
      Portsmouth Point overture. Jean Temperley, ms; Christina Ortiz,
      pno; London Madrigal Singers; LSO; André Previn. Angel S 37001.
      (also HMV ASD 2990 Tape (c) TC ASD 2990).

++Gr 5-74 p2026	+-NYT 6-2-74 pD24
++Gr 6-76 p102 tape	+RR 5-74 p44
/HF 10-74 p117	+RR 12-76 p111 tape
+HFN 7-76 p105 tape	+St 9-74 p124
+-NR 8-74 p2	

      The Rio Grande. cf FRANCAIX: Concerto, piano, D major.
      Songs: Blow the man down; Long time ago; Shenandoah; When Johnny
        comes marching home. cf Department of Defense Bicentennial
        Edition 50-1776.
LaMONTAINE, John
      Birds of paradise. cf HANSON: Concerto, piano, op. 36.
LANDINI, Francisco
      Songs: Se la nimica mie; Adiu adiu. cf 1750 Arch S 1753.
LANE, Richard
      Songs (4). cf HANSON: Elegy in memory of Serge Koussevitzky.
LANG
      Sprintode. cf Hungaroton SLPX 11762.
LANG, C. S.
      Tuba tune, op. 15, D major. cf Audio EAS 16.

LANG, Josephine
    Sie liebt mich. cf Gemini Hall RAP 1010.
LANGFORD, Gordon
    Rhapsody, trombone and brass band. cf Pye TB 3004.
LANGLAIS, Jean
    Incantation pour un jour saint. cf Polydor 2460 262.
1508 Méditations sur L'Apocalypse (5). Marie-Louise Jacquet, org.
    Arion ARN 38312.
        +MU 1-76 p14
    Pasticcio. cf GUILMANT: Sonata, organ, no. 1, op. 42, D minor.
    Suite française: Nazard. cf Wealden WS 142.
    Te Deum. cf Argo ZRG 807.
    Te Deum. cf Wealden WS 142.
    Te Deum, op. 5, no. 3. cf Audio EAS 16.
    Triptyque. cf Vista VPS 1029.
LANG-MULLER, Peter
    Songs: Der var engang: Se natten er svanger; I Wurzburg kling der
        klokker. cf Cantilena 6238.
LANNER, Josef
1509 Dornbacher Ländler, op. 9. Die Romantiker, op. 167. Tyroler
        Ländler, op. 6. STRAUSS, J. II: Wiener Blut, op. 354. Wiener
        Bonbons, op. 307. STRAUSS, Josef: Mein Lebenslauf ist Lieb
        und Lust, op. 263. Alexander Schneider Quintet. CBS 61717.
    Tape (c) 40-61717.
        +-Gr 9-76 p478   +RR 8-76 p63
        +HFN 10-76 p183   +RR 11-76 p109 tape
    Favorit-Polka, op. 201. cf Saga 5421.
    Mitternachtswalzer, op. 8. cf Nonesuch H 71141.
    The parting of the ways. cf Saga 5411.
    Regata-Galopp, op. 134. cf Nonesuch H 71141.
    Die Romantiker, op. 167. cf Dornbacher Ländler, op. 9.
    Summer night's dream. cf Saga 5411.
    Tyroler Ländler, op. 6. cf Dornbacher Ländler, op. 9.
LANSKY, Paul
    Mild und Leise. cf Odyssey Y 34139.
    Modal fantasy. cf CRI SD 342.
LANTINS, Arnold de
    Puisque je voy. cf Telefunken ER 6-35257 (2).
LANTINS, Hugo de
    Gloria. cf Telefunken ER 6-35257 (2).
LAPARRA, Raoul
    Habañera: Et c'est a moi que l'on dit "Chante". cf Club 99-101.
LARA, Agustín
    Granada. cf DG 2530 700.
LARSSON, Lars-Erik
1510 Concertino, double bass and strings, op. 45, no. 11. Little
        serenade, op. 12. ROMAN: Concerto, violin, D minor. Symphony,
        B flat major. Leo Berlin, vln; Luigi Ossoinak, double-bs;
        Philharmonic Chamber Ensemble. Swedish Society SLT 33187.
        +RR 4-76 p52
    The disguised God, op. 24. cf GRIEG: Landkjenning, op. 31.
    Folkvisenatt. cf Swedish Society SLT 33229.
    Little serenade, op. 12. cf Concertino, double bass and strings,
        op. 45, no. 11.
    Orchestral variations, op. 51. cf KARKOFF: Symphony, no. 4,
        op. 69.

1511   Pastoral suite, op. 19. WIREN: Serenade, strings, op. 11.*
           Stockholm Philharmonic Orchestra; Stig Westerberg. Discofil
           SLT 33176. (*Reissue from Decca SEC 5066).
                +-Gr 1-76 p1197   +-RR 7-76 p54
1512   Sonatina, no. 1, op. 16. RANGSTROM: Legends from Lake Mälanen.
           STENHAMMAR: Fantasias, op. 11 (3). WIREN: Ironical minatures,
           op. 19. Staffan Scheja, pno. RCA LSC 3119.
                +RR 7-76 p72

LASCEUX
       Flutes. cf Calliope CAL 1917.
       Noël Lorrain. cf Calliope CAL 1917.
       Récit de tièrce. cf Calliope CAL 1917.
       Symphonie concertante. cf Calliope CAL 1917.
LASERNA, Blas
       Tonadillas. cf HMV RLS 723.
LASSUS, Roland de
1513   Bell Amfitrit altera mass. Psalmus Poenitentalis VII. Christ
           Church Cathedral Choir; Simon Preston. Argo ZRG 735.
                +Gr 8-74 p382   +RR 7-74 p75
                +MQ 1-76 p144   ++St 2-76 p106
                *NR 2-75 p7
       Cathalina, apra finestra; Matona mia cara. cf L'Oiseau-Lyre
           12BB 203/6.
       Je l'ayme bien; Quand mon mari vient de dehors. cf Titanic TI 4.
1514   Les larmes de Saint Pierre. Raphaël Passaquet Vocal Ensemble;
           Raphaël Passaquet. Harmonia Mundi HMU 961.
                +-Gr 11-75 p875   +St 10-76 p126
                +RR 5-76 p70
       Madrigal dell'eterna. cf Argo ZRG 823.
       Matona, mia cara. cf HMV CSD 3756.
       Morescas: O Lucia, miau, miau; Lucia, Celu, ahi, hai; Hai, Lucia;
           Allala, pia calia; Cathalina; Chi chilichi; Canta Giorgia. cf
           Prophetiae Sibyllarum.
1515   Motets: Ave Regina caelorum; Salve Regina; O mors, quam amara est.
           Penitential psalms: Miserere mei, Deus; Domine, ne in furore
           tuo. Roderick Skeaping, Trevor Jones, viol; Pro Cantione
           Antiqua, Early Music Wind Ensemble; Bruno Turner. DG Archive
           2533 290.
                +Gr 10-75 p669   +RR 10-75 p85
                +-HF 5-76 p86   ++St 2-76 p106
                +HFN 11-75 p155
       Penitential psalms: Miserere mei, Deus; Domine, ne in furore tuo.
           cf Motets (DG Archive 2533 290).
1516   Prophetiae Sibyllarum: Carmina chromatico; Sibylla Persica;
           Sibylla Libyca; Sibylla Delphica; Sibylla Cimmeria; Sibylla
           Samia; Sibylla Cumana; Sibylla Hellespontiaca; Sibylla Phrygia;
           Sibylla Europaea; Sibylla Tiburtina; Sibylla Erythraea;
           Sibylla Agrippa. Morescas: O Lucia, miau, miau; Lucia, Celu,
           ahi, hai; Hai, Lucia; Allala, pia calia; Cathalina; Chi
           chilichi; Canta Giorgia. Munich Vocal Soloists and Flute
           Consort; Hans Ludwig Hirsch. Telefunken AW 6-41889.
                ++Gr 1-76 p1225   ++RR 2-76 p59
                +HFN 1-76 p110
       Psalmus Poenitentalis VII. cf Bell Amfitrit altera mass.
1517   Sacrae lectiones ex propheta Job. Raphaël Passaquet Vocal
           Ensemble; Raphaël Passaquet. Harmonia Mundi HMU 761/2.
                +-RR 5-76 p70

1518 Sacrae lectiones ex propheta Job. Prague Madrigal Singers (Prager
Madrigalisten); Miroslav Venhoda. Telefunken 6-41274.
+-Gr 4-76 p1646   +NR 10-75 p6
+HF 5-76 p86   +-RR 5-76 p70
+HFN 5-76 p101
Sacrae lectiones ex propheta Job: Parce mihi, Domine; Taedet
animam meam vitae meae. cf Harmonia Mundi HMU 473.

1519 Songs: Alma redemptoris mater; Omnes de Saba venient; Psalmus
poenitentalis V: Domine exaudi orationem meam; Tui sunt
coeli; Salve regina. Christ Church Cathedral Choir; Simon
Preston. Argo ZRG 795.
+Gr 9-76 p454   ++RR 7-76 p76
+HFN 7-76 p89
Songs: Margo labourez les vignes; La nuit froide et sombre. cf
Argo ZRG 667.

LATOUR
Rule Britannia, a favorite air, with variations for the harpsichord or pianoforte. cf Angel S 36095.

LATVIAN FOLK SONGS. cf Kaibala 50E 01.

LAURO, Antonio
Suite venezelano: Vals. cf Saga 5412.

LAVIGNE, Philippe de
Sonata, recorder, C major. cf Cambridge CRS 2826.

LAVRY, Marc
Jewish dances (3). cf Da Camera Magna SM 93399.

LAW, Andrew
Bunker Hill. cf Nonesuch H 71276.
Bunker Hill. cf Vox SVBX 5350.

LAWES, Henry
Sweet stay awhile. cf National Trust NT 002.

LAWES, William
Courtly masquing ayres. cf Orion ORS 76216.
Gather ye rosebuds. cf National Trust NT 002.
Sonata, no. 7, D minor. cf BYRD: Fantasias, nos. 2 and 3.
Suite, no. 2, F major. cf BYRD: Fantasias, nos. 2 and 3.
Suite, no. 3, B flat major, excerpts. cf BYRD: Fantasias, nos.
2 and 3.

LEAF, Robert
Let the whole creation cry. cf Columbia M 34134.

LEBIC, Lojze
Meditacije za dva. cf KULJERIC: Omaggio à Lukačić.

LECLAIR, Jean-Marie
1520 Sonatas, flute, op. 1, nos. 2, 6; op. 9, nos. 2, 7; op. 2, nos. 1,
3, 5, 8, 11. Christian Lardé, flt; Jean Lamy, vla da gamba;
Huguette Dreyfus, hpd. Telefunken ER 6-48074 (2).
+-Gr 2-76 p1355   ++RR 1-76 p49
+HFN 1-76 p110
Sonatas, flute, op. 2, nos. 1, 3, 5, 8, 11. cf Sonatas, flute,
op. 1, nos. 2, 6.
Sonatas, flute, op. 9, nos. 2, 7. cf Sonatas, flute, op. 1, nos.
2, 6.
1521 Sonatas, 2 saxophones, C major, F major, A flat major. TELEMANN:
Sonatas, 2 saxophones, nos. 1-3, 6 (Canonic). Paul Brodie,
Jean-Marie Londeix, sax. Golden Crest RE 7062.
+IN 5-76 p12
1522 Sonata, violin, no. 3, D minor. NARDINI: Sonata, violin, D major.

VERACINI: Sonata, violin, op. 1, no. 3. VIVALDI: Sonata,
violin, op. 2, no. 2, A major. Arthur Grumiaux, vln; Istvan
Hajdu, pno. Philips 6500 879.
   ++MJ 11-76 p45   +SR 9-18-76 p50
   +NR 11-76 p7
LECUONA, Ernesto
   Malagueña. cf Opus unnumbered.
   Siboney. cf DG 2530 700.
LEEUW, Ton de
1523 Gending. Midare. Music, violin. Night music. Abbie de Quant,
    flt; Michiko Takahashi, marimba; Jos Verkoeyen, vln; Gamelan
    Ensemble; Ton de Leeuw. Donemus CV 7602.
    +RR 11-76 p90
   Midare. cf Gending.
   Music, violin. cf Gending.
   Night music. cf Gending.
   Quartet, strings, no. 2. cf CRUMB: Black angels, electric string
    quartet.
LEGINSKA, Ethel
   Victorian portraits. cf HAUBIEL: Metamorphoses.
LEGRAND
   One day. cf HMV CSD 3766.
LEGRANT, Guillaume
   Entre vous, noviaux mariés. cf Nonesuch H 71326.
LEHAR, Franz
   Beautiful world: Darling, trust in me; Wonderful world. cf CBS
    30070.
   The Count of Luxemburg: Lieber Freund...Bist du's lachendes Glück.
    cf Works, selections (London OSA 26220).
   The Count of Luxembourg: Medley. cf CBS 30070.
   Frasquita: Serenade. cf CBS 30070.
   Friederike: Poet's song. cf KALMAN: Neut nacht hab ich getraumt
    von dir.
   Friederike: Warum hast du mich wachgeküsst. cf Works, selections
    (London OSA 26220).
   Giuditta: Schön wie die blaue Sommernacht. cf HMV CSD 3748.
   Gold and silver waltz. cf Decca SB 322.
   The land of smiles (Das Land des Lächelns): Bei einem Tee en deux;
    Dein ist mein ganzes Herz; Wer hat die Liebe uns ins Herz
    gesenkt. cf Works, selections (London OSA 26220).
   The land of smiles: Dein ist mein ganzes Herz. cf DG 2530 700.
   The land of smiles: Immer nur lacheln; Von Apfelblüten einen
    Kranz; Dein ist mein ganzes Herz. cf Angel S 37108.
   Das Land des Lächelns: Su Tshong's songs. cf KALMAN: Heut nacht
    hab ich getraumt von dir.
   Das Land des Lächelns: Wer hat die Liebe uns ins Herz gesenkt;
    Bei einem Tee en deux. cf HMV CSD 3748.
   The land of smiles: Yours is my heart alone. cf CBS 30070.
1524 Die lustige Witwe (The merry widow): excerpts. Adelaide Singers;
    Adelaide Symphony Orchestra; John Lanchbery. Angel S 37092.
    Tape (c) 4XS 37092.
    +HF 11-76 p97   +MJ 12-76 p28
1525 The merry widow: Ballet music (arr. Lanchberry). Adelaide Sym-
    phony Orchestra; Adelaide Singers; John Lanchbery. HMV
    CSD 3772. (also Angel S 37092 Tape (c) 4XS 37092).
    +HFN 9-76 p112   +-RR 8-76 p41

Die lustige Witwe: Lippen schweigen. cf HMV CSD 3748.
The merry widow: Lippen Schweigen. cf Works, selections (London
    OSA 26220).
1526 Die lustige Witwe: Verehrteste Damen und Herren; So kommen Sie;
    Bitte, meine Herren; Oh, Vaterland; Damenwahl; Es lebt eine
    Vilja Heia, Mädel, aufgeschaut; Wie die Weiber man behandelt;
    Wie eine Rosenknospe; Den Herrschaften hab ich was zu erzählen;
    Ja wir sind es, die Grisetten; Lippen schweigen; Ja das Studium
    der Weiber ist schwer. Elizabeth Harwood, Teresa Stratas, s;
    René Kollo, Werner Hollweg, Donald Grobe, Werner Krenn, t;
    Zoltan Kéléman, bar; German Opera Chorus; BPhO; Herbert von
    Karajan. DG 2530 729. (Reissue from 2707 070).
            +-Gr 11-76 p893   +RR 11-76 p39
Luxembourg march. cf Philips 6308 246.
Paganini: Gern hab ich die Frau'n geküsst; Niemand liebt Dich, du
    Himmel auf Erden. cf Works, selections (London OSA 26220).
Paganini: I have been in love before; Love, you invaded my senses.
    cf CBS 30070.
Schön ist die Welt: Frei und jung dabei; Schön ist die Welt; Ich
    bin verliebt. cf Works, selections (London OSA 26220).
Wiener Frauen: Overture. cf Supraphon 110 1638.
1527 Works, selections: The best of Franz Lehar: The Count of Luxem-
    bourg: Lieber Freund...Bist du's lachendes Glück. Friederike:
    Warum hast du mich wachgeküsst. The land of smiles: Bei einem
    Tee en deux; Dein ist mein ganzes Herz; Wer hat die Liebe uns
    ins Herz gesenkt. The merry widow: Lippen Schweigen. Paganini:
    Gern hab' ich die Frau'n geküsst; Niemand liebt Dich; Liebe
    du Himmel auf Erden. Schön ist die Welt: Frei und jung dabei;
    Schön ist die Welt; Ich bin verliebt. Der Zarewitsch: Wolga-
    lied; Kosende Wellen. Werner Krenn, t; Renate Holm, s; Vienna
    Volksoper Orchestra; Anton Paulik. London OSA 26220. (also
    Decca SXL 6711).
            +Gr 12-75 p1098   ++RR 12-75 p28
            +HFN 1-76 p123   +-SR 3-73 p48
            +NR 5-73 p12   +St 5-73 p117
            +-ON 12-73 p44
Der Zarewitsch: Wolgalied. cf Angel S 37108.
Der Zarewitsch: Wolgalied; Kosende Wellen. cf Works, selections
    (London OSA 26220).
LeHEURTEUR, Guillaume
Troys jeunes bourgeoises. cf HMV SLS 5022 (2).
LEIGHTON, Kenneth
Almighty God. cf Harmonia Mundi HMD 223.
Give me the wings of faith. cf HMV HQS 1350.
Improvisation. cf Wealden WS 137.
Lully, lulla, thou little tiny child. cf HMV CSD 3774.
Scherzo. cf Gaudeamus XSH 101.
LeJEUNE, Claude
Songs: Ce n'est que fiel; La belle Aronde; Comment penser; Revecy
    venir du printemps; Qu'est devenu ce bel ciel; Debat la nostre
    trill'en May. cf Argo ZRG 667.
Songs: Un gentil amoureux; Une puce. cf HMV CSD 3740.
LENOIR
Songs: Parlez-moi d'amour. cf Columbia M 33933.
LENTZ, Daniel
Songs of the sirens. cf CHILDS: Trio, clarinet, violoncello and
    piano.

LEONCAVALLO, Ruggiero
   La bohème: Scuoti o vento. cf RCA LRL 2-7531.
1528 I Pagliacci. Lucine Amara, s; Richard Tucker, Thomas Hayward, t;
     Giuseppe Valdengo, Clifford Harvuot, bar; Metropolitan Opera
     Orchestra and Chorus; Fausto Cleva. CBS 61658. Tape (c)
     40-61658. (Reissue from Philips ABL 3041/2).
       +-Gr 10-75 p679   +-HFN 8-76 p94 tape
       +-HFN 10-75 p153   -RR 10-75 p28
1529 I Pagliacci: Si può; Signore, Signori; Un tal gioco; Don din don
     din; Qual flamma avea nel guardo...Stridono lassu; Recitar...
     Vesti la giubba; Non, Pagliaccio non son. MASCAGNI: Cavalleria
     rusticana: Gli aranci olezzano; Voi lo sapete, o mamma; Oh,
     Il Signore vi manda; Intermezzo; Mamma, quel vino è generoso.
     Joan Carlyle, s; Fiorenza Cossotto, ms; Carlo Bergonzi, Ugo
     Benelli, t; Giuseppe Taddei, Rolando Panerai, Giangiacomo
     Guelfi, bar; Marie Gracia Allegri, con; La Scala Orchestra
     and Chorus; Herbert von Karajan. DG 2535 199. (Reissue
     from SLPM 139205/7).
       ++Gr 12-76 p1046
   I Pagliacci: Bell chorus. cf Decca DPA 525/6.
   I Pagliacci: Hüll dich in Tand. cf Discophilia 13 KGW 1.
   I Pagliacci: Intermezzo. cf HMV SLS 5019.
   I Pagliacci: Prologue. cf Caprice CAP 1062.
   I Pagliacci: Prologue. cf Rubini GV 65.
   I Pagliacci: Un tal gioco, Vesti la giubba. cf Cantilena 6238.
   I Pagliacci: Vesti la giubba. cf Decca SXL 6649.
   I Pagliacci: Vesti la giubba. cf Everest SDBR 3382.
   I Pagliacci: Vesti la giubba. cf RCA CRM 1-1749.
   I Pagliacci: Was gibt's. cf Discophilia DIS 13 KHH 1.
   Songs: Mattinata. cf DG 2530 700.
   Songs: Mattinata. cf Everest SDBR 3382.
   Songs: Mattinata. cf Rococo 5397.
   Zazà: Dir che ci sono al mondo. cf Club 99-100.
   Zazà: Zazà, piccola zingara. cf RCA LRL 2-7531.
LEONIN (12th century France)
   Organa. MACHAUT: Hoquetus David, motet. PEROTIN LE GRAND: Organa.
     ARS ANTIQUA: Motets from the Bamberg and Montpellier Codices.
     Hoqueti: Instrumental pieces. ARS NOVA: Motets from the Roman
     de Fauvel and the Ivrea and Chantilly Codices. Early Music
     Consort; David Munrow. DG Archive 2723 045 (3).
       ++Gr 11-76 p862   ++HFN 10-76 p160
LEONTOVICH
   Songs: Carol of the bells; O come, o come, Emanuel; O come all ye
     faithful; We wish you a merry Christmas. cf CBS 61771.
LeROY, Adrien
   Allemandes (2). cf CBS 76183.
   Basse dance. cf ATTAINGNANT: Basse dance.
   Branle gay. cf ATTAINGNANT: Basse dance.
   Branle gay. cf Saga 5426.
   Destre amoureux. cf ATTAINGNANT: Basse dance.
   Haulberroys. cf ATTAINGNANT: Basse dance.
   Passemeze. cf ATTAINGNANT: Basse dance.
LESCHITZKY, Theodor
   Gavotte. cf Rococo 2049.
LESCUREL, Jehannot de
   A vous douce debonaire. cf HMV SLS 863.

LESLIE
    Memory. cf Transatlantic XTRA 1159.
LESUEUR, Jean François
1531 Coronation march (orch. Thilde). PAISIELLO: Coronation mass. Te
        Deum (orch. Thilde). Mady Mesplé, s; Gerard Dunan, t; Yves
        Bisson, bs; Pierre Cochereau, org; Orchestra and Chorus;
        Armand Birbaum. Philips 6581 012.
            -Gr 2-76 p1368   -MT 8-76 p666
            +-HFN 1-76 p115   +-RR 1-76 p57
LESUR, Daniel
    Elégie. cf Delos DEL FY 008.
LEVEN, Ake
    Psalm, no. 331. cf Swedish Society SLT 33197.
LEVY
    Suite, no. 1. cf Crystal S 371.
LEVY, Burt
    Orbs with flute. cf Nonesuch HB 73028 (2).
LEWANDOWSKI, Louis
    Zocharti loch. cf Serenus SRS 12039.
LEWER
    Fidelia. cf Nonesuch H 71276.
LEWIS, Peter
    Gestes. cf CRI SD 324.
LEWIS, Robert Hall
1532 Symphony, no. 2. PERLE: Three movements, orchestra. RPO; David
        Epstein. CRI SD 331.
            +MJ 7-76 p56
LIDDELL
    Abide with me. cf Prelude PRS 2505.
LIDHOLM, Ingvar
    Laudi. cf BIS LP 14.
LIDL, Vaclav
    Hic homo sum. cf KOHOUTEK: Teatro del mondo.
LIDON, José
    Sonata con trompeta real. cf Audio EAS 16.
LIEBERSON, Peter
1533 Concerto, 4 groups of instruments. Fantasy, piano. LUNDBORG:
        Music forever, no. 2, excerpts. Passacaglia. Ursula Oppens,
        pno; Speculum Musicae, Light Fantastic Players. CRI SD 350.
            +NR 7-76 p4
    Fantasy, piano. cf Concerto, 4 groups of instruments.
LIEURANCE
    By the waters of Minnetonka. cf HMV RLS 719.
LIGETI, György
1534 Concerto, flute, oboe and orchestra. Concerto, 13 instrumental-
        ists. Melodien, orchestra. London Sinfonietta; Aurèle
        Nicolet, flt; Heinz Holliger, ob; David Atherton. Decca HEAD 12.
            ++Gr 8-76 p286   ++RR 7-76 p54
            +-HFN 7-76 p89   +-Te 12-76 p32
    Concerto, 13 instrumentalists. cf Concerto, flute, oboe and
        orchestra.
1535 Etude, no. 1. Lux aeterna. Quartet, strings, no. 2. Volumina.
        La Salle Quartet; North German Radio Chorus; Gerd Zacher, org;
        Helmut Franz. DG 2530 392.
            ++HFN 11-76 p161   +-Te 12-76 p32
            +RR 10-76 p81

Lux aeterna. cf Etude, no. 1.
Melodien, orchestra. cf Concerto, flute, oboe and orchestra.
Quartet, strings, no. 2. cf Etude, no. 1.
Volumina. cf Etude, no. 1.

LIGHT, Edward
Guardian angels. cf Smithsonian Collection N 002.

LINDBERG, Oskar
Leksandssvit. cf ATTERBERG: De favitska jungfrurna.
Pingst. cf Swedish Society SLT 33188.

LINDBLAD, Adolf
Drömmarne. cf Swedish Society SLT 33188.
Songs: The herdsman's mountain song; Upon a summer day. cf
    Library of Congress OMP 101/2.

LINDEMANN
Unter dem Grillenbanner. cf DG 2721 077 (2).

LINKE, Norbert
Violencia. cf BLACHER: Sonata, solo violin, op. 40.

LINLEY, M.
Otello: Air. cf Smithsonian Collection N 002.

LIST
Remember. cf CRI SD 102.

LISZT, Franz
(G refers to Grove's number, 5th edition)
1536 Ad nos, ad salutarem undam, G 259. Adagio, G 263, D flat major.
    Ave Maria (Von Arcadelt), G 659, F major. Prelude and fugue
    on the name B-A-C-H, G 260. Lionel Rogg, org. Connoisseur
    CSQ 2100.
        +NR 8-76 p11
1537 Ad nos, ad salutarem undam, G 259. Introduction, fugue and
    magnificat (arr. Gottschlag). Prelude and fugue on the name
    B-A-C-H, G 260. Peter Le Huray, org. Saga 5401.
        +Gr 8-75 p348   +MT 1-76 p40
        +HFN 8-75 p77   +RR 9-75 p59
    Ad nos, ad salutarem undam, G 259. cf Organ works (Hungaroton
    SLPX 11540).
    Adagio, G 263, D flat major. cf Ad nos, ad salutarem undam, G
    259.
    Adagio, G 263, D flat major. cf Organ works (Argo ZRG 784).
    Am Grabe Richard Wagners, G 267. cf Organ works (Hungaroton
    SLPX 11540.
    Andante maestoso, G 668. cf Organ works (Hungaroton SLPX 11540).
    Andante religioso, G 261a. cf Organ works (Hungaroton SLPX 11540).
    Angelus, G 378. cf Organ works (Hungaroton SLPX 11540).
    Angelus, G 378: Prière aux anges gardiens. cf Organ works (Argo
    ZRG 784).
1538 Années de pèlerinage, G 160, G 161, G 163. Jerome Rose, pno.
    Vox SVBX 5454 (3).
        +HF 8-74 p96   +NYT 1-13-74 pD25
        +MJ 12-76 p28   ++St 5-74 p102
        ++NR 3-74 p12
    Années de pèlerinage, 1st year,G 160: Au bord d'une source. cf
    Piano works (Decca SPA R 371).
    Années de pèlerinage, 1st year, G 160: Au bord d'une source. cf
    HMV HLM 7093.
    Années de pèlerinage, 1st year, G 160: Au lac de Wallenstadt. cf
    DEBUSSY: Estampes: Jardins sous la pluie.

Années de pèlerinage, 2nd year, G 161. cf BEETHOVEN: Sonata,
    piano, no. 27, op. 90, E minor.
1539 Années de pèlerinage, 2nd year, G 161: Sposalizio; I penseroso;
    Canzonetta del Salvator Rosa; Sonnetto del Petrarca, no. 47;
    Sonetto del Petrarca, no. 104; Sonetto del Petrarca, no. 123;
    Gondoliera. Legends, G 175: St. Francis of Assisi; St.
    Francis of Paola. Wilhelm Kempff, pno. DG 2530 560.
        ++Gr 11-75 p860   +NR 3-76 p12
        +HF 4-76 p112   ++RR 11-75 p79
        ++HFN 12-75 p156   +St 11-76 p145
1540 Années de pèlerinage, 2nd year, G 161: Sonetti del Petrarca (3).
    SCHUMANN: Symphonic etudes, op. 13. Alexis Weissenberg, pno.
    Connoisseur Society CS 2109.
        +-MJ 12-76 p28
1541 Années de pèlerinage, 2nd year, G 161: Sonetto de Petrarca, no.
    123. Consolation, no. 5, G 172. Etude de concert, no. 3,
    G 144, D flat: Un sospiro. Mephisto waltz, G 110. Sonata,
    piano, G 178, B minor. Van Cliburn, pno. RCA ARL 1-1173.
    Tape (c) ARK 1-1173 (ct) ARS 1-1173.
        +-HF 11-76 p116   +MJ 11-76 p25
        +HF 12-76 p147 tape   +-St 11-76 p145
    Années de pèlerinage, 2nd year, G 161: Dante sonata. cf Piano
    works (Turnabout Tape (c) KTVS 13432).
    Années de pèlerinage, 2nd year, G 161: Canzonetta del Salvator
    Rosa. cf Piano works (CBS 61680).
    Années de pèlerinage, 2nd year, G 161: Sonetto del Petrarca,
    no. 123. cf Piano works (RCA ARL 1-1173).
    Années de pèlerinage, 3rd year, G 161: Les jeux d'eau a la Villa
    d'Este. cf DEBUSSY: Estampes: Jardins sous la pluie.
    Ave Maria, G 20/2; G 504. cf Organ works (Hungaroton SLPX 11540).
    Ave Maria (Von Arcadelt), G 659, F major. cf Ad nos, ad salutarem
    undam, G 259.
    Ave Maria (Von Arcadelt), G 659, F major. cf Organ works
    (Hungaroton SLPX 11540).
    Ave maris stella, G 669. cf Organ works (Hungaroton SLPX 11540).
    Bagatelle without tonality. cf Piano works (Turnabout Tape (c)
    KTVC 13432).
1542 Ballade, no. 2, G 171, B minor. Paraphrase on Dom Sebastien,
    Funeral march, G 402. THALBERG: Fantasy on Rossini's "Moise",
    op. 33. Fantasy on Rossini's "Barber of Seville", op. 63.
    Raymond Lewenthal, pno. Angel S 36079.
        +NR 11-75 p12   +St 2-76 p108
        +SR 3-6-76 p41
1543 Concerto, piano, no. 1, G 124, E flat major. Concerto, piano,
    no. 2, G 125, A major. Garrick Ohlsson, pno; NPhO; Moshe
    Atzmon. Angel S 37145 (Q) SQ 37145. (also HMV (Q) SQ ASD
    3159).
        ++Gr 4-76 p1603   ++RR 2-76 p31
        +HF 4-76 p111   ++St 3-76 p116 Quad
        +-HFN 2-76 p101   +STL 3-7-76 p37
        +-NR 2-76 p13
1544 Concerto, piano, no. 1, G 124, E flat major. Concerto, piano, no.
    2, G 125, A major. Leonard Pennario, pno; LSO; Rene Leibowitz.
    Camden CCV 5047. (Reissue).
        +HFN 4-76 p123   /RR 4-76 p48
1545 Concerto, piano, no. 1, G 124, E flat major. Concerto, piano,
    no. 2, G 125, A major. György Cziffra, pno; Orchestre de Paris;

György Cziffra, Jr. Connoisseur Society CS 2087 (Q) CSQ 2087.
+-HF 4-76 p111   ++SFC 1-4-76 p27
+-MJ 4-76 p30   ++St 4-76 p113 Quad

1546 Concerto, piano, no. 1, G 124, E flat major. Concerto, piano,
no. 2, G 125, A major. Emil von Sauer, pno; OSCCP; Felix
Weingartner. DaCapo 1C 053 01458M. (Reissue from European
Columbia originals).
+-HF 4-76 p111

1547 Concerto, piano, no. 1, G 124, E flat major. Concerto, piano,
no. 2, G 125, A major. Lazar Berman, pno; VSO; Carlo Maria
Giulini. DG 2530 770. Tape (c) 3300 770.
++Gr 11-76 p792   ++RR 11-76 p68
++HFN 12-76 p144

1548 Concerto, piano, no. 1, G 124, E flat major. Hungarian rhapsody,
no. 2, G 244, C sharp minor. Les préludes, no. 3, G 97.
Richard Kraus, Tamás Vásáry, pno; Berlin Radio Symphony
Orchestra, Bamberg Symphony Orchestra; Ferenc Fricsay, Felix
Prohaska. DG 2548 235. (Reissues from 136 226, 136 020,
138 055).
+Gr 11-76 p792   +RR 10-76 p61

1549 Concerto, piano, no. 1, G 124, E flat major. TCHAIKOVSKY:
Concerto, piano, no. 1, op. 23, B flat minor. Horacio Gutierrez,
pno; LSO; André Previn. HMV SQ ASD 3262. (also Angel S 37177
Tape (c) 4XS 37177).
+-Gr 9-76 p421   ++SFC 11-21-76 p35
++HFN 9-76 p130   ++STL 9-19-76 p36
++RR 9-76 p67

1550 Concerto, piano, no. 1, G 124, E flat major. Totentanz, G 126.
Gyula Kiss, pno; Hungarian State Symphony Orchestra; János
Feréncsik. Hungaroton SLPX 11792.
+ARG 12-76 p35

1551 Concerto, piano, no. 1, G 124, E flat major. Concerto, piano,
no. 2, G 125, A major. Michele Campanella, pno; LPO; Hubert
Soudant. Pye Nixa PCNHX 7.
++RR 12-76 p61

Concerto, piano, no. 1, G 124, E flat major. cf HMV SLS 5068.
Concerto, piano, no. 2, G 125, A major. cf Concerto, piano, no. 1,
G 124, E flat major (Angel 37145).
Concerto, piano, no. 2, G 125, A major. cf Concerto, piano, no. 1,
G 124, E flat major (Camden CCV 5047).
Concerto, piano, no. 2, G 125, A major. cf Concerto, piano, no. 1,
G 124, E flat major (Connoisseur Society CS 2087).
Concerto, piano, no. 2, G 125, A major. cf Concerto, piano, no. 1,
G 124, E flat major (DaCapo 1C 053 01358M).
Concerto, piano, no. 2, G 125, A major. cf Concerto, piano, no. 1,
G 124, E flat major (DG 2530 770).
Concerto, piano, no. 2, G 125, A major. cf Concerto, piano, no. 1,
G 124, E flat major (Nixa 7).
Concerto, piano, no. 2, G 125, A major. cf BEETHOVEN: Sonata,
piano, no. 27, op. 90, E minor.

1552 Consolations, G 172. Harmonies poètiques et réligieuses, G 173.
Legendes, G 175. Weihnachtsbaum, G 186. Jerome Rose, pno.
Vox SVBX 5475 (3).
+HF 5-76 p86   ++NR 2-75 p12

Consolations, G 172, D flat major, E major. cf Organ works
(Hungaroton SLPX 11540).

Consolation, no. 3, G 172, D flat major. cf Piano works (CBS
     61680).
Consolation, no. 3, G 172, D flat major. cf Piano works (Decca
     SPA R 371).
Consolation, no. 3, G 172, D flat major. cf Piano works (Saga
     5405).
Consolation, no. 3, G 172, D flat major. cf RCA ARL 1-1176.
Consolation, no. 5, G 172. cf Années de pèlerinage, 2nd year,
     G 161: Sonetto de Petrarca, no. 123.
Consolation, no. 5, G 172. cf Piano works (RCA ARL 1-1173).
1553 Duo, G 127. Grand duo concertant, G 128. Epithalam, G 129.
     Elegies, nos. 1 and 2, G 130. Romance oubliée, G 132. Endre
     Granat, vln; Françoise Regnat, pno. Orion ORS 76210.
          +NR 11-76 p7
Elegies, nos. 1 and 2, G 130. cf Duo, G 127.
Epithalam, G 129. cf Duo, G 127.
Etudes de concert, nos. 2 and 3, G 144. cf Piano works (CBS
     61680).
Etude de concert, no. 2, G 145: Gnomenreigen. cf Piano works
     (Decca SPA R 371).
Etude de concert, no. 3, G 144, D flat major. cf CBS 30064.
Etude de concert, no. 3, G 144, D flat major: Un sospiro. cf
     Années de pèlerinage, 2nd year, G 161: Sonetto de Petrarca,
     no. 123.
Etude de concert, no. 3, G 144, D flat major: Un sospiro. cf
     Piano works (Decca SPA R 371).
Etude de concert, no. 3, G 144, D flat major: Un sospiro. cf
     Piano works (RCA ARL 1-1173).
1554 Etudes d'exécution transcendente, G 139. Hungarian rhapsody, no. 3,
     G 244, B flat major. Rhapsodie espagnole, G 254. Lazar Berman,
     pno. Columbia/Melodiya M2 33928 (2). (also HMV/Melodiya SLS
     5040).
          +-Gr 2-76 p1359   +NR 4-76 p12
          +-HF 5-76 p75   +RR 1-76 p50
          +-HFN 1-76 p110   +-SFC 2-8-76 p26
          +MJ 10-76 p25
1555 Etudes d'exécution transcendente, G 139. Scherzo and march, G 177.
     Variations on Bach's "Weinen, Klagen, Sorgen, Zagen", G 673.
     Robert Silverman, pno. Orion 76226.
          +NR 7-76 p10   +RR 8-76 p69
Etudes d'exécution transcendente, no. 5, G 139: Feux follets. cf
     Piano works (Decca SPA R 371).
Etudes d'exécution transcendente, no. 5, G 139: Feux follets. cf
     CHOPIN: Etude, op. 10, no. 3, E major.
Etudes d'exécution transcendente, no. 11, G 139: Harmonies du
     soir. cf BACH: Sonata, violin, no. 1, S 1001, G minor.
Etudes d'exécution transcendente, no. 11, G 139: Harmonies du
     soir. cf CHOPIN: Etude, op. 10, no. 3, E major.
Etudes d'exécution transcendente d'après Paganini, no. 3, G 140,
     A flat minor: La campanella. Faust waltz, G 407. Hungarian
     fantasia, G 123. Totentanz, G 126. György Cziffra, pno;
     Orchestre de Paris; György Cziffra, Jr. Connoisseur Society
     CS 2092.
          +St 12-76 p144
Etudes d'exécution transcendente d'après Paganini, no. 3, G 140,
     A flat minor: La campanella. cf Piano works (Decca SPA R 371).

Etudes d'exécution transcendente d'après Paganini, no. 6, G 140,
A minor. cf CHOPIN: Sonata, piano, no. 3, op. 58, B minor.

Evocation à la Chapelle Sixtine, G 658. cf Organ works (Hungaroton
SLPX 11540).

Excelsius, G 666. cf Organ works (Hungaroton SLPX 11540).

Fantasia on Wagner's Rienzi: Sancta Spirita cavaliere, G 439. cf
Piano works (Saga 5405).

1557 A Faust symphony, G 108. Alfonz Bartha, t; Budapest Chorus;
Budapest Symphony Orchestra; János Feréncsik. DG 2535 149.
(Reissue from 138 647/8).

    +Gr 4-76 p1603   /RR 1-76 p36
    +HFN 2-76 p115

Faust waltz, G 407. cf Etudes d'exécution transcendente d'après
Paganini, no. 3, G 140, A flat minor: La campanella.

Gnomenreigen, G 145. cf CHOPIN: Sonata, piano, no. 3, op. 58,
B minor.

Grand duo concertant, G 128. cf Duo, G 127.

Harmonies poètiques et réligieuses, G 173. cf Consolations, G 172.

1558 Harmonies poètiques et réligieuses, G 173: Bénédiction de Dieu
dans la solitude; Funérailles. Liebesträum, G 541. Mephisto
waltz, no. 1, G 514. Garrick Ohlsson, pno. HMV HQS 1361.
(also Angel S 37125).

    +Gr 9-76 p444   +-RR 9-76 p82
    +HF 5-76 p86   ++St 3-76 p116
    +-NR 2-76 p13   +STL 8-8-76 p29

Harmonies poètiques et réligieuses, G 173: Funérailles. cf
Piano works (Saga 5405).

Hosanna, G 677. cf Organ works (Hungaroton SLPX 11540).

Hungarian fantasia, G 123. cf Etudes d'exécution transcendente
d'après Paganini, no. 3, G 140, A flat minor: La campanella.

Hungarian fantasia, G 123. cf CHOPIN: Andante spianato and grande
polonaise, op. 22, E flat major.

Hungarian fantasia, G 123. cf HMV SLS 5033.

1559 Hungarian rhapsodies, nos. 1-16, 19, G 244. György Cziffra, pno.
Connoisseur Society CS 2097/99 (3).

    ++ARG 11-76 p28   +-NR 9-76 p11
    +HF 10-76 p112   ++St 11-76 p145
    +MJ 10-76 p25

1560 Hungarian rhapsodies, nos. 1-19, G 244. Rhapsodie espagnole,
G 254. Robert Szidon, pno. DG 2720 072 (3). Tape (c) 3300
386. (also DG 2709 044 Tape (c) 3371 018).

    +Gr 10-73 p699   ++NR 2-74 p12
    +-Gr 9-74 p592 tape   ++NYT 1-13-74 pD18
    +Gr 12-76 p1066 tape   +RR 10-73 p102
    +HF 5-74 p85   +SFC 1-27-74 p30
    +HF 6-76 p97 tape   ++St 3-74 p114
    ++MJ 5-74 p52

1561 Hungarian rhapsodies, nos. 1-19, G 244. Michele Campanella, pno.
Philips 6747 108.

    ++HF 6-76 p81   +-NR 2-76 p13
    ++MJ 4-76 p30

1562 Hungarian rhapsodies, nos. 2 and 4, G 244. Mazeppa, G 138. Les
préludes, G 97. BPhO; Herbert von Karajan. DG Tape (c) 3335
110.

    +HFN 10-75 p155 tape   ++RR 6-76 p88 tape

Hungarian rhapsodies, nos. 2 and 15, G 244. cf Piano works
(CBS 61680).

Hungarian rhapsody, no. 2, G 244, C sharp minor.   cf Concerto,
    piano, no. 1, G 124, E flat major.
Hungarian rhapsody, no. 2, G 244, C sharp minor.   cf Piano works
    (Decca SPA R 371).
Hungarian rhapsody, no. 2, G 244, C sharp minor.   cf Piano works
    (Saga 5405).
Hungarian rhapsody, no. 2, G 244, G sharp minor.   cf BACH:
    Toccata and fugue, S 565, D minor.
Hungarian rhapsody, no. 2, G 244, C sharp minor.   cf Angel S 37231.
Hungarian rhapsody, no. 2, G 244, C sharp minor.   cf HMV SLS 5019.
Hungarian rhapsody, no. 3, G 244, B flat major.   cf Etudes
    d'exécution transcendente, G 139.
1563 Hungarian rhapsodies, nos. 4 and 5, G 359.   Tasso, lamento e
    trionfo, G 96.   BPhO; Herbert von Karajan.   DG 2530 698.
        ++Gr 8-76 p286   +-RR 8-76 p41
        +HFN 8-76 p83   +SFC 12-26-76 p34
1564 Hungarian rhapsodies, nos. 8-11, 13, G 244.   Claudio Arrau, pno.
    Desmar DSM 1003.   (Reissue from Columbia, 1951-52).
        +-Audio 6-76 p97   ++MJ 12-75 p38
        ++Gr 2-76 p1360   +NR 12-75 p13
        ++HF 1-76 p88   ++RR 1-76 p50
        ++HFN 2-76 p101   ++St 1-76 p106
Hungarian rhapsody, no. 11, G 244.   cf Piano works (Turnabout
    Tape (c) KTVC 13432).
Hungarian rhapsody, no. 12, G 244, C sharp minor.   cf HMV HQS 1354.
Hungarian rhapsody, no. 13, G 244, A minor.   cf Rococo 2049.
Hungarian rhapsody, no. 13, G 244, A minor (abbreviated).   cf
    International Piano Library IPL 104.
Hungarian rhapsody, no. 14, G 244, F minor.   cf Piano works
    (Saga 5405).
Introduction, fugue and magnificat.   cf Ad nos, ad salutarem
    undam, G 259.
Introitus, no. 1, G 268.   cf Organ works (Hungaroton SLPX 11540).
Kirchliche Festouverture, G 675.   cf Organ works (Argo ZRG 784).
Kreuzandachten, G 53.   cf Organ works (Hungaroton SLPX 11540).
1565 Legend of Saint Elizabeth, G 2.   Eva Andor, s; Erzsébet Komlóssy,
    ms; Sándor Nagy, Lajos Miller, György Bordas, bar; Kolos Kováts,
    József Gregor, bs; Czech Radio Children's Choir; Slovak
    Philharmonic Orchestra and Chorus; János Feréncsik.   Hungaroton
    SLPX 11650/52.
        +Gr 12-75 p1089   +NR 11-75 p10
        +-HF 6-76 p82   +RR 10-75 p84
        +HFN 1-76 p110   ++SFC 10-19-75 p33
The legend of St. Elizabeth, G 2, excerpt.   cf Hungaroton SLPX
    11762.
The legend of Saint Elizabeth, G 2: Introduction.   cf Organ works
    (Hungaroton SLPX 11540).
Legendes, G 175.   cf Consolations, G 172.
Legendes, G 175: St. Francis of Assisi; St. Francis of Paola.   cf
    Années de pèlerinage, 2nd year, G 161 (DG 2530 560).
Liebesträum, G 541.   cf Harmonies poétiques et réligieuses, G 173:
    Bénédiction de dieu dans la solitude; Funérailles.
Liebesträum, G 541.   cf Connoisseur Society (Q) CSQ 2065.
Liebesträum, no. 3, G 541, A flat major.   cf Piano works (CBS
    61680).
Liebesträum, no. 3, G 541, A flat major.   cf Piano works (Decca
    SPA R 371).

Liebesträum, no. 3, G 541, A flat major. cf Piano works (Saga 5405).
Liebesträum, no. 3, G 541, A flat major. cf CBS 30064.
Liebesträum, no. 3, G 541, A flat major. cf Rococo 2049.
Mazeppa, G 138. cf Hungarian rhapsodies, nos. 2 and 4, G 244.
Mephisto waltz. cf Piano works (RCS ARL 1-1173).
Mephisto waltz. cf HMV RLS 717.
Mephisto waltz, G 110. cf Années de pèlerinage, 2nd year, G 161: Sonetto de Petrarca, no. 123.

1566 Mephisto waltz, no. 1, G 514. Sonata, piano, G 178, B minor. Venezia e Napoli, G 162. Lazar Berman, pno. Columbia/Melodiya M 33927. (also HMV/Melodiya ASD 3228 Tape (c) TC ASD 3228).

+-Audio 10-76 p149	+-HFN 9-76 p123
+-Gr 9-76 p444	+MJ 10-76 p25
+-Gr 11-76 p887 tape	+-NR 4-76 p12
++HF 5-76 p75	+-RR 8-76 p69
+-HFN 8-76 p81	+STL 9-19-76 p36

Mephisto waltz, no. 1, G 514. cf Harmonies poétiques et religieuses, G 173: Bénédiction de Dieu dans la solitude; Funérailles.
Mephisto waltz, no. 1, G 514. cf Piano works (CBS 61680).
Mephisto waltz, no. 1, G 514. cf Piano works (Decca SPA R 371).
Missa pro organo, G 264. cf Organ works (Hungaroton SLPX 11540).
Nun danket alle Gott, G 61. cf Organ works (Hungaroton SLPX 11540).
Offertorium, G 667. cf Organ works (Hungaroton SLPX 11540).
Ora pro nobis, G 262. cf Organ works (Hungaroton SLPX 11540).

1567 Organ works: Adagio, G 263, D flat major. Angelus, G 378: Prière aux anges gardiens. Kirchliche Festouverture, G 675. Trauerode, no. 2. G 268, E minor. Variationen on Bach's "Weinen, Klagen, Sorgen, Zagen", G 673. Peter Planyavsky, org. Argo ZRG 784.
      +Gr 9-76 p445   ++HFN 8-76 p83

1568 Organ works: Ad nos, ad salutarem undam, G 259. Am Grabe Richard Wagners, G 267. Andante maestoso, G 668. Andante religioso, G 261a. Angelus, G 378. Ave Maria, G 20/2. Ave Maria, G 504. Ave Maria (Von Arcadelt), G 659, F major. Ave maris stella, G 669. Consolations, G 172, D flat major, E major. Evocation à la Chapelle Sixtine, G 658. Excelsius, G 666. Hosanna, G 677. Introitus, no. 1, G 268. Kreuzandachten, G 53. The legend of Saint Elizabeth, G 2: Introduction. Missa pro organo, G 264. Nun danket alle Gott, G 61. Offertorium, G 667. Ora pro nobis, G 262. Orpheus, G 98. Prelude and fugue on the name B-A-C-H, G 260. Requiem, organ, G 266. Resignazione, G 263. Rosario, G 670. Salve Regina, no. 1, G 699. A symphony on Dante's "Divina Commedia", G 109. Trauerode, no. 2, G 268, E minor. Tu es Petrus, G 664. Ungarns Gott, G 674. Variations on Bach's "Weinen, Klagen, Sorgen, Zagen", G 673. Weihnachtsbaum, nos. 1-4, G 186. Weimar's Volkslied, G 672. Zum Haus des Herren ziehen, wir, G 671: Prelude. Zur Trauung, G 60. Sándor Margittay, Gábor Lehotka, Endre Kovacs, org. Hungaroton SLPX 11540 (5).
      +-Gr 4-76 p1639
      Orpheus, G 98. cf Organ works (Hungaroton SLPX 11540).

1569 Paraphrases: Bellini: La sonnambula: Fantaisie, G 393. Norma, G 394. Glinka: Russlan and Ludmila: Tscherkessenmarsch, G 406. Mozart: Don Juan, G 418. Richard and John Contiguglia, pno. Connoisseur Society CS 2039. Tape (c) E 1027.

++Gr 12-75 p1065   +NYT 10-28-73 pD21
+HF 7-72 p84   ++RR 10-75 p78
+HF 1-75 p110 tape   +SR 5-20-72 p50
++MJ 2-75 p39   ++St 10-72 p132
+NR 2-73 p11   ++STL 1-11-76 p36

Paraphrase on Donizetti's Dom Sebastien, Funeral march, G 402.
  cf Ballade, no. 2, G 171, B minor.
Paraphrases on Schubert's Horch, Horch, Die Lerch, G 558/9; Der
  Muller und der Bach; Liebesbotschaft; Das Wandern. cf
  CHOPIN: Sonata, piano, no. 3, op. 58, B minor.
Paraphrase on Tchaikovsky's Eugene Onegin: Polonaise, G 429. cf
  GRAINGER: Paraphrase on Tchaikovsky's Nutcracker: Waltz of the
  flowers.
Paraphrase on Verdi's Rigoletto, G 434. cf Rococo 2049.

1570 Paraphrases (Transcriptions): Wagner: Der fliegende Holländer:
  Spinning chorus, G 440. Lohengrin: Elsa's procession to the
  Minster, G 446, no. 2. Parsifal: Solemi. march to the Holy
  Grail, G 450. Der Ring des Nibelungen: Valhalla, G 449.
  Tannhäuser: Overture, G 442. Tristan und Isolde: Liebestod,
  G 447. Michele Campanella, pno. Pye QS PCNH 2. Tape (c)
  ZCPCNH 2.
      +Gr 3-76 p1482   -RR 9-76 p95 tape

1571 Piano works: Années de pèlerinage, 2nd year, G 161: Canzonetta
  del Salvator Rosa. Etudes de concert, G 144, nos. 2 and 3.
  Consolation, no. 3, G 172, D flat major. Hungarian rhapsodies,
  no. 2 and no. 15, G 244. Liebesträum, no. 3, G 541, A flat
  major. Mephisto waltz, no. 1, G 514. Valse oubliée, no. 1,
  G 215, F sharp minor. Philippe Entremont, pno. CBS 61680.
      -Gr 4-76 p1639   -RR 3-76 p61
      +-HFN 3-76 p109

1572 Piano works: Années de pèlerinage, 1st year, G 160: Au bord d'une
  source. Consolations, no. 3, G 172, D flat major. Etude de
  concert, no. 3, G 144, D flat major: Un sospiro. Etude de
  concert, no. 2, G 145: Gnomenreigen. Etudes d'exécution transcendente, no. 5, G 139: Feux follets. Etudes d'exécution
  transcendente d'après Paganini, no. 3, G 140, A flat minor:
  La campanella . Hungarian rhapsody, no. 2, G 244, C sharp
  minor. Liebesträum, no. 3, G 541, A flat major. Mephisto
  waltz, no. 1, G 514. Valse oubliée, no. 1, G 215, F sharp
  minor. France Clidat, pno. Decca SPA R 371. Tape (c) KCSP
  R 371.
      +-Gr 5-76 p1787   +-RR 4-76 p64
      +-HFN 4-76 p123

1573 Piano works: Années de pèlerinage, 2nd year, G 161: Sonetto del
  Petrarca, no. 123. Consolation, no. 5, G 172, D flat major.
  Etude de concert, no. 3, G 144, D flat major: Un sospiro.
  Mephisto waltz. Sonata, piano, G 178, B minor. Van Cliburn,
  pno. RCA ARL 1-1173.
      +-HF 11-76 p116   +NR 7-76 p10
      +HF 12-76 p147 tape   +-St 11-76 p145
      +MJ 11-76 p25

1574 Piano works: Consolation, no. 3, G 172, D flat major. Fantasia on
  Wagner's Rienzi: Sancta spirito cavaliere, G 439. Harmonies
  poètiques et rèligieuses, G 173: Funérailles. Hungarian
  rhapsody, no. 2, G 244, C sharp minor. Hungarian rhapsody,
  no. 14, G 244, F minor. Liebesträum, no. 3, G 541, A flat

major. David Wilde, pno. Saga 5405.
+–Gr 1-76 p1217   +–RR 12-75 p83
+–HFN 1-76 p110

1575 Piano works: Années de pèlerinage, 2nd year, G 161: Dante sonata.
Bagatelle without tonality. Hungarian rhapsody, no. 11, G 244.
Sonata, piano, G 178, B minor. Unstern, G 208. Alfred Brendel,
pno. Turnabout Tape (c) KTVC 13432.
+–HFN 1-76 p125 tape
Les préludes, G 97. cf Hungarian rhapsodies, nos. 2 and 4, G 244.
Les préludes, G 97. cf BEETHOVEN: Egmont, op. 84: Overture.
Les préludes, G 97. cf Angel S 37231.
Les préludes, G 97. cf HMV RLS 717.
Les préludes, G 97. cf HMV SLS 5019.
Les préludes, no. 3, G 97. cf Concerto, piano, no. 1, G 124, E
flat major.
Prelude and fugue on the name B-A-C-H, G 260. cf Ad nos, ad
salutarem undam, G 259 (Connoisseur CSQ 2100).
Prelude and fugue on the name B-A-C-H, G 260. cf Ad nos, ad
salutarem undam, G 259 (Saga 5401).
Prelude and fugue on the name B-A-C-H, G 260. cf Organ works
(Hungaroton SLPX 11540).
Prelude and fugue on the name B-A-C-H, G 260. cf Argo 5BBA 1013-5.
Prelude and fugue on the name B-A-C-H, G 260. cf Audio 2.
Prelude and fugue on the name B-A-C-H, G 260. cf Decca SDD 499.
Prelude and fugue on the name B-A-C-H, G 260. cf HMV SQ HQS 1356.
Prelude and fugue on the name B-A-C-H, G 260. cf Vista VPS 1030.
Prelude and fugue on the name B-A-C-H, G 260. cf Wealden WS 137.
Requiem, organ, G 266. cf Organ works (Hungaroton SLPX 11540).
Resignazione, G 263. cf Organ works (Hungaroton SLPX 11540).
Rhapsodie espagnole, G 254. cf Etudes d'exécution transcendente,
G 139.
Rhapsodie espagnole, G 254. cf Hungarian rhapsodies, nos. 1-19,
G 244.
Romance oubliée, G 132. cf Duo, G 127.
Rosario, G 670. cf Organ works (Hungaroton SLPX 11540).
Salve Regina, no. 1, G 699. cf Organ works (Hungaroton SLPX
11540).
Scherzo and march, G 177. cf Etudes d'exécution transcendente,
G 139.
Sonata, piano, G 178, B minor. cf Années de pèlerinage, 2nd
year, G 161: Sonetto di Petrarca, no. 123.
Sonata, piano, G 178, B minor. cf Mephisto waltz, no. 1, G 514.
Sonata, piano, G 178, B minor. cf Piano works (RCA ARL 1-1173).
Sonata, piano, G 178, B minor. cf Piano works (Turnabout Tape
(c) KTVC 13432).
Sonata, piano, G 178, B minor. cf BEETHOVEN: Sonata, piano,
no. 23, op. 57, F minor.
Sonata, piano, G 178, B minor. cf CHOPIN: Sonata, piano, no. 3,
op. 58, B minor (Bruno Walter Society AUDAX 761).
Sonata, piano, G 178, B minor. cf CHOPIN: Sonata, piano, no. 3,
op. 58, B minor (Seraphim 60241).
Songs: Angiolin dal biondo crin, G 269; Es rauschen die Winde,
G 294; Es muss ein Wunderbares sein, G 314; In Liebeslust,
G 318; Kling leise, mein Lied, G 301; Schwebe, schwebe, blaues
Auge, G 305; Die Vätergruft, G 281; Wie singt die Lerche schön,
G 312. cf CHOPIN: Songs (Argo ZRG 814).

Songs: Comment disaient-ils, G 276; O, quand je dors, S 282, Uber
    allen Gipfeln ist Ruh', G 306; Vergiftet sind meine Lieder,
    G 289; Mignon's Lied, G 275. cf Desto DC 7118/9.
Songs: Oh, quand je dors, G 282. cf Columbia M 33933.
1576 A symphony on Dante's "Divina Commedia", G 109. Bolshoi Theatre
    Orchestra and Chorus; Boris Khaikin. Columbia/Melodyia
    M 33823.
        ++HF 6-76 p81   +ON 2-7-76 p34
        +NR 2-76 p3   +St 7-76 p110
A symphony on Dante's "Divina Commedia", G 109. cf Organ works
    (Hungaroton SLPX 11540).
Tasso, lamento et trionfo, G 96.   cf Hungarian rhapsodies, nos.
    4 and 5, G 359.
Totentanz, G 126. cf Concerto, piano, no. 1, G 124, E flat major.
Totentanz, G 126. cf Etudes d'exécution transcendente d'après
    Paganini, no. 3, G 140, A flat minor: la campanella.
Totentanz, G 126. cf CHOPIN: Andante spianato and grande polo-
    naise, op. 22, E flat major.
Trauerode, no. 2, G 268, E minor. cf Organ works (Argo ZRG 784).
Trauerode, no. 2, G 268, E minor. cf Organ works (Hungaroton
    SLPX 11540).
Tu es Petrus, G 664. cf Organ works (Hungaroton SLPX 11540).
Ungarns Gott, G 674. cf Organ works (Hungaroton SLPX 11540).
Unstern, G 208. cf Piano works (Turnabout Tape (c) KTVC 13432).
Valses oubliées, no. 1, G 215, F sharp major; no. 2, G 215, A
    flat major. cf CHOPIN: Etude, op. 10, no. 3, E major.
Valse oubliée, no. 1, G 215, F sharp minor. cf Piano works
    (CBS 61680).
Valse oubliée, no. 1, G 215, F sharp minor. cf Piano works
    (Decca SPA R 371).
Variations on Bach's "Weinen, Klagen, Sorgen, Zagen", G 673. cf
    Etudes d'exécution transcendente, G 139.
Variations on Bach's "Weinen, Klagen, Sorgen, Zagen", G 673. cf
    Organ works (Argo ZRG 784).
Variations on Bach's "Weinen, Klagen, Sorgen, Zagen", G 673. cf
    Organ works (Hungaroton SLPX 11540).
Variations on Bach's "Weinen, Klagen, Sorgen, Zagen", G 673. cf
    FRANCK: Chorale, no. 2, B minor.
Venezia e Napoli, G 162. cf Mephisto waltz, no. 1, G 514.
Weihnachtsbaum, G 186. cf Consolations, G 172.
Weihnachtsbaum, nos. 1-4, G 186. cf Organ works (Hungaroton
    SLPX 11540).
Weimar's Volkslied, G 672. cf Organ works (Hungaroton SLPX 11540).
Zum Haus des Herren ziehen, wir, G 671: Prelude. cf Organ works
    (Hungaroton SLPX 11540).
Zur trauung, G 60. cf Organ works (Hungaroton SLPX 11540).
LITAIZE, Gaston
    Toccata sue le veni creator. cf Vista VPS 1029.
LITOLFF, Henri
    Concerto symphonique, no. 4, op. 102, D minor: Scherzo. cf HMV
        SLS 5033.
    Trio, no. 1, op. 47, D minor. cf ALKAN: Trio, no. 1, op. 30,
        G minor.
LOCATELLI, Pietro
1577 Concerti, op. 4, nos. 8, 10-12. Instrumental Ensemble of France;
        Jean-Pierre Wallez. French Decca 7162.
        ++SFC 5-30-76 p24

Concerto grosso, op. 1, no. 8, F minor. cf CORELLI: Concerto
   grosso, op. 6, no. 8, G minor.
Concerto grosso, op. 7, no. 12, F major. cf ALBINONI: Sonata à
   5, op. 2, no. 6, G minor.
Sonata, flute, G minor. cf Saga 5414.

LOCKE, Matthew
   Music for His Majesty's sackbutts and cornetts: Air. cf Argo
      SPA 464.

LOCKWOOD, Larry
   Suite, band. cf IVES: March Mega Lambda Chi.

LODER, Edward
   The diver. cf HMV EMD 5528.

LOEFFLER, Charles
   Memories of my childhood. cf GRIFFES: Bacchanale.
   Music for four string instruments. cf Vox SVBX 5301.
   Poem, orchestra. cf GRIFFES: Bacchanale.

1578 Rhapsodies, oboe, viola and piano (2). MOZART: Quartet, oboe,
   K 370, F major. John Mack, ob; Daniel Majeske, vln; Abraham
   Skernick, vla; S. Geber, vlc; E. Podis, pno. Advent S 5017.
      +NR 6-76 p9

LOEILLET, Jean-Baptiste
   Concerto, trumpet, D major. cf RCA CRL 3-1430.
   Corente. cf DG Archive 2533 172.
   Gigue. cf DG Archive 2533 172.
   Sarabande. cf DG Archive 2533 172.
   Sonata, recorder, G major. cf Cambridge CRS 2826.
   Sonata, recorder, op. 3, no. 3, G minor. cf Transatlantic TRA 292.
   Suite, no. 1. cf Orion ORS 76231.

LOEVENDIE, Theo
   Scaramuccia. cf KEURIS: Concerto, saxophone.

LOEWE
   Get me to the church on time. cf RCA AGL 1-1334.

LOEWE, Karl
   Canzonetta. cf Club 99-99.
   Songs: Der Erlkönig, op. 1, no. 3; Prinz Eugen, op. 92. cf
      Preiser LV 192.
   Songs: Hinkenden Jamben, op. 62, no. 5; Hochzeitlied, op. 20,
      no. 1; Odins Meeres-Ritt, op. 118; Die Wandelnde Glocken, op.
      20, no. 3. cf BRAHMS: Songs (Decca SXL 6738).

LOFFELHOLTZ, Christoph
   Intrada. cf CBS 76183.

LOGES
   Ich schenk dir eine neue Welt. cf DG 2530 700.

LOHET, Simon
   Fuga as der Tabulator von Joh. Woltz (duodecima). cf Pelca
      PRSRK 41017/20.

LOHR
   The marriage market: Little grey home in the west. cf Pye Ember
      GVC 51.
   Songs: When Jack and I were children. cf Argo ZFB 95/6.

LOMAKIN
   Tebe poem. cf Harmonia Mundi HMU 105.

LONQUE, A.
   Sonatina, op. 34, D major. cf Orion ORS 76235.

LONQUE, Georges
   Sonatina, op. 32, D major. cf Orion ORS 76235.

Sonatina, op. 36, G major. cf Orion ORS 76235.
LOPEZ CAPILLAS, Francisco
    Alleluias (2). cf Eldorado S-1.
    Gloria laus et honor a 4. cf Eldorado S-1.
LOQUEVILLE, Richard de
    Sanctus. cf Telefunken ER 6-35257 (2).
LORING
    Nothing in all creation. cf Grosvenor GRS 1039.
LORTZING, Gustav
    Zur und Zimmermann: O sancta justitia. cf Decca GOSE 677/79.
LOTTI, Antonio
    Pur dicesti. cf HMV RLS 719.
LOUDOVA, Ivana
    Air, bass clarinet and piano. cf Panton 110 369.
    Chorale, orchestra. cf KUBIK: Lament of a warrior's wife.
    Spleen: Hommage à Charles Baudelaire. cf KUBIK: Lament of a
        warrior's wife.
    LOUIS DANTO: NONE BUT THE LONELY HEART. cf Musical Heritage
        Society MHS 3276.
    LOVE, LUST, PIETY AND POLITICS. cf BASF 25 22286-1.
LUBBOCK
    The smugglers' song. cf Pearl SHE 528.
LUBIMOV
    Brajen mouj. cf Harmonia Mundi HMU 105.
LUCKY, Stepan
    Pieces (3). cf Panton 110 369.
LUDERS, Gustav
    The Prince of Pilsen, excerpts. cf Everest 3360.
LUENING, Otto
    King Lear: Suite. cf BERGSMA: The fortunate islands.
    A poem in cycles and bells. cf BERGSMA: The fortunate islands.
1579 Quartet, strings, no. 2. Quartet, strings, no. 3. Sonata, solo
      violin, no. 3. Trio, flute, violoncello and piano. Sinnhoffer
      Quartet; Harvey Sollberger, flt; Fred Sherry, vlc; Charles
      Wuorinen, pno; Max Polikoff, vln. CRI SD 303.
        +Gr 9-76 p437   ++MQ 4-74 p321
        ++HF 11-73 p110   ++NR 10-73 p6
        ++MJ 5-74 p47   +RR 8-76 p64
    Quartet, strings, no. 3. cf Quartet, strings, no. 2.
    Sonata, solo violin, no. 3. cf Quartet, strings, no. 2.
    Trio, flute, violoncello and piano. cf Quartet, strings, no. 2.
LULLY, Jean
1580 Alceste. Felicity Palmer, Anne-Marie Rodde, Sonia Nigoghossian,
      s; Bruce Brewer, John Elwes, t; Max von Egmond, Pierre-Yves Le
      Maiget, bs; Rapael Passaquet Vocal Ensemble; Le Grand Ecurie
      et La Chambre du Roy; Jean-Claude Malgoire. CBS 79301 (3).
        +-Gr 1-76 p1235   ++RR 11-75 p32
        +HFN 12-75 p156   +STL 1-11-76 p36
        +-MT 2-76 p142
1581 Alceste, highlights. Phyllis Curtin, s; Anne Ayer, ms; William
      Lewis, Hugues Cuenod, t; Philippe Huttenlocher, bs-bar; Vienna
      Volksoper Orchestra; Willard Straight. Serenus SRS 12057.
        +NR 3-75 p9   +-On 4-10-76 p32
    Alceste: Air de Caron. cf FAURE: Songs (1750 Arch S 1754).
    Amadis: Marche pour le combat. cf Decca GOSD 674/6.
1582 Te deum. Jennifer Smith, Francine Bessac, s; Zeger Vandersteene,

c-t; Louis Devos, t; Philippe Huttenlocher, bs; A Coeur Joi
Vocal Ensemble of Valence; Jean-François Paillard Chamber
Orchestra; Jean-François Paillard. Erato STU 70927.
-Gr 6-76 p71   -RR 6-76 p79
Thésée: Overture. cf Decca GOSD 674/6.
Une noce de village: Dernière entrée. cf DG Archive 2533 172.

LUNA
El niño Judio: De España vengo. cf London OS 26435.

LUNDBORG, Erik
Music forever, no. 2, excerpts. cf LIEBERSON: Concerto, 4
groups of instruments.
Passacaglia. cf LIEBERSON: Concerto, 4 groups of instruments.

LUNDEN, Lennart
Lilltåa och 9 till. cf BIS LP 2.

LUNDVIK, Hildro
Det första vårregnet. cf Swedish Society SLT 33188.
LUTE MUSIC OF THE DUTCH RENAISSANCE. cf DG Archive 2533 302.

LUTOSLAWSKI, Witold
Concerto, violoncello. cf DUTILLEUX: Concerto, violoncello.
Concerto, violoncello. cf FARKAS: Concertino all'antica.

LUTYENS, Elisabeth
O saisons, o chateaux. cf BRITTEN: Prelude and fugue, 18 solo
strings, op. 29.

LVOV, Alexis
To thy heavenly banquet (Communian anthem). cf Vanguard SVD
71212.

LVOVSKY (LVOVSKI)
Gospel chant. cf Ikon IKO 3.
Herouvimska. cf Harmonia Mundi HMU 105.

LYON
Friendship. cf Vox SVBX 5350.

LYSBERG
La fontaine. cf Library of Congress OMP 101/2.

McBETH
Brass. cf Crystal S 206.

McCABE, John
1583 The Chagall windows. Variations on a theme by Karl Amadeus
Hartmann. Hallé Orchestra; James Loughran. HMV (Q) ASD 3096.
+Audio 3-76 p63   +MT 2-76 p143
+Gr 10-75 p621   ++RR 9-75 p39
++HFN 10-75 p144   *Te 3-76 p31
+MM 7-76 p37
Le poisson magique (Meditation after Paul Klee). cf DAVID:
Toccata and fugue, F minor.
Variations on a theme by Karl Amadeus Hartmann. cf The Chagall
windows.

MACARTHY
The bonnie blue flag. cf CBS 61746.

McCAULEY, William
Miniatures, flute and strings (5). cf MATTON: Concerto, 2 pianos.

MacDERMOT
Good morning starshine. cf Stentorian SP 1735.

MacDOUGALL, Robert
Anacoluthon: A confluence. cf BASSETT: Sextet.

MACDOWELL, Edward
1584 Etudes, op. 39, Bk 2, no. 2: Shadow dance. Fantastic pieces,
op. 17, no. 2: Witches' dance. Modern suite, no. 2, op. 14.

Sea pieces, op. 55. Andrea Anderson Swem, pno. Orion ORS
75175. (also Pye ECL 9045).
  -Gr 7-76 p194   +NR 9-75 p12
  +MJ 4-76 p30   +St 12-75 p128
Fantastic pieces, op. 17, no. 2: Witches' dance. cf Etudes,
op. 39, Bk 2, no. 2: Shadow dance.
1585 Modern suite, no. 2, op. 14. Sea pieces, op. 55 (8). Shadow
dance, op. 39. Witches dance, op. 17. Andrea Anderson Swem,
pno. Ember ECL 9045.
  +-HFN 6-76 p89   +RR 6-76 p73
Modern suite, no. 2, op. 14. cf Etudes, op. 39, Bk 2, no. 2:
Shadow dance.
Sea pieces, op. 55. cf Etudes, op. 39, Bk 2, no. 2: Shadow
dance.
Sea pieces, op. 55. cf Modern suite, no. 2, op. 14.
Sea pieces, op. 55, no. 3: A.D. 1620. cf Columbia M 34129.
Shadow dance, op. 39. cf Modern suite, no. 2, op. 14.
1586 Sonata, piano, no. 2, op. 50, G minor. Sonata, piano, no. 3,
op. 57, D minor. Yoriko Takahashi, pno. Orion ORS 75183.
  ++HF 12-75 p98   ++NR 10-75 p10
  +MJ 4-76 p30
1587 Sonata, piano, no. 2, op. 50, G minor. Woodland sketches, op. 51.
Clive Lythgoe, pno. Philips 9500 095.
  ++Gr 11-76 p838   +NYT 7-4-76 pD1
  +-HF 11-76 p118   ++RR 11-76 p91
  +HFN 11-76 p161   +SFC 12-26-76 p34
  +-MJ 12-76 p28   +St 10-76 p126
  +NR 7-76 p12
Sonata, piano, no. 3, op. 57, D minor. cf Sonata, piano, no. 2,
op. 50, G minor.
1588 Suite, no. 1, op. 42. Suite, no. 2, op. 48. Eastman-Rochester
Symphony Orchestra; Howard Hanson.
  *NYT 7-4-76 pD1
Suite, no. 2, op. 48. cf Suite, no. 1, op. 42.
To a wild rose. cf CBS 30062.
To a wild rose. cf Connoisseur Society (Q) CSQ 2066
To a wild rose. cf International Piano Library IPL 102.
To a wild rose. cf Pye TB 3009.
Witches dance, op. 17. cf Modern suite, no. 2, op. 14.
Woodland sketches, op. 51. cf Sonata, piano, no. 2, op. 50,
G minor.
McGIBBON, William
Sonata, no. 5, G major (edit. Elliot). cf HANDEL: Trio sonata,
op. 2, no. 1, C minor.
McKUEN, Rod
1589 Concerto, 4 harpsichords, no. 1. Four statements from three
books. London Arte Orchestra; Arthur Greenslade. Stanyon
SR 10009.
  +-ARG 11-76 p29
1590 Concerto, piano, no. 3. Leslie Pearson, pno; Westminster Sym-
phony Orchestra; Arthur Greenslade. Stanyon SR 9012.
  -ARG 11-76 p29
Four statements from three books. cf Concerto, 4 harpsichords,
no. 1.
McMILLAN, Ann
Carrefours. cf Folkways FTS 33901/4.
Whale I. cf Folkways FTS 33901/4.

MACHAUT, Guillaume
    Amours me fait désirer. cf HMV SLS 863.
    Ballad and plus dure. cf BIS LP 2.
    Dame se vous m'estes. cf HMV SLS 863.
    De bon espoir: Puis que la douce. cf HMV SLS 863.
    De toutes flours. cf HMV SLS 863.
    Douce dame jolie. cf HMV SLS 863.
    Hareu, hareu: Helas, où sera pris confors. cf HMV SLS 863.
    Hoquetus David, motet. cf LEONIN: Organa.
    Ma chière dame. cf DUFAY: Franc cuer gentil.
    Ma fin est mon commencement. cf HMV SLS 863.
    Mes esperis se combat. cf HMV SLS 863.
    Phyton, le merveilleus serpent. cf HMV SLS 863.
    Quant j'ay l'espart. cf HMV SLS 863.
    Quant je sui mis. cf HMV SLS 863.
    Quant Théseus: Ne quier veoir. cf HMV SLS 863.
    Se je souspir. cf HMV SLS 863.
    Trop plus est belle: Biauté paree; Je ne sui. cf HMV SLS 863.
MADERNA, Bruno
    Dedication. cf BLACHER: Sonata, solo violin, op. 40.
    Pièce pour Ivry. cf BLACHER: Sonata, solo violin, op. 40.
MADETOJA, Leevi
    Comedy overture. cf Symphony, no. 3.
1591 Symphony, no. 3. Comedy overture. Helsinki Philharmonic
    Orchestra; Jorma Panula. Finnlevy SFX 20.
      +-RR 7-76 p54
MAHLER, Gustav
1592 Kindertotenlieder. Symphony, no. 10, F sharp minor: Adagio.
    Janet Baker, ms; Israel Philharmonic Orchestra, NYP; Leonard
    Bernstein. Columbia M 33532 (Q) MQ 33532. (also CBS 76475
    Tape (c) 40-76475).
      +Audio 1-76 p70   +NR 11-75 p11
      +Gr 4-76 p1649   +-RR 4-76 p48
      +Gr 11-76 p887 tape   +-RR 6-76 p88 tape
      +HF 10-75 p76   ++SFC 9-21-75 p34
      +HFN 4-76 p107   ++St 11-75 p127
1593 Kindertotenlieder. Lieder eines fahrenden Gesellen. Kirsten
    Flagstad, s; VPO; Adrian Boult. Decca ECS 780. (Reissue
    from SXL 2224).
      +-Gr 8-76 p328   +-RR 6-76 p44
      +HFN 6-76 p102
    Kindertotenlieder. cf Symphony, no. 5, C sharp minor (Columbia
    M2S 698).
    Kindertotenlieder. cf Symphony, no. 5, C sharp minor (DG 2707 081).
    Kindertotenlieder. cf Symphony, no. 5, C sharp minor (Odyssey
    322406/6).
1594 Das klagende Lied (2-part version). Heather Harper, s; Norma
    Proctor, alto; Werner Hollweg, t; Netherlands Radio Chorus;
    COA; Bernard Haitink. Philips 6500 587.
      ++Gr 2-74 p159   +NYT 2-10-74 pD28
      +HF 5-74 p76   +RR 1-74 p61
      +HFN 2-76 p117   /SFC 5-19-74 p30
      ++MJ 3-74 p10   ++St 8-74 p116
      ++NR 3-74 p10
1595 Das klagende Lied: Waldmärchen; Der Speilmann; Hochzeitsstück.
    Symphony, no. 10, F sharp minor: Adagio. Elisabeth Söderström,

Evelyn Lear, s; Grace Hoffman, ms; Ernst Häfliger, t; Stuart
Burrows, t; Gerd Nienstedt, bar; LSO and Chorus; Pierre Boulez.
CBS 77233 (2). (Reissues from 72865, 72773).
+-Gr 3-76 p1494   +-RR 1-76 p58
*HFN 2-76 p117

Des Knaben Wunderhorn. cf Symphony, no. 5, C sharp minor (London
CSA 2228).

Des Knaben Wunderhorn: Das irdische Leben; Rheinlegendchen. cf
Prelude PRS 2505.

1596 Des Knaben Wunderhorn: Revelge; Der Tambourg'sell. Lieder und
Gesange aus der Jugendzeit: Frühlingsmorgen; Erinnering; Hans
und Grethe; Don Juan serenade; Phantasie aus Don Juan; Um
schlimme Kinder artig zu machen; Ich ging mit Lust durch einen
grünen Wald; Aus, aus; Starke Einbildungskraft; Zu Strassburg
auf der Schanz; Ablösung im Sommer; Scheiden und Meiden; Nich
Wiedersehen; Selbstgefühl. Roland Hermann, bar; Geoffrey Parsons, pno. HMV HQS 1346.
+-Gr 10-75 p669   +-MT 2-76 p142
+-HFN 10-75 p143   +-RR 11-75 p86

1597 Das Lied von der Erde (The song of the earth). Alfreda Hodgson,
con; John Mitchinson, t; Scottish National Orchestra; Alexander
Gibson. Classics for Pleasure CFP 40226.
+-Gr 6-76 p72   +-RR 3-76 p68
+-HFN 3-76 p99

1598 Das Lied von der Erde. Yvonne Minton, ms; René Kollo, t; CSO;
Georg Solti. London OS 26292. (also Decca SET 555 Tape (c)
KCET 555).
++Gr 11-72 p939   -NYT 3-11-73 pD30
+-HF 4-73 p84   ++RR 11-72 p104
++HFN 12-76 p155 tape   ++SFC 1-21-73 p41
+-MJ 4-73 p8   -SR 2-73 p54
+NR 6-73 p11   ++St 5-73 p118

1599 Das Lied von der Erde. Nan Merriman, con; Ernst Häfliger, t;
COA; Eugen Jochum. DG Tape (c) 3335 184.
+RR 9-76 p94 tape

1600 Das Lied von der Erde. Christa Ludwig, con; René Kollo, t; BPhO;
Herbert von Karajan. DG Tape (c) 3581 015.
+RR 3-76 p76 tape

1601 Das Lied von der Erde. Rückert Lieder (5). Christa Ludwig, ms;
René Kollo, t; BPhO; Herbert von Karajan. DG 2707 082 (2).
++Gr 12-75 p1089   +-NYT 4-4-76 pREC1
+-HF 4-76 p112   +RR 12-75 p88
+-HFN 1-76 p111   ++SFC 1-4-76 p27
++MJ 3-76 p25   ++St 5-76 p118
+NR 2-76 p13

1602 Das Lied von der Erde. Janet Baker, ms; James King, t; COA;
Bernard Haitink. Philips 6500 831. Tape (c) 7300 362.
+-Gr 10-76 p634   +-RR 10-76 p97
+HFN 10-76 p173

1603 Das Lied von der Erde. Dietrich Fischer-Dieskau, bar; Murray
Dickie, t; PhO; Paul Kletzki. Seraphim S 60260.
+NR 5-76 p12

1604 Lieder eines Fahrenden Gesellen (Songs of a wayfarer). MARTIN:
Monlogues on "Jedermann" (6). Dietrich Fischer-Dieskau, bar;
Bavarian Radio Orchestra, BPhO; Rafael Kubelik, Frank Martin.
DG 2530 630. (Reissues from 2707 056, 138 871).

```
        +Gr 9-76 p459              +RR 9-76 p85
       ++HFN 10-76 p173            +SFC 4-25-76 p30
        +NR 5-76 p12               +-St 7-76 p84
```

Lieder eines fahrenden Gesellen. cf Kindertotenlieder.
Lieder eines fahrenden Gesellen. cf Symphony, no. 5, C sharp
  minor.
Songs of a wayfarer. cf Symphony, no. 6, A minor.
Lieder und Gesange aus der Jugendzeit: Frühlingsmorgen; Erinner-
  ing; Hans und Grethe; Don Juan Serenade; Phantasie aus Don
  Juan; Um schlimme Kinder artig zu machen; Ich ging mit Lust
  durch einen grünen Wald; Aus, Aus; Starke Einbildungskraft;
  Zu Strassburg auf der Schanz; Ablösung im Sommer; Scheiden
  und Meiden; Nicht Wiedersehen; Selbstgefühl. cf Des Knaben
  Wunderhorn: Revelge; Der Tambourg'sell.
Rückert songs. cf Symphony, no. 5, C sharp minor.
Rückert Lieder (5). cf Das Lied von der Erde.
Rückert Lieder: Ich bin der Welt abhanden gekommen. cf Prelude
  PRS 2505.

1605 Symphonies, nos. 1-9. Symphony, no. 10: Adagio. Beverly Sills,
        Natania Davrath, s; Florence Kopleff, con; Utah Symphony
        Orchestra; Maurice Abravanel. Vanguard SRV 330/43 (14).
          +-SFC 4-25-76 p30
1606 Symphony, no. 1, D major. NYP; Leonard Bernstein. Columbia MS
        7069.
          +-St 10-76 p78
1607 Symphony, no. 1, D major. Bavarian Radio Symphony Orchestra;
        Rafael Kubelik. DG 139 331. Tape (c) 923 070 (ct) 88311.
          ++HFN 1-72 p117 tape   ++St 10-76 p78
1608 Symphony, no. 1, D major. Bavarian Radio Symphony Orchestra;
        Rafael Kubelik. DG Tape (c) 3335 172.
          +RR 9-76 p94 tape
1609 Symphony, no. 1, D major. Dresden Staatskapelle Orchestra;
        Otmar Suitner. DG Tape (c) 3348 123.
          -RR 9-76 p92 tape
1610 Symphony, no. 1, D major. LPO; Adrian Boult. Everest 3005.
          +St 10-76 p78
1611 Symphony, no. 1, D major. Israel Philharmonic Orchestra; Zubin
        Mehta. London CS 7004. Tape (c) 5-7004. (also Decca SXL
        6779 Tape (c) KSXC 6779).
          +-Gr 9-76 p422   +-HFN 11-76 p175 tape
          -Gr 11-76 p887 tape   +NR 12-76 p6
          /HF 12-76 p98   ++RR 9-76 p56
          +-HFN 9-76 p125   -SFC 11-14-76 p30
1612 Symphony, no. 1, D major. Columbia Symphony Orchestra; Bruno
        Walter. Odyssey Y 30047. (Reissue from Columbia MS 6394).
          +HF 12-70 p137   ++St 10-76 p78
          +SFC 10-4-70 p28
1613 Symphony, no. 1, D major. COA; Bernard Haitink. Philips 6500
        342. Tape (c) 7300 397 (ct) 7750 093. (Reissue from 6711 003).
          ++Gr 2-73 p1552   +RR 2-73 p60
          ++HF 5-73 p86   +-RR 5-76 p77 tape
          +HFN 2-73 p340   ++St 5-73 p118
          ++NR 4-73 p4
1614 Symphony, no. 1, D major. LSO; James Levine. RCA ARL 1-0894.
        Tape (c) ARK 1-0894 (ct) ARS 1-0894.
          +Gr 5-75 p1967   ++RR 5-75 p33
```
```

                    -HF 6-75 p98   ++SFC 5-11-75 p23
              ++HFN 5-75 p134   +-SR 5-3-75 p33
                    -NR 5-75 p5   +St 5-75 p96
              +NYT 4-4-76 pREC1   +-St 10-76 p78
1615 Symphony, no. 1, D major. Symphony, no. 4, G major. Judith
      Blegen, s; LSO, CSO; James Levine. RCA (Q) CRD 3-1040 (3).
              +Audio 11-75 p89 Quad   +-St 10-76 p78
              +-NR 11-75 p4 Quad
1616 Symphony, no. 1, D major (1893 version). NPhO; Wyn Morris.
      Virtuoso TPLS 13037.
              /Gr 12-70 p994   +St 10-76 p78
1617 Symphony, no. 2, C minor. Janet Baker, Sheila Armstrong, s;
      Edinburgh Festival Chorus; LSO; Leonard Bernstein. Columbia
      M2 32681 (2). Tape (c) M2T 32681 (ct) M2A 32681 (Q) M2Q 32681
      Tape (ct) QMA 32681. (also CBS 78249).
              +-Gr 11-74 p895   +RR 12-74 p31
              +HF 1-75 p71   -SFC 9-29-74 p26
              +-NR 12-74 p3   +-St 1-75 p113
              +NYT 8-4-74 pD20   +-St 10-76 p78
1618 Symphony, no. 2, C minor. Ileana Cotrubas, s; Christa Ludwig,
      con; Vienna State Opera Chorus; VPO; Zubin Mehta. Decca SXL
      6744/5 (2). Tape (c) KSXC 7037. (also London CSA 2242 Tape
      (c) 5-2242).
              ++Gr 12-75 p1037   +NYT 4-4-76 pREC1
              +-HF 7-76 p81   +RR 12-75 p55
              +-HFN 1-76 p111   +RR 1-76 p66 tape
              +-HFN 2-76 p116 tape   ++SFC 4-11-76 p38
              +NR 9-76 p3 tape   +St 8-76 p97, 10-76 p78
1619 Symphony, no. 2, C minor. Edith Mathis, s; Norma Proctor, con;
      Bavarian Radio Symphony Orchestra; Rafael Kubelik. DG 139 332/3.
      Tape (c) 3581 001. (also DG 2707 043).
              ++HFN 1-72 p117 tape   -St 10-76 p78
              /St 10-71 p141 tape
1620 Symphony, no. 2, C minor. Heather Harper, s; Helen Watts, con;
      LSO and Chorus; Georg Solti. London CSA 2217. Tape (c) H 10187.
      (also Decca Tape (c) KCET 27002).
              +HFN 2-76 p116 tape   ++St 10-76 p78
1621 Symphony, no. 2, C minor. Emilia Cundari, s; Maureen Forrester,
      con; Westminster Choir; NYP; Bruno Walter. Odyssey Y2 30848
      (2). (Reissue from Columbia M2S 601). (also CBS 61282/3. Re-
      issue from Philips ABL 3245/6).
              +Gr 11-72 p958   -St 10-76 p78
              ++SFC 4-16-72 p45
1622 Symphony, no. 2, C minor. Elly Ameling, s; COA; Bernard Haitink.
      Philips 802 844/5 (2).
              +St 10-76 p78
1623 Symphony, no. 2, C minor. Brigitte Fassbaender, ms; Margaret
      Price, s; LSO and Chorus; Leopold Stokowski. RCA ARL 2-0852
      (2). Tape (c) CRK 2-0852.
              ++Audio 8-76 p76   +NYT 4-4-76 pREC1
              +-Gr 6-76 p49   +-RR 6-76 p44
              +-HF 4-76 p116   +SFC 1-4-76 p27
              ++HF 7-76 p114 tape   +-SR 3-6-76 p41
              ++HFN 7-76 p90   ++St 4-76 p113
              ++NR 2-76 p2   +-St 10-76 p78
1624 Symphony, no. 2, C minor. Orchestra; Otto Klemperer. Turnabout

34249/50 (2).
-St 10-76 p78
Symphony, no. 2, C minor: Urlicht. cf CBS 30066.
1625 Symphony, no. 3, D minor. Martha Lipton, ms; John Ware, solo
posthorn; Schola Cantorum Women's Chorus, Church of the Trans-
figuration Boys' Choir; NYP; Leonard Bernstein. CBS 77206 (2).
(Reissue from SBRG 72065/6). (also Columbia M2S 675).
++Gr 6-74 p51   ++St 10-76 p78
+RR 6-74 p42
1626 Symphony, no. 3, D minor. Marjorie Thomas, con; Bavarian Radio
Symphony Orchestra; Rafael Kubelik. DG 2707 036 (2).
+St 10-76 p79
1627 Symphony, no. 3, D minor. Helen Watts, con; Ambrosian Singers;
Wandsworth School Boys' Chorus; LSO; Georg Solti. London
2223 (2).
+St 10-76 p79
1628 Symphony, no. 3, D minor. Maureen Forrester; Amsterdam Chorus;
COA; Bernard Haitink. Philips 802711/2 (2).
+St 10-76 p79
1629 Symphony, no. 3, D minor. Marilyn Horne, ms; Glen Ellyn Children's
Chorus; CSO and Chorus; James Levine. RCA ARL 2-1757 (2).
++SFC 11-14-76 p30
1630 Symphony, no. 3, D minor. RAI Torino Orchestra; Bergamo Choir;
Lucrezia West, con; Jascha Horenstein. Rococo 2083 (2).
-NR 4-76 p4
1631 Symphony, no. 3, D minor. Norma Procter, con; Wandsworth School
Boys' Choir; Ambrosian Singers; LSO; Jascha Horenstein.
Unicorn Tape (c) ZCUN 302/2. (also Nonesuch/Advent 73023
Tape (c) E 1009).
+ARG 5-71 p577   ++NR 7-71 p3
+Gr 12-70 p994   +NYT 4-18-71 pD26
+Gr 7-72 p247 tape   +SFC 4-25-71 p32
+HF 7-71 p80   ++St 9-71 p88
+HF 12-74 p146 tape   +St 10-76 p78
1632 Symphony, no. 4, G major. Judith Raskin, s; CO; Georg Szell.
CBS 61056. Tape (c) 40-61056. (also Columbia MS 6833).
+Gr 12-72 p1214 tape   +RR 1-73 p88 tape
+HFN 12-72 p2455 tape   +-St 10-76 p79
1633 Symphony, no. 4, G major. Reri Grist, s; NYP; Leonard Bernstein.
Columbia MS 6152.
+St 10-76 p79
1634 Symphony, no. 4, G major. Elisabeth Schwarzkopf, s; PhO; Otto
Klemperer. HMV ASD 2799. (Reissue from Columbia SAX 2441).
(also Angel 35829).
+Gr 4-73 p1921   +RR 4-73 p60
+HFN 4-73 p782   +-St 10-76 p79
1635 Symphony, no. 4, G major. Sylvia Stahlman, s; COA; Georg Solti.
London 6781.
+-St 10-76 p79
1636 Symphony, no. 4, G major. Heather Harper, s; BRSO; Loren Maazel.
Nonesuch H 71259.
-HF 3-72 p96   +-SR 3-11-72 p18
/NR 1-72 p2   +-St 10-76 p79
/NYT 11-29-71 pD32
1637 Symphony, no. 4, G major. Elly Ameling, s; COA; Bernard Haitink.
Philips 802888. Tape (c) 18230 CAA. (also Tape (c) 7300 408).
+RR 5-76 p77   +St 10-76 p79

1638 Symphony, no. 4, G major. Judith Blegen, s; CSO; James Levine.
      RCA ARL 1-0895. Tape (c) ARK 1-0895 (ct) ARS 1-0895. (also
      RK 11733).
              ++Gr 10-75 p612   +NYT 4-4-76 pREC1
              ++Gr 11-76 p887 tape   ++RR 11-75 p43
              ++HF 6-75 p98   ++St 8-75 p100
               -HFN 11-75 p155   +ST 2-76 p739
              +-HFN 11-76 p175 tape   -St 10-76 p79
               +NR 5-75 p5
1639 Symphony, no. 4, G major. Netania Davrath, s; Utah Symphony
      Orchestra; Maurice Abravanel. Vanguard C 10042. Tape (c)
      GRT 8193 10042E.
               +HF 5-76 p114 tape
      Symphony, no. 4, G major. cf Symphony, no. 1, D major.
1640 Symphony, no. 5, C sharp minor. Kindertotenlieder. Jennie Tourel,
      ms; NYP; Leonard Bernstein. Columbia M2S 698 (2).
               +St 10-76 p79
1641 Symphony, no. 5, C sharp minor. Lieder eines fahrenden Gesellen.
      Dietrich Fischer-Dieskau, bar; Bavarian Radio Symphony Orchestra;
      Rafael Kubelik. DG 2530 189/90. Tape (r) 2056. (also
      DG 2707 056).
              +-Gr 8-72 p380   ++SFC 7-23-72 p33
               +HFN 9-72 p1667   ++St 4-72 p81
              ++NR 3-72 p4   +St 10-76 p79
1642 Symphony, no. 5, C sharp minor. Kindertotenlieder. Christa
      Ludwig, ms; BPhO; Herbert von Karajan. DG 2707 081 (2). Tape
      (c) 3370 006.
               +Gr 6-75 p40   +-NYT 4-4-76 pREC1
              +-HF 1-76 p88   ++RR 6-75 p43
               +HFN 9-75 p100   +-RR 10-75 p97 tape
              +-MT 11-75 p978   ++St 3-76 p116
              ++NR 12-75 p5   +St 10-76 p79
1643 Symphony, no. 5, C sharp minor. LSO; Rudolf Schwarz. Everest
      SDBR 3386.
               +NR 5-76 p3
1644 Symphony, no. 5, C sharp minor. Rückert songs. Janet Baker, s;
      NPhO; John Barbirolli. HMV SLS 785. Tape (c) TC SLS 785.
              ++HFN 4-76 p125 tape   ++RR 5-76 p77 tape
1645 Symphony, no. 5 C sharp minor. Des Knaben Wunderhorn. Yvonne
      Minton, ms; CSO; Georg Solti. London CSA 2228 (2). Tape (r)
      DPK 80232. (also Decca SET 471/2).
              +-ARG 2-71 p346   ++SFC 11-15-70 p30
              ÷-Gr 2-71 p1315   +St 2-71 p108
              ++HF 1-71 p72   +St 10-76 p79
              ++MJ 1-71 p67
1646 Symphony, no. 5, C sharp minor. Kindertotenlieder. NYP; Bruno
      Walter. Odyssey 32260016 (2).
               +St 10-76 p79
1647 Symphony, no. 5, C sharp minor. Symphony, no. 10, F sharp minor:
      Adagio. COA; Bernard Haitink. Philips 6700 048 (2).
              ++Gr 4-72 p1715   /NYT 5-14-72 pD25
              ++HF 6-72 p82   ++SFC 6-25-72 p45
              +-HFN 4-72 p713   +St 6-72 p84
               +HFN 9-72 p1667   +-St 10-76 p79
              ++NR 7-72 p5
1648 Symphony, no. 6, A minor. NPhO; John Barbirolli. Angel S 3760 (2).

```
                  /ARG 11-70 p173          +NYT 8-2-70 pD20
                  +Gr 12-69 p967           +-ON 5-16-70 p30
                  +HF 6-70 p96             +St 6-70 p86
                  +NR 6-70 p3              -St 10-76 p79
```
1649 Symphony, no. 6, A minor. Symphony, no. 9, D major. NYP;
      Leonard Bernstein. Columbia M3S 776 (3).
```
                  +-St 10-76 p79
```
1650 Symphony, no. 6, A minor. Symphony, no. 10, F sharp minor:
      Andante, Purgatorio. CO; Georg Szell. Columbia M2 31313 (2).
      (also CBS 77272).
```
                  +Gr 10-72 p694           -SFC 7-23-72 p33
                  +HF 11-72 p84            ++St 10-72 p82
                  +-HFN 10-72 p1911        +-St 10-76 p79
                  +-RR 10-72 p40
```
1651 Symphony, no. 6, A minor. Symphony, no. 10, F sharp minor:
      Adagio. Bavarian Radio Symphony; Rafael Kubelik. D 139341/2
      (2). (also DG 2707 037).
```
                  /ARG 9-70 p34            /St 10-76 p79
```
1652 Symphony, no. 6, A minor. Songs of a wayfarer. Yvonne Minton,
      ms; CSO; Georg Solti. London CSA 2227 (2). Tape (r) DPK
      80231. (also Decca SET 469/70).
```
                  +-ARG 2-71 p346          ++SFC 11-15-70 p28
                  /Gr 12-70 p997           +St 2-71 p108
                  ++HF 1-71 p72            +St 10-76 p79
                  ++MJ 1-71 p63
```
1653 Symphony, no. 6, A minor. (Includes Horenstein interview with
      Alan Blyth) Stockholm Philharmonic Orchestra; Jascha Horen-
      stein. Nonesuch HB 73029 (2).
```
                  +-HF 10-75 p76           +St 12-75 p129
                  +HFN 9-75 p100           +-St 10-76 p79
                  +NR 9-75 p2
```
1654 Symphony, no. 6, A minor. Stockholm Philharmonic Orchestra;
      Jascha Horenstein. Unicorn RHS 320/1 (2).
```
                  ++Gr 9-75 p465           +-MT 1-76 p41
                  +HFN 9-75 p100           +RR 9-75 p40
```
1655 Symphony, no. 7, E minor. NPhO; Otto Klemperer. Angel S 3740 (2).
```
                  +-St 10-76 p80
```
1656 Symphony, no. 7, E minor. NYP; Leonard Bernstein. Columbia
      M2S 739 (2).
```
                  +-St 10-76 p80
```
1657 Symphony, no. 7, E minor. CSO; Georg Solti. London CSA 2231 (2).
      Tape (c) K 10249 (r) K 80249. (also Decca SET 518/9).
```
                  +Gr 9-71 p448            ++SFC 12-12-71 p32
                  ++HF 2-72 p100           ++SFC 3-5-72 p39
                  ++HF 8-72 p108 tape      -SR 2-26-72 p50
                  +HFN 10-71 p1849         ++St 3-72 p86
                  -NR 3-72 p3              +St 10-76 p80
                  +NYT 11-29-71 pD25
```
1658 Symphony, no. 8, E flat major. LSO; Leonard Bernstein. Columbia
      M2S 751 (2).
```
                  *St 10-76 p80
```
1659 Symphony, no. 8, E flat major. Martina Arroya, Edith Mathis, Erna
      Spoorenberg, s; Julia Hamari, Norma Proctor, con; Donald Grobe,
      t; Dietrich Fischer-Dieskau, bar; Franz Crass, bs; Bavarian
      Radio Symphony Orchestra; Rafael Kubelik. DG 2707 062 (2).
      (Reissue from DG 2720 033).

```
      +ARG 9-72 p659              +ON 1-26-74 p36
      +-HFN 8-72 p1464            +St 12-72 p128
      +NR 12-72 p4               +St 10-76 p80
```

1660 Symphony, no. 8, E flat major. Vienna Festival Orchestra;
      Dimitri Mitropoulos. Everest 3189E (2).
```
      -St 10-76 p80
```

1661 Symphony, no. 8, E flat major. Heather Harper, Lucia Popp, Arleen
      Auger, s; Yvonne Minton, Helen Watts, altos; René Kollo, t;
      John Shirley-Quirk, bar; Martti Talvela, bs; Vienna State
      Opera Chorus, Vienna Singverein, Vienna Boys' Choir; CSO;
      Georg Solti. London OSA 1295 (2). Tape (c) S 31211 (r) 490
      211. (also Decca SET 534/5 Tape (c) KCET 7006).
```
      +ARG 9-72 p659              +RR 3-74 p72 tape
      +Gr 10-72 p699             ++SFC 6-25-72 p45
      ++Gr 9-73 p563 tape        ++SFC 10-14-73 p45 tape
      ++HF 9-72 p87              ++SR 6-17-72 p44
      ++HFN 10-72 p1897          ++St 9-72 p96
      ++NR 7-72 p2               ++St 10-76 p80
      +ON 1-26-74 p36            ++STL 11-19-72 p72
      +RR 10-72 p36
```

1662 Symphony, no. 8, E flat major. Heather Harper, Ileana Cotrubas,
      Hanneke van Bork, s; Birgit Finnila, Marianne Dieleman, con;
      William Cochran, t; Hermann Prey, bar; Hans Sotin, bs; Jo Juda,
      vln; Kees di Wijs, org; Amsterdam Collegium Musicum, Tookunst-
      koor, Amsterdam De Stem des Volks, Kinderkoor St. Willibrord,
      Pius X; COA; Bernard Haitink. Philips 6700 049 (2). (also
      6500 258/9 (2).
```
      +Gr 12-71 p1040            +SFC 12-12-71 p32
      ++HF 2-72 p102             +SFC 6-25-72 p45
      +HFN 12-71 p2304          ++St 4-72 p81
      ++NR 3-72 p3               +St 10-76 p80
      /NYT 11-28-71 pD25
```

1663 Symphony, no. 8, E flat major. Joyce Barker, Elisabeth Simon,
      Norma Burrowes, s; Joyce Blackham, Alfreda Hodgson, ms; John
      Mitchinson, t; Raymond Myers, bar; Gwynne Howell, bs; Orpinton
      Junior Singers, Highgate School Choir, Finchley Children's
      Music Group, Ambrosian Singers, New Philharmonia Chorus,
      Bruchner-Mahler Choir; London Sinfonia; Wyn Morris. RCA CRL
      2-0359 (2). Tape (ct) CRS 2-0359 (c) CRK 2-0359.
```
      +-HF 5-74 p73              +-ON 1-26-74 p36
      +LJ 4-75 p69               -SFC 5-19-74 p29
      ++NR 2-74 p3               +St 8-74 p117
      +-NYT 2-10-74 pD26        ++St 10-76 p80
```

1664 Symphony, no. 9, D major. LSO; Georg Solti. Decca SET 360/1.
      Tape (c) CET 27003. (also London 2220).
```
      ++HFN 2-72 p320 tape       +-St 10-76 p80
```

1665 Symphony, no. 9, D major. NPhO; Otto Klemperer. HMV SXLP 3021/2.
      (Reissue from Columbia SAX 5281/2). (also Angel S 3708).
```
      +Gr 1-76 p1197             +RR 12-75 p56
      ++HFN 12-75 p171
```

1666 Symphony, no. 9, D major. Columbia Symphony Orchestra; Bruno
      Walter. Odyssey Y2 30308 (2).
```
      ++St 10-76 p80
```

1667 Symphony, no. 9, D major. VPO; Bruno Walter. Turnabout THS
      65008/9 (2). (Reissue from Victor M-726).
```
      +HF 10-74 p107             +St 10-76 p80
```

Symphony, no. 9, D major. cf Symphony, no. 6, A minor.
1668 Symphony, no. 10, F sharp minor (revised version by Deryck Cooke).
PO; Eugene Ormandy. Columbia M2S 735 (2).
+-St 10-76 p80
1669 Symphony, no. 10, F sharp minor (finally revised full-length per-
forming version by Deryck Cooke). NPhO; Wyn Morris. Philips
6700 067 (2).

+Gr 3-74 p1700	+-RR 3-74 p40
+HF 5-74 p73	+SFC 1-13-74 p20
+HFN 3-74 p109	+St 5-74 p103
+NR 2-74 p3	++St 10-76 p80
+NYT 2-10-74 pD26	

Symphony, no. 10, F sharp minor: Adagio. cf Kindertotenlieder.
Symphony, no. 10, F sharp minor: Adagio. cf Das klagende Lied:
Waldmärchen; Der Speilmann: Hochzeitsstück.
Symphony, no. 10, F sharp minor: Adagio. cf Symphony, no. 5,
C sharp minor.
Symphony, no. 10, F sharp minor: Adagio. cf Symphony, no. 6,
A minor (DG 139341/2).
Symphony, no. 10, F sharp minor: Andante, Purgatorio. cf Symphony,
no. 6, A minor (Columbia M2 31313).
MAINERIO (17th century Italy)
Schiarazula marazula. cf DG Archive 2533 184.
Ungarescha-Saltarello. cf DG Archive 2533 184.
MAKSIMOVIC, Rajko
Tri Haiku-a. cf KULJERIC: Omaggio à Lukačič.
MALASEK, Jiri
South Bohemian encounter. cf Panton 110 440.
MALATS, Joaquin
Impresiones de España: Serenata española. cf London STS 15306.
MALCOLM, George
Variations on a theme by Mozart for 4 harpsichords. cf BACH,
C.P.E.: Concerto, 4 harpsichords, F major.
MALIBRAN, Maria
Le reveil d'un beau jour. cf Gemini Hall RAP 1010.
MALMFORS, Ake
Songs: Wiegenlied; Scherzlied; Mangard; I fjärren dis. cf
Swedish Society SLT 33188.
MANCHICOURT, Pierre
Doulce memoire. cf Argo ZRG 667.
MANERA (20th century Italy)
Salve regina. cf L'Oiseau-Lyre SOL 343.
MANFREDINI, Vincenzo
1670 Concerto, piano, B flat major. PAISIELLO: Concerto, piano, F
major. PLATTI: Concerto, piano, no. 2, C minor. Felicja
Blumenthal, pno; Salzburg Mozarteum Orchestra, Turin Radio
Orchestra; Albert Zedda, Theodor Guschlbauer. Turnabout
TV 34495.

+-Gr 9-76 p422	+-RR 7-76 p60
+-HFN 8-76 p83	

Concerto grosso, op. 3, no. 12, G major. cf CORELLI: Concerto
grosso, op. 6, no. 8, G minor.
MANOLOV, Hristo
Dostoino est. cf Harmonia Mundi HMU 137.
MANTIA, Simone
Priscilla. cf Washington University Press OLY 104.

MANTLER, Michael
    13, piano and 2 orchestras. cf BLEY: 3/4, piano and orchestra.
MANZIARLY, Marcelle de
    Dialogue. cf Laurel-Protone LP 13.
MARAIS, Marin
1671 Sonnerie de Sainte Geneviève du Mont de Paris. Suite, no. 1,
          C major. Suite, no. 4, D major. Alice Harnoncourt, vln;
          Leopold Stastny, flt; Nikolaus Harnoncourt, vla da gamba;
          Herbert Tachezi, hpd. Oryx Musical Heritage MHS 964.
              +AR 8-76 p78   +Gr 9-73 p502
      Suite, no. 1, C major. cf Sonnerie de Sainte Geneviève du Mont
          de Paris.
      Suite, no. 4, D major. cf Sonnerie de Sainte Geneviève du Mont
          de Paris.
1672 Suites, recorder, B major, G minor. Hotteterre Quartet.
          Telefunken AW 6-41992.
              +Gr 12-76 p1015   ++RR 12-76 p72
              +HFN 12-76 p144
1673 Suite, viola da gamba and continuo, E minor. Catharina Meints,
          James Caldwell, vla da gamba; James Weaver, hpd. Cambridge
          CRS 2201.
              +NR 10-76 p7   +SFC 11-14-76 p30
MARBECK, John
    Credo. cf Argo ZRG 789.
MARCELLO, Alessandro
    Sonata, recorder, D minor. cf Saga 5414.
MARCELLO, Benedetto
    Concerto, oboe, D minor. cf BACH: Concerto, violin, oboe and
        strings, S 1060, D minor.
MARCHAND, Louis
    Basse de cromorne. cf DAQUIN: Noëls.
    Basse de trompette. cf Vista VPS 1030.
    Dialogue (2). cf DAQUIN: Noëls.
    Duo. cf DAQUIN: Noëls.
    Fugue. cf DAQUIN: Noëls.
    Grand jeu. cf DAQUIN: Noëls.
    Plein jeu. cf DAQUIN: Noëls.
    Quatuor. cf DAQUIN: Noëls.
    Suite. cf DUPHLY: La Félix.
    Tierce en taille. cf DAQUIN: Noëls.
MARENZIO, Luca
    O voi che sospirate; Occhi lucenti. cf L'Oiseau-Lyre 12BB 203/6.
    Spring. cf Hungaroton SLPX 11696.
MARESCHAL, Samuel
    Psalms, nos. 8 and 47. cf Pelca PRSRK 41017/20.
MARGETSON
    Tommy lad. cf Pye Ember GVC 51.
MAROS, Miklos
    Descort. cf Caprice RIKS 59.
MARSHALL
    I hear you calling me. cf Pearl SHE 528.
MARTIN
    Serenade to a double Scotch. cf Music for Pleasure SPR 90086.
MARTIN, Frank
    Ballade. cf FARKAS: Concertino all'antica.
1674 Ballade, flute, piano and string orchestra. Ballade, viola, per-

cussion and wind orchestra. Polyptyque. Yehudi Menuhin, vln,
vla; Aurèle Nicolet, flt; Werner Bärtschi, pno; Menuhin Festival
Orchestra, Zurich Chamber Orchestra; Edmond de Stoutz, Michael
Dobson. HMV SQ ASD 3185.

  ++Gr 7-76 p176   ++MT 12-76 p1005
  +HFN 8-76 p83   +RR 7-76 p60

 Ballade, viola, percussion and wind orchestra. cf Ballade, flute,
  piano and string orchestra.

1675 Concerto, 7 wind instruments, percussion and strings. Etudes,
  string orchestra. OSR; Ernest Ansermet. London STS 15270.
   +St 6-76 p105

 Etudes, string orchestra. cf Concerto, 7 wind instruments, percussion and strings.

 Monologues on "Jedermann" (6). cf MAHLER: Lieder eines fahrenden
  Gesellen.

1676 Petite symphonie concertante. ROUSSEL: Sinfonietta, op. 52.
  TORTELIER: Offrande. London Chamber Orchestra; Paul Tortelier.
  Unicorn UNS 233.
   +St 1-76 p108

 Polyptyque. cf Ballade, flute, piano and string orchestra.

 Preludes (8). cf Sterling 1001/2.

MARTIN/COULTER
 Puppet on a string. cf HMV CSD 3766.

MARTINI
 Gavotte, F major. cf Angel S 36095

MARTINI, M.
 Plaire à celui que j'aime. cf Smithsonian Collection N 002.

MARTINI IL TEDESCO, Johann
 Songs: Plaisir d'amour. cf Club 99-101.
 Songs: Plaisir d'amour. cf Columbia M 33933.
 Songs: Plaisir d'amour. cf Decca SXL 6629.

MARTINO, Donald
1677 Paradiso choruses. PINKHAM: For evening draws on. Liturgies.
  Toccatas for the vault of heaven. Soloists; New England
  Conservatory Symphony Orchestra and Chorus; Lorna Cooke
  de Varon. Golden Crest NEC 114.
   +-HF 6-76 p84   +NYT 12-21-75 pD18

MARTINON, Jean
 Symphony no. 2. cf DUTILLEUX: Symphony no. 1.

MARTINU, Bohuslav
1678 Butterflies and birds of paradise. Czech dances (3). Madrigal
  sonata, flute, violin and piano. Les ritournelles. Boris
  Krajný, Jiri Holena, Eva Kramská, pno; Eva Dostalová, flt;
  Maria Motulková, vln. Panton 110 446.
   +Gr 4-76 p1623   ++RR 1-76 p51

 Concertino, violoncello, piano and wind instruments. cf IBERT:
  Concerto, violoncello and wind instruments.

1679 Concerto, flute, violin and orchestra. SCHULHOFF: Concerto,
  flute, piano and orchestra. Jiří Valek, flt; Bohuslav
  Matousek, vln; Josef Hála, pno; Dvořák Chamber Orchestra;
  Vladimir Válek. Panton 110 368.
   +-Gr 11-76 p797   +RR 9-76 p57

 Concerto, 2 orchestras, piano and timpani. cf JANACEK: Idyll,
  string orchestra.

1680 Concerti, violin, nos. 1, 2. Josef Suk, vln; CPhO; Václav Neumann.
  Supraphon 110 1535.

```
        ++Gr 7-75 p187          +NR 11-75 p7
        +HF 3-76 p90            +RR 5-75 p34
```
1681 Concerto, violin, piano and orchestra. Concerto, violoncello,
     no. 1. Josef Chuchro, vlc; Nora Grumliková, vla; Jaroslav
     Kolar, pno; CPhO; Zdeněk Košler. Supraphon 110 1348.
```
        +Gr 3-75 p1654          ++SFC 6-8-75 p23
        +-NR 1-75 p4            +St 4-76 p113
        +RR 2-75 p36
```
     Concerto, violoncello, no. 1. cf Concerto, violin, piano and
     orchestra.
     Czech dances (3). cf Butterflies and birds of paradise.
     Duo, violin and violoncello, no. 2. cf IBERT: Concerto, violon-
     cello and wind instruments.
1682 Impromptus, violin and piano (3). Nonet. Sextet, piano and
     winds. Sonata, violin and piano, D minor. Jiří Tomášek, vln.
     Czech Nonet, Members; Josef Ružička, pno. Panton 110 282.
```
        +-Gr 3-76 p1481         +-NR 1-76 p40
        +HFN 1-76 p111
```
1683 Istar: Suites, nos. 1 and 2. Brno State Philharmonic Orchestra;
     Jiří Waldhans. Supraphon 110 1634.
```
        +-Gr 5-76 p1761         +-RR 4-76 p49
        -NR 5-76 p4
```
     Madrigal sonata, flute, violin and piano. cf Butterflies and
     birds of paradise.
     Nonet. cf Impromptus, violin and piano (3).
     Préludes. cf Supraphon 111 1721/2.
     Les ritournelles. cf Butterflies and birds of paradise.
     Sextet, piano and winds. cf Impromptus, violin and piano (3).
1684 Sinfonietta la jolla. Toccata e due canzoni. PCO; Zdeněk Hnát,
     pno. Supraphon 110 1619.
```
        +ARG 12-76 p36          +NR 11-76 p2
        ++Gr 11-76 p797         +RR 9-76 p57
```
     Sonata, violin and piano. cf BACH: Partita, violin, no. 2, S
     1004, D minor: Chaconne.
     Sonata, violin and piano, D minor. cf Impromptus, violin and
     piano (3).
     Toccata a due canzoni. cf Sinfonietta la jolla.
     Trio, piano, no. 3, C major. cf BEETHOVEN: Trio, piano, no. 3,
     op. 1, no. 3, C minor.
MARTIRANO, Salvatore
     Chansons innocentes. cf CRI SD 324.
MARX, Joseph
     Songs: Hat dich die Liebe berüht; Japanisches Regenlied;
     Nocturne; Selige Nacht; Waldseligkeit; Windräder. cf
     BRITTEN: Winter words, op. 53.
MARZIALS
     Songs: Such merry maids are we; The bud and the bee. cf Trans-
     atlantic XTRA 1159.
MASCAGNI, Pietro
     Amica: Più presso al ciel. cf Rubini GV 58.
     L'Amico Fritz: Intermezzo, Act 3. cf HMV SLS 5019.
     L'Amico Fritz: Non mi resta che il pianto. cf Bongiovanni GB 1.
1685 Cavalleria rusticana. Margaret Harshaw, s; Thelma Votipka,
     Mildred Miller, con; Richard Tucker, t; Frank Guarrera, bar;
     Metropolitan Opera Chorus and Orchestra; Fausto Cleva. CBS
     61640. Tape (c) 40-61640. (Reissue from Philips ABR 4000/1).

```
            -Gr 8-75 p359                +-RR 7-75 p22
            /HFN 8-75 p77                +-RR 1-76 p66 tape
           +-HFN 12-75 p175 tape
```

Cavalleria rusticana: Easter hymn. cf Columbia SCX 6585.

Cavalleria rusticana: Gli aranci olezzano; Voi lo sapete, o
    mamma; Oh, Il Signore vi manda; Intermezzo; Mamma, quel vino
    è generoso. cf LEONCAVALLO: I Pagliacci: Si può; Signore,
    Signori.

Cavalleria rusticana: Intermezzo. cf CBS 30062.

Cavalleria rusticana: Intermezzo. cf Pye TB 3004.

Cavalleria rusticana: Mamma quel vino è generoso. cf Muza SXL
    1170.

Cavalleria rusticana: O Lola. cf Cantilena 6238.

Cavalleria rusticana: Regina coeli. cf HMV SXLP 30205.

Cavalleria rusticana: Siciliana (2 versions). cf Everest SDBR
    3382.

Cavalleria rusticana: Voi lo sapete. cf Club 99-100.

Cavalleria rusticana: Voi lo sapete. cf HMV SLS 5057.

Cavalleria rusticana: Voi lo sapete. cf Rubini GV 58.

Iris, excerpts. cf TCHAIKOVSKY: Mazeppa.

Iris: Serenata. cf Everest SDBR 3382.

Iris: Un dì (ero piccina) al tempio. cf CATALANI: La Wally:
    Ebben, Ne andrò lontana.

Lodoletta: Flammen, perdonami. cf CATALANI: La Wally: Ebben, Ne
    andrò lontana.

Songs: Ave Maria. cf Argo ZFB 95/6.

MASCHERONI, Edoardo
    Lorenza: Susanna al bagno. cf Club 99-100.

MASON, Daniel Gregory
    Quartet on Negro themes, op. 19, G minor. cf Vox SVBX 5301.

MASON, John
    Songs: Quales sumus; Vae nobis miserere. cf APPLEBY: Magnificat.

MASSE, Victor
    Paul et Virginie: Air du tigre. cf Discophilia KIS KGG 3.

    THE MASSED BANDS, PIPES AND DRUMS OF THE WELSH GUARDS AND THE
        ARGYLL AND SUTHERLAND HIGHLANDERS ON TOUR. cf Monitor Records
        MFS 760.

MASSENET, Jules
1686 Ariane: Lamento d'Ariane. Le Cid: Ballet music. MEYERBEER: Les
        patineurs: Ballet suite (arr. Lambert). NPhO; Richard Bonynge.
        Decca SXL 6812.

```
            +Gr 11-76 p797               +RR 11-76 p69
            +HFN 11-76 p161
```

Cendrillon: Enfin, je suis ici. cf CBS 76522.

1687 Le Cid. Grace Bumbry, con; Eleanor Bergquist; Placido Domingo,
        t; Paul Plishka, bs; Arnold Voketaitis; Byrne Camp Chorale;
        New York Opera Orchestra; Eve Queler. Columbia M3 34211 (3).

```
            +ON 12-11-76 p48            ++SFC 11-14-76 p30
```

Le Cid: Ballet music. cf Ariane: Lamento d'Ariane.

El Cid: Ballet music. cf KHACHATURIAN: Gayaneh: Suite.

Le Cid: O souverain. cf Cantilena 6238.

Le Cid: Pleurez pleurez mes yeux. cf Club 99-97.

Le Cid: Pleurez mes yeux. cf HMV RLS 719.

Le Cid: Pleurez mes yeux. cf HMV SLS 5057.

Cleopatre: A-t-il dit vrai..Solitaire sur ma terrasse. cf
    Club 99-101.

Concerto, piano. cf GOUNOD: Fantasy on the Russian national hymn.
Crépuscule. cf HMV ASD 2929.
Don César de Bazan: Sevillana (2). cf HMV RLS 719.
Don Quichotte: Quant apparaissent les étoiles, Mort de Quichotte.
     cf Club 99-101.
Elégie. cf Columbia/Melodiya M 33593.
Elégie. cf Desto DC 7118/9.
Elégie. cf HMV RLS 716.
Elegy. cf Musical Heritage Society MHS 3276.

1688 Esclarmonde. Joan Sutherland, s; Huguette Tourangeau, ms;
     Giacomo Aragall, Ryland Davies, Ian Caley, Graham Clark, t;
     Louis Quilico, bar; Robert Lloyd, Clifford Grant, bs; Finchley
     Children's Music Group, John Alldis Choir; NPhO; Richard
     Bonynge. Decca SET 612/4 (3). (also London 13118 Tape (c)
     5-13118).
          ++Gr 11-76 p867   +RR 11-76 p40
          +HFN 12-76 p144   ++SFC 11-28-76 p45
          +ON 12-11-76 p48

Grisélidis: Il partit au printemps. cf Club 99-97.
Hérodiade: Adieu donc. cf Club 99-98.
Hérodiade: Il est doux. cf Rococo 5397.
Manon: Ancor son io tutt attonita; Addio nostro piccolo desco.
     cf Club 99-100.
Manon: Chiudo gli occhi. cf Everest SDBR 3382.
Manon: En fermant les yieux. cf Decca GOSD 674/6.
Manon: J'ai marqué l'heure du départ. cf Angel S 37143.

1689 La navarraise. Marilyn Horne, ms; Placido Domingo, Ryland Davies,
     Leslie Fyson, t; Sherrill Milnes, Gabriel Bacquier, bar;
     Nicolas Zaccaria, bs; Ambrosian Opera Chorus; LSO; Henry Lewis.
     RCA ARL 1-1114. Tape (c) ARK 1-1114 (ct) ARS 1-1114 (Q) ARD
     1-1114 Tape (ct) ART 1-1114.
          +Gr 11-75 p1894   +-NR 1-76 p11
          +HF 12-75 p99   +ON 1-3-76 p33
          +HFN 12-76 p145   +-RR 11-75 p33
          +MJ 11-75 p21   ++SFC 10-12-75 p22
          +MJ 2-76 p33   ++St 1-76 p108
          +-MT 8-76 p661

La navarraise: O bien aimée. cf Club 99-98.
Panurge: Touraine est un pays. cf Club 99-101.
Phedre: Overture. cf CHABRIER: Works, selections (Mercury SRI
     75078).
Sappho: Ah qu'il est loin de mon pays. cf Club 99-98.
Scènes hongroises. cf CHABRIER: Le Roi malgré lui: Fête polo-
     naise.

1690 Songs: A Colombine; Automne; Elégie; Fleuramye; Madrigal; Oh, si
     les fleurs avaient des yeux; Ouvre tes yeux bleus; Pensée
     d'automne; Poème d'Avril; Nuit d'Espagne; Roses d'Octobre;
     Sérénade de Zanetto; Si tu veux, Mignonne; Souvenir de Venise.
     Bruno Laplante, bar; Janine Lachance, pno. CRD Calliope CAL
     1830.
          +Gr 12-76 p1033   +-RR 11-76 p97

Songs: Chant provençal; Elégie; Nuit d'Espagne; Sérénade d'automne;
     Stances; Un adieu; Vous aimerez demain. cf FAURE: Songs
     (Saga 5419).

1691 Thais. Beverly Sills, Norma Burrowes, Ann-Marie Connors, s; Ann
     Murray, Patricia Kern, ms; Nicolai Gedda, t; Sherrill Milnes,

 bar; Richard Van Allan, bs; John Alldis Choir; NPhO; Lorin
 Maazel. Angel SCLX 3832 (3).
          +ON 12-11-76 p48   +St 12-76 p142
Thais: Méditation. cf CBS 30062.
Thais: Méditation. cf Connoisseur Society (Q) CSQ 2065.
Thais: Méditation. cf HMV SXLP 30188.
Thais: Te souvient-il du lumineux voyage. cf Decca GOSD 674/6.
La vièrge: Le dernier sommeil de la vièrge (The last sleep of the
 virgin). cf BERLIOZ: Les Troyens: Overture, March Troyenne.
Werther: Air des larmes. cf Rubini RS 301.
Werther: Air des lettres. cf Decca GOSD 674/6.
Werther: Des cris joyeux. cf HMV SLS 5057.
Werther: Il faut nous separer...Pourquoi me reveiller. cf Rococo
 5383.
Werther: O nature pleine de grace, Clair de lune, Desolation, Mort
 de Werther. cf Club 99-98.
Werther: Pourquoi me reveiller. cf Cantilena 6239.
Werther: Va, laisse les couler mes larmes. cf CBS 76522.
Werther: Va, laisse couler. cf Discophilia KIS KGG 3.
MATEJ, Josef
 Concerto, trumpet, French horn and trombone. cf JIRASEK: Partita.
MATHIAS, William
1692 Ave Rex, op. 45. Concerto, harp, op. 50. Dance overture, op. 16.
 Invocation and dance, op. 17. Osian Ellis, hp; Welsh National
 Opera Chorus; LSO; David Atherton. L'Oiseau-Lyre SOL 346.
          +NR 3-76 p15
 Celtic dances, op. 60. cf HODDINOTT: Concerto grosso, no. 2, op.
 46.
 Concerto, harp, op. 50. cf Ave Rex, op. 45.
 Dance overture, op. 16. cf Ave Rex, op. 45.
1693 Divertimento, flute, oboe and piano, op. 24. Quartet, strings,
 op. 38. Quintet, winds, op. 22. William Bennett, flt; Anthony
 Camden, ob; Levon Chilingirian, vln; Martin Jones, Clifford
 Benson, pno; Gabrieli Quartet, Nash Ensemble. Argo ZRG 771.
          +Gr 11-76 p830   +RR 6-76 p65
          ++HFN 6-76 p83
 Invocation and dance, op. 17. cf Ave Rex, op. 45.
1694 Invocations, op. 35. Jubilate, op. 67, no. 2. Partita, op. 19:
 Chorale, Postlude, Processional. Toccata giocosa, op. 36,
 no. 2. Variations on a hymn tune, op. 20. Christopher
 Herrick, org. L'Oiseau-Lyre SOL 342.
          +Audio 6-76 p96   +MT 5-76 p409
          +Gr 7-75 p219   +NR 3-76 p11
          +HFN 8-75 p78   +RR 7-75 p48
 Jubilate, op. 67, no. 2. cf Invocations, op. 35.
 Partita, op. 19: Chorale, Postlude, Processional. cf Invocations,
 op. 35.
 Processional. cf Argo ZRG 807.
 Quartet, strings, op. 38. cf Divertimento, flute, oboe and
 piano, op. 24.
 Quintet, winds, op. 22. cf Divertimento, flute, oboe and piano,
 op. 24.
 Sinfonietta, op. 34. cf HODDINOTT: Concerto grosso, no. 2, op.
 46.
 Toccata giocosa, op. 36, no. 2. cf Invocations, op. 35.
 Variations on a hymn tune, op. 20. cf Invocations, op. 35.

1695 The worlde's joie. Janet Price, s; Kenneth Bowen, t; Michael
         Rippon, bar; Bach Choir, St. George's Chapel Choir; NPhO;
         David Willcocks. HMV ASD 3301.
             ++Gr 12-76 p1033
MATTEIS, Nicola
      Suite, recorder and organ, D major. cf Saga 5414.
MATTHYSZ
      Variations from "Der Gooden Fluyt Hemel". cf Transatlantic
         TRA 292.
MATTON, Orger
1696 Concerto, 2 pianos. McCAULEY: Miniatures, flute and strings (5).
         PEPIN: Monade, strings. Renée Morrisset, Victor Bouchard, pno;
         Robert Aitken, flt; Toronto Symphony Orchestra; Walter Susskind,
         William McCauley, Alexander Brott. CRI SD 317.
             *RR 2-76 p31
MAYS, Walter
1697 Invocations to the Svara mandala. WERNICK: A prayer for Jerusalem.
         Jan DeGaetani, ms; Glen Steele, perc; Wichita State University
         Percussion Orchestra; J. C. Combs. CRI SD 344.
             +-NR 8-76 p14   +NR 11-76 p16
MEACHAM
      American patrol. cf RCA AGL 1-1334.
MEAUX, Etienne de
      Trop est mes maris jalos. cf Telefunken 6-41275.
MECHEM
      Make a joyful noise unto the Lord (Psalm 100). cf Columbia M
         34134.
MEDEK, Tilo
      Battaglia alla Turca. cf Desmar DSM 1005.
      A MEDIEVAL CHRISTMAS. cf Nonesuch H 71315.
      MEDIEVAL ENGLISH CAROLS AND ITALIAN DANCES. cf Anonymous works
         (Decca DL 79418).
MEDINS, Janis
1698 Dainas (24 preludes). Arthur Ozolins, pno. Kaibala 60 F 02 (2).
             ++NR 12-76 p13
      Suite concertante. cf BRUCH: Kol Nidrei, op. 47.
MEDTNER, Nicolai
1699 Concerto, piano, no. 3, op. 50. Sonata, piano, op. 22, G minor.
         Sonata tragica, op. 39, C minor. Michael Ponti, pno; Luxem-
         bourg Radio Orchestra; Pierre Cao. Candide CE 31092.
             +NR 10-75 p3   +SFC 1-4-76 p27
      Fairy tale, B flat minor. cf RCA ARM 4-0942/7.
      Sonata, piano, op. 22, G minor. cf Concerto, piano, no. 3, op. 50.
      Sonata tragica, op. 39, C minor. cf Concerto, piano, no. 3, op.
         50.
MEHUL, Etienne
      Overture burlesque. cf Angel S 36080.
MEKEEL, Joyce
1700 Corridors of dream. Planh. MUSGRAVE: Chamber concerto, no. 2.
         SEEGER, R. C.: Two movements for chamber orchestra. Boston
         Musica Viva; Richart Pittman. Delos DEL 25405.
             +HF 1-76 p80   +NYT 4-11-76 pD23
             ++NR 2-76 p6   +-St 5-76 p123
      Planh. cf Corridors of dream.
MELLII, Pietro
      For the Emperor Matthias. cf Saga 5420.

MELLNAS, Arne
   Euphoni. cf Caprice RIKS LP 35.
MENDELSSOHN, Felix
   Andante and rondo capriccioso, op. 14. cf HMV HQS 1354.
   Athalia, op. 74: Overture. cf Overtures (HMV 7003).
   Calm sea and prosperous voyage (Meerstille und Glückliche Fahrt),
      op. 27. cf HMV SEOM 24.
   Calm sea and prosperous voyage, op. 27. cf Overtures (HMV 7003).
   Calm sea and prosperous voyage, op. 27. cf Symphony, no. 3,
      op. 56, A minor.
   Chant populaire. cf HMV SQ ASD 3283.
1701 Concerto, piano, no. 1, op. 25, G minor. Concerto, piano, no. 2,
     op. 40, D minor. Murray Perahia, pno; AMF; Neville Marriner.
     Columbia M 33207. (Q) MQ 33207 Tape (ct) MAQ 33207. (also
     CBS 76376 Tape (c) 40-76376).
        +-Audio 5-76 p76   +HF 5-76 p114 tape
        ++Gr 7-75 p187   +HFN 7-75 p84
        ++Gr 12-75 p1121 tape +-HFN 12-75 p173 tape
        +-Gr 6-76 p102 tape   ++NR 7-75 p6
        +HF 7-75 p79   +St 9-75 p108
1702 Concerto, piano, no. 1, op. 25, G minor. SAINT-SAENS: Concerto,
     piano, no. 2, op. 22, G minor. Daniel Adni, pno; Royal Liver-
     pool Philharmonic Orchestra; Charles Groves. HMV SQ ASD 3208.
        +-Gr 8-76 p291   +-RR 7-76 p60
        +-HFN 7-76 p91
1703 Concerto, piano, no. 1, op. 25, G minor. Concerto, piano, no. 2,
     op. 40, D minor. Rondo brillante, op. 29, E flat major. John
     Ogdon, pno; LSO; Aldo Ceccato. Klavier KS 531.
        +-Audio 5-76 p77   +-NR 11-74 p5
   Concerto, piano, no. 1, op. 25, G minor. cf HMV SLS 5033.
   Concerto, piano, no. 2, op. 40, D minor. cf Concerto, piano, no.
     1, op. 25, G minor (Columbia M 33207).
   Concerto, piano, no. 2, op. 40, D minor. cf Concerto, piano, no.
     1, op. 35, G minor (Klavier KS 531).
1704 Concerto, violin, op. 64, E minor. BRUCH: Concerto, violin, no.
     1, op. 26, G minor. Itzhak Perlman, vln; LSO; André Previn.
     Angel S 36963. Tape (c) 8SX 36963 (c) 4XS 36963. (also HMV
     ASD 2926 Tape (c) TC ASD 2926).
        +Gr 1-74 p1372   +NR 1-74 p8
        ++HF 3-74 p88   +RR 1-74 p42
        +-HF 7-74 p128 tape +St 1-74 p110
        +-HFN 11-76 p175 tape
1705 Concerto, violin, op. 64, E minor. MOZART: Concerto, violin,
     no. 3, K 216, G major. Leonid Kogan, vln; OSCCP; Constantin
     Silvestri. Connoisseur Society CS 2111.
        ++NR 12-76 p7   +SFC 12-76 p34
1706 Concerto, violin, op. 64, E minor. TCHAIKOVSKY: Concerto, violin.
     op. 35, D major. Ruggiero Ricci, vln; Netherlands Radio Orch-
     estra; Jean Fournet. Decca PFS 4345. (also London SPC 2116
     Tape (c) 5-21116).
        +-Gr 1-76 p1198   +NR 12-76 p7
        +-HFN 1-76 p111   +-RR 12-75 p68
1707 Concerto, violin, op. 64, E minor. MOZART: Concerto, violin, no.
     5, K 219, A major. Václav Hudeček, vln; Prague Radio Symphony
     Orchestra; Václav Smetacek. Panton 110 511.
        +RR 10-76 p62

1708 Concerto, violin, op. 64, E minor. PAGANINI: Concerto, violin,
no. 1, op. 6, D major. Eugene Fodor, vln; NPhO; Peter Maag.
RCA ARL 1-1565. Tape (c) ARK 1-1565 (ct) ARS 1-1565.
+-Gr 9-76 p422   +-RR 10-76 p64
+HFN 9-76 p125   +SFC 10-31-76 p35
+-MJ 11-76 p45   +St 11-76 p151
++NR 9-76 p5

1709 Concerto, violin, op. 64, E minor. TCHAIKOVSKY: Concerto, violin,
op. 35, D major. Aaron Rosand, vln; Luxembourg Radio Orchestra;
Louis de Froment. Turnabout TV 34553.
-HFN 5-76 p102   +-RR 5-76 p46

Concerto, violin, op. 64, E minor. cf BRUCH: Scottish fantasia,
op. 46.
Concerto, violin, op. 64, E minor. cf HMV RLS 718.
Concerto, violin, op. 64, E minor. cf HMV SLS 5068.
Concerto, violin, op. 64, E minor: 1st movement. cf CBS 30063.
Concerto, violin, op. 64, E minor: Finale. cf RCA ARM 4-0942/7.

1710 Etudes, op. 104 (3). Fantasy, op. 28, F sharp minor. Rondo
capriccioso, op. 14, E major. Songs without words: Elegy,
Restlessness. Variations sérieuses, op. 54, D minor. Constance
Keene, pno. Laurel-Protone LP 12.
++NR 7-76 p12   ++SFC 10-31-76 p36

Etudes (studies), op. 104 (3). cf FIELD: Nocturne, no. 11, E flat
major.

1711 Fantasie-caprice, op. 16, no. 2: Scherzo. Rondo capriccioso,
op. 14, E major. Song without words, op. 67, no. 4, C major.
RACHMANINOFF: Concerto, piano, no. 2, op. 18, C minor. Alexander Brailowsky, pno; San Francisco Symphony Orchestra; Enrique Jorda. Camden CCV 5037.
-HFN 4-76 p123   -RR 4-76 p51

Fantasy, op. 28, F sharp minor. cf Etudes, op. 104 (3).

1712 Hebrides overture, op. 26. NICOLAI: The merry wives of Windsor:
Overture. WAGNER: The flying Dutchman: Overture. Lohengrin:
Prelude, Act 1. WEBER: Der Freischütz: Overture. BPhO; Herbert
von Karajan. HMV SXLP 30210. Tape (c) TC EXE 205. (Reissue
from Columbia SAX 2439).
+Gr 6-76 p95   +HFN 8-76 p95 tape
+HFN 7-76 p103   +RR 8-76 p61

Hebrides overture, op. 26. cf Overtures (HMV 7003).
Hebrides overture, op. 26. cf Symphony, no. 3, op. 56, A minor.
Hymn of praise, op. 52: I waited for the Lord. cf Abbey LPB 757.

1713 A midsummer night's dream, opp. 21/61. Heather Harper, Janet
Baker, s; NPhO; Otto Klemperer. HMV Tape (c) TC EXE 186.
+-HFN 4-76 p125 tape   +RR 6-76 p88

1714 A midsummer night's dream, opp. 21/61: Incidental music. Jennifer
Vyvyan, Marion Lowe, s; ROHO Women's Chorus; LSO; Peter Maag.
Decca SPA 451. (Reissue from SXL 2060).
+Gr 9-76 p422   +RR 8-76 p41
+HFN 8-76 p93

1715 A midsummer night's dream, opp. 21/71: Incidental music. Heather
Harper, s; Janet Baker, con; Philharmonia Orchestra and Chorus;
Otto Klemperer. HMV SXLP 30196. (Reissue from Columbia SAX
2393).
+Gr 2-76 p1345   ++RR 2-76 p31
+HFN 2-76 p115

A midsummer night's dream, opp. 21/61: Incidental music. cf RCA
CRM 5-1900.

A midsummer night's dream, opp. 21/61: Overture, Scherzo, Nocturne,
   Wedding march. cf Symphony, no. 4, op. 90, A major.
A midsummer night's dream, opp. 21/61: Overture, Incidental music,
   Scherzo, Notturno, Wedding march. cf Symphony, no. 4, op. 90,
   A major.
A midsummer night's dream, op. 61: Scherzo. cf Columbia/Melodiya
   M 33593.
A midsummer night's dream, op. 61: Scherzo. cf RCA LRL 1-5131.
A midsummer night's dream, op. 61: Wedding march and variations.
   cf RCA VH 020.
1716 Octet, strings, op. 20, E flat major. RIMSKY-KORSAKOV: Quintet,
   piano and winds, op. posth., B flat major. Vienna Octet.
   Decca SDD 389. Tape (c) KSDC 389. (also London STS 15308).
          +Gr 12-73 p1218   +RR 9-75 p78 tape
          +HFN 7-75 p90 tape   ++St 9-76 p120
          -NR 7-76 p7   ++STL 1-6-74 p29
          +-RR 1-74 p52
1717 Octet, strings, op. 20, E flat major. Symphony, strings, no. 10,
   B minor. Symphony, strings, no. 12, G minor. I Musici. Philips 6580 103.
          +MJ 10-76 p25   ++St 9-76 p120
          +NR 8-76 p6
1718 Octet, strings, op. 20, E flat major. Janáček Quartet, Smetana
   Quartet. Supraphon SU 4.
          +Audio 6-76 p97
On wings of song, op. 34, no. 2. cf Decca SKL 5208.
On wings of song, op. 34, no. 2 (2). cf RCA ARM 4-0942/7.
On wings of song, op. 34, no. 2. cf Sterling 1001/2.
1719 Overtures: Athalia, op. 74. Hebrides, op. 26. Calm sea and
   prosperous voyage, op. 27. Ruy Blas, op. 95. Son and stranger,
   op. 89. NPhO; Moshe Atzmon. HMV SQ ESD 7003. Tape (c) TC
   ESD 7003.
          +Gr 7-76 p176   +HFN 11-76 p175 tape
          +-Gr 10-76 p658 tape   +RR 7-76 p60
          +HFN 7-76 p91
1720 Quartets, strings, no. 1, op. 12; nos. 3-4, op. 44; op. 80.
   Bartholdy Quartet. BASF 39 2166/7.
          +MT 3-76 p235
Quartet, strings, no. 1, op. 12, E flat major: Canzonetta. cf
   Swedish Society SLT 33189.
Richte mich, Gott, op. 78, no. 2. cf BIS LP 14.
Ronde brillante, op. 29, E flat major. cf Concerto, piano, no. 1,
   op. 25, G minor.
Rondo brillante, op. 29, E flat major. cf HMV SLS 5033.
Rondo capriccioso, op. 14, E major. cf Etudes, op. 104 (3).
Rondo capriccioso, op. 14, E major. cf Fantasie-caprice, op. 16,
   no. 2: Scherzo.
Rondo capriccioso, op. 14, E major. cf Rococo 2049.
Ruy Blas overture, op. 95. cf Overtures (HMV 7003).
Ruy Blas overture, op. 95. cf HMV ESD 7010.
Ruy Blas overture, op. 95. cf Virtuosi VR 7506.
Die schöne Melusine overture, op. 32. cf SCHUBERT: Symphony,
   no. 5, D 485, B flat major.
Shepherd's complaint. cf Coronet LPS 3032.
Son and stranger overture, op. 89. cf Overtures (HMV 7003).
Sonata, organ, op. 65, no. 2, C minor. cf BACH: Chorale prelude:
   Aus tiefer Not schrei ich zu dir, S 687.

1721   Sonata, organ, op. 65, no. 3, A major.   REGER: Pieces, op. 59, no.
        9: Benedictus. Pieces, op. 145, no. 2: Dankpsalm. WESLEY, S. S.:
        Andante, E minor. Andante, G major. Choral song and fugue.
        Introduction and fugue, C sharp minor. Simon Lindley, org.
        Vista VPS 1026.
                +Gr 4-76 p1644   +RR 6-76 p74
        Sonata, organ, op. 65, no. 3, A major. cf Gaudeamus XSH 101.
        Sonata, organ, op. 65, no. 5, D major. cf Pelca PRSRK 41017/20.
1722   Sonata, violoncello and piano, no. 1, op. 45, B flat major. Son-
        ata, violoncello and piano, no. 2, op. 58, D major. Song with-
        out words, op. 109, D major. Variations concertantes, op. 17.
        Friedrich-Jürgen Sellheim, vlc; Eckart Sellheim, pno. CBS
        76547.
                +Gr 11-76 p830   ++RR 12-76 p84
                +-HFN 11-76 p163
1723   Sonata, violoncello and piano, no. 2, op. 58, D major. SCHUBERT:
        Sonata, arpeggione and piano, D 821, A minor. Lynn Harrell,
        vlc; James Levine, pno. RCA ARL 1-1568. Tape (c) ARK 1-1568
        (ct) ARS 1-1568.
                +Gr 9-76 p438   ++NR 9-76 p7
                +-HFN 9-76 p125   +RR 9-76 p83
                +MJ 11-76 p45
        Sonata, violoncello and piano, no. 2, op. 58, D major. cf Sonata,
        violoncello and piano, no. 1, op. 45, B flat major.
        Sonata, violoncello and piano, no. 2, op. 58, D major. cf
        KUPFERMAN: Fantasy concerto, violoncello, piano and tape.
1724   Songs: Part songs, mixed voices, op. 41, nos. 2-4, 6. Partsongs,
        mixed voices, op. 48, nos. 5-6. Partsongs, mixed voices, op.
        59, nos. 2, 4, 6. Partsongs, mixed voices, op. 88, nos. 4, 6.
        Partsongs, mixed voices, op. 100, no. 1. Songs without words,
        nos. 8, 17-18, 27, 30-31, 34. BBC Northern Singers; Keith
        Swallow, pno; Stephen Wilkinson. Abbey LPB 735.
                +Gr 5-76 p1794   +RR 2-76 p52
                +HFN 2-76 p103
1725   Songs: Auf Flügeln des Gesanges, op. 34, no. 2; Das erste Veilchen,
        op. 19, no. 2; Frühlingslied, op. 71, no. 2; Gruss, op. 19,
        no. 5; Hirtenlied, op. 57, no. 2; Neue Liebe, op. 19, no. 4;
        Pagenlied; Venetianisches Gondellied, op. 57, no. 5; Volkslied,
        op. 47, no. 4. WEBER: Songs: Abendsegen, op. 64, no. 5;
        Bettlerlied, op. 25, no. 4; Canzonettas, op. 29; Ah, dove siete,
        Ninfe se liete, Ch'io mai vi possa; Die fromme Magd, op. 54,
        no. 1; Heimlicher Liebe Pein, op. 64, no. 3; Lass mich schlum-
        mern, op. 25, no. 3; Sanftes Licht; Die Schäferstunde, op. 13,
        no. 1; Uber die Berge, op. 25, no. 2; Wiegenlied, op. 13, no.
        2; Die Zeit, op. 13, no. 5. Robert Tear, t; Philip Leger, pno;
        Timothy Walker, gtr. Argo ZRG 827.
                +Gr 6-76 p78   +RR 7-76 p78
                +HFN 6-76 p91   +STL 9-19-76 p36
1726   Songs: Altdeutsches Frühlingslied, op. 86, no. 2: Der trübe Winter
        ist vorbei; Andres Maienlied, op. 8, no. 8: Die Schwalbe
        fliegt, der Frühling siegt; Auf Flügeln des Gesanges, op. 34,
        no. 2; Bei der Wiege, op. 47, no. 6: Schlummre, Schlummre
        und Träume; Erster Verlust, op. 99, no. 1: Ach, wer bringt die
        schönen Tage; Frage, op. 9, no. 1: Ist es wahr, Ist es wahr;
        Frühlingslied, op. 19, no. 1: In dem Walde süsse Töne; Früh-
        lingslied, op. 34, no. 3: Es brechen im schallenden Reigen;

Frühlingslied, op. 47, no. 3: Durch den Wald den Dunkeln;
Gruss, op. 19, no. 5: Leise zieht durch mein Gemüt; Hirtenlied,
op. 57, no. 2: O Winter, schlimmer Winter; Im Herbst, op. 9,
no. 5: Ach, wie schnell die Tage fliehen; Jagdlied, op. 84,
no. 3; Mit Lust tät ich ausreiten; Minnelied, op. 34, no. 1:
Leucht't heller als die Sonne; Der Mond, op. 86, no. 5: Mein
Herz ist wie dunkle Nacht; Neue Liebe, op. 19, no. 4: In dem
Mondenschein im Walde; Pagenlied: Wenn die Sonne lieblich
schiene; Reiselied, op. 34, no. 6: Der Herbstwind rüttelt die
Bäume; Schilflied, op. 71, no. 4: Auf dem Teich dem Regungs-
losen; Venezianisches Gondellied, op. 57, no. 5: Wenn durch
die Piazetta; Wanderlied, op. 57, no. 6: Laue Luft kommt blau
geflossen; Winterlied, op. 19, no. 3: Mein Sohn, wo willst du
hin so spät. Peter Schreier, t; Walter Olbertz, pno. DG
2530 593.

+Gr 4-76 p1649   ++STL 6-6-76 p37

1727 Songs: Altdeutsches Frühlingslied, op. 86, no. 6; Auf Flügeln des
Gesanges, op. 34, no. 2; Bei der Wiege, op. 47, no. 6; Erster
Verlust, op. 99, no. 1; Frage, op. 9, no. 1; Frühlingslied,
op. 19, no. 1; Frühlingslied, op. 34, no. 3; Frühlingslied, op.
47, no. 3; Gruss, op. 19, no. 5; Hexenlied, op. 8, no. 8;
Hirtenlied, op. 57, no. 2; Im Herbst, op. 9, no. 5; Jagdlied,
op. 84, no. 3; Minnelied, op. 34, no. 1; Der Mond, op. 86, no.
5; Neue Liebe, op. 19, no. 4; Pagenlied, op. posth.; Reiselied,
op. 34, no. 6; Schifflied, op. 71, no. 4; Venetianisches Gondellied, op. 57, no. 5; Wanderlied, op. 57, no. 6; Winterlied,
op. 19, no. 3. Peter Schreier, t; Walter Olbertz, pno. DG
2530 596.

+HF 12-76 p104   +ON 12-4-76 p60
+HFN 4-76 p109   ++RR 5-76 p71
+-MT 12-76 p1005   ++SFC 10-24-76 p35
+NR 11-76 p11

1728 Songs: Allnächtlich im Traume, op. 86, no. 4; Altdeutsches Lied,
op. 57, no. 1; An die Entfernte, op. 71, no. 3; Auf der Wand-
erschaft, op. 71, no. 5; Auf Flügeln des Gesanges, op. 34, no.
2; Bei der Wiege, op. 47, no. 6; Da lieg ich unter den Bäumen,
op. 84, no. 1; Der Blumenkranz, op. posth.; Erntelied, op. 8,
no. 4; Das erste Veilchen, op. 19, no. 2; Erster Verlust, op.
99, no. 1; Es lausche das Laub, op. 86, no. 1; Frühlingslied,
op. 19, no. 1; Frühlingslied, op. 34, no. 3; Frühlingslied,
op. 47, no. 3; Gruss, op. 19, no. 5; Hexenlied, op. 8, no. 8;
Hirtenlied, op. 57, no. 2; Jagdlied, op. 84, no. 3; Minnelied,
op. 34, no. 1; Minnelied, op. 47, no. 1; Morgengruss, op. 47,
no. 2; Der Mond, op. 86, no. 5; Nachtlied, op. 71, no. 6;
Neue Liebe, op. 19, no. 4; O Jugend, op. 57, no. 4; Pagenlied,
op. posth.; Reiselied, op. 19, no. 6; Scheidend, op. 9, no. 6;
Schifflied, op. 71, no. 4; Schlafoser Augen Leuchte, op. posth.;
Tröstung, op. 71, no. 1; Venetianisches Gondellied, op. 57,
no. 5; Volkslied, op. 47, no. 4; Das Waldschloss, op. posth.;
Wanderlied, op. 57, no. 6; Warnung vor dem Rhein, op. posth.;
Wenn sich zwei Herzen scheiden, op. 99, no. 5; Winterlied,
op. 19, no. 3. Dietrich Fischer-Dieskau, bar; Wolfgang
Sawallish, pno. EMI Odeon 1C 193 02180/1 (2).

+HF 12-76 p106

Songs: Adeste fideles; Greensleeves; Hark, the herald angels sing.
cf RCA PRL 1-8020.

Songs: Auf Flügeln des Gesanges, op. 34, no. 2. cf HMV ASD 2929.
Songs: Auf flügeln des Gesanges, op. 34, no. 2. cf Rubini GV 70.
Songs: Hark, the herald angels sing. cf CBS 61771.
Songs: Hark, the herald angels sing. cf HMV CSD 3774.
Songs: O for the wings of a dove. cf HMV RLS 719.
Songs without words: Elegy, Restlessness. cf Etudes, op. 104 (3).
Songs without words: Trauermarsch. cf Coronet LPS 3031.
Songs without words, nos. 8, 17-18, 27, 30-31, 34. cf Songs
    (Abbey LPB 735).
Songs without words, op. 30, no. 6. cf FIELD: Nocturne, no. 11,
    E flat major.
Songs without words, op. 38, no. 6. cf FIELD: Nocturne, no. 11,
    E flat major.
Songs without words, op. 53, no. 4. cf FIELD: Nocturne, no. 11,
    E flat major.
Songs without words, op. 62, nos. 5, 6. cf FIELD: Nocturne, no.
    11, E flat major.
Songs without words, op. 62, no. 6; op. 67, no. 5; op. 85, no. 4.
    cf RCA VH 020.
Songs without words, op. 67, no. 4, C major. cf Fantasie-caprice,
    op. 16, no. 2: Scherzo.
Songs without words, op. 67, no. 4, C major. cf International
    Piano Library IPL 5001/2.
Songs without words, op. 67, no. 5. cf FIELD: Nocturne, no. 11,
    E flat major.
Songs without words, op. 109, D major. cf Sonata, violoncello
    and piano, no. 1, op. 45, B flat major.
Songs without words, op. 109, D major. cf HMV RLS 723.
Sweet remembrance. cf RCA ARM 4-0942/7.
Symphonic movement, C minor. cf Symphony, strings, no. 8, op.
    posth., D major.
1729 Symphonies, nos. 1-5. Edith Mathis, s; Liselotte Rebmann, ms;
    Werner Hollweg, t; Berlin German Opera Chorus; BPhO; Herbert
    von Karajan. DG 2720 068 (4). Tape (c) 3371 020.
        ++Gr 12-73 p1198   +HF 12-76 p147
        ++HF 1-74 p74   +-St 12-73 p132
1730 Symphony, no. 1, op. 11, C minor. Symphony, no. 4, op. 90, A
    major. Hamburg Philharmonic Orchestra; Gary Bertini. BASF KBC
    22068. (also BASF BAC 3092).
        +-Gr 6-76 p49   -RR 4-76 p49
        /HFN 4-76 p109   +St 2-76 p106
1731 Symphony, no. 1, op. 11, C minor. Symphony, no. 2, op. 52, B
    flat major. BPhO; Herbert von Karajan. DG 2707 084. (Re-
    issue from 2720 068).
        ++Gr 8-76 p286   +NR 5-76 p2
        +HFN 7-76 p103   +RR 8-76 p42
        /MJ 5-76 p28
Symphony, no. 2, op. 52, B flat major. cf Symphony, no. 1, op.
    11, C minor.
1732 Symphony, no. 3, op. 56, A minor. Calm sea and prosperous voyage,
    op. 27. NPhO; Riccardo Muti. Angel (Q) S 37168. (also HMV
    SQ ASD 3184 Tape (c) TC ASD 3184).
        +Gr 6-76 p49   +-NR 6-76 p5
        -Gr 10-76 p658 tape   +-RR 5-76 p45
        +HFN 6-76 p91   +St 10-76 p127
        -MJ 10-76 p25

1733 Symphony, no. 3, op. 56, A minor. Hebrides overture, op. 26.
        Baltimore Symphony Orchestra; Sergiu Commissiona. Turnabout
        (Q) QTVS 34604.
           ++NR 3-76 p3   +St 10-76 p127
           ++SFC 3-7-76 p27
        Symphony, no. 3, op. 56, A minor: 4th movement. cf HMV SEOM 25.
1734 Symphony, no. 4, op. 90, A major. A midsummer night's dream,
        opp. 21/61: Overture, Scherzo, Nocturne, Wedding march. RPO;
        Hans Vonk. Deccc PFS 4359.
           +-Gr 6-76 p50   +-RR 5-76 p46
           +-HFN 5-76 p102
1735 Symphony, no. 4, op. 90, A major. A midsummer night's dream, opp.
        21/61: Overture, Incidental music, Scherzo, Notturno, Wedding
        march. BSO; Colin Davis. Philips 9500 068. Tape (c) 7300
        480.
              +ARG 12-76 p37   +HFN 11-76 p163
              +Gr 11-76 p797   +RR 12-76 p62
        Symphony, no. 4, op. 90, A major. cf Symphony, no. 1, op. 11,
        C minor.
        Symphony, no. 4, op. 90, A major. cf BEETHOVEN: Symphony, no. 5,
        op. 67, C minor.
1736 Symphony, strings, no. 8, op. posth., D major. Symphonic movement,
        C minor. Leipzig Gewandhaus Orchestra; Kurt Masur. DG Archive
        2533 311. (Reissue from 2722 006).
              +Gr 8-76 p291   ++NR 8-76 p3
              ++HFN 7-76 p103   +-RR 8-76 p41
              +MT 11-76 p914
        Symphony, strings, no. 10, B minor. cf Octet, strings, op. 20,
        E flat major.
        Symphony, strings, no. 12, G minor. cf Octet, strings, op. 20,
        E flat major.
        Trio, piano, no. 1, op. 49, D minor. cf HMV RLS 723.
        Trio, piano, no. 1, op. 49, D minor: Scherzo. cf RCA ARM 4-0942/7.
        Variations concertantes, op. 17. cf Sonata, violoncello and
        piano, no. 1, op. 45, B flat major.
        Variations sérieuses, op. 54, D minor. cf Etudes, op. 104 (3).
        Variations sérieuses, op. 54, D minor. cf RCA VH 013.
        Venetian boat song, op. 19, no. 6. cf Connoisseur Society (Q)
        CSQ 2066.
MENDELSSOHN-HENSEL, Fanny
        Bergeslust. cf Gemini Hall RAP 1010.
        Italien. cf Gemini Hall RAP 1010.
MENNIN, Peter
        Quartet, no. 2. cf Vox SVBX 5305.
        Sonata concertante. cf Desto DC 6435/7.
MERCADANTE, Giuseppe
        Concerto, flute, E minor. cf RCA CRL 3-1429.
        Le sette ultime parole di Nostro Signore sulla croce: Qual'ciglia
        candido (Parola quinta). cf Decca SXL 6781.
        Soirées Italiennes: Evive il galop. cf English National Opera
        ENO 1001.
MERIKANTO, Aare
1737 Juha. Raili Kostia, s; Maiju Kuusoja, con; Hendrik Krumm, t;
        Matti Lehtinen, bar; Finnish National Opera Chorus and Orches-
        tra; Ulf Soderblom. Musical Heritage Society MHS 3079/81 (3).
           ++ON 2-14-76 p49   ++St 9-75 p108

MERKEL, Gustav
    Sonata, organ, op. 30, D minor.   cf Pelca PRS 40581.
MERUCO (14th century)
    De home vray.   cf HMV SLS 863.
MERULO, Claudio
    Canzon françese.   cf L'Oiseau-Lyre 12BB 203/6.
MESSAGER, Andre
    L'Amour masque: J'ai deux amants.   cf London OS 26248.
    Monsieur Beaucaire: Philomel; I do not know; Lightly, lightly;
       What are the names.   cf HMV RLS 716.
MESSIAEN, Oliver
1738 Apparition de l'eglise eternelle.   Messe de la Pentecôte.   Verset
       pour la fête de la dédicace.   Louis Thiry, org.   Calliope CAL
       1927.
          ++HFN 2-76 p103
1739 Apparition de l'eglise eternelle.   L'Ascension.   Le banquet
       céleste.   Charles Krigbaum, org.   Lyrichord LLST 7297.
          ++NR 8-76 p11
    Apparition de l'eglise eternelle.   cf Organ works (CRD CAL 1925/
       30).
1740 L'Ascension.   Les corps glorieux, no. 3.   Louis Thiry, org.
       Calliope CAL 1926.
          ++HFN 1-76 p113
    L'Ascension.   cf Apparition de l'eglise eternelle.
    L'Ascension.   cf Organ works (CRD CAL 1925/30).
    L'Ascension.   cf Argo 5BBA 1013/5.
    L'Ascension: Alléluias sereins; Transports de joie.   cf Decca
       SDD 499.
    Le banquet céleste.   cf Apparition de l'eglise eternelle.
    Le banquet céleste.   cf Organ works (CRD CAL 1925/30).
1741 Catalogue d'oiseau: La rousserolle effarvette.   Vint regards sur
       l'enfant Jesus.   Peter Serkin, pno.   RCA CRL 3-0759 (3).
          ++Gr 9-76 p445   ++NR 9-75 p12
          ++HF 10-75 p80   +-RR 9-76 p83
          +HFN 11-76 p163   ++SFC 8-17-75 p22
    Chronochromie.   cf BOULEZ: Le soleil des eaux.
    Les corps glorieux, nos. 1-7.   cf Organ works (CRD CAL 1925/30).
1742 Le corps glorieux, nos. 1 and 2.   Louis Thiry, org.   Calliope
       CAL 1925.
          ++HFN 1-76 p113
    Le corps glorieux, no. 3: L'Ange aux parfums.   cf L'Ascension.
    Le corps glorieux, no. 6: Joie et clarté des corps glorieux.   cf
       Wealden WS 137.
    Diptyque.   cf Organ works (CRD CAL 1925/30).
    Livre d'orgue: Chants d'oiseaux; Les mains de l'abîme; Pièce de
       en trio; Reprises par interversions; Soixantequatre durées;
       Les yeux dans les roues.   cf Organ works (CRD CAL 1925/30).
    Messe de la Pentecôte.   cf Apparition de l'eglise eternelle.
    Messe de la Pentecôte.   cf Organ works (CRD CAL 1925/30).
    La nativité du Seigneur, nos. 1-9.   cf Organ works (CRD CAL 1925/
       30).
    La nativité du Seigneur: Dieu parmi nous.   cf Gaudeamus XSH 101.
1743 Organ works: Apparition de l'eglise eternelle.   L'Ascension.
       Les corps glorieux, nos. 1-7.   Le banquet céleste.   Diptyque.
       Messe de la Pentecôte.   La nativité du Seigneur, nos. 1-9.
       Livre d'orgue: Chants d'oiseaux; Les mains de l'abîme;

Pièce de en trio (2); Reprises par interversions; Soixante-
quatre durées; Les yeux dans les roues. Verset pour la fête
de la dédicace. Louis Thiry, org. CRD CAL 1925/30 (6).
    ++Gr 11-75 p863   ++STL 10-5-75 p36
    +RR 8-76 p69

1744 Poèmes pour mi. TIPPETT: Songs for Dov. Felicity Palmer, s; BBC
    Symphony Orchestra, Robert Tear, t; London Sinfonietta; David
    Atherton, Pierre Boulez. Argo ZRG 703.
        ++Gr 5-73 p2072   ++NR 6-74 p9
        +-HF 7-74 p94   +NYT 2-17-74 pD33
        +HFN 5-73 p986   +RR 5-73 p25
        +-MJ 12-74 p46   ++SFC 4-7-74 p25
        ++MQ 1-76 p139   ++St 9-74 p9

1745 Quartet for the end of time. Peter Serkin, pno; Ida Kavafian,
    vln; Fred Sherry, vlc; Richard Stoltzman, clt. RCA ARL 1-1567.
        +-MJ 11-76 p45   ++SFC 8-29-76 p29
        +NR 9-76 p6   ++St 10-76 p128

Verset pour la fête de la dédicace. cf Apparition de l'eglise
    eternelle.
Verset pour la fête de la dédicace. cf Organ works (CRD CAL 1925/
    30).
Vint regards ur l'enfant Jesus. cf Catalogue d'oiseau: La
    rousserolle effarvette.

MEYER
    Appalachian echoes. cf Orion ORS 75207.
MEYERBEER, Giacomo
    L'Africana: Di qui si vede il mar...Quai celesti concenti. cf
        Court Opera Classics CO 347.
    L'Africaine: In grembo a me. cf Collectors Guild 611.
    L'Africaine: O paradiso. cf Cantilena 6238.
    L'Africaine: O paradiso. cf RCA CRM 1-1749.
    L'Africaine: Sur mes genoux. cf Rubini GV 58
    Dinorah: Ombra leggiera. cf VERDI: Rigoletto.
    Dinarah: Ombra leggiera. cf HMV SLS 5057.
    Dinorah: Sei vendicata assai. cf Discophilia DIS KGA 2.
    Dinorah: Shadow song. cf BELLINI: I puritani: O rendetemi la
        speme.
    Gli Ugonotti: Bianca al par di neve alpina. cf DONIZETTI: La
        favorita: Spirto gentil; Una vergine, un angel di Dio.
    Gli Ugonotti: Qui sotto it ciel. cf Everest SDBR 3382.
    Les Huguenots: Duet of Marguerite and Raoul. cf Angel S 37143.
    Les Huguenots: Nobil signori salute. cf Collectors Guild 611.
    Les Huguenots: Nobles seigneurs, salut. cf CBS 76522.
    The Huguenots: O beau pays. cf Rubini GV 68.
    Les Huguenots: O glucklich Land. cf Discophilia DIS 13 KHH 1.
    Les Huguenots: Tu l'as dit. cf Decca GOSD 674/6.
    The Huguenots: Tu l'as dit; Oui, tu m'aimes. cf Rubini GV 63.
    Les Huguenots: Welch ein Schreck; Duet, Act 4. cf Rococo 5377.
    Les patineurs: Ballet suite. cf MASSENET: Ariane: Lamento d'Ariane.
    Le prophète: Ah mon fils. cf Collectors Guild 611.
    Le prophète: Coronation march, Act 4. cf Philips 6747 204.
    Robert le diable: Gnadenarie. cf Discophilia DIS 13 KHH 1.
1746 Songs: Cantique du Trappiste; Le chant du dimanche; Der Garten
    des Herzens; Hor' ich das Liedchen klingen; Komm; Menschen-
    feindlich; Mina; Le poète mourant; Die Rose, die Lilie, die
    Taube; Die Rosenblätter; Scirocco; Sicilienne; Sie und ich;

Ständchen. Dietrich Fischer-Dieskau, bar; Karl Engel, pno.
DG Archive 2533 295.
 +Gr 9-75 p500   +ON 5-76 p48
 -HF 1-76 p90   +RR 9-75 p70
 +HFN 9-75 p101   +SR 11-29-75 p50
 +NR 12-75 p12

MEYER-HELMUND, Erik
 Violets. cf Rubini RS 301.

MIASKOVSKY, Nikolai
1747 Concert, violin, op. 44, D minor. YSAYE: Ecstasy, op. 21, E flat
  major. Grigori Feigin, Rosa Fain, vln; MRSO, MPO; Alexander
  Dmitriev, Kyril Kondrashin. HMV Melodiya ASD 3237.
   +Gr 9-76 p422   +RR 9-76 p57
   +HFN 8-76 p83
 Yellowed leaves, op. 31, nos. 1 and 6. cf Odyssey Y 33825.

MICHAEL, David Moritz
 Instrumental suites: Movements (6). cf Columbia MS 6161.

MICKLEM
 Father, we thank you. cf Grosvenor GRS 1039.
 No one has ever seen God. cf Grosvenor GRS 1039.

MIELCZEWSKI
1748 Veni Domine, Deus in nomine tuo. ROZYCKI: Magnificat, Magnificemus
  in cantico, Confitebor. STACHOWICZ: Veni consolator.
  SZARZYNAKI: Ave Regina, Jesu spes mea. Complesso di Musica
  Antica Warsaw; Eugeniusz Sasiadek. Muza SXL 0975.
   +NR 7-76 p9

MIGUEZ, Leopoldo Amérigo
 Nocturne. cf Angel S 37110.

MILAN, Luis
 Fantasias, nos. 10-12, 16. cf Pavanas, nos. 1-6.
1749 Pavanas, nos. 1-6. Fantasias, nos. 10-12, 16. MUDARRA: Difer-
  encias sobre El Conde claros. Fantasia que contrahaza la
  harpa en la manera de Ludovico. Gallarda. O guárdame las
  vacas. Pavana de Alexandre. NARVAEZ: Baxa de contrapunto.
  Diferencias sobre Guárdame las vacas. Fantasia. Mille
  regretz. Konrad Ragossnig, lt. DG Archive 2533 183.
   ++Gr 7-75 p223   ++NR 3-76 p13
   ++HFN 7-75 p87   ++RR 7-75 p48
 Pavana. cf DG Archive 2533 184.
 Pavana. cf National Trust NT 002.
 Pavanas, D major, C major. cf Saga 5426.
 Songs: Aquel caballero, madre; Toda mi vida hos amé. cf
  DG 2530 504.
 Songs: Toda mi vida os amé. cf National Trust NU 002.

MILANO, Francesco da
 Fantasias I-VIII. cf RCA RK 11708.

MILETIC, Miroslav
1750 Cetiri aforizma. Deptih. Folklorne kosacije. Fontana del
  tritone. Koncertantna fantazija. Medjimurska suita. Monolog.
  Ples. Tišina. Pro Arte Quartet. Jugoton LSY 61178.
   *Te 9-76 p33
 Deptih. cf Cetiri aforizma.
 Folklorne kosacije. cf Cetiri aforizma.
 Fontana del tritone. cf Cetiri aforizma.
 Koncertantna fantazija. cf Cetiri aforizma.
 Medjimurska suita. cf Cetiri aforizma.

          Monolog. cf Cetiri aforizma.
          Ples. cf Cetiri aforizma.
          Tišina. cf Cetiri aforizma.
MILHAUD, Darius
1751 L'Automne. Le bal martiniquais. Paris. Le printemps. Scara-
          mouche. Christian Ivaldi, Noel Lee, Michel Béroff, Jean-
          Philippe Collard, pno. Connoisseur Society CS 2101.
              +NR 8-76 p12   ++St 11-76 p148
              ++SFC 8-29-76 p29
          Le bal martiniquais. cf L'Automne.
1752 Cantate de l'enfant et la mère. La muse ménagère. Madeleine
          Milhaud, narrator; Leonid Hambro, Darius Milhaud, pno;
          Juilliard Quartet; Darius Milhaud. Odyssey Y 33790. (Reissue
          from Columbia ML 4305).
              +-ARSC Vol VIII, no. 2-3,   +MJ 10-76 p52
                  p83   +NYT 5-16-76 pD20
1753 Chansons de Ronsard, op. 223 (4). Symphony, no. 6, op. 343.
          Paula Seibel, s; Louisville Orchestra; Jorge Mester. Louisville
          LS 744. (also RCA GL 25020).
              +-Gr 11-76 p797   +RR 10-76 p61
              ++HF 8-75 p90   ++SFC 5-25-75 p17
              +-HFN 12-76 p145
          La muse ménagère. cf Cantate de l'enfant et de la mère.
          Paris. cf L'Automne.
1754 Pastorale, op. 229. Petite suite, op. 348. Preludes, op. 231b.
          Sonata, organ, op. 112. George Baker, org. Delos FY 016.
              +NR 10-76 p15
          Petite suite, op. 348. cf Pastorale, op. 229.
          Preludes, op. 231b. cf Pastorale, op. 229.
          Le printemps. cf L'Automne.
1755 Le printemps, Bks. 1 and 2. Rag-caprices (3). Saudades do Brasil.
          William Bolcom, pno. Nonesuch H 71316.
              ++Gr 1-76 p1217   +SR 5-29-76 p52
              +NR 12-75 p13   +St 2-76 p74
          Quartet, strings, no. 12. cf BONDON: Quartet, strings, no. 1.
          Rag-caprices (3). cf Le printemps, Bks. 1 and 2.
          Saudades do Brasil. cf Le printemps, Bks. 1 and 2.
          Saudades do Brasil, no. 7: Corcovado. cf RCA ARM 4-0942/7.
          Saudades do Brasil, no. 10: Sumaré. cf RCA ARM 4-0942/7.
          Scaramouche. cf L'Automne.
          Scaramouche. cf FRANCAIX: Concerto, piano, D major.
          Sonata, organ, op. 112. cf Pastorale, op. 229.
          Sonatas, viola and piano, nos. 1 and 2. cf FRANCK: Sonata, violin
          and piano, A major.
          Symphony, no. 6, op. 343. cf Chansons de Ronsard, op. 223.
MILLOCKER, Karl
          Die Dubarry: Es lockt die Nacht. cf HMV CSD 3748.
MILLS
          The true beauty. cf CRI SD 102.
MIMAROGLU, Ilhan
1756 La ruche. To kill a sunrise. Various vocalists; composed in the
          studios of the Columbia-Princeton Electronic Music Center;
          Jacques Wiederkehr, vlc; Michel Merlet, hpd; Martine Joste,
          pno; composed in the studios of the Groupe de Recherches
          Musicales, ORTF. Folkways FTQ 33951.
              +-HF 11-76 p121   *NR 9-76 p13

To kill a sunrise. cf La ruche.
1757 Tract. Tuly Sand, speaking and singing with electronic music.
          Folkways FTS 33441.
                  -NR 10-75 p11   +NR 3-76 p13
MINKUS, Ludwig
          Don Quixote: Pas de deux. cf ADAM: Giselle: Pas de deux; Grand
              pas de deux and finale.
MIYAGI, Michio
          Haru no umi (arr. and orch. Gerhardt). cf RCA LRL 1-5094.
MODENA (16th century Spain)
          Recercare a 4. cf L'Oiseau-Lyre 12BB 203/6.
MODERNE, Jacques
          Branle gay nouveau. cf Harmonia Mundi HMU 610.
          Trios branles de Bourgogne. cf Harmonia Mundi HMU 610.
MOERAN, Ernest
          Maltworms. cf Pearl SHE 525.
          Songs: Diaphenia; Sweet o' the year. cf HMV HLM 7093.
MOESCHINGER, Albert
          Fuga mystica. cf Pelca PRSRK 41017/20.
MOLINARO, Simone
          Ballo detto 'Il Conte Orlando'. cf DG Archive 2533 323.
          Saltarello. cf DG Archive 2533 323.
MOLINS, Pierre de
          Amis tout dous. cf HMV SLS 863.
MOLLEDA
          Variations on a theme. cf RCA ARL 1-1323.
MOLLER
          Sonata, organ, D major. cf Columbia MS 6161.
MOLLOY
          Love's old sweet song. cf Prelude PRS 2505.
          Love's old sweet song. cf Transatlantic XTRA 1159.
MOLTER, Johann
          Symphony, 4 trumpets, C major. cf Nonesuch H 71301.
MOLTKE
          Des grossen Kurfürsten Reitermarsch. cf DG 2721 077 (2).
MOMPOU, Federico
1758 Canço i danza, no. 1. Charmes. Fêtes lointaines. Variations
          on a theme by Chopin . Pierre Huybregts, pno. Orion ORS
          76234.
                  +NR 7-76 p11   +St 10-77 p128
          Charmes. cf Canço i danza, no. 1.
          Fêtes lointaines. cf Canço i danza, no. 1.
          Scènes d'enfants. cf HMV HQS 1364.
          Variations on a theme by Chopin. cf Canço i danza, no. 1.
MONDELLO
          Poem, flute and harp. cf CASTELNUOVO-TEDESCO: Sonata, violon-
              cello and harp.
          Siciliana. cf Orion ORS 75207.
MONIUSZKO, Stanislaw
          Halka: Aria, Act 2. cf Muza SX 1144.
1759 Halka: Gdybym rannyn slonkiem, O moj malenki. PUCCINI: La
          bohème: Si mi chiamano Mimi. Tosca: Vissi d'arte. VERDI: Aida:
          O patria mia. Un ballo in maschera: Morrò, ma prima grazia.
          WAGNER: Lohengrin: Einsam in trüben Tagen. Maria Foltyn, s;
          Polish Radio Symphony Orchestra, Warsaw Opera Orchestra; Arnold
          Rezler, Zygmunt Latoszewski. Muza XL 1017.
                  +NR 8-76 p8   ++St 12-76 p154

MONK, William
    Abide with me. cf Sage 5225.
MONOD, Jacques-Louis
1760 Cantus contra cantum I. SHIFRIN: Quartet, strings, no. 4.*
        WEBER: Quartet, strings, no. 2, op. 35.* Merja Sargon, s;
        New Music Quartet, Fine Arts Quartet; Chamber Orchestra;
        Jacques-Louis Monod. CRI SD 358. (*Reissues).
            ++NR 11-76 p8
MONSIGNY, Pierre
    Rose et Colas: La Sagesse est un trésor. cf Desto DC 7118/9.
MONTEVERDI, Claudio
    Amor. cf Il combattimento di Tancredi e Clorinda.
    Baci soavi e cari. cf Harmonia Mundi 204.
1761 Il combattimento di Tancredi et Clorinda. Amor. Con che saovita.
        Tempro la cetra. Elisabeth Speiser, s; Rodolfo Malacarne, t;
        Laerte Malaguti, Karlheinz Peters, bs; Mainz Chamber Orchestra;
        Gunther Kehr. Turnabout Tape (c) KTVC 34018.
            +HFN 5-76 p117 tape   +RR 8-76 p84 tape
    Con che saovita. cf Il combattimento di Tancredi e Clorinda.
    Exultent caeli. cf BASSANO: Hodie Christus natus est.
1762 L'Incoronazione de Poppea. Cathy Berberian, Maria Minetto,
        Margaret Baker, ms; Carlo Gaifa, Kurt Equiluz, t; Enrico
        Fissore, bar; Giancarlo Luccardi, bs; VCM; Nikolaus Harnoncourt.
        Telefunken HS 6-35247 (5).
            +-Audio 10-76 p146   ++NR 2-75 p10
            +-Gr 3-75 p1697   +ON 8-75 p28
            +HF 2-75 p84   +RR 3-75 p16
            +-MT 8-75 p715   +St 6-75 p99
1763 L'Incoronazione di Poppea: Concert suite. L'Orfeo: Concert suite.
        Sue Harmon, s; Michael Sells, t; Robert Rodriguez, hpd; Emanuel
        Gruber, vlc; Orion Chamber Orchestra and Singers. Ember ECL
        9038.
            -Gr 1-76 p1225   -HFN 12-75 p157
    L'Incoronazione di Poppea: Disprezzata Regina; Tu che dagli avi
        miei...Maestade, che prega; A Dio Roma. cf Songs (Telefunken
        AW 6-41930).
    Lamento della Ninfa. cf Harmonia Mundi HMD 204.
    Lamento d'Olimpia. cf L'Oiseau-Lyre 12BB 203/6.
1764 Madrigals, Bk 9: Duets and trios. Accademia Monteverdiana;
        Dennis Stevens. Musical Heritage Society MHS 3104.
            +-St 5-76 p119
1765 Madrigals: Altri canti d'amore; Hor che' el ciel e la terra; Gira
        il nemico; Se vittorie si belle; Armato il cor; Ogni amante
        è guerrier; Ardo avvampo; Il ballo per l'Imperatore Ferdinando.
        Luigi Alva, Ryland Davies, Robert Tear, Alexander Oliver, t;
        Clifford Grant, Stafford Dean, bs; Glyndebourne Festival Chorus;
        Raymond Leppard, Leslie Pearson, Henry Ward, hpd; Joy Hall, vlc;
        Robert Spencer, lt; ECO; Raymond Leppard. Philips 6500 663.
        (From 6799 006).
            +-Gr 1-75 p1379   +-RR 1-75 p56
            +MM 1-76 p43
1766 Madrigals: Madrigali guerrieri, Sinfonia; Altri canti d'amor;
        Hor che' el ciel e la terra; Così sol d'una chiara fonte; Gira
        il nemico insidioso; Ardo avvampo. Madrigali amorosi, Altri
        canti di Marte; Due belli occhi; Lamento della Ninfa; Non
        havea Febo ancora...Amor...Non partir ritrosetta; Dolcissimo

uscignolo; Vago augelletto. Prague Madrigal Singers; Miroslav
Venhoda. Supraphon 112 1306.

   -Gr 6-75 p76   -RR 4-75 p63
   /HFN 5-75 p134   +St 5-76 p119
   +NR 10-75 p6

1767 Madrigali amorosi: Altri canti di Marte; Vago augelletto; Ardo
     e scoprir; O sia tranquillo il mare; Ninfa che scalza il piede;
     Dolcissimo uscignolo; Chi vol haver felice; Non havea Febo
     ancora; Lamento della Ninfa...Si tra sdegnosi pianti; Perchè
     t'en fuggi, O Fillide; Non partir, ritrosetta; Su, su pastorelli
     vezzosi. Yvonne Fuller, Angela Bostok, Lillian Watson, Sheila
     Armstrong, s; Alfred Hodgson, ms; Anne Collins, con; Luigi Alva,
     Ryland Davies, Robert Tear, Alexander Oliver, t; Stafford Dean,
     Clifford Grant, bs; Joy Hall, vlc; Robert Spencer, lt; Osian
     Ellis, hp; Raymond Leppard, Henry Ward, hpd; Glyndebourne
     Festival Chorus, Members; ECO; Raymond Leppard. Philips
     6500 864. (Reissue from 6799 006).

       +Gr 7-75 p228   +MM 1-76 p43
       +HFN 8-75 p91   +-RR 6-75 p80

Magnificat. cf Vespro della beata vergine.

1768 Magnificat a 6 voci (from Vespro, 1610). Magnificat a 7 voci
     (from Vespro, 1610). Mass, In illo tempore (from Vespro,
     1610). Vespro della beata vergine. Paul Esswood, Kevin
     Smith, c-t; Ian Partridge, John Elwes, t; David Thomas,
     Christopher Keyte, bs; Regensburger Domspatzen; Instrumental
     Ensemble; Hanns-Martin Schneidt. DG Archive 2710 017 (3).
     (also 2723 043).

       +-Audio 3-76 p66   +-MR 5-76 p164
       ++HF 5-76 p87   +NR 2-76 p9
       ++MJ 5-76 p28   ++St 3-76 p116

Magnificat a 7 voci (from Vespro, 1610). cf Magnificat a 6 voci.
Mass, In illo tempore. cf Vespro della beata vergine.
Mass, In illo tempore (from Vespro, 1610). cf Magnificat a 6
     voci.
L'Orfeo: Concert suite. cf L'Incoronazione di Poppea: Concert
     suite.
L'Orfeo: Mira, deh, mira, Orfeo...In un fiorito prato. cf Songs
     (Telefunken AW 6-41930).

1769 Songs (vocal works): Con che soavità; Lamento d'Arianna; Lettera
     amorosa. L'Incoronazione di Poppea: Disprezzata Regina; Tu
     che dagli avi miei...Maestade, che prega; A dio Roma.* L'Orfeo:
     Mira, deh, mira, Orfeo...In un fiorito prato.** Cathy Berberian,
     ms; Nigel Rogers, Günther Theuring, Lajos Kozma, t; Paul Esswood,
     c-t; VCM; Nikolaus Harnoncourt. Telefunken AW 6-41930.
     (*Reissue from 6-35247, **Reissue from SKH 21).

       +-Gr 4-76 p1649   +RR 4-76 p73
       +-HFN 11-76 p163

Tempro la cetra. cf Il combattimento di Tancredi e Clorinda.

1770 Vespro della beata vergine. Mass, In illo tempore. Magnificat.
     Paul Esswood, Kevin Smith, c-t; Ian Partridge, John Elwes, t;
     David Thomas, Christopher Keyte, bs; Edward H. Tarr, Ralph
     Bryant, Richard Cook, cor; Fritz Brodersen, Harald Strutz,
     Walfried Kohlert, trom; Sebastian Kelber, Klaus Holsten, flt
     and Renaissance rec; Eduard Melkus, Spiros Rantos, Thomas
     Weaver, vln; Lilo Gabriel, David Becker, vla; Klaus Storck,
     Eugene Eicher, vlc; Laurenzius Strehl, vla da gamba and

violone; Dieter Kirsch, lt; Hubert Gumz, Gerd Kaufmann, org;
Regensburg Domspatzen; Hanns-Martin Schneidt. DG Archive
2723 043 (3).
    +Gr 11-75 p876   +RR 11-75 p86
    ++HFN 11-75 p155   +STL 11-2-75 p38
    +MT 6-76 p495

1771 Vespro della beata vergine. Elly Ameling, Norma Burrowes, s;
Charles Brett, c-t; Anthony Rolfe Johnson, Robert Tear, Martyn
Hill, t; Peter Knapp, John Noble, bs; King's College Chapel
Choir; Early Music Consort; Philip Ledger. HMV SQ SLS 5064
(2). (also Angel S 3837).
    +-Gr 11-76 p857   +-RR 9-76 p85
    +-HFN 9-76 p125   +STL 9-19-76 p36
    +HFN 12-76 p153 tape

1772 Vespro della beata vergine. VCM; Vienna Boys' Choir, Monteverdi
Choir; Nikolaus Harnoncourt. Telefunken 6-35045 (2).
    +MJ 5-76 p28
Vespro della beata vergine. cf Magnificat a 6 voci.

1773 Vespro della beata vergine, excerpts. Exeter University Singers
and Players; Donald James. Exon Audio EAS 14.
    -Gr 11-76 p857

MONTSALVATGE, Xavier
Canciones negras. cf DG 2530 598.

MOODY, James
Quintet, harmonica. cf JACOB: Divertimento, harmonica and string
quartet.

MOORE, Carman
Youth in a merciful house. cf Folkways FTS 33901/4.

MOORE, Douglas
1774 The ballad of Baby Doe. Beverly Sills, s; Walter Cassel, bar;
Frances Bible, ms; Lynn Taussig, Helen Baisley, Grant Williams,
Chester Ludgin, Beatrice Krebs, Jack DeLon, Joshua Hecht,
soloists; New York City Opera Orchestra and Chorus; Emerson
Buckley. DG 2709 061 (3). (Reissue from MGM 3GC-1, Heliodor
H 25035/3).
    *ARG 11-76 p34   +NR 10-76 p11
    +-Gr 7-76 p206   +ON 7-76 p43
    +HFN 7-76 p91   +RR 7-76 p29
    *MJ 11-76 p44

MORALES, Cristóbal de
1775 Andreas Christi famulus. Emendemus in melius. Jubilate Deo omnis
terra. Lamentabatur Jacob. Magnificat secundi toni. Pastores
dicite quidnam vidistis. Pro Cantione Antiqua; Early Music
Consort; Bruno Turner. DG Archive 2533 321.
    ++RR 12-76 p93
Emendemus in melius. cf Andreas Christi famulus.
Jubilate Deo omnis terra. cf Andreas Christi famulus.
Lamentabatur Jacob. cf Andreas Christi famulus.
Magnificat secundi toni. cf Andreas Christi famulus.

1776 Missa mille regretz. Motets: Andreas Christi famulus; Christus
resurgens ex mortuis; Pastores, dicite quidnam vidistis; Per
tuam crucem; Salve regina. Prague Madrigal Singers; Miroslav
Venhoda. Telefunken AW 6-41917.
    +-Gr 3-76 p1499   +RR 4-76 p73
    +-HFN 3-76 p99
Motets: Andreas Christi famulus; Christus resurgens ex mortuis;

Pastores, dicite quidnam vidistis, per tuam crucem; Salve
regina. cf Missa mille regretz.
Pastores dicite quidnam vidistis. cf Andreas Christi famulus.
MORENO
To huey tlahtzin. cf Polydor 2383 389.
MORGAN, Justin
Songs: Amanda; Despair; Montgomery. cf Vox SVBX 5350.
Songs: Judgment anthem; Amanda. cf Nonesuch H 71276.
MORLEY, Thomas
La caccia, a 2. cf Philips 6500 926.
1777 Dances for broken consort. SUSATO: Danserye: Dances (12).
Early Music Consort; David Munrow. HMV Tape (c) TC EXE 104.
(also Angel S 36851 Tape (c) 4XS 36851).
+HF 11-76 p153 tape   +RR 1-75 p72 tape
Fancy. cf BIS LP 22.
First Booke of Consort Lessons: Captaine Piper's pavan and
galliard; La corante; Lachrimae pavan; Lavolto; Mounsier's
almaine; Michill's galliard; My Lord of Oxenford's maske.
cf Angel S 36851.
Hard by a crystal fountain. cf Turnabout TV 34017.
It was a lover and his lass. cf Coronet LPS 3032.
It was a lover and his lass. cf HMV ESD 7002.
Laboravi in gemitu meo. cf Harmonia Mundi HMU 473.
Lamento. cf BIS LP 22.
My bonny lass she smileth. cf Harmonia Mundi 593.
O grief, even on the bud. cf Argo ZRG 833.
La sampogna. cf Philips 6500 926.
Shoot, false love. cf HMV CSD 3766.
Songs: April is my mistress' face; Daemon and Phyllis; Fire,
fire; I love, alas; Leave, alas, this tormenting: My bonny
lass; Now is the month of Maying; O grief, even on the bud;
Those dainty daffadillies; Though Philomela lost her love.
cf Enigma VAR 1017.
Songs: Ay me, the fatal arrow; My bonny lass she smileth; Now
is the month of Maying; Though Philomela lost her love. cf
Prelude PRS 2501.
Songs: Come, sorrow, come; It was a lover and his lass; Thyrsis
and Milla. cf Philips 6500 282.
Songs: Joyne hands; The Lord Zouches maske; O mistress mine;
Sola soletta (Conversi). cf Harmonia Mundi HMD 223.
Songs: Now is the month of Maying; Though Philomela lost her
love. cf HMV CSD 3756.
MORMAN TABERNACLE CHOIR: A JUBILANT SONG. cf Columbia M 34134.
MORNABLE, Antoine de
Je ne scay. cf HMV CSD 3740.
MORRIS, Robert
Phases. cf CURTIS-SMITH: Sonorous inventions.
MORSE
Up the street. cf RCA AGL 1-1334.
MORTENSEN, Finn
Quintet, winds, op. 4. cf CARLSTEDT: Sinfonietta, 5 wind
instruments.
MORTENSON, Jan
Ultra. cf Caprice RIKS LP 35.
MORTON, Robert
La perontina. cf Telefunken ER 6-35257 (2).

MOSCHELES, Ignaz
1778   Grande sonate symphonique, op. 112. PIXIS: Concerto, violin,
           piano and strings. Mary Louise Boehm, Pauline Boehm, pno;
           Kees Kooper, vln; Westphalian Symphony Orchestra; Siegfried
           Landau. Turnabout TV 34590.
                   +Gr 9-76 p445   +NR 12-75 p7
                   -HFN 8-76 p83   +RR 8-76 p49
MOSS
       Songs: The floral dance. cf Argo ZFB 95/6.
MOSS, Lawrence
       Elegy. cf HARRIS: Fantasy, violin and piano.
       Timepiece. cf HARRIS: Fantasy, violin and piano.
MOSZKOWSKI, Moritz
       Caprice espagnole, op. 37. cf International Piano Library IPL
           5001/2.
       Caprice espagnole, op. 37. cf L'Oiseau-Lyre DSLO 7.
1779   Concerto, piano, op. 59, E major. GODOWSKY: Passacaglia, B
           minor. Caren Goodin, Stephen Glover, pno. International
           Piano Library IPL 1001.
                   +NR 6-73 p4   +-RR 6-76 p44
       Etincelles. cf RCA VH 020.
       Etude, A flat major. cf RCA VH 020.
       Serenata. cf Pye Ember GVC 51.
       Stücke, op. 45, no. 2: Guitarre (2). cf RCA ARM 4-0942/7.
       Valse, op. 34. cf International Piano Library IPL 102.
MOURANT
       Elegy, flute and harp. cf CASTELNUOVO-TEDESCO: Sonata, violon-
           cello and harp.
MOURET, Jean Joseph
       Rondeau. cf Stentorian SP 1735.
       Suite of symphonies: 1st suite. cf Klavier KS 551.
MOUTON, Charles
       L'Amant content, canarie. cf Turnabout TV 34137.
       Le dialogue des graces sur Iris, allemande. cf Turnabout TV 34137.
       La, la, la l'oysillon du bois. cf L'Oiseau-Lyre 12BB 203/6.
       La malassis, sarabande. cf Turnabout TV 34137.
       Nesciens Mater virgo virum. cf HMV SLS 5049.
MOUTON, Jean
1780   Noe psallite. SMERT: Nowell: Dieus vous garde. WALTHER: Joseph
           lieber Joseph mein. ANON.: Lux hodie; Orient's partibus,
           Renosemus laudibus, Verbum caro; In hoc anni circulo, Fines
           amourtes, Verbum partis hodie, Lullay lullow, Fulget hodie,
           Now make we merthe, Nowell; The borys hede, Pray for us,
           Verbum patris humanatur, Conditor fut le non-pareil, Nova,
           nova, Riu riu chiu, Verbum caro; Dies est laetitiae. Purcell
           Consort of Voices; Grayston Burgess. Argo ZRG 526.
                   +CJ 3-76 p25
MOYLE, W. E.
       Cornish rock. cf Pye TB 3009.
       Restormel. cf Pye TB 3009.
MOYREAU
       Les cloches d'Orléans. cf Calliope CAL 1917.
MOZART, Leopold
       Concerto, trumpet, D major. cf HUMMEL: Concerto, trumpet, E flat
           major.
1781   Musical sleigh ride, F major. Sinfonia, D major. Sinfonia

burlesca, G major. Eduard Melkus Ensemble. DG Archive 2533
   328.
     +-Gr 12-76 p1015   +RR 11-76 p70
     +HFN 11-76 p163
Sinfonia, D major. cf Musical sleigh ride, F major.
Sinfonia burlesca, G major. cf Musical sleigh ride, F major.
MOZART, Wolfgang Amadeus
1782 Adagio, K 261, E major. Concerto, violin, no. 1, K 207, B flat
     major. Rondos, violin, K 269, B flat major; K 373, C major.
     Wolfgang Schneiderhan, vln; BPhO; Wolfgang Schneiderhan.
     DG 2535 205. (Reissue from SLPM 139 350/2).
       -Gr 12-76 p996   +HFN 12-76 p151
    Adagio, K 356, C major. cf CBS 73487.
    Adagio, K 356, C major. cf Columbia M 33514.
1783 Adagio, K 411, B flat major. Divertimenti, K 439b (5). STADLER:
     Trio, 3 basset horns. Alan Hacker, clt and basset horn;
     Matrix, Members. Prelude PRS 2502.
      /Gr 10-76 p621   +RR 10-76 p25
     +-HFN 11-76 p164
1784 Adagio, K 540, B minor. Sonata, piano, no. 13, K 333, B flat
     major. Sonata, piano, no. 9, K 311, D major. Variations on
     "Salve tu Domine", K 398, F major. Michael Cave, pno. Orion
     ORS 75185.
      +HF 1-76 p91   +NR 9-75 p11
     /MJ 4-76 p31
1785 Adagio, K 540, B minor. Sonata, piano, no. 11, K 331, A major.
     Sonata, piano, no. 13, K 333, B flat major. Alfred Brendel,
     pno. Philips 9500 025 Tape (c) 7300 474.
      ++Gr 6-76 p70   +-RR 6-76 p73
     ++HFN 6-76 p93
    Adagio, K 540, B minor. cf HAYDN: Sonata, piano, no. 46, A flat
     major.
    Adagio, K 540, B minor. cf International Piano Library IPL 5003/4.
    Adagio and allegro, K 594, F minor. cf BACH: Toccata and fugue,
     S 565, D minor.
1786 Adagio and fugue, K 546, C major. Concerto, piano, no. 14, K 449,
     E flat major. Serenade, no. 6, K 239, D major. Rudolf Serkin,
     pno; Busch Chamber Players; Adolf Busch. Turnabout THS 65058.
      +ARSC Vol VIII, no. 2-3   +NR 7-76 p3
       p88   ++SFC 8-1-76 p27
    Adagio and fugue, K 546, C minor. cf Works, selections (HMV SLS
     5048).
    Allegro, K 3. cf Connoisseur Society (Q) CSQ 2065.
1787 Andante, K 315, C major. Concerto, flute, no. 1, K 313, G major.
     Concerto, flute, no. 2, K 314, D major. James Galway, flt;
     Festival Strings; Rudolf Baumgartner. RCA LRL 1-5109. Tape
     (c) RK 11732.
      +-Gr 3-76 p1467   +-HFN 11-76 p175 tape
      +-HFN 4-76 p109   ++RR 3-76 p45
    Andante, K 315, C major. cf RCA LRL 1-5127.
    Andante, K 616, F major. cf HAYDN: Sonata, piano, no. 46, A flat
     major.
    Andante and variations, K 501, G major. cf Piano works (4 hands)
     (Peerless PRCM 213/2).
1788 Arias: Così fan tutte, K 588; Un aura amorosa; Ah lo veggio; In
     qual fiero. Don Giovanni, K 527: Il mio tesoro; Dalla sua

pace. Die Entführung aus dem Serail, K 384; Hier soll ich
dich; Wenn der Freude; Constanze; Ich baue ganz. Idomeneo,
Ré di Creta, K 366: Fuor del mar. Die Zauberflöte, K 620:
Dies Bildnis ist bezaubernd schön; Wie stark ist nicht dein
Zauberton. Stuart Burrows, t; LSO, LPO; John Pritchard.
L'Oiseau-Lyre DSLO 13.

    +-Gr 10-76 p649   +-RR 10-76 p28
    +HFN 11-76 p164

1789 Arias: La clemenza di Tito, K 621: Parto, parto. Don Giovanni,
K 527: In quali accessi...Mi tradi quell' alma ingrata; Crudele,
Ah no, mio bene...Non mi dir. Die Entführung aus dem Serail,
K 384: Martern aller Arten. Idomeneo, Ré di Creta, K 366:
Parto, e l'unico oggetto. Le nozze di Figaro, K 492: Giunse
alfin il momento...Deh vieni non tardar; E Susanna no vien...
Dove sono; Voi che sapete. Il Re pastore, K 208: L'amerò
sarò costante. Margaret Price, s; ECO; James Lockhart. RCA
SER 5675. (also RCA AGL 1-1532).

    +Gr 10-73 p727   +-RR 11-73 p28
    +MJ 12-76 p28   ++SFC 9-12-76 p31
    +NR 9-76 p9   +ST 2-76 p741
    ++ON 9-76 p70   ++St 11-76 p89
    +Op 12-73 p1104

1790 Arias (concert): Männer suchen stets zu naschen, K 433. Don
Giovanni, K 527: Madamina, il catalogo è questo; Ah pietà,
Signori miei; Fin ch'han dal vino; Deh vieni alla finestra;
Metà di voi quà vadano. Le nozze di Figaro, K 492: Bravo,
signor padrone...Se vuol ballare; Non più andrai; Tutto è
disposto...Aprite un po' quegli occhi; Hai già vinta la causa...
Vedrò, mentr'io sospiro. Zaide, K 344: Wer hungrig bei der
Tafel sitzt; Ihr Mächtigen seht ungerührt. Die Zauberflöte,
K 620: Der Vogelfanger bin ich ja; Ein Mädchen oder Weibchen.
Theo Adam, bs-bar; Dresden State Orchestra; Otmar Suitner.
Telefunken AG 6-41946.

    +HFN 5-76 p103   +-RR 4-76 p30

Ave verum corpus, K 618. cf Abbey LPB 754.

1791 Canzonetti, K 549. Divertimenti, K 439b (5). Duos, 2 horns,
K 487 (12). Nocturnes, K 346, 436-439a. Elly Ameling,
Elisabeth Cooymans, s; Peter van der Bilt, bar; NWE members.
Philips 6747 136 (2).

    +HF 7-76 p82   +ON 5-76 p48
    ++MJ 5-76 p28   ++SFC 5-9-76 p35
    ++NR 12-76 p11   ++St 6-76 p105
    *NYT 8-8-76 pD13

Cassation (serenade), no. 1, K 62a, D major (with March, K 62,
D major). cf BACH, J. C.: Sinfonia, op. 6, no. 6, C minor.
Cassation, no. 1, K 63, G major. Adagio. cf Works, selections
(Philips 6775 012).

1792 Cassation, no. 2, K 99, B flat major. Divertimento, no. 2, K 131,
D major. Dresden Philharmonic Orchestra: Günther Herbig.
Philips 6500 703.

    +Gr 6-76 p50   +RR 6-76 p45
    +HFN 6-76 p91

La clemenza di Tito, K 621: Overture. cf Works, selections (HMV
SLS 5048).
La clemenza di Tito, K 621: Parto, parto. cf Arias (RCA SER 5675).
Concert rondo, K 371, E flat major. cf Concerti, horn, nos. 1-4.

1793 Concerto, clarinet, K 622, A major. Quintet, clarinet, K 581, A
      major. Benny Goodman; BSO; Charles Munch. Camden CCV 5006.
      (Reissue).
            +St 2-76 p741
1794 Concerto, clarinet, K 622, A major. Concerto, oboe, K 314, C
      major. Derek Wickens, ob; Thea King, clt; London Little Orch-
      estra; Leslie Jones. Oryx ORPS 21. Tape (c) BRL 21.
            +-Gr 11-76 p798   +RR 1-75 p70
      Concerto, clarinet, K 622, A major. cf Works, selections (Decca
      DPA 521/2).
      Concerto, clarinet, K 622, A major. cf Works, selections (Decca
      DPA 541/2).
      Concerto, clarinet, K 622, A major. cf Decca SPA 395.
1795 Concerto, flute, no. 1, K 313, C major. Concerto, flute, no. 2,
      K 314, D major. James Galway, flt; New Irish Chamber Orchestra;
      André Prieur. New Irish Recording Company NIR 010.
            ++Gr 1-76 p1198
      Concerto, flute, no. 1, K 313, G major. cf Andante, K 315, C
      major.
      Concerto, flute, no. 2, K 314, D major. cf Andante, K 315, C
      major.
      Concerto, flute, no. 2, K 314, D major. cf Concerto, flute, no. 1,
      K 313, C major.
      Concerto, flute, no. 2, K 314, D major. cf Works, selections
      (Decca DPA 521/2).
      Concerto, flute, no. 2, K 314, D major: 3rd movement. cf Decca
      SPA 394.
1796 Concerto, flute and harp, K 299, C major. Sinfonia concertante,
      K 297b, E flat major. Wolfgang Schulz, flt; Nicanor Zabaleta,
      hp; Walter Lehmayer, ob; Peter Schmidl, clt; Günther Högner,
      hn; Fritz Faltl, bsn; VPO; Karl Böhm. DG 2530 715.
            +-Gr 11-76 p798   +RR 10-76 p61
            +HFN 10-76 p175
      Concerto, flute and harp, K 299, C major. cf Works, selections
      (Decca DPA 521/2).
      Concerto, horn, K 494a, E major, excerpt. cf Concerti, horn and
      strings, nos. 1-4.
1797 Concerti, horn and strings, nos. 1-4, K 412, K 417, K 447, K 495.
      Concert rondo, K 371, E flat major. Concerto, K 494a, E major,
      excerpt. Barry Tuckwell, Fr hn; AMF; Neville Marriner. Angel
      S 36840. Tape (c) 4XS 36840. (also HMV ASD 2780 Tape (c) TC
      ASD 2780).
            +Gr 4-72 p1716   ++NR 7-72 p7
            +HF 10-72 p102   ++RR 10-72 p128 tape
            ++HF 10-76 p147 tape   ++St 10-72 p133 tape
1798 Concerti, horn and strings, nos. 1-4, K 412, K 417, K 447, K 495.
      Mason Jones, hn; PO; Eugene Ormandy. CBS 30075. Tape (c) 40-
      30075. (Reissue from 61095). (also Columbia MS 6785).
            +Gr 9-76 p423   +HFN 10-76 p185 tape
            ++HFN 10-76 p181   +-RR 9-76 p58
1799 Concerti, horn and strings, nos. 1-4, K 412, K 417, K 447, K 495.
      James Brown, hn; Virtuosi of England; Arthur Davison. Classics
      for Pleasure CFP 148. Tape (c) TC CFP 148.
            +HFN 12-76 p153 tape
1800 Concerti, horn and strings, nos. 1-4, K 412, K 417, K 447, K 495.
      Dennis Brain, hn; PhO; Herbert von Karajan. HMV ASD 1140.

Tape (c) TC ASD 1140. (Reissue from Columbia 33CX 1140).
    ++Gr 9-73 p549   +RR 9-73 p76
    +-HFN 8-76 p95 tape

1801 Concerti, horn and strings, nos. 1-4, K 412, K 417, K 447, K 495.
      Alan Civil, hn; PhO; Otto Klemperer. HMV SXLP 30207.
      (Reissue from Columbia SAX 2406). (also Angel S 35689).
        ++Gr 2-76 p1345   +RR 2-76 p32
        +HFN 3-76 p111

Concerto, horn and strings, no. 2, K 417, E flat major. cf Works,
    selections (Decca DPA 521/2).

1802 Concerto, horn and strings, no. 4, K 495, E flat major. The
      marriage of Figaro, K 492: Overture. Symphony, no. 40, K 550,
      G minor. Alan Civil, hn; RPO; Lawrence Foster. London SPC
      21093. (also Decca PFS 4314 Tape (c) KPFC 4314).
        +-Gr 7-75 p188   +-RR 7-75 p28
        +-HFN 9-75 p101   +-RR 2-76 p71 tape
        +MJ 3-75 p26

Concerto, horn and strings, no. 4, K 495, E flat major. cf Works,
    selections (CBS 30088).

Concerto, horn and strings, no. 4, K 495, E flat major. cf Works,
    selections (Decca DPA 521/2).

Concerto, horn and strings, no. 4, K 495, E flat major. cf HMV
    SLS 5068.

1803 Concerto, oboe, K 314, C major. Sinfonia concertante, K 297b,
      E flat major. Lothar Koch, Karl Steins, ob; Herbert Stahr, clt;
      Manfred Braun, bsn; Norbert Hauptmann, hn; BPhO; Herbert von
      Karajan. HMV ASD 3191. (Reissue from SLS 817).
        +Gr 8-76 p291   +RR 8-76 p42

Concerto, oboe, K 314, C major. cf Concerto, clarinet, K 622, A
    major.

1804 Concerti, piano, nos. 1-6, 8-9, 11-27. Rondo, piano, K 382, D
      major. Daniel Barenboim, pno; ECO; Daniel Barenboim. HMV SLS
      5031 (12). (Some reissues from ASD 2484, 3032, 3033, 2999,
      2956, 2357, 2434, 2887, 2318, 2465, 2838).
        +Gr 1-76 p1198   +RR 2-76 p32

1805 Concerto, piano, no. 1, K 37, F major. Concerto, piano, no. 2,
      K 39, B flat major. Concerto, piano, no. 26, K 537, D major.
      Karl Engel, pno; Salzburg Mozarteum Orchestra; Leopold Hager.
      Telefunken AW 6-41993.
        +-Gr 12-76 p996   +RR 10-76 p61
        +HFN 11-76 p164

Concerto, piano, no. 2, K 39, B flat major. cf Concerto, piano,
    no. 1, K 37, F major.

1806 Concerto, piano, no. 5, K 175, D major. Concerto, piano, no. 27,
      K 595, B flat major. Rondo, piano, K 382, D major. Karl Engel,
      pno; Salzburg Mozarteum Orchestra; Leopold Hager. Telefunken
      AW 6-41962.
        +-Gr 12-76 p996   +RR 10-76 p61
        +HFN 10-76 p173

1807 Concerto, piano, no. 8, K 246, C major. Concerto, piano, no. 26,
      K 537, D major. Jörg Demus, pno; Collegium Aureum. BASF KHB
      29311. (also Musical Heritage Society MHS 1614, also BASF BAC
      3003).
        +Gr 7-74 p213   +-RR 6-74 p44
        -MQ 4-74 p312   ++SFC 5-26-74 p22
        ++NR 11-73 p5   +-St 2-74 p118
        +NYT 10-14-73 pD33   *St 2-76 p741

1808 Concerto, piano, no. 9, K 271, E major. Concerto, piano, no. 20,
      K 466, D minor. Felicja Blumenthal, pno; Salzburg Mozarteum
      Orchestra; Leopold Hager. Everest SDBR 3381.
          -Gr 5-76 p1761   +NR 11-75 p7
1809 Concerto, piano, no. 9, K 271, E flat major. Concerto, piano,
      no. 21, K 467, C major. Nina Milkina, pno; St. John's Orchestra;
      John Lubbock. Pye Nixa QS PCNH 1. Tape (c) ZCPNH 1.
          +-Gr 4-76 p1604   +RR 3-76 p46
          +-HFN 5-76 p102   +-RR 9-76 p94 tape
      Concerto, piano, no. 9, K 271, E flat major. cf BEETHOVEN: Con-
      certo, piano, no. 1, op. 15, C major.
1810 Concerto, piano, no. 10, K 365, E flat major. Concerto, piano,
      no. 27, K 595, B flat major. Elena Gilels, Emil Gilels, pno;
      BPhO, VPO; Karl Böhm. Rococo 2093.
          *NR 10-76 p5
1811 Concerto, piano, no. 11, K 413, F major. Concerto, piano, no. 12,
      K 414, A major. Rudolf Serkin, pno; Marlboro Festival Orches-
      tra; Alexander Schneider. Columbia M 31728. Tape (c) MT 31728.
      (also CBS 73084)
          +-Gr 2-76 p1345   +-RR 1-76 p36
          +-HF 8-73 p88   ++SFC 7-8-73 p32
          +-HFN 2-76 p105   +St 7-73 p106
          +NR 6-73 p6
1812 Concerto, piano, no. 12, K 414, A major. Concerto, piano, no.
      17, K 453, G major. Alfred Brendel, pno; AMF; Neville Marriner.
      Philips 6500 140. (also Philips 6599 054)
          ++Gr 2-74 p1558   ++NR 5-76 p6
          +HFN 2-74 p338   ++RR 2-74 p30
          *MJ 10-76 p24
1813 Concerto, piano, no. 12, K 414, A major. Concerto, piano, no. 27,
      K 595, B flat major. Myra Hess, pno; American Chamber Orches-
      tra; Robert Scholz. Rococo 2108.
          +NR 10-76 p5
      Concerto, piano, no. 12, K 414, A major. cf Concerto, piano, no.
      11, K 413, F major.
1814 Concerto, piano, no. 13, K 415, C major. Concerto, piano, no. 26,
      K 537, D major. Maria João Pires, pno; Gulbenkian Foundation
      Chamber Orchestra; Theodor Guschlbauer. Erato STU 70887.
          +-Gr 5-76 p1762   +-RR 5-76 p46
          +-HFN 5-76 p102
      Concerto, piano, no. 13, K 415, C major. cf HAYDN: Concerto,
      piano, op. 21, D major.
1815 Concerti, piano, nos. 14-19. Peter Serkin, pno; ECO; Alexander
      Schneider. RCA ARL 3-0732 (3).
          +Gr 11-75 p819   ++SFC 1-26-75 p26
          *HF 5-75 p79   +SR 2-8-75 p37
          +HFN 11-75 p157   +St 4-75 p71
          ++NR 2-75 p6   +ST 2-76 p741
          +RR 11-75 p44
1816 Concerto, piano, no. 14, K 449, E flat major. Concerto, piano,
      no. 24, K 491, C minor. Murray Perahia, pno; ECO; Murray
      Perahia. CBS 76481. Tape (c) 40-76481. (also Columbia M
      34219 Tape (c) MT 34219).
          +Gr 5-76 p1761   +-HFN 10-76 p185 tape
          +Gr 9-76 p497 tape   +RR 5-76 p46
          +HFN 6-76 p93

1817   Concerto, piano, no. 14, K 449, E flat major. Concerto, piano,
       no. 15, K 450, B flat major. Peter Serkin, pno; ECO; Alexan-
       der Schneider. RCA ARL 1-1492. Tape (c) ARK 1-1492 (ct) ARS
       1492.
              +HF 10-76 p147 tape
1818   Concerto, piano, no. 14, K 449, E flat major. Concerto, piano,
       no. 25, K 503, C major. Karl Engel, pno; Salzburg Mozarteum
       Orchestra; Leopold Hager. Telefunken 6-41925. Tape (c)
       4-41925.
              ++HFN 2-76 p105   +-RR 2-76 p32
       Concerto, piano, no. 14, K 449, E flat major. cf Adagio and
       fugue, K 546, C major.
       Concerto, piano, no. 15, K 450, B flat major. cf Concerto, piano,
       no. 14, K 449, E flat major.
       Concerto, piano, no. 15, K 450, B flat major. cf HAYDN: Concerto,
       piano, D major.
1819   Concerto, piano, no. 16, K 451, D major. Concerto, piano, no. 17,
       K 453, G major. Peter Serkin, pno; ECO; Alexander Schneider.
       RCA ARL 1-1943. Tape (c) ARK 1-1943 (ct) ARS 1-1943.
              /NR 12-76 p7
1820   Concerti, piano, nos. 17, 20, 21, 23-24. Rondo, K 511, A minor.
       Arthur Rubinstein, pno; RCA Victor Symphony Orchestra; Alfred
       Wallenstein, Josef Krips. RCA SER 5716/8 (3). (Reissues from
       SB 6578, 6570, 6532, 2117).
              ++Gr 5-75 p1968   ++MT 2-76 p143
              +HFN 7-75 p85   +RR 7-75 p31
1821   Concerto, piano, no. 17, K 453, G major.* Concerto, piano, no.
       26, K 537, D major. Ingrid Haebler, pno; LSO; Witold Rowicki.
       Philips 6580 043. Tape (c) 7317 137. (*Reissue from SAL 3537).
              +Gr 3-74 p1765   +HFN 2-74 p341
              +Gr 9-76 p494 tape   +RR 1-74 p43
1822   Concerto, piano, no. 17, K 453, G major. Concerto, piano, no. 23,
       K 488, A major. Karl Engel, pno; Salzburg Mozarteum Orchestra;
       Leopold Hager. Telefunken AW 6-41888. Tape (c) CX 4-41888.
              /Gr 9-75 p466   /RR 8-75 p40
              /Gr 10-75 p721 tape   /RR 1-76 p66 tape
              +-HFN 8-75 p79
       Concerto, piano, no. 17, K 453, G major. cf Concerto, piano, no.
       12, K 414, A major.
       Concerto, piano, no. 17, K 453, G major. cf Concerto, piano, no.
       16, K 451, D major.
1823   Concerto, piano, no. 18, K 456, B flat major. Concerto, piano,
       no. 24, K 491, C minor. Karl Engel, pno; Salzburg Mozarteum
       Orchestra; Leopold Hager. Telefunken AW 6-41926.
              -Gr 3-76 p1468   +-RR 2-76 p32
              ++HFN 2-76 p105
1824   Concerto, piano, no. 18, K 456, B flat major. Concerto, piano,
       no. 27, K 595, B flat major. Alfred Brendel, pno; AMF; Neville
       Marriner. Philips 6500 948. Tape (c) 7300 383.
              +-Gr 4-75 p1811   ++NR 11-75 p6
              +HF 10-76 p147 tape   ++RR 5-75 p29
              +HFN 7-75 p90   ++SFC 10-5-75 p38
              ++MT 10-75 p885
1825   Concerti, piano, nos. 20-27. Géza Anda, pno; Salzburg Mozarteum
       Orchestra; Géza Anda. DG 2740 138 (4).
              +Gr 10-76 p602   ++HFN 10-76 p181

1826 Concerto, piano, no. 20, K 466, D minor. Concerto, piano, no. 24,
      K 491, C minor. Artur Schnabel, pno; NYP, LAPO; Georg Szell,
      Alfred Wallenstein. Bruno Walter Society BWS 723.
             ++ARSC Vol VIII, no. 2-3 p90
1827 Concerto, piano, no. 20, K 466, D minor. Concerto, piano, no. 23,
      K 488, A major. Alan Schiller, pno; LPO; Charles Mackerras.
      Classics for Pleasure CFP 40249.
             +-RR 12-76 p63
1828 Concerto, piano, no. 20, K 466, D minor. Concerto, piano, no. 21,
      K 467, C major. Friedrich Gulda, pno; VPO; Claudio Abbado.
      DG 2530 548. Tape (c) 3300 492.
             +-Gr 11-75 p820   +-NR 1-76 p6
             +Gr 10-76 p658 tape   ++NYT 1-18-76 pD1
             +-HFN 10-75 p144   /RR 10-75 p45
             +-HFN 12-75 p173 tape +RR 4-76 p80 tape
1829 Concerto, piano, no. 20, K 466, D minor. Deutsche Tänze, K 605
      (3). Serenade, no. 13, K 525, G major. Bruno Walter, pno;
      VPO; Bruno Walter. Turnabout THS 65036. (Reissue).
             +-Audio 3-76 p63   +-NR 7-75 p6
1830 Concerto, piano, no. 20, K 466, D minor. Concerto, piano, no. 24,
      K 491, C minor. Artur Schnabel, pno; PhO; Walter Susskind.
      Turnabout THS 65046.
             +NR 12-75 p6   ++SFC 3-7-76 p27
      Concerto, piano, no. 20, K 466, D minor. cf Concerto, piano, no.
      9, K 271, E major.
1831 Concerti, piano, nos. 21-27. Rondo, piano, K 382, D major.
      Daniel Barenboim, pno; ECO; Daniel Barenboim. Angel SDC 3830
      (4).
             ++NR 7-76 p4   ++St 10-76 p128
             +SFC 8-1-76 p27
1832 Concerto, piano, no. 21, K 467, C major. Concerto, piano, no. 23,
      K 488, A major. Jörg Demus, pno; Collegium Aureum. BASF KHB
      22477. (also BASF BAC 3084 Tape (c) KBACC 3084).
             +-Gr 8-76 p291   +-RR 7-76 p82
             +-HFN 5-76 p117   +St 10-76 p128
1833 Concerto, piano, no. 21, K 467, C major. Concerto, piano, no. 23,
      K 488, A major. Ilana Vered, pno; LPO; Uri Segal. Decca PFS
      4340. Tape (c) KPFC 4340. (also London SPC 21138).
             +Gr 12-75 p1038   +-RR 11-75 p44
             +HFN 12-75 p157   +-RR 7-76 p82 tape
             +-NR 12-76 p7   +St 10-76 p128
1834 Concerto, piano, no. 21, K 467, C major. Concerto, 2 pianos,
      K 365, E flat major. Ingrid Haebler, Ludwig Hoffmann, pno;
      LSO; Witold Rowicki, Alceo Galliera. Philips 6580 083. Tape
      (c) 7317 112. (Reissue from AXS 12000/1).
             +Gr 7-74 p214   +RR 6-74 p44
             +Gr 9-76 p494 tape
      Concerto, piano, no. 21, K 467, C major. Concerto, piano, no. 9,
      K 271, E flat major.
      Concerto, piano, no. 21, K 467, C major. cf Concerto, piano, no.
      20, K 466, D minor (DG 2530 548).
      Concerto, piano, no. 21, K 467, C major. cf Works, selections
      (Decca DPA 541/2).
      Concerto, piano, no. 21, K 467, C major. cf HMV SLS 5068.
1835 Concerto, piano, no. 23, K 488, A major. Symphony, no. 41, K 551,
      C major. Michael Roll, pno; London Mozart Players; Harry Blech.
      Abbey ABY 746.

            +-HFN 3-76 p101   +-RR 3-75 p30
            +-RR 2-76 p35

1836 Concerto, piano, no. 23, K 488, A major. Concerto, piano, no. 24,
      K 491, C minor. Wilhelm Kempff, pno; Bamberg Symphony Orches-
      tra; Ferdinand Leitner. DG 2535 204. (Reissue from SLPM 138645).
          +Gr 12-76 p996   +HFN 12-76 p151
     Concerto, piano, no. 23, K 488, A major. cf Concerto, piano, no.
      17, K 453, G major.
     Concerto, piano, no. 23, K 488, A major. cf Concerto, piano, no.
      20, K 466, D minor.
     Concerto, piano, no. 23, K 488, A major. cf Concerto, piano, no.
      21, K 467, C major (BASF 22477).
     Concerto, piano, no. 23, K 488, A major. cf Concerto, piano, no.
      21, K 467, C major (Decca 4340).
1837 Concerto, piano, no. 24, K 491, C minor. Concerto, 3 pianos,
      K 242, F major. Ingried Haebler, Ludwig Hoffmann, Sas Bunge,
      pno; LSO; Colin Davis, Alceo Galliera. Philips 6580 144.
      (Reissues from SAL 3642, AXL 12000/1-12).
          +-Gr 12-76 p1001   +HFN 11-76 p171
     Concerto, piano, no. 24, K 491, C minor. cf Concerto, piano, no.
      14, K 449, E flat major.
     Concerto, piano, no. 24, K 491, C minor. cf Concerto, piano, no.
      18, K 456, B flat major.
     Concerto, piano, no. 24, K 491, C minor. cf Concerto, piano, no.
      20, K 466, D minor (Bruno Walter Society BWS 723).
     Concerto, piano, no. 24, K 491, C minor. cf Concerto, piano, no.
      20, K 466, D minor (Turnabout 65046).
     Concerto, piano, no. 24, K 491, C minor. cf Concerto, piano, no.
      23, K 488, A major.
     Concerto, piano, no. 24, K 491, C minor: 2nd movement. cf Philips
      6580 114.
1838 Concerto, piano, no. 25, K 503, C major. Concerto, piano, no. 27,
      K 595, B flat major. Friedrich Gulda, pno; VPO; Claudio Abbado.
      DG 2530 642. Tape (c) 3300 642.
          +-Gr 7-76 p181   +-HFN 7-76 p91
          -Gr 10-76 p658 tape   +-RR 6-76 p45
1839 Concerto, piano, no. 25, K 503, C major. Fantasia, K 475, C minor.
      Ivan Moravec, pno; CPhO; Josef Vlach. Vanguard SU 11. (also
      Supraphon 110 1559).
          +Gr 5-76 p1762   ++NR 5-76 p6
          +NR 11-75 p6   ++St 2-76 p107
     Concerto, piano, no. 25, K 503, C major. cf Concerto, piano, no.
      14, K 449, E flat major.
     Concerto, piano, no. 26, K 537, D major. cf Concerto, piano, no.
      1, K 37, F major.
     Concerto, piano, no. 26, K 537, D major. cf Concerto, piano, no.
      8, K 246, C major.
     Concerto, piano, no. 26, K 537, D major. cf Concerto, piano, no.
      13, K 415, C major.
     Concerto, piano, no. 26, K 537, D major. cf Concerto, piano, no.
      17, K 453, G major.
     Concerto, piano, no. 27, K 595, B flat major. cf Concerto, piano,
      no. 5, K 175, D major.
     Concerto, piano, no. 27, K 595, B flat major. cf Concerto, piano,
      no. 10, K 365, E flat major.
     Concerto, piano, no. 27, K 595, B flat major. cf Concerto, piano,
      no. 12, K 414, A major.

Concerto, piano, no. 27, K 595, B flat major. cf Concerto, piano,
    no. 18, K 456, B flat major.

Concerto, piano, no. 27, K 595, B flat major. cf Concerto, piano,
    no. 25, K 503, C major.

1840 Concerto, 2 pianos, K 365, E flat major. Concerto, 3 pianos, K
    242, F major. Vladimir Ashkenazy, Daniel Barenboim, Fou Ts'ong,
    pno; ECO; Daniel Barenboim. Decca SXL 6716. (also London CS
    6937).

            -Gr 6-75 p45   +-RR 6-75 p43
            +-HFN 6-75 p89   +-SFC 2-1-76 p29
            +NR 3-76 p5

Concerto, 2 pianos, K 365, E flat major. cf Concerto, piano, no.
    21, K 467, C major.

Concerto, 3 pianos, K 242, F major. cf Concerto, piano, no. 24,
    K 491, C minor.

Concerto, 3 pianos, K 242, F major. cf Concerto, 2 pianos, K 365,
    E flat major.

Concerto, violin, no. 1, K 207, B flat major. cf Adagio, K 261,
    E major.

Concerto, violin, no. 1, K 207, B flat major. cf Works, selec-
    tions (Philips 6775 012).

1841 Concerto, violin, no. 2, K 211, D major. Concerto, violin, no. 4,
    K 218, D major. Hermann Krebbers, vln; Netherlands Chamber
    Orchestra; David Zinman. Philips 6580 120.

            ++Gr 6-76 p51   +RR 6-76 p46
            /HFN 8-76 p85

Concerto, violin, no. 2, K 211, D major. cf Works, selections
    (Philips 6775 012).

1842 Concerto, violin, no. 3, K 216, G major.* Concerto, violin, no.
    4, K 218, D major. Bronislaw Huberman, vln; VPO, Orchestra;
    Bruno Walter, Issay Dobrowen. Bruno Walter Society BWD 351.
    (*Reissue from Columbia M 258).

            +-NR 11-76 p6

Concerto, violin, no. 3, K 216, G major. cf BACH: Concerto,
    violin and strings, S 1041, A minor.

Concerto, violin, no. 3, K 216, G major. cf MENDELSSOHN: Concerto,
    violin, op. 64, E minor.

Concerto, violin, no. 3, K 216, G major. cf HMV RLS 718.

Concerto, violin, no. 4, K 218, D major. cf Concerto, violin, no.
    2, K 211, D major.

Concerto, violin, no. 4, K 218, D major. cf Concerto, violin,
    no. 3, K 216, G major.

Concerto, violin, no. 5, K 219, A major. cf BERG: Concerto, violin.

Concerto, violin, no. 5, K 219, A major. cf MENDELSSOHN: Concerto,
    violin, op. 64, E minor.

Concerto, violin, no. 5, K 219, A major. cf RCA ARM 4-0942/7.

Concerto, violin, no. 5, K 219, A major: Rondo; Tempo di minuetto.
    cf CBS 30063.

1843 Concerto, violin, no. 7, K 271a, D major. Concertone, 2 violins,
    K 190 (K 166b), C major. Josef Suk, Václav Snítil, vln; Prague
    Chamber Orchestra; Josef Suk. RCA LRL 1-5111.

            +Gr 4-76 p1604   +RR 4-76 p50
            +HFN 4-76 p109

Concerto, no. 7, violin, K 271a, D major. cf Concerto, violin,
    K Anh 294a, D major.

1844 Concerto, violin, K Anh 294a, D major. Concerto, violin, no. 7,

K 271a, D major. Yehudi Menuhin, vln; Menuhin Festival Orches-
tra; Yehudi Menuhin. HMV ASD 3198. (also Angel S 37167).
    +Gr 6-76 p51   +NR 10-76 p5
    +-HFN 6-76 p93   +-RR 6-76 p46
Concertone, 2 violins, K 190 (K166b), C major. cf Concerto, violin,
no. 7, K 271a, D major.
Contredances (Country dances), K 609 (5). cf DG Archive 2533 182.
Country dances, K 609 (5). cf Nonesuch H 71141.

1845 Così fan tutte, K 588. Elisabeth Schwarzkopf, Hanny Steffek, s;
Christa Ludwig, ms; Alfredo Kraus, t; Guiseppe Taddei, bar;
Walter Berry, bs; PhO and Chorus; Karl Böhm. Angel S 3631.
(also HMV SLS 5028).
    +-Gr 10-75 p679   +RR 11-75 p34
    +-HFN 11-75 p159   +St 4-75 p68
    +MT 1-76 p41

1846 Così fan tutte, K 588. Gundula Janowitz, Reri Grist, s; Brigitte
Fassbaender, ms; Peter Schreier, t; Hermann Prey, Rolando
Panerai, bar; Vienna State Opera Chorus; VPO; Karl Böhm. DG
2709 059 (3). Tape (c) 3371 019. (also 2740 118).
    +-Gr 10-75 p679   +-NR 1-76 p11
    +-Gr 12-76 p1066 tape   +ON 12-20-75 p38
    +-HF 2-76 p98   +-RR 10-75 p29
    +HF 12-76 p147 tape   +RR 12-76 p106 tape
    +-HFN 10-75 p145   -SR 1-24-76 p51
    +-MT 1-76 p41   +-St 4-76 p114

1847 Così fan tutte, K 588. Montserrat Caballé, Ileana Cotrubas, s;
Janet Baker, ms; Nicolai Gedda, t; Wladimiro Ganzarolli, bar;
Richard Van Allan, bs; ROHO; Colin Davis. Philips 6707 025 (4).
    +Gr 2-75 p1543   +ON 2-8-75 p33
    +-HF 3-75 p84   +-RR 2-75 p24
    ++MJ 3-75 p25   ++SFC 12-15-74 p33
    +-MT 1-76 p41   +St 4-75 p101
    +-NR 4-75 p8

1848 Così fan tutte, K 588. Elizabeth Schwarzkopf, Lisa Otto, s; Nan
Merriman, ms; Leopold Simoneau, t; Rolando Panerai, Sesto
Bruscantini, bar; PhO and Chorus; Herbert von Karajan. World
Records SOC 195/7 (3).
    +Gr 2-76 p1373   +RR 1-76 p25
    +-HFN 12-75 p157

1849 Così fan tutte, K 588: Bella vita militar...Di scrivermi ogni
giorno...Soave sia il vento; Smanie implacibile; Alla bella
Despinetta...Come scoglio...Non siate ritrosi...E voi ridete...
Un aura amorosa; Una donna a quindici anni...Prenderò quel
brunettino; Secondale, aurette; Il core vi dono; Per pietà;
Fra gli amplessi...Tutti accusan le donne; E nel tuo...Sani e
salve. Pilar Lorengar, Jane Berbié, s; Teresa Berganza, ms;
Ryland Davies, t; Tom Krause, Gabriel Bacquier, bar; ROHO
Chorus; LPO; Georg Solti. Decca SET 595. (Reissue from SET
575/8).
    +-Gr 3-76 p1507   +-RR 2-76 p22
    +HFN 3-76 p111
Così fan tutte, K 588: Un aura amorosa; Ah lo veggio; In qual
fiero. cf Arias (L'Oiseau-Lyre DSLO 13).
Così fan tutte, K 588: Overture. cf Works, selections (CBS 61022).
Così fan tutte, K 588: Overture. cf Works selections (HMV SLS
5048).

Così fan tutte, K 588: Overture. cf BEETHOVEN: Symphony, no. 5,
    op. 67, C minor.
Così fan tutte, K 588: Rivolgete a lui lo sguardo. cf Caprice
    CAP 1062.
Deutsche Tänze (German dances), K 605 (3). cf Concerto, piano,
    no. 20, K 466, D minor.
German dances, K 509 (6). cf Works, selections (CBS 76473).
German dance, no. 3, K 605. cf Works, selections (Decca DPA 541/2).
Divertimento, E flat major (arr. from Serenade, no. 10, K 361,
    B flat major). cf KARKOFF: Figures for wind instruments and
    percussion.
Divertimenti, K 439b (5). cf Adagio, K 411, B flat major.
Divertimenti, K 439b (5). cf Canzonetti, K 549.
1850 Divertimenti, nos. 1-17. New York Philomusica; A. Robert Johnson.
    Vox SVBX 5104/6 (9).
              ++HF 4-76 p91   ++NYT 8-8-76 pD13
              ++NR 2-76 p7   ++SFC 5-9-76 p38
1851 Divertimento, no. 1, K 136, D major. Quintet, clarinet, K 581,
    A major. Thea King, clt; Aeolian Quartet. Saga Tape (c)
    CA 5291.
              +-HFN 10-76 p185 tape   +RR 11-76 p109 tape
    Divertimento, no. 2, K 131, D major. cf Cassation, no. 2, K 99,
    B flat major.
1852 Divertimenti, no. 3, K 166, E flat major; no. 4, K 186, B flat
    major; K Anh 226, E flat major; K Anh 227, B flat major.
    Vienna Philharmonic Wind Ensemble. DG 2530 703. Tape (c)
    3300 703.
              ++Gr 7-76 p181   ++RR 7-76 p61
              +Gr 9-76 p497 tape   ++RR 11-76 p109 tape
              ++HFN 8-76 p85
1853 Divertimento, no. 11, K 251, D major. Symphony, no. 36, K 425,
    C major. London Mozart Players; Harry Blech. Abbey ABY 732.
              +-HFN 2-76 p103   +RR 2-76 p34
    Divertimento, no. 15, K 287, B flat major. cf BACH: Sonata,
    violin and harpsichord, A minor.
    Divertimento, no. 17, K 334, D major: Minuet. cf HMV HLM 7077.
    Divertimento, no. 17, K 334, D major: Minuet (2). cf RCA ARM
    4-0942/7.
1854 Divertimento, string trio, K 563, E flat major. Isaac Stern, vln;
    Pinchas Zukerman, vla; Leonard Rose, vlc. Columbia M 33266.
    (also CBS 76381).
              ++Gr 8-75 p342   +NYT 8-8-76 pD13
              +HF 8-75 p90   +RR 8-75 p52
              +HFN 8-75 p79   +-SFC 5-18-75 p23
              +-NR 6-75 p8   +St 8-75 p101
1855 Don Giovanni, K 527. Luise Helletsgruber, Elisabeth Rethberg,
    Margit Bokor, s; Dino Borgioli, t; Karl Etti, bar; Ezio Pinza,
    Herbert Alsen. Virgilio Lazzari, bs; Orchestra; Bruno Walter.
    Bruno Walter Society BWS 802 (3).
              +NR 12-76 p10
1856 Don Giovanni, K 527. Birgit Nilsson, Leontyne Price, Eugenia
    Ratti, s; Cesare Valletti, t; Heinz Blankenburg, bar; Cesare
    Siepi, Fernando Corena, Arnold van Mill, bs; Vienna State
    Opera Chorus; VPO; Erich Leinsdorf. Decca D10D 4 (4).
    (Reissue from RCA SER 4528/31).
              -Gr 12-76 p1046   +RR 12-76 p44

1857 Don Giovanni, K 527. Elisabeth Schwarzkopf, Irmgard Seefried,
      Ljuba Welitsch, s; Anton Dermota, t; Tito Gobbi, Erich Kunz,
      bar; Josef Greindl, Alfred Poell, bs; VPO; Wilhelm Furtwängler.
      Olympic 9109/4.
            +-NR 1-76 p10
1858 Don Giovanni, K 527: Overture. The magic flute, K 620: Overture.
      The marriage of Figaro, K 492: Overture. Symphony, no. 35,
      K 385, D major. Mozart Festival Orchestra; Antonia Brico.
      Columbia M 33888. (also CBS 61692).
            +-Audio 10-76 p151   ++NR 5-76 p3
            +-Gr 8-76 p292   +RR 6-76 p45
            -HF 5-76 p88   +SFC 2-1-76 p29
            +HFN 6-76 p91   +-SR 5-29-76 p52
            +-MJ 5-76 p28   ++St 8-76 p97
1859 Don Giovanni, K 527: Overture; Notte giorno; Madamina, il catalogo;
      La ci darem la mano; Or sai chi l'onore; Finch'han dal vino;
      Deh vieni; Sola, sola; Il mio tesoro; In quali accessi; Ah
      dove è il perfido. Shiela Armstrong, Rachel Mathes, Ann Murray,
      s; Robert Tear, t; John Shirley-Quirk, bar; Stafford Dean, Don
      Garrard, bs; Scottish Chamber Orchestra; Alexander Gibson.
      Classics for Pleasure CFP 40246.
            +-Gr 6-76 p83   +-RR 6-76 p30
            +-HFN 6-76 p93
      Don Giovanni, K 527: Ah viens à ta fenêtre. cf Club 99-101.
      Don Giovanni, K 527: In quali accessi...Mi tradi quell'alma in-
      grata; Crudele, Ah no, mio bene...Non mi dir. cf Arias (RCA
      SER 5675).
      Don Giovanni, K 527: La ci darem. cf Rubini GV 65.
      Don Giovanni, K 527: Madamina. cf Works, selections (CBS 30088).
      Don Giovanni, K 527: Madamina. cf Hungaroton SLPX 11444.
      Don Giovanni, K 527: Madamina, il catalogo è questo; Ah pietà,
      Signori miei; Finch'han dal vino; Deh vieni all finestra;
      Metà di voi quà vadano. cf Arias (Telefunken AG 6-41946).
      Don Giovanni, K 527: Il mio tesoro. cf Decca SKL 5208.
      Don Giovanni, K 527: Il mio tesoro; Dalla sua pace. cf Arias
      (L'Oiseau-Lyre DSLO 13).
      Don Giovanni, K 527: Overture. cf Symphony, no. 41, K 551, C major.
      Don Giovanni, K 527: Overture. cf Works, selections (HMV SLS 5048).
      Don Giovanni, K 527: Overture. cf Pye PCNHX 6.
      Don Giovanni, K 527: Tränen vom Freund getrocknert. cf Court Opera
      Classics CO 354.
      Duos, 2 horns, K 487 (12). cf Canzonetti, K 549.
1860 Die Entführung aus dem Serail, K 384 (The abduction from the
      seraglio). Rolf Boysen, speaker; Erika Köth, Lotte Schädle, s;
      Fritz Wunderlich, Friedrich Lenz, t; Kurt Böhme, bs; Bavarian
      State Opera Orchestra and Chorus; Eugen Jochum. DG 2726 051 (2).
      (Reissue from SLPM 139 213/5).
            +-Gr 11-76 p868   +RR 11-76 p41
            +HFN 11-76 p173
      Die Entführung aus dem Serail, K 384. cf Fidelio, op. 72.
1861 Die Entführung aus dem Serail, K 384, excerpts. Margaret Price,
      Danièle Perriers, s; Ryland Davies, Kimmo Lapaleinen, t; Noel
      Mangin, bs; LPO: John Pritchard. Classics for Pleasure CFP
      40032. (also Vanguard VSD 71203).
            +Gr 12-72 p1197   +St 1-76 p109
            +-HFN 12-72 p2443

Die Entführung aus dem Serail, K 384: Hier soll ich dich; Wenn der
Freude; Constanze; Ich baue ganz. cf Arias (L'Oiseau-Lyre
DSLO 13).
Die Entführung aus dem Serail, K 384: Martern aller Arten. cf
Arias (RCA SER 5675).
Die Entführung aus dem Serail, K 384: Martern aller Arten. cf
Decca GOSE 677/79.
Die Entführung aus dem Serail, K 384: Overture. cf Works,
selections (HMV SLS 5048).
The abduction from the seraglio, K 384: Wer ein Liebchen hat
gefunden; Ha, wie will ich trimphieren. cf Hungaroton SLPX
11444.
1862 Exsultate jubilate, K 165. Mass, no. 16, K 317, C major. Mass,
no. 18, K 427, C minor: Et incarnatus est. Maria Stader, s
Oralia Dominguez, con; Ernst Häfliger, t; Michel Roux, bs;
Elizabeth Brasseur Chorale; Lamoureux Orchestra, Berlin Radio
Symphony Orchestra; Igor Markevitch, Ferenc Fricsay. DG
2535 148. (Reissues from 133 222, 136 291).
     +Gr 2-76 p1368   +-RR 1-76 p59
     +-HFN 2-76 p117   +-RR 7-76 p82 tape
Exsultate jubilate, K 165. cf Works, selections (CBS 30088).
Exsultate jubilate, K 165. cf BERLIOZ: La mort d'Ophelie, op. 18,
no. 2.
Fantasia, K 396, C minor. cf BACH: Fantasia S 906, C minor.
1863 Fantasia, K 397, D minor. Fantasia, K 475, C minor. Rondo, piano,
K 511, A minor. Sonata, piano, no. 14, K 457, C minor.
Claudio Arrau, pno. Philips 6500 782.
     +Gr 11-75 p863   ++NR 1-76 p6
     +-HF 3-76 p91   +-RR 11-75 p79
     ++HFN 11-75 p157   +-SFC 2-15-76 p38
     +MJ 1-76 p25
Fantasia, K 397, D minor. cf HAYDN: Andante and variations, F
minor.
1864 Fantasia, K 475, C minor. Sonata, pinao, no. 14, K 457, C minor.
Sonata, piano, no. 16, K 570, B flat major. Sonata, piano,
no. 19, K 576, D major. Glenn Gould, pno. Columbia M 33515.
     +-HF 11-75 p108   +-St 3-76 p118
     +NR 11-75 p12
Fantasia, K 475, C minor. cf Concerto, piano, no. 25, K 503,
C major.
Fantasia, K 475, C minor. cf Fantasia, K 397, D minor.
Fantasia, K 594, F minor. cf Piano works (4 hands) (Peerless
PRCM 213/2).
Fantasia, K 594, F minor. cf Argo 5BBA 1013-5.
Fantasia, K 608, F minor. cf Piano works (4 hands) (Peerless
PRCM 213/2).
La finta semplice, K 51: Overture. cf Overtures (HMV SXLP 30213).
1865 Galimathias musicum, K 32. March, K 189, D major. Serenade,
no. 3, K 185, D major. Jürgen Pilz, vln; Hans Otto, hpd;
Dresden Philharmonic Orchestra; Günther Herbig. Philips 6500
702.
     +Gr 5-76 p1765   +RR 5-76 p47
     +HFN 5-76 p103
Idomeneo, Ré di Creta, K 366: Fuor del mar. cf Arias (L'Oiseau-
Lyre DSLO 13).
Idomeneo, Ré di Creta, K 366: Parto, el l'unico oggetto. cf

Arias (RCA SER 5675).

1866 Idomeneo, Rê di Creta, K 367: Ballet music. Les petits riens,
     K Anh 10. Netherlands Chamber Orchestra: David Zinman.
     Philips 6500 681.
          +Gr 3-76 p1468   -MJ 10-76 p24
          +HF 8-76 p92   +NR 4-76 p3
          +HFN 3-76 p101   +RR 3-76 p45

     Eine kleine Gigue, K 574, G major. cf HAYDN: Sonata, piano,
     no. 46, A flat major.
     Eine kleine Gigue, K 574, G major. cf International Piano Library
     IPL 5003/4.
     Ländler, K 606 (6). cf DG Archive 2533 182.
     Ländler, K 606 (6). cf Saga 5411.

1867 Lucio Silla, K 135. Arleen Auger, Julia Varady, Edith Mathis,
     Helen Donath, s; Peter Schreier, Werner Krenn, t; Salzburg
     Radio and Mozarteum Chorus; Salzburg Mozarteum Orchestra;
     Leopold Hager. BASF 78 22472/4 (4).
          +-Gr 1-76 p1236   ++RR 1-76 p25
          ++HFN 2-76 p107   +SLT 1-11-76 p36
          +-MT 7-76 p577

     Lucio Silla, K 135: Overture. cf Overtures (HMV SXLP 30213).
     Manner süchen stets zu naschen, K 433. cf Arias (Telefunken
     AG 6-41946).
     March, K 189, D major. cf Galimathias musicum, K 32.

1868 March, K 215, D major. Serenade, no. 5, K 213a (K 204), D major.
     Uto Ughi, vln; Dresden Staatskapelle; Edo de Waart. Philips
     6500 967.
          ++Gr 3-76 p1467   +-RR 3-76 p46
          +HFN 3-76 p101

     March, K 215, D major. cf Serenade, no. 5, K 213a (K 204),
     D major.

1869 March, K 237 (189c), D major. Serenade, no. 4, K 203, D major.
     Uto Ughi, vln; Dresden State Orchestra; Edo de Waart. Philips
     6500 965.
          +Gr 3-76 p1467   +NR 5-76 p3
          +-HF 6-76 p90   +NYT 8-8-76 pD13
          +HFN 2-76 p103   +RR 2-76 p34
          +MJ 5-76 p28

     March, K 237, D major. cf Serenade, no. 5, K 213a (K 204), D
     major.

1870 March, K 249, D major. Serenade, no. 7, K 250, D major. Uto
     Ughi, vln; Dresden Staatskapelle; Edo de Waart. Philips
     6500 966.
          +Gr 6-76 p50   +NYT 8-8-76 pD13
          +HF 11-76 p122   +-RR 6-76 p45
          +HFN 6-76 p91   ++SFC 8-15-76 p38
          ++NR 11-76 p2   +STL 6-6-76 p37

     March, K 335, no. 1, D major. cf Symphony, no. 32, K 318, G
     major.
     Marches, K 335, no. 10 and 11, D major. cf Serenade, no. 9,
     K 320, D major.
     Marches, K 408 (3). cf Works, selections (CBS 76473).

1871 March, K 408 (K 385a), no. 1, C major. Serenade, no. 13, K 525,
     G major. Symphony, no. 36, K 425, C major. VPO; István
     Kertesz. Decca SDD 480. (Reissue from SXL 6091).
          +-Gr 6-76 p50   +RR 4-76 p50
          +HFN 4-76 p121

March, K 408 (K 385a), no. 2, D major. cf Symphony, no. 35,
    K 385, D major.
Masonic funeral music, K 477. cf Works, selections (CBS 61022).
Masonic funeral music, K 477. cf Works, selections (HMV SLS 5048).
1872 Mass, no. 4, K 139, C minor. Celestina Casapietra, s; Annelies
      Burmeister, con; Peter Schreier, t; Hermann Christian Polster,
      bs; Walter Heinz Bernstein, org; Leipzig Radio Symphony
      Orchestra and Chorus; Herbert Kegel. Philips 6500 867.
          +Gr 1-76 p1225   +NR 3-76 p7
1873 Mass, no. 6, K 192, F major (Missa brevis). Mass, no. 13, K 259,
      C major. Celestina Casapietra, s; Annelies Burmeister, con;
      Peter Schreier, t; Hermann Christian Polster, bs; Walter Heinz
      Bernstein, org; Leipzig Radio Symphony Orchestra and Chorus;
      Herbert Kegel. Philips 6500 867.
          +-Gr 11-75 p881   ++MT 3-76 p235
          ++HF 6-76 p88   +-ON 3-6-76 p42
          +HFN 10-75 p145   ++RR 10-75 p86
          +-MJ 10-76 p24
      Mass, no. 7, K 167, C major. cf HAYDN: Mass, no. 5, B flat major.
      Mass, no. 13, K 259, C major. cf Mass, no. 6, K 192, F major.
      Mass, no. 16, K 317, C major. cf Exsultate jubilate, K 165.
      Mass, no. 16, K 317, C major: Gloria. cf Abbey LPB 754.
1874 Mass, no. 18, K 427, C minor. Ileana Cotrubas, Kiri Te Kanawa,
      s; Werner Krenn, t; Hans Sotin, bs; John Alldis Choir; NPhO;
      Raymond Leppard. HMV ASD 2959. (also Seraphim S 60257).
          +Gr 6-74 p92   +-ON 3-6-76 p42
          -HF 10-76 p114   +RR 5-74 p64
          +MJ 10-76 p24   ++SFC 2-1-76 p29
          ++NR 3-76 p7   /St 9-76 p122
1875 Mass, no. 18, K 427, C minor. Pro Musica Orchestra and Chorus;
      Ferdinand Grossmann. Turnabout 34174. Tape (c) KTVC 34174.
          +-HFN 9-75 p110 tape   +-RR 2-76 p70 tape
1876 Mass, no. 18, K 427, C minor. Carole Bogard, s; Ann Murray, ms;
      Richard Lewis, t; Michael Rippon, bs; Amor Artis Chorale; ECO;
      Johannes Somary. Vanguard VSD 71210.
          +Gr 6-76 p72   -NR 5-76 p8
          -HF 10-76 p114   +-RR 6-76 p79
          +-HFN 8-76 p85   +-St 9-76 p122
          +MJ 10-76 p24
      Mass, no. 18, K 427, C minor: Et incarnatus est. cf Exsultate
      jubilate, K 165.
1877 Mass, no. 19, K 626, D minor. Anna Tomowa-Sintow, s; Agnes
      Baltsa, con; Werner Krenn, t; Jose van Dam, bs; Vienna Singer-
      verein; BPhO: Herbert von Karajan. DG 2530 705.
          +Gr 12-76 p1034   +RR 11-76 p97
          +HFN 11-76 p164
1878 Mass, no. 19, K 626, D minor. Hans Buchhierl, treble; Mario
      Krämer, alto, Werner Krenn, t; Barry McDaniel, bar; Tölzer
      Boys' Choir; Collegium Aureum; Gerhard Schmidt-Gaden. Harmonia
      Mundi KHB 22006.
          +St 11-76 p148
1879 Mass, no. 19, K 626, D minor. Helen Donath, s; Yvonne Minton, con;
      Ryland Davies, t; Gerd Nienstedt, bs; John Alldis Choir; BBC
      Symphony Orchestra; Colin Davis. Philips Tape (c) 7300 393.
          +RR 12-76 p106 tape
1880 Mass, no. 19, K 626, D minor. Elly Ameling, s; Barbara Scherler,

alto; Louis Devos, t; Roger Soyer, bs; Gulbenkian Foundation
Symphony Orchestra and Chorus; Michel Corbóz. RCA AGL 1-1533.
(also Erato STU 70943).
+—Gr 6-76 p72   +—NR 8-76 p10
+MJ 10-76 p24   -RR 6-76 p79
+—MU 10-76 p16   -St 11-76 p148
1881 Mass, no. 19, K 626, D minor. Carole Bogard, s; Ann Murray, ms;
      Richard Lewis, t; Michael Rippon, bs; Amor Artis Chorale; ECO;
      Johannes Sommary. Vanguard VSD 71211.
      ++NR 10-76 p9   ++St 11-76 p148
      Minuet, K 355, D major. cf BACH: The well-tempered clavier,
      Bk II, S 870-893: Prelude, no. 22.
      Minuet, K 355, D major. cf HAYDN: Sonata, piano, no. 46, A flat
      major.
      Minuet, K 409, C major. cf Works, selections (CBS 76473).
      Minuets, K 599 (6). cf Works, selections (CBS 76473).
      Nocturnes, K 346, 436-439. cf Canzonetti, K 549.
1882 Le nozze di Figaro, K 492. Esther Rethy, Aulikki Rautawaara,
      Jarmila Novotna, Angelica Cravcenco, Dora Komarek, s; William
      Wernigk, Giuseppe Nessi, t; Mariano Stabile, bar; Ezio Pinza,
      Virgilio Lazzari, Viktor Madin, bs; Salzburg Festival Orchestra;
      Bruno Walter. Bruno Walter Society RR 801 (3).
      ++NR 10-76 p11
      Le nozze di Figaro, K 492: Bravo, signor padrone...Se vuol ballare;
      Non più andrai; Tutto è disposto...Aprite un po'quegli occhi;
      Hai già vinta la causa...Vedrò, mentr'io sospiro. cf Arias
      (Telefunken 6-41946).
      La nozze di Figaro, K 492: Crudel, perchè finora. cf Rubini GV 65.
      Le nozze di Figaro, K 492: Giunse alfin il momento...Deh vieni
      non tardar; E Susanna no vien...Dove sono; Voi che sapete. cf
      Arias (RCA SER 5675).
      Le nozze di Figaro, K 492: Non so più. cf CBS 76476.
      The marriage of Figaro, K 492: Overture. cf Concerto, horn and
      strings, no. 4, K 495, E flat major.
      The marriage of Figaro, K 492: Overture. cf Don Giovanni, K 527:
      Overture.
      The marriage of Figaro, K 492: Overture. cf Symphony, no. 41,
      K 551, C major.
      Le nozze di Figaro, K 492: Overture. cf Works, selections (CBS
      61022).
      Le nozze di Figaro, K 492: Overture. cf Works, selections (Decca
      DPA 541/2).
      Le nozze di Figaro, K 492: Overture. cf Works, selections (HMV
      SLS 5048).
      The marriage of Figaro, K 492: Overture. cf Creative Record
      Service R 9115.
      Le nozze di Figaro, K 492: Overture. cf Decca SPA 409.
      Le nozze di Figaro, K 492: Porgi amor. cf HMV RLS 719.
      The marriage of Figaro, K 492: Se vuol ballare. cf Works, selec-
      tions (CBS 30088).
      Le nozze di Figaro, K 492: Soli einst das Gräflein. cf Rococo
      5379.
      Le nozze di Figaro, K 492: Voi che sapete. cf Decca SPA 449.
      Le nozze di Figaro, K 492: Voi che sapete. cf Philips 6747 204.
1883 Overtures: La finta semplice, K 51. Lucio Silla, K 135. Il Re
      pastore, K 208. Der Schauspieldirektor, K 486. Ballet music:

Les petits riens, K Anh 10. AMF; Neville Marriner. HMV SXLP
30213. (Reissue from ASD 2834).
    +Gr 9-76 p423   +-RR 8-76 p47
    +HFN 8-76 p93
Les petits riens, K Anh 10. cf Idomeneo, Ré di Creta, K 367:
Ballet music.
Les petits riens, K Anh 10. cf Overtures (HMV SXLP 30213).
Les petits riens, K Anh 10. cf Symphony, no. 31, K 297, D major.
1884 Piano works (4 hands): Andante and variations, K 501, G major.
Fantasia, K 608, F minor. Fantasia, K 594, F minor. Fugue,
K 401, G minor. Sonatas, K 358, B flat major; K 381, D major;
K 497, F major; K 521, C major. Jean and Kenneth Wentworth,
pnos. Peerless PRCM 213/2 (2).
    -Gr 1-76 p1218   +-HFN 1-76 p115
Prelude on "Ave verum", K 580a. cf Works, selections (CBS 30088).
1885 Quartets, flute, K 285, D major; K 285a, G major; K 285b, C major;
K 298, A major. Grumiaux Trio; William Bennett, flt. Philips
6500 034. Tape (c) 7300 401.
    +HF 5-72 p106   ++SFC 2-3-74 p26
    ++HF 3-76 p116 tape   ++St 8-72 p76
    ++NR 4-72 p10
1886 Quartets, flute, K 285, D major; K 285a, G major; K 285b, C major;
K 298, A major. Michel Debost, flt; Trio à Cordes Français.
Seraphim S 60246.
    ++HF 2-76 p98   ++St 12-75 p129
    ++NR 10-75 p5
1887 Quartet, oboe, K 370, F major. Quintet, clarinet, K 581, A major.
George Pieterson, clt; Koji Toyoda, Arthur Grumiaux, vln;
Pierre Pierlot, ob; Max Lesueur, vla; János Scholz, vlc.
Philips 6500 924. Tape (c) 7300 414.
    +-Gr 1-76 p1215   ++HFN 2-76 p116 tape
    ++HF 2-76 p100   +-NR 12-75 p9
    +HF 6-76 p97 tape   ++NYT 8-8-76 pD13
    +-HFN 2-76 p105   +-RR 1-76 p42
1888 Quartet, oboe, K 370, F major. Quintet, horn, K 407, E flat major.
Trio, clarinet, K 498, E flat major. Endres Quartet. Turnabout TV 34035. Tape (c) KTVC 34035.
    +-HFN 1-76 p125 tape   +RR 7-76 p82 tape
Quartet, oboe, K 370, F major. cf LOEFFLER: Rhapsodies, oboe,
viola and piano (2).
1889 Quartets, strings, nos. 20-23. Juilliard Quartet. Columbia MG
33976 (2). (also CBS 79204).
    +Gr 10-76 p621   +RR 10-76 p81
    ++HFN 10-76 p175   ++SFC 7-18-76 p31
    +NR 7-76 p6   +-St 9-76 p122
    +NYT 8-8-76 pD13
1890 Quartet, strings, no. 20, K 499, D major. Quartet, strings, no.
21, K 575, D major. Alban Berg Quartet. Telefunken AW
6-41999.
    +Gr 9-76 p437   +RR 8-76 p64
    ++HFN 9-76 p126
1891 Quartet, strings, no. 21, K 575, D major. Quartet, strings,
no. 23, K 590, F major. Küchl Quartet. Decca SDD 509. Tape
(c) KSDC 509.
    +Gr 11-76 p830   +-HFN 12-76 p155
    +-HFN 10-76 p175   ++RR 11-76 p86

Quartet, strings, no. 21, K 575, D major. cf Quartet, strings,
   no. 20, K 499, D major.
Quartet, strings, no. 23, K 590, F major. cf Quartet, strings,
   no. 21, K 575, D major.
Quintet, clarinet, K 581, A major. cf Concerto, clarinet, K 622,
   A major.
Quintet, clarinet, K 581, A major. cf Divertimento, no. 1, K 136,
   D major.
Quintet, clarinet, K 581, A major. cf Quartet, oboe, K 370, F
   major.
Quintet, clarinet, K 581, A major. cf HMV SLS 5046.
Quintet, horn, K 407, E flat major. cf Quartet, oboe, K 370, F
   major.
1892 Quintets, strings, nos. 1-6, K 174, K 406 (K 516b), K 515, K 516,
   K 593, K 614. Arthur Grumiaux, Arpad Gerécz, vln; Georges
   Janzer, Max Lesueur, vla; Eva Czako, vlc. Philips 6747 107 (3).
      ++Audio 12-76 p91   +RR 1-76 p42
      +Gr 1-76 p1212
1893 Quintet, strings, no. 1, K 174, B flat major. Quintet, strings,
   no. 3, K 515, C major. Grumiaux Trio; Arpad Gerécz, vln;
   Max Lesueur, vla. Philips 6500 619.
      +Audio 4-76 p88   ++NR 8-75 p8
      +HF 11-75 p106   +-NYT 8-8-76 pD13
      +HFN 2-76 p107   +SFC 11-23-75 p26
Quintet, strings, no. 2, K 406, C minor. cf Quintet, strings,
   no. 4, K 516, G minor.
1894 Quintet, strings, no. 3, K 515, C major. Quintet, strings, no. 4,
   K 516, G minor. Griller Quartet; William Primrose, vla. Bach
   Guild HM 29. (Reissue from Philips SGL 5841).
      +Gr 2-76 p1355   +-RR 10-75 p67
      +-HFN 10-75 p153
1895 Quintets, strings, no. 3, K 515, C major; no. 4, K 516, G minor;
   no. 5, K 593, D major; no. 6, K 614, E flat major. William
   Primrose, vla; Griller Quartet. Vanguard GRT Tape E 8184-194,
   15801/2.
      +HF 6-76 p97 tape
Quintet, strings, no. 3, K 515, C major. cf Quintet, strings,
   no. 1, K 174, B flat major.
1896 Quintet, strings, no. 4, K 516, G minor. Quintet, strings, no. 2,
   K 406, C minor. Grumiaux Quintet. Philips 6500 620.
      ++HF 6-76 p90   ++NR 3-76 p5
      +MJ 10-76 p24   +-NYT 8-8-76 pD13
Quintet, strings, no. 4, K 516, G minor. cf Quintet, strings, no. 3,
   K 515, C major.
1897 Quintet, strings, no. 5, K 593, D major. Quintet, strings, no.
   6, K 614, E flat major. Grumiaux Trio; Arpad Gerécz, vln;
   Max Lesueur, vla. Philips 6500 621.
      +MJ 12-76 p44   +-NYT 8-8-76 pD13
      ++NR 9-76 p7
Quintet, strings, no. 5, K 593, D major. cf Quintets (Vanguard
   GRT Tape 8184-194).
Quintet, strings, no. 6, K 614, E flat major. cf Quintet, strings,
   no. 5, K 593, D major.
Quintet, strings, no. 6, K 614, E flat major. cf Quintets
   (Vanguard GRT Tape 8184-194).
1898 Il Re pastore, K 208. Edith Mathis, Arleen Auger, Sona Ghazarian,

s; Peter Schreier, Werner Krenn, t; Salzburg Mozarteum
Orchestra; Leopold Hager. BASF KBL 22043 (3). (also BAC
3072/4).
    +Gr 8-75 p359   +OC 2-76 p4
    +HF 6-75 p100   +ON 4-5-75 p40
    +HFN 8-75 p79   +RR 8-75 p22
    +-MT 3-76 p235   ++St 6-75 p102
    +-NR 6-75 p12

1899 Il Re pastore, K 208, K 208. Reri Grist, Lucia Popp, Arlene
    Saunders, s; Nicola Monti, Luigi Alva, t; Denis Vaughan, hpd;
    Naples Orchestra; Denis Vaughan. RCA PVL 2-9086 (2).
    (Reissue from SER 5567/8).
        +Gr 12-76 p1046
    Il Re pastore, K 208: L'amerò sarò costante. cf Arias (RCA SER
    5675).
    Il Re pastore, K 208: L'amerò, sarò costante. cf BIS LP 45.
    Il Re pastore, K 208: L'amerò, sarò costante. cf HMV RLS 719.
    Il Re pastore, K 208: Overture. cf Overtures (HMV SXLP 30213).
    Rondo, piano, K 382, D major. cf Concerti, piano, nos. 1-6,
    8-9, 11-27.
    Rondo, piano, K 382, D major. cf Concerto, piano, no. 5, K 175,
    D major.
    Rondo, piano, K 382, D major. cf Concerti, piano, nos. 21-27.
    Rondo, piano, K 382, D major. cf BEETHOVEN: Concerto, piano,
    no. 3, op. 37, C minor.
    Rondo, piano, K 386, A major. cf BEETHOVEN: Concerto, piano,
    no. 3, op. 37, C minor.
    Rondo, piano, K 511, A minor. cf Concerti, piano, nos. 17, 20-
    21, 23-24.
    Rondo, piano, K 511, A minor. cf Fantasia, K 397, D minor.
    Rondo, piano, K 511, A minor. cf HAYDN: Sonata, piano, no. 46,
    A flat major.
    Rondo, violin, K 269, B flat major. cf Adagio, K 261, E major.
    Rondo, violin, K 373, C major. cf Adagio, K 261, E major.
    Rondo, violin, K 373, C major. cf CBS 30066.

1900 Die Schauspieldirektor, K 486. Lo sposo deluso, K 430. Ruth
    Welting, Felicity Palmer, Ileana Cotrubas, s; Anthony Rolfe,
    Robert Tear, t; Clifford Grant, bar; LSO; Colin Davis.
    Philips 9500 011.
        +-ARG 11-76 p29   +-ON 12-4-76 p60
        +-MJ 12-76 p28   +SFC 9-12-76 p31
        +NR 12-76 p11
    Der Schauspieldirektor, K 486: Overture. cf Overtures (HMV
    SXLP 30213).
    Der Schauspieldirektor, K 486: Overture. cf Works, selections
    (CBS 61022).
    Serenade, D major: Rondo. cf RCA ARM 4-0942/7.
    Serenade, no. 3, K 185, D major. cf Galimathias musicum, K 32.
    Serenade, no. 3, K 185, D major: Andante, allegro. cf Works,
    selections (Philips 6775 012).
    Serenade, no. 4, K 203, D major. cf March, K 237, D major.
    Serenade, no. 4, K 203, D major: Andante, menuetto, allegro. cf
    Works, selections (Philips 6775 012).

1901 Serenade, no. 5, K 213a (K204), D major. Marches, K 215, D major;
    K 237, D major. Pinchas Zukerman, vln; ECO; Pinchas Zukerman.
    CBS 76489. Tape (c) 40-76489.

```
              ++Gr 5-76 p1765              +-HFN 8-76 p95
              +-HFN 6-76 p91              +RR 5-76 p47
```
Seranade no. 5, K 213a (K 204), D major. cf March, K 215, D
     major.

Serenade, no. 5, K 213a (K 204), D major: Andante moderato,
     menuetto, allegro. cf Works, selections (Philips 6775 012).

Serenade, no. 6, K 239, D major. cf Adagio and fugue, K 546,
     C major.

1902 Serenade, no. 7, K 250, D major. Bavarian Radio Symphony Orch-
     estra; Rafael Kubelik. DG 2535 139. Tape (c) 3335 139.
     (Reissue from SLPM 138 869).
```
              +-Gr 4-76 p1604             +-RR 3-76 p46
              +HFN 3-76 p109              -RR 7-76 p82 tape
```
Serenade, no. 7, K 250, D major. cf March, K 249, D major.

Serenade, no. 7, K 250, D major: Rondo. cf London STS 15239.

1903 Serenade, no. 9, K 320, D major. Marches, K 335, nos. 10 and 11,
     D major. Dresden State Orchestra; Edo de Waart. Philips
     6500 627.
```
              +Gr 4-75 p1808              +RR 4-75 p29
              +-HF 9-75 p87               +SFC 5-18-75 p23
              +NR 7-75 p2                 +STL 5-4-75 p37
              +NYT 8-8-76 pD13
```
1904 Serenade, no. 10, 13 wind instruments, K 361, B flat major.
     Sydney Symphony Orchestra, Soloists; Willem van Otterloo.
     RCA GL 25015.
```
              +Gr 10-76 p602             +-RR 10-76 p62
              +HFN 12-76 p145
```
Serenade, no. 10, 13 wind instruments, K 361, B flat major. cf
     Vanguard VSD 707/8.

1905 Serenade, no. 11, K 375, E flat major. Serenade, no. 12, K 388,
     C minor. New London Wind Ensemble. Classics for Pleasure
     CFP 40211.
```
              ++Gr 11-75 p819             +RR 6-75 p44
              +HFN 2-76 p103
```
Serenade, no. 11, K 375, E flat major. cf BEETHOVEN: Trio, piano,
     clarinet and violoncello, op. 38, E flat major.

Serenade, no. 11, K 375, B flat major. cf GOUNOD: Petite sym-
     phonie, 9 wind instruments.

Serenade, no. 12, K 388, C minor. cf Serenade, no. 11, K 375,
     E flat major.

1906 Serenade, no. 13, K 525, G major (Eine kleine Nachtmusik).
     Symphony no. 32, K 318, D major. Sinfonia concertante, K 364,
     E flat major. Alan Loveday, vln; Stephen Shingles, vla; AMF;
     Neville Marriner. Argo ZRG 679. Tape (c) KZRC 679.
```
              ++ARG 9-72 p663            ++HFN 3-72 p513
              ++HF 11-72 p88             +NR 5-72 p663
              ++HF 12-76 p147 tape       ++St 1-73 p110
```
1907 Serenade, no. 13, K 525, G major. SAINT-SAENS: The carnival of
     the animals.* Alfons and Aloys Kontarsky, pno; VPO; Karl
     Böhm. DG 2530 731. (*Reissue from 2530 588).
```
              +Gr 12-76 p1071            +-RR 12-76 p64
```
Serenade, no. 13, K 525, G major. cf Concerto, piano, no. 20,
     K 466, D minor.

Serenade, no. 13, K 525, G major. cf March, K 408, no. 1, C major.

Serenade, no. 13, K 525, G major. cf Symphony, no. 41, K 551,
     C major.

Serenade, no. 13, K 525, G major. cf Works, selections (CBS 61022).

Serenade, no. 13, K 525, G major. cf Works, selections (Decca DPA
541/2).

Serenade, no. 13, K 525, G major. cf Works, selections (HMV SLS
5048).

Serenade, no. 13, K 525, G major. cf BEETHOVEN: Egmont, op. 84:
Overture.

Sinfonia concertante, K 297b, E flat major. cf Concerto, flute
and harp, K 299, C major.

Sinfonia concertante, K 297b, E flat major. cf Concerto, oboe,
K 314, C major.

Sinfonia concertante, K 297b, E flat major. cf Symphony, no. 30,
K 202, D major.

Sinfonia concertante, K 364, E flat major. cf Serenade, no. 13,
K 525, G major.

1908 Sonatas, flute and harpsichord, nos. 1-6, K 10-K 15. Thomas
Brandis, vln; Karl-Heinz Zöller, flt; Waldemar Döling, hpd;
Wolfgang Boettcher, vlc. DG Archive 2533 135.
        ++HF 12-75 p101   ++SFC 7-18-76 p31
        ++NR 12-75 p9

1909 Sonatas, organ and orchestra, nos. 1-17. Daniel Chorzempa, org;
German Bach Soloists; Helmut Winschermann. Philips 6700 061
(2).
        +Gr 5-76 p1761   +NR 10-74 p15
        +HF 2-75 p98   ++RR 8-76 p71
        +HFN 5-76 p103   ++SFC 1-26-75 p27
        +MJ 11-74 p48   ++St 4-75 p102

1910 Sonatas, piano (17). Christoph Eschenbach, pno. DG 2720 031 (7).
        +HFN 6-76 p102

1911 Sonata, piano, no. 8, K 310, A minor. Sonata, piano, no. 14,
K 457, C minor. Sonata, piano, no. 15, K 545, C major. Christoph Eschenbach, pno. DG 2530 234. (Reissue from 2720 031).
        +HFN 6-72 p1115   +-St 3-76 p119
        +NR 1-76 p6

Sonata, piano, no. 9, K 311, D major. cf Adagio, K 540, B minor.

Sonata, piano, no. 9, K 311, D major. cf HAYDN: Andante and
variations, F minor.

Sonata, piano, no. 10, K 330, C major. cf HAYDN: Andante and
variations, F minor.

1912 Sonata, piano, no. 11, K 331, A major. Sonata, piano, no. 12,
K 332, F major. Valentina Kameniková, pno. Supraphon 111 1417.
        +HFN 8-75 p79   +-RR 9-75 p60
        -NR 1-76 p6

Sonata, piano, no. 11, K 331, A major. cf Adagio, K 540, B minor.

Sonata, piano, no. 11, K 331, A major: Rondo alla turca. cf
CBS 73487.

Sonata, piano, no. 11, K 331, A major: Rondo alla turca. cf
RCA VH 020.

Sonata, piano, no. 12, K 332, F major. cf Sonata, piano, no. 11,
K 331, A major.

Sonata, piano, no. 13, K 333, B flat major. cf Adagio, K 540,
B minor (Orion ORS 75185).

Sonata, piano, no. 13, K 333, B flat major. cf Adagio, K 540,
B minor (Philips 9500 025).

Sonata, piano, no. 14, K 457, C minor. cf Fantasia, K 397, D
minor.

Sonata, piano, no. 14, K 457, C minor. cf Fantasia, K 475, C
   minor.
Sonata, piano, no. 14, K 457, C minor. cf Sonata, piano, no. 8,
   K 310, A minor.
Sonata, piano, no. 15, K 545, C major. cf Sonata, piano, no. 8,
   K 310, A minor.
Sonata, piano, no. 15, K 545, C major. cf Connoisseur Society
   (Q) CSQ 2066.
Sonata, piano, no. 16, K 570, B flat major. cf Fantasia, K 475,
   C minor.
Sonata, piano, no. 17, K 576, D major. cf Fantasia, K 475, C
   minor.
Sonata, piano, no. 17, K 576, D major. cf BACH: Fantasia, S 906,
   C minor.
Sonata, piano, op. posth., D 960, B flat major. cf Fantasia,
   op. 15, D 760, C major.
Sonata, 2 pianos, K 358, B flat major. cf Piano works (4 hands)
   (Peerless PRCM 213/2).
Sonata, 2 pianos, K 381, D major. cf Piano works (4 hands).
Sonata, 2 pianos, K 448 (375a), D major. cf BRAHMS: Variations
   on a theme by Haydn, op. 56a.
Sonata, 2 pianos, K 497, F major. cf Piano works (4 hands).
Sonata, 2 pianos, K 521, C major. cf Piano works (4 hands).
1913 Sonata, violin and piano, no. 19, K 302, E flat major. Sonata,
   violin and piano, no. 20, K 303, C major. Sonata, violin and
   piano, K 547, F major. Variations on "La bergére Célimène",
   K 359, G major (12). Henryk Szeryng, vln; Ingrid Haebler, pno.
   Philips 6500 145.
         ++Gr 3-75 p1674   +NR 7-75 p7
         +NR 1-76 p6   +RR 3-75 p51
1914 Sonata, violin and piano, no. 20, K 303, C major. Sonata, violin
   and piano, no. 23, K 306, D major. Sonata, violin and piano,
   no. 24, K 376, F major. Sonata, violin and piano, no. 32,
   K 454, B flat major. Sonata, violin and piano, no. 33, K 481,
   E flat major. Radu Lupu, pno; Szymon Goldberg, vln. London
   CSA 2243 (2). (also Decca 13BB 207-12)
         ++ARG 12-76 p37   ++St 11-76 p89
         +MT 1-76 p42
Sonata, violin and piano, no. 20, K 303, C major. cf Sonata,
   violin and piano, no. 19, K 302, E flat major.
Sonata, violin and piano, no. 23, K 306, D major. cf Sonata,
   violin and piano, no. 20, K 303, C major.
Sonata, violin and piano, no. 24, K 376, F major. cf Sonata,
   violin and piano, no. 20, K 303, C major.
Sonata, violin and piano, no. 32, K 454, B flat major. cf Sonata,
   violin and piano, no. 20, K 303, C major.
Sonata, violin and piano, no. 33, K 481, E flat major. cf Sonata,
   violin and piano, no. 20, K 303, C major.
Sonata, violin and piano, K 547, F major. cf Sonata, violin and
   piano, no. 19, K 302, E flat major.
1915 Songs: An Chloe, K 524; An die Freude, K 53; Die betrogene Welt,
   K 474; Dans un bois solitaire, K 308; Ich würd auf meinem Pfad,
   K 390; Die ihr des unermesslichen Weltalls Schöpfer ehrt,
   K 619; Komm, liebe Zither, K 351; Lied beim Auszug in das Feld,
   K 552; Lied der Freiheit, K 506; Das Lied der Trennung, K 519;
   Lied zur Gesellenreise, K 468; Lobgesang auf die feierliche
   Johannisloge, K 148; Das Traumbild, K 530; Wie unglücklich bin

ich, K 147; Verdankt sei es dem Glanz der Grossen, K 392;
Warnung, K 433; Die Zufriedenheit, K 473. Hermann Prey, bar;
Takashi Ochi, mand; Bernard Klee, pno. DG 2530 724.
      +Gr 10-76 p639
1916 Songs (Secular and sacred canons and songs): Alleluia, K 553; Auf
      das Wohl aller Freunde, K 508; Ave Maria, K 554; Das Bandel,
      K 441; Bona nox, K 561; Canonic adagio, K 410; Caro bell'idol
      mio, K 562; Difficile lectu, K 559; Essen Trinken, K 234; Gehn
      wir in'n Prater, K 558; G'rechtelt's enk, K 556; Heiterkeit
      und Leichtes Blut, K 507; Kyrie a cinque con diversi canoni,
      K 89; Lacrimoso son'io, K 555; Lasst froh uns sein, K 231;
      Lieber Freistädtler, K 232; Nascosa è il mio sol, K 557; Nichts
      labt mich mehr als Wein, K 233; Nun, liebes Weibchen, K 625;
      O du eselhafter Martin, K 560; Sie, sie ist dahin, K 229;
      V'amo di core, K 348; Wo der perlende Wein, K 347. Berlin
      Soloists, Members; Dietrich Knothe. Philips 6500 917.
          +-Gr 2-76 p1367   +NR 7-76 p8
          ++HF 9-76 p93   +NYT 8-8-76 pD13
          +HFN 2-76 p109   +RR 2-76 p60
          +MJ 10-76 p24   +SFC 9-12-76 p31
      Songs: Abendempfindung, K 523. cf HMV SEOM 25.
      Songs: Due pupille amabile, K 439; Luci care, luci belle, K 346;
      Mi lagnero tacendo, K 437; Piu'non si trovano, K 459. cf RCA
      PRL 1-9034.
      Lo sposo deluso, K 430. cf Die Schauspieldirektor, K 486.
1917 Symphonies, nos. 13-16. AMF; Neville Marriner. Argo ZRG 594.
      Tape (c) KZRC 594.
          +HFN 7-76 p104 tape   -RR 6-76 p88 tape
1918 Symphony, no. 18, K 130, F major. Symphony, no. 19, K 132, E
      flat major. Symphony, no. 24, K 182, B flat major. Mainz
      Chamber Orchestra; Günter Kehr. Turnabout TV 34038. Tape (c)
      KTVC 34038.
          +HFN 7-76 p104 tape
      Symphony, no. 19, K 132, E flat major. cf Symphony, no. 18,
      K 130, F major.
1919 Symphonies, nos. 20-21, 23, 25. Netherlands Philharmonic Orches-
      tra; Otto Ackerman. Discophilia OAA 105.
          +NR 11-76 p3
      Symphony, no. 24, K 182, B flat major. cf Symphony, no. 18,
      K 130, F major.
      Symphonies, nos. 25, 29, 31, 33-36, 38-41. cf Works, selections
      (HMV SLS 5048).
1920 Symphony, no. 25, K 183, G minor. Symphony, no. 29, K 201, A
      major. AMF; Neville Marriner. Argo ZRG 706. Tape (c) ZRC 706.
          +-Gr 4-72 p1716   +HFN 12-72 p2455 tape
          +-Gr 12-72 p1214 tape   +MJ 4-73 p34
          ++HF 5-73 p86   ++NR 5-73 p3
          ++HF 12-76 p147 tape   ++RR 12-72 p107 tape
          ++HFN 5-72 p925   ++SFC 8-12-73 p34
1921 Symphony, no. 26, K 184, E flat major. Thamos, King of Egypt,
      K 345: Incidental music. Karin Eickstädt, s; Gisela Pohl, con;
      Eberhard Büchner, t; Theo Adam, bs; Berlin Radio Chorus; Berlin
      State Orchestra; Bernhard Klee. Philips 6500 840.
          +-Audio 3-76 p63   +NR 1-76 p12
          +/Gr 2-76 p1345   +-ON 3-6-76 p42
          +HF 1-76 p91   +RR 2-76 p34
          +HFN 2-76 p105   ++St 1-76 p76

1922 Symphony, no. 29, K 201, A major. Symphony, no. 33, K 319, B
      flat major. BPhO; Herbert von Karajan. DG 2535 155. Tape (c)
      3335 155.
              +HFN 3-76 p109   +-RR 7-76 p82 tape
              +-RR 3-76 p46
1923 Symphony, no. 29, K 201, A major. Symphony, no. 41, K 551, C
      major. Hallé Orchestra; John Barbirolli. Pye GSGC 15028.
              +-HFN 4-76 p121   +RR 4-76 p49
      Symphony, no. 29, K 201, A major. cf Symphony, no. 25, K 183, G
      minor.
1924 Symphony, no. 30, K 202, D major. Sinfonia concertante, K 297b,
      E flat major. Winterthur Symphony Orchestra; Fritz Rieger.
      Pelca PSR 40007.
              +NR 12-76 p4
1925 Symphony, no. 31, K 297, D major. Symphony, no. 36, K 425, C
      major. Les petits riens, K Anh 10. Bavarian Radio Symphony
      Orchestra; Ferdinand Leitner. DG Tape (c) 3348 220.
              +RR 9-76 p92 tape
1926 Symphony, no. 32, K 318, G major. March, K 335, no. 1, D major.
      Zaide, K 344. Edith Mathis, s; Peter Schreier, Werner Hollweg,
      Armin Ude, Joachim Vogt, Wolfgang Wagner, Gunther Koch, t;
      Ingvar Wixell, Reiner Suss, bs; Friederike Aust, Gerd Grasse,
      Wolfgang Dehler, Manfred Wagner, speakers; Berlin State Orch-
      estra; Bernhard Klee. Philips 6700 097 (2).
              +Gr 10-76 p644   +NR 12-76 p11
      Symphony, no. 32, K 318, G major. cf Serenade, no. 13, K 525,
      G major.
1927 Symphony, no. 33, K 319, B flat major. Symphony, no. 40, K 550,
      G minor. Collegium Aureum; Franzjosef Maier. BASF BAC 3086.
              +Gr 4-76 p1607   +-RR 3-76 p47
              +-HFN 3-76 p99
      Symphony, no. 33, K 319, B flat major. cf Symphony, no. 29, K
      201, A major.
1928 Symphony, no. 35, K 385, D major. Symphony, no. 41, K 551, C
      major. CO; Georg Szell. Columbia MS 6969.
              +-SFC 7-18-76 p31
1929 Symphony, no. 35, K 385, D major. Symphony, no. 41, K 551, C
      major. LPO; Adrian Boult. HMV (Q) ASD 3158. Tape (c) TC ASD
      3158.
              +-Gr 3-76 p1467   +-RR 2-76 p35
              +HFN 3-76 p99   +-RR 7-76 p83 tape
              +HFN 5-76 p117 tape
1930 Symphony, no. 35, K 385, D major. Symphony, no. 40, K 550, G
      minor. March, K 408 (K 385a), no. 2, D major. AMF; Neville
      Marriner. Philips 6500 162. Tape (c) 7300 086. (Reissue
      from Philips 6707 013).
              +HF 12-76 p147 tape   ++NR 7-75 p5
              +HFN 6-72 p1113   ++SFC 7-13-75 p21
1931 Symphony, no. 35, K 385, D major. Symphony, no. 41, K 551, C
      major. COA; Josef Krips. Philips 6500 429. Tape (c) 7300 270.
              +-Gr 4-73 p1876   +-RR 4-73 p61
              +-HFN 4-73 p783   +-RR 2-76 p71 tape
              +-HFN 9-75 p110 tape   +SFC 3-2-75 p25
              ++NR 3-75 p2
      Symphony, no. 35, K 385, D major. cf Don Giovanni, K 527: Over-
      ture.

1932 Symphony, no. 36, K 425, C major. Symphony, no. 41, K 551, C
       major. NYP; Leonard Bernstein. Columbia M 30444.
              +-SFC 7-18-76 p31
       Symphony, no. 36, K 425, C major. cf Divertimento, no. 11, K
       251, D major.
       Symphony, no. 36, K 425, C major. cf March, K 408, no. 1, C
       major.
       Symphony, no. 36, K 425, C major. cf Symphony, no. 31, K 297,
       D major.
1933 Symphony, no. 40, K 550, G minor. Symphony, no. 41, K 551, C
       major. LPO; Charles Mackerras. Classics for Pleasure CFP
       40253. Tape (c) TC CFP 40253.
              +-Gr 11-76 p798   +HFN 12-76 p153
              +-HFN 10-76 p173   +-RR 10-76 p62
1934 Symphony, no. 40, K 550, G minor. Symphony, no. 41, K 551, C
       major. NPhO; Carlo Maria Giulini. Decca SXL 6225. Tape (c)
       KSXC 6225. (also London 6479).
              +-Gr 8-73 p388 tape   +SFC 7-18-76 p31
              +-HFN 6-73 p1189 tape
1935 Symphony, no. 40, K 550, G minor. Symphony, no. 41, K 551, C
       major. VSO; Ferenc Fricsay. DG Tape (c) 3318 036.
              -HFN 2-76 p117 tape
       Symphony, no. 40, K 550, G minor. cf Concerto, horn and strings,
       no. 4, K 495, E flat major.
       Symphony, no. 40, K 550, G minor. cf Symphony, no. 33, K 319,
       B flat major.
       Symphony, no. 40, K 550, G minor. cf Symphony, no. 35, K 385,
       D major.
       Symphony, no. 40, K 550, G minor. cf Works, selections (Decca
       DPA 541/2).
       Symphony, no. 40, K 550, G minor. cf BEETHOVEN: Symphony, no. 5,
       op. 67, C minor (Decca DPA 527/8).
       Symphony, no. 40, K 550, G minor. cf BEETHOVEN: Symphony, no. 5,
       op. 67, C minor (Philips 7317 133).
       Symphony, no. 40, K 550, G minor: 1st movement. cf Works,
       selections (CBS 30088).
       Symphony, no. 40, K 550, G minor: 1st movement. cf Philips
       6580 114.
1936 Symphony, no. 41, K 551, C major. Serenade, no. 13, K 525, G
       major. ECO; Daniel Barenboim. Angel S 36761. Tape (c)
       4SX 36761.
              +-NR 11-76 p2
1937 Symphony, no. 41, K 551, C major. SCHUBERT: Symphony, no. 8,
       D 759, B minor. BSO; Eugen Jochum. DG 2530 357. Tape (c)
       3300 318 (ct) 89468.
              +HFN 2-76 p117 tape   +SFC 7-18-76 p31
1938 Symphony, no. 41, K 551, C major. Overtures: Don Giovanni, K 527;
       The magic flute, K 620; The marriage of Fidago, K 492. Vienna
       State Opera Orchestra; Felix Prohaska. Vanguard S 167. Tape
       (ct) 8184-167.
              +SFC 7-18-76 p31
       Symphony, no. 41, K 551, C major. cf Concerto, piano, no. 23,
       K 488, A major.
       Symphony, no. 41, K 551, C major. cf Symphony, no. 29, K 201,
       A major.
       Symphony, no. 41, K 551, C major. cf Symphony, no. 35, K 385,
       D major (Columbia MS 6969).

Symphony, no. 41, K 551, C major. cf Symphony, no. 35, K 385,
    D major (HMV 3158).
Symphony, no. 41, K 551, C major. cf Symphony, no. 35, K 385,
    D major (Philips 6500 429).
Symphony, no. 41, K 551, C major. cf Symphony, no. 36, K 425,
    C major.
Symphony, no. 41, K 551, C major. cf Symphony, no. 40, K 550,
    G minor (Classics for Pleasure CFP 40253).
Symphony, no. 41, K 551, C major. cf Symphony, no. 40, K 550,
    G minor (Decca SXL 6225).
Symphony, no. 41, K 551, C major. cf Symphony, no. 40, K 550,
    G minor (DG Tape (c) 3318 036).
Thamos, King of Egypt, K 345: Incidental music. cf Symphony,
    no. 26, K 184, E flat major.
Trio, clarinet, K 498, E flat major. cf Quartet, oboe, K 370,
    F major.
Trio, clarinet, K 498, E flat major. cf BRUCH: Trios, op. 83,
    nos. 2, 6, 7.
Variations on a minuet by Duport, K 573. cf BACH: Fantasia, S
    906, C minor.
Variations on "Ah, vous dirai-je, Maman, K 225", C major. cf
    BACH: The well-tempered clavier, Bk II, S 870-893: Prelude, no.
    22.
Variations on "La bergère Célimène", K 359, G major (12). cf
    Sonata, violin and piano, no. 19, K 302, E flat major.
Variations on "Salve tu Domine", K 398, F major. cf Adagio,
    K 540, B minor.
1939 Works, selections: Concerto, horn and strings, no. 4, K 495, E
    flat major.* Exsultate jubilate, K 165. Don Giovanni, K 527:
    Madamina. The marriage of Figaro, K 492: Se vuol ballare.
    Prelude on "Ave verum", K 580a.* Symphony, no. 40, K 550, G
    minor: 1st movement. Judith Raskin, s; Ezio Pinza, bs; E.
    Power Biggs, org; Mason Jones, hn; Marlboro Festival Orchestra,
    CO, Metropolitan Opera Orchestra, PO; Pablo Casals, Georg
    Szell, Bruno Walter, Eugene Ormandy. CBS 30088. (*Reissues
    from SBRG 72477, 61095).
        +Gr 12-76 p1071   +RR 12-76 p62
1940 Works, selections: Masonic funeral music, K 477. Overtures:
    Così fan tutte, K 588; Le nozze di Figaro, K 492; Der Schaus-
    pieldirektor, K 486; Die Zauberflöte, K 620. Serenade, no. 13,
    K 525, G major. Columbia Symphony Orchestra; Bruno Walter.
    CBS 61022. (Reissue from SBRG 72043).
        +Gr 11-76 p798   +-RR 11-76 p69
        +-HFN 11-76 p170
1941 Works, selections: German dances, K 509 (6). Marches, K 408 (3).
    Minuet, K 409, C major. Minuets, K 599 (6). Die Zauberflöte,
    K 620: March of the priests. LSO; Erich Leinsdorf. CBS 76473.
    Tape (c) 40-76473.
        +Gr 4-76 p1604   +-RR 2-76 p32
        +-HFN 2-76 p105   +-RR 7-76 p83 tape
        -HFN 5-76 p117 tape
1942 Works, selections: Concerto, clarinet, K 622, A major. Concerto,
    flute, no. 2, K 314, D major. Concerto, flute and harp, K 299,
    C major. Concerto, horn and strings, no. 2, K 417, E flat major.
    Concerto, horn and strings, no. 4, K 495, E flat major. Alfred
    Prinz, clt; Werner Tripp, Claude Monteux, flt; Hubert Jellinek,

hp; Barry Tuckwell, hn; VPO, LSO; Karl Münchinger, Peter Maag,
Pierre Monteux. Decca DPA 521/2 (2). (Reissues from SXL 6054,
6112, 6108).
    +-Gr 4-76 p1604   +-RR 3-76 p45
    +HFN 3-76 p111
1943 Works, selections: Concerto, clarinet, K 622, A major. Concerto,
piano, no. 21, K 467, C major. German dance, no. 3, K 605.
Le nozze di Figaro, K 492: Overture. Serenade, no. 13, K 525,
G major. Symphony, no. 40, K 550, G minor. Ilana Vered, pno;
Alfred Prinz, clt; VPO, Vienna Mozart Ensemble, LPO, NPhO;
Erich Kleiber, István Kertész, Willi Boskovsky, Karl Münchinger,
Uri Segal, Carlo Maria Giulini. Decca DPA 541/2 (2).
    +Gr 9-76 p483   +HFN 11-76 p175 tape
    +HFN 10-76 p183   +-RR 7-76 p61
1944 Works, selections: Adagio and fugue, K 546, C minor. Così fan
tutte, K 588: Overture. La clemenza di Tito, K 621: Overture.
Don Giovanni, K 527: Overture. Die Entführung aus dem Serail,
K 384: Overture. Masonic funeral music, K 477. Le nozze di
Figaro, K 492: Overture. Serenade, no. 13, K 525, G major.
Symphonies, nos. 25, 29, 31, 33-36, 38-41. Die Zauberflöte,
K 620: Overture. PhO, NPhO; Otto Kelmperer. HMV SLS 5048 (6).
(Reissues from 33CX 1438, SAX 2587, 2436, SAN 137/9, SAX 5252,
2278, 5256, 2546, 2436, 2468, 2278, 2486).
    ++Gr 8-76 p292   +-RR 8-76 p47
    +HFN 8-76 p93
1945 Works, selections: Cassation, no. 1, K 63, G major: Adagio.
Serenade, no. 4, K 203, D major: Andante, menuetto, allegro.
Serenade, no. 3, K 185, D major: Andante, allegro. Serenade,
no. 5, K 213a (K 204), D major: Andante moderato, menuetto,
allegro. Concerto, violin, no. 1, K 207, B flat major. Concerto, violin, no. 2, K 211, D major. Jaap Schröder, vln;
Amsterdam Mozart Ensemble; Frans Brüggen. Philips 6775 012 (2).
    /Gr 7-75 p191   -MT 1-76 p41
    +HFN 7-75 p85   +RR 7-75 p32
1946 Zaide, K 344. Edith Mathis, s; Peter Schreier, Armin Ude, Werner
Hollweg, t; Ingvar Wixell, bar; Reiner Süss, bs; Berlin State
Orchestra; Bernhard Klee. Philips 6700 097 (2).
    +ARG 11-76 p31   +RR 10-76 p28
    +HFN 10-76 p173   ++SFC 9-12-76 p31
    +-MJ 12-76 p28   ++St 12-76 p145
    +ON 12-4-76 p60
Zaide, K 344. cf Symphony, no. 32, K 318, G major.
Zaide, K 344: Wer hungrig bei der Tafel sitzt; Ihr Mächtigen seht
ungerührt. cf Arias (Telefunken AG 6-41946).
1947 Die Zauberflöte (The magic flute), K 620 (sung in Swedish, film
soundtrack). Birgit Nordin, Irma Urrila, Elisabeth Eriksson,
Britt Marie Aruhn, s; Kirsten Vaupel, Birgitta Smiding, ms;
Josef Köstlinger, Ragnar Ulfung, Gösta Prüzelius, t; Erik
Saeden, Hakan Hagegard, bar; Ulrik Cold, Ulf Johanson, bs; Urban
Malmberg, Erland van Heijne, Ansgar Krook, treble; Swedish Radio
Chorus and Symphony Orchestra; Eric Ericson. BBC REK 223 (3).
    /Gr 1-76 p1236
1948 Die Zauberflöte, K 620. Wilma Lipp, Sena Jurinac, Friedl Riegler,
Irmgard Seefried, Hermine Steinmassl, s; Else Schürhof, Eleonore
Dörpinghans, ms; Annelies Stückl, con; Anton Dermota, Peter
Klein, Erich Majkut, t; Erich Kunz, bar; George London, bs-bar;

Harald Pröglhöf, Ljubomir Pantscheff, Ludwig Weber, bs; Vienna
Singverein; VPO; Herbert von Karajan. HMV SLS 5052 (3). (Re-
issue from Columbia 33CS 1013/5).
    +-Gr 6-76 p83   +RR 10-76 p28
    +HFN 7-76 p91
Die Zauberflöte, K 620: Dies Bildnis ist bezaubernd schön. cf
    Decca SKL 5208.
Die Zauberflöte, K 620: Dies Bildnis ist bezaubernd schön. cf
    Rubini GV 70.
Die Zauberflöte, K 620: Dies Bildnis ist bezaubernd schön; Wie
    stark ist nicht dein Zauberton. cf Arias (L'Oiseau-Lyre DSLO
    13).
Die Zauberflöte, K 620: In diesen heil'gen Hallen. cf Preiser LV
    192.
Die Zauberflöte, K 620: In diesen heil'gen Hallen. cf Rococo
    5397.
Die Zauberflöte, K 620: March of the priests. cf Works, selections
    (CBS 76473).
Die Zauberflöte, K 620: O ewige Nacht...Wie stark ist nicht dein
    Zauberton. cf Court Opera Classics CO 354.
Die Zauberflöte, K 620: O Isis und Osiris; In diesen heil'gen
    Hallen. cf Hungaroton SLPX 11444.
The magic flute, K 620: Overture. cf Don Giovanni, K 527: Over-
    ture.
The magic flute, K 620: Overture. cf Symphony, no. 41, K 551,
    C major.
Die Zauberflöte, K 620: Overture. cf Works, selections (CBS 61022).
Die Zauberflöte, K 620: Overture. cf Works, selections (HMV SLS
    5048).
Die Zauberflöte, K 620: Overture. cf Philips 6747 204.
Die Zauberflöte, K 620: Der Vogelfanger; Dies Bildnis. cf Decca
    GOSE 677/79.
Die Zauberflöte, K 620: Der Vogelfanger bin ich ja; Ein Mädchen
    oder Weibchen. cf Arias (Telefunken AG 6-41946).
MUDARRA, Alonso de
    Diferencias sobre El Conde claros. cf MILAN: Pavanas, nos. 1-6.
    Dulces exuviae. cf L'Oiseau-Lyre 12BB 203/7.
    Fantasia. cf Angel S 36093.
    Fantasia, no. 10. cf Saga 5426.
    Fantasia que contrahaza la harpa en la manera de Ludovico. cf
        MILAN: Pavanas, nos. 1-6.
    Gallarda. cf MILAN: Pavanas, nos. 1-6.
    Gallarda. cf Angel S 36093.
    O guárdame las vacas. cf MILAN: Pavanas, nos. 1-6.
    Pavana de Alexandre. cf MILAN: Pavanas, nos. 1-6.
    Songs: Claros y frescos rios; Isabel, perdiste la tua faxa; Triste
        estaba el Rey David. cf DG 2530 504.
MUDGETT, Mark
    Colonial songs. cf DELLA JOIO: Satiric dances.
    Musical history of the national anthem. cf DELLA JOIO: Satiric
        dances.
MUFFAT, George
1949 Apparatus musica: Organasticus. Leena Jacobson, org. Musical
        Heritage Society 3074/6 (3).
        +-Audio 4-76 p90
    Toccata and variations (6). cf Pelca PRS 40577.

MUHLBERGER
     Mir sein die Kaiserjäger.   cf DG 2721 077.
MULET, Henri
     Byzantian sketches: Wienachtslied.   cf BACH: Toccata and fugue,
          S 565, D minor.
     Carillon-Sortie.   cf Grosvenor GRS 1041.
     Du bist der Fels.   cf BACH: Toccata and fugue, S 565, D minor.
     Kirchenfenster.   cf BACH: Toccata and fugue, S 565, D minor.
     Tu es Petra.   cf Wealden WS 149.
MULLEN
     The naggletons.   cf Transatlantic XTRA 1159.
MULLER-ZURICH, Paul
     Capriccio.   cf DUTILLEUX: Sonatine, flute and piano.
MUMMA, Gordon
     Cybersonic cantilevers.   cf Folkways FTS 33901/4.
MUNDY, John
     Robin.   cf National Trust NT 002.
     Sing joyfully.   cf Argo ZRG 789.
     Were I a king.   cf Enigma VAR 1017.
MUNRO
     My lovely Celia.   cf HMV HLM 7093.
MURADELI, Vano
     October: Choruses (2).   cf HMV ASD 3200.
MURRAY
     Songs: I'll walk beside you.   cf Argo ZFB 95/6.
MURRAY, Sunny
     Evocations.   cf Folkways FTS 33901/4.
MURSCHHAUSER, Franz
     Variations super cantilenam.   cf Pelca PRSRK 41017/20.
MURTULA, Giovanni
     Tarantella.   cf DG 2530 561.
MUSET, Colin
     Quant je voi yver retorner.   cf Candide CE 31095.
     Quant je voy yver.   cf Nonesuch H 71326.
MUSGRAVE, Thea
     Chamber concerto, no. 2.   cf MEKEEL: Corridors of dream.
     Concerto, clarinet.   cf BANKS: Concerto, horn.
1950 Concerto, horn.   Concerto, orchestra.   Barry Tuckwell, hn; Keith
          Pearson, clt; Scottish National Orchestra; Thea Musgrave,
          Alexander Gibson.   Decca HEAD 8.   (also London HEAD 8).
               +Gr 6-75 p46   ++RR 5-75 p36
               +-HFN 6-75 p89   +SR 1-24-76 p53
               ++MJ 3-76 p24   ++STL 6-8-76 p36
               ++MT 10-75 p886   ++Te 3-76 p31
               ++NR 2-76 p6
     Concerto, orchestra.   cf Concerto, horn.
1951 Night music, chamber orchestra.   RIEGGER: Dichotomy, chamber
          orchestra, op. 12.   SESSIONS: Rhapsody, orchestra.   Symphony,
          no. 8.   London Sinfonietta, NPhO; Frederick Prausnitz.   Argo
          ZRG 702.
               ++Audio 4-76 p90   +NR 10-74 p3
               +Gr 5-73 p2051   +RR 5-73 p25
               +HF 11-74 p97   +SFC 10-6-74 p26
               +HFN 6-73 p1180   ++St 3-75 p106
               ++MQ 1-76 p139
     Primavera.   cf Caprice RIKS 59.

MUSIC FOR COMPUTERS, ELECTRONIC SOUND AND PLAYERS.   cf CRI SD 300.
MUSIC FOR EVENSONG.   cf Polydor 2460 250.
MUSIC FOR LUTE AND GAMBA.   cf BIS LP 22.
MUSIC FOR THE VYNE.   cf National Trust NT 002.
MUSIC FOR 2 GUITARS.   cf Saga 5412.
MUSIC FROM THE AGE OF JEFFERSON.   cf Smithsonian Collection N 002.
MUSIC OF MEDIEVAL PARIS.   cf Candide CE 31095.
MUSIC OF SPAIN, Zarzuela arias.   cf London OS 26435.
MUSIC OF THE TUDOR COURT.   cf Harmonia Mundi HMD 223.
MUSIC OF VENICE.   cf DG 2548 219.
MUSICKE OF SUNDRE KINDES: Renaissance secular music, 1480-1620.
    cf L'Oiseau-Lyre 12BB 203/7.
MUSSROGSKY, Modest
1952   Boris Godunov. Mark Reizen, M. Mikhailov, bs; Bolshoi Theatre
       Orchestra and Chorus; N. Golovanov. Bruno Walter Society RR
       440 (3).
            +-NR 10-76 p12
       Boris Godunov: Coronation scene; Prologue. cf Decca DPA 525/6.
       Boris Godunov: Fountain duet. cf Rubini GV 63.
       Boris Godunov: Introduction and polonaise. cf Works, selections
         (HMV Melodiya 3101).
       Boris Godunov: Mon coeur est triste, J'ai le pouvoir supreme;
         Scene du carillon, Oh je meurs...Ta soeu a grand besoin. cf
         Club 99-101.
1953   Boris Godunov: Prologue, coronation scene; Boris' monologue;
       Clock scene; Pimen's monologue; Varlaam's song; Farewell and
       death of Boris. Nikolai Ghiuselev, bs; Sofia Philharmonic
       Orchestra and Chorus; Russlan Raychev. Harmonia Mundi HMB 130.
            -Gr 8-76 p332
       By the water. cf RCA VH 020.
       Fair at Sorochinsk: Gopak. cf Connoisseur Society CS 2070.
       Fair at Sorochinsk: Introduction, Gopak. cf Works, selections
         (HMV Melodiya 3101).
       Fair at Sorochinsk: O banish thoughts of sorrow. cf Rubini GV 63.
       Intermezzo, B minor. cf Works, selections (HMV Melodiya 3101).
1954   Khovanschina (edit. Rimsky-Korsakov). Maria Dimchevska, Nadya
       Dobriyanova, s; Alexandrina Milcheva-Noneva, ms; Lyubomir Bod-
       urov, Lyuben Mikhailov, Milen Payunov, Dimiter Dimitrov, Verter
       Vrachovsky, t; Stoyan Popov, bar; Dimiter Petkov, Nikola Guyse-
       lev, bs; Svetoslav Obretenov Chorus; Sofia National Opera Orch-
       estra; Athanas Margaritov. CRD Harmonia Mundi HMB 4-124 (4).
       (also Monitor HS 90104/7 (4)).
            ++Audio 8-76 p75   +-NR 5-76 p10
            +-Gr 11-75 p899   +-RR 12-75 p28
            +-HFN 12-75 p159   ++SFC 5-16-76 p28
            ++MJ 5-76 p28   +St 7-76 p110
1955   Khovanschina. TCHAIKOVSKY: The sleeping beauty, op. 66: Suite.
       Swan Lake, op. 20: Suite. PhO; Herbert von Karajan. HMV SXLP
       30200. Tape (c) TC EXE 183.
            ++Gr 2-76 p1382   +HFN 6-76 p105
       Khovanschina: Dawn on the Moscow River; Dance of the Persian
         slaves; Galitzin's journey. cf Works, selections (HMV Melodiya
         3101).
       Khovanschina: Entr'acte, Act 5. cf Pye QS PCNH 4.
       Khovanschina; Marfa's prophecy. cf Columbia/Melodiya M 33931.
       Mlada: Triumphal march. cf Works, selections (HMV Melodiya 3101).

1956 A night on the bare mountain. Pictures at an exhibition. CSO;
      Fritz Reiner. Camden CCV 5038.
              +HFN 4-76 p121   +-RR 4-76 p50
1957 A night on the bare mountain. Pictures at an exhibition. St.
      Louis Symphony Orchestra; Leonard Slatkin. Turnabout TV 34633.
      Tape (c) CT 2109.
              +MJ 11-76 p45   +NR 8-76 p2
      A night on the bare mountain. cf Works, selections (HMV Melodiya
      3101).
      A night on the bare mountain. cf BORODIN: Prince Igor: Overture.
      A night on the bare mountain. cf BORODIN: Prince Igor: Polovtsian
      dances.
      A night on the bare mountain. cf DEBUSSY: Prelude to the after-
      noon of a faun.
      A night on the bare mountain (orch. Rimsky-Korsakov). cf Decca
      DPA 519/20.
      A night on the bare mountain. cf DG 2584 004.
1958 Pictures at an exhibition. Scherzo, B flat major (arr. Liapunov).
      Turkish march (arr. Chernov). Michel Beroff, pno. Angel S
      37223.
              +HF 12-76 p108   +NR 11-76 p13
1959 Pictures at an exhibition (orch. Ravel). PROKOFIEV: Concerto,
      piano, no. 3, op. 26, C major. Israela Margalit, pno; NPhO;
      Lorin Maazel. London SPC 20179. (also Decca PFS 4255 Tape
      (c) KPFC 4255).
              -Gr 3-73 p1692   +RR 3-73 p54
              /HF 5-73 p87   +-RR 7-76 p83 tape
              +NR 2-73 p5   +-SFC 3-11-73 p30
1960 Pictures at an exhibition (2 versions). LPO; Orchestra; John
      Pritchard, Ralph Burns. Stanyan 9016.
              -ARG 11-76 p34
      Pictures at an exhibition. cf A night on the bare mountain (Cam-
      den CCV 5038).
      Pictures at an exhibition. cf A night on the bare mountain
      (Turnabout TV 34633).
      Pictures at an exhibition (edit. Horowitz). cf BRAHMS: Pieces,
      piano, op. 117, no. 2, B flat minor.
      Pictures at an exhibition. cf CHOPIN: Etude, op. 10, no. 3,
      E major.
      Pictures at an exhibition. cf HANDEL: Royal fireworks music.
      Pictures at an exhibition. cf HMV SLS 5019.
      Scherzo, B flat major (arr. Liapunov). cf Pictures at an
      exhibition.
      Scherzo, B flat major. cf Works selections (HMV Melodiya 3101).
1961 Songs: Darling Savishna; Gopak; Eremushka lullaby; The orphan;
      Peasant's lullaby; Sunless cycle: Between four walls, Thou
      didst not know me in the crowd, The idle noisy day is ended,
      Boredom, Elegy, On the river; Where are you little star.
      SHOSTAKOVICH: Poems, op. 126: Song of Ophelia, Gamayun bird of
      prophecy, We were together, The city sleeps, Strom, Secret
      signs, Music. Satires, op. 109: To a critic, Taste of spring,
      Progeny, Misunderstanding, Kreutzer sonata. TCHAIKOVSKY: Again
      as before alone, op. 73, no. 6; Cradle song, op. 16, no. 1; Do
      not believe, my friend, op. 6, no. 1; The fearful minute, op.
      28, no. 6; If I'd only known, op. 47, no. 1; In this moonlight,
      op. 73, no. 3; It was in the early spring, op. 38, no. 2; Mid

the din of the ball, op. 38, no. 3; Sleep my poor friend, op.
47, no. 4; Was I not a little blade of grass, op. 47, no. 7;
Why, op. 6, no. 5. Galina Vishnevskaya, s; Mstislav Rostropovich, vlc; Olga Rostropovich, Vasso Devetzi, pno; Ulf Hoelscher,
vln. HMV SLS 5055 (3).
     +Gr 9-76 p465   +-RR 9-76 p87
1962 Songs: Eremushka lullaby. Songs and dances of death. Sunless.
Where are you little star. Bernard Kruysen, bar; Noël Lee,
pno. Telefunken AW 6-41998.
     +-Gr 9-76 p459   +RR 8-76 p75
     ++HFN 9-76 p126
Songs: Cradle song; The magpie; The night; On the Dniepr; The
ragamuffin; Where are you little star. cf Philips 6780 751.
Songs: Where are you little star. cf HMV SEOM 25.
1963 Songs and dances of death. RACHMANINOFF: Alfred de Musset, op.
21, no. 6, fragment; Child, thou art fair as a flower, op. 8,
no. 2; A dream, op. 8, no. 5; I wait for thee, op. 14, no. 1;
In the silent night, op. 4, no. 3; Lilacs, op. 21, no. 5; Oh,
never sing to me again, op. 4, no. 4. TCHAIKOVSKY: It was in
early spring, op. 38, no. 2; The lights in the rooms were fad-
ing, op. 63, no. 5; Reconciliation, op. 25, no. 1; Serenade,
op. 63, no. 6; Take my heart away; Why did I dream of you, op.
28, no. 3. Irina Arkhipova, ms; John Wustman, Semyon Stuchevsky, pno. HMV Melodiya ASD 3103.
     +Gr 9-76 p465   ++RR 9-76 p86
     +HFN 9-76 p126
Songs and dances of death. cf Songs (Telefunken AW 6-41998).
Sunless. cf Songs (Telefunken AW 6-41998).
Turkish march (arr. Chernov). cf Pictures at an exhibition.
1964 Works, selections: Boris Godunov: Introduction and polonaise (arr.
Rimsky-Korsakov). Fair at Sorochinsk: Introduction, Gopak.
Intermezzo, B minor (arr. Rimsky-Korsakov). Khovanschina:
Dawn on the Moscow River; Dance of the Persian slaves; Galitzin's
journey (arr. Rimsky-Korsakov). Mlada: Triimphal march (arr.
Rimsky-Korsakov). A night on the bare mountain (arr. Rimsky-
Korsakov). Scherzo, B flat major. USSR Symphony Orchestra;
Yevgeny Svetlanov. HMV Melodiya ASD 3101. (also Angel SR
40273).
     +-Gr 8-75 p326   +RR 8-75 p40
     +-HFN 8-75 p80   +-SFC 10-10-76 p32
     +NR 12-76 p4
MYNYDDOG
Myfanwy. cf Columbia SCX 6585.
NAPRAVNIK, Eduard
Dubrovsky: The forest guards its secret. cf Rubini GV 63.
Dubrovsky: Never to see her. cf Rubini RS 301.
Harold: Lullaby. cf Rubini RS 301.
NARDINI, Pietro
Sonata, violin, D major. cf LECLAIR: Sonata, violin, no. 3, D
minor.
NARVAEZ, Luis de
Baxa de contrapunto. cf MILAN: Pavanas, nos. 1-6.
Con qué la lavaré. cf DG 2530 504.
Diferencias sobre "Guárdame las vacas". cf MILAN: Pavanas, nos.
1-6.
Diferencias (variations) on "Guárdame las vacas". cf Angel S 36093.

Diferencias (variations) on "Guárdame las vacas". cf London
     STS 15306.
Diferencias sobre "Guárdame las vacas". cf Swedish Society SLT
     33189.
Fantasia. cf MILAN: Pavanas, nos. 1-6.
Fantasia. cf L'Oiseau-Lyre 12BB 203/6.
Mille regretz. cf MILAN: Pavanas, nos. 1-6.
Mille regretz. cf L'Oiseau-Lyre 12BB 203/6.

NAUMANN, Siegfried
     Riposte, flauto e percussione, no. 1. cf Caprice RIKS LP 34.

NAYLOR, Peter
     Now the green blade riseth. cf Vista VPS 1023.

NEGRI, Marc Antonio
     Balletto. cf DG Archive 2533 184.

NELHYBEL, Vaclav
     Arco and pizzicato. cf CASTELNUOVO-TEDESCO: Concertino, harp,
          op. 93.
     Miniatures, strings (3). cf CASTELNUOVO-TEDESCO: Concertino,
          harp, op. 93.
     Scherzo concertante. cf Crystal S 371.

NERUDA, Jan Křtitel
     Concerto, trumpet, E flat major. cf HAYDN: Concerto, trumpet,
          E flat major.

NEUENDORFF
     Songs: Der Rattenfänger; Wandern, ach wandern. cf Angel S 37108.

NEUSIDLER, Hans
     The Burgher of Nuremberg. cf Saga 5420.
     Ein guter Venezianer Tantz. cf Hungaroton SLPX 11721.
     Hie folget ein welscher Tantz Wascha Mesa, Der hupff auf. cf
          Hungaroton SLPX 11721.
     Der Judentanz (The Jew's dance). cf DG Archive 2533 302.
     The Jew's dance. cf Saga 5426.
     Der Juden Tantz, Der hupff auf zur Juden Tantz. cf Hungaroton
          SLPX 11721.
     Der polnisch Tantz, Der hupff auf. cf Hungaroton SLPX 11721.
     Preambel. cf DG Archive 2533 302.
     Welscher Tantz Wascha mesa. cf DG Archive 2533 302.
     Welscher Tanz, Wascha mesa: Hupfauff. cf DG Archive 2533 184.

NEVILLE
     Shrewsbury fair. cf Pye TB 3007.

NEVIN
     The rosary. cf Prelude PRS 2505.
     NEW AMERICAN MUSIC. cf Folkways FTS 33901/4.

NICHOLSON, Richard
     The Jew's dance. cf Angel S 36851.
     No more, good herdsman, of thy song. cf National Trust NT 002.

NICOLAI, Otto
     Die lustige Weiber von Windsor (The merry wives of Windsor):
          Frau Fluth's aria. cf Rococo 5377.
     The merry wives of Windsor: Overture. cf MENDELSSOHN: Hebrides
          overture, op. 26.
     The merry wives of Windsor: Overture. cf Decca GOSE 677/79.
     Die lustige Weiber von Windsor: Overture. cf Supraphon 110 1637.

NIELSEN, Carl
1965 At a young artist's bier. Bohemian-Danish folk melody. Helios
          overture, op. 17. Symphonies, nos. 1-3. Danish Radio Symphony

Orchestra; Herbert Blomstedt. Seraphim SIC 6097 (3).
    ++SFC 12-5-76 p58
At a young artist's bier. cf Works, selections (HMV 5027).
1966 At a young artist's bier: Andante lamentoso. The mother, op. 41:
    Incidental music (3). Quintet, woodwinds, op. 43. Serenata
    in vano. West Jutland Chamber Ensemble. DG 2530 515.
        +HF 4-76 p118   +St 5-76 p119
        +NR 2-76 p5
Bohemian-Danish folk melody. cf At a young artist's bier.
Bohemian-Danish folk tunes. cf Works, selections (HMV 5027).
Chaconne, op. 32. cf Piano works (Decca SDD 476).
Concerto, clarinet, op. 57. cf Works, selections (HMV 5027).
Concerto, flute. cf Works, selections (HMV 5027).
1967 Concerto, violin. Beverly Davison, vln; London Schools Symphony
    Orchestra; Frederick Applewhite. Cameo GOCLP 9006.
        -HFN 5-76 p103   +-RR 2-76 p35
Concerto, violin. cf Works, selections (HMV 5027).
Dance of the lady's maids. cf Piano works (Decca SDD 476).
Festival prelude. cf Piano works (Decca SDD 475).
Helios overture, op. 17. cf At a young artist's bier.
Helios overture, op. 17. cf Works, selections (HMV 5027).
Humoresque-bagatelles, op. 11. cf Piano works (Decca SDD 476).
Little suite, op. 1, A minor. cf GRIEG: Elegiac melodies, op. 34.
Little suite, op. 1, A minor. cf HMV SQ ESD 7001.
Little suite, op. 1, A minor: Intermezzo. cf Swedish Society SLT
    33229.
Motetter, op. 55 (3). cf Abbey LPB 757.
The mother, op. 41: Incidental music (3). cf At a young artist's
    bier: Andante lamentoso.
Pan and syrinx, op. 49. cf Works, selections (HMV 5027).
Piano music for young and old, op. 53. cf Piano works (Decca SDD
    475).
1968 Piano works: Festival prelude. Piano music for young and old,
    op. 53. Pieces, piano, op. 3 (5). Pieces, piano, op. 59 (3).
    Symphonic suite, op. 8. John McCabe, pno. Decca SDD 475.
        ++Gr 1-76 p1218   +-MM 12-76 p43
        +HFN 11-75 p161   +-RR 11-75 p81
1969 Piano works: Chaconne, op. 32. Dance of the lady's maids.
    Humoresque-bagatelles, op. 11. Suite, op. 45. Theme and
    variations, op. 40. John McCabe, pno. Decca SDD 476.
        ++Gr 1-76 p1218   +-MM 12-76 p43
        +HFN 11-75 p161   +-RR 11-75 p81
Pieces, piano, op. 3 (5). cf Piano works (Decca SDD 475).
Pieces, piano, op. 59 (3). cf Piano works (Decca SDD 475).
Quintet, woodwinds, op. 43. cf At a young artist's bier: Andante
    lamentoso.
Rhapsodie overture (An imaginary journey to the Faroe Islands).
    cf Works, selections (HMV 5027).
Saga-Drøm, op. 39. cf Symphony, no. 4, op. 29.
Saga-Drøm, op. 39. cf Works, selections (HMV 5027).
1970 Saul and David. Elisabeth Söderström, Sylvia Fisher, Bodil Gøbel,
    s; Willy Hartmann, Alexander Young, t; Boris Christoff, Michael
    Langdon, Kim Borg, bs; John Alldis Choir; Danish Radio Symphony
    Orchestra and Chorus; Jascha Horenstein. Unicorn RHS 343/5 (3).
        +Gr 7-76 p211   +SFC 8-29-76 p28
        ++HF 11-76 p96   ++St 11-76 p150

```
      +HFN 7-76 p93              +STL 8-8-76 p28
      +RR 8-76 p22
```

Serenata in vano. cf At a young artist's bier: Andante lamentoso.
Suite, op. 45. cf Piano works (Decca SDD 476).
Symphonic rhapsody. cf Works, selections (HMV 5027).
Symphonic suite, op. 8. cf Piano works (Decca SDD 475).
Symphonies, nos. 1-3. cf At a young artist's bier.
1971 Symphonies, nos. 1-6. Jill Gomez, s; Brian Rayner Cook, bar; LSO;
     Ole Schmidt. Unicorn RHS 324/330 (6). Tape ZCUNP 324.

```
      +Gr 1-75 p1350             +RR 12-74 p16
      ++HF 4-75 p73              +-RR 3-76 p77 tape
      +MM 2-76 p35               ++SFC 2-16-75 p23
      ++NR 4-75 p2               +-St 6-75 p100
```

Symphonies, nos. 1-6. cf Works, selections (HMV 5027).
1972 Symphony, no. 2, op. 16. LSO; Ole Schmidt. Unicorn RHS 325.
     (Reissue from RHS 324/330).

```
      +Gr 8-76 p292              +-RR 7-76 p61
      +HFN 7-76 p103
```

1973 Symphony, no. 3, op. 27. Jill Gomez, s; Brian Rayner Cook, bar;
     LSO; Ole Schmidt. Unicorn RHS 326. (Reissue from RHS 324/330).

```
      +Gr 9-76 p423              +RR 8-76 p48
      +HFN 8-76 p93
```

1974 Symphony, no. 4, op. 29. Saga-Drøm, op. 39. Royal Danish Orches-
     tra; Igor Markevitch. DG 2548 240. (Reissue from 139 185).

```
      +Gr 11-76 p798             +RR 10-76 p62
      +HFN 11-76 p170
```

1975 Symphony, no. 4, op. 29. Hallé Orchestra; John Barbirolli. Pye
     GSGC 15025.

```
      +-RR 4-76 p50
```

Theme and variations, op. 40. cf Piano works (Decca SDD 476).
1976 Works, selections: At a young artist's bier. Concerto, clarinet,
     op. 57. Concerto, flute. Concerto, violin, op. 33. Bohemian-
     Danish folk tunes. Helios overture, op. 17. Saga-Drøm, op.
     39. Pan and syrinx, op. 49. Rhapsodie overture (An imaginary
     journey to the Faroe Islands). Symphonic rhapsody. Symphonies,
     nos. 1-6. Kjell-Inge Stevensson, clt; Frantz Lemsser, flt;
     Arve Tellefsen, vln; Danish State Radio Symphony Orchestra;
     Herbert Blomstedt. HMV SLS 5027 (8).

```
      +Gr 10-75 p621             ++MT 5-76 p409
      ++HFN 11-75 p161           +RR 11-75 p45
```

NIEMAN, Alfred
     Israeli folk songs (9). cf Serenus SRS 12039.
1977 Sonata, piano, no. 2. POLLACK: Bridgeforms. Robert Pollack,
     Alberto Portugheis, pno. CRI SD 333.

```
      *CL 10-76 p7               +-NR 7-76 p12
      +-MJ 7-76 p57              *St 8-76 p100
```

NIGEL ROGERS, CANTI AMOROSI. cf DG Archive 2533 305.
NILES, J. J.
     I wonder as I wander. cf CBS 30062.
NILSON, Leo
     Viarp I. cf Caprice RIKS LP 35.
NILSSON, Bo
     Gruppen (20). cf Caprice RIKS LP 34.
1978 Quantitaten. PAULSON: Modi, op. 108b. SCHUMANN: Kinderscenen,
     op. 15. STRAVINSKY: Serenade, A major. Hans Palsson, pno.
     Caprice RIKS LP 79.

```
      +RR 10-75 p80              +RR 6-76 p75
```

NIN, Joaquin
    Cantilena asturiana. cf RCA ARM 4-0942/7.
    Granadina. cf HMV SQ ASD 3283.
    Spanish suite. cf Laurel-Protone LP 13.
NOBLE
    Cherokee. cf Pye GH 603.
NOEREN
    Midnight. cf Library of Congress OMP 101/2.
NOLA, Gian Domenico de
    Songs: Chi la gagliarda; Chi chi li chi. cf HMV CSD 3756.
NOORT, van
    Les petit branles. cf BASSANO: Ricercare.
NORDQUIST
    Psalm, no. 23. cf Swedish Society SLT 33197.
NORGARD, Per
1979 Iris. Voyage into the golden screen. Danish Radio Symphony
      Orchestra; Herbert Blomsted, Tamás Veto. Caprice RIKS LP 54.
        *Te 6-76 p50
    Spell. cf CHILDS: Trio, clarinet, violoncello and piano.
    Voyage into the golden screen. cf Iris.
    Wenn die Rose sich selbst schmückt, schmückt Sie auch den Garten.
      cf Caprice RIKS 59.
NOTRE DAME SCHOOL
    Flos filies and motet. cf BIS LP 2.
NOVACEK, Ottokar
    Perpetuum mobile. cf HMV SEOM 22.
    Perpetuum mobile. cf HMV SXLP 30188.
NUNES GARCIA, José Mauricio
    Lauda Sion salvatorem. cf Eldorado S-1.
NUROCK, Kirk
    Battle hymn of the republic. cf Brassworks Unlimited BULP 2.
NYMAN, Michael
1980 Bell set, no. 1. 1-100. Michael Nyman, pno; Nigel Shipway,
      Michael Nyman, perc. Obscure no. 6.
        +-RR 12-76 p85
    1-100. cf Bell set, no. 1.
NYSTROEM, Gösta
    Sinfonia concertante, violoncello and orchestra. cf KOCH: Nordic
      rhapsody, op. 26.
    Sjal och landskap. cf BRITTEN: Winter words, op. 53.
    Songs: Bara hos dem; Vitt land; Onskan. cf FRUMERIE: Songs
      (Swedish Society SLT 33171).
1981 Symphony, no. 4 (Sinfonia del mare). Elisabeth Söderström, s;
      Swedish Radio Symphony Orchestra; Stig Westerberg. Swedish
      Society SLT 33207.
        +RR 7-76 p62
OBRADORS
    Del cabello mas sutil. cf Polydor 2383 389.
OBRECHT, Jacob
    Haec deum caeli Laudemus nunc Dominum. cf HMV SLS 5049.
    Ic draghe de mutze clutze; Mijn morken graf; Pater noster. cf
      L'Oiseau-Lyre 12BB 203/6.
    T saat een Meskin. cf Coronet LPS 3032.
OBROVSKA, Jana
    Preludes (6). cf BARTA: Sonata, guitar.
OCHSENKHUN, Sebastian
    Innsbruck, ich muss dich lassen. cf DG Archive 2533 302.

OCKEGHEM, Johannes
1982 Marienmotetten. Prague Madrigal Singers; Miroslav Venhoda. Tele-
         funken AW 6-41878.
               -Gr 6-75 p79   *NR 10-75 p6
               +-HF 3-76 p92   +RR 6-75 p81
               +-HFN 8-75 p80   +St 3-76 p119
               +-MT 10-75 p886
1983 Missa pro defunctis. Prague Madrigal Singers; Vienna Musica
         Antiqua; Miroslav Venhoda. Telefunken SAWT 9612. (also 6-41265).
               -Gr 9-74 p572   +NR 12-74 p9
               -MQ 7-76 p460   -RR 12-74 p72
         Motets: Intemerata Dei mater. cf HMV SLS 5049.
         Songs: Prenez sur moi; Ma bouche rit. cf HMV SLS 5049.
OFFENBACH, Jacques
         La belle Hélène: Dis-moi Venus. cf London OS 26248.
1984 Les contes d'Hoffmann: Drig, drig, drig; Il était une fois à la
         cour d'Eisenach; Ah, Vivre deux...Une poupée aux yeux d'émail
         ...J'ai des yeux; Les oiseaux dans la charmille; Voici les
         valseurs; Belle nuit, o nuit d'amour; Scintille, diamant; O
         dieu, de quelle ivresse...Aujourd'hui les larmes; Elle a fui,
         la tourterelle; Jour et nuit je me mets en quatre; C'est une
         chanson d'amour; Chère enfant que j'appelle. Joan Sutherland,
         s; Huguette Tourangeau, ms; Placido Domingo, Hugues Cuenod, t;
         Gabriel Bacquier, bs; Lausanne Pro Arte Chorus, Du Brassus
         Choir; ORS and Chorus; Richard Bonynge. Decca SET 569. Tape
         (c) KCET 569. (Reissue from SET 545/7).
               +Gr 7-74 p262   +Op 9-74 p800
               +Gr 12-76 p1066 tape   +RR 5-74 p24
         Les contes d'Hoffmann: Horst du es tonen. cf Rococo 5377.
         Les contes d'Hoffmann: Les oiseaux dans la charmille. cf Court
         Opera Classics CO 342.
1985 Gaité parisiénne (arr. Rosenthal). Hollywood Bowl Symphony Orch-
         estra; Felix Slatkin. Angel S 36086. (Reissue from Capitol
         SP 8405).
               +St 2-76 p107
         Gaité parisiénne, excerpts. cf HMV SLS 5019.
         Gaité parisiénne: Can-can. cf CBS 30073.
         La grande Duchesse de Gérolstein: Dites lui. cf CBS 76522.
         La grande Duchesse de Gérolstein: Portez armes...J'aime les
         militaries. cf London OS 26248.
1986 Orpheus in the underworld, abridged. June Bronhill, s; Eric
         Shilling, bar; Sadler's Wells Orchestra and Chorus. HMV CSD
         1316. Tape (c) TC CSD 1316.
               +-HFN 8-76 p95 tape   +-RR 12-76 p108 tape
         Orphée aux Enfers: Overture. cf Supraphon 110 1638.
         La périchole: Ah, Quel diner je viens de faire. cf CBS 76522.
         La périchole: Laughing song. cf Desto DC 7118/9.
         La périchole: Tu n'est pas beau...Je t'adore, Air de lettre, Ah,
         quel diner. cf London OS 26248.
OGERMAN, Claus
1987 Symbiosis. Bill Evans Trio and Orchestra; Claus Ogerman. BASF
         MC 22094.
               +NR 5-76 p15
OHANA, Maurice
1988 Tres gráficos para guitarra y orquesta. RUIZ-PIPO: Tablas para
         guitarra y orquesta. Narciso Yepes, gtr; LSO; Rafael Frühbeck

de Burgos. DG 2530 585.

+Gr 3-76 p1468	++NR 4-76 p14
+HFN 5-76 p103	++RR 3-76 p47
+MJ 7-76 p56	

OLEY
Jesus meine Zuversicht. cf BACH: Chorale preludes (FAS 16).
Wir Christenleut (2 versions). cf BACH: Chorale preludes (FAS 16).

OLSSON, Otto
Songs: Ave Maria; Psalm, no. 42. cf Swedish Society SLT 33197.

O'NEIL
Regimental march of the Royal Canadian mounted police. cf Citadel
CT 6007.

ORD HUME, J.
The elephant march. cf Pye TC 3006.

ORFF, Carl
1989 Carmina burana. Catulli carmina. Trionfo di Afrodite. Ruth
Margret Pütz, Brigitte Dürrier, Enriqueta Tarrés, Hannelore
Bode, Carol Malone, s; Michael Cousins, Donald Grobe, Horst R.
Laubenthal, André Peysang, Toni Maxen, t; Barry McDaniel,
Roland Hermann, Werner Becker, bar; Marlies Jacobs, Günter
Hess, Friedrich Himmelmann, Hans Günter Nöcker, bs; Tölzer
Boys' Choir; Cologne Radio Symphony Orchestra and Chorus;
Ferdinand Leitner. BASF 21346/3 (3).

+Gr 8-76 p327	++RR 6-76 p79
+HFN 6-76 p93	

1990 Carmina burana. Judith Blegen, s; Kenneth Riegel, t; Peter Bin-
der, bar; Cleveland Orchestra Chorus and Boys' Choir; CO;
Michael Tilson Thomas. Columbia MX 33172. Tape (c) MTX 33172
(ct) MAX 33172 (Q) MQ 33172, M 33172 Tape (c) MAQ 33172. (also
CBS 76372 Tape (c) 40-76372).

+-Audio 10-75 p119	+-NR 9-75 p10
-Gr 5-75 p2006	+-ON 11-75 p70
-Gr 12-75 p1121 tape	+-RR 5-75 p62
+HF 5-75 p84	+-RR 12-75 p99 tape
+-HF 5-76 p114 tape	-SFC 5-11-75 p23
++HFN 5-75 p136	++St 5-75 p74

1991 Carmina burana. Norma Burrowes, s; Louis Devos, t; John Shirley-
Quirk, bar; Brighton Festival Chorus, Southend Boys' Choir;
RPO; Antal Dorati. Decca PFS 4368. (also London 21153 Tape
(c) 5-21153).

+-Gr 12-76 p1034	+-RR 12-76 p94

1992 Carmina burana. Jutta Vulpius, s; Hans Joachim Rotzsch, t;
Kurt Rehm, Kurt Hübenthal, bar; Leipzig Radio Childrens'
Choruses; Leipzig Radio Symphony Orchestra; Herbert Kegel.
DG 2548 194. Tape (c) 3318 051. (Reissue from 89525).

-Gr 9-75 p500	-RR 9-75 p70
+HFN 9-75 p109	-RR 11-75 p94 tape
+HFN 2-76 p117 tape	

1993 Carmina burana. Sheila Armstrong, s; Gerald English, t; Thomas
Allen, bar; St. Clement Danes Grammar School Boys' Choir; LSO
and Chorus; André Previn. HMV (Q) ASD 3117. Tape (c) TC ASD
3117. (also Angel S 37117 Tape (c) 4XS 37117 (ct) 8XS 37117).

+Gr 10-75 p670	+-NR 5-76 p8
+HF 7-76 p83	-ON 4-17-76 p42
+HFN 12-75 p160	++RR 1-76 p60
+-HFN 1-76 p125 tape	+RR 1-76 p67 tape
++MJ 5-76 p29	+St 9-76 p122

1994 Carmina burana. Celestina Casapietra, s; Horst Hiestermann, t;
      Karl-Heinz Stryczek, bar; Dresden Boys' Choir; Leipzig Radio
      Symphony Orchestra and Chorus; Herbert Kegel. Philips 9500
      040. Tape (c) 7300 444 (ct) 7750 096.
          /Gr 10-76 p639   ++HFN 10-76 p161
          ++HF 12-76 p110   ++RR 10-76 p98
1995 Carmina burana. Milada Subrtová, s; Jaroslav Tománek, t; Teodor
      Srubar, bar; CPO; Václav Smetáček. Supraphon SUAST 50409.
      Tape (c) 4-50409.
          +-HFN 2-76 p116 tape
1996 Carmina burana. Catulli carmina. Trionfo di Afrodite. Milada
      Subrtová, Helena Tattermuschova, s; Marta Boháčová, ms; Jaro-
      slav Tománek, Ivo Zidek, Oldřich Lindauer, t; Teodor Srubar,
      bar; Karel Berman, bs; Ludmila Trzcka, Vladimir Topinka,
      Vladimir Mencl, Oldřich Kredba, pno; CPhO and Chorus; PSO;
      Václav Smetáček. Supraphon 112-1461/3 (3).
          +-HFN 10-75 p153   ++RR 10-75 p88
          +NR 5-76 p8   +-St 9-76 p122
1997 Catulli carmina. Judith Blegen, s; Richard Kness, t; Temple
      University Choir; PO; Eugene Ormandy. CBS 61364. Tape (c)
      40-61364. (Reissue from SBRG 72611) (also Columbia MS 7017).
          -Gr 6-76 p77   ++RR 5-76 p71
          +-HFN 7-76 p103   ++RR 10-76 p106 tape
          +-HFN 8-76 p95 tape
1998 Catulli carmina. Ute Mai, s; Eberhard Büchner, t; Leipzig Radio
      Symphony Orchestra and Chorus; Herbert Kegel. Philips 6500
      815.
          ++Gr 1-76 p1226   +NR 9-75 p10
          +HF 11-75 p94   +RR 1-76 p60
          ++HFN 1-76 p115   ++St 11-75 p130
      Catulli carmina. cf Carmina burana (BASF 21346/3).
      Catulli carmina. cf Carmina burana (Supraphon 112 1461/3).
1999 Die Kluge. Gerhard Lenssen, pno. Claves LP D 506.
          ++HFN 2-76 p109
2000 Songs (choral works): Concento di voce: 3, Sermio; Der gute
      Mensch; Nänie und Dithyrambe; Veni creator spiritus; Vom
      Frühjahr, Oeltank, und vom Fliegen. Czech Philharmonic Chorus;
      Instrumental Ensemble; Václav Smetáček. Supraphon 112 1137.
          +-Gr 9-76 p459   +NR 10-76 p7
          +-HF 12-76 p110   ++RR 8-76 p76
          +-HFN 8-76 p85
      Trionfo di Afrodite. cf Carmina burana (BASF 21346/3).
      Trionfo di Afrodite. cf Carmina burana (Supraphon 112 1461/3).
      THE ORGAN IN AMERICA. cf Columbia MS 6161.
      ORGAN MUSIC OF THE FRENCH REVOLUTION. cf Calliope CAL 1917.
ORNSTEIN, Leo
2001 Danse sauvage (Early piano music). Michael Sellers, pno. Orion
      ORS 75194.
          +Audio 12-76 p88   +-NR 11-76 p13
2002 Preludes (3). Sonata, violoncello and piano. Bonnie Hampton,
      vlc; Nathan Schwartz, pno. Orion ORS 76211.
          +Audio 12-76 p88   +NR 7-76 p6
      Quintet, piano and strings, op. 92. cf Three moods.
      Sonata, violoncello and piano. cf Preludes.
2003 Three moods. Quintet, piano and strings, op. 92. William West-
      ney, pno; Daniel Stepner, Michael Strauss, vln; Peter John

    Sacco, vla; Thomas Mansbacher, vlc. CRI SD 339.
        +HF 5-76 p88   +NYT 4-11-76 pD23
        ++MJ 3-76 p24   +SR 9-18-76 p50
        ++NR 1-76 p7   ++St 5-76 p119
ORTHODOX CHURCH MUSIC FROM FINLAND. cf Ikon IKO 4.
ORTHODOX SLAV LITURGY. cf Harmonia Mundi HMU 105.
ORTIZ, Diego
    Dulce memoire; Recercada. cf L'Oiseau-Lyre 12BB 203/6.
    Quinta pars. cf BIS LP 22.
    Recercadas primera y segunda. cf BIS LP 22.
    Recercadas primera y segunda. cf Nonesuch H 71326.
OSBORNE
    Rhapsodie. cf Washington University RAVE 761.
OSGOOD
    Round the clock. cf Transatlantic XTRA 1160.
OSSER
    Bandolero. cf Pye GH 603.
OSTERLING
    Winds on the run. cf Pye GH 603.
    OUR MUSICAL PAST: A concert for brass band, voice and piano. cf
        Library of Congress OMP 101/2.
OWEN, Morfydd
    Madonna songs. cf Argo ZRG 769.
OXENFORD
    The ash grove. cf HMV ESD 7002.
PACHELBEL, Johann
    Ach, was soll ich Sunder machen. cf Pelca PRSRK 41017/20.
    Canon. cf Decca SDD 411.
2004 Canon, D major. Suite, no. 6, B flat major. Suite, G major.
        FASCH: Concerto, trumpet, 2 oboes and strings, D major.
        Symphony, G major. Symphony, A major. Maurice André, tpt;
        Pierre Pierlot, Jacques Chambon, ob; Jean-François Paillard
        Chamber Orchestra; Jean-François Paillard. Erato STU 70468.
            +Gr 4-76 p1607   +NR 3-74 p15
            +HFN 5-76 p105   +RR 4-76 p51
    Canon, D major. cf DG 2548 219.
    Canon, D major. cf Vanguard SRV 344.
    Canon (Kanon). cf CORELLI: Concerto grosso, op. 6, no. 8, G minor.
    Chaconne, A minor. cf Argo ZRG 806.
    Suite, G major. cf Canon, D major.
    Suite, no. 6, B flat major. cf Canon, D major.
    Vom Himmel hoch. cf Grosvenor GRS 1041.
PACOLINI (16th century Italy)
    Padoana commun; Passamezzo commun. cf L'Oiseau-Lyre 12BB 203/6.
    Passamezzo della battaglia. cf DG Archive 2533 323.
    Saltarello della traditoria. cf DG Archive 2533 323.
    Saltarello milanese. cf DG Archive 2533 323.
PADEREWSKI, Ignace Jan
    Fantaisie polonaise, op. 19. cf ARENSKY: Concerto, piano, no. 2.
    Legende, op. 16, no. 1, A flat major. cf International Piano
        Library IPL 102.
2005 Sonata, piano, op. 21, E flat minor. Variations and fugue, op.
        23, E flat minor. Ignace Jan Paderewski, pno. Muza SXL 0570.
            *HF 3-76 p70
    Sonata, violin and piano, op. 13, A minor. cf BUSONI: Sonata,
        violin and piano, no. 2, op. 36a, E minor.

Variations and fugue, op. 23, E flat minor. cf Sonata, piano,
    op. 21, E flat minor.
PAGANINI, Niccolo
    La campanella. cf London STS 15239.
    La campanella. cf Sterling 1001/2.
    Cantabile, op. 17. cf GIULIANI: Sonata, violin and guitar, op. 25.
2006 Caprices, op. 1, nos. 1-24. Daniel Majeske, vln. Advent S 5019-2
    (2).
        +NR 6-76 p9
2007 Caprices, op. 1, nos. 1-24. Michael Rabin, vln. Seraphim SIB
    6096 (2).
        +MJ 11-76 p45
    Caprice, op. 1, no. 9, E major. cf HMV SEOM 22.
    Caprices, op. 1, nos. 13, 20. cf London STS 15239.
    Caprices, op. 1, nos. 13, 20, 24. cf RCA ARM 4-0942/7.
2008 Caprice, op. 1, no. 24, A minor (arr. Callimahos). PROKOFIEV:
    Sonata, flute and piano, op. 94, D major. SANCAN: Sonatine
    pour flute et piano. Louise di Tullio, flt; Virginia di
    Tullio, pno. Crystal S 311.
        +HF 5-76 p89   +NR 12-75 p8
    Caprice, op. 1, no. 24, A minor. cf HMV SEOM 22.
    Caprice, op. 1, no. 24, A minor. cf HMV HLM 7077.
    Centone di sonata, op. 64, no. 1.' cf GIULIANI: Sonata, violin
    and guitar, op. 25.
2009 Centone di sonate, op. 64, no. 2, D major; no. 5, E major. Sonata,
    violin and guitar, A major. Sonatas, violin and guitar, op. 2
    (6). György Terebesi, vln; Sonja Prunnbauer, gtr. Telefunken
    6-41936.
        +HF 7-76 p83   +SR 3-6-76 p41
        ++NR 3-76 p12
2010 Concerti, violin, nos. 1-6. Salvatore Accardo, vln; LPO; Charles
    Dutoit. DG 2740 121 (5). (no. 6 available on 2530 467).
        ++Gr 11-75 p827   +NR 1-76 p6
        ++HF 2-76 p100   ++RR 11-75 p46
        +HFN 12-75 p160   ++SFC 12-7-75 p31
2011 Concerto, violin, no. 1, op. 6, D major.* Le Streghe, variations
    on a theme by Süssmayr, op. 8. Salvatore Accardo, vln; LPO;
    Charles Dutoit. DG 2530 714. (*Reissue from 2740 121).
        +Gr 11-76 p803   +RR 11-76 p70
2012 Concerto, violin, no. 1, op. 6, D major. Concerto, violin, no.
    2, op. 7, B minor. Shmuel Ashkenasi, vln; VSO; Heribert Esser.
    DG 2535 207. (Reissue from SLPM 139424).
        +Gr 12-76 p1001   +RR 12-76 p63
        +HFN 12-76 p151
2013 Concerto, violin, no. 1, op. 6, D major. Concerto, violin, no. 4,
    D minor. Henryk Szeryng, vln; LSO; Alexander Gibson. Philips
    9500 069. Tape (c) 7300 477.
        +Gr 10-76 p602   ++RR 10-76 p64
        ++HFN 10-76 p175
    Concerto, violin, no. 1, op. 6, D major. cf MENDELSSOHN: Concerto,
    violin, op. 64, E minor.
    Concerto, violin, no. 1, op. 6, D major. cf HMV RLS 718.
    Concerto, violin, no. 2, op. 7, B minor. cf Concerto, violin,
    no. 1, op. 6, D major.
2014 Concerto, violin, no. 3, E major.* Sonata, viola, op. 35, C
    minor. Salvatore Accardo, vln; Dino Asciolla, vla; LPO; Charles

Dutoit. DG 2530 629. Tape (c) 3300 629. (*Reissue from
2470 121).
    ++Gr 5-76 p1765   +NR 7-76 p4
    +HFN 4-76 p109   +RR 5-76 p47

Concerto, violin, no. 4, D minor. cf Concerto, violin, no. 1,
op. 6, D major.

Fantasia on "Dal tuo stellato soglio", C minor. cf Sonatas,
violin and guitar, nos. 1-6, opp. 2 and 3.

Moto perpetuo, op. 11. cf HMV SXLP 30188.

Moto perpetuo, op. 11 (arr. and orch. Gerhardt). cf RCA LRL 1-5094.

Moto perpetuo, op. 11. cf RCA LRL 1-5127.

Moto perpetuo, op. 11. cf RCA ARM 4-0942/7.

2015 Sonata, viola, op. 35, C minor. ROLLA: Concerto, violin, A major.
ROSSINI: Duet, violoncello and double bass. Susanne Lauten-
bacher, vln; Ulrich Koch, vla; Georges Mallach, vlc; Jean
Poppe, double-bs; Württemberg Chamber Orchestra, Luxembourg
Radio Orchestra; Jörg Faerber, Pierre Cao. Turnabout QTV
34606.
    +NR 5-76 p6

Sonata, viola, op. 35, C minor. cf Concerto, violin, no. 3,
E major.

Sonata, viola, op. 35, C minor. cf HOFFMEISTER: Concerto, viola,
D major.

Sonata, violin, op. 25, C major. cf DG 2530 561.

2016 Sonatas, violin and guitar, nos. 1-6, opp. 2 and 3. Fantasia
on "Dal tuo stellato soglio", C minor. Tarantella, A minor.
György Terebesi, vln; Sonja Prunnbauer, gtr. Telefunken
6-41995.
    +-NR 7-76 p6

Sonata, violin and guitar, A major. cf Centone di sonate, op.
64, no. 2, D major; no. 5, E major.

Sonata, violin and guitar, A major. cf GIULIANI: Sonata, violin
and guitar, op. 25.

Sonatas, violin and guitar, op. 2 (6). cf Centone di sonate,
op. 64, no. 2, D major; no. 5, E major.

Sonata, violin and guitar, op. 3, no. 6, E minor. cf GIULIANI:
Sonata, violin and guitar, op. 25.

Le Streghe, variations on a theme by Süssmayr, op. 8. cf Concerto,
violin, no. 1, op. 6, D major.

Le Streghe, variations on a theme by Süssmayr, op. 8. cf BACH:
Partita, violin, no. 2, S 1004, D minor: Chaconne.

Tarantella, A minor. cf Sonatas, violin and guitar, nos. 1-6,
opp. 2 and 3.

PAISIELLO, Giovanni

Concerto, piano, F major. cf MANFREDINI: Concerto, piano, B flat
major.

Coronation mass. cf LESUEUR: Coronation march.

La molinara: Nel cor più non mi sento. cf Decca SXL 6629.

Te Deum. cf LESUEUR: Coronation march.

I Zingari in Fiera: Chi vuol la zingarella. cf Decca SXL 6629.

PAIX, Jakob

Mir ist ein feins brauns Maidelein gefallen in mein Sinn. cf
Pelca PRSRK 41017/20.

Phantasia primi toni. cf Grosvenor GRS 1039.

Ungaresca. cf CBS 76183.

Ungaresca. cf Pelca PRSRK 41017/20.

PALADILHE, Emile
    Psyché. cf Polydor 2383 389.
PALESTRINA, Giovanni de
    Alma redemptoris mater. cf Vista VPS 1022.
    Haec dies. cf HMV SLS 5047.
    Missa brevis. cf HMV SLS 5047.
    Sanctus and benedictus (Soriano). cf Vanguard SVC 71212.
2017 Songs (choral works): Hodie beata virgo. Litaniae de beata
        Virgine Maria a 8 vocum. Magnificat a 8 voci (primi toni).
        Senex puerum portabat. Stabat mater. King's College Choir;
        David Willcocks. Argo ZK 4. (Reissue from ZRG 5398).
            +-Gr 12-76 p1034
2018 Songs (choral works): Missa aeterna Christi munera. Oratio
        Jeremiae prophetae. Motets: Sicut cervus desiderat; Super
        flumina Babylonis; O bone Jesu. Pro Cantione Antiqua; Bruno
        Turner. DG Archive 2533 322.
            +-Gr 12-76 p1034   +RR 12-76 p93
    Songs: O beata et gloriosa Trinitas. cf Abbey LPB 750.
2019 Tu es Petrus, mass. VICTORIA: Ave Maria, motet. Mass Pro Vic-
        toria. London Oratory Choir; John Hoban. Discourses ABM 20.
            +-HFN 9-76 p126   +RR 11-76 p99
    Vestiva i colli. cf Hungaroton SLPX 11669/70.
PANELLA
    On the square. cf Michigan University SM 0002.
PANTYSELYN
    Laudamus. cf Columbia SCX 6585.
PANUFNIK, Andrej
2020 Concerto, violin. Sinfonia concertante, flute, harp and strings.
        Yehudi Menuhin, vln; Aurèle Nicolet, flt; Osian Ellis, hp;
        Menuhin Festival Orchestra; Andrej Panufnik. HMV EMD 5525.
            +Gr 12-75 p1038   +RR 12-75 p57
            +HFN 4-76 p111   ++Te 6-76 p48
    Sinfonia concertante, flute, harp and strings. cf Concerto,
        violin.
PAPP, Lajos
    Quintet, cimbalom. cf DUBROVAY: Quartet, strings.
PARADIS, Maria Theresia von
    Sicilienne. cf Gemini Hall RAP 1010.
PARCHAM, Andrew
    Solo, blockflöte. cf Telefunken DT 6-48075.
PARKES
    Old London. cf Philips 6308 246.
PARRY, Hubert
    Elegy, A flat major. cf Vista VPS 1035.
    Jerusalem, op. 208. cf BBC REB 228.
    O mistress mine. cf HMV ASD 2929.
    Old 100th, chorale fantasia. cf Vista VPS 1035.
    Songs: I was glad. cf Audio 3.
    Songs: I was glad; Songs of farewell; My soul, there is a country.
        cf Audio 1.
    Toccata and fugue. cf Wealden WS 149.
PARRY, John
    Flow gently, Deva. cf HMV EMD 5528.
PARSCH, Arnost
    Poetica II. cf Supraphon 111 1390.

PARSLEY, Osbert
2021 Lamentations, 5 voices. SHEPPARD: Playnsong mass for a meane for
      5 voices. TAVERNER: Missa sine nomine, 5 voices. ANON.:
      Kyrie "Orbis factor". Pro Cantione Antiqua; Bruno Turner.
      BASF 22065-6.
             +Gr 8-76 p328   +RR 6-76 p76
             +HFN 6-76 p101
PARSONS, Robert
      Ave Maria. cf Harmonia Mundi HMU 473.
PARTCH, Harry
2022 The bewitched. University of Illinois Musical Ensemble; John
      Garvey. CRI SD 304 (2).
             *NR 4-74 p14   ++SFC 3-3-74 p22
             +-RR 9-76 p33   +St 8-74 p118
2023 The bewitched: Scene 10 and epilogue. Castor and Pollux. The
      letter. Cloud-chamber music. Windsong. Gate 5 Ensemble of
      Sausalito; University of Illinois Ensemble; John Garvey. CRI
      SD 193.
             +-RR 10-76 p88
      Castor and Pollux. cf The bewitched: Scene 10 and epilogue.
      Cloud-chamber music. cf The bewitched: Scene 10 and epilogue.
      The letter. cf The bewitched: Scene 10 and epilogue.
      Windsong. cf The bewitched: Scene 10 and epilogue.
PASSEREAU (16th century France)
      Il est bel et bon. cf Argo SPA 464.
      Il est bel et bon. cf Argo ZRG 823.
      Il est bel et bon. cf HMV CSD 3740.
      He is good and handsome. cf Coronet LPS 3032.
PATACHICH
      Quartet, saxophones. cf Coronet LPS 3030.
PAUER, Jiri
      Concerto, trumpet. cf DREJSL: Concerto, piano.
      Suite, harp. cf Panton 110 380.
2024 Zdravý Nemocný (Le malade imaginaire), excerpts. Miloslava
      Fidlerová, s; Libuše Márová, con; Beno Blachut, t; Jaroslav
      Horáček, bs; CPhO; Bohumir Liška. Panton 110 382.
             ++RR 8-76 p22
PAULSON, Gustaf
      Modi, op. 108b. cf NILSSON: Quantitaten.
PAUMANN, Conrad
      Mit gantzem Willen. cf CBS 76183.
PEARSALL, Robert de
      Songs: Great God of love; Lay a garland; Light of my soul. cf
      Argo ZRG 833.
PEARSON, Leslie
      An Elizabethan fantasy: Now is the month of Maying; The willow
      song; The night watch. cf Argo ZDA 203.
      A medieval pagaent: Agincourt song; Greensleeves; Summer is icumen
      in. cf Argo ZDA 203.
PEETERS, Flor
      Chorale fantasie "Lasst uns erfreuen", op. 81, no. 2. cf KARG-
      ELERT: Chorale improvisations, op. 65.
      Chorale prelude, op. 81, no. 1: Suttgart. cf KARG-ELERT: Chorale
      improvisations, op. 65.
      Sonatina, op. 45. cf Orion ORS 76235.
      Sonatina, op. 46, G major. cf Orion ORS 76235.

Suite modale, op. 43. cf GUILMANT: Sonata, organ, no. 1, op. 42,
    D minor.
Suite modale, op. 43. cf KARG-ELERT: Chorale improvisations, op.
    65.
Toccata, fugue and hymn, op. 28. cf KARG-ELERT: Chorale improvi-
    sations, op. 65.
PENDERECKI, Krzysztof
2025 Canticum canticorum Salomnis. De natura sonoris, no. 1. The
    dream of Jacob. Threnody to the victims of Hiroshima. Cracow
    Philharmonia Chorus; Polish Radio Symphony Orchestra; Krzysztof
    Penderecki. EMI SQ EMD 5529.
        +Gr 6-76 p77   +MT 11-76 p915
        +HFN 6-76 p95   +RR 3-76 p69
    Capriccio per Siegfried Palm. cf DG 2530 562.
    De natura sonoris, no. 1. cf Canticum canticorum Salomnis.
    The dream of Jacob. cf Canticum canticorum Salomnis.
2026 Magnificat. Peter Lagger, bs; Cracow Radio and Children's Chorus;
    Polish National Symphony Orchestra; Krzysztof Penderecki.
    Angel S 37141. (also HMV (Q) EMD 5524).
        +Gr 1-76 p1226   *NYT 12-21-75 pD18
        ++HF 1-76 p92   +ON 2-14-76 p49
        +HFN 7-76 p93   ++RR 12-75 p89
        +MT 4-76 p321   +SFC 9-28-75 p30
        ++NR 11-75 p9   +St 11-76 p152
    Threnody to the victims of Hiroshima. cf Canticum canticorum
        Salomnis.
PENN
    Pansy faces. cf HMV EMD 5528.
PEPIN, Clermont
    Monade, strings. cf MATTON: Concerto, 2 pianos.
PEPPER
    Songs: Over the rolling sea. cf Argo ZFB 95/6.
PEPUSCH, John
    Sonata, no. 4, F major. cf Telefunken DT 6-48075.
PERGAMENT, Moses
    Kol nidre. cf EPHROS: The priestly benediction.
PERGOLESI, Giovanni
    Concerto, flute, no. 1, G major. cf Decca SPA 394.
    Concerto, flute, no. 2, G major. cf RCA CRL 3-1429.
2027 Missa Romana, F major. Escolania Montserrat, Boy soloists;
    Montserrat, Benedictine Monastery, Young Singers; Tölzer Boys'
    Choir; Collegium Aureum; P. Ireneu Segarra. BASF KHB 21230.
        +-Gr 3-76 p1499   +-ON 3-9-74 p25
        +-MT 9-76 p747   +-RR 2-76 p61
        +NR 9-73 p10   +-St 1-74 p110
    La serva padrona: Stizzoso, mio stizzoso. cf Decca SXL 6629.
    Songs: Tre giorni son che Nina. cf Decca SXL 6629.
PERI, Jacopo
    Songs: O durezza di ferro; Tre le donne; Bellissima regina. cf
        DG Archive 2533 305.
PERLE, George
    Three movements, orchestra. cf LEWIS: Symphony, no. 2.
PEROTIN LE GRAND
    Alleluya. cf Candide CE 31095.
    Organa. cf LEONIN: Organa.
PERSICHETTI, Vincent
    Drop, drop slow tears. cf Delos FY 025.

The hollow men. cf Avant AV 1014.
Parable. cf GOLDMAN: Sonata, violin and piano.
2028 Quartets (4). New Art String Quartet. Arizona State University
Unnumbered.
+SR 9-18-76 p50
Songs: The death of a soldier; The grass; Of the surface of
things; The snow man; Thou child so wise. cf Duke University
Press DWR 7306.
PERT, Morris
2029 Chromosphere. Japanese verses (4). Luminos. Georgina Dobrée,
basset hn; Morris Pert, pno; Vernica Hayward, s; Suntreader.
Discourses ABM 21.
+-Gr 8-76 p317   *MT 9-76 p747
+-HFN 5-76 p105   +-RR 10-76 p90
Japanese verses (4). cf Chromosphere.
Luminos. cf Chromosphere.
PERUSIO, Matheus de
Andray soulet. cf HMV SLS 863.
Andray soulet. cf 1750 Arch S 1753.
Le greygnour bien. cf HMV SLS 863.
PETER, J. F.
Songs: I will freely sacrifice to Thee; I will make an everlasting
covenant. cf Vox SVBX 5350.
PETER, S.
Songs: Look ye, how my servants shall be feasting; O, there's a
sight that rends my heart. cf Vox SVBX 5350.
PETERSON-BERGER, Olaf Wilhelm
Om magna ar. cf Swedish Society SLT 33229.
På gräset under lindarna. cf Swedish Society SLT 33229.
PETIT, Pierre
Tarantelle. cf Delos DEL FY 008.
PETTERSSON, Gustaf Allan
2030 Concerto, string orchestra, no. 3: Mesto. SIBELIUS: The tempest,
op. 109: Humoresk; Caliban's song; Scene; Intrada and berceuse;
Chorus of winds; Intermezzo; Nymphs' dance; Prospero; Miranda;
Naiads; Storm. Swedish Radio Symphony Orchestra; Stig Wester-
berg. Swedish Society SLT 33203.
+RR 6-76 p46   ++St 12-76 p146
2031 Symphony, no. 2. Swedish Radio Symphony Orchestra; Stig Wester-
berg. Swedish Society SLT 33219.
+Gr 3-75 p1658   +RR 9-76 p58
PEZEL, Johann
Pieces (3). cf Swedish Society SLT 33200.
Sonatinas, nos. 61-62, 65-66. cf Nonesuch H 71301.
Suite, C major. cf Saga 5417.
PHALESE, Pierre
L'Arboscello ballo furlano. cf Harmonia Mundi HMU 610.
Au joly bois. cf Hungaroton SLPX 11669/70.
La battaglia. cf Hungaroton SLPX 11669/70.
Galliard, Traditore. cf Angel SFO 36895.
Reprise, galliard. cf DG Archive 2533 184.
La roca el Fuso. cf Hungaroton SLPX 11669/70.
PHILE
Hail, Columbia. cf Department of Defense Bicentennial Edition
50-1776.
Hail, Columbia. cf Michigan University SM 0002.

The President's march. cf Columbia MS 6161.
PHILIPS, Peter
    Amarilli di Julio Romano. cf Saga 5402.
    Ascendit Deus. cf Harmonia Mundi HMU 473.
    Galliard to Philips pavan. cf Harmonia Mundi HMD 223.
    Philips pavan. cf Harmonia Mundi HMD 223.
PHILLIPS, Burrill
    Selections from McGuffey's readers. cf CARPENTER: Adventures in
        a perambulator.
    PIANO MUSIC OF SPAIN. cf Opus unnumbered.
PICCININI (16th century Italy)
    Canzona. cf DG Archive 2533 323.
PICCINNI, Niccolo
    Didon: Ah, que je fus bien inspirée. cf Smithsonian Collection
        N 002.
PICK (attrib.)
    March and troop. cf Saga 5417.
    Suite, B flat major. cf Saga 5417.
    PICTURES FROM ISRAEL. cf Da Camera Magna SM 93399.
PIEFKE, G.
    Königgrätzer Marsch. cf DG 2721 077.
    Königgrätzer Marsch. cf Polydor 2489 523.
    Preussens Gloria. cf DG 2721 077.
    Preussens Gloria. cf Polydor 2489 523.
PIERNE, Gabriel
    Canzonetta. cf Grenadilla GS 1006.
2032 Les enfants a Bethlehem. Gerda Hartman, Christiane Chateau, s;
        Nicole Leport, Arlette Durigneux, Martine Bernardi, child
        singers; J. J. Jouineau, t; Michel Piquemal, bar; Pierre
        Fresnay, narrator; ORTF and Children's Chorus; Jacques Jouineau.
        Barclay Inedits 995 029.
            +-Gr 8-73 p383   +St 12-76 p96
    Songs (4). cf Golden Crest CRS 4143.
PIETRI
    Maristella: Io conosco un giardino. cf Rococo 5383.
PILKINGTON, Francis
    Songs: Amyntas with his Phillis fair; Have I found her. cf Har-
        monida Mundi HMV 223.
    Sweet Phillida. cf Harmonia Mundi 593.
PINKHAM, Daniel
    Cantilena. cf COWELL: Works, selections (CRI SD 109).
    Capriccio. cf COWELL: Works, selections (CRI SD 109).
    Concerto, celeste and harpsichord. cf COWELL: Works, selections
        (CRI SD 109).
    For evening draws on. cf MARTINO: Paradiso choruses.
    Liturgies. cf MARTINO: Paradiso choruses.
    Madrigal. cf CRI SD 102.
    Revelations: Litany and toccata. cf WEalden WS 142.
    Songs: Elegy. cf CRI SD 102.
    Songs: Henry was a worthy King; The leaf; Piping Anne and husky
        Paul; Agnus Dei. cf Orion ORS 75205.
    Toccatas for the vault of heaven. cf MARTINO: Paradiso choruses.
PINOS, Alois
    Esca, prepared piano. cf Supraphon 111 1390.
PINTO, George
    Sonata, piano, op. 3, no. 1, E flat minor. cf BENNETT, W.: Jan-
        uary, op. 36, no. 1.

PIPO, Ruiz
    Canción y danza. cf London STS 15306.
    Canción y danza, no. 1. cf Supraphon 111 1230.
PISTON, Walter
2033 The incredible flutist. Louisville Orchestra; Jorge Mester.
      Louisville LS 755.
         +SFC 12-19-76 p50
    Quartet, no. 5. cf Vox SVBX 5305.
2034 Quintet, wind instruments. WEBER: Consort of winds, op. 66.
      Böhm Quintette. Orion ORS 75206.
         +NR 4-76 p7
    Sonata, flute and piano. cf BURTON: Sonatina, flute and piano.
    Sonatina. cf Desto DC 6435/7.
2035 Symphony, no. 7. Symphony, no. 8. Louisville Orchestra; Jorge
      Mester. Louisville LS 746.
         +-HF 5-76 p89   +-St 5-76 p120
    Symphony, no. 8. cf Symphony, no. 7.
    Trio, piano. cf BLOCH: Nocturnes.
PIXIS, Johann
    Concerto, violin, piano and strings. cf MOSCHELES: Grand sonate
      symphonique, op. 112.
PIZZETTI
    Calzare d'Argento: Da vero quanto grande e la misera. cf Rococo
      5383.
PLANQUETTE, Robert
2036 Les cloches de Corneville. Mady Mesplé, Christiane Stutzmann,
      Annie Tallard, Arta Verlen, s; Charles Burles, Jean Giraudeau,
      Jean Bussard, t; Bernard Sinclair, Jean-Christoph Benoit, bar;
      Charles Roeder, bs; Paris Opera Chorus; Opéra-Comique Orches-
      tra; Jean Doussard. Connoisseur Society CS 2-2107 (2).
         +ARG 11-76 p35   +SFC 10-24-76 p35
         +-HF 9-76 p95   +-St 11-76 p152
         *MJ 12-76 p28
PLATTI, Giovanni
    Concerto, piano, no. 2, C minor. cf MANFREDINI: Concerto, piano,
      B flat major.
THE PLAY OF DANIEL. cf Anonymous works.
PLAYFORD, John
    The English dancing master: Country dances. cf DG Archive 2533
      172.
    THE PLEASURE OF THE ROYAL COURTS. cf Nonesuch H 71326.
    PLEASURES OF THE COURT FROM THE TIMES OF HENRY VIII AND ELIZABETH
      I. cf Angel S 36851.
PLESKOW, Raoul
    Motet and madrigal. cf CRI SD 342.
PLOG, Anthony
    Two scenes, soprano, trumpet and organ. cf Avant AV 1014.
PODEST
    Partita, strings, guitar and percussion. cf BOHAC: Elegie in
      memory of Ludvíka Podéstě.
POGLIETTI, Alessandro
    Balletto. cf DG Archive 2533 172.
    Ricercar per lo Rossignolo, D major. cf Pelca PRSRK 41017/20.
    Ricercar secundi toni, G minor. cf Pelca PRSRK 41017/20.
POLDINI, Eduard
    Dancing doll. cf Connoisseur Society (Q) CSQ 2065.

POLDOWSKI, Irene
        Impression Fausse. cf Gemini Hall RAP 1010.
        Tango. cf RCA ARM 4-0942/7.
POLLACK, Robert
        Bridgeforms. cf NIEMAN: Sonata, piano, no. 2.
        Movement and variations. cf EDWARDS: Quartet, strings.
POLLONOIS
        Courante. cf Hungaroton SLPX 11721.   ·
PONCE, Manuel
        Canciones populares mexicanas (3). cf Saga 5412.
        Estrellita. cf RCA ARM 4-0942/7.
        Estrellita. cf Supraphon 111 1230.
        Mexican folksong. cf Supraphon 111 1230.
        Prelude, E major. cf RCA ARL 1-0864.
        Preludes (3). cf Laurel-Protone LP 13.
        Scherzino mexicano. cf Supraphon 111 1230.
        Valse. cf Saga 5412.
PONCHIELLI, Amilcare
        La gioconda: Ah, Pescator altonda l'esca. cf Discophilia DIS
            KGA 2.
        La gioconda: Cielo è mar. cf Everest SDBR 3382.
        La gioconda: Cielo è mar. cf Muza SXL 1170.
        La gioconda: Cielo è mar. cf Rubini GV 70.
        La gioconda: Cielo è mar; Suicidio. cf Rubini GV 63.
        La gioconda: Dance of the hours. cf GOUNOD: Faust: Ballet music;
            Waltz.
        La gioconda: Dance of the hours, Act 3. cf HMV SLS 5019.
        La gioconda: E un anatema...L'amo come il fulgor del creato,
            Cosi mantieni il patto. cf Court Opera Classics CO 347.
        La gioconda: Suicidio. cf Club 99-100.
        La gioconda: Suicidio. cf HMV SLS 5057.
        La gioconda: Voce di donna. cf Collectors Guild 611.
POOT, Marcel
        Concertino, wind quartet. cf BALAI: Divertimento, wind quintet
            and harp, op. 7.
        Sonatina, D major. cf Orion ORS 76235.
POPPER, David
        Dance of the elves, op. 39. cf HMV SQ ASD 3283.
PORCELIJN, David
        Continuations. cf KEURIS: Concerto, saxophone.
        10-5-6-5 (a). cf KEURIS: Concerto, saxophone.
PORPORA, Nicola
2037 Concerto, violoncello, G major. SAMMARTINI: Concerto, viola
        pomposa, C major. VIVALDI: Concerto, viola d'amore, A major.
        Thomas Blees, vlc; Ulrich Koch, vla d'amore, vla pomposa;
        South West German Orchestra; Paul Angerer. Turnabout TV 34574.
            +-Gr 4-76 p1607   ++NR 6-75 p6
            +-HFN 8-76 p85   +RR 3-76 p51
PORTA
        Canzon terza. cf Pelca PRSRK 41017/20.
PORTER, Ambrose
        An Easter meditation. cf Vista VPS 1023.
PORTER, Quincy
        Quartet, strings, no. 8. cf CARTER: Etudes and a fantasy, wood-
            wind quartet (8).
PORTER, Walter
        Thus sang Orpheus. cf L'Oiseau-Lyre 12BB 203/6.

PORTMAN
    Verse for ye double organ. cf Audio EAS 16.
POSTON
    Last night in the open shippen. cf Grosvenor GRS 1034.
POULENC, Francis
    Concert champêtre, harpsichord and orchestra. cf Concerto, organ,
        strings and timpani, G minor.
2038 Concerto, organ, strings and timpani, G minor. Concert champêtre,
        harpsichord and orchestra. Marie-Claire Alain, org; Robert
        Veyron-Lacroix, hpd; ORTF; Jean Martinon. Musical Heritage
        Society MHS 1595. (also Erato STU 70637).
          +Gr 7-76 p181   ++RR 6-76 p51
          +HF 10-73 p109   ++St 2-74 p118
2039 Concerto, piano. Gloria, G major. Norma Burrowes, s; Christina
        Ortiz, pno; Birmingham City Symphony Orchestra and Chorus;
        Louis Frémaux. HMV SQ ASD 3299. (also Angel S 37246).
          ++Gr 12-76 p1041
    Elégie, horn and piano. cf Works, selections (HMV EMSP 553).
    L'Embarquement pour Cythère. cf HMV SXLP 30181.
    Gloria, G major. cf Concerto, piano.
    Mass, G major. cf FRANCK: Songs (Abbey LPB 758).
2040 Motets: Timor et tremor; Vinea mea electa; Tenebrae factae sunt;
        Tristis est anima mea; O magnum mysterium; Quem vidistis past-
        ores dicite; Videntes stellam; Hodie Christus natus est.
        STRAVINSKY: Mass. Christchurch Cathedral Choir; Simon Preston.
        Argo ZRG 720.
          ++Audio 8-76 p75   +RR 12-73 p98
          +Gr 1-74 p1406   ++SFC 6-9-74 p28
          +HFN 12-73 p2614   ++St 8-74 p122
          +NR 8-74 p7
    Motets pour le temps de Noël: O magnum mysterium; Quem vidistis
        pastores dicite. cf Vista VPS 1022.
    Mouvements perpetuels, no. I. cf RCA ARM 4-0942/7.
2041 Mouvements perpetuels, Nos. I, II, III. Nocturne, D major. Suite
        française. SATIE: Avant-dernièrs pensées. Croquis et agac-
        eries d'un gros bonhomme en bois. Descriptions automatiques.
        Gnossienne, no. 3. Gymnopédie, no. 1. Sarabande, no. 2.
        Francis Poulenc, pno. Odyssey Y 33792. (Reissues).
          +ARSC Vol VIII, no. 2-3 +NR 7-76 p11
            p83   ++NYT 5-16-76 pD19
          +MJ 10-76 p52
    Nocturne, D major. cf Mouvements perpetuels, nos. I, II, III.
    Sonata, brass: Allegro. cf Argo SPA 464.
2042 Sonata, clarinet and bassoon. PROKOFIEV: Overture on Hebrew
        themes, op. 34. SPOHR: German songs (6). Alice Howland, s;
        David Weber, clt; Leonard Sharrow, bsn; Instrumental Ensemble.
        Grenadilla GS 1004.
          +NR 8-76 p7
    Sonata, clarinet and bassoon. cf Works, selections (HMV EMSP 553).
    Sonata, clarinet and piano. cf Works, selections (HMV EMSP 553).
    Sonata, 2 clarinets. cf Works, selections (HMV EMSP 553).
    Sonata, flute and piano. cf Works, selections (HMV EMSP 553).
    Sonata, flute and piano. cf BARTOK: Hungarian peasant suite.
    Sonata, horn, trumpet and trombone. cf Works, selections (HMV EMSP
        553).
    Sonata, oboe and piano. cf Works, selections (HMV EMSP 553).

2043 Sonata, 2 pianos. RACHMANINOFF: Suite, 2 pianos, no. 1, op. 5.
         Nadya and Steven Gordon, pno. Klavier KS 549.
            ++NR 2-76 p14   +St 6-76 p106
      Sonata, violin and piano. cf Works, selections (HMV EMSP 553).
      Sonata, violoncello and piano. cf Works, selections (HMV EMSP
         553).
      Songs: L'Anguille; La belle jeunesse; Priez pour paix; Serenade.
         cf FAURE: Songs (1750 Arch S 1754).
      Songs: La belle si nous étions; Clic, clac, dansez sabots; Chan-
         son a boire; Laudes de Saint Antoine de Padoue; Petites prieres
         de Saint François d'Assise (4). cf HMV CSD 3740.
      Songs: Les chemins de l'amour. cf Columbia M 33933.
      Songs: Fêtes galantes; La reine de coeur; Hotel; Sanglots. cf
         Caprice CAP 1107.
2044 The story of Babar, the little elephant (orch. Françaix). SAINT-
         SAENS: The carnival of the animals. Peter Ustinov, narrator;
         Aldo Ciccolini, Alexis Weissenberg, pno; OSCCP; George Prêtre.
         HMV ESD 7020. (Reissues from ASD 2286, 2316).
            +Gr 11-76 p888   +RR 11-76 p70
      Suite française. cf Mouvements perpetuels, nos. I, II, III.
      Suite française. cf Orion ORS 76231.
2045 Works, selections: Elégie, horn and piano. Sonata, clarinet and
         piano. Sonata, clarinet and bassoon. Sonata, 2 clarinets.
         Sonata, flute and piano. Sonata, horn, trumpet and trombone.
         Sonata, oboe and piano. Sonata, violin and piano. Sonata,
         violoncello and piano. Michel Portal, Maurice Gabai, clt;
         Jacques Février, pno; Alan Civil, hn; John Wilbraham, tpt;
         John Iveson, trom; Amaury Wallez, bsn; Yehudi Menuhin, vln;
         Pierre Fournier, vlc; Michael Debost, flt; Michael Borgue, ob.
         HMV EMSP 553 (2).
            ++Gr 3-76 p1481   +MT 8-76 p661
            +HFN 5-76 p105   +-RR 4-76 p64
            +MM 7-76 p30
POULTON
      Aura Lee. cf CBS 61746.
POWELL
      Cardiff Castle. cf Philips 6308 246.
POWELL, Mel
2046 Divertimenti (2). Trio. Herbert Sorkin, vln; Margaret Ross, hp;
         Fairfield Wind Ensemble, Helura Trio. CRI SD 121.
            +NR 7-76 p6
      Trio. cf Divertimenti (2).
PRAETORIUS, Michael
      Ballet. cf Swedish Society SLT 33189.
      Ballet du Roy pour Sonner apres. cf Coronet LPS 3032.
      La bourrée. cf DG Archive 2533 184.
      Galliarde de la guerre. cf DG Archive 2533 184.
      Galliarde de Monsieur Wustron. cf DG Archive 2533 184.
      Gavotte. cf DG Archive 2533 184.
2047 Hosianna dem Sohne Davids. Ein Kind geborn zu Bethlehem. Nun
         komm, der Heiden Heiland. Psallite unigenito Christo. Terpsi-
         chore: Ballet des sorciers; Bransle double; Gaillarde; Sara-
         bande; Ballet des feus; Pavane Spaigne; La rosette; Bransle
         gentil; Volte; Courante. Von Himmel hoch. SCHEIN: Banchetto
         musicale: Suite, no. 1, G major; no. 2, D minor. Hannover
         Niedersächsischer Choir; Ferdinand Conrad Instrumental Ensemble;

Willi Träder. Nonesuch H 71128.
    +HFN 3-76 p112   /RR 12-75 p91
    +-MT 3-76 p236
Ein Kind geborn zu Behtlehem. cf Hosianna dem Sohne Davids.
Motets: Allein Gott in der Höh sei Ehr; Aus tiefer Not schrei
    ich zu dir; Christus, der uns selig macht; Erhalt uns, Herr,
    bei deinem Wort; Gott der Vater wohn uns bei; Resonet in
    laudibus. cf Terpsichore.
Nun komm, der Heiden Heiland. cf Hosianna dem Sohne Davids.
Partita. cf Swedish Society SLT 33200.
Psallite unigenito Christo. cf Hosianna dem Sohne Davids.
Reprinse. cf DG Archive 2533 184.
Spagnoletta. cf DG Archive 2533 184.
Songs: Deck the halls; Es ist ein Ros entsprungen; Gloria, Gott
    in der Höh. cf RCA PRL 1-8020.
2048 Terpsichore (1612): La bourrée; Courante M. M Wustrow; Galliard;
      Reprinse secundam inferiorem; Passameze; Pavane de Spaigne; La
      sarabande; Spagnoletta; Suite de ballets; Suite be voltes.
      Motets: Allein Gott in der Höh sei Ehr; Aus tiefer Not schrei
      ich zu dir; Christus, der uns selig macht; Erhalt uns, Herr,
      bei deinem Wort; Gott der Vater wohn uns bei; Resonet in
      laudibus. Early Music Consort; St. Alban's Abbey Choir,
      Boys' voices; David Munrow. HMV CSD 3761. (also Angel S
      37091).
          +Gr 11-74 p904   +NYT 8-15-76 pD15
          +HF 6-75 p101   +-RR 12-74 p46
          -NR 4-75 p6   +St 11-75 p142
      Terpsichore: Ballet des sorciers; Bransle double; Gaillarde;
      Sarabande; Ballet des feus; Pavane Spaigne; La rosette; Brans-
      le gentil; Volte; Courante. cf Hosianna dem Sohne Davids.
      Terpsichore: Pavane; Spagnoletta; Pavane and galliarde. cf CBS
      76183.
      Volta. cf Swedish Society SLT 33189.
      Von Himmel hoch. cf Hosianna dem Sohne Davids.
PRATT, George
      Songs: By the waters of Babylon; The earth is the Lord's. cf
      Grosvenor GRS 1039.
PRATT, Paul
      Hot house rag. cf Washington University Press OLY 104.
PREIS
      O Du mein Oesterreich. cf DG 2721 077.
PRESTI, Ida
      Danse d'Avila. cf Delos DEL FY 008.
PRESTON, Thomas
      Beatus Laurentius. cf APPLEBY: Magnificat.
PREVIN, André
      André Previn's music night: Signature tune. cf HMV (Q) ASD 3131.
PRICE, Maldwyn
      Heroic march. cf HMV OU 2105.
      A PROCESSION OF VOLUNTARIES. cf Cambridge CRS 2540.
PROKOFIEV, Serge
2049 Alexander Nevsky, op. 78. Lili Chookasian, con; Westminster
      Symphony Choir; NYP; Thomas Schippers. CBS 61769. (Reissue)
      (also Odyssey Y 31014).
          +-RR 12-76 p94
2050 Alexander Nevsky, op. 78. Betty Allen, ms; Mendelssohn Club

Chorus; PO; Eugene Ormandy. RCA ARL 1-1151. Tape (c) ARK
1-1151 (ct) ARS 1-1151 (Q) ARD 1-1151.
+Audio 3-76 p64   +-NR 11-75 p9
+Gr 4-76 p1649   +-ON 1-24-76 p51
+HF 2-76 p102   +-RR 4-76 p73
+-HF 2-76 p102 Quad   ++SFC 11-9-75 p22
+-HFN 4-76 p111   ++St 3-76 p120
+-MM 11-76 p42

2051 Autumnal, op. 8. Concerto, piano, no. 3, op. 26, C major. Sym-
phony, no. 1, op. 25, D major. Vladimir Ashkenazy, pno; LSO;
André Previn, Vladimir Ashkenazy. Decca SXL 6768. Tape (c)
KSXC 6768. (Reissues from 15BB 218).
+-Gr 11-76 p803   +-RR 12-76 p63
+HFN 12-76 p151
Autumnal, op. 8. cf Concerti, piano, nos. 1-5.

2052 Betrothal in the monastery (The duenna), op. 86: Summer night
suite. RIMSKY-KORSAKOV: Le coq d'or: Suite. Bournemouth
Symphony Orchestra; Paavo Berglund. HMV (Q) ASD 3141.
+Gr 12-75 p1045   +-RR 1-76 p37
+HFN 1-76 p115

2053 Cinderella, op. 87. MRSO; Gennady Rozhdestvensky. HMV SXDW 3026
(2). (also Melodiya/Angel S 4102).
+RR 10-76 p65
Cincerella, op. 87: Gavotte. cf CHOPIN: Scherzo, no. 4, op. 54,
E major.
Cinderella, op. 87: Introduction; Quarrel; The dancing lesson;
Spring fairy; Summer fairy; Grasshoppers dance; Winter fairy;
The interrupted departure; Clock scene; Cinderella's arrival
at the ball; Grande valse; Cinderella's waltz; Midnight;
Apotheosis. cf BRITTEN: Young person's guide to the orchestra,
op. 34.

2054 Concerti, piano, nos. 1-5. Autumnal, op. 8. Overture on Hebrew
themes, op. 34. Symphony, no. 1, op. 25, D major. Vladimir
Ashkenazy, pno; Keith Puddy, clt; LSO, Gabrieli Quartet; Vladi-
mir Ashkenazy. Decca 15BB 218/20 (3). (also London CSA 2314
Tape (c) CSA 5-2314).
++Gr 10-75 p622   +NR 9-76 p6
+HF 7-76 p86   +RR 9-75 p45
+-HF 12-76 p147 tape   ++SFC 4-11-76 p36
++HFN 9-75 p102   ++St 6-76 p73
+MJ 12-76 p28   +Te 3-76 p34

2055 Concerto, piano, no. 1, op. 10, D flat major. Concerto, piano,
no. 2, op. 16, G minor. Overture on Hebrew themes, op. 34.
Vladimir Ashkenazy, pno; Keith Puddy, clt; Gabrieli Quartet,
LSO; André Previn. Decca SXL 6767. Tape (c) KSXC 6767.
(Reissue from 15BB 218).
+Gr 6-76 p51   +-RR 5-76 p47
++HFN 6-76 p102   +RR 8-76 p84 tape
+-HFN 7-76 p105 tape

2056 Concerto, piano, no. 2, op. 16, G minor. Concerto, piano, no. 5,
op. 55, G major.* Jorge Bolet, Alfred Brendel, pno; Cincinnati
Symphony Orchestra, VSOO; Thor Johnson, Jonathan Sternberg.
Turnabout TV 34543. (*Reissue from PLP 527).
-Gr 1-76 p1203   +-RR 11-75 p47
+HFN 11-75 p173
Concerto, piano, no. 2, op. 16, G minor. cf Concerto, piano, no.
1, op. 10, D flat major.

Concerto, piano, no. 3, op. 26, C major.   cf Autumnal, op. 8.
Concerto, piano, no. 3, op. 26, C major.   cf MUSSORGSKY: Pictures
   at an exhibition.
Concerto, piano, no. 5, op. 55, G major.   cf Concerto, piano, no.
   2, op. 16, G minor.
2057 Concerto, violin, no. 1, op. 19, D major.   Concerto, violin, no.
      2, op. 63, G minor.   Pierre Amoyal, vln; Strasbourg Philhar-
      monic Orchestra; Alain Lombard.   Erato STU 70866.
            +Gr 12-75 p1038   +-MT 6-76 p495
            ++HFN 2-76 p109   ++RR 12-75 p57
            ++MM 4-76 p29
2058 Concerto, violin, no. 1, op. 19, D major.   SZYMANOWSKI: Concerto,
      violin, no. 1, op. 35.   Shizuka Ishikawa, vln; CPhO; Jan Krenz.
      Supraphon 110 1639.
            +ARG 12-76 p43   ++NR 7-76 p4
            +-Gr 4-76 p1607   +RR 5-76 p58
Concerto, violin, no. 1, op. 19, D major.   cf HMV SLS 5058.
Concerto, violin, no. 2, op. 63, G minor.   cf Concerto, violin,
   no. 1, op. 19, D major.
Concerto, violin, no. 2, op. 63, G minor.   cf RCA ARM 4-0942/7.
Devilish inspiration.   cf Coronet LPS 1722.
Gavotte.   cf RCA ARM 4-0942/7.
Lieutenant Kijé suite, op. 60.   cf KODALY: Háry János: Suite
   (Decca 4355).
Lieutenant Kijé suite, op. 60.   cf KODALY: Háry János: Suite
   (RCA 1-1325).
Lieutenant Kijé suite, op. 60: Wedding.   cf Columbia M 34127.
Love for three oranges, op. 33: March.   cf Columbia M 34127.
Love for three oranges, op. 33: March.   cf RCA CRL 3-2026.
Love for three oranges, op. 33: March; Scherzo.   cf KHACHATURIAN:
   Gayaneh suite.
Love for three oranges, op. 33: Suite.   cf BARTOK: The miraculous
   Mandarin, op. 19: Suite.
March, F minor.   cf RCA ARM 4-0942/7.
Melodies, op. 35 (5).   cf FRANCAIX: Sonatine.
Overture on Hebrew themes, op. 34.   cf Concerti, piano, nos. 1-5.
Overture on Hebrew themes, op. 34.   cf Concerto, piano, no. 1,
   op. 10, D flat major.
Overture on Hebrew themes, op. 34.   cf POULENC: Sonata, clarinet
   and bassoon.
2059 Peter and the wolf, op. 67.   SAINT-SAENS: The carnival of the
      animals.   Hermione Gingold, narrator; Alfons and Aloys Kontar-
      sky, pno; VPO; Karl Böhm.   DG 2530 588.   Tape (c) 3300 588.
            +-Gr 2-76 p1346   *NR 4-76 p3
            +-HF 6-76 p91   +RR 2-76 p36
            +HFN 2-76 p109   +RR 4-76 p80 tape
            +HFN 3-76 p113 tape   +SFC 2-29-76 p25
            +-MJ 5-76 p29   +-St 6-76 p108
Peter and the wolf, op. 67.   cf BRITTEN: The young person's guide
   to the orchestra, op. 34 (Classics for Pleasure CFP 185).
Peter and the wolf, op. 67.   cf BRITTEN: The young person's guide
   to the orchestra (Vanguard 71189).
Poems by Anna Akhmatova, op. 27 (5).   cf Philips 6780 751.
2060 Romeo and Juliet, op. 64, excerpts.   BSO; Erich Leinsdorf.   RCA
      AGL 1-1273.   (Reissue).
            ++NR 1-76 p3

Romeo and Juliet, op. 64, excerpts. cf HMV ESD 7011.
Romeo and Juliet, op. 64: Balcony scene. cf Angel S 37157.
2061 Scythian suite, op. 20. Seven, they are seven, op. 30.* SHOSTA-
    KOVICH: Symphony, no. 2, op. 14, C major. Yuri Elnikov, t;
    MPO, MRSO and Chorus, RSFSR Russian Chorus; Gennady Rozhdest-
    vensky, Kyril Kondrashin. HMV Melodiya ASD 3060. (*Reissue
    from ASD 2669).
         +Gr 5-75 p1968   +-MT 1-76 p42
         +-HFN 5-75 p137   +RR 5-75 p40
    Scythian suite, op. 20. cf BARTOK: The miraculous Mandarin, op.
    19: Suite.
    Seven, they are seven, op. 30. cf Scythian suite, op. 20.
    Sonata, flute and piano, op. 94, D major. cf BARTOK: Hungarian
    peasant suite.
    Sonata, flute and piano, op. 94, D major. cf FRANCK: Sonata,
    violin and piano, A major.
    Sonata, flute and piano, op. 94, D major. cf PAGANINI: Caprice,
    op. 1, no. 24, A minor.
2062 Sonata, piano, no. 2, op. 14, D minor. Sonata, piano, no. 8, op.
    84, B flat major. Tedd Joselson, pno. RCA ARL 1-1570.
         +-Gr 9-76 p446   +-RR 9-76 p83
         +-HF 12-76 p110   +St 11-76 p155
         +HFN 9-76 p127
    Sonata, piano, no. 5, op. 38, C major. cf International Piano
    Library IPL 5003/4.
    Sonata, piano, no. 7, op. 83, B flat major. cf BARBER: Sonata,
    piano, op. 26.
2063 Sonata, piano, no. 8, op. 84, B flat major. Visions fugitives,
    op. 22, nos. 1, 3, 5, 7-8, 10-11, 17. Emil Gilels, pno.
    Columbia/Melodiya M 33824.
         +HF 5-76 p89   +-St 8-76 p98
         +NR 4-76 p12
2064 Sonata, piano, no. 8, op. 84, B flat major. RACHMANINOFF: Mom-
    ents musicaux, op. 16 (6). Lazar Berman, pno. DG 2530 678.
    Tape (c) 3300 678.
         +Gr 3-76 p1487   +-MJ 10-76 p25
         +-Gr 6-76 p102 tape   ++NR 7-76 p11
         +HF 5-76 p75   +-RR 3-76 p62
         +-HF 10-76 p147 tape   +St 8-76 p98
         +HFN 5-76 p105
    Sonata, piano, no. 8, op. 84, B flat major. cf Sonata, piano,
    no. 2, op. 14, D minor.
    Sonata, violin and piano, no. 1, op. 80, F minor. cf BRAHMS:
    Sonata, violin and piano, no. 2, op. 100, A major.
    Sonata, violin and piano, no. 1, op. 80, F minor. cf JANACEK:
    Sonata, violin and piano, op. 21.
    Sonata, violoncello and piano, op. 119. cf DEBUSSY: Sonata,
    violoncello and piano, no. 1, D minor.
2065 The stone flower. Bolshoi Theatre Orchestra; Gennady Rozhdest-
    vensky. Columbia/Melodiya M3 33215 (3). (also HMV Melodiya
    SLS 5024).
         +-Gr 9-75 p466   +NR 4-75 p4
         +HF 5-75 p86   +NYT 3-9-75 pD23
         +HFN 9-75 p102   +RR 8-75 p41
         +MT 7-76 p577   +Te 12-75 p42
2066 The story of a real man, op. 117. Glafira Deomidova, s; Kira

Leonova, ms; Gyorgy Shulpin, Aleksei Maslennikov, t; Yevgeny
Kibkalo, bar; Gennadi Pankov, Mark Reshetin, Artur Eizen, bs;
Bolshoi Theatre Chorus and Orchestra; Mark Ermler. Westminster
WGSO 8317/2 (2).

> +Gr 10-76 p649   +SFC 4-11-76 p38
> +ON 7-75 p30   +St 10-75 p111

Symphony, no. 1, op. 25, D major. cf Autumnal, op. 8.
Symphony, no. 1, op. 25, D major. cf Concerti, piano, nos. 1-5.
Symphony, no. 1, op. 25, D major. cf BIZET: Symphony, C major.
Symphony, no. 1, op. 25, D major. cf KHACHATURIAN: Gayaneh suite.
Symphony, no. 1, op. 25, D major. cf CHABRIER: España.
Symphony, no. 1, op. 25, D major. cf Philips 6780 755.
Symphony, no. 1, op. 25, D major. cf RCA CRL 3-2026.

2067   Symphony, no. 5, op. 100, B flat major. LSO; André Previn. HMV
ASD 3115. Tape (c) TC ASD 3115. (also Angel S 37100)

> +HF 1-76 p92   +-NYT 1-18-76 pD1
> +-HFN 9-75 p102   +RR 9-75 p46
> +-HFN 12-75 p173 tape   +-RR 12-75 p99 tape
> ++NR 1-76 p3   +-St 4-76 p114

2068   Symphony, no. 6, op. 111, E flat minor. LPO; Walter Weller.
Decca SXL 6777.

> +-Gr 9-76 p423   +RR 9-76 p58
> ++HFN 9-76 p126

Tales of an old grandmother, op. 31. cf HMV HQS 1364.
Toccata. cf RCA VH 020.
Visions fugitives, op. 22. cf CHOPIN: Scherzo, no. 4, op. 54,
E major.
Visions fugitives, op. 22, nos. 1, 3, 5, 7-8, 10-11, 17. cf
Sonata, piano, no. 8, op. 84, B flat major.
PROMS FESTIVAL '76. cf HMV SEOM 24.

PROTHEROE
Songs: O mor ben yn y man. cf Columbia SCX 6585.
PRUSSIAN AND AUSTRIAN MARCHES. cf DG 2721 077.

PRYOR, Arthur
The tip topper; The supervisor. cf Washington University Press
OLY 104.

THE PSALMS OF DAVID. cf HMV SQ CSD 3768.

PUCCINI, Giocomo
2069   Arias: La bohème: Sì, mi chiamano Mimi; Donde lieta usci. Gianni
Schicchi: O mio babbino caro. Madama Butterfly: Un bel dì,
vedremo. Manon Lescaut: In quelle trine morbide. Tosca:
Visse d'arte. VERDI: La forza del destino: Son giunta...
Madre, pietosa vergine. Otello: Mia madre aveva...Piangea
cantando...Ave Maria. Il trovatore: Che più t'arresti...Tacea
la notte. Renata Tebaldi, Luisa Maragliano, s; Luisa Ribacchi,
ms; Carlo Bergonzi, t; Santa Cecilia Orchestra and Chorus,
Maggio Musicale Fiorentino Orchestra, Grand Theatre Orchestra,
Geneva; Tullio Serafin, Francesco Molinari-Pradelli, Lamberto
Gardelli, Alberto Erede. Decca SDD 481. (Reissues from SXL
2170/1, 2180/1, 2054/6, 2089/72, 2129/31, LST 2995/7, 5009/11,
SET 236/8).

> +-Gr 5-76 p1801   +RR 5-76 p26
> +HFN 5-76 p115

2070   Arias: La bohème: Che gelida manina...Sì, mi chiamano Mimi...O
soave fanciulla; La commedia e stupenda...Quando me'n vo; In
un coupé...O Mimi, tu più non torni. La fanciulla del West:

Ch'ella mi creda libera. Gianni Schicchi: O mio babbino caro.
Madama Butterfly: Bimba dagli occhi; Un bel di; Scuoti quella
fronda. Manon Lescaut: In quelle trine morbide; Oh, sarò la
più bella...Tu, tu, amore. Tosca: Sante ampolle...Recondita
armonia; Tre sbirri...Te deum; Vissi d'arte; E lucevan le
stelle. Turandot: Signore, ascolta...Non piangere, liù; In
questa reggia; Nessun dorma; Tu che di gel sei cinta. Renata
Tebaldi, Inge Borkh, Gianna d'Angelo, s; Fiorenza Cossoto, ms;
Carlo Bergonzi, Mario del Monaco, Jüssi Bjorling, Piero de
Palma, Mario Carlin, t; Ettore Bastianini, George London,
Renato Cesari, bar; Cesare Siepi, Fernando Corena, Nicola
Zaccaria, bs; Various orchestras and conductors. Decca DPA
553/4 (2).
    +-Gr 6-76 p92   +RR 7-76 p29
    +HFN 6-76 p103

2071 La bohème. Renata Tebaldi, Gianna d'Angelo, s; Carlo Bergonzi,
Piero de Palma, t; Ettore Bastianini, Attilio d'Orazi, bar;
Renato Cesari, Cesare Siepi, Fernando Corena, Giorgio Onesti,
bs; Rome, Santa Cecilia Orchestra and Chorus; Tullio Serafin.
Decca D5D 2 (2). (Reissue from SXL 2170/1).
    +Gr 8-76 p332   +RR 8-76 p23
    +HFN 8-76 p85

2072 La bohème. Victoria de los Angeles, Lucine Amara, s; Jüssi Bjorling, William Nahr, t; Robert Merrill, John Reardon, Thomas
Powell, George del Monte, bar; Giorgio Tozzi, Fernando Corena,
bs; Columbus Boys' Choir; RCA Victor Orchestra and Chorus;
Thomas Beecham. HMV SLS 896. Tape (c) TC SLS 896. (Reissue
from ALP 1409/10). (also Seraphim SIB 6099)
    ++Gr 11-74 p963   +RR 1-76 p67 tape
    +Gr 10-75 p721 tape   +St 7-75 p110
    +RR 11-74 p28

2073 La bohème. Maria Callas, Anna Moffo, s; Giuseppe di Stefano,
Franco Ricciardi, t; Rolando Panerai, Manuel Spatafora, bar;
Nicola Zaccaria, Carlo Badioli, Eraldo Coda, Carlo Forti, bs;
La Scala Opera Orchestra and Chorus; Antonino Votto. HMV SLS
5059 (2). Tape (c) TC SLS 5059. (Reissue from Columbia 33CX
1463/4).
    +Gr 8-76 p332   +-HFN 8-76 p85
    +Gr 12-76 p1066 tape   +-RR 9-76 p34

2074 La bohème: Che gelida manina...Si, mi chiamano Mimi...O soave
fanciulla; Quando me'n vo; Entre'acte...C'e Rodolfo...un ter-
ribil tosse...Donde lieta usci; O Mimi, tu più non torni; Sono
andanti to end of opera. Bidú Sayão, Mimi Benzell, s; Richard
Tucker, Lodovico Oliviero, t; Francesco Valentino, George
Cehanovsky, bar; Salvatore Baccaloni, Nicola Moscona, Lawrence
Davidson, bs; Metropolitan Opera Orchestra and Chorus; Giuseppe
Antonicelli. CBS 30068.
    /-Gr 3-76 p1507   +-RR 3-76 p28
    +-HFN 3-76 p111

La bohème: Addio di Mimi; On m'appelle Mimi; Si mia chiamano Mimi;
Entrate...C'e Rodolfo; Donde lieta usci; Addio dolce svegliare
alla mattina; Gavotta...Minuetto; Sono andanti; Io Musetta...
Oh come e bello e morbido. cf HMV RLS 719.
La bohème: Che gelida manina. cf Cantilena 6238.
La bohème: Che gelida manina. cf Decca SPA 449.
La bohème: Che gelida manina. cf Decca SXL 6649.

La bohème: Che gelida manina. cf Muza SXL 1170.

La bohème: Che gelida manina; Vecchia zimarra. cf Bongiovanni GB 1.

La boheme: Che gelida manina...Sì, mi chiamano Mimì...O soave fanciulla, la commedia è stupenda...Quando me'n vo; In un coupé ...O Mimì, tu più non torni. cf Arias (Decca DPA 553/4).

La bohème: Donde lieta usci. cf HMV SLS 5057.

La bohème: In un coupé; Ah Mimì, tu più non torni. cf RCA LRL 2-7531.

La bohème: Mi chiamano Mimì. cf Club 99-97.

La bohème: Mi chiamano Mimì; Donde lieta. cf Club 99-100.

La bohème: O soave fanciulla. cf Cantilena 6239.

La bohème: O soave fanciulla; O Mimì, tu più non torni. cf Decca DPA 517/8.

La bohème: Quando m'en vo' soletta. cf CATALANI: La Wally: Ebben, Ne andrò lontana.

La bohème: Quando me'n vo. cf Telefunken AG 6-41945.

La bohème: Sì, mi chiamano Mimì. cf MONIUSZKO: Halka: Gdybym rannyn slonkiem, O moj malenki.

La bohème: Sì, mi chiamano Mimì. cf Rubini GV 58.

La bohème: Sì, mi chiamano Mimì; Donde lieta usci. cf Arias (Decca SDD 481).

La bohème: Sì, mi chiamano Mimì. cf BOITO: Mefistofele: L'altra notte in fondo al mare.

La fanciulla del West: Ch'ella mi creda libera. cf Arias (Decca DPA 553/4).

2075 Gianni Schicchi. Suor Angelica. Il tabarro. Margaret Mas, Sylvia Bertona, Victoria de los Angeles, Lidia Marimpietri, Santa Chissari, Anna Marcangeli, Giuliana Raymondi, s; Miriam Pirazzini, Fedora Barbieri, Mina Doro, Corinna Vozza, Teresa Cantarini, Maria Huder, Anna Maria Canali, ms; Giacinto Prandelli, Piero de Palma, Carlo del Monte, Adelio Zagonara, Claudio Cornoldi, Renato Ercolani, t; Tito Gobbi, Fernando Valentini, bar; Plinio Clabassi, Paolo Montarsolo, Alfredo Mariotti, Saturno Meletti, bs; Rome Opera House Orchestra and Chorus; Vicenzo Bellezza, Tullio Serafin, Gabriele Santini, HMV SLS 5066 (3). (Reissues from ALP 1355, 1577, ASD 295).

+Gr 10-76 p650   +-RR 12-76 p44
+HFN 12-76 p151

Gianni Schicchi: O mil babbino caro. cf Arias (Decca SDD 481).

Gianni Schicchi: O mio babbino caro. cf Arias (Decca DPA 553/4).

Gianni Schicchi: O mio babbino caro. cf CATALANI: La Wally: Ebben, Ne andrò lontana.

Gianni Schicchi: O mio babbino caro. cf Decca SPA 449.

2076 Madama Butterfly. Arias: GIORDANO: Andrea Chénier: Vicino a te... La morte nostra. PUCCINI: Madama Butterfly: Ancora un passo, Bimba dagli occhi, E questo...Che tua madre, Un bel di. Manon Lescaut: Tu, tu, Amore...Tentatrice. VERDI: Otello: Gia la notte, Ave Maria. Margaret Sheridan, s; Ida Mannerini, ms; Lionel Cecil, Nello Palai, t; Vittorio Weinberg, bar; A. Gelli, bs; La Scala Orchestra; Carlo Sabajno. Club 99 OP 1001.

+-NR 5-76 p10

2077 Madama Butterfly. Renata Tebaldi, s; Fiorenza Cossotto, Lidia Nerozzi, ms; Carlo Bergonzi, Angelo Mercuriali, t; Enzo Sordello, bar; Michele Cazzato, Paulo Washington, Virgilio Carbonari, Oscar Nanni, bs; Rome Santa Cecilia Orchestra and Chorus; Tullio Serafin. Decca D4D 3 (3). (Reissue from SXL 2054/6).

```
    +-Gr 8-76 p332          +RR 8-76 p23
    +HFN 8-76 p86
```
2078 Madama Butterfly. Maria Callas, s; Lucia Danieli, Luisa Villa, ms;
      Nicolai Gedda, Renato Ercolani, Mario Carlin, t; Mario Borriello,
      bar; Plinio Clabassi, Enrico Campi, bs; La Scala Opera House
      Orchestra and Chorus; Herbert von Karajan. HMV SLS 5015 (3).
      (Reissue from Columbia 33CX 1296/8).
```
          +Gr 5-76 p1796         +-RR 7-76 p30
          +HFN 7-76 p94
```
2079 Madama Butterfly. Licia Albanese, s; James Melton, t; John
      Brownlee, bar; Metropolitan Opera Orchestra; Pietro Cimara.
      Metropolitan Opera MET 2.
```
          +NYT 7-6-75 pD11        +SR 11-29-75 p50
          +ON 1-24-76 p51
```
2080 Madama Butterfly: Bimba, bimba, non piangere. Manon Lescaut: Oh,
      sarò la più bella. VERDI: Un ballo in maschera: Teco io sto.
      Otello: Già nella notte. Leontyne Price, s; Placido Domingo,
      t; NPhO; Nello Santi. RCA ARL 1-0840. Tape (c) ARK 1-0840
      (ct) ARS 1-0840.
```
          +-Gr 4-76 p1659         +-ON 1-3-76 p33
          +-HF 2-76 p117          +RR 4-76 p35
          +HFN 4-76 p119          +SFC 6-13-76 p30
          +NR 1-76 p10            +St 3-76 p124
          +OC 2-76 p4
```
2081 Madama Butterfly: Dovunque al mondo...Ancora un passo or via; Vieni,
      amor mio; Viene la sera; Un bel di, vedremo; Scuoti quella
      fronda; Humming chorus; Addio fiorito asil; Come una mosca
      prigioniera...Con onor muore. Eleanor Steber, s; Jean
      Madeira, Thelma Votipka, ms; Richard Tucker, Alessio de Paolis,
      t; Giuseppe Valdengo, Melchiorre Luise, George Cehanovsky, bar;
      John Baker, bs; Metropolitan Opera Orchestra and Chorus; Max
      Rudolf. CBS 30067. (Reissue from 78246).
```
          +Gr 4-76 p1654         +-RR 3-76 p28
```
2082 Madama Butterfly: Dovunque al mondo...A quanto cielo; Viene la
      sera; Un bel di vedremo; Ah, m'ha scordata; E questo...Che tua
      madre; Una nava da guerra...Scuoti quella fronda; Humming
      chorus; Io so che alle sua pene...Addio fiorito asil; Con onor
      muore...Tu, tu, piccolo iddio. Mirella Freni, bs, Christa
      Ludwig, Elke Schary, ms; Luciano Pavarotti, Michel Sénéchal,
      t; Robert Kerns, Giorgio Stendoro, Siegfried Rudolf Frese, bar;
      Marius Rintzler, Hans Helm, bs; VSOO Chorus; VPO: Herbert von
      Karajan. Decca SET 605. (Reissue from SET 584/6).
```
          +-Gr 11-76 p868        +-RR 11-76 p42
          +HFN 11-76 p173
```
      Madama Butterfly: Ancora un passo or via. cf HMV SXLP 30205.
      Madama Butterfly: Ancora un passo, Bimba dagli occhi, E questo...
          Che tua madre, Un bel di. cf Madama Butterfly (Club 99 OP 1001).
      Madama Butterfly: Bimba dagli occhi pieni di malia; Una nava da
          guerra...Scouti quella fronda. cf Decca DPA 517/8.
      Madama Butterfly: Bimba dagli occhi; Un bel di; Scuoti quella
          fronda. cf Arias (Decca DPA 553/4).
      Madama Butterfly: Con onor muore...Tu, tu, piccolo iddio. cf
          HMV SLS 5057.
      Madama Butterfly: Dicon ch'oltre mare. cf Cantilena 6239.
      Madama Butterfly: Humming chorus. cf Decca DPA 525/6.
      Madama Butterfly: Un bel di, vedremo. cf Arias (Decca SDD 481).

Madama Butterfly: Un bel di. cf Club 99-97.

Madama Butterfly: Un bel di, vedremo. cf Muza SX 1144.

Manon Lescaut: In quelle trine morbide. cf Arias (Decca SDD 481).

Manon Lescaut: In quelle trine morbide. cf Rubini GV 58.

Manon Lescaut: In quelle trine morbide. cf Rubini GV 70.

Manon Lescaut: In quelle trine morbide; Oh, sarò la più bella...
    Tu, tu, amore. cf Arias (Decca DPA 553/4).

Manon Lescaut: In quelle trine morbide; Sola, perduta, abbandonata.
    cf CATALANI: La Wally: Ebben, Ne andrò lontana.

Manon Lescaut: Intermezzo, Act 3. cf HMV SLS 5019.

Manon Lescaut: Oh, sarò la più bella. cf Madama Butterfly:
    Bimba, bimba, non piangere.

Manon Lescaut: Sola perduta, abbandonata. cf HMV SLS 5057.

Manon Lescaut: Sola perduta, abbandonata. cf HMV SXLP 30205.

Manon Lescaut: Sola perduta, abbandonata. cf Muza SX 1144.

Manon Lescaut: Tu, tu, amore...Tentatrice. cf Madama Butterfly.

Manon Lescaut: Tu, tu, amore, tu; Donna non vidi mai; In quelle
    trine morbide. cf BOITO: Mefistofele: Dio de pietà; Dai campi;
    L'altra notte in fondo al mare.

2083   Messa di gloria, A major. William John, t; Philippe Huttenlocher,
       bs; Gulbenkian Foundation Symphony Orchestra and Choir; Michel
       Corbòz. Erato STU 70890. (also RCA FRL 1-5890 Tape (c) FRK
       1-5890 (ct) FRS 1-5890).

       +-Gr 2-75 p1535   /NR 2-76 p9
       +HF 3-76 p92   +-ON 1-3-76 p33
       +HF 8-76 p70 tape   -RR 4-76 p74
       /HFN 5-76 p107   /SFC 12-26-76 p34
       +MJ 2-76 p33   +SR 1-24-76 p53
       +-MT 8-76 p661   ++St 4-75 p114

       La rondine: Ch' il bel sogno di Doretta. cf CATALANI: La Wally:
       Ebben, Ne andrò lontana.

       La rondine: Ch' il bel sogno di Doretta. cf Telefunken AG 6-41947.

       Suor Angelica. cf Gianni Schicchi.

       Suor Angelica: Senza mamma. cf BELLINI: La sonnambula: Ah, se
       una volta sola...Ah, non credea mirarti...Ah, non giunge.

       Suor Angelica: Senza mamma. cf Bongiovanni GB 1.

       Suor Angelica: Senza mamma, o bimbo. cf CATALANI: La Wally:
       Ebben, Ne andrò lontana.

       Il tabarro. cf Gianni Schicchi.

2084   Tosca. Galina Vishnevskaya, s; Franco Bonisolli, Mario Guggia.
       t; Matteo Manuguerra, bar; Guido Mazzini, Antonio Zerbini,
       Domenico Versaci Medici, Giocomo Bertasi, bs; French National
       Orchestra and Chorus; Mstislav Rostropovich. DG 2707 087 (2).
       Tape (c) 3370 008.

       +Gr 12-76 p1051   +-NR 12-76 p10
       -HF 12-76 p112   -SFC 12-26-76 p34
       +MJ 12-76 p28

2085   Tosca. Carmen Melis, Piero Pauli, Apollo Granforte, soloists; La
       Scala Orchestra; Carlo Sabajno. Discophilia DIS 23/KS 10-11 (2).

       +ON 1-3-76 p33

2086   Tosca. Maria Callas, s; Giuseppe di Stefano, Angelo Mercuriali,
       t; Tito Gobbi, bar; Alvaro Cordova, treble; Franco Calabrese,
       Dario Caselli, Melchiorre Luise, bs; La Scala Orchestra and
       Chorus; Victor de Sabata. HMV SLS 825 (2). Tape (c) TC SLS
       825. (Reissue from Columbia CX 1094/5).

<pre>
        +Gr 3-73 p1729   +Op 5-73 p438
        +HFN 3-73 p567   +RR 3-73 p38
        +-HFN 12-75 p173 tape   +RR 1-76 p67 tape
</pre>

Tosca: Amara sol per te m'era il morire. cf Court Opera Classics
    CO 347.
Tosca: E lucevan le stelle. cf Cantilena 6238.
Tosca: E lucevan le stelle. cf Decca SKL 5208.
Tosca: E lucevan le stelle. cf Decca SXL 6649.
Tosca: E lucevan le stelle (3 versions). cf Everest SDBR 3382.
Tosca: Mario, Mario. cf Decca DPA 517/8.
Tosca: Recondita armonia, E lucevan le stelle. cf Muza SXL 1170.
Tosca: Sante ampolle...Recondita armonia; Tre sbirri...Te deum;
    Vissi d'arte; E lucevan le stelle. cf Arias (Decca DPA 553/4).
Tosca: Te deum. cf RCA LRL 2-7531.
Tosca: Tre sbirri...Te Deum. cf Decca SPA 449.
Tosca: Vissi d'arte. cf Arias (Decca SDD 481).
Tosca: Vissi d'arte. cf BOITO: Mefistofele: L'altra notte in
    fondo al mare.
Tosca: Vissi d'arte. cf MONIUSZKO: Halka: Arias, Acts 2, 4.
Tosca: Vissi d'arte. cf Club 99-97.
Tosca: Vissi d'arte; Quanto...Gia mi docon venal. cf Club 99-100.
Tosca: Vissi d'arte. cf HMV RLS 719.
Tosca: Vissi d'arte. cf Rubini GV 70.
Tosca: Vissi d'arte. cf Rubini RS 301.
Tosca: Vissi d'arte. cf Telefunken AG 6-41947.
Turandot: In questa reggia. cf HMV SLS 5057.
Turandot: In questa reggia. cf HMV SXLP 30205.
Turandot: Nessun dorma. cf Bongiovanni GB 1.
Turandot: Nessun dorma. cf Decca SXL 6649.
Turandot: Nessun dorma. cf Rococo 5383.
Turandot: Signore, ascolta...Non piangere, liù; In questa reggia;
    Nessun dorma; Tu che di gel sei cinta. cf Arias (Decca DPA
    553/4).
Turandot: Tu che di gel sei cinta; Signore ascolta. cf BOITO:
    Mefistofele: L'altra notte in fondo al mare.
Le Villi: Non ti scordar di me. cf CATALANI: La Wally: Ebben,
    Ne andrò lontana.
PUJOL VILARRUBI, Emilio
    El abejorro. cf London STS 15306.
    El abejorro. cf Swedish Society SLT 33189.
    Guajira. cf Enigma VAR 1015.
    Tango. cf Enigma VAR 1015.
PURCELL, Daniel
    Psalm 100. cf Smithsonian Collection N 002.
PURCELL, Henry
    Abdelazer: Incidental music. cf Works, selections (L'Oiseau-
    Lyre DSLO 504).
    Abdelazer: Rondeau. cf Columbia M 33514.
    Air and minuet. cf Saga 5426.
    Ayre, D minor. cf CBS 61648.
    Ayre, G major. cf Columbia M 33514.
    Bonduca: Trumpet tune. cf Columbia M 33514.
    Chaconne, F major. cf Orion ORS 76216.
    Chacony, G minor. cf HANDEL: Trio sonata, op. 2, no. 1, C minor.
2087 Dido and Aeneas. Mary Thomas, Honor Sheppard, s; Helen Watts,
    alto; Robert Tear, t; Maurice Bevan, bar; Harald Lister, hpd;

Oriana Concert Orchestra and Choir; Alfred Deller. Vanguard
SRV 279SD. (also Bach Guild HM 46). (Reissue from Philips
SAL 3511).
+Audio 12-76 p90   +HFN 7-76 p103
+-Gr 2-73 p1557   ++RR 12-72 p41
-Gr 8-76 p333   +-RR 6-76 p31
-HFN 1-73 p121

Dido and Aeneas: Dido's recitative and lament. cf Rococo 5380.
Dido and Aeneas: With drooping wings; Sailors' dance. cf Coronet
LPS 3032.
Distressed innocence: Incidental music. cf Works, selections
L'Oiseau-Lyre DSLO 504).
Distressed innovence: Rondeau; Air; Minuet. cf National Trust
NT 002.
2088 The fairy queen. The Deller Consort; Stour Music Festival
Orchestra and Chorus; Alfred Deller. Vanguard SRV 311/12 (2).
+HF 1-76 p92   ++St 11-75 p88
+NR 8-75 p11
Fanfare, C major. cf CBS 61648.
Fantasia. cf BYRD: Fantasias, nos. 2 and 3.
The Gordian knot untied: Incidental music. cf Works, selections
(L'Oiseau-Lyre DSLO 504).
Ground, D minor. cf BYRD: Fantasias, nos. 2 and 3.
The Indian Queen: Trumpet overture. cf Philips 6580 114.
The Indian Queen: Trumpet tune. cf Columbia M 33514.
King Arthur: Fairest isle. cf HMV RLS 716.
The libertine: In these delightful pleasant groves. cf Vista
VPS 1022.
The married beau: Incidental music. cf Works, selections
(L'Oiseau-Lyre DSLO 504).
Martial air. cf CBS 61648.
Musick and ayres: Rondeau, Gavotte, Minuet, Trumpet tune. cf
Argo ZDA 203.
A new ground, E minor. cf BYRD: Fantasias, nos. 2 and 3.
O God, thou art my God: Hallelujah, Christ is made the sure
foundation. cf Abbey LPB 761.
Ode for St. Cecilia's day. cf Ode for the birthday of Queen Mary.
2089 Ode for the birthday of Queen Mary. Ode for St. Cecilia's day.
Christina Clarke, Honor Sheppard, s; Mark Deller, Alfred Deller,
c-t; Neil Jenkins, t; Maurice Bevan, bar; Stour Music Festival
Chamber Orchestra; Alfred Deller. Harmonia Mundi HMD 222.
+-HFN 3-76 p101   +-RR 4-76 74
Overtures, D minor, G minor. cf BYRD: Fantasias, nos. 2 and 3.
Overture and suite, G major. cf BYRD: Fantasias, nos. 2 and 3.
Parts on a ground, D major (3). cf BYRD: Fantasias, nos. 2 and 3.
Pavans, B flat major. A minor. cf BYRD: Fantasias, nos. 2 and 3.
Pavan of four parts, G minor. cf BYRD: Fantasias, nos. 2 and 3.
Sefauchi's farewell, D minor. cf BYRD: Fantasias, nos. 2 and 3.
Sonata, A minor. cf BYRD: Fantasias, nos. 2 and 3.
Sonatas, trumpet, nos. 1 and 2, D major. cf Erato STU 70871.
Sonata, trumpet, no. 1, D major. cf RCA CRL 3-1430.
Sonata, trumpet, no. 2, D major. cf RCA CRL 3-1430.
2090 Songs (verse anthems): Behold, I bring you glad tidings; I was
glad; In Thee, O Lord do I put my trust; O give thanks; O
Lord, God of Hosts. Lynton Atkinson, treble; Paul Esswood,
c-t; Ian Partridge, Antony Dawson, t; Stafford Dean, bs; John

Scott, org; St. John's College Chapel Choir; George Guest.
Argo ZRG 831.
+-Gr 2-76 p1368   +-MT 7-76 p578
+-MM 8-76 p34   +RR 2-76 p61
2091 Songs (verse anthems): Behold, I bring you glad tidings; In thee,
O Lord, do I put my trust; I was glad; O give thanks; O Lord
God of hosts. St. John's College Chapel Choir; Soloists and
strings; George Guest. Argo ZRG 4311.
+HFN 3-76 p101
2092 Songs (Birthday odes for Mary II): Come, ye sons of art; Love's
goddess sure. Norma Burrows, s; James Bowman, Charles Brett,
c-t; Robert Lloyd, bs; Christopher Hogwood, hpd and org; Oliver
Brooks, vla da gamba; Early Music Consort; David Munrow. HMV
ASD 3166.
+-Gr 4-76 p1650   +MT 7-76 p578
+HFN 4-76 p111   +RR 4-76 p74
+-MM 8-76 p34
Songs: Let us wander; Sound the trumpet; Two daughters of this
aged stream. cf BIS LP 17.
Songs: Lucinda is bewitching fair; See where repenting Celia lies.
cf Works, selections (L'Oiseau-Lyre DSLA 504).
Songs: Remember not Lord our offences. cf Abbey LPB 750.
Suite, D major. cf BYRD: Fantasias, nos. 2 and 3.
Te deum. cf HMV SLS 5047.
Three parts upon a ground: Fantasia. cf HMV SLS 5022 (2).
Trio sonata, no. 11, F minor. cf CBS 73487.
Trumpet tune (arr. Trevor). cf Argo 5BBA 1013/5.
Trumpet tune. cf Decca DPA 523/4.
Trumpet tune. cf Klavier KS 551.
Trumpet tune, Bonduca. cf CBS 61648.
Trumpet tune, Cebell. cf CBS 61648.
Trumpet tune (The cebell), D major. cf Philips 6500 926.
Trumpet tune and air. cf BBC REB 228.
Trumpet tune and air. cf Philips 6500 926.
Voluntary, C major. cf CBS 61648.
Voluntary, C major. cf Columbia M 33514.
When the cock begins to crow. cf Decca KCSP 367.
2093 Works, selections: Abdelazer: Incidental music. Distressed
innocence: Incidental music. The Gordian knot untied: Inci-
dental music. The married beau: Incidental music. Songs:
Lucinda is bewitching fair; See where repenting Celia lies.
Joy Roberts, s; Academy of Ancient Music; Christopher Hogwood.
L'Oiseau-Lyre DSLO 504.
+-Gr 6-76 p51   +RR 6-76 p51
+HFN 6-76 p95   +SFC 12-12-76 p55
Yorkshire feast song. cf Philips 6580 114.
PURSWELL, Patrick
It grew and grew. cf CRI SD 324.
PYKINI (14th century)
Plasanche or tost. cf HMV SLS 863.
QUANTZ, Johann
Sonata, flute, op. 1, no. 1, A minor. cf DUTILLEUX: Sonatine,
flute and piano.
Sonata, flute, op. 1, no. 2, B flat major. cf DUTILLEUX: Sonatine,
flute and piano.
QUILTER, Roger
It was a lover and his lass, op. 23, no. 3. cf HMV ASD 2929.

2094   Songs: Dream valley, op. 21, no. 1; Elizabethan lyrics, op. 12
         (7); From lilac; Go, lovely rose, op. 24, no. 3; Love's phil-
         osophy, op. 3, no. 1; Now sleeps the crimson petal, op. 3,
         no. 2; Shakespeare songs, op. 6 (3); Shakespeare songs, op. 30
         (4); Shakespeare songs, op. 23 (5). David Johnston, t;
         Christopher Keyte, bar; Daphne Ibbot, Rae de Lisle, pno.
         Pearl SHE 531.
               +Gr 11-76 p858   +-STL 10-10-76 p37
               +-RR 11-76 p100
       Songs: Now sleeps the crimson petal, op. 3, no. 2. cf Argo
         ZFB 95/6.
       Songs: Now sleeps the crimson petal, op. 3, no. 2. cf HMV RLS
         716.
       Songs: Now sleeps the crimson petal, op. 3, no. 2. cf Pearl SHE
         528.
RABAUD
       Solo de concours. cf Grenadilla GS 1006.
RACHMANINOFF, Sergei
2095   Aleko. Blagovesta Karnobatlova-Dobreva, s; Tony Khristova, con;
         Pavel Kurshumov, t; Nikola Gyuzelev, Dimitre Petkov, bs;
         Bulgarian Radio and Television Vocal Ensemble; Plovdiv
         Symphony Orchestra; Russlan Raychev. Balkanton BOA 1530 (2).
         (also Monitor HS 90102/3, also Harmonia Mundi HMV 135).
               +Gr 9-74 p578   +ON 3-6-76 p42
               +-HF 1-76 p93   +RR 11-75 p34
               +HFN 1-76 p117   +SFC 11-9-75 p22
               +-NR 8-75 p11   /St 10-75 p116
       Barcarolle, op. 10, no. 3, G minor. cf CHOPIN: Piano works
         (Everest SDBR 3377).
2096   The bells, op. 35. Vocalise, op. 34, no. 14. Sheila Armstrong,
         s; Robert Tear, t; John Shirley-Quirk, bar; LSO and Chorus;
         André Previn. HMV SQ ASD 3284. (also Angel S 37169).
               +-Gr 12-76 p1041   ++RR 11-76 p100
               ++HFN 12-76 p145
2097   Concerto, piano, no. 1, op. 1, F sharp minor. Concerto, piano,
         no. 2, op. 18, C minor. Tamás Vásáry, pno; LSO; Yuri Ahrono-
         vitch. DG 2530 717. Tape (c) 3300 717.
               +-Gr 11-76 p803   +-RR 11-76 p76
               +-HFN 11-76 p165
       Concerto, piano, no. 1, op. 1, F sharp minor: Finale. cf Works,
         selections (CBS 30089).
2098   Concerto, piano, no. 2, op. 18, C minor. Rhapsody on a theme by
         Paganini, op. 43. Werner Haas, pno; Franfurt Radio Symphony
         Orchestra; Eliahu Inbal. Philips 6500 920
               +-Gr 3-76 p1468   +-RR 2-76 p35
               -HFN 2-76 p111
2099   Concerto, piano, no. 2, op. 18, C minor. TCHAIKOVSKY: Concerto,
         piano, no. 1, op. 23, B flat minor. Dinorah Varsi, pno;
         Rotterdam Philharmonic Orchestra; Lamberto Gardelli. Philips
         6580 141.
               +Gr 11-76 p803   -RR 11-76 p84
               +-HFN 11-76 p165
       Concerto, piano, no. 2, op. 18, C minor. cf Concerto, piano,
         no. 1, op. 1, F sharp minor.
       Concerto, piano, no. 2, op. 18, C minor. cf Works, selections
         (Decca DPA 565/6).

Concerto, piano, no. 2, op. 18, C minor. cf GRIEG: Concerto,
     piano, op. 16, A minor.
Concerto, piano, no. 2, op. 18, C minor. cf MENDELSSOHN: Fantasie-
     caprice, op. 16, no. 2: Scherzo.
Concerto, piano, no. 2, op. 18, C minor. cf HMV SLS 5033.
Concerto, piano, no. 2, op. 18, C minor. cf HMV SLS 5068.
2100 Concerto, piano, no. 3, op. 30, D minor. Byron Janis, pno; BSO;
     Charles Munch. Camden CCV 5043.
          +HFN 4-76 p123   +RR 5-76 p48
2101 Concerto, piano, no. 3, op. 30, D minor. Alicia de Larrocha, pno;
     LSO; André Previn. Decca SXL 6746. (also London CS 6977
     Tape (c) 5-6977).
          +-ARG 11-76 p37   ++NR 12-76 p6
          +-Gr 3-76 p1471   +-RR 2-76 p35
          +HFN 2-76 p111
2102 Concerto, piano, no. 3, op. 30, D minor. Yevgeny Mogilevsky,
     pno; MPO; Kiril Kondrashin. HMV Melodiya SXLP 30218. (also
     Angel S 40226).
          /Gr 8-76 p297   +RR 8-76 p49
          +-HFN 8-76 p86
2103 Concerto, piano, no. 3, op. 30, D minor. SCHUMANN: Sonata, piano,
     no. 3, op. 14, F minor. Byron Janis, pno; LSO; Antal Dorati.
     Philips Tape (c) 7321 016.
          +Gr 10-76 p658 tape
2104 Concerto, piano, no. 3, op. 30, D minor. Vladimir Ashkenazy, pno;
     PO; Eugene Ormandy. RCA ARL 1-1324. Tape (c) ARK 1-1324 (ct)
     ARS 1-1324 (Q) ARD 1-1324.
          ++Gr 5-76 p1765   ++NR 4-76 p4
          ++HF 5-76 p90   ++RR 5-76 p48
          ++HF 8-76 p70 tape   ++SFC 4-25-76 p30
          +HFN 7-76 p94   ++St 8-76 p98
          +MJ 4-76 p30
Concerto, piano, no. 4, op. 40, G minor: Largo. cf Works, selec-
     tions (CBS 30089).
Daisies, op. 38, no. 3. cf Piano works (Columbia M 33998).
Daisies, op. 38, no. 3. cf RCA ARM 4-0942/7.
Elegie, op. 3, no. 1. cf CHOPIN: Piano works (Everest SDBR 3377).
2105 Etudes tableaux, opp. 33 and 39. Jean-Philippe Collard, pno.
     Connoisseur Society CS 2075.
          +MJ 4-76 p30
Etudes tableaux, op. 33 (8). cf Piano works (Columbia M 33998).
Etudes tableaux, op. 33, no. 3, E flat major. cf BACH: Sonata,
     violin, no. 1, S 1001, G minor.
Etudes tableaux, op. 33, no. 4, B minor. cf CBS 76420.
Etudes tableaux, op. 37, no. 2. cf RCA ARM 4-0942/7.
Etudes tableaux, op. 39, no. 4. cf Columbia M2 33444.
Fantasy pieces, nos. 1, 2. cf Piano works (Columbia M 33998).
2106 The isle of the dead, op. 29. Symphonic dances, op. 45. LSO;
     André Previn. HMV SQ ASD 3259. Tape (c) ASD 3259. (also
     Angel S 37158).
          +Gr 9-76 p423   +RR 9-76 p59
          +-HFN 9-76 p127   ++SFC 10-10-76 p32
          +HFN 11-76 p175 tape
Lilacs, op. 21, no. 5. cf Piano works (Columbia M 33998).
Melodie, op. 3, no. 3, E major. cf CHOPIN: Piano works (Everest
     SDBR 3377).

       Moments musicaux, op. 16 (6). cf PROKOFIEV: Sonata, piano, no. 8,
          op. 84, B flat major.
       Oriental sketch. cf Piano works (Columbia M 33998).
       Oriental sketch. cf RCA ARM 4-0942/7.
2107 Piano works: Daisies, op. 38, no. 3. Etudes tableaux, op. 33 (8).
          Fantasy pieces, nos. 1, 2. Lilacs, op. 21, no. 5. Oriental
          sketch. Variations on a theme by Corelli, op. 42. Ruth Laredo,
          pno. Columbia M 33998.
              +-MJ 12-76 p28   +St 10-76 p129
              +NR 7-76 p11
       Polichinelle, op. 3, no. 4, F sharp minor. cf CHOPIN: Piano works
          (Everest SDBR 3377).
       Polka de W. R. cf CHOPIN: Piano works (Everest SDBR 3377).
       Prelude, C sharp minor. cf CBS 30064.
       Preludes. cf CHOPIN: Scherzo, no. 4, op. 54, E major.
2108 Preludes (24). Vladimir Ashkenazy, pno. Decca Tape (c) KSXC2
          7038.
              +-Gr 6-76 p102 tape   +HFN 5-76 p117 tape
2109 Prelude, op. 3, no. 2, C sharp minor. Preludes, op. 23, nos. 1-10.
          Preludes, op. 32, nos. 1-13. Vladimir Ashkenazy, pno. London
          CSA 2241 (2). Tape (c) 5-2241. (also Decca 5BB 221/2).
              +Gr 2-76 p1360   ++NR 11-76 p12
              +HFN 4-76 p111   ++RR 2-76 p52
              ++MJ 12-76 p28   ++St 10-76 p129
       Prelude, op. 3, no. 2, C sharp minor. cf Works, selections
          (CBS 30089).
       Prelude, op. 3, no. 2, C sharp minor. cf Works, selections
          (Decca DPA 565/6).
       Prelude, op. 3, no. 2, C sharp minor. cf CHOPIN: Piano works
          (Everest SDBR 3377).
       Prelude, op. 3, no. 2, C sharp minor. cf Decca SPA 473.
       Prelude, op. 3, no. 2, C sharp minor. cf Decca PFS 4351.
       Prelude, op. 3, no. 2, C sharp minor. cf Saga 5427.
       Preludes, op. 23, nos. 1-10. cf Prelude, op. 3, no. 2, C sharp
          minor.
       Prelude, op. 23, no. 3, D minor. cf Works, selections (CBS 30089).
       Prelude, op. 23, no. 5, G minor. cf CHOPIN: Piano works (Everest
          SDBR 3377).
       Prelude, op. 23, no. 5, G minor. cf International Piano Library
          IPL 5001/2.
       Prelude, op. 23, no. 6, E flat major. cf Works, selections
          (CBS 30089).
       Prelude, op. 23, no. 10. cf Columbia/Melodiya M 33593.
2110 Preludes, op. 32 (13). Prelude, op. posth, D minor. Ruth Laredo,
          pno. Columbia M 33430.
               +-HF 10-75 p81   +St 10-76 p129
              ++NR 8-75 p12
       Preludes, op. 32, nos. 1-13. cf Prelude, op. 3, no. 2, C sharp
          minor.
       Prelude, op. posth, D minor. cf Preludes, op. 32.
       Rhapsody on a theme by Paganini, op. 43. cf Concerto, piano,
          no. 2, op. 18, C minor.
       Rhapsody on a theme by Paganini, op. 43. cf Works, selections
          (Decca DPA 565/6).
       Rhapsody on a theme by Paganini, op. 43. cf DOHNANYI: Variations
          on a nursery song, op. 25.

Rhapsody on a theme by Paganini, op. 43. cf Philips 6780 755.
Rhapsody on a theme by Paganini, op. 43: 18th variation. cf
   CBS 30072.
The rock, op. 7. cf Symphonies, nos. 1-3. Decca K9K 33.
The rock, op. 7. cf Symphony, no. 3, op. 44, A minor.
Romance, op. 6, no. 1. cf Columbia/Melodiya M 33593.
2111 Sonata, piano, no. 2, op. 36, B flat minor (original version).
      Variations on a theme by Corelli, op. 42. Jean-Philippe Collard,
      pno. Connoisseur Society CS 2082. (Q) CSQ 2082.
           ++HF 12-75 p102   ++SFC 9-28-75 p30
           ++MJ 4-76 p30   ++St 2-76 p107 Quad
           ++NR 11-75 p13
      Sonata, violoncello and piano, op. 19, G minor: Andante. cf
      CBS SQ 79200.
2112 Songs: A-oo, op. 38, no. 6; Daisies, op. 38, no. 3; Dissonance,
      op. 34, no. 13; Dreams, op. 38, no. 5; The harvest of sorrow,
      op. 4, no. 5; How fair this spot, op. 21, no. 7; In my garden
      at night, op. 38, no. 1; The morn of life, op. 34, no. 10; The
      muse, op. 34, no. 1; Oh, never sing to me again, op. 4, no. 4;
      The pied piper, op. 38, no. 4; The poet, op. 34, no. 9; The
      storm, op. 34, no. 3; To her, op. 38, no. 2; Vocalise, op. 34,
      no. 14; What wealth of rapture, op. 34, no. 12. Elisabeth
      Söderström, s; Vladimir Ashkenazy, pno. Decca SXL 6718.
      (also London OS 26428).
           ++Gr 7-75 p228   +NR 11-76 p11
           +-HF 9-76 p95   ++RR 7-75 p59
           ++HFN 7-75 p87   +SR 11-13-76 p52
           +MJ 12-76 p28   +St 10-76 p86
2113 Songs: Arion, op. 34, no. 5; Believe it not, op. 14, no. 7; Day
      to night comparing went the wind her way, op. 34, no. 4; A
      dream, op. 8, no. 5; Fate, op. 21, no. 1; I wait for thee, op.
      14, no. 1; In the silent night, op. 4, no. 3; The little island,
      op. 14, no. 2; Midsummer nights, op. 14, no. 5; Music, op. 34,
      no. 8; The raising of Lazarus, op. 34, no. 6; So dread a fate
      I'll ne'er believe, op. 34, no. 7; So many hours, so many
      fancies, op. 4, no. 6; The soldier's wife, op. 8, no. 4; Spring
      waters, op. 14, no. 11; The world would see thee smile, op. 14,
      no. 6. Elisabeth Söderström, s; Vladimir Ashkenazy, pno.
      Decca SXL 6772.
           +Gr 12-76 p1041   ++RR 12-76 p94
      Songs: Alfred de Musset, op. 21, no. 6, fragment; Child, thou art
      fair as a flower, op. 8, no. 2; A dream, op. 8, no. 5; I wait
      for thee, op. 14, no. 1; In the silent night, op. 4, no. 3;
      Lilacs, op. 21, no. 5; Oh, never sing to me again, op. 4, no.
      4. cf MUSSORGSKY: Songs and dances of death.
      Songs: O cease thy singing, maiden fair; When night descends.
      cf Musical Heritage Society MHS 3276.
2114 Suite, 2 pianos, no. 1, op. 5. Suite, 2 pianos, no. 2, op. 17.
      Vladimir Ashkenazy, André Previn, pno. Decca SXL 6698. Tape
      (c) KSXC 6697. (also London CS 6893).
           +Gr 6-75 p62   ++RR 6-75 p72
           +HFN 8-75 p82   ++RR 2-76 p71 tape
           +MT 10-75 p887   +St 6-76 p106
           +NR 11-75 p13
      Suite, 2 pianos, no. 1, op. 5. cf POULENC: Sonata, 2 pianos.

Suite, 2 pianos, no. 2, op. 17. cf Suite, 2 pianos, no. 1, op. 5.
Suite, 2 pianos, no. 2, op. 17. cf Coronet LPS 1722.
Symphonic dances, op. 45. cf The isle of the dead, op. 29.
Symphonic dance, op. 45, no. 1. cf Works, selections (CBS 30089).
2115   Symphonies, nos. 1-3. The rock, op. 7. LPO, OSR; Walter Weller.
       Decca D9D 3. Tape (c) K9K 33.
           +Gr 9-76 p477   +HFN 10-76 p181
           ++Gr 12-76 p1066 tape   -RR 9-76 p59
2116   Symphony, no. 1, op. 13, D minor. LSO; André Previn. HMV (Q) ASD
       3137. (also Angel (Q) S 37120).
           +Gr 11-75 p827   +NR 11-75 p3
           +HF 12-75 p103   -RR 11-75 p53
           ++HFN 11-75 p163   +St 3-76 p120
           ++HFN 1-76 p125 tape
2117   Symphony, no. 2, op. 27, E minor. LSO; André Previn. Angel S
       36954. Tape (c) 4XS 36954 (ct) 8XS 36954. (also HMV ASD 2889
       Tape (c) TC ASD 2889 (ct) 8X ASD 2889).
           ++Gr 4-73 p1879   +LJ 12-74 p36
           +Gr 7-73 p245 tape   ++NR 12-73 p4
           +Gr 5-74 p2084 tape   +NYT 11-4-73 pD24
           +HF 12-73 p106   -RR 5-73 p64
           +HF 3-76 p116 tape   +St 12-73 p135
           +HFN 4-73 p784
2118   Symphony, no. 2, op. 27, E minor. Hallé Orchestra; James Lough-
       ran. Classics for Pleasure CFP 40065. Tape (c) CT CFP 40065.
           ++Gr 9-74 p511   +RR 9-74 p58
           +-HFN 12-76 p153 tape
2119   Symphony, no. 2, op. 27, E minor. PO; Eugene Ormandy. RCA ARL
       1-1150. Tape (c) ARK 1-1150 (ct) ARS 1-1150.
           +Gr 4-76 p1607   ++NR 11-75 p3
           -HF 12-75 p103   +RR 4-76 p52
           /HF 3-76 p116 tape   +-St 3-76 p120
           +MT 8-76 p661
       Symphony, no. 2, op. 27, E major. cf Works, selections (Decca
       DPA 565/6).
       Symphony, no. 2, op. 27, E minor: Adagio. cf Works, selections
       (CBS 30089).
       Symphony, no. 2, op. 27, E minor: Adagio. cf Angel S 37157.
2120   Symphony, no. 3, op. 44, A minor. The rock, op. 7. LPO; Walter
       Weller. Decca SXL 6720.
           +Gr 10-75 p625   -RR 10-75 p48
           +HFN 10-75 p148   +ST 2-76 p739
2121   Symphony, no. 3, op. 44, A minor. Vocalise, op. 34, no. 14.
       National Philharmonic Orchestra; Leopold Stokowski. Desmar
       DSM 1007.
           ++Gr 4-76 p1608   ++NR 2-76 p2
           +-HF 4-76 p116   ++RR 5-76 p48
           +MJ 3-76 p25   ++St 5-76 p120
2122   Trio elegiaque, op. 9. Andreas Trio. Musical Heritage Society
       MHS 3361.
           +ARG 11-76 p36
       Variations on a theme by Corelli, op. 42. cf Sonata, piano, no.
       2, op. 36, B flat minor.
       Variations on a theme by Corelli, op. 42. cf Piano works (Colum-
       bia M 33998).
       Vocalise, op. 34, no. 14. cf The bells, op. 35.

Vocalise, op. 34, no. 14. cf Symphony, no. 3, op. 44, A minor.
Vocalise, op. 34, no. 14. cf Works, selections (Decca DPA 565/6).
Vocalise, op. 34, no. 14. cf KABALEVSKY: Sonata, violoncello,
   op. 71, B flat major.
Vocalise, op. 34, no. 14. cf CBS 30062.
Vocalise, op. 34, no. 14. cf Command COMS 9006.
Vocalise, op. 34, no. 14. cf HMV SQ ASD 3283.
Vocalise, op. 34, no. 14. cf HMV SXLP 30188.
Vocalise, op. 34, no. 14. cf Odyssey Y 33825.
Vocalise, op. 34, no. 14. cf RCA LRL 1-5131.

2123 Works, selections: Concerto, piano, no. 1, op. 1, F sharp minor:
   Finale. Concerto, piano, no. 4, op. 40, G minor: Largo. Preludes, op. 3, no. 2, C sharp minor; op. 23, no. 3, D minor;
   op. 23, no. 6, E flat major. Symphonic dance, op. 45, no. 1.
   Symphony, no. 2, op. 27, E minor: Adagio. Philippe Entremont,
   pno; PO; Eugene Ormandy. CBS 30089.
      +-RR 12-76 p64

2124 Works, selections: Concerto, piano, no. 2, op. 18, C minor. Prelude, op. 3, no. 2, C sharp minor. Rhapsody on a theme by
   Paganini, op. 43. Symphony, no. 2, op. 27, E major. Vocalise,
   op. 34, no. 14. Julius Katchen, Ilana Vered, Vladimir Ashkenazy, pno; Elisabeth Söderström, s; LSO, LPO; Georg Solti, Hans
   Vonk, Adrian Boult. Decca DPA 565/6.
      +-HFN 12-76 p152   +-RR 12-76 p63

RADECK
   Fridericus-Rex-Grenadiermarsch. cf DG 2721 077.
   Fridericus-Rex. cf Polydor 2489 523.

RADICA, Ruben
   Extensio II. cf KULJERIC: Omaggio à Lukačić.

RAFF, Joachim
   Rigaudon, op. 204, no. 3. cf International Piano Library IPL 102.

RAIMON
   There will come a time for light. cf Hungaroton SLPX 11696.

RAISON, André
   Offertoire "Vive le Roy". cf Wealden WSQ 134.
   Trio en passacaille. cf Argo ZRG 806.

RAMEAU, Jean
   L'Agacante, G major. cf Pièces de clavecin (CRD 1020).
   Les cyclopes, D minor. cf Angel S 36095.
   La Dauphine, G minor. cf Pièces de clavecin (CRD 1020).
   L'Enharmonique, G minor. cf Harmonia Mundi HMU 334.
   L'Entretien des Muses, D minor. cf Agnel S 36095.

2125 Les fetes d'Hebe: Ballet music. Ursula Connors, s; Ambrosian
   Singers; ECO; Raymond Leppard. Angel S 37105. (also HMV ASD
   3084).
      +Gr 6-76 p51   +-ON 10-75 p56
      +HFN 6-76 p95   +RR 6-76 p31
      +MT 11-76 p915   +St 12-75 p129
      ++NR 8-75 p4

   Hippolyte et Aricie: Ah, qu'on daigne du mons; Quelle plaint en
   ces lieux. cf Decca GOSD 674/6.
   L'Indiscrète, B flat major. cf Pièces de clavecin (CRD 1020).
   La livri, C minor. cf Pièces de clavecin (CRD 1020).

2126 Les paladins, excerpts. Anne-Marie Rodde, s; Henri Farge, c-t;
   Jean-Christoph Benoit, bar; Le Grande Ecurie et La Chambre du
   Roy; Jean-Claude Malgoire. Vanguard SRV 318SD.

+–Audio 4-76 p88   +SFC 3-30-75 p16
–NR 4-75 p9   +St 7-75 p105
+–PRO 9/10-76 p19
La pantomime. cf Pièces de clavecin (CRD 1020).
2127 Pièces de clavecin: L'Agacante, G major. La Dauphine, G minor.
L'Indiscrète, B flat major. La livri, C minor. La pantomime.
Suite, A minor. La timide. Trevor Pinnock, hpd. CRD CRD
1020.
+–Gr 6-76 p69   ++RR 6-76 p73
Suite, A minor. cf Pièces de clavecin (CRD 1020).
Suite, E minor. cf Harmonia Mundi HMU 334.
Tambourin. cf L'Oiseau-Lyre DSLO 7.
La timide. cf Pièces de clavecin (CRD 1020).
Zoroastre: Dances. cf BACH, C.P.E.: Minuets, Wq 189.
RANDALL, J. K.
Improvisation. cf ERICKSON: End of the mime.
Music for the film "Eakins". cf CEELY: Elegia.
RANGSTROM, Ture
De fangna. cf Rubini GV 65.
Legends from Lake Mälanen. cf LARSSON: Sonatina, no. 1, op. 16.
Songs: Flickan under nymånen; Pan; Villemo Villemo; Notturno. cf
FRUMERIE: Songs (Swedish Society SLT 33171).
RANKI, György
2128 The tragedy of man, excerpts. Margit Lászlo, s; István Rozsos, t;
György Melis, bar; József Gregor, bs; Debrecen Kodaly Chorus;
Hungarian State Opera Orchestra; János Feréncsik. Hungaroton
SLPX 11714.
+NR 6-76 p12
RAPHLING, Sam
2129 Concerto, piano and percussion. Indiscretions. Movement, piano
and brass quintet. Remembered scene, piano and small dance
band. Sonatina. Gramiston Montague, Sam Raphling, pno; London
Percussion Ensemble, American Brass Quintet, London Melody Group.
Serenus SRS 12061.
–NR 4-76 p6
Indiscretions. cf Concerto, piano and percussion.
Movement, piano and brass quintet. cf Concerto, piano and per-
cussion.
Remembered scene, piano and small dance band. cf Concerto, piano
and percussion.
Sonatina. cf Concerto, piano and percussion.
RASBACH
Songs: Trees. cf Argo ZFB 95/6.
RASI, Francesco
Indarno Febo. cf DG Archive 2533 305.
RATHGEBER, Johann Valentin
Aria, F major. cf Pelca PRSRK 41017/20.
Aria pastorella, G major. cf Pelca PRSRK 41017/20.
Concerto, trumpet, op. 6, no. 15, E flat major. cf Nonesuch H
71301.
RAUH
Ach Elslein. cf BIS LP 22.
RAUTAVAARA, Rinohuhani
2130 Pelimannit, op. 1. SEGERSTAM: Divertimento. SIBELIUS: Canzon-
etta, op. 62, no. 1. Rakastava, op. 14. Suite mignonne, op.
98. Helsinki Chamber Orchestra; Leif Segerstam. BIS LP 19.

```
        +Gr 11-75 p827            +St 5-76 p124
        ++HFN 11-75 p163
    Sonata, flute and guitar.  cf BIS LP 30.
```

RAVEL, Maurice
2131 Boléro. Rapsodie espagnole. Shéhérazade: Ouverture de feerie.
      La valse. Orchestre de Paris; Jean Martinon. Angel S 37147.
      Tape (c) 4XS 31747 (ct) 8XS 37147. (also HMV SQ ASD 3215.
      Reissue from SLS 5016).
```
        +Audio 12-76 p92          +-NR 12-75 p4
        +-Gr 8-76 p297            ++RR 7-76 p62
        +-HF 3-76 p92             ++SFC 11-9-75 p22
        +HF 12-76 p147 tape       ++St 3-76 p120
```
2132 Boléro. Rapsodie espagnole. La valse. BSO; Seiji Ozawa. DG
      2530 475. Tape (c) 3300 459.
```
        +-Audio 8-75 p80          ++NR 6-75 p4
        -Gr 4-75 p1812            +-NYT 1-18-76 pD1
        /HF 6-75 p96              +RR 3-75 p33
        +HF 12-76 p147 tape       ++SFC 4-27-75 p23
        +HFN 6-76 p105 tape       +St 9-75 p111
```
2133 Boléro. TCHAIKOVSKY: Overture, the year 1812, op. 49. BPO;
      Arthur Fiedler. DG 2584 003. (also DG 2535 198).
```
        +-Audio 11-76 p108        +-NR 4-76 p6
        -Gr 5-76 p1807            +RR 6-76 p33
        +-HFN 5-76 p107           ++SFC 3-7-76 p27
        +MJ 5-76 p28
```
2134 Boléro. Ma mère l'oye. Pavane pour une infante défunte. Rap-
      sodie espagnole. Budapest Philharmonic Orchestra; Andras
      Korody. Hungaroton SLPX 11644.
```
        +-NR 4-75 p3              +-St 9-75 p111
        +-RR 3-76 p48
```
      Boléro. cf Works, selections (Decca DPA 561/2).
      Boléro. cf Works, selections (DG 2711 015).
      Boléro. cf Works, selections (DG 2740 120).
      Boléro. cf Works, selections (HMV 5016).
      Boléro. cf CHABRIER: España (HMV 7019).
      Boléro. cf CHABRIER: España (Westminster GRT 8131).
      Boléro. cf Bruno Walter Society RR 443.
2135 Concerto, piano, G major. Concerto, piano, for the left hand,
      D major. Aldo Ciccolini, pno; Orchestre de Paris; Jean Marti-
      non. Angel S 37151. Tape (c) 4XS 37151 (ct) 8XS 37151.
```
        +-HF 8-76 p92             ++NR 6-76 p7
        +HF 12-76 p147 tape
```
2136 Concerto, piano, G major. Concerto, piano, for the left hand,
      D major. Julius Katchen, pno; LSO; István Kertész. Decca SDD
      486. (Reissue from SXL 6209, SXL 6411).
```
        +-Gr 1-76 p1203           /RR 12-75 p58
        +HFN 12-75 p171
```
2137 Concerto, piano, G major. Concerto, piano, for the left hand, D
      major. Anne Queffélec, pno; Strasbourg Philharmonic Orchest-
      ra; Alain Lombard. Erato STU 70928.
```
        +Gr 9-76 p424             +RR 8-76 p49
        +HFN 9-76 p127
```
      Concerto, piano, G major. cf Works, selections (HMV 5016).
      Concerto, piano, G major. cf GRIEG: Concerto, piano, op. 16, A
      minor.
      Concerto, piano, for the left hand, D major. cf Concerto, piano,
      G major (Angel S 37151).

Concerto, piano, for the left hand, D major. cf Concerto, piano,
G major (Decca SDD 486).

Concerto, piano, for the left hand, D major. cf Concerto, piano,
G major (Erato STU 70928).

Concerto, piano, for the left hand, D major. cf Works, selections
(HMV 5016).

2138 Daphnis et Chloe. Orchestre de Paris; Jean Martinon. Angel S
31748. (Q) S 37148 Tape (c) 8XS 37148 (ct) 4XS 37148.

+-HF 3-76 p92	++SFC 12-7-75 p31
++NR 1-76 p4	+St 5-76 p122

2139 Daphnis et Chloe. Camerata Singers; NYP; Pierre Boulez. Columbia
M 33523. (Q) MQ 33523. (also CBS 76425 Tape (c) 40-76425).

+Audio 6-76 p96	+-MJ 11-75 p41
+-Gr 12-75 p1045	++NYT 1-18-76 pD1
+HF 12-75 p104	++ON 3-6-76 p42
+HFN 12-75 p163	++RR 12-75 p58
+HFN 2-76 p116	++St 11-75 p132

2140 Daphnis et Chloe. CO and Chorus; Lorin Maazel. Decca SXL 6703.
Tape (c) KSXC 6703. (also London CS 6898 Tape (c) CS 5 6898).

+Gr 4-75 p1812	+-NYT 1-18-76 pD1
+-HF 9-75 p88	+ON 3-6-76 p42
+-HF 12-76 p147 tape	+RR 3-75 p33
+HFN 8-75 p82	++SFC 7-20-75 p27
+HFN 6-76 p105 tape	++St 11-75 p132
+NR 8-75 p5	

2141 Daphnis et Chloe. Tanglewood Festival Chorus; BSO; Seiji Ozawa.
DG 2530 563. (Reissue from 2740 120).

++Audio 2-76 p94	+-NYT 1-18-76 pD1
++Gr 4-76 p1608	+ON 3-6-76 p42
+-HF 12-75 p104	++RR 3-76 p48
+HFN 4-76 p111	++SFC 10-12-75 p22
+NR 12-75 p4	++St 5-76 p122

Daphnis et Chloe. cf Works, selections (DG 2711 015).

Daphnis et Chloe. cf Works, selections (DG 2740 120).

Daphnis et Chloe. cf Works, selections (HMV 5016).

2142 Daphnis et Chloe: Suites, nos. 1, 2. Mother Goose. COA; Bernard
Haitink. Philips 6500 311. Tape (c) 7300 166.

++Gr 10-72 p707	++NR 11-72 p3
+-HF 12-71 p118	++RR 10-72 p72
+HF 3-75 p66	+RR 12-76 p108 tape
++HFN 10-72 p1917	++SFC 11-19-72 p31
+HFN 8-75 p110 tape	

Daphnis et Chloe: Suite, no. 1. cf Decca DPA 519/20.

2143 Daphnis et Chloe: Suite, no. 2. Miroirs: Alborado del gracioso.
Pavane pour une infante défunte. Rapsodie espagnole. PhO,
NPhO; Carlo Maria Giulini. HMV SXLP 30198. Tape (c) TC EXE
181. (Reissue from Columbia SAX 2476, 5265).

+Gr 4-76 p1608	-HFN 5-76 p119 tape
+HFN 3-76 p109	+RR 2-76 p36

Daphnis et Chloe: Suite, no. 2. cf Works, selections (Decca DPA
561/2).

Daphnis et Chloe: Suite, no. 2. cf CHABRIER: España.

Daphnis et Chloe: Suite, no. 2. cf DEBUSSY: La mer.

Daphnis et Chloe: Suite, no. 2. cf GRIEG: Peer Gynt, op. 46:
Suite, no. 1.

L'Enfant et les sortilèges: Foxtrot. cf Decca GOSD 674/6.

2144 Gaspard de la nuit. Sonatine. Valses nobles et sentimentales.
     Martha Argerich, pno. DG 2530 540.
          +-Gr 8-75 p348   ++NR 1-76 p13
          ++HF 12-75 p104   +RR 8-75 p59
          +HFN 9-75 p103   ++St 7-76 p112
          ++MJ 1-76 p25
     Gaspard de la nuit. cf DEBUSSY: Images, Bk 1.
     Gaspard de la nuit: Ondine. cf DEBUSSY: Estampes: Jardins sous
          la pluie.
     Habañera. cf Odyssey Y 33825.
2145 L'Heure espagnole. Suzanne Danco, s; Paul Derenne, Michel Hamel,
     t; Heinz Rehfuss, bar; André Vessières, bs; OSR; Ernest Anser-
     met. Decca ECS 786. (Reissue from LXT 2828).
          +-Gr 11-76 p873   +RR 11-76 p42
          +HFN 11-76 p170
     Introduction and allegro. cf Panton 110 380.
     Introduction and allegro. cf Works, selections (Decca DPA 561/2).
2146 Jeux d'eau. Ma mère l'oye. Miroirs. Pascal Rogé, Denise Fran-
     çoise Rogé, pno. Decca SXL 6715. (also London CS 6936).
          +Gr 11-75 p864   +NR 2-76 p14
          +-HF 4-76 p120   +RR 11-75 p82
          +HFN 11-75 p163   /St 7-76 p112
          +-MJ 4-76 p30
     Jeux d'eau. cf BACH: Partita, harpsichord, no. 1, S 825, B flat
          major.
     Jeux d'eau. cf CHOPIN: Scherzo, no. 4, op. 54, E major.
     Jeux d'eau. cf DEBUSSY: Estampes: Jardins sous la pluie.
2147 Ma mère l'oye (Mother goose). Tzigane. Valses nobles et senti-
     mentales. Itzhak Perlman, vln; Orchestre de Paris; Jean Marti-
     non. Angel (Q) S 37149. Tape (c) 4XS 37149 (ct) 8XS 37149.
          +-HF 7-76 p88   +-SFC 3-28-76 p28
          ++NR 3-76 p2   ++St 6-76 p106
2148 Ma mère l'oye. Menuet antique. La valse. NYP; Pierre Boulez.
     Columbia M 32838. (Q) MQ 32838 Tape (c) MAQ 32838. (also
     CBS 76306).
          +Gr 5-75 p1968   +MT 10-75 p888
          +HF 6-75 p96   +NR 6-75 p1
          +HF 1-76 p111 Quad tape +-RR 4-75 p30
          +HFN 5-75 p137   ++St 6-75 p104
          -MJ 12-75 p38
     Ma mère l'oye. cf Boléro.
     Mother goose. cf Daphnis et Chloe: Suites, nos. 1 and 2.
     Ma mère l'oye. cf Jeux d'eau.
     Ma mère l'oye. cf Works, selections (Decca DPA 561/2).
     Ma mère l'oye. cf Works, selections (DG 2530 752/3).
     Ma mère l'oye. cf Works, selections (DG 2711 015).
     Ma mère l'oye. cf Works, selections (DG 2740 120).
     Ma mère l'oye. cf Works, selections (HMV 5016).
     Ma mère l'oye. cf BARTOK: Divertimento, string orchestra.
     Ma mère l'oye. cf BARTOK: Rumanian dances, op. 8a.
     Ma mère l'oye. cf BRAHMS: Variations on a theme by Haydn, op. 56a.
     Mélodies hébraiques: Kaddisch. cf HMV HLM 7077.
2149 Menuet antique. Miroirs: Alborada del gracioso; Une barque sur
     l'océan. Pavane pour une infante défunte. Le tombeau de
     Couperin. Orchestra de Paris; Jean Martinon. Angel S 37150.
     Tape (c) 4XS 37150 (ct) 8XS 37150.

2149 Menuet antique. Miroirs: Alborada del gracioso; Une barque sur
         l'océan. Pavane pour une infante défunte. Le tombeau de
         Couperin. Orchestre de Paris; Jean Martinon. Angel S 37150.
         Tape (c) 4XS 37150 (ct) 8XS 37150.
            +HF 8-76 p92   +-SFC 3-28-76 p28
            ++NR 6-76 p2   ++St 6-76 p106

2150 Menuet antique. Miroirs: Une barque sur l'océan. Le tombeau
         de Couperin. Valses nobles et sentimentales. OSCCP; André
         Cluytens. Classics for Pleasure CFP 40093. (Reissue from
         Columbia SAX 2479, 2478).
            +Gr 12-75 p1045   +RR 11-75 p54
            +HFN 2-76 p115

      Menuet antique. cf Ma mère l'oye.
      Menuet antique. cf Works, selections (DG 2530 752/3).
      Menuet antique. cf Works, selections (DG 2711 015).
      Menuet antique. cf Works, selections (DG 2740 120).
      Menuet antique. cf Works, selections (HMV 5016).
      Miroirs. cf Jeux d'eau.
      Miroirs: Alborado del gracioso. cf Daphnis et Chloe: Suite, no.
         2.
      Miroirs: Alborado del gracioso. cf Works, selections (Decca DPA
         561/2).
      Miroirs: Alborado del gracioso. cf FALLA: El amor brujo.
      Miroirs: Alborada del gracioso. cf Philips 6780 030.
      Miroirs: Alborada del gracioso; Une barque sur l'océan. cf
         Menuet antique.
      Miroirs: Alborada del gracioso; Une barque sur l'océan. cf Works,
         selections (DG 2530 752/3).
      Miroirs: Alborada del gracioso; Une barque sur l'océan. cf Works,
         selections (DG 2711 015).
      Miroirs: Alborada del gracioso; Une barque sur l'océan. cf Works,
         selections (DG 2740 120).
      Miroirs: Alborada del gracioso; Une barque sur l'océan. cf Works,
         selections (HMV 5016).
      Miroirs: Une barque sur l'océan. cf Menuet antique.
      Miroirs: Une barque sur l'océan. cf DEBUSSY: Estampes: Jardins
         sous la pluie.
      Miroirs: La vallée des cloches. cf CHOPIN: Scherzo, no. 4, op.
         54, E major.
      Pavane. cf Stentorian SC 1724.
      Pavane pour une infante défunte. cf Boléro.
      Pavane pour une infante défunte. cf Daphnis et Chloe: Suite, no.
         2.
      Pavane pour une infante défunte. cf Menuet antique.
      Pavane pour une infante défunte. cf Works, selections (Decca DPA
         561/2).
      Pavane pour une infante défunte. cf Works, selections (DG 2530
         752/3).
      Pavane pour une infante défunte. cf Works, selections (DG 2711
         015).
      Pavane pour une infante défunte. cf Works, selections (DG 2740
         120).
      Pavane pour une infante défunte. cf Works, selections (HMV 5016).
      Pavane pour une infante défunte. cf BARTOK: Divertimento, string
         orchestra.
      Pavane pour une infante défunte. cf DEBUSSY: Prelude à l'après-
         midi d'un faune.

Pavane pour une infante défunte. cf FALLA: El amor brujo.
Pavane pour une infante défunte. cf CBS 30072.
Pavane pour une infante défunte. cf Claves LP 30-406.
Pavane pour une infante défunte. cf RCA SB 6891.
2151 Piano works, complete. Janine Dacosta, Maria Antoinette Pictet,
       pno. Musical Heritage Society MHS 3083/5 (3).
            /St 7-76 p112
2152 Piano works, 2 hands, complete. Kun Woo Paik, pno. ACM 10002-4
       (3). (from Ruth Uebel, 205 E 63rd St, NY 10021).
            ++St 7-76 p112
Piece en forme de habañera. cf Claves LP 30-406.
Quartet, strings, F major. cf DEBUSSY: Quartet, strings, op. 10,
     G minor.
Rapsodie espagnole. cf Boléro (Angel 37147).
Rapsodie espagnole. cf Boléro (Hungaroton 11644).
Rapsodie espagnole. cf Bolero (DG 2530 475).
Rapsodie espagnole. cf Daphnis et Chloe: Suite, no. 2.
Rapsodie espagnole. cf Works, selections (Decca DPA 561/2).
Rapsodie espagnole. cf Works, selections (DG 2711 015).
Rapsodie espagnole. cf Works, selections (DG 2740 120).
Rapsodie espagnole. cf Works, selections (HMV 5016).
Rapsodie espagnole. cf DEBUSSY: La mer.
Shéhérazade: Ouverture de féerie. cf Bolero (Angel 37147).
Shéhérazade: Ouverture de féerie. cf Works, selections (HMV 5016).
Sonata, violin and piano. cf DEBUSSY: Sonata, violin and piano.
Sonata, violin and piano. cf FRANCAIX: Sonatine.
2153 Sonatine. Le tombeau de Couperin. SCHUMANN: Kreisleriana, op.
       16. WEBER: Konzertstück, op. 79, F minor. Robert Casadesus,
       Lazare Levy, pno. Rococo 2054.
            +-NR 10-76 p13
Sonatine. cf Gaspard de la nuit.
2154 Songs: Chansons madécasses; Histoires naturelles; Mélodies popu-
       laires grecques (6); Poèmes de Stephane Mallarmé (3). Felicity
       Palmer, s; Judith Pearse, flt; Christopher van Kempen, vlc;
       Clifford Benson, John Constable, pno; Nash Ensemble; Simon
       Rattle. Argo ZRG 834.
            +-Gr 3-76 p1494   +RR 2-76 p62
            +-HFN 2-76 p111
Songs: D'Anne jouant delespinette; D'Anne que me jecta de la
     neige; Shéhérazade. cf HMV RLS 716.
Le tombeau de Couperin. cf Menuet antique (Angel 37150).
Le tombeau de Couperin. cf Menuet antique (Classics for Pleasure
     CFP 40093).
Le tombeau de Couperin. cf Sonatine.
Le tombeau de Couperin. cf Works, selections (DG 2530 752/3).
Le tombeau de Couperin. cf Works, selections (DG 2711 015).
Le tombeau de Couperin. cf Works, selections (DG 2740 120).
Le tombeau de Couperin: Suite. cf Works, selections (HMV 5016).
Tzigane. cf Ma mère l'oye.
Tzigane. cf Works, selections (HMV 5016).
Tzigane. cf CHAUSSON: Poème, op. 25.
Tzigane. cf CBS 76420.
Tzigane. cf Columbia M2 33444.
Tzigane. cf RCA ARM 4-0942/7.
La valse. cf Boléro (Angel 37147).
La valse. cf Bolero (DG 2530 475).
La valse. cf Ma mère l'oye.

La valse. cf Works, selections (Decca DPA 561/2).
La valse. cf Works, selections (DG 2711 015).
La valse. cf Works, selections (DG 2740 120).
La valse. cf Works, selections (HMV 5016).
La valse. cf DEBUSSY: La mer.
La valse. cf Angel S 37157.
La valse. cf HMV (Q) ASD 3131.

2155 Valses nobles et sentimentales. SCHUMANN: Faschingsschwank aus
      Wien, op. 26. Arturo Benedetti Michelangeli, pno. Rococo
      2112.
          +NR 10-76 p15
      Valses nobles et sentimentales. cf Gaspard de la nuit.
      Valses nobles et sentimentales. cf Ma mère l'oye.
      Valses nobles et sentimentales. cf Menuet antique.
      Valses nobles et sentimentales. cf Works, selections (DG 2530
      752/3).
      Valses nobles et sentimentales. cf Works, selections (DG 2711
      015).
      Valses nobles et sentimentales. cf Works, selections (DG 2740
      120).
      Valses nobles et sentimentales. cf Works, selections (HMV 5016).
      Valses nobles et sentimentales, nos. 6 and 7. cf RCA ARM 4-0942/7.

2156 Works, selections: Boléro. Daphnis et Chloe: Suite, no. 2.
      Introduction and allegro. Ma mère l'oye. Miroirs: Alborado
      del gracioso. Pavane pour une infante défunte. Rapsodie es-
      pagnole. La valse. Melos Ensemble, OSR; Ossian Ellis, hp;
      Ernest Ansermet. Decca DPA 561/2.
          +HFN 12-76 p152

2157 Works, selections: Miroirs: Alborada del gracioso; Une barque sur
      l'océan. Ma mère l'oye. Menuet antique. Pavane pour une
      infante défunte. Le tombeau de Couperin. Valses nobles et
      sentimentales. BSO; Seiji Ozawa. DG 2530 752/3 (2). (Reissue
      from 2740 120).
          -Gr 11-76 p804   +RR 10-76 p65
          +HFN 11-76 p170   +RR 11-76 p76

2158 Works, selections: Boléro. Menuet antique. Daphnis et Chloe. Ma
      mère l'oye. Miroirs: Alborada del gracioso; Une barque sur
      l'océan. Pavane pour une infante défunte. Rapsodie espagnole.
      Le tombeau de Couperin. La valse. Valses nobles et sentimen-
      tales. Tanglewood Festival Chorus; BSO; Seiji Ozawa. DG
      2711 015 (4).
          +-HF 7-76 p88   +-SFC 3-28-76 p28
          ++NR 4-76 p6   +-St 6-76 p106

2159 Works, selections: Boléro.* Daphnis et Chloe. Ma mère l'oye.
      Menuet antique. Miroirs: Alborada del gracioso; Une barque
      sur l'océan. Pavane pour une infante défunte. Rapsodie
      espagnole.* Le tombeau de Couperin. La valse.* Valses nobles
      et sentimentales. Tanglewood Festival Chorus; BSO; Seiji
      Ozawa. DG 2740 120 (4). (*available on 2530 475).
          +-Gr 11-75 p828   ++RR 11-75 p53
          +HFN 1-76 p117

2160 Works, selections: Boléro. Concerto, piano, G major. Concerto,
      piano, for the left hand, D major. Daphnis et Chloe. Ma mère
      l'oye. Menuet antique. Miroirs: Alborada del gracioso; Une
      barque sur l'océan. Pavane pour une infante défunte. Rapsodie
      espagnole. Shéhérazade: Ouverture de féerie. Le tombeau de

Couperin: Suite. Tzigane. La valse. Valses nobles et sentimentales. Aldo Ciccolini, pno; Orchestre de Paris; Jean Martinon. HMV (Q) SLS 5016 (5).
  +Gr 10-75 p626   +R 11-75 p54
  +HFN 1-76 p117

RAVENSCROFT, Thomas
  Remember, O thou man. cf HMV CSD 3774.
  Songs: By a bank as I lay; He that will an alehouse keep; Jinkin
    the jester; The maid she went a-milking; Malt's come down;
    What I hap had to marry a shrow. cf Pearl SHE 525.

RAVINA, Jean Henri
  Etude de style. cf International Piano Library IPL 102.

RAXACH, Enrique
  Imaginary landscape. cf EISMA: Le gibet.
  Paraphrase. cf EISMA: Le gibet.
  Quartet, strings, no. 2. cf CRUMB: Black angels, electric string
    quartet.

RAWSTHORNE, Alan
  Elegy, guitar. cf BENNETT: Concerto, guitar and chamber orchestra.

READ
  Down steers the bass; Russia. cf Vox SVBX 5350.

READ, Daniel
  Newport. cf Nonesuch H 71276.

REBEL, Jean-Fery
  Sonata, recorder, no. 3, D major. cf DORNEL: Suite, no. 1, C
    minor.
  Sonata, recorder, no. 6, B minor. cf DORNEL: Suite, no. 1, C
    minor.

THE RED ARMY CHOIR IN CONCERT. cf Everest SDBR 3388.

REDHEAD, Richard
  Shout salvation: Suite. cf HMV SXLP 50017.

REED, Alfred
  Festive overture. cf Pye GH 603.

REEVES, D. W.
  Second Connecticut Regiment. cf Columbia 33513.
  Second Regiment Connecticut National Guard march. cf Mercury SRI
    75055.

REGER, Max
  Aus meinem Tagebuche: Adagio and vivace, op. 82. cf International
    Piano Library IPL 102.
  Benedictus, op. 59, no. 9. cf Vista VPS 1022.
  Fantasia and fugue on the chorale "Hallelujah, Gott zu loben
    bleibe meine Seelen freund", op. 52, no. 3.
  Eine feste Burg, op. 67, no. 6. cf Vista VPS 1035.
  Introduction and passacaglia, D minor. cf Vista VPS 1035.
  O Tod, wie bitter bist du, op. 110, no. 3. cf BIS LP 14.
  Pieces, op. 59, no. 9: Benedictus. cf MENDELSSOHN: Sonata, organ,
    op. 65, no. 3, A major.
  Pieces, op. 59, no. 9: Benedictus. cf Decca SDD 499.
  Pieces, op. 145, no. 2: Dankpsalm. cf FRANCK: Chorale, no. 2, B
    minor.
  Pieces, op. 145, no. 2: Dankpsalm. cf HOLLER: Ciacona, op. 54.
  Pieces, op. 145, no. 2: Dankpsalm. cf MENDELSSOHN: Sonata, organ,
    op. 65, no. 3, A major.
2161 Quartet, piano, op. 133, A minor. STRAUSS, R.: Quartet, piano,
    op. 13, C minor. Cardiff Festival Ensemble. Argo ZRG 809.

```
                +-Gr 9-75 p485              +MT 1-76 p43
                +-HFN 7-75 p87             +RR 7-75 p44
```
2162 Quintet, clarinet and strings, op. 146, A major. Karl Leister,
        clt; Drolc Quartet. DG 2530 303.
```
                +Gr 6-73 p68               +RR 6-73 p66
                ++HF 2-76 p102             +St 4-76 p114
                +HFN 6-73 p1181            +STL 7-8-73 p36
                ++NR 12-75 p8
```
2163 Quintet, clarinet and strings, op. 146, A major. Rudolf Gall,
        clt; Keller Quartet. Oryx 1832.
```
                +St 4-76 p115
```
      Quintet, clarinet and strings, op. 146, A major: 2nd movement.
        cf HMV SLS 5046.

      Scherzino, horn and orchestra. cf BASF BAC 3085.

      Sonata, solo viola, op. 31, G minor. cf HUMMEL: Sonata, viola,
        E flat major.

      Toccata and fugue, op. 59. cf Argo 5BB 1013/5.

REICHA (Rejcha), Antonin

      Fugue, A major. cf Panton 110 418/9.

2164 Quartet, flute and strings, op. 98, no. 1, G minor. KRAMAR-KROMMER:
        Quartet, flute and strings, op. 75, D major. Peter Brock, flt;
        Josef Vlach, vln; Josef Kodousek, vla; Viktor Moucka, vlc.
        Supraphon 111 1450.
```
                ++Gr 1-76 p1212            ++RR 10-74 p71
                +NR 3-75 p6               ++RR 12-75 p73
```
REICHE, Gottfried

      Abblasen. cf BACH: Organ works (Philips 6500 925).

REIMANN, Aribert

2165 Concerto, piano and 19 players. Engführung, tenor and piano.
        Ernst Häfliger, t; Klaus Billing, pno; Heinz Holliger, ob, Eng
        hn; Aurèle Nicolet, flt; Hanzheinz Schneeberger, vln; Basler
        Solisten Ensemble; Francis Travis. Wergo WER 60072.
```
                +ARG 12-76 p37
```
      Engführung, tenor and piano. cf Concerto, piano and 19 players.

REINECKE, Carl

      Toy symphony, C major. cf Angel S 36080.

REINER

      Trio, flute, bass clarinet and percussion. cf ISTVAN: Ritmi ed
        Antiritmi.

REINHOLD

      Impromptu, C sharp minor. cf Connoisseur Society (Q) CSQ 2066.

RENARD

      Le temps de cerises. cf Club 99-101.

RENWICK

      Dance. cf Crystal S 206.

RESPIGHI, Ottorino

2166 Ancient airs and dances: Suites, nos. 1-3. Los Angeles Chamber
        Orchestra; Neville Marriner. Angel S 37301. Tape (c) 4XS
        37301 (ct) 8XS 37301. (also HMV SQ ASD 3188).
```
                +Gr 6-76 p52               ++NR 6-76 p3
                ++HF 7-76 p88              ++RR 6-76 p52
                +HF 10-76 p147 tape        +-SFC 4-25-76 p30
                +HFN 6-76 p97              +-St 9-76 p123
```
2167 Ancient airs and dances: Suites, nos. 1-3. PH; Antal Dorati.
        Philips 6582 010. (Reissue from Mercury AMS 16028).
```
                +Gr 4-76 p1608            +-HFN 3-76 p109
```

2168   The birds (Gli uccelli). Trittico botticelliano. PCO, "prepared
          for recording" by Josef Vlach and Petr Skvor. Supraphon 110
          1769.
                   +-Gr 7-76 p181   +-RR 6-76 p52
                   +-HF 7-76 p89   +St 9-76 p123
                   ++NR 6-76 p3
       The birds. cf Decca KCSP 367.
       La boutique fantasque. cf DUKAS: The sorcerer's apprentice.
2169   Feste Romane. The fountains of Rome. The pines of Rome. PO;
          Eugene Ormandy. CBS 30077. Tape (c) 40-30077. (Reissues
          from Philips SABL 113, SBRG 72026).
                   +Gr 9-76 p424   +HFN 10-76 p185 tape
                   +-HFN 10-76 p181   +-RR 9-76 p59
       Feste Romane. cf RCA CRM 5-1900.
2170   The fountains of Rome. The pines of Rome. NPhO; Charles Munch.
          Decca SDD 494. (Reissue from PRS 4131).
                   +-Gr 8-76 p297   +RR 8-76 p52
                   +HFN 8-76 p93
2171   The fountains of Rome. The pines of Rome. Minneapolis Symphony
          Orchestra; Antal Dorati. Philips 6582 015.
                   +Gr 12-76 p1001   ++RR 11-76 p76
                   +HFN 11-76 p170
       The fountains of Rome. cf Feste Romane.
       The pines of Rome. cf Feste Romane.
       The pines of Rome. cf The fountains of Rome (Decca SDD 494).
       The pines of Rome. cf The fountains of Rome (Philips 6582 015).
       The pines of Rome. cf BERLIOZ: Le carnaval romain, op. 9.
       The pines of Rome. cf HMV SLS 5019.
       Songs: Il tramonto. cf DEBUSSY: Songs: Le promenoir des deux
          amants.
       Stornellactrice, 1st and 2nd edition. cf Rubini RS 301.
       Trittico botticelliano. cf The birds.
REUSNER, Esaias
       German dances. cf DG Archive 2533 172.
REVER
       Sigurd: J'ai gardé mon ame, Esprits gardiens, Un souvenir poignant.
          cf Club 99-98
REYNOLDS, Roger
       Ambages. cf Nonesuch HB 73028.
REZAC, Ivan
       Sisyfova nedele (The Sunday of Sisyfos). cf DEBUSSY: Images, Bk 1.
       Songs on Nezval (4). cf Panton 110 385.
       Torso of a Schumann statue. cf HLOBIL: Concerto, double bass,
          op. 70.
REZNICEK, Emil
       Donna Diana: Overture. cf HMV ESD 7010.
       Donna Diana: Overture. cf Suolphon 110 1637.
RHEINBERGER, Josef
2172   Sonata, organ, no. 1, op. 27, C minor. Sonata, organ, no. 20,
          op. 196, F major. Conrad Eden, Timothy Farrell, org. Vista
          VPS 1011.
                   +-Gr 5-75 p1996   +-MT 7-76 p578
                   +HFN 5-75 p138   +-RR 4-75 p53
2173   Sonata, organ, no. 2, op. 63, A flat major. Sonata, organ, no. 9,
          op. 142, B flat minor. Conrad Eden, Robert Munns, org. Vista
          VPS 1013.

```
                    /Gr 8-75 p348              +-MT 7-76 p578
                    +-HFN 8-75 p82            +-RR 9-75 p61
```
2174 Sonata, organ, no. 3, op. 88, G major. Sonata, organ, no. 19,
        op. 192, G minor. Conrad Eden, Timothy Farrell, org. Vista
        VPS 1015.
```
                    +-Gr 3-76 p1487           +-MT 7-76 p578
                    +HFN 4-76 p113            +RR 3-76 p63
```
2175 Sonatas, organ, nos. 4-5, 7-8, 10, 13-16, 18. Robert Munns,
        Roger Fisher, Timothy Farrell, Conrad Eden, org. Vista VPS
        1016/20 (5).
```
                    +Gr 4-76 p1639            +-MT 7-76 p578
                    +HFN 4-76 p113            +-RR 6-76 p74
```
2176 Sonata, organ, no. 6, op. 119, E flat minor. Sonata, organ, no.
        11, op. 148, D minor. Robert Munns, Roger Fisher, org.
        Vista VPS 1012.
```
                    +Gr 6-75 p69             ++MT 7-76 p578
                    ++HFN 6-75 p91           +RR 7-75 p53
```
        Sonata, organ, no. 9, op. 142, B flat minor. cf Sonata, organ,
          no. 2, op. 63, A flat major.
        Sonata, organ, no. 11, op. 148, D minor. cf Sonata, organ, no.
          6, op. 119, E flat minor.
2177 Sonata, organ, no. 12, op. 154, D flat major. Sonata, organ, no.
        17, op. 181, B major. Roger Fisher, Timothy Farrell, org.
        Vista VPS 1014.
```
                    +Gr 10-75 p658           ++MT 7-76 p578
                    +HFN 12-75 p163          +RR 12-75 p84
```
        Sonata, organ, no. 17, op. 181, B major. cf Sonata, organ, no.
          12, op. 154, D flat major.
        Sonata, organ, no. 19, op. 192, G minor. cf Sonata, organ, no.
          3, op. 88, G major.
        Sonata, organ, no. 20, op. 196, A major. cf Sonata, organ, no. 1,
          op. 27, C minor.
RHENE-BATON (Baton, Rene)
      Heures d'été. cf HMV RLS 716.
RIBARI, Antal
      Pezzi, violoncello and harp (4). cf CASTELNUOVO-TEDESCO: Sonata,
        violoncello and harp.
RICCIOTTI, Carlo
      Concertino, no. 2, G major. cf CORELLI: Concerto grosso, op. 6,
        no. 8, G minor.
      Concertino, no. 2, G major. cf Decca SDD 411.
RICHARD, Coeur de Lion
      Ja nun nons pris. cf Vanguard VCS 10049.
RIDOUT, Alan
      The history of the flood. cf HMV CSD 3766.
      Spiritus domini. cf Abbey LPB 739.
RIEGGER, Wallingford
      Canon and fugue, op. 33, D minor. cf BECKER: Concerto arabesque,
        piano.
      Dichotomy, chamber orchestra, op. 12. cf MUSGRAVE: Night music,
        chamber orchestra.
      Quartet, strings, no. 2. cf HARRIS: Fantasy, violin and piano.
      Sonatine, op. 39. cf Desto DC 6435/7.
RIETI, Vittorio
2178 Capriccio. Pastorale e fughetta. Quartet, strings, no. 3. Lir-
        iche saffiche (7). Silografie. Benedetta Pecchioli, ms;

Oscar Ravina, vln; Gleb Gzhashvili, pno; Phoenix String Quartet,
Ariel Wind Quintet, Casa Serena Trio, Chamber Orchestra; Mario
Rossi. Serenus SRS 12063.
          ++NR 4-76 p15
Incisioni. cf Crystal S 204.
Liriche saffiche (7). cf Capriccio.
Pastorale e fughetta. cf Capriccio.
Quartet, strings, no. 3. cf Capriccio.
Silografie. cf Capriccio.
RIETZ, Julius
Concert overture, op. 7. cf BRUCH: Symphony, no. 2, op. 36, F
     minor.
RILEY, Dennis
Variations II: Trio. cf CRI SD 324.
RIMMER, Drake
Golden rain. cf Pye TB 3009.
King Lear. cf Pye TB 3007.
RIMSKY-KORSAKOV, Nikolai
2179 Capriccio espagnol, op. 34. Mlada: Procession of the nobles.
     TCHAIKOVSKY: Capriccio italien, op. 45. Marche slav, op. 31.
     Mazeppa: Gopak. LPO; Adrian Boult. HMV ASD 3903. Tape (c)
     TC ASD 3093. (also Angel S 37227 Tape (c) 4XS 37227).
          +Gr 7-75 p245   +MJ 11-76 p45
          +-Gr 7-76 p230 tape   +NR 10-76 p3
          ++HFN 7-75 p87   +RR 8-75 p42
          +HFN 6-76 p105 tape
     Capriccio espagnol, op. 34. cf BORODIN: Prince Igor: Polovtsian
     dances.
     Capriccio espagnol, op. 34. cf DVORAK: Slavonic dance, op. 72,
     no. 2, E minor.
     Capriccio espagnol, op. 34. cf GLINKA: Jota aragonesa.
2180 Concerto, piano, op. 30, C sharp minor. Russian Easter festival
     overture, op. 36. Symphony, no. 3, op. 32, C major. Sviatos-
     lav Richter, pno; MRSO; Alexander Gauk, Kiril Kondrashin.
     Everest SDBR 3393.
          +-NR 5-76 p2
     Le coq d'or (The golden cockerel): March. cf Classics for pleasure
     CFP 40254.
2181 Le coq d'or: Suite. Tale of the Tsar Sultan: Suite. MRSO; Kon-
     stantin Ivanov. Angel SR 40259.
          ++NR 3-76 p2
     Le coq d'or: Suite. cf BORODIN: Prince Igor: Overture; Polovtsian
     dances.
     Le coq d'or: Suite. cf PROKOFIEV: Betrothal in the monastery, op.
     86: Summer night suite.
     Le coq d'or: Wedding march. cf BERLIOZ: Les Troyens: Overture,
     March Troyenne.
     Le coq d'or: Wedding march (Bridal procession). cf RCA CRL 3-2026.
     Fantasy on Russian themes, op. 33. cf ARENSKY: Concerto, violin,
     op. 54, A minor.
     Kashchei, the deathless: The night descends. cf Columbia/Melodiya
     M 33931.
     The legend of Sadko, op. 5. cf Philips 6780 755.
     The legend of Sadko, op. 5: Chanson hindoue. cf HMV RLS 719.
2182 May night. Olga Pastushenko, Tamara Antipova, s; Lyudmilla Sapeg-
     ina, Nina Derbina, ms; Anna Matyushina, Lyuziya Rashkovets, con;

Konstantin Lisovsky, Yuri Yelnikov, t; Ivan Budrin, bs-bar; Alexei Krivchenya, Gennady Troitsky, bs; MRSO and Chorus; Vladimir Fedoseyev. DG 2709 063 (3).
    +-Gr 11-76 p873   +-RR 10-76 p30
    +-HFN 11-76 p165
May night: Song about the village elder. cf HMV ASD 3200.
Mlada: Procession of the nobles. cf Capriccio espagnol, op. 34.
Quintet, piano and winds, op. posth., B flat major. cf MENDEL-SSOHN: Octet, strings, op. 20, E flat major.
Russian Easter festival overture, op. 36. cf Concerto, piano, op. 30, C sharp minor.
Russian Easter festival overture, op. 36. cf BORODIN: Prince Igor: Polovtsian dances.
Russian Easter festival overture, op. 36. cf Philips 6747 204.
2183 Scheherazade, op. 35. Ruben Yordanoff, vln; Orchestre de Paris; Mstislav Rostropovich. Angel S 37061. (also HMV ASD 3047 (Q) Q4 ASD 3047).
    +-Gr 2-75 p1499   +-RR 2-75 p37
    +-HF 5-75 p89   +-SFC 4-13-75 p23
    +-HF 10-76 p147 tape   ++St 6-75 p104
    -NR 3-75 p2
2184 Scheherazade, op. 35. CSO; Fritz Reiner. Camden CCV 5010. (Reissue).
    +ST 2-76 p741
2185 Scheherazade, op. 35. PhO; Lovro von Matačíc. Classics for Pleasure SIT 60042. Tape (c) TC SIT 60042.
    +HFN 12-76 p153 tape
2186 Scheherazade, op. 35. LAPO; Sydney Harth, vln; Zubin Mehta. Decca SXL 6731. Tape (c) KSXC 6731. (also London 6950 Tape (c) 56590 (ct) 6950).
    -Gr 11-75 p828   +-NR 1-76 p2
    -HF 2-76 p103   +-NYT 1-18-76 pD1
    +-HF 7-76 p114 tape   +-RR 10-75 p49
    +HFN 10-75 p148   +RR 3-76 p77 tape
    +MJ 12-76 p28   +SFC 9-14-75 p28
2187 Scheherazade, op. 35. Erich Gruenberg, vln; RPO; Leopold Stokowski. RCA ARL 1-1182. Tape (c) ARK 1-1182 (ct) ARS 1-1182.
    +ARG 12-76 p39   +MJ 12-76 p28
    +-HF 12-76 p104   +-NR 10-76 p3
Scheherazade, op. 35, excerpts. cf Stentorian SP 1735.
Scheherazade, op. 35: The sea and Sinbad's ship. cf HMV SEOM 25.
Skazka, op. 29. cf BALAKIREV: Russia.
The snow maiden: Dance of the tumblers. cf Columbia M 34127.
Songs: The rose and the nightingale. cf Musical Heritage Society MHS 3276.
Symphony, no. 3, op. 32, C major. cf Concerto, piano, op. 30, C sharp minor.
The tale of the Tsar Sultan: Flight of the bumblebee. cf Creative Record Service R 9115.
The tale of the Tsar Sultan: Flight of the bumblebee. cf HMV RLS 723.
The tale of the Tsar Sultan: Flight of the bumblebee. cf HMV SQ ASD 3283.
The tale of the Tsar Sultan: Flight of the bumblebee. cf Odyssey Y 33825.
The tale of the Tsar Sultan: Flight of the bumblebee (arr. and orch. Gerhardt). cf RCA LRL 1-5094.

409   RIMSKY-KORSAKOV (cont.)

The tale of the Tsar Sultan: Flight of the bumblebee (2). cf
    RCA ARM 4-0942/7.
The tale of the Tsar Sultan: Suite. cf Le coq d'or: Suite.
The tale of the Tsar Sultan: Suite; Flight of the bumblebee. cf
    BORODIN: Symphony, no. 2, B minor.
2188 Trio, C minor. Sviatoslav Knushevitzky, vlc; David Oistrakh, vln;
    Lev Oborin, pno. Westminster WGM 8321.
        +NR 9-76 p7
The Tsar's bride: Lyubasha's aria. cf Columbia/Melodiya M 33931.
RITCHIE
    City of the sticks. cf Philips 6308 246.
RIVERS, Samuel
    Shadows. cf Folkways FTS 33901/4.
ROBERTON
    The old woman. cf Columbia SCX 6585.
ROBERTS
    Prelude and trumpetings. cf Delos FY 025.
ROBERTS, Myron
    Homage to Perotin. cf Grosvenor GRS 1039.
ROBINSON
    The Queenes good night. cf BIS LP 22.
ROBINSON, McNeil
    Improvisation on a submitted theme. cf L'Oiseau-Lyre SOL 343.
ROBINSON, Thomas
    Lessons for the lute. cf Saga 5420.
    A toy. cf DG Archive 2533 323.
ROBISON
    Fiducia. cf Nonesuch H 71276.
ROCHBERG, George
    Blake songs. cf Nonesuch H 71302/3.
2189 Chamber symphony, 9 instruments. Music for the magic theater.
    Oberlin Orchestra; Kenneth Moore. Desto DC 6444.
        +HF 5-76 p90   +St 7-76 p111
2190 Duo concertante. Quartet, strings, no. 1. Ricordanza, soliloquy
    for piano and violoncello. Mark Sokol, vln; Norman Fischer,
    vlc; George Rochberg, pno; Concord String Quartet. CRI SD 337.
        +HF 5-76 p90   ++NR 1-76 p7
        ++MQ 10-76 p337   +St 7-76 p111
    Music for the magic theater. cf Chamber symphony, 9 instruments.
    Quartet, strings, no. 1. cf Duo concertante.
    Ricordanza, soliloquy for piano and violoncello. cf Duo concer-
        tante.
RODGERS
    Lagoon. cf Columbia MG 33728.
    Victory at sea: Guadalcanal march. cf Mercury SRI 75055.
RODGERS, Richard
    Lover. cf Personal Touch 88.
RODRIGO, Joaquin
    Adagio. cf Supraphon 111 1230.
    Bolero. cf Supraphon 111 1230.
2191 Concierto de Aranjuez, guitar and orchestra. VILLA-LOBOS: Concerto,
    guitar and small orchestra. John Williams, gtr; James Brown,
    cor anglais; ECO; Daniel Barenboim. CBS 76369. Tape (c)
    40-76369. (also Columbia M 33208 Tape (c) MT 33208 (Q) MQ
    33208 Tape (ct) MAQ 33208).

```
                ++Gr 1-75 p1354              +NR 4-75 p5
                +Gr 5-75 p2032               ++RR 1-75 p33
                +HF 2-76 p130 tape           ++RR 6-75 p93 tape
                +MM 8-76 p35                 ++St 7-75 p104
```

Concierto de Aranjuez, guitar and orchestra. cf BERKELEY: Concerto,
    guitar (RCA 11734).
Concierto de Aranjuez, guitar and orchestra. cf BERKELEY: Concerto,
    guitar (RCA 1-1181).
Concierto de Aranjeuz, guitar and orchestra. cf HMV SLS 5068.
Concierto madrigal, 2 guitars and orchestra. cf GIULIANI: Concerto,
    guitar, op. 30, A major.
En los trigales. cf RCA SB 6891.
Fandango. cf Angel S 36094.
Fandango. cf Enigma VAR 1015.
Fantasia para un gentilhombre. cf CASTELNUOVO-TEDESCO: Concertino,
    harp, op. 93.
Fantasia para un gentilhombre. cf GIULIANI: Introduction, theme
    with variations and polonaise, op. 65.
Pequeñas sevillanas. cf Enigma VAR 1015.
Sonata giacosa. cf ALBENIZ: Suite española, no. 3: Sevillanas;
    no. 4: Cadiz.
Ya se van los pastores. cf Enigma VAR 1015.

ROGAN, Mackenzie
    Milanollo. cf Pye GH 603.
ROGER-DUCASSE, Jean
    Pastorale, F major. cf Creative Record Service R 9115.
ROGERS, James H.
    The time for making songs has come. cf Cambridge CRS 2715.
ROLLA, Alessandro
    Concerto, violin, A major. cf PAGANINI: Sonata, viola, op. 35,
        C minor.
ROMAN, Johann Helmich
    Concerto, violin, D minor. cf LARSSON: Concertino, double bass
        and strings, op. 45, no. 11.
2192 Drottningholmsmusiken. Symphony, no. 16, D major. Symphony, no.
        20, E minor. Drottningholm Chamber Orchestra; Stig Westerberg.
        Swedish Society SLT 33140.
            +RR 8-76 p52
    Symphony, B flat major. cf LARSSON: Concertino, double bass and
        strings, op. 45, no. 11.
    Symphony, no. 16, D major. cf Drottningholmsmusiken.
    Symphony, no. 20, E minor. cf Drottningholmsmusiken.
RONALD, Landon
    Songs: Away on the hill; Down in the forest; Sound of earth. cf
        HMV RLS 719.
RONCALLI, Lodovico
    Suite, G major. cf DG 2530 561.
RONTANI, Raffaello
    Nerinda bella. cf L'Oiseau-Lyre 12BB 203/6.
ROOT, George
    The battle cry of freedom. cf CBS 61746.
    The battle cry of freedom. cf Department of Defense Bicentennial
        Edition 50-1776.
    Tramp, tramp, tramp. cf CBS 61746.
ROPARTZ, Joseph Guy
    Pièce, E flat minor. cf Boston Brass BB 1001.

ROPEK, Jiří
    Variations on "Victimae Paschali Laudes". cf London STS 15222.
RORE, Cyprien
    De la belle contrade. cf L'Oiseau-Lyre 12BB 203/6.
ROREM, Ned
        King Midas. cf ARGENTO: To be sung upon the water.
        Songs: A Christmas carol; For Susan; Clouds; Guilt; What sparks
            and wiry cries. cf Duke University Press DWR 7306.
        Songs: Sing, my soul. cf Orion ORS 75205.
ROSENBERG, Hilding
        Quartet, strings, no. 4. cf BARTOK: Quartet, strings, no. 4,
            C major.
        Voyage to America: Intermezzo; The railway fugue. cf BERWALD:
            Symphony, no. 2, D major.
ROSENBLUTH, Leo
        Y'hi ratson; R'tseh; Psalm 116; Ravo l'Fanecha; M'loch; V'Hagen.
            cf EPHROS: The priestly benediction.
ROSETTI, Francisco Antonio (Franz Anton Rössler)
        Concerto, flute, G major. cf DEVIENNE: Concerto, flute, no. 4,
            G major.
ROSS, Walter
        Concerto, tuba and orchestra. cf IVES: March Mega Lambda Chi.
ROSSETER, Philip
        Songs: Sweet come again; What then is love but mourning; Whether
            men do laugh or weep. cf Philips 6500 282.
        Songs: When Laura smiles. cf Pearl SHE 525.
ROSSI, Salomone
        Kaddish. cf Serenus SRS 12039.
ROSSINI, Gioacchino
        Il barbiere di Siviglia (The barber of Seville): Air de Rosina.
            cf Discophilia DIS 13 KHH 1.
        Il barbiere di Siviglia: All'idea di quel metallo. cf Rococo
            5383.
        Il barbiere di Siviglia: La calunnia. cf Hungaroton SLPX 11444.
        Il barbiere di Siviglia: Contro un cor. cf Telefunken AG 6-41945.
        Il barbiere di Siviglia: Dunque io son. cf Rubini GV 68.
        Il barbiere di Siviglia: Ecco ridente in cielo. cf Rubini GV 70.
        Il barbiere di Siviglia: Largo al factotum. cf Columbia/Melodiya
            M 33593.
        Il barbiere di Siviglia: Largo al factotum; Dunque io son. cf
            Rubini GV 65.
        The barber of Seville: Largo al factotum; Una voce poco fa. cf
            Works, selections (Decca SPA 445).
        The barber of Seville: Overture. cf Overtures (Decca PFS 4386).
        Il barbiere di Siviglia: Overture. cf Overtures (DG 2530 559).
        Il barbiere di Siviglia: Overture. cf Overtures (HMV SXLP 30205).
        Il barbiere di Siviglia: Overture. cf Overtures (Philips 6500
            878).
        Il barbiere di Siviglia: Overture. cf Decca SPA 449.
        Il barbiere di Siviglia: Overture. cf HMV SLS 5019.
        Il barbiere di Siviglia: Una voce poco fa. cf BELLINI: Capuletti
            ed i Montecchi: O, quante volte, O quante.
        Il barbiere di Siviglia: Una voce poco fa. cf VERDI: Rigoletto.
        The barber of Seville: Una voce poco fa. cf Caprice CAP 1107.
        Il barbiere di Siviglia: Una voce poca fa. cf Court Opera Classics
            CO 342.
        Il barbiere di Siviglia: Una voce poca fa. cf HMV SLS 5057.

La cambiale di matrimonio: Overture. cf Overtures (Philips 6500
    878).
La cenerentola: Overture. cf Overtures (DG 2530 559).
Cinderella: Overture. cf Overtures (Decca PFS 4386).
Cinderella: Zitto, zitto; Piano, piano; Non più mesta. cf Works,
    selections (Decca SPA 445).
Duet, violoncello and double bass. cf DITTERSDORF: Sinfonia con-
    certante, viola, double bass and orchestra.
Duet, violoncello and double bass. cf PAGANINI: Sonata, viola,
    op. 35, C minor.
2193 Elisabetta Regina d'Inghilterra. Montserrat Caballé, Valerie
      Masterson, s; Rosanne Creffield, ms; José Carreras, Ugo Benelli,
      Neil Jenkins, t; Ambrosian Singers; LSO; Gianfranco Masini.
      Philips 6703 067 (3).
            +Gr 9-76 p467   +RR 9-76 p22
            +HFN 9-76 p115   +-SFC 11-21-76 p35
La gazza ladra (The thieving magpie): Di piacer mi balza il cor.
    cf BELLINI: Capuletti ed i Montecchi: O, quante volte, O quante.
La gazza ladra: Overture. cf Overtures (Classics for Pleasure
    CFP 40077).
The thieving magpie: Overture. cf Overtures (Decca PFS 4386).
La gazza ladra: Overture. cf Overtures (DG 2530 559).
La gazza ladra: Overture. cf Overtures (HMV SXLP 30205).
The thieving magpie: Overture. cf Works, selections (Decca SPA
    445).
La gazza ladra: Overture. cf BACH: Toccata and fugue, S 565, D
    minor.
The thieving magpie: Overture. cf Decca SB 322.
The thieving magpie: Overture. cf Decca KCSP 367.
The thieving magpie: Overture. cf Decca SXL 6782.
La gazza ladra: Overture. cf HMV SLS 5019.
Giovanna d'Arco: La danza. cf BELLINI: Songs (RCA AGL 1-1341).
Guglielmo Tell (William Tell): Ah Matilde; Troncar suoi, O muto
    asil. cf Cantilena 6239.
2194 Guillaume Tell: Ballet music. Moise: Ballet music. Otello:
      Ballet music. Le siège de Corinthe: Ballet music. Monte Carlo
      Opera Orchestra; Antonio de Almeida. Philips 6780 027 (2).
            -Gr 12-76 p1001   ++RR 11-76 p77
            +HFN 11-76 p167   +-SFC 11-21-76 p35
Guglielmo Tell: Duet. cf Rococo 5379.
Guglielmo Tell: O muto asil del pianto. cf DONIZETTI: La favorita:
    Spirto gentil; Una vergine, un angel di Dio.
Guillaume Tell: Overture. cf Overtures (Classics for Pleasure
    CFP 40077).
William Tell: Overture. cf Overtures (Decca PFS 4386).
Guglielmo Tell: Overture. cf Overtures (HMV SXLP 30205).
William Tell: Overture. cf Works, selections (Decca SPA 445).
Guglielmo Tell: Overture. cf Angel S 37231.
William Tell: Overture. cf CBS 61748.
William Tell: Overture. cf Pye PCNHX 6.
L'Inganno felice: Overture. cf Overtures (Philips 6500 878).
L'Invito. cf L'Oiseau-Lyre SOL 345.
L'Italiana in Algeri (The Italian girl in Algiers): Finale, Act 1.
    cf Works, selections (Decca SPA 445).
L'Italiana in Algeri: Overture. cf Overtures (Classics for Plea-
    sure CFP 40077).

Italian girl in Algiers: Overture. cf Overtures (Decca PFS 4386).
L'Italiana in Algeri: Overture. cf Overtures (DG 2530 559).
L'Italiana in Algeri: Overture. cf Overtures (HMV SXLP 30205).
L'Italiana in Algeri: Overture. cf Overtures (Philips 6500 878).
Moise: Ballet music. cf Guillaume Tell: Ballet music.
Otello: Ballet music. cf Guillaume Tell: Ballet music.
2195 Overtures: La gazza ladra. Guillaume Tell. L'Italiana in Algeri.
      Semiramide. Il Signor Bruschino. RPO; Colin Davis. Classics
      for Pleasure CFP 40077. Tape (c) TC CFP 40077. (Reissue).
            +HFN 12-76 p153 tape   +RR 5-74 p39
2196 Overtures: The barber of Seville. Cinderella. Italian girl in
      Algiers. Semiramide. The thieving magpie. William Tell.
      RPO; Carlos Paita. Decca PFS 4386.
            ++Gr 12-76 p1071
2197 Overtures: L'Assedio di Corinto. Il barbiere di Siviglia. La
      cenerentola. La gazza ladra. L'Italiana in Algeri. Il
      Signor Bruschino. LSO; Claudio Abbado. DG 2530 559. Tape (c)
      3300 497.
            +Gr 1-76 p1244   +RR 1-76 p37
            +HFN 1-76 p117   +-RR 9-76 p95 tape
            +HFN 2-76 p116 tape   ++SFC 4-4-76 p34
            ++MJ 5-76 p28   +St 7-76 p111
            -NR 5-76 p1
2198 Overtures: Il barbiere di Siviglia. La gazza ladra. Guglielmo
      Tell. L'Italiana in Algeri. La scala di seta. Semiramide.
      PhO; Herbert von Karajan. HMV SXLP 30203. Tape (c) TC EXE 194.
      (Reissue from Columbia SAX 2378).
            ++Gr 5-76 p1807   +HFN 8-76 p95
            +-HFN 7-76 p103   +-RR 6-76 p52
2199 Overtures: Il barbiere di Siviglia. La cambiale di matrimonio.
      L'Inganno felice. L'Italiana in Algeri. La scala di seta.
      Il Signor Bruschino. Tancredi. Il Turco in Italia. AMF;
      Neville Marriner. Philips 6500 878. Tape (c) 7300 368.
            +Audio 12-76 p93   +RR 11-75 p54
            ++HF 2-76 p103   +-RR 9-76 p95 tape
            +HF 5-76 p114 tape   ++SFC 12-28-75 p30
            +HFN 10-75 p148   +St 7-76 p111
            +NR 1-76 p6
      La pietra del paragone: Pubblico fu l'oltraggio. cf English
      National Opera ENO 1001.
      La scala di seta: Overture. cf Overtures (HMV SXLP 30205).
      La scala di seta: Overture. cf Overtures (Philips 6500 878).
      Semiramide: Bel raggio lusinghier. cf Works, selections (Decca
      SPA 445).
      Semiramide: Overtures (Classics for Pleasure CFP 40077).
      Semiramide: Overture. cf Overtures (Decca PFS 4386).
      Semiramide: Overture. cf Overtures (HMV SXLP 30205).
      Semiramide: Serbami ognor. cf Decca DPA 517/8.
2200 Serenade, small orchestra, E flat major. Sonatas, strings (6).
      Variations, clarinet and small orchestra, C major. Jacques
      Lancelot, clt; I Solisti Veneti; Claudio Scimone. RCA AGL
      2-1339 (2).
            +SFC 6-27-76 p29   ++St 8-76 p99
2201 The siege of Corinth (L'Assedio di Corinto). Beverly Sills, Delia
      Wallis, s; Shirley Verrett, ms; Harry Theyard, Gaetano Scano,
      t; Justino Diaz, Gwynne Howell, Robert Lloyd, bs; Ambrosian

Opera Chorus; LSO; Thomas Schippers. Angel SLCX 3819 (3).
(also HMV SLS 981).

+-Gr 6-75 p83	+NYT 4-6-75 pD18
+-HF 6-75 p75	+ON 4-19-75 p54
+HFN 7-75 p87	+RR 6-75 p24
+-MQ 10-75 p626	++SFC 3-30-75 p16
+MT 1-76 p43	+-SR 5-3-75 p33
+NR 5-75 p11	+St 5-75 p98

Le siège de Corinthe: Ballet music. cf Guillaume Tell: Ballet
    music.
L'Assedio di Corinto: Overture. cf Overtures (DG 2530 559).
Il Signor Bruschino: Overture. cf Overtures (Classics for
    Pleasure CFP 40077).
Il Signor Bruschino: Overture. cf Overtures (DG 2530 559).
Il Signor Bruschino: Overture. cf Overtures (Philips 6500 878).
Sonatas, strings (6). cf Serenade, small orchestra, E flat major.

2202 Sonatas, strings, nos. 1-6. I Solisti Veneti; Claudio Scimone.
        RCA AGL 2-1339 (2).
            ++NR 5-76 p1

2203 Sonatas, strings, nos. 1-3, 6. BPhO; Herbert von Karajan. DG
        2535 187. (Reissue from 139041).
            +-Gr 9-76 p424   +-RR 8-76 p64

2204 Sonatas, strings, nos. 1-4. ECO; Antonio Ros-Marba. Pye NEL 2017.
            +HFN 4-76 p113   +RR 5-76 p50
      Sonatas, strings, nos. 2 and 4. cf DONIZETTI: Quartet, strings,
          D major.
      Sonata, 2 violins, violoncello and double bass, no. 3, C major.
          cf DG 2548 219.
      Songs: Duetto buffo di due gatti; La pesca; La regata Veneziana.
          cf BIS LP 17.
      Songs: La danza; L'orgia; La promessa. cf BELLINI: Songs (Cetra
          LPO 2003).

2205 Stabat mater. Sung-Sook Lee, s; Florence Quivar, ms; Kenneth
        Riegel, t; Paul Plishka, bs; Cincinnati May Festival Chorus;
        CnSO; Thomas Schippers. Turnabout (Q) QTV S 34634.
            /NR 10-76 p9   ++St 12-76 p146
            ++SFC 6-27-76 p29

      Stabat mater: Cuius aninam. cf Bongiovanni GB 1.
      Stabat mater: Inflammatus. cf Works, selections (Decca SPA 445).
      Tancredi: Overture. cf Overtures (Philips 6500 878).
      Il Turco in Italia: Overture. cf Overtures (Philips 6500 878).
      Variations, clarinet and small orchestra, C major. cf Serenade,
          small orchestra, E flat major.

2206 Works, selections: The barber of Seville: Largo al factotum; Una
        voce poco fa. Cinderella: Zitto, zitto; Piano, piano; Non più
        mesta. The Italian girl in Algiers: Finale, Act 1. Stabat
        mater: Inflammatus. Semiramide: Bel raggio lusinghier. The
        thieving magpie: Overture. William Tell: Overture. Pilar
        Lorengar, Giuliana Tavolaccini, s; Teresa Berganza, ms; Luigi
        Alva, Ugo Benelli, t; Rolando Panerai, Paolo Montarsolo, Sesto
        Bruscantini, Ettore Bastianini, bar; Fernando Corena, bs; LSO
        and Chorus, Maggio Musicale Fiorentino Orchestra and Chorus;
        Pierino Gamba, Alberto Erede, Alexander Gibson, Silvio Varviso,
        Olivero de Fabritiis, Istváñ Kertész. Decca SPA 445. Tape (c)
        KCSP 445. (Reissue).
            +/HFN 8-76 p93   +RR 8-76 p23
            +HFN 10-76 p185 tape

ROSTAL, Peter
    Anglo-American fantasy. cf HMV HQS 1360.
    The entertainer: Variations on a theme by Scott Joplin. cf HMV
        HQS 1360.
ROTOLI
    Fior che langue. cf Rubini RS 301.
ROUSSAKIS, Nicolas
    Short pieces, 2 flutes (6). cf Nonesuch HB 73028 (2).
ROUSSEL, Albert
    L'Accueil des muses. cf DUKAS: La plainte au loin du faune.
    Bacchus et Ariane, op. 43: Suite, no. 2. cf DEBUSSY: Prélude à
        l'après-midi d'un faune.
    Impromptu, harp, op. 21. cf Panton 110 380.
    Poèmes de Ronsard (2). cf BIS LP 45.
    Prélude et fugue sur le nom de Bach, op. 46. cf DUKAS: La plainte
        au loin du faune.
    Segovia, op. 29. cf RCA SB 6891.
    Sinfonietta, op. 52. cf MARTIN: Petite symphonie concertante.
    Songs: Adieu; A flower given to my daughter; Jazz dans la nuit;
        Light; Mélodies, op. 20; Poèmes chinois, op. 12; Poèmes chinois,
        op. 35; Odes anacréontiques, nos. 1, 5; Odelette. cf ENESCO:
        Songs, op. 15.
    Symphony, no. 3, op. 42, G minor. cf DUKAS: La péri.
ROUTLEY, Erik
    Songs: God who spoke in the beginning; You we praise as God. cf
        Grosvenor GRS 1039.
ROZSA, Miklós
    Notturno Ungherese, op. 25. cf Hungarian sketches, op. 14.
2207 Hungarian sketches, op. 14 (3). Notturno Ungherese, op. 25. Over-
        ture to a symphony concert, op. 26a. Theme, variations and
        finale, op. 13. RCA Italiana Orchestra; Miklós Rózsa. RCA
        GL 25010.
            +-Gr 10-76 p607   +RR 10-76 p66
            +HFN 12-76 p147
2208 Little suite, op. 5. Duo, violin and piano, op. 7. Variations
        on a Hungarian peasant song, op. 4. Edwin Herbst, Leonard
        Pennario, pno; Endre Granat, vln. Orion ORS 73127. (also
        Ember ECL 9043).
            ++Gr 5-76 p1781   ++NR 12-73 p12
            ++HF 1-74 p79
    Overture to a symphony concert, op. 26a. cf Hungarian sketches,
        op. 14.
2209 Quintet, op. 2. Trio, op. 1. Leonard Pennario, pno; Sheldon
        Sanov, Endre Granat, vln; Milton Thomas, vla; Nathaniel Rosen,
        vlc. Orion OR 75191.
            +NR 1-76 p8
    Theme, variations and finale, op. 13. cf Hungarian sketches, op.
        14.
    Trio, op. 1. cf Quintet, op. 2.
    Variations on a Hungarian peasant song, op. 4. cf Little suite,
        op. 5.
ROZSAVOLGYI, Mark
    Czardas. cf CSERMAK: Hungarian dances.
    First Hungarian round dance. cf Csermak: Hungarian dances.
ROZYCKI, Ludomir
    Magnificat, Magnificemus in cantico, Confitebor. cf MIELCZEWSKI:

Veni Domine, Deus in nomine tuo.
Orchestral prelude, op. 31. cf BORODIN: Prince Igor: Polovtsian
     dances.

RUBBRA, Edmund
Dormi Jesu, op. 3. cf Songs (RCA LRL 1-5119).
Missa Cantuariensis, op. 59. cf Songs (RCA LRL 1-5119).
Missa in honorem Sancti Dominici, op. 66. cf Songs (RCA LRL
     1-5119).
2210 Songs (choral works): Dormi Jesu, op. 3. Missa Cantuariensis,
     op. 59. Missa in honorem Sancti Dominici, op. 66. That Virgin's
     child most meek, op. 114, no. 2. St. Margaret's Westminster
     Singers; Ian Watson, org; Richard Hickox. RCA LRL 1-5119.
          +Gr 5-76 p1795   +RR 5-76 p72
          +-HFN 5-76 p107   +STL 6-6-76 p37
          +-MT 9-76 p747
Songs: I care not for these ladies; It fell on a summer's day.
     cf Argo ZRG 833.
That Virgin's child most meek, op. 114, no. 2. cf Songs (RCA LRL
     1-5119).

RUBINSTEIN, Anton
Concerto, piano, no. 4, op. 70, D minor. cf International Piano
     Library IPL 5001/2.
The demon: The night. cf HMV ASD 3200.
Es blinkt der Tah. cf Discophilia 13 KGW 1.
Melody, op. 3, no. 1, F major. cf HMV SXLP 30188.
Melody, op. 3, no. 1, F major. cf Odyssey Y 33825.
Melody, op. 3, no. 1, F major. cf L'Oiseau-Lyre DSLO 7.
Prelude and fugue, op. 53, no. 2. cf International Piano Library
     IPL 102.
Sonata, violoncello and piano, no. 2, op. 39, G major. cf
     GERNSHEIM: Sonata, violoncello and piano, no. 1, op. 12, D
     minor.
Songs: Night. cf Rubini RS 301.

RUE, Pierre de la
Missa Ave santissima Maria: Sanctus. cf HMV SLS 5049.
Pour ung jamais. cf L'Oiseau-Lyre 12BB 203/6.

RUIZ-PIPO, Antonio
Danza, no. 1. cf DG 3335 182.
Tablas para guitarra y orquesta. cf OHANA: Tres gráficos para
     guitarra y orquesta.

RUSSELL
By appointment: White roses. cf HMV RLS 716.
Honey. cf Personal Touch 88.
The ship on fire. cf Transatlantic XTRA 1159.
RUSSIAN AND BULGARIAN ORTHODOX CHANTS. cf Harmonia Mundi HMU 137.
RUSSIAN CHORAL WORKS OF THE SEVENTEENTH AND EIGHTEENTH CENTURIES.
     cf HMV Melodiya ASD 3102.
RUSSIAN FOLK SONGS. cf Columbia M 33822.
RUSSIAN ORTHODOX CHURCH MUSIC (CHRISTMAS). cf Ikon IKO 3.
RUSSIAN SONGS. cf Angel SR 40269.
RUSSIAN WEDDING, FESTIVAL AND SEASON SONGS. cf Westminster WGM
     8320.

RUZDJAK, Marko
Piste. cf BOZIC: Audiogemi I-IV.

RYBAR, Jaroslav
Sonata, 12 wind instruments. cf KUBIK: Lament of a warrior's wife.

RYCHLIK
     Relazioni, alto flute, English horn and bassoon. cf ISTVAN: Ritmi
          ed Antiritmi.
RZEWSKI, Frederic
     Songs: Apolitical intellectuals; God to a hungry child; Struggle.
          cf Folkways FTS 33901/4.
SACCHI
     Lungi dal caro bene. cf Club 99-99.
LE SAGE DE RICHEE
     Ouverture. cf Turnabout TV 34137.
SAGRERAS
     El colibri. cf London CS 7015.
SAHL, Michael
     Quartet, strings. cf Desto DC 6435/7.
SAINT-MARTIN, Leonce de
     Toccata de la libération. cf Vista VPS 1029.
SAINT-SAENS, Camille
     Africa, op. 89. cf GOUNOD: Fantasy on the Russian national hymn.
     Allegro appassionato, op. 43. cf Concerto, violoncello, no. 1,
          op. 33, A minor.
     Allegro appassionato, op. 43. cf Works, selections (Vox 5134).
     Ascanio: Ballet music, Adagio and variation. cf RCA LRL 1-5094.
     Le bonheur est chose legère. cf CBS 76476.
     Caprice, violin and orchestra, op. 122. cf Works, selections
          (Vox 5134).
2211 Caprice, violin and orchestra, D major. Concerto, violin, no. 2,
          op. 58, C major. Concerto, violin, no. 4, op. 62, G major.
          Ivry Gitlis, vln; Monte Carlo Opera Orchestra; Eduard van
          Remoortel. Philips 6581 011.
               +-Gr 1-76 p1203   +RR 1-76 p37
               /HFN 2-76 p111
2212 The carnival of the animals (text by Ogden Nash). WALTON: Façade.
          Noel Coward, Edith Sitwell, David Horner, narrators; Leonid
          Hambro, Jascha Zayde, pno; Frank Miller, vlc; Orchestras;
          André Kostelanetz, Frederick Prausnitz. CBS 61693. (Reissue
          from Philips NBR 6001).
               -Gr 8-76 p297   +-RR 6-76 p52
               +-HFN 7-76 p103
     The carnival of the animals. cf CHABRIER: Trois valse romantiques.
     The carnival of the animals. cf MOZART: Serenade, no. 13, K 525,
          G major.
     The carnival of the animals. cf POULENC: The story of Babar, the
          little elephant.
     The carnival of the animals. cf PROKOFIEV: Peter and the wolf,
          op. 67.
     Le carnaval des animaux. cf HMV SLS 5033.
     The carnival of the animals. cf HMV SXLP 30181.
     The carnival of the animals. cf Philips 6780 030.
     The carnival of the animals: Le cygne. cf Columbia/Melodiya
          M 33593.
     The carnival of the animals: The swan. cf Decca KCSP 367.
     Le carnaval des animaux: The swan. cf L'Oiseau-Lyre DSLO 7.
     Cavatine, op. 144. cf Boston Brass BB 1001.
     Concerto, piano, no. 2, op. 22, G minor. cf MENDELSSOHN: Concerto,
          piano, no. 1, op. 25, G minor.
     Concerti, violin (3). cf Works, selections (Vox 5134).

Concerto, violin, no. 2, op. 58, C major. cf Caprice, violin and orchestra, D major.

2213 Concerto, violin, no. 3, op. 61, B minor. VIEUXTEMPS: Concerto, violin, no. 5, op. 37, A minor. Kyung-Wha Chung, vln; LSO; Lawrence Foster. Decca SXL 6759. Tape KSXC 6759. (also London 6992 Tape (c) 5-6992).

+Gr 9-76 p424   +-HFN 11-76 p175
++HFN 9-76 p129   +RR 9-76 p68

2214 Concerto, violin, no. 3, op. 61, B minor. Havanaise, op. 83, Louis Kaufman, vln; Netherlands Philharmonic Orchestra; Maurits Van Den Berg. Orion ORS 75177. (also Ember ECL 9046).

+Audio 6-76 p96   +IN 2-76 p8
-Gr 8-76 p297   -NR 6-75 p7
+-HF 9-75 p89   -RR 5-76 p50
-HFN 9-76 p129   +-St 9-75 p111

Concerto, violin, no. 4, op. 62, G major. cf Caprice, violin and orchestra, D major.

Concerti, violoncello (2). cf Works, selections (Vox 5134).

2215 Concerto, violoncello, no. 1, op. 33, A minor. Concerto, violoncello, no. 2, op. 119, D minor. Suite, violoncello, op. 16. Allegro appassionato, op. 43. Christine Walevska, vlc; Monte Carlo Opera Orchestra; Eliahu Inbal. Philips 6500 459. Tape (c) 7300 343.

+-HF 9-75 p89   +-RR 8-76 p84 tape
+HFN 7-75 p91   ++St 6-75 p106
+NR 3-75 p4

Concerto, violoncello, no. 1, op. 33, A minor. cf BRUCH: Kol Nidrei, op. 47.

Concerto, violoncello, no. 2, op. 119, D minor. cf Concerto, violoncello, no. 1, op. 33, A minor.

Danse macabre, op. 40. cf BIZET: Jeux d'enfants, op. 22.

Danse macabre, op. 40. cf Decca DPA 519/20.

Danse macabre, op. 40. cf DG 2584 004.

Danse macabre, op. 40. cf Pye QS PCNH 4.

Danse macabre, op. 40. cf RCA VH 020.

Danse macabre, op. 40. cf Sterling 1001/2.

Fantasie, E flat major. cf Wealden WS 149.

Havanaise, op. 83. cf Concerto, violin, no. 3, op. 61, B minor.

Havanaise, op. 83. cf Works, selections (Vox 5134).

Havanaise, op. 83. cf CHAUSSON: Poème, op. 25.

Havanaise, op. 83. cf RCA ARM 4-0942/7.

Introduction and rondo capriccioso, op. 28. cf Works, selections (Vox 5134) .

Introduction and rondo capriccioso, op. 28. cf CHAUSSON: Poème, op. 25.

Introduction and rondo capriccioso, op. 28. cf CBS 30063.

Introduction and rondo capriccioso, op. 28. cf RCA ARM 4-0942/7.

Marche militaire française. cf HMV SXLP 50017.

Marche militaire française: Africa, Valse mignonne, Reverie à Blidah, Suite Algérienne. cf Rococo 2049.

Morceau de concert, op. 62. cf Works, selections (Vox 5134).

Prelude and fugue, op. 99, no. 3, E flat major. cf FRANCK: Chorale, no. 2, B minor.

Romance, violin and orchestra, op. 48. cf Works, selections (Vox 5134).

Le rouet d'omphale, op. 31. cf BIZET: Jeux d'enfants, op. 22.

Samson et Dalila: Arrêtz, o mes frères. cf Rubini GV 63.
Samson et Dalila: Mon coeur s'ouvre à ta voix. cf Collectors
  Guild 611.
Samson et Dalila: Mon coeur s'ouvre à ta voix. cf Decca GOSD
  674/6.
Samson et Dalila: Mon coeur s'ouvre à ta voix; Amour, viens aider.
  cf Columbia/Melodiya M 33931.
Samson et Dalila: Mon coeur s'ouvre, Printemps qui commence,
  Amour viens aider, Mon coeur s'ouvre. cf Discophilia KIS KGG 3.
Sonata, violin, no. 1, op. 75, D minor. cf RCA ARM 4-0942/7.
Sonata, violoncello and piano, no. 1, op. 32, A minor. cf DEBUSSY:
  Sonata, violoncello and piano, no. 1, D minor.
Suite, violoncello, op. 16. cf Concerto, violoncello, no. 1,
  op. 33, A minor.
2216 Symphony, no. 1, op. 2, E flat major. Symphony, no. 2, op. 55, A
  minor. ORTF; Jean Martinon. HMV ASD 2946. Tape (c) TC ASD
  2946. (also Angel S 36995).
        ++Gr 1-74 p1387   ++NYT 5-19-74 pD27
        +Gr 10-75 p721 tape   +RR 12-73 p74
        +HF 7-74 p100   +RR 8-76 p85 tape
        +HFN 10-75 p155 tape   ++St 9-74 p128
        +NR 5-74 p5
2217 Symphony, no. 1, op. 2, E flat major.** Symphony, no. 2, op. 55,
  A minor.* Symphony, no. 3, op. 78, C minor. Symphony, A
  major.* Symphony, F major.* Bernard Gavoty, org; French
  National Radio Orchestra; Jean Martinon. HHM SQ SLS 5035 (3).
  (*Reissue from ASD 3138, **ASD 2946).
        +Gr 5-76 p1766
2218 Symphony, no. 1, op. 2, E flat major. Symphony, no. 2, op. 55,
  A minor. Frankfurt Radio Symphony Orchestra; Eliahu Inbal.
  Philips 9500 079.
        +Gr 10-76 p607   +-RR 11-76 p77
        ++HFN 10-76 p175
Symphony, no. 2, op. 55, A minor. cf Symphony, no. 1, op. 2,
  op. 2, E flat major (HMV ASD 2946).
Symphony, no. 2, op. 55, A minor. cf Symphony, no. 1, op. 2, E
  flat major (HMV 5035).
Symphony, no. 2, op. 55, A minor. cf Symphony, no. 1, op. 2, E
  flat major (Philips 9500 079).
2219 Symphony, no. 3, C minor. Bernard Gavoty, org; French National
  Radio Orchestra; Jean Martinon. Angel S 37122. Tape (c) 4XS
  37122 (ct) 8XS 37122.
        ++HF 2-76 p104   +-NR 1-76 p5
        -MJ 1-76 p39   ++St 3-76 p120
2220 Symphony, no. 3, op. 78, C minor. Gaston Litaize, org; CSO; Daniel
  Barenboim. DG 2530 619. Tape (c) 3300 619.
        ++Gr 4-76 p1608   +NR 8-76 p3
        ++Gr 7-76 p230 tape   +-RR 4-76 p53
        +HF 10-76 p118   +SFC 5-23-76 p36
        +HFN 4-76 p113   ++St 10-76 p131
        +HFN 6-76 p105 tape   ++STL 4-11-76 p36
2221 Symphony, no. 3, op. 78, C minor. LAPO; Zubin Mehta. London
  C 6680. (also Decca SXL 6482 Tape (c) KSXC 6482).
        +Gr 1-71 p1160   -NR 8-71 p5
        +HF 9-71 p104   ++SFC 10-3-71 p37
        +HFN 11-76 p175 tape   +St 12-71 p94

Symphony, no. 3, op. 78, C minor. cf Symphony, no. 1, op. 2,
    E flat major.
2222 Symphonies, A major, F major. ORTF; Jean Martinon. Angel S 37089.
    (also HMV (Q) ASD 3138).
          ++Gr 11-75 p828 Quad   +NR 8-75 p3
          +HF 8-75 p75   +RR 12-75 p61
          +HFN 12-75 p163   ++St 11-75 p135
          +MT 8-76 p662
Symphony, A major. cf Symphony, no. 1, op. 2, E flat major.
Symphony, F major. cf Symphony, no. 1, op. 2, E flat major.
Toccata, op. 111, no. 6. cf Desmar DSM 1005.
2223 Works, selections: Allegro appassionato, op. 43. Caprice, violin,
    op. 122. Concerti, violoncello (2). Concerti, violin (3).
    Havanaise, op. 83. Introduction and rondo capriccioso, op. 28.
    Morceau de concert, op. 62. Romance, violin and orchestra,
    op. 48. Ruggiero Ricci, vln; Laszlo Varga, vlc; Luxembourg
    Radio Orchestra, Westphalian Symphony Orchestra, PH; Pierre
    Cao, Reinhard Peters, Siegfried Landau. Vox (Q) SVBX 5134 (3).
          +-HF 1-76 p94   ++SFC 9-28-75 p30
          ++NR 10-75 p3
SAINZ DE LA MAZA, Eduardo
    Campanas del Alba. cf London CS 7015.
    Habañera. cf London STS 15306.
SAINZ DE LA MAZA, Regino
    Rondeña (Andaluzza). cf Supraphon 111 1230.
SAKAC, Branimir
    Turm-Musik. cf KULJERIC: Omaggio à Lukačić.
SALIERI, Antonio
    Minuet. cf DG Archive 2533 182.
SALLINEN, Aulis
    Chaconne. cf BACH: Prelude and fugue, S 546, C minor.
SALMENHAARA, Erkki
    Quintet, winds. cf CARLSTEDT: Sinfonietta, 5 wind instruments.
SALOME, Théodor
    Grand choeur. cf Wealden WS 142.
SALZEDO, Carlos
    Pièce concertante, op. 27. cf Boston Brass BB 1001.
SALZEDO, Leonard
    Divertimento: Prelude. cf Argo SPA 464.
SAMARA, Spiro
    Histoire d'amour. cf Rococo 5397.
SAMAZEUILH, Gustave
    Serenade. cf RCA ARL 1-1323.
SAMMARTINI, Giovanni
    Concerto, viola pomposa, C major. cf PORPORA: Concerto, violon-
        cello, G major.
SAMMARTINI, Giuseppe
    Concerto, flute, F major. cf RCA CRL 3-1429.
SAN SEBASTIAN
    Preludios vascos: Dolor. cf RCA ARL 1-1323.
SANCAN, Pierre
    Sonatine pour flute et piano. cf PAGANINI: Caprice, op. 1, no.
        24, A minor.
SANDERSON
    Songs: Friend o' mine. cf Argo ZFB 95/6.
SANDRIN, Pierre
    Doulce memoire (2 settings). cf Argo ZRG 667.

Doulce memoire. cf L'Oiseau-Lyre 12BB 203/6
SANDSTROM, Sven-David
    In the meantime. cf Caprice RIKS LP 34.
SANTOLIQUIDO, Francesco
    Poesi persiane (3). cf DEBUSSY: Songs: Le promenoir des deux
        amants.
SANTORSOLA, Guido
    Prelude, no. 1. cf Supraphon 111 1230
SANZ, Gaspar
    Canarios. cf DG Archive 2533 172.
    Canarios. cf Enigma VAR 1015.
    Castillian dances (4). cf Swedish Society SLT 33189.
    Españoleta. cf DG Archive 2533 172.
    Españoleta. cf Enigma VAR 1015.
    Gallard y villano. cf DG Archive 2533 172.
    Passacalle de la Cavalleria de Napoles. cf DG Archive 2533 172.
    Suite española. cf Angel S 36093.
    Suite española. cf London STS 15306.
SARACINI, Claudio
    Songs: Da te parto; Deh, come invan chiedete; Giovinetta vezzo-
        setta; Io moro; Quest'amore, quest'arsura. cf DG Archive 2533
        305.
SARASATE, Pablo
    Caprice basque, op. 24. cf HMV HLM 7077.
    Carmen fantasy, op. 25. cf HMV SEOM 22.
    Carmen fantasy, op. 25. cf RCA ARM 4-0942/7.
    Danzas españolas, op. 21, no. 1: Malagueña; no. 2: Habañera. cf
        RCA ARM 4-0942/7.
    Danzas españolas, op. 22, no. 1: Romanza andaluza. cf Connoisseur
        Society CS 2070.
    Danzas españolas, op. 23, no. 2: Zapateado. cf RCA ARM 4-0942/7.
    Introduction and tarantelle. cf RCA ARM 4-0942/7.
    Zigeunerweisen, op. 20, no. 1. cf Columbia/Melodiya M 33593.
    Zigeunerweisen, op. 20, no. 1. cf Odyssey Y 33825.
    Zigeunerweisen, op. 20, no. 1. cf RCA ARM 4-0942/7.
SARTI, Guiseppé
    Songs: Lungi dal caro bene. cf Decca SXL 6629.
SATIE, Erik
    Avant-dernières pensées. cf POULENC: Mouvements perpetuels, nos.
        I, II, III.
    Croquis et agaceries d'un gros bonhomme en bois. cf POULENC: Mouve-
        ments perpetuels, nos. I, II, III.
    Descriptions automatiques. cf POULENC: Mouvements perpetuels, nos.
        I, II, III.
    Gnossiennes, nos. 1-6. Gymnopédies, nos. 1-3. Nouvelles pièces
        froides (3). Ogives (2). Pièces froides: Airs à faire fuir
        (3). Rêveries nocturnes (2). Sarabandes, nos. 1-3. Songe
        creux. Peter Kraus, Mark Bird, gtr. Orion ORS 74163. (also
        Ember ECL 9048).
            +Audio 12-75 p105   +NR 11-74 p12
            +Gr 5-76 p1781   +RR 6-76 p75
            +HFN 8-76 p86
    Gnossiennes, nos. 1, 4, 5. cf Piano works (Saga 5387).
    Gnossienne, no. 2. cf Works, selections (Unicorn RHS 338).
    Gnossienne, no. 3. cf POULENC: Mouvements perpetuels, nos. I, II,
        III.

Gymnopédies, nos. 1-3. cf Gnossiennes, nos. 1-6.
Gymnopédies, nos. 1-3. cf Piano works (Saga 5387).
Gymnopédie, no. 1. cf Works, selections (Unicorn RHS 338).
Gymnopédie, no. 1. cf POULENC: Mouvements perpetuels, nos. I, II,
    III.
Gymnopédie, no. 1. cf Connoisseur Society (Q) CSQ 2065.
Gymnopédie, no. 1. cf Stentorian SC 1724.
Jack in the box. cf Supraphon 111 1721/2.
Nocturne, no. 1. cf Piano works (Saga 5387).
Nouvelles pièces froides (3). cf Gnossiennes, nos. 1-6.
Ogives (2). cf Gnossiennes, nos. 1-6.
Parade: Rag-time. cf Piano works (Saga 5387).
Passacaille. cf Piano works (Saga 5387).
2225 Piano works: Gnossiennes, nos. 1, 4, 5. Gymnopédies (3). Nocturne,
    no. 1. Parade: Rag-time (arr. Ourdine). Passacaille. Pièces:
    Désespoir agréable; Effronterie; Poésie; Prélude canin; Pro-
    fondeur; Songe creux. Sarabandes, nos. 1, 3. Sonatine bureau-
    cratique. Sports et divertissements. Veritable préludes
    flasques. Vieux sequins et vieilles cuirasses. John McCabe,
    pno. Saga 5387. Tape (c) CA 5387.
        +-Gr 12-74 p1182   +RR 12-74 p63
        +-HFN 10-76 p185 tape
Le piccadilly. cf Works, selections (Unicorn RHS 338).
Pièces: Désespoir agréable; Effronterie; Poésie; Prélude canin;
    Profondeur; Songe creux. cf Piano works (Saga 5387).
Pièces froides: Airs à faire fuir (3). cf Gnossiennes, nos. 1-6.
Pièces froides: Airs à faire fuir. cf Works, selections (Unicorn
    RHS 338).
Poudre d'or. cf Works, selections (Unicorn RHS 338).
Rêveries nocturnes (2). cf Gnossiennes, nos. 1-6.
Sarabandes, nos. 1-3. cf Gnossiennes, nos. 1-6.
Sarabandes, nos. 1, 3. cf Piano works (Saga 5387).
Sarabande, no. 2. cf POULENC: Mouvements perpetuels, nos. I, II,
    III.
Sonatine bureaucratique. cf Piano works (Saga 5387).
Songe creux. cf Gnossiennes, nos. 1-6.
Songs: Chanson; Chanson médiévale; La diva de l'Empire; Elégie;
    Hymne: Salut drapeau; Geneviève de Brabant: Air de Geneviève;
    Je te veux; Petit air; Les anges, Les fleurs, Sylvie; Tendre-
    ment. cf Works, selections (Unicorn RHS 338).
Sports et divertissements. cf Piano works (Saga 5387).
Véritable préludes flasques. cf Piano works (Saga 5387).
Vexations. cf Works, selections (Unicorn RHS 338).
Vieux sequins et vieilles cuirasses. cf Piano works (Saga 5387).
2226 Works, selections: Pièces froides: Airs à faire fuir. Gnossienne,
    no. 2. Gymnopédie, no. 1. Poudre d'or. Le piccadilly. Vexa-
    tions. Songs: Chanson; Chanson médiévale; Le diva de l'Empire;
    Elégie; Hymne: Salut drapeau; Geneviève de Brabant; Air de
    Geneviève; Je te veux; Petit air. Songs: Les anges, Les fleurs,
    Sylvie; Tendrement. Peter Dickinson, pno; Meriel Dickinson, ms.
    Unicorn RHS 338.
        +Gr 11-76 p858   ++RR 11-76 p101
        +-HFN 11-76 p167
SAVIO (20th century Brazil)
    Batucada. cf Supraphon 111 1230.
    Escenas brasilenas. cf London STS 15306.
SCARLATTI
    Tu lo sai. cf Club 99-101.

SCARLATTI, Alessandro
  Christmas cantata: O humble city of Bethlehem...Of a virgin pure;
    Thus he taketh upon him...Our hope of life undying. cf Abbey
    LPB 761.
  Endimione e Cintia: Se geloso e il mio core. cf CBS 76476.
2227 Madrigals: Arsi un tempo; Cor mio, deh non languire; Intenerite
    voi, lagrime mie; Mori, mi dici; O morte, agl'altri fosca; O
    selce, o tigre, o ninfa; Or che da te, mio bene; Sdegno la
    fiamma estinse. Monteverdi Choir, Hamburg; Jürgen Jürgens.
    DG Archive 2533 300.
        +Gr 8-76 p328   ++NR 7-76 p7
        ++HF 8-76 p93   +RR 7-76 p78
        +HFN 8-76 p86   +St 11-76 p155
  Minuet and gavotte. cf Saga 5426.
  Pastorale and capriccio. cf BEETHOVEN: Sonata, piano, no. 14,
    op. 27, no. 2, C sharp minor.
2228 Sinfonias, nos. 1-2, 4-5, 8, 12. Paris Instrumental Ensemble;
    Charles Ravier. Oryx 3C 313.
        +-St 11-76 p155
  Songs: Le violette. cf Decca SXL 6629.
2229 Stabat mater. Mirella Freni, s; Teresa Berganza, ms; Kuentz
    Chamber Orchestra; Charles Mackerras. DG Archive 2533 324.
    Tape (c) 3310 324.
        ++HFN 12-76 p147
  Su le sponde del Tebro. cf BACH: Cantata, no. 51, Jauchzet Gott
    in allen Landon.
SCARLATTI, Domenico
  Sonatas, guitar (2). cf RCA ARL 1-0864.
  Sonata, guitar, A minor. cf London CS 7015.
  Sonata, guitar, A major. cf London CS 7015.
  Sonatas, guitar, A major, A minor. cf Philips 6581 017.
2230 Sonata, guitar, K 2, E major. Sonata, K 11, G minor. Sonata, K 9,
    A minor. Sonata, K 430, G major. Sonata, K 380, A major.
    Sonata, K 446, A major. SOLER: Fandango, M. 1a. Sonatas, B
    minor, G major. Peter Kraus, Mark Bird, gtr. Orion ORS 75187.
        +NR 3-76 p13
  Sonata, guitar, K 9, A minor. cf Sonata, guitar, K 2, E major.
  Sonata, guitar, K 11, G minor. cf Sonata, guitar, K 2, E major.
2231 Sonatas, guitar, L 23, E major; L 238, A major; L 429, A minor;
    L 485, A major; L 108, D minor; L 104, D major (trans. Williams).
    VILLA-LOBOS: Preludes, nos. 1-5. John Williams, gtr. CBS
    73545. (Reissues from 72979, 73350).
        +Gr 7-76 p194   ++RR 7-76 p73
  Sonatas, guitar, L 33, E minor; L 352, E minor. cf RCA SB 6891.
  Sonata, guitar, L 23 (K 380), E major. cf Swedish Society SLT
    33189.
  Sonatas, guitar, L 83, A major; L 352, E minor; L 423, A minor;
    L 483, D major. cf Angel S 36093.
  Sonata, guitar, K 380, A major. cf Sonata, guitar, K 2, E major.
  Sonata, guitar, K 430, G major. cf Sonata, guitar, K 2, E major.
  Sonata, guitar, K 446, A major. cf Sonata, guitar, K 2, E major.
2232 Sonatas, harpsichord (6). VILLA-LOBOS: Preludes (5). John
    Williams, gtr. Columbia M 34198. Tape (c) MT 34198. (also
    CBS 73545 Tape (c) 40-73545). (Reissue).
        +ARG 12-76 p39   ++HFN 10-76 p185 tape
        +HFN 10-76 p183
2233 Sonatas, harpsichord, K 1-2, 6, 8-10, 14, 18, 21, 27, 54, 123,
    314. Richard Lester, hpd. Sutton SSLP 109.

　　　　　　+HFN 4-76 p113　　　　　　+RR 4-76 p65
　　　　　　+MT 8-76 p662
2234　Sonatas, harpsichord, L 3 (K 502), C major; L 10 (K 84), C minor;
　　　L 14 (K 492), D major; L 189 (K 184), F minor; L 198 (K 296),
　　　F major; L 204 (K 105), G major; L 209 (K 455), G major; L 223
　　　(K 532), A minor; L 238 (K 208), A major; L 281 (K 239), F
　　　minor; L 389 (K 375), G major; L 422 (K 141), D minor; L 483
　　　(K 322), A major; L 23 (K 380), E major; L 31 (K 318), F sharp
　　　major; L 35 (K 319), F sharp major; L 93 (K 149), A minor; L
　　　148 (K 261), B major; L 205 (K 487), C major; L 225 (K 381), E
　　　major; L 256 (K 247), C sharp minor; L 260 (K 246), C sharp
　　　minor; L 446 (K 262), B major; L 457 (K 132), C major; L supp.
　　　31 (K 83), A major. Kenneth Cooper, hpd. Vanguard VSD 71201/2.
　　　　　++AR 5-76 p27　　　　　　　+NR 8-75 p14
　　　　　+Audio 12-75 p104　　　　++SFC 11-30-75 p34
　　　　　+HF 11-75 p115
2235　Sonatas, harpsichord, K 206, 212, 222, 364, 365, 370, 371, 481,
　　　501, 502, 513, 524, 525, 532. Colin Tilney, hpd. Argo ZK 5.
　　　　　+Gr 12-76 p1022　　　　　+RR 12-76 p86
2236　Sonatas, harpsichord, Kk 495, 496, E major; Kk 211, 212, A major;
　　　Kk 386, 387, F minor; Kk 418, 419, F major; Kk 234, G minor;
　　　Kk 235, G major; Kk 534, 535, D major. Gilbert Rowland, hpd.
　　　Keyboard KGR 1002.
　　　　　+Gr 9-76 p446　　　　　　+RR 6-76 p75
　　　　　+-MT 8-76 p662
2237　Sonatas, harpsichord, K 115, C minor; K 132, C major; K 133, C
　　　major; K 238, F minor; K 239, F minor; K 513, C major; K 481,
　　　F minor; K 208, A major; K 209, A major; K 215, E major; K 216,
　　　E major; K 124, G major; K 490, D major; K 491, D major; K 492,
　　　D major. Blandine Verlet, hpd. Philips 6581 015.
　　　　　++Gr 2-76 p1360　　　　　++RR 2-76 p52
　　　　　+HFN 3-76 p103　　　　　++STL 3-7-76 p37
　　　Sonatas, harpsichord, K 87, B minor; K 201, G major; K 370, E flat
　　　major; K 371, E flat major. cf Saga 5402.
　　　Sonatas, harpsichord, L 352, C minor; L 384, F major; L 387, G
　　　major. cf Angel S 36095.
SCHAEFER
2238　Symphony, large orchestra, op. 25. Brno Philharmonic Orchestra;
　　　Jiří Waldhans. Panton 110 371.
　　　　　++RR 9-76 p59
SCHAEFER, Paul
　　　Anglo-American fantasy. cf HMV HQS 1360.
　　　The entertainer: Variations on a theme by Scott Joplin. cf HMV
　　　HQS 1360.
SCHARWENKA, Xaver
　　　Polish dance, op. 31, no. 1. cf Rococo 2049.
SCHAT, Peter
2239　Canto general. To you. Lucia Kerstens, ms; Vera Beths, vln;
　　　Reinbert de Leeuw, Maarten Bon, Stanley Hoogland, Bart Berman,
　　　pno; Bert van Dijk, Christian Ingelse, org; Hans Bredenbeek,
　　　Ton Burmanje, Louis Ignatius Gall, Harmoed Greef, Dick Hoogeveen,
　　　Franck Noya, Jorge Oraison, Frank Wesstein, Bob Zimmermann, gtr.
　　　Peter Schat. Donemus Audio-Visual DAVS 7475/1.
　　　　　+-RR 10-75 p88　　　　　　+Te 9-76 p28
　　　Thema. cf KEURIS: Concerto, saxophone. Donemus 7374/4.
　　　To you. cf Canto general. Donemus Audio-Visual 7475/1.

SCHEIDEMANN, Heinrich
    Toccata Mixolydisch, G major. cf BACH: Chorale preludes (Pelca
        PSR 40597).
SCHEIDT, Samuel
    Canzon cornetto. cf Nonesuch H 71301.
    Galliard (9 variationen). cf Hungaroton SLPX 11669/70.
    Galliard (John Dowland, 10 variations). cf Hungaroton SLPX
        11669/70.
    Galliard battaglia. cf Argo SPA 464.
    Warum betrübst du dich. cf BACH: Chorale prelude: Sei gegrüsset,
        Jesus gütig, S 768.
SCHEIN, Johann
    Allemande-Tripla. cf DG Archive 2533 184.
    Banchetto musicale: Allemande, gaillarde and courante. cf CBS
        76183.
    Banchetto musicale: Suite, no. 1, G major; no. 2, D minor. cf
        PRAETORIUS: Hosianna dem Sohne Davids.
SCHEU
    Sleepy Sidney two-step. cf Everest 3360.
SCHINDLER, Alan
2240 Sextet, strings, in six movements. THORNE: Six set pieces for
        thirteen players. Contemporary Chamber Players of the Univer-
        sity of Chicago; Ralph Shapey. Owl ORLP 20.
            +-St 9-76 p129
SCHLICK, Johann
    Divertimento, 2 mandolins and continuo, D major. cf BEETHOVEN:
        Adagio, E flat major.
SCHMELZER, Johann
    Sonata à 7 flauti. cf HMV SLS 5022 (2).
SCHMIDT, Franz
2241 Das Buch mit sieben Siegeln (The book of the seven seals). Hanny
        Steffek, s; Hertha Töpper, ms; Julius Patzak, Erich Majkut, t;
        Otto Wiener, bar; Frederick Guthrie, bs; Franz Illenberger, org;
        Graz Cathedral Choir; Munich Philharmonic Orchestra; Anton Lippe.
        Musical Heritage Society MHS 33501 (2). (Reissue from Amadeo
        5004/5).
            +-HF 11-76 p122
    Notre Dame: Intermezzo. cf HMV SLS 5019.
2242 Quintet, piano, G major. Variations on a theme by Josef Labor.
        Alfred Prinz, clt; Vienna Philharmonia Quintet. Decca SDD 491.
            +-Gr 5-76 p1781   +RR 5-76 p62
            +HFN 5-76 p107
    Variations on a Hussar's song. cf ARRIAGA Y BALZOLA: Symphony,
        D major.
    Variations on a theme by Josef Labor. cf Quintet, piano, G major.
SCHMITT, Florent
    Chaine brisée, op. 87: La tragique chevauchée. cf DUKAS: La
        plainte au loin du faune.
    Mirages, op. 70. cf DUKAS: La plainte au loin du faune.
SCHNEIDER
    Tower music. cf Swedish Society SLT 33200.
SCHOBERT, Johann
    Sinfonia, op. 10, no. 1, E flat major. cf Smithsonian Collection
        N 002.
SCHOENBERG, Arnold
2243 Das Buch der Hängenden Gärten (The book of hanging gardens), op. 15.

SCHUBERT: Songs: An mein Herz, D 860; Blondel zu Marien, D 626; Ganymed, D 544; Heidenröslein, D 257; Der Musensohn, D 764; Nur wer die Sehnsucht kennt, D 877/4; Schäfers Klagelied, D 121; Sprache der Liebe, D 412; Rastlose Liebe, D 138. Jan DeGaetani, ms; Gilbert Kalish, pno. Nonesuch H 71320.

+Gr 9-76 p459   +-NR 2-76 p12
++HF 3-76 p75   +-ON 2-7-76 p34
+-HFN 9-76 p129   ++St 10-76 p132
-MQ 7-76 p456

2244 Chamber symphony, no. 1, op. 9, E major (arrangement for orchestra, op. 9b). Chamber symphony, no. 2, op. 38, E flat major. Frankfurt Radio Symphony Orchestra; Eliahu Inbal. Philips 6500 923.

+-Gr 12-75 p1045   ++NR 7-76 p2
+HFN 12-75 p163   +RR 11-75 p55

Chamber symphony, no. 1, op. 9, E major. cf Works, selections (Decca SXLK 6660/4).

Chamber symphony, no. 2, op. 38, E flat major. cf Chamber symphony, no. 1, op. 9, E major (arr. for orchestra, op. 9b).

Die eiserne Brigade. cf Works, selections (Decca SXLK 6660/4).

Fantasia, violin and piano, op. 47. cf Works, selections (Decca SXLK 6660/4).

2245 Gurrelieder. Marita Napier, s; Yvonne Minton, con; Jess Thomas, Kenneth Bowen, t; Siegmund Nimsgern, Günter Reich, bs; BBC Singers, BBC Chorus and Choral Society, Goldsmiths Choral Union, London Philharmonic Chorus, Mens voices; BBC Symphony Orchestra; Pierre Boulez. CBS 78264 (2). (also Columbia M2 33303 (2) (Q) M2Q 33303, also X 1398 (2).

+Audio 6-76 p96   ++NR 7-75 p7
+Gr 4-75 p1855   +-NYT 9-21-75 pD18
+-HF 8-75 p96   +ON 9-75 p60
++MJ 2-76 p32   +-RR 4-75 p64
++MQ 1-76 p147   ++SFC 11-9-75 p22
+MT 7-75 p631   +St 9-75 p120

Herzgewächse, op. 20. cf Works, selections (Decca SXLK 6660/4).

Klavierstück, op. 33a and b. cf BARTOK: Out of doors.

Lied der Waldtaube. cf Works, selections (Decca SXLK 6660/4).

Little pieces, piano, op. 19 (6). cf Piano works (DG 2530 531).

Little pieces, piano, op. 19 (6). cf Piano works (Nonesuch H 71309).

2246 Moses und Aron. Felicity Palmer, Jane Manning, s; Gillian Knight, ms; Richard Cassilly, John Winfield, t; John Noble, Roland Hermann, bar; Richard Angas, Michael Rippon, bs; Günter Reich, speaker; BBC Singers, Orpheus Boys Choir; BBC Symphony Orchestra; Pierre Boulez. CBS 79201 (2). (also Columbia M2 33594 (2)).

+Gr 11-75 p900   +ON 4-3-76 p56
+-HF 6-76 p92   +-RR 11-75 p35
++HFN 11-75 p167   +SFC 8-29-76 p28
+NR 4-76 p10   ++St 10-76 p131

Nachtwandler. cf Works, selections (Decca SXLK 6660/4).

Der neue Klassizismus, op. 28, no. 3. cf Works, selections (Decca SXLK 6660/4).

Ode to Napoleon, op. 41. cf Works, selections (Decca SXLK 6660/4).

2247 Pelleas und Melisande, op. 5. BPhO; Herbert von Karajan. DG 2530

485. (Reissue from 2711 014).
    +Audio 9-75 p70   +NR 5-75 p3
    +Gr 10-76 p607   +SR 5-3-75 p33
Pelleas und Melisande, op. 5. cf BERG: Lyric suite: Pieces.
2248 Piano works: Pieces, op. 11 (3). Pieces, op. 23 (5). Pieces,
    opp. 33a and 33b. Little pieces, op. 19 (6). Suite, op. 25.
    Maurizio Pollini, pno. DG 2530 531.
        ++Gr 5-75 p1999   +NR 12-75 p13
        +HF 11-75 p116   ++RR 6-75 p76
        ++HFN 6-75 p92   ++St 3-76 p121
        +MT 7-76 p578
2249 Piano works: Pieces, op. 11 (3). Little pieces, op. 19 (6).
    Pieces, op. 23 (5). Pieces, opp. 33a and 33b. Suite, op. 25.
    Paul Jacobs, pno. Nonesuch H 71309.
        +HF 6-75 p86   +NYT 4-27-75 pD19
        +HFN 7-75 p88   ++SFC 6-15-75 p24
        +MT 7-76 p578   +St 9-75 p120
        ++NR 6-75 p13
Pieces, chamber orchestra (3). cf Works, selections (Decca SXLK
    6660/4).
Pieces, piano, op. 11 (3). cf Piano works (DG 2530 531).
Pieces, piano, op. 11 (3). cf Piano works (Nonesuch H 71309).
Pieces, piano, op. 23 (5). cf Piano works (DG 2530 531).
Pieces, piano, op. 23 (5). cf Piano works (Nonesuch H 71309).
Pieces, piano, opp. 33a and 33b. cf Piano works (DG 2530 531).
Pieces, piano, opp. 33a and 33b. cf Piano works (Nonesuch H 71309).
2250 Pierrot Lunaire, op. 21. Erika Stiedry-Wagner, speaker; Rudolf
    Kolisch, vln and vla; Stefan Auber, vlc; Leonard Posella, flt
    and pic; Kalman Bloch, clt; Edward Steuermann, pno; Arnold
    Schoenberg. Odyssey Y 33791. (Reissue from Columbia ML 4471).
        +ARSC Vol VIII, no. 2-3 +-HF 9-76 p96
              p83   +NYT 5-16-76 pD19
Pierrot Lunaire, op. 21. cf Works, selections (Decca SXLK 6660/4).
Quintet, wind instruments, op. 26. cf Works, selections (Decca
    SXLK 6660/4).
Rondo. cf Works, selections (Decca SXLK 6660/4).
Serenade, op. 24. cf Works, selections (Decca SXLK 6660/4).
2251 Songs: Cabaret songs (8); Early songs (9). Marni Nixon, s;
    Leonard Stein, pno. RCA ARL 1-1231.
        +HF 3-76 p74   +ON 2-7-76 p34
        ++MJ 2-76 p32   ++SFC 12-14-75 p39
        +-MQ 7-76 p456   ++SR 1-24-76 p53
        +NR 2-76 p12   ++St 4-76 p115
Ein Stelldichein. cf Works, selections (Decca SXLK 6660/4).
Suite, op. 25. cf Piano works (DG 2530 531).
Suite, op. 25. cf Piano works (Nonesuch H 71309).
Suite, op. 29. cf Works, selections (Decca SXLK 6660/4).
Suite, string orchestra. cf BRITTEN: Prelude and fugue, 18 solo
    strings, op. 29.
A survivor from Warsaw, op. 46. cf RCA LRL 2-7531.
2252 Variations, orchestra, op. 31. Verklärte Nacht, op. 4. BPhO;
    Herbert von Karajan. DG 2530 627. Tape (c) 3300 627. (Reissue
    from 2711 014).
        +Gr 5-76 p1766   +-MT 9-76 p748
        +Gr 6-76 p102 tape   ++RR 6-76 p53
        +HFN 5-76 p115

SCHOENBERG (cont.)   428

    Variations, orchestra, op. 31. cf BERG: Lyric suite: Pieces.
    Verklärte Nacht, op. 4. cf Variations, orchestra, op. 31.
    Verklärte Nacht, op. 4. cf Works, selections (Decca SXLK 6660/4).
    Verklärte Nacht, op. 4. cf BERG: Lyric suite: Pieces.
    Weihnachtsmusik. cf Works, selections (Decca SXLK 6660/4).
2253 Works, selections: Chamber symphony, no. 1, op. 9, E major. Die
    eiserne Brigade. Fantasia, violin and piano, op. 47. Herz-
    gewächse, op. 20. Lied der Waldtaube. Nachtwandler. Der
    neue Klassizimus, op. 28, no. 3. Ode to Napoleon, op. 41.
    Pieces, chamber orchestra (3). Pierrot Lunaire, op. 21.
    Quintet, wind instruments, op. 26. Rondo. Serenade, op. 24.
    Ein Stelldichein. Suite, op. 29. Verklärte Nacht, op. 4.
    Weihnachtsmusik. Der wunsch des Liebhabers, op. 27, no. 4.
    Nona Liddell, vln; John Constable, pno; Mary Thomas, June
    Barton, s; Anna Reynolds, ms; John Shirley-Quirk, bar; Gerald
    English, speaker; London Sinfonietta and Chorus; David Atherton.
    Decca SXL 6660/4 (5). (also London SXLK 6660/4).
         +Gr 9-74 p535   +NR 2-76 p4
         +-HF 7-76 p89   +RR 9-74 p34
         ++MJ 2-76 p32   ++SFC 12-14-75 p39
         +-MQ 7-76 p456   ++Te 9-75 p42
    Der wunsch des Liebhabers, op. 27, no. 4. cf Works, selections
    (Decca SXLK 6660/4).
SCHOENBERG, Stig Gustav
    Tower music. cf Swedish Society SLT 33200.
SCHOLZ
    Torgauer Marsch. cf DG 2721 077 (2).
SCHRAMMEL
    Wien bleibt Wien. cf DG 2721 077 (2).
    Wien bleibt Wien. cf Philips 6308 246.
    Wien bleibt Wien. cf Polydor 2489 523.
SCHREINER
    General Lee's grand march. cf Michigan University SM 0002.
SCHROEDER, Hermann
    Praeludium, Kanzone and Rondo. cf BADINGS: Intermezzo.
SCHUBERT, Franz
    Alfonso und Estrella, D 732: Overture. cf Symphonies, nos. 1-6,
    8-9.
    Allegretto, D 915, C minor. cf Piano works (Philips 6500 928/9).
    Allegretto, D 915, C minor. cf Piano works (Philips 6747 175).
    Ave Maria, D 839. cf Decca SXL 6781.
    Ave Maria, D 839. cf HMV HLM 7077.
    The bee, op. 13, no. 9. cf HMV SQ ASD 3283.
    Dances, D 365, nos. 1-2, 6, 22, 26, 29, 34-36. cf Works, selec-
    tions (Telefunken DX 6-35266).
    Deutsche Tanze (German dances), D 336 (4). cf Sonata, piano, no.
    17, op. 53, K 850, D major.
    German dance, D 643, C sharp minor. cf Works, selections (Tele-
    funken DX 6-35266).
    Deutsche Tanze, nos. 1-16, D 783. cf Piano works (Philips 6747
    175).
    German dances, nos. 1-16, D 783. cf Sonata, no. 17, op. 53,
    D 850, D major.
    German dances, D 973 (3); D 769 (2); D 820 (6). cf Piano works
    (Saga 5407).
    German dances with coda and 7 trios. cf Saga 5411.

Ecossaises, D 145, nos. 1-3. cf Works, selections (Telefunken
   DX 6-35266).
Ecossaises, D 299, nos. 11-12. cf Works, selections (Telefunken
   DX 6-35266).
Ecossaises, D 421, nos. 1,2. cf Works, selections (Telefunken DX
   6-35266).
Ecossaises, D 529, nos. 1-3, 5; D 783, nos. 1 and 2. cf Saga 5421.
Ecossaises, D 618, b, c (variants of Ecossaises D 146, nos. 4, 8).
   cf Works, selections (Telefunken DX 6-35266).
Ecossaises, D 735, nos. 7, 8. cf Works, selections (Telefunken
   DX 6-35266).
Ecossaises, D 781. cf Piano works (Philips 6500 928/9).
Ecossaises, D 781. cf Piano works (Philips 6747 175).
Ecossaises, D 783. cf Works, selections (Telefunken DX 6-35266).
2254 Fantasia, op. 15, D 760, C major. SCHUMANN: Sonata, piano, no. 2,
     op. 22, G minor. Bruno-Leonardo Gelber, pno. Connoisseur
     Society (Q) CSQ 2085.
          ++Audio 7-76 p71   +-NR 3-76 p12
          ++HF 6-76 p93   ++SFC 11-16-75 p32
          ++MJ 4-76 p30   +St 10-76 p132
2255 Fantasia, op. 15, D 760, C major. Sonata, piano, no. 16, op. 42,
     D 845, A minor. Maurizio Pollini, pno. DG 2530 473. Tape (c)
     3300 504.
          ++Gr 1-75 p1372   -RR 12-74 p64
          ++HF 5-75 p86   ++RR 3-76 p77 tape
          ++HF 10-76 p147 tape   ++SFC 3-2-75 p24
          ++NR 4-75 p12
2256 Fantasia, op. 15, D 760, C major. Moments musicaux, op. 94, D 780.
     Anthony Goldstone, pno. Oryx ORPS 49. (Reissue from RBL 49).
          -Gr 9-76 p446
2257 Fantasia, op. 15, D 760, C major. Sonata, piano, op. posth.,
     D 960, B flat major. Alfred Brendel, pno. Philips 6500 285.
     Tape (c) 7300 396.
          +Gr 11-72 p932   +-RR 11-72 p92
          +HF 10-76 p147 tape
     Fantasia, op. 15, D 760, C major. cf Piano works (Philips 6747
        175).
2258 Fantasia, op. 159, D 934, C major. Rondo brillant, op. 70, D 895,
     B minor. Sonata, violin and piano, op. 162, D 574, A major.
     Gerald Tarack, vln; David Hancock, pno. Monitor MCA 2146.
          ++Audio 9-76 p85   +NR 6-76 p9
          +MJ 10-76 p25
     Fantasia, op. 159, D 934, C major. cf Works, selections (DG
        2734 004).
     Fantasia, D 993, C minor. cf Piano works (Saga 5407).
     Fugue, op. posth., E minor. cf Pelca PRS 40581.
2259 Gesang der Geister über den Wassern, male chorus and strings.
     Mirjam's Siegesgesang, soprano, chorus and piano. Nacht-
     gesang im Walde, D 913. Ursula Buckel, s; Gerd Lohmeyer, pno;
     South German Madrigal Choir; Wolfgang Gönnenwein. Candide
     (Q) QCE 31087.
          /HF 11-75 p119   ++SFC 7-6-75 p16
          +NR 3-76 p6
     Hungarian melody, D 817, B minor. cf Piano works (Philips 6500
        928/9).
     Hungarian melody, D 817, B minor. cf Piano works (Philips 6747
        175).

2260 Impromptus, op. 90, D 899. Sonata, piano, no. 19, op. posth.,
      D 958, C minor. Sándor Falvai, pno. Hungaroton SLPX 11747.
      +NR 5-76 p12   +-RR 8-76 p71
2261 Impromptus, op. 90, D 899 and op. 142, D 935. Christoph Eschenbach,
      pno. DG 2530 633.
      ++Gr 4-76 p1643   +RR 4-76 p66
      +HFN 4-76 p113
2262 Impromptus, op. 90, D 899 and op. 142, D 935. Ingrid Haebler,
      pno. Philips 6580 075.
      +Gr 6-76 p70   +RR 8-76 p71
      +HFN 5-76 p114
      Impromptus, opp. 90 and 142, D 899, D 935, D 946. cf Piano works
      (Philips 6747 175).
      Impromptus, op. 90, D 935, D 946. cf Piano works (Philips 6500
      928/9).
      Impromptus, op. 90, nos. 2, 4, D 899. cf BEETHOVEN: Sonata, piano,
      no. 14, op. 27, no. 2, C sharp minor.
      Impromptu, op. 90, no. 2, D 899, E flat major. cf CHOPIN: Etude,
      op. 10, no. 3, E major.
      Impromptu, op. 90, no. 3, D 899, G flat major. cf Connoisseur
      Society (Q) CSQ 2065.
      Impromptu, op. 90, no. 3, D 899, G flat major. cf RCA ARM 4-0942/7.
      Impromptu, op. 90, no. 4, D 899, A flat major. cf Works, selec-
      tions (Decca SPA 426).
      Impromptu, op. 90, no. 4, D 899, A flat major. cf CHOPIN: Etude,
      op. 10, no. 3, E major.
      Impromptu, op. 90, no. 4, D 899, A flat major. cf HMV HQS 1354.
      Impromptus, op. 142, nos. 1, 2, D 935. cf BEETHOVEN: Sonata,
      piano, no. 14, op. 27, no. 2, C sharp minor.
      Impromptu, op. 142, no. 2, D 935, A flat major. cf CHOPIN: Ballade,
      no. 2, op. 38, F major.
      Impromptu, op. 142, no. 3, D 935, B flat major. cf Works, selec-
      tions (Decca DPA 545/6).
      Impromptu, op. 142, no. 3, D 935, B flat major. cf BEETHOVEN:
      Concerto, piano, no. 4, op. 58, G major.
      In the Italian style, D 590/1, C major and D major. cf Symphonies,
      nos. 1-6, 8-9.
      Introduction and variations on "Ihr Blumlein alle", D 802, E minor.
      cf Works, selections (DG 2734 004).
      Ländler (Komsiche), D 354, D major. cf Works, selections (Tele-
      funken DX 6-35266).
      Ländler, D 734, nos. 3, 5, 9-11, 14-15. cf Works, selections
      (Telefunken DX 6-35266).
      Ländler, D 790. cf Piano works (Philips 6747 175).
      Marche militaire, no. 1, D 733, D major. cf HMV SXLP 30181.
2263 Mass, no. 2, G 167, G major. Mass, no. 3, D 324, B flat major.
      Emilia Maksimova, s; Reni Pentchova, con; Christo Kamenov,
      Naiden Borodjiev, t; Ivan Dobrev, Pavel Gherdjikov, bs; Rousse
      Rodina Choir, Rodna Pessen Choir; Sofia Soloists Instrumental
      Ensemble; Vassil Kazandjikiev. Harmonia Mundi HMU 111.
      +-Gr 8-76 p328
      Mass, no. 3, D 324, B flat major. cf Mass, no. 2, G 167, G major.
2264 Mass, no. 6, D 950, E flat major. Felicity Palmer, s; Helen Watts,
      con; Kenneth Bowen, Wynford Evans, t; Christopher Keyte, bs;
      St. John's College Chapel Choir; AMF; George Guest. Argo ZRG
      825.

```
        +Gr 11-75 p881              +MT 4-76 p321
        +HF 5-76 p92                +NR 2-76 p9
        +HFN 11-75 p165             +-RR 12-75 p92
        +MJ 10-76 p25               +St 5-76 p122
```

2265 Mass, no. 6, D 950, E flat major. Pilar Lorengar, s; Betty Allen,
     ms; Fritz Wunderlich, Manfred Schmidt, t; Josef Greindl, bs;
     St. Hedwig's Cathedral Choir; BPhO; Erich Leinsdorf. Seraphim
     S 60243. (Reissue from Capital SP 8579).
```
        -HF 11-75 p119              ++St 5-76 p122
```
     Minuet, D 995, F major. cf Piano works (Saga 5407).
     Minuets with 6 trios, D 89 (5). cf Nonesuch H 71141.
     Mirjam's Siegesgesang, soprano, chorus and piano. cf Gesang der
        Geister über den Wassern, male chorus and strings.
2266 Moments musicaux, op. 94, D 780. Sonata, piano, no. 19, op.
     posth., D 958, C minor. Ingrid Haebler, pno. Philips 6580
     128. (Reissue from SAL 3647, 6741 002).
```
        +Gr 10-76 p628             +HFN 11-76 p171
```
     Moments musicaux, op. 94, D 780. cf Fantasia, op. 15, D 760, C
     major.
     Moments musicaux, op. 94, D 780. cf Piano works (Philips 6747
     175).
     Moments musicaux, op. 94, nos. 1, 3, 6, D 780. cf CHOPIN:
        Ballade, no. 2, op. 38, F major.
     Moments musicaux, op. 94, no. 1, D 780, C major. cf CHOPIN:
        Etude, op. 10, no. 3, E major.
     Moment musical, op. 94, no. 3, D 780, F minor. cf Works, selec-
        tions (Decca SPA 426).
     Moment musical, op. 94, no. 3, D 780, F minor. cf Works, selec-
        tions (Decca DPA 545/6).
     Moment musical, op. 94, no. 3, D 780, F minor. cf ALBENIZ: España,
        op. 165, no. 2: Tango.
     Moment musical, op. 94, no. 3, D 780, F minor. cf Decca PFS 4351.
     Moment musical, op. 94, no. 3, D 780, F minor. cf International
        Piano Archives IPA 5007/8.
     Moments musicaux, op. 94, no. 3, D 780, F minor. cf L'Oiseau-
        Lyre DSLO 7.
     Moment musical, op. 94, no. 3, D 780, F minor. cf Saga 5427.
     Nachtgesang im Walde, D 913. cf Gesang der Geister über den
        Wassern, male chorus and strings.
     Nocturne, piano, violin and violoncello, op. 148, D 897, E flat
        major. cf Works, selections (DG 2734 004).
2267 Octet, op. 166, D 803, F major. Berlin Philharmonic Octet.
     Philips 6580 110.
```
        +-Gr 2-76 p1356            +-RR 11-75 p60
        +HFN 12-75 p164
```
2268 Octet, op. 166, D 803, F major. Cleveland Quartet; Thomas Martin,
     double bs; Jack Brymer, clt; Barry Tuckwell, hn. RCA ARL
     1-1047.
```
        +-Gr 2-76 p1356            ++NR 11-75 p7
        +HF 11-75 p120            +-RR 2-76 p46
        +HFN 2-76 p111            ++SR 9-6-75 p40
        ++MJ 1-76 p39
```
2269 Octet, op. 166, D 803, F major. Munich Octet. Turnabout TV
     34152. Tape (c) KTVC 34152.
```
        +-HFN 6-76 p105 tape       +-RR 7-76 p83 tape
```
     Octet, op. 166, D 803, F major. cf HMV SLS 5046.

Octet, op. 166, D 803, F major: Scherzo.   cf Works, selections
(Decca SPA 426).

2270 Piano works: Allegretto, D 915, C minor. Ecossaises, D 781 (11).
Hungarian melody, D 817, B minor. Impromptus, op. 90, D 935,
D 946. Sonata, piano, no. 16, op. 42, D 845, A minor. Alfred
Brendel, pno. Philips 6500 928/9 (2).

+Gr 9-75 p491   +-MJ 10-76 p52
+-HF 11-76 p122   +NR 5-76 p12
++HFN 9-75 p104   +RR 9-75 p62

2271 Piano works: Allegretto, D 915, C minor. Deutsche Tanze, nos.
1-16, D 783. Ecossaises, D 781 (11). Fantasia, op. 15, D 760,
C major. Hungarian melody, D 817, B minor. Impromptus, opp.
90, 142, D 899, D 935, D 946. Ländler, D 790 (12). Moments
musicaux, op. 94, D 780. Sonatas, piano, nos. 14-21. Alfred
Brendel, pno. Philips 6747 175 (8). (Reissues from 6500 418,
6500 929, 6500 763, 6500 416, 6500 415, 6500 284, 6500 285,
6500 928).

+Gr 9-75 p491   +MT 6-76 p495
++HFN 9-75 p109   +RR 11-75 p82

2272 Piano works: Fantasia, D 993, C minor. German dances, D 973 (3);
D 769 (2); D 820 (6). Minuet, D 995, F major. Sonata, piano,
no. 5, D 557, A flat major. Variation on a waltz by Anton
Diabelli, D 718, C minor. Waltz, G flat major (Kupelwieser).
Rosario Marciano, pno. Saga 5407.

+-Gr 2-76 p1360   +-RR 11-75 p82
+-HFN 11-75 p165

Quartet, guitar, flute, viola and violoncello, D 96, G major.   cf
HAYDN: Cassation, no. 6, C major.

2273 Quartets, strings, nos. 1-3, D 18, D 32, D 36. Melos Quartet.
DG 2530 322.

+Gr 8-73 p344   +RR 8-73 p57
+NR 1-76 p8   +SFC 12-28-75 p30

2274 Quartets, strings, nos. 1-15. Melos Quartet. DG 2740 123 (7).
(Reissues from 2530 322, 2530 533).

+Gr 10-75 p651   ++MT 8-76 p662
+HF 7-76 p90   +NR 4-76 p8
+HFN 10-75 p148   +RR 10-75 p67

2275 Quartet, strings, no. 9, D 173, G minor. Quartet, strings, no.
13, op. 29, D 804, A minor. Alban Berg Quartet. Telefunken
AW 6-41882. Tape (c) 4-41882.

+Gr 6-75 p62   +NR 6-76 p9
++HF 8-76 p94   +RR 5-75 p50
++HFN 6-75 p92   ++RR 10-75 p97 tape
+HFN 9-75 p110 tape   ++St 11-76 p157

2276 Quartet, strings, no. 12, D 703, C minor. Quartet, strings, no.
14, D 810, D minor. Melos Quartet. DG 2530 533.

+Gr 5-75 p1991   +-NR 12-75 p9
+-HF 2-76 p104   +-RR 5-75 p51
+-HFN 6-75 p92

2277 Quartet, strings, no. 12, D 703, C minor. Quintet, piano, op. 114,
D 667, A major. Emil Gilels, pno; Norbert Brainin, vln; Peter
Schidlof, vla; Martin Lovett, vlc; Rainer Zepperitz, double-bs;
Amadeus Quartet. DG 2530 646.

+Gr 8-76 p317   +-RR 6-76 p65
++HFN 6-76 p97

Quartet, strings, no. 13, op. 29, D 804, A minor. cf Quartet, strings, no. 9, D 173, G minor.

2278 Quartet, strings, no. 14, D 810, D minor. Collegium Aureum Quartet. BASF KHC 22059.

    +Gr 10-75 p651   +-MT 7-76 p579
    +HF 11-75 p120   ++NR 3-75 p6
    +HFN 10-75 p148   ++RR 10-75 p68
    +MM 5-76 p33

Quartet, strings, no. 14, D 810, D minor. cf Quartet, strings, no. 12, D 703, C minor.

2279 Quintet, piano, op. 114, D 667, A major. Stuart Sankey, double-bs; Festival Quartet. Camden CCV 5046.

    +-RR 5-76 p62

2280 Quintet, piano, op. 114, D 667, A major. Rudolf Serkin, pno; Jaime Laredo, vln; Philipp Naegele, vla; Leslie Parnas, vlc; Julius Levine, double-bs. CBS 61623. (Reissue from 72640).

    +Gr 9-76 p438   +RR 10-76 p82
    +HFN 11-76 p171

2281 Quintet, piano, op. 114, D 667, A major. Maura Lympany, pno; LSO. Classics for Pleasure CFP 40085. Tape (c) TC CFP 40085.

    +Gr 9-74 p536   -RR 9-74 p69
    +-HFN 12-76 p153 tape

2282 Quintet, piano, op. 114, D 667, A major. Beaux Arts Trio; Samuel Rhodes, vla; Georg Hortängel, double-bs. Philips 9500 071. Tape (c) 7300 481.

    +Gr 8-76 p317   ++NR 12-76 p8
    +HFN 7-76 p95   ++RR 7-76 p68
    +MJ 12-76 p44

Quintet, piano, op. 114, D 667, A major. cf Quartet, strings, no. 12, D 703, C minor.

Quintet, piano, op. 114, D 667, A major. cf Works, selections (Decca DPA 545/6).

Quintet, piano, op. 114, D 667, A major: Theme and variations. cf Works, selections (Decca SPA 426).

2283 Quintet, strings, op. 163, D 956, C major. Isaac Stern, Alexander Schneider, vln; Milton Katims, vla; Pablo Casals, Paul Torte-lier, vlc. CBS 61043. (Reissue from Philips ABL 3100).

    +Gr 6-76 p62   +RR 5-76 p62
    +-HFN 7-76 p103

2284 Quintet, strings, op. 163, D 956, C major. Alberni Quartet; Thomas Igloi, vlc. CRD CRD 1018.

    +Gr 11-75 p854   ++RR 11-75 p65
    +HFN 12-75 p164   ++St 3-76 p80

2285 Quintet, strings, op. 163, D 956, C major. Guarneri Quartet; Leonard Rose, vlc. RCA ARL 1-1154. Tape (c) ARS 1-1154 (ct) ARK 1-1154.

    ++HF 2-76 p104   +NR 11-75 p7

Rondo brillant, op. 70, D 895, B minor. cf Fantasia, op. 159, D 934, C major.

Rondo brillant, op. 70, D 895, B minor. cf Works, selections (DG 2734 004).

Rosamunde, op. 26, D 797: Ballet music, G major. cf Works, selec-tions (Decca SPA 426).

2286 Rosamunde, op. 26, D 797: Incidental music. Rohangiz Yachmi, con; Vienna State Opera Chorus; VPO; Karl Münchinger. London OS 26444. (also Decca SXL 6748 Tape (c) KSXC 6748).

```
            +ARG 12-76 p39              +HFN 5-76 p117 tape
            +Gr 3-76 p1471             +RR 2-76 p36
            +Gr 9-76 p497 tape         +-RR 4-76 p80 tape
            +HFN 2-76 p111
```

Rosamunde, op. 26, D 797: Overture. cf Pye PCNHX 6.

Rosamunde, op. 26, D 797: Overture; Ballet music, nos. 1 and 2.
    cf Symphonies, nos. 1-6, 8-9.

Rosamunde, op. 26, D 797: Overture (Die Zauberharfe, D 644);
    Entre'acte, B flat; Ballet, B minor and G major. cf Works,
    selections (Decca DPA 545/6).

Rosamunde, op. 26, D 797: Romance. cf Songs (Philips 6500 704).

Scherzi, D 593 (2). cf Sonata, piano, no. 18, op. 78, D 894, G
    major.

2287 Die schöne Müllerin, op. 25, D 795. Dietrich Fischer-Dieskau,
    bar; Gerald Moore, pno. DG 2530 544. (Reissue from 2720 059).

```
            ++Gr 8-75 p355              *ON 5-76 p48
            +HFN 9-75 p109             ++RR 10-75 p88
            +NR 2-76 p12               ++St 4-76 p116
```

2288 Die schöne Müllerin, op. 25, D 795. Werner Krenn, t; Rudolf Buch-
    binder, pno. Oryx EXP 20.

```
            +-St 4-76 p116
```

Die schöne Müllerin, op. 25, D 795. cf Schwanengesang, D 957.

Die schöne Müllerin, op. 25, D 795. cf Works, selections (Tele-
    funken DX 6-35266).

Die schöne Müllerin, op. 25, D 795: Morgengruss. cf Songs
    (Philips 6580 111).

Die schöne Müllerin, op. 25, D 795: Wohin. cf Works, selections
    (Decca SPA 426).

2289 Schwanengesang, D 957. Die schöne Müllerin, op. 25, D 795.
    Winterreise, op. 89, D 911. Dietrich Fischer-Dieskau, bar;
    Gerald Moore, pno. HMV SLS 840 (3). (Riessues from AFS 481,
    551, 552, 544).

```
            ++Gr 2-76 p1368            +RR 1-76 p61
            +HF 1-76 p123
```

Schwanengesang, D 957, excerpts (4). cf Seraphim S 60251.

2290 Schwanengesang, D 957: Abschied; Der Atlas; Das Fischermädchen;
    Frühlingssehnsucht; Kriegers Ahnung; Ihr Bild; Der Atlas;
    Liebesbotschaft; Die Taubenpost. Winterreise, op. 89, D 911.
    Gérard Souzay, bar; Dalton Baldwin, pno. Philips 6780 028 (2).
    (Reissue from SAL 3428/9).

```
            +-Gr 10-76 p639           +RR 12-76 p95
            ++HFN 11-76 p173
```

Schwanengesang, D 957: Der Döppelganger; Die Stadt. cf Rococo 5370.

Schwanengesang, D 957: Ständchen. cf Works, selections (Decca SPA
    426).

Schwanengesang, D 957: Ständchen. cf Works, selections (Decca
    DPA 545/6).

Serenade. cf CBS 30062.

Serenade. cf HMV HQS 1360.

Sonata, arpeggione and piano, D 821, A minor. cf MENDELSSOHN:
    Sonata, violoncello and piano, no. 2, op. 58, D major.

Sonata, arpeggione and piano, D 821, A minor. cf Works, selections
    (DG 2734 004).

2291 Sonatas, piano, complete. Wilhelm Kempff, pno. DG 2740 132 (9).
    (Reissue from 2720 024).

```
            +Gr 4-76 p1640            +RR 2-76 p53
            +-HFN 12-75 p171
```

2292 Sonata, piano, no. 2, D 279, C major. Sonata, piano, no. 20,
        op. posth., D 959, A major. Wilhelm Kempff, pno. DG 2530 237.
            ++NR 1-76 p8
      Sonata, piano, no. 5, D 557, A flat major. cf Piano works (Saga
        5407).
2293 Sonata, piano, no. 6, D 566, E minor. Sonata, piano, no. 11,
        D 625, F minor. Wilhelm Kempff, pno. DG 2530 354.
            +NR 5-76 p12
      Sonata, piano, no. 11, D 625, F minor. cf Sonata, piano, no. 6,
        D 566, E minor.
      Sonatas, piano, nos. 14-21. cf Piano works (Philips 6747 175).
2294 Sonata, piano, no. 14, op. 143, D 784, A minor. Sonata, piano,
        no. 21, op. posth., D 960, B flat major. Ingrid Haebler, pno.
        Philips 6580 133. (Reissue from SAL 3756).
            +-Gr 11-76 p383   +HFN 11-76 p171
      Sonata, piano, no. 16, op. 42, D 845, A minor. cf Fantasia, op.
        15, D 760, C major.
      Sonata, piano, no. 16, op. 42, D 845, A minor. cf Piano works
        (Philips 6500 928/9).
      Sonata, piano, no. 16, op. 42, D 845, A minor. cf International
        Piano Library IPL 5003/4.
      Sonata, piano, no. 16, op. 42, D 845, A minor: Finale. cf BEET-
        HOVEN: Andante favori, F major.
2295 Sonata, piano, no. 17, op. 53, D 850, D major. German dances,
        D 336 (4). Vladimir Ashkenazy, pno. London CS 6961.
            ++HF 12-76 p116
2296 Sonata, piano, no. 17, op. 53, D 850, D major. German dances,
        nos. 1-16, D 783. Alfred Brendel, pno. Philips 6500 763.
            ++Gr 7-75 p219   +NR 2-76 p13
            +-HF 2-76 p107   +RR 7-75 p53
            ++HFN 7-75 p88
2297 Sonata, piano, no. 18, op. 78, D 894, G major. Scherzi, D 593 (2).
        Radu Lupu, pno. London CS 6966. (also Decca SXL 6741).
            +Gr 5-76 p1787   +HFN 5-76 p107
            +-HF 12-76 p116   +RR 5-76 p66
      Sonata, piano, no. 18, op. 78, D 894, G major. cf Works, selections
        (Decca SPA 426).
      Sonata, piano, no. 19, op. posth., D 958, C minor. cf Impromptus,
        op. 90, D 899.
      Sonata, piano, no. 19, op. posth., D 958, C minor. cf Moments
        musicaux, op. 94, D 780.
2298 Sonata, piano, no. 20, op. posth., D 959, A major. Rudolf Serkin,
        pno. CBS 61645. (Reissue from SBRG 72432).
            ++Gr 12-76 p1022
2299 Sonata, piano, no. 20, op. posth., D 959, A major. Delia Calapai,
        pno. Orion ORS 76218.
            +NR 9-76 p11
      Sonata, piano, no. 20, op. posth., D 959, A major. cf Sonata,
        piano, no. 2, D 279, C major.
2300 Sonata, piano, no. 21, op. posth., D 960, B flat major. Rudolf
        Serkin, pno. Columbia M 33932. (also CBS 76501).
            +Gr 8-76 p319   +RR 8-76 p71
            +-HF 6-76 p93   ++SFC 2-29-76 p25
            +HFN 9-76 p129   +St 9-76 p124
            +NR 4-76 p12
2301 Sonata, piano, no. 21, op. posth., D 960, B flat major. Christoph

Eschenbach, pno. DG 2530 477.
        +-Gr 3-75 p1682   +RR 4-75 p54
        /HF 8-75 p99   ++SFC 6-29-75 p26
        +NR 7-75 p14   +St 9-76 p124
2302 Sonata, piano, no. 21, op. posth., D 960, B flat major. Daniel
        Adni, pno. HMV SQ HQS 1355.
        +-Gr 4-76 p1640   +-RR 3-76 p63
        +HFN 3-76 p103
2303 Sonata, piano, no. 21, op. posth., D 960, B flat major. Gabriel
        Chodos, pno. Orion ORS 75179.
        ++HF 3-76 p94   +-NR 2-76 p13
2304 Sonata, piano, no. 21, op. posth., D 960, B flat major. SCHUMANN:
        Kinderscene, op. 15. Vladimir Horowitz, pno. RCA VH 016.
        (Reissues from HMV ALP 1430, 1469).
        +-Gr 12-75 p1082   +-RR 12-75 p84
        +-HFN 2-76 p117
     Sonata, piano, no. 21, op. posth., D 960, B flat major. cf Son-
        ata, piano, no. 14, op. 143, D 784, A minor.
     Sonata, piano trio, D 28, B flat major. cf Works, selections (DG
        2734 004).
2305 Sonata, violin and piano, op. 162, D 574, A major. Sonatinas,
        violin and piano, nos. 1-3, D 384, 385, 408. Henryk Szeryng,
        vln; Ingrid Haebler, pno. Philips 6500 885.
        ++Gr 3-76 p1481   ++RR 3-76 p63
        +HFN 3-76 p103
     Sonata, violin and piano, op. 162, D 574, A major. cf Fantasia,
        op. 159, D 934, C major.
     Sonata, violin and piano, op. 162, D 574, A major. cf Works,
        selections (DG 2734 004).
     Sonatinas, violin and piano, nos. 1-3, D 384, 385, 408. cf Sonata,
        violin and piano, op. 162, D 574, A major.
     Sonatinas, violin and piano, nos. 1-3, D 384, 385, 408. cf Works,
        selections (DG 2734 004).
     Sonatina, violin and piano, no. 1, op. 137, D 384, D major: Rondo.
        cf RCA ARM 4-0942/7.
     Sonatina, violin and piano, no. 3, op. 137, D 408, G minor. cf
        RCA ARM 4-0942/7.
     Sonatina, violin and piano, no. 3, op. 137, D 408, G minor: Finale.
        cf Pearl SHE 528.
2306 Songs: Abendstern, D 806; An die Entfernte, D 765; Atys, D 585;
        Auf dem Wasser zu singen, D 774; Auflösung, D 807; Der Einsame,
        D 800; Das Fischermädchen, D 957; Der Geistertanz, D 116; Im
        Frühling, D 882; Lachen und Weinen, D 777; Nacht und Träume,
        D 827; Nachtstuck, D 672; Sprache der Liebe, D 410. Peter Pears,
        t; Benjamin Britten, pno. Decca SXL 6722.
        ++Gr 6-75 p79   +-MT 1-76 p43
        ++HFN 6-75 p92
2307 Songs: An den Mond, D 296; Bertha's Lied in der Nacht, D 653; Dass
        sie hier gewesen, D 775; Klärchens Lied, D 210; Lied der Anna
        Lyle, D 830; Lied der Mignon I and II, D 877/2-3; Das Mädchen,
        D 652; Mignons Gesang, D 321; Lilla an die Morgenröte, D 273;
        Sehnsucht, D 636b; Ständchen, D 921; Wehmut, D 772; Der Zwerg,
        D 771. Christa Ludwig, ms; Irwin Gage, pno. DG 2530 528.
        +Gr 6-75 p80   +NR 2-76 p12
        +HF 12-75 p105   ++SFC 2-15-76 p38
        +HFN 7-75 p88   +St 2-76 p109

2308   Songs: An Sylvia, D 891; Gretchen am Spinnrade, D 118; Erntelied,
       D 434; Meeres Stille, D 216; Wehmut, D 772. SCHUMANN: Songs:
       Leis rudern hier, op. 25, no. 17*; Der Nussbaum, op. 25, no. 3;
       Wenn durch die Piazza, op. 25, no. 18*; Widmung, op. 25, no. 1*.
       WOLF: Songs: An den Schlaf; An eine Aeolsharfe; Auf einer Wand-
       erung; Auftrag; Begegnung; Denk es, o Seele; Der Gärtner;
       Keine gleicht von allen Schönen; Sonne der Schlummerlosen.
       Elisabeth Schwarzkopf, s; Geoffrey Parsons, Gerald Moore, pno.
       HMV ASD 3124. (*Reissues from Columbia SAX 5268).
            ++Gr 3-76 p1504   +-RR 2-76 p63
2309   Songs: An die Laute; D 905; An Sylvia, D 891; Der Blumenbrief,
       D 622; Du liebst mich nicht, D 756; Der Einsame, D 800; Im
       Abendrot, D 799; Die Liebe hat gelogen, D 751; Der liebliche
       Stern, D 861; Das Mädchen, D 652; Die Männer sind mechant, D
       866; Minnelied, D 429; Nacht und Träume, D 827; Rosamunde,
       op. 26, D 797: Romance; Schlummerlied, D 527; Seligkeit, D 443;
       Die Sterne, D 939. Elly Ameling, s; Dalton Baldwin, pno.
       Philips 6500 704.
            +-Gr 1-76 p1226   +-ON 5-76 p48
            ++HF 2-76 p94   +RR 1-76 p61
            +HFN 1-76 p119   ++St 5-76 p122
            +MJ 1-76 p24   +STL 5-9-76 p38
            +NR 3-76 p10
2310   Songs: An die Nachtigall, D 196; Fischerweise, D 869; Die Gebüsche,
       D 646; Im Freien, D 880; Im Haine, D 738; Das Lied im Grünen,
       D 917; Der Schmetterling, D 633; Die Vogel, D 691; Der Wachtel-
       schlag, D 742. SCHUMANN: Frauenliebe und Leben, op. 42. Elly
       Ameling, s; Dalton Baldwin, pno. Philips 6500 706.
            +-Gr 10-75 p670   ++NR 11-75 p11
            ++HF 12-75 p106   +RR 1-75 p88
            +HFN 10-75 p149   ++SFC 9-14-75 p28
            +-MT 3-76 p236   +St 12-75 p83
2311   Songs: An Sylvia, D 891; Die Forelle, D 550; Ganymed, D 544; Im
       Abendrot, D 799; Der Jungling an der Quelle, D 300; Nacht und
       Träume, D 827; Normans Gesang, D 846; Die schöne Müllerin, op.
       25, D 795: Morgengruss. Winterreise, op. 89, D 911: Die Wetter-
       fahne; Der Lindenbaum; Frühlingstraum; Einsamkeit. Gérard
       Souzay, bar; Dalton Baldwin, pno. Philips 6580 111. (Reissues
       from SABL 214, SAL 3501, 3651, 3248/9).
            +-Gr 6-76 p77   +RR 5-76 p73
            ++HFN 5-76 p115
       Songs: An die Laute, D 405; Erlkönig, D 328; Gruppe aus dem Tar-
       tarus, D 583; Kreuzzug; Der Musensohn, D 764. cf Rococo 5370.
       Songs: An die Musik, D 547; Du bist die Ruhe, D 776; Litanie, D
       343; Mädchens Klage, D 6. cf Club 99-99.
       Songs: An mein Herz; Blondel zu Marien, D 626; Ganymed, D 544;
       Heidenroslein; Der Musensohn, D 764; Nur wer die Sehnsucht
       kennt, D 877/4; Schäfers Klagelied, D 121; Sprache der Liebe,
       D 410; Rastlose Liebe, D 138. cf SCHOENBERG: Das Buch der
       hangenden Gärten, op. 15.
       Songs: An Sylvia, D 891; Die Forelle, D 550; Heidenröslein, D 257.
       cf Works, selections (Decca DPA 545/6).
       Songs: Auf dem Strom, op. 119. cf Crystal S 371.
       Songs: Ave Maria, D 839. cf London STS 15239.
       Songs: Ave Maria, D 839 (2). cf RCA ARM 4-0942/7.
       Songs: Ave Maria, D 839; An Sylvia, D 891; Heidenröslein, D 257.
       cf Works, selections (Decca SPA 426).

Songs: Ave Maria, D 938; Die Forelle, D 550. cf Philips 6747 204.

Songs: Christ ist erstanden, S 440; Gebet, D 815; Gott im Unge-
witter, D 985; Psalm 23, Gott ist mein Hirt, D 706. cf BRAHMS:
Songs (HMV KASD 3091).

Songs: Du bist die Ruh, D 776; Die Forelle, D 550; Gretchen am
Spinnrade, D 118; Heidenröslein, D 257. cf L'Oiseau-Lyre SOL
345.

Songs: Die Forelle, D 550. cf Club 99-101.

Songs: Die Forelle, D 550. cf HMV HLM 7093.

Songs: Gott, meine Zuversicht, D 706; Die Nachtigall, D 724; La
Pastorella, D 513; Ständchen, D 921; Widerspruch, D 865. cf
RCA PRL 1-9034.

Songs: Heidenröslein, D 257. cf Decca SKL 5208.

Songs: Heidenröslein, D 257. cf HMV ASD 2929.

Songs: Der Hirt auf dem Felsen, D 965. cf Westminster WGS 8268.

Songs: Night music. cf Hungaroton SLPX 11696.

Songs: Ständchen, D 889. cf Polydor 2383 389.

2312 Symphonies, nos. 1-6, 8-9. Rosamunde, op. 26, D 979: Overture;
Ballet music, nos. 1 and 2. BPhO; Karl Böhm. DG 2740 127 (5).
+-Gr 1-76 p1204   +-RR 10-75 p49
+-HFN 10-75 p152

2313 Symphonies, nos. 1-6, 8-9. Overtures: Alfonso und Estrella, D
732. In the Italian style, D 590/1, C major and D major. Die
Zwillingsbrüder. Bath Festival Orchestra; Yehudi Menuhin. HMV
SLS 5007 (5).
+-Gr 1-76 p1204   +-RR 7-75 p33

2314 Symphony, no. 3, D 200, D major. Symphony, no. 5, D 485, B flat
major. RPO; Thomas Beecham. HMV SXLP 30204. Tape (c) TC EXE
184. (Reissue from ASD 345).
++Gr 2-76 p346   +RR 2-76 p41
+HFN 2-76 p115   +RR 4-76 p80 tape
+HFN 5-76 p117 tape

2315 Symphony, no. 4, D 417, C minor. Symphony, no. 5, D 485, B flat
major. Orchestra; Karl Böhm. DG Tape (c) 3300 484.
+HFN 12-75 p173 tape   +RR 1-76 p67 tape

2316 Symphony, no. 5, D 485, B flat major. Symphony, no. 8, D 759,
B minor. Columbia Symphony Orchestra, NYP; Bruno Walter. CBS
61033. (Reissue from Philips SABL 209).
+-Gr 10-76 p607   +-RR 11-76 p78
-HFN 11-76 p170

2317 Symphony, no. 5, D 485, B flat major. Symphony, no. 8, D 759,
B minor. LPO; John Pritchard. Classics for Pleasure CFP
40245.
+Gr 12-76 p1001   +RR 12-76 p64
+-HFN 12-76 p147

2318 Symphony, no. 5, D 485, B flat major. MENDELSSOHN: Die schöne
Melusine overture, op. 32. Winterthur Symphony Orchestra;
Fritz Busch. Discophilia GBE 141.
++NR 10-76 p4

2319 Symphony, no. 5, D 485, B flat major. Symphony, no. 8, D 759, B
minor. COA; Bernard Haitink. Philips 9500 099.
-Gr 11-76 p804   +-RR 11-76 p78
/HFN 11-76 p167

Symphony, no. 5, D 485, B flat major. cf Symphony, no. 3, D 200,
D major.

Symphony, no. 5, D 485, B flat major. cf Symphony, no. 4, D 417,
C minor.

Symphony, no. 8, D 759, B minor. cf Symphony, no. 5, D 485, B
flat major (CBS 61033).
Symphony, no. 8, D 759, B minor. cf Symphony, no. 5, D 485, B
flat major (Classics for Pleasure CFP 40245).
Symphony, no. 8, D 759, B minor. cf Symphony, no. 5, D 485, B
flat major (Philips 9500 099).
Symphony, no. 8, D 759, B minor. cf Works, selections (Decca DPA
545/6).
Symphony, no. 8, D 759, B minor. cf BEETHOVEN: Symphony, no. 5,
op. 67, C minor (DG 3335 103).
Symphony, no. 8, D 759, B minor. cf BEETHOVEN: Symphony, no. 5,
op. 67, C minor (RCA 25002).
Symphony, no. 8, D 759, B minor. cf HAYDN: Symphony, no. 104, D
major.
Symphony, no. 8, D 759, B minor. cf MOZART: Symphony, no. 41,
K 551, C major.
Symphony, no. 8, D 759, B minor: 1st movement. cf Works, selec-
tions (Decca SPA 426).
2320 Symphony, no. 9, D 944, C major. LPO; John Pritchard. Classics
for Pleasure CFP 40233.
          ++Gr 8-76 p297   +-RR 5-76 p50
          -HFN 5-76 p107
2321 Symphony, no. 9, D 944, C major. BSO; William Steinberg. RCA GL
25008.
          +-Gr 11-76 p804   -RR 11-76 p78
          +HFN 12-76 p147
Symphony, no. 9, D 944, C major. cf RCA CRM 5-1900.
Trio, piano, E flat major: Andante con moto. cf CBS 61684.
Trios, piano, nos. 1 and 2. cf Works, selections (DG 2734 004).
Variation on a waltz by Anton Diabelli, D 718, C minor. cf Piano
works (Saga 5407).
Die Verschworenen, D 787: Ja, wir schwören. cf CBS 76476.
Waltz, G flat major (Kupelweiser). cf Piano works (Saga 5407).
Waltz, D 146, no. 11. cf Works, selections (Telefunken DX 6-35266).
Waltz (Valses nobles), D 969, nos. 8, 9. cf Works, selections
Telefunken DX 6-35266).
Winterreise, op. 89, D 911. cf Schwanengesang, D 957 (HMV 840).
Winterreise, op. 89, D 911. cf Schwanengesang, D 957 (Philips
6780 028).
Winterreise, op. 89, D 911: Die Wetterfahne; Der Lindenbaum;
Frühlingstraum; Einsamkeit. cf Songs (Philips 6580 111).
Winterreise, op. 89, D 911: Le Tilleul. cf Club 99-101.
2322 Works, selections: Impromptu, op. 90, no. 4, D 899, A flat major.
Moment musical, op. 94, no. 3, D 780, F minor. Octet, op. 166,
D 803, F major: Scherzo. Quintet, piano, op. 114, D 667, A
major: Theme and variations. Rosamunde, op. 26, D 797: Ballet
music, G major. Die schöne Müllerin, op. 25, D 795: Wohin.
Schwanengesang, D 957: Ständchen. Sonata, piano, no. 18, op.
78, D 894, G major: Menuetto and trio. Songs: Ave Maria, D
839; An Sylvia, D 891; Heidenröslein, D 257. Symphony, no. 8,
D 759, B minor: 1st movement. Joan Sutherland, s; Stuart Bur-
rows, Peter Pears, t; Tom Krause, Hermann Prey, bar; John
Constable, Benjamin Britten, Clifford Curzon, Karl Engel,
Vladimir Ashkenazy, pno; LPO, Vienna Octet, OSR, NPhO; Ambrosian
Singers; Leopold Stokowski, Richard Bonynge. Decca SPA 426.
          +Gr 4-76 p1665   +-RR 4-76 p53
          +HFN 4-76 p123

2323 Works, selections: Impromptu, op. 142, no. 3, D 935, B flat major.
      Moment musical, op. 94, no. 3, D 780, F minor. Quintet, piano,
      op. 114, D 667, A major. Rosamunde, op. 26, D 797: Overture
      (Die Zauberharfe, D 644): Entr'acte, B flat; Ballet, B minor and
      G major. Schwanengesang, D 957: Ständchen. Songs: An Sylvia,
      D 891; Die Forelle, D 550; Heidenröslein, D 257. Symphony, no.
      8, D 759, B minor. Margaret Price, s; Tom Krause, Hermann
      Prey, bar; Wilhelm Backhaus, Irwin Gage, James Lockhart, Karl
      Engel, Clifford Curzon, pno; Vienna Octet, Members, VPO; Carl
      Schuricht, Pierre Monteux. Decca DPA 545/6. Tape (c) KDPC
      545/6.
          +Gr 9-76 p483   +HFN 11-76 p175 tape
          +HFN 10-76 p183   +-RR 8-76 p52
2324 Works, selections: Sonata, arpeggione and piano, D 821, A minor.
      Fantasia, op. 159, D 934, C major. Introduction and variations
      on "Ihr Blumlein alle", D 802, E minor. Nocturne, piano,
      violin and violoncello, op. 148, D 897, E flat major. Rondo
      brillant, op. 70, D 895, B minor. Sonata, piano trio, D 28,
      B flat major. Sonata, violin and piano, op. 162, D 574, A
      major. Sonatinas, violin and piano, nos. 1-3, D 384, 385, 408.
      Trios, piano, nos. 1 and 2. Wolfgang Schneiderhan, Rudolf
      Koeckert, vln; Walter Klien, Jean Fonda, Karl Engel, Christoph
      Eschenbach, pno; Pierre Fournier, Josef Merz, vlc; Aurèle
      Nicolet, flt; Trieste Trio. DG 2734 004 (4). (Reissues from
      SLPM 139101, 139164, 139368, 138053, 139106, 139434, SLPEM
      136488, 2538 067).
          +-Gr 4-76 p1624   +-RR 4-76 p66
          +HFN 4-76 p123
2325 Works, selections: Dances, D 365, nos. 1-2, 6, 22, 26, 29, 34-36.
      Ecossaises, D 145, nos. 1-3. Ecossaises, D 299, nos. 11-12.
      Ecossaises, D 421, nos. 1, 2. Ecossaises, D 618, b, c (vari-
      ants of Ecossaises, D 145, nos. 4, 8). Ecossaises, D 735, nos.
      7, 8. Ecossaises, D 783. Ländler (Komische), D 354, D major.
      Ländler, D 734, nos. 3, 5, 9-11, 14-15. German dance, D 643,
      C sharp minor. Valses nobles, D 969, nos. 8, 9. Waltz, D
      146, no. 11. Die schöne Müllerin, op. 25, D 795.* Nigel Rogers,
      t; Richard Burnett, pno. Telefunken DX 6-35266 (2). (*also
      CX 4-41892).
          +-Gr 1-76 p1226
2326 Die Zwillingsbrüder. Helen Donath, s; Nicolai Gedda, t; Dietrich
      Fischer-Dieskau, bar; Kurt Moll, Hans-Joachim Gallus, bs;
      Bavarian Radio Orchestra and Chorus; Wolfgang Sawallisch. EMI
      Electrola 065 28833. (also HMV ASD 3300).
          -Gr 12-76 p1051   ++RR 12-76 p45
          +HFN 12-76 p133   ++St 9-76 p124
      Die Zwillingsbrüder. cf Symphonies, nos. 1-6, 8-9.
SCHULHOFF, Erwin
      Concerto, flute, piano and orchestra. cf MARTINU: Concerto, flute,
      violin and orchestra.
      Esquisses de jazze. cf Supraphon 111 1721/2.
      Rag music. cf Supraphon 111 1721/2.
SCHULLER, Gunther
      Duets, solo horns, nos. 1 and 3. cf Crystal S 371.
SCHULZ
      Ihr Kinderlein kommet. cf RCA PRL 1-8020.
SCHUMAN, William
      Chester overture. cf DELLA JOIO: Satiric dances.

Quartet, no. 3. cf Vox SVBX 5305.
2327 Symphony, no. 8. SUDERBURG: Concerto "Within the memory of time".
      Bela Siki, pno; Seattle Symphony Orchestra, NYP; Milton Katims,
      Leonard Bernstein. Odyssey Y 34140. (Reissue from Columbia
      MS 6512).
            ++ARG 11-76 p39   +NR 12-76 p7
            +-HF 12-76 p126
      Variations on "Ameria" (after Charles Ives). cf London CSA 2246.
SCHUMANN, Clara
      Das ist ein Tag. cf Gemini Hall RAP 1010.
      Impromptu. cf Gemini Hall RAP 1010.
      Liebst du um Schönheit. cf Gemini Hall RAP 1010.
      Mazurka, op. 6, G major. cf BENNETT, W. S.: Sonata, piano, no. 1,
      op. 13, F minor.
      Romances, op. 21, nos. 2-3. cf BENNETT, W. S.: Sonata, piano, no.
      1, op. 13, F minor.
      Variations on a theme by Robert Schumann. cf BENNETT, W. S.:
      Sonata, piano, no. 1, op. 13, F minor.
SCHUMANN, Robert
      Adagio and allegro, op. 70, A flat major. cf BASF BAC 3085.
2328 Album fur die Jügend, op. 68. Kinderscenen, op. 15. Alexis
      Weissenberg, pno. Connoisseur Society CS 2-2110 (2).
            ++SFC 10-31-76 p35
      Album fur die Jügend, op. 68. cf Piano works (Vox SVBX 5470).
2329 Album fur die Jügend, op. 68, nos. 1-43. Alexis Weissenberg, pno.
      HMV SQ ASD 3202.
            +Gr 7-76 p194   +RR 7-76 p72
            +-HFN 7-76 p95
2330 Album fur die Jügend, op. 68: Sailor's songs; Wintertime. Carna-
      val, op. 9. Arturo Benedetti Michelangeli, pno. HMV ASD 3129.
      (also Angel S 37137).
            ++Gr 10-75 p658   +-NR 11-75 p5
            +-HF 12-75 p106   +RR 10-75 p80
            +-HFN 11-75 p165   -SFC 9-21-75 p34
            +-MT 2-76 p143   +-SR 11-29-75 p50
      Album for the young, op. 68: Knight Rupert. cf Connoisseur Soci-
      ety (Q) CSQ 2065.
      Album for the young, op. 68: New year's eve. cf Coronet LPS 3032.
      Albumblätter, op. 124. cf Piano works (Telefunken FK 6-35287).
      Andante and variations, op. 46, B flat major. cf DUSSEK: Concerto,
      2 pianos, no. 10, op. 63, B flat major.
      Ballscenen, op. 109. cf Piano works (4 hands) (arion ARN 236005).
      Bilder aus Osten, op. 66. cf Piano works (4 hands) (Arion ARN
      236005).
      Bunte Blätter, op. 99. cf Piano works (Vox SVBX 5470).
      Bunte Blätter, op. 99, nos. 1-14. cf Piano works (Telefunken FK
      6-35287).
      Canon, B minor. cf Westminster WGS 8116.
      Canons, D major, B minor. cf Wealden WSQ 134.
      Canon, op. 56, no. 5, B minor. cf Vista VPS 1035.
      Carnaval, op. 9. cf Album für die Jugend, op. 68: Sailor's songs;
      Wintertime.
2331 Concerto, piano, op. 54, A minor. Introduction and allegro, op.
      134, D minor. Introduction and allegro, op. 92, G major.
      Peter Frankl, pno; Bamberg Symphony Orchestra; Janos Fürst.
      Turnabout TV 34559.
            +-Gr 1-76 p1203   +-HFN 1-76 p119

Concerto, piano, op. 54, A minor. cf BEETHOVEN: Andante favori,
F major.
Concerto, piano, op. 54, A minor. cf GRIEG: Concerto, piano, op.
16, A minor (Classics for Pleasure CFP 40255).
Concerto, piano, op. 54, A minor. cf GRIEG: Concerto, piano, op.
16, A minor (Decca SXL 6624).
Concerto, piano, op. 54, A minor. cf GRIEG: Concerto, piano, op.
16, A minor (HMV 3133).
Concerto, piano, op. 54, A minor. cf HMV SLS 5033.
2332 Concerto, violoncello, op. 129, A minor. TCHAIKOVSKY: Pezzo
capriccioso, op. 62. Variations on a rococo theme, op. 33.
Maurice Gendron, vlc; VSO; Christoph von Dohnányi. Philips
6580 131. (Reissue from 835130AY).
        +-Gr 8-76 p298   +-RR 7-76 p62
Concerto, violoncello, op. 129, A minor. cf CHOPIN: Andante
spianato and grande polonaise, op. 22, E flat major.
Concerto, violoncello, op. 129, A minor. cf LALO: Concerto,
violoncello, D minor.
2333 Davidsbündlertänze, op. 6. Kriesleriana, op. 16. Géza Anda, pno.
DG 2535 145. (Reissue from 139199).
        +-Gr 3-76 p1487   +-RR 3-76 p64
        +HFN 3-76 p111
2334 Davidsbündlertänze, op. 6. Walter Gieseking, pno. Everest SDBR
3389.
        +-NR 10-76 p15
2335 Dichterliebe, op. 48. Songs, op. 35. Gérard Souzay, bar; Piano
accompaniments. Rococo 5372.
        +NR 9-76 p9
Dichterliebe, op. 48. cf CBS SQ 79200.
Dichterliebe, op. 48: Ich grolle nicht. cf Discophilia 13 KGW 1.
Etudes on caprices by Paganini, op. 10. cf Piano works (Vox SVBX
5470).
Etudes on caprices by Paganini, op. 10 (6). cf Piano works
(Telefunken FK 6-35287).
Fantasiestücke. cf Piano works (Vox SVBX 5470).
2336 Fantasiestücke, op. 12, nos. 2-4. Kinderscenen, op. 15. Romances,
op. 28, nos. 2 and 3. Waldscenen, op. 82, nos. 1, 3-7. John
McCabe, pno. Oryx ORPS 59.
        +-Gr 9-76 p446
Fantasiestücke, op. 12, no. 2: Aufschwung. cf HMV HQS 1354.
Fantasiestücke, op. 12, no. 3: Warum. cf Connoisseur Society
(Q) CSQ 2066.
Fantasy, violin and orchestra, op. 131, C major. cf BRAHMS:
Concerto, violin and violoncello, op. 102, A minor.
Faschingsschwank aus Wien, op. 26. cf RAVEL: Valses nobles et
sentimentales.
2337 Frauenliebe und Leben, op. 42. Liederkries, op. 39. Janet Baker,
ms; Daniel Barenboim, pno. Angel S 37222. (also HMV ASD 3217).
        +Gr 7-76 p205   ++NR 11-76 p11
        +-HF 11-76 p124   ++RR 7-76 p78
        ++HFN 8-76 p86   +-STL 9-19-76 p36
Frauenliebe und Leben, op. 42. cf SCHUBERT: Songs (Philips
6500 706).
Fugues (4). cf Piano works (Vox SVBX 5470).
Fugues, op. 72 (4). cf Piano works (Telefunken FK 6-35287).
Genoveva overture, op. 81. cf Symphony, no. 1, op. 38, B flat
major.

Humoreske, op. 20, B flat major. cf Piano works (Telefunken FK
    6-35287).
Humoreske, op. 20, B flat major. cf Piano works (Vox SVBX 5470).
2338 Impromptus on a theme by Clara Wieck, op. 5. Sonata, piano, no.
    no. 3, op. 14, F minor. Jean-Philippe Collard, pno. Connois-
    seur Society CS 2081.
        +Audio 12-76 p86   +-MJ 4-76 p30
        +HF 12-75 p106   +NR 11-75 p14
Introduction and allegro, op. 92, G major. cf Concerto, piano,
    op. 54, A minor.
Introduction and allegro, op. 134, D minor. cf Concerto, piano,
    op. 54, A minor.
Julius Caesar, op. 128: Overture. cf Symphony, no. 2, op. 61,
    C major.
Kinderball, op. 130. cf Piano works (4 hands) (Arion ARN 236005).
Kinderscenen, op. 15. cf Album für die Jugend, op. 68.
Kinderscenen, op. 15. cf Fantasiestücke, op. 12, nos. 2-4.
Kinderscenen, op. 15. cf NILSSON: Quantitaten.
Kinderscenen, op. 15. cf SCHUBERT: Sonata, piano, no. 21, op.
    posth., D 960, B flat major.
Kinderscenen, op. 15: Hasche-Mann; Bittendes Kind; Gluckes genüg;
    Wichtige Begebenheit. cf Vista VPS 1022.
Kinderscenen, op. 15, no. 7: Träumerei. cf CBS 30064.
Kinderscenen, op. 15, no. 7: Träumerei. cf HMV RLS 723.
Kinderscenen, op. 15, no. 7: Träumerei. cf RCA LRL 1-5131.
Klavierstücke, op. 32 (4). cf Piano works (Telefunken FK 6-35287).
Kreisleriana, op. 16. cf Davidsbündlertänze, op. 6.
Kreisleriana, op. 16. cf Piano works (Vox SVBX 5470).
Kreisleriana, op. 16. cf RAVEL: Sonatine.
Kreisleriana, op. 16. cf International Piano Archives IPA 5007/8.
Liederkreis, op. 24. cf Songs (DG 2530 543).
Liederkreis, op. 39. cf Frauenliebe und Leben, op. 42.
Manfred overture, op. 115. cf Symphonies, nos. 1-4.
Manfred overture, op. 115. cf Symphony, no. 3, op. 97, E flat
    major.
Manfred overture, op. 115. cf BORODIN: Prince Igor: Polovtsian
    dances.
Märchenerzählungen, op. 132. cf BRUCH: Trios, op. 83, nos. 2, 6,
    7.
Märchenerzählungen, op. 132. cf GLINKA: Trio pathétique, D minor.
Myrthen, op. 25. cf Songs (DG 2530 543).
Myrthen, op. 25, no. 1: Widmung. cf RCA ARM 4-0942/7.
Nachtstücke, op. 23. cf Piano works (Telefunken FK 6-35287).
Nachtstücke, op. 23. cf Piano works (Vox SVBX 5470).
Overture, scherzo and finale, op. 52, E major. cf Symphonies,
    nos. 1-4 (DG 2740 129).
Overture, scherzo and finale, op. 52, E major. cf Symphonies,
    nos. 1-4 (HMV SLS 867).
Overture, scherzo and finale, op. 52, E major. cf Symphonies,
    nos. 1-4 (RCA 3-5114).
Overture, scherzo and finale, op. 52, E major. cf Symphony, no.
    2, op. 61, C major.
Overture to Goethe's "Hermann und Dorothea", op. 136, B minor. cf
    Symphonies, nos. 1-4.
Papillons, op. 2. cf Piano works (Vox SVBX 5470).
2339 Piano works (4 hands): Ballscenen, op. 109. Bilder aus Osten, op.
    66. Kinderball, op. 130. Vierhändige Klavierstücke, op. 85 (12).

Jacqueline and Otto Delfino, pno. Arion ARN 236005 (2).
        -Gr 1-76 p1218   +-RR 1-76 p51
2340 Piano works: Albumblätter, op. 124. Bunte Blätter, op. 99, nos.
        1-14. Etudes on caprices by Paganini, op. 10 (6). Fugues,
        op. 72 (4). Humoreske, op. 20, B flat major. Klavierstücke,
        op. 32 (4). Nachtstücke, op. 23. Toccata, op. 7, C major.
        Variationen über eigenes Thema. Waldscenen, op. 82. Karl
        Engel, pno. Telefunken FK 6-35287 (4).
          +-Gr 5-76 p1787
2341 Piano works: Album für die Jugend, op. 68. Bunte Blätter, op. 99.
        Fantasiestücke. Fugues (4). Humoreske, op. 20, B flat major.
        Kreisleriana, op. 16. Nachtstücke, op. 23. Papillons, op. 2.
        Etudes on caprices by Paganini, op. 10. Peter Frankl, pno.
        Vox SVBX 5470 (3).
          +MJ 12-76 p29   +SFC 10-3-76 p33
2342 Quartet, piano, op. 47, E flat major. Quintet, piano, op. 44,
        E flat major. Thomas Rajna, pno; Alberni Quartet. CRD 1024.
          +Gr 8-76 p317   +-RR 7-76 p68
          +-HFN 7-76 p95   ++St 11-76 p158
2343 Quartet, piano, op. 47, E flat major. Quintet, piano, op. 44, E
        flat major. Beaux Arts Trio; Samuel Rhodes, vla; Dolf Bettel-
        heim, vln. Philiops 9500 065.
          +-Gr 5-76 p1781   +RR 5-76 p62
          /HFN 5-76 p109   ++STL 8-8-76 p29
      Quintet, piano, op. 44, E flat major. cf Quartet, piano, op. 47,
        E flat major (CRD 1024).
      Quintet, piano, op. 44, E flat major. cf Quartet, piano, op. 47,
        E flat major (Philips 9500 065).
      Romances, op. 28, nos. 2 and 3. cf Fantasiestücke, op. 12, nos.
        2-4.
      Sonata, piano, no. 2, op. 22, G minor. cf SCHUBERT: Fantasia,
        op. 15, D 760, C major.
2344 Sonata, piano, no. 3, op. 14, F minor. SCRIABIN: Sonata, piano,
        no. 5, op. 53, F sharp major. Vladimir Horowitz, pno. RCA
        ARL 1-1766. Tape (c) ARK 1-1766 (ct) ARS 1-1766.
          +Gr 12-76 p1022   ++SFC 10-10-76 p32
          ++RR 12-76 p87
      Sonata, piano, no. 3, op. 14, F minor. cf Impromptus on a theme
        by Clara Wieck, op. 5.
      Sonata, piano, no. 3, op. 14, F minor. cf RACHMANINOFF: Concerto,
        piano, no. 3, op. 30, D minor.
      Sonata, piano, no. 3, op. 14, F minor. cf International Piano
        Library IPL 5003/4.
2345 Sonata, violin, no. 1, op. 105, A minor. Sonata, violin, no. 2,
        op. 121, D minor. Jean-Pierre Wallez, vln; Bruno Rigutto, pno.
        French Decca QS 7292.
          +Gr 8-76 p318   +RR 6-76 p75
          ++HFN 7-76 p95
      Sonata, violin, no. 2, op. 121, D minor. cf Sonata, violin, no.
        1, op. 105, A minor.
2346 Songs: Liederkreis, op. 24. Myrthen, op. 25. Dietrich Fischer-
        Dieskau, bar; Christoph Eschenbach, pno. DG 2530 543.
          +Gr 8-75 p355   *ON 5-76 p48
          +HF 12-75 p106   +-RR 1-76 p61
          ++HFN 3-76 p103   ++SFC 11-16-76 p32
          +NR 12-75 p12   ++St 3-76 p121

Songs, op. 35. cf Dichterliebe, op. 48.
Songs: Botschaft, op. 74, no. 8; Das Glück, op. 79, no. 16. cf
    CBS 76476.
Songs: Frühlingsnacht, op. 39, no. 12; Der Schatzgraber, op. 45,
    no. 1; Der Soldat, op. 40, no. 3. cf Rococo 5370.
Songs: Lust der Sturmnacht, op. 35, no. 1; Mein schöner Stern;
    Stille Liebe, op. 35, no. 8; Stille Tränen, op. 35, no. 10;
    Widmung, op. 25, no. 1. cf Seraphim S 60251.
Songs: Der Nussbaum, op. 25, no. 3. cf HMV ASD 2929.
Stücke im Volkston, op. 102 (5). cf CBS Classics 61579.
Symphonic etudes, op. 13. cf ALBENIZ: España, op. 165, no. 2:
    Tango.
Symphonic etudes, op. 13. cf LISZT: Années de pelerinage, 2nd
    year, G 161: Sonetti del Petrarca (3).
2347 Symphonies, nos. 1-4. Overture, scherzo and finale, op. 52, E
    major. BPhO; Herbert von Karajan. DG 2740 129.
        +HFN 3-76 p109   +-RR 3-76 p48
2348 Symphonies, nos. 1-4. Manfred overture, op. 115. Overture,
    scherzo and finale, op. 52, E major. Dresden Staatskapelle
    Orchestra; Wolfgang Sawallisch. HMV SLS 867 (3). Tape (c)
    TC SLS 867.
        ++Gr 2-74 p1565   +-RR 2-74 p34
        +HFN 5-76 p117 tape   ++RR 3-76 p77 tape
2349 Symphonies, nos. 1-4. Overture to Goethe's "Hermann und Dorothea",
    op. 136, B minor. Overture, scherzo and finale, op. 52, E major.
    Leipzig Gewandhaus Orchestra; Kurt Masur. RCA LRL 3-5114 (3).
        +-Gr 9-76 p431   +-RR 9-76 p60
        +-HFN 9-76 p129
2350 Symphony, no. 1, op. 38, B flat major. Symphony, no. 3, op. 97,
    E flat major. CO; Georg Szell. CBS 61595. Tape (c) 40-61595.
    (Reissues from Columbia SAX 2475, 2506).
        +Gr 3-75 p1663   ++RR 3-75 p34
        +-HFN 2-76 p116 tape   +-RR 2-76 p71
        +-HFN 8-76 p95 tape
2351 Symphony, no. 1, op. 38, B flat major. Symphony, no. 4, op. 120,
    D minor. BPhO; Rafael Kubelik. DG 2535 116. Tape (c) 3335 116.
    (also DG 2530 169 Tape (ct) 89479 (c) 3300 419).
        +RR 6-75 p47   +RR 2-76 p71 tape
2352 Symphony, no. 1, op. 38, B flat major. Genoveva overture, op. 81.
    Pécs Philharmonic Orchestra; Tamás Breitner. Hungaroton SLPX
    11785.
        /NR 8-76 p4
Symphony, no. 1, op. 38, B flat major. cf BEETHOVEN: Symphony,
    no. 5, op. 67, C minor.
2353 Symphony, no. 2, op. 61, C major. Julius Caesar, op. 128: Over-
    ture. VPO; Georg Solti. Decca SXL 6487. Tape (c) KSXC 6487.
        ++Gr 12-70 p998   +STL 1-10-71 p36
        +-RR 3-76 p77 tape
2354 Symphony, no. 2, op. 61, C major. Overture, scherzo and finale,
    op. 52, E major. BPhO; Herbert von Karajan. DG 2530 170.
    Tape (c) 3300 482. (Reissue from 2720 046).
        +Gr 10-75 p626   +RR 8-75 p42
        +HFN 8-75 p89   +RR 10-75 p97 tape
        +NR 1-76 p5
2355 Symphony, no. 2, op. 61, C major. BPhO; Rafael Kubelik. DG Tape
    (c) 3335 117.
        +RR 2-76 p71 tape

SCHUMANN (cont.) 446

2356 Symphony, no. 3, op. 97, E flat major. Manfred overture, op. 115.
      BPhO; Rafael Kubelik. DG 2535 118. Tape (c) 3335 118. (also
      DG 138 908).
          +RR 6-75 p48   +RR 2-76 p71 tape
      Symphony, no. 3, op. 97, E flat major. cf Symphony, no. 1, op. 38,
      B flat major.
      Symphony, no. 4, op. 120, D minor. cf Symphony, no. 1, op. 38,
      B flat major.
      Toccata, op. 7, C major. cf Piano works (Telefunken FK 6-35287).
      Trio, piano, no. 1, op. 63, D minor. cf HMV RLS 723.
      Variationen über eigenes Thema. cf Piano works (Telefunken FK
      6-35287).
      Vierhändige Klavierstücke, op. 85 (12). cf Piano works (4 hands)
      (Arion ARN 236005).
      Waldscenen, op. 82. cf Piano works (Telefunken FK 6-35287).
      Waldscenen, op. 82, nos. 1, 3-7. cf Fantasiestücke, op. 12, nos.
      2-4.
      Waldscenen, op. 82, no. 7: The prophet bird. cf Decca KCSP 367.
      Waldscenen, op. 82, no. 7: The prophet bird. cf Saga 5427.
      Ziguenerleben, op. 29, no. 3. cf RCA PRL 1-9034.
SCHUTZ, Heinrich
      Domini est terra. cf Harmonia Mundi HMU 473.
      Erbarm dich mein, o Herre Gott. cf Harmonia Mundi HMU 473.
      Herr, nun lässest du deinen Diener in Frieden fahren. cf Harmonia
      Mundi HMU 473.
      Heute ist Christus der Herr geboren. cf Harmonia Mundi HMU 473.
2357 Music at the Court of Dresden: Domini est terra, S 476 (from Psalm
      24). Erbarm dich mein, S 447. Heute ist Christus der Herr
      geboren, S 439. O bone Jesu, S 471. Song of Simeon, S 433.
      Vater Abraham, S 477. Ars Europea Choeur National; Jacques
      Grimbert. Harmonia Mundi HMU 958.
          +-HFN 5-76 p109   +RR 3-76 p71
      Psalm, no. 150. cf HMV SLS 5047.
2358 Songs (Italian madrigals): Alma afflitta; Così morir debb'io;
      d'Orrida selce alpina; Di marmo siete voi; Dunque addio, care
      selve; Feriteve, viperette mordaci; Fiamma ch'allaccia; Fuggio
      o mio core; Giunto è pur; Io moro; Mi saluta costei; O dolcezze
      amarissime; O primavera; Quella damma son io; Ride la primavera;
      Selve beaté; Sospir che del bel petto; Tornate a cari baci;
      Vasto mar. Monteverdi Choir; Jürgen Jürgens. DG Archive 2708
      033 (2).
          +-Gr 11-76 p858   +HFN 10-76 p175
      Songs: Also hat Gott die Welt geliebt, S 380; Ich weiss, dass mein
      Erlöser lebt, S 393; Jauchzet dem Herren alle Welt, S 396. cf
      BIS LP 14.
      Songs: The heavens are telling. cf Abbey LPB 750.
SCHWANTNER, Joseph
      Autumn canticles. cf BLOCH: Nocturnes.
      Consortium I. cf Delos DEL 25406.
      In aeternum. cf Delos DEL 25406.
SCOTT
      Annie Laurie. cf HMV RLS 719.
      The gentle maiden. cf RCA ARM 4-0942/7.
      Goodnight. cf HMV RLS 719.
      Tallahassee suite. cf RCA ARM 4-0942/7.

SCOTT, Cyril
2359 Concerto, piano, no. 1, C major. John Ogdon, pno; LPO; Bernard
        Hermann. Lyrita SRCS 81.
                ++Gr 9-75 p471   +HFN 10-75 p149
                +HF 6-76 p94   ++RR 9-75 p46
SCOTTO (15th century Italy)
        O fallace speranza. cf Nonesuch H 71326.
SCRIABIN, Alexander
2360 Concerto, piano, op. 20, F sharp minor. Prometheus, Poem of fire,
        op. 60. Vladimir Ashkenazy, pno; LPO; Lorin Maazel. London
        CS 6732. Tape (c) M 10251 (r) L 80251. (also Decca SXL 6527
        Tape (c) KSXC 6527).
                +Gr 1-72 p1220   +NR 4-72 p3
                ++Ha 11-72 p126   ++RR 10-76 p107 tape
                ++HF 6-72 p92   ++SFC 2-20-72 p32
                 +HF 11-72 p132 tape   /SR 4-22-72 p52
                ++HFN 1-72 p113   +St 6-72 p92
                ++HFN 8-75 p110 tape   +STL 2-13-72 p29
     Etude, op. 8, no. 7, B flat minor. cf BARBER: Sonata, piano, op.
        26.
     Etude, op. 42, no. 5, C sharp minor. cf BARBER: Sonata, piano,
        op. 26.
     Poème de l'extase, op. 54. cf DVORAK: Slavonic dance, op. 72,
        no. 2, E minor.
     Poème de l'extase, op. 54. cf RCA CRL 3-2026.
     Prelude, op. 11, no. 5, D major. cf BARBER: Sonata, piano, op. 26.
     Prelude, op. 22, no. 1, G sharp minor. cf BARBER: Sonata, piano,
        op. 26.
     Prelude and nocturne for the left hand, op. 9, D flat major. cf
        HMV HQS 1354.
     Prometheus, Poem of fire, op. 60. cf Concerto, piano, op. 20,
        F sharp minor.
2361 Sonatas, piano, nos. 3-5, 9. Vladimir Ashkenazy, pno. Decca
        SXL 6705. Tape (c) KSXC 6705. (also London CS 6920).
                +Gr 11-75 p864   ++NR 11-76 p12
                +HFN 12-75 p164   ++RR 11-75 p84
                +HFN 8-76 p95 tape   ++St 11-76 p158
                +MJ 12-76 p28
     Sonata, piano, no. 5, op. 53, F sharp major. cf SCHUMANN: Sonata,
        piano, no. 3, op. 14, F minor.
     Sonata, piano, no. 9, op. 68, F major. cf BARBER: Sonata, piano,
        op. 26.
2362 Universe (Nemtin). Aleksei Lyubimov, pno; Irina Orlova, org;
        RSFSR Yurlov Chorus; MPO; Kiril Kondrashin. Angel/Melodiya
        SR 40260. (also HMV/Melodiya ASD 3201).
                +Gr 5-76 p1795   *NR 5-76 p8
                +-HF 6-76 p94   +-RR 7-76 p79
                +HFN 7-76 p95   *SFC 3-14-76 p27
                -MJ 5-76 p28   +-St 10-76 p132
                +MT 8-76 p662   +STL 9-76 p38
SEARLE, Humphrey
        Aubade, op. 28. cf BANKS: Concerto, horn.
2363 Symphony, no. 1, op. 23. Symphony, no. 2, op. 33. LPO; Adrian
        Boult, Josef Krips. Lyrita SRCS 72. (Reissue from Decca SXL
        2232).
                +Gr 6-75 p49   +RR 5-75 p39
                ++HFN 5-75 p139   +Te 3-76 p32

Symphony, no. 2, op. 33. cf Symphony, no. 1, op. 23.

SEEGER, Ruth Crawford

Two movements for chamber orchestra. cf MEKEEL: Corridors of
dream.

SEGER, Joseph

Fugue, F minor. cf Panton 110 418/9.
Prelude, A minor. cf Panton 110 418/9.

SEGERSTAM, Leif

Divertimento. cf RAUTAVAARA: Pelimannit, op. 1.

SEIBER, Matyas

Concertino, clarinet and strings. cf ADDISON: Concerto, trumpet,
strings and percussion.
Morgenstern Lieder. cf Hungaroton SLPX 11713.

SEIFERT

Kärtner Liedermarsch. cf DG 2721 077 (2).

SEITZ

March grandioso. cf Mercury SRI 75055.

SEJAN, Nicholas

Fugue, no. 3. cf Calliope CAL 1917.
Noël Suisse. cf Calliope CAL 1917.

SELBY

Fugue or voluntary, D major. cf Columbia MS 6161.
Songs: O be joyful to the Lord; Ode for the New Year. cf VOX
SVBX 5350.

SENAILLE, Jean

Entrée et cotillon. cf HMV HLM 7093.

SENFL, Ludwig

Non wöllt ihr hören neue Mär. cf Nonesuch H 71326.

SERMISY, Claude de

Allez souspirs. cf HMV SLS 5022 (2).
Amour me voyant. cf HMV SLS 5022 (2).
Au pres de vous (2 settings). cf Argo ZRG 667.
C'est a grant tort. cf Hungaroton SLPX 11669/70.
Content désir; La, je m'y plains. cf L'Oiseau-Lyre 12BB 203/6.
Tant que vivray. cf Harmonia Mundi 204.

SEROV, Alexander

The hostile force: Yeryomka's song. cf HMV ASD 3200.
Judith: Judith's aria, Act 1. cf Rubini GV 63.
Rogneda: Groaned the blue sea. cf Rubini GV 63.

SERRANO Y RUIZ, Emilio

El carro del sol: Canción veneciana. cf London OS 26435.

SESSIONS, Roger

2364 Concerto, violin. Paul Zukofsky, vln; ORTF; Gunther Schuller.
CRI SD 220.
+RR 12-76 p64
Duo, cf Desto DC 6435/7.
Quartet, no. 2. cf Vox SVBX 5305.
Rhapsody, orchestra. cf MUSGRAVE: Night music, chamber orchestra.
Symphony, no. 8. cf MUSGRAVE: Night music, chamber orchestra.

SEVERAC, Deodat de

Ma poupee cherie. cf Club 99-101.
Le soldat de plomb. cf CHABRIER: Trois valse romantiques.

SHAPEY, Ralph

Evocation. cf Desto DC 6435/7.

2365 Praise. Paul Geiger, bs-bar; University of Chicago Contemporary
Chamber Players; Ralph Shapey. CRI SD 355.
+NR 10-76 p7

2366   Quartet, strings, no. 6. Rituals. SHIFRIN: Pieces for orchestra
          (3). Lexington Quartet, London Sinfonietta; Ralph Shapey,
       Jacques Monod. CRI SD 275.
                +Gr 9-76 p431   *NYT 10-31-71 pD26
                +HF 10-72 p106   +-RR 9-76 p60
                ++NR 3-72 p6   +St 6-72 p90
       Rituals. cf Quartet, strings, no. 6.

SHARPE, Trevor
       Three blades of Toledo. cf Pye GH 603.

SHAW
       Trip to Pawtucket. cf Columbia MS 6161.
       Trip to Pawtucket. cf Columbia M 34129.

SHAW, Christopher
       A lesson from Ecclesiastes. cf DALLAPICCOLA: Tempus destruendi,
          Tempus aedificandi.
       Music when soft voices die. cf DALLAPICCOLA: Tempus destruendi,
          Tempus aedificandi.
       Peter and the lame man. cf DALLAPICCOLA: Tempus destruendi,
          Tempus aedificandi.
       To the Bandusian spring. cf DALLAPICCOLA: Tempus destruendi,
          Tempus aedificandi.

SHAW, Geoffrey
       Anglican folk mass: The Creed. cf HMV HQS 1350.

SHCHEDRIN, Rodion
       Humoresque. cf Desmar DSM 1005.
       In the style of Albéniz. cf Odyssey Y 33825.

SHELDON
       Saraband. cf Orion ORS 76231.

SHEPPARD, John
       Our Father. cf Abbey LPB 739.
       Playnsong mass for a meane for 5 voices. cf PARSLEY: Lamentations,
          5 voices.

SHIELD, Alice
       Farewell to a hill. cf Finnadar QD 9010.

SHIFRIN, Seymour
       Pieces for orchestra. cf SHAPEY: Quartet, strings, no. 6.
       Quartet, strings, no. 4. cf MONOD: Cantus contra cantum I.
       Serenade for five instruments. cf FOSS: Behold, I build an house.

SHOSTAKOVICH, Dmitri
       Age of gold, op. 22. cf Columbia M 34127.
       Concertino, 2 pianos, op. 94. cf Works, selections (HMV RLS 721).
       Concerto, piano, no. 2, op. 101, F major. cf Concerto, piano and
          trumpet, no. 1, op. 35, C minor.
       Concerto, piano, no. 2, op. 101, F major. cf Works, selections
          (HMV RLS 721).
2367   Concerto, piano and trumpet, no. 1, op. 35, C minor. Concerto,
       piano, no. 2, op. 102, F major. André Previn, Leonard Bernstein,
       pno; William Vacchiano, tpt; NYP; Leonard Bernstein. CBS
       73400. Tape (c) 40-73400. (Reissues from CBS SBRG 72349/50,
       Philips SABL 134).
                +-Gr 5-75 p1974   +-HFN 8-76 p95 tape
                +HFN 6-75 p93   +-RR 5-75 p39
                +-HFN 2-76 p116 tape   +RR 3-76 p78 tape
       Concerto, violin, no. 1, op. 99, A minor. cf HMV SLS 5058.
       Concerto, violin, no. 2, op. 129, C sharp minor. cf HMV SLS 5058.
       Concerto, violoncello, no. 2, op. 126, G major. cf GLAZUNOV: Song
          of the troubadour, op. 71.

Fantastic dances. cf Odyssey Y 33825.

From Jewish folk poetry, op. 79. cf Works, selections (HMV RLS 721).

The gadfly, op. 72a: Romance. cf HMV SXLP 30188.

2368   Katerina Ismailova, op. 29. Eleanora Andreyeva. Eduard Bulavin, Vyacheslav Radzievsky, Gennady Yefimov, Dina Potapovskaya, Vyacheslav Fedorkin, Vasily Shtefutza, Lev Yeliseyev, Vladimir Popov, Yevgeny Maksimenko, Mikhail Turemnov, Vladimir Generalov, Konstantin Mogilevsky, Matvei Matveyev, Georgy Dudarev, Yevgeny Korenev, Nina Isakova, Olga Borisova; Stanislavsky-Nemirovich-Danchenko Musical Drama Theatre Orchestra and Chorus; Gennady Provatorov. HMV Melodiya SLS 5050 (4).

      +Gr 8-76 p333   +RR 8-76 p24
      +-HFN 7-76 p95   +STL 7-4-76 p36
      +MT 11-76 p915

Novorossüsk Chimes. cf HMV ASD 3200.

2369   Preludes, op. 34 (24). Sonata, piano, no. 2, op. 64, B minor. Inger Wikström, pno. RCA GL 25003.

      +-Gr 11-76 p843   +RR 10-76 p91
      +HFN 12-76 p147

Preludes, piano, op. 34, nos. 10, 15-16, 24. cf Works, selections (HMV RLS 721).

2370   Preludes and fugues, op. 87 (24). Roger Woodward, pno. RCA LRL 2-5100 (2).

      +Gr 12-75 p1082   +MM 5-76 p32
      ++HF 5-76 p92   +NR 4-76 p12
      +-HFN 1-76 p119   +RR 1-76 p52
      ++MJ 4-76 p31   +-Te 9-76 p26

Preludes and fugues, op. 87, nos. 2-3, 5-7, 16, 20, 23. cf Works, selections (HMV RLS 721).

2371   Quartet, strings, no. 7, op. 108, F sharp minor. Quartet, strings, no. 13, op. 138, B flat minor. Quartet, strings, no. 14, op. 142, F sharp major. Fitzwilliam Quartet. L'Oiseau-Lyre DSLO 9.

      +Audio 6-76 p96   ++NR 2-76 p6
      +Gr 12-75 p1065   +RR 11-75 p65
      ++HF 5-76 p94   ++St 8-76 p68
      +HFN 12-75 p164   +STL 4-11-76 p36
      +MJ 12-76 p44   ++Te 9-76 p26
      ++MT 12-76 p1005

2372   Quartet, strings, no. 8, op. 110, C minor. Quartet, strings, no. 15, op. 144, E flat minor. Fitzwilliam Quartet. L'Oiseau-Lyre DSLO 11.

      ++ARG 12-76 p40   ++RR 4-76 p61
      ++Gr 4-76 p1624   +SR 11-13-76 p52
      ++HFN 4-76 p113   +STL 4-11-76 p36
      ++MT 9-76 p748   +Te 9-76 p26

Quartet, strings, no. 10, op. 118. cf BERG: Quartet, strings, op. 3.

Quartet, strings, no. 13, op. 138, B flat minor. cf Quartet, strings, no. 7, op. 108, F sharp minor.

2373   Quartet, strings, no. 14, op. 142, F sharp major. Quartet, strings, no. 15, op. 144, E flat minor. Beethoven Quartet. HMV HQS 1362.

      +Gr 11-76 p830   +RR 9-76 p77
      +HFN 9-76 p129

Quartet, strings, no. 14, op. 142, F sharp major. cf Quartet, strings, no. 7, op. 108, F sharp minor.
Quartet, strings, no. 15, op. 144, E flat minor. cf Quartet, strings, no. 8, op. 110, C minor.
Quartet, strings, no. 15, op. 144, E flat minor. cf Quartet, strings, no. 14, op. 142, F sharp major.
Quintet, piano, op. 57, G minor. cf Works, selections (HMV RLS 721).
Satires, op. 109: To a critic, Taste of spring, Progeny, Misunderstanding, Kreutzer sonata. cf MUSSORGSKY: Songs (HMV SLS 5055).
Sonata, piano, no. 2, op. 64, B minor. cf Preludes, op. 34.
Sonata, violoncello, op. 40, D minor. cf Works, selections (HMV RLS 721).
Sonata, violoncello, op. 40, D minor. cf CHOPIN: Sonata, violoncello, op. 65, G minor.
Songs: Poems, op. 126: Song of Ophelia, Gamayun bird of prophecy, We were together, The city sleeps, Strom, Secret signs, Music. cf MUSSORGSKY: Songs (HMV SLS 5055).
Symphony, no. 2, op. 14, C major. cf PROKOFIEV: Scythian suite, op. 20.

2374 Symphony, no. 4, op. 43, C minor. PO; Eugene Ormandy. CBS 61696. (Reissue from SBRG 72129).
+Gr 8-76 p298   +RR 7-76 p63
+HFN 8-76 p93

2375 Symphony, no. 5, op. 47, D minor. Washington National Orchestra; Howard Mitchell. Camden CCV 5045.
-HFN 4-76 p121   /-RR 4-76 p53

2376 Symphony, no. 5, op. 47, D minor. Symphony, no. 10, op. 93, E minor. Bournemouth Symphony Orchestra; Paavo Berglund. HMV SQ SLS 5044 (2). Tape (c) TC SLS 5044.
+-Gr 3-76 p1471   +HFN 6-76 p105 tape
+Gr 6-76 p102 tape   +-RR 3-76 p49
+-HFN 3-76 p105   -RR 8-76 p85 tape

2377 Symphony, no. 5, op. 47. PO; Eugene Ormandy. RCA ARL 1-1149. Tape (c) ARK 1-1149 (ct) ARS 1-1149 (Q) ARD 1-1149 Tape (ct) ART 1-1149.
++Audio 6-76 p96   +-HFN 8-76 p87
+-Gr 8-76 p298   +MJ 11-75 p20
+HF 1-76 p95   ++NR 11-75 p2
+-HF 1-76 p95 Quad   +RR 8-76 p52
++HF 4-76 p148 tape   ++SFC 9-14-75 p28

2378 Symphony, no. 7, op. 60, C major. Symphony, no. 9, op. 70, E flat major. CPhO; Václav Neumann. Supraphon 110 1771/2 (2).
-Gr 8-76 p298   -RR 6-76 p53
+NR 11-76 p3

2379 Symphony, no. 8, op. 65, C minor. LSO; André Previn. HMV ASD 2917. Tape (c) TC ASD 2917. (also Angel 36980).
++Gr 10-73 p689   +NR 11-74 p2
+-Gr 9-76 p497 tape   ++RR 11-73 p50
++HF 10-74 p84   ++St 1-75 p75
+HFN 8-76 p95 tape
Symphony, no. 9, op. 70, E flat major. cf Symphony, no. 7, op. 60, C major.

2380 Symphony, no. 10, op. 93, E minor. LPO; Andrew Davis. Classics for Pleasure CFP 40216. (also Seraphim S-60255).

```
        +Gr 5-75 p1973              +-NR 1-76 p3
        +-HF 5-76 p95               +RR 6-75 p53
        +-HFN 6-75 p93              +St 4-76 p116
```

Symphony, no. 10, op. 93, E minor. cf Symphony, no. 5, op. 47, D minor.

2381 Symphony, no. 12, op. 112. Bulgarian Radio and Television Orchestra; Russlan Raychev. Monitor MCS 2148.

```
        +-NR 5-76 p2
```

2382 Symphony, no. 12, op. 112. Leipzig Gewandhaus Orchestra; Ogan Durjan. Philips 6580 012. Tape (c) 7317 120.

```
        ++Gr 8-72 p341             +RR 8-72 p64
        ++HF 1-72 p102             +RR 7-76 p83 tape
        +HFN 8-72 p1468            -SFC 2-20-72 p33
        -NR 2-72 p2
```

2383 Symphony, no. 14, op. 135. Galina Vishnevskaya, s; Mark Reshetin, bs; MPO; Mstislav Rostropovich. HMV Melodiya ASD 3090.

```
        +Gr 12-75 p1046           ++RR 1-76 p38
        ++HFN 1-76 p119
```

Trio, piano, no. 2, op. 67, E minor. cf FAURE: Trio, piano, op. 120, D minor.

Trio, piano, no. 2, op. 67, E minor. cf IVES: Trio, violin, violoncello and piano.

2384 Works, selections: Concerto, piano, no. 2, op. 101, F major. Concertino, 2 pianos, op. 94. From Jewish folk poetry, op. 79. Quintet, piano, op. 57, G minor.* Sonata, violoncello, op. 40, D minor.** Preludes and fugues, op. 87, nos. 2-3, 5-7, 16, 20, 23.*** Preludes, piano, op. 34, nos. 10, 15-16, 24. Dmitri Shostakovich, Maxim Shostakovich, pno; Leonid Kogan, vln; Mstislav Rostropovich, vlc; Nina Dorliak, s; Zara Dolukhanova, ms; Alexei Maslennikov, t; MRSO, Beethoven Quartet; Alexander Gauk. HMV RLS (3). (*Reissue from Parlophone PMA 1040 **PMA 1043 ***nos. 6, 7, 20 from PMC 1056).

```
        +Gr 10-76 p607            +-RR 10-76 p66
        +-HFN 11-76 p167
```

SIBELIUS, Jean

Andante festivo, strings. cf Works, selections (HMV SQ ASD 3287).

Belshazzar's feast, op. 51. cf Symphony, no. 1, op. 39, E minor.

Canzonetta, op. 62a. cf Works, selections (HMV SQ ASD 3287).

Canzonetta, op. 62, no. 1. cf RAUTAVAARA: Pelimannit, op. 1.

2385 Concerto, violin, op. 47, D minor. Finlandia, op. 26. Karelia suite, op. 11. Kuolema, op. 44: Valse triste. Legends, op. 22: The swan of Tuonela. Symphony, no. 2, op. 43, D major. Ruggiero Ricci, vln; LSO; Alexander Gibson, Pierre Monteux. Decca DPA 531/2. Tape (c) KDPC 531/2.

```
        +-Gr 6-76 p95             +-RR 6-76 p53
        +-HFN 6-76 p103
```

2386 Concerto, violin, op. 47, D minor. Tapiola, op. 112. Tossy Spivakovsky, vln; LSO; Tauno Hannikainen. Everest Tape (c) GRT 8059 3045.

```
        +HF 7-76 p114 tape
```

2387 Concerto, violin, op. 47, D minor. Humoreske, no. 5, op. 89, no. 3, E flat major. Serenade, no. 1, op. 69a, D major. Serenade, no. 2, op. 69b, G minor. Ida Haendel, vln; Bournemouth Symphony Orchestra; Paavo Berglund. HMV SQ ADS 3199. Tape (c) TC ASD 3199.

```
    +Gr 6-76 p52              +HFN 8-76 p95
    ++Gr 9-76 p497 tape       ++RR 5-76 p51
    +HFN 7-76 p97
```

Concerto, violin, op. 47, D minor. cf BEETHOVEN: Concerto, violin,
    op. 61, D major.
Concerto, violin, op. 47, D minor. cf BEETHOVEN: Romances, nos.
    1 and 2, opp. 40, 50.
Concerto, violin, op. 47, D minor. cf DVORAK: Concerto, violin,
    op. 53, A minor.
Dance intermezzo, op. 45, no. 2. cf Works, selections (HMV SQ
    ASD 3287).
The Dryad, op. 45, no. 1. cf Works, selections (HMV SQ ASD 3287).
2388 Finlandia, op. 26. Karelia suite, op. 11. Legends, op. 22: The
    swan of Tuonela. En saga, op. 9. VPO; Malcolm Sargent.
    Classics for Pleasure CFP 40247. Tape (c) TC CFP 40247.
        +HFN 6-76 p102   +-RR 5-76 p50
        -HFN 12-76 p153 tape
2389 Finlandia, op. 26. Kuolema, opp. 44 and 62. Scènes historiques,
    opp. 25 and 66. HSO; Jussi Jalas. London CS 6956.
        +NR 12-76 p3   ++SFC 8-29-76 p29
Finlandia, op. 26. cf Concerto, violin, op. 47, D minor.
Finlandia, op. 26. cf Angel S 37232.
Finlandia, op. 26. cf CBS 30073.
Finlandia, op. 26. cf HMV SLS 5019.
Finlandia, op. 26. cf Philips 6747 204.
Humoreske, no. 5, op. 89, no. 3, E flat major. cf Concerto, violin,
    op. 47, D minor.
Humoreske, no. 5, op. 89, no. 3, E flat major. cf HMV SEOM 24.
In memoriam, op. 59. cf Legends, op. 22.
In memoriam, op. 59. cf The tempest, op. 109.
2390 Karelia suite, op. 11. Legends, op. 22. Helsinki Radio Symphony
    Orchestra; Okko Kamu. DG 2530 656.
        ++Gr 10-76 p608   ++RR 11-76 p78
        +HFN 10-76 p175
Karelia suite, op. 11. cf Concerto, violin, op. 47, D minor.
Karelia suite, op. 11. cf Finlandia, op. 26.
Karelia suite, op. 11, no. 3: March. cf BERLIOZ: Les Troyens:
    Overture, March Troyenne.
Kuolema, opp. 44 and 62. cf Finlandia, op. 26.
Kuolema, op. 44: Valse triste. cf Concerto, violin, op. 47, D
    minor.
Kuolema, op. 44: Valse triste. cf BEETHOVEN: Symphony, no. 3,
    op. 55, E flat major.
Kuolema, op. 44: Valse triste. cf HMV SLS 5019.
Kuolema, op. 44: Valse triste. cf HMV SXLP 30188.
Legends, op. 22. cf Karelia suite, op. 11.
2391 Legends, op. 22: Lemminkäinen and the maidens of the island; The
    swan of Tuonela; Lemminkäinen in Tuonela; Lemminkäinen's return.
    In memoriam, op. 59. Hungarian State Symphony Orchestra;
    Jussi Jalas. Decca SDD 488. Tape (c) KSDC 488. (*Reissue
    from DPA 531/2). (also London CS 6955).
        +-Gr 9-76 p431   +HFN 10-76 p185 tape
        -HF 11-76 p126   +-NR 9-76 p5
        +-HFN 9-76 p129   +-RR 8-76 p53
Legends, op. 22: The swan of Tuonela. cf Concerto, violin, op.
    47, D minor.

Legends, op. 22: The swan of Tuonela. cf Finlandia, op. 26.
Legends, op. 22: The swan of Tuonela. cf Symphonies, nos. 1, 5-7.
Luonnotar, op. 70. cf Symphony, no. 6, op. 104, D minor.

2392 The oceanides, op. 73. Pelléas and Mélisande, op. 46: Incidental
      music. Tapiola, op. 112. RPO; Thomas Beecham. HMV SXLP 30197.
      Tape (c) TC EXE 180. (Reissue from ASD 468, 518).
          ++Gr 2-76 p1349   +-HFN 5-76 p117 tape
          +-HFN 2-76 p115   +RR 2-76 p41

Pan and echo, op. 53a. cf Works, selections (HMV SQ ASD 3287).
Pelléas and Mélisande, op. 46: Incidental music. cf The oceanides,
    op. 73.
Pohjola's daughter, op. 49. cf Symphony, no. 6, op. 104, D minor.
Rakastava, op. 14. cf GRIEG: Elegiac melodies, op. 34.
Rakastava, op. 14. cf RAUTAVAARA: Pelimannit, op. 1.
Romance, op. 42, C major. cf Works, selections (HMV SQ ASD 3287).
En saga, op. 9. cf Finlandia, op. 26.
En saga, op. 9. cf Symphony, no. 5, op. 82, E flat major (Clas-
    sics for Pleasure 40218).
En saga, op. 9. cf Symphony, no. 5, op. 82, E flat major (HMV
    3038).
En saga, op. 9. cf BEETHOVEN: Symphony, no. 3, op. 55, E flat
    major.
Scènes historiques, opp. 25 and 66. cf Finlandia, op. 26.
Scènes historiques, op. 25: All overtura, Scena, Festivo. cf
    Symphony, no. 1, op. 39, E minor.
Serenade, no. 1, op. 69a, D major. cf Concerto, violin, op. 47,
    D minor.
Serenade, no. 2, op. 69b, G minor. cf Concerto, violin, op. 47,
    D minor.
Spring song, op. 16. cf Works, selections (HMV SQ ASD 3287).
Suite champêtre, op. 98b. cf Works, selections (HMV SQ ASD 3287).
Suite mignonne, op. 98. cf Works, selections (HMV SQ ASD 3287).
Suite mignonne, op. 98. cf RAUTAVAARA: Pelimannit, op. 1.
Suite mignonne, op. 98. cf Swedish Society SLT 33229.

2393 Symphonies, nos. 1-7. VPO; Lorin Maazel. Decca D7D 4. Tape
      (c) KE 9. (Reissue).
          +-Gr 9-76 p477   +RR 4-76 p81 tape
          +-HFN 10-76 p181   +-RR 9-76 p66

2394 Symphonies, nos. 1, 5-7. Legends, op. 22: The swan of Tuonela.
      BSO; Serge Koussevitzky. Rococo 2103 (2).
          +-NR 4-76 p3

2395 Symphony, no. 1, op. 39, E minor. Scenes historiques, op. 25:
      All overtura, Scena, Festivo. Bournemouth Symphony Orchestra;
      Paavo Berglund. HMV SQ ASD 3216.
          +-Gr 12-76 p1002

2396 Symphony, no. 1, op. 39, E minor. Belshazzar's feast, op. 51.
      Orchestra; Robert Kajanus. Turnabout THS 65045. (Reissues
      from Columbia and HMV originals).
          +-Audio 6-76 p96   +NR 8-75 p6
          +HF 1-76 p95

2397 Symphony, no. 2, op. 43, D major. LSO; Pierre Monteux. Decca
      ECS 789. Tape (c) KECC 789. (Reissue from RCA SB 2070).
          +Gr 10-76 p611   +RR 11-76 p83
          +HFN 11-76 p170   +RR 12-76 p108 tape
          +HFN 12-76 p155

2398 Symphony, no. 2, op. 43, D major. COA; Georg Szell. Philips

6580 051. Tape (c) 7317 099. (also Philips 835506. Tape (c)
PCR 4-900092 (r) PC 8-900092). (Reissue from SAL 3515).
    +Gr 11-72 p965   -RR 11-72 p61
    +-RR 12-76 p108 tape

2399 Symphony, no. 2, op. 43, D major. LPO; John Pritchard. Pye GSGC
    15003. Tape (c) ZCCCB 15003. (Reissue from Virtuosi TPLS
    13032/3).
        -Gr 8-75 p335   -RR 12-76 p108 tape
        +-RR 8-75 p43

2400 Symphony, no. 2, op. 43, D major. RPO; John Barbirolli. RCA GL
    25011. Tape (c) GK 25011. (Previously issued by Reader's
    Digest).
        ++Gr 10-76 p608   +-RR 11-76 p83
        +-HFN 12-76 p147   +RR 12-76 p104

2401 Symphony, no. 2, op. 43, D major. PO; Eugene Ormandy. RCA (Q)
    ARD 1-0018. Tape (c) ARS 1-0018 (c) ARK 1-0018.
        +Gr 6-73 p58   +-NR 7-76 p2
        +HF 9-73 p114   +RR 7-73 p57
        /HFN 6-73 p1183   ++SFC 10-14-73 p34
        +MJ 1973 annual p14   +St 8-73 p111
        +NR 5-73 p5

Symphony, no. 2, op. 43, D major. cf Concerto, violin, op. 47,
    D minor.

2402 Symphony, no. 4, op. 63, A minor. Symphony, no. 5, op. 82, E flat
    major. RPO; Loris Tjeknavorian. RCA LRL 1-5134.
        +Gr 11-76 p804   +RR 11-76 p16

2403 Symphony, no. 5, op. 82, E flat major. En saga, op. 9. Scottish
    National Orchestra; Alexander Gibson. Classics for Pleasure
    CFP 40218. Tape (c) TC CFP 40218.
        +Gr 9-75 p466   +-HFN 12-76 p153
        /HFN 8-75 p83   +RR 9-75 p47

2404 Symphony, no. 5, op. 82, E flat major. En saga, op. 9. Bourne-
    mouth Symphony Orchestra; Paavo Berglund. HMV ASD 3038. Tape
    (c) TC ASD 3038. (also Angel S 37104).
        +Gr 1-75 p1354   /NR 8-75 p2
        -Gr 9-76 p497 tape   +-RR 1-75 p33
        /HF 9-75 p91   +St 8-75 p105
        ++HFN 8-76 p95 tape

2505 Symphony, no. 5, op. 82, E flat major. Symphony, no. 7, op. 105,
    C major. BSO; Colin Davis. Philips 6500 959. Tape (c) 7300
    415.
        +-Audio 6-76 p98   +-MT 7-76 p579
        +-Gr 11-75 p833   +-NR 12-75 p3
        +HF 12-75 p107   ++NYT 1-18-76 pD1
        ++HF 4-76 p148 tape   +RR 11-75 p55
        +-HFN 11-75 p165   ++SFC 9-28-75 p30
        +-HFN 2-76 p116   +St 3-76 p121

Symphony, no. 5, op. 82, E flat major. cf Symphony, no. 4, op.
    63, A minor.

2406 Symphony, no. 6, op. 104, D minor. Luonnotar, op. 70. Pohjola's
    daughter, op. 49. Taru Valjakka, s; Bournemouth Symphony
    Orchestra; Paavo Berglund. HMV (Q) ASD 3155.
        +Gr 2-76 p1346   +RR 2-76 p41
        +HFN 3-76 p105

Symphony, no. 7, op. 105, C major. cf Symphony, no. 5, op. 82,
    E flat major.

Tapiola, op. 112. cf Concerto, violin, op. 47, D minor.

Tapiola, op. 112. cf The oceanides, op. 73.

2407 The tempest, op. 109. In memoriam, op. 59. Royal Liverpool Philharmonic Orchestra; Charles Groves. HMV ASD 2916. Tape (c)
TC ASD 2961.
              ++Gr 2-74 p1566   +-HFN 1-76 p125 tape
              +HF 2-75 p101   +RR 2-74 p35

The tempest, op. 109: Humoresk; Caliban's song; Scene; Intrada and
berceuse; Chorus of winds; Intermezzo; Nymphs' dance; Prospero;
Miranda; Naiads; Storm. cf PETTERSSON: Concerto, string orchestra, no. 3: Mesto.

The tempest, op. 109: Miranda, The Naiads, The storm. cf BERLIOZ:
Les Troyens: Overture, March Troyenne.

Valse romantique, op. 62b. cf Works, selections (HMV SQ ASD 3287).

2408 Works, selections: Andante festivo, strings. Canzonetta, op. 62a.
Dance intermezzo, op. 45, no. 2. The Dryad, op. 45, no. 1.
Pan and echo, op. 53a. Romance, op. 42, C major. Spring song,
op. 16. Suite champêtre, op. 98b. Suite mignonne, op. 98.
Valse romantique, op. 62b. Royal Liverpool Philharmonic Orch-
estra; Charles Groves. HMV SQ ASD 3287.
              +Gr 11-76 p811   +RR 11-76 p78
              +-HFN 12-76 p148

SIECZYNSKI, Richard
Vienna, city of my dreams. cf CBS 30070.

SIEGMEISTER, Elie
American harp. cf Orion ORS 75207.

SIFLER, Paul
2409 Autumnal song. The despair and agony of Dachau. Fantasia. Gloria
in Excelsis Deo. Joseph's vigil. The last supper. Prelude on
"God of might". Shepherd pipers before the manger. Toccata
on "Ein Feste Burg". Paul Sifler, org. Fredonia FD 2.
              +NR 4-76 p13

The despair and agony of Dachau. cf Autumnal song.

Fantasia. cf Autumnal song.

Gloria in Excelsis Deo. cf Autumnal song.

Joseph's vigil. cf Autumnal song.

The last supper. cf Autumnal song.

Prelude on "God of might". cf Autumnal song.

Shepherd pipers before the manger. cf Autumnal song.

Toccata on "Ein Fest Burg". cf Autumnal song.

SILVERMAN, Stanley
Planh. cf Folkways FTS 33901/4.

SIMON
Kurassiermarsch "Grosser Kurfürst". cf Polydor 2489 523.

SIMONS, Gardell
Atlantic zephyrs. cf Washington University Press OLY 104.

SIMONS
Marta. cf DG 2530 700.

SIMPSON, Dudley
Exposé. cf BRINDLE: Orion 42.

SINDING, Christian
Rustle of spring, op. 32, no. 3. cf Decca SPA 473.

SKALKOTTAS, Nikos
2410 Octet. Quartet, strings, no. 3. Variations on a Greek folk tune
(8). Robert Masters, vln; Derek Simpson, vlc; Marcel Gazelle,
pno; Dartington String Quartet. Argo ZRG 753. (Reissue from
HMV ASD 2289).

```
      +Gr 11-74 p921              ++RR 11-74 p73
      ++MQ 1-76 p139              ++St 6-75 p107
      ++NR 2-75 p6
```
Variations on a Greek folk tune. cf Octet.

SKULTE, Bruno
2411 Behold a bright star. Latvian folk songs of the winter solstice:
        Blow, northwind, blow (2); God came ever so gently; A silver
        rain was falling; Winterfest has arrived (2); Light the kindling;
        Fan the fire; Open the door, mother; What thunders, what rum-
        bles; Pray, let me in; As winterfest departs. Silvija Erdmane,
        s; Alfred Genovese, ob; Ilze Akerberga, org; Latvian Folk Sin-
        gers, Members; Andrejs Jansons. Kaibala 50E 02.
          +NR 12-76 p1
      Latvian folk songs of the winter solstice: Blow, northwind, blow
        (2); God came ever so gently; A silver rain was falling; Winter-
        fest has arrived (2); Light the kindling; Fan the fire; Open
        the door, mother; What thunders, what rumbles; Pray, let me in;
        As winterfest departs. cf Behold a bright star.

SLONIMSKY, Sergei
2412 Concerto buffo. TISHCHENKO: Symphony, no. 3. Musici de Praga;
        Eduard Fischer. Supraphon 110 1433.
          +RR 4-76 p54

SLUKA, Lubos
      Sonata, bass clarinet and piano. cf Panton 110 369.
      Sonata, violin and piano. cf BACH: Partita, violin no. 2, S 1004,
        D minor: Chaconne.

SMART, Henry
      Hymn to Cynthia. cf Prelude PRS 2501.
      Postlude, D major. cf Vista VPS 1035.

SMERT, Richard
      Nowell: Dieus vous garde. cf MOUTON: Noe psallite.

SMETANA, Bedrich
2413 The bartered bride (in German). Teresa Stratas, Janet Perry,
        Margarethe Bence, s; Gudrun Wewezow, ms; René Kollo, Heinz
        Zednik, t; Jörn Wilsing, Walter Berry, Karl Dönch, bar; Alex-
        ander Malta, Theodor Nicolai, bs; Bavarian Radio Orchestra and
        Chorus; Jaroslav Krombholc. Eurodisc 89036 XGR (3).
          +-HF 3-76 p94   +-ON 3-20-76 p56
      The bartered bride: Ah, bitterness. cf HMV SXLP 30205.
      The bartered bride: Overture. cf Decca SPA 409.
2414 The bartered bride: Overture and dances. Má Vlast (My fatherland).
        St. Louis Symphony Orchestra; Walter Susskind. Turnabout (Q)
        QTVS 34619/20 (2).
          +-HF 7-76 p93 Quad   +NR 6-76 p4
          +MJ 5-76 p29 Quad   +St 7-76 p112 Quad
2415 The bartered bride: Overture, Polka, Furiant, Dance of the comed-
        ians. Má Vlast: Vltava. TCHAIKOVSKY: Romeo and Juliet: Fantasy
        overture. LPO; Adrian Boult. HMV SXLP 30199. Tape (c) TC EXE
        182. (Reissues from World Record Club ST 665, 683).
          +Gr 1-76 p1209   ++RR 1-76 p38
          +HFN 1-76 p123   +RR 10-76 p107 tape
          +HFN 6-76 p105 tape
      The bartered bride: Overture, Polka, Furiant, Dance of the comed-
        ians. cf DVORAK: Slavonic dances, opp. 46, 72.
      The bartered bride: Polka. cf Decca SB 322.
      The bartered bride: This girl I've found you. cf English National
        Opera ENO 1001.

The bartered bride: Wedding chorus. cf Hungaroton SLPX 11762.
The bartered bride: Wer in Lieb entbrannt. cf Preiser LV 192.
2416 Czech dances and polkas. Brno Philharmonic Orchestra; Frantisek
      Jílek. Supraphon 110 1225. Tape (c) 4-01225.
            +HFN 2-76 p116 tape   +RR 4-76 p81 tape
      Dalibor: Blickst du mein Freund. cf Discophilia 13 KGW 1.
      Dalibor: Do I live. cf HMV SXLP 30205.
2417 The devils wall. Soloists; Prague National Theatre Orchestra and
      Chorus; Zdenek Chalabala. Supraphon 50361/3 (3).
            ++SFC 6-27-76 p29
2418 Haakon Jarl, op. 16. Richard III, op. 11. Solemn march for
      Shakespeare celebrations. Wallenstein's camp, op. 14. CPhO;
      Václav Neumann. Supraphon 110 1584.
            +Gr 4-76 p1611   +NR 9-76 p4
            +-HFN 4-76 p115   +-RR 4-76 p53
2419 Libuše. Nadežda Kniplová, s; Věra Soukupová, con; Ivo Zidek, t;
      Václav Bednár, bar; Karel Berman, bs-bar; Zdenek Kroupa, bs;
      Prague National Theatre Orchestra and Chorus; Jaroslav Kromb-
      holc. Supraphon SUAST 50701/4 (4).
            +-Gr 3-74 p1744   +RR 2-74 p20
            +-HFN 2-74 p345   +SFC 12-2-73 p30
            +NR 11-76 p11
2420 Má Vlast. CSO; Rafael Kubelik. Mercury SRI 2-77006 (2). (Re-
      issue from OL 3103).
            +-HF 7-76 p93   +-NR 2-76 p2
2421 Má Vlast. CPhO; Karel Ancerl. Supraphon GS 50521. Tape (c)
      4-50521.
            +HFN 2-76 p116 tape   ++RR 3-76 p78 tape
2422 Má Vlast. CPhO; Karel Ancerl. Vanguard/Supraphon SU 9/10 (2).
      (Reissue from Crossroads 22 26 0002).
            +HF 7-76 p93
      Má Vlast. cf The bartered bride: Overture and dances.
      Má Vlast: Vltava. cf The bartered bride: Overture, Polka, Furiant,
      Dance of the comedians.
      Má Vlast: The Moldau. cf BIZET: L'Arlesienne: Suite, no. 2.
      Má Vlast: Vltava. cf BEETHOVEN: Egmont, op. 84: Overture.
      Má Vlast: Vltava. cf DVORAK: Carnival overture, op. 92.
      Má Vlast: Vltava. cf DVORAK: Slavonic dances, opp. 46 and 72.
      Má Vlast: Vltava. cf DVORAK: Slavonic dances, op. 46, nos. 1-6.
      Má Vlast: Vltava. cf DVORAK: Symphony, no. 9, op. 95, E minor.
      Má Vlast: The Moldau. cf Angel S 37232.
      Má Vlast: Vltava. cf Decca DPA 519/20.
      On the seashore, op. 17. cf DEBUSSY: Estampes: Jardins sous la
      pluie.
      Richard III, op. 11. cf Haakon Jarl, op. 16.
      Solemn march for Shakespeare celebrations. cf Haakon Jarl, op. 16.
2423 Songs: Choruses for female voices (3); The dedication; Festive
      chorus; Our song; The peasant; The prayer; The renegade (2
      versions); Slogans (2); Song of the sea; The three riders.
      Miroslav Svejda, t; Jindrich Jindrák, bar; Jaroslav Horácek,
      bs; Czech Philharmonic Chorus; Josef Veselka. Supraphon 112
      1143.
            +Gr 2-74 p1595   +NR 3-74 p8
            +HF 6-74 p101   +RR 2-74 p58
            +HFN 2-74 p345   ++SFC 2-22-76 p29
            +-LJ 5-75 p44   +St 8-74 p122

Trio, piano, op. 15, G minor. cf DVORAK: Trio, piano, op. 90,
    E minor.
Wallenstein's camp, op. 14. cf Haakon Jarl, op. 16.
SMITH, Gregg
2424 Beware of the soldier. Rosalind Rees, s; Chuck Garretson, boy
        soprano; Douglas Perry, t; Charles Greenwell, bs; Texas Boys'
        Choir, Columbia University Men's Glee Club; Orchestra Ensemble;
        Gregg Smith. CRI SD 341.
            +NR 11-75 p9   ++St 1-76 p109
SMITH, Seymore
    The spider and the fly. cf Transatlantic XTRA 1159.
SMITH, William O.
    Straws. cf Crystal S 351.
SMOLKA
    The heathcock sits above the cloud. cf Panton 110 385.
SODERLUNDH, Lille Bror
    Land o' Bella: Suite. cf KARKOFF: Figures for wind instruments
        and percussion.
SODERMAN, Johan
    Kung Heimar och Aslög. cf Rubini GV 65.
SOKOLOV
    Russian folk song suite. cf Hungaroton SLPX 11762.
SOLAGE (14th century France)
    Fumeux, fume. cf HMV SLS 863.
    Helas, je voy mon cuer. cf HMV SLS 863.
SOLER, Antonio
    The emperor's fanfare. cf Columbia M 33514.
    Fandango, M. la. cf SCARLATTI, D.: Sonata, guitar, K 2, E major.
    Fandango, D minor. cf BACH, C.P.E.: Concerto, piano, A minor.
    Sonatas, guitar, B minor, G major. cf SCARLATTI, D.: Sonata,
        guitar, K 2, E major.
2425 Sonatas, harpsichord (1). Fernando Valenti, hpd. Desmar 1001
            +-Gr 2-76 p1363   +RR 1-76 p52
            +-MJ 12-75 p38   +St 1-76 p106
            ++NR 3-76 p11
    Sonata, harpsichord, D major. cf BACH, C.P.E.: Concerto, piano,
        A major.
    Sonata, piano, G minor. cf London CS 6953.
    Sonata, piano, D major. cf London CS 6953.
SOLLBERGER, Harvey
2426 Riding the wind I. WYNER: Intermedio. Susan Davenny Wyner, s;
        Patricia Spencer, flt; String Orchestra, Da Capo Chamber Play-
        ers; Yehudi Wyner, Harvey Sollberger. CRI SD 352.
            +ARG 12-76 p44   +NR 11-76 p8
    Solos for 6 players. cf Desto DC 6435/7.
SOMERVELL, Arthur
    Maud. cf BUTTERWORTH: A Shropshire lad.
    SONGS AND OPERATIC ARIAS, RENATA TEBALDI. cf Decca SXL 6629.
SOPRONI, Jozsef
2427 Concerto, violoncello. Eklypsis. Quartet, strings, no. 4. Son-
        ata, flute and piano. László Mező, vlc; Erzsébet Csik, flt;
        Zoltán Kocsis, pno; Hungarian Radio and Television Orchestra;
        György Lehel. Hungaroton SLPX 11743.
            +NR 9-76 p14
    Eklypsis. cf Concerto, violoncello.
    Quartet, strings, no. 4. cf Concerto, violoncello.

Sonata, flute and piano. cf Concerto, violoncello.
Songs: Ejszaka, Dicseret, Sötet lett. cf Hungaroton SLPX 11713.
SOR, Fernando
Divertimento, op. 55, G major. cf GIULIANI: Rondo, no. 1, C major.
L'Encouragement, op. 34, E major. cf GIULIANI: Rondo, no. 1, C major.
Fantasía elegiaca, op. 59. cf Enigma VAR 1015.
Introduction and variations on "Malbrough a'en va-t-en guerre". cf RCA ARL 1-1323.
Minuet, D major. cf London STS 15306.
Sicilienne, D minor. cf RCA ARL 1-1323.
Variations on a theme by Mozart, op. 9. cf Angel S 36093.
Variations on "Folies d'Espagne". cf Supraphon 111 1230.
SOUSA, John Philip
American marching songs. cf BBC REB 228.
The army goes rolling along. cf Department of Defense, Bicentennial Edition 50-1776.
The blue and the gray. cf BBC REB 228.
The bride elect. cf Works, selections (Antilles AN 7015).
El Capitán. cf Works, selections (Delos 25413).
El Capitán. cf Columbia 33513.
El Capitán. cf Michigan University SM 0002.
The charlatan. cf Works, selections (Antilles AN 7015).
Coquette. cf Works, selections (Antilles AN 7015).
The diplomat. cf Works, selections (Delos 25413).
The fairest of the fair. cf Works, selections (Delos 25413).
The free lance. cf Works, selections (Delos 25413).
Free lance. cf Everest 3360.
The gallant Seventh. cf Columbia 33513.
The gladiator. cf Works, selections (Delos 25413).
Hands across the sea. cf BBC REB 228.
High school cadets. cf Works, selections (Delos 25413).
The liberty bell. cf Columbia 33513.
Liberty loan. cf Works, selections (Delos 25413).
Manhattan Beach. cf Everest 3360.
2428 Marches: El Capitán. The dauntless battalion. Golden jubilee. King Cotton. Liberty loan. March of the mitten men, or power and the glory. Nobles of the mystic shrine. Sabre and spurs. Semper fidelis. Solid men to the front. The Stars and Stripes forever. The thunderer. U.S. Field artillery. The Washington post. Sousa Band, Special Band, Philadelphia Rapid Transit Company Band, Pryor Band, Victor Military Band; John Philip Sousa, Arthur Pryor, Rosario Bourdon. Pelican LP 135. (Reissues from Victor 78s).
      +HF 5-76 p95
2429 Marches: The belle of Chicago; El Capitán; George Washington bicentennial; The liberty bell; The New York hippodrome; Our flirtations; The pathfinder of Panama; The stars and stripes forever; The thunderer; U.S field artillery; The Washington post; We are coming. Detroit Concert Band; Leonard B. Smith. Sousa American Bicentennial Collection 1.
      -HF 9-75 p92   +IN 6-76 p25
Mother Hubbard. cf Works, selections (Antilles AN 7015).
The national game. cf Works, selections (Delos 25413).
Nymphalin. cf Works, selections (Antilles AN 7015).
The pathfinder of Panama. cf Everest 3360.

President Garfield's inauguration march. cf Columbia M 33838.
Presidential polonaise. cf Works, selections (Delos 25413).
The red man. cf Works, selections (Antilles AN 7015).
La reine de la mer. cf Works, selections (Antilles AN 7015).
Royal Welsh fusiliers. cf Works, selections (Delos 25413).
Semper fidelis. cf Works, selections (Delos 25413).
Semper fidelis. cf Columbia 33513.
Semper fidelis. cf RCA AGL 1-1334.
The stars and stripes forever. cf Works, selections (Delos 25413).
The stars and stripes forever. cf CBS 30073.
The stars and stripes forever. cf Columbia 33513.
The stars and stripes forever. cf Columbia M 34129.
The stars and stripes forever. cf Department of Defense Bicenten-
    nial Edition 50-1776.
The stars and stripes forever. cf Personal Touch 88.
The stars and stripes forever. cf Pye QS PCNH 4.
The stars and stripes forever. cf RCA VH 020.
The stars and stripes forever. cf RCA AGL 1-1334.
The summer girl. cf Works, selections (Antilles AN 7015).
The thunderer. cf Columbia 33513.
The triumph of time. cf Works, selections (Antilles AN 7015).
Washington post. cf Works, selections (Delos 25413).
Washington post. cf Philips 6308 246.
2430 Works, selections: The bride elect. Coquette. The charlatan.
      Mother hubbard. Nymphalin. The red man. The summer girl. La
      reine de la mer. The triumph of time. Antonín Kubalek. pno.
      Antilles AN 7015.
            +NR 9-76 p13
2431 Works, selections: El Capitán. The diplomat. Fairest of the fair.
      The gladiator. The free lance. High school cadets. Liberty
      loan. The national game. Presidential polonaise. Royal Welsh
      fusiliers. Semper fidelis. The stars and stripes forever.
      Washington post. Sousa Band, West German Army Band; narrated
      by Tony Thomas. Delos 25413.
            +MJ 7-76 p59   +NR 9-76 p12
    SOVIET ARMY CHORUS AND BAND: CELEBRATION. cf Columbia/Melodiya
      M 33592.
SOWANDE, Fela
2432 African suite: Akinla; Joyful day; Nostalgia. STILL: Sahdji.
      WALKER: Lyric for strings. Morgan State College Choir; LSO;
      Paul Freeman. Columbia M 33433.
            ++Audio 12-75 p106   +NYT 4-11-76 pD23
            +HF 10-75 p65   ++St 12-75 p132
            ++NR 9-75 p5
SOWERBY, Leo
2433 Sonata, organ. Gordon Wilson, org. Century Advent GW 31-734.
            +MJ 7-76 p57
    SPANISH SONGS OF THE MIDDLE AGES AND RENAISSANCE. cf DG 2530 504.
SPEAKS, Oley
    On the road to Mandalay. cf Cambridge CRS 2715.
    SPECTRUM: NEW AMERICAN MUSIC. cf Nonesuch H 71302/3.
SPEER, Daniel
    Sonatas (3). cf Crystal S 206.
    Sonata, brass, G major/C major. cf Crystal S 204.
SPENDIAROV, Alexander
2434 Crimean sketches, op. 9: Elegiac song; Drinking song. Almast.

SPENDIAROV (cont.)   462

TANEIEV: Symphony, op. 12, C minor. MRSO; Bolshoi Theatre
Orchestra; Gennady Rozhdestvensky. HMV Melodiya ASD 3106.
+Gr 9-75 p472   +-MM 1-76 p42
+HFN 9-75 p104   +RR 9-75 p50
SPIRITUALS. cf RCA AVM 1-1735.

SPOHR, Ludwig
2435 Concertante, violin, harp and orchestra, no. 1, G major. Sonata,
flute and harp, E major. Hansheinz Schneeberger, vln; Ursula
Holliger, hp; Peter-Lukas Graf, flt; ECO; Peter-Lukas Graf.
CRD Claves 30-407.
+Gr 8-76 p298   +RR 6-76 p54
German songs (6). cf POULENC: Sonata, clarinet and bassoon.
Notturno, op. 34, C major. cf HUMMEL: Partita, E flat major.
Sonata, flute and harp, E major. cf Concertante, violin, harp
and orchestra, no. 1, G major.

SPONTINI, Gasparo
La vestale: Tu che invoco. cf HMV SLS 5057.

STACHOWICZ
Veni consolator. cf MIELCZEWSKI: Veni Domine, Deus in nomine tuo.

STADLER, Anton
Trio, 3 basset horns. cf MOZART: Adagio, K 411, B flat major.

STAFFORD
Watchman, what of the night. cf HMV EMD 5528.

STAINER, John
I saw the Lord. cf Abbey LPB 739.
2436 On a bass. Prelude and fughetta. Songs: Drop down ye heavens;
Lead kindly light; I saw the Lord; Magnificat and nunc dimittis,
B flat major; Let Christ the King. Magdalen College Choir,
Oxford; Bernard Rose. Argo ZRG 811.
+Audio 10-76 p150   +MU 10-76 p16
+Gr 7-75 p233   +NR 6-76 p13
+HFN 8-75 p84   +RR 7-75 p60
+MT 4-76 p322
Prelude and fughetta. cf On a bass.
Songs: Drop down ye heavens; Lead kindly light; I saw the Lord;
Magnificat and nunc dimittis, B flat major; Let Christ the
King. cf On a bass.

STALDER
Sonata quarta. cf Pelca PRSRK 41017/20.

STAMITZ, Johann
2437 Concerto, clarinet, B flat major. Symphony, op. 4, no. 2, D major.
Symphony, G major. Symphony, op. 3, no. 2, D major. Alan
Hacker, clt; Academy of Ancient Music; Christopher Hogwood.
L'Oiseau-Lyre DSLO 505.
++Gr 2-76 p1349   +NR 6-76 p5
+HFN 5-76 p109   +RR 1-76 p38
+MJ 10-76 p24   +St 6-76 p110
Symphony, G major. cf Concerto, clarinet, B flat major.
Symphony, op. 3, no. 2, D major. cf Concerto, clarinet, B flat
major.
Symphony, op. 4, no. 2, D major. cf Concerto, clarinet, B flat
major.

STAMITZ, Karl
Concerto, viola, D major. cf HANDEL: Concerto, viola, B minor.
Concerto, viola, D major. cf HOFFMEISTER: Concerto, viola, D
major.

STANFORD, Charles
    Beati quorum via, op. 51, no. 3. cf HMV HQS 1350.
    Irish dances: Reel; Leprechaun's dance. cf GRAINGER: Piano works
        (Everest X 913).
    Postlude, D minor. cf Wealden WSQ 134.
    Psalm, no. 23: The Lord is my shepherd. cf Polydor 2460 250.
    Psalm, no. 53: The foolish body. cf HMV SQ CSD 3768.
2438 Songs (Motets): English motets, op. 135: Ye holy angels bright;
        Eternal Father; Glorious and powerful God. Latin motets, op.
        51: Justorum animae; Coelos ascendit; Beati quorum via. WOOD:
        Expectans expectavi; God omnipotent reigneth; Glory and honour
        and laud; Hail, gladdening light; Tis the day of resurrection;
        O thou the central orb. Magdalen College Choir; Ian Crabbe,
        org; Bernard Rose. Argo ZRG 852.
            +Gr 11-76 p858   +RR 11-76 p102
            +-HFN 10-76 p177
    Songs: The Lord is my shepherd; Te Deum, C major. cf Audio 1.
    Songs: Magnificat and nunc dimittis, A major. cf Audio 3.
    Songs: When Mary through the garden went, op. 127, no. 3. cf
        Vista VPS 1023.
    When in man's music God is glorified. cf Grosvenor GRS 1039.
STANLEY
    Sonata, recorder, F major. cf Orion ORS 76216.
STANLEY, John
    Trumpet tunes, nos. 1 and 2. cf Argo ZDA 203.
    Voluntaries, D major, A minor. cf Cambridge CRS 2540.
    Voluntary, A minor. cf Grosvenor GRS 1041.
    Voluntary, C major. cf Argo ZRG 807.
    Voluntary, D minor: Adagio. cf Argo ZDA 203.
STARER, Robert
    Prelude. cf Orion ORS 75207
STARZER, Josef
    Diane et Endimione: Dances. cf BACH, C. P. E.: Minuets, Wq 189.
    Gli Orazi e Gli Curiazi: Dances. cf BACH, C. P. E.: Minuets,
        Wq 189.
    Roger et Bradamante: Dances. cf BACH, C. P. E.: Minuets, Wq 189.
STEDRON, Milos
    Leich on a theme by Heinrich von Meissen. cf Supraphon 111 1390.
STEELE, Jan
    All day. cf CAGE: Experiences, nos. 1 and 2.
    Distant saxophones. cf CAGE: Experiences, nos. 1 and 2.
    Rhapsody spaniel. cf CAGE: Experiences, nos. 1 and 2.
STEIBELT, Daniel
    Bacchanales, op. 53 (3). cf Angel S 36080.
STELLA
    Riso. cf Rubini GV 68.
STELZMULLER
    Dance. cf Saga 5421.
STENHAMMAR, Wilhelm
    Fantasias, op. 11 (3). cf LARSSON: Sonatina, no. 1, op. 16.
    Serenade, op. 31, F major. cf KOCH: Oxberg variations.
    Songs: Det far att skepp; Flickan knyter i Johannenatten; Flickan
        Kom ifrån sin älsklings möte; I skogen. cf FRUMERIE: Songs
        (Swedish Society SLT 33171).
2439 Symphony, no. 2, op. 17, G minor. Stockholm Philharmonic Orches-
        tra; Tor Mann. Swedish Society SLT 33198. (Reissue).

     +RR 1-75 p29   /St 6-75 p52
     +-RR 10-76 p67

STENIUS, Torsten
 Partita on a Finnish folk melody. cf BACH: Prelude and fugue,
  S 546, C minor.
STERNWALD, Jiri
 Symphonic picture. cf Panton 110 440.
STEVENS
 Sigh no more, ladies. cf HMV ESD 7002.
STEVENS, Halsey
 Songs: Go, lovely rose; Like as the culver; Weepe, O mine eyes.
  cf Orion ORS 75205.
 Songs: Like as the culver on the bared bough. cf CRI SD 102.
 Symphony, no. 1. cf COPLAND: Dance symphony.
STEVENSON
 Behold I bring you glad tidings. cf Vox SVBX 5350.
STEWART, Richard
 Prelude, organ and tape. cf Abbey LPB 752.
STILL, William Grant
 Darker America. cf KAY: Dances for string orchestra.
 From the black belt. cf KAY: Dances for string orchestra.
 Sahdji. cf SOWANDE: African suite.
STOCK, Dave
 Quintet, clarinet and strings. cf BLANK: Rotation.
STOCKHAUSEN, Karlheinz
2440 Bird of passage. Ceylon. Harald Bojé, electronium; Peter Eötvös,
  camel bells, triangles, synthesizer; Aloys Kontarsky, pno;
  Joachim Krist, tam tam; Tim Souster, sound projections; Markus
  Stockhausen, tpt, electric tpt and flugel horn; John Miller,
  tpt; Karlheinz Stockhausen, chromatic rin, lotus flute, Indian
  bells, bird whistle, voice, kandy drum. Crysalis CHR 1110.
   +-Gr 9-76 p438   +-RR 8-76 p64
 Ceylon. cf Bird of passage.
2441 Mikrophonie I and II. Aloys Kontarsky, Fred Alings, tam tam;
  Johannes Fritsch, Harald Bojé, microphones; Hugh Davies, Jaap
  Spek, Karlheinz Stockhausen, electronics; WDR Choir, Cologne
  Studio Choir; Alfons Kontarsky, org; Herbert Schernus. DG 2530
  583.
   ++HFN 11-76 p167   +RR 9-76 p77
2442 Momente (Europa version 1972). Gloria Davy, s; West German Radio
  Chorus; Musique Vivante Ensemble Instrumentalists; Karlheinz
  Stockhausen. DG 2709 055 (3).
   ++RR 10-76 p67
2443 Prozession (1971 version). Harald Bojé, electronium; Christoph
  Caskel, tam tam; Joachim Krist, microphone; Peter Eötvös, elec-
  trochord with synthesizer; Aloys Kontarsky, pno; Karlheinz
  Stockhausen, electronics. DG 2530 582.
   +HFN 11-76 p167   +-RR 9-76 p77
2444 Set sail for the sun (Setz die Segel zur Sonne). Short wave (Kurz-
  wellen). The Negative Band. Finnadar SR 9009. Tape (c) CS
  9009 (ct) TP 9009.
   *HF 1-76 p96   +St 5-76 p125
   -NR 3-76 p15
 Short wave. cf Set sail for the sun.
2445 Stop. Ylem. London Sinfonietta; Karlheinz Stockhausen. DG 2530
  422.

```
        +Gr 6-75 p50            +-NR 6-75 p2
        +HF 6-75 p103           +-RR 4-75 p32
        ++HFN 5-75 p139         +St 5-76 p125
        +-MT 12-75 p1072
```
    Ylem. cf Stop.
    Zyklus. cf BRINDLE: Orion 42.
STOCKS
    Come, living God. cf Grosvenor GRS 1039.
STOKOWSKI, Leopold
    Caprice oriental. cf International Piano Archives IPA 5007/8 (2).
STOLZ, Robert
    Frühling im Prater: Du bist auf dieser Welt. cf HMV CSD 3748.
    Zwei Herzen im Dreivierteltakt: Zwei Herzen im Dreivierteltakt.
        cf HMV CSD 3748.
STRADELLA, Allessandro
    Songs: Per Pietà. cf Desto DC 7118/9.
    Songs: Pietà Signore. cf Decca SXL 6781.
STRAESSER
    Ramassasin. cf KUNST: No time.
STRAIGHT, Willard
    Development. cf HOVHANESS: Triptych.
STRAUSS, Eduard
    Bahn frei, op. 45. cf STRAUSS, J. II: Works, selections (Classics
        for Pleasure CFP 40256).
    Unter der Enns, op. 121. cf Saga 5421.
STRAUSS, Johann I
    Radetzky march, op. 228. cf STRAUSS, J. II; Works, selections
        (Decca SXL 6740).
    Radetzky march, op. 228. cf STRAUSS, J, II: Works, selections
        (DG 3318 062).
    Radetzky march, op. 228. cf STRAUSS, J. II: Works, selections
        (Pye GSGC 15024).
    Radetzky march, op. 228. cf STRAUSS, J. II: Works, selections
        (Saga 5408).
    Radetzky march, op. 228. cf Angel S 37232.
    Radetzky march, op. 228. cf Classics for Pleasure CFP 40254.
    Vienna carnival. cf Saga 5411.
STRAUSS, Johann II
    Accelerations waltz, op. 234. cf Works, selections (Telefunken
        4-41306).
2446 An der schönen blauen Donau (The beautiful blue Danube), op. 314.
        Frühlingstimmen, op. 410. Geschichten aus dem Wienerwald, op.
        325. Künstlerleben, op. 316. Rosen aus dem Suden, op. 388.
        Boston Pops Orchestra; Arthur Fiedler. London SPC 21144.
        Tape (c) SP5 21144. (also Decca PFS 4353).
```
              +-Gr 12-76 p1072        +-MJ 12-76 p28
              ++HF 8-76 p94           +NR 8-76 p2
```
2447 The beautiful blue Danube, op. 314. Auf der Jagd, op. 373.
        Egyptian march, op. 335. Frühlingsstimmen, op. 410. Perpe-
        tuum Mobile, op. 257. STRAUSS, J.II/Josef: Pizzicato Polka.
        STRAUSS, Josef: Delirien, op. 212. Transaktionen, op. 184.
        Ohne Sorgen, op. 271. VPO; Willi Boskovsky. London STS 15269.
```
                 +NR 4-76 p3
```
    An der schönen blauen Donau, op. 314. cf Works, selections (Decca
        DKPA 513/4).
    The beautiful blue Danube, op. 314. cf Works, selections (Decca
        DPA 549/50).
```
```

The beautiful blue Danube, op. 314. cf Works, selections (DG 3318 062).

The beautiful blue Danube, op. 314. cf Works, selections (HMV ASD 3132).

An der schönen blauen Donau, op. 314. cf Works, selections (HMV ESD 7025).

The beautiful blue Danube, op. 314. cf Works, selections (Pye GSGC 15024).

An der schönen blauen Donau, op. 314. cf Works, selections (RCA SMA 7012).

An der schönen blauen Donau, op. 314. cf Works, selections (RCA GL 25019).

The beautiful blue Danube, op. 314. cf Works, selections (Saga 5408).

The beautiful blue Danube, op. 314. cf CBS 30073.

The beautiful blue Danube, op. 314. cf Everest 3360.

Annen Polka, op. 117. cf Works, selections (Classics for Pleasure CFP 40048).

Annen polka, op. 117. cf Works, selections (Decca SXL 6740).

Annen polka, op. 117. cf Works, selections (DG 3318 062).

Annen polka, op. 117. cf Works, selections (HMV ASD 3132).

Annen polka, op. 117. cf Works, selections (Pye GSGC 15024).

Annen polka, op. 117, cf Works, selections (Saga 5408).

Annen polka, op. 117, cf Works, selections (Telefunken 4-41306).

Auf der Jagd, op. 373. cf The beautiful blue Danube, op. 314.

Auf der Jagd, op. 373. cf Works, selections (Decca SXL 6740).

2448 Banditen, op. 378. Champagne polka, op. 211. Egyptian march, op. 335. Künstlerleben, op. 316. Wiener Bonbons, op. 307. STRAUSS, Josef: Delirien, op. 212. Heiterer Mut, op. 281. Mein Lebenslauf ist Lieb und Lust, op. 263. Ohne Sorgen, op. 271. Plappermäulchen, op. 245. VPO; Willi Boskovsky. Decca SDD 474. Tape (c) KSDC 474. (Reissues from SXL 2163, 2082, 2198).

+Gr 6-76 p96   +RR 5-76 p52
+-HFN 5-76 p115   ++RR 11-76 p110 tape
+HFN 7-76 p104 tape

Banditen, op. 378. cf Works, selections (Classics for Pleasure CFP 40256).

Bei uns zu Haus, op. 361. cf Works, selections (Decca SXL 6740).

Bijouterie Quadrille, op. 169. cf Works, selections (RCA SMA 7012).

Bouquet Quadrille, op. 135. cf Works, selections (RCA SMA 7012).

Casanova: Ich steh' zu dir; Ich hab' dich lieb. cf HMV CSD 3748.

Champagne polka, op. 211. cf Banditen, op. 378.

Champagne polka, op. 211. cf Works, selections (Classics for Pleasure CFP 40048).

Du und du, op. 367. cf Works, selections (Decca DPA 549/50).

Egyptian march, op. 335. cf Banditen, op. 378.

Egyptian march, op. 335. cf The beautiful blue Danube, op. 314.

Egyptian march, op. 335. cf Works, selections (Classics for Pleasure CFP 40256).

Egyptian march, op. 335. cf Works, selections (Decca DPA 549/50).

Egyptian march, op. 335. cf Works, selections (Telefunken 4-41306).

Eljen a Magyar, op. 332. cf Works, selections (Classics for Pleasure CFP 40048).

Eljen a Magyar, op. 332. cf Works, selections (DG 3318 062).

Es gibt nur a Kaiserstadt, es gibt nur ein Wien, op. 291. cf
Works, selections (RCA SMA 7012).

Explosions, op. 43. cf Works, selections (Decca DPA 549/50).

Explosions, op. 43. cf Works, selections (Decca SXL 6740).

2449 Die Fledermaus. Julia Varady, Lucia Popp, Evi List, s; René
Kollo, Perry Gruber, t; Hermann Prey, Benno Kusche, Bernd
Weikl, Nikolai Lugovoi, bar; Ivan Rebroff, Franz Muxeneder,
bs; Bavarian State Opera Chorus; Bavarian State Orchestra;
Carlos Kleiber. DG 2707 088 (2). Tape (c) 3370 009.
        +Gr 10-76 p650   +RR 10-76 p30
       ++HFN 10-76 p159   ++SFC 11-7-76 p33
        -ON 12-4-76 p60

Die Fledermaus, op. 363: Du und du. cf Works, selections (Decca
DKPA 513/4).

Die Fledermaus, op. 363: Mein Herr was dachten Sie von mir. cf
Discophilia DIS 13 KHH 1.

2450 Die Fledermaus, op. 363: Overture. Geschichten aus dem Wienerwald, op. 325. Rosen aus dem Süden, op. 388. Der Zigeunerbaron, op. 420: Overture. NYP; Leonard Bernstein. Columbia
M 34125.
        +-HF 12-76 p118   +NR 7-76 p3
        +-MJ 12-76 p28

2451 Die Fledermaus, op. 363: Overture. Le beau Danube, op. 314 (arr.
Desormier). STRAUSS, J. II and J. I: Bal de Vienne (arr.
Gamley). National Philharmonic Orchestra; Richard Bonynge.
Decca SXL 6701. Tape (c) KSXC 6701. (also London CS 6896
Tape (c) CS5 6896).
        +Gr 12-75 p1098   +HFN 2-76 p116 tape
        +HF 11-76 p97   ++RR 12-75 p62
        +HFN 12-75 p165   +RR 5-76 p78

Die Fledermaus: Overture. cf Works, selections (Classics for
Pleasure CFP 40048).

Die Fledermaus, op. 363: Overture. cf Works, selections (Decca
DPA 549/50).

Die Fledermaus, op. 363: Overture. cf Works, selections (DG 3318
062).

Die Fledermaus, op. 363: Overture. cf Works, selections (HMV
ASD 3132).

Die Fledermaus, op. 363: Overture. cf Works, selections (Pye
GSGC 15024).

Die Fledermaus, op. 363: Overture. cf Works, selections (Saga
5408).

Die Fledermaus, op. 363: Overture. cf Supraphon 110 1638.

2452 Die Fledermaus, op. 363: Overture and ballet music. Graduation
ball, op. 97 (arr. Dorati). WEBER: Invitation to the dance,
op. 65 (arr. Berlioz). VPO; Willi Boskovsky, Herbert von
Karajan. Decca SPA 406. (Reissue).
        +-HFN 8-76 p93   +RR 8-76 p54

2453 Die Fledermaus, op. 363: Overture: Täubchen, das entflattert ist;
So soll allein ich bleiben; Trinke, Liebchen, trinke, schnell;
Ich lade gern mir Gäste ein; Klänge der Heimat; Spiel ich die
Unschuld. Der Zigeunerbaron, op. 420: Als flotter Geist; Ja,
das Schreiben und das Lesen; Doch treu und wahr; O habet Acht;
Ei, ei, er lacht; Wer uns getraut; Her die Hand; Nach Wien;
Entr'act, Act 3. Hilde Gueden, Anneliese Rothenberger,
Margarethe Sjöstedt, s; Hilde Rössl-Majdan, ms; Karl Terkal,

Kurt Equiluz, t; Erich Kunz, Walter Berry, bar; Claude Heater,
bs; Vienna Musikfreunde Chorus; VPO; Heinrich Hollreiser.
Classics for Pleasure CFP 40251. (Reissue from HMV ASD 444,
394/5).
+-Gr 11-76 p893   +RR 10-76 p32
+-HFN 11-76 p173
2454 Die Fledermaus, op. 363: Overture; Täubchen, das Entflattert ist;
Komm mit mir Souper; Trinke liebchen, trinke schnell; Mein
Herr, was dächten sie von mir; Mein Schönes, grosses Vogel-
haus; Ich lade gern mir Gäste ein; Mein Herr Marquis; Die
Klänge der Heimat; Im feuerstrom der Reben; Brüderlein, Brüder-
lein und Schwesterlein; Ich stehe voll Zagen; O Fledermaus;
Champagner hat's verschuldet; Die Majestät wird anerkannt.
Anneliese Rothenberger, Renate Holm, s; Nicolai Gedda, Adolf
Dallapozza, t; Brigitte Fassbaender, con; Dietrich Fischer-
Dieskau, bar; Walter Berry, bs-bar; Vienna State Opera Chorus;
VSO; Willi Boskovsky. HMV ASD 2891. (Q) Q4ASD 2891.
+-Gr 11-74 p963   +RR 7-74 p31
+Gr 11-74 p970 Quad   +RR 11-76 p42
+-Gr 11-76 p893
Freikugeln, op. 326. cf Works, selections (RCA SMA 7012).
Frühlingsstimmen (Voices of spring), op. 410. cf The beautiful
blue Danube, op. 314 (London STS 15269).
Voices of spring, op. 410. cf Works, selections (Decca DKPA 513/4).
Voices of spring, op. 410. cf Works, selections (Decca DPA 549/
50).
Frühlingsstimmen, op. 410. cf An der schönen blauen Donau, op.
314 (London SPC 21144).
Frühlingsstimmen, op. 410. cf Court Opera Classics CO 342.
Voices of spring, op. 410. cf HMV RLS 717.
Geschichten aus dem Wienerwald (Tales from the Vienna Woods), op.
325. cf An der schönen blauen Donau, op. 314.
Geschichten aus dem Wienerwald, op. 325. cf Works, selections
(Classics for Pleasure CFP 40256).
Tales from the Vienna Woods, op. 325. cf Works, selections
(Decca DKPA 513/4).
Tales from the Vienna Woods, op. 325. cf Works, selections
(Decca DPA 549/50).
Tales from the Vienna Woods, op. 325. cf Works, selections (DG
3318 062).
Geschichten aus dem Wienerwald, op. 325. cf Works, selections
(HMV ESD 7025).
Tales from the Vienna Woods, op. 325. cf Works, selections (Pye
GSGC 15024).
Geschichten aus dem Wienerwald, op. 325. cf Works, selections
(RCA SMA 7012).
Geschichten aus dem Wienerwald, op. 325. cf Works, selections
(RCA GL 25019).
Geschichten aus dem Wiener Wald, op. 325. cf Pye QS PCNH 4.
Graduation ball, op. 97. cf Die Fledermaus, op. 363: Overture
and ballet music.
The hunt. cf Works, selections (Decca DPA 549/50).
I long for Vienna. cf CBS 30070.
Im Krapfenwald, op. 336. cf Works, selections (Classics for
Pleasure CFP 40256).
Jubilee waltz. cf Works, selections (RCA SMA 7012).

Kaiser Walzer (Emperor waltz), op. 437. cf Works, selections (Decca
    DPA 549/50).
Emperor waltz, op. 437. cf Works, selections (Decca DKPA 513/4).
Emperor waltz, op. 437. cf Works, selections (DG 3318 062).
Emperor waltz, op. 437. cf Works, selections (HMV ASD 3132).
Kaiser Walzer, op. 437. cf Works, selections (HMV EDS 7025).
Kaiser Walzer, op. 437. cf Works, selections (RCA GL 25019).
Emperor waltz, op. 437. cf Works, selections (Saga 5408).
Kreuzfidel, op. 301. cf Works, selections (RCA SMA 7012).
Künstlerleben (Artist's life), op. 316. cf An der schönen blauen
    Donau, op. 314.
Künstlerleben, op. 316. cf Banditen, op. 378.
Artist's life, op. 316. cf Works, selections (Decca DKPA 513/4).
Artist's life, op. 316. cf Works, selections (Decca DPA 549/50).
Leichtes Blut, op. 319. cf Works, selections (Classics for Plea-
    sure CFP 40048).
Leichtes Blut, op. 319. cf Works, selections (Decca SXL 6740).
Leichtes Blut, op. 319. cf Works, selections (Telefunken 4-41306).
Liebeslieder, op. 114. cf Works, selections (Decca SXL 6740).
Loreley-Rheinklange, op. 154. cf Saga 5421.
Man lebt nur einmal, op. 167. cf Saga 5421.
Morgenblätter (Morning papers), op. 279. cf Works, selections
    (Decca DKPA 513/4).
Morgenblätter, op. 279. cf Works, selections (Classics for Plea-
    sure CFP 40256).
Morgenblätter, op. 279. cf Works, selections (RCA SMA 7012).
Eine Nacht in Venedig (A night in Venice): Ach, wie so herrlich
    zu shau'n. cf Angel S 37108.
Neue pizzicato polka, op. 49. cf Works, selections (Decca DPA
    549/50).
Perpetuum Mobile, op. 257. cf The beautiful blue Danube, op. 314.
Perpetuum Mobile, op. 257. cf Works, selections (Classics for
    Pleasure CFP 40048).
Perpetuum Mobile, op. 257. cf Works, selections (Decca DPA 549/50).
Perpetuum Mobile, op. 257. cf Works, selections (Decca SXL 6740).
Perpetuum Mobile, op. 257. cf Works, selections (Pye GSGC 15024).
Perpetuum Mobile, op. 257. cf Works, selections (Telefunken
    4-41306).
Persian march, op. 289. cf Works, selections (Classics for Plea-
    sure CFP 40048).
Persian march, op. 289. cf Works, selections (Telefunken 4-41306).
Ritter Pasman, op. 441: Csardas. cf Works, selections (Decca SXL
    6740).
Rosen aus dem Süden (Roses from the South), op. 388. cf An der
    schönen blauen Donau, op. 314.
Rosen aus dem Süden, op. 388. cf Works, selections (Classics for
    Pleasure CFP 40256).
Roses from the South, op. 388. cf Works, selections (Decca DKPA
    513/4).
Roses from the South, op. 388. cf Works, selections (Decca DPA
    549/50).
Rosen aus dem Süden, op. 388. cf Works, selections (HMV ESD 7025).
Roses from the South, op. 388. cf Works, selections (Telefunken
    4-41306).
Sinngedichte, op. 1. cf Saga 5421.
Stadt und Land, op. 322. cf Works, selections (Decca SXL 6740).

Ein Tausend und eine Nacht (A thousand and one nights), op. 346.
cf Works, selections (Decca DKPA 513/4).

A thousand and one nights, op. 346. cf Works, selections (Decca
DPA 549/50).

A thousand and one nights, op. 346: Intermezzo. cf Works, selections (Saga 5408).

Tritsch-Tratsch, op. 214. cf Works, selections (Classics for
Pleasure CFP 40048).

Tritsch-Tratsch, op. 214. cf Works, selections (DG 3318 062).

Tritsch-Tratsch, op. 214. cf Works, selections (HMV ASD 3132).

Tritsch-Tratsch, op. 214. cf Works, selections (Saga 5408).

Tritsch-Tratsch, op. 214. cf Works, selections (Telefunken
4-41306).

Unter Donner und Blitz (Thunder and lightning), op. 324. cf
Works, selections (Classics for Pleasure CFP 40048).

Thunder and lightning, op. 324. cf Works, selections (Decca
DPA 549/50).

Thunder and lightning, op. 324. cf Works, selections (Saga
5408).

Thunder and lightning, op. 324. cf Angel S 37231.

2455 Valses de Vienne (arr. Korngold, Bittner, Cools). Mady Mesplé,
Christiane Stutzmann, Arta Verlen, s; Philippe Gaudin, t;
Bernard Sinclair, Jacques Loreau, bar; René Duclos Choir;
Paris Opéra-Comique Orchestra; Jean Doussard. Connoisseur
Society CS 2-2106 (2).

+ARG 11-76 p38   +ON 12-4-76 p60
+HF 10-76 p118   +SFC 11-7-76 p33
+-MJ 12-76 p28   +St 12-76 p147

Vergnugungzug, op. 281. cf Works, selections (Decca SXL 6740).

Waldmeister overture. cf Works, selections (Decca SXL 6740).

Waldmeister overture. cf Supraphon 110 1638.

Wein, Weib und Gesang (Wine, women and song), op. 333. cf Works,
selections (Decca DKPA 513/4).

Wine, women and song, op. 333. cf Works, selections (Decca DPA
549/50).

Wein, Weib und Gesang, op. 333. cf Works, selections (HMV ESD
7025).

Wein, Weib und Gesang, op. 333. cf Works, selections (RCA GL
25019).

Wein, Weib und Gesang, op. 333. cf HMV RLS 717.

Wine, women and song, op. 333. cf L'Oiseau-Lyre DSLO 7.

Wiener Blut (Vienna blood), op. 354. cf Works, selections (Decca
DKPA 513/4).

Vienna blood, op. 354. cf Works, selections (Decca DPA 549/50).

Wiener Blut, op. 354. cf Works, selections (HMV ASD 7025).

Wiener Blut, op. 354. cf Works, selections (RCA GL 25019).

Wiener Blut, op. 354. cf Works, selections (Telefunken 4-41306).

Wiener Blut, op. 354. cf LANNER: Dornbacher Ländler, op. 9.

Wiener Blut, op. 354: Ich bin ein echtes Wiener Blut. cf HMV CSD
3748.

Wiener Bonbons (Vienna bonbons), op. 307. cf Works, selections
(Decca DKPA 513/4).

Wiener Bonbons, op. 307. cf Banditen, op. 378.

Wiener Bonbons, op. 307. cf LANNER: Dornbacher Ländler, op. 9.

2456 Works, selections: Annen polka, op. 117. Champagne polka, op.
211. Eljen a Magyar, op. 332. Die Fledermaus: Overture.

Leichtes Blut, op. 319. Persian march, op. 289. Perpetuum
Mobile, op. 257. Tritsch-Tratsch, op. 214. Unter Donner und
Blitz, op. 324. Der Zigeunerbaron: Overture. LPO; Theodor
Guschlbauer. Classics for Pleasure CFP 40048. Tape (c) TC
CFP 40048.
    ++HFN 12-76 p153 tape   ++RR 8-73 p54

2457 Works, selections: Banditen, op. 378. Egyptian march, op. 335.
Geschichten aus dem Wienerwald, op. 325. Im Krapfenwald, op.
336. Morgenblätter, op. 279. Rosen aus dem Süden, op. 388.
Der Zigeunerbaron, op. 420: March. STRAUSS, J.II/Josef:
Pizzicato polka. STRAUSS, Josef: Eingesendent, op. 240.
STRAUSS, Eduard: Bahn frei, op. 45. Hallé Orchestra; James
Loughran. Classics for Pleasure CFP 40256.
    +Gr 11-76 p894   +-RR 7-76 p63
    +HFN 7-76 p101

2458 Works, selections: An der schönen blauen Donau, op. 314. Artist's
life, op. 316. Emperor waltz, op. 437. Die Fledermaus, op.
363: Du und du. Morning papers, op. 279. Tales from the
Vienna Woods, op. 325. A thousand and one nights, op. 346.
Roses from the South, op. 388. Vienna bonbons, op. 307.
Vienna blood, op. 354. Voices of spring, op. 410. Wein, Weib
und Gesang, op. 333. VPO; Willi Boskovsky. Decca DKPA 513/4.
Tape (c) KSDC 513/4.
    ++Gr 1-76 p1244   +RR 10-75 p57

2459 Works, selections: Artist's life, op. 316. The beautiful blue
Danube, op. 314. Du und du, op. 367. Egyptian march, op. 335.
Emperor waltz, op. 437. Explosions, op. 43. Die Fledermaus,
op. 363: Overture. The hunt. Neue pizzicato polka, op. 49.
Perpetuum Mobile, op. 257. Roses from the South, op. 388.
Tales from the Vienna Woods, op. 325. Thunder and lightning,
op. 324. Vienna blood, op. 354. Voices of spring, op. 410.
Wine, women and song, op. 333. A thousand and one nights, op.
346. STRAUSS, J. II/Josef: Pizzicato polka. VPO; Willi
Boskovsky. Decca DPA 549/50 (2). Tape (c) KDPC 549/50. (Re-
issues).
    +HFN 11-76 p173   +RR 10-76 p68
    +HFN 12-76 p155 tape

2460 Works, selections: Annen polka, op. 117. Auf der Jagd, op. 373.
Bei uns zu Haus, op. 361. Leichtes Blut, op. 319. Liebes-
lieder, op. 114. Explosions, op. 43. Ritter Pasman, op. 441:
Csardas. Perpetuum Mobile, op. 257. Stadt und Land, op. 322.
Vergnugungzug, op. 281. Waldmeister overture. STRAUSS, J. I:
Radetzky march, op. 228. VPO; Willi Boskovsky. Decca SXL
6740. Tape (c) KSXC 6740.
    ++Audio 12-76 p93   ++HFN 1-76 p125 tape
    ++Gr 1-76 p1253   ++RR 12-75 p62
    ++HFN 1-76 p121   ++RR 4-76 p82 tape

2461 Works, selections: Annen polka, op. 117. The beautiful blue
Danube, op. 314. Eljen a Magyar, op. 332. Emperor waltz,
op. 437. Die Fledermaus, op. 363: Overture. Tales from the
Vienna Woods, op. 325. Tritsch-Tratsch, op. 214. STRAUSS, J.
I: Radetzky march, op. 228. Berlin Radio Symphony Orchestra;
Ferenc Fricsay. DG Tape (c) 3318 062.
    +-RR 4-76 p81 tape

2462 Works, selections: Annen polka, op. 117. The beautiful blue
Danube, op. 314. Emperor waltz, op. 437. The gipsy baron,

op. 420: Overture. Die Fledermaus, op. 363: Overture. Tritsch-
Tratsch, op. 214. BPhO; Herbert von Karajan. HMV ASD 3132.
(also Angel S 37144).
+HFN 8-76 p87   ++RR 8-76 p53
+NR 11-76 p4

2463 Works, selections: An der schönen blauen Donau, op. 314. Geschichten aus dem Wienerwald, op. 325. Kaiser Walzer, op. 437, Rosen
aus dem Süden, op. 388. Wein, Weib und Gesang, op. 333.
Weiner Blut, op. 354. Johann Strauss Orchestra, Vienna; Willi
Boskovsky. HMV ESD 7025. (Reissues from TWO 368, 389, SLS
5017).
+Gr 12-76 p1072   +RR 12-76 p65

2464 Works, selections: An der schönen Blauen Donau, op. 314. Frühlingsstimmen, op. 410. Geschichten aus dem Wienerwald, op.
325. Kaiserwälzer, op. 437. Künstlerleben, op. 316. Morgenblätter, op. 279. Rosen aus dem Süden, op. 388. Wein, Weib
und Gesang, op. 333. Wiener Blut, op. 354. VSO; Robert Stolz.
Olympic 8132. Tape (c) 8132.
+HF 11-76 p153 tape   +NR 6-75 p3

2465 Works, selections: Annen polka, op. 117. The beautiful blue
Danube, op. 314. Die Fledermaus, op. 363: Overture. Perpe-
tuum Mobile, op. 257. Tales from the Vienna Woods, op. 325.
Der Zigeunerbaron, op. 420: Overture. STRAUSS, J.I.: Radetzky
march, op. 228. STRAUSS, J. II/Josef: Pizzicato polka. Hallé
Orchestra; John Barbirolli. Pye GSGC 15024. Tape (c) ZCCCB
15024. (Reissue from CCL 30130).
+-Gr 2-76 p1382   +RR 7-76 p83 tape
+RR 12-75 p67

2466 Works, selections: An der schönen blauen Donau, op. 314. Bouquet
Quadrille, op. 135. Bijouterie Quadrille, op. 169. Es gibt
nur a Kaiserstadt, es gibt nur ein Wien, op. 291. Freikugeln,
op. 326. Geschichten aus dem Wienerwald, op. 325. Kreuzfidel,
op. 301. Jubilee waltz. Morgenblätter, op. 279. STRAUSS, J.
II/Josef: Pizzicato polka. BPO; Arthur Fiedler. RCA SMA
7012. (Reissue from HMV CLP 1040).
+Gr 1-76 p1253   +RR 12-75 p62
+HFN 12-75 p165

2467 Works, selections: An der schönen blauen Donau, op. 314. Geschich-
ten aus dem Wienerwald, op. 325. Kaiser Walzer, op. 437. Wein,
Weib und Gesang, op. 333. Wiener Blut, op. 354. VSO; Jascha
Horenstein. RCA GL 25019. Tape (c) GK 25019.
+-Gr 10-76 p661   /RR 10-76 p68
+HFN 12-76 p148   +RR 12-76 p104 tape
+-HFN 12-76 p153

2468 Works, selections: Annen polka, op. 117. The beautiful blue
Danube, op. 314. Emperor waltz, op. 437. Die Fledermaus,
op. 363: Overture. A thousand and one nights, op. 346: Intermezzo. Thunder and lightning, op, 324. Tritsch-Tratsch, op.
214. Der Zigeunerbaron, op. 420: Overture. STRAUSS, J. II/
Josef: Pizzicato polka. STRAUSS, J. I: Radetzky march, op.
228. VSO; Robert Stolz. Saga 5408.
+-Gr 1-76 p1253   +RR 12-75 p52
+HFN 12-75 p165

2469 Works, selections: Accelerations waltz, op. 234. Annen polka, op.
117. Egyptian march, op. 335. Leichtes Blut, op. 319. Roses
from the South, op. 388. Persian march, op. 289. Perpetuum

Mobile, op. 257. Tritsch-Tratsch. op. 214. Wiener Blut, op.
354. Bamberg Symphony Orchestra; Joseph Keilberth. Tele-
funken Tape (c) 4-41306.
+RR 4-76 p81 tape
Der Zigeunerbaron (The gipsy baron), op. 420: Als flotter Geist;
Ja, das Schreiben und das Lesen; Doch treu und wahr; O habet
Acht; Ei, ei, er lacht; Wer uns getraut; Her die Hand; Nach
Wien; Entr'act, Act 3. cf Die Fledermaus, op. 363 (Classics
for Pleasure CFP 40251).
Der Zigeunerbaron (The gypsy baron), op. 420: Einzugsmarsch. cf
DG 2721 077 (2).
Der Zigeunerbaron, op. 420: March. cf Works, selections (Classics
for Pleasure CFP 40256).
Der Zigeunerbaron: Overture. cf Works, selections (Classics for
Pleasure CFP 40048).
The gipsy baron, op. 420: Overture. cf Works, selections (HMV
ASD 3132).
Der Zigeunerbaron, op. 420: Overture. cf Works, selections (Pye
GSGC 15024).
Der Zigeunerbaron, op. 420: Overture. cf Works, selections (Saga
5408).
STRAUSS, Johann II/Josef
Pizzicato polka. cf STRAUSS, J. II: The beautiful blue Danube,
op. 314.
Pizzicato polka. cf STRAUSS, J. II: Works, selections (Classics
for Pleasure CFP 40256).
Pizzicato polka. cf STRAUSS, J. II: Works, selections (Decca
DPA 549/50).
Pizzicato polka. cf STRAUSS, J. II: Works, selections (Pye
GSGC 15024).
Pizzicato polka. cf STRAUSS, J. II: Works, selections (RCA SMA
7012).
Pizzicato polka. cf STRAUSS, J. II: Works, selections (Saga 5408).
STRAUSS, Josef
Delirien, op. 212. cf Works, selections (Decca SXL 6817).
Delirien, op. 212. cf STRAUSS, J. II: Banditen.
Delirien, op. 212. cf STRAUSS, J. II: The beautiful blue Danube,
op. 314.
Dorfschwalben aus Oesterreich (Village swallows from Austria), op.
164. cf Works, selections (Decca SXL 6817).
Dorfschwalben aus Osterreich, op. 164. cf Court Opera Classics
CO 342.
Village swallows from Austria, op. 164. cf Decca KCSP 367.
Eingesendet, op. 240. cf STRAUSS, J. II: Works, selections
(Classics for Pleasure CFP 40256).
Feuerfest, op. 269. cf Works, selections (Decca SXL 6817).
Galoppin-polka, op. 237. cf Saga 5421.
Heiterer Mut, op. 281. cf Works, selections (Decca SXL 6817).
Heiterer Mut, op. 281. cf STRAUSS, J. II: Banditen, op. 378.
Jokey, op. 278. cf Works, selections (Decca SXL 6817).
Im Fluge, op. 230. cf Works, selections (Decca SXL 6817).
Mein Lebenslauf ist Lieb und Lust, op. 263. cf Works, selections
(Decca SXL 6817).
Mein Lebenslauf ist Lieb und Lust, op. 263. cf LANNER: Dornbacher
Ländler, op. 9.
Mein Lebenslauf ist Lieb und Lust, op. 263. cf STRAUSS, J. II:
Banditen, op 378.

Ohne Sorgen, op. 271. cf STRAUSS, J. II: The beautiful blue
  Danube, op. 314.
Ohne Sorgen, op. 271. cf Works, selections (Decca SXL 6817).
Ohne Sorgen, op. 271. cf STRAUSS, J. II: Banditen, op. 378.
Plappermäulchen, op. 245. cf Works, selections (Decca SXL 6817).
Plappermäulchen, op. 245. cf STRAUSS, J. II: Banditen, op. 378.
Transaktionen, op. 184. cf Works, selections (Decca SXL 6817).
Transaktionen, op. 184. cf STRAUSS, J. II: The beautiful blue
  Danube, op. 314.
2470 Works, selections: Delirien, op. 212. Dorfschwalben aus Oester-
  reich, op. 164. Feuerfest, op. 269. Heiterer Mut, op. 281.
  Jokey, op. 278. Im Fluge, op. 230. Mein Lebenslauf ist Lieb
  und Lust, op. 263. Plappermäulchen, op. 245. Ohne Sorgen, op.
  271. Transaktionen, op. 184. VPO; Willi Boskovsky. Decca
  SXL 6817. Tape (c) KSXC 6817.
          ++Gr 12-76 p1075   +RR 12-76 p65
          +HFN 12-76 p148

STRAUSS, Oscar
Rund und die Liebe: Hans' songs. cf KALMAN: Heut nacht hab ich
  getraumt von dir.
Valses: Saison d'amour, Je ne suis pas, Je t'aime. cf London OS
  26248.
A waltz dream: Love song of May. cf CBS 30070.

STRAUSS, Richard
2471 Der Abend, op. 34, no. 1. Deutsche Motette, op. 62. Hymne, op. 34,
  no. 2. Jessica Cash, s; Jean Temperley, ms; Wynford Evans, t;
  Stephen Varcoe, bar; Schütz Choir; Roger Norrington. Argo
  ZRG 803.
          ++Audio 2-76 p96   ++NYT 9-21-75 pD18
          +Gr 4-75 p1856   ++RR 5-75 p67
          +-HF 1-76 p97   ++SFC 8-24-75 p28
          +-MT 12-75 p1072   ++STL 5-4-75 p37
          +NR 10-75 p6
2472 An Alpine symphony, op. 64. Dresden Staatskapelle; Rudolf Kempe.
  HMV SQ ASD 3173. Tape TC ASD 3173. (Reissue from SLS 861).
          +Gr 4-76 p1611   ++HFN 6-76 p105 tape
          +Gr 6-76 p102 tape   ++RR 3-76 p49
          +HFN 4-76 p121 Quad
2473 An Alpine symphony, op. 64. LAPO; Zubin Mehta. London CS 6981.
  Tape (c) CS5 6981. (also Decca SXL 6752 Tape (c) KSXC 6752).
          +-Gr 4-76 p1611   +-MT 12-76 p1006
          +-Gr 6-76 p102 tape   +NR 8-76 p8
          +HF 11-76 p126   +RR 5-76 p51
          +HFN 4-76 p115   ++SFC 6-20-76 p26
          ++HFN 7-76 p104 tape   +St 11-76 p158
2474 Also sprach Zarathustra, op. 30. CSO; Fritz Reiner. Camden CCV
  5040.
          +HFN 4-76 p121   +RR 5-76 p51
2475 Also sprach Zarathustra, op. 30. Till Eulenspiegels lustige
  Streiche, op. 28. Salome, op. 54: Dance of the seven veils.
  Don Juan, op. 20. Der Rosenkavalier, op. 59: Waltz. VPO;
  Herbert von Karajan. Decca DPA 543/4. Tape (c) KDPC 543/4.
          +Gr 9-76 p483   +RR 8-76 p53
          +-HFN 8-76 p93
2476 Also sprach Zarathustra, op. 30. Don Juan, op. 20.* Till Eulenspiegels lustige Streiche, op. 28. Samuel Magad, vln; CSO;

Georg Solti. Decca SXL 6749. Tape (c) KSXC 6749. (also London CS 6978 Tape (c) 5-6978 (ct) 8-6978) (Reissue from CS 6800).

+Gr 12-75 p1053	+MJ 12-76 p28
+Gr 6-76 p102 tape	+NR 7-76 p2
+HF 8-76 p95	+RR 12-75 p62
+HFN 1-76 p121	++SFC 3-21-76 p28
++HFN 2-76 p116 tape	++St 8-76 p99

2477 Also sprach Zarathustra, op. 30. Till Eulenspiegels lustige Streiche, op. 28. VPO; Clemens Krauss. London R 23208.

+-NR 12-76 p6

2478 Also sprach Zarathustra, op. 30. PO; Eugene Ormandy. RCA ARL 1-1120. Tape (c) ARK 1-1120 (ct) ARS 1-1120 (Q) ARD 1-1120 Tape (c) ART 1-1120.

+-HF 3-76 p96	++SFC 12-7-75 p31
++NR 1-76 p5	++St 6-76 p110

2479 Also sprach Zarathustra, op. 30. St. Louis Symphony Orchestra; Walter Susskind. Turnabout (Q) QTVS 34584. Tape (c) SMG 8T 145.

+HF 9-75 p92	+-SFC 8-3-75 p30
+HF 1-76 p111	+St 9-75 p114
+NR 8-75 p2	

Also sprach Zarathustra, op. 30. cf Works, selections (Vanguard SRV 325/9).

Arabella, op. 79: Der Richtige. cf Decca GOSE 677/79.

2480 Ariadne auf Naxos, op. 60. Erna Berger, Viorica Ursuleac, Meliza Korjus, s; Helge Roswänge, t; Stuttgart Radio Orchestra; Clemens Krauss. BASF KBF 21806 (2).

+-HF 6-75 p104	+SFC 3-30-75 p16
+NR 6-75 p11	+-St 8-75 p105
+ON 1-10-76 p33	

2481 Ariadne auf Naxos, op. 60: Es gibt ein Reich. Capriccio, op. 85: Wo ist mein Bruder. Songs: Einerlei, op. 69, no. 3; Befreit, op. 39, no. 4; Ich wollt ein Sträusslein binden, op. 68, no. 2; Schlechtes Wetter, op. 69, no. 5; Vier letze Lieder, op. posth. Lisa Della Casa, s; Franz Bierbach, bs; Karl Hudez, pno; VPO; Heinrich Hollreiser, Karl Böhm. Decca ECM 778. (Reissues from LXT 5017, 5258, LW 5056).

+-Gr 3-76 p1507	+RR 2-76 p22
+-HFN 3-76 p112	

Le bourgeois gentilhomme, op. 60. cf Works, selections (Vanguard SRV 325/9).

Burleske, D minor. cf Works, selections (HMV SQ SLS 5067).

2482 Capriccio, op. 85: Monolog des Theaterdirektors. Guntram, op. 25: Prelude, Act 1. Der Rosenkavalier, op. 59: Waltzes. Die schweigsame Frau, op. 80: Schlusszene. TCHAIKOVSKY: Eugene Onegin, op. 24: Gremins aria. Liebtrabant: Overture. The Voyevode: Overture. Kurt Böhme, bs; Southwest German Radio Symphony Orchestra; Otto Ackermann. Discophilia OAA 102.

+NR 11-76 p3

Capriccio, op. 85: Wo ist mein Bruder. cf Ariadne auf Naxos, op. 60: Es gibt ein Reich.

Concertino, clarinet, bassoon, harp and string orchestra. cf Works, selections (HMV SQ SLS 5067).

Concertino, clarinet, bassoon, harp and string orchestra. cf Hungaroton SLPX 11748.

2483 Concerto, horn, no. 1, op. 11, E flat major. Concerto, horn, no.

2, op. 11, E flat major. Peter Damm, hn; Dresden Staatskapelle;
   Rudolf Kempe. Angel S 37004.
         +ARG 12-76 p42   +NR 11-76 p5
         +HF 11-76 p128

Concerto, horn, no. 1, op. 11, E flat major. cf Works, selections
   (HMV SQ SLS 5067).

Concerto, horn, no. 2, op. 11, E flat major. cf Concerto, horn,
   no. 1, op. 11, E flat major.

Concerto, horn, no. 2, op. 11, E flat major. cf Works, selections
   (HMV SQ SLS 5067).

Concerto, oboe, D major. cf Works, selections (HMV SQ SLS 5067).

Concerto, violin, op. 8, D minor. cf Works, selections (HMV SQ
   SLS 5067).

2484 Daphne, op. 82. Rose Bampton, s; Anton Dermota, Set Svanholm, t;
         Ludwig Weber, bs; Teatro Colon Orchestra and Chorus; Erich
         Kleiber. Bruno Walter Society IGI 295 (2).
              +-NR 10-76 p10

Deutsche Motette, op. 62. cf Der Abend, op. 34, no. 1.

2485 Don Juan, op. 20. Festival prelude, op. 61. Salome, op. 54:
         Salome's dance. Till Eulenspiegels lustige Streich, op. 28.
         BPhO; Karl Böhm. DG 2535 208. (Reissue from SLPM 138866).
              +Gr 12-76 p1002   +-RR 12-76 p65
              +HFN 12-76 p151

2486 Don Juan, op. 20. Der Rosenkavalier, op. 59: First and second
         waltz sequence. Till Eulenspiegels lustige Streiche, op. 28.
         COA; Eugen Jochum. Philips 6580 129. (Reissue from SABL 201).
              +-Gr 8-76 p299   +-RR 7-76 p63

2487 Don Juan, op. 20. Der Rosenkavalier, op. 59: Suite. Till Eulen-
         spiegels lustige Streiche, op. 28. PO; Eugene Ormandy. RCA
         ARL 1-1408. Tape (c) ARK 1-1408 (ct) ARS 1-1408.
              +-NR 7-76 p2   /SFC 6-6-76 p33

Don Juan, op. 20. cf Also sprach Zarathustra, op. 30 (Decca 543/4).

Don Juan, op. 20. cf Also sprach Zarathustra, op. 30 (Decca SXL
   6749).

Don Juan, op. 20. cf Works, selections (Vanguard SRV 325/9).

Don Juan, op. 20. cf GRIEG: Peer Gynt, op. 46: Suite, no. 1.

2488 Don Quixote, op. 35. Ulrich Koch, vla; Mstislav Rostropovich, vlc;
         BPhO; Herbert von Karajan. Angel S 37057. Tape (c) 4XS 37057.
         (also HMV SQ ASD 3118 Tape (c) TC ASD 3118).
              ++Gr 6-76 p52   +-NR 6-76 p5
              +HF 8-76 p96   +-RR 6-76 p54
              +-HFN 6-76 p97   ++SFC 6-6-76 p33
              ++HFN 8-76 p95

2489 Don Quixote, op. 35. Pierre Fournier, vlc; Giusto Cappone, vla;
         BPhO; Herbert von Karajan. DG 2535 195. (Reissue from 139009).
              ++Gr 8-76 p299   +RR 8-76 p53

Fanfare für die Wien Philharmoniker. cf IVES: March Mega Lambda
   Chi.

Festival prelude, op. 61. cf Don Juan, op. 20.

Die Frau ohen Schatten, op. 65: Ich, mein Gebieter. cf Decca
   GOSE 677/79.

Guntram, op. 25: Prelude, Act 1. cf Caprice, op. 85: Monolog des
   Theaterdirektors.

2490 Ein Heldenleben, op. 40. Michel Schwalbé, vln; BPhO; Herbert von
         Karajan. DG 2535 194. (Reissue from 138025).
              ++Gr 8-76 p299   +RR 8-76 p53

2491 Ein Heldenleben, op. 40. BPhO; Herbert von Karajan. HMV (Q) ASD
      3126. (also Angel (Q) S 37060).
          ++Gr 10-75 p633   -NR 12-75 p4
          +-HF 5-76 p95   +-RR 9-75 p48
          +HFN 9-75 p105
2492 Ein Heldenleben, op. 40. NYP; Willem Mengelberg. RCA SMA 7001.
      (Reissue from HMV D 1711/5).
          +-Gr 10-75 p634   +St 2-76 p737
          +RR 8-75 p43
2493 Ein Heldenleben, op. 40. VPO; Clemens Krauss. Richmond R 23209.
      (Reissue).
          +-ARG 12-76 p41
      Ein Heldenleben, op. 40. cf Works, selections (Vanguard SRV 325/9).
      Hymne, op. 34, no. 2. cf Der Abend, op. 34, no. 1.
      Panathenäenzug, op. 74. cf Works, selections (HMV SQ SLS 5067).
      Parergon to Symphonia domestica, op. 73. cf Works, selections
      (HMV SQ SLS 5067).
2494 Quartet, piano, op. 13, C minor. Los Angeles String Trio; Irma
      Vallecillo, pno. Desmar DSM 1002.
          -Gr 2-76 p1356   +RR 2-76 p46
          -HF 5-76 p96   +SR 3-6-76 p41
          ++MJ 12-75 p38   +St 1-76 p106
          ++NR 12-75 p9
      Quartet, piano, op. 13, C minor. cf REGER: Quartet, piano, op.
      133, A minor.
2495 Der Rosenkavalier, op. 59. Elisabeth Schwarzkopf, Ljuba Welitsch,
      s; Christa Ludwig, ms; Teresa Stich-Randall, con; PhO; Herbert
      von Karajan. Angel S 3563 (4). (also HMV SLS 810 Tape (c)
      TC SLS 810).
          +HFN 8-76 p94 tape   +RR 11-76 p110 tape
          +Op 12-71 p1088
2496 Der Rosenkavalier, op. 59. Regine Crespin, Helen Donath, Emmy
      Loose, s; Yvonne Minton, Anne Howells, ms; Murray Dickie,
      Luciano Pavarotti, t; Otto Wiener, bar; Manfred Jungwirth,
      Herbert Lachner, bs; VPO; Vienna Staatsoper Chorus; Georg
      Solti. London OSA 1435. Tape (c) 131165 (r) 90165. (also
      Decca Tape (c) K3N 23).
          +Gr 8-76 p341 tape   +-RR 8-76 p82 tape
          +HF 4-71 p68   ++St 6-72 p109 tape
          ++HFN 8-76 p94 tape
      Der Rosenkavalier, op. 59: Finale, Act 2. cf Hungaroton SLPX
      11444.
      Der Rosenkavalier, op. 59: First and second waltz sequence. cf
      Don Juan, op. 20.
      Der Rosenkavalier, op. 59: Presentation of the rose; Final trio
      and duet. cf Decca GOSE 677/79.
      Der Rosenkavalier, op. 59: Suite. cf Don Juan, op. 20.
      Der Rosenkavalier, op. 59: Waltz. cf Also sprach Zarathustra,
      op. 30.
      Der Rosenkavalier, op. 59: Waltzes. cf Capriccio, op. 85: Mono-
      log des Theaterdirektors.
2497 Salome, op. 54. Margarete Klose, Astrid Varnay, s; Julius Patzak,
      Hans Hopf, t; Hans Braun, bar; Bavarian Radio Orchestra;
      Hermann Weigert. Bruno Walter Society IGI 289 (2).
          ++NR 10-76 p10

Salome, op. 54: Closing scene. cf Decca GOSE 677/9.

Salome, op. 54: Dance of the seven veils. cf Also sprach Zarathustra, op. 30.

Salome, op. 54: Final scene. cf BERG: Lulu: Suite.

Salome, op. 54: Salome's dance. cf Don Juan, op. 20.

Schlagobers, op. 70: Waltz. cf Works, selections (Vanguard SRV 325/9).

Die schweigsame Frau, op. 80: Schlusszene. cf Capriccio, op. 85: Monolog des Theaterdirektors.

Sonata, violin and piano, op. 18, E flat major. cf FRANCK: Sonata, violin and piano, A major.

Sonata, violin and piano, op. 18, E flat major. cf Columbia M2 33444.

Sonata, violin and piano, op. 18, E flat major. cf RCA ARM 4-0942/7.

Sonata, violoncello and piano, op. 6. cf BRUCH: Kol Nidrei, op. 47.

2498   Songs: Als mir dein Lied erklang, op. 68, no. 4; Befreit, op. 39, no. 4; Einerlei, op. 69, no. 3; Freundliche Vision, op. 48, no. 1; Heimkehr, op. 15, no. 5; Ich wollt ein Sträusslein binden, op. 68, no. 2; Mienem Kinde, op. 37, no. 3; Die Nacht, op. 10, no. 3; Säusle, liebe Myrte, op. 68, no. 3; Schlagende Herzen, op. 29, no. 2; Schlechtes Wetter, op. 69, no. 5; Der Stern, op. 69, no. 1; Wie sollten wir geheim sie halten, op. 19, no. 4. Hilde Gueden, s; Friedrich Gulda, pno. London R 23212.
+NR 11-76 p11

2499   Songs: Ach weh mir unglückhaftem Mann, op. 21, no. 4; All mein Gedanken, op. 21, no. 1; Breit über mein Haupt, op. 19, no. 2; Freundliche Vision, op. 48, no. 1; Heimliche Aufforderung, op. 27, no. 3; Ich liebe dich, op. 37, no. 2; Mein Auge, op. 37, no. 4; Morgan, op. 27, no. 4; Die Nacht, op. 10, no. 3; Nachtgang, op. 29, no. 3; Nichts, op. 10, no. 2; Ruhe meine Seele, op. 27, no. 1; Ständchen, op. 17, no. 2; Traum durch die Dämmerung, op. 29, no. 1; Wie sollten wir gehiem sie halten, op. 19, no. 4; Wozu nach Mädchen, op. 19, no. 1; Zueignung, op. 10, no. 1. Gérard Souzay, bar; Dalton Baldwin, pno. Philips 6580 143. (Reissue from SAL 3483).
+Gr 10-76 p639   +RR 12-76 p95
+HFN 11-76 p173

2500   Songs: Amor, op. 68, no. 5; Einkehr, op. 47, no. 2; Heimkehr, op. 15, no. 5; Ich schwebe, op. 48, no. 2; Ich wollt ein Sträusslein binden, op. 68, no. 2; Säusle liebe Myrte, op. 68, no. 3; Schlagende Herzen, op. 29, no. 2; Der Stern, op. 69, no. 1. WOLF: Songs: Eichendorff Lieder: Waldmädchen; Verschwiegene Liebe. Goethe Lieder: Die Bekehrte; Epiphanias; Die Spröde. Morike Lieder: Schlafendes Jesuskind; Zum neuen Jahre. Spanish songbook: Ach des Knaben Augen; Die ihr schwebet; Nun wandre, Maria. Judith Blegen, s; Martin Katz, pno. RCA ARL 1-1571.
+Gr 9-76 p466   +RR 10-76 p99
+HFN 10-76 p177   ++St 11-76 p164
+ON 12-4-76 p60

Songs: Ach weh mir unglückhaftem Mann, op. 21, no. 4; All mein Gedanken, mein Herz und mein Sinn, op. 21, no. 1; Du meines Herzens Krönelein, op. 21, no. 2; Gefunden, op. 56, no. 1; Himmelsboten, op. 32, no. 5; Nachtgang, op. 29, no. 3. cf BRAHMS: Songs (Decca SXL 6738).

Songs: Allerseelen, op. 10, no. 8; Befreit, op. 39, no. 4;
    Cäcilie, op. 27, no. 2. cf Rococo 5380.
Songs: Einerlei, op. 69, no. 3; Befreit, op. 39, no. 4; Ich wollt
    ein Sträusslein binden, op. 68, no. 2; Schlechtes Wetter, op.
    69, no. 5; Vier letze Lieder, op. posth. cf Ariadne auf Naxos,
    op. 60: Es gibt ein Reich.
Songs: Liebeshymnus, op. 32, no. 3; Muttertändelie, op. 43, no. 2;
    Das Rosenband, op. 36, no. 1; Ruhe, meine Seele, op. 27, no. 1.
    cf BRAHMS: Alto rhapsody, op. 53.
Songs: Morgen, op. 27, no. 4. cf Club 99-99.
Songs: Ständchen, op. 17, no. 2. cf HMV ASD 2929.
Stadt Wien: Fanfare. cf Argo SPA 464.
Stimmungsbilder: An einsamer Quelle. cf RCA ARM 4-0942/7.
Symphonia domestica, op. 53. cf Works, selections (Vanguard SRV
    325/9).
Till Eulenspiegels lustige Streiche, op. 28. cf Also sprach
    Zarathustra, op. 30 (Decca DPA 543/4).
Till Eulenspiegels lustige Streiche, op. 28. cf Also sprach
    Zarathustra, op. 30 (Decca SXL 6749).
Till Eulenspiegels lustige Streiche, op. 28. cf Also sprach
    Zarathustra, op. 30 (RCA R 23208).
Till Eulenspiegels lustige Streiche, op. 28. cf Don Juan, op. 20
    (DG 2535 208).
Till Eulenspiegels lustige Streiche, op. 28. cf Don Juan, op. 20
    (Philips 6580 129).
Till Eulenspiegels lustige Streiche, op. 28. cf Don Juan, op. 20
    (RCA ARL 1-1408).
Till Eulenspiegels lustige Streiche, op. 28. cf Works, selections
    (Vanguard SRV 325/9).
Tod und Verklärung (Death and transfiguration), op. 24. cf Works,
    selections (Vanguard SRV 325/9).
Death and transfiguration, op. 24. cf HINDEMITH: Mathis der Maler
    symphony.
Tod und Verklärung, op. 24. cf RCA CRM 5-1900.
Vier letze Lieder: Beim Schlafengehen; Frühling; Im Abendrot;
    September. cf Rococo 5380.
2501 Works, selections: Burleske, D minor. Concertino, clarinet,
    bassoon, harp and string orchestra. Concerto, horn, no. 1,
    op. 11, E flat major. Concerto, horn, no. 2, op. 11, E flat
    major. Concerto, oboe, D major. Concerto, violin, op. 8, D
    minor. Panathenäenzug, op. 74. Parergon to Sinfonia domestica,
    op. 73. Peter Damm, hn; Ulf Hoescher, vln; Manfred Clement,
    ob; Manfred Weise, clt; Wolfgang Liebschner, bsn; Peter Rösel,
    Malcolm Frager, pno; Dresden Staatskapelle; Rudolf Kempe.
    HMV SQ SLS 5067 (4).
            +Gr 10-76 p611   +MT 12-76 p1006
            +HFN 11-76 p168   +RR 10-76 p67
2502 Works, selections: Also sprach Zarathustra, op. 30. Le bourgeois
    gentilhomme, op. 60. Don Juan, op. 20. Ein Heldenleben, op.
    40. Schlagobers, op. 70: Waltz. Symphonia domestica, op. 53.
    Teill Eulenspiegels lustige Streiche, op. 28. Tod und Verk-
    lärung, op. 24. VPO; Richard Strauss. Vanguard SRV 325/9 (5).
    (also Eterna 82620/8; Clavier CT 1501/5).
            +ARSC Vol VIII, no. 2-3 +-MJ 11-76 p44
            p123   +NR 8-76 p2
            +-Audio 10-76 p145   +SFC 7-25-76 p29
            +-HF 10-76 p118   +St 11-76 p134

STRAVINSKY, Igor
2503 Apollon musagète. Le sacre du printemps. OSR; Ernest Ansermet.
       London STS 15265.
           +-SFC 7-25-76 p29
      Apollon musagète: Variation. cf Works, selections (Crystal S 302).
2504 Le baiser de la fée (The fairy's kiss). Duo concertante. Pulci-
       nella: Suite italienne. Itzhak Perlman, vln; Bruno Canino, pno.
       Angel S 37115.
           ++Gr 9-76 p438   +NR 3-76 p6
           ++HF 5-76 p101   ++RR 9-76 p84
            +HFN 9-76 p130   ++SFC 7-25-76 p29
      The fairy's kiss. cf Works, selections (Crysal S 302).
2505 Canticum sacrum. Symphony of psalms. Christ Church Cathedral
       Choir, Oxford; Philip Jones Ensemble; Simon Preston. Argo ZRG
       799.
           ++Audio 6-76 p97   ++NR 5-76 p7
            +-HF 10-76 p120   +-St 9-76 p125
            +-MT 8-76 p663   +-Te 3-76 p28
      Canticum sacraum. cf Symphony of psalms.
      Circus polka. cf CHABRIER: España.
      Concertino, 12 instruments. cf Works, selections (DG 2530 551).
2506 Concerto, piano and wind instruments. Ebony concerto. Octet,
       wind instruments. Symphonies, wind instruments. Theo Bruins,
       pno; George Pieterson, clt; NWE; Edo de Waart. Philips 6500
       841.
            +Audio 6-76 p97   ++MT 6-76 p496
           ++Gr 3-76 p1472   ++NR 12-75 p7
           ++HF 3-76 p96   +-RR 2-76 p42
            +HFN 2-76 p113   ++SFC 12-14-75 p40
           ++MJ 2-76 p33   ++St 6-76 p111
      Concerto, 2 pianos. cf Coronet LPS 1722.
2507 Concerto, 16 wind instruments, E flat major. Octet, wind instru-
       ments. The soldier's tale: Suite. Nash Ensemble; Elgar
       Howarth. Classics for Pleasure CFP 40098.
            +Gr 2-75 p1510   ++RR 3-75 p35
           ++HFN 5-75 p140   +-ST 2-76 p739
      Duo concertante. cf Le baiser de la fée.
      Ebony concerto. cf Concerto, piano and wind instruments.
2508 The firebird (L'Oiseau de fue). Petrouchka (1947 version). The
       rite of spring: Ballet. Apropos Le sacre, recorded commentary
       by Igor Stravinsky. Columbia Symphony Orchestra; Igor Stra-
       vinsky. CBS 78307 (3). (Reissues from Philips SABL 174, 175,
       SBRG 72046).
           ++Gr 11-76 p811   ++RR 12-76 p65
           ++HFN 11-76 p170
2509 The firebird. NYP; Pierre Boulez. Columbia M 33508. (Q) MQ
       33508. (also CBS 76418 Tape (c) 40-76418).
           ++Audio 2-76 p94   ++NR 11-75 p2
            +-Gr 12-75 p1053   ++NYT 1-18-76 pD1
            +-HF 11-75 p126   +RR 12-75 p67
            +HFN 12-75 p165   -SFC 9-28-75 p30
            +HFN 1-76 p125 tape   ++St 2-76 p110
            -MJ 12-75 p38
2510 The firebird. Igor Stravkinsky, pno. Klavier KS 126. (Reissue
       from Duo-Art piano rolls).
            +-Audio 2-76 p94   *NR 10-75 p8
            /-HF 12-75 p109

2511 The firebird. LPO; Bernard Haitink. Philips 6500 483. Tape
        (c) 7300 278. (Reissue from 6747 094).
                ++Gr 11-75 p834   ++MJ 2-76 p33
                +HF 2-76 p130 tape   +NR 1-76 p4
                ++HF 9-76 p84 tape   ++RR 11-75 p55
                ++HF 6-76 p95
      The firebird. cf DEBUSSY: Prelude to the afternoon of a faun.
      L'Oiseau de feu: Berceuse, Scherzo. cf Works, selections (Crys-
        tal S 302).
      The firebird: Danse infernale. cf DG 2584 004.
2512 L'Oiseau de feu: Suite. Petrouchka. NYP; Leonard Bernstein.
        CBS 61122. Tape (c) 40-61122.
                +RR 4-76 p82 tape
2513 The firebird: Suite. Jeu de cartes. LSO; Claudio Abbado. DG
        2530 537. Tape (c) 3300 483.
                +Gr 8-75 p331   ++NYT 1-18-76 pD1
                +HFN 8-75 p84   +RR 8-75 p44
                +HFN 10-75 p155 tape   +-RR 4-76 p82 tape
                /NR 1-76 p4   ++SFC 12-14-75 p40
      L'Oiseau de feu: Suite. cf RCA CRL 3-2026.
      Fireworks, op. 4. cf CHABRIER: España.
2514 L'Histoire du soldat (The soldier's tale) (English version by
        Michael Flanders and Kitty Black). Sir John Gielgud, narrator;
        Tom Courtenay, soldier, Ron Moody, the devil; Boston Symphony
        Chamber Players. DG 2530 609. Tape (c) 3300 609.
                +-Gr 3-76 p1471   +RR 3-76 p49
                +-HF 8-76 p96   +-RR 4-76 p82
                ++HFN 3-76 p105   ++SFC 5-16-76 p28
                ++HFN 3-76 p113 tape   +-SR 11-13-76 p52
                ++NR 7-76 p2   +St 11-76 p159
2515 The soldier's tale. Jean Cocteau, narrator; Peter Ustinov, Jean-
        Marie Fertey, Anne Tonietta; Instrumental Ensemble; Igor Marke-
        vitch. Philips 6580 136. (Reissue).
                +Gr 9-76 p432   +RR 9-76 p66
                +-HFN 10-76 p183
2516 The soldier's tale. Octet, wind instruments. BSO; Leonard
        Bernstein. RCA SMA 7014. (Recorded 1947).
                -Gr 8-76 p299   -RR 6-76 p56
      L'Histoire du soldat: Suite. cf Vanguard VSD 707/8.
      The soldier's tale: Suite. cf Concerto, 16 wind instruments,
        E flat major.
      Jeu de cartes. cf The firebird: Suite.
      Mass. cf POULENC: Motets (Argo ZRG 720).
      Mavra: Chanson Russe. cf Works, selections (Crystal S 302).
      Octet, wind instruments. cf Concerto, piano and wind instruments.
      Octet, wind instruments. cf Concerto, 16 wind instruments, E
        flat major.
      Octet, wind instruments. cf The soldier's tale.
      Octet, wind instruments. cf Works, selections (DG 2530 551).
2517 Oedipus Rex. Tatiana Troyanos, ms; René Kollo, Frank Hoffmeister,
        t; Tom Krause, David Evitts, bar; Ezio Flagello, bs; Michael
        Wager, narrator; Harvard Glee Club; BSO; Leonard Bernstein.
        Columbia M 33999. (Reissue from M4X 33032). (also CBS 76380).
                +Gr 6-76 p83   +NR 6-76 p12
                +-HF 9-76 p97   +-RR 6-76 p31
                ++HFN 7-76 p97   ++SFC 5-2-76 p38
                ++MT 12-76 p1006   +-St 10-76 p133

2518 Oedipus Rex. Martha Mödl, ms; Peter Pears, Helmut Krebs, t;
      Heinz Rehfuss, bar; Otto von Rohr, bs; Jean Cocteau, narrator;
      Cologne Radio Symphony Orchestra and Chorus; Igor Stravinsky.
      Odyssey Y 33789. (Reissue from Columbia ML 4644).
              ++ARSC Vol VIII, no. 2-3   +NYT 5-16-76 pD19
                 p83   ++SFC 5-2-76 p38
              ++HF 9-76 p97   ++St 10-76 p133
      Pastorale. cf Works, selections (DG 2530 551).
      Pater noster. cf Abbey LPB 739.
2519 Petrouchka (1911 version). LSO; Charles Mackerras. Vanguard
      VSD 71177. Tape (c) ZCVSM 71177 (Q) VSQ 30021 Tape (r) VSS 23.
              +-Gr 2-74 p1566   ++RR 11-73 p53
              +-HF 8-73 p104   +-RR 2-75 p76
               +HF 5-76 p114 Quad tape   ++St 6-73 p123
              +-HFN 12-73 p2617
      Petrouchka (1947 version). cf The firebird.
      Petrouchka (1947 version). cf Philips 6780 755.
      Petrouchka. cf L'Oiseau de feu: Suite.
      Petrouchka. cf BARTOK: Out of doors.
2520 Petrouchka: 3 movements. TCHAIKOVSKY: Concerto, piano, no. 1,
      op. 23, B flat minor. Ilana Vered, pno; LSO; Kazimierz Kord.
      London SPC 21148. Tape (c) 5-21148 (ct) 8-21148. (also Decca
      PFS 4362 Tape (c) KPFC 4362).
              +Gr 4-76 p1611   ++NR 8-76 p5
              -HF 11-76 p130   +-RR 4-76 p54
              +HF 10-76 p147 tape   +-RR 12-76 p108 tape
              +-HFN 4-76 p117   ++SFC 3-7-76 p27
              +HFN 6-76 p105 tape   ++St 9-76 p128
      Petrouchka: Danse Russe. cf Works, selections (Crystal S 302).
      Pieces, solo clarinet (3). cf Hungaroton SLPX 11748.
2521 Pulcinella. Song of the nightingale. OSR; Ernest Ansermet.
      Decca ECS 776. (Reissue from SXL 2188).
              +Gr 10-76 p711   +-RR 12-76 p66
              +-HFN 11-76 p170
2522 Pulcinella. OSR; Ernest Ansermet. London STS 15218. (Reissue).
              +-NR 5-76 p2
      Pulcinella: Suite. cf BARTOK: Rumanian dances, op. 8a.
      Pulcinella: Suite italienne. cf Le baiser de la fée.
      Ragtime, 11 instruments. cf Works, selections (DG 2530 551).
2523 The rite of spring (Le sacre du printemps). CO; Pierre Boulez.
      CBS 72807. Tape (c) 40-72807.
              +HFN 1-76 p125 tape   -RR 11-75 p94 tape
2524 The rite of spring. OSCCP; Pierre Monteux. Decca ECS 750. (Re-
      issue from RCA RB 16007). (also London STS 15318).
              +-Gr 9-74 p518   +-RR 11-74 p61
              ++MJ 5-76 p28
2525 The rite of spring. VPO; Lorin Maazel. Decca SXL 6735. (also
      London 6954 Tape (c) 5-6954).
              -ARG 12-76 p42   +NR 12-76 p3
              -Gr 8-76 p299   +-RR 6-76 p56
              +HFN 6-76 p97
2526 The rite of spring. LSO; Claudio Abbado. DG 2530 635.
              +ARG 12-76 p42   +NR 11-76 p3
              +Gr 5-76 p1766   +-RR 5-76 p52
2527 Le sacre du printemps. LPO; Erich Leinsdorf. London SPC 2114.
      Tape (c) 52114 (ct) 82114.

                +--HF 5-76 p101   +--NR 3-76 p2
                +--MJ 5-76 p28   +St 5-76 p126
2528   The rite of spring. LSO; Colin Davis. Philips Tape (c) 7317 119.
                +--HFN 2-76 p116 tape
       Le sacre du printemps. cf Apollon musagète.
       The rite of spring: Ballet. cf The firebird.
       Septet. cf Works, selections (DG 2530 551).
       Serenade, A major. cf NILSSON: Quantitaten.
2529   Sonata, piano, F minor. TCHAIKOVSKY: Sonata, piano, op. 37, G
          major. Paul Crossley, pno. Philips 6500 884.
                +Gr 2-76 p1363   +NR 12-75 p13
                +HF 3-76 p97   ++RR 2-76 p53
                +HFN 3-76 p105   +SR 11-29-75 p50
                +MT 8-76 p663   +SFC 12-21-75 p42
       Song of the nightingale (Chant du rossignol). cf Pulcinella.
       Song of the nightingale. cf Symphony in three movements.
       Songs: Ave Maria; Pater noster. cf Abbey LPB 750.
       Songs: Podblioudnuia. cf Hungaroton SLPX 11762.
       Songs: La rosée sainte. cf HMV RLS 716.
       Suite italienne. cf DVORAK: Trio, piano, op. 65, F minor.
       Suite italienne. cf KABALEVSKY: Sonata, violoncello, op. 71, B
          flat major.
       Suite on themes by Pergolesi. cf Works, selections (Crystal S
          302).
       Symphonies, wind instruments. cf Concerto, piano and wind instru-
          ments.
2530   Symphony in three movements. Song of the nightingale. PhO; Con-
          stantin Silvestri. Classics for Pleasure CFP 40094. (Reissue
          from HMV ASD 401).
                +--Gr 6-75 p50   +ST 2-76 p739
                +--RR 6-75 p54
2531   Symphony of psalms. Canticum sacrum. Richard Morton, t; Marcus
          Creed, bar; Philip Jones Ensemble; Simon Preston. Argo ZRG
          799.
                ++Audio 6-76 p97   ++NR 5-76 p7
                 +Gr 10-75 p675   +RR 10-75 p90
                +--HF 10-76 p120   +--St 9-76 p125
                +--HFN 10-75 p150   +ST 2-76 p739
                +--MT 8-76 p663   +--Te 3-76 p28
2532   Works, selections: Apollon musagète: Variation. The fairy's kiss.
          Mavra: Chanson Russe. Petrouchka: Danse Russe. L'Oiseau de
          feu: Berceuse, Scherzo. Suite on themes by Pergolesi. Eudice
          Shapiro, vln; Ralph Berkowitz, pno. Crystal S 302.
                ++Audio 11-76 p108   ++NR 6-76 p7
                 +IN 9-76 p20   ++SFC 5-2-76 p38
2533   Works, selections: Concertino, 12 instruments. Octet, wind
          instruments. Pastorale. Ragtime, 11 instruments. Septet.
          Boston Symphony Chamber Players. DG 2530 551.
                ++Gr 1-76 p1215   ++NR 1-76 p8
                 +HF 3-76 p96   ++RR 11-75 p66
                ++HFN 12-75 p165   ++SFC 12-14-75 p40
                ++MT 6-76 p496   ++St 6-76 p111
STROBAEUS, Johann
       Alia Chorea polonica. cf Hungaroton SLPX 11721.
STUDER, Hans
       Invicatio, fugue and epilog. cf Pelca PRSRK 41017/20.

SUBOTNIK, Morton
2534 Until spring (created on the electric music box). Odyssey Y 34158.
          +HF 12-76 p126
SUGAR
      Choral studies: 2 songs. cf Hungaroton SLPX 11762.
SUK, Josef
      Love song. cf Seraphim S 60259.
2535 Ripening (Zrani), op. 34. Serenade, strings, op. 6, E flat major.
          Symphony, op. 27, C minor. CPhO; Václav Talich. Supraphon
          010 1441/3 (3). (Reissues from LPV 343, 5, 269/70)
              ++Gr 12-76 p1002   ++RR 6-76 p56
      Serenade, strings, op. 6, E flat major. cf Ripening, op. 34.
      Serenade, strings, op. 6, E flat major. cf DVORAK: Serenade,
          strings, op. 22, E major.
      Study, E minor. cf London STS 15306.
      Symphony, op. 27, C minor. cf Ripening, op. 34.
SULLIVAN, Arthur
2536 Cox and box. Trial by jury. Soloists; Gilbert and Sullivan Fes-
          tival Orchestra and Chorus; Peter Murray. Pye NSPH 15. Tape
          (c) ZCP 15 (ct) Y8P 15.
              +-Gr 1-76 p1248   +-RR 12-75 p31
              +-HFN 12-75 p171
      Day dreams, op. 14, no. 5: Elle et lui. cf Pearl SHE 528.
2537 Festival Te Deum. The golden legend: Prologue. The tempest:
          Dance. Soloists; Nova Singers; New Westminster Philharmonic
          Orchestra. Rare Recorded Editions SRRE 164.
              +-HFN 7-76 p97
      The golden legend: Prologue. cf Festival Te Deum.
2538 The gondoliers (without dialogue). Edna Graham, Elsie Morison,
          s; Monica Sinclair, Helen Watts, Marjorie Thomas, con; Alex-
          ander Young, Richard Lewis, t; Geraint Evans, John Cameron,
          bar; James Milligan, bs-bar; Owen Brannigan, bs; Glyndebourne
          Festival Chorus; Pro Arte Orchestra; Malcolm Sargent. HMV SXCW
          3027 (2). Tape (c) SXDW 3027. (Reissue from ASD 265/6).
              +Gr 9-76 p478   +HFN 12-76 p155
              +HFN 11-76 p173   +-RR 9-76 p34
2539 The gondoliers, excerpts. HMS Pinafore, excerpts. The Mikado,
          excerpts. The pirates of Penzance, excerpts. Marion Studholme,
          s; Jean Allister, con; Edmund Bohan, t; Ian Wallace, bs; London
          Concert Orchestra; Marcus Dods. Polydor 2383 366. Tape (c)
          3170 266.
              +Gr 1-76 p1247   +HFN 5-76 p111
      The gondoliers, excerpts. cf Works, selections (Pye NSPH 16).
2540 The gondoliers: Overture; For the merriest fellows are we; We're
          called gondolieri; Thank you, gallant gondolieri...Duke of
          Plaza Toro's song; I stole the prince; Then one of us will be
          a queen; Come let's away and Act 1 finale; Rising early in the
          morning...Take a pair of sparkling eyes; Dance a cachucha;
          Duchess's song; Gavotte song...Here is a case unprecedented and
          Act 2 finale. Soloists; Gilbert and Sullivan Festival Orches-
          tra and Chorus; Peter Murray. Pye NSPH 8. Tape (c) ZCP 8 (ct)
          Y8P 8.
              +-Gr 1-76 p1248   +RR 12-75 p31
              +-HFN 12-75 p171
      The gondoliers: I am a courtier, grave and serious. cf English
          National Opera ENO 1001.

The gondoliers: In enterprise of martial kind; I stole the prince.
    cf Works, selections (Pearl GEM 135).
The gondoliers: Overture. cf Overtures (Pye NSPH 7).
The gondoliers: Take a pair of sparkling eyes; I am a courtier,
    grave and serious...Gavotte; Dance a cachucha. cf Works,
    selections (Classics for Pleasure CFP 40238).
2541 The grand Duke. Barbara Lilley, Julia Goss, Anne Eggleston,
    Glynis Prendergast, s; Jane Metcalfe, Patricia Leonard, ms;
    Lyndsie Holland, Beti Lloyd-Jones, con; Meston Reid, t; John
    Reed, Kenneth Sandford, Michael Rayner, John Ayldon, James
    Conroy-Ward, bar; Jon Ellison, bs; D'Oyly Carte Opera Company;
    RPO; Royston Nash. Decca SKL 5239 /40 (2). Tape (c) K17K 22.
        +Gr 12-76 p1071   ++HFN 12-76 p153
        +HFN 12-76 p149   +RR 12-76 p45
Henry VIII: March; Graceful dance. cf Trial by jury.
2542 HMS Pinafore, highlights. Soloists; Gilbert and Sullivan Festival
    Orchestra and Chorus; Peter Murray. Pye NSPH 9. Tape (c)
    ZCP 9 (ct) Y8P 9.
        +-Gr 1-76 p1248   +RR 12-75 p31
        +-HFN 12-75 p171
HMS Pinafore, excerpts. cf The gondoliers, excerpts.
HMS Pinafore, excerpts. cf Works, selections (Pye NSPH 16).
HMS Pinafore: Opening chorus...Little Buttercup's song; Captain's
    song; Finale. cf Works, selections (Classics for Pleasure CFP
    40238).
HMS Pinafore: Overture. cf Overtures (Pye NSPH 7).
HMS Pinafore: When I was a lad. cf Works, selections (Pearl GEM
    135).
Imperial march. cf Utopia limited.
2543 Iolanthe. Pamela Field, Marjorie Williams, Rosalind Griffiths, s;
    Lyndsie Holland, con; Judi Merri, Patricia Leonard, ms; Mal-
    colm Williams, t; John Reed, Michael Rayner, bar; John Ayldon,
    Kenneth Sandford, bs; D'Oyly Carte Opera Chorus; RPO; Royston
    Nash. Decca SKL 5188/9 (2). Tape (c) KSKC 5188/9 (ct) ESKC
    5188/9. (also London OSA 12104 (2)).
        +Gr 12-74 p1211   +-RR 10-74 p27
        +Gr 12-74 p1237 tape   ++St 11-76 p142
        +NR 10-76 p13
2544 Iolanthe, highlights. Soloists; Gilbert and Sullivan Festival
    Orchestra and Chorus; Peter Murray. Pye NPHS 11. Tape (c)
    ZCP 11 (ct) Y8P 11.
        +-Gr 1-76 p1248   +-RR 12-75 p31
        +-HFN 12-75 p171
Iolanthe, excerpts. cf Works, selections (Pye NSPH 16).
Iolanthe: Entrance and march of peers; If we're weak enough to
    tarry; Final chorus. cf Works, selections (Classics for
    Pleasure CFP 40238).
Iolanthe: The law is the true embodiment; When I went to the bar;
    Love unrequited...When you're lying awake. cf Works, selec-
    tions (Pearl GEM 135).
Iolanthe: Overture. cf Overtures (Pye NSPH 7).
Macbeth: Overture. cf Trial by jury.
2545 The Mikado. Valerie Masterson, Pauline Wales, s; Peggy Ann Jones,
    ms; Lyndsie Holland, con; Colin Wright, t; Michael Rayner,
    John Reed, bar; John Ayldon, Kenneth Sandford, John Broad, bs;
    D'Oyly Carte Opera Chorus; RPO; Royston Nash. Decca SKL 5158/9
    (2). Tape (c) KSKC 5158/9 (ct) ESKC 5158/9. (also London OSA

12103).

+-Audio 2-76 p95   +MJ 10-75 p45
+-Gr 1-74 p1409   +NR 5-75 p12
+Gr 4-74 p1919 tape   ++RR 4-74 p93 tape
+HF 3-75 p74   ++SFC 2-23-75 p23
+HFN 2-74 p345   +St 7-75 p99

2546 The Mikado (without dialogue). Elsie Morison, Jeannette Sinclair,
      s; Monica Sinclair, Marjorie Thomas, con; Richard Lewis, t;
      Geraint Evans, John Cameron, bar; Owen Brannigan, Ian Wallace,
      bs; Glyndebourne Festival Chorus; Pro Arte Orchestra; Malcolm
      Sargent. HMV SXDW 3019 (2). Tape (c) TC 2 EXE 1021. (Reissue
      from ASD 256/7).
               +Gr 1-76 p1248   +RR 1-76 p26
               +-HFN 1-76 p123   +RR 8-76 p85 tape
               +HFN 5-76 p117 tape

2547 The Mikado, highlights. Soloists; Gilbert and Sullivan Festival
      Orchestra and Chorus; Peter Murray. Pye NSPH 13. Tape (c)
      ZCP 13 (ct) Y8P 13.
               +-Gr 1-76 p1248   +RR 12-75 p31
               +-HFN 12-75 p171

      The Mikado, excerpts. cf The gondoliers, excerpts.
      The Mikado, excerpts. cf Works, selections (Pye NSPH 16).
      The Mikado: Opening chorus...A wand'ring minstrel I; The sun whose
         rays; Final chorus. cf Works, selections (Classics for Pleas-
         ure CFP 40238).
      The Mikado: Overture. cf Overtures (Pye NSPH 7).
      The Mikado: Tit-willow. cf Works, selections (Pearl GEM 135).
2548 Overtures: The gondoliers. HMS Pinafore. Iolanthe. The Mikado.
      The pirates of Penzance. Ruddigore. The yeomen of the guard.
      Soloists; Gilbert and Sullivan Festival Orchestra; Peter Murray.
      Pye NSPH 7. Tape (c) ZCP 7 (ct) Y8P 7.
               +-Gr 1-76 p1248   +RR 12-75 p31
               +-HFN 12-75 p171

      Patience: Bunthorne's song. cf Works, selections (Pearl GEM 135).
2549 Pineappe Poll (arr. Mackerras). Pro Arte Orchestra; John Hollings-
      worth. Pye GSGC 15023. Tape (c) ZCCCB 15023. (Reissue from
      CML 33000).
               +-Gr 2-76 p1382   +RR 12-75 p67
               +-HFN 1-76 p123   +-RR 6-76 p89 tape

2550 The pirates of Penzance, highlights. Soloists; Gilbert and Sulli-
      van Festival Orchestra and Chorus; Peter Murray. Pye NSPH 14.
      Tape (c) ZCP 14 (ct) Y8P 14.
               +-Gr 1-76 p1248   +RR 12-75 p31
               +-HFN 12-75 p171

      The pirates of Penzance, excerpts. cf The gondoliers, excerpts.
      The pirates of Penzance, excerpts. cf Works, selections (Pye NSPH
         16).
      The pirates of Penzance: Major General's song; Sighing softly to
         the river. cf Works, selections (Pearl GEM 135).
      The pirates of Penzance: Major General's song; Oh, is there not
         one maiden breast...Poor wand'ring one; Sergeant of police's
         song. cf Works, selections (Classics for Pleasure CFP 40238).
      The pirates of Penzance: Overture. cf Overtures (Pye NSPH 7).
      Princess Ida: If you give me your attention; Whene'er I spoke
         sarcastic joke. cf Works, selections (Pearl GEM 135).
      The rose of Persia: There was once a small street Arab. cf Works,
         selections (Pearl GEM 135).

2551 Ruddigore. Elsie Griffin, s; Eileen Sharp, ms; Derek Oldham, t;
        Bertha Lewis, con; George Baker, Darrell Fancourt, bar; Leo
        Sheffield, bs-bar; Edward Halland, bs; D'Oyly Carte Opera
        Company, Members; Orchestra. Pearl GEM 133/4 (2). (Reissue
        from HMV D 878-886).
            +-Gr 2-76 p1382   +-RR 3-76 p29
      Ruddigore, excerpts. cf Works, selections (Pye NSPH 16).
2552 Ruddigore: Overture; Sir Rupert Murgatroyd his leisure; I know a
        youth...(entire score to, and including) You understand; I
        think you do; When the buds are blossoming; Within this breast
        there beats a heart; Happily coupled are we; Painted emblems...
        (entire score to, and including) My eyes are fully open; Oh,
        happy the lily. Soloists; Gilbert and Sullivan Festival Orch-
        estra and Chorus; Peter Murray. Pye NSPH 12. Tape (c) ZCP 12
        (ct) Y8P 12.
            +-Gr 1-76 p1248   +RR 12-75 p31
            +-HFN 12-75 p171
      Ruddigore: Overture. cf Overtures (Pye NSPH 7).
2553 Songs: Oh, ma charmante; Shakespeare songs; O mistress mine;
        Willow song; Rosalind; Sigh no more, ladies; Orpheus with
        his lute; Sweet day so cool; Where the bee sucks; The window.
        Francis Loring, bar; Paul Hamburger, pno. Rare Recorded Edi-
        tions SRRE 163.
            +-Gr 8-76 p328   +HFN 7-76 p98
      Songs: The frost is here. cf Transatlantic XTRA 1159.
      Songs: The lost chord. cf Prelude PRS 2505.
      Songs: Once again. cf HMV EMD 5528.
      Songs: Orpheus with his lute. cf HMV ASD 2929.
      Songs: Orpheus with his lute. cf Polydor 2383 389.
      The sorcerer: My name is John Wellington Wells. cf Works, selec-
        tions (Pearl GEM 135).
      The tempest: Dance. cf Festival Te Deum.
2554 Trial by jury. Macbeth: Overture. Henry VIII: March; Graceful
        dance. Julia Goss, Colin Wright, s; John Reed, Michael Rayner,
        John Ayldon, Kenneth Sandford, bar; D'Oyly Carte Opera Chorus,
        Members; RPO; Royston Nash. Decca TXS 113. Tape (c) KTXC 113
        (ct) EDXC 113. (also London OSA 1167).
            +-Gr 5-75 p2014   +RR 5-75 p24
            /HFN 7-75 p91 tape   +St 10-76 p121
            +NR 10-76 p13
      Trial by jury. cf Cox and box.
      Trial by jury: Judge's song. cf Works, selections (Pearl GEM 135).
2555 Utopia limited. Imperial march. Pamela Field, Julia Goss, Rosa-
        lind Griffiths, s; Judi Merri, ms; Lyndsie Holland, con; Meston
        Reid, Colin Wright, t; Kenneth Sandford, John Reed, John Ayldon,
        Jon Ellison, Michael Buchan, James Conroy-Ward, Michael Rayner,
        David Porter, bar; John Broad, bs; D'Oyly Carte Opera Company
        Chorus; RPO; Royston Nash. Decca SKL 5225/6 (2). Tape (c)
        K2C 17.
            +-Gr 4-76 p1664   +-MT 6-76 p496
            +HFN 4-76 p115   +-RR 4-76 p30
            ++HFN 6-76 p105 tape
2556 Utopia limited. D'Oyly Carte Opera Company; RPO; Royston Nash.
        London SA 12105 (2). Tape (c) 5-12105.
            +NR 10-76 p13   +SFC 11-7-76 p33.
2557 Utopia limited: O make way...In every mental lore; Let all your
        doubts take wing; Quaff the nectar; How fair, how modest;

Although of native maids the cream; This morning we propose...
Bold-faced ranger; First you're born; Finale, Increase your
army; O Zara...A tenor, all singers above; Words of love too
loudly spoken; Society has quite forsaken; This ceremonial...
Eagle high; With wily brain; A wonderful joy; Upon our sea-girt
land; There's a little group of isles. Marion Scodari, s;
Susan Hoagland, Grace Boave, ms; Carroll Mattoon, con; Barry
Morley, t; Peter Kline, Thomas Jones, Gregory Wise, Philip
Graneto, Jerry Holloway, bar; Lyric Theatre Company Orchestra
and Chorus, Washington; John Landis. Pearl SHE 529.
    -Gr 7-76 p221

Utopia limited: First you're born; Some seven men form an association. cf Works, selections (Pearl GEM 135).

2558 Works, selections: The gondoliers: Take a pair of sparkling eyes;
I am a courtier, grave and serious...Gavotte; Dance a cachucha.
Iolanthe: Entrance and march of peers; If we're weak enough to
tarry; Final chorus. HMS Pinafore: Opening chorus...Little
Buttercup's song; Captain's song; Finale. The Mikado: Opening
chorus...A wand'ring minstrel I; The sun whose rays; Final
chorus. The pirates of Penzance: Major General's song; Oh,
is there not one maiden breast...Poor wand'ring one; Sergeant
of police's song. The yeomen of the guard: When maiden loves;
I have a song to sing. Soloists; Glyndebourne Festival Chorus;
Pro Arte Orchestra; Malcolm Sargent. Classics for Pleasure
CFP 40238. Tape (c) TC CFP 40238.
     +HFN 3-76 p111   +HFN 12-76 p153 tape
     +HFN 5-76 p115   +RR 3-76 p29

2559 Works, selections: The gondoliers: In enterprise of martial kind;
I stole the prince. HMS Pinafore: When I was a lad. Iolanthe:
The law is the true embodiment; When I went to the bar; Love
unrequited...When you're lying awake. The Mikado: Tit-willow.
The pirates of Penzance: Major General's song; Sighing softly
to the river. Patience: Bunthorne's song. Princess Ida: If
you give me your attention; Whene'er I spoke sarcastic joke.
The rose of Persia: There was once a small street Arab. The
sorcerer: My name is John Wellington Wells. Trial by jury:
Judge's song. Utopia limited: First you're born; Some seven
men form an association. The yeomen of the guard: I have a
song to sing; I've jibe and joke; A private buffoon. C. H.
Workman, bar; Herman Finck Orchestra; Orchestra; Herman Finck.
Pearl GEM 135.
    +-RR 7-76 p30

2560 Works, selections: The gondoliers, excerpts. HMS Pinafore, ex-
cerpts. The Mikado, excerpts. The pirates of Penzance, excerpts
Ruddigore, excerpts. Iolanthe, excerpts. The yeomen of the
guard, excerpts. Soloists; Gilbert and Sullivan Festival
Orchestra; Peter Murray. Pye NSPH 16. Tape (c) ZCP 16 (ct)
Y8P 16.
    +-Gr 1-76 p1248   +-RR 12-75 p31
    +-HFN 12-75 p171

2561 The yeomen of the guard, highlights. Soloists; Gilbert and Sullivan Festival Orchestra and Chorus; Peter Murray. Pye NSPH 10.
Tape (c) ZCP 10 (ct) Y8P 10.
    +-Gr 1-76 p1248   +-RR 12-75 p31
    +-HFN 12-75 p171

The yeomen of the guard, excerpts. cf Works, selections (Pye
NSPH 16).

The yeomen of the guard: I have a song to sing; I've jibe and
    joke; A private buffoon. cf Works, selections (Pearl GEM 135).
The yeomen of the guard: Overture. cf Overtures (Pye NSPH 7).
The yeomen of the guard: Overture. cf BBC REB 228.
The yeomen of the guard: When maiden loves; I have a song to
    sing. cf Works, selections (Classics for Plesure CFP 40238).

SUPPE, Franz von
    Boccaccio: Medley. cf CBS 30070.
    Flotte Bursche overture. cf Overtures (Philips 6580 113).
    The jolly robbers overture. cf Overtures (Philips 6580 113).
    Light cavalry: Overture. cf Overtures (Decca 374).
    Light cavalry: Overture. cf Overtures (Philips 6580 113).
    Light cavalry: Overture. cf Angel S 37231.
    Light cavalry: Overture. cf Decca SPA 409.
    Light cavalry: Overture. cf Everest 3360.
    Light cavalry: Overture. cf HMV ESD 7010.
    Morning, noon and night in Vienna: Overture. cf Overtures (Decca
        374).
2562 Overtures: Poet and peasant. Light cavalry. Pique Dame. Morning,
    noon and night in Vienna. VPO; Georg Solti. Decca SPA 374.
    Tape (c) KSDC 194. (Reissue from SXL 2174; SDD 194).
        +HFN 9-75 p108   +-RR 10-75 p57
        +HFN 5-76 p117 tape
2563 Overtures: The beautiful Galatea. Flotte Bursche. The jolly
    robbers. Light cavalry. Poet and peasant. The queen of
    spades. Dresden Staatskapelle; Otmar Suitner. Philips 6580
    113.
        ++Gr 8-76 p342   +RR 8-76 p54
        +HFN 8-76 p93
    Poet and peasant: Overture. cf Overtures (Decca 374).
    Poet and peasant: Overture. cf Overtures (Philips 6580 113).
    Poet and peasant: Overture. cf BACH: Toccata and fugue, S 565,
        D minor.
    Poet and peasant: Overture. cf CBS 61748.
    The queen of spades (Pique Dame): Overture. cf Overtures
        (Philips 6580 113).
    Pique Dame: Overture. cf Overtures (Decca 374).
2564 Die schöne Galathea (The beautiful Galathea). Elisabeth Roon, s;
    Waldemar Kmentt, t; Kurt Preger, bar; Otto Wiener, bs; VSOO
    and Chorus; Anton Paulik. Saga 5418. (also Urania).
        +Gr 5-76 p1811   +RR 5-76 p26
        +HFN 5-76 p109
    The beautiful Galatea: Overture. cf Overtures (Philips 6580 113).
    Die schöne Galathea: Overture. cf Supraphon 110 1638.

SUSATO, Tielman
    La bataille. cf Argo ZRG 823.
    La battaglia. cf Hungaroton SLPX 11669/70.
    Bergeret sans roch. cf Argo ZRG 823.
    Branle quatre branles. cf Argo ZRG 823.
    C'est a grand tort. cf Hungaroton SLPX 11669/70.
    The Danserye, dances (12). cf Angel S 36851.
    Il estoit une fillette, ronde. cf Harmonia Mundi HMU 610.
    Hoboecken dans. cf Harmonia Mundi HMU 610.
    Mille regretz, pavane. cf JOSQUIN DES PRES: Songs (Argo ZRG 793).
    Mille regretz, pavane. cf Harmonia Mundi HMU 610.
    Mon amy, ronde. cf Argo ZRG 823.

La Mourisque. cf Angel SFO 36895.
La Mourisque. cf Argo SPA 464.
La Mourisque. cf Argo ZRG 823.
La Mourisque. cf CBS 76183.
Ronde. cf Argo ZRG 823.
Ronde. cf DG Archive 2533 184.
Ronde. cf Harmonia Mundi HMU 610.
Ronde et salterelle. cf Harmonia Mundi HMU 610.
Si pas soffrir, pavane. cf Harmonia Munid HMU 610.
SVENDSEN, Johan
        Kom Carina. cf Cantilena 6238.
SVENSSON, Sven
        Dar bor en konung. cf Swedish Scoeity SLT 33197.
        SWABIAN WIND MUSIC IN HUNGARY. cf Hungaroton SLPX 16570.
SWAN
        China. cf Vox SVBX 5350.
SWANSON, Howard
        Trio, flute, clarinet and piano. cf Folkways FTS 33901/4.
SWEELINCK, Jan
        Balleto del granduca. cf Hungaroton SLPX 11669/70.
        Hodie Christus natus est. cf Vanguard SVC 71212.
        Mein Junges Leben hat ein End. cf BACH: Toccata and fugue, S 565,
        D minor.
        Mein Junges Leben hat ein End, variations. cf Grosvenor GRS 1041.
        Orsus, serviteurs deu Seigneur. cf Abbey LPB 757.
        Paduana lachrimae. cf Hungaroton SLPX 11669/70.
        Psalms, nos. 5, 23. cf DG Archive 2533 302.
SZABADOS, Gyorgy
2565   Duo piano and violin. The interrogation of Irma Szabo. Miracle.
        The wedding. Szabados Quartet. Hungaroton SLPX 17475.
            +NR 1-76 p8
        The interrogation of Irma Szabo. cf Duo, piano and violin.
        Miracle. cf Duo, piano and violin.
        The wedding. cf Duo, piano and violin.
SZARZYNAKI
        Ave Regina, Jesu spes mea. cf MIELCZEWSKI: Veni Domine, Deus in
        nomine tuo.
SZERVANSZKY, Endre
2566   Concerto, clarinet. Serenade, clarinet. Petofi choruses: Song of
        the dogs. Variations, orchestra. Béla Kovács, clt; Veszprém
        City Mixed Choir; Hungarian Radio and Television Orchestra;
        Adám Medveczky. Hungaroton SLPX 11716.
            +NR 3-76 p3   +RR 5-76 p52
        Petofi choruses: Song of the dogs. cf Concerto, clarinet.
        The river has overflown. cf Hungaroton SLPX 11696.
        Serenade, clarinet. cf Concerto, clarinet.
        Variations, orchestra. cf Concerto, clarinet.
SZOKOLAY, Sándor
        Quartet, strings, no. 1. cf DUBROVAY: Quartet, strings.
2567   Samson. Erzsébet Hazy, s; György Melis, bar; Sándor Palcsó,
        Janos Nagy, t; Hungarian State Opera Orchestra and Chorus;
        Andras Kórodi. Hungaroton SLPX 11738/9 (2).
            +HFN 7-76 p98   +RR 7-76 p35
            +NR 8-76 p10   +SFC 8-29-76 p28
SZOLLOSY
2568   Music for orchestra. Transfigurations, orchestra. Hungarian

Radio and Television Orchestra; György Lehel. Hungaroton SLPX
11733.
+NR 5-76 p4
Transfigurations, orchestra. cf Music for orchestra.
SZULC, Jozsef
Clair de lune, op. 81, no. 1. cf HMV RLS 716.
Clair de lune, op. 81, no. 1. cf HMV RLS 719.
SZYMANOWSKA, Maria
Etudes, F major, C major, E major. cf Avant AV 1012.
Nocturne, B flat major. cf Avant AV 1012.
SZYMANOWSKI, Karol
2569 Concerti, violin (2). Wanda Wilkomirska, Charles Treger, vln.
Muza SXL 0383.
*HF 3-76 p70
Concerto, violin, no. 1, op. 35. cf PROKOFIEV: Concerto, violin,
no. 1, op. 19, D major.
Concerto, violin, no. 1, op. 35. cf HMV SLS 5058.
La Fontaine d'Areethuse, op. 30, no. 1. cf FRANCAIX: Sonatine.
2570 Harnasie, op. 55. Kazimierz Pustelak, t; Warsaw National Philhar-
monic Orchestra and Chorus; Witold Rowicki. Muza SX 1317.
+NR 9-76 p3   ++St 12-76 p147
Le Roi Roger (King Roger), op. 46: Chant do Roxane. cf RCA ARM
4-0942/7.
King Roger, op. 46: Roxane's song. cf Connoisseur Society CS 2070.
2571 Sonata, piano, no. 1, op. 8, C minor. Sonata, piano, no. 3, op.
36. Daniel Graham, pno. Musical Heritage Society MHS 3136.
+MJ 11-76 p45   +St 3-76 p121
Sonata, piano, no. 3, op. 36. cf Sonata, piano, no. 1, op. 8, C
minor.
TAILLEFERRE, Germaine
Quatuor. cf Gemini Hall RAP 1010.
Quartet, strings. cf BONDON: Quartet, strings, no. 1.
TALLIS, Thomas
2572 Motets: Ece tempus idoneum; Gaude gloriosa; Hear the voice and
prayer; If ye love me; Lamentations I; Loquebantur variis lin-
guis; O nata lux de lumine; Spem in alium. Clerkes of Oxen-
ford; David Wulstan. Seraphim S 60256.
++HF 11-76 p130   +NR 11-76 p10
Salvator mundo. cf Harmonia Mundi HMU 473.
Songs: O Lord, give Thy holy spirit; O nata lux de lumine. cf
Abbey LPB 750.
Songs: O nata lux de lumine. cf Vista VPS 1022.
TALMA, Louise
Alleluia in form of toccata. cf Avant AV 1012.
TANSMAN, Alexandre
Mouvement perpetuel. cf RCA ARM 4-0942/7.
Suite, bassoon. cf Washington University RAVE 761.
TARREGA, Francisco
Adelita. cf Angel S 36094.
Capricho arabe. cf Enigma VAR 1015.
Danza mora. cf ALBENIZ: Suite española, no. 3: Sevillanas; no.
4: Cadiz.
Estudio brillante. cf Angel S 36094.
Gran jota. cf Enigma VAR 1015.
Gran jota. cf London STS 15306.
Maria. cf Angel S 36094.

Maria. cf Enigma VAR 1015.
Marieta. cf Angel S 36094.
Mazurka. cf Angel S 36094.
Prelude, G major. cf London STS 15306.
Preludio. cf ALBENIZ: Suite española, no. 3: Sevillanas; no. 4:
    Cadiz.
Preludes, nos. 2, 5. cf Angel S 36094.
Recuerdos de la Alhambra. cf DG 3335 182.
Sueño. cf ALBENIZ: Suite española, no. 3: Sevillanas; no. 4:
    Cadiz.
Sueño. cf Enigma VAR 1015.
Tango. cf DG 3335 182.

TARTINI, Giuseppe
Concerto, flute, D major. cf RCA CRL 3-1430.
Concerto, flute, G major. cf RCA CRL 3-1429.
Concerto, trumpet, D major. cf Erato STU 70871.
2573 Concerti, violin, D major, E minor, E major. Pierre Amoyal, vln;
    I Solisti Veneti; Claudio Scimone. Erato STU 70972.
        +-Gr 8-76 p300   +-RR 8-76 p54
          +HFN 10-76 p177
Concerto, violoncello, D major: Grave ed espressivo. cf HMV RLS
    723.

TAUSINGER, Jan
Concertino, viola and chamber orchestra. cf ISTVAN: Ritmi ed
    Antiritmi.
Contemplations (2). cf Panton 110 369.
Duetti compatibili. cf HUMMEL: Sonata, viola, E flat major.

TAVENER, John
2574 Canciones españolas. Requiem for Father Malachy. James Bowman,
    Kevin Smith, c-t; King's Singers; Nash Ensemble; John Tavener.
    RCA LRL 1-5104.
        +Gr 10-76 p640   +RR 9-76 p86
        +HFN 11-76 p168
Missa sine nomine, 5 voices. cf PARSLEY: Lamentations, 5 voices.
Requiem for Father Malachy. cf Canciones españolas.

TAYLOR
Toy symphony: Adagio and finale. cf Angel S 36080.

TAYLOR, Deems
2575 Through the looking glass. Eastman-Rochester Symphony Orchestra;
    Howard Hanson. Era 1008.
        *NYT 7-4-76 pD1

TCHAIKOVSKY, Peter
2576 Andante and finale, op. 70. Concert fantasia, op. 56. Werner
    Haas, pno; Monte Carlo Opera Orchestra; Eliahu Inbal. Philips
    6500 316. (Reissue from 6703 033).
        +Gr 7-76 p181   ++RR 6-76 p57
        +HFN 7-76 p103
2577 Capriccio italien, op. 45. The nutcracker, op. 71: Suite. Eugene
    Onegin, op. 24: Waltz and polonaise. LPO; Leopold Stokowski.
    Philips 6500 766. Tape (c) 7300 332.
        +Gr 5-75 p2037 tape   +NR 12-75 p2
        +Gr 7-75 p245   +RR 7-75 p36
        +HFN 6-75 p109   +RR 4-76 p82 tape
        ++HFN 7-75 p89   -SFC 11-9-75 p22
Capriccio italien, op. 45. cf BACH: Toccata and fugue, S 565,
    D minor.

Capriccio italien, op. 45. cf GLINKA: Jota aragonesa.
Capriccio italien, op. 45. cf RIMSKY-KORSAKOV: Capriccio espag-
     nol, op. 34.
Capriccio italien, op. 45. cf RCA CRL 3-2026.
Capriccioso, op. 19, no. 5. cf Piano works (CBS 61718).
Capriccioso, op. 62. cf Command COMS 9006.
Chanson triste, op. 40, no. 2. cf Piano works (CBS 61718).
Concert fantasia, op. 56. cf Andante and finale, op. 70.
2578 Concerto, piano, no. 1, op. 23, B flat minor. Concerto, piano,
     no. 2, op. 44, G major. Concerto, piano, no. 3, op. 75/79,
     E flat major. Emil Gilels, pno; NPhO; Lorin Maazel. Angel
     S 3798 (2). (also HMV SLS 865 Tape (c) TC SLS 865).
          +Gr 2-74 p1575   ++RR 2-74 p16
          +-HF 1-74 p81   ++RR 12-76 p108 tape
          +HFN 3-76 p113 tape   ++St 1-74 p112
          ++NR 2-74 p4
2579 Concerto, piano, no. 1, op. 23, B flat minor. Peter Katin, pno;
     LPO; John Pritchard. Classics for Pleasure CFP 115. Tape
     (c) TC CFP 115.
          +-HFN 12-76 p153 tape
2580 Concerto, piano, no. 1, op. 23, B flat minor. Solomon, pno; PhO;
     Issay Dobrowen. Da Capo 1C 053 01412. (Reissue from HMV
     originals).
          +HF 11-76 p130
2581 Concerto, piano, no. 1, op. 23, B flat minor. Lazar Berman, pno;
     BPhO; Herbert von Karajan. DG 2530 677. Tape (c) 3300 677.
          ++Gr 3-76 p1477   +-MJ 4-76 p30
          ++HF 5-76 p75   +-NR 4-76 p4
          +-HF 10-76 p147 tape   -RR 3-76 p50
          +HFN 4-76 p117   ++STL 5-9-76 p38
          +-HFN 6-76 p105
2582 Concerto, piano, no. 1, op. 23, B flat minor. Vladimir Horowitz,
     pno; NBC Symphony Orchestra; Arturo Toscanini. RCA VH 015.
          +Gr 1-76 p1204   +RR 12-75 p68
          +-HFN 2-76 p117
2583 Concerto, piano, no. 1, op. 23, B flat minor. Van Cliburn, pno;
     Symphony Orchestra; Kyril Kondrashin. RCA Tape (c) RK 11704.
          -Gr 7-76 p230 tape   -RR 10-76 p107 tape
          -HFN 7-76 p104 tape
2584 Concerto, piano, no. 1, op. 23, B flat minor. Earl Wild, pno;
     RPO; Anatole Fistoulari. RCA GL 25013. Tape (c) GK 25013.
     (Previously issued by Reader's Digest).
          +Gr 10-76 p612   +-RR 10-76 p68
          +HFN 12-76 p149   +RR 12-76 p104
          +-HFN 12-76 p153 tape
Concerto, piano, no. 1, op. 23, B flat minor. cf Works, selec-
     tions (Decca DPA 547/8).
Concerto, piano, no. 1, op. 23, B flat minor. cf LISZT: Concerto,
     piano, no. 1, G 124, E flat major.
Concerto, piano, no. 1, op. 23, B flat minor. cf RACHMANINOFF:
     Concerto, piano, no. 2, op. 18, C minor.
Concerto, piano, no. 1, op. 23, B flat minor. cf STRAVINSKY:
     Petrouchka: 3 movements.
Concerto, piano, no. 1, op. 23, B flat minor. cf HMV SLS 5033.
Concerto, piano, no. 1, op. 23, B flat minor. cf HMV SLS 5068.
2585 Concerto, piano, no. 2, op. 44, G major. Sylvia Kersenbaum, pno;

ORTF; Jean Martinon. Connoisseur Society CS 2076. (Q) CSQ
2076.
+-HF 1-76 p97   ++SFC 11-9-75 p22
+MJ 4-76 p30   +St 10-75 p118
++NR 12-75 p2
Concerto, piano, no. 2, op. 44, G major. cf Concerto, piano, no.
1, op. 23, B flat minor.
Concerto, piano, no. 2, op. 44, G major. cf GRAINGER: Paraphrase
on Tchaikovsky: Nutcracker: Waltz of the flowers.
Concerto, piano, no. 3, op. 75/79, E flat major. cf Concerto,
piano, no. 1, op. 23, B flat minor.
2586 Concerto, violin, op. 35, D major. Sérénade mélancolique, op.
26. Arthur Grumiaux, vln; NPhO; Jan Krenz. Philips 9500 086.
Tape (c) 7300 490.
++Gr 6-76 p52   ++RR 6-76 p58
++HFN 6-76 p101
Concerto, violin, op. 35, D major. cf MENDELSSOHN: Concerto,
violin, op. 64, E minor (Decca 4345).
Concerto, violin, op. 35, D major. cf MENDELSSOHN: Concerto,
violin, op. 64, E minor (Turnabout TV 34553).
Concerto, violin, op. 35, D major. cf RCA ARM 4-0942/7.
Elegy, G major. cf HMV SQ ESD 7001.
The enchantress: Nastasya's aria; Where are you my beloved. cf
Rubini GV 63.
2587 Eugene Onegin, op. 24. Teresa Kubiak, s; Anna Reynolds, Julia
Hamari, Enid Hartle, ms; Stuart Burrows, Michel Sénéchal, t;
Bernd Weikl, bar; William Mason, Nicolai Ghiaurov, Richard Van
Allan, bs; John Alldis Choir; ROHO; Georg Solti. Decca SET
596/8 (3). (also London OSA 13112).
+-Gr 6-75 p84   +ON 12-6-75 p52
+-HF 2-76 p107   +RR 5-75 p18
++HFN 6-75 p96   ++SFC 11-2-75 p28
+MJ 3-76 p25   +SR 3-6-76 p38
++NR 2-76 p10   ++St 1-76 p75
Eugene Onegin, op. 24, excerpts. cf GLINKA: Jota aragonesa.
Eugene Onegin, op. 24: Gremins' aria. cf STRAUSS, R: Capriccio:
Monolog des Theaterdirektors.
Eugene Onegin, op. 24: Lensky's aria. cf Muza SXL 1170.
2588 Eugene Onegin, op. 24: Polonaise. Sleeping beauty, op. 66a: Suite.
Swan Lake, op. 20: Suite. Warsaw National Philharmonic Orch-
estra; Witold Rowicki. DG 2548 125. Tape (c) 3348 125.
(Reissue from 136 036).
+Gr 4-75 p1867   +-RR 9-76 p92 tape
-RR 4-75 p36
2589 Eugene Onegin, op. 24: Waltz. The nutcracker, op. 71: Waltz.
Serenade, strings, op. 48, C major. Sleeping beauty, op. 66:
Waltz. Swan Lake, op. 20: Waltz. PO; Eugene Ormandy. CBS
30076. Tape (c) 40-30076.
/HFN 10-76 p181   +-RR 9-76 p68
+-HFN 10-76 p185 tape
Eugene Onegin, op. 24: Waltz, Act 2. cf Decca DPA 525/6.
Eugene Onegin, op. 24: Waltz and polonaise. cf GOUNOD: Faust:
Ballet music; Waltz.
Eugene Onegin, op. 24: Waltz, Polonaise. cf Columbia M 34127.
Eugene Onegin, op. 24: Written words. cf Rubini GV 65.
2590 Fate, op. 77. Romeo and Juliet: Fantasy overture. The tempest,

op. 18. Washington National Symphony Orchestra; Antal Dorati.
Decca SXL 6694. (also London CS 6891).
    +Gr 2-75 p1500   +RR 2-75 p40
    +NR 9-75 p6   +St 9-76 p125
2591 Fatum, op. 77. The tempest, op. 18. The storm, op. 76. The
    Voyevode, op. 78. Frankfurt Radio Symphony Orchestra; Eliahu
    Inbal. Philips 6500 467.
        ++Audio 11-76 p108   +NR 1-76 p2
        +Gr 2-76 p1349   ++RR 2-76 p42
        +HFN 3-76 p107   ++SFC 3-7-76 p27
        ++MJ 1-76 p39   +St 9-76 p125
2592 Fatum, op. 77. The storm, op. 76. The Voyevoda, op. 78. Bochum
    Orchestra; Othmar Maga. Vox STPL 513460.
        +NR 4-76 p4
2593 Francesca da Rimini, op. 32. Hamlet overture, op. 67a. LPO;
    Vernon Handley. Classics for Pleasure CFP 40223.
        +-Gr 1-76 p1203   +RR 9-75 p50
        +-HFN 9-75 p106
2594 Francesca da Rimini, op. 32. Hamlet overture, op. 67a. Utah
    Symphony Orchestra; Maurice Abravanel. Turnabout QTV 34601.
        ++NR 2-76 p4
    Hamlet overture, op. 67a. cf Francesca da Rimini, op. 32 (Classics
    for Pleasure CFP 40223).
    Hamlet overture, op. 67a. cf Francesca da Rimini, op. 32 (Turna-
    bout QTV 34601).
    Humoresque, op. 10, no. 2. cf Piano works (CBS 61718).
    Liebtrabant: Overture. cf STRAUSS, R.: Capriccio, op. 85: Monolog
    des Theaterdirektors.
    The maid of Orleans: Adieu, forêts. cf Caprice CAP 1107.
    The maid of Orleans: Joan's aria. cf Columbia/Melodiya M 33931.
    The maid of Orleans: Joan's prayer. cf Rubini GV 63.
2595 Manfred symphony, op. 58. MRSO: Gennady Rozhdestvensky. Angel/
    Melodiya S 40267.
        /HF 7-75 p66   -SFC 4-13-75 p23
        +NR 3-75 p2   +St 1-76 p110
2596 Manfred symphony, op. 58. LSO; André Previn. HMV ASD 3018.
    (also Angel S 37018).
        +Gr 12-74 p1150   +-NYT 1-18-76 pD1
        ++HF 11-75 p126   +-RR 12-74 p40
        +NR 8-75 p2   +-St 1-76 p110
    Manfred symphony, op. 58. cf Symphonies, nos. 1-6.
2597 Marche slav, op. 31. Overture, the year 1812, op. 49. Romeo and
    Juliet: Fantasy overture. NPhO; Norman Del Mar. Pickwick CN
    2021. (Reissue from Contour 2870 419).
        +Gr 9-76 p432
    Marche slav, op. 31. cf Symphonies, nos. 4-6 (Pye GGCD 303).
    Marche slav, op. 31. cf Symphonies, nos. 4-6 (Sine Qua Non SQN
    140).
    Marche slav, op. 31. cf RIMSKY-KORSAKOV: Capriccio espagnol,
    op. 34.
    Marche slav, op. 31. cf Classics for Pleasure CFP 40254.
    Marche slav, op. 31. cf HMV ESD 7011.
    Marche slav, op. 31. cf RCA CRL 3-2026.
2598 Mazeppa. MASCAGNI: Iris, excerpts. Magda Olivero, s; Marianna
    Radev, ms; David Poleri, Piero de Palma, t; Ettore Bastianini,
    bar; Boris Christoff, bs; Orchestra and Chorus; Jonel Perlea.
    Rococo 1016 (3).
        +NR 10-76 p11

Mazeppa: Gopak. cf RIMSKY-KORSAKOV: Capriccio espagnol, op. 34.

2599 Mazeppa: Introduction; Cossack dance. The Oprichnik: Overture;
     Dance. The Voyevoda, op. 78: Overture; Entr'acte; Dances of
     the maids. Bamberg Symphony Orchestra; Janos Fürst. Turna-
     bout TV 34548. (Q) QTV 34548.
        +Gr 1-76 p1204   +NR 12-74 p4
        +-HFN 12-75 p167   +RR 1-76 p39

Mazeppa: Sleep, my pretty baby. cf Rubini GV 63.

Mazurka, op. 9, no. 3. cf Piano works (CBS 61718).

2600 The months (The seasons), op. 37b (original piano version). Alexei
     Cherkassov, pno; USSR Symphony Orchestra; Yevgeny Svetlanov.
     HMV Melodiya SXDW 3025 (2).
        +Gr 8-76 p300   +RR 8-76 p60
        ++HFN 8-76 p87

The months, op. 37b: Barcarolle. cf Piano works (CBS 61718).

The seasons, op. 37, no. 5: May. cf BACH: Sonata, violin, no. 1,
     S 1001, G minor.

Morning prayer. cf Coronet LPS 3032.

Nocturne, op. 19, no. 4. cf Piano works (CBS 61718).

Nocturne, op. 19, no. 4. cf Command COMS 9006.

Nocturne, op. 19, no. 4. cf L'Oiseau-Lyre DSLO 7.

2601 The nutcracker, op. 71, complete ballet. LSO; Ambrosian Singers;
     André Previn. Angel SB 3788 (2). Tape (c) 4X2S 3788. (also
     HMV SLS 834 Tape (c) TC SLS 834).
        ++Gr 1-73 p1333   +-NR 4-73 p2
        ++HF 3-73 p96   +RR 1-73 p52
        +HFN 1-73 p123   +RR 5-76 p78 tape
        +HFN 3-76 p113 tape   ++SFC 11-19-72 p31

2602 The nutcracker, op. 71. Bolshoi Theatre Orchestra; Gennady Rozh-
     destvensky. Columbia/Melodiya M2 33116 (2). (also HMV Melo-
     diya SXDW 3028) (Reissue from Artia).
        +Gr 11-76 p812   +-NR 2-75 p2
        +-HF 3-75 p72   -RR 11-76 p84
        +-HFN 11-76 p170   +-St 3-75 p109

2603 The nutcracker, op. 71. National Philharmonic Orchestra; Richard
     Bonynge. London CSA 2239 (2). (also Decca SXL 6688/9 Tape
     (c) KSXC 6688/9).
        +Gr 11-75 p833   ++NR 2-75 p2
        +HF 3-75 p71   ++NYT 3-9-75 pD23
        +HFN 11-75 p169   +RR 11-75 p55
        +HFN 1-76 p125 tape   +St 3-75 p109

2604 The nutcracker, op. 71. COA; St. Bavo Cathedral Boys' Choir;
     Antal Dorati. Philips 6747 257 (2).
        ++NR 12-76 p5   ++SFC 10-24-76 p35

The nutcracker, op. 71. cf GRIEG: Peer Gynt, op. 46: Suite, no.
     1; op. 55: Suite, no. 2: Solveig's song.

2605 The nutcracker, op. 71, excerpts. Ambrosian singers; LSO; André
     Previn. HMV ASD 3051. Tape (c) TC ASD 3051.
        +Gr 2-75 p1556   ++RR 4-76 p82 tape
        ++RR 3-75 p36

The nutcracker, op. 71, excerpts. cf Works, selections (Decca
     DPA 547/8).

The nutcracker, op. 71, excerpts. cf ADAM: Giselle.

2606 The nutcracker, op. 71: Overture, March. Romeo and Juliet: Fan-
     tasy overture. Swan Lake, op. 20, excerpts. PO; Eugene Ormandy.
     CBS 30086.
        +-HFN 11-76 p173   -RR 11-76 p84

The nutcracker, op. 71: Pas de deux, Act 2. cf ADAM: Giselle:
     Pas de deux; Grand pas de deux and finale.
The nutcracker, op. 71: Suite. cf Capriccio italien, op. 45.
The nutcracker, op. 71: Waltz. cf Eugene Onegin, op. 24: Waltz.
The Oprichnik: Overture; Dance. cf Mazeppa: Introduction; Cossack
     dance.
Overture, the year 1812, op. 49. cf Marche slav, op. 31.
Overture, the year 1812, op. 49. cf Serenade, strings, op. 48,
     C major.
Overture, the year 1812, op. 49. cf Works, selections (Decca
     DPA 547/8).
Overture, the year 1812, op. 49. cf RAVEL: Bolero.
Overture, the year 1812, op. 49. cf Angel S 37232.
Overture, the year 1812, op. 49. cf Bruno Walter Society RR 443.
Pater noster. cf CBS SQ 79200.
Pezzo capriccioso, op. 62. cf SCHUMANN: Concerto, violoncello,
     op. 129, A minor.
2607 Piano works: Capriccioso, op. 19, no. 5. Chanson triste, op. 40,
     no. 2. Humoresque, op. 10, no. 2. Mazurka, op. 9, no. 3. The
     months (The seasons), op. 37b: Barcarolle. Nocturne, op. 19,
     no. 4. Romance, op. 5, F minor. Scherzo, op. 40, no. 11.
     Scherzo humoristique, op. 19, no. 2. Song without words, op.
     2, no. 3, F major. Song without words, op. 40, no. 6, A minor.
     Valse scherzo, op. 7, A major. Philippe Entremont, pno. CBS
     61718. (Reissue from SBRG 7213).
          +Gr 9-76 p446   +RR 8-76 p72
          +-HFN 10-76 p183
2608 Pique Dame (Queen of spades), op. 68. Tamara Milashkina, s; Val-
     entina Levko, ms; Galina Borisova, con; Vladimir Atlantov, t;
     Vladimir Valaitis, bar; Andrei Fedoseyev, bs; Bolshoi Theatre
     Orchestra; Mark Ermler. Columbia/Melodiya M 33328 (3).
          +-HF 4-76 p120   +SR 3-6-76 p38
          +NR 2-76 p10   +St 4-76 p116
          +ON 12-6-75 p52
Pique Dame, op. 68: Lisa's aria, Act 3. cf Muza SX 1144.
Queen of spades, op. 68: Twill soon be midnight. cf Rubini GV 63.
Queen of spades, op. 68: Twill soon be midnight. cf Rubini RS 301.
Pique Dame, op. 68: Yeletsky's aria. cf Caprice CAP 1062.
Quartet, strings, no. 1, op. 11, D major: Andante cantabile. cf
     Symphonies, nos. 4-6.
Romance, op. 5, F minor. cf Piano works (CBS 61718).
Romeo and Juliet: Fantasy overture. cf Fate, op. 77.
Romeo and Juliet: Fantasy overture. cf Marche slav, op. 31.
Romeo and Juliet: Fantasy overture. cf The nutcracker, op. 71:
     Overture, March.
Romeo and Juliet: Fantasy overture. cf SMETANA: The bartered
     bride: Overture, Polka, Furiant, Dance of the comedians.
Romeo and Juliet: Overture. cf BORODIN: In the steppes of central
     Asia.
Romeo and Juliet: Overture. cf Symphonies, nos. 4-6.
Romeo and Juliet: Overture. cf BERLIOZ: Romeo and Juliet, op. 17:
     Romeo alone; Festival of the Capulets; Love scene; Queen Mab:
     Scherzo.
Scherzo, op. 40, no. 11. cf Piano works (CBS 61718).
Scherzo humoristique, op. 19, no. 2. cf Piano works (CBS 61718).
The seasons. cf The months.

2609 Serenade, strings, op. 48, C major. Souvenir de Florence, op. 70,
      D minor. AMF; Neville Marriner. Argo ZRG 584. Tape (c) 584.
          -ARG 3-72 p325   /MJ 11-76 p44
2610 Serenade, strings, op. 48, C major. Hamlet, op. 67. NYP; Leonard
      Bernstein. Columbia M 34128. Tape (c) MT 34128. (also CBS
      76506 Tape (c) 40-76506).
          +-Gr 9-76 p432   +MJ 11-76 p44
          +-HF 11-76 p132   +NR 7-76 p3
          +-HFN 9-76 p130   +RR 9-76 p67
          +-HFN 12-76 p155 tape
2611 Serenade, strings, op. 48, C major. Overture, the year 1812, op.
      49. VSO; Michael Gielen. Everest SDBR 3394.
          /NR 4-76 p3
2612 Serenade, strings, op. 48, C major. Francesca da Rimini, op. 32.
      LSO; Leopold Stokowski. Philips 6500 921. Tape (c) 7300 364.
          +Gr 5-76 p1766   +-NR 6-76 p4
          -HF 11-76 p132   +RR 5-76 p58
          +-HFN 4-76 p117   +RR 11-76 p110 tape
          /MJ 11-76 p44
      Serenade, strings, op. 48, C major. cf Eugene Onegin, op. 24:
      Waltz.
      Serenade, strings, op. 48, C major. cf DVORAK: Serenade, strings,
      op. 22, E major (Argo 848).
      Serenade, strings, op. 48, C major. cf DVORAK: Serenade, strings,
      op. 22, E major (DG 2548 121).
      Serenade, strings, op. 48, C major. cf DVORAK: Serenade, strings,
      op. 22, E major (HMV ASD 3036).
      Serenade, strings, op. 48, C major. cf GRIEG: Holberg suite,
      op. 40.
      Serenade, strings, op. 48, C major: Valse. cf RCA ARM 4-0942/7.
      Sérénade mélancolique, op. 26. cf Concerto, violin, op. 35, D
      major.
      Sérénade mélancolique, op. 26. cf RCA ARM 4-0942/7.
2613 Sleeping beauty, op. 66. LSO; André Previn. Angel SCLX 3812 (3).
      Tape (c) 4XS 3812. (also HMV SLS 5001 Tape (c) TC SLS 5001).
          +Gr 12-74 p1150   ++RR 12-74 p41
          ++HF 3-75 p71   +RR 10-76 p104 tape
          +HFN 12-76 p155 tape   /SFC 11-24-74 p31
          +-NR 2-75 p3   ++St 5-75 p103
          ++NYT 3-9-75 pD23
      Sleeping beauty, op. 66, excerpts. cf ADAM: Giselle.
      Sleeping beauty, op. 66: Blue bird pas de deux. cf ADAM: Giselle:
      Pas de deux; Grand pas de duex and finale.
      Sleeping beauty, op. 66: Coda prologue. cf BEETHOVEN: Sonata,
      piano, no. 14, op. 27, no. 2, C sharp minor.
      Sleeping beauty, op. 66: Suite. cf Eugene Onegin, op. 24: Polo-
      naise.
      Sleeping beauty, op. 66: Suite. cf MUSSORGSKY: Khovanschina.
      Sleeping beauty, op. 66: Waltz. cf Eugene Onegin, op. 24: Waltz.
      Sleeping beauty, op. 66: Waltz. cf Columbia M 34127.
      Sonata, piano, op. 37, G major. cf STRAVINSKY: Sonata, piano,
      F minor.
      Song without words (Chant sans paroles), op. 2, no. 3, F major.
      cf Piano works (CBS 61718).
      Song without words, op. 40, no. 6, A minor. cf Piano works (CBS
      61718).

Chant sans paroles, op. 40, no. 6, A minor. cf Decca PFS 4351.

2614 Songs: Again as before alone, op. 73, no. 6; At the ball, op. 38,
      no. 3; Believe me not, my friend, op. 6, no. 1; Cradle song,
      op. 16, no. 1; The fearful minute, op. 28, no. 6; If I'd
      only known, op. 47, no. 1; In this moonlit night, op. 73, no.
      3; It was in the early spring, op. 38, no. 2; Sleep, my wistful
      friend, op. 47, no. 4; Was I not a little blade of grass, op.
      47, no. 7; Why, op. 6, no. 5. Galina Vishnevskaya, s; Mstislav
      Rostropovich, pno. Angel S 37166.
            ++HF 9-76 p97   +ON 5-76 p48
            +-NR 6-76 p14   ++St 6-76 p111

2615 Songs (choral): Blessed is he that smiles; Evening; The golden
      cloud had slept; Hymn to St. Cyril and St. Methodius; Juris-
      prudence students' song; Legend, op. 54, no. 5; The merry
      voice grew silent; Morning prayer, op. 39, no. 1; The nightin-
      gale; Tis not the cuckoo in the damp woods; To sleep; Without
      time or season. Vadim Korshunov, t; USSR Russian Chorus;
      Aleksander Sveshnikov. HMV Melodiya ASD 3165.
            +Gr 3-76 p1499   +RR 2-76 p63
            +HFN 3-76 p107

Songs: Again as before, op. 73, no. 6. cf Pye QS PCNH 4.

Songs: Again as before alone, op. 73, no. 6; Cradle song, op. 16,
      no. 1; Do not believe, my friend, op. 6, no. 1; The fearful
      minute, op. 28, no. 6; If I'd only known, op. 47, no. 1; In
      this moonlight, op. 73, no. 3; It was in the early spring, op.
      38, no. 2; Mid the din of the ball, op. 38, no. 3; Sleep my
      poor friend, op. 47, no. 4; Was I not a little blade of grass,
      op. 47, no. 7; Why, op. 6, no. 5. cf MUSSORGSKY: Songs (HMV
      SLS 5055).

Songs: At the ball, op. 38, no. 3. cf Rubini RS 301.

Songs: Au jardin, pres du ruisseau, op. 46, no. 4; L'Aube, op. 46,
      no. 6; Larmes humaines, op. 46, no. 3.

Songs: Do not believe me, my friend, op. 6, no. 1; None but the
      lonely heart, op. 6, no. 6; Not a word, my friend, op. 6, no.
      2. cf Philips 6780 751.

Songs: It was in early spring, op. 38, no. 2; The lights in the
      rooms were fading, op. 63, no. 5; Reconciliation, op. 25, no.
      1; Serenade, op. 63, no. 6; Take my heart away; Why did I
      dream of you, op. 28, no. 3. cf MUSSORGSKY: Songs and dances
      of death.

Songs: The merry voices grew silent. cf HMV ASD 3200.

Songs: None but the lonely heart, op. 6, no. 6. cf Desto DC 7118/9.

Songs: None but the lonely heart, op. 6, no. 6. cf Columbia/
      Melodiya M 33593.

Songs: None but the lonely heart, op. 6, no. 6. cf Musical Heri-
      tage Society MHS 3276.

Souvenir de Florence, op. 70, D minor. cf Serenade, strings, op.
      48, C major.

Souvenir d'un lieu cher, op. 42: Scherzo. cf RCA ARM 4-0942/7.

The storm, op. 76. cf Fatum, op. 77 (Philips 6500 467).

The storm, op. 76. cf Fatum, op. 77 (Vox 513460).

2616 Suite, no. 3, op. 55, G major. OSCCP; Adrian Boult. Decca ECS
      766. (Reissue from LXT 5099).
            +-Gr 3-76 p1472   -RR 2-76 p42
            +HFN 8-75 p89

2617 Suite, no. 3, op. 55, G major. LPO; Adrian Boult. HMV (Q) ASD

3135. Tape (c) TC ASD 3135.
    ++Gr 11-75 p834   ++HFN 8-76 p59 tape
    -Gr 9-76 p497 tape   ++RR 10-75 p58
    ++HFN 10-75 p150

2618 Swan Lake, op. 20. Ida Haendel, vln; Douglas Cummings, vlc; LSO;
     André Previn. HMV SQ SLS 5070 (3). (also Angel SX 3834 Tape
     (c) 4X3S 3834).
       +-Gr 12-86 p1009   +RR 12-76 p66
       ++HFN 12-76 p149

     Swan Lake, op. 20, excerpts. cf The nutcracker, op. 71: Overture,
     March.
     Swan Lake, op. 20, excerpts. cf Works, selections (Decca DPA 547/8).
     Swan Lake, op. 20, excerpts. cf ADAM: Giselle.
2619 Swan Lake, op. 20: Ballet suite. Royal Swedish Orchestra; Leif
     Segerstam. Everest SDBR 3390.
       +-NR 6-76 p4   +-SR 5-29-76 p52

     Swan Lake, op. 20: Dance of the swans, excerpt. cf CBS 30072.
     Swan Lake, op. 20: Pas de deux, Act 3. cf ADAM: Giselle: Pas
     de deux; Grand pas de deux and finale.
     Swan Lake, op. 20: Suite. cf Eugene Onegin, op. 24: Polonaise.
     Swan Lake, op. 20: Suite. cf MUSSORGSKY: Khovanschina.
     Swan Lake, op. 20: Waltz. cf Eugene Onegin, op. 24: Waltz.
2620 Symphonies, nos. 1-3. NYP; Leonard Bernstein. CBS 78300. (Re-
     issues from 72949, 73047, 73098). (also Columbia D3M 32996).
       +Gr 7-76 p182   +RR 7-76 p64
       +-HFN 8-76 p93
2621 Symphonies, nos. 1-6. Manfred symphony, op. 58. VPO; Lorin
     Maazel. Decca D8D 6 (6). (Reissue)
       +Gr 9-76 p477   +RR 9-76 p67
       ++HFN 10-76 p181
2622 Symphony, no. 1, op. 13, G minor. NPhO; Riccardo Muti. HMV SQ
     ASD 3213. Tape (c) TC ASD 3213. (also Angel S 37114).
       +Gr 7-76 p181   -RR 7-76 p64
       +-HFN 7-76 p98   +STL 6-6-76 p37
       /HFN 12-76 p155 tape
2623 Symphony, no. 1, op. 13, G minor. LSO; Antal Dorati. Philips
     6582 016.
       +-Gr 12-76 p1002   +-RR 11-76 p84
       +-HFN 11-76 p168
2624 Symphony, no. 1, op. 13, G minor. USSR Symphony Orchestra;
     Konstantin Ivanov. Westminster WGS 8319.
       +NR 10-76 p3
2625 Symphonies, nos. 4-6. Quartet, strings, no. 1, op. 11, D major:
     Andante cantabile. Marche slav, op. 31. Hallé Orchestra;
     John Barbirolli. Pye GGCD 303 (2). (Reissues from CCL 30154,
     30116, 30146, GSGC 14076).
       ++Gr 8-76 p300   ++RR 8-76 p54
2626 Symphonies, nos. 4-6. BSO; Pierre Monteux. RCA LVL 3-7530 (3).
     (Reissues from SB 2093, 2045, HMV ALP 1356).
       +-Gr 12-75 p1054   /-RR 11-75 p56
       +-HFN 11-75 p173   +ST 2-76 p737
2627 Symphonies, nos. 4-6. Marche slav, op. 31. Romeo and Juliet:
     Overture. Utah Symphony Orchestra; Maurice Abravanel. Sine
     Qua Non SQN 140 (4).
       ++IN 2-76 p8
2628 Symphony, no. 4, op. 36, F minor. Symphony, no. 6, op. 74, B

        minor. Symphony Orchestra; Serge Koussevitzky. Bruno Walter
        Society SID 730 (2).
              -ARSC Vol VIII, no. 2-3 p90
2629 Symphony, no. 4, op. 36, F minor. Scottish National Orchestra;
        Alexander Gibson. Classics for Pleasure CFP 40228.
              +-Gr 1-76 p1204   -RR 11-75 p56
              +-HFN 11-75 p169
2630 Symphony, no. 4, op. 36, F minor. NYP; Leonard Bernstein. Col-
        umbia XM 33886. Tape (c) XMT 33886 (ct) XMA 33886 (Q) XMQ 33886
        Tape (c) QAX 33886. (also CBS 76482 Tape (c) 40-76482).
              -Gr 7-76 p182   +MJ 5-76 p29
              -Gr 9-76 p497   +NR 4-76 p4
              -HF 11-76 p132   -RR 6-76 p57
              -HFN 6-76 p101   ++SFC 5-30-76 p24
              /-HFN 10-76 p185 tape   ++St 9-76 p128
2631 Symphony, no. 4, op. 36, F minor. VPO; Claudio Abbado. DG 2530
        651.
              ++Gr 11-76 p812   +-RR 11-76 p84
              ++HFN 11-76 p168
2632 Symphony, no. 4, op. 36, E minor. BPhO; Lorin Maazel. DG 2548
        176. Tape (c) 3348 176. (Reissue from 138789).
              +-Gr 11-75 p840   +RR 10-75 p58
              +HFN 10-75 p152   +-RR 9-76 p93 tape
2633 Symphony, no. 4, op. 36, F minor. Symphony, no. 6, op. 74, B
        minor. BPhO; Herbert von Karajan. HMV ASD 2814, 2816 (4).
        Tape (c) TC ASD 2814, TC 2816. (Reissues from SLS 833).
              +Gr 6-75 p50   ++RR 4-75 p36
              +Gr 5-75 p2032 tape   +-RR 3-76 p78
2634 Symphony, no. 5, op. 64, E minor. NPhO; Leopold Stokowski. Decca
        SDD 493.
              +HFN 12-76 p151   +-RR 12-77 p66
2635 Symphony, no. 5, op. 64, E minor. BPhO; Herbert von Karajan.
        DG 2530 699. Tape (c) 3300 699.
              +Gr 8-76 p300   +HFN 10-76 p177
              ++Gr 10-76 p658 tape   +-SFC 10-24-76 p35
2636 Symphony, no. 5, op. 64, E minor. BPhO; Herbert von Karajan.
        HMV Q4ASD 2815. Tape (c) TC ASD 2815. (Reissue from SLS 833).
        (also Angel S 36885 Tape (c) 8XS 36885 (c) 4XS 36885).
              ++Gr 6-74 p112 Quad   ++RR 2-74 p36 Quad
              +Gr 1-75 p1357   +-RR 4-76 p82 tape
              +Gr 5-75 p2032 tape
2637 Symphony, no. 5, op. 64, E minor. BPhO; Rudolf Kempe. HMV SXLP
        30216. (Reissue from ASD 379).
              +Gr 11-76 p812
2638 Symphony, no. 5, op. 64, E minor. CSO; Georg Solti. London CS
        6983. Tape (c) CS 5-6983. (also Decca SXL 6754 Tape (c)
        KSXC 6754).
              -Gr 4-76 p1612   +-RR 4-76 p54
              +HF 10-76 p120   +-RR 8-76 p85 tape
              +-HFN 4-76 p117   -SFC 5-30-76 p24
              +HFN 7-76 p104 tape   +-St 9-76 p128
              +NR 12-76 p6
2639 Symphony, no. 5, op. 64, E minor. COA; Bernard Haitink. Philips
        6500 922. Tape (c) 7300 365.
              ++Audio 12-76 p93   +MJ 5-76 p29
              +Gr 2-76 p1349   ++NR 6-76 p4

```
                     +HF 10-76 p120              -RR 2-76 p42
                    ++HFN 2-76 p113             ++SFC 5-30-76 p24
                     +HFN 3-76 p113              +-St 9-76 p128
                     +MJ 5-76 p29              ++STL 2-8-76 p36, 5-9-76 p38
                    ++NR 6-76 p4
```

2640 Symphony, no. 5, op. 64, E minor. LSO; Antal Dorati. Philips
      6582 013. (Reissue from EMI Mercury AMS 16125).
```
                     +Gr 10-76 p611              -RR 9-76 p68
                     +HFN 10-76 p181
```

2641 Symphony, no. 5, op. 64, E minor. NPhO; Jascha Horenstein. RCA
      GL 25007. Tape (c) GK 25007. (Previously issued by Reader's
      Digest).
```
                     +Gr 10-76 p611              -RR 10-76 p74
                     +HFN 12-76 p148             -RR 12-76 p104 tape
                     +-HFN 12-76 p153 tape
```

      Symphony, no. 5, op. 64, E minor: Themes. cf Decca SB 322.

2642 Symphony, no. 6, op. 74, B minor. MRSO; Gennady Rozhdestvensky.
      Angel/Melodiya SR 40266. (also HMV Melodiya ASD 3226).
```
                     +-Gr 8-76 p300             +-RR 8-76 p60
                     +-HFN 8-76 p87              +St 9-75 p115
                      -NR 8-75 p4
```

2643 Symphony, no. 6, op. 74, B minor. PhO; Paul Kletzki. Classics
      for Pleasure CFP 40220.
```
                    ++HFN 12-76 p151             -RR 12-76 p66
```

2644 Symphony, no. 6, op. 74, B minor. OSCCP; Erich Kleiber. Decca
      ECS 787. (Reissue from LXT 2888).
```
                      -Gr 9-76 p432              -RR 6-76 p58
                      -HFN 6-76 p102
```

2645 Symphony, no. 6, op. 74, B minor. BPhO; Wilhelm Furtwängler.
      DG 2535 165.
```
                     +-Gr 5-76 p1772            +RR 5-76 p22
```

2646 Symphony, no. 6, op. 74, B minor. PhO; Carlo Maria Giulini. HMV
      SXLP 30208. Tape (c) TC EXE 191. (Reissue from Columbia
      SAX 2368).
```
                    ++Gr 6-76 p57              +-RR 6-76 p57
                     +-HFN 6-76 p105           +-RR 12-76 p108 tape
```

2647 Symphony, no. 6, op. 74, B minor. LSO; Loris Tjeknavorian. RCA
      LRL 1-5129.
```
                     +Gr 11-76 p812            +RR 11-76 p16
```

      Symphony, no. 6, op. 74, B minor. cf Symphony, no. 4, op. 36,
      F minor (Bruno Walter Society SID 730).
      Symphony, no. 6, op. 74, B minor. cf Symphony, no. 4, op. 36,
      F minor (HMV 2814, 2816).
      Symphony, no. 6, op. 74, B minor. cf Works, selections (Decca
      DPA 547/8).
      Symphony, no. 6, op. 74, B minor. cf BRAHMS: Symphony, no. 1, op.
      68, C minor.
      Symphony, no. 6, op. 74, B minor. cf RCA CRM 5-1900.
      The tempest, op. 18. cf Fate, op. 77 (Decca 6694).
      The tempest, op. 18. cf Fatum, op. 77 (Philips 6500 467).

2648 Trio, piano, op. 50, A minor. Temianko Trio. Saga 5415.
```
                     +-HFN 7-76 p101           +-RR 2-76 p47
                      /-MT 10-76 p832
```

      Trio, piano, op. 50, A minor: Pezzo elegiaco. cf CBS SQ 79200.
      Valse scherzo. cf Seraphim S 60259.
      Valse scherzo, op. 7, A major. cf Piano works (CBS 61718).

503 TCHAIKOVSKY (cont.)

Valse scherzo, violin and piano, op. 34. cf BACH: Sonata, violin,
no. 1, S 1001, G minor.
Variations on a rococo theme, op. 33. cf CHOPIN: Andante spianato
and grande polonaise, op. 22, E flat major.
Variations on a rococo theme, op. 33. cf SCHUMANN: Concerto,
violoncello, op. 129, A minor.
The Voyevode, op. 78. cf Fatum, op. 77 (Philips 6500 467).
The Voyevoda, op. 78. cf Fatum, op. 77 (Vox 513460).
The Voyevode: Overture. cf STRAUSS, R.: Capriccio, op. 85: Mono-
log des Theaterdirektors.
The Voyevoda, op. 78: Overture; Entr'acte; Dance of the maids.
cf Mazeppa: Introduction; Cossack dance.
2649 Works, selections: Concerto, piano, no. 1, op. 23, B flat minor.
The nutcracker, op. 71, excerpts. Overture, the year 1812,
op. 49. Swan Lake, op. 20, excerpts. Symphony, no. 6, op. 74,
B minor. Peter Katin, pno; RPO, LSO, OSR, Grenadier Guards
Band; Henry Lewis, Edric Kundell, Ernest Ansermet, Kenneth
Alwyn. Decca DPA 547/8 (2). Tape (c) KDPC 547/8.
+Gr 9-76 p483   /HFN 12-76 p155 tape
+HFN 10-76 p183   +-RR 7-76 p64
TCHEREPNIN, Alexander
Fanfare. cf Unicorn RHS 339.
TCHESNOKOV
Hvalite imia Gospodnie. cf Harmonia Humdi HMU 137.
TEIKE
Alte Kameraden. cf DG 2721 077 (2).
Old comrades. cf Pye GH 603.
TELEMANN, Georg Philipp
Bourrée alla polacca. cf CBS 73487.
Concerto, B flat major. cf Concerto, flute, oboe d'amore, violin
and strings, E major.
2650 Concerto, flute, oboe d'amore, violin and strings, E major. Con-
certo, B flat major. Overture, C major. Sinfonia, F major.
German Bach Soloists; Helmut Winschermann. Oryx C3 306.
++ARG 11-76 p39   +St 7-76 p113
Concerto, horn, D major. cf CORELLI: Sonata, violin and continuo,
op. 5, no. 5, G minor.
Concerto, trumpet, D major. cf ALBINONI: Concerto, trumpet, C
major.
Concerto, trumpet, D major. cf HUMMEL: Concerto, trumpet, E flat
major.
Concerto, trumpet, D major. cf Erato STU 70871.
Concerto, trumpet, D major. cf RCA CRL 3-1430.
2651 Concerto, viola, G major. Don Quichotte. Overture, D major.
Stephen Shingles, vla; AMF; Neville Marriner. Argo ZRG 836.
Tape (c) KZRC 836.
++Gr 2-76 p1350   +HFN 6-76 p105
+Gr 6-76 p102 tape   +NR 6-76 p5
++HF 10-76 p122   +RR 2-76 p42
+HFN 3-76 p107   ++RR 5-76 p78 tape
Concerto, viola, G major. cf HANDEL: Concerto, viola, B minor.
2652 Conclusion, D major (Tafelmusik). Overture, D major. TORELLI:
Concerti, trumpet, D major (2). Maurice André, tpt; Concerto
Amsterdam; Frans Brüggen. Telefunken AG 6-41982. (Reissues
from SAWT 9451/2, 9499).
++Gr 9-76 p432   ++RR 8-76 p60
+HFN 9-76 p131

Don Quichotte. cf Concerto, viola, G major.
Don Quichotte. cf BACH, J. C.: Sinfonias, op. 18, nos. 2, 4, 6.
Fantasias. cf BASSANO: Ricercare.
Fantasia, E major. cf London STS 15222.
2653 Fantasias, flute (12). Jean-Pierre Rampal, flt. Odyssey Y 33200.
           ++Audio 10-76 p150   ++St 8-76 p99
           ++NR 1-76 p15
Der getreue Musikmeister: Overture; Heldenmusik. cf CBS 61648.
Heroic music. cf Klavier KS 551.
Melante: Heroic music. cf FANTINI: Sonatas, organ.
2654 Musique de table. Frans Brüggen, rec; Concerto Amsterdam. Tele-
           funken 6-35298. (also 2648 006/8).
           ++SFC 8-15-76 p38
Overture, C major. cf Concerto, flute, oboe d'amore, violin and
     strings, E major.
Overture, D major. cf Conclusion, D major.
Overture, D major. cf Concerto, viola, G major.
Partita, no. 5, E minor. cf Orion ORS 76216.
2655 Pimpinone (incorporating Tessarini's Violin concerto, op. 1, no.
           7, B flat major; Albinoni's Oboe concerto, op. 9, no. 8, B
           flat major; Vivaldi's Violin concerto, op. 7, no. 2, C major).
           Uta Spreckelsen, s; Siegmund Nimsgern, bar; Florilegium Musi-
           cum Ensemble; Hans Ludwig Hirsch. Telefunken 6-35285ER (2).
           +-HF 7-76 p94   ++SFC 1-18-76 p38
           +-HFN 5-76 p111   ++St 6-76 p114
           +NR 2-76 p11
Pimpinone. cf ALBINONI: Concerto, oboe and strings, op. 9, no.
     8, G minor.
Sinfonia, F major. cf Concerto, flute, oboe d'amore, violin and
     strings, E major.
Sonatas, bassoon, E minor, D major, F minor. Washington University
     RAVE 761.
Sonata, recorder, F major. cf Cambridge CRS 2826.
Sonatas, 2 saxophones, nos. 1-3, 6. cf LECLAIR: Sonatas, 2 saxo-
     phones, C major, F major, A flat major.
Songs: Ach Herr, strafe mich nicht. cf Telefunken 6-41929.
Suite, recorder and strings, A minor. cf Suite, viola da gamba
     and strings, D major.
2656 Suite, viola da gamba and strings, D major. Suite, recorder and
           strings, A minor. Johannes Koch, vla da gamba; Hans-Martin
           Linde, rec; Collegium Aureum; Rolf Reinhardt. BASF 20 29015-8.
           +Gr 8-76 p305   +RR 5-76 p58
           +HFN 5-76 p111
Suite, solo violin, F major. cf Telefunken DT 6-48075.
Wie schön leuchtet der Morgenstern (2 settings). cf BACH: Chorale
     preludes (FAS 16).
TEMPLETON
     Bach goes to town. cf Angel S 36095.
TESSARINI, Carlo
     Concerto, violin, no. 8, op. 1, no. 7, B major. cf ALBINONI:
          Concerto, oboe and strings, op. 9, no. 8, G minor.
     Sonata, trumpet, D major. cf RCA CRL 3-1430.
THALBERG, Sigismond
     Les capricieuses, op. 64. cf Piano works (Candide 31084).
     Concerto, piano, op. 5, F minor. cf Piano works (Candide 31084).
     Fantasy on Meyerbeer's "Les Huguenots", op. 20. cf Piano works
          (Candide 31084).

Fantasy on Rossini's "Barber of Seville", op. 63. cf LISZT:
    Ballade, no. 2, G 171, B minor
Fantasy on Rossini's "Moise", op. 33. cf LISZT: Ballade, no. 2,
    G 171, B minor.
2657 Piano works: Concerto, piano, op. 5, F minor. Les capricieuses,
    op. 64. Fantasy on Meyerbeer's "Les Huguenots", op. 20.
    Variations on "Home, sweet home", op. 72. Variations on "The
    last rose of summer", op. 73. Michael Ponti, pno; Westpahlian
    Orchestra; Richard Kapp. Candide CE 31084.
            +HF 9-75 p93   +St 2-76 p108
            +-NR 4-75 p6
Trio, op. 69, A major. cf ALKAN: Trio, no. 1, op. 30, G minor.
Variations on "Home, sweet home", op. 72. cf Piano works (Candide
    31084).
Variations on "The last rose of summer", op. 73. cf Piano works
    (Candide 31084).
THIMMIG, Leslie
    Seven profiles. cf EDWARDS: Quartet, strings.
THOMAS, Ambroise
    Hamlet: A vos jeux, mes amis. cf Telefunken AG 6-41945.
    Hamlet: A vos jeux...Partagez-vous me fleurs...Et maintenant
        écoutez ma chanson. cf HMV SLS 5057.
    Hamlet: J'ai pu frapper le miserable...Etre ou non etre. cf Club
        99-101.
    Hamlet: Mad scene. cf BELLINI: I puritani: O rendetemi la speme.
    Hamlet: Mad scene. cf HMV RLS 719.
    Mignon: Connais-tu le pays. cf CBS 76522.
    Mignon: Connais-tu le pays. cf Decca GOSD 674/6.
    Mignon: Elle ne croyait pas. cf Rubini GV 70.
    Mignon: Je suis Titania. cf Rubini GV 68.
    Mignon: Legères hirondelles. cf Rubini GV 65.
    Mignon: Non consoci il bel suol. cf Club 99-100.
    Mignon: Oui, pour ce soir je suis reine des fees. cf Telefunken
        AG 6-41945.
    Mignon: Overture. cf BERLIOZ: Le carnaval romain, op. 9.
    Mignon: Overture. cf CBS 61748.
    Mignon: Overture. cf HMV ESD 7010.
    Mignon: Styrienne. cf Rococo 5379.
    Raymond: Overture. cf CBS 61748.
THOMAS, Mansel
    Prayers from the Gaelic (4). cf Argo ZRG 769.
THOME, Francis
    Rogodon, op. 97. cf Angel S 36095.
THOMPSON, Randall
    Songs: Felices ter; The paper reeds. cf Orion ORS 75205.
    The testament of freedom. cf HANSON: Drum taps: Songs.
THOMSON, Virgil
2658 Autumn (Concertino, harp, strings and percussion). The plow that
    broke the plains. The river. Ann Mason Stockton, hp; Los
    Angeles Chamber Orchestra; Neville Marriner. Angel (Q) S
    37300. (also HMV ASD 3294).
            -HF 9-76 p98   ++RR 12-76 p67
            +HFN 12-76 p132   +SFC 6-13-76 p30
            ++MJ 7-76 p57   ++St 7-76 p83 Quad
            ++NR 5-76 p4
Feast of love. cf HANSON: Psalms.

2659   The plow that broke the plains. The river. Symphony of the Air;
          Leopold Stokowski. Vanguard VSD 2095.
             +St 7-76 p74
       The plow that broke the plains. cf Autumn.
       The plow that broke the plains. cf Vanguard VSD 707/8.
       Quartet, no. 2. cf Vox SVBX 5305.
       The river. cf Autumn.
       The river. cf The plow that broke the plains.
       Symphony on a hymn tune. cf HANSON: Psalms.
THORNE, Francis
       Burlesque overture. cf DIAMOND: Romeo and Juliet.
2660   Concerto, piano and chamber orchestra. Lyric variations II, wind
          quintet and percussion. Sonatina, solo flute. Dennis Russell
          Davies, pno; Harvey Sollberger, flt; St. Paul Chamber Orchestra;
          Böhm Quintet. Serenus 12058.
             ++MJ 7-76 p57   +NR 4-76 p6
       Lyric variations II, wind quintet and percussion. cf Concerto,
          piano and chamber orchestra.
       Rhapsodic variations. cf DIAMOND: Romeo and Juliet.
       Six set pieces for thirteen players. cf SCHINDLER: Sextet, strings,
          in six movements.
       Sonatina, solo flute. cf Concerto, piano and chamber orchestra.
TIERSOT, Julien (Jean Baptiste Elesee)
       Chants de la vieille France (4). cf FAURE: Songs (1750 Arch S
          1754).
TILLIS, Frederick
       Quintet, brass. cf HUSTON: Sounds at night.
TINCTORIS, Johannes de
       Missa 3 vocum: Kyrie. cf HMV SLS 5049.
TIPPETT, Michael
2661   A child of our time. Jessye Norman, s; Janet Baker, ms; Richard
          Cassilly, t; John Shirley-Quirk, bs; BBC Singers, BBC Choral
          Society; BBC Symphony Orchestra; Colin Davis. Philips 6500
          985. Tape (c) 7300 458.
             +Gr 11-75 p882   ++MT 12-76 p1006
             ++HF 2-76 p109   ++NR 2-76 p8
             +HF 8-76 p70 tape   ++NYT 12-21-75 pD18
             +-HFN 12-75 p167   +RR 11-75 p88
             +-MJ 1-76 p24   +St 4-76 p117
             +MM 8-76 p31   +STL 1-4-76 p36
       A child of our time: Negro spirituals (2). cf Abbey LPB 739.
       A child of our time: Negro spirituals (5). cf Abbey LPB 754.
       A child of our time: Negro spirituals (5). cf Grosvenor GRS 1034.
2662   Concerto, 2 string orchestra. Variations on a theme by Corelli.
          Little music, strings. Alan Loveday, Karmel Caine, vln; Ken-
          neth Heath, vlc; AMF; Neville Marriner. Argo ZRG 680.
             ++Gr 1-72 p1225   ++NR 5-72 p2
             ++HF 8-72 p88 .   ++RR 5-76 p58
             +HFN 2-72 p317
       Fantasia concertante on a theme by Corelli. cf ELGAR: Introduction
          and allegro, op. 47.
       Little music, strings. cf Concerto, 2 string orchestras.
       Little music, strings. cf ELGAR: Introduction and allegro, op. 47.
       Preludio al Vespro di Monteverdi. cf Audio 2.
2663   Quartets, strings, nos. 1-3. Lindsay Quartet. L'Oiseau-Lyre
          DSLO 10.

```
        ++Gr 12-75 p1065              ++NR 7-76 p6
        +HFN 12-75 p167              ++RR 12-75 p74
        +MJ 12-76 p144                +St 10-76 p136
        +MM 8-76 p31                 ++STL 1-4-76 p36
        +MT 11-76 p915               +Te 3-76 p31
```

2664 Songs: Boyhood's end; Songs for Achilles; Songs for Ariel; The
      heart's assurance. Philip Langridge, t; John Constable, pno;
      Timothy Walker, gtr. L'Oiseau-Lyre DSLO 14.

```
        +-Gr 11-76 p858              +RR 11-76 p103
        +HFN 10-76 p177
```

Songs for Dov. cf MESSIAEN: Poèmes pour mi.

2665 Suite for the birthday of Prince Charles. Symphony, no. 1. LSO;
      Colin Davis. Philips 9500 107.

```
        ++ARG 11-76 p40              +-NR 12-76 p5
        +Gr 10-76 p612              +RR 10-76 p74
        ++HFN 10-76 p179            *SFC 12-19-76 p50
        ++MJ 12-76 p44             ++St 12-76 p148
```

Symphony, no. 1. cf Suite for the birthday of Prince Charles.
Variations on a theme by Corelli. cf Concerto, 2 string orch-
     estras.

TISHCHENKO, Boris
      Symphony, no. 3. cf SLONIMSKY: Concerto buffo.

TJEKNAVORIAN, Loris
2666 Armenian bagatelles. Requiem for the massacred. Howard Snell,
      tpt; London Percussion Virtuosi; LSO Members; Loris Tjekna-
      vorian. Unicorn RHS 334.

```
        +/Gr 1-76 p1209             +-MT 2-76 p143
        +HFN 1-76 p121             ++RR 12-75 p92
```

Requiem for the massacred. cf Armenian bagatelles.

2667 Simorgh. Roudaki Hall Soloists; Loris Tjeknavorian. Unicorn
      RHS 333.

```
        +HFN 12-75 p167             +RR 2-76 p43
```

TOCH, Ernest
2668 Capricetti. Geographical fugue. Little dances, piano (3).
      Sonata, violin and piano, no. 1. Valse, spoken chorus. Eud-
      ice Shapiro, vln; Ralph Berkowitz, Armen Guzelimian, pno; Los
      Angeles Camerata Chorus; H. Vincent Mitzelfelt. Crystal S 502.

```
        +MJ 7-76 p57                +NR 6-76 p8
```

Geographical fugue. cf Capricetti.
Impromptu, op. 90c. cf Command COMS 9006.
Little dances, piano (3). cf Capricetti.
Sonata, violin and piano, no. 1. cf Capricetti.
Valse, spoken chorus. cf Capricetti.

TOMASEK, Václav Jan
2669 Dithyrambs, piano, op. 65 (3). Goethe songs: Am Flusse; Erster
      Verlust, op. 56: Erlkönig; Das Geheimnis; Der König in Thule;
      Der Fischer, op. 59; Rastlose Liebe; Die Sorge; Die Spinnerin,
      op. 55; Das Veilchen, op. 57; Wanderers Nachtlied, op. 58.
      Libuse Márová, s; Jindrich Jindrák, bar; Dagmar Simonkova,
      Alfréd Holecek, pno. Panton 110 516.

```
        +HFN 12-76 p149             +RR 12-76 p96
```

Goethe songs: Am Flusse; Erster Verlust, op. 56; Erlkönig; Der
      Fischer, op. 59; Das Geheimnis; Der König in Thule; Rastlose
      Liebe; Die Sorge; Die Spinnerin, op. 55; Das Veilchen, op. 57;
      Wanderers Nachtlied, op. 58. cf Dithyrambs, piano, op. 65.

TOMASI, Henri
      Le muletier des Andes. cf London CS 7015.

TOMKINS, Thomas
      Barafostus' dream. cf BASF BAC 3075.
      A fancy for two to play. cf Pelca PRS 40581.
      The Lady Folliott's galliard. cf Turnabout TV 34017.
      A sad pavan for these distracted times. cf BYRD: Fantasias, nos.
         2 and 3.
      Songs: My beloved spake; O sing unto the Lord. cf Abbey LPB 757.
      Songs: O let me live; O let me die. cf Harmonia Mundi HMD 223.
      Songs: See, see the shepherd's queen; Weep no more, thou sorry
         boy. cf Harmonia Mundi 593.
      Songs: Too much I once lamented. cf Argo ZRG 833.
      Songs: When David heard. cf Abbey LPB 750.
TOOP, David
      The chairs story. cf EASTLEY: The centriphone.
      The divination of the bowhead whale. cf EASTLEY: The centriphone.
      Do the bathosphere. cf EASTLEY: The centriphone.
TOPLADY
      Rock of ages. cf BBC REB 228.
TORELLI, Giuseppe
      Concerti, trumpet, D major (2). cf TELEMANN: Conclusion, D major.
      Concerto grosso, op. 8, no. 6, G minor. cf CORELLI: Concerto
         grosso, op. 6, no. 8, G minor.
      Sonata a 5, no. 7, D major. cf Philips 6580 114.
TORRE, Francisco de la
      Dime, triste corazón. cf DG 2530 504.
      Pámpano verde, racimo albar. cf DG 2530 598.
TORRES
      Saeta, no. 4. cf BACH: Chorale prelude: Aus tiefer Not schrei
         ich zu dir, S 687.
TORROBA, Federico
      Burgalesa. cf Enigma VAR 1015.
      Madronos. cf Angel S 36094.
TORTELIER, Paul
      Miniatures, 2 violoncellos (3). cf HMV SQ ASD 3283.
      Offrande. cf MARTIN: Petite symphonie concertante.
      Valse, no. 1. cf HMV SQ ASD 3283.
TOSTI, Francesco
      Aprile. cf Collectors Guild 611.
      Dopo. cf Club 99-100.
      La mia canzone. cf Everest SDBR 3382.
      Ninon. cf Club 99-101.
      Penso. cf Rubini RS 301.
      Songs: Goodbye; Mattinata; La serenata. cf HMV RLS 719.
      Songs: Parted. cf Argo ZFB 95/6.
      Songs: My dreams; The Venetian song. cf Transatlantic XTRA 1159.
TOURNEMIRE, Charles
      Chorale sur le victimae Paschali. cf Vista VPS 1029.
      Chorale sur le victimae Paschali. cf Westminster WGS 8116.
TOURS
      Songs: Mother o' mine. cf Argo ZFB 95/6.
TOYE, Geoffrey
      The haunted ballroom: Waltz. cf RCA GL 25006.
TRIANA (17th century Spain)
      Dinos, madre del donsel. cf DG 2530 504.
      A TRIBUTE TO DAVID OISTRAKH. cf Seraphim S 60259.
TRIMARCHI
      Un bacio ancora. cf Everest SDBR 3382.

TRIMBLE, Lester
    Songs: Love seeketh not itself to please; Tell me where is fancy
        bred. cf Duke University Press DWR 7306.
TROJAN
    The nightingale's concerto. cf Rediffusion 15-46.
TROMBLY, Preston
    Kinetics III. cf Nonesuch HB 73028 (2).
TRUHLAR
    Impromptu. cf BARTA: Sonata, guitar.
TURCHANINOV
    Litany at the Lity. cf Ikon IKO 3.
TURCO, Giovanni del
    Songs: Occhi belli. cf DG Archive 2533 305.
TURINA, Joaquin
    Danzas fantásticas. cf ALBENIZ: Iberia.
    Fandanguillo. cf Angel S 36094.
2670 Fandanguillo, op. 36. Homenage a Tárrega, op. 69: Garrotín;
        Soleares. Ráfaga, op. 53. Sonata, guitar, op. 71, C major.
        Sevillana, op. 29. VILLA-LOBOS: Preludes (5) (arr. Carlevaro/
        Costanzo). Irma Costanzo, gtr. HMV SXLP 30215.
            +-Gr 8-76 p319   +RR 6-76 p76
            +HFN 6-76 p101
    Garrotín. cf Angel S 36094.
    Homenage a Tárrega, op. 69: Garrotín; Soleares. cf Fandanguillo,
        op. 36.
    Poema en forma de canciones, op. 19. cf FALLA: Mélodies.
    Ráfaga. cf Angel S 36094.
    Ráfaga, op. 53. cf Fandanguillo, op. 36.
    Sacro-monte, op. 55, no. 5. cf London CS 6953.
    Sacro-monte, op. 55, no. 5. cf Opus unnumbered.
    Sevillana, op. 29. cf Fandanguillo, op. 36.
    Soleares. cf Angel S 36094.
    Sonata, guitar, op. 61, D major. cf Fandanguillo, op. 36.
    Sonata, guitar, op. 61, D major. cf Supraphon 111 1230.
    Songs: Saeta en forma de Salve a la Virgen de la Esperanza; El
        fantasma; Cantares. cf DG 2530 598.
    Zapateado, op. 8, no. 3. cf London CS 6953.
    TWO HUNDRED YEARS OF AMERICAN MARCHES. cf Michigan University
        SM 0002.
URBANNER, Erich
    Quartet, strings, no. 3. cf HAUBENSTOCK-RAMATI: Quartet, strings,
        no. 1.
USSACHEVSKY, Vladimir
    Conflict. cf Folkways FTS 33901/4.
2671 The creation, 3 scenes. Missa brevis. Jo Ann Otley, s; Macalester
        College Chamber Chorus, University of Utah Chorus; Utah Symph-
        ony Orchestra Members; Ian Morton, Newell Weight. CRI SD 297.
            +HF 5-74 p101   -RR 6-76 p80
            +NR 8-74 p8   +-St 12-74 p146
    Missa brevis. cf The creation, 3 scenes.
    Piece, tape recorder. cf BERGSMA: The fortunate island.
VACEK, Milos
    Indian summer. cf Panton 110 440.
2672 Poem of fallen heroes. VALEK: Symphony, no. 11. Věra Soukupová,
        ms; Jiří Tomášek, vln; Hubert Simáček, vla; Josef Ruzička, pno;
        Jaroslav Josífko, flt; Pavel Verner, ob; Josef Vocatý, flt;

VACEK (cont.)   510

Lumir Vaněk, bsn; František Langweil, hn; Czech Radio Symphony
         Orchestra, Prague Symphony Orchestra; Jaromil Nohejil, Eduard
         Fischer. Panton 110 528.
                   -RR 12-76 p67
VAILLANT, Jean
         Par maintes foys. cf 1750 Arch S 1753.
         Tres doulz amis: Ma dame; Cent mille fois. cf HMV SLS 863.
VAINBERG, Moisei
         Concerto, trumpet and orchestra, op. 94, B flat major. cf
            ARUTYUNIAN: Concerto, trumpet and orchestra, A flat major.
VALDERRABANO, Enriquez de
         De dónde venís, amore. cf DG 2530 504.
VALEK, Jiří
         Symphony, no. 7. cf BARTA: Symphony, no. 3.
         Symphony, no. 11. cf VACEK: Poem of fallen heroes.
VALENTI, Antonio
         Gavotte. cf HMV RLS 723.
VALLET, Nicolas
         Galliarde. cf DG Archive 2533 302.
         Prelude. cf DG Archive 2533 302.
         Slaep, soete, slaep. cf DG Archive 2533 302.
VAN DER HOVE
         Lieto godea. cf DG Archive 2533 323.
VAN VACTOR, David
         Sonatina, flute and piano. cf BURTON: Sonatina, flute and piano.
VANHAL, Jan
         Fugues (6). cf Panton 110 418/9.
VANHAL, Johann Baptist
         Cantabile. cf Philips 6581 017.
VANTOURA, Suzanne Haik
2673 La Musique de la Bible Révélée. Psaume 23; Bénédiction sacer-
             dotale; Psaume 24; Cantique des cantiques; Psaume 6; Les
             lamenatations; Psaume 133; Ecoute Israël; Psaume 150; Elégie de
             David; Psaume 122; Psaume 123; Esther; Le buisson ardent.
             Adolphe Attia, t; Michel Scherb, bar; Emile Kaçmann, bs; Mar-
             tine Geliot, Celtic harp; Raymond Couste, lt; Gérard Perrotin,
             perc; Pierre Pollin, tpt; Chorus; Maurice Benhamou, cond.
             Harmonia Mundi HMU 989.
                   +-St 11-76 p162
VAQUEIRAS, Raimbault de
         Kalenda maya. cf Vanguard VCS 10049.
VARESE, Edgard
2674 Amériques. Ecuatorial. Nocturnal. Utah Symphony Orchestra;
             Maurice Abravanel. Vanguard SRV 308 SD. (Reissue).
                   ++MJ 7-74 p30   +St 7-76 p74
                   ++SFC 3-24-74 p27
         Denisty 21.5. cf Nonesuch HB 73028.
         Ecuatorial. cf Amériques.
         Nocturnal. cf Amériques.
         Octandre. cf Caprice RIKS LP 34.
VARLAMOV, Alexander
         Romance. cf Rubini GV 63.
VASARHELYI
         Folk song suite, no. 2, excerpt. cf Hungaroton SLPX 11762.
VASQUEZ, Juan
         En la fuente del rosel. cf DG 2530 504.
         Vos me matastes (arr. Fuenllano). cf DG 2530 504.

VASS, Lajos
    Nocturne. cf Hungaroton SLPX 11762.
VAUGHAN THOMAS, David
    Saith O Ganeuon. cf Argo ZRG 769.
VAUGAN WILLIAMS, Ralph
2675 Concerto grosso. Fantasia on a theme by Thomas Tallis. Partita,
    2 string orchestras. LPO; Adrian Boulr. HMV SQ ASD 3286.
    (also Angel S 37211).
        +Gr 11-76 p812   ++RR 11-76 p85
        +HFN 12-76 p149   ++SFC 12-26-76 p34
2676 English folk song suite (arr. Jacob). Fantasia on "Greensleeves".
    Fantasia on a theme by Thomas Tallis. VSOO; Adrian Boult.
    ABC Westminster WGS 8111. (Reissue from WST 14111).
        +-Gr 12-76 p1009
    English folk song suite. cf Works, selections (Pye GSGC 15019).
    Evening hymn and last post. cf BBC REB 228.
2677 Fantasia on a theme by Thomas Tallis. Fantasia on "Greensleeves".
    Variants on "Dives and Lazarus" (5). The lark ascending. AMF;
    Neville Marriner. Argo ZRG 696. Tape (c) KZRC 15696.
        ++Gr 10-72 p708   ++HFN 2-73 p348 tape
        +Gr 2-73 p1565 tape   +NR 4-73 p4
        ++HF 5-73 p97   ++RR 10-72 p86
        +HF 10-76 p147 tape   -SFC 5-20-73 p34
        ++HFN 10-72 p1919   ++St 6-73 p124
    Fantasia on a theme by Thomas Tallis. cf Concerto grosso.
    Fantasia on a theme by Thomas Tallis. cf English folk song suite.
    Fantasia on a theme by Thomas Tallis. cf Works, selections (Pye
    GSGC 15019).
    Fantasia on a theme by Thomas Tallis. cf ELGAR: In the south
    overture, op. 50.
    Fantasia on Christmas carols. cf HELY-HUTCHINSON: Carol symphony.
    Fantasia on "Greensleeves". cf English folk song suite.
    Fantasia on "Greensleeves". cf Fantasia on a theme by Thomas
    Tallis.
    Fantasia on "Greensleeves". cf Works, selections (Pye GSGC 15019).
    Fantasia on "Greensleeves". cf DELIUS: Aquarelles.
    Fantasia on "Greensleeves". cf CBS 30062.
    Fantasia on "Greensleeves". cf CBS 30072.
    Fantasia on "Greensleeves". cf CBS 30073.
    Fantasia on "Greensleeves". cf HMV ESD 7011.
2678 Folksong arrangements: Ca' the yowes; Greensleeves; The seeds of
    love; The unquiet grave; Ward, the pirate; and ten others.
    London Madrigal Singers; Christopher Bishop. Seraphim S 60249.
        +HF 11-75 p126   ++St 4-76 p117
    Greensleeves. cf HMV RLS 716.
2679 Hugh the drover, abridged. Mary Lewis, s; Constance Willis, Nellie
    Walker, con; Tudor Davies, t; Frederic Collier, William Ander-
    son, William Michael, Peter Dawson, Robert Gwynne, bar; Keith
    Faulkner, bs; Orchestra; Malcolm Sargent. Pearl GEM 128.
    (Reissue from HMV D 922-6).
        +Gr 3-86 p1504
    Hymn tune preludes, no. 1, Eventide; no. 2, Dominus regit me. cf
    ELGAR: Works, selections (Polydor 2383 359).
    In the Fen country. cf Symphony, no. 3.
    The lark ascending. cf Fantasia on a theme by Thomas Tallis.
    The lark ascending. cf DELIUS: Aquarelles.

Mass, G minor. cf HMV SLS 5047.
O taste and see. cf HMV SQS 1350.
Old King Cole: Ballet. cf Works, selections (Pye GSGC 15019).
On Wenlock edge: Is my team ploughing. cf HMV HLM 7093.
Partita, 2 string orchestras. cf Concerto grosso.
The posoned kiss: Overture. cf ELGAR: Works, selections (Poly-
    dor 2383 359).
2680 Preludes on Welsh folk songs: Romance, Toccata. Preludes on Welsh
    hymn tunes: Bryn Calfaria, Rhosymedre, Hyfrydol. Songs: A
    choral flourish; The hundredth psalm; O vos omnes; A vision of
    aeroplanes; A voice out of the whirlwind. Exultate Singers;
    Timothy Farrell, org; Garrett O'Brien. RCA GL 25016. Tape
    (c) GK 25016.
              +Gr 10-76 p640   +-RR 10-76 p99
              +-HFN 12-76 p149   +RR 12-76 p104 tape
              +HFN 12-76 p153   ++STL 10-10-76 p37
Preludes on Welsh hymn tunes: Bryn Calfaria, Rhosymedre, Hyfrydol.
    cf Preludes on Welsh folk songs: Romance, Toccata.
Rhosymedre. cf HMV SQ HQS 1356.
The running set. cf ELGAR: Works, selections (Polydor 2383 359).
Sea songs: Quick march. cf ELGAR: Works, selections (Polydor
    2383 359).
2681 Sir John in love. Felicity Palmer, Wendy Eathorne, s; Elizabeth
    Bainbridge, Helen Watts, ms; Robert Tear, Gerald English, t;
    Raimund Herincx, John Noble, Rowland Jones, bar; Robert Lloyd,
    bs; John Alldis Choir; NPhO; Meredith Davies. Angel SCLX 3822
    (2). (also HMV SLS 980).
              +Gr 9-75 p513   ++NR 7-75 p9
              +-HF 9-75 p76   +NYT 8-10-75 pD14
              ++HFN 9-75 p106   +ON 7-75 p30
              ++MM 1-76 p43   ++RR 9-75 p16
              +MT 12-75 p1071   ++St 9-75 p83
Songs: A choral flourish; The hundredth psalm; O vos omnes; A
    vision of aeroplanes; A voice out of the whirlwind. cf Pre-
    ludes on Welsh folk songs: Romance, Toccata.
Songs: Linden Lea. cf Pearl SHE 528.
Songs: O clap your hands. cf Audio 3.
Songs: Mystical songs: Antiphon. cf Argo ZRG 789.
Songs: Mystical songs: The call. cf Saga 5213.
Songs of travel: Youth and love. cf Saga 5213.
2682 Symphony, no. 1. LPO; Adrian Boult. Decca Tape (c) KECC 583.
              +HFN 6-76 p105 tape
2683 Symphony, no. 3. Symphony, no. 5. LPO; Adrian Boult. Decca
    Tape (c) KECC 607.
              +HFN 6-76 p105 tape   +-RR 6-76 p89 tape
2684 Symphony, no. 3. In the Fen country. NPhO; Adrian Boult. HMV
    ASD 2393. Tape (c) TC ASD 2393. (also Angel S 36532).
              +Gr 7-76 p230 tape   ++RR 8-76 p85 tape
              +HFN 6-76 p105 tape
Symphony, no. 5. cf Symphony, no. 3.
2685 Symphony, no. 7. LPO; Adrian Boult. Decca Tape (c) KECC 577.
              +HFN 6-76 p105 tape   +-RR 6-76 p89
Variants on "Dives and Lazaurs" (5). cf Fantasia on a theme by
    Thomas Tallis.
The wasps: March-past of the kitchen utensils. cf BBC REB 228.
The wasps: Overture. cf Works, selections (Pye GSGC 15019).

The wasps: Overture. cf ELGAR: In the south overture, op. 50.
2686 Works, selections: English folk song suite (orch. Jacob).* Fan-
        tasia on "Greensleeves".* Fantasia on a theme by Thomas Tallis.*
        Old King Cole: Ballet. The wasps: Overture. LPO; Adrian Boult.
        Pye GSGC 15019. (*Reissues from Nixa NLP 905).
            +Gr 2-86 p1350   +HFN 1-76 p123
VAUTOR, Thomas
        Mother, I will have a husband. cf Harmonia Mundi HMU 204.
        Shepherds and nymphs. cf Harmonia Mundi HMU 593.
        Sweet Suffolk owl. cf Enigma VAR 1017.
VECCHI, Orazio
        Saltarello. cf Argo ZRG 823.
        This song out for you. cf Hungaroton SLPX 11696.
VEDEL, Artemii
        How long, O Lord, how long. cf HMV Melodiya ASD 3102.
VERACINI, Francesco
        Sonata, violin, op. 1, no. 3. cf LECLAIR: Sonata, violin, no. 3,
        D minor.
VERDELOT, Philippe
        Ave sanctissima Maria. cf HMV SLS 5049.
VERDI, Giuseppe
2687 Aida. Renata Tebaldi, s; Giuletta Simionato, ms; Carlo Bergonzi,
        t; Cornell MacNeil, bar; Vienna Singverein; VPO; Herbert von
        Karajan. Decca Tape (c) K2A 20.
            +Gr 8-76 p341 tape   +RR 8-76 p82 tape
            ++HFN 8-76 p94 tape
2688 Aida, excerpts. Leontyne Price, s; Grace Bumbry, con; Sherrill
        Milnes, bar; Placido Domingo, t; Orchestra; Erich Leinsdorf.
        RCA Tape (c) RK 11706
            +HFN 12-76 p155 tape
2689 Aida: Se que guerrier io fossi...Celeste Aida; Or, di Vulcano al
        tempio muovi...Ritorna vincitor; Gloria all'egitto...O Re;
        Pei sacri numi; Que Radames verra...O patria mia...Rivedrai
        le foreste imbalsamate; Vorrei salvario...Gia i sacerdote
        adunansi; Qual gemito; Morir, si pura e bella. Montserrat
        Caballé, Esther Casas, s; Fiorenza Cossotto, ms Placido Domingo,
        Nicola Martinucci, t; Piero Cappuccilli, bar; Nicolai Ghiaurov,
        Luigi Roni, bs; Royal Opera House Chorus; Royal Military School
        of Music Trumpeters; NPhO; Riccardo Mut. HMV SQ ASD 3292.
        (Reissue from SLS 977).
            +Gr 12-76 p1052   ++RR 12-76 p46
            +HFN 12-76 p151
    Aida: Ballet music. cf GOUNOD: Faust: Ballet music; Waltz.
    Aida: Celeste Aida. cf Cantilena 6238.
    Aida: Celeste Aida. cf Decca SXL 6649.
    Aida: Celeste Aida. cf Everest SDBR 3382.
    Aida: Celeste Aida; Su, del nilo...Ritorna vincitor; Gloria all'
        egitto; Gia i sacerdoti...Ah, tu dei vivre. cf Arias (Decca
        DPA 555/6).
    Aida: Ciel mio padre, A te grave cagion m'adduce, Aida, Pur ti
        riveggo. cf Court Opera Classics CO 347.
    Aida: Final duet. cf Rococo 5377.
    Aida: Gloria all'egitto. cf Works, selections (DG 2530 549).
    Aida: Grand march. cf Classics for Pleasure CFP 40254.
    Aida: Grand march, Act 2. cf Decca DPA 525/6.
    Aida: O ciel d'azur. cf Rococo 5397.

Aida: O cieli azzuri. cf CATALANI: La Wally: Ebben, ne andrò
   lontana.
Aida: O patria mia. cf MONIUSZKO: Halka: Gdybym rannyn slonkiem,
   O moj malenki.
Aida: O patria mia; Radames so che qui attendi...Su dunque; La
   tra foreste; Fuggiam gli ardori inospiti. cf Rubini GV 63.
Aida: Prelude. cf Overtures (DG 2707 090).
Aida: Pur ti reveggo...La tra foreste; La fatal pietra...O terra
   addio. cf Cantilena 6239.
Aida: Pur ti riveggo, mia dolce Aida; La fatal pietra; Se guerrier
   ...Celeste Aida; Ritorna vincitor; O patria mia. cf BOITO:
   Mefistofele: Dio de pieta; Dai campi; L'altra notte in fonde
   al mare.
Aida: Ritorna vincitor. cf Club 99-100.
Aida: Ritorna vincitor. cf Rococo 5386.
Aida: Ritorna vincitor. cf Rubini GV 58.
Aida: Ritorna vincitor. cf Telefunken AG 6-41947.
Aida: Schon sind die Priester all'vereint. cf Court Opera Classics
   CO 354.
Aida: Se quel guerrier fossi...Celeste Aida. cf Arias (Philips
   6747 193).
Alzira: Miserandi avanzi...Irne lungi. cf Arias (Philips 6747
   193).

2690 Arias: La battaglia di Legnano: Voi lo diceste...Quante volte
      come un dono. I Lombardi: Se vano è il pregare. Nabucco:
      Ben io t'invenni...Anch'io dischiuso. Otello: Willow song; Ave
      Maria. La traviata: Addio del passato. I vespri siciliani:
      Arrigo; Ah, Parli a un core; Mercé dilette amiche. Renata
      Scotto, s; Elizabeth Bainbridge, ms; William Elvin, bar; LPO;
      Gianandrea Gavazzeni. Columbia M 33516. (also CBS 76426 Tape
      (c) 40-76426).
          +-Gr 8-76 p334   +-NR 12-75 p11
           +Gr 9-76 p494 tape   +-ON 12-6-75 p52
          +-HF 10-75 p87   ++RR 9-76 p35
          ++HFN 7-76 p101   +St 9-75 p116
          ++HFN 10-76 p185 tape

2691 Arias: Aida: Celeste Aida (Bergonzi); Su, del nilo...Ritorna
      vinctor (Tebaldi); Gloria all'egito; Già i sacerdoti...Ah, tu
      dei vivre (Simionato, Bergonzi). Un ballo in maschera: Morrò,
      ma prima in grazia (Regine Crespin). Don Carlo: O don fatale
      (Grace Bumbry). La forza del destino: Overture; Morir, Tremenda
      cosa...Urna fatale (Leonard Warren). Falstaff: Ehi, Paggio...
      L'onore (Geraint Evans). Luisa Miller: Oh, fede negar potessi
      ...Quando le sere al placido (Carlo Bergonzi). Macbeth: Ap-
      paritions scnee (Dietrich Fischer-Dieskau). Nabucco: Va,
      pensiero (Dietrich Fischer-Dieskau). Otello: Willow song...
      Ave Maria (Crespin). Rigoletto: Gualtier Maldè...Caro nome
      (Joan Sutherland); La donna è mobile (Luciano Pavarotti). La
      traviata: Prelude, Act 1; Libiamo (Sutherland, Bergonzi); Un
      di felice (Sutherland, Bergonzi). Il trovatore: Vedi, le
      fosche...Stride la vampa (Giulietta Simionata); Di quella pira
      (Mario del Monaco); Miserere (Del Monaco, Renata Tebaldi).
      Various choruses and orchestras. Decca DPA 555/6. Tape (c)
      KDPC 555/6.
          +HFN 10-76 p183   +-RR 9-76 p34
          +HFN 12-76 p155 tape

2692 Arias: Un ballo in maschera: Morrò, ma prima in grazia. Ernani:
Surta e la notte...Ernani, Involami. La forza del destino:
Son giunta, grazie o Dio. Macbeth: Una macchia e qui tuttora.
Nabucco: Ben io t'invenni...Ach'io dischiuso un giorno. Otello:
Era piu calmo...Piangea cantando...Ave Maria. I vespri sicili-
ani: Merce, dilette amiche. Gabriella Déry, s; Hungarian State
Opera Orchestra; András Kórodi. Hungaroton SLPX 11715.
            -NR 5-76 p9

2693 Arias: Aida: Se quel guerrier fossi...Celeste Aida. Alzira:
Miserandi avanzi...Irne lungi. Aroldo: Sotto il sol di Siria
ardent. Attila: Qual notte...Qui, qui sostiamo...Ella in
poter del barbaro...Cara patria; Que del convegno e il loco...
Che non avrebbe il misero. Un ballo in maschera: Di tu se
fedele; Forse la soglia attinse...Ma se m'e forza perderti.
La battaglia di Legnano: O magnanima...La pia materna mano.
Il Corsaro: Eccomi prigioniero...Al mio stanco cadavere. Don
Carlo: Fontainebleau...Io la vidi e al suo sorriso. I due
Foscari: Notte, perpetua notte...Non maledirmi, O prode.
Ernani: Merce, diletti amici...Come rugiado al cespite. Fal-
staff: Dal labbro il canto estasiato vola. La forza del des-
tino: La vita e inferno...O tu che in seno agli angeli. Un
giorno di regno: Pietoso al lungo pianto. Giovanna d'Arco:
Nel suo bel volto...Sotto una guercia. I Lombardi: La mia
Letizia infondere...Sien miei sensi...Come poteva un angelo.
Luisa Miller: Oh, fede negar potessi...Quando le sere al placido.
Macbeth: O figli...Ah, la paterna mano. I Masnadieri: Quando
io leggo in Plutarcho...O mio castel paterna...Ecco un foglio
a te diretto...Fiere umane, umane fieri...Senti O Moor...Nell'
argilla maledetta; Come splendido...Di ladroni attorniato.
Oberto, Conte di San Bonifacio: Ciel che feci...Ciel pietosa.
Otello: Dio, mi potevi scagliar; Niun mi tema. Rigoletto:
Questa o quella; Ella mi fu rapita...Parmi veder le lagrime;
La donna e mobile. Simon Boccanegra: O inferno...Sento avvam-
par nel l'amina. La traviata: Lunge a lei...De miei bollenti
spiriti. Il trovatore: Il presagio funesto...Ah si, ben mio
...Di quella pira. I vespri siciliani: E di monforte il cenno
...Giorno di pianto. Carlo Bergonzi, t; NPhO; Nello Santi,
Lamberto Gardelli. Philips 6747 193 (3).
        +-ARG 11-76 p41   +ON 10-76 p72
        +Gr 9-75 p520   +RR 10-75 p35
        +HFN 9-75 p107   ++SFC 11-14-76 p30
        +NR 10-76 p9   ++St 9-76 p85
        +OC 2-76 p4

Aroldo: Overture. cf Overtures (DG 2707 090).
Aroldo: Sotto il sol di Siria ardent. cf Arias (Philips 6747 193).
Attila: Oh nel fuggente nuvolo. cf HMV SLS 5057.
Attila: Overture. cf Overtures (DG 2707 090).
Attila: Qual notte...Qui, qui sostiamo...Ella in poter del bar-
baro...Cara patria; Qui del convegno e il loco...Che non avrebbe
il misero. cf Arias (Philips 6747 193).
Attila: Tregua e cogli'unni...Dagl'immortali vertici e gettat la
mia sorte. cf RCA LRL 2-7531.
Attila: Tregua e cogl'unni...Dagli immortali vertici. cf Rubini
GV 70.
Ave Maria. cf L'Oiseau-Lyre SOL 345.

2694 Un ballo in maschera. Martina Arroyo, Reri Grist, s; Fiorenza

Cossotto, ms; Placido Domingo, Kenneth Collins, t; Giorgio
Giorgetti, Piero Cappuccilli, bar; Gwynne Howell, bs-bar;
Richard Van Allan, bs; ROHO Chorus; NPhO; Riccardo Muti. HMV
SLS 894. Tape (c) TC ASD 984. (also Angel SX 3762).

   ++Gr 12-75 p1094   +NR 1-76 p9
   +-Gr 12-76 p1066 tape   +-ON 1-17-76 p32
   +-HF 2-76 p111   ++RR 12-75 p32
   +-HFN 1-76 p122   ++RR 12-76 p110 tape
   +-HFN 12-76 p153 tape   ++St 3-76 p80
   +-MT 8-76 p663

2695 Un ballo in maschera, excerpts. Cristina Deutekom, Patricia Hay,
s; Charles Craig, t; Scottish Opera Chorus; Scottish National
Orchestra; Alexander Gibson. Classics for Pleasure CFP 40252.
   +HFN 12-76 p149

Un ballo in maschera: Di tu se fedele; Forse a soglia attinse...
Ma se m'e forza perderti. cf Arias (Philips 6747 193).
Un ballo in maschera: Ecco l'orrido campo. cf HMV SLS 5057.
Un ballo in maschera: Eri tu. cf Discophilia DIS KGA 2.
Un ballo in maschera: Fors'e la soglia...Ma se m'e forza perderti.
cf Rococo 5383.
The masked ball: Ma dell'arido stelo. cf Muza SX 1144.
Un ballo in maschera: Morrò, ma prima in grazia. cf Arias (Decca
DPA 555/6).
Un ballo in maschera: Morrò, ma prima in grazia. cf Arias (Hunga-
roton SLPX 11715).
Un ballo in maschera: Morrò, ma prima in grazia. cf BELLINI: La
sonnambula: Ah, se una volta sola...Ah, non credea mirarti...
Ah, non giunge.
Un ballo in maschera: Morrò, ma prima grazia. cf MONIUSZKO:
Halka: Gdybym rannyn slonkiem, O moj malenki.
Un ballo in maschera: Prelude. cf Overtures (DG 2707 090).
Un ballo in maschera: Teco io sto. cf PUCCINI: Madama Butterfly:
Bimba, bimba, non piangere.
La battaglia di Legnano: O magnanima...La pia materna mano. cf
Arias (Philips 6747 193).
La battaglia di Legnano: Overture. cf Overtures (DG 2707 090).
La battaglia di Legnano: Voi lo diceste...Quante volte come un
dono. cf Arias (Columbia M 33516).

2696 Il Corsaro. Jessye Norman, Montserrat Caballé, s; José Carreras,
Alexander Oliver, t; Gian-Piero Mastromei, John Noble, bar;
Clifford Grant, bs; Ambrosian Singers; NPhO; Lamberto Gardelli.
Philips 6700 098 (2).

   ++Gr 6-76 p84   +-NYT 6-20-76 pD27
   ++HF 9-76 p77   +ON 10-76 p72
   ++HFN 6-76 p99   +RR 6-76 p21
   ++MJ 11-76 p44   ++SFC 6-13-76 p30
   ++MM 12-76 p42   ++St 9-76 p86
   ++MT 12-76 p1007   ++STL 8-8-76 p29
   +-NR 8-76 p8

Il Corsaro: Eccomi prigioniero...Al mio stanco cadavere. cf
Arias (Philips 6747 193).
Il Corsaro: Overture. cf Overtures (DG 2707 090).

2697 Don Carlo: Ballo della regina. Otello: Ballet music, Act 3.
I vespri siciliani: Le quattro stagioni. CO; Lorin Maazel.
Decca SXL 6727. (also London CS 6945 Tape (c) 5-6945).
   ++Gr 8-75 p331   +-ON 2-14-76 p49

```
        -HF  4-76 p124              ++RR  8-75 p45
        +HFN 8-75 p84              ++SFC 1-18-76 p38
        -NR  3-76 p2               +SR  4-17-76 p51
        +NYT 1-18-76 pD1
```

Don Carlo: Ascolta; Le porte dell'asil s'apron già. cf Decca DPA
    517/8.
Don Carlo: Ella giammai m'amo. cf Hungaroton SLPX 11444.
Don Carlo: Fontainebleau...Io la vidi e al suo sorriso. cf Arias
    (Philips 6747 193).
Don Carlo: Io vengo a domandar, E dessa; Fontainebleau; Non pian-
    ger, mia, compagna; Tu che le vanità. cf BOITO: Mefistofele:
    Dio de pietà; Dai campi; L'altra notte in fondo al mare.
Don Carlo: O don fatale. cf Arias (Decca DPA 555/6).
Don Carlo: O don fatale. cf Caprice CAP 1107.
Don Carlo: O don fatale. cf Collectors Guild 611.
Don Carlo: O don fatale. cf Columbia/Melodiya M 33931.
Don Carlo: O don fatale. cf Decca SPA 449.
Don Carlo: Per me giunto. cf Caprice CAP 1062.
Don Carlo: Spuntato ecco il di. cf Works, selections (DG 2530 549).
Don Carlo: Tu che le vanità. cf HMV SLS 5057.
I due Foscari: Notte perpetua notte...Non maledirmi, O prode. cf
    Arias (Philips 6747 193).
Ernani: Infelice...e tuo credevi. cf Bongiovanni GB 1.
Ernani: Infelice...e tuo credevi. cf English National Opera ENO
    1001.
Ernani: Merce, diletti amici...Come rugiada al cespite. cf Arias
    (Philips 6747 193).
Ernani: Prelude. cf Overtures (DG 2707 090).
Ernani: Si ridesti il Leon di Castiglia. cf Works, selections
    (DG 2530 549).
Ernani: Surta è la notte. cf HMV SLS 5057.
Ernani: Surta è la notte...Ernani, involami. cf Arias (Hungaroton
    SLPX 11715).
Ernani: Surta è la notte...Ernani, involami. cf Telefunken AG
    6-41947.
Ernani: Surta è la notte...Ernani, Ernani, involami. cf DONIZETTI:
    Linda di Chamonix: Ah, tardai troppo...O luce di quest'anima.
2698 Falstaff. Elisabeth Schwarzkopf, Anna Moffo, s; Nan Merriman,
    Fedora Barbieri, ms; Luigi Alva, t; Tito Gobbi, Rolando Pan-
    erai, bar; PhO and Chorus; Herbert von Karajan. Angel CL 3552
    (3). (Also SLS 5037 Tape (c) TC SLS 5037 (Reissue from Colum-
    bia SAX 2254/6)).
        ++Gr 1-76 p1239   +RR 1-76 p26
        ++HFN 2-76 p113   +RR 11-76 p110 tape
        +HFN 6-76 p109 tape   +St 4-75 p69
2699 Falstaff: Ehi, paggio...L'onore ladri; Udrai quanta egli stoggia
    to end of Act 1; Signore, v'assista il cielo to end of Scene 1,
    Act 2; Quando ero paggio del Duca di Norfolk to end of Act 2;
    Ehi, Taverniere; Cavaliero, Reverenza to end of opera. Henny
    Neumann-Knapp, Martina Wulf, s; Elsie Tegetthoff, Hedwig Ficht-
    müller, con; Philipp Rasp, Peter Markwort, Wilhelm Ulbricht, t;
    Hans Hotter, Arno Schellenberg, bar; Gottlieb Zeithammer, bs;
    Leipzig Radio Orchestra and Chorus; Hans Weisbach. BASF 10
    22137. (Recorded, 1939).
        +Gr 1-76 p1239
Fallstaff: Dal labbro il canto estasiato vola. cf Arias (Philips
    6747 193).

Falstaff: Ehi, Paggio...L'onore. cf Arias (Decca DPA 555/6).
La forza del destino (The force of destiny): Overture. cf
    Overtures (DG 2707 090).
The force of destiny: Overture. cf Decca SXL 6782.
La forza del destino: Overture; Morir, Tremenda cosa...Urna fatale.
    cf Arias (Decca DPA 555/6).
La forza del destino: Pace, pace, mio Dio. cf Rococo 5386.
La forza del destino: Pace, pace, mio Dio. cf CATALANI: La Wally:
    Ebben, ne andrò lontana.
La forza del destino: Solenne in quest'ora. cf Cantilena 6238.
La forza del destino: Solenne in quest'ora. cf Decca DPA 517/8.
La forza del destino: Son giunta, grazie o Dio. cf Arias (Hunga-
    roton SLPX 11715).
La forza del destino: Son giunta...Madre, pietosa vergine. cf
    PUCCINI: Arias (Decca SDD 481).
La forza del destino: Unsonst Alvarez. cf DELIBES: Lakmé: Fanta-
    sien nebelhafte Träume.
La forza del destino: La vita è inferno...O tu che in seno agli
    angeli. cf Arias (Philips 6747 193).
Un giorno di regno: Overture. cf Overtures (DG 2707 090).
Un giorno di regno: Pietoso al lungo pianto. cf Arias (Philips
    6747 193).
Giovanna d'Arco: Nel suo bel volto...Sotto un guercia. cf Arias
    (Philips 6747 193).
Giovanna d'Arco: Overture. cf Overtures (DG 2707 090).
I Lombardi: Jerusalem, Jerusalem, O Signore. cf Works, selections
    (DG 2530 549).
I Lombardi: La mia Letizia infondere. cf DONIZETTI: La favorita:
    Spirto gentil; Una vergine, un angel di Dio.
I Lombardi: La mia Letizia infondere...Sien miei sensi...Come
    poteva un angelo. cf Arias (Philips 6747 193).
I Lombardi: O signore dal tetto natio. cf Decca DPA 525/6.
I Lombardi: Se vano è il pregare. cf Arias (Columbia M 33516).
2700 Luisa Miller. Montserrat Caballé, Annette Celine, s; Anna Rey-
    nolds, ms; Luciano Pavarotti, t; Sherrill Milnes, bar; Richard
    Van Allan, Bonaldo Giaiotti, bs; London Opera Chorus; National
    Philharmonic Orchestra; Peter Maag. London OSA 13114 (3).
    Tape (c) 5-13114. (also Decca SET 606/8 Tape (c) K2L 25).
        +Gr 5-76 p1796   +ON 10-76 p72
        +Gr 8-76 p341 tape   +-RR 5-76 p27
       ++HF 9-76 p98   +RR 8-76 p82
        +HFN 5-76 p111   +SFC 6-27-76 p29
        +HFN 8-76 p94 tape   +St 10-76 p136
        +MT 11-76 p916   +STL 6-6-76 p37
        +NR 11-76 p8
Luisa Miller: Oh, fede negar potessi...Quando le sere al placido.
    cf Arias (Decca DPA 555/6).
Luisa Miller: Oh, fede negar potessi...Quando le sere al placido.
    cf Arias (Philips 6747 193).
Luisa Miller: Overture. cf Overtures (DG 2707 090).
2701 MacBeth. Shirley Verrett, Stefania Malagu, ms; Placido Domingo,
    Antonio Savastano, t; Piero Cappuccilli, bar; Nicolai Ghiaurov,
    Carlo Zardo, Giovanni Foiani, Alfredo Mariotti, Sergio Fontana,
    bs; La Scala Orchestra and Chorus; Claudio Abbado. DG 2709
    062 (3). Tape (c) 3371 022.
       ++Gr 10-76 p653   ++NR 12-76 p9

```
        +Gr 12-76 p1066          ++RR 10-76 p32
        +HFN 10-76 p179          ++SFC 11-14-76 p30
        ++MJ 12-76 p28
```

2702 Macbeth. Fiorenza Cossotto, s; Maria Borgato, ms; José Carreras,
        Giuliano Bernardi, Leslie Fyson, t; Sherrill Milnes, John
        Noble, Neilson Taylor, bar; Ruggero Raimondi, Carlo Del Bosco,
        bs; Ambrosian Opera Chorus; NPhO; Riccardo Muti. HMV SQ SLS
        992 (3). (also Angel SX 3833 Tape (c) 4X35 3833)
        ++Gr 12-76 p1052
    Macbeth: Apparitions scene. cf Arias (Decca DPA 555/6).
    Macbeth: Chorus of refugees. cf Decca DPA 525/6.
    Macbeth: O figli...Ah, la paterna mano. cf Arias (Philips 6747
        193).
    Macbeth: Una macchia e qui tuttora. cf Arias (Hungaroton SLPX
        11715).
    Macbeth: Nel di dilla vittoria. cf HMV SLS 5057.
    Macbeth: Patria oppressa. cf Works, selections (DG 2530 549).
    Macbeth: Prelude. cf Overtures (DG 2707 090).
2703 I Masnadieri (The brigands). Montserrat Caballé, s; Carlo Ber-
        gonzi, John Sandor, t; Piero Cappuccilli, William Elvin, bar;
        Ruggero Raimondi, Maurizio Mazzieri, bs; Ambrosian Singers;
        NPhO; Lamberto Gardelli. Philips 6703 064 (3). Tape (c) 7699
        010.
```
        +Gr 9-75 p514            +ON 10-75 p56
        ++Gr 12-76 p1066         +RR 9-75 p24
        +HF 12-75 p80            +RR 11-76 p108 tape
        +HFN 9-75 p106           +SFC 8-31-75 p20
        ++MT 3-76 p236           +SR 10-4-75 p50
        +NR 10-75 p7             ++St 11-75 p87
        ++OC 12-75 p49           +STL 11-2-75 p38
```
    I Masnadieri: Prelude. cf Overtures (DG 2707 090).
    I Masnadieri: Quando io leggo in Plutarcho...O mio castel paterno
        ...Ecco un foglio a te diretto...Fiere umane, umane fieri...
        Senti O Moor...Nell'argilla maledetta; Come splendido...Di
        ladroni attorniato. cf Arias (Philips 6747 193).
    Nabucco: Ben io t'invenni. cf HMV SLS 5057.
    Nabucco: Ben io t'invenni...Anch'io dischiuso. cf Arias (Columbia
        M 33516).
    Nabucco: Ben io t'invenni...Anch'io dischiuso un giorno. cf Arias
        (Hungaroton SLPX 11715).
    Nabucco: Chorus of Hebrew slaves. cf Decca DPA 525/6.
    Nabucco: Gli arredi festivi; Va pensiero. cf Works, selections
        (DG 2530 549).
    Nabucco: Overture. cf Overtures (DG 2707 090).
    Nabucco: Overture. cf HMV ESD 7010.
    Nabucco: Va, pensiero. cf Arias (Decca DPA 555/6).
    Oberto, Conte di San Bonifacio: Ciel che feci...Ciel pietosa. cf
        Arias (Philips 6747 193).
    Oberto, Conte de San Bonifacio: Overture. cf Overtures (DG 2707
        090).
2704 Otello. Renata Tebaldi, s; Mario del Monaco, t; Aldo Protti, bar;
        Vienna State Opera Chorus; VPO; Herbert von Karajan. Decca
        Tape (c) K2A 21 (2).
```
        +Gr 8-76 p341 tape       +RR 8-76 p82 tape
        +-HFN 8-76 p94 tape
```
2705 Otello. Leonie Rysanek, s; Miriam Pirazzini, ms; Jon Vickers,

Florindo Andreolli, Mario Carlin, t; Tito Gobbi, Robert Kerns,
bar; Ferruccio Mazzoli, Franco Calabrese, bs; Rome Opera Orch-
estra and Chorus; Tullio Serafin. RCA SER 5646/8. (Reissue
from LDS 6155). (also RCA AGL 3-1969. Reissue).
    +-Gr 11-72 p966   +-RR 11-72 p36
    +-Op 1-73 p54   +SFC 11-14-76 p30
Otello: Ballet music. cf GOUNOD: Faust: Ballet music; Waltz.
Otello: Ballet music, Act 3. cf Don Carlo: Ballo della regina.
Otello: Dio, mi potevi scagliar; Niun mi tema. cf Arias (Philips
    6747 193).
Otello: Era piu calmo...Piangea cantando...Ave Maria. cf Arias
    (Hungaroton SLPX 11715).
Otello: Fuoco di gioia. cf Works, selections (DG 2530 549).
Otello: Già nella notte. cf PUCCINI: Madama Butterfly: Bimba,
    bimba, non piangere.
Otello: Già la notte, Ave Maria. cf PUCCINI: Madama Butterfly.
Otello: Mia madre aveva...Piangea cantando...Ave Maria. cf
    PUCCINI: Arias (Decca SDD 481).
Otello: Piangea cantando; Ave Maria piena di grazia. cf HMV
    RLS 719.
Otello: Una vela, Una vela; Fuoco di gioia. cf Decca DPA 525/6.
Otello: Willow song; Ave Maria. cf Arias (Columbia M 33516).
Otello: Willow song...Ave Maria. cf Arias (Decca DPA 555/6).
2706 Overtures: Aroldo. Attila. La battaglia di Legnano. Il Corsaro.
    La forza del destino. Giovanna d'Arco. Un giorno di regno.
    Luisa Miller. Nabucco. Oberto, Conte de San Bonifacio.
    I vespri siciliani. Preludes: Aida. Un ballo in maschera.
    Ernani. Macbeth. I Masnadieri. Rigoletto. La traviata.
    BPhO; Herbert von Karajan. DG 2707 090 (2). Tape (c) 3370
    010.
    +Gr 8-76 p305   +SFC 12-26-76 p34
    +HFN 9-76 p131   ++SLT 8-8-76 p29
    +RR 9-76 p68
Pezzi sacri, no. 4: Te Deum. cf BOITO: Mefistofele: Prologue.
2707 Requiem. Martina Arroyo, s; Josephine Veasey, ms; Placido Domingo,
    t; Ruggero Raimondi, bs; LSO and Chorus; Leonard Bernstein.
    CBS 77231 (2). (Reissue from 72873/4). (also Columbia M2-
    30060).
    +-Gr 6-76 p77   +-RR 8-76 p77
    ++HFN 6-76 p102
2708 Requiem. Heather Harper, s; Josephine Veasey, ms; Carlo Bini, t;
    Hans Sotin, bs; London Philharmonic Chorus; RPO; John Alldis,
    Carlos Paita. Decca OPFS 5-6. (also London 21140/1 Tape (c)
    SPC 5-21140/1 (ct) SPC8-21140/1).
    +-Gr 12-75 p1090   -MJ 11-76 p45
    -HF 6-76 p95   -NR 8-76 p10
    +-HFN 12-75 p167   /RR 12-75 p94
2709 Requiem. Elisabeth Schwarzkopf, s; Christa Ludwig, ms; Nicolai
    Gedda, t; Nicolai Ghiaurov, bar; PhO and Chorus; Carlo Maria
    Giulini. HMV SLS 909. Tape (c) TC SLS 909. (also Angel
    S 3649).
    ++HFN 6-76 p109 tape   +-RR 10-76 p107 tape
2710 Requiem. Julia Wiener-Chenisheva, s; Alexandrina Milcheva-Nonova,
    ms; Lyubomir Bodurov, t; Nikola Gyuzelev, bs; Svetoslav Obretenov
    Chorus; Sofia Philharmonic Orchestra; Ivan Marinov. Monitor
    HS 90108/9 (2).
    +MJ 11-76 p45   +-NR 11-76 p10

711 Rigoletto. Operatic arias: DELIBES: Lakmé: Dov è l'Indiana bruna.
     MEYERBEER: Dinorah: Ombra leggiera. ROSSINI: Il barbiere di
     Siviglia: Una voce poco fa. VERDI: I vespri siciliani: Merce,
     dilette amiche. Maria Callas, Elvira Galassi, s; Adriana Lazza-
     rini, Luisa Mandelli, ms; Giuseppe di Stefano, Renato Ercolani,
     t; Tito Gobbi, William Dickie, bar; Nicola Zaccaria, Plinio
     Clabassi, Carlo Forti, bs; La Scala Opera House Orchestra and
     Chorus; PhO; Tullio Serafin. HMV SLS 5018 (3). (Reissues
     from 33CXS 1324, 33CX 1325/6, 33CX 1231).
          +-Gr 3-76 p1507   +-RR 2-76 p23
          +-HFN 3-76 p107

712 Rigoletto: Questa o quella; Pari siamo...Figlia, mio padre; E il
     sol dell'anima; Gualtier Malde...Caro nome; Parmi veder le
     lagrime; Povero Rigoletto...Cortigiani; Parla, siam soli...
     Tutte le feste al tempio; La donna è mobile...Bella figlia dell'
     amore; Chi à mai. Joan Sutherland, Gillian Knight, Kiri Te
     Kanawa, Josephte Clément, s; Huguette Tourangeau, ms; Luciano
     Pavarotti, Ricardo Cassinelli, t; Sherrill Milnes, Christian
     du Plessis, John Gibbs, bar; Martti Talvela, Clifford Grant,
     bs; Ambrosian Opera Chorus; LSO; Richard Bonynge. Decca SET
     580. Tape (c) KSXC 580. (Reissue from SET 542/4).
          +-Gr 12-74 p1219   +RR 5-76 p78 tape
          +RR 11-74 p32
     Rigoletto: Bella figlia dell'amore. cf Decca SPA 449.
     Rigoletto: Caro nome. cf BELLINI: Capuletti ed i Montecchi: O,
     quante volte, O, quante.
     Rigoletto: Caro nome. cf BELLINI: La sonnambula: Ah, se una volta
     sola...Ah, non credea mirarti...Ah, non giunge.
     Rigoletto: Caro nome; Quartet. cf HMV RLS 719.
     Rigoletto: Cortigiani, vil razza. cf Caprice CAP 1062.
     Rigoletto: La donna è mobile. cf Decca SXL 6649.
     Rigoletto: La donna è mobile. cf Muza SXL 1170.
     Rigoletto: La donna è mobile, Questa o quella. cf Everest SDBR
     3382.
     Rigoletto: E il sol dell'anima. cf Court Opera Classics CO 342.
     Rigoletto: E il sol dell'anima. cf Rubini GV 70.
     Rigoletto: Gualtier Maldé...Caro nome. cf HMV SLS 5057.
     Rigoletto: Gualtier Maldé...Caro nome; La donna è mobile. cf
     Arias (Decca DPA 555/6).
     Rigoletto: Liebe ist Seligkeit. cf Rococo 5379.
     Rigoletto: Mio padre; Povero Rigoletto; Cortigiani vil razza
     danata. cf Discophilia DIS KGA 2.
     Rigoletto: Prelude. cf Overtures (DG 2707 090).
     Rigoletto: Questa o quella; Ella mi fu rapita...Parmi veder le
     lagrime; La donna è mobile. cf Arias (Philips 6747 193).
     Simon Boccanegra: Il lacerato spirito. cf Hungaroton SLPX 11444.
     Simon Boccanegra: O inferno...Sento avvampar nel l'anima. cf
     Arias (Philips 6747 193).
     Songs: Brindisi; Lo spazzacamino; Stornello. cf BELLINI: Songs
     (RCA AGL 1-1341).

2713 La traviata. Mirella Freni, s; Hania Kovicz, ms; Franco Bonisolli,
     Peter Bindszus, t; Sesto Bruscantini, Rudolf Jedlicka, bar;
     Berlin State Opera Chorus and Orchestra; Lamberto Gardelli.
     BASF KBL 21644 (3). (also BASF BAC 3101/2).
          +Gr 5-76 p1796   +-NR 3-75 p7
          +HF 6-75 p105   -ON 1-18-75 p32
          +-HFN 5-76 p111   +RR 5-76 p28
          +-MT 9-76 p748   +-St 6-75 p107

2714 La traviata. Joan Sutherland, s; Carlo Bergonzi, Piero de Palma,
t; Robert Merrill, bar; Maggio Musicale Orchestra and Chorus;
John Pritchard. Decca Tape (c) K19K 32.
        +HFN 12-76 p153 tape

2715 La traviata. Renata Scotto, Giuliana Tavolaccini, s; Armanda
Bonato, ms; Gianni Raimondi, t; Ettore Bastianini, bar; Silvio
Maionica, bs; La Scala Orchestra and Chorus; Antonino Votto.
DG 2726 049 (2). (Reissue from SLPM 138 832/4).
        +-Gr 8-76 p333   +RR 8-76 p24
        +-HFN 8-76 p91

2716 La traviata. Maria Callas, Ines Marietti, s; Ede Marietti
Gandolfo, ms; Francesco Albanese, Mariano Caruso, Tommaso Saley
t; Ugo Savarese, Alberto Albertini, bar; Mario Zorgniotti, bs;
Cetra Chorus; Radio Italiana Orchestra, Turin; Gabriele Santini
Turnabout THS 65047/8 (2). (Reissue from Cetra LPC 1246).
        +-HF 10-76 p122

La traviata: Addio del passato. cf Arias (Columbia M33516).
La traviata: Addio del passato. cf CATALANI: La Wally: Ebben,
ne andrò lontana.
La traviata: Addio del passato; Alfredo, di questo core. cf
Rubini GV 68.
La traviata: Addio del passato; Madimagella Valery...Pura siccome
un angelo...E grave il sacrifizio...Diet alla giovane. cf
Rubini GV 58.
La traviata: Ah fors'è lui...Follie. cf Court Opera Classics
CO 342.
La traviata: Ah fors'è lui; Sempre libera; Dite alla giovine. cf
HMV RLS 719.
La traviata: Di provenza. cf Rubini GV 65.
La traviata: Di provenza il mar...Dunque invano. cf RCA LRL
2-7531.
La traviata: Drinking song; Matadors' chorus. cf DG 2548 212.
La traviata: E strano...Ah, fors'è lui...Sempre libera. cf
Telefunken AG 6-41945.
La traviata: Lunge a lei...Dei miei bollenti spiriti. cf Arias
(Philips 6747 193).
La traviata: Prelude. cf Overtures (DG 2707 090).
La traviata: Prelude, Act 1. cf Philips 6747 204.
La traviata: Prelude, Act 1; Libiamo; Un dì felice. cf Arias
(Decca DPA 555/6).
La traviata: Prelude, Act 3. cf HMV SLS 5019.
La traviata: Un dì felice. cf Cantilena 6238.
La traviata: Un dì felice. cf Decca DPA 517/8.

2717 Il trovatore. Antonietta Stella, Armanda Bonato, s; Fiorenza
Cossotto, ms; Carlo Bergonzi, Franco Ricciardi, Angelo
Mercuriali, t; Ettore Bastianini, bar; Ivo Vinco, Giuseppe
Morresi, bs; La Scala Orchestra and Chorus; Tullio Serafin.
DG 2728 008 (3). (Reissue from SLPM 138 835/7).
        +Gr 8-76 p334   +-RR 7-76 p35
        /HFN 7-76 p101

2718 Il trovatore. BOITO: Mefistofele: L'altra notte. CATALANI: La
Wally: Ebben, ne andrò lontana. CILEA: Adriana Lecouvreur: Io
son l'umile; Poveri fiori. GIORDANO: Andrea Chénier: La mamma
morta. Maria Callas, s; Fedora Barbieri, ms; Giuseppe di
Stefano, Renato Ercolani, t; Rolando Panerai, bar; Nicolas
Zaccaria, bs; La Scala Orchestra and Chorus; Herbert von Karajan

Tullio Serafin. HMV SLS 869 (3). Tape (c) TC SLS 869. (Reissue from Columbia and Columbia 33CX 1321).
    +Gr 1-74 p1422   +-Op 3-74 p224
    +-HFN 8-76 p94 tape   +-RR 1-74 p34

2719 Il trovatore. Caterina Mancini, Graziella Sciutti, s; Miriam
Pirazzini, ms; Giacomo Lauri-Volpi, t; Carlo Tagliabue, bar;
Alfredo Colella, bs; Italian Radio and Television Symphony
Orchestra and Chorus; Fernando Previtali. Turnabout THS
65037/9 (3). (Reissue from Cetra LPC 1226).
    -HF 10-76 p122   +-NR 10-75 p7

Il trovatore: Ai nostri monti. cf Decca DPA 517/8.

Il trovatore: Anvil chorus. cf Decca DPA 525/6.

Il trovatore: Anvil chorus. cf DG 2548 212.

Il trovatore: Il balen. cf Rubini GV 65.

Il trovatore: Il balen del suo sorriso...Per me ora fatale. cf
RCA LRL 2-7531.

Il trovatore: Che più t'arresti...Tacea la notte. cf PUCCINI:
Arias (Decca SDD 481).

Il trovatore: D'amor sull'ali rosee. cf BELLINI: La sonnambula:
Ah, se una volta sola...Ah, non credea mirarti...Ah, non
giunge.

Il trovatore: D'amore sull'ali. cf Rococo 5397.

Il trovatore: Di quella pira. cf Decca SXL 6649.

Il trovatore: Di quella pira; Ah, si ben mio coll'essere; Miserere
...Ah, Che la morte. cf DONIZETTI: La favorita: Spirto gentil;
Una vergine, un angel di Dio.

Il trovatore: Mira d'acerbe lagrime; Vivra, Contenda il giubilo.
cf Discophilia DIS KGA 2.

Il trovatore: Il presagio funesto...Ah si, ben mio...Di quella
pira. cf Arias (Philips 6747 193).

Il trovatore: Stride la vampa. cf Rubini GV 70.

Il trovatore: Tacea la notte placida...Di tale amor. cf Telefunken AG 6-41947.

Il trovatore: Timor di me...D'amor sull'ali rosee. cf HMV SLS
5057.

Il trovatore: Timor di me...D'amor sull'ali rosee. cf CATALANI:
La Wally: Ebben, ne andrò lontana.

Il trovatore: Vedi, le fosche. cf Works, selections (DG 2530 549).

Il trovatore: Vedi, le fosche. cf Decca SPA 449.

Il trovatore: Vedi, le fosche...Stride la vampa; Di quella pira;
Miserere. cf Arias (Decca DPA 555/6).

I vespri siciliani: Arrigo; Ah, parli a un core; Mercè dilette
amiche. cf Arias (Columbia M33516).

I vespri siciliani: Arrigo, Ah parli a un core; Mercè, dilette
amiche. cf BELLINI: La sonnambula: Ah, se una volta sola...Ah,
non credea mirarte...Ah, non giunge.

I vespri siciliani: E' di monforte il cenno...Giorno di pianto.
cf Arias (Philips 6747 193).

I vespri siciliani: Mercè, dilette amiche. cf Arias (Hungaroton
SLPX 11715).

I vespri siciliani: Mercè, dilette amiche. cf Rigoletto.

I vespri siciliani: Mercè, diletti amiche. cf DONIZETTI: Linda di
Chamonix: Ah, tardai troppo...O luce di quest'anima.

I vespri siciliani: Mercè dilette amiche. cf Telefunken AG
6-41947.

I vespri siciliani: Mercè, dilette amiche; Arrigo, Ah parli a un
core. cf HMV SLS 5057.

I vespri siciliani: Overture. cf Overtures (DG 2707 090).
I vespri siciliani: Overture. cf BORODIN: Prince Igor: Polovtsiar dances.
I vespri siciliani: Le quattro stagioni. cf Don Carlo: Ballo della regina.
I vespri siciliani: Si, mi'abboriva...in braccio alle dovizie. cf RCA LRL 2-7531.

2720 Works, selections: Aida: Gloria all'egitto. Don Carlo: Spuntato ecco il di. Ernani: Si ridesti il Leon di Castiglia. I Lombardi: Jerusalem, Jerusalem; O Signore. Macbeth: Patria oppressa. Nabucco: Gli arredi festivi; Va pensiero. Otello: Fuoco di gioia. Il trovatore: Vedi, le fosche. La Scala Orchestra and Chorus; Claudio Abbado. DG 2530 549. Tape (c) 3300 485.

+-Gr 11-75 p903	++NYT 1-18-76 pD1
+HFN 11-75 p171	+ON 1-17-76 p32
++HFN 2-76 p116	+RR 11-75 p35
+HFN 8-76 p95	

VIANNA, Fructuoso
      Dansa de negros. cf Angel S 37110.
      Jogos pueris. cf Angel S 37110.
VIARDOT-GARCIA, Pauline
      Dites, que faut-il faire. cf Gemini Hall RAP 1010.
      Fluestern athemscheues Lauschen. cf Gemini Hall RAP 1010.
      Die Sterne. cf Gemini Hall RAP 1010.
VICTORIA, Tomas de
      Ave Maria, motet. cf PALESTRINA: Tu es Petrus, mass.
      Ave Maria à 8. cf Abbey LPB 754.
      Mass Pro Victoria. cf PALESTRINA: Tu es Petrus, mass.
      O Magnum mysterium. cf Vanguard SVD 71212.
      Salve regina à 5. cf Eldorado S-1.
      Songs: Popule meus, quid feci tibi. cf Abbey LPB 750.
VICTORY, Gerard
      Jonathan Swift. cf BOYDELL: Symphonic inscapes.
      VIENNESE DANCE MUSIC FROM THE CLASSICAL PERIOD. cf DG Archive 2533 182.
VIERNE, Luis
      Divertissement. cf Polydor 2460 262.
      Pastorale. cf Wealden WS 142.
      Pièces de fantaisie, op. 55: Naiades. cf Audio 2.
      Pieces in free style, op. 19, Berceuse. cf HMV SQ HQS 1356.
      Symphony, no. 1, op. 14, D minor: Final. cf Argo 5BBA 1013/5.
      Symphony, no. 1, op. 14, D minor: Final. cf Polydor 2460 262.
VIEUXTEMPS, Henri
      Concerto, violin, no. 4, op. 31, D minor. cf RCA ARM 4-0942/7.
      Concerto, violin, no. 5, op. 37, A minor. cf SAINT-SAENS: Concerto, violin, no. 3, op. 61, B minor.
      Concerto, violin, no. 5, op. 37, A minor. cf RCA ARM 4-0942/7.
VILLA-LOBOS, Heitor
      Carnaval das criancas. cf Piano works (DG 2530 634).
      Chôros, no. 1, E major. cf Saga 5412.
      Chôros, no. 4. cf BALAI: Divertimento, wind quintet and harp, op. 7.
      Chôros, no. 5 (Alma brasileira). cf Angel S 37110.
      Ciclo brasileiro. cf Angel S 37110.
      Cirandinha, no. 1: Therezinha de Jesus. cf Saga 5412.

Cirandinha, no. 10: A canoa virou. cf Saga 5412.
Concerto, guitar and small orchestra. cf RODRIGO: Concierto de
    Aranjuez, guitar and orchestra.
Concerto, harp. cf CASTELNUOVO-TEDESCO: Concertino, harp, op. 93.
Etude, no. 1. cf London STS 15306.
Etude, no. 11. cf Supraphon 111 1230.
Festa no sertão. cf Angel S 37110.
A fiandeira. cf Piano works (DG 2530 634).
Impressões seresteiras. cf Angel S 37110.
A lenda do caboclo. cf Piano works (DG 2530 634).
New York skyline (1957 version). cf Piano works (DG 2530 634).
2721 Piano works: A fiandeira. A lenda do caboclo. Carnaval das
    criancas. New York skyline (1957 version). Rudepoêma.
    Saudades das selvas brasileiras. Suite floral, op. 97 (1949
    revision). Roberto Szidon, Richard Metzler, pno. DG 2530 634.
        +-Gr 10-76 p628   ++RR 9-76 p84
        +HFN 10-76 p179
Preludes (5). cf SCARLATTI, D.: Sonatas, harpsichord.
Preludes (5). cf TURINA: Fandanguillo, op. 36.
Preludes, nos. 1-5. cf SCARLATTI, D.: Sonatas, guitar (CBS 73545).
Preludes, no. 1, E minor; no. 3, A minor. cf DG 3335 182.
Prelude, no. 3. cf London STS 15306.
A próle do bêbê, no. 1. cf Angel S 37110.
Rudepoêma. cf Piano works (DG 2530 634).
Saudades das selvas brasileiras. cf Piano works (DG 2530 634).
Song of the black swan. cf Laurel-Protone LP 13.
Suite floral, op. 97 (1949 revision). cf Piano works (DG 2530 634).
VILLETTE, Pierre
    Hymne à la vierge. cf FRANCK: Songs (Abbey LPB 758).
VIRGILIO
    Jana: Morte di Jane. cf Club 99-100.
VISEE, Robert de
    Suite, D minor. cf d'ANGLETERRA: Carillon, G major.
    Suite, D minor. cf Turnabout TV 34137.
VIVALDI, Antonio
2722 Beatus vir. Credo. Gloria. Lauda Jerusalem. Melinda Lugosi,
    s; Katalin Szökefalvi-Nagy, Maria Zempléni, s; Klára Takács,
    con; János Rolla, vln; Budapest Madrigal Choir; Liszt Academy
    Chamber Orchestra; Ferenc Szekeres. Hungaroton LSPX 11695.
        +NR 10-76 p9
2723 Beatus vir (Psalm III). Priscilla Salgo, s; Carmel Bach Festival
    Chorale and Orchestra; Sandor Salgo. Orion ORS 75208.
        +HF 8-76 p77   ++SFC 4-18-76 p23
        +NR 4-76 p9
    Concerto, no. 11, D minor: Largo. cf HMV RLS 723.
2724 Concerti, op. 4, nos. 1-12 (La stravaganza). AMF; Neville
    Marriner. Argo ZRG 800/1 (2).
        ++SFC 9-26-76 p29
2725 Concerti, op. 4, nos. 1-12 (La stravaganza). French Instrumental
    Ensemble; Jean-Pierre Wallez, vln. French Decca 7303/4 (2).
        +Gr 8-76 p305   +-RR 6-76 p58
    Concerto, op. 7, no. 2. cf ALBINONI: Concerto, oboe and strings,
    op. 9, no. 8, G minor.
2726 Concerto, bassoon and strings, E minor. Concerto, recorder and
    strings, C minor (ed. Giegling). Concerto, flute and strings,
    D major (ed. Negri). Concerto, oboe, A minor (ed. Negri). Leo
    Driehuys, ob; Marco Constantini, bsn; Severino Gazzelloni, flt;

I Musici. Philips 6580 152.
                +HFN 11-76 p171   +RR 11-76 p85
2727 Concerti, bassoon and strings, P 137, E minor; P 70, A minor;
      P 305, F major; P 382, B flat major. Klaus Thunemann, bsn;
      I Musici. Philips 6500 919.
                    +Audio 12-76 p93   +MM 4-76 p28
                    ++Gr 11-75 p834   +RR 11-75 p56
                    +HFN 12-75 p169
2728 Concerto, bassoon and strings, P 137, E minor. Concerto, oboe,
      P 89, A minor. Concerto, bassoon, flute, oboe and violin,
      P 360, G minor. Concerto, harpsichord and strings, P 361, G
      minor. Jürg Schaeftlein, ob; Milan Turkovic, bsn; Leopold
      Stastny, flt; Alice Harnoncourt, vln; VCM; Nikolaus Harnoncourt
      Telefunken AW 6-41961.
                    ++Gr 12-76 p1009   +NR 6-76 p6
                    ++HFN 10-76 p179   ++RR 10-76 p75
      Concerto, bassoon, flute, oboe and violin, P 360, G minor. cf
      Concerto, bassoon and strings, P 137, E minor.
      Concerto, flute, A minor. cf RCA CRL 3-1429.
2729 Concerti, flute, P 140, G major; P 203, D major; P 80, A minor;
      P 205, D major. Concerto, flute and bassoon, op. 10, no. 2,
      G minor. Severino Gazzelloni, flt; Jiri Stavicek, bsn; I
      Musici. Philips 6500 707.
                    +Gr 4-75 p1817   ++NR 8-75 p6
                    ++HF 10-75 p86   ++RR 4-75 p38
                    +HFN 8-75 p85   ++St 1-76 p110
      Concerto, flute and bassoon, op. 10, no. 2, G minor. cf Concerti,
      flute, P 140, G major; P 203, D major; P 80, A minor; P 205,
      D major.
      Concerto, flute and bassoon, op. 10, no. 2, G minor. cf RCA LRL
      1-5127.
      Concerto, flute and strings, D major. cf Concerto, bassoon and
      strings, E minor.
2730 Concerto, flute and strings, P 155, D major. Concerti, piccolo
      and strings, P 78, C major; P 79, C major; P 83, A minor.
      Julius Baker, flt and pic; Herbert Tachezi, hpd; I Solisti
      di Zagreb, VSOO; Antonio Janigro, Felix Prohaska. Pye Van-
      guard Bach Guild HM 45. (Reissue from Philips Vanguard VSL
      11022, VSL 11013).
                    +-Gr 11-76 p823   ++NR 5-76 p6
                    +-HFN 7-76 p103   +RR 7-76 p65
2731 Concerto, 4 flutes, 4 violins, 2 organs and strings, P 226, A
      major. Concerto, oboe, violin, organ, harpsichord and
      strings, P 36, C major. Concerto, violin, organ, harpsichord
      and strings, P 274, F major. Concerto, violin, organ, harpsi-
      chord and strings, P 311, D minor. Monique Frasca-Colombier,
      vln; Andre Isoir, positive organ; Michel Giboureau, ob; Paul
      Kuentz Chamber Orchestra; Paul Kuentz. DG 2530 652. Tape (c)
      3300 652.
                    +-Gr 8-76 p305   +RR 7-76 p65
                    +-Gr 10-76 p658 tape   +RR 12-76 p110 tape
                    +HFN 8-76 p91
      Concerto, guitar and strings, D major. cf FASCH: Concerto,
      guitar and strings, D minor.
      Concerto, 2 guitars, G major. cf CBS 73487.
      Concerto, harpsichord and strings, P 361, G minor. cf Concerto,
      bassoon and strings, P 137, E minor.

Concerto, lute, D major. cf HANDEL: Concerto, 2 lutes, strings
    and recorders, op. 4, no. 6, B flat major.
Concerto, 2 mandolins, G major: Andante. cf Saga 5426.
2732 Concerti, oboe, P 42, A minor; P 41, C major; P 50, C major;
    op. 11, no. 6, G minor. Heinz Holliger, ob; I Musici. Philips
    9500 044. Tape (c) 7300 443.
        ++Gr 6-76 p57   ++RR 5-76 p59
        ++HFN 5-76 p113   +STL 7-4-76 p36
Concerto, oboe, A minor. cf Concerto, bassoon and strings, E
    minor.
Concerto, oboe, A minor. cf BACH: Concerto, violin, oboe and
    strings, S 1060, D minor.
Concerto, oboe, P 89, A minor. cf Concerto, bassoon and strings,
    P 137, E minor.
Concerto, oboe, op. 8, no. 9, D minor. cf ALBINONI: Concerto,
    oboe, op. 7, no. 3, B flat major.
Concerto, oboe, op. 9, no. 2, A minor. cf ALBINONI: Concerto,
    oboe, op. 7, no. 3, B flat major.
Concerto, oboe, violin, organ, harpsichord and strings, P 36,
    C major. cf Concerto, 4 flutes, 4 violins, 2 organs and
    strings, P 226, A major.
2733 Concerto, orchestra, P 84, C major. Concerto, orchestra, P 383,
    G minor. Concerto, orchestra, P 267, F major. Concerto,
    orchestra, P 359, G minor. I Solisti Veneti; Claudio Scimone.
    Erato STU 70818.
        /Gr 12-75 p1054   +MM 4-76 p28
        +HFN 12-75 p169   +RR 12-75 p69
Concerto, piccolo and strings, P 78, C major; P 79, C major; P 83,
    A minor. cf Concerto, flute and strings, P 155, D major.
Concerto, piccolo and strings, P 83, A minor. cf HAYDN: Concerto,
    trumpet, E flat major.
Concerto, recorder, op. 44, no. 11, C major. cf Transatlantic
    TRA 292.
Concerto, recorder and strings, C minor. cf Concerto, bassoon
    and strings, E minor.
Concerti, alto recorder and strings, A minor (2), F major. cf
    Concerto, soprano recorder and strings, D major.
Concerto, sopranino recorder and strings, C major. cf Concerto,
    soprano recorder and strings, D major.
2734 Concerto, sopranino recorder and strings, P 79, C major. Con-
    certo, viola d'amore, lute and strings, P 266, D major. Con-
    certo, 2 violins and strings, P 222, A major. Concerto, violon-
    cello and strings, E minor. Pierre Fournier, vlc; Walter
    Prystawski, Herbert Höver, vln; Hans-Martin Linde, sopranino
    rec; Monique Frasca-Colombier, vla d'amore; Narciso Yepes, gtr;
    Lucerne Festival Strings, Emil Seiler Chamber Orchestra, Paul
    Kuentz Chamber Orchestra; Rudolf Baumgartner, Wolfgang Hofmann,
    Paul Kuentz. DG 2535 200. (Reissues).
        +-HFN 12-76 p151   +RR 12-76 p68
2735 Concerto, soprano recorder and strings, D major. Concerti, alto
    recorder and strings, A minor (2), F major. Concerto, sopranino
    recorder and strings, C major. Laszlo Czidra, rec; Franz
    Liszt Academy Chamber Orchestra; Frigyes Sándor. Hungaroton
    SLPX 11671.
        +RR 3-75 p36   ++SFC 1-25-76 p30
Concerto, trumpet, A flat major. cf HUMMEL: Concerto, trumpet,
    E flat major.

Concerto, viola d'amore, A major. cf PORPORA: Concerto, violon-
cello, G major.
Concerto, viola d'amore, lute and strings, P 266, D major. cf
Concerto, sopranino recorder and strings, P 79, C major.
Concerto, violin, E flat major. cf ALBINONI: Sonata à 5, op. 2,
no. 6, G minor.

2736 Concerti, violin (oboe) and strings, op. 7 (12). Heinz Holliger,
ob; Salvatore Accardo, vln; I Musici. Philips 6700 100 (2).

    +ARG 11-76 p44   +HFN 9-76 p131
    ++Gr 11-76 p823   ++RR 9-76 p75
    +-HF 12-76 p118   ++SFC 9-26-76 p29

2737 Concerti, violin and strings, opp. 11-12, nos. 1-12. Salvatore
Accardo, vln; I Musici. Philips 6747 189 (3).

    ++HF 10-76 p126   ++SFC 9-26-76 p29
    ++MJ 10-76 p24   ++St 8-76 p99
    ++NR 5-76 p6

2738 Concerti, violin and strings, op. 11, nos. 1-4. Salvatore Accardo,
vln; I Musici. Philips 6500 933.

    ++Gr 3-76 p1477   ++RR 3-76 p51
    ++HFN 3-76 p107

2739 Concerti, violin and strings, op. 11, nos. 5 and 6. Concerti,
violin and strings, op. 12, nos. 1 and 2. Salvatore Accardo,
vln; I Musici. Philips 6500 934.

    +Gr 6-76 p57   ++RR 4-76 p59
    +HFN 4-76 p119

Concerti, violin and strings, op. 12, nos. 1 and 2. cf Concerti,
violin and strings, op. 11, nos. 5 and 6.

2740 Concerti, violin, op. 12, nos. 3-6. Salvatore Accardo, vln; I
Musici. Philips 6500 937.

    +Gr 1-76 p1209   +RR 1-76 p39
    ++HFN 1-76 p122   ++STL 2-8-76 p36

Concerto, violin, organ, harpsichord and strings, P 274, F major.
cf Concerto, 4 flutes, 4 violins, 2 organs and strings, P 226,
A major.
Concerto, violin, organ, harpsichord and strings, P 311, D minor.
cf Concerto, 4 flutes, 4 violins, 2 organs and strings, P 226,
A major.
Concerto, 2 violins and strings, G minor. cf CBS 61684.
Concerto, 2 violins and strings, P 222, A major. cf Concerto,
sopranino recorder and strings, P 79, C major.

2741 Concerti, 2 violins, P 281, D minor; P 436, C minor; P 366, G
minor; P 189, D major. Isaac Stern, David Oistrakh, vln; PO
Members; William R. Smith, hpd; Eugene Ormandy. CBS 61629.
(Reissue from Fontana SCFL 136).

    +-Gr 10-75 p639   +ST 2-76 p739
    +-HFN 9-75 p109   ++STL 10-5-75 p36
    ++RR 9-75 p51

Concerto, violoncello and strings, E minor. cf Concerto, sopranino
recorder and strings, P 79, C major.

2742 Credo. Gloria. Kyrie. Jennifer Smith, Wally Stämpfli, s; Nicole
Rossier, Hanna Schaer, altos; Lausanne Vocal and Instrumental
Ensemble; Michel Corboz. RCA AGL 1-1340. (also Erato 70910).

    +Gr 5-76 p1795   +HFN 5-76 p113
    +HF 8-76 p77   -NR 5-76 p8
    +-HFN 5-76 p113   ++RR 5-76 p73
    +HFN 5-76 p113   +SFC 9-26-76 p29

Credo. cf Beatus vir.
L'Estro armonico, op. 3, no. 11, D minor. cf Vanguard VSD 707/8.
2743 The four seasons, op. 8, nos. 1-4. Simon Standage, baroque violin;
        English Concert; Trevor Pinnock, hpd and cond. CRD CRD 1025.
              +Gr 11-76 p823   +RR 10-76 p75
2744 The four seasons, op. 8, nos. 1-4. Monique Frasca-Colombier, vln;
        Paul Kuentz Chamber Orchestra; Paul Kuentz. DG 2548 005. Tape
        (c) 3348 005.
              +Gr 11-75 p839   -RR 10-75 p61
              ++HFN 10-75 p153   +RR 9-76 p92 tape
2745 Il cimento dell'armonia e dell'invenzione (The four seasons), op.
        8. Piero Toso, vln; Pierre Pierlot, ob; I Solisti Veneti;
        Claudio Scimone. Musical Heritage Society MHS 1727/9 (3).
        (also Erato STU 70679. Reissue from STU 70679-71).
              +-Gr 7-76 p182   +-RR 6-76 p6
              +HFN 7-76 p101   +-St 10-74 p139
2746 The four seasons, op. 8, nos. 1-4. Mikhail Vaiman, vln; Leningrad
        Chamber Orchestra; Lev Shinder. HMV Melodiya SXLP 30195.
              -GR 2-76 p1350   +-RR 2-76 p43
              +-HFN 3-76 p109
2747 The four seasons, op. 8, nos. 1-4. Igor Kipnis, hpd; Konstanty
        Kulka, vln; Stuttgart Chamber Orchestra; Karl Münchinger.
        London CS 6809.
              +NR 2-76 p4   ++St 4-76 p117
              +SFC 1-25-76 p30
Gloria. cf Beatus vir.
2748 Juditha triumphans. Elly Ameling, s; Ingeborg Springer, Julia
        Hamari, ms; Birgit Finnilä, Annelies Burmeister, con; Berlin
        Chamber Orchestra; Vittorio Negri.   Philips 6747 173 (3).
              +-Gr 10-75 p676   ++NR 2-76 p9
              ++HF 3-76 p97   +-ON 2-28-76 p53
              +-HFN 11-75 p171   +RR 9-75 p72
              ++MJ 1-76 p24   +SFC 1-25-76 p30
              +-MT 2-76 p146   ++St 4-76 p117
Lauda Jerusalem. cf Beatus vir.
2749 Laudate pueri (Psalm no. 112). Motets: In furore; Nulla in mundo
        pax sincera. Magda Kalmár, s; Ferenc Liszt Academy Chamber
        Orchestra; Frigyes Sándor. Hungaroton SLPX 11632.
              +-HF 8-76 p77   +-SFC 4-18-76 p23
              +-NR 6-76 p13
Motets: In furore; Nulla in mundo pax sincera. cf Laudate pueri
        (Psalm no. 112).
2750 Nisi Dominus (Psalm 126). Motets: Invicti bellate; Longe mala
        umbrae terrores. Teresa Berganza, ms; ECO; Antonio Ros-Marba.
        Pye NEL 2018.
              +Gr 3-76 p1500   +RR 4-76 p75
              ++HFN 5-76 p113
Sonata, violin, op. 2, no. 2, A major. cf LECLAIR: Sonata, violin,
        no. 3, D minor.
Sonata, violin, op. 2, no. 2, A major. cf RCA ARM 4-0942/7.
Songs: Piango gemo sospiro. cf Decca SXL 6629.
VIVIANI, Giovanni
      Sonata prima per tromba e organ. cf Klavier KS 551.
VOIS, Alewijnsz de
      Pavanne de Spanje. cf BASSANO: Ricercare.
THE VOLGA BOATMEN AND OTHER RUSSIAN FAVORITES. cf London CS 26398.

VOLKMANN, Robert
  Fantasie, op. 25a, C major. cf KIRCHNER: Piano works (Genesis
    GS 1032).
  Sonata, piano, no. 12, C minor. cf KIRCHNER: Piano works (Gene-
    sis GS 1032).
VON HAGEN
  Funeral dirge on the death of General Washington. cf Vox SVBX
    5350.
VON HESSEN, Moritz
  Pavane. cf DG Archive 2533 302.
VORISEK, Jan Hugo
  Sonata, piano, op. 20, B flat minor. cf DUSSEK: La consolation,
    op. 62.
VOSTRAK, Zbynek
  Scale of light. cf ISTVAN: Isle of toys.
VRIEND, Jan
  Huantan. cf KUNST: No time.
VUILDRE, Philippe van
  Je fille quant Dieu. cf HMV CSD 3740.
WAGNER, J. F.
  Tiroler Holzhackerbaub'n. cf DG 2721 077 (2).
  Unter dem Doppeladler. cf DG 2721 077 (2).
WAGNER, Richard
  Adagio, clarinet and strings. cf Decca SPA 395.
2751 Arias: Der fliegende Holländer: Senta's ballad; Mögst du, mein
    Kind; Wie aus der Ferne. Götterdämmerung: Seit er von dir
    geschieden; Hier sitz ich zur Wacht. Die Meistersinger von
    Nürnberg: Wahn, Wahn; Morgenlich leuchtend. Parsifal: Ich
    sah das Kind; Das ist Karfreitagzüber, Herr. Das Rheingold:
    Weiche, Wotan, weiche. Siegfried: Ewig war ich. Tannhäuser:
    Wie Todesahnung...O du, mein holder abendstern; Inbrunst im
    Herzen; Dich teure Halle. Tristan und Isolde: Mild und leise.
    Die Walküre: Ein Schwert verhiess mir der Vater. Emmy Destinn,
    Frida Leider, Astrid Varnay, Birgit Nilsson, Leonie Rysanek,
    s; Sigrid Onegin, Karin Branzell, ms; Lauritz Melchior, Max
    Lorenz, Wolfgang Windgassen, t; Walter Soomer, Friedrich Schorr,
    Hans Hotter, bar; Paul Knüpfer, Richard Mayr, Josef Greindl,
    bs; Orchestra, Munich Philharmonic Orchestra, Würtemberg State
    Orchestra, Berlin State Opera Orchestra, Bavarian Radio Sym-
    phony Orchestra, Bayreuth Festival Orchestra; Bruno Seidler-
    Winkler, Ferdinand Leitner, Manfred Gurlitt, Arthur Rother,
    Robert Heger, Hermann Weigert, Karl Böhm. DG 2721 115 (2).
    (Reissues from 043 064, 002 416, 73940, 65598, 72863, LPM
    19069, 18097, 66853, 72977, 30025, LPM 19259, 67973, LPM 19047,
    19045, SLPM 139221/5).
        +ARG 11-76 p10   +-ON 8-76 p39
        +Gr 8-76 p337   +-RR 9-76 p35
        +HFN 8-76 p88   +STL 9-19-76 p36
2752 Der fliegende Holländer (The flying Dutchman). Gwyneth Jones, s;
    Sieglinde Wagner, ms; Hermin Esser, Harald Ek, t; Thomas
    Stewart, bar; Karl Ridderbusch, bs; Bayreuth Festival Orchestra
    and Chorus; Karl Böhm. DG 2709 040 (3). Tape (r) 47040.
    (also DGG 2520 052).
        +-Gr 10-72 p747   +-ON 3-24-73 p36
        +-HF 11-72 p106   +-Op 11-72 p1000
        +-HFN 10-72 p1919   +-RR 10-72 p52

```
        +LJ 4-75 p70 tape          +-SR 7-24-76 p36
        +-MJ 2-73 p36              +-St 12-72 p141
        +-NR 11-72 p12
```

2753 Der fliegende Holländer: Mit Gewitter und Sturm; Die Frist ist um;
     Traft ihr das Schiff im Meere an; Senta, willst Du mich verder-
     ben; Mögst Du, mein Kind, den fremden Mann; Wohl kenn ich
     Weibes heil'ge Pflichten; Senta, oh Senta, leugnest Du...Willst
     jenes Tag's Du nich mehr Dich entsinnen; Verloren, ach ver-
     loren. Viorica Ursuleac, s; Luise Willer, alto; Karl Ostertag,
     Franz Klarwein, t; Hans Hotter, bar; George Hann, bs; Bavarian
     State Opera Orchestra and Chorus; Clemens Krauss. BASF KBF
     21538. (Recorded 1944).

```
        +Gr 1-76 p1239             +-ON 4-12-75 p36
        +HF 7-75 p86               +-RR 1-76 p29
        +NR 6-75 p11               +St 8-75 p107
```

     Der fliegende Holländer, excerpts. cf Works, selections (DG 2721
     109).
     Der fliegende Holländer, excerpts. cf Works, selections (DG 2721
     110).
     Der fliegende Holländer, excerpts. cf Works, selections (DG 2721
     112).
     Der fliegende Holländer, excerpts. cf Works, selections (DG 2721
     113).
     The flying Dutchman: Die Frist ist um. cf Decca GOSE 677/79.
2754 The flying Dutchman: Overture. Lohengrin: Preludes, Acts 1 and
     3. Parsifal: Preludes, Acts 1, 3. Die Meistersinger von Nürn-
     berg: Prelude. BPhO; Herbert von Karajan. Angel S 37098.
     Tape (c) 4XS 37098 (ct) 8XS 37098. (also HMV SQ ASD 3160).

```
        ++Gr 3-76 p1477            +NR 3-76 p3
        +HF 7-76 p114 tape         ++RR 3-76 p51
        +HFN 4-76 p119             +SFC 12-21-75 p39
```

     The flying Dutchman: Overture. cf MENDELSSOHN: Hebrides overture,
     op. 26.
     Der fliegende Holländer: Overture. cf Decca SPA 409.
     Der fliegende Holländer: Sailors' chorus. cf Decca DPA 525/6.
     Der fliegende Holländer: Sailors' chorus. cf DG 2548 212.
     Der fliegende Holländer: Senta's ballad. cf Rococo 5377.
     Der fliegende Holländer: Senta's ballad; Mögst du, mein Kind; Wie
     aus der Ferne. cf Arias (DG 2721 115).
2755 Götterdämmerung. Hilde Konetzni, Kirsten Flagstad, Elisabeth
     Höngen, Margret Weth-Falke, s; Margarite Kenney, con; Max Lorenz,
     t; Josef Hermann, Alois Pernerstorfer, bar; Ludwig Weber, bs;
     La Scala Orchestra and Choir; Wilhelm Furtwängler. Bruno
     Walter Society RR 420/4 (5).

```
        +NR 3-76 p8
```

     Götterdämmerung, excerpts. cf Works, selections (DG 2721 109).
     Götterdämmerung, excerpts. cf Works, selections (DG 2721 111).
     Götterdämmerung, excerpts. cf Works, selections (DG 2721 113).
2756 Götterdämmerung, Act 3, scenes 2 and 3 (Andrew Porter's English
     translation). Rita Hunter, s; Alberto Remedios, t; Clifford
     Grant, bs; Norman Bailey, bs-bar; Sadler's Wells Orchestra
     and Chorus; Reginald Goodall. Unicorn UNS 245/6. Tape (c)
     ZCUND 245 (ct) Y8UND 245.

```
        +Gr 7-73 p228              +-ON 8-76 p39
        +Gr 1-74 p1430 tape        +Op 8-73 p718
        +HF 2-74 p106              +RR 7-73 p26
```

Götterdämmerung: Ha weisst du was er mir ist. cf Rococo 5377.
Götterdämmerung: Hagens Ruf. cf Rococo 5397.
2757 Götterdämmerung: Immolation scene. Tristan und Isolde: Prelude
       and Liebestod. Eileen Farrell, s; BSO; Charles Munch. RCA AGL
       1-1274. (Reissue from LSC 2255).
            /HF 10-76 p127
2758 Götterdämmerung: Rhine journey. Tannhäuser: Overture and Venusberg
       music. Die Walküre: Fire music. BSO; Charles Munch. Camden
       CCV 5044.
            +HFN 4-76 p121   -RR 5-76 p59
       Götterdämmerung: Seit er von dir geschieden; Hier sitz ich zur
       Wacht. cf Arias (DG 2721 115).
       Götterdämmerung: Siegfried's Rhine journey; Siegfried's funeral
       march; Immolation of the Gods. cf Der Ring des Nibelungen
       (Decca SXL 6743).
       Götterdämmerung: Zu neuen Taten, Starke Scheite. cf Rococo 5380.
2759 Lohengrin. Eleanor Steber, Astrid Varnay, s; Wolfgang Windgassen,
       t; Hermann Uhde, Hans Braun, bar; Josef Greindl, bs; Bayreuth
       Festival Orchestra and Chorus, 1953; Joseph Keilberth. Decca
       D12D 5 (5). (Reissue from LXT 2880/4).
            +Gr 8-76 p334   +-RR 8-76 p25
            +HFN 8-76 p88
2760 Lohengrin. Elisabeth Grümmer, s; Christa Ludwig, ms; Jess Thomas,
       t; Dietrich Fischer-Dieskau, bar; Gottlob Frick, Otto Wiener,
       bs; Vienna State Opera Chorus; VPO; Rudolf Kempe. HMV SQ SLS
       5071 (5). (Reissue from Angel SAN 121/5).
            +Gr 12-76 p1057   ++HFN 12-76 p151
2761 Lohengrin. Anja Silja, Astrid Várnay, s; Jess Thomas, t; Ramón
       Vinay, Tom Krause, bar; Franz Crass, bs; Bayreuth Festival
       Orchestra and Chorus; Wolfgang Sawallisch. Philips 6747 241
       (4).
            /HF 6-76 p96   +NYT 3-14-76 pD15
            +HFN 7-76 p99   +ON 8-76 p39
            +MJ 5-76 p28   +RR 7-76 p36
            +-MT 12-76 p1007   ++SFC 3-21-76 p28
            /NR 5-76 p10   +St 7-76 p113
2762 Lohengrin.* Die Meistersinger von Nürnberg.** Tristan und Isolde.***
       Anja Silja, Astrid Varnay, Hannelore Bode, Birgit Nilsson, s;
       Anna Reynolds, Christa Ludwig, ms; Jess Thomas, Niels Möller,
       Gerhard Stolze, Heribert Steinbach, Robert Licha, Wolf Appel,
       Norbert Orth, Jean Cox, Frieder Stricker, Wolfgang Windgassen,
       Claude Heater, Erwin Wohlfahrt, Peter Schreier, t; Franz Crass,
       Ramón Vinay, Eberhard Wächter, bar; Karl Ridderbusch, bs-bar;
       Tom Krause, Klaus Kirchner, Zoltán Kelemen, Hans Sotin, József
       Dene, Klaus Hirte, Gerd Nienstedt, Heinz Feldhoff, Harmut Bauer,
       Nikolaus Hillebrand, Bernd Weikl, Martti Talvela, bs; Bayreuth
       Festival Orchestras and Choruses of 1962, 1966, 1974; Wolfgang
       Sawallisch, Silvio Varviso, Karl Böhm. Philips 6747 243 (14).
       (*also 6747 241, **Reissue from 6747 167, ***Reissue from DG
       SLPM 139 221-4).
            +Gr 7-76 p217   +-RR 7-76 p36
            +HFN 7-76 p99
2763 Lohengrin. Lucine Amara, s; Rita Gorr, ms; Sándor Kónya, t; Williar
       Dooley, Calvin Marsh, bar; Jerome Hines, bs; Boston Pro Musica
       Chorus; BSO; Erich Leinsdorf. RCA PVL 5-9046 (5). (Reissue
       from SER 5544/8).

          +-Gr 11-76 p874   +-RR 11-76 p47
          +-HFN 11-76 p173
2764 Lohengrin, excerpts. Siegfried, excerpts. Tannhäuser, excerpts.
        Max Lorenz, Franz Volker, t; Various Orchestra; Heinz Tietjen,
        Hans Schmidt-Isserstedt. Telefunken AJ 6-42019.
          +-HFN 11-76 p169
      Lohengrin, excerpts. cf Works, selections (DG 2721 109).
      Lohengrin, excerpts. cf Works, selections (DG 2721 110).
      Lohengrin, excerpts. cf Works, selections (DG 2721 113).
      Lohengrin: Du Armste; Entweihte Götter. cf Rococo 5379.
      Lohengrin: Einsam in trüben Tagen. cf MONIUSZKO: Halka: Gdybym
        rannyn slonkiem, O moj malenki.
      Lohengrin: Einsam in trüben Tagen. cf Decca GOSE 677/79.
      Lohengrin: Einsam in trüben Tagen. cf Muza SX 1144.
      Lohengrin: Elsa's dream. cf HMV RLS 719.
      Lohengrin: Gott grüss Euch. cf Rococo 5397.
      Lohengrin: Hochstes Vertrauen (2); In fernem Land. cf Discophilia
        13 KGW 1.
      Lohengrin: In fernem Land. cf Cantilena 6238.
      Lohengrin: In fernem Land, Mein lieber Schwan. cf Court Opera
        Classics CO 354.
      Lohengrin: Mein Herr und Gott nun ruf ich dich. cf Preiser LV
        192.
2765 Lohengrin: Prelude, Act 1. Tannhäuser: Overture and bacchanale.
        Tristan und Isolde: Prelude and Liebestod. BPhO; Herbert von
        Karajan. Angel S 37097. Tape (c) 4XS 37097 (ct) 8XS 37097.
        (also HMV (Q) ASD 3130 Tape (c) TC ASD 3130).
          +Gr 12-75 p1054   +HFN 12-76 p155 tape
          +HF 12-75 p109   ++NR 9-75 p2
          +-HFN 12-75 p169   +-RR 12-75 p70
      Lohengrin: Prelude, Act 1. cf MENDELSSOHN: Hebrides overture, op.
        26.
2766 Lohengrin: Preludes, Acts 1 and 3. Parsifal: Prelude; Good Friday
        music. Tannhäuser: Overture. Lamoureux Orchestra, Bavarian
        Radio Symphony Orchestra; Igor Markevitch, Eugen Jochum. DG
        2548 221. (Reissues from SLPM 133010, 138005).
          +Gr 4-76 p1617   +-RR 5-76 p59
          +HFN 4-76 p121
2767 Lohengrin: Preludes, Acts 1 and 3. Die Meistersinger von Nürn-
        berg: Prelude, Act 1. Tristan und Isolde: Prelude and Liebe-
        stod. Parsifal: Prelude. COA; Bernard Haitink. Philips 6500
        932. Tape (c) 7300 391. (also 6500 032)
          +-Gr 11-75 p839   /MJ 7-76 p31
          +HF 7-76 p114   +NR 3-76 p3
          +HFN 12-75 p169   ++RR 11-75 p58
          +HFN 2-76 p116 tape   ++SFC 12-21-75 p39
      Lohengrin: Preludes, Acts 1 and 3. cf The flying Dutchman: Over-
        ture.
2768 Lohengrin: Prelude, Act 3. Tannhäuser: Overture, Grand march.
        Die Walküre: Magic fire music; Ride of the Valkyries. Tristan
        und Isolde: Liebstod. PO; Eugene Ormandy. CBS 30065. Tape
        40-30065.
          +-Gr 3-76 p1513   +-RR 12-75 p70
          +HFN 1-76 p123
      Lohengrin: Das susse Lied erhalt. cf Rococo 5377.
      Lohengrin: Wedding march. cf DG 2548 212.

2769 Die Meistersinger von Nürnberg. Hannelore Bode, s; Julia Hamari,
      ms; Adalbert Kraus, Martin Schomberg, Wolfgang Appel, Michel
      Sénéchal, René Kollo, Adolf Dallapozza, t; Norman Bailey, Bernd
      Weikl, bar; Kurt Moll, Martin Egel, Gerd Nienstedt, Helmut
      Berger-Tuna, Kurt Rydl, Rudolf Hartmann, bs; Gumpoldskirchener
      Spatzen; Vienna State Opera Chorus; VPO; Georg Solti. Decca
      D13D 5 (5). Tape (c) K13K 54. (also London 1512 Tape (c)
      5-1512).

              +Gr 9-76 p468   +-ON 12-18-76 p76
             ++HFN 9-76 p113   +-RR 9-76 p36
             ++HFN 12-76 p155   ++SFC 12-5-76 p58
              +-MT 12-76 p1007

2770 Die Meistersinger von Nürnberg. Catarina Ligendza, s; Christa
      Ludwig, ms; Peter Maus, Loren Driscoll, Karl-Ernst Mercker,
      Martin Vantin, Placido Domingo, Horst Laubenthal, t; Dietrich
      Fischer-Dieskau, Klaus Lang, bar; Peter Lagger, Roland Hermann,
      Gerd Feldhoff, Ivan Sardi, bs; Berlin Opera Orchestra and
      Chorus; Eugen Jochum. DG 2740 149. (also DG 2713 011).

              +-Gr 12-76 p1057   +-RR 12-76 p46
              +ON 12-18-76 p76

2771 Die Meistersinger von Nürnberg (nearly complete recording). Maria
      Muller, s; Camilla Kallab, ms; Max Lorenz, Erich Zimmermann,
      Benno Arnold, Gerhard Witting, Gustaf Rodin, Karl Krollmann, t;
      Jaro Prohaska, Eugen Fuchs, bar; Josef Greindl, Fritz Krenn,
      Helmut Fehn, Herbert Gosebruch, Franz Sauer, Alfred Dome, Erich
      Pina, bs; Bayreuth Festival Orchestra and Chorus; Wilhelm
      Furtwängler. EMI Odeon 1C 181 01797/801 (5). (also Da Capo).

              +Gr 12-76 p1057   +-ON 12-18-76 p76
              +-HF 9-76 p100   +-St 11-76 p162
              +-NYT 6-20-76 pD27

2772 Die Meistersinger von Nürnberg. Hannelore Bode, s; Anna Reynolds,
      ms; Heribert Steinbach, Robert Licha, Wolf Appel, Norbert Orth,
      Jean Cox, Frieder Stricker, t; Karl Ridderbusch, bar; Jozsef
      Dene, Klaus Hirte, Gerd Nienstedt, Heinz Feldhoff, Hartmut
      Bauer, Nikolaus Hillebrand, Bernd Weikl, bs; Bayreuth Festival
      Orchestra and Chorus; Silvio Varviso. Philips 6747 167 (5).

              +Gr 9-75 p519   ++ON 4-17-76 p42
              +-HF 4-76 p124   +-RR 9-75 p25
              /HFN 9-75 p107   ++SFC 1-18-76 p38
              +HFN 7-76 p99   +-SR 1-24-76 p51
              +MJ 3-76 p24   +-SR 7-24-76 p36
              +-NR 3-76 p8   +-St 5-76 p126
              +-NYT 3-14-76 pD15

      Die Meistersinger von Nürnberg. cf Lohengrin.
      Die Meistersinger von Nürnberg, excerpts. cf Works, selections
         (DG 2721 109).
      Die Meistersinger von Nürnberg, excerpts. cf Works, selections
         (DG 2721 110).
      Die Meistersinger von Nürnberg, excerpts. cf Works, selections
         (DG 2721 111).
      Die Meistersinger von Nürnberg, excerpts. cr Works, selections
         (DG 2721 112).
      Die Meistersinger von Nürnberg, excerpts. cf Works, selections
         (DG 2721 113).
      Die Meistersinger von Nürnberg: Ansprache des Pogner. cf Rococo
         5397.

Die Meistersinger von Nürnberg: Fanget an. cf Rococo 5379.

Die Meistersinger von Nürnberg: Jerum, Jerum, Oh, ihr boshafter
    Geselle. cf Preiser LV 192.

Die Meistersinger von Nürnberg: Morgendlich leuchtend im roseigen
    Schein. cf Court Opera Classics CO 354.

Die Meistersinger von Nürnberg: Overture. cf Bruno Walter Society
    RR 443.

Die Meistersinger von Nürnberg: Preislied; Fanget an Orchester-
    begleitung. cf Discophilia 13 KGW 1.

Die Meistersinger von Nürnberg: Prelude. cf The flying Dutchman:
    Overture.

Die Meistersinger von Nürnberg: Prelude, Act 1. cf Lohengrin:
    Preludes, Acts 1 and 3.

2773 Die Meistersinger von Nürnberg: Preludes, Acts 1 and 3; Dance of
      the apprentices; Procession of the Meistersinger. Tannhäuser:
      Overture and Venusberg music. PO; Eugene Ormandy. RCA ARL
      1-1868. Tape (c) ARK 1-1868 (ct) ARS 1-1868.
          +NR 11-76 p4

Die Meistersinger von Nürnberg: Wahn, Wahn; Morgenlich leuchtend.
    cf Arias (DG 2721 115).

Die Meistersinger von Nürnberg: Was duftet. cf Decca GOSE 677/79.

Nibelungen-Marsch (arr. Sonntag). cf DG 2721 077.

2774 Parsifal. Gundula Janowitz, Anja Silja, Else-Margrete Gardelli,
      Dorothea Siebert, Rita Bartos, s; Irene Dalis, Sona Cervena,
      ms; Jesse Thomas, Niels Möller, Gerhard Stolze, Georg Paskuda,
      t; Ursula Boese, con; George London, Hans Hotter, bs-bar; Martti
      Talvela, Gustav Neidlinger, Gerd Nienstedt, bs; Bayreuth Festi-
      val Orchestra and Chorus of 1962; Hans Knappertsbusch. Philips
      6747 250 (5). (Reissue from SAL 3475/9). (also Philips 835
      220/4).
          +Gr 7-76 p217   ++RR 7-76 p40
          ++HFN 7-76 p99

Parsifal, excerpts. cf Works, selections (DG 2721 109).

Parsifal, excerpts. cf Works, selections (DG 2721 112).

Parsifal, excerpts. cf Works, selections (DG 2721 113).

Parsifal: Amfortas, Die Wunde, Nur eine Waffe taugt. cf Court
    Opera Classics CO 354.

Parsifal: Ich sah das Kind; Das ist Karfreitagzüber, Herr. cf
    Arias (DG 2721 115).

Parsifal: Prelude. cf Lohengrin: Preludes, Acts 1 and 3.

Parsifal: Preludes, Acts 1 and 3. cf The flying Dutchman: Over-
    ture.

Parsifal: Prelude; Good Friday music. cf Lohengrin: Preludes,
    Acts 1 and 3.

2775 Das Rheingold. Elisabeth Höngen, s; Margaret Weth-Falke, ms;
      Günther Treptow, Joachim Sattler, Peter Markwort, t; Ferdinand
      Frantz, Angelo Mattiello, Alois Pernerstorfer, bar; Ludwig
      Weber, Albert Emmerich, bs; La Scala Orchestra and Chorus;
      Wilhelm Furtwängler. Bruno Walter Society RR 420/1 (3).
          +NR 3-76 p8

2776 The Rheingold (sung in English). Valerie Masterson, Shelagh
      Squires, Lois McDonall, s; Helen Attfield, Anne Collins, ms;
      Katherine Pring, con; Robert Ferguson, Emile Belcourt, Gregory
      Dempsey, t; Derek Hammond-Stroud, Norman Bailey, Norman Welsby,
      bar; Robert Lloyd, Clifford Grant, bs; English Opera Group
      Orchestra; Reginald Goodall. HMV (Q) SLS 5032 (4). (also
      Angel SDC 3825).

```
        +-Gr 11-75 p903           +-ON 8-76 p39
        +-HFN 12-75 p161          ++RR 12-75 p33
        +-MT 7-76 p579            +-SR 4-17-76 p51
        +-NYT 3-14-76 pD15
```

Das Rheingold, excerpts. cf Works, selections (DG 2721 109).

Das Rheingold, excerpts. cf Works, selections (DG 2721 110).

Das Rheingold: Abendlich strahit der Sonne Auge. cf Preiser LV 192.

Das Rheingold: Entrance of the Gods into Valhalla. cf Der Ring des Nibelungen (Decca SXL 6743).

Das Rheingold: Immer ist Undank Loges Lohn, uber Stock und Stein zu Tal. cf Court Opera Classics CO 354.

Das Rheingold: Weiche, Wotan, weiche. cf Arias (DG 2721 115).

2777 Rienzi. Siv Wennberg, Janis Martin, Ingeborg Springer, s; René Kollo, Peter Schreier, t; Theo Adam, Nikolaus Hillebrand, Siegfried Vogel, Günther Leib, bs; Leipzig Radio Chorus, Dresden State Opera Chorus; Dresden Staatskapelle; Heinrich Hollreiser. HMV SQ SLS 990 (5). (also Angel SC 3818).

```
        +-Gr 11-76 p873           +-RR 11-76 p47
        +-HFN 11-76 p151          +SFC 11-14-76 p30
```

Rienzi: Gesang des Friedensboten. cf Discophilia DIS 13 KHH 1.

Rienzi: Overture. cf HMV RLS 717.

Rienzi: Overture. cf Supraphon 110 1637.

2778 Der Ring des Nibelungen. Claire Watson, Birgit Nilsson, Regine Crespin, Joan Sutherland, s; Jean Madeira, ms; Christa Ludwig, Marga Höffgen, con; Set Svanholm, Paul Kuen, Waldemar Kmentt, James King, Wolfgang Windgassen, Gerhard Stolze, t; Eberhard Wächter, Dietrich Fischer-Dieskau, bar; George London, Hans Hotter, bs-bar; Gustav Neidlinger, Walther Kreppel, Gottlob Frick, bs; VPO; Georg Solti. Decca Tape (c) K2W 29 (2), K3W 30 (3), K3W 31 (3), K4W 32 (4).

```
        ++Gr 9-76 p494 tape       +-RR 12-76 p110 tape
        ++HFN 9-76 p133 tape
```

2779 Der Ring des Nibelungen. Walburga Wegner, Magda Gabory, Margarita Kenney, Kirsten Flagstad, Hilde Konetzni, Ilona Steingruber, Karen Marie Cerkal, Julia Moor, s; Elisabeth Höngen, Dagmar Schmedes, Margret Weth-Falke, Sieglinde Wagner, Polly Batic, con; Joachim Sattler, Gunther Treptow, Peter Markwort, Set Svanholm, Max Lorenz, t; Ferdinand Frantz, Angelo Mattiello, Josef Hermann, bar; Alois Pernerstorfer, Ludwig Weber, Albert Emmerich, bs; La Scala Orchestra and Chorus; Wilhelm Furtwängler. Murray Hill 940477 (11).

```
        +Gr 10-76 p654            +-RR 10-76 p37
        +HFN 8-76 p89
```

2780 Der Ring des Nibelungen: Gotterdämmerung. Das Rheingold. Siegfried. Die Walküre. Birgit Nilsson, Ludmilla Dvořáková, Dorothea Siebert, Helga Dernesch, Anja Silja, Erika Köth, Leonie Rysanek, Daniza Mastilovic, Liane Synek, s; Martha Mödl, Annelies Burmeister, Věra Soukupová, Ruth Hesse, Gertrud Hopf, Sona Cervena, ms; Sieglinde Wagner, Marga Höffgen, Elisabeth Schartel, con; Wolfgang Windgassen, Hermin Esser, Irwin Wohlfahrt, James King, t; Thomas Stewart, bar; Gustav Neidlinger, Josef Greindl, Theo Adam, Gerd Nienstedt, Martti Talvela, Kurt Böhme, bs; Bayreuth Festival Orchestra and Chorus; Karl Böhm. Philips 6747 037 (16).

```
        +-Gr 9-73 p535            -NYT 9-9-73 pD32
```

```
        +-Gr 7-76 p211              +RR 9-73 p38
        ++HFN 7-76 p99             +-RR 7-76 p40
        +-MJ 10-73 p11            +SFC 8-26-73 p28
```

2781 Der Ring des Nibelungen, orchestral excerpts: Götterdämmerung:
      Siegfried's Rhine journey; Siegfried's funeral march; Immola-
      tion of the Gods. Das Rheingold: Entrance of the Gods into
      Valhalla. Siegfried: Forest murmurs. Die Walküre: Ride of
      the Valkyries; Wotan's farewell and magic fire music. National
      Philharmonic Orchestra; Antal Dorati. Decca SXL 6743. Tape
      (c) KSXC 6743. (also London CS 6970 Tape (c) 5-6970).

```
        +Gr 4-76 p1612            -RR 12-76 p111 tape
        -RR 5-76 p59             ++SFC 6-27-76 p29
```

2782 Siegfried. Elisabeth Höngen, Kirsten Flagstad, Julia Moor, s;
      Set Svanholm, Peter Markwort, t; Josef Hermann, Alois Perner-
      storfer, bar; Ludwig Weber, bs; La Scala Orchestra and Chorus;
      Wilhelm Furtwängler. Bruno Walter Society RR 420/3 (5).

```
        +NR 3-76 p8
```

2783 Siegfried (sung in English). Maurine London, Rita Hunter, s;
      Anne Collins, ms; Gregory Dempsey, Alberto Remedios, t; Nor-
      man Bailey, Derek Hammond-Stroud, bar; Clifford Grant, bs;
      Sadler's Wells Orchestra; Barry Tuckwell, hn; Reginald Goodall.
      HMV SLS 875 (5). Tape (c) TC SLS 875. (also EMI Odeon
      SLS 873).

```
        +Gr 4-74 p1893            +Op 6-74 p516
        +HF 4-76 p93             ++RR 4-74 p32
        +-HFN 12-76 p155          +RR 10-76 p104 tape
        ++NYT 6-16-74 pD34
```

      Siegfried, excerpts. cf Lohengrin, excerpts.
      Siegfried, excerpts. cf Works, selections (DG 2721 110).
      Siegfried, excerpts. cf Works, selections (DG 2721 111).
      Siegfried: Ewig war ich. cf Arias (DG 2721 115).

2784 Siegfried: Final scene. Wesendonck Lieder. Eileen Farrell, s;
      Set Svanholm, t; Stokowski Symphony Orchestra, Rochester
      Philharmonic Orchestra; Leopold Stokowski, Erich Leinsdorf.
      RCA AVM 1-1413. (Reissue from LM 1066, 1000).

```
        +-ARSC Vol VIII, nos. 2-3 +-HF 10-76 p127
          p98                     +-NR 7-76 p9
```

      Siegfried: Forest murmurs. cf Der Ring des Nibelungen (Decca SXL
      6743).
      Siegfried: Schmiedelied. cf Discophilia 13 KGW 1.
      Tannhäuser, excerpts. cf Lohengrin, excerpts.
      Tannhäuser, excerpts. cf Works, selections (DG 2721 109).
      Tannhäuser, excerpts. cf Works, selections (DG 2721 111).
      Tannhäuser, excerpts. cf Works, selections (DG 2721 112).
      Tannhäuser, excerpts. cf Works, selections (DG 2721 113).
      Tannhäuser: Dich, teure Halle; Allmächt'ge Jungfrau. cf Rubini
      GV 63.
      Tannhäuser: Dich teure Halle, Frubitte der Elisabeth. cf Rococo
      5377.
      Tannhäuser: Dir töne Lob, Stets soll nur dir. cf Rococo 5397.
      Tannhäuser: Grand march; Pilgrims' chorus. cf DG 2548 212.
      Tannhäuser: Lied des Hirtenknaben; Als du in kuhnem Sange. cf
      Rococo 5379.
      Tannhäuser: Loblied an die Venus. cf Discophilia 13 KGW 1.
      Tannhäuser: O, du mein holder Abendstern. cf Caprice CAP 1062.
      Tannhäuser: O, du mein holder Abendstern. cf English National
      Opera ENO 1001.

Tannhäuser: Overture. cf Lohengrin: Preludes, Acts 1 and 3.
Tannhäuser: Overture. cf BEETHOVEN: Symphony, no. 8, op. 93, F
     major.
Tannhäuser: Overture. cf Bruno Walter Society RR 443.
Tannhäuser: Overture and bacchanale. cf Lohengrin: Prelude, Act 1.
Tannhäuser: Overture, Grand march. cf Lohengrin: Prelude, Act 3.
Tannhäuser: Overture and Venusberg music. cf Götterdämmerung:
     Rhine journey.
Tannhäuser: Overture and Venusberg music. cf Die Meistersinger
     von Nürnberg: Preludes, Acts 1 and 3; Dance of the apprentices;
     Procession of the Meistersinger.
Tannhäuser: Pilgrims' chorus. cf Decca DPA 523/4.
Tannhäuser: Pilgrims' chorus. cf Decca DPA 525/6.
Tannhäuser: Prelude, Act 3. cf HMV RLS 717.
Tannhäuser: Wie Todesahnung...O du, mein holder Abendstern; In-
     brunst im Herzen; Dich teure Halle. cf Arias (DG 2721 115).
2785 Tristan und Isolde. Kirsten Flagstad, s; Kerstin Thorborg, ms;
     Lauritz Melchior, Emery Darcy, Karl Laufkoetter, t; Julius
     Huehn, bar; Alexander Kipnis, John Gurney, bs; Metropolitan
     Opera Orchestra and Chorus; Erich Leinsdorf. MET 3 (4).
     (available with donation to Metropolitan Opera Fund).
          +-HF 10-76 p91   +ON 8-76 p39
          ++NYT 6-20-76 pD27
Tristan und Isolde. cf Lohengrin.
Tristan und Isolde, excerpts. cf Works, selections (DG 2721 109).
Tristan und Isolde, excerpts. cf Works, selections (DG 2721 110).
Tristan und Isolde, excerpts. cf Works, selections (DG 2721 112).
Tristan und Isolde, excerpts. cf Works, selections (DG 2721 113).
Tristan und Isolde: Liebestod. cf Lohengrin: Prelude, Act 3.
Tristan und Isolde: Mild und leise. cf Arias (DG 2721 115).
Tristan und Isolde: Mild und leise. cf Decca GOSE 677/9.
Tristan und Isolde: O sink hernieder. cf Rococo 5377.
Tristan und Isolde: O sink hernieder, Liebestod. cf Rococo 5380.
Tristan und Isolde: Prelude, Act 3. cf HMV RLS 717.
2786 Tristan und Isolde: Prelude and Liebestod. Wesendonk Lieder.
     Jessye Norman, s; LSO; Colin Davis. Philips 9500 031.
          +-Audio 7-76 p71   +-NR 5-76 p9
          +-Gr 7-76 p205   +-RR 7-76 p41
          +-HF 7-76 p94   ++SFC 5-16-76 p28
          +-HFN 8-76 p91
Tristan und Isolde: Prelude and Liebestod. cf Götterdämmerung:
     Immolation scene.
Tristan und Isolde: Prelude und Liebestod. cf Lohengrin: Prelude,
     Act 1.
Tristan und Isolde: Prelude and Liebestod. cf Lohengrin: Preludes,
     Acts 1 and 3.
2787 Die Walküre: Hilde Konetzni, Kirsten Flagstad, Elisabeth Höngen,
     Walburga Wegner, Ilona Steingruber, Dagmar Schmedes, Karen
     Marie Cerkal, Margaret Weth-Falke, Sieglinde Wagner, s; Polly
     Batic, Margarita Kenney, con; Gunther Treptow, t; Ferdinand
     Frantz, bar; Ludwig Weber, bs; La Scala Orchestra and Chorus;
     Wilhelm Furtwängler. Bruno Walter Society RR 420/2 (5).
          +NR 3-76 p8
2788 The Valkyrie (sung in English). Margaret Curphey, Rita Hunter,
     Katie Clarke, Helen Attfield, Anne Evans, s; Ann Howard, Eliz-
     abeth Connell, Sarah Walker, Selagh Squires, Ann Collins, ms;

Alberto Remedios, s; Norman Bailey, bs-bar; Clifford Grant,
bs; English Opera Orchestra; Reginald Goodall. HMV SQ SLS
5063 (5). (also Angel SX 3826).
          ++Gr 9-76 p473   +-MT 12-76 p1008
          +HFN 9-76 p111   +RR 9-76 p41
Die Walküre, excerpts. cf Works, selections (DG 2721 109).
Die Walküre, excerpts. cf Works, selections (DG 2721 110).
Die Walküre, excerpts. cf Works, selections (DG 2721 111).
Die Walküre: Ein Schwert verhiess mir der Vater. cf Arias (DG
    2721 115).
Die Walküre: Fire music. cf Götterdämmerung: Rhine journey.
Die Walküre: Fort denn eile. cf Collectors Guild 611.
Die Walküre: Ho-jo-to-ho. cf Rococo 5379.
Die Walküre: Ho-jo-to-ho...So fliehe denn; Magic fire music. cf
    Decca GOSE 677/9.
Die Walküre: Magic fire music; Ride of the Valkyries. cf
    Lohengrin: Prelude, Act 3.
Die Walküre: Nun zaume...Ho-jo-to-ho, Siegmund, Sieh auf mich,
    War est so schmählich. cf Rococo 5380.
Die Walküre: Ride of the Valkyries. cf BEETHOVEN: Symphony, no.
    3, op. 55, E flat major.
Die Walküre: Ride of the Valkyries. cf Decca DPA 519/20.
Die Walküre: Ride of the Valkyries; Wotan's farewell and magic
    fire music. cf Der Ring des Nibelungen (Decca SXL 6743).
Die Walküre: Wintersturme wichen dem Wonnemond (2). cf Disco-
    philia 13 KGW 1.
Wesendonck Lieder. cf Siegfried: Final scene.
Wesendonck Lieder. cf BRAHMS: Alto rhapsody, op. 53.
2789 Works, selections (Bayreuth Festival, 1900-1930): Der fliegende
    Holländer, excerpts. Götterdämmerung, excerpts. Lohengrin,
    excerpts. Die Meistersinger von Nürnberg, excerpts. Parsifal,
    excerpts. Das Rheingold, excerpts. Tannhäuser, excerpts.
    Tristan und Isolde, excerpts. Die Walküre, excerpts. Emmy
    Destinn, Fride Leider, s; Sigrid Onegin, con; Heinrich Hensel,
    Lauritz Melchior, t; Friedrich Schorr, Walter Soomer, bar;
    Richard Mayr, Paul Knupfer, bs; Bayreuth Festival Orchestra.
    DG 2721 109 (2).
          +-RR 12-76 p25
2790 Works, selections (Bayreuth Festival, 1930-1944): Der fliegende
    Holländer, excerpts. Lohengrin, excerpts. Die Meistersinger
    von Nürnberg, excerpts. Das Rheingold, excerpts. Siegfried,
    excerpts. Tristan und Isolde, excerpts. Die Walküre, excerpts.
    Elisabeth Ohms, Margarete Klose, Frida Leider, Maria Muller, s;
    Franz Volker, Max Lorenz, Lauritz Melchior, t; Heinrich Schlus-
    nus, bar; Josef von Manowarda, bs; Bayreuth Festival Orchestra.
    DG 2721 110 (2).
          +-RR 12-76 p25
2791 Works, selections (Bayreuth Festival, 1951-1960): Götterdämmerung,
    excerpts. Dei Meistersinger von Nürnberg, excerpts. Siegfried,
    excerpts. Tannhäuser, excerpts. Die Walküre, excerpts. Leon-
    ie Rysanek, Annelies Kupper, Astrid Varnay, s; Wolfgang Wind-
    gassen, t; Eberhard Wächter, bar; Hans Hotter, bs-bar; Josef
    Greindl, bs; Bayreuth Festival Orchestra and Chorus; Wilhelm
    Pitz. DG 2721 111 (2).
          +RR 12-76 p25
2792 Works, selections (Bayreuth Festival, 1960-): Der fliegende

Holländer, excerpts. Die Meistersinger von Nürnberg, excerpts.
Parsifal, excerpts. Tannhäuser, excerpts. Tristan und Isolde,
excerpts. Gwyneth Jones, Birgit Nilsson, s; Christa Ludwig,
Sieglinde Wagner, con; Jess Thomas, Wolfgang Windgassen, James
King, t; Dietrich Fischer-Dieskau, Eberhard Wächter, bar; Karl
Ridderbusch, Martti Talvela, bs; Bayreuth Festival Orchestra.
DG 2721 112 (2).
+-RR 12-76 p25

2793 Works, selections (orchestral): Der fliegende Holländer, excerpts.
Götterdämmerung, excerpts. Lohengrin, excerpts. Die Meister-
singer von Nürnberg, excerpts. Parsifal, excerpts. Tannhäuser,
excerpts. Tristan und Isolde, excerpts. Bayreuth Festival
Orchestra; Hans Knappertsbusch, Richard Strauss, Karl Elmen-
dorff, Wilhelm Furtwängler, Victor de Sabata, Eugen Jochum,
Pierre Boulez, Karl Böhm. DG 2721 113 (2).
+-RR 12-76 p25

WAGNES
Die Bosniaken kommen. cf DG 2721 077.

WAISSELIUS (also WAISSEL), Matthaus
La battaglia, phantasia. cf Hungaroton SLPX 11669/70.
Deutscher Tanzt. cf DG Archive 2533 302.
Fantasia. cf DG Archive 2533 302.
Gagliarda: Chi passa; La gamba; La rocca del Fuso; La traditora.
cf Hungaroton SLPX 11669/70.
Un gay bergier, phantasia. cf Hungaroton SLPX 11669/70.
Je prens en gré, phantasia. cf Hungaroton SLPX 11669/70.
Polonischer Tantz. cf Hungaroton SLPX 11721.

WALDTEUFEL, Emile
Acclamations, op. 223. cf Works, selections (HMV SQ ESD 7012).
Bella bocca, op. 163. cf Works, selections (HMV SQ ESD 7012).

2794 Dolores. España, op. 236. The Grenadiers. Mon rêve. Pomone.
Les patineurs, op. 183. Toujours ou jamais. National Phil-
harmonic Orchestra; Douglas Gamley. Decca SXL 6704. Tape
(c) KSXC 6704. (also London CS 6899).
+Audio 12-76 p93   +NR 1-76 p5
+-Gr 1-76 p1248   +RR 12-75 p71
+-HF 10-76 p128   +-SR 3-6-76 p41
+HFN 12-75 p169   +St 8-76 p102
España, op. 236. cf Dolores.
España, op. 236. cf Works, selections (HMV SQ ESD 7012).
L'Esprit français, op. 182. cf Works, selections (HMV SQ ESD 7012).
Estudiantina, op. 191. cf Works, selections (HMV SQ ESD 7012).
The Grenadiers. cf Dolores.
Minuit, op. 168. cf Works, selections (HMV SQ ESD 7012).
Mon rêve. cf Dolores.
Les patineurs, op. 183. cf Dolores.
Les patinuers, op. 183. cf Works, selections (HMV SQ ESD 7012).
Pomone. cf Dolores.
Prestissimo, op. 152. cf Works, selections (HMV SQ ESD 7012).
Toujours ou jamais. cf Dolores.

2795 Works, selections: Acclamations, op. 223. Bella bocca, op. 163.
España, op. 236. L'Esprit français, op. 182. Estudiantina,
op. 191. Les patineurs, op. 183. Prestissimo, op. 152.
Minuit, op. 168. Monte Carlo Opera Orchestra; Willi Boskovsky.
HMV SQ ESD 7012.
+Gr 11-76 p894   +RR 11-76 p85
+HFN 11-76 p169

WALKER, Ernest
    I will lift up mine eyes. cf Argo ZRG 789.
WALKER, George
    Lyric for strings. cf SOWANDE: African suite.
WALKER, Louise
    Small variations on a Catalonian folksong. cf Supraphon 111 1230.
WALKER, Robert
    The sun of the Celandines. cf Prelude PRS 2501.
WALLACE
    Scots wha hae variations. cf Library of Congress OMP 101/2.
WALLACE, Vincent
    Maritana: Yes, let me like a soldier fall. cf HMV EMD 5528.
WALLEBOM
    Hasta la vista. cf Pye TB 3006.
WALLNOFER, Adolf
    Woher die Liebe. cf Discophilia 13 KGW 1.
WALMISLEY, Thomas
    Psalm, no. 49, O hear ye this, all ye people. cf HMV SQ CSD 3768.
WALOND, William
    Voluntary, D minor. cf Cambridge CRS 2540.
WALTERS
    Trumpets wild. cf HMV OU 2105.
WALTHER, Johann
    Concerto del Signore Torelli, appropriato all'organo. cf BOHM:
        Ach wie nichtig, ach wie flüchtig, chorale partita.
    Jesu, meine Freude. cf BACH: Chorale prelude: Sei gegrüsset,
        Jesus gütig, S 768.
    Jesu, meine Freude, partita. cf BOHM: Ach wie nichtig, ach wie
        flüchtig, chorale partita.
    Joseph lieber Joseph mein. cf MOUTON: Noe psallite.
    Sonata, G major. cf BADINGS: Intermezzo.
WALTON, William
    Bagatelles, guitar (5). cf BENNETT: Concerto, guitar and chamber
        orchestra.
    Concerto, violin, B minor. cf RCA ARM 4-0942/7.
    Crown imperial. cf BRITTEN: Simple symphony, strings, op. 4.
    Crown imperial. cf Decca 419.
2796 Façade. Tony Randall, reader; Columbia Chamber Ensemble; Arthur
        Fiedler. Columbia M 33980.
                +HF 9-76 p101   +St 10-76 p137
            ++NR 8-76 p4
    Façade. cf SAINT-SAENS: The carnival of the animals.
    Façade. cf HMV HQS 1360.
    Façade. cf Polydor 2383 391.
    Henry V: Death of Falstaff; Passacaglia; Touch her soft lips and
        part. cf DELIUS: Aquarelles.
    Henry V: Touch her soft lips and part. cf RCA GL 25006.
    Orb and sceptre. cf HMV (Q) ASD 3131.
    Portsmouth Point overture. cf LAMBERT: The Rio Grande.
    Portsmouth Point overture. cf HMV ESD 7011.
    Scapino overture. cf LAMBERT: The Rio Grande.
    Sonata, violin. cf ELGAR: Sonata, violin, op. 82, E minor.
    Songs: Set me as a seal upon thine heart. cf HMV HQS 1350.
    Songs: Set me as a seal upon thine heart. cf Wealden WS 137.
    Symphony, no. 2. cf LAMBERT: The Rio Grande.
WARD
    America the beautiful. cf Department of Defense Bicentennial
        Edition 50-1776.

WARLOCK, Peter
      Adam lay y bounden. cf HELY-HUTCHINSON: Carol symphony.
      Bethlehem down. cf HELY-HUTCHINSON: Carol symphony.
      Motets. cf DELIUS: A late lark.
      Songs: Away to Twivver; Fill the cup, Philip; Capriol suite:
         Tordion; Mattachins; Hey troly loly lo; I asked a thief to
         steal me a peach; In an arbour green; Jillian of Berry; My
         ghostly fader; Piggesine; Sweet content; Peter Warlock's fancy.
         cf Pearl SHE 525.
      Songs: The Bayley beareth the bell away; Lullaby. cf HMV RLS 716.
      Songs: Pretty ring-time; My own country; Passing by; A prayer to
         St. Anthony of Padua; The sick heart. cf BUTTERWORTH: A Shrop-
         shire lad.
      Songs: Where be ye, my love. cf Argo ZRG 833.
WATSON
      Songs: Anchored. cf Argo ZFG 95/6.
WEBB
      The shepherd's song. cf HMV SXLP 50017.
WEBBER, Amherst
      Vieille chanson. cf HMV RLS 716.
WEBER, Ben
2797 Concerto, piano, op. 52. WUORINEN: Concerto, piano. William
         Masselos, Charles Wuorinen, pno; RPO; Gerhard Samuel, James
         Dixon. CRI SD 239.
              *RR 2-76 p44
      Consort of winds, op. 66. cf PISTON: Quintet, wind instruments.
      Quartet, strings, no. 2, op. 35. cf MONOD: Cantus contra cantum
         I.
WEBER, Carl Maria von
2798 Abu Hassan. Siegfried Göhler, Kurt Kachlicki, Gerd Biewer, August
         Hütten, Dorothea Garlin, speakers; Ingeborg Hallstein, s;
         Peter Schreier, t; Theo Adam, bs-bar; Gerhard Wüstner Student
         Chorale, Dresden State Opera Chorus; Dresden Staatskapelle;
         Heinz Rögner. RCA LRL 1-5125.
             +-Gr 8-76 p337   +RR 8-76 p25
             +-HFN 8-76 p91
      Abu Hassan: Overture. cf Supraphon 110 1637.
      Adagio and rondo. cf HMV SQ ASD 3283.
      Andante and rondo, op. 35, C minor. cf HANDEL: Concerto, viola,
         B minor.
      Concertino, clarinet, op. 26, C minor. cf HAYDN: Concerto, trumpet,
         E flat major.
2799 Concerto, piano, no. 1, C major. Concerto, piano, no. 2, E flat
         major. Friedrich Wührer, pno; Vienna Pro Musica Orchestra;
         Hans Swarowsky. Turnabout TV 34155. Tape (c) KTVC 34155.
            -HFN 7-76 p105 tape   +-RR 10-76 p107 tape
      Concerto, piano, no. 2, E flat major. cf Concerto, piano, no. 1,
         C major.
      Duo concertant, E flat major. cf Decca SPA 395.
      Duo concertant, op. 48. cf GLINKA: Trio pathetique, D minor.
2800 Euryanthe. Jessye Norman, Rita Hunter, Renate Krahmer, s; Nicolai
         Gedda, Harald Neukirch, t; Tom Krause, bar; Siegfried Vogel,
         bs; Leipzig Radio Chorus; Dresden Staatskapelle; Marek Janowski.
         HMV (Q) SLS 983. (also Angel SDL 3764).
            +Gr 10-75 p687   +NYT 11-16-75 pD1
            ++HF 2-76 p81   +ON 1-10-76 p33

```
          +-HFN 11-75 p171          +RR 11-75 p22
          +-MM 3-76 p41             +SFC 12-7-75 p31
          +MT 7-76 p579             +SR 1-24-76 p52
          +NR 12-75 p11             ++St 2-76 p110
```

Euryanthe: Overture. cf Invitation to the dance, op. 65
Euryanthe: Overture. cf BEETHOVEN: Symphony, no. 8, op. 93, F
    major.

2801 Der Freischütz. Gundula Janowitz, Edith Mathis, s; Peter Schreier,
    t; Bernd Weikl, bar; Theo Adam, bs-bar; Franz Crass, Siegfried
    Vogel, bs; Leipzig Radio Chorus; Dresden State Orchestra;
    Carlos Kleiber. DG 2720 071 (3). Tape (c) 3371 008. (also
    DG 2709 045).

```
          +Gr 11-73 p985            +ON 10-74 p64
          +HF 3-74 p100             +-Op 11-73 p1006
          +HF 4-76 p148 tape        +-RR 10-73 p54
          +MJ 7-74 p50              +RR 4-74 p93 tape
          +NR 7-74 p10              +St 4-74 p77
```

2802 Der Freischütz. Irmgard Seefried, Rita Streich, Margot Laminet,
    Gisela Ohrt, s; Richard Holm, Paul Kuen, t; Albrecht Peter,
    bar; Eberhard Wächter, Kurt Böhme, Walter Kreppel, bs; Bav
    arian Radio Symphony Orchestra and Chorus; Eugen Jochum. DG
    2726 061 (2). (Reissue from SLPM 138639/40).

```
          +-Gr 11-76 p874           +-RR 11-76 p48
          +-HFN 10-76 p179
```

2803 Der Freischütz: Overture; Nein, länger trag ich nicht die Qualen;
    Schweig, schweig, damit dich niemand warnt; Durch die Wälder;
    Schelm, Malt fest; Kommt ein Schlanker Bursch gegangen; Wie was;
    Entsetzen; Und ob die Wolke sie verhülle; Wir winden dir den
    Jungfernkranz; Was gleicht wohl auf Erden. Gundula Janowitz,
    Edith Mathis, s; Peter Schreier, t; Theo Adam, bs; Leipzig
    Radio Chorus; Dresden Staatskapelle; Carlos Kleiber. DG 2537
    020. (Reissue from 2720 071).

```
          +Gr 1-76 p1239
```

Der Freischütz: Chorus of huntsmen; Bridesmaids' chorus. cf
    DG 2548 212.
Der Freischütz: Durch die Walder. cf Rococo 5379.
Der Freischütz: Hier im irdischen Jammertal. cf Preiser LV 192.
Der Freischütz: Nein langer trag ich nicht die Qualen. cf Court
    Opera Classics CO 354.
Der Freischütz: Overture. cf Invitation to the dance, op. 65.
Der Freischütz: Overture. cf MENDELSSOHN: Hebrides overture, op.
    26.
Der Freischütz: Trube Augen. cf Discophilia DIS 13 KHH 1.
Der Freischütz: Wie nahte...Leise, leise. cf Decca GOSE 677/9.

2804 Invitation to the dance, op. 65 (orch. Berlioz). Overtures:
    Euryanthe. Der Freischütz. Oberon. NYP; Leonard Bernstein.
    CBS 61685. (also Columbia M 33585 Tape (c) MT 33585).

```
          +Gr 9-76 p435             ++MJ 11-76 p44
          +-HFN 8-76 p91            +NR 7-76 p3
          ++MJ 11-76 p44            +-RR 8-76 p61
```

Invitation to the dance, op. 65. cf STRAUSS, J. II: Die Fleder-
    maus, op. 363: Overture and ballet music.
Invitation to the dance, op. 65. cf HMV SLS 5019.
Konzertstück, op. 79, F minor. cf RAVEL: Sonatine.

2805 Oberon. Birgit Nilsson, Marga Schiml, Arleen Auger, s; Julia
    Hamari, ms; Donald Grobe, Placido Domingo, t; Hermann Prey,
    bar; Bavarian Radio Symphony Orchestra and Chorus; Rafael

Kubelik. DG 2726 052 (2). (Reissue from 2709 035).
    +Gr 11-76 p874   ++RR 12-76 p50
    +HFN 10-76 p179
Oberon: Overture. cf Invitation to the dance, op. 65.
Oberon: Overture. cf Bruno Walter Society RR 443.
Oberon: Ozean du Ungeheuer. cf Decca GOSE 677/79.
2806   Overtures: Abu Hassan. Euryanthe. Der Freischütz. Jubel.
       Oberon. Preciosa. Bavarian Radio Symphony Orchestra; Rafael
       Kubelik. DG 2535 136. (Reissues from SLPEM 136463).
           +-Gr 3-76 p1477   +RR 3-76 p52
           +HFN 3-76 p109
2807   Overtures: Abu Hassan. Euryanthe. Der Freischütz. Jubel.
       Oberon. Peter Schmoll und seine Nachbarn. Bamburg Symphony
       Orchestra; Theodor Guschlbauer. Erato STU 70568.
           +-Gr 6-76 p57   +MM 12-76 p43
           +-HFN 8-76 p91   /RR 6-76 p61
       Quintet, clarinet, op. 34, B flat major. cf HMV SLS 5046.
2808   Sonatas, violin, op. 10a, nos. 1-6. Hanno Haag, vln; Anneliese
       Schlicker, pno. Oryx ORYX 1834.
           -Gr 9-76 p438
       Sonata, violoncello, A major. cf Command COMS 9006.
       Songs: Abendsegen, op. 64, no. 5; Bettlelied, op. 25, no. 4; Can-
           zonettas, op. 29; Ah, dove siete, Ninfe se liete, Ch'io mai vi
           possa; Die fromme Magd, op. 54, no. 1; Heimlicher Liebe Pein,
           op. 64, no. 3; Lass mich schlummern, op. 25, no. 3; Sanftes
           Licht; Die Schäferstunde, op. 13, no. 1; Uber die Berge, op. 25,
           no. 2; Wiegenlied, op. 13, no. 2; Die Zeit, op. 13, no. 5. cf
           MENDELSSOHN: Songs (Argo ZRG 827).
2809   Symphony, no. 1, op. 19, C major. Symphony, no. 2, op. 51, C
       major. Turandot: Overture and march. LSO; Hans-Hubert
       Schönzler. RCA LRL 1-5106.
           +-Gr 4-76 p1612   +MT 7-76 p580
           +HFN 4-76 p119   +RR 4-76 p59
           +-MM 12-76 p43   +STL 4-11-76 p36
       Trio, piano, op. 63, G minor. cf HAYDN: Cassation, no. 6, C major.
WEBERN, Anton
2810   Bagatelles, string quartet, op. 9 (6). Movements, string quartet,
       op. 5 (5). Quartet, strings (1905). Quartet, strings, op. 28.
       LaSalle Quartet. DG 2530 284. (Reissue from 2720 029).
           +Gr 10-76 p621   ++RR 9-76 p78
           +HFN 11-76 p169   ++STL 10-10-76 p37
       Bagatelles, string quartet, op. 9 (6). cf HAUBENSTOCK-RAMATI:
           Quartet, strings, no. 1.
       Five Movements for string quartet, op. 5. cf HAUBENSTOCK-RAMATI:
           Quartet, strings, no. 1.
       Kleine Stücke, violoncello, op. 11 (3). cf DG 2530 562.
       Kleine Stücke, violoncello, op. 11. cf DEBUSSY: Sonata, violon-
           cello and piano, no. 1, D minor.
       Konzert, op. 24. cf Caprice RIKS LP 34.
       Movements, string quartet, op. 5 (5). cf Bagatelles, string
           quartet, op. 9.
       Movements, string quartet, op. 5 (5). cf BERG: Lyric suite:
           Pieces.
       Passacaglia, op. 1. cf BERG: Lyric suite: Pieces.
       Pieces, op. 6 (6). cf BERG: Lyric suite: Pieces.
       Quartet, strings (1905). cf Bagatelles, string quartet, op. 9.

Quartet, strings, op. 28. cf Bagatelles, string quartet, op. 9.
Quartet, strings, op. 28. cf HAUBENSTOCK-RAMATI: Quartet, strings,
    no. 1.
Sonata, violoncello. cf DG 2530 562.
Sonata, violoncello. cf Laurel-Protone LP 13.
Songs: Lieder, op. 12 and op. 25. cf Hungaroton SLPX 11713.
Symphony, op. 21. cf BERG: Lyric suite: Pieces.

WEBSTER
    Lorena. cf CBS 61746.

WECK (15th century Germany)
    Spanyöler Tanz and Hopper dancz. cf Nonesuch H 71326.

WEELKES, Thomas
    Songs: As Vestas was from Latmos Hill descending. cf Turnabout
        TV 34017.
    Songs: Cease sorrows now. cf Harmonia Mundi 204.
    Songs: Cease sorrows now; Come sirrah, Jack ho; Since Robin Hood.
        cf Enigma VAR 1017.
    Songs: Come, sirrah Jack ho; Four arms, two necks, one wreathing;
        Hark, all ye lovely saints; The nightingale, the organ of
        delight; Since Robin Hood. cf HMV CSD 3756.
    Songs: Hark, all ye lovely saints above; Say, dear, when will your
        frowning leave. cf Harmonia Mundi 593.

WEIGL, Karl
    Songs, contralto and string quartet (3). cf CASTELNUOVO-TEDESCO:
        Concertino, harp, op. 93.

WEILL, Kurt
2811 Aufstieg und Fall der Stadt Mahagonny. Gisela Litz, ms; Lotte
        Lenya, s; Heinz Sauerbaum, t; Horst Günter, bs; Richard Munch,
        speaker; NDR Orchestra and Chorus; Wilhelm Brückner-Rüggeberg.
        CBS 77341. (Reissue from Philips LO 9418/20).
            +Gr 11-73 p1004   +Op 3-74 p225
            +HFN 11-73 p2328   ++RR 12-73 p52
            +-HFN 2-76 p117
      Berlin requiem. cf Works, selections (DG 2740 153).
      Concerto, violin, wind orchestra and percussion, op. 12. cf
        Works, selections (DG 2740 153).
      Death in the forest, op. 23. cf Works, selections (DG 2740 153).
2812 Die Dreigroschenoper (The threepenny opera). Mahagonny: Havana-
        Lied; Alabamasong; Wie mann sich bettet. Der Silbersee: Lied
        der Fennimore; Cäsars Tod. Songs: Das Berliner Requiem:
        Ballade vom ertrunken Mädchen; Happy end: Bilbao-song; Suraba-
        ya-Johnny; Matrosen-tango. Lotte Lenya, s; Berlin Radio Free
        Orchestra; Orchestra; Roger Bean, Wilhelm Brückner-Rüggeberg.
        CBS 78279 (2). (Reissue).
            +Gr 4-75 p1864   +-RR 4-75 p19
            +-HFN 5-75 p141   +STL 5-4-75 p37
            +MM 11-76 p43   +-Te 6-75 p51
      Die Dreigroschenoper: Suite, wind orchestra. cf Works, selections
        (DG 2740 153).
2813 Happy end. Lotte Lenya, s; Orchestra and Chorus; Wilhelm Brückner-
        Rüggeberg. CBS 73463. (Reissue from Philips SABL 193).
            +Gr 1-76 p1240   ++RR 1-76 p29
            +HFN 5-76 p115
      Happy end. cf Works, selections (DG 2740 153).
2814 Kleine Dreigroschenemusik (Three penny opera music). Kurka: The
        good soldier Schweik. Westchester Symphony Orchestra; Sieg-

fried Landau. Candide CE 31089.
            +Audio 8-76 p75
    Kurka: The good soldier Schweik. cf Kleine Dreigroschenemusik.
    Mahagonny songspiel. cf Works, selections (DG 2740 153).
    Protagonist, op. 14: Pantomime, no. 1. cf Works, selections (DG
        2740 153).
    Quodlibet, op. 9. cf KORNGOLD: Suite, op. 11.
2815 Symphony, no. 1. Symphony, no. 2. BBS Symphony Orchestra; Gary
      Bertini. Argo ZRG 755. (Reissue from HMV ASD 2390, Angel S
      36506).
            +-Gr 10-74 p713   ++NR 1-75 p4
             +HF 9-75 p75   +RR 9-74 p64
            +-MM 11-76 p43   +SFC 2-16-75 p24
            ++MQ 1-76 p139   +Te 9-75 p41
2816 Symphony, no. 1. Symphony, no. 2. Leipzig Gewandhaus Orchestra;
      Edo de Waart. Philips 6500 642.
            +-Gr 7-75 p196   +RR 6-75 p56
             +HF 9-75 p75   +SFC 5-18-75 p23
             +HFN 6-75 p96   +SR 6-14-75 p46
            +-MM 11-76 p43   +-Te 9-75 p41
            ++NR 6-75 p5
    Symphony, no. 2. cf Symphony, no. 1 (Argo ZRG 755).
2817 Works, selections: Berlin requiem. Concerto, violin, wind
      orchestra and percussion, op. 12. Die Dreigroschenoper: Suite,
      wind orchestra. Death in the forest, op. 23. Happy end.
      Mahagonny songspiel. Protagonist, op. 14: Pantomime, no. 1.
      Meriel Dickinson, Mary Thomas, ms; Philip Langridge, Ian
      Partridge, t; Benjamin Luxon, bar; Michael Rippon, bs; Nona
      Liddell, vln; London Sinfonietta; David Atherton. DG 2640
      153 (3).
            +Gr 11-76 p823   ++RR 11-76 p103
            ++HFN 11-76 p152
WEINBERGER, Jaromir
    Schwanda the bagpiper: Polka. cf HMV SLS 5019.
WEINER, Leo
    Ballade, clarinet and piano, op. 8. cf Hungaroton SLPX 11748.
WEINGARTNER, Felix
    Schafers Sonntaglied thou art a child. cf Rococo 5370.
WEINZWEIG
    Round dance. cf Citadel CT 6007.
WEISGALL, Hugo
    End of summer. cf GIDEON: The condemned playground.
WEISMANN, Julius
    Conzertino, op. 118, E flat major. cf BASF BAC 3085.
WEISS, Sylvius
    Chaconne, E flat major. cf d'ANGLETERRA: Carillon, G major.
    Fantasia. cf Swedish Society SLT 33189.
    Fantasia, E minor. cf Philips 6581 017.
    Suite, C minor. cf d'ANGLETERRA: Carillon, G major.
WELDON
    Gate city. cf Michigan University SM 0002.
WELLESZ, Egon
    Sonata, solo violoncello, op. 30. cf CRUMB: Sonata, solo violon-
        cello.
WENNERBERG
    Songs: Flickorna; Marketentersksorna. cf BIS LP 17.

WERLE, Lars Johan
 Now all the fingers of this tree, op. 9. cf Caprice RIKS 59.
 Tintomara: Crispin's monologue. cf Caprice CAP 1062.
WERNICK, Richard
 Kaddish-requiem. cf Nonesuch H 71302/3.
 A prayer for Jerusalem. cf MAYS: Invocations to the Svara
  mandala.
WERT, Giaches de
 Songs: Vezzosi augelli; Valle, che de' lamenti. cf HMV CSD 3756.
WESLEY, Samuel
 Symphony, D major. cf ARNE: Symphony, no. 1, C major.
WESLEY, Samuel S.
 Andante, E minor, G major. cf MENDELSSOHN: Sonata, organ, op.
  65, no. 3, A major.
 Choral song and fugue. cf Songs (RCA LRL 1-5129).
 Choral song and fugue. cf MENDELSSOHN: Sonata, organ, op. 65,
  no. 3, A major.
 Introduction and fugue, C sharp minor. cf MENDELSSOHN: Sonata,
  organ, op. 65, no. 3, A major.
 Largetto, F minor. cf Songs (RCA LRL 1-5129).
 Psalm, no. 94, O Lord God, to whom vengeance belongeth. cf HMV
  SQ CSD 3768.
2818 Songs (Anthems and hymns): Anthems, Blessed be God the Father;
  Brightest and best of the songs of the morning (Ephiphany);
  I am thine, o save me; The Lord is my shepherd; O Lord my God;
  O Lord, Thou art my God; For this mortal must put on immortal-
  ity; Psalm, no. 126, B flat major; Psalm, no. 127, F major;
  Wash me thoroughly from my wickedness. Hymns, the church's
  one foundation; O help us Lord; Each hour of need; O thou, who
  camest from above. Choral song and fugue. Largetto, F minor.
  Roy Massey, Robert Green, org; Hereford Cathedral Choir. RCA
  LRL 1-5129.
    +Gr 9-76 460   +RR 8-76 p77
    +HFN 9-76 p131
 Songs: Magnificat and nunc dimittis, E major; Thou wilt keep
  him. cf Polydor 2460 250.
WESTERGAARD, Peter
 Divertimento on Discobbolic fragments. cf Nonesuch HB 73028 (2).
WESTON, P. G.
 Venamair. cf Pye TB 3007.
WHITE
 Power. cf Nonesuch H 71276
WHITE, José Silvestre de los Dolores
 Concerto, violin, F sharp minor. cf BAKER: Sonata, violoncello
  and piano.
WHITLOCK, Percy
 Folk tune and scherzo. cf Polydor 2460 262.
 Plymouth suite: Toccata. cf Vista VPS 1035.
 Plymouth suite: Toccata. cf Wealden WS 142.
WHITTENBERG, Charles
 Set for 2, viola and piano. cf Serenus SRS 12064.
WHYTHORNE, Thomas
 As thy shadow itself apply'th. cf Pearl SHE 525.
WIDMANN, Erasmus
 Agatha. cf CG Archive 2533 184.
 Magdalena. cf DG Archive 2533 184.

Regina. cf DG Archive 2533 184.
WIDOR, Charles Marie
    Mass, 2 choirs and 2 organs. cf FRANCK: Songs (Abbey LPB 758).
    Symphony, organ, no. 5, op. 42, no. 1, F minor: Toccata. cf
       Argo 5BBA 1013/5.
    Symphony, organ, no. 5, op. 42, no. 1, F minor: Toccata. cf
       Audio 2.
    Symphony, organ, no. 5, op. 42, no. 1, F minor: Toccata. cf
       Decca DPA 523/4.
    Symphony, organ, no. 5, op. 42, no. 1, F sharp major: Toccata.
       cf HMV SQ HQS 1356.
    Symphony, organ, no. 5, op. 42, no. 1, F minor: Toccata. cf
       HMV HQS 1360.
    Symphony, organ, no. 6, op. 42, no. 2: Allegro. cf Decca SDD
       499.
2819 Symphony, no. 9, op. 70. Symphony, op. 73. Rollin Smith, org.
      Repertoire Recording Society RRS 17.
          +MU 10-76 p12
    Symphony, no. 73. cf Symphony, no. 9, op. 70.
WIENIAWSKI, Henryk
    Concerto, violin, no. 2, op. 22, D minor. cf RCA ARM 4-0942/7.
    Concerto, violin, no. 2, op. 22, D minor: Romance. RCA ARM
       4-0942/7.
    Légende, op. 17. cf HMV RLS 718.
    Légende, op. 17. cf Seraphim S 60259.
    Mazurka, op. 19, no. 1, G major. cf Connoisseur Society CS 2070.
    Polonaise, violin. cf ARENSKY: Concerto, violin, op. 54, A minor.
    Polonaise, op. 4, D major. cf HAYDN: Concerto, trumpet, E flat
       major.
    Polonaise, op. 4, D major. cf RCA ARM 4-0942/7.
    Polonaise brillante, no. 2, op. 21, A major. cf Connoisseur
       Society CS 2070.
    Scherzo tarantelle (2). cf RCA ARM 4-0942/7.
    Scherzo tarantelle, op. 16. cf HAYDN: Concerto, trumpet, E flat
       major.
WILBYE, John
    Songs: Adieu sweet Amarillis; Thus saith my Cloris bright. cf
       Enigma VAR 1017.
    Songs: Cruel, behold my heavy ending; O wretched man. cf HMV CSD
       3756.
    Songs: Draw on sweet night; Softly, softly. cf Argo ZRG 833.
    Songs: Lady, when I behold the roses. cf Harmonia Mundi 204.
WILDER, Alec
    Duets, horns, nos. 3, 15, 17, 19. cf Crystal S 371.
WILLAERT, Adrian
    Allons, allons gay. cf Hungaroton SLPX 11669/70.
    Allons, allons gay; Faulte d'Argent. cf HMV CSD 3740.
    O magnum mysterium. cf Vista VPS 1022.
WILLCOCKS, David
    Psalm, no. 13, Lord, I am not high-minded. cf HMV SQ CSD 3768.
WILLIAMS
    Lord of the boundless curves of space. cf Grosvenor GRS 1039.
WILLIAMS, Gerrard
    Raguette extra sec. cf Polydor 2383 391.
    Valsette brut. cf Polydor 2383 391.
WILLIAMS, Mansel
    Cwn pennant; Y Blodau Ger y Drws. cf Argo ZRG 769.

WILLIAMS, Mary Lou
    Gloria. cf Folkways FTS 33901/4.
WILLIAMSON, Malcolm
2820 Concerto, organ. Concerto, piano, no. 3, E flat major. Malcolm
        Williamson, org, pno; LPO; Adrian Boult, Leonard Dommett.
        Lyrita SRCS 79.
            +-Gr 5-75 p1979   +RR 5-75 p46
            +HFN 6-75 p96   +-Te 3-76 p31
        Concerto, piano, no. 3, E flat major. cf Concerto, organ.
WILLS, Arthur
    Carillon on Orientis Partibus. cf Vista VPS 1030.
WILLSON
    76 trombones. cf RCA AGL 1-1334.
WILSON
    There's a spirit in the air. cf Grosvenor GRS 1039.
WILSON, Olly Woodrow
    Akwan. cf ANDERSON: Squares.
WIREN, Dag
    Ironical miniatures, op. 19. cf LARSSON: Sonatina, no. 1, op. 16.
    Serenade, strings, op. 11. cf LARSSON: pastoral suite, op. 19.
    Serenade, strings, op. 11. cf HMV SQ ESD 7001.
    Sinfonietta, op. 7a, C major. cf ATTERBERG: Suite, violin, viola
        and orchestra, op. 19, no. 1.
WISHART, Peter
    Alleluya, a new work is come on hand. cf HMV CSD 3774.
WODIZKA
    Sonata, violin and continuo, op. 1, no. 3, D minor. cf Smithson-
        ian Collection N 002.
WOLF, Hugo
2821 Songs (Mörike): Abschied; An die Geliebte; Auf einer Wanderung;
        Bei einer Trauung; Begegnung; Der Feuerreiter; Fussreise; Der
        Genesene an die Hoffnung; Im Frühling; In der Frühe; Der Jäger;
        Jägerlied; Lebe wohl; Neue Liebe; Peregrina I and II; Storchen-
        botschaft; Verborgenheit. Dietrich Fischer-Dieskau, bar;
        Sviatoslav Richter, pno. DG 2530 584.
            +-Gr 3-76 p1500   *ON 5-76 p48
            +HF 7-76 p95   ++RR 3-76 p72
            +HFN 4-76 p119   +-SFC 2-15-76 p38
            ++MT 6-76 p497   +St 6-76 p114
            +NR 3-76 p10   +STL 4-11-76 p36
2822 Songs: Abendbilder; Anakreons Grab; Beherzigung, 1 and 2; Blumen-
        gruss; Cophtisches Lied, 1 and 2; Dank des Paria; Dies zu dueten
        bin erbötig; Du bist wie eine Blume; Epiphanias; Erschaffen
        und Beleben; Frage nicht; Frech und Froh, 1 and 2; Frühling
        übers Jahr; Ganymed; Genialisch Treiben; Gleich und gleich;
        Grenzen der Menschheit; Guttmann und Gutweib; Der Harfenspieler,
        1-3; Hatt ich irgend wohl bedenken; Herbst; Herbstentschluss;
        Komm Liebchen, komm; Königlich Gebet; Locken haltet mich
        gefangen; Mädchen mit dem roten Mündchen; Mit schwarzen Segelin;
        Nicht Gelegenheit macht Diebe; Der neue Amadis; Ob der Koran
        von Ewigkeit sei; Phänomen; Prometheus; Der Rattenfänger;
        Ritter Kurts Brautfahrt; St. Nepomuks Vorabend; Der Sänger;
        Der Schäfer; Solang man nüchtern ist; Spätherbstnebel; Spott-
        lied; Trunken mussen wir alle sein; Wanderers Nachtlied; Was
        in her Schenke waren heute; Wenn ich dein gedenke; Wenn ich in
        deine Augen seh; Wie des Mondes Abbild zittert; Wie sollt ich

WOLF (cont.) 550

Songs continued; Wo wird einst. Dietrich Fischer-Dieskau, bar;
        Daniel Barenboim, pno. DG 2740 156 (3).
                +Gr 11-76 p861   +RR 11-76 p104
                +HFN 11-76 p169
    Songs: Anakreons Grab; Der Musikant; Der Soldat I; Der verzwei-
        felte Liebhafer; Wenn du zu Blumen gehts; Wer sein holdes Lieb
        verloren. cf BRAHMS: Songs (Decca SXL 6738).
    Songs: Eichendorff Lieder: Waldmädchen; Verschwiegene Liebe.
        Goethe Lieder: Die Bekehrte; Epiphanias; Die Spröde. Mörike
        Lieder: Schlafendes Jesuskind; Zum neuen Jahre. Spanish
        songbook: Ach des Knaben Augen; Die ihr Schwebet; Nun wandre,
        Maria. cf STRAUSS, R.: Songs (RCA ARL 1-1571).
    Songs: Heimweh; Verschwiegene Liebe: Waldwanderung. cf DELIBES:
        Lakmé: Fantasien nebelhafte Träume.
    Songs: Verborgenheit; Verlassene Mägdlein. cf Club 99-99.
WOLFF, Christian
    Accompaniments. cf Lines.
2823 Lines. Accompaniments. Nathan Rubin, Thomas Halpin, vln; Nancy
        Ellis, vla; Judiyaba, vlc; Frederick Rzewski, pno. CRI SD 357.
                +ARG 11-76 p44   -NR 11-76 p8
WOLF-FERRARI, Ermanno
    Il campiello, excerpts. cf Works, selections (London STS 15362).
    La dama boba, excerpts. cf Works, selections (London STS 15362).
    I gioielli della Madonna (Jewels of the Madonna), excerpts. cf
        Works, selections (London STS 15362).
    Jewels of the Madonna: Aprila o belle. cf Discophilia DIS KGA 2.
    Quattro rispetti, op. 11, no. 4. cf DEBUSSY: Songs: Le promenoir
        des deux amants.
    I quattro Rusteghi, excerpts. cf Works, selections (London STS
        15362).
2824 Il segreto di Susanna. Maria Chiara, s; Bernd Weikl, bar; ROHO;
        Lamberto Gardelli. Decca SET 617. Tape (c) KCET 617. (also
        London 1169 Tape (c) 5-1169).
                ++Gr 11-76 p874   ++RR 11-76 p49
                +HFN 11-76 p169
    Il segreto di Susanna, excerpts. cf Works, selections (London
        STS 15362).
2825 Works, selections: Il campiello: Intermezzo, Act 2; Ritornello.
        La dama boba: Overture. I gioielli della Madonna: Orchestral
        suite. I quattro Rusteghi: Prelude; Intermezzo, Act 2. Il
        segreto di Susanna: Overture. OSCCP; Nello Santi. London STS
        15362. (Reissue). (also Decca SDD 452. Reissue).
                +ARG 12-76 p44   +-RR 1-76 p40
                +-HFN 1-76 p123
WOLKENSTEIN, Oswald von
2826 Songs: Ach senliches leiden; Der mai mit lieber zal; Durch barberei
        arabia; Her wiert uns durstet; Froleichen so well wir; In
        Suria; Ir alten weib Stand auff, Maredel; Sag an herzlieb Ain
        graserin; Wol auf und wacht. Walther von der Vogelweide
        Chamber Choir, Members; Othmar Costa. Telefunken AW 6-41139.
                +HFN 5-76 p113   +RR 2-76 p64
WOLPE, Stefan
    Piece in two parts, 6 players. cf BUSONI: Berceuse elégiaque,
        op. 42.
    Quartet, trumpet, tenor saxophone, percussion and piano. cf
        Nonesuch H 71302/3.

WOOD, Charles
> Nunc dimittis, E flat major: O Thou the central orb. cf Audio 1.
> Short communion service in the Phrygianmode: Sanctus and bene-
>   dictus. cf HMV HQS 1350.
> Songs (anthems): Expectans expectavi; God omnipotent reighneth;
>   Glory and honour and laud; Hail, gladdening light; Tis the day
>   of resurrection; O thou the central orb. cf STANFORD: Songs
>   (Argo ZRG 852).
> Songs: Hail, gladdening light. cf Audio 3.

WOOD, Hugh
> Pieces, piano, op. 6 (3). cf BIRTWISTLE: Tragoedia.

WOOD, Joseph
> Poem, orchestra. cf HOVHANESS: Meditation on Orpheus.

WOODFIELD
> Amsterdam. cf Philips 6308 246.

WORK, Henry Clay
2827 Songs of the Civil War era: Agnes by the river; The buckskin bag
>   of gold; Come home, father; Crossing the grand Sierras (chor.
>   only); Grandfather's clock; Drafted into the army; Kingdom
>   coming; The picture on the wall; Now, Moses; Poor Kitty pop-
>   corn (or the soldier's pet); The silver horn; Take them away,
>   they'll drive me crazy; Uncle Joe's "Hail Columbia"; When the
>   evening star went down; Who shall rule this American nation.
>   Joan Morris, ms; Clifford Jackson, bar; Washington Camerata
>   Chorus; William Bolcom, pno. Nonesuch H 71317.
>   ++HF 2-76 p114   +St 1-76 p79
>   +NR 11-75 p10
> WORLD OF THE CLARINET. cf Decca SPA 395.

WRANITZKY, Pavel (Paul)
> German dances (10). cf DG Archive 2533 182.
> Quodlibet. cf DG Archive 2533 182.

WUORINEN, Charles
2828 Bearbeitungen über das Glogauer Liederbuch. Grand bamboula,
>   string orchestra. Trio, strings. Speculum Musicae Members,
>   The Light Fantastic Players; Daniel Shulman. Nonesuch H 71319.
>   +NR 2-76 p5   ++St 8-76 p102
> Concerto, piano. cf WEBER, B.: Concerto, piano, op. 52.
> Grand bamboula, string orchestra. cf Bearbeitungen über das
>   Glogauer Liederbuch.
> Quartet, strings. cf BABBITT: Quartet, no. 3.
> Trio, strings. cf Bearbeitungen über das Glogauer Liederbuch.
> Variations, flute, I and II. cf Nonesuch HB 73028 (2).

WYLIE, Ruth
> Psychogram. cf GOLDMAN: Sonata, violin and piano.

WYNER, Yehudi
> Intermedio. cf SOLLBERGER: Riding the wind I.

WYNNE, David
> Evening prayers. cf Argo ZRG 769.

XENAKIS, Iannis
2829 Antikhthon. Aroura. Synaphai. NPhO; Geoffrey Douglas Madge, pno;
>   Elgar Howarth. Decca HEAD 13.
>   +Gr 9-76 p435   +RR 7-76 p66
>   ++HFN 7-76 p101
> Aroura. cf Antikhthon.
> Nomos alpha. cf DG 2530 562.
> Synaphai. cf Antikhthon.

YAMADA
    Akatonbo. cf London STS 15239
YARDUMIAN, Richard
2830 Armenian suite. Cantus animae et cordis. Symphony, no. 1.
      Bournemouth Symphony Orchestra; Anshel Brusilow. HMV SQ EMD
      5527.
          +Gr 6-76 p57   +RR 4-76 p59
          +HFN 6-76 p101
    Cantus animae et cordis. cf Armenian suite.
    Symphony, no. 1. cf Armenian suite.
YARNOLD
    March, D major. cf Columbia MS 6161.
YEPES, Narciso
    Danza inca. cf London STS 15306.
YON, Pietro
    Humoresque. cf Wealden WSQ 134.
    Humoresque l'organo primitivo. cf Polydor 2460 262.
YRADIER, Sebastián
    La paloma. cf Everest 3360.
YSAYE, Eugene
    Ecstasy, op. 21, E flat major. cf MIASKOVSKY: Concerto, violin,
      op. 44, D minor.
    Extase, op. 21, E flat major. cf Seraphim S 60259.
    Sonata, solo violoncello, op. 28. cf CRUMB: Sonata, solo violon-
      cello.
YUN, Tsang
    Glissées. cf DG 2530 562.
ZACZ, Jan
    Fugue, A minor. cf Panton 110 418/9.
ZARZYCKI, Aleksander
    Mazurka. cf Seraphim S 60259.
ZEHLE, William
    Wellington march. cf Transatlantic XTRA 1160.
ZELENSKI, Wladyslaw
    Janek: Gdy slub wezmiersz z twoim Stachem. cf Muza SXL 1170.
ZELLER, Karl
2831 Der Vogelhändler (The bird seller), excerpts. Anneliese Rothen-
      berger, Renate Holm, s; Gisela Litz, con; Gerhard Unger, Adolf
      Dallapozza, t; Walter Berry, bs; VSOO and Chorus; Willi
      Boskovsky. Angel S 37165.
          +HF 11-76 p134   ++SFC 6-27-76 p29
          +NR 8-76 p10   +St 10-76 p140
    Der Vogelhändler: Adam's song. cf KALMAN: Heut nacht hab ich
      getraumt von dir.
    Der Vogelhändler: Nightingale song. cf Club 99-99.
    The bird seller: Roses from Tyrol. cf CBS 30070.
ZESSO (15th century Italy)
    E quando andarete al monte. cf Nonesuch H 71326.
ZIANI, Marc
    Alma redemptoris mater. cf Telefunken 6-41929.
ZIMMERMAN
    Anchors aweigh. cf Department of Defense Bicentennial Edition
      50-1776.
    Fugue, E minor. cf Panton 110 418/9.
ZIMMERMAN, Bernd
    Vier kurze Studien. cf CG 2530 562.

MMERMAN, Heinz Werner
    Songs: Praise the Lord; Praise you servants of the Lord.   cf
        Grosvenor GRS 1039.
    Uns ist ein Kind geboren, Gelobt sei der Herr taglich.   cf Abbey
        LPB 754.
POLI, Domenico
    Missa: Gloria.   cf Eldorado S-1.
    Pastorale.   cf Pelca PRSRK 41017/20.
CKERMAN, Mark
    Paraphrases, solo flute.   cf CRI SD 342.
TANO
    La mantella.   cf Pye TB 3006.

MUSIC IN COLLECTIONS

ABBEY

PB 739
432 BAIRSTOW: The lamentation. BOYCE: Voluntary, D major. DAVIES:
     God be in my head. RIDOUT: Spiritus domini. SHEPPARD: Our
     Father. STAINER: I saw the Lord. STRAVINSKY: Pater noster.
     TIPPETT: A child of our time: Negro spirituals (2). ANON.
     (arr. Woodward): The God of love. Philip Moore, org; Canter-
     bury Cathedral Choir; Allan Wicks.
          +HFN 2-76 p111   +-RR 2-76 p62

PB 750
433 BROCKLESS: Christ is now rysen agayne. BYRD: Make ye joy to God;
     Justorum animae. PALESTRINA: O beata et glorioso Trinitas.
     PURCELL: Remember not Lord our offences. SCHUTZ: The heavens
     are telling. TALLIS: O Lord, give Thy holy spirit; O nata lux
     de lumine. TOMKINS: When David heard. STRAVINSKY: Ave Maria;
     Pater noster. VICTORIA: Popule meus, quid feci tibi. St.
     Bartholomew-the-Great Choir; Andrew Morris, Brian Brockless,
     org.
          +-Gr 11-76 p867

PB 752
434 DAQUIN: Noel no. VII en trio et en dialogue. BACH: Trio sonata,
     no. 5, S 529, C major. BYRD: Ut Re mee fa sol la. REGER:
     Fantasia and fugue on the chorale "Hallelujah, Gott zu loben
     bleibe meine Seelen freund", op. 52, no. 3. STEWART: Prelude,
     organ and tape. Murray Somerville, org.
          +Gr 10-76 p628   +-HFN 11-76 p164

PB 754
435 ALBINONI: Adagio, organ and strings. BACH: Cantata, no. 147,
     Jesu, joy of man's desiring. CASALS: Nigra sum, sed formosa.
     HANDEL: Messiah: Hallelujah chorus. MOZART: Ave verum corpus,
     K 618. Mass, no. 16, K 317, C major: Gloria. TIPPETT: A
     child of our time: Negro spirituals (5). VICTORIA: Ave Maria
     a 8. ZIMMERMANN: Uns ist ein Kind geboren, Gelobt sei der
     Herr taglich. Anthony Langford, org; Yorkshire Sinfonia;
     Leeds Parish Church Choir; Donald Hunt.
          +-HFN 2-76 p95   ++RR 2-76 p58
          +-HFN 3-76 p95

PB 757
436 BAIRSTOW: Blessed city, heavenly Salem. BRAHMS: Ein deutsches
     Requiem, op. 45: How lovely are thy dwellings fair. DAY: When
     I survey the wondrous cross. MENDELSSOHN: Hymn of praise, op.
     52: I waited for the Lord. NIELSEN: Motetter, op. 55. SWEEL-
     INCK: Orsus, serviteurs du Seigneur. TOMKINS: My beloved spake;

O sing unto the Lord. John Davies, Jonathan Nott, trebles;
John Southall, alto; Alan Fairs, bs; Harry Bramma, org; Worcester Cathedral Choir; Donald Hunt.
       +RR 2-76 p54

LPB 761
2837   BALLET: Sweet was the song the virgin sang. DAVIES: The Lord is
       my shepherd, Psalm no. 23. FAURE: Requiem, op. 48: Pie Jesu.
       HANDEL: Messiah: How beautiful are the feet; I know that my
       redeemer liveth. HORSLEY: There is a green hill. KIRKPATRICK
       Away in a manger. PURCELL, H.: O God, thou art my God: Halle-
       lujah, Christ is made the sure foundation. SCARLATTI, A.:
       Christmas cantata: O humble city of Bethlehem...Of a virgin
       pure; Thus he taketh upon him...Our hope of life undying.
       TRAD. (German): Quem pastores laudavere. (Irish): St. Columba
       The King of love my shepherd is. (Scottish): Balulalow.
       (Polish): Infant holy, infant lowly. (Czech): Rocking. Mich-
       ael Criswell, treble; David Lumsden, org, hpd.
       +-RR 11-76 p100

                              ANGEL

S 36080
2838   GURLITT: Toy symphony, op. 169, C major. KLING: Kitchen symphony
       op. 445. MEHUL: Overture burlesque. REINICKE: Toy symphony,
       C major. STEIBELT: Bacchanales, op. 53 (3). TAYLOR: Toy
       symphony: Adagio and finale. Instrumental Ensemble; Raymond
       Lewenthal, pno.
       +NR 10-75 p3   +St 2-76 p108

S 36093
2839   GIULIANI: Grand overture, op. 61. MUDARRA: Fantasia. Gallarda.
       NARVAEZ: Variations on "Guardame las vacas". SANZ: Suite
       española. SCARLATTI, D.: Sonatas, guitar, L 83, A major; L
       352, E minor; L 423, A minor; L 483, D major (arr. Romero).
       SOR: Variations on a theme by Mozart, op. 9. Angel Romero,
       gtr.
       ++NR 4-76 p14

S 36094
2840   ALBENIZ: España, op. 165, no. 2: Tango. Cantos de España, op.
       232: Cordoba. GRANADOS: Tonadillas al estilo antiguo: La
       maja de Goya. RODRIGO: Fandango. TARREGA: Adelita. Estudio
       brillante. Maria. Marieta. Mazurka. Preludes, nos. 2, 5.
       TORROBA: Madronos. TURINA: Fandanguillo. Garrotin. Rafaga.
       Soleares. Angel Romero, gtr.
       *NR 7-76 p13

S 36095
2841   BACH: Inventions, 2 parts, no. 1, S 772, C major. Das wohltemp-
       erierte Klavier: Prelude, no. 1, S 846, C major; Prelude and
       fugue, no. 2, S 847, C minor. COUPERIN, F.: Livres de clave-
       cin, Bk I, Ordre, no. 3: La favorite; Bk I, Ordre, no. 4: Le
       reveilmatin. DELIUS: Dance rhapsody, harpsichord. FROBERGER
       Tombeau de M. Blancrocher. LATOUR: Rule Britannia, a favorite
       air, with variations for harpsichord or pianoforte. MARTINI:
       Gavotte, F major. RAMEAU: Les cyclopes, D minor. L'Entretien
       des Muses, D minor. SCARLATTI, D.: Sonatas, harpsichord, L
       352, C minor; L 384, F major; L 387, G major. TEMPLETON: Bach
       goes to town. THOME: Rogodon, op. 97. ANON.: Chi passa per

questa strada. Igor Kipnis, hpd.
+NR 7-76 p13

S 26851. Tape (c) 4XS 36851
2842 Pleasures of the Court from the times of Henry VIII and Elizabeth
I: DOWLAND: My Lord Chamberlain's galliard. MORLEY: First
Booke of Consort Lessons: Captaine Piper's pavan and galliard;
La corante; Lachrimae pavan; Lavolto; Mounsier's almaine;
Michill's galliard; My Lord of Oxenford's maske. NICHOLSON:
The Jew's dance. SUSATO: The Danserye, dances (2). Early
Music Consort, Morley Consort; David Munrow.
++AR 11-72 p125   +NYT 8-15-76 pD15
++HF 10-76 p118   ++SFC 5-14-72 p42
+HF 11-76 p153 tape   ++St 8-72 p86
+NR 5-72 p5

SFO 36895
2843 Henry VIII and his six wives: Music from the film soundtrack,
arranged, composed and directed by David Munrow. Music per-
formed includes the following 16th century pieces: ARBEAU:
Basse danse "Jouyssance vous donneray". GERVAISE: Galliards
(2). HENRY VIII: Ballad, Pastime with good company. PHALESE:
Galliard, Traditore. SUSATO: La Mourisque. ANON.: Galliard;
King Harry VIII pavan; My Lady Carey's dompe; O death rock
me asleep; La pastorella; Le petit gentilhomme; The short
mesure off my Lady Wynkfyld's rownde. Early Music Consort;
David Munrow.
+AR 2-74 p20   +NYT 8-15-76 pD15

S 37108
2844 FALL: Der Fidele Bauer: O frag mich nicht. Die Rose von Stambul:
Zwei Augen; Ihr stillen, süssen Frau'n. KALMAN: Die Zirkus-
prinzessin: Zwei Märchenaugen. Countess Maritza: Komm Zigany;
Grüss mir mein Wien. LEHAR: The land of smiles: Immer nur
lacheln; Von Apfelblüten einen Kranz; Dein ist mein ganzes
Herz. Der Zarewitsch: Wolgalied. NEUENDORFF: Der Ratten-
fänger: Wandern, ach wandern. STRAUSS, J. II: A night in
Venice: Ach, wie so herrlich zu shau'n. Fritz Wunderlich, t;
Various orchestras and conductors.
+NR 8-76 p10   +St 12-76 p154

S 37110 (also HMV HQS 1339)
2845 FERNANDEZ: Brasileira, no. 2: Ponteio, Moda, Cataretè. GUARNIERI:
Dansa brasileira. Dansa negra. Ponteios, nos. 24, 30. MIGUEZ:
Nocturne. VIANNA: Dansa de negros. Jogos pueris. VILLA-LOBOS:
A próle do bebê, no. 1. Ciclo brasileiro. Festa no sertão.
Impressões seresteiras. Chôros, no. 5 (Alma brasileira).
Cristina Ortiz, pno.
+Gr 1-75 p1372   ++NR 8-75 p13
+HF 9-75 p95   +-RR 1-75 p52
++HFN 6-75 p97   +SR 9-18-76 p50
+MM 4-76 p29   +St 9-75 p115

S 37143
2846 BIZET: Les pêcheurs de perles: Leila, Leila, Dieu puissant le
voilà. GLUCK: Orphée et Eruydice: Viens, viens, Eurydice,
suis-moi. GOUNOD: Mireille: Vincenette a votre âge. Roméo
et Juliette: Madrigal of Juliet and Roméo. LALO: Le Roy d'Ys:
Cher mylio. MASSENET: Manon: J'ai marqué l'heure du départ.
MEYERBERR: Les Huguenots: Duet of Marguerite and Raoul. Mady
Mesplé, s; Nicola Gedda, t; Paris Opera Orchestra; Pierre Der-
vaux.

　　　　　　　+-HF 4-76 p128　　　　　　+ON 12-13-75 p48
　　　　　　　/NR 12-75 p11　　　　　　　+SFC 11-2-75 p28
　　　　　　　+OC 2-76 p4　　　　　　　　+St 4-76 p118

S 37157
2847　ALBINONI: Adagio, G minor (arr. Giazotto). HOLST: The planets,
　　　　op. 32: Venus. PROKOFIEV: Romeo and Juliet, op. 64: Balcony
　　　　scene. RACHMANINOFF: Symphony, no. 2, op. 27, E minor: Adagio.
　　　　RAVEL: La valse. LSO; André Previn.
　　　　　　　+NR 1-76 p2

S 37231
2848　BERLIOZ: La dmanation de Faust, op. 24: Hungarian march. LISZT:
　　　　Hungarian rhapsody, no. 2, G 244, C sharp minor. Les préludes,
　　　　G 97. ROSSINI: Guglielmo Tell: Overture. STRAUSS, J. II:
　　　　Thunder and lightning, op. 324. SUPPE: Light cavalry: Over-
　　　　ture. PhO; Herbert von Karajan.
　　　　　　　-NR 12-76 p4

S 37232
2849　BORODIN: Prince Igor: Dance of the Polovtsian maidens, Polovtsian
　　　　dances. SIBELIUS: Finlandia, op. 26. SMETANA: Má Vlast: The
　　　　Moldau. STRAUSS, J. I: Radetzky march, op. 228. TCHAIKOVSKY:
　　　　Overture, the year 1812, op. 49. PhO; Herbert von Karajan.
　　　　　　　-NR 12-76 p4

SR 40269
2850　Russian songs: Fantasies on Russian folk melodies; I shall go; I
　　　　will sow goose-foot; In the evening; Korobeyniky; The little
　　　　duck; The Orenburg shawl; The slender Rowan tree; The Steppe
　　　　lands; The swift post-troika; Under the troika's arched yoke;
　　　　Variations on Saratov melodies; The volga flows on. Ludmila
　　　　Zykina, s; Russian Folk Instruments Ensemble; Vladimir Pitelin.
　　　　　　　+NR 8-76 p11

　　　　　　　　　　　　　　ARGO
ZFB 95/6 (2)
2851　ADAMS: The holy city. BENNETT: The carol singers. BRAHE: Bless
　　　　this house. BOND: A perfect day. CLARKE: The blind plough-
　　　　man. DAVIS: God will watch over you. GLOVER: Rose of Tralee.
　　　　GOULD: The curfew. HARRISON: Give me a ticket to heaven.
　　　　HUHN: Invictus. KNIGHT: Rocked in the cradle of the deep.
　　　　LAMB: The volunteer organist. LOHR: When Jack and I were chil-
　　　　dren. MASCAGNI: Ave Maria (adapted from the Intermezzo from
　　　　Cavalleria Rusticana). MOSS: The floral dance. MURRAY: I'll
　　　　walk beside you. PEPPER: Over the rolling sea. QUILTER: Now
　　　　sleeps the crimson petal, op. 3, no. 2. RASBACH: Trees.
　　　　SANDERSON: Friend o' mine. TOSTI: Parted. TOURS: Mother o'
　　　　mine. WATSON: Anchored. ANON.: Mr. Shadowman (arr. E. Kaye).
　　　　Benjamin Luxon, bar; David Willison, pno.
　　　　　　　+Gr 10-76 p661　　　　　　+RR 10-76 p96
　　　　　　　+HFN 11-76 p160

ZDA 203
2852　The baroque sound of the trumpet: CLARKE: Suite, D major. HANDEL:
　　　　Suite, D major. PEARSON: An Elizabethan fantasy: Now is the
　　　　month of Maying; The willow song; The night watch. A medieval
　　　　pageant: Agincourt song; Greensleeves; Summer is icumen in.
　　　　PURCELL: Musick and ayres: Rondeau, Gavotte, Minuet, Trumpet
　　　　tune. STANLEY: Trumpet tunes, nos. 1 and 2. Voluntary, D

minor: Adagio. John Wilbraham, tpt; Leslie Pearson, org.
+Audio 6-76 p97   ++HFN 7-75 p73
+Gr 8-75 p342   ++NR 1-76 p15
++HF 10-75 p136   +-RR 7-75 p50

'A 464 (also Decca SPA 464) (Reissues from ZRG 823, 717, 731, 655, 813,
         Decca SDD 363, 274)

!53 ADDISON: Divertimento, op. 9: Valse. ALTENBURG: Concerto, 7
     tumpets and timpani: Allegro. ARNOLD: Quintet, brass: Con
     brio. ASTON: Hornpype (arr. Howarth). BEETHOVEN: Equali, 4
     trombones, no. 1: Andante (arr. Jones). BLISS: Antiphonal
     fanfare, 3 brass choirs. BRITTEN: Fanfare for St. Edmunds-
     bury. BYRD: Earle of Oxford's march (arr. Howarth). DODGSON:
     Sonata, brass: Poco adagio. DUKAS: La Péri: Fanfare. FARNABY:
     A toye (arr. Howarth). GABRIELI: G.: Sonata, pian'e forte (arr.
     Jones). GRIEG: Funeral marche (arr. Emerson). HOLBORNE: Heigh-
     ho holiday (arr. Howarth). LOCKE: Music for His Majesty's
     sackbutts and cornetts: Air. PASSEREAU: Il est bel et bon
     (arr. Reeve). POULENC: Sonata, brass: Allegro. SALZEDO:
     Divertimento: Prelude. SCHEIDT: Galliard battaglia (arr.
     Jones). SUSATO: La Mourisque (arr. Iveson). STRAUSS, R.:
     Stadt Wien: Fanfare (arr. Emerson). John Wilbraham, Michael
     Laird, tpt; Philip Jones Brass Ensemble.
          ++Gr 12-76 p1009   ++RR 11-76 p83
          +HFN 11-76 p173

RG 667
!54 Doulce memoire: ARCADELT: Margot labourez les vignes. DE BOIS:
     Je suis deshéritée. BERTRAND: Je suis un demi-dieu. BONNET:
     Francion vint l'autre jour. CERTON: Que n'est elle aupres de
     moi. COSTELEY: Mignonne, allons voir. GERVAISE: M'amye est
     tant honneste. JANNEQUIN: La plus belle de la ville. LASSUS:
     Margot labourez les vignes; La nuit froide et sombre. JEUNE:
     Ce n'est que fiel; La belle Aronde; Comment penser; Revecy
     venir du printemps; Qu'est devenu ce bel ciel; Debat la nostre
     trill'en May. MANCHICOURT: Doulce memoire. SANDRIN: Doulce
     memoire. SANDRIN (Cabezon): Doulce memoire. SERMISY: Au pres
     de vous (2 settings). ANON.: Aupres de vous. Purcell Consort
     of Voices; Elizabethan Consort of Viols; Andrew David, hpd;
     David Munrow, Richard Lee, flt and rec; James Tyler, lt.
          ++AR 2-74 p20   ++SFC 10-27-74 p6
          ++HF 6-73 p116   +SFC 2-22-76 p28
          ++NR 3-73 p21

G 769
!55 ELWYN-EDWARDS: Caneuom y tri aderun. OWEN: Madonna songs (2).
     THOMAS: Prayers from the Gaelic (4). VAUGHAN THOMAS: Saith O
     Ganeuon. WILLIAMS: Cwn pennant; Y Blodau Ger y Drws. WYNNE:
     Evening prayers. Janet Price, s; Kenneth Bowen, t; Elinor
     Bennett, hp; Anthony Saunders, pno.
          +Gr 5-75 p2009   +MT 2-76 p146
          ++HFN 6-75 p98   +RR 6-75 p81

G 789
!56 Five centuries at St. George's. BAINTON: And I saw a new heaven.
     BATTEN: O praise the Lord. BRITTEN: Festival Te Deum. BYRD:
     Exsurge Domine. CAMPBELL: Jubilate Deo. FARRANT: Call to
     remembrance. GIBBONS: O clap your hands. GREENE: Lord let me
     know mine end. HARRIS: Behold now praise the Lord. MARBECK:
     Credo. MUNDY: Sing joyfully. VAUGHAN WILLIAMS: Mystical songs:
     Antiphon. WALKER: I will lift up mine eyes. Simon Morris,

Lester Gray, trebles; Timothy Rowe, bar; St. George's Chapel
Choir; John Porter, org; Sidney Campbell.
>+Gr 4-75 p1856   +MT 3-76 p237
>+HFN 6-75 p100   +RR 3-75 p61

ZRG 806
2857 BACH: Passacaglia and fugue, S 582, C minor. BUXTEHUDE: Passa-
      caglia, D minor. Prelude, fugue and chaconne, C major. BYRD:
      Ut, Re. CABANILLES: Passacalles du 1er mode. CHAMBONNIERES:
      Chaconne, G major. FRESCOBALDI: Cento. Partite sopra passa-
      cagli. PACHELBEL: Chaconne, F minor. RAISON: Trio en passa-
      caille. Peter Hurford, org.
          ++Gr 4-76 p1644   ++RR 3-86 p59
          +HFN 3-76 p95

ZRG 807
2858 BOSSI: Etude symphonique. FRANCK: Chorale, no. 1. GIGOUT:
      Scherzo. HURFORD: Laudate Dominum. LANGLAIS: Te Deum.
      MATHIAS: Processional. STANLEY: Voluntary, C major. Peter
      Hurford, org.
          +-Gr 7-75 p220   +MT 3-76 p237
          ++HFN 6-75 101   +RR 7-75 p46

ZRG 823. Tape (c) KZRC 823
2859 AGRICOLA: Oublier veul. BYRD: Earle of Oxford's march (arr. How-
      arth). FARNABY: Giles Farnaby's dream. His rest. The new
      Sa-Hoo. The old spagnoletta. Tell me, Daphne. A toye.
      FRANCHOS: Trumpet intrada (ed. Herbert). GIBBONS: In nomine
      (arr. Howarth). Royal pavane (ed. Jones). LASSUS: Madrigal
      dell'eterna. PASSEREAU: Il est bel et bon (arr. Reeve). SUS-
      ATO: La bataille. Bergeret sans roch. Branle quatre branles.
      Mon amy, ronde. Ronde. La Mourisque. VECCHI: Saltarello.
      Philip Jones Brass Ensemble.
          +Audio 12-76 p91   +RR 6-76 p63
          +Gr 6-76 p61   +RR 10-76 p106 tape
          +HFN 6-76 p97

ZRG 833
2860 CORNYSHE: Part songs (2). DELIUS: To be sung on a summer night
      on the water. DOWLAND: Can she excuse my wrongs; Come again
      sweet love. GARDNER: The old man and young wife; Sandgate
      girl's lament. MORLEY: O grief, even on the bud. PEARSALL:
      Great God of love; Lay a garland; Light of my soul. RUBBRA:
      I care not for these ladies; It fell on a summer's day. TOM-
      KINS: Too much I once lamented. WARLOCK: Where be ye, my love.
      WILBYE: Draw on sweet night; Softly, softly. Alban Singers;
      Peter Purford.
          +-Gr 4-76 p1653   +RR 3-76 p69
          +-HFN 4-76 p115

ZRG 845. Tape (c) KZRC 845
2861 BARBER: Quartet, strings, op. 11, B minor: Adagio. COPLAND: Quiet
      city. COWELL: Hymn and fuguing tune, no. 10. CRESTON: A
      rumor. IVES: Symphony, no. 3. Celia Nicklin, ob, cor anglais;
      Michael Laird, tpt; AMF; Neville Marriner.
          ++Gr 7-76 p182   ++NR 6-76 p5
          ++Gr 11-76 p887 tape   +NYT 7-4-76 pD1
          +-HF 10-76 p132   +RR 7-76 p42
          +HFN 7-76 p84   ++St 10-76 p123
          +-MJ 7-76 p57

5BBA 1013/5 (3). (also Decca 5BBA 1013/5) (Reissues from ZRG 5419, 5420

5448, 571, 663, 5339, 503, 5237, 528, ZFA 68, 47)
2862   BACH: Chorale preludes (Schübler), S 645-50. BRITTEN: Prelude
       and fugue on a theme by Vittoria. BRAHMS: Chorale preludes,
       op. 122, nos. 1, 4, 8, 10. DAVIES: Organ fantasia on "O mag-
       num mysterium". FRANCK: Piece heroïque. Prelude, fugue and
       variations, op. 18. ELGAR: Sonata, organ, op. 28, G major.
       HINDEMITH: Sonata, organ, no. 3. LISZT: Prelude and fugue
       on the name B-A-C-H, G 260. MESSIAEN: L'Ascension. MOZART:
       Fantasia, F 594, F major. PURCELL: Trumpet tune (arr. Trevor).
       REGER: Toccata and fugue, op. 59. VIERNE: Symphony, no. 1,
       op. 14, D minor: Final. WIDOR: Symphony, organ, no. 5, op. 42,
       no. 1, F minor: Toccata. Simon Preston, org.
            +Gr 10-75 p661   ++MU 10-76 p16
            +-HFN 9-75 p109   +-NR 3-76 p11
            +MJ 3-76 p25   +-RR 1-76 p43
            +MT 4-76 p322   +SFC 3-7-76 p27

                                AUDIO

WELCATH 1
2863   BAIRSTOW: Jesu, the very thought of Thee; Lord, I call upon Thee.
       GRAY: Evening service, F minor. PARRY: I was glad; Songs of
       farewell: My soul, there is a country. STANFORD: The Lord is
       my shepherd; Te Deum, C major. WOOD: Nunc dimittis, E flat
       major; O Thou the central orb. David Ponsford, org; Wells
       Cathedral Choir; Anthony Crossland.
            +-Gr 9-76 p460

EXCATH 2
2864   ALAIN: Litanies, op. 79. BACH: Concerto, organ, no. 1, S 592,
       G major. HAYDN: Pieces, mechanical clock, no. 17, F major;
       no. 21, C major; no. 23, C major. JACKSON: Fanfare. HOWELLS:
       Pieces, organ: Paean. LISZT: Prelude and fugue on the name
       B-A-C-H, S 260. TIPPETT: Preludio al Vespro di Monteverdi.
       VIERNE: Pièces de fantaisie, op. 55: Naiades. WIDOR: Symphony,
       organ, no. 5, op. 42, no. 1, F minor: Toccata. Paul Morgan,
       org.
            +Gr 9-76 p460

EXCATH 3
2865   DAVIES: God be in my head; Psalm, no. 121. ELGAR: Sonata, organ,
       G major: Last movement (organ solo). IRELAND: Greater love
       hath no man. PARRY: I was glad. STANFORD: Magnificat and nunc
       dimittis, A major. VAUGHAN WILLIAMS: O clap your hands. WOOD:
       Hail, gladdening light. Paul Morgan, org; Exeter Cathedral
       Choir; Lucian Nethsingha.
            +-Gr 9-76 p460

EAS 16
2866   BACH: Chorale prelude: Liebster, Jesu, wir sind hier. Prelude
       and fugue, S 545, C major. BRAHMS: Chorale prelude: Es ist ein
       Ros entsprungen, op. 122, no. 8. BUXTEHUDE: Chorale prelude:
       Nun komm, der Heiden Heiland. FERGUSON: Festival march. Pre-
       lude on the hymn tune "Durness". Toccata. GIBBONS: Verse for
       ye single organ. LANG: Tuba tune, op. 15, D major. LANGLAIS:
       Te Deum, op. 5, no. 3. LIDON: Sonata con trompeta real. PORT-
       MAN: Verse for ye double organ. Barry Ferguson, org.
            +Gr 9-76 p460

AVANT

AV 1012
2867 BACEWICZ: Sonata, piano, no. 2. BOULANGER: Cortege. D'un vieux
        jardin. JACQUET DE LA GUERRE: Suite, D minor. SZYMANOWSKA:
        Etudes, F major, C major, E major. Nocturne, B flat major.
        TALMA: Alleluia in form of toccata. Nancy Fierro, pno.
              ++NR 3-75 p11   +-St 5-76 p124
AV 1014
2868 HOVAHNESS: Prayer of St. Gregory. PERSICHETTI: The hollow men.
        PLOG: Two scenes, soprano, trumpet and organ. SOUTHERS: Evo-
        lutions. WUENSCH: Suite, trumpet and organ. Anthony Plog, tpt;
        Madolyn Swearingen, org; Barbara Bing, s.
              ++IN 8-76 p22

BASF

BAC 3075
2869 English virginalists. BULL: English toy. Fantasia, D minor. The
        King's hunt. BYRD: Pavan and galliard of Mr. Peter. Walsing-
        ham variations. FARNABY: Maske, G minor. GIBBONS: Fantasia,
        D minor (2). Fancy, D minor. Pavan, G minor. TOMKINS: Bara-
        fostus' dream. Gustav Leonhardt. hpd.
              ++Gr 4-75 p1843   +MT 1-76 p44
              +HFN 6-75 p99   +-RR 4-75 p54
BAC 3085. Tape (c) KBACC 3085
2870 CHERUBINI: Sonata, horn, no. 1, F major. Sonata, horn, no. 2, F
        major. KALLIWODA: Introduction and rondo, op. 51, F major.
        REGER: Scherzino, horn and orchestra. SCHUMANN: Adagio and
        allegro, op. 70, A flat major (orch. Ansermet). WEISMANN:
        Conzertino, op. 118, E flat major. Hermann Baumann, hn; Munich
        Philharmonic Orchestra; Marinus Voorberg.
              ++Gr 4-76 p1617   ++RR 4-76 p39
              +HFN 5-76 p111   +RR 10-76 p105 tape
              +HFN 8-76 p95 tape
25 22286-1
2871 Love, lust, piety and politics: BROWNE: Woefully array'd. CORNYSSH:
        Ah, Robin. Blow thy horn, hunter. Hoyda, jolly Rutterkin.
        HENRY VIII, King: Pastime with good company. NEWARK: The far-
        ther I go, the more behind. TURGES: Enforce yourself as Goddës
        knight. ANON.: Deo gracias Anglia, Agincourt carol. Alas, de-
        parting is ground of woe. And I were a maiden. Synge we to
        this mery cumpane. Goday, my Lord, Syr Christenmasse. Tapp-
        ster, dryngker. Pro Cantione Antiqua; Early Music Consort;
        Bruno Turner.
              +-Gr 4-76 p1650   +MT 4-76 p322
              +-HFN 6-76 p89   +RR 12-75 p86

BBC

REB 228. Tape (c) RMC 4036
2872 ELGAR: Pomp and circumstance march, op. 39, no. 1, D major.
        PARRY: Jerusalem, op. 208. PURCELL: Trumpet tune and air.
        SULLIVAN: The yeomen of the guard: Overture. SOUSA: American
        marching songs (arr. Peter Smith). The blue and gray (arr.
        Grundman). Hands across the sea. TOPLADY: Rock of ages.
        TRAD.: Taps (arr. Philip Lang and Kenneth Force). VAUGHAN

WILLIAMS: Evening hymn and last post. The wasps: March-past
of the kitchen utensils. Stewart Gaudion, cor; HM Welsh Guards
Band; J. W. T. A. Malcolm.
+Gr 6-76 p97

BIS

LP 2
2873   BURKHART: Tre advetntssanger. CERTON: Psalms and nunc dimittis
(3). DUFAY: Vergine bella. KUKUCK: Die Brücke. LUNDEN:
Lilltåa och 9 till. MACHAUT: Ballad and plus dure. NOTRE DAME
SCHOOL: Flos filies and motet. Musica Intima.
+ON 5-76 p48   -RR 6-75 p78

LP 14
2874   LIDHOLM: Laudi. MENDELSSOHN: Richte mich, Gott, op. 78, no. 2.
REGER: O Tod, wie bitter bist du, op. 110, no. 3. SCHUTZ:
Also hat Gott die Welt geliebt, S 380. Ich weiss, dass mein
Erlöser lebt, S 393. Jauchzet dem Herren alle Welt, S 396.
ANON.: I himmelen, i himmelen (arr. Aberg). Säg mig den vägen
(arr. Olsson). Det blir magot i himlen för barnen att få (arr.
Svedlund). Stockholm Motet Choir; Dan-Olof Stenlund.
+-HFN 12-76 p166   +-RR 3-76 p70

LP 17
2875   DVORAK: Die Bescheidene, op. 32, no. 8; Die Gefangene, op. 32,
no. 11; Scheiden ohne Leiden, op. 32, no. 4; Die verlassene,
op. 32, no. 6; Die Zuversicht, op. 32, no. 10. GEIJER: Dansen.
KODALY: Csillagoknak teremtöje; Kiolvaso. PURCELL: Let us
wander; Sound the trumpet; Two daughters of this aged stream.
ROSSINI: Duetto buffo di due gatti; La pesca; La regata Vene-
ziana. TCHAIKOVSKY: Au jardin, près du ruisseau, op. 46, no.
4; L'Aube, op. 46, no. 6; Larmes humaines, op. 46, no. 3.
WENNERBERG: Flickorna; Marketenterskorna. Elisabeth Söderström,
s; Kerstin Meyer, ms; Jan Eyron, pno.
+-Gr 6-75 p80   +RR 7-75 p58
+ON 5-76 p48

LP 22
2876   Music for lute and gamba. ABEL: Sonata, G major. BALLARD: Alle-
mande. Courante. Prelude. Rocantins. CAIX d'HERVELOIS:
Suite, A major. DOWLAND: Resolution. MORLEY: Fancy. Lamento.
ORTIZ: Quinta pars. Recercadas primera y segunda. RAUH/
NEWSIDLER: Ach Elslein. ROBINSON: The Queenes good night.
Bengt Ericson, vla da gamba; Rolf La Fluer, lt.
++St 8-76 p106

LP 30
2877   Works for flute and guitar: CASTELNUOVO-TEDESCO: Sonatina, op.
205. GIULIANI: Sonata, flute and guitar, op. 85, A major.
IBERT: Entr'acte. RAUTAVAARA: Sonata, flute and guitar. KOCH:
Canto a danza. Gunilla von Bahr, flt; Diego Blanco, gtr.
-Gr 6-76 p62   +RR 6-76 p71

LP 45
2878   ALABIEFF (Alabiev): The Russian nightingale. ADAM: Variations on
a theme of Mozart's "Ah vous dirai-je, Maman". BISHOP: Lo,
here the gentle lark. BENEDICT: La capinera. DELL'ACQUA:
Villanelle. DOROW: Dream; Pastourelles, pastoureux. MOZART:
Il Re pastore, K 208: L'amerò sarò costante. ROUSSEL: Poèmes
de Ronsard (2). Dorothy Dorow, s; Gunilla von Bahr, flt;
Lucia Negro, pno.
+St 12-76 p153

BONGIOVANNI

GB 1 (Available from Thomas, Michael G. 54 Lymington Road, London,
     NW6)
2879 Operatic and oratorio arias: BIZET: Carmen: Micaëla'a aria. DONI-
        ZETTI: La figlia del reggimento: Convien partir. MASCAGNI:
        L'Amico Fritz: Non mi resta che il pianto. PUCCINI: La bohème:
        Che gelida manina; Vecchia zimarra. Turandot: Nessun dorma.
        Suor Angelica: Senza mamma. ROSSINI: Stabat mater: Cuius
        aninam. VERDI: Ernani: Infelice...e tuo credevi. Mirella
        Freni, s; Giuliano Cianella, t; Antonio Zerbini, bs; Leone
        Magiera, pno.
          +-Gr 3-76 p1508

BOSTON BRASS

BB 1001
2880 BOUTRY: Capriccio. BERGHMANS: La femme à barbe. DEFAYE: Danses
        (2). GUILMANT: Morceau symphonique, op. 88. ROPARTZ: Pièce,
        E flat minor. SALZEDO: Pièce concertante, op. 27. SAINT-SAENS:
        Cavatine, op. 144. Ronald Barron, trom; Fredrik Wanger, pno.
          +-HF 8-76 p96

BRASSWORKS UNLIMITED

BULP 2
2881 BACH (Leidzen): Suite, orchestra, S 1068, D major: Air on the G
        string. BROUGHTON: My country 'tis of thee. HIMES: America
        the beautiful. NUROCK: Battle hymn of the republic. TURRIN:
        March and choral. Various students from Juilliard, Eastman
        School of Music, Manhattan School of Music, Mannes College,
        Berklee College.
          +-IN 5-76 p12

BRUNO WALTER SOCIETY

RR 443 (2) (Reissues from Columbia and Telefunken)
2882 BACH: Suite, orchestra, S 1068, D major. BEETHOVEN: Coriolan
        overture, op. 62. BRAHMS: Tragic overture, op. 81. RAVEL:
        Boléro. TCHAIKOVSKY: Overture, the year 1812, op. 49. WAGNER:
        Die Meistersinger von Nürnberg: Overture. Tannhäuser: Over-
        ture. WEBER: Oberon: Overture. COA; Willem Mengelberg.
          +-ARSC Vol VIII, no. 2-3, p91

CALLIOPE

CAL 1917
2883 Organ music of the French revolution: BALBASTRE: Fugue et duo.
        CALVIERE: Pièce. CORRETTE: Magnificat du 8ème ton. LASCEUX:
        Flûtes. Récit de tièrce. Noël Lorrain. Symphonie concertante.
        MOYREAU: Les cloches d'Orléans. SEJAN: Fugue, no. 3. Noël
        Suisse. André Isoir, org.
          +-Gr 9-76 p453   +HFN 11-76 p152

CAMBRIDGE

CRS 2540
2884 A procession of voluntaries: ALCOCK: Voluntary, D major. BOYCE:
         Voluntaries, D major, G minor. GREENE: Voluntaries, G major,
         C minor. STANLEY: Voluntaries, D major, A minor. WALOND:
         Voluntary, D minor. Lawrence Moe, org.
              +CL 11-76 p10   +NR 11-76 p11

CRS 2715
2885 Favorite American concert songs: BARBER: The daisies; Monks and
         raisins; Nocturne; Songs to poems from chamber music (by James
         Joyce) (3); Sure on this shining night; With rue my heart is
         laden. CHARLES: My lady walks in loveliness. DUKE: Luke
         Havergal. GRIFFES: The lament of Ian the Proud. HAGEMAN: Do
         not go, my love. HOMER: The sick rose. KELLEY: Eldorado.
         ROGERS: The time for making songs has come. SPEAKS: On the
         road to Mandalay. Dale Moore, bar; Betty Ruth Tomfohrde, pno.
              +-ON 6-76 p52   /St 4-73 p126

CRS 2826
2886 BERTOLI: Sonata prima. BOISMORTIER: Rondeau, A minor. HANDEL:
         Sonata, flute, op. 1, no. 11, F major. LOEILLET: Sonata, re-
         corder, G major. LAVIGNE: Sonata, recorder, C major. TELE-
         MANN: Sonata, recorder, F major. Trio Primavera.
              +NR 10-76 p7

CANDIDE

CE 31095
2887 Music of Medieval Paris. ADAM DE LA HALLE: De cueur pensieu; En
         mai quant rosier. CROIX: S'amours eust point de poer.
         l'ESCURIEL: Amours, cent mille merciz. MUSET: Quant je voi
         yver retorner. PEROTIN LE GRAND: Alleluya. ANON.: Amores
         dont je sui espris; Ave virgo regia; Chanconnette; Dieus qui
         porroit; Danse real; Ductia; Estampe royal, Hocquet (2); O
         Maria virgo; Quant voi l'aloete; Veris ad imperia. Purcell
         Consort of Voices, Praetorius Consort; Grayston Burgess,
         Christopher Ball.
              +NR 7-76 p7

CANON

VAR 5968
2888 ANDROZZO: If I can help somebody (arr. F. Wright). BARSOTTI: Sun-
         down. CACAVAS: Burnished brass. ELMS: Wembley Way. FINLAYSON:
         Bright eyes. KNIPPER: Cavalry of the Steppes (arr. Woodfield).
         RICHARDSON: The White company. SIEBERT: Cucurumba. SKORNIKA:
         Instrumentalist march. STREET: Doone Valley. TROMBEY: Eye-
         level (arr. Richardson). Morris Concert Band; Harry Mortimer.
              ++Gr 1-76 p1247

CANTILENA

6238
2889 BIZET: Les pêcheurs de perles: Au fond du temple. Carmen: La fleur
         que vous. GRIEG: Og jeg vil ha'. LANGE-MULLER: Der var engang:
         Se natten er svanger; I Wurzburg kling de klokker. LEONCAVALLO:
         I Pagliacci: Un tal gioco, Vesti la giubba. MASCAGNI: Caval-

leria rusticana: O Lola. MASSENET: Le Cid: O souverain.
MEYERBEER: L'Africaine: O paradis. PUCCINI: La bohème: Che
gelida manina. Tosca: E lucevan le stelle. SVENDSEN: Kom
Carina. VERDI: Aida: Celeste Aida. La forza del destino:
Solenne in quest'ora. La traviata: Un di felice. WAGNER:
Lohengrin: In fernem Land. Vilhelm Herold, t; Instrumental
accompaniment.
  +NR 3-76 p7

6239
2890 AUBER: Manon Lescaut: Donna non vidi mai. BIZET: Ivan le Terrible:
      Ouvre ton coeur. DONIZETTI: Lucia di Lammermoor: Fra poco a
      me, Tu che a Dio. MASSENET: Werther: Pourquoi me reveiller.
      PUCCINI: La bohème: O soave fanciulla. Madama Butterfly: Dicon
      ch'oltre mare. ROSSINI: Guglielmo Tell: Ah Matilde; Troncar
      suoi, O muto asil. VERDI: Aida: Pur ti reveggo...La tra for-
      este; La fatal pietra...O terra addio. Giovanni Martinelli, t;
      Rosa Ponselle, Frances Alda, s; Giuseppe de Luca, bar; Marcel
      Journet, Jose Mardones, bs; Instrumental accompaniments.
        +NR 9-76 p9

                              CAPRICE
RIKS LP 34
2891 HAUBENSTOCK-RAMATI: Ständchen sur le nom de Heinrich Strobel.
      KONTONSKI: Cano für kammerorchester. NAUMANN: Riposte, flauto
      e percussione, no. 1. NILSSON: Gruppen (20). SANDSTROM: In
      the meantime. VARESE: Octandre. WEBERN: Konzert, op. 24.
      Musica Nova; Siegfried Naumann.
        +-RR 12-76 p74   /Te 6-76 p50
RIKS LP 35
2892 BODIN: Place of plays. Dedicated to you II. HANSON: L'inferno
      de Strindberg. JOHNSON: Through the mirror of thirst (second
      passage). MELLNAS: Euphoni. MORTENSON: Ultra. NILSON: Viarp
      I.
        *Te 6-76 p50
RIKS 59 (also CAP 1059)
2893 BACK: Neither nor. BELL: Grass. MAROS: Descort. MUSGRAVE: Prima-
      vera. NORGARD: Wenn die Rose sich selbst schmückt, schmückt
      Sie auch den Garten. WERLE: Now all the fingers of this tree,
      op. 9. Dorothy Dorow, s; Ulf Bergström, flt; Martin Bergstrand,
      double-bs; Seppo Asikainen, perc; Ragnar Dahl, pno; Daniel Bell.
        +Gr 8-76 p331   +ON 5-76 p48
        +MT 12-76 p1008   +-Te 6-76 p50
CAP 1062
2894 GLUCK: Iphigénie en Tauride: Dieux qui me poursuivez; Dieux pro-
      tecteurs. GOUNOD: Faust: Avant de quitter ces lieux. LEON-
      CAVALLO: I Pagliacci: Prologue. MOZART: Così fan tutte, K 588:
      Rivolgete a lui lo sguardo. TCHAIKOVSKY: Pique Dame, op. 68:
      Yeletsky's aria. VERDI: Don Carlo: Per me giunto. Rigoletto:
      Cortigiani, vil razza. WAGNER: Tannhäuser: O du mein holder
      Abendstern. WERLE: Tintomara: Cirspin's monologue. Håkan
      Hagegård, bar; Stockholm Royal Court Orchestra; Carlo Felice
      Cillario.
        +Gr 6-76 p84   ++RR 6-76 p30
        +-HF 8-76 p98   +St 9-76 p130

CAP 1107
2895 ALNAES: Songs (4). FALLA: La vida breve: Vivan los que rien.
POULENC: Songs: C; Fêtes galantes; La reine de coeur; Hotel;
Sanglots. ROSSINI: The barber of Seville: Una voce poco fa.
TCHAIKOVSKY: The maid of Orleans: Adieu, forêts. VERDI: Don
Carlo: O don fatale. Edith Thallaug, ms; Jan Eyron, pno;
Stockholm Royal Court Orchestra; Carlo Felice Cillario.
+ON 5-76 p48   +St 9-76 p132
+-RR 5-76 p27

CBS
30062. Tape (c) 40-30062
2896 GRIEG: Elegiac melodies, op. 34: Heart's wounds; The last spring.
MacDOWELL (arr. Frost): To a wild rose. MASSENET: Thais: Medi-
tation. MASCAGNI: Cavalleria rusticana: Intermezzo. NILES:
I wonder as I wander. RACHMANINOFF: Vocalise, op. 34, no. 14.
SCHUBERT: Serenade. VAUGHAN WILLIAMS: Fantasia on "Green-
sleeves". TRAD. (arr. Arthur Harris): Londonderry air. PhO;
Eugene Ormandy.
+Gr 3-76 p1513   -HFN 2-76 p115
30063. Tape (c) 40-30063
2897 ALBENIZ (arr. Kreisler): España, op. 165, no. 2: Tango. CHAUSSON:
Poème, op. 25. KREISLER: Liebesfreud. Liebesleid. MENDELS-
SOHN: Concerto, violin, op. 64, E minor: 1st movement. MOZART:
Concerto, violin, no. 5, K 219, A major: Rondo; Tempo di men-
uetto. SAINT-SAENS: Introduction and rondo capriccioso, op. 28.
Pinchas Zukerman, vln; Various accompaniments.
+Gr 3-76 p1513   +-HFN 3-76 p112
30064. Tape (c) 40-30064
2898 BEETHOVEN: Bagatelle, no. 25, A minor (Für Elise). Sonata, piano,
no. 14, op. 27, no. 2, C sharp minor: 1st movement. DEBUSSY:
Suite bergamasque: Clair de lune. CHOPIN: Etude, C minor.
Fantasie-Impromptu. Polonaise, A flat major. Waltz, A flat
major. Waltz, C sharp minor. LISZT: Etude de concert, no. 3,
G 144, D flat major. Liebesträum, no. 3, G 541, A flat major.
RACHMANINOFF: Prelude, C sharp minor. SCHUMANN: Kinderscenen,
op. 15, no. 7: Träumerei. Philippe Entremont, pno.
+Gr 3-76 p1513   -HFN 3-76 p112
30066
2899 BACH: Suite, lute, S 1006a, E major (trans. Williams). CHOPIN:
Sonata, piano, no. 3, op. 58, B minor: 2nd and 3rd movements.
DEBUSSY: Prélude à l'après-midi d'un faune. HANDEL: Water
music: Suite, G major; Rigaudons and minuets. MAHLER: Sym-
phony, no. 2, C minor: Urlicht. MOZART: Rondo, violin, K 373,
C major. Janet Baker, con; John Williams, gtr; Murray Perahia,
pno; Pinchas Zukerman, vln; LSO, ECO, NPhO, La Grande Ecurie;
Leonard Bernstein, Daniel Barenboim, Pierre Boulez, Jean-
Claude Malgoire.
+Gr 1-76 p1244   +RR 1-76 p44
30070. Tape (c) 40-30070
2900 HEUBERGER: The opera ball: In our secluded Rendezvous. LEHAR:
Beautiful world: Darling, trust in me; Wonderful world. The
Count of Luxembourg: Medley. Frasquita: Serenade. The land
of smiles: Yours is my heart alone. Paganini: I have been in
love before; Love, you invaded my senses. SIECZYNSKI: Vienna,

city of my dreams. STRAUSS, J. II: I long for Vienna (arr. Korn-
gold). STRAUSS, O.: A waltz dream: Love song of May. SUPPE:
Boccaccio: Medley. ZELLER: The bird seller: Roses from Tyrol.
Richard Tucker, t; Columbia Symphony Orchestra; Franz Allers.
        +-Gr 4-76 p1664   +-HFN 8-76 p95

30072
2901 BIZET: Carmen: Micaëla's aria. BORODIN: Quartet, strings, no. 2,
      D major: Nocturne (arr. Sargent). Prince Igor: Polovtsian
      dance, no. 2. CHOPIN: Les sylphides: Waltz (arr. Desormiere).
      DEBUSSY: Rêverie (arr. Smith). Suite bergamasque: Clair de
      lune (arr. Caillet). GRIEG: Peer Gynt, op. 46: Dawn. RACHMAN-
      INOFF: Rhapsody on a theme by Paganini, op. 43: 18th variation.
      RAVEL: Pavane pour une infante défunte. TCHAIKOVSKY: Swan Lake:
      Dance of the swans, excerpt. VAUGHAN WILLIAMS: Fantasia on
      "Greensleeves". Philippe Entremont, pno; PO; Eugene Ormandy.
        +-HFN 10-76 p183   +RR 8-76 p30

30073
2902 BENJAMIN: Jamaican rumba (arr. Harris). DEBUSSY: Suite bergamasque:
      Clair de lune (arr. Caillet). DELIBES: Sylvia: Pizzicato polka.
      FALLA: El amor brujo: Ritual fire dance. KHACHATURIAN: Gayaneh:
      Sabre dance. OFFENBACH: Gaîté parisiénne: Can-can (arr. Rosen-
      thal). SIBELIUS: Finlandia, op. 26. SOUSA: The stars and
      stripes forever. STRAUSS, J. II: The beautiful blue Danube,
      op. 314. VAUGHAN WILLIAMS: Fantasia on "Greensleeves". Mormon
      Tabernacle Choir; PO; Eugene Ormandy.
        /RR 8-76 p48

61579. Tape (c) 40-61579 (Reissues)
2903 BACH: Concerto, organ, no. 3, S 594, C major: Recitative (trans.
      Rosanoff). Pastorale, S 590, F major: Aria. FALLA: Spanish
      popular songs: Nana. HAYDN: Sonata, piano, no. 9, D major:
      Adagio. SCHUMANN: Stücke im Volkston, op. 102 (5). TRAD. (arr.
      Casals): Cant dels Ocells. Sant Marti del Canigo. Pablo
      Casals, vlc; Eugene Istomin, Leopold Mannes, pno; Prades
      Festival Orchestra; Perpignan Festival Orchestra.
        +Gr 1-75 p1361   ++RR 9-74 p63
        +-HFN 1-76 p125 tape

61648 (Reissue from SBRG 72077)
2904 CLARKE: Interlude. King William's march. Prince of Denmark's
      march. CROFT: Trumpet tune, D major. Voluntary, organ and
      trumpets. HANDEL: Ode for St. Cecilia's day: A trumpet volun-
      tary. Samson: Awake the trumpet's lofty sound. PURCELL: Ayre,
      D minor. Fanfare, C major. Martial air. Trumpet tune, Bon-
      duca. Trumpet tune, Cebell. Trumpet tune. Voluntary, C
      major (attrib. Purcell). TELEMANN: Der getreue Musikmeister:
      Overture; Heldenmusik. E. Power Biggs, org; New England
      Brass Ensemble.
        +Gr 6-76 p61   +RR 5-76 p39
        +HFN 6-76 p103

61684
2905 BACH: Concerto, 2 harpsichords and strings, C minor: Adagio. Con-
      certo, violin and strings, C minor. HANDEL: Belshazzar: Mar-
      tial symphony. Saul: Dead march. Suite, harpsichord, no. 11,
      D minor: Sarabande (2). SCHUBERT: Trio, piano, E flat major:
      Andante con moto. VIVALDI: Concerto, 2 violins and strings,
      G minor. TRAD.: The British Grenadiers, Greensleeves, Lilli-
      burlero. Igor Kipnis, hpd; Isaac Stern, David Oistrakh, vln;

Leonard Rose, vlc; Robert and Gaby Casadesus, Eugene Istomin,
pno; Marcel Tabuteau, ob; LPO, RPO, PO, Members, Zurich Chamber
Orchestra, Prades Chamber Orchestra; Adrian Boult, Charles
Groves, Pablo Casals.
+RR 4-76 p63

61746
2906 CROUCH: Kathleeen Mavourneen. EMMETT: Dixie. GILMORE: When
Johnny comes marching home. KITTREDGE: Tenting on the old
camp ground. MACARTHY: The bonnie blue flag. POULTON: Aura
Lee. ROOT: The battle cry of freedom. Tramp, tramp, tramp.
WEBSTER: Lorena. TRAD.: The battle hymn of the republic. He's
gone away. Sometimes I feel like a motherless child. Sweet
Evelina. Mormon Tabernacle Choir; Richard Condie.
+RR 11-76 p101

61748 (Reissue from 61748)
2907 GLAZUNOV: Raymonda, op. 57: Overture. HEROLD: Zampa: Overture.
ROSSINI: William Tell: Overture. SUPPE: Poet and peasant: Over-
ture. THOMAS: Mignon: Overture. Raymond: Overture. NYP;
Leonard Bernstein.
++Gr 11-76 p888   ++RR 11-76 p67
+-HFN 11-76 p173

61771
2908 ADAM: O holy night. GRUBER: Silent night. LEONTOVICH (arr.
Wilkhousky): Carol of the bells; O come, O come, Emanuel;
O come all ye faithful; We wish you a merry Christmas. MENDELS-
SOHN: Hark, the herald angels sing. HANDEL: Joy to the world.
TRAD.: The first nowell. Deck the hall with boughs of holly.
O Tannenbaum. Philadelphia Brass Ensemble; Mormon Tabernacle
Choir; Richard Condie.
+-RR 12-76 p92

73487. Tape (c) 40-73487
2909 BACH: Cantata, no. 147, Jesu, joy of man's desiring. Suite, solo
violoncello, no. 3, S 1009, C major: Bourrée. Sonata, 2 violins
and harpsichord, S 1037, C major: Gigue. DAQUIN: Le coucou.
MOZART: Adagio, K 356, C major. Sonata, piano, no. 11, K 331,
A major: Rondo alla turca. PURCELL: Trio sonata, no. 11, F
minor. TELEMANN: Bourrée alla polacca. VIVALDI: Concerto, 2
guitars, G major. (All arranged by Brian Gascoigne). John
Williams, Carlos Bonell, gtr; Brian Gascoigne, Morris Pert,
marimbas and vibraphone; Keith Marjoram, double-bs.
+Gr 5-76 p1807   +-RR 4-76 p66
+HFN 4-76 p119

76183
2910 Dance music of the Renaissance: ATTAINGNANT: Bransle. BALLARD:
Tourdion. DUTERTRE: Pavane and galliarde. GERVAISE: Danceries
à quatre. LOFFELHOLTZ: Intrada. LEROY: Allemandes (2). PAIX:
Ungaresca. PRAETORIUS: Terpsichore: Pavane; Spagnoletta;
Pavane and galliarde. PAUMANN: Mit gantzem Willen. SCHEIN:
Banchetto musicale: Allemande, gaillarde and courante. SUSATO:
La Mourisque. ANON.: Ballet des coqs. Dit le Bourgignon.
Saltarello. La Grand Ecurie et la Chambre du Roy, Florilegium
Musicum de Paris; Jean-Claude Malgoire.
+Gr 9-76 p435   +RR 9-76 p53
+HFN 9-76 p130

76420. Tape (c) 40-76420
2911 BACH: Partita, violin, no. 3, S 1006, E major: Prelude; Loure;

Gigue. BLOCH: Baal Shem: Ningun. CASTELNUOVO-TEDESCO: Etudes
d'ondes: Sea murmurs (trans. Heifetz). DEBUSSY: La plus que
lente (trans Roques). FALLA: Spanish popular songs: Nana
(trans. Kochanski). KREISLER: La chasse (in the style of
Cartier). RACHMANINOFF: Etudes tableaux, op. 33, no. 4, B
minor (trans. Heifetz). RAVEL: Tzigane. Jascha Heifetz, vln;
Brooks Smith, pno.
    +Gr 12-75 p1075   +HFN 3-76 p113 tape
    +HFN 12-75 p159   +-RR 12-75 p78

76476. Tape (c) 40-76476
2912 BRAHMS: Songs: Klänge II, op. 66, no. 2; Klosterfräulein, op. 61,
no. 2; Phänomen, op. 61, no. 3; Walpurgisnacht, op. 75, no. 4;
Weg der Liebe I and II, op. 20, nos. 1 and 2. CHAUSSON: Chan-
son perpetuelle, op. 37. MOZART: Le nozze di Figaro, K 492:
Non so più. SAINT-SAENS: Le bonheur est chose legère. SCAR-
LATTI, A.: Endimione e Cintia: Se geloso e il mio core.
SCHUBERT: Die Verschworenen, D 787: Ja, wir schwören. SCHUMANN:
Botschaft, op. 74, no. 8; Das Glück, op. 79, no. 16. Judith
Blegen, s; Frederica von Stade, ms; Instrumental accompani-
ment.
    ++Gr 2-76 p1373   +HFN 7-76 p105 tape
    +Gr 6-76 p102 tape   +RR 2-76 p63
    +HFN 3-76 p103   +STL 3-7-76 p37

76522. Tape (c) 40-76522
2913 BERLIOZ: Beatrice et Benedict: Dieu, Que viens-je d'entendre...Il
m'en souvient. La damnation de Faust, op. 24: D'amour l'ardente
flamme. GOUNOD: Romeo et Juliette: Depuis hier je cherche en
vain. MASSENET: Cendrillon: Enfin, je suis ici. Werther: Va,
laisse les couler mes larmes. MEYERBEER: Les Huguenots: Nobles
seigneurs, salut. OFFENBACH: La grande Duchesse de Gerolstein:
Dites lui. La perichole: Ah, Quel diner je viens de faire.
THOMAS: Mignon: Connais-tu le pays. Frederica von Stade. ms;
LPO; John Pritchard.
    ++Gr 7-76 p212   +RR 7-76 p28
    +HFN 7-76 p83   +STL 9-19-76 p36
    ++HFN 10-76 p185 tape

SQ 79200 (2)
2914 BACH: Concerto, 2 violins and strings, S 1043, D minor. BEETHOVEN:
Leonore overture, no. 3, op. 72. HANDEL: Messiah: Hallelujah
chorus. RACHMANINOFF: Sonata, violoncello and piano, op. 19,
G minor: Andante. SCHUMANN: Dichterliebe, op. 48. TCHAIKOVSKY:
Trio, piano, op. 50, A minor: Pezzo elegiaco. Yehudi Menuhin,
Isaac Stern, vln; Vladimir Horowitz, pno; Mstislav Rostropovich,
vlc; Dietrich Fischer-Dieskau, bar; Leonard Bernstein, hpd;
Oratorio Society; NYP; Leonard Bernstein.
    +-Gr 12-76 p1009   +-RR 12-76 p51

CITADEL

CT 6007
2915 APPLEBAUM: All's well that ends well: The Stratford fanfares:
Suite of dances. CABLE: Newfoundland rhapsody. CAMPBELL:
Capital City suite: River by night and confusion square. GAY-
FER: Royal visit. O'NEIL: Regimental march of the Royal Canad-
ian mounted police. WEINZWEIG: Round dance. Howard Cable
Symphonic Band.
    +NR 12-76 p15

CLASSICS FOR PLEASURE

CFP 40254
2916 BEETHOVEN: The ruins of Athens, op. 113: Turkish march. BERLIOZ:
      La damnation de Faust, op. 24: Hungarian march. Les Troyens:
      Trojan march. CHABRIER: Marche joyeuse. RIMSKY-KORSAKOV: The
      golden cockerel: March. STRAUSS, J. I: Radetzky march, op. 228.
      TCHAIKOVSKY: Marche slav, op. 31. VERDI: Aida: Grand march.
      LPO; Arthur Davison.
              +HFN 7-76 p87   +RR 7-76 p63

CLAVES

LP 30-406
2917 BURKHARD: Serenade, op. 71, no. 3. CARULLI: Serenade, C major.
      GIULIANI: Sonata, violin and guitar, op. 25, E minor. IBERT:
      Entr'acte. RAVEL: Pièce en forme de habañera. Pavane pour
      une infante défunte. Peter-Lukas Graf, flt; Konrad Ragossnig,
      gtr.
              +HFN 2-76 p93   +RR 6-76 p72

THOMAS L. CLEAR

TLC 2580 (3)
2918 Augmented history of the violin on records, 1920-1950. Portions of
      violin concertos by Beethoven, Hindemith, Mendelssohn, Paganini,
      Tchaikovsky; sonatas by Bach, Beethoven, Grieg; miscellaneous
      concert pieces. Jacques Thibaud, Jenö Hubay, Carl Flesch,
      Henri Marteau, Cecilia Hansen, Mischa Elman, Duci de Kerekjarto,
      László Szentgyörgyi, Harry Soloway, Renee Chemet, Max Strub,
      Louis Zimmerman, Alexander Moguilewsky, Franz von Vecsey,
      Ibolyka Zilzer, Toscha Samaroff, Eddy Brown, Albert Spalding,
      Albert Dubois, Gerhard Taschner, Alberto Bachmann, Richard
      Czerwonky, Heinz Stanske, Henri Merckel, Juan Manen, Alfredo
      San Malo, Manuel Quiroga, Gregoras Dinicu, Georges Enesco,
      Miguel Candela, René Bendetti, Samuel Gardner, Karl Freund,
      Joseph Hassid, Hugo Kolberg, Tossy Spivakovsky, vln; various
      accompaniments.
              +ARSC Vol VII, no. 2-3   ++St 5-75 p105
                p94

CLUB

99-97
2919 DELIBES: Arioso. CHAUSSON: Le colibri, op. 2. GODARD: Le tasse:
      Les regrets. GOUNOD: Cinq-mars: Nuit resplendissante. Faust:
      Seigneur, daignez permettre; Prison scene. La Reine de Saba:
      Plus grande dans son obscurité. LALO: Le Roi d'Ys: En silence
      pourquoi souffrir. MASSENET: Le Cid: Pleurez pleurez mes yeux.
      Grisélidis: Il partit au printemps. PUCCINI: La bohème: Mi
      chiamano Mimi. Madama Butterfly: Un bel di. Tosca: Vissi
      d'arte. Charlotte Tirard, s; Instrumental accompaniment.
              +NR 5-76 p9

99-98
2920 ADAM: Si j'etais roi: Un regard de ses yeux. BIZET: Carmen: Finale.
      BRUNEAU: L'attaque du Moulin: Adieu forêt profonde. DELIBES:
      Lakmé: Fantaisie aux divins mensonges. GOUNOD: Faust: Duet,

Act 1. Mireille: Anges du paradis. MASSENET: Hérodiade:
Adieu donc. La navarraise: O bien aimée. Sappho: Ah qu'il
est loin de mon pays. Werther: O nature pleine de grace, Clair
de lune, Desolation, Mort de Werther. REVER: Sigurd: J'ai
gardé mon âme, Esprits gardiens, Un souvenir poignant. Leon
Campagnola, t; Instrumental accompaniment.
　　+NR 5-76 p9

99-99
2921　BACH: Komm süsser Tod, S 42. BIZET: Carmen: Je dis que rien ne
m'épouvante. BRAHMS: Auf dem Kirchhofe, op. 105, no. 4;
Mainacht, op. 43, no. 2; Sappische ode, op. 94, no. 4. BAYLY:
Long long ago. CHARPENTIER: Louise: Depuis le jour. FRANZ:
Widmung, op. 14, no. 1. GLUCK: Paride ed Elena: Spiagge amate.
LOEWE: Canzonetta. SACCHI: Lungi dal caro bene. SCHUBERT: An
die Musik, D 547; Du bist die Ruhe, D 776; Litanei, D 343;
Mädchens Klage, D 6. STRAUSS, R.: Morgen, op. 27, no. 4.
WOLF: Verborgenheit; Verlassene Mägdlein. ZELLER: Der Vogel-
händler: Nightingale song. Hulda Lashanska, s; Instrumental
accompaniments.
　　+NR 5-76 p9　　　　　　+-ON 5-76 p48

99-100
2922　BOITO: Mefistofele: L'altra notte. GIORDANO: Adriana Lecouvreur:
Io sono l'umile ancella, Poveri fiori. Siberia: Nel suo amor,
Non odi la il martis. Fedora: Morte di Fedora. LEONCAVALLO:
Zaza: Dir che ci sono al mondo. MASCAGNI: Cavalleria rusti-
cana: Voi lo sapete. MASCHERONI: Lorenza: Susanna al bagno.
MASSENET: Manon: Ancor son io tutt attonita; Addio nostro
piccolo desco. PONCHIELLI: La gioconda: Suicidio. PUCCINI:
Tosca: Vissi d'arte; Quanto...Gia mi docon venal. La bohème:
Mi chiamano Mimi, Donde lieta. THOMAS: Mignon: Non consoci
il bel suol. TOSTI: Dopo. VERDI: Aida: Ritorna vincitor.
VIRGILIO: Jana: Morte di Jane. Emma Carelli, s; Instrumental
accompaniment.
　　*NR 8-76 p9

99-101 (2)
2923　BERLIOZ: La damnation de Faust, op. 24: Serenade de Mephistopheles
CARPENTER: Jazz boys; Crying' blues. CHARPENTIER: Louise: Berceuse. FEVRIER: Monna Vanna: Ce n'est pas un vieillard. HAHN:
Je me mets en votre mercy; Offrande. LAPARRA: Habañera: Et
c'est a moi que l'on dit "Chante". LARBEY: Le chant des Char-
pentiers; Le chant des demoiselles de Magasin. MARTINI IL
TEDESCO: Plaisir d'amour. MASSENET: Cleopatre: A-t-il dit
vrai...Solitaire sur ma terrasse. Don Quichotte: Quant apparaissent les etoiles, Mort de Quichotte. Panurge: Touraine est
un pays. MOZART: Don Giovanni, K 527: Ah viens a ta fenetre.
MUSSORGSKY: Boris Godunov: Mon coeur est triste, J'ai le pou-
voir supreme, Scene du carillon, Oh je meurs...Ta soeur a grand
besoin. RENARD: Le temps de cerises. SCARLATTI: Tu lo sai.
SCHUBERT: Die Forelle, D 550; Die Winterreise, op. 89, D 911:
Le Tilleul. SEVERAC: Ma poupee cherie. THOMAS: Hamlet: J'ai
pu frapper le miserable...Etre ou non etre. TOSTI: Ninon.
ANON.: L'amour du mois de Mai; Beau sejour; Cantatille dedi-
cated to Mme. de Pompadour; Chanson de Marie; Chanson de Nor-
mande; Etoile du matin; Menuet d'exaudet; La peche de Moules;
Roussignolet qui cantos; Trois princesses; Vivrons heureux.
Vanni Marcoux, bs-bar; Orchestral accompaniments.
　　*NR 7-76 p8

COLLECTORS GUILD

COLLECTORS GUILD

611
2924   BISHOP: Home sweet home. DONIZETTI: Lucrezia Borgia: Brindisi.
       GOUNOD: Faust: Faites-lui mes aveux. MEYERBEER: L'Africaine:
       In grembo a me. Les Huguenots: Nobil signori salute. Le pro-
       phète: Ah mon fils. PONCHIELLI: La gioconda: Voce di donna.
       SAINT-SAENS: Samson et Dalila: Mon coeur s'ouvre à ta voix.
       TOSTI: Aprile. VERDI: Don Carlo: O don fatale. WAGNER: Die
       Walküre: Fort denn eile. Margarethe Matzenauer, ms; Instru-
       mental accompaniment.
               ++NR 8-76 p7

COLUMBIA

MS 6161
2925   The organ in America. BILLINGS: Chester. BROWN: Rondo, G major.
       HEWITT: The battle of Trenton. IVES: Variations on "America".
       MICHAEL: Instrumental suites: Movements (6). MOLLER: Sonata,
       D major. PHILE: The President's march. SELBY: Fugue or vol-
       untary, D major. SHAW: Trip to Pawtucket. YARNOLD: March, D
       major. ANON.: Captain Sargent's (Light Infantry Company's)
       quick march. The London march. The unknown. E. Power Biggs,
       org.
               +St 7-76 p72

SCX 6585
2926   ADAM (Linley): Comrades in arms. ARNOLD (Davies): Sara (arr.
       Jones). HOWE (Steffe): Battle hymn of the republic (arr.
       Ringwald). JONES (Watts): Morte Christe. ISALOW: Sanctus
       (arr. Hughes). MASCAGNI: Cavalleria rusticana: Easter hymn
       (arr. Mansfield). MYNYDDOG (Parry): Myfanwy. PANTYSELYN (Owen):
       Laudamus (arr. Protheroe). PROTHEROE (Gwyllt): O mor ben yn y
       man. ROBERTON: The old woman. TRAD.: Counting the goats (arr.
       Roberts). Sospan Fach (arr. Davies). O Mary, don't you weep
       (arr. Rhea). Treochy Male Choir; David Bell, org; John Cynan
       Jones.
               ++Gr 6-76 p96

M2 33444 (2)
2927   Jascha Heifetz, in concert. BACH: Partita, violin, no. 3, S 1006,
       E major: Prelude, Louré, Gigue. BLOCH: Baal Shem: Ningun.
       CASTELNUOVO-TEDESCO: Etudes d'ondes: Sea murmurs. DEBUSSY: La
       plus que lente. FALLA: Spanish popular songs: Nana. FRANCK:
       Sonata, violin and piano, A major. KREISLER: La chasse. RACH-
       MANINOFF: Etudes tableaux, op. 39, no. 4 (arr. Heifetz). RAVEL:
       Tzigane. STRAUSS, R.: Sonata, violin and piano, op. 18, E flat
       major. Jascha Heifetz, vln; Brooks Smith, pno.
               ++HF 2-76 p116   ++SFC 12-28-75 p30
               ++NR 3-76 p5   ++St 2-76 p110

XM 33513.   Tape (c) XMT 33513 (ct) XMA 33513 (Q) XMQ 33513 Tape (ct)
               XQA 33513 (also CBS 73478)
2928   Footlifter: A century of American marches in authentic versions.
       ALFORD: Purple carnival. EVANS: Symphonia. FILLMORE: the
       footlifter. HUFFINE: Them basses. IVES: March intercol-
       legiate. March Omega Lambda Chi. JOPLIN: Combination march
       (arr. Schuller). REEVES: Second Connecticut Regiment. SOUSA:
       El Capitán; The gallant Seventh; The liberty bell; Semper
       Fidelis; The stars and stripes forever; The thunderer. Col-

COLUMBIA (cont.)   574

umbia All-Star Band; Gunther Schuller.
            +Gr 4-76 p1665   +MJ 11-75 p21
            +HF 10-75 p90   +NR 10-75 p13
            +-HF 5-76 p114 tape   +RR 4-76 p40
            +IN 1-76 p14   +St 12-75 p134
M 33514. (Q) MQ 33514
2929 The antiphonal organs of the Cathedral of Freiburg. BANCHIERI:
      Dialogo per organo. BUXTEHUDE: Toccata and fugue, F major.
      CAMPRA: Rigaudon, A major. HANDEL: Aylesford pieces: Fugue,
      G major; Sarabande; Impertinence. Samson: Awake the trumpet's
      lofty sound. Water music: Pomposo. KREBS: Fugue on B-A-C-H.
      MOZART: Adagio, K 356, C major. PURCELL: Abdelazer: Rondeau.
      Ayre, G major. Bonduca: Trumpet tune. The Indian Queen:
      Trumpet tune. Voluntary, C major (attrib.). SOLER: The empe-
      ror's fanfare. E. Power Biggs, org.
            +NR 12-75 p14   +St 3-76 p122
Melodiya M 33592
2930 Soviet Army Chorus and Band: Celebration. Ballad about a soldier;
      Evening at the pier; The Guardsmen in Berlin; In the Bryansk
      Forest; Nightingales; Rostov-Town; Soviet Army song; Sacred
      war; Vasya-Vasilyok. Soviet Army Band and Chorus: Boris Alek-
      sandrov.
            +NR 2-76 p8   +St 5-76 p129
Melodiya M 33593
2931 BACH: Partita, violin, no. 2, S 1004, D minor: Chaconne. GLIERE:
      Tarantella, op. 9, no. 2. MASSENET: Elegie. MENDELSSOHN: A
      midsummer night's dream, op. 61: Scherzo. RACHMANINOFF: Pre-
      lude, op. 23, no. 10. Romance, op. 6, no. 1. ROSSINI: Il
      barbiere di Siviglia: Largo al factotum. SAINT-SAENS: The
      carnival of the animals: Le cygne. SARASATE: Zigeunerweisen,
      op. 20, no. 1. TCHAIKOVSKY: None but the lonely heart, op. 6,
      no. 6. Rodion Azarkhin, double bs; Piano accompaniments.
            +-NR 3-76 p15   +-SR 1-24-76 p53
MG 33728 (2)
2932 André Kostelanetz: Spirit of '76. BARBER: Vanessa: Intermezzo.
      COWELL: Twilight: Texas. CRESTON: Midnight: Mexico. GERSHWIN:
      Concerto, F major. GRIFFES: The pleasure dome of Kubla Khan.
      The white peacock. GROFE: Trick or treat. HOVHANESS: Medita-
      tion on Orpheus. IVES/SCHUMAN: Variations on "America".
      RODGERS: Lagoon. André Previn, pno; Orchestra; André Kostel-
      anetz.
            ++MJ 7-76 p57   ++SFC 8-8-76 p38
            +NR 5-76 p2   +-St 7-76 p114
M 33822
2933 Russian folk songs: Ah, what was the use; Along the Peterskoi
      road; The ballad of Stenka Razin; Beautiful sea, Holy Lake
      Baikal; Bridal song; Dark eyes; From beyond the island; Let's
      go; Little cranberry tree; Meadowlands; Monotonously rings
      the little bell; Oh, how long the night; Oh, Nastasia; On the
      hill; Over the fields; The red Sarafan; Sasedka; Song of the
      coachman. USSR State Academic Chorus, Soviet Army Chorus,
      Piatnitsky Chorus; Osipov Russian Folk Orchestra.
            +NR 9-76 p8
M 33838
2934 A Bicentennial celebration, 200 years of American music. CENNICK:
      Happy in the Lord (arr. Parker). BILLINGS: America. DELLO

JOIO: Notes from Tom Paine. HERBERT: The gold bug. HEWITT:
The bottle of Trenton. SOUSA: President Garfield's inaugura-
tion march (and 7 others). Goldman Band; Leonard dePaur,
Richard Franko Goldman, Ainslee Cox.
    +NR 4-76 p13   +St 4-76 p111

Melodiya M 33931
2935 DONIZETTI: La favorita: O mio Fernando. MUSSORGSKY: Khovanschina:
      Marfa's prophecy. RIMSKY-KORSAKOV: The Tsar's bride: Lyuba-
      sha's aria. Kashchei, the deathless: The night descends.
      SAINT-SAENS: Samson et Dalila: Mon coeur s'ouvre à ta voix;
      Amour, viens aider. TCHAIKOVSKY: The maid of Orleans: Joan's
      aria. VERDI: Don Carlo: O don fatale. Elena Obraztsova, ms;
      Bolshoi Theatre Orchestra; Boris Khaikin, Odyssei Dimitriadi.
          +-Audio 11-76 p106   +-SFC 4-4-76 p34
          +-HF 7-76 p102   +SR 5-29-76 p52
          +NR 5-76 p9   +-St 8-76 p106
M33933. Tape (c) MT 33933 (ct) MA 33933 (also CBS 76502 Tape (c) 40-
         76502)

2936 BIZET: Ouvre ton coeur. DELIBES: Les filles de Cadix. DELL'
      ACQUA: Villanelle. GOUNOD: Mireille: Waltz. KOECHLIN: Si tu
      le veux. LENOIR: Parlez-moi d'amour. LISZT: Oh, quand je
      dors, G 282. MARTINI: Plaisir d'amour. POULENC: Les chemins
      de l'amour. Beverly Sills, s; Page Brook, flt; Columbia
      Symphony Orchestra; André Kostelanetz.
          +Audio 11-76 p106   +NR 4-76 p10
          +-HFN 7-76 p93   +ON 4-10-76 p32
          +-HFN 10-76 p185   +RR 7-76 p77
          *MJ 5-76 p28   +St 7-76 p115

M 34127
2937 Age of gold. BORODIN: Prince Igor: Polovtsian dances. In the
      steppes of Central Asia. GLIERE: The red poppy, op. 70:
      Sailors' dance. IPPOLITOV-IVANOFF: Caucasian sketches, op. 10:
      Procession of the Sardar. PROKOFIEV: Love for three oranges,
      op. 33: March. Lt. Kijé suite, op. 60: Wedding. RIMSKY-
      KORSAKOV: The snow maiden: Dance of the tumblers. SHOSTAKOVICH:
      Age of gold, op. 22. TCHAIKOVSKY: Eugene Onegin, op. 24:
      Waltz, Polonaise. Sleeping beauty, op. 66: Waltz. NYP;
      Leonard Bernstein.
          +NR 7-76 p3   +SFC 10-31-76 p35

M 34129
2938 BETHUNE: The battle of Manassas. BILLINGS: Chester. BUCK: Con-
      cert variations on "The star-spangled banner". HEWITT: The
      battle of Trenton. HOLDEN: Donshier quickstep. JOPLIN: March-
      ing onward. Treemonisha: Finale. MACDOWELL: Sea pieces, op.
      55, no. 3: A.D. 1620. SHAW: Trip to Pawtucket. SOUSA: The
      stars and stripes forever. ANON.: Brandywine quickstep. Cap-
      tain Sargent's quick march. The Duke of York's march. General
      Burgoyne's march. General Washington's march. The London
      march. The unknown. E. Power Biggs, org.
          +NR 12-76 p12

M 34134
2939 Mormon Tabernacle Choir: A jubilant song. BERGER, J.: I lift up
      my eyes (Psalm 121). BRIGHT: Rainsong. CUNDICK: The west
      wind. DELLO JOIO: A jubilant song. GATES: Oh, my luve's like
      a red, red rose. HANSON: Psalm, no. 150. LEAF: Let the whole
      creation cry. MECHEM: Make a joyful noise unto the Lord
      (Psalm 100). THOMPSON: Glory to God in the highest. Jo Ann

Ottley, s; Robert Cundick, pno; Alexander Schreiner, org; Mormon Tabernacle Choir; Jerold D. Ottley.
<pre>
      +ARG 11-76 p49   +NR 9-76 p8
      +HF 11-76 p140   *ON 11-76 p98
</pre>

COMMAND

COMS 9006
2940 DEBUSSY: Songs: Ariettes oubliées, no. 2: Il pleure dans mon coeur (trans. Hartmann). FAURE: Elégie, op. 24, C minor. Papillon, op. 77. Sicilienne, op. 78. RACHMANINOFF: Vocalise, op. 34, no. 14. TCHAIKOVSKY: Capriccioso, op. 62. Nocturne, op. 19, no. 4. TOCH: Impromptu, op. 90c (arr. Solow). WEBER: Sonata, violoncello, A major. Jeffrey Solow, vlc; Doris Stevenson, pno.
<pre>
      +Audio 4-76 p89   +SFC 4-11-76 p38
      +HF 1-76 p100   +St 6-76 p116
      +NR 12-75 p9
</pre>

CONNOISSEUR SOCIETY

(Q) CSQ 2065
2941 Great hits you played when you were young, vol. 3: BACH: Prelude, C minor. BEETHOVEN: Sonata, piano, no. 20, op. 49, no. 2, G major. CHOPIN: Waltz, op. 69, no. 1, A flat major. CHAMINADE: Scarf dance. DEBUSSY: Children's corner suite: Golliwog's cakewalk. GRIEG: Lyric pieces, op. 43, no. 6: To the spring. LISZT: Liebesträum, G 541. MASSENET: Thais: Méditation. MOZART: Allegro, K 3. POLDINI: Dancing doll. SATIE: Gymnopédie, no. 1. SCHUBERT: Impromptu, op. 90, no. 3, G flat major. SCHUMANN: Album for the young, op. 68: Knight Rupert. Morton Estrin, pno.
<pre>
      +MJ 4-76 p31   +St 2-75 p115
      +NR 4-76 p11
</pre>

(Q) CSQ 2066
2942 Great hits you played when you were young, vol. 4: BACH: Musette, D major. Minuet, G major. BEETHOVEN: Sonatina, G major. CHOPIN: Waltz, op. 34 no. 2, A minor. DEBUSSY: Suite bergamasque: Clair de lune. DVORAK: Humoresque, op. 101. FALLA: El amore brujo: Ritual fire dance. HELLER: L'Avalanche. MACDOWELL: To a wild rose. MENDELSSOHN: Venetian boat song, op. 19, no. 6. MOZART: Sonata, piano, no. 15, K 545, C major. REINHOLD: Impromptu, C sharp minor. SCHUMANN: Fantasiestücke, op. 12, no. 3: Warum. Morton Estrin, pno.
<pre>
      +MJ 4-76 p31   +St 2-75 p115
      +NR 4-76 p11
</pre>

CS 2070. (Q) CSQ 2070
2943 Wanda Wilkomirska, recital. BARTOK: Rumanian folk dances. DEBUSSY: Petite suite: En bateau. La plus que lente. KREISLER: Londonderry air (arr.). MUSSORGSKY: Fair at Sorochinsk: Gopak. SARASATE: Danzas españolas, op. 22, no. 1: Romanza andaluza. SZYMANOWSKI: King Roger, op. 46: Roxane's song. WIENIAWSKI: Mazurka, op. 19, no. 1, G major. Polonaise brillante, no. 2, op. 21, A major. Wanda Wilkomirska, vln; David Garvey, pno.
<pre>
      ++Gr 8-76 p342   ++SFC 12-22-74 p24
      +-HF 5-75 p94   +St 4-75 p99
      +NR 10-75 p4
</pre>

CORONET

LPS 1722
2944 BACH: In thee is joy. CHOPIN: Waltz, op. 64, no. 1, D flat major.
      KERN: All the things you are. KEY: Star spangled banner.
      PROKOFIEV: Devilish inspiration. RACHMANINOFF: Suite, 2 pianos,
      no. 2, op. 17. STRAVINSKY: Concerto, 2 pianos. Wallace
      Hornibrook, Charles Webb, pno.
          +-NR 10-76 p14
LPS 3030
2945 BOREL: Fugato, F major. GLASER: Kleine Stücke, 4 saxophones, op.
      8a. HLOBIL: Quartet, saxophones, op. 93. HINDEMITH: Konzert-
      stück, 2 saxophones. LAMB: Madrigal, 3 saxophones. PATACHICH:
      Quartet, saxophones. Rascher Saxophone Quartet.
          +NR 10-76 p16
LPS 3031
2946 BACH: Cantata, no. 99: Opening chorus. BEETHOVEN: The heavens re-
      sound. GRIEG: Holberg suite, op. 40. HANDEL: Concerto grosso,
      C major: 1st movement. HARTLEY: Octet, saxophones. KASTNER:
      Sextet, saxophones. MENDELSSOHN: Song without words: Trauer-
      marsch. ANON.: The turtle dove. Ye banks and braes o' Bonnie
      Doon. Rascher Saxophone Ensemble.
          +NR 10-76 p16
LPS 3032
2947 BACH: In dulci jubili in canone doppio al ottava. Das wöhltemper-
      ierte Klavier: Fugues, nos. 2, 7, 17. FARMER: Fair Phyllis I
      saw sitting all alone. FRESCOBALDI: Canzona. HANDEL: Terpsi-
      chore: Sarabande; Gigue. HAYDN: Minuet and trio. JACOBI:
      Skizze. Niederdeutscher Tanz. MENDELSSOHN: Shepherd's com-
      plaint. MORLEY: It was a lover and his lass. PASSEREAU: He
      is good and handsome. PRAETORIUS: Ballet du Roy pour Sonner
      apres. PURCELL: Dido and Aeneas: With drooping wings; Sailors'
      dance. OBRECHT: T saat een Meskin. SCHUMANN: Album for the
      young, op. 68: New Year's eve. TCHAIKOVSKY: Morning prayer.
      ANON.: I believe; Pisu Beog. TRAD.: Psalm, no. 55. Rascher
      Saxophone Quartet.
          +NR 10-76 p16

COURT OPERA CLASSICS

CO 342
2948 ARDITI: Parla. DELL'ACQUA: Villanelle. GOUNOD: Romeo et Juliette:
      Je veux vivre dans ce reve. GRIEG: Peer Gynt, op. 46: Der
      Winger mag scheiden. OFFENBACH: Les contes d'Hoffmann: Les
      oiseux dans la charmille. ROSSINI: Il barbiere di Siviglia:
      Una voce poca fa. STRAUSS, Josef: Dorfschwalben aus Oster-
      reich, op. 164. STRAUSS, J. II: Frühlingstimmen, op. 410.
      VERDI: Rigoletto: E il sol dell'anima. La traviata: Ah fors'
      è lui...Follie. Melitta Heim, s; Instrumental accompaniment.
          +-NR 8-76 p9
CO 347
2949 BELLINI: Norma: Dormono entrambi...Teneri, teneri figlia, In mia
      man alfin tu sei. CATALANI: Loreley: Non fui da um padre mai
      bendetta, Dove son, d'onde vengo...O forze recondite. La Wally:
      Ebben, Ne andrò lontana. GIORDANO: Siberia: Nel suo amore
      rianimata la coscienza. MEYERBEER: L'Africana: Di qui si vede
      il mar...Quai celesti concenti. PONCHIELLI: La gioconda: E un

anatema...L'amo come il fulgor del creato, Cosi mantieni il
patto. PUCCINI: Tosca: Amara sol per te m'era il morire.
VERDI: Aida: Ciel mio padre, A te grave cagion m'adduce, Aida,
Pur ti riveggo. Ester Mazzoleni, s; Instrumental accompaniments.
        +-NR 8-76 p9

CO 354
2950 HUMPERDINCK: Königskinder: Ei, ist das schwer, ein Bettler sein.
MOZART: Don Giovanni, K 527: Tranen vom Freund getrocknert.
Die Zauberflöte, K 620: O ew'ge Nacht...Wie stark ist nicht
dein Zauberton. VERDI: Aida: Schon sind die Priester all'
vereint. WAGNER: Lohengrin: In fernem Land, Mein lieber Schwan.
Die Meistersinger von Nürnberg: Morgenlich leuchtend im roseig-
en Schein. Das Rheingold: Immer ist Undank Loges Lohn, Uber
Stock und Stein zu Tal. Parsifal: Amfortas, Die Wunde, Nur
eine Waffe taugt. WEBER: Der Freischütz: Nein langer trag ich
nicht die Qualen. Karl Jörn, t; Instrumental accompaniments.
        *NR 7-76 p8

                    CREATIVE RECORD SERVICE
R 9115
2951 A computer-performed organ recital. BACH: Concerto, organ, no. 2,
S 593, A minor (after Vivaldi, S 593). IVES: Variations on
"America". JOPLIN: Maple leaf rag. MOZART: The marriage of
Figaro, K 492: Overture. RIMSKY-KORSAKOV: The tale of the
Tsar Sultan: Flight of the bumblebee. ROGER-DUCASSE: Pastorale,
F major. Digital Equipment Corporation PDP-8; ninety-rank
Schliker pipe organ.
        -St 12-76 p154

                              CRI
SD 102
2952 CANBY: The interminable farewell. CLAFLIN: Lament for April 15;
The quangle wangle's hat; Design for the atomic age. DVORKIN:
Maurice. HARMAN: A hymn to the Virgin. KAY: How stands the
glass around; What's in a name. LIST: Remember. MILLS: The
true beauty. PINKHAM: Elegy. Madrigal. STEVENS: Like as the
culver on the bared bough. Randolph Singers; David Randolph.
        +NR 7-76 p9
SD 300
2953 Music for computers, electronic sound and players. AREL: Mimiana
II: Frieze. BORETZ: Group variations. DODGE: Folia, chamber
orchestra. Extensions, trumpet and tape. Ronald Anderson,
tpt; Instrumental Ensemble; Jacques-Louis Monod.
        +MQ 1-76 p152   *NR 8-74 p12
SD 324
2954 HIBBARD: Trio. JENNI: Cucumber music. LEWIS: Gestes. MARTIRANO:
Chansons innocentes. PURSWELL: It grew and grew. RILEY: Variations II: Trio. Candace Nightbay, s; Jon English, bs trombone; Charles West, bs clarinet; Motter Forman, hp; Patrick
Purswell, flt; Andreas Marchand, pno; Betty Bang Mather, alto
flt, pic; William Hibbard, vla, toy pno; James Avery, pno,
celesta; William Parsons, perc; Robert Strava, vln; Carolyn
Berdahl, vlc.

```
        +Gr 12-75 p1066          +NR 2-75 p14
        +MQ 1-76 p152            -RR 1-76 p40
```

SD 342
2955 BAZELON: Duo, viola and piano. CROSS: Etudes, magnetic tape.
         LANSKY: Modal fantasy. PLESKOW: Motet and madrigal. ZUCKER-
         MAN: Paraphrases, solo flute. Judith Allen, s; Paul J. Sperry,
         t; Patricia Spencer, James Winn, flt; Linda Quan, vln; Allen
         Blustine, clt; Fred Sherry, vlc; Karen Philips, vla; Glenn
         Jacobsen, Ursula Oppens, Robert Miller, pno; Charles Wuorinen.
         Realized at the University of Toronto Electronic Music Studio.
                 -NR 10-75 p14   +NYT 4-11-76 pD23

SD 345
2956 BURGE: Sources IV. CHAITKIN: Etudes, piano (3). CURTIS-SMITH:
         Rhapsodies, piano. FRANK: Orpheum (Night music I). HUDSON:
         Reflexives, piano and tape. David and Lois Burge, pno.
                 +CL 10-76 p7   +NR 7-76 p12
                 +MJ 7-76 p57

                              CRYSTAL

S 204
2957 BERNSTEIN: Fanfare for Bima. BOZZA: Sonatina, brass quintet.
         HUGGLER: Quintet, no. 1. RIETI: Incisioni. SPEER: Sonata,
         brass, G major/C major. Cambridge Brass Quintet.
                 ++NR 11-76 p7

S 206
2958 BACH: Contrapunctus VII (trans. Posten). EAST: Peccavi (trans.
         Cran). FELD: Quintet. HARTLEY: Orpheus. McBETH: Brass.
         RENWICH: Dance. SPEER: Sonatas (3) (trans. Fetter). Annapolis
         Brass Quintet.
                 +NR 11-76 p7

S 351
2959 AITKEN: Montages, solo bassoon. BOZZA: Sonatina, flute and bas-
         soon. GABAYE: Sonatine, flute and bassoon. GERSTER: Bird in
         the spirit. GOODMAN: Jadis III. SMITH: Straws. Felix
         Skowronek, flt; Arthur Grossmann, bsn.
                 +MJ 7-76 p57

S 371
2960 HARTLEY: Sonorities II. HEIDEN: Canons, 2 horns, nos. 2 and 5.
         LEVY: Suite, no. 1. NELHYBEL: Scherzo concertante. SCHUBERT:
         Auf dem Strom, op. 119. SCHULLER: Duets, solo horns, nos. 1
         and 3. WILDER: Duets, horns, nos. 3, 15, 17, 19. Calvin
         Smith, hn; William Zsembery, Fr hn; Linda Ogden, s; John
         Dressler, pno.
                 +-NR 6-76 p7

                          DA CAMERA MAGNA

SM 93399
2961 Pictures from Israel: ACHRON: Hebrew lullaby. Scher. BEN HAIM:
         Sephardic lullaby. BLOCH: Baal Shem. CHAJES: Hechassid, op.
         24, no. 1. ENGEL: Chabad melody and Freilachs, op. 20, nos.
         1, 2. KAMINSKI: Recitative and dance. KOUGUELL: Berceuse.
         LAVRY: Jewish dances (3). Theodore Mamlock, vln; Richard Laugs,
         pno.
                 +-RR 6-76 p68   ++St 3-75 p114
```
```

DECCA

SB 322. Tape (c) KBSC 322
2962 BANTOCK: The frogs of Aristophanes overture. GODARD: Jocelyn:
      Berceuse (arr. Ball). DOUGHTY: Grandfather's clock. LEHAR:
      Gold and silver waltz. ROSSINI: The thieving magpie: Overture
      (arr. Wright). SMETANA: The bartered bride: Polka (arr. Wright).
      TCHAIKOVSKY: Symphony, no. 5, op. 64, E minor: Themes (arr.
      Ball). TRAD.: Hava nagila (arr. Siebert). Stanshaw (Bristol)
      Band; Walter Hargreaves.
            ++Gr 5-76 p1808
Tape (c) KCSP 367
2963 Birds in music: BENEDICT: The gypsy and the bird. DAQUIN: The
      cuckoo. GRANADOS: Goyescas: The maiden and the nightingale.
      PURCELL, H.: When the cock begins to crow. RESPIGHI: The
      birds. ROSSINI: The thieving magpie: Overture. SAINT-SAENS:
      The carnival of the animals: The swan. SCHUMANN: Waldscenen,
      op. 82, no. 7: The prophet bird. STRAUSS, Josef: Village
      swallows from Austria, op. 164. Various soloists, orchestras
      and conductors.
            +RR 7-76 p83 tape

SPA 394
2964 BACH: Suite, orchestra, S 1067, B minor: Rondeau; Minuet, Badin-
      erie. BEETHOVEN: Serenade, flute, violin and viola, op. 25,
      D major: Adagio; Allegro. BENEDICT: The gypsy and the moth.
      CIMAROSA: Concerto, 2 flutes and orchestra, G major: Allegro.
      DEBUSSY: Sonata, flute, viola and harp. GLUCK: Orfeo ed Euri-
      dice: Dance of the blessed spirits. HANDEL: Sonata, flute,
      op. 1, no. 5, G major. MOZART: Concerto, flute, no. 2, K 314,
      D major: 3rd movement. PERGOLESI: Concerto, flute, no. 1,
      G major. Claude Monteux, André Pepin, Alexander Murray, Jean-
      Pierre Rampal, Aurèle Nicolet, Christiane Nicolet, Richard
      Adeney, flt; Raymond Leppard, hpd; Claude Viala, vlc; Emanuel
      Hurwitz, vln; Cecil Aronowitz, vla; Joan Sutherland, s; LSO,
      Stuttgart Chamber Orchestra; Pierre Monteux, Karl Münchinger,
      Richard Bonynge.
            +Gr 1-76 p1244   ++RR 12-75 p49
            +HFN 12-75 p171
SPA 395 (Reissues from SXL 6054, 2297, L'Oiseau-Lyre SOL 60028)
2965 World of the clarinet. BRAHMS: Quintet, clarinet, op. 115, B
      minor: 3rd movement. DEBUSSY: Petite pièce, B flat major.
      HOROWITZ: Majorcan pieces (2). MOZART: Concerto, clarinet, K
      622, A major. WAGNER: Adagio, clarinet and strings. WEBER:
      Duo concertant, E flat major. Alfred Prinz, Gervase de Peyer,
      Alfred Boskovsky, clt; Cyril Preedy, pno; Anton Fietz, Philipp
      Matheis, vln; Günther Breitenbach, vla; Nikolaus Hübner, vlc;
      Johann Krump, double bass; VPO; Karl Münchinger.
            +Gr 1-76 p1210   +-RR 1-76 p36
            +-HFN 12-75 p171

SPA 409
2966 BEETHOVEN: Egmont, op. 84: Overture. GLINKA: Russlan and Ludmila:
      Overture. MOZART: Le nozze di Figaro, K 492: Overture. SMET-
      ANA: The bartered bride: Overture. SUPPE: Light cavalry:
      Overture. WAGNER: Der fliegende Holländer: Overture. LSO,
      London Festival Orchestra, LPO, NPhO, RPO; Hans Vonk, Robert
      Sharples, Lawrence Foster, Leopold Stokowski, Carlos Paita.
            +Gr 4-76 p1664   +-RR 4-76 p41
            +HFN 4-76 p121

SDD 411. Tape (c) KSDC 411 (Reissues from SET 346/8, SXL 2265)
2967 BACH: Christmas oratorio, S 248: Sinfonia. CORELLI: Concerto
          grosso, op. 6, no. 8, G minor. GLUCK: Chaconne. PACHELBEL:
          Canon (arr. Münchinger). RICCIOTTI: Concertino, no. 2, G ma-
          jor. SCO; Karl Münchinger.
              +Gr 3-75 p1667   +RR 6-76 p88 tape
              +RR 12-74 p29
SPA 419. Tape (c) KCSP 419
2968 BAX: Fanfare for the wedding of Princess Elizabeth, 1948. BLISS:
          Antiphonal fanfare, 3 brass choirs. BRITTEN: Fanfare for St.
          Edmondsbury. COATES: The dambusters march. ELGAR: Imperial
          march, op. 32. Pomp and circumstance marches, op. 39. WALTON:
          Crown imperial. ANON.: God save the Queen (arr. Britten). LSO
          and Chorus; Grenadier Guards Band; Philip Jones Brass Ensemble;
          Arthur Bliss, Rodney Bashford, Benjamin Britten.
              +HFN 1-76 p125 tape   +RR 10-75 p43
              +-HFN 9-75 p109
SPA 449
2969 BEETHOVEN: Fidelio, op. 72: Ha, Welch ein Augenblick. DONIZETTI:
          L'Elisir d'amore: Una furtiva lagrima. DVORAK: Rusalka, op.
          114: O silver moon. GOUNOD: Faust: O Dieu, que les bijoux.
          MOZART: Le nozze di Figaro, K 492: Voi che sapete. PUCCINI:
          La bohème: Che gelida manina. Gianni Schicchi: O mio babbino
          caro. Tosca: Tre sbirri...Te Deum. ROSSINI: Il barbiere di
          Siviglia: Overture. VERDI: Don Carlo: O don fatale. Rigo-
          letto: Bella figlia dell'amore. Il trovatore: Vedi, le fosche.
          Joan Sutherland, Pilar Lorengar, Renata Tebaldi, s; Theresa
          Berganza, ms; Grace Bumbry, con; Giuseppe di Stefano, Carlo
          Bergonzi, t; Geraint Evans, bar; George London, bs-bar; Various
          orchestras and conductors.
              +-Gr 4-76 p1664   +RR 7-76 p30
              +-HFN 5-76 p115
SPA 473
2970 BEETHOVEN: Sonata, piano, no. 8, op. 13, C minor: Adagio cantabile.
          BADARZEWSKA-BARANOWSKA: The maiden's prayer (arr. Cooper).
          CHOPIN: Chant polonais, op. 74, no. 5 (trans. Liszt). DEBUSSY:
          Prelude, Bk 1, no. 8: La fille aux cheveux de lin. DOHNANYI:
          Rhapsody, op. 11, no. 3, C major. RACHMANINOFF: Prelude, op.
          3, no. 2, C sharp minor. SINDING: Rustle of spring, op. 32,
          no. 3. Hidden melodies: I've got you under my skin (in the
          style of Mendelssohn). Three blind mice (Bach). Waltzing
          Matilda (Scarlatti). I saw three ships come sailing by (Schu-
          mann). When Johnny comes marching home (Schubert). For he's
          a jolly good fellow (Chopin). The Lambeth walk (Rachmaninoff).
          The Londonderry air (Brahms). Three blind mice (Debussy).
          Joseph Cooper, pno.
              +Gr 12-76 p1072   +RR 12-76 p87
SDD 499
2971 ALAIN: Litanies, op. 79. BACH: Toccata and fugue, S 565, D minor.
          LISZT: Prelude and fugue on the name B-A-C-H, G 260. MESSIAEN:
          L'Ascension: Alléluias sereins; Transports de joie. REGER:
          Pieces, op. 59, no. 9: Benedictus. WIDOR: Symphony, organ,
          no. 6, op. 42, no. 2: Allegro. Allan Wicks, org.
              +Gr 9-76 p453   +HFN 9-76 p126
DPA 517/8. Tape (c) KDPC 517/8
2972 BERLIOZ: Béatrice et Bénédict: Vous soupirez, madame. BIZET: Car-
          men: C'est toi, c'est moi. Les pêcheurs de perles: C'est toi..

Au fond du temple saint. PUCCINI: La bohème: O soave fanciulla;
O Mimì, tu più non torni. Madama Butterfly: Bimba dagli occhi
pieni di malia; Una nava da guerra...Scuoti quella fronda.
Tosca: Mario, Mario. ROSSINI: Semiramide: Serbami ognor. VER-
DI: La forza del destino: Solemne in quest'ora. La traviata:
Un dì felice. Il trovatore: Ai nostri monti. Don Carlos:
Ascolta; Le porte dell'asil s'apron gia. Joan Sutherland, Ren-
ata Tebaldi, April Cantelo, s; Marilyn Horne, Giuletta Simio-
nato, Fiorenza Cossotto, Helen Watts, Regina Resnik, ms; Mario
del Monaco, Carlo Bergonzi, Giuseppe di Stefano, Libero de
Luca, t; Dietrich Fischer-Dieskau, Leonard Warren, Ettore Bas-
tianini, bar; Various orchestras and conductors.
          +—HFN 11-75 p173   +—RR 11-75 p35
          +HFN 1-76 p125 tape
DPA 519/20 (2). Tape (c) KDPC 519/20
2973 BERLIOZ: Damnation of Faust, op. 24: Dance of the sylphs. CHAB-
      RIER: España. DEBUSSY: Prélude à l'après-midi d'un faune.
      DVORAK: Slavonic dance, op. 46, no. 1, C major. DUKAS: The
      sorcerer's apprentice. FAURE: Pavane, op. 50. MUSSORGSKY: A
      night on the bare mountain (orch. Rimsky-Korsakov). RAVEL:
      Daphnis et Chlöe: Suite, no. 1. SAINT-SAENS: Danse macabre,
      op. 40. SMETANA: Ma Vlast: Vltava. WAGNER: Die Walküre:
      Ride of the Valkyries. Various orchestras; Leopold Stokowski,
      Stanley Black, Bernard Herrmann.
          +Gr 11-75 p915   +HFN 1-76 p125 tape
          +HFN 11-75 p173   +RR 1-76 p39
DPA 523/4 (2). (Reissues from ZRG 371, SXL 2219, SDD 463, SXL 2115,
          2201, ZRG 448, ZRG 5339, ZRG 5448, ZRG 5371)
2974 BACH: Chorale prelude: Wachet auf, S 645. Fantasia and fugue, S
      542, G minor. Toccata and fugue, S 565, D minor. BOELLMANN:
      Suite gothique, op. 25: Toccata. CLARKE: Trumpet voluntary
      (arr. Ratcliffe). DAVIES: Solemn melody. FRANCK: Pièce hero-
      ique. HANDEL: Concerto, organ, op. 4, no. 2, B flat major.
      Concerto, organ, op. 7, no. 16, D minor. Water music: Air;
      Largo (arr. Blake, Martin). KARG-ELERT: Nun danket alle Gott.
      PURCELL: Trumpet tune (arr. Trevor). WAGNER: Tannhäuser: Pil-
      grims' chorus (arr. Lemaire). WIDOR: Symphony, organ, no. 5,
      op. 42, no. 1, F minor: Toccata. Karl Richter, Michael Nicho-
      las, Simon Preston, org.
          +—Gr 4-76 p1644   +RR 3-76 p58
          +HFN 3-76 p112
DPA 525/6 (2)
2975 BIZET: Carmen: Toreador chorus. BORODIN: Prince Igor: Polovtsian
      dances. GOUNOD: Faust: Soldiers' chorus. LEONCAVALLO: I Pag-
      liacci: Bell chorus. MUSSORGSKY: Boris Godunov: Coronation
      scene; Prologue. PUCCINI: Madama Butterfly: Humming chorus.
      TCHAIKOVSKY: Eugene Onegin, op. 24: Waltz, Act 2. VERDI: Aida:
      Grand march, Act 2. I Lombardi: O signore dal tetto natio.
      MacBeth: Chorus of refugees. Nabucco: Chorus of Hebrew slaves.
      Otello: Una vela, Una vela; Fuoco di gioia. Il trovatore:
      Anvil chorus. WAGNER: Der fliegende Holländer: Sailors' cho-
      rus. Tannhäuser: Pilgrims' chorus. Grand Theatre Chorus,
      Geneva, Ambrosian Opera Chorus, Kingsway Symphony Orchestra and
      Chorus, Santa Cecilia Orchestra and Chorus, LPO, ROHO and Cho-
      rus, LSO, OSR, Belgrade Opera Orchestra and Chorus, Lausanne
      Radio Chorus and Children's Chorus.
          +HFN 3-76 p112   +—RR 3-76 p30

GOSD 674/6 (3)
2976   ADAM: Si j'étais roi: Overture. AUBER: Manon Lescaut: C'est
        l'histoire amoureuse. BERLIOZ: Béatrice et Bénédict: Vous
        soupirez, madame. Les Troyens: Je vais mourir. BIZET: Car-
        men: L'amour est un oiseau rebelle (Habañera). The pearl
        fishers: C'est toi...Au fond du temple saint. BOIELDIEU: La
        dame blanche: Maintemant, observons...Viens gentille dame.
        CHARPENTIER: Louise: Depuis le jour. DEBUSSY: Pelléas et
        Mélisande: C'est le dernier soir. DELIBES: Lakmé: Prendre le
        dessin d'un bijou...Fantasie aux divins mesonges; Ou va la
        jeune Indoue. GOUNOD: Faust: Avant de quitter ces lieux;
        Que vois-je là...Ah, je ris; Alerte, alerte. Sappho: Où suis-
        je...O ma lyre immortelle. LULLY: Amadis: Marche pour le com-
        bat. Thésée: Overture. MASSENET: Manon: En fermant les yieux.
        Thais: Te souvient-il du lumineux voyage. Werther: Air des
        lettres. MEYERBEER: Les Huguenots: Tu l'as dit. RAMEAU:
        Hippolyte et Aricie: Ah, qu'on daigne du mons; Quelle plaint
        en ces lieux. RAVEL: L'Enfant et les sortilèges: Foxtrot.
        SAINT-SAENS: Samson et Dalila: Mon coeur s'ouvre à ta voix.
        THOMAS: Mignon: Connais-tu le pays. Various choirs, soloists
        and orchestras.
                +-Gr 10-76 p661   +-RR 11-76 p40
                +-HFN 11-76 p173

GOSE 677/79 (3)
2977   BEETHOVEN: Fidelio, op. 72: Ha, Welch ein Augenblick; Abscheu-
        licher. LORTZING: Zur und Zimmermann: O sancta justitia.
        MOZART: Die Entführung aus dem Serail, K 384: Martern aller
        Arten. Die Zauberflöte, K 620: Der Vogelfanger; Dies Bildnis.
        NICOLAI: The merry wives of Windsor: Overture. STRAUSS: Ara-
        bella, op. 79: Der Richtige. Die Frau ohne Schatten, op. 65:
        Ich, mein Gebieter. Der Rosenkavalier, op. 59: Presentation
        of the rose; Final trio and duet. Salome, op. 54: Closing
        scene. WAGNER: The flying Dutchman: Die Frist ist um. Die
        Meistersinger von Nürnberg: Was duftet. Tristan und Isolde:
        Mild und leise. Lohengrin: Einsam in trüben Tagen. Die Wal-
        küre: Ho-jo-to-ho...So fliehe denn; Magic fire music. WEBER:
        Der Freischütz: Wie nahte...Leise, leise. Oberon: Ozean du
        Ungeheuer. Joan Sutherland, Birgit Nilsson, Pilar Lorengar,
        Kirsten Flagstad, Marianne Schech, Lisa della Casa, Elisabeth
        Söderström, Hilde Gueden, Régine Crespin, Leonie Rysanek,
        Judith Hellwig, Inge Borkh, s; Werner Krenn, t; Tom Krause,
        Geraint Evans, bar; George London, bs-bar; Oskar Czerwenka,
        Otto Edelmann, bs; OSCCP; Various orchestras, choruses and
        conductors.
                +Gr 11-76 p893   +-RR 11-76 p41
                +-HFN 11-76 p173

PFS 4351. (Reissues from SDD N 436/8, except *)
2978   BYRD: The Earl of Salisbury pavan (orch. Stokowski). Galliard
        (after Francis Tregian)*. CHOPIN: Mazurka, op. 17, no. 4, A
        minor (orch. Stokowski)*. CLARKE: Trumpet voluntary (arr.
        Stokowski). DUPARC: Extase (orch. Stokowski)*. DVORAK: Sla-
        vonic dance, op. 72, no. 2, E minor. ELGAR: Enigma variations,
        op. 36: Nimrod. RACHMANINOFF: Prelude, op. 3, no. 2, C sharp
        minor (orch. Stokowski). SCHUBERT: Moment musical, op. 94, no.
        3, D 780, F minor (orch. Stokowski). TCHAIKOVSKY: Chant sans
        paroles, op. 40, no. 6, A minor (orch. Stokowski). Howard

DECCA (cont.)   584

Snell, tpt; David Gray, hn; CPhO, LSO; Leopold Stokowski.
        +Gr 6-76 p95   +-RR 6-76 p52
        +HFN 7-76 p97

SKL 5208
2979 BROUGHTON: The immortal hour: Faery song. COLERIDGE-TAYLOR: Hia-
      watha's wedding feast: Onaway, awake, beloved. DONIZETTI: L'
      Elisir d'amore: Una furtiva lagrima. Don Pasquale: Com'e gen-
      til. HAHN: Si mes vers avaient des ailes. HAYDN: The creation:
      In native worth. LALO: Le Roi d'Ys: Vainement ma bien-aimée.
      MENDELSSOHN: On wings of song, op. 34, no. 2. MOZART: Don
      Giovanni, K 527: Il mio tesoro. Die Zauberflöte, K 620: Dies
      Bildnis ist bezaubernd schön. PUCCINI: Tosca: E lucevan le
      stelle. SCHUBERT: Heidenröslein, D 257. STRAUSS, R.: Der
      Rosenkavalier, op. 59: Di rigori armato. Kenneth McKellar, t;
      Alfred Furnish, pno; Orchestra; Peter Knight.
        +Gr 1-76 p1253

SXL 6629. (also London OS 26376)
2980 Songs and operatic arias, Renata Tebaldi. BONONCINI: Deh più a
      me non vascondete. GLUCK: Alceste: Divinités du Styx. Paride
      ed Elena: O del mio dolce ardor. HANDEL: Alcina: Verdi prati.
      Xerzes: Ombra mai fù. MARTINI IL TEDESCO: Plaisir d'amour
      PAISIELLO: La Molinara: Nel cor più non me sento. I Zingari
      in Fiera: Chi vuol la zingarella. PERGOLESI: La serva padrona:
      Stizzoso, mio stizzoso. Songs: Tre giorni son che Nina.
      SARTI: Lungi dal caro bene. SCARLATTI, A.: Le violette. VIV-
      ALDI: Piango gemo sospiro (Songs arr. by Gamley). Renata
      Tebaldi, s; NPhO; Richard Bonynge.
        -Gr 8-75 p360   -NYT 6-8-75 pD19
        -HF 9-75 p103   +-ON 8-75 p28
        '+HFN 9-75 p103   +RR 10-75 p85
        +-HFN 4-76 p117   +-St 9-75 p123
        ++NR 8-75 p9

SXL 6649. Tape (c) KSXC 6649 (also London OS 26384)
2981 BIZET: Carmen: Flower song. FLOTOW: Martha: M'appari. GOUNOD:
      Faust: Salut, demeure. LEONCAVALLO: I pagliacci: Vesti la
      giubba. PUCCINI: La bohème: Che gelida manina. Tosca: E
      lucevan le stelle. Turandot: Nessun dorma. VERDI: Aida: Cel-
      este Aida. Rigoletto: La donna è mobile. Il trovatore: Di
      quella pira. Luciano Pavarotti, t; John Alldis Choir; LPO,
      NPhO, Vienna Volksoper Orchestra and Chorus, BPhO, LSO, RPO;
      Leone Magiera, Richard Bonynge, Herbert von Karajan, Zubin
      Mehta, Nicola Rescigno.
        +-Gr 7-75 p241   +OC 5-76 p49
        +HF 1-76 p99   +-ON 1-17-76 p32
        +-HFN 8-75 p91   +RR 7-75 p16
        +HFN 7-76 p105 tape   ++SFC 10-12-75 p22
        ++MJ 12-76 p28   ++St 4-76 p118
        ++NR 12-75 p11

SXL 6781
2982 ADAM: Cantique Noël. BACH (Gounod): Ave Maria. BERLIOZ: Requiem,
      op. 5: Sanctus. BIZET: Agnus Dei (arr. Guiraud). FRANCK:
      Panis Angelicus. GOUNOD: Ave Maria (Bach). MERCADANTE: Le
      sette ultime parole di Nostro Signore sulla croce: Qual'ciglia
      candido (Parola quinta). SCHUBERT: Ave Maria, D 839. STRA-
      DELLA: Pietà Signore. ANON.: Adeste fideles. Luciano Pava-
      rotti, t; NPhO; Wandsworth Boys' Choir, London Voices; Kurt

Herbert Adler.
        ++Gr 11-76 p888   +-RR 12-76 p87
        +HFN 11-76 p165
SXL 6782. Tape (c) KSXC 6782
2983 BEETHOVEN: The creatures of Prometheus, op. 43: Overture. BER-
        LIOZ: Roman carnival, op. 9. BRAHMS: Academic festival over-
        ture, op. 80. GLINKA: Russlan and Ludmilla: Overture. ROSSINI:
        The thieving magpie: Overture. VERDI: The force of destiny:
        Overture. CO; Lorin Maazel.
        ++Gr 12-76 p1071   +RR 12-76 p68

DELOS

DEL FY 008
2984 BROUWER: Micro-piezas (4). DUARTE: Friendships (6). JOLIVET:
        Sérénade. LESUR: Elégie. PETIT: Tarantelle. PRESTI: Danse
        d'avila. Ako Ito, Henri Dorigny, gtr.
        ++NR 8-76 p13

FY 025
2985 ANDERSON: Canticle of praise: Te deum. BAKER: Far-West toccata.
        BERLINSKI: The burning bush. IVES: Variations on "America".
        PERSICHETTI: Drop, drop slow tears. ROBERTS: Prelude and
        trumpetings. George Baker, org.
        +-MU 10-76 p16

DEL 25406
2986 BERIO: O King. DAVIDOVSKY: Synchronisms, no. 3. HARRIS: Ludis
        II. IVES: Largo, violin, clarinet and piano (1901). SCH-
        WANTNER: Consortium I. In aeternum. Boston Musica Viva;
        Richard Pittman.
        +NR 2-76 p6   +NYT 4-11-76 pD23

DEPARTMENT OF DEFENSE

50-1776 (Bicentennial Edition. Available from Radio Shack stores)
2987 Broad stripes/bright stars. BAGLEY: National emblem. BERLIN:
        God bless America. BILLINGS: Chester. BOSKERCK: Semper para-
        tus. COHAN: Over there. You're a grand old flag. Yankee
        doodle boy. CRAWFORD: The Air Force song. DYKES: Eternal
        father. The Marine's hymn. EMMETT: Dixie. FILLMORE: Ameri-
        cans we. GERSHWIN: Strike up the band. GRAFULLA: Washington
        greys. HEWITT: The battle of Trenton. HOWE: Battle hymn of
        the Republic. LAMBERT: Blow the man down; Long time ago;
        Shenandoah; When Johnny comes marching home. PHILE: Hail
        Columbia. ROOT: The battle cry of freedom. SOUSA: The army
        goes rolling along. The stars and stripes forever. WARD:
        America the beautiful. ZIMMERMANN: Anchors aweigh. U.S. Army
        Band and Chorus; U.S. Navy Band and Sea Chanters; U. S. Marine
        Band; U.S. Air Force Band; Samuel Loboda, Ned E. Muffley, Jack
        Kline, Arnald Gabriel.
        +-St 7-76 p107

DESMAR

DSM 1005
2988 International piano library gala concert. ALBENIZ (arr. Godowsky):

DESMAR (cont.)   586

Tango (Shura Cherkassky). BACH, W. F. E.: Das Dreyblatt (Gina
Bachauer, Alicia de Larrocha, Garrick Ohlsson). BARTOK: Mikrokosmos, nos. 1, 2, 5-6: Bulgarian dances (Stephen Bishop-Kovacevich). BEETHOVEN (arr. Blackford): The ruins of Athens, op.
113: Turkish march (Gina Bachauer, Jorge Bolt, Jeanne-Marie
Darré, Alicia de Larrocha, John Lill, Radu Lapu, Garrick Ohlsson, Bálint Vázsonyi). CHOPIN (arr. Balakirev): Concerto,
piano, no. 1, op. 11, E minor: Romanza (Garrick Ohlsson).
DEBUSSY (arr. Ravel): Nocturnes: Fetes (3) (Bálint Vázsonyi,
Tamás Vásáry). FRIEDMAN (arr. Gaertner): Viennese dance, no.
1 (Victor Borge). MEDEK: Battaglia alla Turca (John Lill,
John Ogdon). SAINT-SAENS: Toccata, op. 111, no. 6 (Jeanne-
Marie Darré). SHCHEDRIN: Humoresque (Radu Lupu). TRAD.: God
save the Queen (John Lill).
          +-Audio 12-76 p90   +MJ 3-76 p25
          +-Gr 2-76 p1356   +-RR 1-76 p47
          +HFN 3-76 p93   +St 1-76 p106

                              DESTO
DC 7118/9
2989 Jennie Tourel at Alice Tully Hall: BEETHOVEN: Songs: An die
      Hoffnung, op. 94; Ich liebe dich, G 235. BERLIOZ: Absence.
      DARGOMIJSKY: Romance. DEBUSSY: Chansons de bilitis (3).
      GLINKA: Songs: Doubt; Vain temptation. HAHN: Si mes vers
      avaient des ailes. LISZT: Songs: Comment disaient-ils, G 276;
      O, quand je dors, G 282; Uber allen Gipfeln ist Ruh', G 306;
      Vergiftet sind meine Lieder, G 289; Mignon's Lied, G 275.
      MASSENET: Elégie. MONSIGNY: Rose et Colas: La Sagesse est un
      trésor. OFFENBACH: La périchole: Laughing song. STRADELLA:
      Per Pietà. TCHAIKOVSKY: Songs: None but the lonely heart, op.
      6, no. 6. Jennie Tourel, ms.
          +Gr 3-75 p1692   +NYT 9-22-74 pD32
          +HF 12-71 p115   +ON 2-12-72 p34
          +HFN 1-76 p121   ++St 2-72 p100
DC 6435/7 (3)
2990 Twentieth century American violin music. BABBITT: Sextets. BER-
      GER: Duo, no. 2. BRANT: Quombex, viola d'amore, music box and
      organ. CAGE: Nocturne. CRUMB: Night music, II. FELDMAN:
      Vertical thoughts. MENNIN: Sonata concertante. PISTON: Sona-
      tina. RIEGGER: Sonatina, op. 39. SAHL: Quartet, strings.
      SESSIONS: Duo. SHAPEY: Evocation. SOLLBERGER: Solos for 6
      players. WOLPE: Piece, solo violin, no. 2. Paul Zukofsky,
      vln, vla d'amore; Gilbert Kalish, pno; Sophie Sollberger, flt;
      Allen Blustine, clt; David Jolley, hn; Raymond DesRoches, perc;
      Alvin Brehm, double-bs; New York String Quartet; Harvey Soll-
      berger.
          +Gr 4-76 p1627   +NYT 4-27-75 pD19
          +HF 11-75 p132

                        DEUTSCHE GRAMMOPHON

2530 504. Tape (c) 3300 477
2991 Spanish songs from the Middle Ages and Renaissance: ALFONSO X, El
      Sabio: Rosa das rosas. Santa Maria. ENCINA: Romerico. FUEN-

LLANA: Perdida de Antequera. MILAN: Toda mi vida hos amé.
Aquel caballero, madre. MUDARRA: Si me llaman a mi. Claros y
frescos ríos. Triste estaba el rey David. Isabel, perdiste la
tu faxa. NARVAEZ: Con qué la lavaré. TRIANA: Dinos, madre del
donsel. TORRE: Dime, triste corazón. VALDERRABANO: De dónde
venís, amore. VASQUEZ: Vos me matastes (arr. Fuenllana). En
la fuente del rosel (arr. Pisador). ANON.: Dindirindin. Los
hombres con gran plazer. Nuevas te traygo, Carillo. Teresa
Berganza, ms; Narico Yepes, gtr.

    +-Audio 6-76 p97   ++RR 3-75 p66
    +-Gr 3-75 p1692   ++RR 12-76 p108 tape
    +-HF 8-75 p102   +SR 6-14-75 p46
    ++NR 5-75 p11   ++St 7-75 p106
    +NYT 6-8-75 pD19

2530 561
2992 BUSSOTTI: Ultima rara. CAROSO: Laura soave: Balletto, Gagliarda,
      Saltarello (Balleto). CASTELNUOVO-TEDESCO: La guarda cuyda-
      dosa. Tarantella. GUILIANI: Grande ouverture, op. 61. MUR-
      TULA: Tarantella. PAGANINI: Sonata, violin, op. 25, C major.
      RONCALLI: Suite, G major. ANON.: Suite, lute. Pieces, lute
      (5). Siegfried Behrend, gtr; Claudia Brodzinska Behrend,
      vocalist.
          +Gr 11-75 p869   +RR 11-75 p75
          +HFN 12-75 p155   ++SFC 2-8-76 p26
          +NR 3-76 p12

2530 562
2993 BROWN: Music, violoncello and piano. KAGEL: Unguis incarnatus
      est. PENDERECKI: Capriccio per Siegfried Palm. WEBERN: Son-
      ata, violoncello. Kleine Stücke, violoncello, op. 11 (3).
      XENAKIS: Nomos alpha. YUN: Glissées. ZIMMERMAN: Vier kurze
      Studien. Siegfried Palm, vlc; Aloys Kontarsky, pno.
          ++Gr 1-76 p1215   ++RR 11-75 p84

2530 598
2994 ANCHIETA: Con amores, la mi madre. ESTEVE Y GRIMAU: Alma, sin-
      tamos; Ojos, llorar. GRANADOS: Tonadillas: La maja dolorosa,
      nos. 1-3; El majo discreto; El tra-la-la y el punteado; El
      majo timido. GURIDI: Llámle con el pañuelo; No quiero tus
      avellanas; Como quiere que advine. MONTSALVATGE: Canciones
      negras. TORRE: Pámpano verde, racimo albar. TURINA: Saeta en
      forma de Salve a la Virgen de la Esperanza; El fantasma; Can-
      tares. Teresa Berganza, ms; Felix Lavilla, pno.
          +Gr 12-75 p1093   +ON 12-4-76 p60
          ++HF 8-76 p97   +RR 12-75 p85
          +HFN 12-75 p149   ++St 8-76 p106
          ++NR 5-76 p11

2530 700
2995 BRODSZKY: Be my love. CARDILLO: Core 'ngrato. CURTIS: Non ti
      scordar di me. d'HARDELOT: Because. FREIRE: Ay-ay-ay. GREVER:
      Magic is the moonlight; Jurame. LACALLE: Amapola. LARA: Gra-
      nada. LEHAR: The land of smiles: Dein ist mein ganzes Herz.
      LEONCAVALLO: Mattinata. LECUONA: Siboney. LOGES: Ich schenk
      dir eine neue Welt. SIMONS: Marta. Placido Domingo, t; LSO;
      Karl-Heinz Loges, Marcel Peeters.
          *NR 8-76 p11   +-St 12-76 p153
          +SFC 6-13-76 p30

Archive 2533 172
2996 Dance music of the high baroque. BOUIN: La montauban. CHEDVILLE:
      Musette. CORRETTE: Menuets, nos. 1, 2. DESMARETS: Menuet.
      Passepied. FISCHER: Bourrée. Gigue. HOTTETERRE: Bourrée.
      LOEILLET: Corente. Gigue. Sarabande. LULLY: Unce noce de
      village: Dernière entrée. PLAYFORD: The English dancing mas-
      ter: Country dances. POGLIETTI: Balletto. REUSNER: German
      dances (arr. Stanley). SANZ: Canarios. Españoleta. Gallard
      y Villano. Passacalle de la Cavalleria de Napoles. Konrad
      Ragossnig, gtr; Eduard Melkus, Spiros Rantos, baroque vln;
      Alfred Sous, Helmut Hucke, Baroque ob; René Zosso, hurdy-gurdy;
      Ulsamer Collegium.
            +Gr 3-75 p1667   +RR 3-75 p37
            ++MT 12-76 p1008   ++St 11-75 p142
            +NR 3-75 p6

Archive 2533 182
2997 Viennese dance music from the classical period: BEETHOVEN: Contra-
      dances, Wo014. EYBLER: Polonaise. GLUCK: Orfeo ed Euridice:
      Ballet. Don Juan: Allegretto. HAYDN: Minuets (2). MOZART:
      Ländler, K 606 (6). Contredances, K 609 (5). SALIERI: Minuet.
      WRANITZKY: German dances (10). Quodlibet. Eduard Melkus
      Ensemble.
            +Audio 4-76 p88   ++NR 11-75 p4
            +Gr 8-75 p332   ++RR 7-75 p37
            +HF 11-75 p130   +St 8-76 p108
            +HFN 9-75 p107

Archive 2533 184
2998 Golden dance hits of 1600. BEHREND: Tanz im Aicholdinger Schloss.
      Eichstatter Hofmühltanz. Riedenburger Tanz. CAROUBEL: Courante
      (2). Volte (2). DALZA: Calata ala Spagnola. GERVAISE: Branle
      de Bourgogne. Branle de Champaigne. HAUSSMANN: Catkanei.
      Galliard. Tantz. MAINERIO: Schiarazula marazula. Ungarescha
      Saltarello. MILAN: Pavana. NEGRI: Balletto. NEUSIDLER:
      Welscher Tanz, Wascha mesa: Hupfauff. PHALESE: Reprise, gall-
      iard. PRAETORIUS: La bourrée. Galliarde de la guerre. Gall-
      iarde de Monsieur Wustron. Gavotte. Reprinse. Spagnoletta.
      SCHEIN: Allemande-Tripla. SUSATO: Ronde. WIDMANN: Agatha.
      Magdalena. Regina. ANON.: Basse danse la Magdalena. Branle
      de Bourgogne. Istampita ghaetta. Italiana. Lamento di Tris-
      tano: Rotta. Saltarello. Siegfried Behrend, gtr; Siegfried
      Finck, perc; Das Ulsamer-Collegium; Collegium Terpsichore.
            +Audio 6-76 p97   ++SFC 2-22-76 p28
            +-NR 8-75 p4   ++St 11-75 p142

Archive 2533 302
2999 Lute music of the Dutch Renaissance: ADRIAENSSEN: Branle Englese.
      Branle simple de Poictou. Courante. Fantasia. HOWET: Fan-
      tasie. JUDENKUNIG: Ellend bringt peyn. Hoff dantz. NEU-
      SIDLER: Der Judentanz. Preambel. Welscher Tantz Wascha mesa.
      OCHSENKHUN: Innsbruck, ich muss dich lassen. SWEELINCK:
      Psalms, nos. 5, 23. HOVE: Galliarde. VALLET: Prelude. Gal-
      liarde. Slaep, soete, slaep. VON HESSEN: Pavane. WAISSELIUS:
      Deutscher Tanzt. Fantasia. ANON.: Der gestraifft Danntz.
      Konrad Ragossnig, Renaissance lute.
            +Gr 7-76 p194   +RR 6-76 p72
            +HFN 7-76 p91   ++SFC 8-22-76 p38
            ++NR 4-76 p14

2533 305
3000 Nigel Rogers, Canti amorosi. CACCINI: Amarilli mia bella; Belle
      rose porporine; Perfidissimo volto; Udite amante. CALESTANI:
      Damigella tutta bella. GAGLIANO: Valli profonde. d'INDIA:
      Cruda amarilli; Intenerite voi, lagrime mie. PERI: O durezza
      di ferro; Tra le donne; Bellissima regina. RASI: Indarno Febo.
      SARACINI: Da te parto; Deh, come invan chiedete; Giovinetta
      vezzosetta; Io moro; Quest'amore, quest'arsura. TURCO: Occhi
      belli. Nigel Roger, t; Colin Tilney, hpd, positive organ;
      Anthony Bailes, chitarrone; Jordi Savalo, vla da gamba; Pere
      Ros, violone.
              +Gr 4-76 p1650   +NR 4-76 p10
              +HFN 5-76 p97   +RR 5-76 p68
              +MJ 11-76 p45   ++St 6-76 p72
2533 323
3001 ADRIAENSSEN: Madonna mia pietà. Io vo gridando. DOWLAND: Mr.
      George Whitehead his almand. My Lord Willoughby's welcome
      home. GALILEI, V.: Fuga a l'unisono. HASSLER: Canzon. JOHN-
      SON: Treble to a ground. MOLINARO: Ballo detto 'Il Conte
      Orlando'. Saltarello. PACOLINI: Passamezzo della battaglia.
      Saltarello della traditoria. Saltarello milanese. PICCININI:
      Canzona. ROBINSON: A toy. VAN DER HOVE: Lieto godea. ANON.:
      De la trumba. Drewries accords. La rosignoll. Konrad
      Ragossnig, Jürgen Hübscher, Dieter Kirsch, lt.
             ++HFN 12-76 p145   +RR 12-76 p74
2548 212 (Reissues from SLPM 138 639/40, 138 835/7, 138 832/4, 138 099/
          100, 138 890/1 SLPEM 136 006)
3002 BERLIOZ: La damnation de Faust, op. 24, Soldiers' chorus. BEETHO-
      VEN: Fidelio, op. 72: Prisoners' chorus. VERDI: La traviata:
      Drinking song; Matadors' chorus. Il trovatore: Anvil chorus.
      WAGNER: Der fliegende Höllander: Sailors' chorus. Lohengrin:
      Wedding march. Tannhäuser: Grand march; Pilgrims' chorus.
      WEBER: Der Freischütz: Chorus of huntsmen: Bridesmaids' cho-
      rus. Margot Laminet, Gisela Ort, s; Bavarian Radio Orchestra
      and Chorus, Bayreuth Festival Orchestra and Chorus, La Scala
      Orchestra and Chorus, Elisabeth Brasseur Choir, Lamoureux
      Orchestra, Bavarian State Opera Chorus, Bavarian State Orches-
      tra; Eugen Jochum, Wilhelm Pitz, Tullio Serafin, Antonio Votto,
      Igor Markevitch, Ferenc Fricsay.
             +-Gr 4-76 p1664   +-RR 5-76 p28
             +HFN 4-76 p123
2548 219
3003 Music of Venice: ALBINONI (Giazotti): Adagio, G minor. Sonata a
      5, op. 5, no. 9, E minor. GEMINIANI: Concerto grosso, D minor
      (from Corelli's Sonata, violin, op. 5, no. 12). PACHELBEL:
      Canon, D major. ROSSINI: Sonata, 2 violins, violoncello and
      double bass, no. 3, C major. YSAYE: Paganini variations. En-
      semble d'Archets Eugène Ysaye; Lola Bobesco, vln.
             +-Gr 5-76 p1771   +RR 4-76 p52
             ++HFN 4-76 p109
2584 004
3004 DUKAS: The sorcerer's apprentice. GINASTERA: Estancia: Danza
      final. KHACHATURIAN: Gayaneh: Sabre dance. MUSSORGSKY: A
      night on the bare mountain (arr. Rimsky-Korsakov). SAINT-SAENS:
      Danse macabre, op. 40. STRAVINSKY: The firebird: Danse in-
      fernale. BPO; Arthur Fiedler.

                ++Gr 6-76 p95   +-RR 6-76 p33
                +HFN 5-76 p99
2721 077 (2)
3005 Prussian and Austrian marches: BEETHOVEN: Pariser Einzugsmarsch
      (attrib). York'scher Marsch. ERTL: Hoch-und-Deutschmeister-
      Marsch. FUCIK: Florentiner Marsch. Regimentskinder. HAYDN,
      M.: Pappenheimer-Marsch. Coburger-Marsch. HENRION: Kreuz-
      ritter-Fanfare. Fehrbelliner Reitermarsch. KOMZAK: Vindobona-
      Marsch. Erzherzog-Albrecht-Marsch. LINDEMANN: Unter dem Gril-
      lenbanner. MOLTKE: Des grossen Kurfürsten Reitermarsch. MUHL-
      BERGER: Mir sein die Kaiserjäger. PIEFKE: Königgrätzer Marsch.
      Preussens Gloria. PREIS: O Du mein Oesterreich. RADECK:
      Fridericus-Rex-Grenadiermarsch. SCHOLZ: Torgauer Marsch.
      SCHRAMMEL: Wien bleibt Wien. SEIFERT: Kärtner Liedermarsch.
      STRAUSS, J. II: Der Zigeunerbaron, op. 420: Einzugsmarsch.
      TEIKE: Alte Kameraden. WAGNER, J. F.: Unter dem Doppeladler.
      Tiroler Holzhackerbaub'n. WAGNER, R.: Nibelungen-Marsch (arr.
      Sonntag). WAGNES: Die Bosniaken kommen. ANON.: Marsch der
      Finnländischen Reiterei. Petersburger Marsch. Hofenfried-
      berger Marsch. BPhO Wind Ensemble; Herbert von Karajan.
              +Audio 9-76 p76   +HF 3-76 p102
              ++Gr 2-75 p1504   +NR 12-75 p15
Tape (c) 3335 182 (Reissues)
3006 ALBENIZ: España, op. 165, no. 3: Malagueña. Piezas caracterís-
      ticas, no. 12: Torre bermeja. Recuerdas de viaje, op. 71, no.
      6: Rumores de la caleta. Suite española, no. 5: Asturias.
      FALLA: El amor brujo: El círculo magico; Canción del fuego
      fatuo. El sombrero de tres picos: Danza del molinero. GRAN-
      ADOS: Danza española, op. 37: Villanesco. RUIZ-PIPO: Danza,
      no. 1. TARREGA: Recuerdos de la Alhambra. Tango. VILLA-
      LOBOS: Preludes, no. 1, E minor; no. 3, A minor. TRAD. (arr.
      Llobet): La canço del Llandre. La filla del marxant. Narciso
      Yepes, gtr.
              +-RR 12-76 p105 tape

                              DISCOPHILIA
DIS 13 KHH 1
3007 LEONCAVALLO: I Pagliacci: Hüll dich in Tand. RUBINSTEIN: Es
      blinkt der Tan. SCHUMANN: Dichterliebe, op. 48: Ich grolle
      nicht. SMETANA: Dalibor: Blickst du mein Freund. WAGNER:
      Lohengrin: Hochstes Vertrauen (2); In fernem Land. Die Mei-
      stersinger von Nürnberg: Preislied; Fanget an Orchesterbeglei-
      tung. Siegfried: Schmiedelied. Tannhäuser: Loblied an die
      Venus. Die Walküre: Winterstürme wichen dem Wonnemond (2).
      WALLNOFER: Woher die Liebe. Hermann Winkelmann, Adolf
      Wallnofer, Karel Burian, t.
              *NR 7-76 p8
DIS 13 KHH 1
3008 BIZET: Carmen: Micaëla's air. DONIZETTI: La figlia del reggimento:
      Heil dir mein Vaterland. LEONCAVALLO: I Pagliacci: Was gibt's.
      MEYERBEER: Les Huguenots: O glucklich Land. Robert le diable:
      Gnadenarie. ROSSINI: Il barbiere di Siviglia: Air de Rosina.
      STRAUSS: Die Fledermaus, op. 363: Mein Herr was dachten Sie
      von mir. WAGNER: Rienzi: Gesang des Friedensboten. WEBER:
      Der Freischütz: Trube Augen. Emilie Herzog, s; Instrumental

accompaniments.
        +NR 11-76 p9

DIS KGA 2
3009 BIZET: Carmen: Con vol ber; Si tu m'aimes. DONIZETTI: La favorita:
        Ah l'alto ardor. MEYERBEER: Dinorah: Sei vendicata assai.
        PONCHIELLI: La gioconda: Ah, Pescator altonda l'esca. VERDI:
        Un ballo in maschera: Eri tu. Rigoletto: Mio padre; Povero
        Rigoletto; Cortigiani vil razza danata. Il trovatore: Mira
        d'acerbe lagrime; Vivra, Contenda il giubilo. WOLF-FERRARI:
        Jewels of the Madonna: Aprila o belle. Pasquale Amato, bar;
        Frieda Hempel, Johanna Gadski, s; Margarete Matzenauer, ms;
        Instrumental accompaniment.
            +NR 10-76 p12

KIS KGG 3
3010 BIZET: Carmen: Habañera, Kartenarie. CHAMINADE: Chanson slave.
        DEBUSSY: L'Enfant prodigue: Air de Lia. GODARD: Vivandière:
        Viens avec nous. GOUNOD: Königin von Saba: Plus grand dans
        son obscurite. Sappho: O ma lyre immortelle. HAHN: D'une
        prison. MASSE: Paul et Virginie: Air du tigre. MASSENET:
        Werther: Va laisse couler. SAINT-SAENS: Samson et Dalila: Mon
        coeur s'ouvre, Printemps que commence, Amour viens aider, Mon
        coeur s'ouvre. Jean Gerville-Reache, ms; Instrumental accom-
        paniments.
            +NR 11-76 p9

                            DUKE UNIVERSITY PRESS

DWR 7306 AX
3011 The art song in America, volume II. CUMMING: Go lovely rose;
        The little black boy; Memory, hither come. DUKE: I carry
        your heart; In just spring; The mountains are dancing. EARLS:
        Arise my love; Entreat me not to leave you. PERSICHETTI: The
        death of a soldier; The grass; Of the surface of things; The
        snow man; Thou child so wise. ROREM: A Christmas carol; For
        Susan; Clouds; Guilt; What sparks and wiry cries. TRIMBLE:
        Love seeketh not itself to please; Tell me where is fancy bred.
        John Kennedy Hanks, t; Ruth Friedberg-Erickson, pno.
            -HF 8-74 p111   +-St 6-75 p107
            /ON 6-76 p52

                                ELDORADO

S-1 (UCLA Latin American Center, U. of California, Los Angeles, Calif-
        ornia 90024)
3012 Festival of early Latin American music. BELSAYAGA: Magnificat
        a 8. BLASCO: Versos con duo para chirimias. DURAN DE LA MOTA:
        Laudate pueri. FERNANDES: Eso rigor e repente. FERNANDEZ
        HIDALGO (Victoria): Salve regina a 5. LOPEZ CAPILLAS: Gloria
        laus et honor a 4. Alleluias (2). NUNES GARCIA: Lauda Sion
        salvatorem. ZIPOLI: Missa: Gloria. ANON.: Sa qui turo zente
        pleta. Roger Wagner Chorale; Sinfonia Chamber Orchestra;
        Roger Wagner.
            ++St 3-76 p125

ENGLISH NATIONAL OPERA

ENO 1001

3013 BELLINI: Adelson e Salvini: Io provo un palpito per quel dimora.
     Norma: Casta diva. La sonnambula: Lisa, mendace anch essa.
     DONIZETTI: Anna Bolena: Cielo a miei lunghi spasimi. L'Elisir
     d'amore: Una furtiva lagrima. Lucrezia Borgia: T'amo qual'
     s'ama un angelo. FLOTOW: Martha: Last rose of summer; Twelve
     o'clock, twelve o'clock. GLUCK: Alceste: Divinités du Styx.
     MERCANDANTE: Soirées Italiennes: Evive il galop. ROSSINI: La
     pietra del paragone: Pubblico fu l'oltraggio. SMETANA: The
     bartered bride: This girl I've found you. SULLIVAN: The gon-
     doliers: I am a courtier, grave and serious. VERDI: Ernani:
     Infelice, e tuo credevi. WAGNER: Tannhäuser: O du mein holder
     Abendstern. Ava June, Rita Hunter, Margaret Haggart, Anne
     Conoley, Josephine Barstow, Lois McDonall, Margaret Curphey, s;
     Shelagh Squires, Sandra Dugdale, Sara Walker, Ann Hood, ms;
     Alberto Remedios, John Brecknock, Keith Erwen, Terry Jenkins,
     Robert Ferguson, Tom Swift, t; David Ashton-Smith, John Kitch-
     iner, Christian Du Plessis, Norman Bailey, Alan Opie, Denis
     Dowling, Neil Howlett, Geoffrey Chard, bar; Dennis Wicks, Eric
     Shilling, Clifford Grant, Harold Blackburn, bs; Tom Hammond,
     Hazel Vivienne, John Barker, Noël Davies, Mark Elder, Noël
     Barker, pno.
           +-Gr 4-76 p1660   +RR 3-76 p28

ENIGMA

VAR 1015

3014 PUJOL VILARRUBI: Guajira. Tango. RODRIGO: Fandango. Pequeñas
     sevillanas. Ya se van los pastores. SANZ: Españoleta. Canar-
     ios. SOR: Fantasia elegiaca, op. 59. TARREGA: Capricho arabe.
     Gran jota. Maria. Sueño. TORROBA: Burgalesa. TRAD.: Brin-
     can y bailan. Don Gato. La serrana. Ya se van la paloma (all
     arr. Bonell). Carlos Bonell, gtr.
           +Gr 12-76 p1027   +RR 12-76 p84

VAR 1017

3015 Elizabethan and Jacobean madrigals: DOWLAND: Fine knacks for lad-
     ies. GIBBONS: The silver swan. MORLEY: April is my mistress'
     face; Daemon and Phyllis; Fire, fire; I love, alas; Leave, alas,
     this tormenting; My bonny lass; Now is the month of Maying; O
     grief, even on the bud; Those dainty daffadillies; Though
     Philomela lost her love. MUNDY: Were I a king. VAUTOR: Sweet
     Suffolk owl. WEELKES: Cease sorrows now; Come sirrah, Jack
     ho; Since Robin Hood. WILBYE: Adieu sweet Amarillis; Thus
     saith my Cloris bright. The Scholars.
           +-Gr 12-76 p1042

ERATO

STU 70871

3016 ALBINONI: Concerto, trumpet, B flat major. HANDEL: Concerto,
     trumpet, G minor. PURCELL: Sonatas, trumpet, nos. 1 and 2, D
     major. TARTINI: Concerto, trumpet, D major. TELEMANN: Con-
     certo, trumpet, D major. Maurice André, tpt; AMF; Neville
     Marriner.
           +Gr 11-74 p915   ++RR 11-74 p61

++Gr 2-76 p1355   ++RR 12-75 p67
+HFN 12-75 p148   +STL 1-11-76 p36

EVEREST

3360 (recorded 1908-15)
3017 John Philip Sousa conducts band music of the world: DE LISLE: La
     Marseillaise. LUDERS: The Prince of Pilsen, excerpts. SCHEU:
     Sleepy Sidney two-step. SOUSA: Free lance. Manhattan Beach.
     The pathfinder of Panama. STRAUSS, J. II: The beautiful blue
     Danube, op. 314. SUPPE: Light cavalry: Overture. YRADIER: La
     poloma. Sousa Band; various conductors.
          +HF 9-74 p109   +-SFC 2-15-76 p38
          +NR 7-74 p12

SDBR 3382
3018 BIZET: I pescatori di perle: Mi par d'udir ancora. BOITO: Luna
     fedel. Mefistofele: Giunto sul passo estremo, Dai campi dai
     prati. CARBONETTI: Tu non vi vuoi piu ben. CILEA: Adriana
     Lecouvreur: No piu nobile. DENZA: Non t'amo piu. DONIZETTI:
     L'Elisir d'amore: Una furtiva lagrima (2 versions). FRANCHETTI:
     Germania: No, non chiuder gli occhi yaghi (2 versions); Stu-
     denti, Udite. GIORDANO: Fedora: Amor ti vieta. LEONCAVALLO:
     I Pagliacci: Vesti la giubba. Mattinata. MASCAGNI: Cavalleria
     rusticana: Siciliana (2 versions). Iris: Serenata. MASSENET:
     Manon: Chiudo gli occhi. MEYERBEER: Gli Ugonotti: Qui sotto
     it ciel. PONCHIELLI: La gioconda: Cielo è mar. PUCCINI: Tosca:
     E lucevan le stelle (3 versions). TOSTI: La mia canzone. TRI-
     MARCHI: Un bacio ancora. VERDI: Aida: Celeste Aida. Rigoletto:
     La donne è mobile, Questa o quella. Enrico Caruso, t; Instru-
     mental accompaniments.
          +-NR 3-76 p7

SDBR 3388
3019 The Red Army Choir in concert: A village lies yonder; Along the
     Peterskaya road; The cossack; Fair tresses; I see a village;
     I shall walk outside the gates and high river banks; Little
     onion; Look to the sky; Old bachelor; Rise, beautiful sun; The
     sea has spread wide and afar; Song of peace; Song of the barge
     haulers; Uncle Nimra; Young heroes. Soviet Army Chorus.
          +-NR 9-76 p8

FINNADAR

QD 9010. Tape (ct) QT 9010F
3020 AREL: Stereo electronic music, no. 2. BABBIT: Ensembles for syn-
     thesizer. DAVIDOVSKY: Electronic study, no. 3. SHIELD: Fare-
     well to a hill. SMILEY: Eclipse. USSACHEVSKY: Piece for tape
     recorder. Columbia-Princeton Electronic Music Center.
          +HF 11-76 p121   *NR 9-76 p13

FOLKWAYS

FTS 33901/4 (4)
3021 New American music. AIN: Used to call me sadness. BOLCOM: Whis-
     per moon. CHADABE: Echoes. DLUGOSZEWSKI: Angels of the ut-
     most heavens. EVANS: Bluefish. FULKERSON: Patterns, no. 7.
     GLASS: Two pages. GRAVES: Transmutations. HAKIM: Placements.

McMILLAN: Carrefours. Whale I. MOORE: Youth in a merciful
house. MUMMA: Cybersonic cantilevers. MURRAY: Evocations.
RIVERS: Shadows. RZEWSKI: Songs: Apolitical intellectuals;
God to a hungry child; Struggle. SILVERMAN: Planh. SWANSON:
Trio, flute, clarinet and piano. USSACHEVSKY: Conflict.
WILLIAMS: Gloria. Zoning fungus II.
    +MJ 3-76 p24   +St 9-76 p126
    +NYT 4-11-76 pD23

GAUDEAMUS

XSH 101
3022 DUPRE: Triptyque, op. 51. FRANCK: Chorale, no. 2, B minor.
    LEIGHTON: Scherzo. MENDELSSOHN: Sonata, organ, op. 65, no.
    3, A major. MESSIAEN: La nativite du Seigneur: Dieu parmi
    nous. Jonathan Rennert, org.
        +-RR 8-76 p70

GEMINI HALL

RAP 1010 (2) (From The Bookers, 200 W. 51st St., New York)
3023 Woman's work. AMALIA, Princess of Prussia: Regimental marches
    (3). ANNA AMALIA, Duchess of Saxe-Weimar: Erwin and Elmire,
    excerpt. ANDREE: Quintet, E major: Allegro molto vivace.
    BOULANGER, L.: Clairieres dans le ciel, excerpts. Nocturne.
    BRONSART: Jery und Bately: Lied and duet. CACCINI, F.: La
    liberazione di Ruggiero dall'Isola d'Alcina, excerpts. CHAM-
    INADE: Caprice espagnole. FARRENC: Quintet: Scherzo. JACQUET
    DE LA GUERRE: Jacob et Rachel: Air. Susanne: Recitative and
    air. HERITTE-VIARDOT: Quartet: Serenade. LANG: Sie liebt
    mich. MALIBRAN: Le reveil d'un beau jour. MENDELSSOHN-HENSEL:
    Bergeslust. Italien. PARADIS: Sicilienne. POLDOWSKI: Im-
    pression fausse. SCHUMANN, C.: Das ist ein Tag. Impromptu.
    Liebst du um Schonheit. TAILLEFERRE: Quatuor. VIARDOT-GARCIA:
    Dites, que faut-il faire. Fluestern athemscheues Lauschen.
    Die Sterne. Berenice Bramson, s; Mertine Johns, ms; Thomas
    Theis, bs; Michael May, pno, hpd; Vieuxtemps String Quartet;
    Roger Rundle, pno; Yvonne Cable, vlc.
        *MJ 11-76 p44   +St 5-76 p124
        +NR 4-76 p15

GOLDEN CREST

CRS 4143
3024 ABSIL: Suite on popular Rumanian themes. BACH: Suite, orchestra,
    S 1068, D major: Air on the G string. BAUER: Sokasodik.
    Willy rag. DEBUSSY: Le petite negre. JOPLIN: Something doing
    rag. PIERNE: Songs (4). TURPIN: Pan-Am rag. TCHAIKOVSKY:
    Quartet, strings, no. 1, op. 11, D major: Andante cantabile.
    Paul Brodie Saxophone Quartet.
        +IN 4-76 p16

GRENADILLA

GS 1006
3025 CAHUZAC: Cantilene. DEBUSSY: Rhapsody, no. 1. HONNEGER: Sona-

tina. JEANJEAN: Arabesques. PIERNE: Canzonetta. RABAUD: Solo
de concours. Louis Cahuzac, Gaston Hamelin, Auguste Perier,
clt; Perre Vilbert, Folmer Jensen, pno; Orchestra; Piero Coppola.
+NR 1-76 p8

GROSVENOR

GRS 1034
3026 BAX: A Christmas carol song. BRITTEN: A shepherd's carol. DAW-
SON: Ain' a that good news. HENDRIE: As I outrode this end'
res night. HOLST: Partsongs, op. 34: Lullay my liking. POS-
TON: Last night in the open shippen. TIPPETT: A child of our
time: Negro spirituals (5). ANON.: Fisherman Peter; Baby born
today: Tomorrow shall be my dancing day (arr. Poston). Sans
day carol (arr. Rutter). Ding don merrily on high (arr.
Williamson). ANON.: (Spanish, 1556): Riu, riu, chiu. Stephen
Darlington, org; Canterbury Cathedral Choir; Allan Wicks.
+RR 1-76 p56

GRS 1039
3027 BARTLETT (Wilson): Praise the Lord with joyful cry. BUXTEHUDE:
Prelude on 'Ein feste Burg". Prelude on the Te Deum. CLARKE:
Come, my way, my truth, my life. CUTTS: As the bridegroom to
his chosen. DUNCALF: My God, my King, thy various praise.
LORING: Nothing in all creation. MICKLEM: Father, we thank you.
No one has ever seen God. PAIX: Phantasia primi toni. PRATT:
By the waters of Babylon; The earth is the Lord's. ROUTLEY:
God who spoke in the beginning. You we praise as God. ROBERTS:
Homage to Perotin. STANFORD: When in man's music God is glori-
fied. STOCKS: Come, living God. WILLIAMS: Lord of the bound-
less curves of space. WILSON: There's a spirit in the air.
ZIMMERMAN: Praise the Lord; Praise you servants of the Lord.
David Strong, org; Keele University Chapel Choir; George Pratt.
+HFN 1-76 p105   +-RR 3-76 p66

GRS 1041
3028 BACH: Chorale preludes: In dulci jubilo, S 608; Liebster Jesu, wir
sind hier, S 731. Fantasia, S 572, G major. FRANCK: Prelude,
fugue and variations. MULET: Carillon-Sortie. PACHELBEL: Vom
Himmel hoch. STANDLEY: Voluntary, A minor. SWEELINCK: Mein
Junges Leben hat ein End, variations. Malcolm Archer, org.
/RR 11-76 p92

HARMONIA MUNDI

HMU 105
3029 Orthodox Slav liturgy: ARKANGELSKII: Outoli bolzni. BORTNIANSKY:
Dostoino est; Slava vo vichnih Bogou. CHRISTOV: Hvalite imia
Gospodne. DUBENSKI: Otche nach. LOMAKIN: Tebe poem. LUBIMOV:
Brajen mouj. LVOVSKI: Herouvimska. VEDEL: Pokayania. Bulgar-
ian Radio and Television Men's Choir; Mikhail Milkov.
+-Gr 1-76 p122
HMU 137 (Reissue from Balkanton BX1 1091)
3030 Russian and Bulgarian orthodox chants. ARKHANGELSKII: Brajen
mouj; Pomichliaiou dien strachniy. CHRISTOV: Ije heruvimi,
lako da tsaria; Tebe poem; Vo tsarstiviy tvoiem. DEGTIAREV:
Preslavnaia dnies. GRETCHANINOV: Otpoutchtchaiechi (9).

MANOLOV: Dostoino est. TCHESNOKOV: Hvalite imia Gospodnie.
ANON.: Blagoobrajniy Iossif. Christo Kamenov, t; Sofia Chorale;
Dimiti Rouskov.
+Gr 9-76 p465

HMD 204
3031 CORNYSHE: Ah, Robin: Hoyda jolly rutterkin. GENTIAN: Je suis Robert. GIBBONS: The cries of London. JOHNSON: Care-charming
sleep. MONTEVERDI: Baci soavi e cari; Lamento della Ninfa.
SERMISY: Tant que vivray. VAUTOR: Mother, I will have a hus-
band. WEELKES: Cease sorrows now. WILBYE: Lady, when I behold the roses. Deller Consort; Bulgarian Quartet; René
Saorgin, hpd, org; Raphael Perulli, vla da gamba.
+-Gr 11-75 p888   +RR 2-76 p59

HMD 223
3032 Music of the Tudor Court: DOWLAND: If that a sinner's sighs; In
this trembling shadow cast; My Lord Willoughby's welcome home;
Now, o now I needs must part. FORD: Almighty God. JOHNSON, E.:
Eliza is the fairest queen. LEIGHTON: Almighty God. MORLEY:
Joyne hands; The Lord Zouches maske; O mistress mine; Sola
soletta (Conversi). PHILIPS: Philips pavan. Galliard to
Philips pavan. PILKINGTON: Amyntas with his Phillis fair; Have
I found her. TOMKINS: O let me live; O let me die. Alfred
Deller, c-t; Desmond Dupre, Robert Spencer, lt; Deller Consort,
Morley Consort.
+-HFN 8-76 p85   +RR 2-76 p60

HMU 334
3033 d'ANGLEBERT: Chaconne, D major. Le tombeau de M. de Chambonnieres.
CHAMBONNIERES: Chaconne, F major. Rondeau. COUPERIN, L.: La
Piémontaise, A minor. Passacaille, C major. DUMONT: Pavane,
D minor. RAMEAU: Suite, E minor. L'Enharmonique, G minor.
Kenneth Gilbert, hpd.
++Gr 1-76 p1221   ++RR 11-75 p76
+HFN 2-76 p103

HMU 335
3034 Carmina Burana: Songs of drinking and eating; Songs of unhappy
love. Bacche, bene venies; Virent prata hiemata; Nomen a sol-
lempnibus; Alte clamat Epicurus; Vite perdite (2 versions)
Vacillantis trutine; In taberna quando sumus; Iste mundus furi-
bundus; Axe Phebus aureo; Dulce solum natalis patrie; Pro-
curans odium; Sic mea fata canendo solo; Ich was ein chint so
wolgetan. Clemencic Consort; René Clemencic.
+St 9-76 p129

HMU 337
3035 Carmina Burana: Officium lusorum; Olim sudor Herculis; Virent
prata hiemata. Clemencic Consort; René Clemencic.
+St 12-76 p152

HMU 473 (2)
3036 BYRD: Haec dies, a 5. DERING: Jesu dulcis memoria. GESUALDO:
Sacrae cantiones a 5: Ave, dulcissima Maria; Ave, Regina coelorum; Hei mihi, Domine; O crux benedicta; O vos omnes. LASSUS:
Sacrae lectiones es propheta Job: Parce mihi, Domine; Teadet
aninam meam vitae meae. MORLEY: Laboravi in gemitu meo. PAR-
SONS: Ave Maria. PHILIPS: Ascendit Deus. SCHUTZ: Erbarm dich
mein, o Herre Gott. Herr, nun lässest du deinen Diener in
Frieden fahren. Domini est terra. Heute ist Christus der Herr
geboren. TALLIS: Salvator mundi. Raphaël Passaquet Vocal

Emsemble; Deller Consort, Ars Europea Choeur National and
Instrumental Ensemble; Raphael Passaquet, Alfred Deller, Jac-
ques Grimbert.
+-RR 6-76 p78

HM 593
3037 English madrigals and folksongs: CAVENDISH: Sly thief, if so you
will believe. FARMER: A little pretty bonny lass. MORLEY:
My bonny lass she smileth. PILKINGTON: Sweet Phillida. TOM-
KINS: See, see, the shepherd's queen; Weep no more, thou sorry
boy. VAUTOR: Shephers and nymphs. WEELKES: Hark, all ye
lovely saints above; Say, dear, when will you frowning leave.
ANON.: The cuckoo; Cold blows the wind (arr. Morris); The
jolly carter; O' 'twas on a Monday morning (arr. Holst); O
waly, waly (arr. Cashmore); The sailor and young Nancy (arr.
Moeran); The sheep sharing (arr. Sharp); The turtle dove
(arr. Morris). Deller Consort.
/Gr 11-75 p887   +RR 2-76 p63

HMU 610
3038 ATTAINGNANT: Tordion. Pavane et galliarde. DEMANTIUS: Polnischer
Tanz und Galliarda. FRANCK, M.: Pavane et galliarde. GER-
VAISE: Branle. HASSLER: Intraden (3). MODERNE: Branle gay
nouveau. Trios branles de Bourgogne. PHALESE: L'Arboscello
ballo furlano. SUSATO: Hoboecken dans. Il estoit une fillette,
ronde. Mille regretz, pavane. Ronde et salterelle. Si pas
souffrir, pavane. Ronde. Clemencic Consort; Rene Clemencic.
-HFN 1-76 p105   +-RR 8-76 p42

HMV

SEOM 22 (Reissues from ASD 2785, 3075, 3001, SLS 832)
3039 BACH: Concerto, violin and strings, S 1042, E major. JOPLIN:
The entertainer. KREISLER: Variations on a theme by Corelli
in the style of Tartini. NOVACEK: Perpetuum mobile. PAGANINI:
Caprices, op. 1, no. 9, E major; no. 24, A minor. SARASATE:
Carmen fantasy, op. 25. Itzhak Perlman, vln; ECO, RPO; Samuel
Sanders, André Previn, pno; Daniel Barenboim, Lawrence Foster,
cond.
+Gr 1-76 p1209   +RR 1-76 p29
++HFN 3-76 p112

SEOM 24
3040 HMV Proms Festival '76: BEETHOVEN: Concerto, piano, no. 5, op. 73,
E flat major: 1st movement. Symphony, no. 9, op. 125, D minor:
4th movement excerpt. BERLIOZ: Les nuits d'ete, op. 7: Villanelle. CHOPIN: Concerto, piano, no. 2, op. 21, F minor: 2nd
movement. DELIUS: North country sketches, no. 2. ELGAR:
Elegy, strings, op. 58. Pomp and circumstance march, op. 39,
no. 1, D major. MENDELSSOHN: Calm sea and prosperous voyage,
op. 27. SIBELIUS: Humoreske, no. 5, op. 89, no. 3, E flat
major. Ursula Koszut, s; Janet Baker, ms; Brigitte Fassbaender,
on; Nicolai Gedda, t; Donald McIntyre, bs; Bruno-Leonardo
Gelber, Garrick Ohlsson, pno; Ida Haendel, vln; Munich Philharmonic Chorus, Munich Motet Choir; Munich Philharmonic Orchestra, NPhO, RPO, PhO; Polish Radio Symphony Orchestra, Bournemouth Symphony Orchestra, LPO; Rudolf Kempe, Riccardo Muti,
Ferdinand Leitner, John Barbirolli, Charles Groves, Jerzy

Maksymuik, Paavo Berglund, Adrian Boult.
+HFN 10-76 p183   +-RR 8-76 p41

SEOM 25
3041 Edinburg festival sampler: BEETHOVEN: Mass, op. 123, D major:
      Credo. BIZET: Carmen: Prelude, Act 1. BRAHMS: Sandmännchen.
      FALLA: The three-cornered hat: Jota. MENDELSSOHN: Symphony,
      no. 3, op. 56, A minor: 4th movement. MOZART: Abendempfindung,
      K 523. MUSSORGSKY: Where are you little star. RIMSKY-KORSAKOV:
      Scheherazade, op. 35: The sea and Sinbad's ship. TRAD.: The
      mermaid (arr. Whitworth). Elisabeth Schwarzkopf, Galina Vish-
      nevskaya, Heather Harper, s; Janet Baker, con; Robert Tear,
      t; Dietrich Fischer-Dieskau, bar; Hans Sotin, bs; Orchestre
      de Paris, LPO, PhO, NPhO and Chorus; King's Singers; Geoffrey
      Parsons, Mstislav Rostropovitch, Daniel Barenboim, pno; Daniel
      Barenboim, Carlo Maria Giulini, Rafael Frühbeck de Burgos,
      Riccardo Muti.
           +HFN 10-76 p183   +-RR 9-76 p45

RLS 716 (4) (Reissues)
3042 Opera, operetta and song recital: BERLIOZ: Les nuits d'été,
      op. 7: Le spectre de la rose; Absence. CHAUSSON: Le colibri,
      op. 2, no. 7; Les papillons, op. 2, no. 3; Poème de l'amour
      et la mer, op. 19; Les temps des lilas. DEBUSSY: Ballade des
      femmes de Paris; Chanson de Bilitis (3); Fêtes galantes, I
      and II; Green. Pelléas et Mélisande: Voici ce qu'il ecrit;
      Tu ne sais pas pourquoi. DUPARC: Chanson triste; Extase;
      L'invitation au voyage; Phidyle. GODARD: Chanson d'Estelle.
      GOETZE: Still as the night. HAHN: En sourdine; L'Heure ex-
      quise; L'Offrande; Si mes vers avaient des ailes. Mozart:
      Etre adoré; L'adieu. FAURE: L'absent, op. 5, no. 3; Apres
      un rêve, op. 7, no. 1; Clair de lune, op. 46, no. 2; Dans
      les ruines d'une abbaye, op. 2, no. 1; Ici-bas, op. 8, no. 3;
      Nell, op. 18, no. 1; Les roses d'Ispahan, op. 39, no. 4; Le
      secret, op. 23, no. 3; Soir, op. 83, no. 2. FRANZ: O thank me
      not, op. 14, no. 1. d'HARDELOT: Because. KENNEDY (Frazer):
      Land of hearts desire. MASSENET: Elégie. MESSAGER: Monsieur
      Beaucaire: Philomel; I do not know; Lightly, lightly; What
      are the names. QUILTER: Now sleeps the crimson petal, op. 3,
      no. 2. PURCELL: King Arthur: Fairest isle. RAVEL: D'Anne
      jouant delespinette; D'Anne qui me jecta de la neige; Shé-
      hérazade. RHENE-BATON: Heures d'été. RUSSELL: By appoint-
      ment: White roses. STRAVINSKY: La rosee sainte. SZULC: Clair
      de lune, op. 81, no. 1. VAUGHAN WILLIAMS: Greensleeves.
      WARLOCK: The Bayley beareth the bell away; Lullaby. WEBBER:
      Vieille chanson. TRAD.: Coming thro the rye; Oft the stilly
      night; Vieille chanson de chasse. Maggie Teyte, s; Gerald
      Moore, Alfred Cortot, pno; Various orchestras and conductors.
           +-ARSC Vol VIII, no. 2-3   +HFN 11-76 p169
                p99   +RR 10-76 p94
           +Gr 10-76 p657

RLS 717 (Reissues from Columbia LX 532/7, 712/3, 899/903, 918, 909,
      861, 877/8, 897/8, 860/1, 868, 866, 898, CLX 2189/90, 2197/
      8, 2187, 2188, 2165/6, 2167/8, CAS 8184, 7, 8717/26, 8733/6,
      8737/9, 8742, DX 226, WAX 6050/1)
3043 BEETHOVEN: Leonore overture, no. 2, op. 72. Die Ruinen von

599 HMV (cont.)

Athens, op. 113: Overture. Symphony, no. 3, op. 55, E flat
major. BERLIOZ: Les Troyens: Trojan march. BRAHMS: Symphony,
no. 2, op. 73, D major. HANDEL: Alcina: Dream music (arr.
Whittaker). LISZT: Les préludes, G 97. Mephisto waltz.
STRAUSS, J. II: Voices of spring, op. 410. Wein, Weib und Gesange, op. 333. WAGNER: Rienzi: Overture. Tannhäuser: Prelude, Act 3. Tristan und Isolde: Prelude, Act 3. VPO, LSO,
LPO, OSCCP, British Symphony Orchestra; Felix Weingartner.
        +Gr 12-75 p1061   +RR 12-75 p37
        +HFN 2-76 p113

RLS 718 (3) (Reissues from DB 2911/2, 1718/9, 2729/31, 3653, 2279/83,
            1961/3, 3555/8, 1611/3)
3044 BACH: Concerto, violin and strings, S 1041, A minor. Concerto,
      2 violins and strings, S 1043, D minor. BRUCH: Concerto, vio-
      lin, no. 1, op. 26, G minor. CHAUSSON: Poème, op. 25. MEN-
      DELSSOHN: Concerto, violin, op. 64, E minor. MOZART: Concerto,
      violin, no. 3, K 216, G major. PAGANINI: Concerto, violin,
      no. 1, op. 6, D major (ed. Kreisler). WIENIAWSKI: Légende,
      op. 17. Yehudi Menuhin, Georges Enesco, vln; Paris Symphony
      Orchestra, Colonne Concerts Orchestra, LSO; Georges Enesco,
      Pierre Monteux, Landon Ronald.
        +-Gr 4-76 p1617   +MT 8-76 p663
        +HFN 5-76 p102   +RR 5-76 p33

RLS 719 (5)
3045 Dame Nelli Melba, The London recordings 1904-1926: ARDITI: Se
      saran rose. BACH (Gounod): Ave Maria (1904, 1906, 1913).
      BEMBERG: Les anges pleurent; Un ange est venu; Chant hindou;
      Chant venétien; Elaine: L'amour est pur; Nymphs et Sylvains:
      Sur le lac. BISHOP: Bid me discourse; Home, sweet home; Lo,
      hear the gentle lark. BIZET: Pastorale. CHAUSSON: Le temps
      des lilas. BARNARD: Come back to Erin. DONIZETTI: Lucia
      di Lammermoor: Mad scene. DUPARC: Chanson triste. FOSTER: Old
      folks at home. GOUNOD: Faust: Jewel song; Final trio. Romeo
      et Juliette: Waltz song. HAHN: Si mes vers avaient des ailes.
      HANDEL: Il pensieroso: Sweet bird (1904, 1926). d'HARDELOT:
      Three green bonnets. HENSCHEL: Spring. HUE: Soir paien.
      LALO: Le Roi d'Ys: Aubade. LIEURANCE: By the waters of Minne-
      tonka. LOTTI: Pur dicesti. MASSENET: Le Cid: Pleurez mes
      yeux. Don César de Bazan: Sevillana (2). MENDELSSOHN: O for
      the wings of a dove. MOZART: Le nozze di Figaro, K 492: Porgi
      amor. Il Re pastore, K 208: L'amerò, sarò costante. PUCCINI:
      La bohème: Addio di Mimi (2); On m'appelle Mimi; Sì mia chia-
      mano Mimi; Entrate...C'e Rodolfo; Donde lieta usci; Addio
      dolce svegliare alla mattina; Gavotta...Minuetto; Sono andati;
      Io Musetta...Oh come è bello e morbido. Tosca: Vissi d'arte.
      RIMSKY-KORSAKOV: The legend of Sadko, op. 5: Chanson hindoue.
      RONALD: Away on the hill; Down in the forest; Sounds of earth.
      SCOTT (Gatty): Goodnight. SCOTT: Annie Laurie. SZULC: Clair
      de lune, op. 81, no. 1. THOMAS: Hamlet: Mad scene (1904, 1910).
      TOSTI: Goodbye (two stanzas); Mattinata; La serenata. VERDI:
      La traviata: Ah fors'è lui; Sempre libera; Dite alla giovine.
      Otello: Piangea cantando; Ave Maria piena di grazia. Rigo-
      letto: Caro nome; Quartet. WAGNER: Lohengrin: Elsa's dream.
      TRAD.: Auld lang syne; Coming thro the rye; God save the King;
      Swing low, sweet chariot (arr. Burleigh). Lord Stanley's ad-
      dress; Nellie Melba's farewell speech. Nelli Melba, s; various

other artists; Landon Ronald, pno; Orchestral accompaniment;
Landon Ronald.

+Gr 11-76 p877   +RR 11-76 p23
+HFN 12-76 p133

RLS 723 (3)
3046 BACH: Sonata, violin, no. 2, S 1003, A minor: Andante. Suite, no.
      3: Aria. BEETHOVEN: Variations on Mozart's "Bei Männern" (7).
      BOCCHERINI: Concerto, violoncello, B flat major (ed. Grütz-
      macher). Sonata, violoncello, no. 6, A major: Adagio and
      allegro. BRAHMS: Concerto, violin and violoncello, op. 102,
      A minor. DVORAK: Songs my mother taught me, op. 55. HAYDN:
      Sonata, no. 1: Tempo di menuetto. LASERNA: Tonadillas.
      MENDELSSOHN: Song without words, op. 109, D major. Trio, piano,
      no. 1, op. 49, D minor. RIMSKY-KORSAKOV: The tale of the Tsar
      Sultan: Flight of the bumblebee. SCHUMANN: Kinderscenen, op.
      15, no. 7: Träumerei. Trio, piano, no. 1, op. 63, D minor.
      TARTINI: Concerto, violoncello, D major: Grave ed espressivo.
      VALENTI: Gavotte. VIVALDI: Concerto, no. 11, D minor: Largo.
      Pablo Casals, vlc; Jacques Thibaud, vln; Alfred Cortot, Blas-
      Net, Otto Schulhof, pno; LSO, Casals Barcelona Orchestra;
      Landon Ronald, Alfred Cortot.
                      ++RR 12-76 p52
SLS 863 (3) (also EMI OC 191-05410/2, also Seraphim SIC 6092)
3047 The art of courtly love: BINCHOIS: Amoreux suy. Bien puist.
      Files a marier. Je ne fai toujours. Jeloymors. Votre très
      doulz regart. BORLET: Hé tres doulz roussignol. Ma tredol
      rosignol. CASERTA: Amour m'a le cuer mis. DANDRIEU: Armes,
      amours: O flour des flours. DUFAY: La belle se siet. Ce moys
      de may. Donnés l'assault. Hélas mon dueil. Lamentatio
      Sanctae matris ecclesiae. Navre je suis. Par droit je puis.
      Vergine bella. FRANCISCUS (FRANCISQUE): Phiton, Phiton.
      GRIMACE: A l'arme, à l'arme. HASPROIS: Ma douce amour.
      LESCUREL: A vous douce debonaire. MACHAUT: Amours me fait
      désirer. Dame se vous m'estes. De bon espoir: Puis que la
      douce. De toutes flours. Douce dame jolie. Hareu, hareu:
      Helas, où sera pris confors. Ma fin est mon commencement.
      Mes esperis se combat. Quant je sui mis. Quant j'ay l'espart.
      Quant Théseus: Ne quier veoir. Phyton, le merveilleus serpent.
      Se je souspir. Trop plus est belle: Biauté paree; Je ne sui.
      MERUCO: De home vray. MOLINS: Amis tous dous. PERUSIO: Andray
      soulet. Le gregnour bien. PYKINI: Plasanche or tost. SOLAGE:
      Fumeux, fume. Helas, je voy mon cuer. VAILLANT: Trés doulz
      amis: Ma dame; Cent mille fois. ANON.: La septime estampie
      real. Istampitta tre fontaine. Contre le temps. Restoés,
      restoés. Basse danses, I and II. Early Music Consort; David
      Munrow.

          +-Gr 12-73 p1238   +PRO 3/4-76 p17
          +HFN 12-73 p2619   ++RR 1-74 p60
          *NR 3-76 p9   ++SFC 8-22-76 p38
          +NYT 8-15-76 pD15   +St 5-76 p78

SLS 988 (2) (also Angel SBZ 3810)
3048 Instruments of the Middle Ages and Renaissance (Includes 96 page
      booklet). Early Music Consort; David Munrow.

          ++Gr 6-76 p58   +RR 7-76 p71
          ++HFN 6-76 p89   ++SFC 8-22-76 p38
          ++NR 9-76 p15   +STL 5-9-76 p38
           +NYT 8-15-76 pD15

601   HMV (cont.)

HQS 1350
3049 Chichester Cathedral, 900 years: BAIRSTOW: Let all mortal flesh
      keep silence. DAVIES: Ave Maria. HOLST: Festival chorus, op.
      36, no. 2: Turn back, O man. HOWELLS: Magnificat. IRELAND:
      Greater love hath no man. LEIGHTON: Give me the wings of
      faith. SHAW: Anglican folk mass: Creed. STANFORD: Beati quo-
      rum via, op. 51, no. 3. VAUGHAN WILLIAMS: O taste and see.
      WALTON: Set me as a seal upon thine heart. WOOD: Short com-
      munion service in the Phrygianmode: Sanctus and benedictus.
      Chicester Cathedral Choir; John Birch, org; Richard Seal.
           +Gr 9-75 p506   +RR 8-75 p63
           +MT 3-76 p237

SQS 1354
3050 BRAHMS: Pieces, piano, op. 117, no. 2, B flat minor. GLINKA: The
      lark (arr. Balakirev). GRANADOS: Goyescas: The maiden and the
      nightingale. LISZT: Hungarian rhapsody, no. 12, G 244, C
      sharp minor. MENDELSSOHN: Andante and rondo capriccioso, op.
      14. SCHUBERT: Impromptu, op. 90, no. 4, D 899, A flat major.
      SCHUMANN: Fantasiestücke, op. 12, no. 2: Aufschwung. SCRIABIN:
      Prelude and nocturne for the left hand, op. 9, D major. Sylvia
      Kersenbaum, pno.
           +-Gr 3-76 p1488   +RR 2-76 p47
           +HFN 2-76 p101
SQ HQS 1356. Tape (c) TC HQS 1356
3051 BACH: Toccata and fugue, S 565, D minor. BRAHMS: Chorale prelude,
      op. 122, no. 8: Es ist ein Ros entsprungen. FRANCK: Chorale,
      no. 3, A minor. LISZT: Prelude and fugue on the name B-A-C-H,
      G 260. VAUGHAN WILLIAMS: Rhosymedre. VIERNE: Pieces in free
      style, no. 19: Berceuse. WIDOR: Symphony, organ, no. 5, op.
      42, no. 1, F sharp major: Toccata. Philip Ledger, org.
           +Gr 7-76 p199   +HFN 9-76 p133 tape
           -Gr 9-76 p497 tape   +RR 12-76 p105 tape
           +HFN 7-76 p93

HQS 1360
3052 BACH: Sonata, flute and harpsichord, S 1031, E flat major: Sicil-
      iano (arr. Rostal/Schaefer). BIZET: Carmen fantasy, op. 25
      (arr. Rostal/Schaefer). BORODIN: Quartet, strings, no. 2, D
      major: Nocturne (arr. Rostal/Schaefer). BRAHMS: Hungarian
      dance, no. 5, G minor. GRANADOS: Goyescas: The maiden and the
      nightingale. GRIEG: Lyric pieces, op. 65, no. 6: Wedding day
      at Troldhaugen (arr. Ruthardt). HANDEL: Suite, harpsichord,
      no. 5, E major (The harmonious blacksmith). ROSTAL/SCHAEFER:
      The entertainer: Variations on a theme by Scott Joplin. Anglo-
      American fantasy. SCHUBERT: Serenade (arr. Rostal/Schaefer).
      WALTON: Façade (arr. Seiber). WIDOR: Symphony, organ, no. 5,
      op. 42, no. 1, F minor: Toccata (arr. Rostal/Schaefer). Peter
      Rostal, Paul Schaefer, pno.
           +Gr 9-76 p478   +RR 9-76 p79
           +HFN 9-76 p127

HQS 1364
3053 DEBUSSY: Children's corner suite. IBERT: Histoires. KHACHATUR-
      IAN: Gayaneh: Adagio. Pictures of childhood (4). MOMPOU:
      Scenes d'enfants. PROKOFIEV: Tales of an old grandmother, op.
      31. Christina Ortiz, pno.
           +Gr 11-76 p843   ++RR 11-76 p89
           +-HFN 11-76 p157

OU 2105
3054 BALL: Journey into freedom. GEEHL: Romanza. GERMAN (arr. D.
        Wright): Men of Harlech. GOODWIN (arr. Brand): The headless
        horseman. GREGSON: Prelude for an occasion. HUGHES (arr.
        Rayner): Cwm Rhondda. JONES, M.(arr. Ball): Rhondda rhapsody.
        PRICE: Heroic march. WALTERS (trans. Brush): Trumpets wild.
        TRAD. (arr. H. Bebb): David of the white rock; The hunting of
        the hare. Phillip Morgan, trom; Gareth Pritchard, Derek Holvey,
        Alun Williams, cor; Parc and Dare Band; Ieuan Morgan.
               +Gr 3-76 p1513
ASD 2929. Tape (c) TC ASD 2929
3055 Lieder and song recital: CHABRIER: Villanelle des petits canards.
        FINZI: Let us garlands bring, op. 18: It was a lover and his
        lass. GOUNOD: Serenade. HAHN: L'heure exquise. HOWELLS:
        Gavotte. IRELAND: The Salley gardens. MASSENET: Crépuscule.
        MENDELSSOHN: Auf Flügeln des Gesanges, op. 34, no. 2. PARRY:
        O mistress mine. QUILTER: It was a lover and his lass, op. 23,
        no. 3. SCHUBERT: Songs: Heidenröslein, D 257. SCHUMANN: Der
        Nussbaum, op. 25, no. 3. STRAUSS, R.: Songs: Ständchen, op.
        17, no. 2. SULLIVAN: Orpheus with his lute. TRAD.: Me suis
        mis en danse (arr. Bax); Bushes and briars (arr. Vaughan
        Williams); Drink to me only (arr. Kinloch Anderson); I know
        where I'm going (arr. Hughes). Janet Baker, ms; Gerald Moore,
        pno.
               +Gr 2-74 p1595   ++RR 4-76 p81 tape
               +RR 4-74 p81
Melodiya ASD 3102
3056 Russian choral works of the seventeenth and eighteenth centuries:
        BEREZOVSKY: Do not reject me in my old age. BORTNYANSKY: Cheru-
        bim hymn, no. 7. I will lift up my eyes to the hills. DIL-
        ETZKY: Glorify the name of the Lord. KALASHNIKOV: Concerto,
        12 voice choir: Cherubic hymn. KRESTYANIN: Befittingly.
        VEDEL: How long, O Lord, how long. USSR Russian Chorus; Alek-
        sander Yurlov.
               +Gr 9-75 p506   +MT 1-76 p44
               +HFN 8-75 p83   +RR 8-75 p62
(Q) ASD 3131. Tape (c) TC ASD 3131 (ct) 8XASD 3131.
3057 ALBINONI (Giozotto): Adagio, organ and strings, G minor. DUKAS:
        L'apprenti sorcier. DVORAK: Slavonic dance, op. 72, no. 1,
        B major. HUMPERDINCK: Hänsel and Gretel: Overture. PREVIN:
        André Previn's music night: Signature tune. RAVEL: La valse.
        WALTON: Orb and sceptre. LSO; André Previn.
               +Gr 9-75 p479   +HFN 1-76 p125 tape
               ++HFN 10-75 p135   +-RR 10-75 p61
ASD 3200
3058 ALEXANDROV: Volga bargehaulers song. GLINKA: Travelling song.
        KNIPPER: Cavalry song. MURADELI: October: Choruses (2).
        KABALEVSKY: Song about Russia. RIMSKY-KORSAKOV: May night:
        Song about the village elder. RUBINSTEIN: The demon: The
        night. SEROV: The hostile force: Yeryomka's song. SHOSTAKO-
        VITCH: Novorossüsk chimes. TCHAIKOVSKY: The merry voices grew
        silent. TRAD.: Cottage cheese pies; Dubinushka; The neighbour;
        Song of Stenka Razin. A. Sergeyev, bs; N. Gres, E. Belyaev, t;
        Soviet Army Ensemble; Boris Aleksandrov.
               +HFN 6-76 p102   +RR 7-76 p76
               +-RR 5-76 p69

SQ ASD 3283
3059 BACH: Toccata, organ, C major: Adagio (arr. Siloti/Casals).
         DEBUSSY: Preludes, Bk I, no. 12: Minstrels. FAURE: Sicilienne,
         op. 78. HEKKING: Villageoise. KARJINSKY: Esquisse.
         MENSELSSOHN: Chant populaire (arr. de Hartmann). NIN: Granadina
         (arr. Kochanski). POPPER: Dance of the elves, op. 39 (ed.
         Fournier). RACHMANINOFF: Vaocalise, op. 34, no. 14. RIMSKY-
         KORSAKOV: The tale of the Tsar Sultan: Flight of the bumblebee.
         SCHUBERT: The bee, op. 13, no. 9 (arr. Casals). TORTELIER:
         Miniatures, 2 violoncellos (3). Valse, no. 1. WEBER: Adagio
         and rondo (arr. Piatigorsky). Paul Tortelier, Maud Tortelier,
         vlc; Maria de la Pau, pno.
              ++Gr 10-76 p661   +RR 10-76 p85
              +HFN 10-76 p177
CSD 3740. Tape (c) TC CSD 3740
3060 French songs: ARBEAU: Belle qui tient ma vie. CERTON: Là, là, la
         je ne l'oise dire. JACOTIN: A Paris à trois fillettes.
         JANNEQUIN (VERDELOT): La guerre (parts 1, 2). LE JEUNE: Un
         gentil amoureux; Une puce. MORNABLE: Je ne scay. POULENC: La
         belle si nous étions; Clic, clac, dansez sabots; Chanson a
         boire; Laudes de Saint Antoine de Padoue; Petites prières de
         Saint Francois d'Assise (4). PASSEREAU: Il est bel et bon.
         VUILDRE: Je fille quant Dieu. WILLAERT: Allons, allons gay;
         Faulte d'Argent. The King's Singers; Early Music Consort;
         David Munrow.
              +Gr 7-73 p222   ++RR 8-73 p67
              ++Gr 7-76 p230 tape   ++RR 12-76 p106 tape
              ++HFN 7-76 p105 tape
CSD 3748. Tape (c) TC CSD 3748
3061 Favourite operetta duets: KALMAN: Die Czardasfürstin: Tanzen mocht'
         ich; Tausend kleine Engel singen; Sich verlieben kann man öfters;
         Mädchen gibt es wunderfeine. Grafin Maritza: Mein lieber
         Schatz. LEHAR: Das Land des Lächelns: Wer hat die Liebe uns
         ins Herz gesenkt; Bei einem Tee en deux. Giuditta: Schön wie
         die blaue Sommernacht. Die lustige Witwe: Lippen schweigen.
         MILLOCKER: Die Dubarry: Es lockt die Nacht. STOLZ: Frühling
         im Prater: Du bist auf dieser Welt. Zwei Herzen im Dreivier-
         teltakt: Zwei Herzen im Dreivierteltakt. STRAUSS, J. II:
         Casanova: Ich steh' zu dir; Ich hab' dich lieb. Wiener Blut:
         Ich bin ein echtes Wiener Blut. Anneliese Rothenberger, s;
         Nicolai Gedda, t; Graunke Symphony Orchestra; Willy Mattes,
         Robert Stolz.
              +HFN 8-76 p95   +RR 12-73 p45
              +-Op 5-74 p424
CSD 3756. Tape (c) TC CSD 3756
3062 BANCHIERI: Contrapunto bestiale. BENNET: Weep, o mine eyes.
         CAIMO: Mentre il cuculo. FARMER: Fair Phyllis I saw sitting
         all alone. FESTA: L'ultimo di mi maggio. GASTOLDI: Mascherata
         di cacciatori. LASSUS: Matona, mia cara. MORLEY: Now is the
         month of maying; Though Philomela lost her love. NOLA: Chi la
         gagliarda; Chi chi li chi. WEELKES: Come, sirrah Jack ho; Four
         arms, two necks, one wreathing; Hark, all ye lovely saints; The
         nightingale, the organ of delight; Since Robin Hood. WERT:
         Vezzosi augelli; Valle, che de' lamenti. WILBYE: Cruel, behold
         My heavy ending; O wretched man. ANON.: Un cavalier di Spagna.
         King's Singers.
              +Gr 11-74 p946   ++RR 10-74 p86
              ++HFN 3-76 p113 tape   ++RR 6-76 p88 tape

(Q) CSD 3766. Tape (c) TC CSD 3766
3063 CLEMENS NON PAPA: La belle margarite. DAGGERE: Downberry down.
        FARMER: A little pretty bonny lass. GIACOBBI: Exultate Deo.
        GRIEG: I laid me down to slumber, op. 30, no. 1; Kvaalin's
        Halling, op. 30, no. 4; Little Thora, op. 30, no. 3; When I
        take a stroll, op. 30, no. 6. HANDL: Jesu dulcis memoria.
        HENRY VIII, King: Pastime with good company. JANNEQUIN: Au
        joly jeu. JOSQUIN DES PRES: Baisez moi; Petite camusette.
        LEGRAND: One day (arr. Bennett). MARTIN/COULTER: Puppet on a
        string (arr. Overton). MORLEY: Shoot, false love. RIDOUT:
        The history of the flood. TOMKINS: Weep no more thou sorry
        boy. VICTORIA: Nigra sum. ANON.: Mon coeur en vous. TRAD.:
        Billy boy (arr. Langford); The mermaid (arr. Whitworth). The
        King's Singers.
              +-Audio 12-76 p93   ++HFN 6-76 p109 tape
              +-Gr 2-76 p1373   +-RR 2-76 p58
              +HFN 3-76 p91   ++RR 11-76 p109 tape
SQ CSD 3768
3064 The psalms of David: ATKINS: No. 107, O give thanks unto the Lord.
        DAVIES: No. 130, Out of the deep. GARRETT: No. 93, The Lord
        is King. GOSS (Day): No. 37, Fret not thyself: Gloria. HAWES:
        No. 45, My heart is inditing. STANFORD: No. 53: The foolish
        body. WALMISLEY: No. 49, O hear ye this, all ye people.
        WESLEY: No. 94, O Lord God, to whom vengeance belongeth.
        WILLCOCKS: No. 13, Lord, I am not high-minded. King's College
        Chapel Choir; Philip Ledger.
              +Gr 5-76 p1795   +RR 6-76 p77
              +HFN 7-76 p93
CSD 3774
3065 BRITTEN: A hymn to the virgin. HADLEY: I sing of a maiden.
        HOWELLS: A spotless rose. KIRKPATRICK: Away in a manger.
        LEIGHTON: Lully, lulla, thou little tiny child. MENDELSSOHN:
        Hark, the herald angels sing. RAVENSCROFT: Rmember, O thou
        man. WISHART: Alleluya, a new work is come on hand. TRAD.:
        Up, good Christen folk and listen. I saw three ships; O little
        town of Bethlehem; The first Nowell; O come all ye faithful
        (English). In dulci jubilo (German). Sans day carol (Cornish).
        Quelle est cette odeur agréable; Quittez pasteurs (French).
        Francis Grier, org; King's College Chapel Choir; Philip Ledger.
              +HFN 12-76 p140   +RR 12-76 p96
SLS 5019 (5)
3066 BERLIOZ: Le carnaval romain, op. 9. La damnation de Faust, op.
        24: Hungarian march. Les Troyens: Royal hunt and storm.
        BIZET: L'Arlésienne: Suites, nos. 1 and 2: Intermezzo; Faran-
        dole. BORODIN: Prince Igor: Polovtsian dances, Act 2.
        CHABRIER: Marche joyeuse. GRANADOS: Goyescas: Intermezzo.
        LISZT: Hungarian rhapsody, no. 2, G 244, C sharp minor (orch.
        Muller-Berghaus). Les préludes, G 97. LEONCAVALLO: I
        Pagliacci: Intermezzo. MASCAGNI: L'Amico Fritz: Intermezzo,
        Act 3. MUSSORGSKY: Pictures at an exhibition (orch. Ravel).
        OFFENBACH: Gaité parisiènne, excerpts (orch. Rosenthal).
        PONCHIELLI: La gioconda: Dance of the hours, Act 3. PUCCINI:
        Manon Lescaut: Intermezzo, Act 3. RESPIGHI: The pines of Rome.
        ROSSINI: Il barbiere di Siviglia: Overture. La gazza ladra:
        Overture. SCHMIDT: Notre Dame: Intermezzo. SIBELIUS: Finlandia,
        op. 26. Kuolema, op. 44: Valse triste. VERDI: La traviata:

Prelude, Act 3. WEBER: Invitation to the dance. op. 65 (orch.
Berlioz). WEINBERGER: Schwanda the bagpiper: Polka. PhO;
Herbert von Karajan.
+Gr 12-75 p1097   ++RR 12-75 p60
++HFN 2-76 p115

SLS 5022 (2). Tape (c) TC SLS 5022
3067 The art of the recorder: ARNE: As you like it: Under the green-
wood tree. BACH: Cantata, no. 208, Schafe können sicher weiden.
Cantata, no. 106: Sonatina. Magnificat, S 243, D major:
Esurientes. BARBIREAU: Een vrolic Wesen. BASTON: Concerto,
D major. BRITTEN: Scherzo. BUTTERLY: The white throated
warbler. BYRD: The leaves be green. COUPERIN, F.: Musête de
choisi. Musête de taverni. DICKINSON: Recorder music. HANDEL:
Acis and Galatea: O ruddier than the cherry. LE HEURTEUR:
Troys jeunes bourgeoises. HINDEMITH: Trio, soprano and 2 alto
recorders. HOLBORNE: The choice. Muylinda. Pavan and gal-
liard. Sic semper soleo. JACOTIN: Voyant souffrir. PURCELL:
Three parts upon a ground: Fantasia. SCHMELZER: Sonata à 7
flauti. SERMISY: Allez souspirs. Amour me voyant. VIVALDI:
Concerto, P 77, A minor. ANON.: English dance. Saltarello.
Early Music Consort; David Munrow.
+Gr 9-75 p479   +HFN 8-76 p95 tape
+Gr 12-76 p1066 tape   +MM 3-76 p43
+HFN 9-75 p91   +RR 9-75 p51

SLS 5033 (4)
3068 FRANCK: Symphonic variations, piano and orchestra. GRIEG: Concerto, piano, op. 16, A minor. LISZT: Hungarian fantasia, G
123. LITOLFF: Concerto symphonique, no. 4, op. 102, D minor:
Scherzo. MENDELSSOHN: Concerto, piano, no. 1, op. 25, G minor.
Rondo brillante, op. 29, E flat major. RACHMANINOFF: Concerto,
piano, no. 2, op. 18, C minor. SAINT-SAENS: Le carnaval des
animaux. SCHUMANN: Concerto, piano, op. 54, A minor.
TCHAIKOVSKY: Concerto, piano, no. 1, op. 23, B flat minor.
John Ogdon, Brenda Lucas, pno; PhO, NPhO, LSO, Birmingham City
Symphony Orchestra; John Barbirolli, John Pritchard, Paavo
Berglund, Aldo Ceccato, Louis Fremaux.
+-Gr 1-76 p1244   +-RR 12-75 p68
+-HFN 1-76 p123

SLS 5046 (4) (Reissues from ASD 605, 2374, 2417, 620, HQS 1286)
3069 BEETHOVEN: Septet, op. 20, E flat major. Duo, clarinet and bassoon,
no. 1, C major. BRAHMS: Quintet, clarinet, op. 115, B minor.
MOZART: Quintet, clarinet, K 581, A major. REGER: Quintet,
clarinet and strings, op. 146, A major: 2nd movement. SCHUBERT:
Octet, op. 166, D 803, F major. WEBER: Quintet, clarinet, op.
34, B flat major. Gervase de Peyer, clt; Melos Ensemble,
Members.
+Gr 6-76 p62   +RR 6-76 p62

SLS 5047 (3) (Reissues from HQS 1237, 1144, 1205, ASD 641, 2340, 2548,
CSD 3755)
3070 BACH: Chorale prelude: Singet dem Herrn, S 225. BYRD: Haec dies.
CHARPENTIER: Messe de minuet. BRITTEN: Hymn to St. Cecilia,
op. 27. Missa brevis, op. 63, D major. PALESTRINA: Haec dies.
Missa brevis. PURCELL: Te deum. SCHUTZ: Psalm, no. 150.
VAUGHAN WILLIAMS: Mass, G minor. April Cantelo, Helen Gelmar,
s; John Eaton, treble; James Bowman, c-t; Nigel Perrin, alto;
Ian Partridge, Robin Doveton, t; Christopher Keyte, David van
Asch, bs; Andrew Davis, org; King's College Chapel Choir,

Cambridge University Musical Society Chorus, Bach Choir;
Wilbraham Brass Soloists; ECO; David Willcocks.
        +Gr 8-76 p331   +RR 8-76 p76

SLS 5049 (3)
3071 The art of the Netherlands: AGRICOLA: Comme femme. BARBIREAU:
Song: Ein fröhlich wesen. BRUMEL: Songs: Du tout plongiet;
Fors seulement, l'attente. Missa et ecce terrae motus: Gloria.
Vray dieu d'amours. BUSNOIS: Songs: Fortuna desperata.
COMPERE: O bone Jesu, motet. GHISELIN: Songs: Ghy syt die
werste boven al (Verbonnet). HAYNE VON GHIZEGHEM: Songs: De
tous biens plaine; A la audienche. ISAAC, H.: Songs: Donna di
dentro di tua casa; Missa la bassadanza: Agnus Dei; A la bat-
taglia. JOSQUIN DES PRES: Songs: Allegez moy, doulce plaisant
brunette; Adieu mes amours; El grillo e buon cantore; Guillaume
se va; Scaramella va alla guerra. Motets: Benedicta es caelorum
Regina; De profundis; Inviolata, integra et casta es, Maria.
La Bernadina. La Spagna. Vive le roy. MOUTON: Nesciens
Mater virgo virum. OBRECHT: Haec deum caeli Laudemus nunc
Dominum. OCKEGHEM: Songs: Prenez sur moi; ma bouche rit.
Motets: Intemerata Dei mater. RUE: Missa Ave sanctissima
Maria: Sanctus. TINCTORIS: Missa 3 vocum: Kyrie. VERDELOT:
Ave sanctissima Maria. ANON.: La guercia; Est-il conclu par
un arrêt d'amour; Heth sold ein meisken garn om win; Lute
dances (2); Mijm morken gaf mij een jonck wifjj; Andernaken.
Early Music Consort, David Munrow.
        +Gr 11-76 p861

SLS 5057 (4)
3072 La divina: The art of Maria Callas. BELLINI: Norma: Sediziose
voci...Casta diva. Il pirata: Oh, s'io potessi...Col sorriso
d'innocenza. I puritani: O rendetemi la speme...Qui la voce.
La sonnambula: Care compagne...Come per me sereno. CHERUBINI:
Medea: Dei tuoi figli. CILEA: Adriana Lecouvreur: Respiro
appena...Io son l'umile ancella. DELIBES: Lakmé: Dov' è
l'Indiana bruna. DONIZETTI: Lucia di Lammermoor: Oh giusto
cielo; Ardon gl'incensi. MASCAGNI: Cavalleria rusticana: Voi
lo sapete. MASSENET: Le Cid: Pleurez, mes yeux. Werther: Des
cris joyeux. MEYERBEER: Dinorah: Ombra leggiera. PONCHIELLI:
La gioconda: Suicidio. PUCCINI: La bohème: Donde lieta usci.
Madama Butterfly: Con onor muore...Tu, tu, piccolo iddio.
Manon Lesaut: Sola perduta, abbandonata. Turandot: In questa
reggia. ROSSINI: Il barbiere di Siviglia: Una voce poca fa.
SPONTINI: La vestale: Tu che invoco. THOMAS: Hamlet: A vos
jeux...Partagez-vous mes fleurs...Et maintenant écoutez ma
chanson. VERDI: Attila: Oh nel fuggente nuvolo. Un ballo in
maschera: Ecco l'orrido campo. Ernani: Surta e la notte. Don
Carlo: Tu che le vanita. MacBeth: Nel di della vittoria.
Nabucco: Ben io t'invenni. Rigoletto: Gualtier Maldé...Caro
nome. Il trovatore: Timor di me...D'amor sull'ali rosee. I
vespri siciliani: Mercè, dilette amiche; Arrigo, Ah parli a un
core. Maria Callas, s; with various artists, orchestras and
conductors.
        +Gr 11-76 p878   +RR 12-76 p42

Melodiya SLS 5058 (4) (Reissues from MK D 010977, ASD 2472, Saga SIX
        5160, Parlophone PMB 1014, ASD 2447, Artia ALP 156)
3073 BARTOK: Concerto, violin, no. 1. HINDEMITH: Concerto, violin, D
flat major. KHACHATURIAN: Concerto, violin, D minor. PROKOFIEV:

Concerto, violin, no. 1, op. 19, D major. SHOSTAKOVICH: Concerto, violin, no. 1, op. 99, A minor. Concerto, violin, no. 2, op. 129, C sharp minor; SZYMANOWSKI: Concerto, violin, no. 1, op. 35. David Oistrakh, vln; MRSO, USSR Symphony Orchestra, Leningrad Philharmonic Orchestra, MPO; Gennady Rozhdestvensky, Aram Khachaturian, Kyril Kondrashin, Yevgeny Mravinsky, Kurt Sanderling.
      ++Gr 10-76 p615   ++RR 10-76 p39
      +-HFN 9-76 p117

SLS 5068 (5) (Reissues from ASD 2938, 2465, 655, 2802, 334, 2363, 2361, 3262, 33CX 1140)

3074 BRUCH: Concerto, violin, no. 1, op. 26, G minor. ELGAR: Concerto, violoncello, op. 85, E minor. GRIEG: Concerto, piano, op. 16, A minor. HAYDN: Concerto, trumpet, E flat major. LISZT: Concerto, piano, no. 1, G 124, E flat major. MENDELSSOHN: Concerto, violin, op. 64, E minor. MOZART: Concerto, horn and strings, no. 4, K 495, E flat major. Concerto, piano, no. 21, K 467, C major. RACHMANINOFF: Concerto, piano, no. 2, op. 18, C minor. RODRIGO: Concierto de Aranjuez, guitar and orchestra. TCHAIKOVSKY: Concerto, piano, no. 1, op. 23, B flat minor. John Wilbraham, tpt; Dennis Brain, hn; Daniel Barenboim, John Ogdon, Agustin Anievas, Horacio Gutierrez, pno; Jacqueline du Pré, vlc; Yehudi Menuhin, vln; Alirio Diaz, gtr; AMF, PhO, ECO, LSO, NPhO, Spanish National Orchestra; Neville Marriner, Herbert von Karajan, Daniel Barenboim, John Barbirolli, Paavo Berglund, Walter Susskind, Rafael Frühbeck de Burgos, Moshe Atzmon, André Previn.
      +-Gr 12-76 p1072   +RR 12-76 p56

EMD 5528
3075 ALLITSEN: The lute player. BLOCKLEY: List to the convent bells. BISHOP: Home, sweet home. CHERRY: Will-o-the-wisp. CLAY: I'll sing three songs of Araby. DAVIES: Creep-mouse. LODER: The diver. PARRY, J.: Flow gently, Deva. PENN: Pansy faces. STAFFORD: Watchman, what of the night. SULLIVAN: Once again. WALLACE: Maritana: Yes, let me like a soldier fall. ANON. (arr. Moffat): My snowy-breasted pearl; The wee cooper o' Fyfe. Robert Tear, t; Benjamin Luxon, bar; André Previn, pno.
      +HFN 6-76 p87   +-RR 6-76 p80

SQ ESD 7001
3076 DVORAK: Nocturne, op. 40, B major. GRIEG: Norwegian melodies, op. 63: In popular folk style; Cowkeeper's tune and country dance. NIELSEN: Little suite, op. 1, A minor. TCHAIKOVSKY: Elegy, G major. WIREN: Serenade, strings, op. 11. Bournemouth Sinfonietta; Kenneth Montgomery.
      +Gr 7-76 p187   +RR 7-76 p61
      +HFN 7-76 p87

ESD 7002. Tape (c) TC ESD 7002 (Reissue from CSD 1542)
3077 ARNE: Where the bee sucks. BOYCE: Heart of oak. DAVY: The Bay of Biscay. HORN: Cherry ripe. MORLEY: It was a lover and his lass. OXENFORD: The ash grove. STEVENS: Sigh no more, ladies. TRAD.: A hunting we will go. The bailiff's daughter of Islington. Charlie is my darling. Early one morning. John Peel. The keel row. The miller of Dee. Oh, the oak and the ash. The vicar of Bray. Ye banks and braes. Elizabeth Harwood, s; Owen Brannigan, bs; Hendon Grammar School Choir; Pro Arte Orchestra; Charles Mackerras.

```
            ++Gr 9-76 p483              +RR 7-76 p74
            +HFN 7-76 p103
```
ESD 7010 (Reissue from Columbia Studio 2 TWO 190)
3078 Overtures: HEROLD: Zampa. MENDELSSOHN: Ruy blas, op. 95. REZNICEK:
      Donna Diana. SUPPE: Light cavalry. THOMAS: Mignon. VERDI:
      Nabucco. Royal Liverpool Philharmonic Orchestra; Charles
      Groves.
```
            ++Gr 11-76 p888            ++RR 11-76 p83
            +HFN 11-76 p173
```
ESD 7011. Tape (c) TC ESD 7011 (Reissues from ASD 3131, 2960, 2784,
      3002, 2894, 2990, 2784, SLS 864)
3079 ALBINONI: Adagio, G minor. BEETHOVEN: The creatures of Prometheus,
      op. 43: Overture. ENESCO: Rumanian rhapsody, op. 11, no. 1,
      A minor. HOLST: The planets, op. 32: Jupiter, the bringer of
      jollity. PROKOFIEV: Romeo and Juliet, op. 64, excerpts.
      TCHAIKOVSKY: Marche slav, op. 31. VAUGHAN WILLIAMS: Fantasia
      on "Greensleeves". WALTON: Portsmouth Point overture. LSO;
      André Previn.
```
            +-Gr 11-76 p888            +RR 11-76 p50
```
HLM 7077 (Reissues from DB 3506, 6158, 2856, 6139, 6178, 1501, DA 1861,
      1281, 1499)
3080 BARTOK: Rumanian folk dances, nos. 1-6 (arr. Székely). BEETHOVEN:
      Rondo, G major. CORELLI: Sonata, violin, no. 12, D minor.
      DEBUSSY: Préludes, Bk I, no. 8: La fille aux cheveux de lin
      (arr. Hartmann). DVORAK: Gypsy songs, op. 55, no. 4: Songs my
      mother taught me (arr. Persinger). KREISLER: Tambourin chinois,
      op. 3. MOZART: Divertimento, no. 17, K 334, D major: Minuet.
      PAGANINI: Caprice, op. 1, no. 24, A minor (arr. Kreisler).
      RAVEL: Mélodies hébraïques: Kaddisch (arr. Garban). SARASATE:
      Caprice basque, op. 24. SCHUBERT: Ave Maria, D 839 (arr.
      Menuhin). Yehudi Menuhin, vln; Marcel Gazelle, Hubert Giesen,
      Gerald Moore, Hephzibah Menuhin, pno.
```
            +-Gr 4-76 p1617           +RR 6-76 p70
            +HFN 5-76 p102
```
HLM 7093 (Reissues from DB 1936, 1083, Columbia 7146/50, 2EA 4573,
      DB 3711, OEA 2182, DB 1851, ALP 1532, HQS 1091, Columbia
      DX 1912, Columbia LX 57, B 9412, B 9035, DA 1627, CSD 1419,
      DB 6371)
3081 BACH: Cantata, no. 147: Jesu, joy of man's desiring (arr. Hess).
      Partita, violin, no. 3, S 1006, E major: Gavotte and rondo.
      BLISS: Welcome the Queen. BRUCH: Kol Nidrei, op. 47. CHOPIN:
      Nocturne, op. 15, no. 2, F sharp major. DELIUS: Irmelin:
      Prelude. DUNHILL: The cloths of heaven. ELGAR: Pomp and cir-
      cumstance march, op. 39, no. 4, G major. KREISLER: Schön
      Rosmarin. LISZT: Années de pelerinage, 1st year, G 160: Au
      bord d'une source. MOERAN: Diaphenia; Sweet o' the year.
      MUNRO: My lovely Celia. SCHUBERT: Die Forelle, D 550.
      SENAILLE: Entrée et cotillon (arr. Moffat). VAUGHAN WILLIAMS:
      On Wenlock edge: Is my team ploughing. Isobel Baillie, s;
      Janet Baker, ms; Gervase Elwes, John McCormack, Heddle Nash, t;
      Guilhermina Suggia, vlc; Frederick Kiddle, Ignace Jan Paderewski,
      Edwin Schneider, Solomon, Gerald Moore, Myra Hess, Franz Rupp,
      Ivor Newton, pno; Yehudi Menuhin, Fritz Kreisler, vln; Martin
      Isepp, hpd; Ambrose Gauntlett, vla da gamba; Leon Goossens, ob;
      BBC Symphony Orchestra, London String Quartet, PhO; RPO; Edward
      Elgar, Arthur Bliss, Thomas Beecham.

```
          +Gr 8-76 p306              +RR 8-76 p35
          +HFN 10-76 p183
```

SXLP 30181 (Reissues from CSD 1624, 3641, 3563)
3082 ARENSKY (arr. Tillett): Suite, 2 pianos, op. 15: Waltz. BACH
      (arr. Howe/Miles): Cantata, no. 208, Sheep may safely graze.
      BENJAMIN: Jamaican rumba (arr. White). Mattie rag (arr. Lane).
      BIZET (arr. Rachmaninoff/Lane): L'Arlésienne: Suite, no. 1:
      Minuet. BRAHMS (arr. Lane): Hungarian dance, no. 1, G minor.
      GRAINGER (arr. Farrar): Handel in the Strand. GRIEG (arr.
      Odom): Lyric pieces, op. 65, no. 6: Wedding day at Troldhaugen.
      POULENC (arr. Smithe): L'Embarquement pour Cythère. SAINT-SAENS:
      The carnival of the animals. SCHUBERT (arr. Odom): Marche
      militaire, no. 1, D 733, D major. Cyril Smith, Phyllis
      Sellick, pno.
          +-Gr 2-76 p1359   +-RR 2-76 p36
          +HFN 2-76 p117

SXLP 30188
3083 DVORAK: Humoresque, op. 101, no. 7, G flat major. FIBICH: Poem,
      no. 14. GLUCK: Orfeo ed Euridice: Mélodie. HANDEL: Largo (arr.
      Bezrukov). KREISLER: Variations on a theme by Corelli in the
      style of Tartini. MASSENET: Thais: Méditation. NOVACEK: Per-
      petuum mobile. PAGANINI: Moto perpetuo, op. 11. RACHMANINOFF:
      Vocalise, op. 34, no. 14. RUBINSTEIN: Melody, op. 3, no. 1,
      F major. SHOSTAKOVICH: The gadfly, op. 97a: Romance. SIBELIUS:
      Kuolema, op. 44: Valse triste. Bolshoi Theatre Orchestra,
      Violins; Yuli Reyentovich.
          +Gr 1-76 p1248   -RR 1-76 p36
          +-HFN 1-76 p123

SXLP 30205 (Reissue from ASD 302)
3084 DVORAK: Rusalka, op. 114: Gods of the lake. GIORDANO: Andrea
      Chénier: La mamma reggia. MASCAGNI: Cavalleria rusticana:
      Regina coeli. PUCCINI: Madama Butterfly: Ancora un passo or
      via. Manon Lescaut: Sola perduta, abbandonata. Turandot: In
      questa reggia. SMETANA: The bartered bride: Ah, bitterness.
      Dalibor: Do I live. TCHAIKOVSKY: Eugene Onegin, op. 24: Oh,
      what shall I do now. The Queen of Spades, op. 68: Twill soon
      be midnight now. Joan Hammond, s; PhO; Walter Susskind.
          +-Gr 2-76 p1374   +-RR 1-76 p26
          +-HFN 1-76 p123

SXLP 50017
3085 BRAHMS: German requiem, op. 45: How lovely are thy dwellings fair
      (ed West). CAMSEY: Benediction; Sing hallelujah, shout halle-
      lujah. COLE: The reason. FEATHERSTONE-CATELINET: My Jesus I
      love Thee. FRENCH-COLES: Why hang your harp on the willow.
      GUILMANT: Concert piece, trombone (arr. Steadman-Allen).
      HERIKSTAD: Heaven came down: March. HAYDN: The creation: The
      heavens are telling. REDHEAD: Shout salvation: Suite. SAINT-
      SAENS: Marche militaire française (arr. Kenyon). WEBB: The
      shepherds' song. International Staff Band; Salvation Army
      South London Chorus; Bernard Adams, cond; Gordon Hill, trom;
      C. Cole, cor.
          +Gr 1-76 p1247
```

<div align="center">HUNGAROTON</div>

SLPX 11444
3086  BORODIN: Prince Igor: How goes it, Prince.  HAYDN: La infedeltà
```

delusa: Non v'e rimedio. MOZART: The abduction from the
seraglio, K 384: Wer ein Liebchen hat gefunden; Ha, wie will
ich triumphieren. Don Giovanni, K 527: Madamina. Die Zauber-
flöte, K 620: O Isis und Osiris; In diesen heil'gen Hallen.
ROSSINI: Il barbiere di Siviglia: La calunnia. STRAUSS, R.:
Der Rosenkavalier, op. 59: Finale, Act 2. VERDI: Don Carlo:
Ella giammai m'amo. Simon Boccanegra: Il lacerato spirito.
Mihály Székely, bs; Various Orchestras; János Ferencsik,
Lamberto Gardelli, Pál Varga, Vilmos Komor.
    +NR 8-76 p8   +St 9-76 p130

SLPX 11669/70
3087 ATTAINGNANT: Au joly bois. Content desire. C'est grand plaisir,
phantasia. AZZAIOLO: Al di dolce. Chi passa. BANCHIERI:
Rostiva i corni. CABEZON: La gamba, pavan. CASTRO: Enfans a
laborder. CAVALIERI: O che nuovo. CLEMENS NON PAPA: Au joly
bois. Je prens en gré, à 3. CRECQUILLON: Content desire. Un
gay bergier. DOWLAND: Flow my teares. The King of Denmark's
galliard. GABRIELI, A.: Pour ung plaisir, phantasia. GOMBERT:
Je suis trop ionette. JANNEQUIN: De son amour. J'ai trop
soudainement aymé. PALESTRINA: Vestiva i colli. PHALESE: Au
joly bois. La battaglia. La rocca el Fuso. SCHEIDT: Galliard
(9 variationen). Galliard (John Dowland, 10 variations).
SERMISY: C'est a grand tort. SUSATO: La battaglia. C'est a
grand tort. SWEELINCK: Balleto del granduca. Paduana lachrimae. WAISSELIUS: La battaglia, phantasia. Gagliarda: Chi
passa; La gamba; La rocca del Fuso; La traditora. Je prens
en gré, phantasia. Un gay bergier, phantasia. WILLAERT: Allons,
allons gay. ANON.: C'est grand plaisir, chanson à 3. Gagliarda: La rocca el Fuso; La traditora. Doulce memoir. Doulce
memoir, phantasia. Pour ung plaisir. Camerata Hungarica;
Primavera Quintet; Laszlo Czidra.
    +HFN 1-76 p104   +RR 7-76 p79
    +NR 11-75 p11

SLPX 11696
3088 BALAZS: Borsod song; The song about melody. BARDOS: Gay melodies;
Uszküdárá; The youth of March. BARTOK: Girls mocking song; Lads
mocking song; Letter to the folks at home. COPLAND: Ching-a-ring
chaw. KOCSAR: The dawn awakens. KODALY: Dance song; Egyetem
Begyetem; Italian madrigal, no. 3; László Lengyel; Peal of bells;
Stork song. MARENZIO: Spring. RAIMON: There will come a time
for light. SCHUBERT: Night music. SZERVANSZKY: The river has
overflown. VECCHI: This song out for you. ANON.: Russian folk
songs (2). Singing Youth Movement Prize-winning Choirs, 1973.
    +-NR 9-76 p8

SLPX 11713
3089 BERG: Songs: Frühe Lieder (7). Lieder, op. 2 (4). KAPR: Studies,
soprano, flute and harp. KADOSA: Verseire, op. 68. SEIBER:
Morgenstern Lieder. SOPRONI: Songs: Ejszaka, Dicseret, Sötet
lett. WEBERN: Songs: Lieder, op. 12 and op. 25. Erika Sziklay,
s; Instrumental accompaniments.
    +NR 9-76 p10

SLPX 11721
3090 Central European lute music, 16th and 17th centuries. BAKFARK:
Fantasia (after "D'amour me plains" by Roger). CATO: Fantasia.
Favorito. Villanella. CRAUS: Tantz, Hupff auff. Chorea, Auff
und nider. Die trunke pinter. DLUGORAI: Chorea polonica.

Fantasia. Finale. Villanella polonica. HECKEL: Ein ungar-
scher Tantz, Proportz auff den ungarischen Tantz. NEUSIDLER:
Ein guter Venezianer Tantz. Hie folget ein welscher Tantz
Wascha Mesa, Der hupff auf. Der Juden Tantz, Der hupff auf zur
Juden Tantz. Der polnisch Tantz, Der hupff auf. POLLONOIS:
Courante. STOBAEUS: Alia Chorea polonica. WAISSELIUS: Polon-
ischer Tantz. ANON.: Danza (ed. Chilesotti). Almande de
Ungrie (ed. Phalese). Batori Tantz, Proportio. Tantz, Pro-
portion (2). Paduana Hispanica. Lengyel tánc. Psalmus CXXX.
Andras Kecskes, lt; Anges Meth, tabor, timbrel.
    ++NR 4-76 p14   ++SFC 8-22-76 p38
    +RR 3-76 p56

SXLP 11741
3091 BULL: Fantasia, D minor. COUPERIN, L.: Suite, A minor. FARNABY:
     Fantasia. FRESCOBALDI: Toccata IX. Galliards (5). FROBERGER:
     Suite, no. 19, C minor. Zsuzsa Pertis, hpd.
     +NR 7-76 p13   +-RR 3-76 p60

SXLP 11748
3092 DEBUSSY: Petite pièce, B flat major. Rhapsody, clarinet and
     orchestra. HONEGGER: Sonatina, clarinet and piano, A major.
     STRAUSS, R.: Concertino, clarinet, bassoon, harp and string
     orchestra. STRAVINSKY: Pieces, solo clarinet (3). WEINER:
     Ballade, clarinet and piano, op. 8. Kálmán Berkes, clt; Tibor
     Fülemile, bsn; Zoltán Kocsis, pno; Budapest Philharmonic Orch-
     estra; András Kórodi.
     +-Gr 4-76 p1612   +-RR 3-76 p60
     ++NR 3-76 p6   +SFC 4-25-76 p30

SXLP 11762
3093 BALAZS: Deceptive enticing; Song about the song. BARDOS: Jocose
     marrying off; Hey put it right. BARTOK: Bread breaking; Pillow
     dance. DEBUSSY: Noël des enfants qui n'ont plus de maisons.
     DOWLAND: Precious moment. FARKAS: Tillio-lio. KARAI: Summer
     night. KODALY: Egyetem Begyeterm; Gömör District folk songs
     (3). LANG: Come Sprintode. LISZT: The legend of St. Eliza-
     beth, G 2, excerpt. SMETANA: The bartered bride: Wedding
     chorus. SOKOLOV: Russian folk song suite. STRAVINSKY: Podb-
     lioudnuia. SUGAR (SZABO): Choral studies: 2 songs. VASARHELYI:
     Folk song suite, no. 2, excerpt. VASS: Nocturne. Singing
     Youth.
     +NR 4-76 p9   +RR 3-76 p67

SXLP 16570
3094 Swabian wind music in Hungary: Auf Wiederseh'n Polka; Dorfs Mädl
     Ländler; Gruss und Kuss Polka; Gruss aus Dunaharaszti Polka;
     Gut Morgen Polka; Immer Frisch Schnell Polka; Langweile Mazurka;
     Maria Walzer Potpourri; Schneeglöcklein Mazurka; Schöne Mietzl
     Polka; Stieglitz Walzer; Tanzabend Ländler. Budapest Blas-
     orchester; Rudolf Borst.
     +NR 12-76 p15

IKON

IKO 2
3095 Russian orthodox church music. ASTAFEV: Paschal Troparion (Christ
     is risen). BORTNIANSKY: I shall praise the name of God; O
     come, let us bless the ever-memorable Joseph. CHESNOKOV:
     Paschal canon (Hymns to the Mother of God); Troparia; Great

compline (God is with us). KASTAISKY: From the rising of the
sun unto the going down thereof. Vespers (O joyful light).
NIKOLSKY: Paschal canon: Selected irmoi. SMOLENSKY: Great
Prokimenon (Who is so great a God as our God); Paschal Mattins:
Stikhera; Znamenny chant (arr. Smolensky). TCHAIKOVSKY: Lit-
urgy of St. John Chrysostom: Hymn to the Mother of God.
TURCHANINOV: Arise, O God; Let all mortal flesh keep silence.
TRAD.: Easter Mattins processional hymn. Ludmila Shishkova,
Vera Arkhangelskaia, s; Alevtina Filatova, ms; A. Rudkovsky,
t; Moscow Conservatoire Mixed Choir; Nicholas V. Matveev.
     +-Gr 3-76 p1504   +RR 11-76 p101

IKO 3
3096 Russian orthodox church music (Christmas): ARKANGELSKII: Nunc
    Dimittis; Polveleos. BORTNIANSKY: Nativity Kontakion. LVOVSKY:
    Gospel chant. TURCHANINOV: Litany at the Lity. ANON. (Christ-
    mas vigil): Gradual chant; God is with us (arr. Kastalskiy).
    ANON. (Ukrainian carols): Across the wide fields; The bells
    of Jerusalem; Chedrick; Early, very early; A falcon; Good
    evening to you; Heaven and earth; O God sees; O the new joy;
    On Jordan's banks; On the Jordan; Oh yes, here is a splendid
    host. TRAD.: From the Lity. USSR Russian Chorus; Feodor
    Potorjinsky.
       +-Gr 12-76 p1045   +RR 11-76 p101

IKO 4
3097 Orthodox church music from Finland: Excerpts from the Moleben to
    the holy mother of God and SS Sergius and Herman of Valamo;
    selected hymns and chants. Hymnodia Choir; Archbishop Paul of
    Karelia and All Finland.
       +RR 11-76 p101

INTERNATIONAL PIANO LIBRARY

IPL 102
3098 BUSONI: Sonatina, no. 2. CASELLA: Contrasts, op. 31 (2). CHAB-
    RIER: Bourrée fantasque. DUSSEK: La chasse. FIELD: Nocturne,
    no. 9, E minor. GODOWSKY: The gardens of Buitenzorg. JENSEN:
    Erotikon, op. 44: Eros. MACDOWELL: To a wild rose. MOSZKOWSKI:
    Valse, op. 34. PADEREWSKI: Legende, op. 16, no. 1, A flat
    major. RAFF: Rigaudon, op. 204, no. 3. RAVINA: Etude de style.
    REGER: Aus meinem Tagebuche: Adagio and vivace, op. 82. RUBIN-
    STEIN: Prelude and fugue, op. 53, no. 2. Arthur Loesser, pno.
       +-Gr 10-72 p764   ++St 12-76 p62
       +NR 6-73 p12

IPL 104
3099 BACH: Prelude and fugue, S 545, C major. Chorale prelude: Rejoice,
    beloved Christians, S 734 (arr. Busoni). Partita, violin, no.
    2, S 1004, D minor: Chaconne (arr. Busoni). BEETHOVEN: Eccos-
    saisses (arr. Busoni). BIZET: Carmen fantasy, op. 25 (arr.
    Busoni). BUSONI: Indian diary, Bk 1. CHOPIN: Etude, op. 10,
    no. 5, G flat major (2 versions). Etude, op. 25, no. 5, E
    minor. Nocturne, op. 15, no. 2, F sharp major. Prelude, op.
    28, no. 7, A major. LISZT: Hungarian rhapsody, no. 13, G 244,
    A minor (abbreviated). Ferruccio Busoni, Michael von Zadora,
    Egon Petri, Edward Weiss, pno.
       +-Gr 10-72 p764   +St 12-76 p62
       +NR 5-73 p12

IPL 5001/2 (2)
3100 BEETHOVEN (Rubinstein): The ruins of Athens, op. 113: Turkish
      march. BRAHMS: Academic festival overture, op. 80. CHOPIN:
      Andante spianato and grande polonaise, op. 22, E flat major.
      Ballade, no. 1, op. 23, G minor. Berceuse, op. 57, D flat
      major. Etude, op. 25, no. 9, G flat major. Nocturne, op. 9,
      no. 2, E flat major. Nocturne, op. 15, no. 2, F sharp major.
      Waltz, op. 42, A flat major. Waltz, op. 64, no. 2, C sharp
      minor. HOFMANN: Chromaticon, pno and orchestra. MENDELSSOHN:
      Song without words, op. 67, no. 4, C major: Spinning song.
      MOSZKOWSKI: Caprice espagnole, op. 37. RACHMANINOFF: Prelude,
      op. 23, no. 5, G minor. RUBINSTEIN: Concerto, piano, no. 4,
      op. 70, D minor. Josef Hofmann, pno; Curtis Institute Symphony
      Orchestra; Fritz Reiner.
            ++Gr 9-76 p453   ++St 12-71 p91
            ++NR 8-73 p14

IPL 5003/4 (2)
3101 BEETHOVEN: Variations on "Kind, willst du ruhig schlafen". CHOPIN:
      Mazurka, op. 59, no. 3, F sharp minor. Nocturne, op. 9, no.
      3, B major. Variations brillantes on a melody from the opera
      Ludovic, op. 12. HAYDN: Sonata, piano, no. 42, D major.
      MOZART: Adagio, K 540, B minor. Eine kleine Gigue, K 574, G
      major. PROKOFIEV: Sonata, piano, no. 5, op. 38, C major.
      SCHUBERT: Sonata, piano, no. 16, op. 42, D 845, A minor.
      SCHUMANN: Sonata, piano, no. 3, op. 14, F minor. Arthur Loes-
      ser, pno.
            +-ARSC Vol 6, no. 2   *NR 5-74 p12
                  p71   +-RR 9-76 p82
            +-Gr 9-74 p558   +St 3-74 p125
            +HF 4-74 p114

IPA 5007/8 (2)
3102 BEETHOVEN: Sonata, piano, no. 21, op. 53, C major. CHOPIN: Bal-
      lade, no. 4, op. 52, F minor. Nocturne, op. 9, no. 3, B major.
      Polonaise, op. 26, no. 2, E flat minor. Waltz, op. 18, E flat
      major. Waltz, op. 64, no. 1, D flat major. HOFMANN: Kaleide-
      skop. SCHUBERT (Godowsky): Moment musical, op. 94, no. 3, D
      780, F minor. SCHUMANN: Kreisleriana, op. 16. STOKOWSKI:
      Caprice oriental. Josef Hofmann, pno.
            +-HF 3-76 p99   +St 12-76 p62
            +St 12-75 p130

                              KAIBALA

50E 01
3103 Latvian folk songs: At the edge of the garden; Come to me; The
      cuckoo is calling; A dance from Merdzene; Dressed in white;
      Evening song; Go on to God, sun; God went into a field; Herd-
      ing song; I sang, I exulted; Laima teased me; The nightingale;
      Other girls; Rise early, dear sun; The sun and the bee; Wake
      at the cowherder, mother; What a lad; When I went dancing; Where
      did you find her; The white goat; Who goes there. Silvija
      Erdmane, s; Martins Aldins, t.
            +NR 12-76 p12

KLAVIER   614

KLAVIER

KS 551
3104 BACH: Cantata, no. 156: Sinfonia. Concerto, organ, A minor.
      Prelude and fugue, S 541, G major. CLARKE: Prince of Den-
      mark's march. GERVAISE: Dances from the French Renaissance (4)
      MOURET: Suite of symphonies: 1st suite. PURCELL: Trumpet tune.
      TELEMANN: Heroic music. VIVIANI: Sonata prima per tromba e
      organ. ANON.: Hornpipe. Richard Morris, org; Martin Berin-
      baum, tpt.
           +NR 9-76 p7

LAUREL-PROTONE

LP L3
3105 BOND: Sonata, violoncello. CASSADO: Requiebros. MANZIARLY:
      Dialogue. NIN: Spanish suite. PONCE: Preludes (3). VILLA-
      LOBOS: Song of the black swan. WEBERN: Sonata, violoncello.
      Gilberto Munguia, vlc; Thomas Hrynkiv, pno.
           +NR 11-76 p14   ++SFC 12-5-76 p58

LIBRARY OF CONGRESS

OMP 101/2 (2)
3106 Our musical past: A concert for brass band, voice and piano.
      BALFE: The bohemian girl; The heart bowed down. BENEDICT: The
      rose of Erin; Hungers' chorus. DONIZETTI: Don Pasquale: O
      summer night. DOWNING (arr.): Free and easy. FARMER: Moon-
      beam waltzes (arr. Downing). FOSTER: Ah, may the red rose live
      alway; Old memories; Why, no one to love. FRIEDERICH: Lilly
      Belle quickstep. GOODWIN: Door latch quickstep. GRAFULLA:
      Captain Finch's quickstep; Captain Shepherd's quickstep.
      JAEGER: Indiana polka (arr. Schatzman). LINDBLAD: The herds-
      man's mountain song; Upon a summer day. LYSBERG: La fontaine.
      KNAEBLE: General Taylor storming Monterey. NOEREN: Midnight
      (arr. Schatzman). WALLACE: Scots wha hae variations. Merja
      Sargon, s; Bernard Rose, pno; Brass Band; Frederick Fennell.
           +-HF 10-76 p132   +-St 11-76 p137

LONDON

CSA 2246 (2)
3107 BERNSTEIN: Candide: Overture. COPLAND: Appalachian spring.
      GERSHWIN: An American in Paris. IVES: Symphony, no. 2. Holi-
      days: Decoration day. Variations on "America". SCHUMAN:
      Variations on "America" (after Charles Ives). LAPO; Zubin
      Mehta.
           +NR 9-76 p2   ++SFC 8-8-76 p38
CS 6953. (also Decca SXL 6734 Tape (c) KSXC 6734)
3108 ALBENIZ, I.: España, op. 165, no. 2: Tango; no. 3: Malagueña.
      Pavana-capricho, op. 12. Recuerdos de viaje, op. 71, no. 5:
      Puerta de tierra; no. 6: Rumores de la caleta. Suite española,
      no. 3: Sevillanas. ALBENIZ, M.: Sonata, piano, D major.
      GRANADOS: Danza española, op. 37: Andaluza; Valenciana. SOLER:
      Sonata, piano, G minor. Sonata, piano, D major. TURINA: Sacro-
      monte, op. 55, no. 5. Zapateado, op. 8, no. 3. Alicia de
      Larrocha, pno.

```
    +-Gr 6-76 p70           +NR 11-76 p14
    ++HF 10-76 p130         +RR 6-76 p66
    ++HFN 6-76 p85          +SFC 7-11-76 p13
    +MJ 10-76 p52
```

CS 7015
3109 ALBENIZ: Recuerdos de viaje, op. 71: Rumores de la caleta (arr.
        Azpiazu). Zambra granadina (trans. Segovia). BACH: Prelude,
        A minor. Gavottes, nos. 1 and 2. Presto, A minor (arr. Kry-
        tiuk). CALLEJA: Canción triste. DEBUSSY: Preludes, Bk 1, no.
        8: La fille aux cheveux de lin (trans. Bream). GUIMARAES:
        Sounds of bells. SAGRERAS: El colibri. SAINZ DE LA MAZA:
        Campanas del Alba. SCARLATTI: Sonata, guitar, A minor (trans.
        Krytiuk). Sonata, A major (trans. Lima). TOMASI: Le muletier
        des Andes. Liona Boyd, gtr.
            ++NR 12-76 p14   +SFC 9-19-76 p33
STS 15222 (also Decca ECS 631)
3110 BACH: Chorale preludes: Wachet auf, ruft uns die Stimme, S 645;
        Ach bleib bei uns, S 649. Fantasia and fugue, S 542, G minor.
        BUXTEHUDE: Prelude and fugue, D minor. ROPEK: Variations on
        "Victimae Paschali Laudes". TELEMANN: Fantasia, E major.
        Jiří Ropek, org.
            +Gr 1-72 p1245   +-NR 3-76 p11
            ++HFN 1-72 p102
STS 15239
3111 ALBENIZ: España, op. 165, no. 2: Tango. BAZZINI: La ronde des
        lutins, op. 25. BRAHMS: Waltz, op. 39, no. 15, A flat major.
        BACH: Arioso. PAGANINI: La campanella. Caprices, op. 1, nos.
        13, 20 (arr. Kreisler). MOZART: Serenade, no. 7, K 250, D
        major: Rondo. SCHUBERT: Ave Maria, D 839. YAMADA: Akatonbo
        (arr. Campoli). Jogashima no ame (arr. Campoli). Alfredo
        Campoli, vln; Norhiko Wada, pno.
            +NR 8-76 p13
STS 15306
3112 MALATS: Impresiones de España: Serenata española. NARVAEZ: Vari-
        ations on "Guardame las vacas". PUJOL VILARRUBI: El abejorro.
        PIPO: Canción y danza. SAINZ DE LA MAZA: Habañera. SANZ: Suite
        española. SAVIO: Escenas brasilenas. SOR: Minuet, D major.
        Study, E minor. TARREGA: Gran jota. Prelude, G major. VILLA-
        LOBOS: Prelude, no. 3. Etude, no. 1. YEPES: Danza inca. Nar-
        ciso Yepes, gtr.
            +NR 7-76 p13
OS 26248 (Reissue from OSA 1292)
3113 CHRISTINE: Phi-Phi: Ah, cher monsieur, excusez-moi. HAHN: Cibou-
        lette: Moi je m'appelle, Y'a des arbres...C'est sa banlieue.
        MESSAGER: L'Amour masque: J'ai deux amants. OFFENBACH: La
        belle Hélène: Dis-moi Venus. La grande Duchesse de Gérolstein:
        Portez armes...J'aime les militaries. La périchole: Tu n'est
        pas beau...Je t'adore, Air de lettre, Ah, quel diner. STRAUSS,
        O.: Valses: Saison d'amour, Je ne suis pas, Je t'aime. Regine
        Crespin, s; OSR, Vienna Volksoper; Alain Lombard, Georges
        Sebastian.
            +NR 1-76 p10   ++SFC 3-7-76 p27
CS 26398
3114 The Volga boatmen and other Russian favorites: Along Petersburg
        Street; The cliff; Bandura; Dark eyes; Dear little night;
        Farewell, joy; In the dark forest; The little oak cudgel; Song

of the Volga boatmen; Styen'ka razin; The twelve brigands.
Nicolai Ghiaurov, bs; Kaval Balalaika Orchestra and Chorus;
Atanas Margaritov.
+HF 3-76 p99   +-ON 1-24-76 p51
+NR 2-76 p13

OS 26435. Tape (c) OS 5-26435 (also Decca SXLR 6792)
3115 Music of Spain, Zarzuela arias. BARBIERI: El barberillo de
lavapiés; Canción de paloma; Jugar con fuego; Romanza de la
Duquesa. CABALLERO: Chateau Margaux: Romanza de Angelita.
Gigantes y cabezudas: Romanza de Pilar. El señor Joaquin:
Balada y alborada. CHAPI Y LORENTE: Las hijas del Zebedeo:
Carceleras. La patria chica: Canción de pastora. GIMENEZ:
El barbero de Sevilla: Me llaman la primorosa. LUNA: El niño
Judio: De España vengo. SERRANO Y RUIZ: El carro del sol:
Canción veneciana. Montserrat Caballé, s; Barcelona Orquesta
Sinfonica; Eugenio Marco.
+ARG 11-76 p49   +RR 11-76 p40
+-Gr 11-76 p877   +SR 11-13-76 p52
+HFN 11-76 p165   ++St 11-76 p88
+NR 12-76 p11

LOUISVILLE

LS 753/4
3116 BIRD: Carnival scene. CHADWICK: Euterpe. CONVERSE: Endymion's
narrative, op. 10. Flivver ten million. FOOTE: Francesca da
Rimini. ORNSTEIN: Nocturne and dance of the fates. Louisville Orchestra; Jorge Mester.
+HF 12-76 p120   +MJ 11-76 p44

MERCURY

SRI 75055. (Reissue from SR 90170)
3117 March time: ALFORD: The mad major. FILLMORE: American we. GOLD-
MAN: Bugles and drums. Boy Scouts of America. Children's
march. Illinois march. The Interlochen bowl. Onward-upward.
HALL: Officer of the day. REEVES: Second Regiment Connecticut
National Guard march. RODGERS: Victory at sea: Guadalcanal
march. SEITZ: March grandioso. Eastman Wind Ensemble; Frederick Fennell.
+NR 5-76 p14

MICHIGAN UNIVERSITY

SM 0002
3118 200 years of American marches: BAGLEY: National emblem. BELSTER-
LING: March of the steel men. BILLINGS: Chester. CHAMBERS:
The boys of the old brigade. FARRAR: Bombasto. FILLMORE:
American we. GRAFULLA (Reeves): Washington greys. HUFFER:
Black Jack. HOLLOWAY: Wood-up quick-step. KLOHR: The bill-
board. KING: Hosts of freedom. PANELLA: On the square. PHILE:
Hail, Columbia. SCHREINER: General Lee's grand march. SOUSA:
El Capitan. WELDON: Gate city. TRAD.: The rose tree. The
roving sailor. The white cockade. The world turned upside

down. Yankee Doodle. Michigan School of Music Winds and
Percussion; Clifford P. Lillya.
       +IN 2-76 p8   +St 12-75 p135
       ++MJ 3-76 p24

MONITOR RECORDS

MFS 760
3119 The massed bands, pipes and drums of the Welsh Guards and the
       Argyll and Sutherland Highlanders on tour. Desmond Walker,
       cond.
       ++IN 10-76 p18

MUSIC FOR PLEASURE

SPR 90086
3120 BINGE: Elizabethan serenade. DEBUSSY: Suite bergamasque: Clair
       de lune. GOODWIN: London serenade. Prairie serenade. Puppet
       serenade. MARTIN: Serenade to a double Scotch. WIREN: Sere-
       nade, strings. Ron Goodwin and His Orchestra.
       +Gr 2-76 p1382

MUSICAL HERITAGE SOCIETY

MHS 3276
3121 Louis Danto: None but the lonely heart. BORODIN: Listen,
       maidens, to my song. DENZA: Si vous l'aviez compris. GLINKA:
       Doubt. GODARD: Jocelyn: Berceuse. GOUNOD: Serenade. HANDEL:
       Jehova, to my words give ear. MASSENET: Elegy. RACHMANINOFF:
       O cease thy singing, maiden fair; When night descends. RIMSKY-
       KORSAKOV: The rose and the nightingale. TCHAIKOVSKY: None but
       the lonely heart, op. 6, no. 6. Louis Danto, t; Jascha Silber-
       stein, vlc; Artur Balsam, pno.
       ++St 6-76 p117

MUZA

SX 1144
3122 DVORAK: Rusalka, op. 114: O lovely moon. MONIUSZKO: Halka: Aria,
       Act 2. PUCCINI: Madama Butterfly: Un bel di vedremò. Manon
       Lescaut: Sola, perduta, abbandonata. TCHAIKOVSKY: Pique Dame,
       op. 68: Lisa's aria, Act 3. VERDI: The masked ball: Ma dell'
       arido stelo. WAGNER: Lohengrin: Einsam in trüben Tagen.
       Teresa Kubiak, s; Lodz Philharmonic Orchestra; Henryk Czyż.
       +NR 8-76 p8   +-St 8-76 p106

SXL 1170
3123 DONIZETTI: Lucia di Lammermoor: Tombe degiavi miei. GIORDANO:
       Fedoro: Amor ti vieta. MASCAGNI: Cavalleria rusticana: Mamma
       quel vino e generoso. PUCCINI: La bohème: Che gelida manina.
       Tosca: Recondita armonia, E lucevan le stelle. PONCHIELLI: La
       gioconda: Cielo e mar. TCHAIKOVSKY: Eugene Onegin, op. 24:
       Lensky's aria. VERDI: Rigoletto: La donna è mobile. ZELENSKI:
       Janek: Gdy slub wezmiersz z twoim Stachem. Paulos Raptis, t;
       Orchestra accompaniments.
       /NR 8-76 p8

NATIONAL TRUST

NT 002
3124 Music for the Vyne: ARNE: Blow, blow thou winter wind. Come away
       death. BOYCE: Trio sonata, D major. FARNABY: The old spagno-
       letta. FORD: The pill to purge milancholy. HENRY VIII, King:
       Taunder naken. HOLBORNE: The image of melancholly. JENKINS:
       Newark siege. LAWES, H.: Sweet stay awhile. LAWES, W.:
       Gather ye rosebuds. MILAN: Toda mi vida os ame. Pavana.
       MUNDY: Robin. NICHOLSON: No more, good herdsman, of thy song.
       PURCELL, H.: Distressed innocence: Rondeau; Air; Minuet. ANON
       The Dalling alman; Come holy ghost; Dances: My robbin; Tickle
       my toe; Hollis berrie. Paul Elliott, t; King's Musick.
              +-MT 10-76 p832   +RR 8-76 p74

                              NONESUCH

H 71141 (Reissue)
3125 BEETHOVEN: Country dances, op. 141 (12). HAYDN: Raccolta de
       menuetti ballabili, nos. 1, 14. LANNER: Mitternachtswalzer,
       op. 8. Regata-Galopp, op. 134. MOZART: Country dances, K
       609 (5). SCHUBERT: Minuets with 6 trios, D 89 (5). Vienna
       State Opera Orchestra; Paul Angerer.
                  +Gr 1-76 p1248   +RR 12-75 p50
                  +HFN 12-75 p171
H 71276. (Q) HQ 1276
3126 Early American vocal music--New England anthems and Southern folk
       hymns: CHAPIN: Rockbridge. BILLINGS: I am come into my gar-
       den; I am the rose of Sharon; I charge you; An anthem for
       Thanksgiving: O praise the Lord. DARE: Babylonian captivity.
       INGALLS: Northfield. LAW: Bunker Hill. LEWER: Fidelia.
       MORGAN: Judgment anthem; Amanda. READ: Newport. ROBISON:
       Fiducia. WHITE: Power. ANON.: Animation; Canaan; Concert;
       Lonsdale; Messiah; Pilgrim's farewell; Springhill; Triumph;
       Washington. The Western Vocal Wind; Mary Lesnick, ms; Raymond
       Murcell, bs-bar; Stuart Schulman, vln; Bonney McDowell, bs viol
       Paul Fleischer pic; Allen Herman, snare drum.
                  +Gr 3-75 p1711   +NR 7-73 p9
                  +HF 3-73 p100   +RR 3-75 p14
                  +HF 1-74 p84 Quad   ++SFC 2-18-73 p34
                  +LJ 5-73 p34   ++St 4-73 p80
                  +MT 8-76 p663   +St 7-76 p71
H 71301. Tape (c) D 1039
3127 BIBER, C. H.: Sonata, trumpet, C major. Sonata, 2 choirs. BIBER,
       H.: Sonata a 7, C major. GABRIELI: Sonata, trumpet. MOLTER:
       Symphony, 4 trumpets, C major. PEZEL: Sonatinas, nos. 61-62,
       65-66. RATHGEBER: Concerto, trumpet, op. 6, no. 15, E flat
       major. SCHEIDT: Canzon cornetto. Gerard Schwarz, tpt; New
       York Trumpet Ensemble; Kenneth Cooper, hpd and org.
                  ++HF 2-75 p108   ++NR 4-75 p14
                   +HF 6-76 p75 tape   +RR 4-75 p22
                  ++HFN 6-75 p99   ++SFC 3-16-75 p25
                  ++MT 9-75 p797   ++St 4-76 p118
H 71302/3
3128 Spectrum: New American music. ANDERSON: Variations on a theme by
       M. B. Tolson. BABBITT: All set. JONES: Ambiance. ROCHBERG:
       Blake songs. WERNICK: Kaddish-requiem. WOLPE: Quartet, trum-
       pet, tenor saxophone, percussion and piano. Phyllis Bryn-Julso

s; Jan DeGaetani, ms; Ramon Gilbert, cantor; Contemporary
Chamber Ensemble; Arthur Weisberg.
```
     +-Gr 4-76 p1627           +MT 3-76 p237
     ++HF 4-75 p98             ++NR 1-75 p6
     +HFN 5-76 p91             -RR 12-75 p71
     +HFN 5-76 p101            +St 6-75 p110
     +MM 7-76 p32
```
71315
129 A Medieval Christmas: Isaiah's prophecy; The Sybil's prophecy
     (Iudicii Signum); Conductus (Adest sponsuo); Gabriel's prophecy
     (Oiet, virgines); Clausula (Domino); Hymn (conditor alme sid-
     erum); Lauda (Verbum caro factum est); Reading (On frymde waes
     work); Organum (Judea et Iherusalem); Conductus (Gedonis area);
     Reading (O moder mayde); Hymn (Ave maris stella); Sacred song
     (O Maria, Deu maire); Prosa (Ave Maria); Responsory (Gaude
     Maria); Hymn (Joseph, Liber nefe min); Conductus with refrain
     (Conguadet hodie, In dulci jubilo); Reading (Hand by hand we
     shule us take); Conductus (Lux hodie...Orientis partibus).
     Boston Camerata; Joel Cohen.
```
     +HF 2-76 p115              +NR 12-75 p1
     +HFN 12-75 p157            +RR 1-76 p58
```
71326
130 The pleasures of the royal courts. Courtly art of the trouveres:
     ADAM DE LA HALLE: Fines amouretes ai; Tant con je vivrai.
     MUSET: Quant je voy yver. ANON.: Ductia; La sexte estampie
     real; Souvent souspire mon cuer. Burgungian Court of Philip
     the Good: DUFAY: Vergine bella. GULIELMUS: Falla con misuras.
     LEGRANT: Entre vous, noviaux maries. German Court of Emperor
     Maximilian I: ISAAC: Innsbruck, ich muss dich lassen. SENFL:
     Nun wöllt ihr hören neue Mar. WECK: Spanyoler Tanz und Hopper
     dancz. Italian music of the Medici Court: CARA: Non e tempo.
     FO: Tua voisi esser sempre mai. SCOTTO: O fallace speranza.
     ZESSO: E quando andarete al monte. ANON.: Polyphonic dances
     (6). Spanish Courts in the early Sixteenth Century: ALONSO:
     La tricotea Samartin. CABEZON: Diferencias sobre "La dama le
     demanda". ENCINA: Ay triste que vengo. ORTIZ: Recercadas
     primera y segunda. ANON.: Pase el agoa, ma Julieta; Rodrigo
     Martines. Early Music Consort; David Munrow.
```
     +Gr 12-76 p1042           +-RR 10-76 p91
     +HFN 11-76 p165           ++SFC 8-22-76 p38
     +NYT 8-15-76 pD15
```
B 73028 (2)
131 BERIO: Sequenza. DAVIDOVSKY: Junctures. FUKUSHIMA: Pieces from
     Chu-u. LEVY, B.: Orbs with flute. REYNOLDS: Ambages. ROUS-
     SAKIS: Short pieces, 2 flutes (6). TROMBLY: Kinetics III.
     VARESE: Density 21.5. WESTERGAARD: Divertimento on Discobbolic
     fragments. WUORINEN: Variations, flute, I and II. Harvey
     Sollberger, Sophie Sollberger, flt; Allen Blustine, clt; Jeanne
     Benjamin, vln; Charles Wuorinen, pno.
```
     ++HF 9-75 p102            +NR 9-75 p9
     +HFN 11-75 p169           +NYT 4-27-75 pD19
     +MT 1-76 p44
```

ODYSSEY

33825
132 ARENSKY: Concerto waltz. DEBUSSY: La plus que lente. DINICU

(Heifetz): Hora staccato. KREISLER: Liebesfreud. Liebesleid.
Schön Rosmarin. MIASKOVSKY: Yellowed leaves, op. 31, nos. 1
and 6. RACHMANINOFF: Vocalise, op. 34, no. 14. RAVEL: Habañera. RIMSKY-KORSAKOV: The tale of the Tsar Sultan: Flight of
the bumblebee. RUBINSTEIN: Melody, op. 3, no. 1, F major.
SARASATE: Zigeunerweise, op. 20, no. 1. SHCHEDRIN: In the
style of Albeniz. SHOSTAKOVICH: Fantastic dances. Timofey
Dokschitzer, tpt; Abram Zhak, Arnold Kaplan, pno.
      ++NR 2-76 p15

Y 34139
3133   CANN: Bonnylee. GRESSEL: Points in time. KRIEGER: Short piece.
       LANSKY: Mild und Leise. SEMEGEN: Electronic composition, no.
       1. WRIGHT: Electronic composition. ZUR: Chants.
          +HF 12-76 p126

                           L'OISEAU-LYRE
DSLO 7
3134   ALBENIZ: España, op. 165, no. 2: Tango (arr. Godowsky). CHAMINAD
       Autrefois, op. 87, A minor. GLAZUNOV: Waltz, op. 42, no. 3,
       D major. GODOWSKY: Waltz-poem, no. 4, for the left hand. Alt
       Wien. HOFMANN: Kaleideskop, op. 40. MOSZKOWSKI: Caprice
       espagnol, op. 37. RAMEAU: Tambourin (arr. Godowsky). RUBIN-
       STEIN: Melody, op. 3, no. 1, F major. SAINT-SAENS: Le carnava
       des animaux: The swan (arr. Godowsky). SCHUBERT: Moments musi
       caux, op. 94, no. 3, D 780, F minor (arr. Godowsky). STRAUSS,
       J. II: Wine, women and song, op. 333 (arr. Godowsky). TCHAI-
       KOVSKY: Nocturne, op. 19, no. 4. Shura Cherkassky, pno.
          +Gr 6-75 p75   +RR 6-75 p60
          +HF 3-76 p98   +SFC 11-16-75 p32
          ++NR 4-76 p10   +SR 1-24-76 p53

12BB 203/6 (also Decca 12BB 203/6)
3135   Musicke of sundre kindes: Renaissance secular music, 1480-1620.
       ALISON: Dolorosa pavan. ATTAINGNANT: Content desir basse dans
       AZZAIOLO: Quando le sera; Sentomi la formicula. BARBERIIS:
       Madonna, qual certezza. BOTTEGARI: Mi stare pone Totesche.
       CAURROY: Fantasia; Prince la France te veut. CAVENDISH: Wand'
       ring in this place. COMPERE: Virgo celesti. COSTELEY: Helas,
       helas, que de mal. DALZA: Recercar; Suite ferrarese; Tastar
       de corde. DUNSTABLE (Anon.): O rosa bella; Hastu mir. FAYRFAX
       I love, loved; Thatt was my woo. FONTANA: Sonata, violin.
       FORSTER: Vitrum nostrum gloriosum. FRESCOBALDI: Toccata.
       GABRIELI, A.: Canzona francese. GABRIELI, G.: Sanctus Dominus
       Deus. GESUALDO: Canzona francese; Mille volte il dir moro.
       GIBBONS: Now each flowery bank. GOMBERT: Caeciliam cantate.
       GUAMI: La brillantina. HECKEL: Mille regretz; Nach willen dei
       HOFHEIMER: Nach willen dein. HUME: Musick and mirth. ISAAC,
       A.: Ne piu bella di queste; Palle, palle; Quis dabit pacem.
       ISAAC, H.: La la hö hö. JANNEQUIN: Les cris de Paris. JEUNE:
       Fiere cruelle. JOSQUIN DES PRES: Mille regretz. LASSUS:
       Cathalina, apra finestra; Matona mia cara. MARENZIO: O voi che
       sospirate; Occhi lucenti. MERULO: Canzona françese. MODENA:
       Ricercare a 4. MONTEVERDI: Lamento d'Olimpia. MOUTON: La, la,
       la l'oysillon du bois. MUDARRA: Dulces exuyiae. NARVAEZ:
       Fantasia; Mille regret. OBRECHT: Ic draghe die mutze clutze;
       Mign morken gaf; Pater noster. ORTIZ: Dulce memoire; Recercada

PACOLINI: Padoana commun; Passamezzo commun. PORTER: Thus
sang Orpheus. RONTANI: Nerinda bella. RORE: De la belle
contrade. RUE: Pour ung jamais. SANDRIN: Doulce memoire.
SERMISY: Content desir; La, je m'y plains. TERZI: Canzona
française. TIBURTINO: Ricercare a 3. TORRE: Adormas te,
Señor; La Spagna. TROMBONCINO: Ave Maria; Hor ch'el ciel e
la terra; Ostinato vo seguire. TYE: In nomine. VASQUEZ:
Lindos ojos aveys, señora. VERDELOT: Madonna, qual certezza.
WILLAERT (arr. Cabezon): E qui, la dira; Madonna qual certezza.
ANON.: Belle, tenés mo; Calata, Celle qui m'a demande; Chui
dicese e non l'amare; Der Katzenflöte; Lady Wynkfyldes rownde;
Las, je n'ecusse; L'e pur morto Feragù; Mignonne, allons;
Pavana de la morte de la ragione; Pavana el tedescho; Pavane
venetiana; Der rather Schwanz; Le rossignol; Saltarello de la
morte de la ragione; Se mai per maraveglia; Shooting the guns
pavan; Sta notte; Suite regina; La triquottée. ANON/AZZAIOLO:
Girometa. ANON/EDWARDS: Where griping griefs. ANON/ISAAC:
Bruder Konrad. ANON/JUDENKUNIG: Christ der ist erstanden.
ANON/PACOLINI: La bella Franceschina. ANON/SPINACINO: Je ne
fay. ANON/WYATT: Blame not my lute. Consort of Musicke;
Anthony Rooley.

+-Gr 12-75 p1066	+NYT 8-15-76 pD15
++HFN 12-75 p159	+RR 12-75 p89
+MJ 11-76 p45	++SFC 9-26-76 p29
+-MM 7-76 p37	++STL 2-8-76 p36
+NR 8-76 p15	

SOL 343
3136 BEAUVARLET-CHARPENTIER: Fugue, G minor. FRANCK: Chorale, no. 2,
      B minor. JONGEN: Toccata. MANERA: Salve regina. ROBINSON:
      Improvisation on a submitted theme. McNeil Robinson, org.
          +Gr 8-76 p323   +NR 12-76 p12
          +HFN 7-76 p93

SOL 345 (Reissue from Cambrian SCLP 591)
3137 BELLINI: Almen se non poss'io. DONIZETTI: La conocchia. ROSSINI:
      L'Invito. SCHUBERT: Songs: Du bist die Ruh, D 776; Die Forelle,
      D 550; Gretchen am Spinnrade, D 118; Heidenroslein, D 257.
      VERDI: Ave Maria. TRAD.: Bugeilio'r Gwenith Gwyn; Dafydd y
      Garreg Wen; Wrth Fynd Efo Deio I Dowyn (arr. E. T. Davies); Y
      Bore Glas (arr. G. Williams); Y Deryn Pur. Margaret Price, s;
      James Lockhart, pno.
          +-Gr 4-76 p1653   +RR 8-76 p73
          +HFN 6-76 p102

                              OPUS

Unnumbered (Available from Box 95, Cathedral Station, New York 10025)
3138 Piano music of Spain: ALBENIZ: Suite española, excerpts. FALLA:
      El amor brujo: Ritual fire dance. GRANADOS: Goyescas: The maja
      and the nightingale. LECUONA: Malagueña. TURINA: Sacro-monte,
      op. 55, no. 5. Olga Llano Kuehl, pno.
          *CL 11-76 p10

                              ORION

ORS 75205
3139 American songs for A Capella Choir. ADLER: A kiss; Strings in

the earth. BARBER: Reincarnation, op. 16. BERGER, J.: Snake
baked a hoe-cake; The Frisco whale. CHORBAJIAN: Bitter for
sweet. HENNAGIN: Crossing the Han River; Walking on the green
grass. PINKHAM: Henry was a worthy King; The leaf; Piping
Anne and husky Paul; Agnus Dei. ROREM: Sing, my soul. STEVENS
Go, lovely rose; Like as the culver; Weepe, O mine eyes.
THOMPSON, R.: Felices ter; The paper reeds. The King Chorale;
Gordon King.

+–Audio 7-76 p71   +NR 4-76 p8
+HF 11-76 p136   +St 7-76 p113

ORS 75207
3140 HAINES: Sonata, harp. MEYER: Appalachian echoes. MONDELLO:
Siciliana. SIEGMEISTER: American harp. STARER: Prelude.
Pearl Chertok, hp.
++NR 11-76 p15

ORS 76216
3141 Le festin: COUPERIN, F.: Livres de clavecin, Bk III, Order no.
14: La linote efarouchee; Le rossignol vainqueur. CROFT:
Sonata, recorder, G major. HANDEL: Sonata, recorder, op. 1,
no. 4, A minor. HASSE: Sonata, recorder, F major. LAWES:
Courtly masquing ayres. PURCELL: Chaconne, F major. STANLEY:
Sonata, recorder, F major. TELEMANN: Partita, no. 5, E minor.
ANON.: Greensleeves to a ground. Carl Dolmetsch, rec; Joseph
Saxby, hpd.
+NR 7-76 p6

ORS 76231
3142 CHERTOK: Around the clock. LOEILLET: Suite, no. 1. KLEINSINGER:
Pavane for Seskia. POULENC: Suite française. SHELDON: Sara-
band. Pearl Chertok, hp.
+NR 11-76 p15

ORS 76235
3143 ABSIL: Sonata, op. 37. DE BO: Sonatina, D major. LONQUE, A.:
Sonatina, op. 34, D major. LONQUE, G.: Sonatina, op. 32, D
major. Sonatina, op. 36, G major. PEETERS: Sonatina, op. 45.
Sonatina, op. 46, G major. POOT: Sonatina, D major. Pierre
Huybregts, pno.
+–NR 11-76 p13

PANTON

110 369
3144 HINDEMITH: Sonata, bass clarinet and piano. LOUDOVA: Air, bass
clarinet and piano. LUCKY: Pieces (3). SLUKA: Sonata, bass
clarinet and piano. TAUSINGER: Contemplations (2). Due Boemi
di Praga.
+RR 4-76 p64

110 380
3145 DVORACEK: Music, harp. GRANDJANY: Fantasia on a theme by Haydn.
PAUER: Suite, harp. RAVEL: Introduction and allegro. ROUSSEL:
Impromptu, harp, op. 21. ANON.: Partita of old dances and airs
(arr. Suriano). Renata Kodadová, Magdalena Spitzerová, hp;
František Cech, flt; Karel Dlouhý, clt; Suk Quartet.
+–RR 4-76 p61

110 385
3146 BORKOVEC: Songs on poems by V. Nezval, op. 15 (7). JEZEK: Songs
(6). Children's choruses on poems by V. Nezval. KREJCI: Songs

on texts by V. Nezval (5). REZAC: Songs on Nezval (4). SMOLKA:
The heathcock sits above the cloud. Jana Jonášová, s; Jiří
Pokorný, Alfréd Holeček, František Vrána, pno; Věra Soukupova,
con; Kühn Female Choir; Pavel Kühn.
        +-RR 4-76 p72

.10 418/9
8147 KUCHAR: Fantasia, D minor. REICHA: Fugue, A major. SEGER: Pre-
      lude, A minor. Fugue, F minor. VANHAL: Fugues (6). ZACZ:
      Fugue, A minor. ZIMMERMAN: Fugue, E minor. ANON.: Fugues, D
      major, E minor (3), E major, A major (2), C major. Fugues (8)
      (arr. Smolka). Jiri Reinberger, Bohumír Rabas, Jaroslav Vod-
      rážka, org.
        +-RR 5-76 p67

.10 440
8148 CEREMUGA: Lasské pastorale. DVORACEK: Dialogue. FELIX: Living
      earth. MALASEK: South Bohemian encounter. STERNWALD: Sym-
      phonic picture. VACEK: Indian summer. Soloists; Brno Radio
      and Moravian Philharmonic Orchestras.
        +RR 12-76 p54

                              PEARL

SHE 525 (Reissue from Unicorn UNS 249) (*Readings)
8149 ARISTOTLE: An observation on beer drinkers.* BELDAMINIS, P. de,
      Jr. (pseud. of Warlock): Prosdorinus de Beldaminis Senior.*
      BLUNT: The drunken wizard.* BEAUMONT/FLETCHER: The night of
      the burning pestle, excerpt.* DOWLAND (trans Warlock): Mrs
      White's nothinge; My Lady Hunsdon's puffe. MOERAN (Warlock):
      Maltworms. OINOPHILUS: In good company. NASH: Eight kinds
      of drunkenness.* RAB NOOLAS (pseud. of Warlock): Drunken song
      in the Saurian mode; Mother's ruin.* RAVENSCROFT (ed. Warlock):
      By a bank as I lay; He that will an alehouse keep; Jinkin the
      jester; The maid she went a-milking; Malt's come down; What
      I hap had to marry a shrow. ROSSETER (ed. Warlock): When Laura
      smiles. WARLOCK: Away to Twivver; Fill the cup, Philip; Cap-
      riol suite: Tordion; Mattachins; Hey troly loly lo; I asked a
      thief to steal me a peach; In an arbour green; Jillian of Berry;
      My ghostly fader; Piggesine (arr. Tomlinson); Sweet content.
      Peter Warlock's fancy. WHYTHORNE (ed. Warlock): As thy shadow
      itself apply'th. ANON.: Have you seen but a white lily grow;
      The lady's birthday; One more river; Wine versus women.* Ian
      Partridge, t; Neilson Taylor, bar; Jennifer Partridge, s and
      pno; Peter Gray, speaker; Fred Tomlinson, bar, pno and cond;
      Singers.
        +-Gr 3-76 p1503   +-RR 5-76 p73

SHE 528
8150 An Edwardian musical evening: CAPEL: Love, could I only tell thee.
      ELGAR: Salut d'amour, op. 12. FAURE: Sicilienne, op. 78. Clair
      de lune, op. 46. GENIN: The carnival of Venice variations.
      LUBBOCK: The smugglers' song. MARSHALL: I hear you calling me.
      QUILTER: Now sleeps the crimson petal, op. 3, no. 2. SCHUBERT:
      Sonatina, violin and piano, no. 3, op. 137, D 408, G minor:
      Finale. VAUGHAN WILLIAMS: Linden Lea. SULLIVAN: Day dreams,
      op. 14, no. 5: Elle et lui. Thomas Round, Ian Partridge, John
      McCormack, t; Trevor Locke, bar; Clive Timms, Clara Taylor,

John Parry, Jennifer Partridge, Clifford Benson, Charles
Marshall, pno; Levon Chilingirian, John Georgiadis, vln; Elmer
Cole, flt.
+-Gr 5-76 p1808   +-RR 5-76 p72

PELCA

PRS 40577
3151 ALBRECHTSBERGER: Fugue, op. 16, no. 5, G major. FUX: Sonata,
      organ, no. 7. HAYDN: Stücke für die Flötenuhr. HERZOGENBERG:
      Choralvorspiele. MUFFAT: Toccata and variations (6). Irmenga
      Knitl, org.
            ++NR 6-76 p14

PRS 40581
3152 ALBRECHTSBERGER: Prelude and fugue, B flat major. HESSE: Fantasy
      op. 87, D minor. LACHNER: Introduction and fugue, op. 62, D
      minor. MERKEL: Sonata, organ, op. 30, D minor. SCHUBERT:
      Fugue, op. posth., E minor. TOMKINS: A fancy for two to play.
      Hermann Busch, Wolfgang Metzler, org.
            +NR 6-76 p14

PRSRK 41017/20 (4)
3153 Historic and new organs of the Swiss countryside. ALBRECHTSBERGF
      Fugue on B-A-C-H, G minor. BACH, C. P. E.: Sonata, organ, F
      major. BACH, J. C.: Aria with 15 variations, A minor. BACH,
      J. S.: Chorale prelude: Jesu Christus, unser Heiland. BENN:
      Ricercar a 3. COUPERIN, F.: Messe pour les paroisses. ERBACH
      Canzon a 4 del quarto tono. FROBERGER: Capriccio, C major.
      HASSLER: Canzon, G minor. HOMILIUS: Wer nur den lieben Gott
      lasst walten. HONEGGER: Choral. KREBS: Warum betrubst du dic
      mein Herz. Sei Lob und Ehr dem höchsten Gut. KUHNAU: Sonata,
      organ, no. 1. LOHET: Fuga as der Tabulator von Joh. Woltz
      (duodecima). MARESCHAL: Psalms, nos. 8 and 47. MENDELSSOHN:
      Sonata, organ, op. 65, no. 5, D major. MOESCHINGER: Fuga
      mystica. MURSCHHAUSER: Variations super cantilenam. PACHELBF
      Ach, was soll ich Sunder machen. PAIX: Mir ist ein feins brau
      Maidelein gefallen in mein Sinn. Ungaresca. POGLIETTI: Ricer
      per lo Rossignolo, D major. Ricercar secundi toni, G minor.
      PORTA: Canzon terza. RATHGEBER: Aria, F major. Aria pastorel
      G major. STALDER: Sonada quarta. STUDER: Invicatio, fugue an
      epilog. ZIPOLI: Pastorale. Rudolf Scheidegger, Jean-Claude
      Zehnder, Bernhard Billeter, Heinrich Gurtner, org.
            ++NR 8-76 p12

PERSONAL TOUCH

88WL (from Personal Touch, Inc., 10 Columbus Circle, New York, 10019)
3154 American sampler. BARBER: Souvenirs: Pas de deux. COPLAND:
      Billy the kid: The open prairie; Celebration dance; Billy's
      demise; The open prairie again. GOTTSCHALK: The banjo, op. 15
      GOULD: Interplay: Blues. Latin American symphonette: Guaracha
      Party rag. GRIFFES: The white peacock. IVES: Variations on
      "America". JOPLIN: The easy winners. RODGERS: Lover. RUSSEL
      Honey. SOUSA: The stars and stripes forever. ANON.: Lorena.
      Sometimes I feel like a motherless child. Arthur Whittemore,
      Jach Lowe, pno.
            *MJ 7-76 p57   +St 8-76 p93

6308 246
3155 BEAT: Royal mile. BRIGHAM: City of Madrid. COLLIER: Arc de
      Triomphe. COATES: Knightsbridge. CROSSE: Unter den Linden.
      FREDERIKSEN: Copenhagen (arr. Mackenzie). GRAFULLA: Washington
      greys (arr. Reeves). LEHAR: Luxembourg march. PARKES: Old
      London. POWELL: Cardiff Castle. RITCHIE: City of the sticks.
      SCHRAMMEL: Wien bleibt Wien. SOUSA: Washington post. WOODFIELD:
      Amsterdam. Scots Guards Band; Duncan Beat.
            +Gr 6-76 p96

6500 282
3156 Elizabethan lute songs and solos: CAMPIAN: It fell on a summer's
      day; The cypress curtain of the night; Shall I come, sweet love,
      to thee. CUTTING: Galliard, G minor. DOWLAND: Awake, sweet
      love thou art return'd; Away with these self-loving lads; Come
      again, sweet love doth now invite; Galliard, D major; Fine knacks
      for ladies; In darkness let me dwell; I saw my lady weep; Shall
      I sue; Tarletones risurrectione; What if I never speed. MORLEY:
      Come, sorrow, come; It was a lover and his lass; Thyrsis and
      Milla. ROSSETER: Sweet, come again; What then is love but
      mourning; Whether men do laugh or weep. Frank Patterson, t;
      Robert Spencer, lute.
            +Gr 5-72 p1922   ++NR 3-73 p12
            +HF 6-73 p76   ++SFC 2-22-76 p28
            +HFN 5-72 p919   +St 4-73 p125
            +MJ 4-73 p30

6500 926
3157 BIBER: Suite, 2 clarino trumpets. BLOW: Fugue, F major: Vers.
      BULL: Variations on the Dutch chorale "Laet ons met herten
      reijne". CAMPIAN: Never weather-beaten sail. DOWLAND: Lacri-
      mae: Antiquae pavan. FANTINI: Sonata, 2 trumpets, B flat major.
      FRESCOBALDI: Capriccio sopra un soggetto. HANDEL: Concerto,
      trumpet, B flat major. MORLEY: La caccia a 2. La sampogna.
      PURCELL: Trumpet tune (The cebell), D major. Trumpet tune and
      air. STANLEY: Trumpet voluntary, D major. ANON.: Hejnat
      Krakowska (2). Clarion Consort.
            +-Gr 2-76 p1350   ++RR 3-76 p52
            +HFN 3-76 p107

6580 114
3158 BACH, J. C.: Symphony, op. 3, no. 1, D major. BACH, J. S.: Branden-
      burg concerto, no. 3, S 1048, G major. MOZART: Concerto, piano,
      no. 24, K 491, C minor: 2nd movement. Symphony, no. 40, K 550,
      G minor: 1st movement. PURCELL: The Indian Queen: Trumpet over-
      ture. Yorkshire feast song. TORELLI: Sonata a 5, no. 7, D
      major. Don Smithers, Michael Laird, tpt; AMF; Neville Marriner.
            +-Gr 6-76 p95   ++NR 3-76 p15
            +HFN 7-76 p103   +-RR 6-76 p35

6581 017
3159 BACH: Anna Magdalena notebook, S 508: 3 pieces (trans Lagoya).
      Suite, lute, S 997, C minor: Prelude and fugue. HANDEL: Sara-
      bande, D major (trans. Lagoya). SCARLATTI, D.: Sonatas, guitar,
      A major, A minor. VANHAL: Cantabile. WEISS: Fantasia, E minor.
      Alexandre Lagoya, gtr.
            +Gr 8-76 p320   +RR 6-76 p66
            +HFN 6-76 p89

6747 204 (2)
3160 BEETHOVEN: Bagatelle, no. 25, G 173, A minor. Egmont, op. 84:

Overture. Symphony, no. 9, op. 125, D minor: 3rd movement.
BIZET: Carmen: Suite, no. 1. CHOPIN: Waltz, op. 18, E flat
major. CLARKE: Trumpet voluntary. GOUNOD: Faust: Soldiers'
chorus. HANDEL: Water music: Suite, no. 3. MEYERBEER: Le pro-
phète: Coronation march, Act 4. MOZART: Le nozze di Figaro,
K 492: Voi che sapete. Die Zauberflöte, K 620: Overture.
RIMSKY-KORSAKOV: Russian Easter festival overture, op. 36.
SCHUBERT: Songs: Ave Maria, D 839; Die Forelle, D 550.
SIBELIUS: Finalndia, op. 26. VERDI: La traviata: Prelude,
Act 1. Elly Ameling, Elizabeth Ebert, s; Gerard Souzay, bar;
Dalton Baldwin, Hans Richter-Haaser, Adam Harasiewicz, pno;
John Wilbraham, tpt; BBC Symphony Orchestra, ECO, Leipzig
Radio Symphony Orchestra, Leipzig Gewandhaus Orchestra, VSO,
NPhO, Lamoureux Concerts Orchestra, Limberg Symphony Orchestra
and Chorus, COA, LSO, AMF; Colin Davis, Edo de Waart, Robert
Hanell, Franz Konwitschny, Raymond Leppard, Igor Markevitch,
Martin Koekelkoren, Eduard van Beinum, Charles Mackerras,
Neville Marriner, Christoph von Dohnanyi.
+Gr 1-76 p1244   +-RR 12-75 p56
+-HFN 12-75 p171

6780 030 (2) (Reissues from SAL 3573, 835, 108AY, 6580 047, Fontana
SFL 14048, 6580 107, 6580 085, 6500 968, SAL 3798)
3161 A festival of French music: BERLIOZ: Les Francs Juges, op. 3.
Rêverie et caprice, op. 8. BIZET: L'Arlésienne: Suite, no. 2,
excerpts. DEBUSSY: La mer. FAURE: Pavane, op. 50. FRANCK:
Symphonic variations, piano and orchestra. GOUNOD: Funeral
march of the marionette. RAVEL: Miroirs: Alborada del gracioso.
SAINT-SAENS: The carnival of the animals. Marie-Françoise
Bucquet, pno; Arthur Grumiaux, vln; LSO, Monte Carlo Opera
Orchestra, COA, NPhO, Lamoureux Orchestra, Hague Philharmonic
Orchestra, Rotterdam Philharmonic Orchestra; Colin Davis, Paul
Capolongo, Bernard Haitink, Edo de Waart, Igor Markevitch,
Willem van Otterloo, Roberto Benzi, Jean Fournet, Eliahu Inbal.
+-Gr 8-76 p342   +RR 7-76 p47
+HFN 10-76 p183

6780 751 (2) (*Reissues from SAL 3453/4, 835 134AY, 3430)
3162 BEETHOVEN: Sonata, violoncello, no. 3, op. 69, A major.*
MUSSORGSKY: Songs: Cradle song; The magpie; The night; On the
Dniepr; The ragamuffin; Where are you, little star.* PROKOFIEV:
Poems by Anna Akhmatova, op. 27 (5).* TCHAIKOVSKY: Songs: Do
not believe me, my friend, op. 6, no. 1; None but the lonely
heart, op. 6, no. 6; Not a word, my friend, op. 6, no. 2.*
TRAD.: Tati-Tati (orch. N. Tcherepnin). Mstislav Rostropovich,
vlc; Sviatoslav Richter, Mstislav Rostropovich, Olga Rostropo-
vich, pno; Galina Vishnevskaya, s; USSR Symphony Orchestra;
Igor Markevitch.
+Gr 12-76 p1015   +RR 12-76 p89
+HFN 12-76 p152

6780 755 (2) (*Reissues from 6585 012, 6580 099, 6580 053, EMI AMS
16056, SABL 144)
3163 BORODIN: In the steppes of Central Asia.* GLINKA: Russlan and
Ludmilla: Overture.* KHATCHATURIAN: Spartacus: Adagio.*
PROKOFIEV: Symphony, no. 1, op. 25, D major. RACHMANINOFF:
Rhapsody on a theme by Paganini, op. 43. RIMSKY-KORSAKOV: The
legend of Sadko, op. 5.* STRAVINSKY: Petrouchka (1947 version).*
Rafael Orozco, pno; Zagreb Philharmonic Orchestra, Monte Carlo

Opera Orchestra, RPO, Minneapolis Symphony Orchestra, COA,
LPO; Milan Horvat, Edouard van Remoortel, Edo de Waart, Antal
Dorati, Bernard Haitink, Jean Fournet, David Lloyd-Jones.
   +Gr 11-76 p888   ++RR 11-76 p70
   +-HFN 12-76 p152

PLEIADES

P 257
3164 Late 16th century music, vol. 3. University of Chicago Collegium,
University of Kentucky Collegium Musicum; Arnold Blackburn,
org; James Bonn, hpd; Howard Brown, Wesley Morgan.
   ++NR 3-76 p16

POLYDOR

2383 389
3165 GINASTERA: Canción al arbol. GRIEG: A dream, op. 48, no. 6.
GUASTAVINO: La rose y el sauce. HAHN: If my songs were only
winged. HALFFTER: Al que linda moca. MORENO: To huey tlahtzin.
PALADILHE: Psyché. OBRADORS: Del cabello mas sutil. SCHUBERT:
Ständchen, D 889. SULLIVAN: Orpheus with his lute. TCHAIKOVSKY:
None but the lonely heart, op. 6, no. 6. TRAD.: Blow the wind
southerly (arr. Moore); El molondron (arr. Obradors); Sakura,
sakura (arr. Shinohara); Sometimes I feel like a motherless
child (arr. Burleigh); La vera sorrentina. Victoria de los
Angeles, s; Geoffrey Parsons, pno.
   +-Gr 10-76 p640
2383 391
3166 BLISS: Bliss. The rout trot. GOOSSENS: Folk-tune. LAMBERT:
Concerto, piano. Elegiac blues. Elegy. WALTON: Façade.
WILLIAMS: Raguette extra sec. Valsette brut. Richard Rodney
Bennett, pno; English Sinfonia, Members; Neville Dykes.
   ++Gr 8-76 p320   +RR 9-76 p56
   +-HFN 9-76 p123
2460 250. Tape (c) 3170 241
3167 Music for evensong: BRIDGES (Howells): All my hope on God is
founded. DAY, E.: Psalm 84: O how amiable. GARDINER: Evening
hymn. HOWELLS: Magnificat and nunc dimittis, G major. STANFORD:
Psalm no. 23: The Lord is my Shepherd. WESLEY, S. S.: Magnificat
and nunc dimittis, E major; Thou wilt keep him. Worcester
Cathedral Choir; Harry Bramma, org; Christopher Robinson.
   +Gr 11-75 p887   +-HFN 4-76 p109
2460 262
3168 BACH: Cantanta, no. 29, Wir danken die, Gott: Sinfonia. CHARPENTIER:
Te deum: Prelude. CHARKE, J.: Trumpet voluntary. FIOCCO:
Andante. LANGLAIS: Incantation pour un jour saint. VIERNE:
Symphony, no. 1, op. 14, D minor: Final. Divertissement.
WHITLOCK: Folk tune and scherzo. YON: Humoresque l'organo
primitivo. Noël Rawsthorne, org.
   ++Gr 8-76 p323   +-HFN 8-76 p81
2489 523
3169 FREDERICK II, King of Prussia: Torgauer Marsch (arr. Breuer).
HENRION: Kreuzritter Fanfare (arr. Manneke). HERZER: Hoch
Heidecksburg (arr. Hubert). JUREK: Deutschmeister Regiments
Marsch. KOMZAK: Erzherzog-Albrecht-Marsch, op. 136 (arr. Breuer).

PIEFKE: Königgrätzer Märsch (arr. Breuer). Preussens Gloria
(arr. Gutzeit). RADECK: Fridericus-Rex (arr. Friess). SCHRAMMEL:
Wien bliebt Wien (arr. Heyer). SIMON: Kurassiermarsch "Grosser
Kurfürst" (arr. Friess). TRAD.: Marsch I Bataillon Garde (arr.
Friess). Petersburger Marsch (arr. Breuer). 11th Panzer
Grenadier Division, Music Corps; Hans Friess.
          +-Gr 1-76 p1247

PREISER

LV 192
3170 BEETHOVEN: Fidelio, op. 72: Hat man nicht auch Gold beineban; Ha
      Welch ein Augenblick. BIZET: Carmen: Euren Toast kann ich wohl
      erwidern. GOUNOD: Faust: Le veau d'or est toujours debout, Il
      etait tems, Vous qui faites l'endormie. HALEVY: La Juive: Wenn
      ew'ger Hass gluhende Rache. LOEWE: Der Erlkönig, op. 1, no. 3;
      Prinz Eugen, op. 92. MOZART: Die Zauberflöte, K 620: In diesen
      heil'gen Hallen. SMETANA: The bartered bride: Wer in Lieb ent-
      brannt. WAGNER: Lohengrin: Mein Herr und Gott nun ruf ich dich.
      Die Meistersinger von Nürnberg: Jerum, Jerum, Oh, ihr boshafter
      Geselle. Das Rheingold: Abendlich strahit der Sonne Auge.
      WEBER: Der Freischütz: Hier im irdischen Jammertal. Michael
      Bohene, bs-bar; Orchestral accompaniment.
          +NR 7-76 p8

PRELUDE

PRS 2501
3171 BARNBY: Sweet and low. BISHOP: Forrester sound the cheerful horn.
      BYRD: Constant Penelope; What pleasures have great princes.
      GIBBONS: The silver swan. HATTON: The Indian maid. MORLEY:
      Ay me, the fatal arrow; My bonny lass she smileth; Now is the
      month of Maying; Though Philomela lost her love. SMART: Hymn
      to Cynthia. WALKER: The sun of the Celandines. TRAD.: Barbara
      Allen (arr. Willcocks). Bobby Shaftoe. Brigg fair (arr.
      Grainger). Dance to your Daddy. Rue and thyme (arr. Doveton).
      The Scholars.
          +-Gr 11-76 p867   +RR 10-76 p97

PRS 2505
3172 BISHOP: Home, sweet home. BOHN: Still as the night. BRAHMS: An
      eine Aeolscharfe, op. 19, no. 5; Meine Liebe ist grün, op. 63,
      no. 5; O wüsst ich doch den Weg Zurück, op. 63; Der Tod, das
      ist die kühle Nacht, op. 96, no. 1; Verzagen, op. 72, no. 4.
      COLERIDGE-TAYLOR: Big lady moon. HORN: Cherry ripe. LIDDELL:
      Abide with me. MAHLER: Des Knaben Wunderhorn: Das irdische
      Leben; Rheinlegendchen. Ruckert Lieder: Ich bin der Welt
      abhanden gekommen. MOLLOY: Love's old sweet song. NEVIN: The
      rosary. SULLIVAN: The lost chord. Norma Procter, con; Paul
      Hamburger, pno.
          +Gr 12-76 p1042   -HFN 12-76 p139

PYE

QS PCNH 4. Tape (c) ZCPCNH 4 (c) Y8PCNH 4
3173 BERLIOZ: La damnation de Faust, op. 24: Hungarian march. BRAHMS:

Hungarian dance, no. 1, G minor (arr. Stokowski). CHABRIER:
España: Rhapsody. HAYDN: Quartet, strings, op. 3, no. 5, F
major: Andante cantabile (attrib. arr. Stokowski). IPPOLITOV-
IVANOV: Caucasian sketches, op. 10: Procession of the Sardar.
MUSSORGSKY: Khovanschina: Entr'acte, Act 5 (arr. Stokowski).
SAINT-SAENS: Danse macabre, op. 40. SOUSA: The stars and
stripes forever (arr. Stokowski). STRAUSS, J. II: Geschichten
aus dem Wiener Wald, op. 325. TCHAIKOVSKY: Again as before,
op. 73, no. 6 (arr. Stokowski). National Philharmonic
Orchestra; Leopold Stokowski.
+-Gr 3-76 p1478

PCNHX 6
3174 BEETHOVEN: Leonore overture, no. 3, op. 72. BERLIOZ: Roman
carnival, op. 9. MOZART: Don Giovanni, K 527: Overture.
ROSSINI: William Tell: Overture. SCHUBERT: Rosamunde, op. 26,
D 797: Overture. National Philharmonic Orchestra; Leopold
Stokowski.
-HFN 12-76 p148

Ember GVC 51 (Reissues)
3175 Ballads and songs: BIMBONI: Sospiri miei adnant ove vi mando.
BURLEIGH: Little mother of mine. BUTTERFIELD: When you and I
were young, Maggie. CADMAN: At dawning. DANKS: Silver threads
among the gold. DEL RIEGO: Thank God for a garden. FEARIS:
Beautiful isle of somewhere. GODARD: Jocelyn: Angels guard
thee. KENNEDY: Say "Au revoir" but not "Goodbye". KREISLER:
Cradle song. LOHR: The marriage market: Little grey home in
the west. MARGETSON: Tommy lad. MOSZKOWSKI: Serenata. TATE:
Somewhere a voice is calling. WALDROP: Sweet Peggy O'Neill.
John McCormack; Other soloists; Orchestral accompaniment.
+Gr 3-76 p1503

GH 603. Tape (c) ZCGH 603 (ct) Y8GH 603
3176 ANDERSON: Clarinet candy. BAYCO: Royal Windsor (arr. Richardson).
BLISS: An age of kings. FARNON: State occasion (arr. Sharpe).
FOSTER: Jeanie with the light brown hair (arr. Hatch).
GERSHWIN: Strike up the band, medley. NOBLE: Cherokee. OSSER:
Bandolero. OSTERLING: Winds on the run. ROGAN: Milanollo.
REED: Festive overture. SHARPE: Three blades of Toledo.
TEIKE: Old comrades. TRAD.: Northumbrian airs; Deep river
(arr. T. Sharpe). Coldstream Guards Band; Trevor L. Sharpe.
+Gr 2-76 p1385

TB 3004. Tape (c) ZCTB 3004 (ct) Y8TB 3004
3177 ANDERSON: Belle of the ball (arr. Tomlinson). BERLIOZ: The Trojans:
Trojan march (arr. S. Robinson). DEBUSSY: Preludes, Bk I, no.
8: The girl with the flaxen hair (arr. M. Brand). FARNON:
Colditz march (arr. G Langford). Concorde march (arr. Street).
Westminster waltz (arr. G. Brand). LANGFORD, G.: Rhapsody,
trombone and brass band. MASCAGNI: Cavalleria rusticana:
Intermezzo. WOOD, G.: Tombstone, Arizona overture. Brighouse
and Rastrick, Cory, Fairey, Yorkshire Imperial Metals Bands;
Don Lusher, trom; Robert Farnon, Geoffrey Brand, H. Arthur
Kenney.
+Gr 2-76 p1382

TB 3006
3178 ALLISON: Mancini magic (arr. Street). Silver threads among the
gold. BACH: Prelude and fugue (arr. Cess Brugman). CATELINET:
Jolson memories (arr. Young). The star. ORD HUME: The elephant

march. WALLEBOM: Hasta la vista (arr. Langstrand and Howe).
ZUTANO: La mantella (arr. D. Rimmer). TRAD.: The girl I left
behind me (arr. G. Langford). Ken Vernon, euphonium; Ralph
Blackett, tenor hn; Burton Constructional Band; Ernest Woodhouse.
    +Gr 7-76 p221

TB 3007
3179 BALL: Holiday overture. GOUNOD: Mors et vita: Judex (arr. Frank
     Wright). HOROBIN: Star of Erin. KENNEY: Concerto, horn.
     NEVILLE: Shrewsbury fair. RIMMER: King Lear. WESTON: Venamair.
     Brian Rostron, tenor hn; James Layland, flugel hn; Carlton Main
     Frickley Colliery Band; H. A. Kenney.
        +Gr 7-76 p221

TB 3009
3180 BALL: Cornish festival overture. BROADBENT: Centaur. DVORAK:
     Symphony, no. 9, op. 95, E minor (arr. Rimmer). HEATH: Frolic
     for trombones. Angel voices (arr. Jordan). MacDOWELL: To a
     wild rose (arr. Ball). MOYLE: Cornish rock. Restormel.
     RIMMER: Golden rain. P. W. Minear, S. Mennear, cor; M. Faro,
     euphonium; S. B. Goudge, S. M. Bazely, M. J. Bazely, trom;
     St. Dennis Band; E. J. Williams.
        +Gr 9-76 p483

RCA

VH 013 (Reissues from RB 6554, RB 16019, HMV BLP 1014)
3181 CLEMENTI: Sonata, piano, op. 47, no. 2, B flat major: Rondo.
     CZERNY: Variations on La ricordanza, op. 33. MENDELSSOHN:
     Variations sérieuses, op. 54, D minor. MOZART: Sonata, piano,
     no. 12, K 332, F major. SCARLATTI, D.: Sonata, piano, L 23,
     E major. SCHUMANN: Sonata, piano, no. 3, op. 14, F minor:
     Variations on a theme by Clara Wieck. Vladimir Horowitz, pno.
        +Gr 2-76 p1363   +RR 2-76 p48
        +HFN 2-76 p117

VH 020
3182 BACH: Chorale prelude: Nun Komm, der Heiden Heiland (Busoni),
     S 659. DEBUSSY: Children's corner suite: Serenade for the
     doll. HOROWITZ: Variations on themes from Bizet's Carmen.
     MENDELSSOHN: Songs without words, op. 62, no. 6; op. 67, no.
     5; op. 85, no. 4. A midsummer night's dream, op. 61: Wedding
     march and variations (arr. Liszt, Horowitz). MUSSORGSKY: By
     the water (arr. Horowitz). PROKOFIEV: Toccata. MOZART: Sonata,
     piano, no. 11, K 331, A major: Alla turca. MOSZKOWSKI: Etude,
     A flat major. Etincelles. SAINT-SAENS: Danse macabre, op. 40
     (arr. Liszt, Horowitz). SOUSA: The stars and stripes forever
     (arr. Horowitz). Vladimir Horowitz, pno.
        +Gr 12-75 p1085   +-RR 12-75 p81
        +-HFN 2-76 p117

SB 6891 (Reissue from RB 16239)
3183 ALBENIZ, M.: Sonata, guitar (arr. Pujol). BERKELEY: Sonatina, op.
     51. CIMAROSA: Sonatas, guitar, C sharp minor, A major (arr.
     Bream). FRESCOBALDI: Aria detta "La frescobaldi" (arr. Segovia).
     RAVEL: Pavane pour une infante défunte (arr. Bream). RODRIGO:
     En los trigales. ROUSSEL: Segovia, op. 29. SCARLATTI, D.:
     Sonatas, guitar, L 33, E minor (arr. Bream); L 352, E minor
     (arr. Segovia). Julian Bream, gtr.

    +Gr 8-76 p320   +RR 8-76 p68
    +HFN 10-76 p181

Tape (c) RK 11708
3184 BYRD: The woods so wild (Cutting). CUTTING: Packington's pound.
      DOWLAND: Bonnie sweet robin. Go from my window. Loth to depart.
      Walsingham. HOLBORNE: The fairy round. Heart's ease. Heigh-
      ho holiday. MILANO: Fantasias I-VIII. ANON.: Greensleeves
      (Cutting). Julian Bream, lt.
        +HFN 7-76 p105 tape   +-RR 9-76 p93 tape

GL 25006. Tape (c) GK 25006 (Previously issued by Reader's Digest)
3185 ARNOLD: Scottish dances, op. 59 (4). BINGE: Elizabethan serenade.
      BRITTEN: Variations and fugue on a theme by Purcell, op. 34.
      ELGAR: Enigma variations, op. 36: Nimrod. Chanson de matin,
      op. 15, no. 2. TOYE: The haunted ballroom: Waltz. WALTON:
      Henry V: Touch her soft lips and part. NPhO; Charles Gerhardt.
        +Gr 10-76 p612   +RR 10-76 p45
        +HFN 12-76 p136   +RR 12-76 p104 tape
        +HFN 12-76 p153 tape

ARL 1-0864. Tape (c) ARK 1-0864 (ct) ARS 1-0864
3186 ASENCIO: Dipso. BACH: Suite, solo violoncello, no. 1, S 1007, G
      major: 3 movements. BENDA: Sonatina, D major. Sonatina, D
      minor. PONCE: Prelude, E major. SCARLATTI: Sonatas, guitar
      (2). SOR: Sandante, C minor. Minuets, C major, A major, C
      major. WEISS: Bourrée. Andrés Segovia, guitar.
        +HF 11-76 p153   ++St 10-75 p119
        +NR 10-75 p12

ARL 1-1172
3187 BAZZINI: La ronde des lutins, op. 25. BLOCH: Baal Shem: Ningun.
      PAGANINI: Sonata appassionata, op. 38. SARASATE: Danzas
      españolas, no. 2, op. 22: Romanza andaluza; no. 6, op. 23:
      Zapateado. SCRIABIN: Etude, op. 8, no. 10, D flat major (arr.
      Szigeti). VITALI: Chaconne, G minor (arr. Charlier). Eugene
      Fodor, vln; Judith Olson, pno; Joseph Payne, org.
        +HF 4-76 p126   ++NR 3-76 p13
        +-MJ 1-76 p39

ARL 1-1176. Tape (c) ARK 1-1176 (ct) ARS 1-1176
3188 CHOPIN: Barcarolle, op. 60, F sharp major. DEBUSSY: L'isle
      joyeuse. GRANADOS: Goyescas: Quejas o la maja y el ruiseñor.
      LISZT: Consolation, no. 3, G 172, D flat major. RACHMANINOFF:
      Prelude, op. 23, no. 5, G minor. Prelude, op. 32, no. 12, G
      sharp minor. RAVEL: Pavane pour une infante défunte. Le
      tombeau de Couperin: Toccata. SCHUMANN: Romance, op. 28, no.
      2, F sharp major. TCHAIKOVSKY: The months, op. 37a, no. 3:
      Song of the lark. Van Cliburn, pno.
        -HF 11-76 p116   +-MJ 10-76 p25
        +-HF 12-76 p147 tape   ++NR 7-76 p10

ARL 1-1323
3189 ALBENIZ: Capricho catalan (trans. Lorimer). BACH: Anna Magdalena
      notebook, S 508, excerpts. MOLLEDA: Variations on a theme.
      SAMAZEUILH: Serenade. SAN SEBASTIAN: Preludios vascos: Dolor
      (trans. Segovia). SOR: Introduction and variations on "Malbrough
      s'en va-t-en guerre". Sicilienne, D minor. Andrés Segovia,
      gtr.
        +NR 3-76 p13   +-SFC 2-29-76 p25

AGL 1-1334
3190 Great American marches: BAGLEY: National emblem march. EMMETT:

Dixie. GERSHWIN: Strike up the band. HANDY: St. Louis blues
march (arr. Hayman). HERBERT: March of the toys. LOEWE: Get
me to the church on time (arr. Hayman). MEACHAM: American
patrol (arr. Hayman). MORSE: Up the street. SOUSA: Semper
fidelis. The stars and stripes forever. WILLSON: 76 trombones.
TRAD.: Yankee Doodle (arr. Gould). BPO; Arthur Fiedler.
+NR 5-76 p14

ARL 1-1403
3191 Hymns: A mighty fortress; Abide with me; The church in the wild-
wood; Come, thou Almighty King; Eternal life; The fifth psalm;
I walked today where Jesus walked; I wonder as I wander; The
penitent; Rock of ages; This is my father's world; Were you
there. Sherrill Milnes, bar; Jon Spong, org.
*NR 8-76 p11

AVM 1-1735 (Reissue from LM 2032)
3192 Spirituals: Crucifixion; De gospel train; Deep river; Everytime I
feel de spirit; Go down, Moses; Hear de lam's a crying'; He's
got the whole world in His hands; Honor, honor; If He change
my name; Let us break bread together; My Lord, what a morning;
Nobody knows the trouble I see; O what a beautiful city; On ma
journey; Plenty good room; Ride on, King Jesus; Roll Jerdn'n
roll; Sinner, please; Sometime I feel like a motherless child;
Soon'a will be down; Were you there. Marian Anderson, s; Franz
Rupp, pno.

+ARSC Vol VIII, no. 2-3   +NR 9-76 p10
p97   ++SFC 8-8-76 p38
+-HF 10-76 p128   +St 12-76 p149
+MJ 11-76 p44

CRM 1-1749 (Reissues from RCA originals, 1906-20)
3193 BIZET: Carmen: La fleur que tu m'avais jetée. DONIZETTI: L'Elisir
d'amore: Una furtiva lagrima. FLOTOW: Martha: Ach, so fromm
(in Italian). GOUNOD: Faust: Salut, demeure chaste et pure.
HALEVY: La Juive: Rachel, quand du Seigneur. HANDEL: Serse:
Ombra mai fù. LEONCAVALLO: I Pagliacci: Vesti la giubba.
MEYERBEER: L'Africaine: O paradiso (in Italian). PONCHIELLI:
La gioconda: Cielo e mar. PUCCINI: La bohème: Che gelida
manina. Tosca: Recondita armonia; E lucevan le stelle.
VERDI: Rigoletto: Questa o quella; La donna è mobile. Il
trovatore: Di quella pira. Aida: Celeste Aida. Enrico Caruso,
t; Orchestral accompaniment.

+ARSC Vol VIII, no. 2-3   +MJ 11-76 p44
p106   +-NR 10-76 p12
+-HF 11-76 p98   +-NR 11-76 p9

LRL 1-5094. Tape (c) RK 11719
3194 BACH: Suite, orchestra, S 1067, B minor: Minuet; Badinerie.
CHOPIN: Waltz, op. 64, no. 1, D flat major (arr. and orch.
Gerhardt). DINICU: Hora staccato (arr. and orch. Gerhardt).
DOPPLER: Fantaisie pastorale hongroise, op. 26. DRIGO: Les
millions d'Arlequin: Serenade (arr. and orch. Gamley). GLUCK:
Orfeo ed Euridice: Dance of the blessed spirits. GODARD: Pieces,
op. 116: Waltz. MIYAGI: Haru no umi (arr. and orch. Gerhardt).
PAGANINI: Moto perpetuo, op. 11 (arr. and orch. Gerhardt).
RIMSKY-KORSAKOV: The tale of the Tsar Sultan: Flight of the
bumblebee (arr. and orch. Gerhardt). SAINT-SAENS: Ascanio:
Ballet music, Adagio and variation. James Galway, flt; National
Philharmonic Orchestra; Charles Gerhardt.

```
      ++Gr 11-75 p915        +MJ 11-76 p60
      +Gr 1-76 p1244         ++NR 11-76 p16
      +Gr 7-76 p230 tape     +-RR 10-75 p38
      +-HFN 12-75 p164
```

LRL 1-5127 (Reissues from LRL 1-5094, LRL 1-5085, LRL 1-5109)
3195 BACH: Suite, orchestra, S 1067, B minor: Minuet; Badinerie.
      BERKELEY: Sonatina, op. 13.* DEBUSSY: Fêtes galantes, Set I:
      Clair de lune.* Syrinx, flute.* The little shepherd.*
      GLUCK: Orpheus and Eurydice: Dance of the blessed spirits.
      MOZART: Andante, K 315, C major. PAGANINI: Moto perpetuo, op.
      11. VIVALDI: Concerto, flute and bassoon, op. 10, no. 2, G
      minor. James Galway, flt; Anthony Goldstone, pno; National
      Philharmonic Orchestra, Lucerne Festival Strings; Charles
      Gerhardt, Rudolf Baumgartner. (*New to UK).
          ++Gr 9-76 p478   ++RR 9-76 p43
          +-HFN 9-76 p125

LRL 1-5131
3196 BACH: Sonata, flute and harpsichord, S 1033, C minor: Allegro.
      BRISCIALDI: Carnival of Venice. CHOPIN: Variations on a theme
      from Rossini's "La Cenerentola". DVORAK: Humoresque, op. 101,
      no. 7, G flat major. GOSSEC: Tambourin. HANDEL: Solomon:
      Arrival of the Queen of Sheba. KREISLER: Schön Rosmarin.
      MENDELSSOHN: A midsummer night's dream, op. 61: Scherzo.
      RACHMANINOFF: Vocalise, op. 34, no. 14. SCHUMANN: Kinderscenen,
      op. 15, no. 7: Träumerei. James Galway, flt; National Philhar-
      monic Orchestra; Charles Gerhardt.
          +Gr 11-76 p824   +RR 11-76 p67
          +HFN 11-76 p161

PRL 1-8020
3197 DAVIS: Little drummer boy. GRUBER: Stille Nacht. HANDEL: Joy to
      the world. HAYDN, M.: Anima nostra; In dulci jubilo. HERBECK:
      Angels we have heard on high; Kommet ihr Hirten; Pueri concinite.
      KODALY: Die Engel und die Hirten. MENDELSSOHN: Adeste fideles;
      Greensleeves; Hark, the herald angels sing. PRAETORIUS (arr.):
      Deck the halls; Es ist ein Ros entsprungen; Gloria, Gott in der
      Höh. SCHULZ: Ihr Kinderlein kommet. Anton Neyder, pno;
      Johannes Sonnleitner, org; Vienna Boys' Choir; Vienna Chamber
      Orchestra; Hans Gillesberger.
          +ARG 12-76 p55   +NR 12-76 p1
PRL 1-9034. Tape (c) PRK 1-9034 (ct) 1-9034
3198 BRAHMS: Am Wildbach die Weiden; Die Berge sind spitz; Nun stehn
      die Rosen in Blüte; Und gehst du über den Kirchhof. DRECHSLER:
      Der Bauer als Millionär: Brüderlein fein. MOZART: Due pupille
      amabile, K 439; Luci care, luci belle, K 346; Mi lagnero tacendo,
      K 437; Piu non si trovano, K 459. SCHUBERT: Gott, meine Zuver-
      sicht, D 706; Die Nachtigall, D 724; La Pastorella, D 513;
      Ständchen, D 921; Widerspruch, D 865. SCHUMANN: Zigeunerleben,
      op. 29, no. 3. Vienna Boys' Choir; Hans Gillesberger.
          +HF 8-76 p98   +NR 4-76 p8
LRL 2-7531 (2) (Reissues from SER 5593, 6825, 5584, 5636, 5564/6, ARL
          4-0370, 2-0105. *New to UK)
3199 The art of Sherrill Milnes, Operatic arias, duets and songs.
      BERNSTEIN: West side story: Maria.* Mass: A simple song.*
      BIZET: Les pêcheurs de perles: Au fond du temple saint.
      BRAHMS: Vier ernste Gesänge, op. 121. HANDEL: Joshua: See the
      raging flames arise. KERN: Showboat: Ol' man river.*
```
```

LEONCAVALLO: La bohème: Scuoti o vento. Zazà: Zazà, piccola
Zingara. PUCCINI: La bohème: In un coupé; Ah Mimi, tu più non
torni. Tosca: Te deum. SCHONBERG: A survivor from Warsaw, op.
46. VERDI: Attila: Tregua e cogli'unni...Dagl'immortali
vertici e gettata la mia sorte. La traviata: Di provenza il
mar...Dunque invano. Il trovatore: Il balen del suo sorriso...
Per me ora fatale. I vespri siciliani: Si, m'abboriva...in
braccio alle dovizie. TRAD.: Shenandoah.* Stephen Foster
medley.* Sherrill Milnes, bar; John Mitchinson, Carlo Bergonzi,
Francis Egerton, Placido Domingo, t; Terence Sharpe, Bonaldo
Giaiotti, bs; Ambrosian Opera Chorus, John Alldis Choir, Wands-
worth School Boys' Choir, New England Conservatory Chorus; NPhO,
RCA Italiana Opera Orchestra, LSO, BSO, National Philharmonic
Orchestra; Anton Guadagno, James Levine, Zubin Mehta, Georges
Prêtre, Nello Santi, Erich Leinsdorf, Charles Gerhardt, Marcus
Dods.
        +-Gr 7-76 p218

ARL 3-0997 (3) (Reissues from RB 16252, SB 6635, SB 6852, SB 6876)
3200 Guitar and lute concertos: ARNOLD: Concerto, guitar, op. 67.
      BRITTEN: Gloriana, op. 53: The courtly dances. BENNETT:
      Concerto, guitar and chamber ensemble. GIULIANI: Concerto,
      guitar, op. 30, A major. RODRIGO: Concierto de Aranjuez,
      guitar and orchestra. VILLA-LOBOS: Concerto, guitar and small
      orchestra. VIVALDI: Concerto, lute and strings, D major (ed.
      Bream). Julian Bream, gtr and lt; Melos Ensemble, Melos Chamber
      Orchestra, LSO, Julian Bream Consort; André Previn.
            +Gr 2-76 p1350   +RR 2-76 p29
            +HFN 2-76 p93

CRL 3-1429
3201 GIANELLA: Concerto, flute, no. 1, D minor. Concerto, flute, no.
      3, C major. Concerto lugubre, C minor. JOLIVET: Concerto,
      flute. Suite en concert, flute and percussion. MERCADANTE:
      Concerto, flute, E minor. PERGOLESI: Concerto, flute, no. 2,
      G major. SAMMARTINI: Concerto, flute, F major. TARTINI:
      Concerto, flute, G major. VIVALDI: Concerto, flute, A minor.
      Jean-Pierre Rampal, flt; Various orchestras.
            ++NR 6-76 p7   ++SFC 5-30-76 p24

CRL 3-1430
3202 ALBINONI: Concerto, trumpet, B flat major. BACH: Cantata, no.
      140: Wachet auf, ruft uns die Stimme. Concerto, trumpet, D
      minor. Suite, flute, no. 2, S 1067, B minor. JOLIVET: Concer-
      tino. Concerto, trumpet, no. 2. LOEILLET: Concerto, trumpet,
      D major. HANDEL: Concerto, trumpet, G minor. PURCELL: Sonata,
      trumpet, no. 1, D major. Sonata, trumpet, no. 2, D major.
      TESSARINI: Sonata, trumpet, D major. TARTINI: Concerto, flute,
      D major. TELEMANN: Concerto, trumpet, D major. Maurice André,
      tpt; Instrumental accompaniments.
            ++NR 6-76 p6

CRL 3-2026 (3)
3203 BORODIN: Quartet, strings, no. 2, D major: Nocturne (arr. Sargent).
      IPPOLITOV-IVANOV: Caucasian sketches, op. 10: Procession of the
      Sardar. PROKOFIEV: Love for three oranges, op. 33: March.
      Symphony, no. 1, op. 25, D major. RIMSKY-KORSAKOV: Le coq d'or:
      Bridal procession. SCRIABIN: Poème de l'extase, op. 54.
      STRAVINSKY: L'Oiseau de feu: Suite. TCHAIKOVSKY: Capriccio
      italien, op. 45. Marche slav, op. 31. Romeo and Juliet:

Overture. PO; Eugene Ormandy.
+NR 12-76 p5

ARM 4-0942/7
3204 The Heifetz collection, Vol. 1, Acoustic recordings complete:
ACHRON: Hebrew dance. Hebrew lullaby. Hebrew melody, op. 33.
Stimmung. d'AMBROSIO: Serenade. BAZZINI: La ronde des lutins,
op. 25. BEETHOVEN: The ruins of Athens, op. 113: Chorus of
dervishes Turkish march. BOULANGER: Cortege. Nocturne, F
major. BRAHMS: Hungarian dance, no. 1, G minor. CHOPIN:
Nocturne, D flat major. Nocturne, E flat major. DRIGO: Valse
bluette. DVORAK: Slavonic dance, no. 2. Slavonic dance, op.
72, no. 2, E minor. Slavonic dance, op. 72, no. 8. ELGAR:
La capricieuse, op. 17. GLAZUNOV: Meditation, op. 32. Raymonda, op. 57: Valse grande adagio. GODOWSKY: Waltz, D major.
GOLDMARK: Concerto, violin, A minor: Andante. GRANADOS: Danza
española, op. 37: Andaluza. HAYDN: Quartet, strings, D major:
Vivace. JUON: Berceuse. KREISLER: Minuet. Sicilienne et
rigaudon. LALO: Symphonie espagnole, op. 21: Andante. MENDEL-
SSOHN: Concerto, violin, op. 64, E minor: Finale. On wings of
song, op. 34, no. 2. MOSZKOWSKI: Stücke, op. 45, no. 2:
Guitarre. MOZART: Divertimento, no. 17, K 334, D major: Minuet.
Serenade, D major: Rondo. PAGANINI: Caprices, op. 1, nos. 13,
20. Moto perpetuo, op. 11. SARASATE: Carmen fantasy, op. 25.
Danzas españolas, op. 21, no. 1: Malagueña; no. 2: Habañera;
op. 23, no. 2: Zapateado. Introduction and tarantelle.
Zigeunerweisen, op. 20, no. 1. SCHUBERT: Ave Maria, D 839.
SCHUMANN: Myrthen, op. 25, no. 1: Widmung. SCOTT: The gentle
maiden. TCHAIKOVSKY: Concerto, violin, op. 35, D major: Can-
zonetta. Sérénade melancolique, op. 26. Souvenir d'un lieu
cher, op. 42: Scherzo. Serenade, strings, op. 48, C major:
Valse. WIENIAWSKI: Concerto, violin, no. 2, op. 22, D minor:
Romance. Scherzo tarantelle. Vol. 2, The first electrical
recordings: ACHRON: Hebrew melody, op. 33. ALBENIZ: Suite
española, no. 3: Sevillanas. BACH: Partita, violin, no. 3,
S 1006, E minor: Minuets 1 and 2. English suite, no. 3, S 808,
G minor: Sarabande, Gavotte, Musette. CASTELNUOVO-TEDESCO:
Etudes d'ondes: Sea murmurs. Valse. CLERAMBAULT: Largo on
the G string. COUPERIN: Livres de clavecin, Bk IV, Ordre, no.
17: Les petits Moulins a vent. DEBUSSY: L'Enfant prodigue:
Prelude. Prelude, Bk I, no. 8: La fille aux cheveux de lin.
La plus que lente. DOHNANYI: Ruralia Hungarica: Gypsy andante.
DRIGO: Valse bluette. ELGAR: La capricieuse, op. 17. FALLA:
Spanish popular songs: Jota. GLAZUNOV: Concerto, violin, op.
82, A minor. Meditation, op. 32. GODOWSKY: Alt Wien. GRIEG:
Lyric piece, op. 54, no. 6: Scherzo. Lyric piece, op. 71, no.
3: Puck. Sonata, violin and piano, no. 3, op. 45, C minor.
HUMMEL: Rondo, op. 11, E flat major. KORNGOLD: Much ado about
nothing, op. 11: Holzapfel und Schlehwein. MENDELSSOHN: On
wings of song, op. 34, no. 2. MILHAUD: Saudades do Brasil,
no. 10: Sumaré. MOSZKOWSKI: Stücke, op. 45, no. 2: Guitarre.
MOZART: Concerto, violin, no. 5, K 219, A major. PAGANINI:
Caprices, op. 1, nos. 13, 24, 20. PONCE: Estrellita. RAVEL:
Tzigane. RIMSKY-KORSAKOV: The tale of the Tsar Sultan: Flight
of the bumblebee. SARASATE: Danza española, op. 23, no. 2:
Zapateado. SCHUBERT: Ave Maria, D 839. Impromptu, op. 90, no.
3, D 899, G flat major. Sonatina, violin and piano, op. 137,

D 384, D major: Rondo. STRAUSS: Sonata, violin and piano, op. 18, E flat major. Stimmungsbilder: An einsamer Quelle. VIVALDI: Sonata, violin, op. 2, no. 2, A major. WIENIAWSKI: Scherzo tarantelle. Vol. 3, 1935-1937: BACH: Sonata, violin, no. 1, S 1001, G minor. Sonata, violin, no. 3, S 1005, C major. BAZZINI: La ronde des lutins, op. 25. BRAHMS: Sonata, violin, no. 2, op. 100, A major. DINICU: Hora staccato. FAURE: Sonata, violin and piano, no. 1, op. 13, A major. FALLA: La vida breve: Danza, no. 1. GRIEG: Sonata, violin, no. 2, op. 13, G minor. POULENC: Mouvements perpetuels, no. 1. SAINT-SAENS: Introduction and rondo capriccioso, op. 28. SCOTT: Tallahassee suite. SZYMANOWSKI: Le Roi Roger, op. 46: Chant do Roxane. VIEUXTEMPS: Concerto, violin, no. 4, op. 31, D minor. WIENIAWSKI: Concerto, violin, no. 2, op. 22, D minor. Polonaise, op. 4, D major. Vol. 4, 1937-1941: BEETHOVEN: Concerto, violin, op. 61, D major. BRAHMS: Concerto, violin, op. 77, D major. Concerto, violin and violoncello, op. 102, A minor. CHAUSSON: Concerto, violin, piano and string quartet, op. 21, D major. PROKOFIEV: Concerto, violin, no. 2, op. 63, G minor. SAINT-SAENS: Havanaise, op. 83. SARASATE: Zigeuner-weisen, op. 20, no. 1. WALTON: Concerto, violin, B minor. Vol. 5, 1946-1949: ARENSKY: Concerto, violin, A minor: Tempo di valse. BACH: Concerto, 2 violins and strings, S 1043, D minor. English suite, no. 6, S 811: Gavottes, nos. 1 and 2. BAX: Mediterranean. BEETHOVEN: German dance, no. 6. BRUCH: Scottish fantasia, op. 46. CASTELNUOVO-TEDESCO: Etudes d'ondes: Sea murmurs. Tango. CHOPIN: Nocturne, E minor. DEBUSSY: Chanson de Bilitis: La chevelure. Song: Ariettes oubliées, no. 2: Il pleure dans mon coeur. Preludes, Bk I, no. 8: La fille aux cheveaux de lin. ELGAR: Concerto, violin, op. 61, B minor. FALLA: El amor brujo: Pantomime. Spanish popular songs: Jota. HALFFTER: Danza de la gitana (Escriche). KORN-GOLD: Much ado about nothing, op. 11: Holzapfel und Schlehwein, Garden scene. MEDTNER: Fairy tale, B flat minor. MENDELSSOHN: Trio, piano, no. 1, op. 49, D minor: Scherzo. Sweet remem-brance. MILHAUD: Saudades do Brasil, no. 7: Corcovado. MOZART: Divertimento, no. 17, K 334, D major: Minuet. NIN: Cantilena asturiana. POLDOWSKI: Tango. PROKOFIEV: Gavotte. March, F minor. RACHMANINOFF: Daisies, op. 38, no. 3. Etudes tab-leaux, op. 37, no. 2. Oriental sketch. RAVEL: Valses nobles et sentimentales, nos. 6 and 7. RIMSKY-KORSAKOV: The tale of the Tsar Sultan: Flight of the bumblebee. TANSMAN: Mouvement perpetuel. VIEUXTEMPS: Concerto, violin, no. 5, op. 37, A minor. Vol. 6, 1950-1955: BEETHOVEN: Romance, no. 1, op. 40, G major. Romance, no. 2, op. 50, F major. Sonata, violin and piano, no. 9, op. 47, A major. BLOCH: Sonatas, violin, nos. 1 and 2. BRAHMS: Sonata, violin, no. 3, op. 108, D minor. BRUCH: Concerto, violin, no. 1, op. 26, G minor. HANDEL: Sonata, violin, op. 1, no. 13, D major. RAVEL: Tzigane. SAINT-SAENS: Sonata, violin, no. 1, op. 75, D minor. SCHUBERT: Sonatina, violin and piano, no. 3, G minor. TCHAIKOVSKY: Concerto, violin, op. 35, D major. WIENIAWSKI: Polonaise, op. 4, D major. Jascha Heifetz, vln; with assisting artists.

+HF 8-75 p71
++NR 7-75 p12
+NYT 6-29-75 pD17

+SR 10-4-75 p51
+St 1-76 p112

CRM 5-1900 (5)
3205 BERLIOZ: Roméo and Juliet, op. 17: Queen Mab scherzo. DEBUSSY:
      Images pour orchestra: Ibéria. La mer. MENDELSSOHN: A mid-
      summer night's dream, op. 21/61: Incidental music. RESPIGHI:
      Feste Romane. SCHUBERT: Symphony, no. 9, D 944, C major.
      STRAUSS, R.: Tod und Verklärung, op. 24. TCHAIKOVSKY: Symphony,
      no. 6, op. 74, B minor. PO; Arturo Toscanini.
          ++NR 12-76 p2   +-SFC 10-3-76 p33

                        REDIFFUSION

15-46
3206 ADDINSELL: Warsaw concerto. BARGONI: Autumn concerto. FIBICH:
      Poem. FISCHER: South of the Alps. TROJAN: The nightingale's
      concerto. Jan Tesik, pno; Vera Zegzulková, vln; Brno Radio
      Orchestra; Jiři Hudec.
          +-Gr 6-76 p96

                        RICHMOND

R 23197 (Reissue)
3207 ADAM: Variations on a theme of Mozart's "Ah, vous dirai-je Maman".
      ALABIEFF: The nightingale. BELLINI: I puritani: Qui la voce...
      Vien, diletto. La sonnambula: Come per me sereno...Sovra il
      sen. DONIZETTI: Lucia di Lammermoor: Mad scene. GOUNOD:
      Mireille: O légère hirondelle; Heureux petit berger. Mado
      Robin, s; Various orchestras; Anatole Fistoulari.
          +St 1-76 p111

                        ROCOCO

2049
3208 DIEMER: Valse de concerto. CHOPIN: Valse, op. 34, no. 1, A flat
      major. KIENZL: Kahn Szene, neuer Walzer. LESCHITZKY: Gavotte.
      LISZT: Hungarian rhapsody, no. 13, G 244, A minor. Liebesträum,
      no. 3, G 541, A flat major. Paraphrase on VERDI: Rigoletto,
      G 434. MENDELSSOHN: Rondo capriccioso, op. 14, E major. SAINT-
      SAENS: Marche militaire française: Africa, Valse mignonne,
      Reverie a Blidah, Suite Algérienne. SCHARWENKA: Polish dance,
      op. 31, no. 1. Camille Saint-Saens, Louis Diemer, Wilhelm
      Kienzl, Ethel Leginska, Anna Esipova, Xavier Scharwenka, pno.
          +-ARSC Vol VIII, no. 2-3 +-NH 2-75 p11
                  p93

5370
3209 d'ALBERT: Zur Drossel sprach der Fink. BRAHMS: Liebestreu, op. 3,
      no. 1; Nicht mehr zu dir zu gehen. SCHUBERT: An die Laute,
      D 905; Erlkönig, D 328: Gruppe aus dem Tartarus, D 583: Kreuz-
      zug; Der Musensohn, D 764. Schwanengesang, D 957: Der Doppel-
      gänger; Die Stadt. SCHUMANN: Frühlingsnacht, op. 39, no. 12;
      Der Schatzgraber, op. 45, no. 1; Der Soldat, op. 40, no. 3.
      WEINGARTNER: Schafers Sonntaglied thou art a child. Therese
      Behr, con; Lucille Marcel, s; Instrumental accompaniments.
          +-ARSC Vol VIII, no. 2-3 +-NR 3-76 p9
                  p88

5377
3210 BIZET: Carmen: Seguidilla. BOIELDIEU: Weisse Dame: Duet.

MEYERBEER: Les Huguenots: Welch ein Schreck; Duet, Act 4.
OFFENBACH: Les contes d'Hoffmann: Horst du es tonen. NICOLAI:
Die lustige Weiber von Windsor: Frau Fluth's aria. WAGNER:
Götterdämmerung: Ha weisst du was er mir ist. Der fliegende
Holländer: Senta's ballad. Lohengrin: Das susse Lied erhalt.
Tannhäuser: Dich teure Halle, Frubitte der Elisabeth. Tristan
und Isolde: O sink hernieder. VERDI: Aida: Final duet. Sophie
Sedlmair, Marie Gutheil-Schoder, s; Instrumental accompaniments.
　　　*NR 3-76 p7

5379
3211 ALABIEFF: Die Nachtigall. AUBER: Fra Diavolo: Romanze. DONIZETTI:
Figlia del Reggimento: Weiss night die Welt. EDWARDES: My
colleague Caruso. GRUNFELD: Schonen von Fogaras: Ganselied.
GOLDMARK: Königin von Saba: Magische tone. KIENZL: Evangeli-
mann: Selig song. MOZART: Le nozze di Figaro, K 492: Soli einst
das Gräflein. ROSSINI: Guglielmo Tell: Duet. THOMAS: Mignon:
Styrienne. VERDI: Rigoletto: Liebe ist Seligkeit. WAGNER:
Lohengrin: Du Armste; Entweihte Götter. Die Meistersinger
von Nürnberg: Fanget an. Tannhäuser: Lied des Hirtenknaben;
Als du in kuhnem Sange. Die Walküre: Ho-jo-to-ho. WEBER: Der
Freischütz: Durch die Walder. Erika Wedekind, Gertrude Forstel,
s; Frieda Langendorf, ms; Dezso Aranyi, Karel Burrian, t;
Dezso Zador, bar; Instrumental accompaniments.
　　　*NR 9-76 p9

5380 (3)
3212 Kirsten Flagstad memorial album. GRIEG: Songs: The first meeting;
The goal; A hope; I love thee. PURCELL: Dido and Aeneas: Dido's
recitative and lament. STRAUSS, R.: Songs: Allerseelen, op. 10,
no. 8; Befreit, op. 39, no. 4; Cäcilie, op. 27, no. 2. Vier
letzte Lieder: Beim Schlafengehen; Frühling; Im Abendrot;
September. WAGNER: Götterdämmerung; Zu neuen Taten, Starke
Scheite. Tristan und Isolde: O sink hernieder, Liebestod.
Die Walküre: Nun zaume...Ho-jo-to-ho, Siegmund, Sieh auf mich,
War est so schmählich. Interview with Flagstad, excerpts;
Flagstad's farewell words after the last Dido and Aeneas per-
formance. Kirsten Flagstad, s; Instrumental accompaniment.
　　　+NR 8-76 p7

5383
3213 DONIZETTI: Lucia di Lammermoor: Tombe degli avi miei...Fra poco
a me. GIORDANO: Andrea Chénier: Un di all'azzuro spazio.
MASSENET: Werther: Il faut nous separer...Pourquoi me reveiller.
PIETRI: Maristella: Io conosco un giardino. PIZZETTI: Calzare
d'Argento: Da vero quanto grande e la misera. PUCCINI: Turandot;
Nessun dorma. ROSSINI: Il barbiere di Siviglia: All'idea di
quel metallo. VERDI: Un ballo in maschera: Fors'e la soglia...
Ma se m'e forza perderti. Giuseppe di Stefano, t; Instrumental
accompaniment.
　　　+NR 9-76 p9

5386
3214 BELLINI: I puritani: O amato zio...Sai com arde. BOITO: Mefistofele:
L'altra notte. CILEA: Adriana Lecouvreur: Poveri fiori.
DONIZETTI: Anna Bolena: Piangete voi...Al dolce guidami
L'Elisir d'amore: Con se va contento...Quanto amore. Maria di
Rohan: Rival, se tu spaersi...Cupa fatal mestizia. VERDI:
Aida: Ritorna vincitor. La forza del destino: Pace, pace, mio
Dio. Virginia Zeani, s; Instrumental accompaniment.
　　　+NR 8-76 p9

5397
3215 BALFE: Killarney. BOWERS: Because I love you. BRAGA: La sere-
      nata. BISHOP: Coming though the rye; Should he upbraid.
      DELL-ACQUA: Villanelle. d'HARDELOT: I know a lovely garden.
      HALEVY: La juive: Wenn ew'ger Hass. LEONCAVALLO: Mattinata.
      MASSENET: Hérodiade: Il est doux. MOZART: Die Zauberflöte,
      K 620: In diesn heil'gen Hallen. SAMARA: Histoire d'amour.
      VERDI: Aida; O ciel d'azur. Il trovatore: D'amore sull ali.
      WAGNER: Götterdämmerung: Hagens Ruf. Lohengrin: Gott grüss
      Euch. Die Meistersinger von Nürnberg: Ansprache des Pogner.
      Tannhäuser: Dir töne Lob, Stets soll nur dir. Ada Adini,
      Elizabeth Parkina, s; Francis MacLennan, t; Allen Hinckley,
      Eduard Landow, bs; Instrumental accompaniment.
          *NR 8-76 p9

                          RUBINI

GV 58
3216 BOITO: Mefistofele: Spunta l'aurora palida. GIORDANO: Siberia:
      Nel suo amore. MASCAGNI: Amica: Più presso al ciel. Caval-
      leria rusticana: Voi lo sapete. MEYERBEER: L'Africaine: Sur
      mes genoux. PUCCINI: La bohème: Si, mi chiamano Mimi. Manon
      Lescaut: In quelle trine morbide. VERDI: Aida: Ritorna
      vincitor. La traviata: Addio del passato; Madimagella Valery
      ...Pura siccome un angelo...E grave il sacrifizio...Diet alla
      giovane. (All selections sung in Italian). Giannina Russ, s;
          +-RR 11-76 p24
GV 63 (2)
3217 BELLINI: Norma: Casta diva...Ah, bello, a me ritorno. BOITO:
      Mefistofele: Lontano, lontano. BORODIN: Prince Igor: Ah, I
      bitterly weep (Jaroslavna's aria, Act 4). DARGOMIZHSKY:
      Russalka: Natasha's aria, Act 4. GLINKA: Do not tempt me
      needlessly. GRODSKI: The night hovered over the earth. MEYER-
      BEER: The Huguenots: Tu l'as dit; Oui, tu m'aimes. MUSSORGSKY:
      Boris Godunov: Fountain duet. Fair at Sorochinsk: O banish
      thoughts of sorrow (Parassia's daydream, Act 3). NAPRAVNIK:
      Duvrobsky: The forest guards its secret. PONCHIELLI: La
      gioconda: Cielo e mar; Suicidio. SAINT-SAENS: Samson et Dalila:
      Arrêtz, ô mes frères. SEROV: Judith: Judith's aria, Act 1.
      Rogneda: Groaned the blue sea (The Viking ballad). TCHAIKOVSKY:
      The enchantress: Nastasya's aria; Where are you my beloved.
      Maid of Orelans: Joan's prayer. Mazeppa: Sleep, my pretty
      baby. Queen of spades, op. 68: Twill soon be midnight. VARLA-
      MOV: Romance. WAGNER: Tannhäuser: Dich, teure Halle; Allmacht'ge
      Jungfrau. VERDI: Aida: O patria mia; Radames so che qui
      attendi...Su dunque; La tra foreste; Fuggiam gli ardori inos-
      piti. (All selections sung in Russian). Natalya Yermolenko-
      Juzhina, s.
          +-RR 11-76 p24
GV 65
3218 DONIZETTI: La favorita: Pour tant d'amour. GRIEG: Song: En
      svane. KORLING: Aftonstamning. LEONCAVALLO: I Pagliacci:
      Prologue. MOZART: Don Giovanni, K 527: La ci darem. Le nozze
      di Figaro, K 492: Crudel, perchè finora. RANGSTROM: De fangna.
      ROSSINI: Il barbiere di Siviglia: Largo al factotum; Dunque io

son. SODERMAN: Kung Heimar och Aslög. TCHAIKOVSKY: Eugene
Onegin, op. 24: Written words. THOMAS: Mignon: Legères hiron-
delles. VERDI: La traviata: Di provenza. Il trovatore: Il
balen. (All selections sung in Swedish). John Forsell, bar.
        +-RR 11-76 p24

RS 301 (2)
3219 BICCI-BILLI: E canta il grillo. BIZET: Carmen: Habañera; Chanson
bohème; Seguidille. BOITO: Mefistofele: Lontano, lontano.
DAVIDOV: Night, love and moon. FAURE: The crucifix. GLINKA:
Do not tempt me needlessly. GRODSKI: The seagull's cry.
HUMPERDINCK: Hänsel und Gretel: Ein Männlein steht im Walde.
MASSENET: Werther: Air des larmes. MEYER-HELMUND: Violets.
NAPRAVNIK: Dubrovsky: Never to see her. Harold: Lullaby.
PUCCINI: Tosca: Vissi d'arte (2). RESPIGHI: Stornellatrice,
1st and 2nd edition. ROTOLI: Fior che langue. RUBINSTEIN:
Night. TCHAIKOVSKY: At the ball, op. 38, no. 3. Queen of
spades, op. 68: Twill soon be midnight. TOSTI: Penso. ANON.:
Russian song. Interview with Medea Mei-Figner, Paris, 1949.
Medea Mei-Figner, s.
        +HFN 3-76 p99   +RR 11-76 p24

GV 68
3220 ARDITI: Il bacio. BELLINI: Norma: Casta diva. I puritani: Son
vergin vezzosa. DONIZETTI: Don Pasquale: So anch'io la virtu;
Pronto io son. Linda di Chamounix: O luce di quest anima;
Rondo. Lucia di Lammermoor: Regnava nel silenzio; Sulla tomba;
Verrano a te. MEYERBEER: The Huguenots: O beau pays. ROSSINI:
Il barbiere di Siviglia: Dunque io son. STELLA: Riso. THOMAS:
Mignon: Je suis Titania. VERDI: La traviata: Addio del passato;
Alfredo, di questo core. (All selections sung in Italian).
Giuseppina Huguet, s.
        +-RR 11-76 p24

GV 70
3221 DELIBES: Lakmé: Où va la jeune Indoue (Yvonne de Treville, s).
MENDELSSOHN: Auf flügeln des Gesanges, op. 34, no. 2 (Heinrich
Hensel, t). PONCHIELLI: La gioconda: Cielo e mar (Giovanni
Martinelli, t). MOZART: Die Zauberflöte, K 620: Dies Bildnis
ist bezaubernd schön (Jacques Urlus, t). PUCCINI: Manon
Lescaut: In quelle trine morbide (Maria Rappold, s). Tosca:
Vissa d'arte (Emmy Destinn, s). ROSSINI: Il barbiere di
Seviglia: Ecco ridente in cielo (Jose Mojica, t). THOMAS:
Mignon: Elle ne croyait pas (Pierre Asselin, t). VERDI: Attila:
Tregua è cogl'unni...Dagli immortali vertici (Mario Laurenti,
bar). Rigoletto: E il sol dell'anima (Anna Case, s; Ralph
Errolle). Il trovatore: Stride la vampa (Maria Duchêne, ms);
Mira d'acerbe lagrime (Marie Rappold, s; Taurino Parvis, bar).
        +-RR 11-76 p24

                           SAGA

5213
3222 Anthology of English song: DUNHILL: The cloths of heaven; To the
Queen of heaven. FINZI: Let us garlands bring, op. 18: Come
away, come away, death; It was a lover and his lass. GIBBS:
Love is a sickness; This is a sacred city. GURNEY: I will go
with my father a'ploughing; Sleep. HEAD: A piper. HOWELLS:
Come sing and dance; King David. IRELAND: Her song; A thanks-

giving. VAUGHAN WILLIAMS: Mystical songs: The call. Songs of travel: Youth and love. WARLOCK: Balulalow; Youth. Janet Baker, con; Martin Isepp, pno.
+-ARG 12-76 p56   +St 11-76 p165

5225
3223 CROFT: O worship the King. DYKES: The King of love my shepherd is; Eternal Father, strong to save. DARWALL: Ye holy angels bright. GIBBONS: Jesu, grant me this, I pray. GOSS: Praise, my soul, the king of heaven. HANDEL: Royal fireworks music: Bourrée. MONK: Abide with me. PURCELL: Abdelazer: Introduction and rondeau. SHEELES: The spacious firmament on high. STAINER: My God I love Thee. TALLIS: When rising from the bed of death. WESLEY, S.: Air, F major. WESLEY, S. S.: O thou who camest from above. WILKES (arr.): Let us, with gladsome mind. Salisbury Cathedral Choir; Christopher Dearnley, Richard Lloyd, org.
+Gr 1-76 p1229   /RR 2-76 p56

5402
3224 BACH: Toccata, S 912, D major. BOHM: Präludium, Fuge und Postludium, G minor. BULL: The King's hunt. BYRD: Wolsey's wilde. COUPERIN, L.: Le tombeau de M. Blancrocher. FARNABY: Loath to depart. PHILIPS, P.: Amarilli di Julio Romano. SCARLATTI, D.: Sonatas, harpsichord, K 87, B minor; K 201, G major; K 370, E flat major; K 371, E flat major. Elizabeth de la Porte, hpd.
+Gr 10-75 p661   +MT 2-76 p146
+HFN 12-75 p153   +RR 8-75 p55

5411
3225 BEETHOVEN: Mödlinger dances, nos. 1-4, 6, 8. HAYDN: German dances, nos. 4, 10-11. LANNER: The parting of the ways. Summer night's dream. MOZART: Ländler, K 606 (6). SCHUBERT: German dances with coda and 7 trios. STRAUSS, J. I: Vienna carnival. Vienna Volksoper Orchestra; Paul Angerer.
+Gr 1-76 p1248   +RR 12-75 p50
+HFN 2-76 p115

5412. Tape (c) CA 5412
3226 Music for 2 guitars. BARRIOS: Danza paraguaya. CRESPO: Norteña, Homenaje a Julian Aguirre. LAURO: Suite venezelano: Vals. PONCE: Canciones populares mexicanas (3). Valse. VILLA-LOBOS: Chôros, no. 1, E minor. Cirandinha, no. 1: Therezinha de Jesus. Cirandinha, no. 10: A canõa virou. TRAD.: Boleras sevillanas. Buenos reyes. Cantar montañes. Cubana. De blanca tierra. Linda amiga. El pano moruno. El puerto. Salamanca. Tutu maramba. Villancico. Walter Reybli, Konrad Ragossnig, gtr.
+Gr 1-76 p1215   +RR 3-76 p64
+HFN 12-75 p159   +RR 12-76 p108 tape
++HFN 10-76 p185 tape

5414
3227 Works for flute, recorder and organ. LOCATELLI: Sonata, flute, G minor. MATTEIS: Suite, recorder and organ, D major. MARCELLO: Sonata, recorder, D minor. POGLIETTI: Ricercar septimi toni. PURCELL, D.: Sonata, recorder, no. 2, D minor. Prelude, F major. PURCELL, H.: Chaconne, F major. Prelude, D major. STANLEY: Sonata, flute and organ, D minor. Voluntary, op. 7, no. 9, G major. Hans-Martin Linde, rec and flt; Kurt Rapf, org.

SAGA (cont.) 642

```
              +Gr 3-76 p1482              +RR 3-76 p63
              +MT 8-76 p661
```

<u>5417</u>. Tape (c) CA 5417
3228 BISHOP: Grand march, E major. CLARKE: Prince of Denmark's march.
      HAYDN: Feldpartita, B flat major. PEZEL: Suite, C major.
      PICK (attrib): March and troop. Suite, B flat major. London
      Bach Ensemble; Trevor Sharpe.

```
              +Gr 3-76 p1478             +-RR 12-75 p44
              +HFN 1-76 p123             +-RR 12-76 p105 tape
              +-HFN 10-76 p185 tape
```

<u>5420</u>
3229 CALVI: The Medici court. FOSCARINI: Il furioso. GALILEI: For
      the Duke of Bavaria. MELLII: For the Emperor Matthias.
      NEUSIDLER: The Burgher of Nuremberg. ROBINSON: Lessons for
      the lute. ANON.: At an English ale house. James Tyler,
      renaissance lute, archlute, baroque guitar.

```
              +Gr 4-76 p1643             +RR 3-76 p56
              +MT 8-76 p666              +-SFC 8-22-76 p38
```

<u>5421</u>
3230 BEETHOVEN: Mödlinger dances: Waltzes, nos. 3, 10, 11. DEBIASY:
      Dance. LANNER: Favorit-polka, op. 201. SCHUBERT: Ecossaises,
      D 529, nos. 1-3, 5; D 783, nos. 1 and 2. STELZMULLER: Dance.
      STRAUSS, E.: Unter der Enns, op. 121. STRAUSS, J. I: Loreley-
      Rheinklänge, op. 154. STRAUSS, J. II: Man lebt nur einmal, op.
      167. Sinngedichte, op. 1. STRAUSS, Josef: Galoppin-polka, op.
      237. ANON.: Dance, D major. Schellerl-Tanz. Vienna Baroque
      Ensemble; Hans Totzauer.

```
              +Gr 7-76 p221              +-RR 7-76 p53
              ++HFN 7-76 p85
```

<u>5426</u>
3231 BACH: Anna Magdalena notebook, S 508: Musette, March. Sonatas
      and partitas, violin, S 1002: Sarabande; S 1003: Andante.
      Prelude, S 999, C minor. CUTTING: Greensleeves. DOWLAND:
      Melancholy galliard. Mistress Winter's jump. My Lady Hundson's
      puffe. HANDEL: Aylesford pieces: Minuets, A minor (2). LeROY:
      Branle gay. MILAN: Pavanas, D major, C major. MUDARRA: Fan-
      tasia, no. 10. NEUSIDLER: The Jew's dance. PURCELL: Air and
      minuet. SCARLATTI, A.: Minuet and gavotte. VIVALDI: Concerto,
      2 mandolins, G major: Andante. ANON.: Kemp's jig. Timothy
      Walker, gtr.

```
              +-Gr 7-76 p194             +RR 5-76 p66
```

<u>5427</u>
3232 BRAHMS: Pieces, piano, op. 117, no. 1, E flat major. CHOPIN:
      Nocturne, op. 9, no. 2, E flat major. Etude, op. 10, no. 3,
      E major. Waltz, op. 64, no. 2, C sharp minor. DEBUSSY:
      Children's corner suite: Golliwog's cakewalk. GRIEG: Lyric
      pieces, op. 54, no. 4: Nocturne. Lyric pieces, op. 65, no. 6:
      Wedding day at Troldhaugen. RACHMANINOFF: Prelude, op. 3, no.
      2, C sharp minor. SCHUBERT: Moment musical, op. 94, no. 3,
      D 780, F minor. SCHUMANN: Waldscenen, op. 82, no. 7: The
      prophet bird. Edward Moore, pno.

```
              +Gr 8-76 p342              -RR 8-76 p68
              +HFN 8-76 p76
```

SERAPHIM

S 60251
3233 DEBUSSY: Ariettes oubliées; Beau soir. DUPARC: L'Invitation au
      voyage; Phidylé. FAURE: Les berceaux, op. 23, no. 1; La
      chanson du pêcheur, op. 4, no. 1; Mai, op. 1, no. 2. GOUNOD:
      Aimons-nous; Ou voulez-vous aller. SCHUBERT: Schwanengesang,
      D 957, excerpts (4). SCHUMANN: Lust der Sturmnacht, op. 55,
      no. 1; Mein schöner Stern; Stille Liebe, op. 35, no. 8: Stille
      Tränen, op. 35, no. 10; Widmung, op. 25, no. 1. Gerard Souzay,
      bar; Dalton Baldwin, pno.
           +-HF 9-76 p106   *ON 5-76 p48
           ++NR 12-75 p12   ++St 2-76 p112
S 60259 (Reissue from Angel 35354)
3234 A tribute to David Oistrakh. DEBUSSY: Suite bergamasque: Clair
      de lune. FALLA: Canciones populares españolas: Jota. KODALY:
      Hungarian dances (3). SUK: Love song. TCHAIKOVSKY: Valse
      scherzo. WIENIAWSKI: Legende, op. 17. YSAYE: Extase, op. 21,
      E major. ZARZYCKI: Mazurka. David Oistrakh, vln; Vladimir
      Yampolsky, pno.
           +/NR 6-76 p15

SERENUS

SRS 12039
3235 BRAUN: Psalm 98. BERLINSKI: Kol Nidre. HANDEL: Esther: Arias.
      Israel in Egypt: Arias. Judas Maccabaeus: Arias. LEWANDOWSKI:
      Zocharti loch. NIEMAN: Israeli folk songs (9). ROSSI: Kad-
      dish. Nathan Lam, bar; David Tilman Choir; Max Walmer, pno.
           +-NR 5-76 p8   +-St 10-76 p140
SRS 12064
3236 BALADA: Cuatris, 4 instruments. Musica en cuatro tiempos, piano.
      FARBERMAN: Alea, 6 percussion. HUSTON: Lifestyles, clarinet,
      violoncello and piano. KAPR: Dialogues, flute and harp (5).
      WHITTENBERG: Set for 2, viola and piano. Samuel Baron, flt;
      Dagmar Platilova, hp; David Sackson, vla; George Plasko, clt;
      Ken Miller, vlc; Dwight Peltzer, Elizabeth Marshall, Donna
      Hallen Lowry, pno; Pugwash Percussion Ensemble; Barcelona
      Conjunto Cameristico.
           +ARG 12-76 p56   ++NR 4-76 p7

1750 ARCH

S 1753
3237 DUFAY: Songs: C'Est bien raison de devoir essaucier; Je me com-
      plains piteusement; Invidia nimica; Malheureux, cuer que veux
      to faire; Par droit je puis bien complaindre. GRIMACE:
      A l'arme a l'arme. LANDINI: Se la nimica mie; Adiu adiu.
      PERUSIO: Andray soulet. VAILLANT: Par maintes foys. ANON.:
      Istampita ghaetta. Music for a While.
           ++AR 2-76 p133   +NR 11-75 p11

SMITHSONIAN COLLECTION

N 002 (from Smithsonian Collection, P.O. Box 5734, Terra Haute, Ind)
3238 Music from the age of Jefferson. ARNE: Artaxerxes: Water parted

from the sea. BACH, J. C.: Sonata, harpsichord, op. 18, no. 2, A major. CLEMENTI: Sonata, flute and piano, op. 2, no. 3, G major. HOOK: The Caledonian laddy. LIGHT: Guardian angels. LINLEY: Otello: Air. MARTINI, M.: Plaire a celui que j'aime. PICCINI: Didon: Ah, que je fus bien inspiree. PURCELL, D.: Psalm 100. SCHOBERT: Sinfonia, op. 10, no. 1, E flat major. WODIZKA: Sonata, violin and continuo, op. 1, no. 3, D minor. Carole Bogard, s; John Solum, one-keyed flute; Sonya Monosoff, vln; Robert Sheldon, Thomas Murray, hn; John Hsu, vlc; Howard Bass, English guitar; Albert Fuller, hpd, pno; James Weaver, hpd, pno, org.
+HF 7-76 p69   +St 5-76 p113

STENTORIAN

SC 1724
3239 ARNE: Flute tune. BACH: Anna Magdalena notebook, S 508: Bist du bei mir. DEBUSSY: Arabesques, nos. 1 and 2. Cortège et air de danse. RAVEL: Pavane. SATIE: Gymnópedie, no. 1. Keith Chapman, org.
++NR 5-76 p13

SP 1735
3240 ANDERSON: Summer skies. BERLIOZ: Rakoczy march. BOELLMAN: Ronde français. CONFREY: Kitten on the keys. FALLA: El amor brujo: Ritual fire dance. MacDERMOT: Good morning starshine. MOURET: Rondeau. RIMSKY-KORSAKOV: Scheherazade, op. 35, excerpts. Keith Chapman, org.
-NR 8-76 p11

STERLING

1001/2 (2)
3241 BEETHOVEN: Fantasia, op. 77, G minor. BRUCKNER: Quadrilles. HUMMEL: Sonata, piano, op. 92, A flat major. MARTIN: Preludes (8). MENDELSSOHN (Liszt): On wings of song, op. 34, no. 2. PAGANINI (Liszt, Busoni): La campanella. SAINT-SAENS (Liszt rev. Wikman): Danse macabre, op. 40. Bertil Wikman, Solveig Wikman, pno.
+-RR 10-76 p92

SUPRAPHON

110 1637
3242 FLOTOW: Allesandro stradella: Overture. KREUTZER: Das Nachtlager von Granada: Overture. NICOLAI: Die lustige Weiber von Windsor: Overture. REZNICEK: Donna Diana: Overture. WAGNER: Rienzi: Overture. WEBER: Abu Hassan: Overture. Brno State Philharmonic Overture; Zeljko Straka.
+NR 11-76 p4   +-RR 12-76 p68

110 1638
3243 HEUBERGER: Der Opernball: Overture. LEHAR: Wiener Frauen: Overture. OFFENBACH: Orphée aux Enfers: Overtures. STRAUSS, J. II: Die Fledermaus, op. 363: Overture. Waldmeister overture. SUPPE: Die schöne Galathea: Overture. CPhO; Václav Neumann.
+NR 11-76 p4

111 1230. Tape (c) 4-11230
3244 PONCE: Estrellita. Mexican folksong. Scherzino mexicano.
      SAINZ DE LA MAZA: Rondeña (Andaluza). RODRIGO: Adagio.
      Bolero. PIPO: Cancion y danza, no. 1. SANTORSOLO: Prelude,
      no. 1. SAVIO: Batucada. SOR: Variations on "Folies d'Espagne".
      TURINA: Sonata, guitar, op. 61, D major. VILLA-LOBOS: Etude,
      no. 11. WALKER: Small variations on a Catalonian folksong.
      Louis Walker, gtr.
           +HFN 3-76 p113 tape   +RR 9-74 p79
           +NR 11-74 p12   +RR 4-76 p81 tape

111 1390
3245 HABA: Suite, bass clarinet, op. 96. KLUSAK: Reydowak. KUCERA:
      Invariant, bass clarinet, piano and stereo tape recorder.
      PARSCH: Poetica II. PINOS: Esca, prepared piano. STEDRON:
      Leich on a theme by Heinrich von Meissen. Josef Horak, bs
      clt; Emma Kovarnova, pno, hpd.
           +NR 8-74 p6   +RR 1-76 p53

111 1721/2 (2)
3246 AURIC: Adieu, New York. BURIAN: American suite. COPLAND: Piano
      blues (4). DEBUSSY: Children's corner suite: Golliwog's cake-
      walk. GERSHWIN: Preludes. HINDEMITH: Suite "1922", op. 26.
      MARTINU: Preludes. SATIE: Jack in the box. SCHULHOFF: Es-
      quisses de jazze. Rag music. Petr Toperczer, Jan Vrana,
      Emil Leichner, Jan Marcol, Milos Mikula, pno.
           +Gr 12-76 p1022

                         SWEDISH SOCIETY

SLT 33188
3247 AFLVEN: Aftonen. HEDAR: Musik. LINDBERG: Pingst. LINDBLAD:
      Drömmarne. LUNDVIK: Det första vårregnet. MALMFORS: Wiegen-
      lied; Scherzlied; Mangard; I fjärren dis. Jan Eyron, pno;
      Swedish Radio Chorus; Eric Ericson.
           +RR 12-76 p92

SLT 33189
3248 BACH: Suite, lute, S 996, E minor. MENDELSSOHN (arr. Blanco):
      Quartet, strings, no. 1, op. 12, E flat major: Canzonetta.
      NARVAEZ: Diferencias sobre "Guardame las vacas.". PRAETORIUS
      (arr. Williams): Ballet. Volta. PUJOL VILARRUBI: El abejorro.
      SANZ: Castillian dances (4). SCARLATTI, D.: Sonata, guitar,
      L 23/K 380, E major. WEISS: Fantasia. Diego Blanco, gtr.
           +-HFN 1-76 p103   -RR 11-75 p81

SLT 33197
3249 ASMUSSEN: So sorry. GRIEG: Ave maris stella. JARNEFELT: Söndag.
      JONSSON: Du käre gud fader i himmelrik. LEVEN: Psalm, no.
      331. NORDQUIST: Psalm, no. 23. OLSSON: Ave Maria; Psalm, no.
      42. SVENSSON: Där bor en konung. Alice Babs, s; Ake Leven,
      org; Leo Berlin, vln.
           +-RR 2-76 p61

SLT 33200
3250 BARTOK: Bagatelles (3). HOLMBOE: Quintet, 2 trumpets, horn, trom-
      bone and tuba, op. 79. KABALEVSKY: Variations on a Ukrainian
      folksong. PRAETORIUS: Partita. PEZEL: Pieces (3). SCHNEIDER:
      Tower music. SCHOENBERG, S. G.: Tower music. Stockholm Phil-
      harmonic Brass Ensemble.
           +RR 3-76 p55

SLT 33229
3251 BJORKANDER: Sketches from the Archipelago: Idyll. BULL: The herd
      girl's Sunday. GRIEG: Elegiac melodies, op. 34 (2). JARNE-
      FELT: Berceuse. KOCH: Dance, no. 2. LARSSON: Folkvisenatt.
      NIELSEN: Little suite, op. 1, A minor: Intermezzo. PETERSON-
      BERGER: Om magna ar. Pa gräset under lindarna. SIBELIUS:
      Suite mignonne, op. 98. Orebro Chamber Orchestra; Lennart
      Hedwall.
           +-RR 3-76 p40

                         TELEFUNKEN
(Teldec) (2)
3252 Teldec Informationsplatten Quadraphonie. SQ, CD-4.
           +Audio 4-76 p88
ER 6-35257 (2)
3253 DANDRIEU: Armes, amours. BINCHOIS: Dueil angoisseus. Gloria,
      laud et honor. BRASSART: O flos fragrans. CESARIS: Bonté
      biauté. DUFAY: L'alta bellezza. Ave virgo. Bien veignes
      vous. Bon jour, bon mois. C'est bien raison. Credo. La
      dolce vista. Dona i ardente ray. Ecclesiae militantis.
      Gloria. Helas mon dueil. J'ay mis mon cuer. Je vous pri.
      Kyrie. Lamentatio Sanctae matris ecclesiae Constantinopoli-
      tanae. Mon chier amy. Moribus et genere Christo. Qui latuit.
      Sanctus. Veni creator spiritus. DUNSTABLE: Beat mater.
      GRENON: La plus jolie. HASPROIS: Puisque je voy. LANTINS, A.:
      Puisque je voy. LANTINS, H.: Gloria. LOQUEVILLE: Sanctus.
      MORTON: La perontina. SOLAGE: Femeux fume. ANON.: Kere dame.
      Musicorum decus et species. Syntagma Musicum; Kees Otten.
           +Audio 9-75 p71   +MQ 7-76 p453
           +Gr 10-75 p662   ++NR 6-75 p12
           +HF 10-75 p74   +RR 6-75 p78
           ++HFN 7-75 p80
6-41275
3254 Chansons der trouveres, 13th century northern France: BERNEVILLE:
      De moi doleros vos chant. BRULE: Biaus m'est estez. CAMBRAI:
      Retrowange novelle. DIJON: Chanterai por mon coraige. MEAUX:
      Trop est mes maris jalos. ANON.: Lasse, pour quoi refusai,
      Li joliz temps d'estey. Studio der Frühen Musik; Thomas Bink-
      ley.
           +MJ 11-76 p45   ++NR 7-76 p7
AF 6-41872 (Some reissues from EK 6-35082, BC 25099, 25100, SAWT 9594)
3255 BACH: Chorale prelude, Wachet auf, ruft uns die Stimme, S 645.
      Prelude and fugue, S 543, A minor. Toccata and fugue, S 565,
      D minor. Trio, S 583, D minor. BUXTEHUDE: Chorale preludes,
      Nun bitten wir den heiligen Gott; Jesus Christus, unser Heiland.
      Fugue, C major. Toccata, F major. CLERAMBAULT: Basse et
      dessus de trompette. Dialogue sur les grande jeux. Récits
      de cornet séparé. DANDRIEU: Allons voir de divin cage. DAQUIN:
      Noël étranger. Michel Chapuis, org.
           +Gr 1-76 p1221   +RR 1-76 p43
           +HFN 3-76 p112
6-41928
3256 Music of the minstrels: Chose tassin; Ductia, Estampie (2);
      Chominciamento di gioia; La quinte, sexte, septime, ultime
      estampie real; Retrove; Saltarello; La Tierche estampie royal.
      Early Music Quartet; Thomas Binkley.

```
          ++Gr 4-76 p1653        +NR 3-76 p9
          +HF 7-76 p96          +St 5-76 p128
          +HFN 3-76 p92
```

6-41929
3257 Paul Esswood, countertenor recital. BACH, J. C.: Ach, dass ich
      Wassers genug hätte. BERNHARD: Was betrübst du dich, meine
      Seele. BUXTEHUDE: Jubilate Domino. TELEMANN: Ach Herr,
      strafe mich nicht. ZIANI: Alma redemptoris mater. Paul
      Esswood, c-t; VCM; Nikolaus Harnoncourt.

```
          ++Gr 4-76 p1653        +NR 3-76 p9
          +HF 7-76 p96          +St 5-76 p128
          +HFN 3-76 p92
```

AG 6-41945
3258 GOUNOD: Faust: Il était un Roi de Thulé...O Dieu, que de bijoux.
      Roméo et Juliette: Je veux vivre. PUCCINI: La bohème: Quando
      me'n vo. ROSSINI: Il barbiere di Siviglia: Contro un cor.
      THOMAS: Hamlet: A vos jeux, mes amis. Mignon: Oui, pour ce
      soir je suis reine des fées. VERDI: La traviata: E strano...
      Ah, fors è lui...Sempre libera. Sylvia Geszty, s; Dresden
      Philharmonic Orchestra; Kurt Masur.

```
          +-HFN 5-76 p100        +RR 4-76 p27
```

AG 6-41947
3259 BOITO: Mefistofele: L'altra notte in fondo al mare. CILEA:
      Adriana Lecouvreur: Ecco, respiro appena...Io son l'umile
      ancella. GIORDANO: Andrea Chénier: La mamma morta. PUCCINI:
      La rondine: Ch' il bel sogno di Doretta. Tosca: Vissa d'arte.
      VERDI: Aida: Ritorna vincitor. Ernani: Surta è la notte...
      Ernani, involami. Il trovatore: Tacea la notte placida...Di
      tale amor. I vespri siciliani: Mercè dilette amiche. Felicia
      Weathers, s; Munich Radio Orchestra; Kurt Eichhorn.

```
          +-Gr 10-76 p653        +-RR 9-76 p34
```

DT 6-48075 (2)
3260 BIGAGLIA: Sonata, recorder, A minor. Sonata, recorder, B flat
      major. COUPERIN: Les nations: L'Espagnola. EYCK: Engels
      Nachtgaeltje. HOTTETERRE: Suite, 2 blockflöten. PARCHAM:
      Solo, blockflöte. PEPUSCH: Sonata, no. 4, F major. TELEMANN:
      Suite, violo violin, F major. Frans Brüggen, rec and flt;
      Jaap Schröder, vln; Frans Vester, Joost Tromp, flt; Kees Boeke,
      rec; Nikolaus Harnoncourt, vla da gamba; Hermann Baumann,
      Adrian van Woudenberg, hn; Gustav Leonhardt, hpd; Concerto
      Amsterdam; Frans Brüggen.

```
          +HFN 5-76 p97            +-RR 3-76 p62
```

TITANIC

TI 4 (Available from Titanic Records, 43 Rice St., Cambridge, Mass.)
3261 Courts and chapels of Renaissance France. DUFAY: J'attendray
      tant qu'il vous playra; J'ay grant doleur; Gloria ad modum
      tubae. GOUDIMEL: Psalms 77, 86, 137; Priere avant le repas.
      LASSUS: Je l'ayme bien; Quand mon mari vient de dehors. (and
      17 others). Boston Camerata; Joel Cohen.

```
          +NR 3-76 p9            +St 6-76 p114
```

TRANSATLANTIC

TRA 292. Tape (c) ZCTRA 292 (ct) Y8TRA 292
3262 Divisions on a ground: An introduction to the recorder and its
          music. FINGER: Divisions on a ground. HANDEL: Sonata, record-
          er, op. 1, no. 11, F major. LOEILLET: Sonata, recorder, op.
          3, no. 3, G minor. MATTHYSZ: Variations from "Der Gooden
          Fluyt Hemel". EYCK: Variations on "Amarilli mia bella".
          VIVALDI: Concerto, recorder, op. 44, no. 11, C major. Richard
          Harvey, rec; Andrew Parrott, hpd; Monica Hugett, Eleanor
          Sloan, vln; Trevor Jones, vla; Catherine Finnis, vlc; Adam
          Skeaping, violone.
                +Gr 10-75 p652   +HFN 2-76 p95
                ++HFN 7-75 p80   +STL 6-8-75 p36
TRA 308
3263 ABSIL: Suite on a Rumanian folk tune, op. 90. CORDELL: Patterns.
          DUBOIS: Quatuor pour saxophones. JACOB: Quartet, saxophone.
          TRAD.: The Agincourt song (arr. P. Harvey). London Saxophone
          Quartet.
                ++Gr 4-76 p1624   +RR 6-76 p64
                +HFN 5-76 p101
XTRA 1159
3264 BEHREND: Auntie. BLEWITT: They don't propose. BOONE: A grand
          fantasy, pianoforte. GLOVER: The gipsy countess. LESLIE:
          Memory. MARZIALS: Such merry maids are we; The bud and the
          bee. MOLLOY: Love's old sweet song. MULLEN: The nattletons.
          RUSSELL: The ship on fire. SULLIVAN: The frost is here.
          SMITH: The spider and the fly. TOSTI: My dreams; The Venetian
          song. Sylvia Eaves, Maureen Keetch, s; Robert Carpenter Turner,
          bar; Kenneth Barclay, pno.
                +Gr 5-76 p1808   +RR 5-76 p71
XTRA 1160
3265 AZEVEDO: Delicada. BLISS: Checkmate: The ceremony of the red
          bishops; Finale checkmate. DEBUSSY: Preludes, Bk I, no. 8:
          La fille aux cheveux de lin. FARNON: Une vie de matelot.
          GERSHWIN: Summertime. HOLST: The planets, op. 32: Mars.
          OSGOOD: Round the clock. ZEHLE: Wellington march. Brighouse
          and Rastrick Band; James Scott.
                ++Gr 5-76 p1808

TURNABOUT

TV 34017. Tape (c) KTVC 34017
3266 BULL: Prelude and In nomine. BYRD: Elegy on the death of Thomas
          Tallis. In nomine a 5. Lord Salisbury's pavan. O Lord, how
          vain. FERRABOSCO: Pavane. GIBBONS: Great Lord of Lords.
          Lord Salisbury's pavan. O God, the King of glory. MORLEY:
          Hard by a crystal fountain. TOMKINS: The Lady Folliott's
          galliard. WEELKES: As Vestas was from Latmos Hill descending.
          Purcell Consort of Voices; Jaye Consort of Viols; Grayston
          Burgess.
                +RR 1-76 p65 tape
TV 34137. Tape (c) KTVC 34137
3267 ATTAINGNANT: Basse danse La brosse and recoupe. Chanson. Gail-
          larde. Pavane. Tordion. Tant que vivray. BITTNER: Allemande.
          Courante. Passacaglia. Praeludium. Sarabande. MOUTON: Le

dialogue des graces sur Iris, allemande. La malassis, sarabande. L'Amant content, canarie. LE SAGE DE RICHEE: Ouverture. VISSEE: Suite, D minor. Michael Schäffer, lt.
+RR 10-76 p105 tape

UNICORN

RHS 339
3268 Contrasts in brass: BARBER: Mutations from Bach. BUXTEHUDE:
Fanfare and chorus. CARR: Prism for brass. GRIEG: Funeral
march. HAYDN: March for the Prince of Wales. KAUFFMANN:
Music for brass. TCHEREPNIN: Fanfare. Locke Brass Consort;
James Stobart.
+Gr 6-76 p58   ++RR 6-76 p40
+HFN 6-76 p84

VANGUARD

SRV 344 SD
3269 Classic favorites for strings. ALBINONI: Adagio, G minor.
BORODIN: Quartet, strings, no. 2, D major: Nocturne. BRITTEN:
Simple strings symphony, op. 4. HANDEL: Solomon: Entrance
of the Queen of Sheba. PACHELBEL: Canon, D major. ECO;
Johannes Somary.
++NR 3-76 p2   ++St 5-76 p128
707/8. Tape (c) 5707/8 (also Precision ZCVBP Tape (c) 707/8)
3270 The best of Leopold Stokowski: BACH: Cantata, no. 147, Jesu,
joy of man's desiring. Cantata, no. 208, Sheep may safely
graze. MOZART: Serenade, no. 10, 13 wind instruments, K 361,
B flat major. STRAVINSKY: L'Histoire du soldat: Suite.
THOMSON: The plow that broke the plains. VIVALDI: L'Estro
armonico, op. 3, no. 11, D minor. Various instrumentalists
and orchestra; Leopold Stokowski.
+-Gr 12-74 p1157   +SFC 5-27-73 p27
-HF 9-71 p118   +St 2-72 p119 tape
+-RR 1-76 p66 tape

VCS 10049
3271 The jolly minstrels: Minstrel tunes, songs and dances of the
Middle Ages on authentic instruments. ANON. 12th Century:
The song of the ass. ANON. 13th century: Ductia, English
dance, Rege mentem; 2 estampies; C'est la fin; Ductia; Novus
miles sequitur; Moulin de Paris; In seculum artifex; Sol ovi-
tur; Worldes blis; Pour mon coeur; Estampie royale. ANON.
14th Century: Saltarello; Estampie; Lamento di Tristan; Trotto.
ANON. 15th Century: Vierhundert Jar uff Diser Erde; Die Suss
Nachtigall; Alta. ADAM DE LA HALLE: Li mans d'amer. VAQUEIRAS:
Kalenda maya. RICHARD, COEUR DE LION: Ja nun nons pris. Jaye
Consort of Viols.
++AR 8-71 p101   ++SFC 2-22-76 p28

SVC 71212
3272 Christmas Eve at the Cathedral of St. John the Divine: LVOV: To
thy heavenly banquet (Communian anthem). PALESTRINA: Sanctus
and benedictus (Soriano). SWEELINCK: Hodie Christus natus est.
VICTORIA: O magnum mysterium. ANON.: Adeste fideles; The
beatitudes; The first noel; Hark, the herald angels sing; Ninefold alleluia and announcement of the holy gospel; The shepherd's

carol; Trisagion; While shepherds watched their flocks. David
Pizarro, org; Cathedral Choir and Cathedral Boys' Choir;
Richard Westernburg.
     +NR 12-76 p1

VIRTUOSI

VR 7506
3273 BINGE: The watermill. BLISS: Kenilworth. GOLLAND: Relay.
      MENDELSSOHN: Ruy Blas overture, op. 95 (arr. G. Thompson).
      SHOSTAKOVICH: Festival overture, op. 96 (arr. Cornthwaite).
      VERDI: Aida: Grand march (arr. D. Wright). La traviata: Pre-
      lude, Act 1 (arr. D. Rimmer). James Shepherd, cor; Virtuosi
      Brass Band; Harry Mortimer.
          ++Gr 1-76 p1247

VISTA

VPS 1022
3274 BACH: Chorale prelude: In dulci jubilo. DOWLAND: Awake, sweet
      love. GIBBONS: The silver swan. PALESTRINA: Alma redemptoris
      mater. POULENC: Motets pour le temps de Noel: O magnum mys-
      terium; Quem vidistis pastores dicite. PURCELL: The libertine:
      In these delightful, pleasant groves. REGER: Benedictus, op.
      59, no. 9. SCHUMANN: Kinderscenen, op. 15: Hasche-Mann;
      Bittendes Kind; Gluckes genug; Wichtige Begebenheit. TALLIS:
      O nata lux de lumine. WILLAERT: O magnum mysterium. TRAD.:
      My love's an arbutus (arr. Stanford). David of the white rock;
      The bells of Aberdovey; Gabriel's message (arr. R. Bevan).
      The Taunton carol (arr. Mack). Little David play on your harp
      (arr. Sargent). I will give my love an apple (arr. Wilkinson).
      PLAINSONG: Puer natus est nobis. David Bevan, org; Anthony
      Bevan, pno; Bevan Family Choir; Roger Bevan.
          +HFN 1-76 p115   +-RR 1-76 p60
VPS 1023
3275 ARMSTRONG: Christ whose glory fills the skies. BACH: God liveth
      still (Turner). Chorale preludes (Orgelbuchlein): Erstanden
      ist der Heilige Christ, S 628. Cantata, no. 67: Halt im
      Gedachtnis Jesum Christ (Turner). Chorale prelude: Christ lag
      in Todesbanden, S 695. BYRD: Haec dies a 6. HANDEL: Messiah:
      Hallelujah chorus. NAYLOR: Now the green blade riseth. PORTER:
      An Easter meditation. STANFORD: When Mary through the garden
      went, op. 127, no. 3. ANON.: Come ye faithful, raise the
      strain (16th century). Christ the Lord is risen (arr. Turner).
      This joyful Eastertide (arr. Wood). Now glad of heart be
      everyone (arr. Turner). TRAD.: Cheer up, friends and neigh-
      bours. John Turner, org; Glasgow Cathedral Choir; John Turner.
          +-Gr 3-76 p1503   +RR 3-76 p65
          +HFN 3-76 p115
VPS 1029
3276 DUPRE: Fileuse. DURUFLE: Suite, op. 5: Toccata, B minor. LANG-
      LAIS: Triptyque. LITAIZE: Toccata sur le veni creator. SAINT-
      MARTIN: Toccata de la libération. TOURNEMIRE: Chorale sur le
      victimae Paschali (arr. Duruflé). Jane Parker-Smith, org.
          +Gr 8-76 p320

VPS 1030
3277 BACH: Prelude and fugue, S 532, D major. COUPERIN: Chaconne, C
         major. FRANCK: Chorale, no. 2, B minor. LISZT: Prelude and
         fugue on the name B-A-C-H, G 260. MARCHAND: Basse de trompette.
         WILLS: Carillon on Orientis Partibus. Arthur Wills, org.
              +Gr 8-76 p323   +RR 8-76 p67
VPS 1035
3278 ANDRIESSEN: Theme and variations. BOSSI: Divertimento in forma
         de giga. HOLLINS: Concert overture, C minor. PARRY: Old
         100th, chorale fantasia. Elegy, A flat major. REGER: Ein feste
         Burg, op. 67, no. 6. Introduction and passacaglia, D minor.
         SCHUMANN: Canon, op. 56, no. 5, B minor. SMART: Postlude, D
         major. WHITLOCK: Plymouth suite: Toccata. Roy Massey, org.
              +Gr 10-76 p633   +HFN 11-76 p164

                              VOX

SVBX 5301 (3)
3279 Early String Quartet in U. S. A. CHADWICK: Quartet, strings, no.
         4, E minor. FOOTE: Quartet, strings, op. 70, D major.
         FRANKLIN: Quartet, strings. GRIFFES: Indian sketches (2).
         HADLEY: Quintet, piano, op. 50, A minor. LOEFFLER: Music for
         four string instruments. MASON: Quartet on Negro themes, op.
         19, G minor. Kohon Quartet; Isabelle Byman, pno.
              +HF 1-72 p106   ++St 6-72 p92
              ++NR 6-71 p5   +St 7-76 p73
              +NYT 9-15-74 pD32
SVBX 5304 (3)
3280 America sings, the great sentimental age--Stephen Foster to Charles
         Ives: FOSTER: We are coming Father Abraham, 300,000 more;
         Willie has gone to war; Jenny June; Wilt thou be true; Katy
         Bell. HANBY: Nelly Gray. HAWTHORNE: Listen to the mocking
         bird. IVES: An old flame; Circus band; A Civil War memory; In
         the alley; Karen; Romanzo di Central Park; A son of a gambolier.
         KITTRIDGE: Tenting on the old campground. Forty-six other
         American songs. Gregg Smith Singers; New York Vocal Arts
         Ensemble; Instrumental accompaniment.
              +HF 7-76 p69   +St 3-75 p86
SVBX 5305 (3)
3281 The American String Quartet in U. S. A., 1900-1950. COPLAND:
         Pieces, string orchestra (2). GERSHWIN: Lullaby, string
         quartet. HANSON: Quartet, op. 23. IVES: Scherzo, string
         quartet. MENNIN: Quartet, no. 2. PISTON: Quartet, no. 5.
         SCHUMAN: Quartet, no. 3. SESSIONS: Quartet, no. 2. THOMSON:
         Quartet, no. 2. Kohon Quartet.
              +HF 12-74 p122   +SFC 11-3-74 p22
              ++NR 10-74 p7   *St 7-76 p75
              +NYT 9-15-74 pD32
SVBX 5350 (3)
3282 America sings: The founding years, 1620-1800. Religious vocal
         music from The Ainsworth Psalter; The Bay Psalm Book; Tuft's
         Introduction; Walter's Grounds and Rules; Lyon's Urania.
         ANTES: How beautiful upon the mountains. BELKNAP: The seasons.
         BILLINGS: As the hart panteth; Consonance; Thus saith the high,
         the lofty one. BROWNSON: Salisbury. DENCKE: O, be glad, ye
         daughters of His people. HERBST: God was in Jesus. HOPKINSON:

VOX (cont.)   652

A toast; Beneath a weeping willow's shade; Come fair Rosina;
Enraptur'd I gaze; My days have been so wond'rous free; My
gen'rous heart disdains; My love is gone to sea; O'er the hills
far away; See, down Maria's blushing cheek; The traveler be-
nighted and lost. INGALLS: Northfield. LAW: Bunker Hill.
KELLY: Last week I took a wife; The mischievous bee. LYON:
Friendship. MORGAN: Amanda; Despair; Montgomery. PETER, J. F.:
I will freely sacrifice to Thee; I will make an everlasting
covenant. PETER, S.: Look ye, how my servants shall be feasting; O, there's a sight that rends my heart. READ: Down steers
the bass; Russia. SELBY: O be joyful to the Lord; Ode for the
New Year. STEVENSON: Behold I bring you glad tidings. SWAN:
China. VON HAGEN: Funeral dirge on the death of General
Washington. ANON.: Washington's march. Gregg Smith Singers
and Chamber Orchestra; Gregg Smith.
    +HF 7-76 p69   +NR 10-76 p8
    +MJ 11-76 p44   ++St 9-76 p117

                    WASHINGTON UNIVERSITY
(Q) OLY 104 (2)
3283 American sampler. BENSHOOF: The cow; Dinky; The fox; John Brown's
      body; The waking. BOND: I love you truly; A perfect day.
      CARPENTER: The player queen. DETT: Juba dance. FILLMORE:
      Miss trombone. FOSTER: I cannot sing tonight; Some folks;
      Summer longings; Why, no one to love. GRIFFES: Early morning
      in London; The lament of Ian the Proud. HELD: Chromatic rag.
      IVES: Walt Whitman; The white gulls. JOPLIN: Fig leaf rag.
      LAMB: American beauty rag; Ragtime nightingale. MANTIA:
      Priscilla. PRATT: Hot house rag. PRYOR: The tip topper; The
      supervisor. SIMONS: Atlantic zephyrs. Elizabeth Suderburg,
      s; Robert Suderburg, Victor Steinhardt, pno; Stuart Dempster,
      trom, euphonium.
          +IN 6-76 p25   +ON 11-76 p98
          ++MJ 7-76 p57   +St 6-76 p101
RAVE 761
3284 ELGAR: Romance, bassoon, op. 62. JACOB: Partita. OSBORNE:
      Rhapsodie. TANSMAN: Suite, bassoon. TELEMANN: Sonatas,
      bassoon, E minor, D major, F minor. Arthur Grossman, bsn;
      Silvia Kind, hpd; Randolph Hokanson, pno.
          +NR 12-76 p8

                    WEALDEN
WS 121
3285 BACH: Prelude and fugue, S 533, E minor. Chorale prelude, Liebster
      Jesus, wir sind hier, S 731. BLOW: Verse for cornet and single
      organ. Voluntary for a single organ. BOYCE: Voluntary, A
      minor. BRAHMS: Chorale preludes, op. 122, no. 8: Es ist ein
      Ros entsprungen; no. 9, Herzlich tut mich verlangen; no. 11,
      O Welt, ich muss dich lassen. CAMIDGE: Concerto, organ, op.
      13, no. 3, A minor: Gavotte. CLERAMBAULT: Suite du deuxième
      ton. WESLEY, C.: Concerto, organ, Set 2, no. 1, G minor:
      Siciliano. WESLEY, S.: Air. Richard Townend, org.
          /Gr 2-76 p1363   /RR 3-76 p62
          /-HFN 1-76 p104

WSQ 134
3286 BACH: Chorale prelude: Kommst du Nun, Jesu, S 650. Fugue, S 578,
        G minor. BRAHMS: Chorale preludes: Est ist ein Ros entsprungen.
        op. 122, no. 8. Schmücke dich, op. 122, no. 5. GUILMANT:
        Sonata, organ, op. 42, no. 1, D minor: Finale. HOWELLS: Rhap-
        sody, no. 3, C sharp minor. RAISON: Offertoire "Vive le Roy".
        SCHUMANN: Canons, D major, B minor. STANFORD: Postlude, D
        minor. YON: Humoresque. David Petit, org.
              +-HFN 6-76 p95   +-RR 8-76 p66

WS 137
3287 BACH: Chorale prelude: Das alte Jahr vergangen ist, S 614.
        COCKER: Tuba tune. GARDINER: Evening hymn. HOWELLS: Siciliano
        for a high ceremony. LEIGHTON: Improvisation. MESSIAEN: Les
        corps glorieux: Joie et clarté des corps glorieux. LISZT:
        Prelude and fugue on the name B-A-C-H, G 260. WALTON: Set me
        as a seal upon thine heart. Christopher Rathbone, org;
        Marlborough College Choir; Royal Wilkinson.
              -Gr 11-76 p867   +-RR 7-76 p77

WS 142
3288 ARNE: Concerto, organ: Con spirito. BACH: Chorale preludes:
        Liebster Jesu, wir sind hier, S 731; Valet will ich dir geben,
        S 736. JOHNSON: Trumpet tune. JONGEN: Petit prélude. LANGLAIS:
        Suite française: Nazard. Te deum. PINKHAM: Revelations: Litany
        and toccata. SALOME: Grand choeur. VIERNE: Pastorale.
        WHITLOCK: Plymouth suite: Toccata. Jack Hindmarsh, org.
              +Gr 8-76 p323   +-RR 8-76 p66

WS 149
3289 BERKELEY: Aria. Aubade. Toccata. BRIDGE: Allegro con spirito,
        B flat major. Adagio, E major. GINASTERA: Toccata, villancico
        y fuga. MULET: Tu es Petra. PARRY: Toccata and fugue. SAINT-
        SAENS: Fantasie, E flat major. Eric Suddick, Steven Hollas,
        org.
              +-HFN 6-76 p95   +RR 8-76 p66

                    WESTMINSTER

WGS 8116
3290 BACH: Chorale prelude: Herr Gott nun Schleuss den Himmel auf.
        BUXTEHUDE: Fugue à la gigue, C major. DURUFLE: Prelude and
        fugue on the name Alain, op. 7. HANDEL, Concerto, organ, op.
        4, no. 8, A major. SCHUMANN: Canon, B minor. TOURNEMIRE:
        Chorale sur le victimae Paschali. Maurice Duruflé, Marie-
        Madeleine Duruflé-Chevalier, org.
              /RR 12-76 p83

WGS 8268 (also ABC ATS 20011)
3291 ADAM: Variations on a theme of Mozart's, "Ah, vous dirai-je Maman".
        ARNE: Artaxerxe: The soldier tir'd of war's alarms. CALDARA:
        La rosa. BISHOP: Lo, here the gentle lark. HANDEL: Meine Seele
        hor im Sehen. SCHUBERT: Der Hirt auf dem Felsen, D 965.
        Beverly Sills, s; Gervase de Peyer, clt; Leslie Parnas, vlc;
        Paula Robison, flt; Charles Treger, vln; Charles Wadsworth,
        pno and hpd.
              -HF 11-72 p111   +ON 6-72 p34
              +MJ 10-72 p58   +RR 12-76 p88
              ++NR 7-72 p12   +-St 8-72 p86
              +NYT 6-11-72 pD24

<u>WGM 8320</u> (Reissue from Melodiya D 08467/8)
3292 Russian wedding, festival and season songs: Danube my Danube;
        Do not be joyous aspen tree; Dunya the spinner; Carols; The
        grapes are blooming in the garden; How the mistress met Ivan;
        I hide gold; I walk in the meadow; I walk with a loach; In
        the meadows; In the field a birch tree stood; The moon has
        golden horns; Oh in the evening at midnight; Our faithful
        well; The seven sons-in-law; Spring fortune telling; What
        is this burning; You my dawn. Siberian State Chorus,
        Piatnitsky Chorus, Fydorov Sisters, Orenburg Chorus, Russian
        Song Chorus, USSR Academic Russian Chorus, Choir of the
        North, Alexandrov Song and Dance Ensemble; Sergei Lemeshev,
        Ivan Skobtsov.
            *NR 10-76 p8

ANONYMOUS WORKS

A hunting we will go. cf HMV ESD 7002.
Adeste fideles. cf Decca SXL 6781.
Adeste fideles. cf Vanguard SVD 71212.
The Agincourt song (arr. P. Harvey). cf Transatlantic TRA 308.
Airs and dances of Renaissance Scotland. cf HANDEL: Trio sonata, op.
     2, no. 1, C minor.
Alas, departing is ground of woe. cf BASF 25 22286-1.
Almande de Ungrie. cf Hungaroton SLPX 11721.
Alta. cf Vanguard VCS 10049.
Ambrosian chant. Capella musicale del Duomi di Milano; Luigi Benedetti,
     Luciano Migliavacca. DG 2533 284-1.
              +-Gr 7-76 p199   +NR 4-76 p9
              ++HF 7-76 p96   ++RR 8-76 p76
              +HFN 5-76 p91   ++SFC 4-25-76 p30
Amores dont je sui espris. cf Candide CE 31095.
L'Amour du mois de Mai. cf Club 99-101.
And all in the morning (arr. Vaughan Williams). cf HELY-HUTCHINSON:
     Carol symphony.
And I were a maiden. cf BASF 25 22286-1.
Andernaken. cf HMV SLS 5049.
Animation. cf Nonesuch H 71276.
At an English ale house. cf Saga 5420.
Auld lang syne. cf HMV RLS 719.
Aupres de vous. cf Argo ZRG 667.
Ave virgo regia. cf Candide CE 31095.
Baby born today. cf Grosvenor GRS 1034.
Bagpipes solo. cf DUFAY: Franc cuer gentil.
The bailiff's daughter of Islington. cf HMV ESD 7002.
Ballet des coqs. cf CBS 76183.
Balulalow. cf Abbey LPB 761.
Barbara Allen. cf Prelude PRS 2501.
Basse danse la Magdalena. cf DG Archive 2533 184.
Basse danses I and II. cf HMV SLS 863.
Batori Tantz, Proportio. cf Hungaroton SLPX 11721.
The battle hymn of the republic. cf CBS 61746.
The beatitudes. cf Vanguard SVD 71212.
Beau sejour. cf Club 99-101.
La bella Franceschina. cf L'Oiseau-Lyre 12BB 203/6.
Belle, tenés mo. cf L'Oiseau-Lyre 12BB 203/6.
The bells of Aberdovey. cf Vista VPS 1022.
Ben ch'io. cf DUFAY: Franc cuer gentil.
Benedicamus domino. cf DUFAY: Franc cuer gentil.
Blame not my lute. cf L'Oiseau-Lyre 12BB 203/6.

Blow the wind southerly. cf Polydor 2383 389.
Billy boy. cf HMV CSD 3766.
Blagoobrajniy Iossif. cf Harmonia Mundi HMU 137.
Bobby Shaftoe. cf Prelude PRS 2501.
Boleras sevillanas. cf Saga 5412.
The borys hede, Pray for us, Verbum patris humanatur, Confitor fut le
    non-pareil, Nova nova, Riu riu chiu, Verbum caro. cf MOUTON: Noe
    psallite.
Brandywine quickstep. cf Columbia M 34129.
Branle de Bourgogne. cf DG Archive 2533 184.
Brigg fair (arr. Grainger). cf Prelude PRS 2501.
Brincan y bailan. cf Enigma VAR 1015.
The British Grenadiers. cf CBS 61684.
Bruder Konrad. cf L'Oiseau-Lyre 12BB 203/6.
Buenos reyes. cf Saga 5412.
Bugeilio'r Gwenith Gwyn. cf L'Oiseau-Lyre SOL 345.
Bushes and briars (arr. Vaughan Williams). cf HMV ASD 2929.
Calata. cf L'Oiseau-Lyre 12BB 203/6.
Canaan. cf Nonesuch H 71276.
La cançó del Llandre (arr. Llobet). cf DG 3335 182.
Cant dels Ocells. cf CBS Classics 61579.
Cantar montañes. cf Saga 5412.
Cantatille dedicated to Mme. de Pompadour. cf Club 99-101.
Captain Sargent's quick march. cf Columbia M 34129.
Captain Sargent's (Light Infantry Company's) quick march. cf Columbia
    MS 6161.
Un cavalier di Spagna. cf HMV CSD 3756.
Celle qui m'a demande. cf L'Oiseau-Lyre 12BB 203/6.
C'est grand plaisir, chanson à 3. cf Hungaroton SLPX 11669/70.
C'est la fin. cf Vanguard VCS 10049.
Chanconnette. cf Candide CE 13095.
Chanson de Marie. cf Club 99-101.
Chanson de Normande. cf Club 99-101.
Charlie is my darling. cf HMV ESD 7002.
Cheer up, friends and neighbours. cf Vista VPS 1023.
Christ der ist erstanden. cf L'Oiseau-Lyre 12BB 203/6.
Christ the Lord is risen. cf Vista VPS 1023.
Christmas Vigil: Gradual chant; God is with us. cf Ikon IKO 3.
Chui dicese e non l'amare. cf L'Oiseau-Lyre 12BB 203/6.
Cold blows the wind. cf Harmonia Mundi HM 593.
Come holy ghost. cf National Trust NT 002.
Come ye faithful, raise the strain. cf Vista VPS 1023.
Coming thro the rye. cf HMV RLS 716.
Coming thro the rye. cf HMV RLS 719.
Concert. cf Nonesuch H 71276.
Contre le temps. cf HMV SLS 863.
Cottage cheese pies. cf HMV ASD 3200.
Counting the goats. cf Columbia SCX 6585.
Cubana. cf Saga 5412.
The cuckoo. cf Harmonia Mundi HM 593.
Dafydd y Garreg Wen. cf L'Oiseau-Lyre SOL 345.
The Dalling alman. cf National Trust NT 002.
Dance to your Daddy. cf Prelude PRS 2501.
Dances: My robbin; Tickle my toe; Hollis berrie. cf National Trust NT
    002.
Dances "In te Domine speravi". cf JOSQUIN DES PRES: Songs (Argo ZRG
    793).

Dance, D major. cf Saga 5421.
La danse de clèves. cf DUFAY: Franc cuer gentil.
Danse real. cf Candide CE 31095.
Danza. cf Hungaroton SLPX 11721.
David of the white rock. cf HMV OU 2105.
David of the white rock. cf Vista VPS 1022.
De blanca tierra. cf Saga 5412.
De la trumba. cf DG Archive 2533 323.
Deck the hall with boughs of holly. cf CBS 61771.
Deep river. cf Pye GH 603.
Deo gracias Anglia, Agincourt carol. cf BASF 25 22286-1.
Det blir magot i himlen för barnen att få. cf BIS LP 14.
Dies est laetitiae. cf MOUTON: Noe psallite.
Dieus qui porroit. cf Candide CE 31095.
Dindirindin. cf DG 2530 504.
Ding don merrily on high. cf Grosvenor GRS 1034.
Dit le Bourgignon. cf CBS 76183.
Don Gato. cf Enigma VAR 1015.
Donc le rieu de la fontaine. cf DUFAY: Franc cuer gentil.
Doulce memoir. cf Hungaroton SLPX 11669/70.
Doulce memoir, phantasia. cf Hungaroton SLPX 11669/70.
Drewries accords. cf DG Archive 5233 323.
Drink to me only (arr. Kinloch Anderson). cf HMV ASD 2929.
Dubinushka. cf HMV ASD 3200.
Ductia. cf Candide CE 31095.
Ductia. cf Nonesuch H 71326.
Ductia. cf Vanguard VCS 10049.
Ductia, English dance, Rege mentem. cf Vanguard VCS 10049.
The Duke of York's march. cf Columbia M 34129.
Early one morning. cf HMV ESD 7002.
Easter Mattins processional hymn. cf CRD Ikon IKO 2.
English dance. cf HMV SLS 5022 (2).
English folk songs: (arranged Grainger): British waterside or The jolly
     sailor; Died for love; The pretty maid milkin' her cow; The lost
     lady found; Shallow brown; Six Dukes went afishin'; The sprig of
     thyme; Willow Willow. (Arr. Holst): Abroad as I was walking; Our
     ship she lies in the harbour; The willow tree. (Arr. Vaughan
     Williams): The maid of Islington; O who is that that raps at my
     window; The seeds of love; Lovely Joan; The unquiet grave or How
     cold the wind doth blow. Robin Doveton, t; Victoria Hartung, pno.
     Prelude PMS 1502.
                +Gr 12-76 p1042   +HFN 12-76 p141
L'Espérance de Bourbon. cf DUFAY: Franc cuer gentil.
Estampe royal. cf Candide CE 31095.
Estampie. cf Vanguard VCS 10049.
Estampies (2). cf Vanguard VCS 10049.
Estampie royale. cf Vanguard VCS 10049.
Est'il conclu par un arrêt d'amour. cf HMV SLS 5049.
Etoile du matin. cf Club 99-101.
Filles à marier. cf DUFAY: Franc cuer gentil.
La filla del marxant. cf DG 3335 182.
The first noel. cf Vanguard SVD 71212.
The first nowell. cf CBS 61771.
The first Nowell. cf HMV CSD 3774.
Fisherman Peter. cf Grosvenor GRS 1034.
From the Lity. cf Ikon IKO 3.

Fugues (8). cf Panton 110 418/9.
Fugues, D major, E minor (3), E major, A major (2), C major. cf Panton
    110 418/9.
Gabriel's message. cf Vista VPS 1022.
Galliard. cf Angel SFO 36895.
General Burgoyne's march. cf Columbia M 34129.
General Washington's march. cf Columbia M 34129.
Der gestraifft Danntz. cf DG Archive 2533 302.
The girl I left behind me. cf Pye TB 3006.
Girometa. cf L'Oiseau-Lyre 12BB 203/6.
The God of love. cf Abbey LPB 739.
God save the Queen (arr. Britten). cf Decca 419.
God save the Queen. cf Desmar DSM 1005.
God save the King. cf HMV RLS 719.
Goday, my Lord, Syr Christenmasse. cf BASF 25 22286-1.
Greensleeves: An acre of land; Bushes and briars; Ca' the yowes; The
    dark-eyed sailor; Early in the spring; Greensleeves; John Dory;
    Just as the tide was flowing; Loch Lomand; The lover's ghost;
    The seeds of love; The springtime of the year; The turtle dove;
    The unquiet grave; Ward, the pirate; Wassail song. London Madri-
    gal Singers; Christopher Bishop. Seraphim S 60249.
              +NR 2-76 p8
Greensleeves. cf CBS 61684.
Greensleeves. cf RCA RK 11708.
Greensleeves to a ground. cf Orion ORS 76216.
Gregorian chant. Schola Cantorum of Amsterdam Students; Wim van Gerven.
    Columbia M3X 32329 (3).
              +MQ 10-74 p673   +SFC 2-22-76 p28
              +-NYT 10-13-74 pD34
Gregorian chant: Chants for feasts of Mary: Propia Missarum; Antiphonae
    Mariae. Schola Cantorum Francesco Coradini; Fosco Corti. DG
    Archive 2533 310.
              +Gr 12-76 p1058   +RR 12-76 p91
              +-HFN 12-76 p143
Gregorian chant: The complete mass for the seventh Sunday of Easter.
    Abbey of Notre Dame d'Argentan Benedictine Nuns Choir. French
    Decca 7554.
              +-Gr 7-76 p218   +HFN 8-76 p80
Gregorian chant: Le jour octave de Noël: Solennité de Sainte Marie Mère
    de Dieu. Notre Dame d'Argentan Abbaye Nuns Choir. French Decca
    7553.
              +GR 12-76 p1058   +RR 12-76 p91
Gregorian chant: Rameaux; Jeudi Saint; Vendredi Saint; Samedi Saint;
    Christ lag in Todesbanden. Gaston Litaize, org; Saint Pierre
    Abbaye Monks Choir; Dom Joseph Gajard. French Decca 278 054/5 (2).
              ++Gr 4-76 p1653   +HFN 3-76 p93
Gregorian chant: Responsories and monodies from the Gallicad rite.
    Deller Consort; Alfred Deller. Harmonia Mundi HMD 234.
              +HFN 12-75 p152   +St 7-76 p114
Gregorian chant: Salut du Saint-Sacrement. Vêpres de la Sainte-Trinité.
    Abbey Saint Pierre de Solesmes, Monks' Choir; Dom Jean Claire.
    French Decca 7545. (also London OS 26431).
              -Audio 8-76 p76   +-NR 6-76 p12
              +Gr 11-73 p986   ++SFC 4-18-76 p23
              +HF 7-76 p96   +St 7-76 p114
              ++MJ 5-76 p28

Gregorian Chant: Salva Festa Dies. Benedictine Monks of the Abbey St.
     Maurice and St. Maur of Clairvaux. Philips 6580 061.
          +Gr 11-73 p986   +NYT 10-13-74 pD34
          +HF 4-73 p108   +-RR 10-73 p114
          +NR 5-73 p8   +SFC 2-22-76 p28
La guercia. cf HMV SLS 5049.
Hark, the herald angels sing. cf Vanguard SVD 71212.
Hava nagila. cf Decca SB 322.
Hejnat Krakowska (2). cf Philips 6500 926.
He's gone away. cf CBS 61746.
Heth sold ein meisken garn om win. cf HMV SLS 5049.
Hocquet (2). cf Candide CE 31095.
Hohenfriedberger Marsch. cf DG 2721 077 (2).
Los hombres con gran plazer. cf DG 2530 504.
Hornpipe. cf Klavier KS 551.
The hunting of the hare. cf HMV OU 2105.
I believe. cf Coronet LPS 3032.
I Himmelen, i himmelen. cf BIS LP 14.
I know where I'm going (arr. Hughes). cf HMV ASD 2929.
I saw three ships. cf HMV CSD 3774.
I will give my love an apple. cf Vista VPS 1022.
In dulci jubilo. cf HMV CSD 3774.
In hoc anni circulo, Fines amourtes, Verbum partis hodie, Lullay lullow,
     Fulget hodie, Now make we merthe, Nowell. cf MOUTON: Noe psallite.
In seculum artifex. cf Vanguard VCS 10049.
Infant holy, infant lowly. cf Abbey LPB 761.
Istampita ghaetta. cf DG Archive 2533 184.
Istampita ghaetta. cf 1750 Arch S 1753.
Istampitta tre fontaine. cf HMV SLS 863.
Italiana. cf DG Archive 2533 184.
Je ne fay. cf L'Oiseau-Lyre 12BB 203/6.
John Peel. cf HMV ESD 7002.
The jolly carter. cf Harmonia Mundi HM 593.
Der Katzenföte. cf L'Oiseau-Lyre 12BB 203/6.
The keel row. cf HMV ESD 7002.
Kemp's jig. cf Saga 5426.
Kere dame. cf Telefunken ER 6-35257 (2).
King Harry VIII pavan. cf Angel SFO 36895.
Kyrie "Orbis factor". cf PARSLEY: Lamentations, 5 voices.
Lady Wynkfyldes rownde. cf L'Oiseau-Lyre 12BB 203/6.
Lamento di Tristan. cf Vanguard VCS 10049.
Lamento di Tristano: Rotta. cf DG Archive 2533 184.
Las, je n'ecusse. cf L'Oiseau-Lyre 12BB 203/6.
Lasse, pour quoi refusai, Le joliz temps d'estey. cf Telefunken 6-41275.
L'e pur morto Feragù. cf L'Oiseau-Lyre 12BB 203/6.
Lengyel tánc. cf Hungaroton SLPX 11721.
Lilliburlero. cf CBS 61684.
Linda amiga. cf Saga 5412.
Little David, play on your harp. cf Vista VPS 1022.
The London march. cf Columbia MS 6161.
The London march. cf Columbia M 34129.
Londonderry air. cf CBS 30062.
Lonsdale. cf Nonesuch H 71276.
Lorena. cf Personal Touch 88.
Lute dances (2). cf HMV SLS 5049.
Lux hodie. cf MOUTON: Noe psallite.

Marches from the music of the Royal Södermanland Regiment (6).   cf
    KARKOFF: Figures for wind instruments and percussion.
Marsch der Finnländischen Reiterei.   cf DG 2721 077 (2).
Marsch I Bataillon Garde.   cf Polydor 2489 523.
Me suis mis en danse (arr. Bax).   cf HMV ASD 2929.
Medieval English carols and Italian dances: Anonymous carols: Nowel syng
    we bothe al and som; Lullay lullow, Ave Maria, Ther is no rose of
    swych vertu, Ave rex angelorum, Nova, nova, Make we joye nowe in
    this fest, Hayl Mary ful of grace, Mervele noght, Nowell, nowell,
    Salve sancta parens, Deo gracias anglia. Anonymous 14th century
    Italian monophonic dances: Saltarello (3), Istampita Palamento.
    Decca DL 79418.
           +CJ 3-76 p25
Medieval German Plainchant and polyphony: Advent/Christmas: Ad te levavi;
    Congaudeat turba; Puer natus est; Viderunt omnes; Johannes post-
    quam senuit; Hodie progeditur. Passiontide: Gloria, laus et honor;
    Dominus Jesus; Christus factus est; Hely, hely. Easter: Confite-
    mini...Laudate; Unicornis captivatur; Haec dies; Ad regnum...
    Noster cetus. Pentecost: Spiritus Domini; Factus est repente;
    Catholicorum concio. Parousia: Cum natus est Jesus; Gloria in
    excelsis. Schola Antiqua; R. John Blackley. Nonesuch H 71312.
           +-Gr 11-75 p888   ++RR 12-75 p90
           +HFN 11-75 p163   +-St 7-76 p114
           +NR 10-75 p6
Menuet d'exaudet. cf Club 99-101.
The mermaid. cf HMV SEOM 25.
The mermaid. cf HMV CSD 3766.
Messiah. cf Nonesuch H 71276.
Mignonne, allons. cf L'Oiseau-Lyre 12BB 203/6.
Mijm morken gaf mij een jonck wifjj. cf HMV SLS 5049.
The miller of Dee. cf HMV ESD 7002.
Missa Salisburgensis. Plaudit tympana, Hymnus. James Griffet, James
    Lewington, Heinrich Weber, Erwin Abel, t; Brian Etheridge, David
    Thomas, Heinz Haggenmüller, Eberhard Wiederhut, bs; Tölzer Boys'
    Choir; Escolania de Montserrat; Collegium Aureum; Ireneu Segarra.
    BASF 25 22073/7.
           +-Gr 11-75 p869   ++MT 5-76 p409
           +HFN 11-75 p148   ++RR 12-75 p86
Mr. Shadowman. cf Argo ZFB 95/6.
El molondrón. cf Polydor 2383 389.
Mon coeur en vous. cf HMV CSD 3766.
Moulin de Paris. cf Vanguard VCS 10049.
Musicorum decus et species. cf Telefunken ER 6-35257 (2).
My Lady Carey's dompe. cf Angel SFO 36895.
My love's an arbutus. cf Vista VPS 1022.
My snowy-breasted pearl. cf HMV EMD 5528.
The neighbour. cf HMV ASD 3200.
Ninefold alleluia and announcement of the holy gospel. cf Vanguard SVD
    71212.
Northumbrian airs. cf Pye GH 603.
Novus miles sequitur. cf Vanguard VCS 10049.
Now glad of heart be everyone. cf Vista VPS 1023.
Nuevas te traygo, Carillo. cf DG 2530 504.
O come all ye faithful. cf HMV CSD 3774.
O death rock me asleep. cf Angel SFO 36895.
O little town of Bethlehem. cf HMV CSD 3774.

O Maria virgo. cf Candide CE 31095.
O Mary, don't you weep. cf Columbia SCX 6585.
O Tannenbaum. cf CBS 61771.
O' 'twas on a Monday morning. cf Harmonia Mundi HM 593.
O waly, waly. cf Harmonia Mundi HM 593.
Oft the stilly night. cf HMV RLS 716.
Oh, the oak and the ash. cf HMV ESD 7002.
Orient's partibus, Renosemus laudibus, Verbum caro. cf MOUTON: Noe
     psallite.
Paduana Hispanica. cf Hungaroton SLPX 11721.
El pano moruno. cf Saga 5412.
Partita of old dances and airs. cf Panton 110 380.
Pase el agoa, ma Julieta. cf Nonesuch H 71326.
La pastorella. cf Angel SFO 36895.
Pavana de la morte de la ragione. cf L'Oiseau-Lyre 12BB 203/6.
Pavana el tedesco. cf L'Oiseau-Lyre 12BB 203/6.
Pavane Venetiana. cf L'Oiseau-Lyre 12BB 203/6.
La peche de Moules. cf Club 99-101.
Petersburger Marsch. cf DG 2721 077 (2).
Petersburger Marsch. cf Polydor 2489 523.
Le petit gentilhomme. cf Angel SFO 36895.
Pieces, lute (5). cf DG 2530 561.
Pilgrim's farewell. cf Nonesuch H 71276.
Pisu Beog. cf Coronet LPS 3032.
Plaudite tympani. cf Missa Salisburgensis. BASF 22073/7.
The play of Daniel. The Clerkes of Oxenford; David Wulstan. Calliope
     CAL 1848.
            +Gr 11-76 p844
Polyphonic dances. cf Nonesuch H 71326.
Pour mon coeur. cf Vanguard VCS 10049.
Pour ung plaisir. cf Hungaroton SLPX 11669/70.
Psalm, no. 55. cf Coronet LPS 3032.
Psalmus CXXX. cf Hungaroton SLPX 11721.
Puer natus est nobis. cf Vista VPS 1022.
El puerto. cf Saga 5412.
Quant voi l'aloete. cf Candide CE 31095.
Quelle est cette odeur agréable. cf HMV CSD 3774.
Quem pastores laudavere. cf Abbey LPB 761.
Quittez pasteurs. cf HMV CSD 3774.
Der rather Schwanz. cf L'Oiseau-Lyre 12BB 203/6.
Restoés, restoés. cf HMV SLS 863.
Riu, riu, chiu. cf Grosvenor GRS 1034.
La rocca el Fuso. cf Hungaroton SLPX 11669/70.
Rocking. cf Abbey LPB 761.
Rodrigo Martines. cf Nonesuch H 71326.
The rose tree. cf Michigan University SM 0002.
Le rosignol. cf DG Archive 2533 323.
Le rossignol. cf L'Oiseau-Lyre 12BB 203/6.
Roussignolet qui cantos. cf Club 99-101.
The roving sailor. cf Michigan University SM 0002.
Rue and thyme. cf Prelude PRS 2501.
Russian folk songs (2). cf Hungaroton SLPX 11696.
Russian song. cf Rubini RS 301.
Sa qui turo zente pleta. cf Eldorado S-1.
Säg mig den vägen. cf BIS LP 14.
The sailor and young Nancy. cf Harmonia Mundi HM 593.

St. Columba, The King of love my shepherd is. cf Abbey LPB 761.
Sakura, sakura. cf Polydor 2383 389.
Salamanca. cf Saga 5412.
Saltarello. cf CBS 76183.
Saltarello. cf DG Archive 2533 184.
Saltarello. cf HMV SLS 5022 (2).
Saltarello. cf Vanguard VCS 10049.
Saltarello (14th century). cf DUFAY: Franc cuer gentil.
Sans day carol. cf Grosvenor GRS 1034.
Sans day carol. cf HMV CSD 3774.
Sans faire. cf DUFAY: Franc cuer gentil.
Sant Marti del Canigo. cf CBS Classics 61579.
Schellerl-Tanz. cf Saga 5421.
Se mai per maraveglia. cf l'Oiseau-Lyre 12BB 203/6.
La septime estampie real. cf HMV SLS 863.
La serrana. cf Enigma VAR 1015.
La sexte estampie real. cf Nonesuch H 71326.
The sheep sharing. cf Harmonia Mundi HM 593.
Shenandoah. cf RCA LRL 2-7531.
The shepherd's carol. cf Vanguard SVD 71212.
Shooting the guns pavan. cf L'Oiseau-Lyre 12BB 203/6.
The short mesure off my Lady Wynkfyld's Rownde. cf Angel SFO 36895.
Sol ovitur. cf Vanguard VCS 10049.
Sometimes I feel like a motherless child. cf CBS 61746.
Sometimes I feel like a motherless child. cf Personal Touch 88.
Sometimes I feel like a motherless child. cf Polydor 2383 389.
Song of Stenka Razin. cf HMV ASD 3200.
The song of the ass. cf Vanguard VCS 10049.
Sospan Fach. cf Columbia SCX 6585.
Souvent souspire mon cuer. cf Nonesuch H 71326.
Springhill. cf Nonesuch H 71276.
Sta notte. cf L'Oiseau-Lyre 12BB 203/6.
Suite, lute. cf DG 2530 561.
Suite regina. cf L'Oiseau-Lyre 12BB 203/6.
Die Suss Nachtigall. cf Vanguard VCS 10049.
Sweet Avelina. cf CBS 61746.
Swing low, sweet chariot. cf HMV RLS 719.
Synge we to this mery cumpane. cf BASF 25 22286-1.
Tantz, Proportio. cf Hungaroton SLPX 11721.
Tappster, dryngker. cf BASF 25 22286-1.
Taps. cf BBC REB 228.
Tati-Tati. cf Philips 6780 751.
The Taunton carol. cf Vista VPS 1022.
This joyful Eastertide. cf Vista VPS 1023.
Tomorrow shall be my dancing day. cf Grosvenor GRS 1034.
La traditora. cf Hungaroton SLPX 11669/70.
La triquotée. cf L'Oiseau-Lyre 12BB 203/6.
Trisagion. cf Vanguard SVD 71212.
Triumph. cf Nonesuch H 71276.
Trois princesses. cf Club 99-101.
Trotto. cf DUFAY: Franc cuer gentil.
Trotto. cf Vanguard VCS 10049.
The turtle dove. cf Coronet LPS 3031.
The turtle dove. cf Harmonia Mundi HM 593.
Tutu maramba. cf Saga 5412.
Ukrainian carols: Across the wide fields; The bells of Jerusalem;

Chedrick; Early, very early; A falcon; Good evening to you; Heaven
and earth; O God sees; O the new joy; On Jordan's banks; On the
Jordan; O yes, here is a splendid host. cf Ikon IKO 3.
The unknown. cf Columbia MS 6161.
The unknown. cf Columbia M 34129.
Up, good Christen folk and listen. cf HMV CSD 3774.
La vera sorrentina. cf Polydor 2383 389.
Veris ad imperia. cf Candide CE 31095.
The vicar of Bray. cf HMV ESD 7002.
Vieille chanson de chasse. cf HMV RLS 716.
Vierhundert Jar uff Diser Erde. cf Vanguard VCS 10049.
Vallancico. cf Saga 5412.
Vivrons heureux. cf Club 99-101.
Washington. cf Nonesuch H 71276.
Washington's march. cf Vox SVBX 5350.
Wassail song (arr. Vaughan Williams). cf HELY-HUTCHINSON: Carol sym-
    phony.
The wee cooper o'Fyfe. cf HMV EMD 5528.
Where griping griefs. cf L'Oiseau-Lyre 12BB 203/6.
While shepherds watched their flocks. cf Vanguard SVD 71212.
The white cockade. cf Michigan University SM 0002.
The world turned upside down. cf Michigan University SM 0002.
Worldes blis. cf Vanguard VCS 10049.
Wrth Fynd Efo Deio I Dowyn. cf L'Oiseau-Lyre SOL 345.
Y Bore Glas. cf L'Oiseau-Lyre SOL 345.
Y Deryn Pur. cf L'Oiseau-Lyre SOL 345.
Ya se van la paloma. cf Enigma VAR 1015.
Yankee Doodle. cf Michigan University S 0002.
Yankee Doodle. cf RCA AGL 1-1334.
Ye banks and braes o' Bonnie Doon. cf Coronet LPS 3031.
Ye banks and braes. cf HMV ESD 7002.

PERFORMER INDEX

Abbado, Claudio, conductor   596, 1828, 1838, 2197, 2513, 2526, 2631,
        2701
Abdoun, Georges, bass-baritone   527
Abravanel, Maurice, conductor   845, 1160, 1195, 1221, 1237, 1301, 1605,
        1639, 2594, 2627, 2674
Academy of Ancient Music   36, 2093, 2437
Academy of St. Martin-in-the-Fields   57, 94, 122, 134, 155, 216, 401,
        416, 548, 567, 693, 850, 851, 929, 941, 998, 999, 1046,
        1069, 1093, 1172, 1173, 1244, 1245, 1246, 1247, 1251, 1256,
        1265, 1270, 1277, 1292, 1293, 1294, 1305, 1318, 1701, 1797,
        1812, 1824, 1883, 1906, 1917, 1920, 1930, 2199, 2264, 2609,
        2651, 2662, 2677, 2724, 2861, 3016, 3074, 3158, 3160
Accademia Monteverdiana   1213, 1764
Accardo, Salvatore, violin   2010, 2011, 2014, 2736, 2737, 2738, 2739,
        2740
Acezantez Ensemble   595, 925
Achucarro, Joaquin, piano   1085
Ackerman, Otto, conductor   1919, 2482
Ackerman, Manfred, bass   178
Adam, Theo, bass-baritone   79, 81, 92, 96, 181, 467, 1267, 1790, 1921,
        2777, 2780, 2798, 2801, 2803
Adams, Bernard, conductor   3085
Adams, Carl, flute   1436
Adams, John, conductor   730
The Adelaide Singers   1524, 1525
Adelaide Symphony Orchestra   1524, 1525
Adeney, Richard, flute   2964
Adini, Ada, soprano   3215
Adler, Kurt, conductor   2982
Adler, Samuel, conductor   1194
Adni, Daniel, piano   1096, 1702, 2302
Aebersold, Jamey, saxophone   220
Aeolian Quartet   1331, 1333, 1336, 1851
Agay, Karola, soprano   939
Ahrans, Allan, bass   178
Ahronovitch, Yuri, conductor   2097
Aitkin, Robert, flute   1696
Akerberga, Ilze, organ   2411
Alain, Marie-Claire, organ   2038
Alarie, Pierette, soprano   1274
The Alban Singers   108, 110, 2860
Albanese, Francesco, tenor   2716
Albanese, Licia, soprano   2079

Bassett, Michael, piano 1459
Bastianini, Ettore, baritone 2070, 2071, 2206, 2598, 2715, 2717, 2972
Bath Festival Orchestra 2313
Batic, Polly, contralto 2779, 2787
Baudo, Serge, conductor 970, 1094, 1409
Bauer, Frieda, piano 1450
Bauer, Hans, conductor 39
Bauer, Hartmut, bass 2762, 2772
Bauer, Kurt, conductor 1262
Baumann, Hermann, horn 855, 2870, 3260
Baumgartner, Rudolf, conductor 1787, 2734, 3195
Bavarian Radio Chorus 392, 711, 1186, 1326, 1496, 2326, 2413, 2802,
        2805, 3002
Bavarian Radio (Symphony) Orchestra 335, 392, 400, 711, 716, 983, 987,
        988, 996, 1326, 1604, 1607, 1608, 1619, 1626, 1641, 1651,
        1659, 1902, 1924, 2326, 2413, 2497, 2751, 2766, 2802, 2805,
        2806, 3002
Bavarian State Opera Chorus 186, 2449, 2753, 3002
Bavarian State Opera Orchestra 1860, 2753
Bavarian State Orchestra 293, 2449, 3002
Bavarian State Orchestra Chamber Ensemble 763
Bayreuth Festival Chorus 2752, 2759, 2761, 2762, 2771, 2772, 2774,
        2780, 2790, 3002
Bayreuth Festival Orchestra 2751, 2752, 2759, 2761, 2762, 2771, 2772,
        2774, 2780, 2789, 2790, 2792, 2793, 3002
Bazelon, Irwin, conductor 268
Bazely, S. M., trombone 3180
B.B.C. Chorus and Choral Society 2245, 2261
B.B.C. Northern Singers 1406, 1724
B.B.C. Singers 877, 2246, 2661
B.B.C. Symphony Chorus 1404
B.B.C. Symphony Orchestra 240, 254, 279, 289, 409, 415, 435, 487, 535,
        593, 1150, 1404, 1744, 1879, 2245, 2246, 2661, 2815, 3081,
        3160
Bean, Hugh, violin 140
Bean, Roger, conducto 2812
Beardslee, Bethany, soprano 1078
Beat, Duncan, conductor 3155
Beaux Arts Trio 1373, 1374, 1444, 2282, 2343
Becker, David, viola 1770
Becker, Werner, baritone 1989
Bedford, David, keyboards, recorders, percussion 270
Bedford, Steuart, conductor 682
Bednář, Václav, baritone 2419
Beecham Choral Society 224
Beecham, Thomas, conductor 224, 518, 541, 1038, 1275, 2072, 2314, 2392,
        3081
Beethoven Quartet 2373, 2384
Behr, Therese, contralto 3209
Behrend, Claudia Brodzinska, singer 2992
Behrend, Siegfried, guitar 2992, 2998
Beissel, Heribert, conductor 791
Belcourt, Emile, tenor 2776
Belgian Radio Chorus 1130
Belgrade Opera Chorus 2975
Belgrade Opera Orchestra 2975
Bell, Daniel, conductor 2893

Boston Pops Orchestra   214, 1233, 2133, 2446, 2466, 3190
Boston Pro Musica Chorus   2763
Boston Symphony Chamber Players   2514, 2533
Boston Symphony Orchestra   72, 392, 428, 451, 498, 505, 508, 611, 645,
          723, 840, 890, 908, 1178, 1735, 1793, 1937, 2060, 2100,
          2132, 2141, 2157, 2158, 2159, 2321, 2394, 2405, 2516, 2517,
          2626, 2757, 2758, 2763, 3199
Bottone, Robert, piano   679
Bouchard, Victor, piano   1696
Boucher, Gene, baritone   1298
Boulanger, Nadia, conductor   1123
Boulez, Pierre, conductor   240, 487, 535, 593, 956, 1083, 1595, 1744,
          2139, 2148, 2245, 2246, 2509, 2523, 2793, 2899
Boult, Adrian, conductor   267, 309, 313, 597, 598, 643, 700, 707, 737,
          1032, 1037, 1040, 1044, 1051, 1055, 1066, 1069, 1139, 1217,
          1253, 1290. 1292, 1293, 1398, 1593, 1610, 1929, 2124, 2179,
          2363, 2415, 2616, 2617, 2675, 2676, 2682, 2683, 2684, 2685,
          2686, 2905, 3040
Bour, Ernest, conductor   1501
Bourdon, Rosario, conductor   2428
Bournemouth Municipal Choir   1401
Bournemouth Sinfonietta Orchestra   37, 1072, 1073, 3076
Bournemouth Symphony Orchestra   223, 1053, 1401, 2052, 2376, 2387, 2395,
          2404, 2406, 2830, 3040
Bowen, Kenneth, tenor   682, 1695, 2245, 2264, 2855
Bowers-Broadbent, Christopher, organ   1492
Bowman, James, countertenor (alto)   678, 682, 1247, 1264, 1270, 2092,
          2574, 3070
Boyd, Liona, guitar   3109
Bozzi, Guido, conductor   1308
Bradbury, Colin, clarinet   733
Bradley, Desmond, violin   737
Brahms, Antony, baritone   480
Brailowsky, Alexander, piano   1711
Brain, Dennis, horn   1800, 3074
Brainin, Norbert, violin   2277
Bramma, Henry, organ   1134, 2836, 3167
Bramson, Berenice, soprano   3023
Brand, Geoffrey, conductor   556
Brandis, Thomas, violin   857, 1908
Brandt, Birger, bass   916
Brannigan, Owen, bass-baritone   685, 2538, 2546, 3077
Branzell, Karin, contralto   2751
Brass Band   3106
Brasseur, Elisabeth, Chorale   1100, 1862, 3002
Braun, Hans, bass-baritone   2497, 2759
Braun, Manfred, bassoon   1803
Bream, Julian, Consort   3200
Bream, Julian, guitar   476, 490, 491, 3183, 3200
Bream, Julian, lute   946, 1248, 3184, 3200
Brecknock, John, tenor   3013
Bredenbeek, Hans, guitar   2239
Brehm, Alvin double bass   2990
Breitenbach, Günther, viola   2965
Breitner, Tamás, conductor   2352
Brendel, Alfred, piano   274, 356, 359, 364, 365, 1007, 1575, 1785, 1812,
          1824, 2056, 2257, 2270, 2271, 2296

Buckel, Ursula, soprano 2259
Buckley, Emerson, conductor 1774
Bucquet, Marie-Françoise, piano 1091, 3161
Budapest Blasorchester 3094
Budapest Chamber Ensemble 226
Budapest Chorus (Choir) 454, 1557
Budapest Madrigal Choir 2722
Budapest Philharmonic Orchestra 246, 538, 1026, 2134, 3092
Budapest Symphony Orchestra 226, 514, 965, 1089, 1135, 1557
Budrin, Ivan, bass-baritone 2182
Buffalo Philharmonic Orchestra 746
Buhl-Møller, Kirsted, soprano 916
Buketoff, Igor, conductor 1194
Bulavin, Eduard 2368
Bulgarian Quartet 3031
Bulgarian Radio and Television Men's Choir 3029
Bulgarian Radio and Television Symphony Orchestra 2381
Bulgarian Radio and Television Vocal Ensemble 2095
Bumbry, Grace, mezzo-soprano (contralto) 1687, 2688, 2691, 2969
Bunge, Sas, piano 1837
Bunger, Richard, piano 28
Burge, David, piano 2956
Burge, Lois, piano 2956
Burgess, Grayston, conductor 1780, 2887, 3266
Burian, Karel (Karl Burrian), tenor 3007, 3211
Burke, John, organ 219
Burles, Charles, tenor 2036
Burmanje, Ton, guitar 2239
Burmeister, Annelies, mezzo-soprano (contralto) 181, 1872, 1873, 2748,
          2780
Burnett, Richard, piano 1118, 2325
Burns, Ralph, conductor 1960
Burrowes, Norma, soprano 542, 1407, 1663, 1691, 1771, 1991, 2039, 2092
Burrows, Stuart, tenor 450, 502, 1266, 1278, 1595, 1788, 2322, 2587
Burton Constructional Band 3178
Busch, Adolf, violin 388
Busch Chamber Players 1786
Busch, Fritz, conductor 418, 1786, 2318
Busch, Herrmann, organ 3152
Busch, Nicholas, horn 681
Busch Quartet 348
Busoni, Ferruccio, piano 3099
Bussard, Jean, tenor 2036
Busser, Henri, conductor 1198
Butler, Douglas, organ 1088
Byman, Isabelle, piano 3279
Byrne Campe Chorale 1687

Caballé, Montserrat, soprano 577, 934, 1847, 2193, 2689, 2696, 2700,
          2703, 3115
Cable, Howard, Symphonic Band 2915
Cable, Yvonne, violoncello 3023
Cahuzac, Louis, clarinet 3025
Calabrese, Franco, bass 2086, 2705
Calapai, Delia, piano 2299
Caldwell, James, viola da gamba 1673

Cales, Claude, baritone 549
Caley, Ian, tenor 1492, 1688
California State University Northridge Chamber Singers 1498
Callas, Maria, soprano 2073, 2078, 2086, 2711, 2716, 2718, 3072
Calloway, Cab, piano 1162
Calusdian, George, piano 1494
Cambridge Brass Quintet 2957
Cambridge, King's College Chapel Choir 94, 212, 567, 637, 920, 1245,
        1246, 1247, 1270, 1277, 1293, 1771, 2017, 3064, 3065, 3070
Cambridge University Musical Society Chorus 920, 3070
Camden, Anthony, oboe 1693
Camden Wind Ensemble 64
Camerata Hungarica 3087
Camerata Singers 2139
Cameron, John, bass-baritone 2538, 2546
Caminada, Anita, mezzo-soprano 470
Campagnola, Leon, tenor 2920
Campanella, Michele, piano 1551, 1561, 1570
Campbell, Sidney, conductor 2856
Campi, Enrico, bass 2078
Campoli, Alfredo, violin 557, 707, 3111
Canali, Anna Maria, mezzo-soprano 2075
Candela, Miguel, violin 2918
Canino, Bruno, piano 487, 2504
Cantarini, Teresa, mezzo-soprano 2075
Cantelo, April, soprano 2972, 3070
Canterbury Cathedral Choir 2832, 3026
Cao, Pierre, conductor 768, 1699, 2015, 2223
Capella Cordina 949, 952
Capolongo, Paul, conductor 1091, 3161
Capon, Frédéric, piano 692
Cappone, Giusto, viola 2489
Cappuccilli, Piero, baritone 470, 2689, 2694, 2701, 2703
Carbonari, Virgilio, tenor 2077
Cardiff Festival Ensemble 2161
Carelli, Emma, soprano 2922
Carelli, Gabor, tenor 876
Caridis, Miltiades, conductor 132
Carlin, Mario, tenor 2070, 2078, 2705
Carlton Main Frickley Colliery Band 3179
Carlyle, Joan, soprano 1529
Carmel Bach Festival Chorale 2723
Carmel Bach Festival Orchestra 2723
Carminata, Maria Rosa, soprano 540
Carreras, José, tenor 2193, 2696, 2702
Caruso, Enrico, tenor 3018, 3193
Caruso, Mariano, tenor 2716
Casa Serena Trio 2178
Casadesus, Gaby, piano 2905
Casadesus, Robert, piano 2153, 2905
Casals Barcelona Orchestra 3046
Casals, Pablo, conductor 144, 437, 1939, 2905
Casals, Pablo, violoncello 390, 459, 705, 2283, 2903, 3046
Casapietra, Celestina, soprano 1872, 1873, 1994
Casas, Esther, soprano 2689
Case, Anna, soprano 3221

Chissari, Santa, soprano 2075
Chodos, Gabriel, piano 2303
Choir of the North 3292
Chookasian, Lili, soprano (contralto) 452, 2049
Choral Art Society 558
Chorzempa, Daniel, organ 207, 1102, 1252, 1909
Chovanec, Jaroslav, violoncello 570
Christ Church Cathedral Choir 986, 1264, 1513, 1519, 2040, 2505, 2531
Christensen, Roy, violoncello 869
Christoff, Boris, bass 1970, 2598
Chuchro, Josef, violoncello 316, 1681
Chung, Kyung-Wha, violin 170, 2213
Church of the Transfiguration Boys' Choir 1625
Cianella, Giuliano, tenor 2879
Ciani, Dino, piano 286
Ciccolini, Aldo, piano 829, 897, 2044, 2160
Cillario, Carlo Felice, conductor 934, 2894, 2895
Cimara, Pietro, conductor 2079
Cincinnati May Festival Chorus 2205
Cincinnati Symphony Orchestra 2056, 2205
Cioni, Renato, tenor 937
Civil, Alan, conductor 543
Civil, Alan, horn 678, 684, 1801, 1802, 2045
Clabassi, Plinio, bass 2075, 2078, 2711
Clarion Consort 164, 3157
Clarion Wind Quintet 211
Clark, Graham, tenor 1688
Clarke, Christina, soprano 2089
Clarke, Katie, soprano 2788
Clatworthy, David, baritone 1298
Clayton, John, bass 220
Clemenčíc Consort 950, 953, 3034, 3035, 3038
Clemenčíc, René, conductor 950, 953, 3034, 3035, 3038
Clément, Josephte, soprano 2712
Clement, Manfred, oboe 2501
Cleobury, Stephen, organ 966, 1093
The Clerkes, Oxenford 1167, 2572
Cleva, Fausto, conductor 1528, 1685
Cleveland Orchestra 231, 304, 320, 392, 536, 647, 915, 973, 974, 1004,
          1005, 1015, 1153, 1158, 1632, 1650, 1928, 1939, 1990, 2140,
          2350, 2523, 2697, 2983
Cleveland Orchestra Boys' Choir 1990
Cleveland Orchestra Children's Chorus 1158
Cleveland Orchestra Chorus 1158, 1990, 2140
Cleveland Quaret 234, 1334, 2268
Cliburn, Van, piano 1541, 1573, 2583, 3188
Clidat, France, piano 1572
Cluytens, André, conductor 310, 419, 1100, 2150
Cochereau, Pierre, organ 964, 1129, 1531
Cochran, William, tenor 1662
Cockpit Ensemble 730, 1076
Cocteau, Jean, narrator 2515, 2518
COD Vocal Ensemble 1497
Coda, Eraldo, bass 2073
A Coeur Joi Vocal Ensemble of Valence 1582
Cohen, Frank, clarinet 1203
Cohen, Joel, conductor 3129, 3261

Cuckston, Alan, piano 478
Cuénod, Hugues, tenor 1581, 1984
Cummings, Douglas, violoncello 2618
Cundari, Emilia, soprano 1621
Cundick, Robert, organ 2939
Curphey, Margaret, soprano 2788, 3013
Curtin, Phyllis, soprano 1581
Curtis, Alan, harpsichord 860
Curtis Institute Symphony Orchestra 3100
Curtis, Jan, mezzo-soprano 580
Curtis-Smith, Curtis, piano 873
Curzon, Clifford, piano 297, 627, 1217, 2322, 2323
Custer, Arthur, conductor 874
Custer, Arthur, piano 874
Cutner, Solomon, piano 357, 798, 1216, 2580, 3081
Czako, Eva, violoncello 1892
Czech Chamber Orchestra 997
Czech Nonet 1682
Czech Philharmonic Chorus 1452, 1996, 2000, 2423
Czech Philharmonic Orchestra 249, 316, 405, 447, 570, 571, 585, 705,
        930, 947, 957, 1011, 1014, 1018, 1019, 1028, 1048, 1409,
        1448, 1452, 1680, 1681, 1839, 1995, 1996, 2024, 2058, 2378,
        2418, 2421, 2422, 2535, 2978, 3243
Czech Philharmonic Studio Orchestra 943
Czech Philharmonic Wind Ensemble 1453, 1454, 1455, 1499
Czech Radio Children's Choir 1565
Czech Radio (Symphony) Orchestra 930, 947, 2672
Czech Wind Quintet 930
Czerwenka, Oskar, bass-baritone 2977
Czerwonky, Richard, violin 2918
Czidra, Laszlo, recorder 2735, 3087
Cziffra, György, Jr., conductor 1135, 1545, 1556
Cziffra, György, piano 1135, 1545, 1556, 1559
Czyź, Henryk, conductor 3122

Da Capo Chamber Players 1116, 2426
Dacosta, Janine, piano 2151
Dael, Lucy van, violin 859
Dales, Richard, baritone 1412
Dalis, Irene, mezzo-soprano 2774
Dallapozza, Adolf, tenor 2454, 2769, 2831
Dalton, James, organ 106
Damm, Peter, horn 2483, 2501
Danby, Nicholas, organ 1127
Danco, Suzanne, soprano 1101, 2145
Danieli, Lucia, contralto (mezzo-soprano) 2078
Danish (State) Radio Symphony Chorus 418, 916, 1970
Danish (State) Radio Symphony Orchestra 418, 658, 916, 1965, 1970,
        1976, 1979
Danto, Louis, tenor 3121
DaPaur, Leonard, conductor 2934
Darasse, Xavier, organ 53
Darcy, Emery, tenor 2785
Dare, John, piano 1498
Daris, Yannis, conductor 907
Darlington, Stephen, organ 3026

Darré, Jeanne-Marie, piano 2988
Dartington String Quartet 1177, 2410
Da Silva, Howard 559
Datyner, Henry, violin 733
Davidson, Lawrence, baritone 2074
Davies, Dennis Russell, conductor 60, 579
Davies, Dennis Russell, piano 2660
Davies, Hugh, electronics 2441
Davies, John, treble 2836
Davies, Meredith, conductor 916, 2681
Davies, Noel, piano 3013
Davies, Peter Maxwell, conductor 881, 882, 883
Davies, Ryland, tenor 938, 1268, 1688, 1689, 1765, 1767, 1849, 1861,
        1879
Davies, Tudor, tenor 2679
Davis, Andrew, conductor 318, 697, 1050, 1099, 1230, 2380
Davis, Andrew, harpsichord 129, 1246, 2854
Davis, Andrew, organ 3070
Davis, Colin, conductor 244, 279, 289, 415, 426, 435, 494, 496, 501,
        504, 515, 520, 690, 1272, 1294, 1735, 1837, 1847, 1879,
        1900, 2195, 2405, 2528, 2661 2665, 2786, 3160, 3161
Davis, Ivan, piano 1152, 1193
Davison, Arthur, conductor 70, 140, 1258, 1287, 1394, 1799, 2916
Davison, Beverly, violin 1967
Davrath, Natania, soprano 1605, 1639
Davy, Gloria, soprano 2442
Dawkes, Hubert, piano 907
Dawson, Antony, tenor 2090
Dawson, Peter, bass-baritone 2679
Dean, Stafford, bass 1765, 1767, 1859, 2090
Dearnley, Christopher, organ 3223
Debost, Michel, flute 1886, 2045
Debrecen Kodály Chorus 2128
Dediukhin, Aleksander, piano 825
DeGaetani, Jan, mezzo-soprano (soprano) 864, 1083, 1121, 1170, 1440,
        1697, 2243, 3128
Del Bosco, Carlo, bass 2702
Delfino, Jacqueline, piano 2339
Deflino, Otto, piano 2339
della Casa, Lisa, soprano 2481, 2977
Deller, Alfred, conductor 1242, 2087, 2088, 2089, 3036
Deller, Alfred, countertenor 2089, 3032
Deller Consort 2088, 3031, 3032, 3036, 3037
Deller, Mark, countertenor 2089
DeLon, Jack 1774
de los Angeles, Victoria, soprano 498, 541, 749, 1087, 1100, 2070,
        2075, 3163
Del Mar, Norman, conductor 228, 685, 688, 1049, 1073, 2597
de Luca, Giuseppe, baritone 2890
Dempsey, Gregory, tenor 2776, 2783
Dempster, Stuart, euphonium 3283
Dempster, Stuart, trombone 3283
Demus, Jörg, piano 322, 672, 1807, 1832
Dene, József, baritone 2762, 2772
Denize, Nadine, soprano 540
Deomidova, Glafira, soprano 2066

Donderer, Georg, violoncello 672
Dooley, William, baritone 2763
Doráti, Antal, conductor 248, 254, 422, 450, 483, 529, 588, 593, 690,
            876, 889, 909, 1006, 1008, 1010, 1150, 1311, 1324, 1329,
            1349, 1350, 1354, 1363, 1364, 1487, 1489, 1490, 1991, 2103,
            2167, 2171, 2590, 2604, 2623, 2640, 2781, 3163
Dorigny, Henri, guitar 2984
Dorliak, Nina, soprano 2384
Doro, Mina, mezzo-soprano 2075
Dorow, Dorothy, soprano 731, 2878, 2893
Dorpinghans, Eleonore, mezzo-soprano 1948
Dosse, Marylene, piano 769, 1197
Dostalova, Eva, flute 1678
Doussard, Jean, conductor 2036, 2455
Doveton, Robin, tenor 3070
Dowis, Jeanane, piano 1304
Dowling, Denis, baritone 3013
Downe House School Choir 1037
Downes, Edward, conductor 534
Downes, Herbert, viola 737
D'Oyly Carte Opera Chorus 2543, 2554, 2555
D'Oyly Carte Opera Company 2541, 2551, 2556
Dresden Boys Choir 1994
Dresden Cathedral Choir 1262
Dresden Cathedral Orchestra 1262
Dresden Kreuzchor 181
Dresden Philharmonic Orchestea 1792, 1865, 3258
Dresden Staatskapelle Orchestra 334, 586, 596, 1001, 1609, 1868, 1869,
            1870, 2472, 2483, 2501, 2562, 2777, 2798, 2800, 2803
Dresden State (Opera) Chorus 2777, 2798
Dresden State (Opera) Orchestra 1790, 1903, 2348, 2801
Dressler, John, piano 2960
Dreyfus, Huguette, fortepiano 51
Dreyfus, Huguette, harpsichord 146, 154, 190, 854, 858, 1520
Driehuys, Leo, oboe 2726
Driscoll, Loren, tenor 2770
Drolc, Eduard, violin 672
Drolc Quartet 2162
Drottingholm Baroque Ensemble 575
Drottingholm Chamber Orchestra 2192
Drower, Meryl, soprano 480
Drucker, Stanley, clarinet 1083
Dubois, Alfred, violin 2918
Dubost, Michel, flute 460
Dubow, Marilyn, violin 1438
Du Brassus Choir 1984
Duchâble, François, piano 818
Duchêne, Maria, mezzo-soprano 3221
Duclos, Rene, Choir 1202, 2455
Dudarev, Georgy 2368
Due Boemi di Praga 3144
Dugdale, Sandra, mezzo-soprano 3013
Dunan, Gérard, tenor 517, 1531
Duncan, Todd, baritone 1163
Dunkel, Paul, flute 1203
Du Plessis, Christian, baritone 2712, 3013

Elliott, Robert, harpsichord 862
Ellis, Nancy, viola 2823
Ellis, Osian, harp 677, 1692, 1767, 2020, 2156
Ellison, Jon, bass 2541, 2555
Elman, Mischa, violin 2918
Elmendorff, Karl, conductor 2793
Elms, Lauris, contralto 1206
Elnikov, Yuri, tenor 2061
Elvin, William, baritone 2690, 2703
Elwes, Gervase, tenor 3081
Elwes, John, tenor 750, 1580, 1768, 1770
Emile, Romano, tenor 876
Emmerich, Albert, bass 2775, 2779
Empire Brass Quintet 1141
Endres Quartet 1888
Enesco (Enescu), George, conductor 3044
Enesco, Georges, violin 2918, 3044
Engel, Karl, piano 636, 1746, 1805, 1806, 1818, 1822, 1823, 2322,
            2323, 2324, 2340
Engel, Lehman, conductor 1159
Engen, Keith, bass 179
English Chamber Orchestra 48, 49, 66, 69, 71, 123, 129, 138, 682, 699,
            713, 733, 912, 926, 1002, 1032, 1067, 1185, 1228, 1268, 1269,
            1271, 1273, 1279, 1293, 1294, 1355, 1382, 1407, 1765, 1767,
            1789, 1804, 1815, 1816, 1817, 1819, 1831, 1840, 1876, 1881,
            1901, 1936, 2125, 2191, 2204, 2750, 2899, 3039, 3070, 3074,
            3160, 3269
English Chamber Orchestra Chorus 1268
English Concert 2743
English, Gerald, speaker 2253
English, Gerald, tenor 489, 1993, 2681
English, Jon, trombone 2954
English Opera Group Chorus 682
English Opera Group Orchestra 685, 2776, 2788
English Sinfonia Orchestra 1408, 3166
Ensemble d'Archets Eugene Ysaye 3003
Entremont, Philippe, piano 785, 1571, 2123, 2607, 2898, 2901
Eötvös, Peter, camel bells, triangles, synthesizer 2440
Eötvös, Peter, electrochord with synthesizer 2443
Ephrikian, Angelo, conductor 833
Epsilon Quartet 907
Epstein, David, conductor 1532
Equiluz, Kurt, tenor 84, 85, 87, 1762, 2453
Erb, Karl, tenor 180
Ercolani, Renato, tenor 2075, 2078, 2711, 2718
Erdélyi, Csaba, viola 1393
Erdmane, Silvija, soprano 2411, 3103
Erede, Alberto, conductor 2069, 2206
Ericson, Barbro, contralto (mezzo-soprano) 566
Ericson, Bengt, viola da gamba 2876
Ericson, Eric, Choir 566
Ericson, Eric, conductor 1077, 1947, 3247
Eriksson, Elisabeth, soprano 1947
Ermler, Mark, conductor 2066
Errolle, Ralph 3221
Erwen, Keith, tenor 3013

Flagello, Nicolas, conductor 759, 2517
Flagstad, Kirsten, soprano 1593, 2755, 2779, 2782, 2785, 2787, 2977,
        3212
Fleet, Edgar, tenor 1213
Fleischer, Eva, mezzo-soprano (contralto) 81
Fleischer, Paul, piccolo 3126
Flesch, Carl, violin 2918
Flook, A., xylophone 18
Florilegium Musicum See Paris Florilegium Musicum
Fodor, Eugene, violin 1708, 3187
Fog, Peter, baritone 916
Foiani, Giovanni, bass 2701
Földes, Andor, piano 189, 303, 362
Foltyn, Maria, soprano 1759
Fonda, Jean, piano 2324
Fontana, Sergio, bass 2701
Forbes, Colin, piano 1206
Forchert, Walter, violin 1314
Forman, Motter, harp 2954
Forrester, Maureen, contralto (alto) 759, 1621, 1628
Forsell, John, baritone 3218
Forsmann, Folke, organ 177
Forstel, Gertrude, soprano 3211
Forti, Carlo, bass 2073, 2711
Foss, Lukas, conductor 746
Foss, Lukas, piano 1120
Foster, Lawrence, conductor 534, 1802, 2213, 2966, 3039
Fournet, Jean, conductor 1102, 1706, 3161, 3163
Fournier, Pierre violoncello 315, 457, 466, 563, 633, 672, 703, 2045,
        2324, 2489, 2734
Frager, Malcolm, piano 2501
Francaix, Jean, piano 1123
Francis, Hannah, soprano 1492
Francis, Sarah, oboe 687
Frankfurt Radio Symphony Orchestra 2098, 2218, 2244, 2591
Frankl, Peter, piano 2331, 2341
Fransson, Göran, singer 1080
Frantz, Ferdinand, baritone 2775, 2779, 2787
Frantz, Justus, piano 975
Franz, Helmut, conductor 1535
Frasca-Colombier, Monique, viola d'amore 2734
Frasca-Colombier, Monique, violin 2731, 2744
Freeman, James, piano 867
Freeman, Paul, conductor 28, 50, 221, 1473, 2432
Fremaux, Louis, conductor 503, 519, 546, 1423, 1503, 2039, 3068
French Garde Republicaine Band 205
French Instrumental Ensemble 2725
French (National) Radio Chorus 502, 541, 2084
French (National) Radio Orchestra 310, 502, 513, 541, 884, 976, 2084,
        2217, 2219
Freni, Mirella, soprano 2082, 2229, 2713, 2879
Frese, Siegfried Rudolf, bass 2082
Fresk Quartet 256
Fresnay, Pierre, narrator 2032
Freund, Karl, violin 2918
Frick, Gottlob, bass 2760, 2778, 2787

Ghione, Franco, conductor 567
Ghitalla, Armando, trumpet 88
Ghiuselev, Nikolai 1953, 1954
Giaiotti, Bonaldo, bass 2700, 3199
Gibbons, John, harpsichord 1410
Gibbs, John, bass-baritone 2712
Giboureau, Michel, oboe 2731
Gibson, Alexander, conductor 701, 1064, 1597, 1859, 1950, 2013, 2206,
          2385, 2403, 2629, 2695
Gibson, Barbara, soprano 1188
Gielen, Michael, conductor 245, 2611
Gielgud, John, narrator 2514
Gieseking, Walter, piano 169, 281, 2334
Giesen, Hubert, piano 3080
Gilbert, David, conductor 1078
Gilbert, Kenneth, harpsichord 29, 151, 962, 3033
Gilbert, Kenneth, organ 572
Gilbert and Sullivan Festival Chorus 2536, 2540, 2542, 2544, 2547,
          2548, 2550, 2552, 2560, 2561
Gilbert and Sullivan Festival Orchestra 2536, 2540, 2542, 2544, 2547,
          2548, 2550, 2552, 2560, 2561
Gilchrist, Alden, harpsichord 1111
Gilchrist, Alden, piano 1111
Gilels, Elena, piano 1810
Gilels, Emil, piano 304, 366, 374, 599, 604, 1227, 1810, 2063, 2277,
          2578
Giles, Allen, harpsichord 971
Giles, Allen, piano 971
Giles, Anne Diener, flute 971
Gillesberger, Hans, conductor 84, 3197, 3198
Gilvan, Raimund, tenor 692, 742
Gingold, Hermione, narrator 2059
Giorgetti, Giorgio, bass 2694
Giorini, Gino, piano 735
Giraudeau, Jean, tenor 2036
Gitlis, Ivry, violin 2211
Giulini, Carlo Maria, conductor 327, 420, 577, 717, 893, 1032, 1547,
          1934, 1943, 2143, 2646, 2709, 3041
Glasgow Cathedral Choir 3275
Glazer, David, clarinet 456
Glazer, Frank, piano 456, 1445
Gleeson, Patrick, conductor 1402
Glen Ellyn Children's Chorus 1629
Glick, Jacob, viola 1171
Glover, Stephen, piano 1779
Glyndebourne Festival Chorus 1765, 1767, 2538, 2546, 2558
Gobbi, Tito, baritone 1857, 2075, 2086, 2698, 2705, 2711
Gøbel, Bodil, soprano 1970
Goberman Chamber Orchestra 189
Goberman, Max, conductor 189, 1351
Gold, Ernest, conductor 1190, 1411
Goldberg, Laurette, harpsichord 209
Goldberg, Szymon, violin 1914
Goldman Band 2934
Goldman, Richard Franko, conductor 2934
Goldsmiths' Choral Union 2245
Goldstein, Martha, piano 168

Gyuzelev, Nikola, bass   2095, 2710

Haag, Hanno, violin   2808
Haas, Werner, piano   623, 2098, 2576
Hacker, Alan, basset horn   1781
Hacker, Alan, clarinet   881, 1781, 2437
Haebler, Ingrid, fortepiano   58, 63
Haebler, Ingrid, piano   1821, 1834, 1837, 1913, 2262, 2266, 2294, 2305
Haendel, Ida, violin   2387, 2618, 3040
Haerle, Dan, piano   220
Häfliger, Ernst, tenor   79, 81, 179, 394, 1595, 1599, 1862, 2165
Hagegård, Håkan, baritone   1947, 2894
Hager, Leopold, conductor   427, 446, 1805, 1806, 1808, 1818, 1822,
          1823, 1867, 1898
Haggart, Margaret, soprano   3013
Hague Philharmonic Orchestra   790, 3161
Hahn, Reynaldo, conductor   1123
Haitink, Bernard, conductor   285, 290, 311, 657, 673, 690, 1009, 1020,
          1594, 1602, 1613, 1622, 1628, 1637, 1647, 1662, 2142, 2319,
          2511, 2639, 2767, 3161, 3163
Hajdu, Istvan, piano   1522
Haken, Eduard, baritone   995
Hála, Josef, piano   1679
Hall, Joy, violoncello   1765, 1767
Hall, William, Chorale   694
Hall, William, conductor   694
Halland, Edward, bass   2551
Hallé Orchestra   500, 646, 653, 659, 798, 921, 1063, 1070, 1123, 1400,
          1583, 1923, 1975, 2118, 2457, 2465, 2625
Hallé Womens Chorus   1400
Hallin, Margareta, soprano   566
Hallstein, Ingeborg, soprano   2798
Halpin, Thomas, violin   2823
Halsey, Louis, conductor   112, 751, 917
Halsey, Louis, Singers   112, 751
Hamari, Julia, mezzo-soprano (alto, contralto)   178, 505, 1659, 2587,
          2748, 2769, 2805
Hambro, Leonid, piano   1752, 2212
Hamburg Philharmonic Orchestra   1730
Hamburg Radio Orchestra   1001
Hamburg Symphony Orchestra   791
Hamburg Wind Ensemble   954
Hamburger, Paul, piano   680, 2553, 3172
Hamel, Michel, tenor   541, 2145
Hamelin, Gaston, clarinet   3025
Hammond, Joan, soprano   3084
Hammond, Tom, piano   3013
Hammond-Stroud, Derek, bass-baritone   1061, 2776, 2783
Hancock, David, piano   2258
Handel Opera Society Chorus   1291
Handel Opera Society Orchestra   1291
Handford, Maurice, conductor   1045
Handley, Vernon, conductor   891, 1061, 2593
Hanell, Robert, conductor   3160
Hanks, John Kennedy, tenor   3011
Hann, Georg, bass-baritone   1200, 2753

Hannikainen, Tauno, conductor 1856
Hannover Niedersächsischer Singkreis 2047
Hanover Boys' Choir 84, 85, 87
Hansen, Cecilia, violin 2918
Hansen, Hans Christian, baritone 916
Hansen, Michael, tenor 916
Hanson, Howard, conductor 229, 753, 1238, 1295, 1296, 1297, 1298, 1441,
      1588, 2575
Harand, Richard, violoncello 967
Harasiewicz, Adam, piano 3160
Hargreaves, Glenville, baritone 863
Hargreaves, Walter, conductor 2962
Harmon, Sue, soprano 1763
Harnoncourt, Alice, violin 186, 191, 1671, 2728
Harnoncourt, Nikolaus, conductor 80, 84, 85, 86, 1249, 1762, 1769,
      1772, 2728, 3257
Harnoncourt, Nikolaus, viola da gamba 191, 1671, 3260
Harnoncourt, Nikolaus, violoncello 186
Harper, Heather, soprano 327, 684, 733, 1247, 1272, 1294, 1594, 1620,
      1636, 1661, 1662, 1713, 1715, 2708, 3041
Harrell, Lynn, violoncello 906, 981, 1723
Harrell, Ray, tenor 876
Harris, Alan, violoncello 1445
Harrison, Beatrice, violoncello 915
Harshaw, Margaret, soprano 1685
Hartemann, Jean-Claude, conductor 1202
Harth, Sidney, violin 2186
Hartle, Enid, mezzo-soprano 2587
Hartman, Gerda, soprano 2032
Hartmann, Rudolf, bass 2769
Hartmann, Willy, tenor 1970
Harty, Hamilton, conductor 798
Harty, Hamilton, piano 1123
Harvard Glee Club 2517
Harvey, Richard, recorder 3262
Harvuot, Clifford, baritone 1528
Harwood, Elizabeth, soprano 1269, 1526, 3077
Haskil, Clara, piano 734
Hassard, Donald, piano 34
Hassbecker, Eva, soprano 181
Hassid, Joseph, violin 2918
Hasson, Maurice, violin 701, 903
Hasty, Stanley, clarinet 1445
Haublein, Hans, violoncello 1314
Hauptmann, Norbert, horn 1803
Hay, Patricia, soprano 2695
Haynes, Bruce, oboe 859
Hayward, Thoms, tenor 1528
Hayward, Vernica, soprano 2029
Haywood, Lorna, soprano 325, 464
Heater, Claude, bass 2453, 2762
Heath, Kenneth, violoncello 567, 929, 2662
Hecht, Joshua, singer 1774
Hechtl, Gottfried, flute 1307
Hediguer, Max, double bass 582
Hedwall, Lennart, conductor 3251

Kalmar, Magda, soprano 1261, 2749
Kamenikova, Valentina, piano 1912
Kamenov, Christo, tenor 2263, 3030
Kamu, Okko, conductor 731, 2390
Kamu, Okko, violin 143
Kanawa, Kiri Te, contralto (soprano) 542, 1874, 2712
Kansas City Philharmonic Orchestra 271
Kaplan, Arnold, piano 3132
Kaplan, Melvin, oboe 1120
Kapp, Richard, conductor 2657
Karajan, Herbert von, conductor 183, 199, 241, 308, 319, 326, 329, 394,
        404, 484, 510, 610, 641, 720, 727, 728, 827, 1025, 1320,
        1372, 1417, 1526, 1529, 1562, 1563, 1600, 1601, 1642, 1712,
        1729, 1731, 1800, 1803, 1848, 1877, 1922, 1948, 1955, 2078,
        2082, 2198, 2203, 2247, 2252, 2347, 2354, 2462, 2475, 2488,
        2489, 2490, 2491, 2495, 2581, 2633, 2635, 2636, 2687, 2698,
        2704, 2706, 2754, 2765, 2848, 2849, 2981, 3005, 3066, 3074
Karnobatlova-Dobreva, Blagovesta, soprano 2095
Karp, Bess, harpsichord 569
Karr, Gary, double bass 1382
Kasandjiev, Vassil, conductor 1145
Katchen, Julius, piano 282, 465, 603, 670, 1235, 2124, 2136
Katims, Milton, conductor 933, 2327
Katims, Milton, viola 564, 2283
Katin, Peter, piano 828, 2579, 2649
Katz, Martin, piano 2500
Kaufman, Annette, piano 477
Kaufman, Louis, violin 477, 2214
Kaufmann, Gerd, organ 1770
Kaval Balalaika Orchestra and Chorus 3114
Kayser, Jan Henrik, piano 269
Kazandjikiev, Vassil, conductor 2263
Kecskes, Andras, lute 3090
Keele University Chapel Choir 3027
Keene, Constance, piano 1710
Keetch, Maureen, soprano 3264
Keeting, Otto, conductor 1474
Kegel, Herbert, conductor 1872, 1873, 1992, 1994, 1998
Kehr, Gunter, conductor 1918
Keilberth, Joseph, conductor 1200, 2469, 2759
Kejmar, Miroslav, trumpet 1455
Kelber, Sebastian, flute 1770
Kelber, Sebastian, recorder 1770
Kéléman, Zoltan, bass-baritone 1526, 2762
Keller (String) Quartet 2163
Keltsch, Werner, Instrumental Ensemble 715
Kempe, Rudolf, conductor 305, 397, 432, 605, 644, 1447, 2472, 2483,
        2501, 2637, 2760, 3040
Kempen, Christopher van, violoncello 2154
Kempff, Wilhelm, piano 213, 277, 288, 457, 466, 802, 1539, 1836, 2291,
        2292, 2293
Kennedy Center Theater Orchestra 526
Kenney, H. Arthur, conductor 3177, 3179
Kenney, Margarita, soprano (contralto) 2755, 2779, 2787
Kerekjarto, Duci de, violin 2918
Kern, Patricia, mezzo-soprano 520, 1420, 1691

Perry, Janet, soprano 2413
Pert, Morris, marimba 2909
Pert, Morris, piano 2909
Pert, Morris, vibraphone 2909
Pertis, Szuzsa, harpsichord 1339, 3091
Perulli, Raphael, viola da gamba 3031
Peter, Albrecht, baritone 2802
Peters, Karlheinz, bass 1761
Peters, Reinhard, conductor 1176, 1396, 2223
Petit, Annie, piano 769
Petit, David, organ 3286
Petit, Françoise, harpsichord 1308
Petite, Jean-Louis, harpsichord 554
Les Petits Chanteurs de Versailles 541
Petkov, Dimiter, bass 1904, 2095
Petr, Milos, horn 1455
Petri, Egon, piano 375, 3099
Pettersson, Claes-Merithz, conductor 1468
Petz, Ferenc, percussion 259
Pevecky Choir 1456
Peyer, Gervase de, clarinet 228, 2965, 3069, 3291
Peysang, André, tenor 1989
Pezzullo, Louis, trombone 874
Philadelphia Brass Ensemble 1391, 2908
Philadelphia Composers Forum Ensemble 1114
Philadelphia Orchestra 74, 204, 452, 608, 913, 1042, 1300, 1403, 1428,
          1429, 1491, 1668, 1798, 1939, 1997, 2050, 2104, 2119, 2123,
          2169, 2374, 2377, 2401, 2478, 2487, 2589, 2606, 2741, 2768,
          2773, 2901, 2902, 2905, 3203, 3205
Philadelphia Rapid Transit Company Band 2428
Philharmonia Chorus 393, 1715, 1845, 1848, 2698, 2709
Philharmonia Hungarica 248, 317, 330, 1176, 1329, 1349, 1350, 1354,
          1363, 1364, 1396, 1487, 1490, 2167, 2223
Philharmonia Orchestra, London 324, 393, 394, 612, 650, 654, 663, 915,
          980, 1022, 1032, 1047, 1070, 1087, 1216, 1232, 1603, 1634,
          1715, 1800, 1801, 1830, 1845, 1848, 1944, 1955, 2143, 2185,
          2198, 2495, 2530, 2580, 2643, 2646, 2698, 2709, 2711, 2848,
          2849, 2896, 3040, 3041, 3066, 3068, 3074, 3081, 3084
Philharmonic Chamber Ensemble 1510
Phillabaum, Katja, piano 1075
Phillips, Karen, viola 1115, 2955
Phoenix String Quartet 2178
Phoenix String Trio 759
Piasetzki, Erich, 207
Piatigorsky, Gregor, violoncello 351, 1029
Piatnitsky Chorus 2933, 3292
Pictet, Maria, Antoinette, piano 2151
Pierlot, Pierre, oboe 1887, 2004, 2745
Pieterson, George, clarinet 1887, 2506
Pignotti, Alfio, violin 1422
Pikaizen, Viktor, violin 1464
Pilarczyk, Helga, soprano 483
Pilgram, Neva, soprano 580, 1078
Pilou, Jeannette, soprano 540
Pilz, Jurgen, violin 1865
Pina, Erich, bass 2771

741

Saunders, Iris, soprano 682
Sautereau, Nadine, soprano 936
Saval, Jordi, viola da gamba 185, 3000
Savarese, Ugo, baritone 2716
Savastano, Antonio, tenor 2701
Sawallisch, Wolfgang, conductor 1022, 1728, 2326, 2348, 2761, 2762
Sax, Manfred, bassoon 857
Saxby, Joseph, harpsichord 3141
Saxon State (Staatskapelle) Orchestra 169
Sayão, Bidú, soprano 2074
La Scala Chorus 1187, 1529, 2073, 2078, 2086, 2701, 1711, 2717, 2718,
        2720, 2755, 2775, 2779, 2782, 2787, 3002
La Scala Orchestra 1187, 1529, 2073, 2076, 2078, 2085, 2086, 2701,
        2711, 2717, 2718, 2720, 2755, 2775, 2779, 2782, 2787, 3002
Scano, Gaetano, tenor 2201
Scarpino, Pietro, piano 295
Schädle, Lotte, soprano 79, 1860
Schaefer, Paul, piano 3052
Schaeftlein, Jürg, oboe 859, 2728
Schaer, Hanna, contralto 750, 2742
Schäffer, Michael, lute 3267
Schärtel, Elisabeth, contralto (mezzo-soprano) 2780
Scharwenka, Xavier, piano 3208
Schary, Elke, mezzo-soprano 2082
Schat, Peter, conductor 1476, 2239
Schech, Marianne, soprano 2977
Scheele-Müller, Ida von, alto 1199
Scheidegger, Rudolf, organ 3153
Scheit, Karl, lute 854
Schein, Ann, piano 34, 533
Scheja, Staffan, piano 1512
Schellenberg, Arno, bass-baritone 2699
Scherb, Michel, baritone 2673
Scherchen, Hermann, conductor 1181, 1274
Scherchen, Hermann, piano 767
Scherler, Barbara, alto (contralto) 1880
Schernus, Herbert, conductor 2441
Schey, Hermann, bass 180
Schidlof, Peter, viola 2277
Schiff, Andras, piano 630
Schiller, Allan, piano 1827
Schiml, Marga, contralto 711, 2805
Schippers, Thomas, conductor 230, 2049, 2201, 2205
Schlicker, Anneliese, piano 2802
Schlusnus, Heinrich, baritone 2790
Schmedes, Dagmar, contralto 2779, 2787
Schmidl, Peter, clarinet 1796
Schmidt, Manfred, tenor 2265
Schmidt, Ole, conductor 1971, 1972, 1973
Schmidt, Trudeliese, mezzo-soprano 1186
Schmidt-Gaden, Gerhard, conductor 116, 1878
Schmidt-Isserstedt, Hans, conductor 283, 291, 296, 301, 353, 528, 1001,
        2764
Schmohl, Traugott, bass-baritone 742
Schnabel, Artur, piano 273, 379, 1826, 1830
Schneeberger, Hanzheinz, violin 2165, 2435

Stockhausen, Karlheinz, conductor 2442, 2445
Stockhausen, Karlheinz, electronics, 2441, 2443
Stockhausen, Markus, flugel horn 2440
Stockhausen, Markus, trumpet 2440
Stockholm Motet Choir 2874
Stockholm Philharmonic Brass Ensemble 3250
Stockholm Philharmonic Orchestra 19, 24, 25, 26, 42, 528, 529, 566,
          1486, 1511, 1653, 1654, 2439
Stockholm Radio Symphony Orchestra 1226, 1415, 1485
Stockholm Royal Conservatory Chamber Choir 1077
Stockholm Royal Court Orchestra 2894, 2895
Stockholm Studentsangarforbund 1226
Stockton, Ann Mason, harp 2658
Stojanovic, J., violoncello 1500
Stokes, Sheridon, flute 569
Stokowski, Leopold, conductor 74, 217, 413, 438, 509, 561, 665, 1011,
          1048, 1280, 1428, 1463, 1623, 2121, 2187, 2322, 2577, 2612,
          2634, 2659,2784, 2966, 2973, 2978, 3173, 3174, 3270
Stokowski Symphony Orchestra 2784
Stolte, Adele, soprano 81, 181
Stoltzman, Richard, clarinet 460, 1203, 1745
Stolz, Robert, conductor 2464, 2468, 3061
Stolze, Gerhard, tenor 2762, 2774, 2778
Stone, David, violin 919
Storck, Klaus, violoncello 1770
Stour Music Festival Chamber Orchestra 2089
Stour Music Festival Chorus 2088
Stour Music Festival Orchestra 2088
Stoutz, Edmond de 1310, 1312, 1674
Stradivari Quartet 1116
Straight, Willard, conductor 1581
Straka, Zeljko, conductor 3242
Strasbourg Philharmonic Orchestra 540, 900, 2057, 2137
Strasbourg Radio Symphony Orchestra 242
Stratas, Teresa, soprano 1526, 2413
Strauss, Johann, Orchestra, Vienna 2463
Strauss, Michael, violin 2003
Strauss, Paul, conductor 1130, 1215
Strauss, Richard, conductor 2502
Strava, Robert, violin 2954
Stravinsky, Igor, conductor 2508, 2518
Stravinsky, Igor, piano 2510
Strehl, Laurenzius, viola da gamba 857, 1770
Strehl, Laurenzius, violone 1770
Streich, Rita, soprano 2802
Stricker, Frieder, tenor 2762, 2772
Strickland, William, conductor 232, 269, 928, 1413, 1431
Stringer, Alan, trumpet 1293, 1305
Strong, David, organ 3027
Strub, Max, violin 2918
Strutz, Harald, trombone 1770
Stryczek, Karl-Heinz, baritone 1994
Stryja, Karol, conductor 590
Stuchevsky, Semyon, piano 1963
Stuckl, Annelies, contralto 1948
Studholme, Marion, soprano 2539

Studio der Frühen Musik (Studio of Early Music)   3254
Studt, Richard, violin   122
Stupka, Frantisek, conductor   1019
Stutch, Nathan, violoncello   760
Stuttgart Chamber Orchestra   61, 178, 212, 852, 1223, 1235, 2747, 2964,
          2967
Stuttgart Hymnus Boys Choir   178
Stuttgart Radio Chorus   1200
Stuttgart Radio Orchestra   1200, 2480
Stutzmann, Christiane, soprano   2036, 2455
Subrtova, Milada, soprano   995, 1995, 1996
Suddick, Eric, organ   3289
Suderburg, Elizabeth, soprano   3283
Suderburg, Robert, piano   3283
Suffolk Children's Orchestra   685
Suggia, Guilhermina, violoncello   3081
Suitner, Otmar, conductor   1001, 1369, 1609, 1790, 2563
Suk, Josef, conductor   1843
Suk, Josef, violin   141, 305, 313, 316, 382, 1424, 1680, 1843
Suk Quartet   3145
Sulcova, Brigita, soprano   1419, 1456, 1499
Suntrader   2029
Suss, Reiner, bass   1926, 1946
Susskind, Walter, conductor   982, 1033, 1232, 1405, 1696, 1830, 2414,
          2479, 3074, 3084
Sutherland, Joan, soprano   353, 470, 471, 472, 936, 937, 938, 940,
          1271, 1292, 1293, 1688, 1984, 2322, 2691, 2712, 2714, 2778,
          2964, 2969, 2972, 2977
Svanholm, Set, tenor   2484, 2778, 2779, 2782, 2784
Svejda, Miroslav, tenor   1493, 2423
Sveshnikov, Aleksander, conductor   2615
Svetlanov, Yevgeny, conductor   1481, 1483, 1964, 2600
Svetoslav Obretenov Chorus   1954, 2710
Swallow, Keith, piano   1406, 1724
Swan, Sylvia, contralto   712
Swarowsky, Hans, conductor   2779
Swearingen, Madolyn, organ   2868
Swedish Radio Chorus   1949, 3247
Swedish Radio Symphony Orchestra   20, 22, 23, 43, 44, 45, 731, 1469,
          1949, 1981, 2030, 2031
Swedish Radio Women's Chorus   731
Swem, Andrea Anderson, piano   1584, 1585
Swift, Tom, tenor   3013
Swingle II   488
Sydney Symphony Orchestra   516, 1206, 1904
Sykes, James, piano   479
Sylvester, Robert, violoncello   868
Symphony of the Air   561, 2659
Synek, Liane, soprano   2780
Syntagma Musicum   3253
Szabados Quartet   2565
Szebenyi, Janos, flute   16, 927
Szekely, Mihaly, bass   3086
Szekely, Zoltan, violin   250
Szekeres, Ferenc, conductor   2722
Szell, Georg, conductor   231, 304, 320, 426, 467, 536, 705, 973, 974,
          1004, 1005, 1015, 1293, 1632, 1650, 1826, 1928, 1939, 2350,
          2398

Walter, Bruno, piano 1829
Walther von der Vogelweide Chamber Choir 2826
Wandsworth School Boys' Choir 1627, 1631, 2982, 3199
Wanger, Fredrik, piano 2880
Ward, David, bass 1293
Ward, Henry, harpsichord 1765, 1767
Ware, John, posthorn 1625
Warfield, William, baritone 1165
Warren, Leonard, baritone 2691, 2972
Warsaw Opera Orchestra 1759
Warsaw (National) Philharmonic Chorus 560, 2570
Warsaw (National) Philharmonic Orchestra 560, 1470, 1471, 1472, 2570,
          2588
Washington Camerata Chorus 2827
Washington Natiional Symphony Orchestra 139, 876, 889, 2375, 2590
Washington Oratorio Society 889
Washington, Paolo, bass-baritone 2077
Watanabe, Akeo, conductor 843
Watson, Claire, soprano 2778
Watson, David, bass 558
Watson, Ian, organ 2210
Watson, Lilian, soprano 1264, 1767
Watt, Alan, bass 1264
Watts, Andre, piano 1156
Watts, Helen, contralto (alto) 89, 670, 1037, 1044, 1247, 1268, 1272,
          1620, 1627, 1661, 2087, 2264, 2538, 2681, 2972
WDR Radio Choir 2441
Weathers, Felicia, soprano 3259
Weaver, James, harpsichord 1673, 3238
Weaver, James, organ 3238
Weaver, James, piano 3238
Weaver, Thomas, violin 1770
Webb, Charles, piano 2944
Webb, Richard, violoncello 1149
Weber, David, clarinet 2042
Weber, Ludwig, bass 1948, 2484, 2755, 2775, 2779, 2782, 2787
Wedekind, Erika, soprano 3211
Weede, Robert, baritone 544
Wegner, Walburga, soprano 2779, 2787
Wehrung, Herrad, soprano 715
Weide, Genevieve, soprano 1498
Weigert, Hermann, conductor 2497, 2751
Weight, Newell, conductor 2671
Weikl, Bernd, baritone 1186, 2449, 2587, 2762, 2769, 2772, 2801, 2824
Weinberg, Vittorio, baritone 2076
Weiner, Susan, oboe 13
Weingartner, Felix, conductor 1546, 3043
Weisbach, Hans, conductor 182, 2699
Weiberg, Arthur, conductor 266, 754, 864, 1191, 3128
Weise, Manfred, clarinet 2501
Weiss, Edward, piano 3099
Weiss, Gunther, violoncello 967
Weiss, Jeanne, piano 1058
Weissenberg, Alexis, piano 1540, 2044, 2328, 2329
Weldon, George, conductor 1068
Welitsch, Alexander, baritone 1200